# Lecture Notes in Computer Science 4653

*Commenced Publication in 1973*
Founding and Former Series Editors:
Gerhard Goos, Juris Hartmanis, and Jan van Leeuwe

Roland Wagner Norman Revell
Günther Pernul (Eds.)

# Database and Expert Systems Applications

18th International Conference, DEXA 2007
Regensburg, Germany, September 3-7, 2007
Proceedings

Volume Editors

Roland Wagner
University of Linz
Institute of FAW
Altenbergerstrasse 69
4040 Linz, Austria
E-mail: rrwagner@faw.uni-linz.ac.at

Norman Revell †
Middlesex University
United Kingdom

Günther Pernul
University of Regensburg
Universitätsstrasse 31
D-93053 Regensburg, Germany
E-mail: guenther.pernul@wiwi.uni-regensburg.de

Library of Congress Control Number: 2007933508

CR Subject Classification (1998): H.2, H.4, H.3, H.5, I.2, J.1

LNCS Sublibrary: SL 3 – Information Systems and Application, incl. Internet/Web and HCI

ISSN 0302-9743
ISBN-10 3-540-74467-3 Springer Berlin Heidelberg New York
ISBN-13 978-3-540-74467-2 Springer Berlin Heidelberg New York



Springer is a part of Springer Science+Business Media

springer.com



Typesetting: Camera-ready by author, data conversion by Scientific Publishing Services, Chennai, India
Printed on acid-free paper SPIN: 12112913 06/3180 5 4 3 2 1 0

We would like to dedicate this volume of DEXA proceedings to our dear friend

**Norman Revell**

(14.5.47 – 7.2.07)

We miss you.

The DEXA Society

# Preface

The annual international conference on Database and Expert Systems Applications (DEXA) is now well established as a reference scientific event. The reader will find in this volume a collection of scientific papers that represent the state of the art of research in the domain of data, information and knowledge management, intelligent systems, and their applications.

The 18th instance of the series of DEXA conferences was held at the University of Regensburg, Germany, September 3–7, 2007.

Several collocated conferences and workshops covered specialized and complementary topics to the main conference topic. Six conferences—the Eighth International Conference on Data Warehousing and Knowledge Discovery (DaWaK), the Seventh International Conference on Electronic Commerce and Web Technologies (EC-Web), the Fifth International Conference on Electronic Government (EGOV), the 3rd International Conference on Trust, Privacy, and Security in Digital Business (TrustBus), the 3rd International Conference on Industrial Applications of Holonic and Multi-Agent Systems (HoloMAS), and the 1st International Conference on Network-Based Information Systems (NBiS)—and 19 workshops were collocated with DEXA.

The conference is a unique international event with a balanced depth and breadth of topics. Its much appreciated conviviality fosters unmatched opportunities to meet, share the latest scientific results and discuss the latest technological advances in the area of information technologies with young scientists and engineers as well as senior world-renown experts.

This volume contains the papers selected for presentation at the conference. Each submitted paper was reviewed by three or four reviewers, members of the Program Committee or external reviewers appointed by members of the Program Committee. Based on the reviews, the Program Committee accepted 86 of the 267 originally submitted papers.

The excellence brought to you in these proceedings would not have been possible without the efforts of numerous individuals and the support of several organizations.

First and foremost, we thank the authors for their hard work and for the quality of their submissions.

We also thank Josef Küng, A Min Tjoa, Gerald Quirchmayr, Gabriela Wagner, the members of the Program Committee, the reviewers, and the many others who assisted in the DEXA organization for their contribution to the success and high standard of DEXA 2007 and of these proceedings.

Finally we thank the DEXA Association, the Austrian Computer Society, the Research Institute for Applied Knowledge Processing (FAW), and the University of Regensburg, especially Günther Pernul and his team, for making DEXA 2007 happen.

June 2007 Roland Wagner

# Organization

## External Reviewers

Yu Cao
Weihai Yu
Wei Ni
Bo Chen
Sebastian Obermeier
Rita Steinmetz
Brigitte Mathiak
Luciano Caroprese
Andrea Tagarelli
Massimo Mazzeo
Yiping Ke
James Cheng
Stardas Pakalnis
Stephan Vornholt
Ingolf Geist
Nasreddine Aoumeur
Eike Schallehn
Somchai Chatvichienchai
Amgoud Leila
Yu Suzuki
Ludwig Fuchs
Rolf Schillinger
Christian Schläger
Nele Dexters
Roel Vercammen
Jan Hidders
Marco A. Casanova
Alfio Ferrara
Stefano Montanelli
Michele Melchiori
Sander Evers
Domenico Famularo
Pasquale Legato
Andrea Pugliese
Francesco Scarcello
Antonio Sala
Lipyeow Lim
Jinsoo Lee
Wooseong Kwak
Suyun Chen
Antonio Cisternino
Laura Semini
Stefano Chessa
Gregory Leighton
Eddy Dragut
Ozgul Unal
Simon Msanjila
Cristóbal Costa
Jennifer Pérez
José Hilario Canós
Umberto Straccia
Fabrizio Falchi
Carlo Meghini
Andrzej Bassara
Konstanty Haniewicz
Michael Rys
Philip Groth
Timo Glaesser
Bastian Quilitz
Satoshi Oyama
Shun Hattori
Adam Jatowt
Satoshi Nakamura
Shinsuke Nakajima
Taro Tezuka
Wee Hyong Tok
Derry Wijaya
Le Dzung
Jarogniew Rykowski
Sergiusz Strykowski
Miroslaw Stawniak
Grant Weddell
Jarek Gryz
Qihong Shao
Ziyang Liu
Michel Kinsy

# Program Committee

## General Chair

Günther Pernul, University of Regensburg, Germany

## Conference Program Chairpersons

Roland R. Wagner, FAW, University of Linz, Austria
Norman Revell, Middlesex University, UK †

## Workshop Chairpersons

A Min Tjoa, Technical University of Vienna, Austria
Roland R. Wagner, FAW, University of Linz, Austria

## Publication Chairperson

Vladimir Marik, Czech Technical University, Czech Republic

## Program Committee

Witold Abramowicz, The Poznan University of Economics, Poland
Hamideh Afsarmanesh, University of Amsterdam, The Netherlands
Fuat Akal, ETH Zürich, Switzerland
Toshiyuki Amagasa, University of Tsukuba, Japan
Bernd Amann, LIP6 - UPMC, France
Vasco Amaral, New University of Lisbon, Portugal
Stanislaw Ambroszkiewicz, Polish Academy of Sciences, Poland
Ira Assent, Aachen University, Germany
Ramazan S. Aygun, University of Alabama in Huntsville, USA
Torben Bach Pedersen, Aalborg University, Denmark
Denilson Barbosa, University of Calgary, Canada
Leonard Barolli, Fukuoka Institute of Technology (FIT), Japan
Kurt Bauknecht, Universität Zürich, Switzerland
Peter Baumann, University of Bremen, Germany
Bishwaranjan Bhattacharjee, IBM Thomas J. Watson Research Center, USA
Sourav S Bhowmick, Nanyang Technological University, Singapore
Stephen Blott, Dublin City University, Ireland
Peter Boncz, Centrum voor Wiskunde en Informatica, The Netherlands
Angela Bonifati, ICAR-CNR, Italy
Stefan Böttcher, University of Paderborn, Germany
Zizette Boufaida, Mentouri University Constantine, Algeria

Kjell Bratbergsengen, Norwegian University of Science and Technology, Norway
Stephane Bressan, National University of Singapore, Singapore
Martin Breunig, University of Osnabrück, Germany
Ahmet Bulut, University of California Santa Barbara, USA
Ioana Burcea, University of Toronto, Camada
Luis M. Camarinha-Matos, Universidade Nova de Lisboa and Uninova, Portugal
Antonio Cammelli, ITTIG-CNR, Italy
K. Selcuk Candan, Arizona State University, USA
Silvana Castano, Università degli Studi di Milano, Italy
Barbara Catania, Università di Genova, Italy
Wojciech Cellary, University of Economics at Poznan, Poland
Elizabeth Chang, Curtin University, Australia
Sudarshan S. Chawathe, University of Maryland, USA
Yi Chen, Arizona State University, USA
Rosine Cicchetti, IUT, University of Marseille, France
Cindy Chen, University of Massachusetts Lowel, USA
Henning Christiansen, Roskilde University, Denmark
Chris Clifton, Purdue University, USA
Frans Coenen, The University of Liverpool, UK
Bin Cui, Peking University, China
Tran Khanh Dang, Ho Chi Minh City University of Technology, Vietnam
John Debenham, University of Technology, Sydney, Australia
Elisabetta Di Nitto, Politecnico di Milano, Italy
Gillian Dobbie, University of Auckland, New Zealand
Dirk Draheim, Software Competence Center Hagenberg, Austria
Silke Eckstein, Technical University of Braunschweig, Germany
Johann Eder, University of Vienna, Austria
Suzanne M. Embury, The University of Manchester, UK
Tomoya Enokido, Rissho University, Japan
Leonidas Fegaras, The University of Texas at Arlington, USA
Ling Feng, University of Twente, The Netherlands
Alvaro A.A. Fernandes, University of Manchester, UK
Eduardo Fernandez, Florida Atlantic University, USA
Simon Field, Office for National Statistics, UK
Mariagrazia Fugini, Politecnico di Milano, Italy
Antonio L. Furtado, Pontificia Universidade Catolica do R.J., Brazil
Manolo Garcia-Solaco, IS Consultant, USA
Mary Garvey, University of Wolverhampton, UK
Alexander Gelbukh, Centro de Investigacion en Computacion (CIC),
Instituto Politecnico Nacional (IPN), Mexico
Giorgio Ghelli, University of Pisa, Italy
Jan Goossenaerts, Eindhoven University of Technology, The Netherlands
William Grosky, University of Michigan, USA
Le Gruenwald, University of Oklahoma, USA
Francesco Guerra, Università degli Studi Di Modena e Reggio Emilia, Italy
Hele-Mai Haav, Tallinn University of Technology, Estonia
Abdelkader Hameurlain, University of Toulouse, France

Beda Christoph Hammerschmidt, Oracle Corporation, USA
Wook-Shin Han, Kyungpook National University, Korea
Kenji Hatano, Doshisha University, Japan
Igor T. Hawryszkiewycz, University of Technology, Sydney, Australia
Alexander Hinneburg, University of Halle, Gemany
Wynne Hsu, National University of Singapore, Singapore
Ela Hunt, ETH Zürich, Switzerland
Mohamed Ibrahim, University of Greenwich, UK
Mirjana Ivanovic, University of Novi Sad, Serbia
Mizuho Iwaihara, Kyoto University, Japan
Dimitris Karagiannis, University of Vienna, Austria
Randi Karlsen, University of Tromsö, Norway
Rudolf Keller, Zühlke Engineering AG, Switzerland
Anastasios Kementsietsidis, University of Edinburgh, UK
Myoung Ho Kim, KAIST, Korea
Stephen Kimani, University of Rome "La Sapienza", Italy
Gary J. Koehler, University of Florida, USA
Christian König, Microsoft Research, USA
Hanna Kozankiewicz, Polish Academy of Sciences, Poland
Michal Krátký, VSB-Technical University of Ostrava, Czech Republic
John Krogstie, SINTEF, Norway
Petr Kroha, Technische Universität Chemnitz-Zwickau, Germany
Josef Küng, University of Linz, Austria
Sergey Kuznetcov, Russian Academy of Sciences, Russia
Lotfi Lakhal, University of Marseille, France
Christian Lang, IBM T.J. Watson Research Center, USA
Paul Larson, Microsoft Corporation, USA
Young-Koo Lee, Kyung Hee University, Korea
Mong Li Lee, National University of Singapore, Singapore
Dongwon Lee, Pennsylvania State University, USA
Ulf Leser, Humboldt University of Berlin, Germany
Changqing Li, Duke University, USA
Xuemin Lin, University of New South Wales, Sydney, Australia
Tok Wang Ling, National University of Singapore, Singapore
Volker Linnemann, University of Lübeck, Germany
Mengchi Liu, Wuhan University, China
Jørgen Løland, Norwegian University of Science and Technology, Norway
Peri Loucopoulos, The University of Manchester, UK
Andras Lukacs, Hungarian Academy of Sciences, Hungary
Qiong Luo, The Hong Kong University of Science and Technology, Hong Kong
Sanjai Kumar Madria, University of Missouri-Rolla, USA
Stefan Manegold, Centrum voor Wiskunde en Informatica, The Netherlands
Vladimir Marik, Czech Technical University, Czech Republic
Simone Marinai, University of Florence, Italy
Holger Märtens, Univ. of Applied Sciences Braunschweig/Wolfenbüttel, Germany
Elio Masciari, University of Southern California, Italy
Heinrich C. Mayr, University of Klagenfurt, Austria

Subhasish Mazumdar, New Mexico Tech, USA
Dennis McLeod, University of Southern California, USA
Xiaofeng Meng, Renmin University, China
Elisabeth Metais, CNAM, France
Klaus Meyer-Wegener, University of Erlangen and Nuremberg, Germany
Philippe Michiels, Univesity of Antwerpen, Belgium
Dezso Miklos, Hungarian Academy of Sciences, Hungary
Yang-Sae Moon, Kangwon National University, Korea
Reagan Moore, San Diego Supercomputer Center, USA
Tadeusz Morzy, Poznan University of Technology, Poland
Günter Müller, Universität Freiburg, Germany
Wolfgang Nejdl, University of Hannover, Germany
Wilfred Ng, University of Science and Technology, Hong Kong
Daniela Nicklas, University of Stuttgart, Germany
Matthias Nicola, IBM Silicon Valley Laboratory, USA
Silvia Nittel, University of Maine, USA
Gultekin Ozsoyoglu, University Case Western Research, USA
Oscar Pastor, Universidad Politecnica de Valencia, Spain
Verónica Peralta, Universidad de la Republica, Uruguay
Dimitris Plexousakis, FORTH and Univeristy of Crete, Greece
Jaroslav Pokorny, Charles University in Prague, Czech Republic
Philippe Pucheral, INRIA, Université de Versailles, France
Magdalena Punceva, CERN, Switzerland
Gerald Quirchmayr, University of Vienna, Austria and University of South Australia, Australia
Fausto Rabitti, ISTI, CNR Pisa, Italy
Wenny Rahayu, La Trobe University, Australia
Rajugan Rajagopalapillai, DEBII, Curtin University of Technology, Australia
Isidro Ramos, Technical University of Valencia, Spain
Ralf Rantzau, IBM Silicon Valley Laboratory, USA
P. Krishna Reddy, International Institute of Information Technology, India
Colette Rolland, University Paris I, Sorbonne, France
Gunter Saake, University of Magdeburg, Germany
Domenico Sacca, University of Calabria, Italy
Simonas Saltenis, Aalborg University, Denmark
Marýa Luýsa Sapino, Università degli Studi di Torino, Italy
Kai-Uwe Sattler, Technical University of Ilmenau, Germany
Marinette Savonnet, Université de Bourgogne, France
Ralf Schenkel, Max Planck Institute, Germany
Stefanie Scherzinger, Saarland University, Germany
Ingo Schmitt, University of Magdeburg, Germany
Harald Schöning, Software AG, Germany
Holger Schwarz, University of Stuttgart, Germany
Erich Schweighofer, University of Vienna, Austria
Sergej Sizov, University of Koblenz, Germany
Darunee Smavatkul, Chiangmai University, Thailand
Giovanni Soda, University of Florence, Italy

Dmitri Soshnikov, Moscow Aviation Technical University, Microsoft Russia, Russia
Srinath Srinivasa, IIIT-B, India
Bala Srinivasan, Monash University, Australia
Uma Srinivasan, University of Western Sydney, Australia
Zbigniew Struzik, The University of Tokyo, Japan
Julius Stuller, Academy of Sciences of the Czech Republic, Czech Republic
Makoto Takizawa, Tokyo Denki University, Japan
Katsumi Tanaka, Kyoto University, Japan
Yufei Tao, City University of Hong Kong, Hong Kong
E. Nesime Tatbul, Brown University, USA
Wei-Guang Teng, National Cheng Kung University, Taiwan
Stephanie Teufel, University of Fribourg, Switzerland
Jukka Teuhola, University of Turku, Finland
Bernhard Thalheim, University of Kiel, Germany
J.M. Thevenin, University of Toulouse, France
Helmut Thoma, University of Basel, Switzerland
A Min Tjoa, Technical University of Vienna, Austria
Frank Tompa, University of Waterloo, Canada
Roland Traunmüller, University of Linz, Austria
Peter Triantafillou, University of Patras, Greece
Maurice van Keulen, University of Twente, The Netherlands
Genoveva Vargas-Solar, LSR-IMAG, France
Yannis Vassiliou, National Technical University of Athens, Greece
Krishnamurthy Vidyasankar, Memorial Univ. of Newfoundland, Canada
Stratis Viglas, University of Edinburgh, UK
Jesus Vilares Ferro, University of Coruna, Spain
Peter Vojtas, Charles University in Prague, Czech Republic
Matthias Wagner, DoCoMo Communications Laboratories Europe GmbH, Germany
John Wilson, University of Strathclyde, UK
Marek Wojciechowski, Poznan University of Technology, Poland
Viacheslav Wolfengagen, Institute for Contemporary Education, Russia
Ming-Chuan Wu, Microsoft Corporation, USA
Vilas Wuwongse, Asian Institute of Technology, Thailand
Liang Huai Yang, National University of Singapore, Singapore
Clement Yu, University of Illinios at Chicago, USA
Hailing Yu, Oracle, USA
Yidong Yuan, University of New South Wales, Sydney, Australia
Maciej Zakrzewicz, Poznan University of Technology, Poland
Gian Piero Zarri, University Paris IV, Sorbonne, France
Arkady Zaslavsky, Monash University, Australia
Baihua Zheng, Singapore Management University, Singapore
Yifeng Zheng, University of Pennsylvania, USA
Aoying Zhou, Fudan University, China
Yongluan Zhou, National University of Singapore, Singapore
Qiang Zhu, The University of Michigan, USA
Ester Zumpano, University of Calabria, Italy
Sergej Sizov, University of Koblenz, Germany

# Table of Contents

## XML and Databases I

On the Efficient Processing Regular Path Expressions of an Enormous Volume of XML Data ..... 1
*Michal Krátký, Radim Bača, and Václav Snášel*

Improving XML Instances Comparison with Preprocessing Algorithms ..... 13
*Rodrigo Gonçalves and Ronaldo dos Santos Mello*

Storing Multidimensional XML Documents in Relational Databases ..... 23
*N. Fousteris, M. Gergatsoulis, and Y. Stavrakas*

## Expert Systems and Semantics

On Constructing Semantic Decision Tables ..... 34
*Yan Tang and Robert Meersman*

Artificial Immune Recognition System Based Classifier Ensemble on the Different Feature Subsets for Detecting the Cardiac Disorders from SPECT Images ..... 45
*Kemal Polat, Ramazan Şekerci, and Salih Güneş*

A Multisource Context-Dependent Semantic Distance Between Concepts ..... 54
*Ahmad El Sayed, Hakim Hacid, and Djamel Zighed*

Self-healing Information Systems (Invited Talk) ..... 64
*Barbara Pernici*

## XML and Databases II

A Faceted Taxonomy of Semantic Integrity Constraints for the XML Data Model ..... 65
*Khaue Rezende Rodrigues and Ronaldo dos Santos Mello*

Beyond Lazy XML Parsing ..... 75
*Fernando Farfán, Vagelis Hristidis, and Raju Rangaswami*

Efficient Processing of XML Twig Pattern: A Novel One-Phase Holistic Solution ..... 87
*Zhewei Jiang, Cheng Luo, Wen-Chi Hou, Qiang Zhu, and Dunren Che*

## Database and Information Systems Architecture and Performance I

Indexing Set-Valued Attributes with a Multi-level Extendible Hashing Scheme ..... 98
*Sven Helmer, Robin Aly, Thomas Neumann, and Guido Moerkotte*

Adaptive Tuple Differential Coding ..... 109
*Jean-Paul Deveaux, Andrew Rau-Chaplin, and Norbert Zeh*

Space-Efficient Structures for Detecting Port Scans ..... 120
*Ali Şaman Tosun*

## XML and Databases III

A Dynamic Labeling Scheme Using Vectors ..... 130
*Liang Xu, Zhifeng Bao, and Tok Wang Ling*

A New Approach to Replication of XML Data ..... 141
*Flávio R.C. Sousa, Heraldo J.A. Carneiro Filho, and Javam C. Machado*

An Efficient Encoding and Labeling Scheme for Dynamic XML Data ..... 151
*Xu Juan, Li Zhanhuai, Wang Yanlong, and Yao Rugui*

## Database and Information Systems Architecture and Performance II

Distributed Semantic Caching in Grid Middleware ..... 162
*Laurent d'Orazio, Fabrice Jouanot, Yves Denneulin, Cyril Labbé, Claudia Roncancio, and Olivier Valentin*

Multiversion Concurrency Control for Multidimensional Index Structures ..... 172
*Walter Binder, Samuel Spycher, Ion Constantinescu, and Boi Faltings*

Using an Object Reference Approach to Distributed Updates ..... 182
*Dalen Kambur, Mark Roantree, and John Murphy*

## Applications of Database Systems and Information Systems I

Towards a Novel Desktop Search Technique ..... 192
*Sujeet Pradhan*

An Original Usage-Based Metrics for Building a Unified View of Corporate Documents ........ 202
*Guillaume Cabanac, Max Chevalier, Claude Chrisment, and Christine Julien*

Exploring Knowledge Management with a Social Semantic Desktop Architecture: The Case of Professional Business Services Firms ........ 213
*Niki Papailiou, Dimitris Apostolou, and Gregoris Mentzas*

Classifying and Ranking: The First Step Towards Mining Inside Vertical Search Engines ........ 223
*Hang Guo, Jun Zhang, and Lizhu Zhou*

## Query Processing and Optimisation I

Progressive High-Dimensional Similarity Join ........ 233
*Wee Hyong Tok, Stéphane Bressan, and Mong-Li Lee*

Decomposing DAGs into Disjoint Chains ........ 243
*Yangjun Chen*

Evaluating Top-k Skyline Queries over Relational Databases ........ 254
*Carmen Brando, Marlene Goncalves, and Vanessa González*

A P2P Technique for Continuous k-Nearest-Neighbor Query in Road Networks ........ 264
*Fuyu Liu, Kien A. Hua, and Tai T. Do*

Information Life Cycle, Information Value and Data Management (Invited Talk) ........ 277
*Rudolf Bayer*

## XML and Databases IV

Vague Queries on Peer-to-Peer XML Databases ........ 287
*Bettina Fazzinga, Sergio Flesca, and Andrea Pugliese*

Proximity Search of XML Data Using Ontology and XPath Edit Similarity ........ 298
*Toshiyuki Amagasa, Lianzi Wen, and Hiroyuki Kitagawa*

Cooperative Data Management for XML Data ........ 308
*Katja Hose and Kai-Uwe Sattler*

## Query Processing and Optimisation II

C-ARIES: A Multi-threaded Version of the ARIES Recovery Algorithm ........ 319
*Jayson Speer and Markus Kirchberg*

Optimizing Ranked Retrieval ..... 329
*Thomas Neumann*

Similarity Search over Incomplete Symbolic Sequences ..... 339
*Jie Gu and Xiaoming Jin*

## Applications of Database Systems and Information Systems II

Random Multiclass Classification: Generalizing Random Forests to Random MNL and Random NB ..... 349
*Anita Prinzie and Dirk Van den Poel*

Related Terms Clustering for Enhancing the Comprehensibility of Web Search Results ..... 359
*Michiko Yasukawa and Hidetoshi Yokoo*

Event Specification and Processing for Advanced Applications: Generalization and Formalization ..... 369
*Raman Adaikkalavan and Sharma Chakravarthy*

An Evaluation of a Cluster-Based Architecture for Peer-to-Peer Information Retrieval ..... 380
*Iraklis A. Klampanos and Joemon M. Jose*

## Query Processing and Optimisation III

A Conceptual Framework for Automatic Text-Based Indexing and Retrieval in Digital Video Collections ..... 392
*Mohammed Belkhatir and Mbarek Charhad*

Dimensionality Reduction in High-Dimensional Space for Multimedia Information Retrieval ..... 404
*Seungdo Jeong, Sang-Wook Kim, and Byung-Uk Choi*

Integrating a Stream Processing Engine and Databases for Persistent Streaming Data Management ..... 414
*Yousuke Watanabe, Shinichi Yamada, Hiroyuki Kitagawa, and Toshiyuki Amagasa*

## Applications of Database Systems and Information Systems III

Data Management for Mobile Ajax Web 2.0 Applications ..... 424
*Stefan Böttcher and Rita Steinmetz*

Data Management in RFID Applications ..... 434
*Dan Lin, Hicham G. Elmongui, Elisa Bertino, and Beng Chin Ooi*

When Mobile Objects' Energy Is Not So Tight: A New Perspective on Scalability Issues of Continuous Spatial Query Systems ............... 445
*Tai T. Do, Fuyu Liu, and Kien A. Hua*

Sequence Alignment as a Database Technology Challenge ............. 459
*Hans Philippi*

## Query Processing and Optimisation IV

Fuzzy Dominance Skyline Queries ................................. 469
*Marlene Goncalves and Leonid Tineo*

Pruning Search Space of Physical Database Design .................. 479
*Ladjel Bellatreche, Kamel Boukhalfa, and Mukesh Mohania*

A Two-Phased Visual Query Interface for Relational Databases ........ 489
*Sami El-Mahgary and Eljas Soisalon-Soininen*

Wavelet Synopsis: Setting Unselected Coefficients to Zero Is Not Optimal ................................................ 499
*Chong Sun, Yan Sheng Lu, Chong Zhou, and Jun Liu*

## Data and Information Modelling I

A Logic Framework to Support Database Refactoring................ 509
*Shi-Kuo Chang, Vincenzo Deufemia, Giuseppe Polese, and Mario Vacca*

An Iterative Process for Adaptive Meta- and Instance Modeling ....... 519
*Melanie Himsl, Daniel Jabornig, Werner Leithner, Peter Regner, Thomas Wiesinger, Josef Küng, and Dirk Draheim*

Compiling Declarative Specifications of Parsing Algorithms............ 529
*Carlos Gómez-Rodríguez, Jesús Vilares, and Miguel A. Alonso*

## XML Query Processing and Optimisation I

Efficient Fragmentation of Large XML Documents .................. 539
*Angela Bonifati and Alfredo Cuzzocrea*

Locating and Ranking XML Documents Based on Content and Structure Synopses ............................................... 551
*Weimin He, Leonidas Fegaras, and David Levine*

MQTree Based Query Rewriting over Multiple XML Views............ 562
*Jun Gao, Tengjiao Wang, and Dongqing Yang*

## Data and Information Modelling II

Convex Cube: Towards a Unified Structure for Multidimensional Databases ........ 572
*Alain Casali, Sébastien Nedjar, Rosine Cicchetti, and Lotfi Lakhal*

Dependency Management for the Preservation of Digital Information ... 582
*Yannis Tzitzikas*

Constraints Checking in UML Class Diagrams: SQL vs OCL .......... 593
*D. Berrabah and F. Boufarès*

## XML Query Processing and Optimisation II

XML-to-SQL Query Mapping in the Presence of Multi-valued Schema Mappings and Recursive XML Schemas ........ 603
*Mustafa Atay, Artem Chebotko, Shiyong Lu, and Farshad Fotouhi*

Efficient Evaluation of Nearest Common Ancestor in XML Twig Queries Using Tree-Unaware RDBMS ........ 617
*Klarinda G. Widjanarko, Erwin Leonardi, and Sourav S. Bhowmick*

## Data Mining I

Exclusive and Complete Clustering of Streams ........ 629
*Vasudha Bhatnagar and Sharanjit Kaur*

Clustering Quality Evaluation Based on Fuzzy FCA ........ 639
*Minyar Sassi, Amel Grissa Touzi, and Habib Ounelli*

Comparing Clustering Algorithms and Their Influence on the Evolution of Labeled Clusters ........ 650
*Rene Schult*

Journey to the Centre of the Star: Various Ways of Finding Star Centers in Star Clustering ........ 660
*Derry Tanti Wijaya and Stéphane Bressan*

## Semantic Web and Ontologies I

Improving Semantic Query Answering ........ 671
*Norbert Kottmann and Thomas Studer*

A Method for Determining Ontology-Based Semantic Relevance ........ 680
*Tuukka Ruotsalo and Eero Hyvönen*

Semantic Grouping of Social Networks in P2P Database Settings ....... 689
*Verena Kantere, Dimitrios Tsoumakos, and Timos Sellis*

Benchmarking RDF Production Tools ..... 700
*Martin Svihla and Ivan Jelinek*

## Semantic Web and Ontologies II

Creating Learning Objects and Learning Sequence on the Basis of Semantic Networks ..... 710
*Przemysław Korytkowski and Katarzyna Sikora*

SQORE-Based Ontology Retrieval System ..... 720
*Rachanee Ungrangsi, Chutiporn Anutariya, and Vilas Wuwongse*

Crawling the Web with OntoDir ..... 730
*Antonio Picariello and Antonio M. Rinaldi*

## Data Mining II

Extracting Sequential Nuggets of Knowledge ..... 740
*Froidevaux Christine, Lisacek Frédérique, and Rance Bastien*

Identifying Rare Classes with Sparse Training Data ..... 751
*Mingwu Zhang, Wei Jiang, Chris Clifton, and Sunil Prabhakar*

Clustering-Based K-Anonymisation Algorithms ..... 761
*Grigorios Loukides and Jianhua Shao*

Investigation of Semantic Similarity as a Tool for Comparative Genomics ..... 772
*Danielle Welter, W. Alexander Gray, and Peter Kille*

## WWW and Databases

On Estimating the Scale of National Deep Web ..... 780
*Denis Shestakov and Tapio Salakoski*

Mining the Web for Appearance Description ..... 790
*Shun Hattori, Taro Tezuka, and Katsumi Tanaka*

Rerank-by-Example: Efficient Browsing of Web Search Results ..... 801
*Takehiro Yamamoto, Satoshi Nakamura, and Katsumi Tanaka*

Computing Geographical Serving Area Based on Search Logs and Website Categorization ..... 811
*Qi Zhang, Xing Xie, Lee Wang, Lihua Yue, and Wei-Ying Ma*

## Temporal and Spatial Databases

A General Framework to Implement Topological Relations on Composite Regions ... 823
*Magali Duboisset, François Pinet, Myoung-Ah Kang, and Michel Schneider*

Active Adjustment: An Approach for Improving the Performance of the TPR*-Tree ... 834
*Sang-Wook Kim, Min-Hee Jang, and Sungchae Lim*

## Data and Information Semantics

Performance Oriented Schema Matching ... 844
*Khalid Saleem, Zohra Bellahsene, and Ela Hunt*

Preference-Based Integration of Relational Databases into a Description Logic ... 854
*Olivier Curé and Florent Jochaud*

A Context-Based Approach for the Discovery of Complex Matches Between Database Sources ... 864
*Youssef Bououlid Idrissi and Julie Vachon*

## Closing Session: Knowledge and Design

Ontology Modularization for Knowledge Selection: Experiments and Evaluations ... 874
*Mathieu d'Aquin, Anne Schlicht, Heiner Stuckenschmidt, and Marta Sabou*

The Role of Knowledge in Design Problems (Invited Talk) ... 884
*Zdenek Zdrahal*

e-Infrastructures (Invited Talk) ... 895
*Wolfgang Gentzsch*

**Author Index** ... 905

# On the Efficient Processing Regular Path Expressions of an Enormous Volume of XML Data*

Michal Krátký, Radim Bača, and Václav Snášel

Department of Computer Science, VŠB – Technical University of Ostrava
17. listopadu 15, 708 33 Ostrava–Poruba, Czech Republic
{michal.kratky,radim.baca,vaclav.snasel}@vsb.cz

**Abstract.** XML (Extensible Mark-up Language) has recently been embraced as a new approach to data modeling. Nowadays, more and more information is formatted as semi-structured data, i.e. articles in a digital library, documents on the web and so on. Implementation of an efficient system enabling storage and querying of XML documents requires development of new techniques. The indexing of an XML document is enabled by providing an efficient evaluation of a user query. XML query languages, like XPath or XQuery, apply a form of path expressions for composing more general queries. The evaluation process of regular path expressions is not efficient enough using the current approaches to indexing XML data. Most approaches index single elements and the query statement is processed by joining individual expressions. In this article we will introduce an approach which makes it possible to efficiently process a query defined by regular path expressions. This approach indexes all root-to-leaf paths and stores them in multi-dimensional data structures, allowing the indexing and efficient querying of an enormous volume of XML data.

**Keywords:** indexing XML data, regular path expression, multi-dimensional data structures.

## 1 Introduction

The mark-up language, XML (*Extensible Mark-up Language*) [21], has recently been embraced as a new approach to data modeling. A *well-formed* XML document or a set of documents is an XML database and the scheme is its database schema.

An XML document is usually modelled as a graph of the nodes which correspond to XML elements and attributes. The graph is usually a tree (without `IDREF` or `IDREFS` attributes). A number of special query languages like *XPath* [23] and *XQuery* [22] have been developed to obtain specified data from an XML database. Most XML query languages are based on the XPath language. The

* Work is partially supported by Grant of GACR No. 201/03/0912.

R. Wagner, N. Revell, and G. Pernul (Eds.): DEXA 2007, LNCS 4653, pp. 1–12, 2007.

language applies *regular path expressions* (*RPEs*) for composing paths in the XML tree. This path is a sequence of steps describing how to get a set of result nodes from a set of input nodes (a set of *context nodes*). The language allows us to generate common used queries like `//books/book[author='John Smith']`, `//book//author` or `/books/book[title='The XML Book']/author`.

In recent years, many approaches to indexing XML data have been developed. One way to index XML data is to summarize all nodes with the same labelled path (e.g. *DataGuide* [17] or *APEX* [5]). We call this one the *summary index*. DataGuide is an early work which indexes only paths starting in the root node. It only supports total matching simple path queries which do not contain wildcards and descendant axes. *APEX* utilizes a data-mining algorithm retrieving *refined paths*. APEX is able to proceed unprepared queries more efficiently. However, branching and content-based queries require additional processing.

A simple way to index an XML document is to store each node as a tuple with a *document order* value [21]. We call these approaches *element-based* approaches. The proposed element-based approaches, *XPA* [7], *XISS* [15], Zhang et al. [25], can easily solve ancestor-descendant and parent-child relationships between two sets of nodes. For improved query support, XPA can utilize the multi-dimensional data structures and Dietz labeling scheme, so it can find all nodes in relation with one node using a single range query. Works proposed by Zhang et al. and *XISS* apply the inverted list for retrieving sets of ancestors and descendants according to a node name and resolve their relationship (make a *structural join*). Also, other approaches [1,10] improve the structural join since performance of query processing in element-based approaches relies on the effectiveness of the structural join. An algorithm proposed in [1] introduces *in-memory stack* and improves the structural join in that each node in joined sets is handled only once. Their algorithm has better results outside RDBMS, as well. *XR-tree* [10] is a special data structure which can even skip the nodes having no matches in an ancestor-descendant relationship. The structural join is an extensive operation because the size of the intermediate result sets can be much larger then the size of the result set in the end. The advantage of element-based approaches is that they can process branching queries, partial matching queries and content-based queries without altering the algorithm's performance.

Other approaches try to decrease the number of structural joins; this is also the goal of path-based approaches, e.g. *Blas* [3], [9], *XRel* [16], *ViST* [24]. Blas indexes suffix paths of XML documents and allows us to retrieve results for a simple path query without time-consuming joins. The XML query is split into paths with a 1 or $> 1$ length, and results for each path are joined together as in element-based approaches. A disadvantage of the Blas method is that it indexes only paths beginning with `descendant-or-self` axis (abbreviated by `//`) and containing a child axis in every location step of the path. The approach described in [9] works like the Blas method. This is a combination of the element-based and path-based methods, where path indexing is used for data set preprocessing before structural joins are applied. The best results are obtained when processing simple path queries. The ViST method converts an XML file into a sequence of pairs, where each pair contains a path in the XML tree. Therefore, the ViST method is sometimes referred to as a *sequence-based* approach. Although no join is necessary the result may contain false hits, resulting in the need for additional processing of the result set. XRel is work which applies labelled paths for decreasing the number of structural joins. In [4] authors propose a path-based approach as well, however, each path of all subtrees is stored.

Recently, a lot of works have been introduced for improving the efficiency of XML-to-relational mapping [6,14]. They use an XML-shredding [20] system which decomposes XML data into relations and processes queries in RDBMS. Relational approaches translate XML queries into SQL queries which usually apply self-joins (edge XML-shredding) or even compare unindexed values in a relation. When an XML schema is available, XML-shredding based on inlining [18] may be used for improved performance of SQL queries. However, time-consuming structural joins are still applied and the number of relations grows rapidly with the number of different XML schemas. More complex content-based or branching queries cannot be easily handled without joining many tables together and this issue limits them when processing large collections.

The main characteristics of our *multi-dimensional approach to indexing XML data* (*MDX*), are as follows: We introduce a new, easy to understand, model of XML data. The index for XML data is described based on this model. Paths in an XML tree are represented as tuples and multi-dimensional data structures are utilized for the best performance without an application of structural join. The novel formalization of the MDX approach is depicted in Section 2. In Section 3, we introduce the approach for efficient processing of RPEs for an enormous volume of XML data based on the formalization. Algorithms are based only on the range queries which retrieve result nodes in the index. In Section 4, we compare our approach with other approaches. The last Section summarises the paper contribution and depicts a future work.

## 2 Multi-dimensional Approach to Indexing XML Data

MDX was first mentioned in [11]. Now, we are depicting a formalization to be a foundation for a novel approach to processing RPE. Due to the formalization purpose, the approach is broken into three layers: model, indexing scheme and implementation scheme. Each of them is depicted.

### 2.1 A Model

An XML document $\mathbb{X}$ may be modelled as a tree $\mathbb{X}.Tree$, having nodes corresponding to elements, attributes, and *string values* (an element content and attribute value). An attribute is modelled as a child of the related element. A string value is a node (*string node*) as well and its parent is a special node to be tagged by a `PCDATA` tag or a `CDATA` tag. The node is added especially when indexing XML documents with *mixed content*. Consequently, an XML document $\mathbb{X}$ may be modelled as a set of paths from the root node to all leaf nodes (see [19]). In Figure 2, we see the $\mathbb{X}.Tree$ of XML document $\mathbb{X}$ in Figure 1.

```
<!DOCTYPE books [
  <!ELEMENT books(book)>
  <!ELEMENT book(title,author)>
  <!ATTLIST book id CDATA #REQUIRED>
  <!ELEMENT title(#PCDATA)>
  <!ELEMENT author(#PCDATA)>
]>
```

```
<?xml version="1.0" ?>
<books>
 <book id="003-04312">
   <title>The XML Book</title>
   <author>John Smyth</author>
 </book>
 <book id="045-00012">
   <title>The XQuery Book</title>
   <author>Frank Nash</author>
 </book>
</books>
```

**Fig. 1.** (a) DTD of documents containing information about books and authors. (b) Well-formed XML document valid w.r.t DTD.

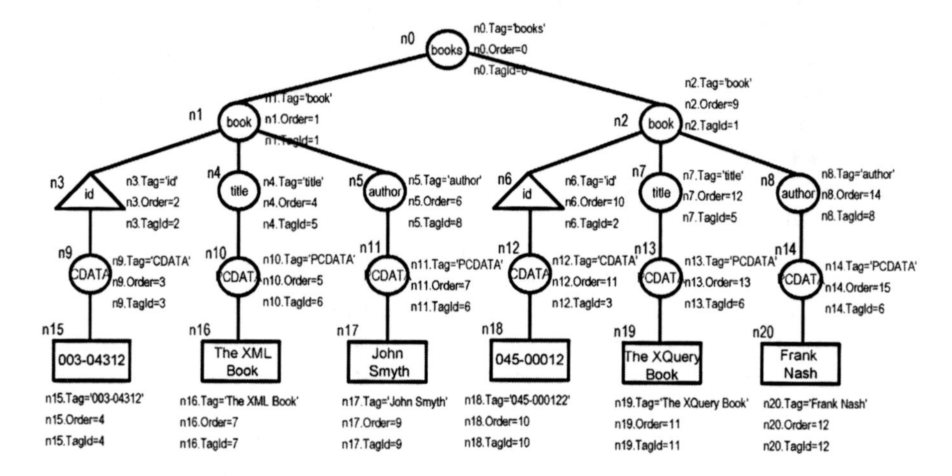


**Fig. 2.** Example of an XML tree $\mathbb{X}.Tree$ of XML document $\mathbb{X}$ in Figure 1

Let $n$ be a node in $\mathbb{X}.Tree$. The node is an object with various attributes. In Table 1, all defined attributes are shown. The unique number $n.Order$ is obtained by incrementing the counter according to the *document order*. The unique number may be generated by another numbering scheme. From an update and insert operations point of view, if we increment the counter by a number $> 1$, we create gaps for future insert and update operations.

**Definition 1 (path $p$ in $\mathbb{X}.Tree$)**
*The path $p$ is a sequence $n_0, n_1, \ldots, n_{|p|-1}, n_{|p|}$ where $|p|$ is the length of the path $p$. Let $\mathcal{P}(\mathbb{X}.Tree)$ be a set of all paths in $\mathbb{X}.Tree$. Let us note that $n_0$ is the root node and only $n_{|p|}$ can be the string node.* ∎

We can extend the node attribute to the path attribute. Path attributes are shown in Table 2. An important abstraction in our path-based approach is a differentiation of the *path* (the sequence $p.Order$) and *path type* (or *labelled path*, the sequence $p.Tag$).

**Table 1.** Attributes of a node $n$ in $\mathbb{X}.Tree$

| Node attribute | Description |
|---|---|
| $n.Tag$ | Tag of a node, or the string value if the $n$ is the string node, we insert it especially for content-based queries |
| $n.Order$ | The unique number of a node defined by a numbering scheme, if the $n$ is the string node, $n.Order = n.IdTag$ |
| $n.IdTag$ | The unique number of $n.Tag$ |

**Table 2.** Attributes of a path $p$ in $\mathbb{X}.Tree$

| Path attribute | Description |
|---|---|
| $p.Tag$ | $n_0.Tag, n_1.Tag, \ldots, n_{\vert p\vert-1}.Tag, n_{\vert p\vert}.Tag$ |
| $p.Order$ | $n_0.Order, n_1.Order, \ldots, n_{\vert p\vert-1}.Order, n_{\vert p\vert}.Order$ |
| $p.IdTag$ | $n_0.IdTag, n_1.IdTag, \ldots, n_{\vert p\vert-1}.IdTag$ |
| $p.pId$ | The unique number of the path |
| $p.lpId$ | The unique number of the $p.IdTag$ (labelled path) |

*Example 1 ($\mathcal{P}(\mathbb{X}.Tree)$ of the XML document $\mathbb{X}$ in Figure 1)*
In Figure 2, the $\mathbb{X}.Tree$ of the XML document in Figure 1 is shown. The tree includes nodes $n_0, \ldots, n_{20}$. Each of them is labelled by its attribute values, e.g. $n_0.Tag$ ='`book`', $n_0.Order = 0$ and $n_0.IdTag = 0$. Let us consider the path $p$ to the element content '`John Smyth`'. The path is the sequence: $n_0$, $n_1$, $n_5$, $n_{11}$, $n_{17}$. Each path is labelled by its attribute values. Attribute values of the path are as follows:

$p.Tag$ = '`books`','`book`','`author`','`PCDATA`','`John Smyth`'
$p.Order = 0, 1, 6, 7, 9$ $\quad p.IdTag = 0, 1, 8, 6$ $\quad p.pId = 2$ $\quad p.lpId = 2$
The number of $p.Order$ and $p.IdTag$ is 6 and 3, respectively, in this document. ∎

## 2.2 An Indexing Scheme

It is necessary to support this model with a data structure. In this approach, we generate multi-dimensional tuples and we need to query each attribute value of the tuple. Consequently, we need to index each attribute of the tuple. Multi-dimensional data structures (see the next Section) match the requirements. Now the process of converting a path to multi-dimensional points is described.

Let $\mathbb{X}.Height$ be the height of $\mathbb{X}.Tree$. Multi-dimensional points $\bar{p}.Order$ and $\bar{p}.IdTag$ are defined for each $p \in \mathcal{P}(\mathbb{X}.Tree)$. Naturally, one unique $\bar{p}.Order$ is generated for each path, but every path has one $\bar{p}.IdTag$, which is also used by many other paths. These points contain extra information: $p.lpId$ is added in the first coordinate of $\bar{p}.Order$, $docId$ (the unique number of XML collection) and $attrFlag \in \{true, false\}$ are added in the first two coordinates of $\bar{p}.IdTag$; `true` indicates the path belonging to an attribute, `false` indicates the path belonging to an element. Let us note that the $\bar{p}.IdTag$ includes $p.lpId$, however, the value is not indexed.

Spaces are defined as follows: $\Omega_{Order} = \Omega_{IdTag} = D_1 \times D_2 \times \ldots \times D_d$, where dimension $d = \mathbb{X}.Height + 2$, $D_i = 0, 1, \ldots, 2^\tau - 2$ is a domain, $\tau$ is the bit length of the domain. $\bar{p}.Order \in \Omega_{Order}, \bar{p}.IdTag \in \Omega_{IdTag}$. Let us distinguish some important items of the domain: $min(D_i)$ and $max(D_i)$ are assigned for a definition of range queries. The value $b_D = 2^\tau - 1$ is assigned as a *blank value*. Due to the fact that the dimension of all indexed points must be the same, points $\bar{p}.Order$ and $\bar{p}.IdTag$ are filled in by $b_D$ in coordinates $|p| + 2, \ldots, d$.

*Example 2 (Multi-dimensional points $\bar{p}.Order$ and $\bar{p}.IdTag$)*
Let us take the path with $p.pId = 2$ in Example 1. Multi-dimensional points follow: $\bar{p}.Order = (2, 0, 1, 6, 7, 9)$ $\bar{p}.IdTag = (0, false, 0, 1, 8, 6)$ ■

### 2.3 An Implementation Scheme

We have defined three indexes: *Index.Order*, *Index.IdTag* and *Index.Term* including vectors $\bar{p}.Order$, $\bar{p}.IdTag$, and couples $\langle n.IdTag, n.Tag \rangle$, respectively. Due to the fact that each path must be indexed for its efficient retrieval, it is suitable to apply multi-dimensional data structures for indexing all tuples $\bar{p}.Order$ and $\bar{p}.IdTag$. The attribute *Size* of indices contains the number of stored objects.

We apply paged and balanced multi-dimensional data structures like (B)UB-tree [2] and R-tree [8]. Vectors of different dimensionalities are indexed by a multi-dimensional forest [12]. The range queries processed in MDX are called *narrow range queries*. In [13], we proposed a novel multi-dimensional data structure for efficient query processing.

## 3 Efficient Processing of Regular Path Expressions

### 3.1 A Model

An RPE query $\mathbb{Q}$ may be modelled as a tree $\mathbb{Q}.Tree$, whose nodes correspond to element and attribute names, and string values (an element content and attribute value). Let us note that the attribute is marked by the prefix `@` in a query. It is obvious the tree contains two types of edges. These edges mean `/` and `//` axes, respectively. We can define attributes $n^Q.Tag$ and $n^Q.IdTag$ of a node $n^Q \in \mathbb{Q}.Tree$, as well.

Consequently, the $\mathbb{Q}.Tree$ is modelled as a set of paths as well as the XML tree. The path $p^Q$ in $\mathbb{Q}.Tree$ is a sequence $n_0^Q a_0, n_1^Q a_1 \ldots a_{|p^Q|-1}, n_{|p^Q|-1}^Q, a_{|p^Q|}, n_{|p^Q|}^Q$ where $|p^Q|$ is the length of the path $p^Q$, $a_i \in \{/, //\}$. Let $\mathcal{P}^Q(\mathbb{Q}.Tree)$ be a set of all paths in the $\mathbb{Q}.Tree$. Let us note that only $n_{|p^Q|}^Q.Tag$ can correspond to the string value and $n_0^Q$ and $n_{|p^Q|}^Q$ can be $\epsilon$, since the path can be started or finished by an axis. Obviously, we can define attributes of $p^Q$: $p^Q.Tag$, $p^Q.IdTag$ and $p^Q.Axes$.

### 3.2 Processing a Simple Path Query

The query evaluation is implemented in two steps. In the first step, we are searching for $\bar{p}.IdTag$ in $Index.IdTag$. In the second, we are searching for $\bar{r}.Order$ in $Index.Order$ such $r.lpId = p.lpId$.

$p.IdTag$ is searched by a sequence of range queries in $Index.IdTag$. Naturally, if the query contains no // axes then the query is the point query and the number of queries is exactly 1. Otherwise, we must filter the output of the algorithm. The number of unique $p.IdTag$ in real XML collections is rather low, e.g. 462 in *XMark*[1] or 110 in *Protein Sequence DB*[2], although such collections contain millions of paths. Consequently, the complexity of searching $p.IdTag$ is not a fundamental problem. In Listing 1.1, an algorithm for searching in $Index.Order$ is shown.

**Listing 1.1.** Searching a set of $p.Order$ for a set of $p.IdTag$

```
// sets of p.IdTag, p.lpId and string values
Input: Tuples IdTag[]; int lpid [], stringId []; Index.Order; p^Q;

Tuple QB_l[lpid.length()], QB_h[lpid.length()]; // two query boxes
int min, max;
// go through all query boxes
for (int i = 0 ; i < lpid.length() ; i++) {
  for (int j = 0 ; j < d ; j++) { // go through coordinates
    if (j == 0)
      { min = max = lpid[i]; } // set the id of labelled path
    else if (j == IdTag[i].length())
      { min = max = stringId[i]; } // set the id of the string
    else if (j < IdTag[i].length()) // set the whole domain
      { min = min(D); max = max(D); }
    else
      { min = max = b_D;} // path is shorter - clear values
    // set query box values
    QB_l[i].setValue(j, min); QB_h[i].setValue(j, max);
}}
// process a sequence of range queries
return Index.Order.RangeQuery(QB_l, QB_h, lpid.length());
```

*Example 3 (The evaluation plan of the query* `//book[author= "John Smyth"]`*)*

1. In $Index.Term$ find $IdTag$ of terms `'book'` ($IdTag = 1$), `author` ($IdTag = 8$), `'PCDATA'` ($IdTag = 6$) and `'John Smyth'` ($IdTag = 9$).
2. In $Index.IdTag$ find $p.lpId$ of the $p.Tag = books, book, author, PCDATA$ ($p.lpId = 2$) by a sequence of range queries: $(0, false, 1, 8, 6)$, $(0, false, *, 1, 8, 6)$. The number of range queries of the sequence is 2 due to the fact the maximal length of a path is 4 and the last range query matches $\bar{p}.Tag$ for such

[1] `http://monetdb.cwi.nl/xml/`
[2] `http://www.cs.washington.edu/research/xmldatasets/`

path. We have applied the abbreviation $*$: $(0, false, *, 1, 8, 6)$ which means the range query is $(0, false, min(D), 1, 8, 6)$:$(0, false, max(D), 1, 8, 6)$.
3. In *Index.Order* find $\bar{p}.Order$ by the range query to be defined by two points $(2, min(D), min(D), min(D), min(D), 9)$ and $(2, max(D), max(D), max(D), max(D), 9)$. The result is $\bar{p}.Order = (2, 0, 1, 6, 7, 9)$, therefore, the result element $n$ is as follows: $n.Tag =$ `'book'`, $n.IdTag = 1$, $n.Order = 1$.

The complexity of the complex range query is $\sum_{i=1}^{m} O(r_i \times \log(Index.Order.Size))$ as well. $r_i$ is proportional to the result size $N_{RS}$ of the whole XML query. As long as the structural join approaches are considered, we have to retrieve $N_{N_i}$ elements in the step $i$. $\sum_{i=1}^{N_L} N_{N_i} >> N_{RS}$, where $N_L$ is the number of location steps. This issue, along with the results of our experiments, prove the path-based approach is more efficient than element based approaches. Now, other path-based approaches are considered.

*Blas* [3] indexes only paths beginning with // axis and containing the child axis in every location step of the path. Others paths are omitted. In [9], authors process value predicates by a kind of join, therefore the evaluation of a query containing string values is less efficient than in the case of MDX. Labeled path searching can also be found in other works like [16,6]. They store labelled paths as a string in some RDBMS and utilize regular path expressions for searching paths corresponding to the XPath query. Labeled paths are only applied for the reduction of a number of structural joins.

### 3.3 Processing a Branch Query

A $\mathbb{Q}$ branch query is processed as follows:

1. Get a set $P^Q(\mathbb{Q})$.
2. For each $p^Q$ find sets of $p.IdTag$ and $p.Order$.
3. Process a *path join* – the join of results (sets of $p.Order$) for two branches.
4. The result of the last path join is the result of the query.

It is important to understand the difference between the structural and path join. The structural join is processed in each location step, the path join is only processed for two branches. Consequently, for a simple path query, a lot of structural joins are processed, however no path join is processed.

## 4 Experimental Results

First, we compare element-based approaches (XPA and XISS) with our path-based approach. We show the element-based approaches are less efficient than MDX. XPA and MDX are based on multi-dimensional data structures (R-tree [8] and Signature R-tree [13], respectively). Although compared approaches are very different we can compare the same parameters (e.g. disk access cost – DAC). In our experiments[3] we test the XMark collection[4]. The collection contains one file,

[3] The experiments were performed on an Intel Pentium ®4 2.4Ghz, 1GB DDR400, using Windows XP.

[4] `http://monetdb.cwi.nl/xml/`

111 MB in size. It includes 2,082,854 elements. In Table 3, tested queries are put forward. The size of indices is approximately 150 MB in all cases.

**Table 3.** XPath queries evaluated in our experiments

| Query | XPath query | Result Size |
|---|---|---|
| $\mathbb{Q}_1$ | `/site/people/person[@id='person0']/name` | 1 |
| $\mathbb{Q}_2$ | `//open_auction//description` | 12,000 |
| $\mathbb{Q}_3$ | `/site/closed_auctions/closed_auction/annotation/description/parlist/listitem/parlist/listitem/text/ emph/keyword/` | 180 |
| $\mathbb{Q}_4$ | `/site/regions/africa/item[location = 'United States']` | 398 |
| $\mathbb{Q}_5$ | `/site//closed auction//description/parlist/listitem/parlist/listitem//emph/ keyword/` | 180 |
| $\mathbb{Q}_6$ | `/site/regions/africa/item[@id = 'item1']` | 1 |

Let us define parameters for the purpose of quality measurement:

$N_N$ - Number of nodes in the result set after evaluation of one location step
$N_U$ - Number of nodes which leads to at least one node in the next step
$\delta_{eff}$ - $\delta_{eff}$ = Result Size / $\sum_{i=1}^{N_L} N_{N_i}$, the efficiency rate for a query, where $N_L$ is the number of location steps

The value $\delta_{eff}$ argues the inefficiency of element-based approaches. Results of an evaluation of query $\mathbb{Q}_5$ for XPA follows.

| $N_N$ | $N_U$ | Time [s] | DAC | |
|---|---|---|---|---|
| 55,383 | 36,003 | 28.27 | 283,324 | $\delta_{eff} = 0.0033$ |

During query processing, 55,383 nodes are handled but the result contains only 180 elements; that is $\delta_{eff} = 0.0033$. In MDX, no insignificant nodes are held, therefore, $\delta_{eff} = 1$. In Table 4, and Figure 3, we can see a comparison of time and DAC for MDX, XPA and XISS. The time and DAC is much lower in our approach than in the case of XPA and XISS. The average time of query

**Table 4.** Comparison of XISS, XPA and MDX approaches

| Query | Method | $N_N$ | $\delta_{eff}$ | Time [s] | DAC [MB] |
|---|---|---|---|---|---|
| **Avg.** | XISS | 29,072.5 | 0.073 | 24.6 | 602.35 |
| | XPA | 29,072.5 | 0.073 | 12.5 | 251.5 |
| | MDX | – | 1 | 0.7 | 4.06 |

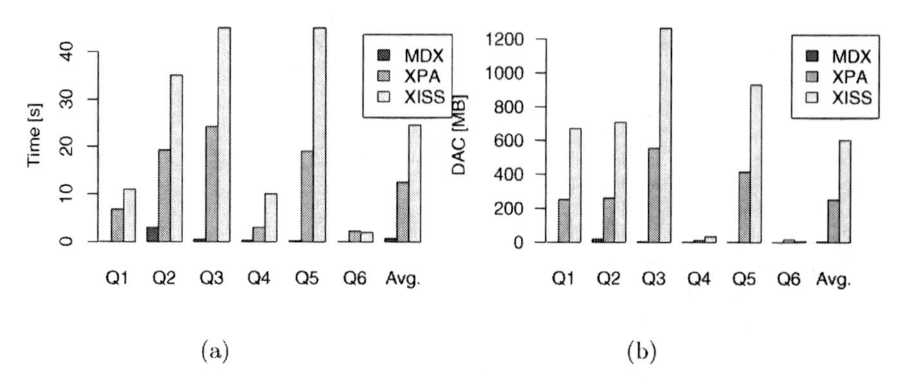


**Fig. 3.** The evaluation of queries (a) Time (b) DAC

processing is 17.9× faster than XPA and 35.1× faster than in the case of XISS. The advantage of the approach without structural joins is obvious.

We have compared our approach with other existing approaches. Query `//people/person/profile[/age=18]/education` has been evaluated for the XMark collection with a factor of 0.5, 64 MB in size. The number of accessed elements is 13,000 for the path-based approach [9]. Due to the fact that the result size is 2,336, the same number of elements was accessed in our approach. The approach is path-based as well, however, value predicates are processed by a kind of join. Therefore, the evaluation of a query containing string values is less efficient than in the case of MDX.

Since our approach applies persistent multi-dimensional data structures, we can index an enormous volume of XML data. We have generated an XMark collection with a factor of 10, 1.1 GB in size. The evaluation time is 1 s, DAC is 7.3 MB for the $\mathbb{Q}_3$ query. An index for the well-known MonetDB/XQuery[5] and eXist[6] engines was not successfully created for this XML document. MS SQL Server 2005[7] indexed the document, however the query evaluation time is 83 s. Obviously, the server applies an element-based approach.

## 5 Conclusion

In this paper, the formalization of the MDX method and efficient implementation of RPEs are described. The implementation is based on the formalization. An XML document is represented as multi-dimensional points and these points are indexed by multi-dimensional data structures. The structural join is omitted in the approach and the result proves the accuracy of the concept when compared with both element and path-based approaches. In our future work, we would like to develop an efficient algorithm for the path join to be enforced in the case of branch queries.

[5] `http://monetdb.cwi.nl/`

[6] `http://exist-db.org/`

[7] `http://www.microsoft.com/sql/`

## References

1. Al-Khalifa, S., Jagadish, H.V., Koudas, N.: Structural Joins: A Primitive for Efficient XML Query Pattern Matching. In: Proceedings of ICDE 2002, The IEEE International Conference on Data Engineering, San Jose, IEEE Computer Society Press, Los Alamitos (2002)
2. Bayer, R.: The Universal B-Tree for multidimensional indexing: General Concepts. In: Masuda, T., Tsukamoto, M., Masunaga, Y. (eds.) WWCA 1997. LNCS, vol. 1274, Springer, Heidelberg (1997)
3. Chen, Y., Davidson, S.B., Zheng, Y.: Blas: an efficient xpath processing system. In: Proceedings of the 2004 ACM SIGMOD International Conference on Management of Data, Paris, France, pp. 47–58. ACM Press, New York (2004)
4. Chen, Z., Korn, G., Koudas, F., Shanmugasundaram, N., Srivastava, J.: Index Structures for Matching XML Twigs Using Relational Query Processors. In: Proceedings of ICDE 2005, The IEEE International Conference on Data Engineering, Tokyo, Japan, pp. 1273–1273. IEEE Computer Society Press, Los Alamitos (2005)
5. Chung, C.-W., Min, J.-K., Shim, K.: Apex: an adaptive path index for xml data. In: Proceedings of the 2002 ACM SIGMOD International Conference on Management of Data, Madison, pp. 121–132. ACM Press, New York (2002)
6. Georgiadis, H., Vassalos, V.: Improving the Efficiency of XPath Execution on Relational Systems. In: Ioannidis, Y., Scholl, M.H., Schmidt, J.W., Matthes, F., Hatzopoulos, M., Boehm, K., Kemper, A., Grust, T., Boehm, C. (eds.) EDBT 2006. LNCS, vol. 3896, p. 570. Springer, Heidelberg (2006)
7. Grust, T.: Accelerating XPath Location Steps. In: Proceedings of the 2002 ACM SIGMOD International Conference on Management of Data, Madison, ACM Press, New York (2002)
8. Guttman, A.: R-Trees: A Dynamic Index Structure for Spatial Searching. In: Proceedings of the 1984 ACM SIGMOD International Conference on Management of Data, Boston, pp. 47–57. ACM Press, New York (1984)
9. Li, W.H.H., Lee, M.L.: A path-based labeling scheme for efficient structural join. In: Bressan, S., Ceri, S., Hunt, E., Ives, Z.G., Bellahsène, Z., Rys, M., Unland, R. (eds.) XSym 2005. LNCS, vol. 3671, pp. 34–48. Springer, Heidelberg (2005)
10. Jiang, H., Lu, H., Wang, W., Ooi, B.: XR-Tree: Indexing XML Data for Efficient Structural Join. In: Proceedings of ICDE 2003, The IEEE International Conference on Data Engineering, India, IEEE Computer Society Press, Los Alamitos (2003)
11. Krátký, M., Pokorný, J., Snášel, V.: Implementation of XPath Axes in the Multidimensional Approach to Indexing XML Data. In: Bertino, E., Christodoulakis, S., Plexousakis, D., Christophides, V., Koubarakis, M., Böhm, K., Ferrari, E. (eds.) EDBT 2004. LNCS, vol. 2992, Springer, Heidelberg (2004)
12. Krátký, M., Skopal, T., Snášel, V.: Multidimensional Term Indexing for Efficient Processing of Complex Queries. Kybernetika, Journal 40(3), 381–396 (2004)
13. Krátký, M., Snášel, V., Zezula, P., Pokorný, J.: Efficient Processing of Narrow Range Queries in the R-Tree. In: Proceedings of International Database Engineering & Applications Symposium, IDEAS 2006, IEEE Computer Society Press, Los Alamitos (2006)
14. Krishnamurthy, R., Kaushik, R., Naughton, J.F.: Efficient XML-to-SQL Query Translation: Where to Add the Intelligence?. In: Proceedings of the 30th International Conference on Very Large Data Bases, VLDB 2004 (2004)
15. Li, Q., Moon, B.: Indexing and Querying XML Data for Regular Path Expressions. In: Proceedings of 27th International Conference on Very Large Data Bases, VLDB 2001 (2001)

16. Shimura, T., Yoshikawa, M., Amagasa, T., Uemura, S.: Xrel: a path-based approach to storage and retrieval of xml documents using relational databases. ACM Trans. Inter. Tech. 1(1), 110–141 (2001)
17. Widom, J., Goldman, R.: DataGuides: Enabling Query Formulation and Optimization in Semistructured Databases. In: Proceedings of International Conference on Very Large Data Bases, VLDB 1997, pp. 436–445 (1997)
18. Shanmugasundaram, J., et al.: A general technique for querying XML documents using a relational database system. SIGMOD Rec. 30, 20–26 (2001)
19. Shasha, D.: Algorithmics and Applications of Tree and Graph Searching, tutorial. In: Proceedings of ACM Symposium on Principles of Database Systems, PODS 2002, ACM Press, New York (2002)
20. Tatarinov, I., et al.: Storing and querying ordered XML using a relational database system. In: Proceedings of the 2002 ACM SIGMOD International Conference on Management of Data, Madison, pp. 204–215. ACM Press, New York (2002)
21. W3 Consortium. Extensible Markup Language (XML) 1.0, W3C Recommendation, February 10, (1998), `http://www.w3.org/TR/REC-xml`
22. W3 Consortium. XQuery 1.0: An XML Query Language, W3C Working Draft (November 12, 2003), `http://www.w3.org/TR/xquery/`
23. W3 Consortium. XML Path Language (XPath) Version 2.0, W3C Working Draft (November 15, 2002), `http://www.w3.org/TR/xpath20/`
24. Wang, H., Park, S., Fan, W., Yu, P.S.: ViST: a dynamic index method for querying XML data by tree structures. In: Proceedings of the 2003 ACM SIGMOD International Conference on Management of Data, San Diego, pp. 110–121. ACM Press, New York (2003)
25. Zhang, C., Naughton, J., DeWitt, D., Luo, Q., Lohman, G.: On supporting containment queries in relational database management systems. In: Proceedings of the 2001 ACM SIGMOD International Conference on Management of Data, Santa Barbara, pp. 425–436. ACM Press, New York, USA (2001)

# Improving XML Instances Comparison with Preprocessing Algorithms*

Rodrigo Gonçalves and Ronaldo dos Santos Mello

Universidade Federal de Santa Catarina
Florianópolis, Santa Catarina, 88045-360, Brazil
{rodrigog,ronaldo}@inf.ufsc.br
http://www.inf.ufsc.br/~{rodrigog,ronaldo}/

**Abstract.** Data instances integration, specially on the web, involves analyzing and matching data from two or more sources, including XML sources. XML sources, in particular, introduce new challenges to the integration process, given their dynamic and irregular structure. In this context, one of the hardest steps is to find out which XML instances are similar. This paper presents a group of algorithms to prepare XML instances for comparison. We analyse the benefit of these algorithms over existing XML comparison approaches.

## 1 Introduction

For most areas of knowledge, there is more than one source of information about it. With the Web, these sources became even more numerous, given the fact that the Web provides a place to store and publish them with reasonable low cost, when compared to previous ways, like books, reports or centralized databases[1].

However, each source usually defines its own way to represent data. This raises the need for data integration approaches to provide global access over distributed repositories[2], integrating their data.

A typical data integration system accomplishes two main tasks: data comparison and data matching. The comparison step faces, besides the syntatic problem regarding the publishing of the data, a semantic problem, i.e., which data, among the sources, are the same[2,3].

This problem becomes harder when XML data sources are considered. XML, as a markup language for semistructured data representation[4], makes possible the definition of self-describing data instances[5]. With XML, each author in a data source may specify their own way of representing data, giving more importance to specific aspects of the data.

To solve this problem, many approaches have been proposed to compare XML instances and determine a similarity score[1,5,6,7], which can be used to deduce if the instances should be integrated. Despite of reducing the complexity of

* This work is partially supported by the DIGITEX Project of CNPq Foundation. CTInfo Process Nr.: 550.845/2005-4.

R. Wagner, N. Revell, and G. Pernul (Eds.): DEXA 2007, LNCS 4653, pp. 13–22, 2007.

determining equivalent XML data, these approaches face some problems related to differences in the structure of the XML instances being compared.

Because of these drawbacks, we propose preprocessing algorithms that execute changes in the structure of XML instances. These algorithms have the objective of reducing the structural differences between XML instances, aiding the further analysis executed by a comparison process in recognizing instances to be integrated.

This paper is organized as follows. Section 2 gives an overview of the approaches related to XML instances comparison, which aims at defining a similarity score. To deal with their limitations, we present five preprocessing algorithms in Section 3, as well as the best sequence to execute them. Some tests are presented in Section 4, showing the increase of quality in the comparison process through the application of these algorithms. Section 5 is dedicated to the conclusion.

## 2 Approaches for XML Data Similarity Definition

The research about similarity between semistructured data is based on the research about similarity between structured data. Initial efforts have established metrics to compute the similarity between data organized in tree structures[8,9]. These researches evolved to what we call today *tree edit distance*[10,11,12].

The need to compare XML documents has been motivating the research for adaptations of the similarity metrics for structured trees in order to be applied to semistructured data[13,14,15]. Some related work are presented in the following.

### 2.1 Related Work

The work of Carvalho et al. [1] identifies similar objects in a vectorial space. Four methods to establish and compare the objects are introduced. They take into consideration how the properties of the objects are represented (by a single or multiple vectors) and which properties are considered (all or a subset of them). They apply such method to XML instances comparison.

In the work of Jagadish et al. [7], the *tree edit distance* between trees representing XML instances is used as a metric to estimate the similarity. It considers some aspects of XML structures, like repeated and optional elements, and also restricts the sequence of allowed operations, reducing the cost to obtain the *edit distance* score.

Buttler [16] uses the concept of *shingles* to compare the structure of XML documents. *Shingles*[17] are parts of documents that can be extracted based on specific rules and used to compare the documents through set operations (intersection, union, etc.). It adopts the paths of elements as the *shingles* used to compare the documents.

In the approach proposed by Weis et al. [5], it is treated the problem of detecting duplicate objects in an XML document. The objective is to eliminate duplicated data in the XML document. Their solution applies an iterative *top-down* analysis of the elements hierarchy in the XML document, identifying and eliminating duplicates objects (complex elements) on each level. The similarity between the objects is calculated and the similar objects are clusterized. Each cluster originates a single object which replaces all the other objects of the cluster in the XML document.

### 2.2 Limitations of the Approaches

Several problems limitate the performance of the related work. Regarding the tags used in XML instances, they can be written in upper, lower and mixedcase. They also can be composed by two or more words (separated by a character or not separated at all). Similar words can also be used for the same concept, like synonyms in a same language or in different languages.

We also have problems related to the way XML instances are structured. A concept may be represented as a simple or as a complex element. Another issue regarding XML representation is when two complex elements do not share the same structure. In this case, we have to decide what to do about the data with heterogeneous structures. A last problem with the structure of XML instances is when their hierarchies of elements are inverted. It happens when an element $e_x$ is an ancestor of another element $e_y$ in one instance, and in other instance $e_x$ is a descendant of $e_y$. This is a reasonably complex issue and may cause a great negative impact on the similarity scores.

## 3 Preprocessing Algorithms

This section presents the set of proposed preprocessing algorithms to be previously applied to XML instances in order to improve the comparison quality of existing similarity metrics. For each algorithm, we give its definition and an example of its application.

### 3.1 Lexical Preparation

As *lexical preparation* we understand small changes in the way the names of elements (*tags*) are written. It aims at solving irrelevant differences in their writing. This algorithm iterates over each tag in the document to execute lexical modifications, in the following order:

1. Inserting the "_" character before uppercase letters if the previous letter is in lowercase. This is justified by the fact that usually uppercase letters are used to distinguish words in tags with composite words;
2. If there are "-" or other characters used as word separators, they are replaced with "_". With this, we uniformize word separation with "_";
3. All uppercase letters are changed to lowercase letter.

Table 1 exemplifies the application of this algorithm.

**Table 1.** Lexical Preparation Examples

| Original tag | Transformed tag |
|---|---|
| dateOfBirth, date-of-birth | date_of_birth |
| NAME, maritalStatus | name, marital_status |

### 3.2 Terms Uniformization

The intention here is to solve the problem of multiple terms used to represent a single concept in the structures of the XML instances. The basic idea is to use a common term for similar concepts in the XML documents, through the identification of tags for the same concept with the aid of semantic dictionaries (thesauri and terminological databases) and *string similarity* metrics[18].

The algorithm works as follows: the tags found in the documents are organized in clusters, where all the tags in a cluster either have a high *string similarity* or share the same meaning. For each cluster, a tag is elected to replace the other tags from the cluster in the documents. Table 2 exemplifies its execution.

**Table 2.** Terms Uniformization Examples

| Original tags | Unified tag |
|---|---|
| vehicle, vehice, cars, car , automobile, vehicle | automobile |
| title, caption | title |

The algorithm does not deal with the issue of choosing the best tag to represent the concept. A random tag is chosen among the set of equivalent tags. But the algorithm can be easily adapted to choose a tag based on an specific set of rules. For example, a word contained in the thesaurus or in a specific language, when dealing with documents' structures in different languages.

### 3.3 Hierarchy Reestructuring

It is possible that two XML instances differ in the way they structure hierarchies of elements. Table 3 exemplifies such a case.

Due to it, most similarity metrics may fail to detect equivalences, because the structural differences misguide the comparison. The proposed algorithm performs structural transformations on the trees of the XML documents in order to solve this problem.

First, we have to verify if such a problem occurs. For each document, we list once the tags it contains and, for each tag, we identify the elements that appear *before* and *after* it in the document's hierarchy. Then, we compare these *before/after* sets between the common tags for the two documents searching for conflicts. A conflict occurs when a *before-element* of a tag in a document appears as an *after-element* of the same tag in the other document.

**Table 3.** XML Documents with Heterogeneous Hierarchical Structures

| Document 1 | Document 2 |
|---|---|
| <collection><br><author><br><name>Paul Jason</name><br><book><br><name>Similarity guide</name><br><edition>1st</edition><br></book><br><country>England</country><br></author><br></collection> | <collection><br><book><br><name>Similarity guide</name><br><edition>1st</edition><br><author><br><name>Paul Jason</name><br><country>England</country><br></author><br></book><br></collection> |

If we identify conflicts, we execute the second stage of the algorithm, which basically moves one of the conflicting elements to a position in the hierarchy so that the conflict does not ocurr anymore. In the current version of the algorithm, the document elected to have its structure modified is randomly chosen. For the given example, we could move the `author` element of Document 2 to be the direct ancestor of the `book` element.

### 3.4 Complex to Simple Element Transformation

The same information in different XML instances may be represented in a structured, semistructured or atomic (simple) way. In this case, comparing two elements where one of them is a simple element and the other is a complex element (structured or semistructured) may compromise the similarity analysis. Table 4 illustrates such heterogeneity.

**Table 4.** XML Documents with Simple/Complex Elements Conflict

| Document 1 | Document 2 |
|---|---|
| <author><br><name><br><firstName>Paul</firstName><br><lastName>Simon</lastName><br></name><br><book><br><name>Similarity Search</name><br><press>J.J. Press</press><br></book><br></author> | <author><br><name>Paul Simon</name><br><book>Similarity Search -<br>J.J. Press</book><br></author> |

The algorithm that deals with this problem basically identifies, for each class (tag) of element, if it appears as a simple or complex element. After that, for the complex version of those elements which appear as simple and complex, it extracts the data values from the complex structure and replaces them with a simple version, containing only the extracted values. In the given example, we would transform `author` element of Document 1 to an `author` element similar to the one in Document 2. With this, we increase the chances of a similarity metric to find out an equivalence between the elements, since the structural differences do not exist anymore.

### 3.5 Complex Elements Compatibilization

Complex elements to be compared may not share the same structure, as shown in Table 5. In this case, approaches based on structural similarity may fail to find out similar instances. To solve this problem, we suggest three possible non-exclusive actions:

1. To elect uncommon subelements and put their contents into a subelement structured as a list;
2. To match uncommon subelements by comparing their contents;
3. To remove subelements which are not common between the elements. This is the approach used in this work.

**Table 5.** XML documents with uncommon elements issue

| Document 1 | Document 2 |
|---|---|
| <author><br><name><br><firstName>Paul</firstName><br><lastName>Simon</lastName><br></name><br><book><br><name>Similarity Search</name><br></book><br></author> | <author><br><name><br><firstName>Paul</firstName><br><middleName>J.</middleName><br><lastName>Simon</lastName><br><name><br><book><br><edition>1st</edition><br><name>Similarity Search</name><br></book><br></author> |

The proposed algorithm identifies, for each complex element in the instances, their direct child elements. In a second step, those child elements which have not been found in both instances are removed. This leaves the complex elements with a compatible structure, which can be better compared. In the example of Table 5, we would let `author` instance of Document 2 equal to the `author` instance of Document 1.

### 3.6 Algorithms Execution

The purpose of each preprocessing algorithm is to improve the application of similarity metrics for XML instances. However, we obtain real advantages when we execute them together in such a way that each algorithm helps the following to achieve higher quality results. With this objective in mind, we execute the algorithms in the same order they were presented in the previous sections.

The first two algorithms (*lexical preparation* and *terms uniformization*) guarantee better results to the following other ones by optimizing the names of the tags. The support of semantic dictionaries is very important at this point, because they allow the validation of terms (synonyms or even false positive terms) for string similarity metrics, and help to determine the most compatible term in a set of equivelent terms.

In the sequence, we apply *hierarchy reestructuring*, due to the fact that the remaining algorithms will fail if there are very heterogeneous hierarchies in the XML documents. Finally, we execute the *complex to simple conversion* and *complex compatibilization* algorithms. We follow such order here because the first one can accelerate the last one, by avoiding it to compare some complex elements which will be further turned into simple elements.

## 4 Algorithms Validation

In this section we present some experiments for evaluating the efficiency of the preprocessing algorithms on pair-wise instances comparison. For string comparison, we use the implementation of the Levenstein metric in the *SecondString*[1] library. For tree-related manipulation, we use the JDSL[2] library. A custom *Portuguese to English* Thesaurus was also used.

In order to evaluate how much the performance of the similarity metrics improves when comparing XML instances, we test our algorithms on XML instances that were further used as input to the metrics of Weis[5], Jagadish[7] and Carvalho[1]. We chose these metrics based on the fact they represent recent work on XML comparison and also illustrate three types of XML comparison: content-based (Weis), structure-based (Jagadish) and both structure- and content-based (Carvalho).

*Precision* and *recall* were used as measures for evaluating the algorithms quality, since they are standards in the Information Retrieval area[1].

We select 141 XML instances from the DBLP[3] database. The instances represent published articles at VLDB 2005 conference.

### 4.1 Scenarios

Four scenarios were defined to the experiments. For each scenario, one or more of the issues identified in this work were tested, considering instances coming from two different sources. The first one (*Scenario1*) supposes instances written in different languages: Portuguese and English. In *Scenario2*, we have instances that organize their data in different hierarchies. In *Scenario3*, we have instances with a very different set of attributes. Finally, in *Scenario4* we define some similar properties that are structured in one instance and textual in the other one.

---

[1] `http://secondstring.sourceforge.net/`

[2] `http://www.cs.brown.edu/cgc/jdsl/`

[3] `http://www.informatik.uni-trier.de/~ley/db/`

For each scenario, 282 instances were generated, based on the mentioned data from VLDB articles, supposing that each set of 141 instances comes from two different sources.

### 4.2 Experiments

The experiments were performed in three stages: in the first stage, no preprocessing was considered; in the second stage, only *tag-naming* algorithms were used (the first two algorithms); and in the third stage, all algorithms were applied, according to the order described in Section 3.6.

**Table 6.** Quantitative results on the selected sample from two XML sources

| Scenario | Metric | None | | Tag-naming | | All | |
|---|---|---|---|---|---|---|---|
| | | Recall | Precision | Recall | Precision | Recall | Precision |
| Scenario1 | Carvalho | 0% | 0% | 100% | 100% | 100% | 100% |
| | Weis | 0% | 0% | 100% | 100% | 100% | 100% |
| | Jagadish[4] | 6 | | 0 | | 0 | |
| Scenario2 | Carvalho | 0% | 0% | 0% | 0% | 100% | 100% |
| | Weis | 0% | 0% | 0% | 0% | 100% | 100% |
| | Jagadish | 3 | | 3 | | 0 | |
| Scenario3 | Carvalho | 100% | 100% | 100% | 100% | 100% | 100% |
| | Weis | 30% | 100% | 100% | 100% | 100% | 100% |
| | Jagadish | 2 | | 2 | | 0 | |
| Scenario4 | Carvalho | 100% | 0% [5] | 100% | 100% | 100% | 100% |
| | Weis | 100% | 100% | 100% | 100% | 100% | 100% |
| | Jagadish | 3 | | 3 | | 0 | |

Table 6 summarizes the results of the experiments. It illustrates how our preprocessing was able to improve mainly the *recall* of the selected similarity metrics in the given scenarios. As shown in Table 6, the execution of the Tag-naming algorithms already produces very good results in some cases, which denotes that they can be applied separetely.

It is also important to note that our algorithms were able to improve the *recall* without loss of *precision.* We consider these results very positive because in the context of data similarity determination for integration purposes we are interested in high *recall,* i.e., we want to retrieve the most number of true correct matchings between data. We assume for these tests a fully automated similarity analysis, where the minimum *precision* allowed was 100%.

The *tag-naming* algorithms by themselves improved the *recall,* specially for the metric of Carvalho, since it uses the tags as the basis for the comparison process. Weis metric got better results for some cases, since one of the requisites

[4] For Jagadish work, we show the edit distance values, since we consider only structured-related information.

[5] The precision was around 0,3%. Thus, we assume a 0% precision.

for two instances to be compared in their method is that the name of the elements representing the instances match.

When all algorithms are applied, all of the three metrics improved considerably the *recall*, with all algorithms improving considerably the *recall* for mosts cases compared to the situation which no preprocessing algorithms are executed.

## 5 Conclusion

In this paper we introduce some relevant work on comparison of XML instances for similarity matching purposes. We have described some major problems they face due to the dynamic and irregular nature of most XML instances, and we propose a set of preprocessing algorithms to be applied on pairs of XML documents in order to deal with these problems and improve the result quality of these work.

The first version of these algorithms were developed with a focus on XML instances and does not consider any schema analysis. Although we have many XML data with associated schemas, there are still many XML data sources where there is no an XML schema or the schema is incomplete or innacurate.

We have shown, through experiments based on related work, how the proposed algorithms increase the *recall* and *precision* of XML instances pair-wise comparison in these work. Finally, in order to maximize their application, we have elected a better sequence of execution for the algorithms.

As future work, we suggest to add the support of domain ontologies. These may help in the *terms uniformization* algorithm for detecting tags meaning the same concept and also support the *hierarchy reestructuring* algorithm. We also believe our algorithms can be used in the context of *record-linkage*[19] when no specialist is available to guide the process.

Due to the lack of paper space, we have not detailed the steps and the complexity of the algorithms. We intend to analyze them to find out lower cost approaches. For the *hierarchy reestructuring* algorithm, for example, we may decide which is the best transformation to be done in order to obtain a lower cost and/or a more suitable structure for further comparison.

Executing more than one algorithm at the same time is another topic of research. By applying them together, we may be able to reduce significantly the running time. Once we have decided on lower cost approaches to apply the algorithms, the running time will also be a topic of analysis. We didn't evaluate such aspect in this article due to the lack of space and the *proof of concept* implementation used.

The use of a *caching system* or a *catalog* to optimize comparison of instances whose structures are similar to previous compared ones is another topic of future research. This system could store either the structure of transformed instances or the rules applied to modify them. However, we must carefully analyse if we will really achieve performance improvements with such additional management of a cache/catalog.

## References

1. Carvalho, J.C.P., da Silva, A.S.: Finding similar identities among objects from multiple web sources. In: Chiang, R.H.L., Laender, A.H.F., Lim, E.-P. (eds.) WIDM, pp. 90–93. ACM Press, New York (2003)
2. Wiederhold, G.: Intelligent integration of information. In: Buneman, P., Jajodia, S. (eds.) Proceedings of the 1993 ACM SIGMOD International Conference on Management of Data, SIGMOD '93, SIGMOD Record (ACM Special Interest Group on Management of Data), Washington, May 26–28, 1993, vol. 22(2), pp. 434–437. ACM Press, New York (1993)
3. Manolescu, I., Florescu, D., Kossmann, D.K.: Answering XML queries over heterogeneous data sources. In: Proceedings of the 27th International Conference on Very Large Data Bases(VLDB '01), Orlando, pp. 241–250. Morgan Kaufmann, San Francisco (2001)
4. Consortium, W.W.W.: Extensible markup language (XML) 1.0, W3C recommendation. 2nd edn. (2000), Available at `http://www.w3.org/TR/2000/WD-xml-2e-20000814`
5. Weis, M., Naumann, F.: Detecting duplicate objects in XML documents. In: Naumann, F., Scannapieco, M. (eds.) IQIS, pp. 10–19. ACM Press, New York (2004)
6. Flesca, S., Manco, G., Masciari, E., Pontieri, L., Pugliese, A.: Fast detection of XML structural similarity. IEEE Trans. Knowl. Data Eng. 17(2), 160–175 (2005)
7. Nierman, A., Jagadish, H.V.: Evaluating structural similarity in XML documents. In: WebDB, pp. 61–66 (2002)
8. Tai, K.-C.: The tree-to-tree correction problem. J. ACM 26(3), 422–433 (1979)
9. Lu, S.-Y.: A tree-to-tree distance and its application to cluster analysis. IEEE Trans. Pattern Anal. Mach. Intell. 1(2), 219–224 (1979)
10. Shasha, D., Zhang, K.: Fast algorithms for the unit cost editing distance between trees. J. Algorithms 11(4), 581–621 (1990)
11. Wang, J.T.-L., Zhang, K., Jeong, K., Shasha, D.: A system for approximate tree matching. IEEE Trans. Knowl. Data Eng. 6(4), 559–571 (1994)
12. Shasha, D., Zhang, K.: Approximate tree pattern matching. In: Pattern Matching Algorithms, pp. 341–371. Oxford University Press, Oxford (1997)
13. Chen, J., DeWitt, D.J., Tian, F., Wang, Y.: NiagaraCQ: A scalable continuous query system for Internet databases. SIGMOD Record (ACM Special Interest Group on Management of Data) 29(2), 379–390 (2000)
14. Wang, Y., DeWitt, D.J., yi Cai, J.: X-diff: An effective change detection algorithm for XML documents. In: ICDE, pp. 519–530 (2003)
15. Marian, A., Abiteboul, S., Cobéna, G., Mignet, L.: Change-centric management of versions in an XML warehouse. In: Proceedings of the 27th International Conference on Very Large Data Bases(VLDB '01), Orlando, pp. 581–590. Morgan Kaufmann, San Francisco (2001)
16. Buttler, D.: A short survey of document structure similarity algorithms. In: International Conference on Internet Computing, pp. 3–9 (2004)
17. Broder, A.: On the resemblance and containment of documents. In: SEQS: Sequences '91 (1998)
18. Navarro, G.: A guided tour to approximate string matching. ACM Computing Surveys 33(1), 31–88 (2001)
19. Winkler, W.: The state of record linkage and current research problems (1999), `http://citeseer.ist.psu.edu/article/winkler99state.html`

# Storing Multidimensional XML Documents in Relational Databases*

N. Fousteris, M.Gergatsoulis, and Y. Stavrakas

Department of Archive and Library Sciences, Ionian University,
Palea Anaktora, Plateia Eleftherias, 49100 Corfu, Greece
{nfouster,manolis}@ionio.gr, ys@dblab.ntua.gr

**Abstract.** The problem of storing and querying XML data using relational databases has been considered a lot and many techniques have been developed. MXML is an extension of XML suitable for representing data that assume different facets, having different value and structure under different contexts, which are determined by assigning values to a number of dimensions. In this paper, we explore techniques for storing MXML documents in relational databases, based on techniques previously proposed for conventional XML documents. Essential characteristics of the proposed techniques are the capabilities a) to reconstruct the original MXML document from its relational representation and b) to express MXML context-aware queries in SQL.

## 1 Introduction

The problem of storing XML data in relational databases has been intensively investigated [4,10,11,13] during the past 10 years. The objective is to use an RDBMS in order to store and query XML data. First, a relational schema is chosen for storing the XML data, and then XML queries, produced by applications, are translated to SQL for evaluation. After the execution of SQL queries, the results are translated back to XML and returned to the application.

Multidimensional XML (MXML) is an extension of XML which allows context specifiers to qualify element and attribute values, and specify the contexts under which the document components have meaning. MXML is therefore suitable for representing data that assume different facets, having different value or structure, under different contexts. Contexts are specified by giving values to one or more user defined dimensions. In MXML, dimensions may be applied to elements and attributes (their values depend on the dimensions). An alternative solution would be to create a different XML document for every possible combination, but such an approach involves excessive duplication of information.

In this paper, we present two approaches for storing MXML in relational databases, based on XML storage approaches. We use MXML-graphs, which

* This research was partially co-funded by the European Social Fund (75%) and National Resources (25%) - Operational Program for Educational and Vocational Training (EPEAEK II) and particularly by the Research Program "PYTHAGORAS II".

R. Wagner, N. Revell, and G. Pernul (Eds.): DEXA 2007, LNCS 4653, pp. 23–33, 2007.

are graphs using appropriate types of nodes and edges, to represent MXML documents. In the first (naive) approach, a single relational table is used to store all information about the nodes and edges of the MXML-graph. Although simple, this approach presents some drawbacks, like the large number of expensive self-joins when evaluating queries. In the second approach we use several tables, each of them storing a different type of nodes of the MXML-graph. In this way the size of the tables involved in joins is reduced and consequently the efficiency of query evaluation is enhanced. Both approaches use additional tables to represent context in a way that it can be used and manipulated by SQL queries.

## 2 Preliminaries

### 2.1 Mutidimensional XML

In MXML, data assume different facets, having different value or structure, under different contexts according to a number of *dimensions* which may be applied to elements and attributes [7,8]. The notion of "world" is fundamental in MXML. A world represents an environment under which data obtain a meaning. A *world* is determined by assigning to every dimension a single value, taken from the domain of the dimension. In MXML we use syntactic constructs called *context specifiers* that specify sets of worlds by imposing constraints on the values that dimensions can take. The elements/attributes that have different facets under different contexts are called *multidimensional elements/attributes*). Each multidimensional element/attribute contains one or more facets, called *context elements/attributes*), accompanied with the corresponding context specifier which denotes the set of worlds under which this facet is the holding facet of the element/attribute. The syntax of MXML is shown in Example 1, where a MXML document containing information about a book is presented.

*Example 1.* The MXML document shown below represents a book in a book store. Two dimensions are used namely **edition** whose domain is {**greek, english**}, and **customer_type** whose domain is {**student, library**}.

```
<book isbn=[edition=english]"0-13-110362-8"[/]
           [edition=greek]"0-13-110370-9"[/]>
  <title>The C programming language</title>
  <authors>
     <author>Brian W. Kernighan</author>
     <author>Dennis M. Ritchie</author>
  </authors>
  <@publisher>
     [edition = english] <publisher>Prentice Hall</publisher>[/]
     [edition = greek] <publisher>Klidarithmos</publisher>[/]
  </@publisher>
  <@translator>
     [edition = greek] <translator>Thomas Moraitis</translator>[/]
  </@translator>
  <@price>
```

```
    [edition=english]<price>15</price>[/]
    [edition=greek,customer_type=student]<price>9</price>[/]
    [edition=greek,customer_type=library]<price>12</price>[/]
  </@price>
  <@cover>
    [edition=english]<cover><material>leather</material></cover>[/]
    [edition=greek]
       <cover>
          <material>paper</material >
          <@picture>
            [customer_type=student]<picture>student.bmp</picture>[/]
            [customer_type=library]<picture>library.bmp</picture>[/]
          </@picture>
        </cover>
     [/]
  </@cover>
</book>
```

Notice that multidimensional elements (see for example the element `price`) are the elements whose name is preceded by the symbol `@` while the corresponding context elements have the same element name but without the symbol `@`.

A MXML document can be considered as a compact representation of a set of (conventional) XML documents, each of them holding under a specific world. In Subsection 3.3 we will present a process called *reduction* which extracts XML documents from a MXML document.

### 2.2 Storing XML Data in Relational Databases

Many researchers have investigated how an RDBMS can be used to store and query XML data. Work has also been directed towards the storage of temporal extensions of XML [15,1,2]. The techniques proposed for XML storage can be divided in two categories, depending on the presence or absence of a schema:

1. *Schema-Based XML Storage techniques*: the objective here is to find a relational schema for storing an XML document, guided by the structure of a schema for that document [9,13,5,14,10,3,11].
2. *Schema-Oblivious XML Storage techniques*: the objective is to find a relational schema for storing XML documents independent of the presence or absence of a schema [13,5,14,16,10,6,4].

The approaches that we propose in this paper do not take schema information into account, and therefore belong to the Schema-Oblivious category.

## 3 Properties of MXML Documents

### 3.1 A Graphical Model for MXML

In this section we present a graphical model for MXML called *MXML-graph*. The proposed model is node-based and each node is characterized by a unique "id".

In MXML-graph, except from a special node called *root node*, there are the following node types: *multidimensional element nodes*, *context element nodes*, *multidimensional attribute nodes*, *context attribute nodes*, and *value nodes*. The *context element nodes*, *context attribute nodes*, and *value nodes* correspond to the element nodes, attribute nodes and value nodes in a conventional XML graph. Each multidimensional/context element node is labelled with the corresponding element name, while each multidimensional/context attribute node is labelled with the corresponding attribute name. As in conventional XML, value nodes are leaf nodes and carry the corresponding value. The facets (context element/attribute nodes) of a multidimensional node are connected to that node by edges labelled with context specifiers denoting the conditions under which each facet holds. These edges are called *element / attribute context edges* respectively. Context elements/attributes are connected to their child elements/attribute or value nodes by edges called *element/attribute/value edges* respectively. Finally, the context attributes of type IDREF(S) are connected to the element nodes that they point to by edges called *attribute reference edges*.

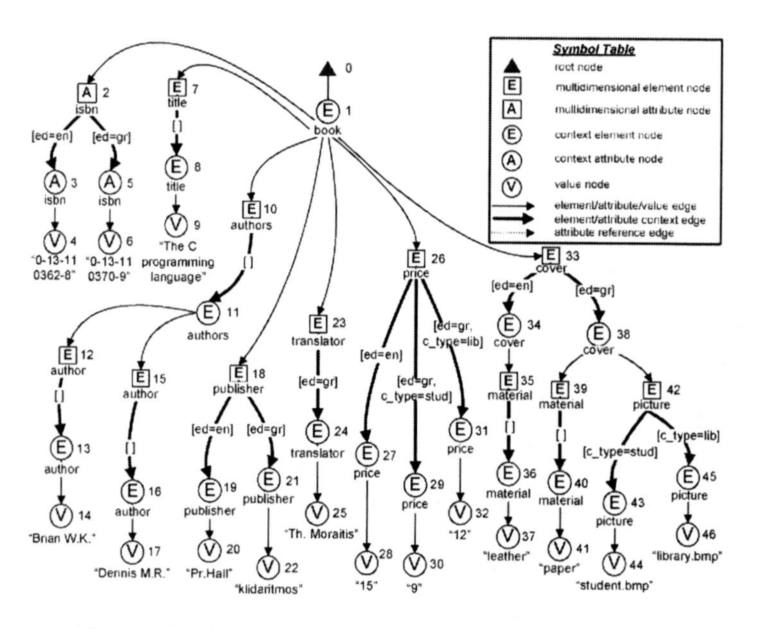


**Fig. 1.** Graphical representation of MXML (MXML tree)

*Example 2.* In Fig. 1, we see the representation of the MXML document of Example 1 as a MXML-graph. Note that some additional multidimensional nodes (e.g. nodes 7 and 10) have been added to ensure that the types of the edges alternate consistently in every path of the graph. This does not affect the information contained in the document, but facilitates the navigation in the graph and the formulation of queries. For saving space, in Fig. 1 we use obvious abbreviations for dimension names and values that appear in the MXML document.

### 3.2 Properties of Contexts

Context specifiers qualifying element/attribute context edges give the *explicit contexts* of the nodes to which these edges lead. The explicit context of all the other nodes of the MXML-graph is considered to be the *universal context* [ ], denoting the set of all possible worlds. The explicit context can be considered as the true context only within the boundaries of a single multidimensional element/attribute. When elements and attributes are combined to form a MXML document, the explicit context of each element/attribute does not alone determine the worlds under which that element/attribute holds, since when an element/attribute $e_2$ is part of another element $e_1$, then $e_2$ have substance only under the worlds that $e_1$ has substance. This can be conceived as if the context under which $e_1$ holds is inherited to $e_2$. The context propagated in that way is combined with (constraint by) the explicit context of a node to give the *inherited context* for that node. Formally, the inherited context $ic(q)$ of a node $q$ is defined as $ic(q) = ic(p) \cap^c ec(q)$, where $ic(p)$ is the inherited context of its parent node $p$. $\cap^c$ is an operator called *context intersection* defined in [12] which combines two context specifiers and computes a new context specifier which represents the intersection of the worlds specified by the original context specifiers. The evaluation of the inherited context starts from the root of the MXML-graph. By definition, the inherited context of the root of the graph is the universal context [ ]. Note that contexts are not inherited through attribute reference edges.

As in conventional XML, the leaf nodes of MXML-graphs must be value nodes. The *inherited context coverage* of a node further constraints its inherited context, so as to contain only the worlds under which the node has access to some value node. This property is important for navigation and querying, but also for the reduction process presented in the next section. The inherited context coverage $icc(n)$ of a node $n$ is defined as follows: if $n$ is a leaf node then $icc(n) = ic(n)$; otherwise $icc(n) = icc(n_1) \cup^c icc(n_2) \cup^c ... \cup^c icc(n_k)$, where $n_1, \ldots, n_k$ are the child element nodes of $n$. $\cup^c$ is an operator called *context union* defined in [12] which combines two context specifiers and computes a new one which represents the union of the worlds specified by the original context specifiers. The inherited context coverage gives the true context of a node in a MXML-graph.

### 3.3 Reduction of MXML to XML

*Reduction* is a process that given a world $w$, and a MXML document (MXML-graph) $G$, we can obtain a conventional XML document (XML-Graph) $G'$ which is the facet of $G$ under $w$. Reduction is based on the idea that we should eliminate all subtrees of $G$ for which the world $w$ does not belong to the worlds specified by the inherited context coverage of their roots. Then, we eliminate each element context edge (resp. attribute context edge) $(p, C, q)$ of the graph $G_1$ obtained from $G$ in this way, as follows: Let $(s, p)$ be the element edge (resp. attribute edge) leading to the node $p$. Then a) add a new element edge (resp. attribute edge) $(s, q)$, and b) remove the edges $(p, C, q)$ and $(s, p)$ and the node $p$.

The XML document (XML-graph) $G'$ obtained in this way is the holding facet of the MXML document $G$ under the world $w$.

# 4 Storing MXML in Relational Databases

In this section we present two approaches for storing MXML documents using relational databases.

## 4.1 Naive Approach

The first approach, called *naive approach*, uses a single table (*Node Table*), to store all information contained in a MXML document. Node Table contains all the information which is necessary to reconstruct the MXML document(graph). Each row of the table represents a MXML node. The attributes of Node Table are: `node_id` stores the id of the node, `parent_id` stores the id of the parent node, `ordinal` stores a number denoting the order of the node among its siblings, `tag` stores the label (tag) of the node or NULL (denoted by "-") if it is a value node, `value` stores the value of the node if it is a value node or NULL otherwise, `type` stores a code denoting the node type (CE for context element, CA for context attribute, ME for multidimensional element, MA for multidimensional attribute, and VN for value node), and `explicit_context` stores the explicit context of the node (as a string). Noted that the explicit context is kept here for completeness, and does not serve any retrieval purposes. In the following we will see how the correspondence of nodes to the worlds under which they hold is encoded.

*Example 3.* Fig. 2 shows how the MXML Graph of Fig. 1 is stored in the Node Table. Some of the nodes have been omitted, denoted by (...), for brevity.

| Node Table | | | | | | |
|---|---|---|---|---|---|---|
| node_id | parent_id | ordinal | tag | value | type | explicit_context |
| 1 | 0 | 1 | book | - | CE | - |
| 2 | 1 | 1 | isbn | - | MA | - |
| 3 | 2 | 1 | isbn | - | CA | [ed=en] |
| 4 | 3 | 1 | - | 0-13-110362-8 | VN | - |
| 5 | 2 | 2 | isbn | - | CA | [ed=gr] |
| 6 | 5 | 1 | - | 0-13-110370-9 | VN | - |
| 7 | 1 | 2 | title | - | ME | - |
| 8 | 7 | 1 | title | - | CE | [ ] |
| 9 | 8 | 1 | - | The C progr. lang. | VN | - |
| .... | .... | .... | .... | .... | .... | .... |
| 43 | 42 | 1 | picture | - | CE | [c_type=stud] |
| .... | .... | .... | .... | .... | .... | .... |

**Fig. 2.** Storing the MXML-graph of Fig. 1 in a Node Table

We now explain how context is stored in such a way so as to facilitate the formulation of context-aware queries. We introduce three additional tables, as shown in Fig. 3. The *Possible Worlds Table* which assigns a unique ID (attribute `word_id`) to each possible combination of dimension values. Each dimension in the MXML document has a corresponding attribute in this table. The *Explicit Context Table* keeps the correspondence of each node with the worlds represented by its explicit context. Finally, the *Inherited Coverage Table* keeps the correspondence of each node with the worlds represented by its inherited context coverage.

*Example 4.* Fig. 3, depicts (parts of) the Possible Worlds Table, the Explicit Context Table, and the Inherited Coverage Table obtained by encoding the context information appearing in the MXML-graph of Fig. 1. For example,

| Possible Worlds Table | | |
|---|---|---|
| world_id | edition | customer_type |
| 1 | gr | stud |
| 2 | gr | lib |
| 3 | en | stud |
| 4 | en | lib |

| Explicit Context Table | |
|---|---|
| node_id | world_id |
| 1 | 1 |
| 1 | 2 |
| 1 | 3 |
| 1 | 4 |
| .... | .... |
| 5 | 1 |
| 5 | 2 |
| 6 | 1 |
| 6 | 2 |
| 6 | 3 |
| 6 | 4 |
| .... | .... |

| Inherited Coverage Table | |
|---|---|
| node_id | world_id |
| 1 | 1 |
| 1 | 2 |
| 1 | 3 |
| 1 | 4 |
| .... | .... |
| 5 | 1 |
| 5 | 2 |
| 6 | 1 |
| 6 | 2 |
| .... | .... |

**Fig. 3.** Mapping MXML nodes to worlds

the inherited context coverage of the node with `node_id=6` includes the worlds { `(edition, greek)`, `(customer_type, student)`} and {`(edition, greek)`, `(customer_type, library)`}. This is encoded in the Inherited Coverage Table as two rows with `node_id=6` and the world ids 1 and 2. In the Explicit Context Table the same node corresponds to all possible worlds (ids 1, 2, 3 and 4).

Representing in this way the context information of MXML-graphs facilitates the construction of SQL queries referring to context. Moreover, it makes possible the translation of queries expressed in a language called MXPath, which is a multidimensional extension of XPath, into equivalent SQL queries. Encoding both the explicit context and the inherited context coverage as above allows us to construct queries which use both the explicit context and the inherited context coverage of nodes. As an example consider the following query given in natural language: *`Find the ISBN of the greek edition of the book with title ''The C Programming Language''`*. This query is encoded in SQL as:

```
select N4.value
from Node as N1, Node as N2, Node as N3,..., Node as N7
```

```
where N7.type="VN" and N7.value="The C Programming language" and
      N7.parent_id=N6.id and
   N6.type="CE" and N6.tag="title" and N6.parent_id=N5.id and
   N5.type="ME" and N5.tag="title" and N5.parent_id=N1.id and
   N1.type="CE" and N1.tag="book" and N1.id=N2.parent_id and
   N2.type="MA" and N2.tag="isbn" and N2.id=N3.parent_id and
   N3.type="CA" and N3.tag="isbn" and N3.id=N4.parent_id and
   N4.type="VN" and N4.id in (select IC1.node_id
     from Inherited_Coverage as IC1, Inherited_Coverage as IC2
     where IC1.world_id=1 and IC2.world_id=2 and IC1.node_id=IC2.node_id)
```

The "where" clause implements the navigation on the tree of Fig. 1, while the nested query implements the constraints related to context, in order to finally return node 6 but not node 4. Note that to make the query more readable we have named the table variables after corresponding node ids, and we have included in the query some conditions, which are redundant as they are deduced from the properties of the MXML graph. Observe that, the "greek edition" context contains both the worlds with ids 1 and 2 according to the Possible Worlds table, which has not been used in the SQL query for brevity. Finally, notice the large number of self-joins which is proportional to the depth of the navigation path.

### 4.2 Limitations of the Naive Approach

The naive approach is straightforward, but it has some drawbacks mainly because of the use of a single table. As the different types of nodes are stored in the table, many NULL values appear in the fields **explicit_context**, **tag**, and **value**. Those NULL values could be avoided if we used different tables for different node types. Moreover, as we showed in Subsection 4.1, queries on MXML documents involve a large number of self-joins of the Node Table, which is anticipated to be a very long table since it contains the whole tree. Splitting the Node Table would reduce the size of the tables involved in joins, and enhance the overall performance of queries. Finally, notice that the context representation scheme we introduced leads to a number of joins in the nested query. Probably a better scheme could be introduced that reduces the number of joins.

### 4.3 A Better Approach

In the *Type Approach* presented here, MXML nodes are divided into groups according to their type. Each group is stored in a separate table named after the type of the nodes. In particular *ME Table* stores multidimensional element nodes, *CE Table* stores context element nodes, *MA Table* stores multidimensional attribute nodes, *CA Table* stores context attribute nodes, and *Value Table* stores value nodes. The schema of these tables is shown in Fig. 4. Each row in these tables represents a MXML node. The attributes in the tables have the same meaning as the respective attributes of the Node Table. Using this approach we tackle some of the problems identified in the previous section. Namely, we eliminate NULL

| ME Table | | | |
|---|---|---|---|
| node_id | parent_id | ordinal | tag |
| 7 | 1 | 2 | title |
| 10 | 1 | 3 | authors |
| .... | .... | .... | .... |

| CE Table | | | | |
|---|---|---|---|---|
| node_id | parent_id | ordinal | tag | explicit_ context |
| 1 | 0 | 1 | book | - |
| 8 | 7 | 1 | title | [ ] |
| .... | .... | .... | .... | .... |
| 19 | 18 | 1 | publisher | [ed=en] |
| 21 | 18 | 2 | publisher | [ed=gr] |
| .... | .... | .... | .... | .... |

| MA Table | | | |
|---|---|---|---|
| node_id | parent_id | ordinal | tag |
| 2 | 1 | 1 | isbn |

| CA Table | | | | |
|---|---|---|---|---|
| node_d | parent_id | ordinal | tag | explicit_context |
| 3 | 2 | 1 | isbn | [ed=en] |
| 5 | 2 | 2 | isbn | [ed=gr] |

| Value Table | | |
|---|---|---|
| node_id | parent_id | value |
| 4 | 3 | 0-13-110362-8 |
| 6 | 5 | 0-13-110362-9 |
| 9 | 8 | The C programming language |
| .... | .... | .... |

**Fig. 4.** The Type tables

| edition | |
|---|---|
| id | value |
| 0 | * |
| 1 | en |
| 2 | gr |

| customer_type | |
|---|---|
| id | value |
| 0 | * |
| 1 | stud |
| 2 | lib |

| Inherited Coverage Table | |
|---|---|
| node_id | world_id |
| 1 | 0.0 |
| 2 | 0.0 |
| 3 | 1.0 |
| 4 | 1.0 |
| 5 | 2.0 |
| 6 | 2.0 |
| .... | .... |

| Explicit Context Table | |
|---|---|
| node_id | world_id |
| 1 | 0.0 |
| 2 | 0.0 |
| 3 | 1.0 |
| .... | .... |
| 31 | 2.2 |
| .... | .... |
| 43 | 0.1 |
| .... | .... |

**Fig. 5.** Dimension Tables

values and irrelevant attributes, while at the same time we reduce the size of the tables involved in joins when navigating the MXML-Graph.

To represent context, we propose a scheme that reduces the size of tables and the number of joins in context-driven queries. First, we use one table for each dimension (in our example `edition` and `customer_type`) to assign an id (`id` column) to each possible value (`value` column). Additionally, id "0" represents all possible values of the dimension (for id = 0 we use the value "*"). Then, we assume a fixed order of the dimension names, which will eventually be taken into account in the formulation of queries. Finally, in the Inherited Coverage and Explicit Context tables we use world ids of the form "$a_1.a_2 \ldots a_n$", where $a_1$, $a_2$, …, $a_n$ are ids of dimension values (Fig. 5). For example, the inherited context coverage of the node with id 6 in Fig. 1 is encoded as "2.0" in Fig. 5.

## 5 Discussion and Motivation for Future Work

Two techniques to store MXML documents in relational databases are presented in this paper. The first one is straightforward and uses a single table to store MXML. The second divides MXML information according to node types in the MXML-graph and, although it is more complex than the first one, it performs better during querying. We are currently working on an extension of XPath for MXML and its translation to SQL. Our plans for future work include the investigation of techniques to update MXML data stored in relational databases.

## References

1. Amagasa, T., Yoshikawa, M., Uemura, S.: A Data Model for Temporal XML Documents. In: Ibrahim, M., Küng, J., Revell, N. (eds.) DEXA 2000. LNCS, vol. 1873, pp. 334–344. Springer, Heidelberg (2000)
2. Amagasa, T., Yoshikawa, M., Uemura, S.: Realizing Temporal XML Repositories using Temporal Relational Databases. In: Proc. of the 3rd Int. Symp. on Cooperative Database Systems and Applications, Beijing, pp. 63–68 (2001)
3. Bohannon, P., Freire, J., Roy, P., Siméon, J.: From XML Schema to Relations: A Cost-Based Approach to XML Storage. In: Proc. of ICDE 2002 (2002)
4. Deutsch, A., Fernandez, M.F., Suciu, D.: Storing Semistructured Data with STORED. In: Proc. of ACM SIGMOD Int. Conf. on Management of Data, pp. 431–442. ACM Press, New York (1999)
5. Du, F., Amer-Yahia, S., Freire, J.: ShreX: Managing XML Documents in Relational Databases. In: Proc. of VLDB' 04, pp. 1297–1300. Morgan Kaufmann, San Francisco (2004)
6. Florescu, D., Kossmann, D.: Storing and Querying XML Data using an RDBMS. Bulletin of the IEEE Comp. Soc. Tech. Com. on Data Eng. 22(3), 27–34 (1999)
7. Gergatsoulis, M., Stavrakas, Y., Karteris, D.: Incorporating Dimensions in XML and DTD. In: Mayr, H.C., Lazanský, J., Quirchmayr, G., Vogel, P. (eds.) DEXA 2001. LNCS, vol. 2113, pp. 646–656. Springer, Heidelberg (2001)
8. Gergatsoulis, M., Stavrakas, Y., Karteris, D., Mouzaki, A., Sterpis, D.: A Web-based System for Handling Multidimensional Information through MXML. In: Caplinskas, A., Eder, J. (eds.) ADBIS 2001. LNCS, vol. 2151, pp. 352–365. Springer, Heidelberg (2001)
9. Ramanath, M., Freire, J., Haritsa, J.R., Roy, P.: Searching for Efficient XML-to-Relational Mappings. In: Proc. of XSym 2003, pp. 19–36. Springer, Heidelberg (2003)
10. Shanmugasundaram, J., Shekita, E.J., Kiernan, J., Krishnamurthy, R., Viglas, S., Naughton, J.F., Tatarinov, I.: A General Technique for Querying XML Documents using a Relational Database System. SIGMOD Record 30(3), 20–26 (2001)
11. Shanmugasundaram, J., Tufte, K., Zhang, C., He, G., DeWitt, D.J., Naughton, J.F.: Relational Databases for Querying XML Documents: Limitations and Opportunities. In: Proc. of VLDB'99, pp. 302–314. Morgan Kaufmann, San Francisco (1999)
12. Stavrakas, Y., Gergatsoulis, M.: Multidimensional Semistructured Data: Representing Context-Dependent Information on the Web. In: Pidduck, A.B., Mylopoulos, J., Woo, C.C., Ozsu, M.T. (eds.) CAiSE 2002. LNCS, vol. 2348, pp. 183–199. Springer, Heidelberg (2002)

13. Tatarinov, I., Viglas, S., Beyer, K.S., Shanmugasundaram, J., Shekita, E.J., Zhang, C.: Storing and querying ordered XML using a relational database system. In: Proc. of the 2002 ACM SIGMOD Int. Conf. on Management of Data, pp. 204–215. ACM Press, New York (2002)
14. Tian, F., DeWitt, D.J., Chen, J., Zhang, C.: The Design and Performance Evaluation of Alternative XML Storage Strategies. SIGMOD Record 31(1), 5–10 (2002)
15. Wang, F., Zhou, X., Zaniolo, C.: Using XML to Build Efficient Transaction-Time Temporal Database Systems on Relational Databases. In: Proc. of ICDE 2006 (2006)
16. Yoshikawa, M., Amagasa, T., Shimura, T., Uemura, S.: XRel: a path-based approach to storage and retrieval of XML documents using relational databases. ACM Transactions on Internet Technology 1(1), 110–141 (2001)

# On Constructing Semantic Decision Tables

Yan Tang and Robert Meersman

VUB STAR Lab,
Department of Computer Science,
Vrije Universiteit Brussels
Pleinlaan 2, B-1050 BRUSSEL 5, Belgium
{Yan.Tang,Robert.Meersman}@vub.ac.be

**Abstract.** Decision tables are a widely used knowledge management tool in the decision making process. Ambiguity and conceptual reasoning difficulties arise while designing large decision tables in a collaborative environment. We introduce the notion of Semantic Decision Table (SDT), which enhances a decision table with explicit *decision semantics* by annotating it properly with a *domain ontology*. In this paper, we focus on the SDT construction process. First, we map decision items to the ontology by building a rooted tree of *decision binary facts* and visualize it in a scalable manner. Formal ontological roles are used during this mapping process. Then, we *commit* the decision rules to the mapping results with a high level pseudo-natural language to ground their semantic. We illustrate with an SDT example from the domain of human resource management.

**Keywords:** semantics, decision table, ontologies, DOGMA.

## 1 Introduction

As one aspect of knowledge management, decision support systems are often used to supplement, complement or amplify the knowledge resource and/or knowledge processing capabilities of a user (or users) engaged in making decisions. Decision tables are widely used in many decision support domains, such as business decision supporting systems, software engineering or system analysis (or evaluation).

Decision tables are a simple yet important powerful tool to provide reasoning in a compact form. However, traditional decision tables diverge widely in their representation and decision semantics are often hidden. Thus, to draw a large decision table with a considerable number of conditions in general is a rather time consuming and difficult task. The situation gets naturally worse when many decision makers are involved in collaborative decision making circumstances. In this paper, we introduce Semantic Decision Tables (SDT), which are annotated properly with a domain ontology, to ground decision semantics explicitly.

We focus on the SDT construction problem in this paper. The paper is organized as follows: in Section 2, we introduce the background and motivation of SDT. We present the notion of SDT in Section 3. An SDT construction method is described in

R. Wagner, N. Revell, and G. Pernul (Eds.): DEXA 2007, LNCS 4653, pp. 34–44,2007.

Section 4. We present related work and open a discussion in Section 5. We summarize and illustrate our future work in Section 6.

## 2 Background and Motivation

We first look at a decision table example of hiring and training a driver in the problem domain of human resource management. Table 1 is a traditional decision table designed by following *de facto* international standard [3].

**Table 1.** A decision table example of hiring and training a driver[1]

| | 1 | 2 | 3 | 4 | 5 | 6 |
|---|---|---|---|---|---|---|
| **Condition** | | | | | | |
| Previous relevant occupation | Bus driver | Taxi driver | Railway driver | Bus driver | Taxi driver | Taxi driver |
| Driving experience | 2 years | 7 years | 5 years | 3 years | 1 year | 5 years |
| License type | C | C | Special | C | B | B |
| Nbr. Of accidents | 0 | 5 | 0 | 1 | 2 | 2 |
| Skill of survival | Unknown | High | Unknown | Medium | High | High |
| Speak required language(s) | Yes | No | Yes | Yes | Yes | No |
| Already hired | No | Yes | No | No | Yes | No |
| **Action** | | | | | | |
| To hire | * | | | | | |
| To hire and to train | | | * | | | |
| To train | | | | | * | |
| To fire | | * | | | | |
| To reject | | | | * | | * |

A decision *condition* is a combination of a *condition stub* and a *condition entry*. For example, a condition stub "Previous relevant occupation" together with a condition entry "Bus driver" constructs a condition as "his *previous relevant occupation* is *bus driver*" in Table 1. A decision *action* contains an *action stub* and an *action entry*. For example, an action stub "To hire" and an action entry "*" construct an action as "the action is *to hire*" in Table 1.

Traditional decision tables diverge widely in their representation. For example, the condition "his *previous relevant occupation* is *bus driver*" in Table 1 can have another representation as a condition stub "*previous relevant occupation* is *bus driver*" and a condition entry "yes". Similarly, the action "the action is *to hire*" in Table 1 might have another representation as an action stub "take an action" together with an action entry "to hire".

We observe that there is a problem of ambiguity during the phase of building a large decision table especially in the cooperative decision making process. If we don't know the background of Table 1, the lexical representation of 'driving' is ambiguous. For instance, *driving* can refer to the concepts of "drive a vehicle", "drive a golf ball" or "drive into a discourse direction when talking in a meeting". This problem gets naturally worse when multiple languages are involved.

[1] We acquired the data from the PoCeHRMOM project – an ongoing project about ontology based human resource management. http://www.starlab.vub.ac.be/website/PoCehrMOM

Moreover, we observe that decision semantics are hidden in a traditional decision table. For instance, the condition "previous relevant occupation is bus driver" and the condition "license type is C" are related because a bus driver *must at least have* a license with type C. Thus, conceptual reasoning difficulties arise when we need to check *conceptual consistency* of a decision table with a considerable number of conditions and actions.

We try to tackle the mentioned problems by bringing the notion of *Semantic Decision Table* in the next section.

## 3 Semantic Decision Table

Ontology was first introduced by T. Gruber as an *explicit specification* of a *conceptualization* [7]. N. Guarino later achieved the explicit specification by defining an ontology as a logical theory accounting for the *intended meaning* of a formal vocabulary, which indirectly reflects the *commitment* (and the underlying conceptualization) by approximating these intended models [8].

In the DOGMA (Developing Ontology-Grounded Methods and Applications, [13]) framework, one constructs (or converts) ontologies into two layers: the *lexon base* layer that contains a vocabulary of simple binary facts called *lexons*, and the *ontological commitment* layer that formally defines the axiomatized rules and constraints through which applications may make use of these lexons.

A lexon is a quintuple $< \gamma, t_1, r_1, r_2, t_2 >$ where $t_1$ and $t_2$ are terms that represent two concepts in some language. $r_1$ and $r_2$ are roles referring to the relationships between $t_1$ and $t_2$. $\gamma$ is a context identifier to disambiguate $t_1$ and $t_2$ and to make $r_1$ and $r_2$ meaningful. A lexon represents a *binary fact* that reflects the relations between two concepts within a context. E.g. the lexon *<γ*[2]*, driver, has, is issued to, drivers license>* means: a driver has drivers license(s); and a driver's license is issued to a driver.

A particular application view of reality, such as the use by the application of the (meta-) lexons in the ontology base, is described by an ontological *commitment.* The ontological commitment, which corresponds to an explicit instance of an intentional logical theory interpretation of applications, contains a set of rules in a given syntax. This describing process is also called '*to commit ontologically*'. The commitments need to be expressed in a *commitment language* that can be interpreted. We explicate a commitment by using the notion of *semantic path* that are restricted by relational database constraints.

The *semantic path* provides the *construction direction* of a lexon. For instance, if we want to apply the constraint "*one* driver has *at most one* driver's license" to the lexon *<γ, driver, has, is issued to, drivers license>*, we use the uniqueness constraint *UNIQ*[3] on the path *p1 = [γ, driver, has, is issued by, drivers license]*. The following commitment statement indicates this constraint:

$$p1 = [\gamma, \text{driver, has, is issued by, drivers license}]: \text{UNIQ (p1)}. \quad (1)$$

[2] As the discussion of context identifier is out of the scope of this paper, we use γ to indicate the context identifier of a lexon.

[3] To formulate each ontological semantic constraint, which is far too trivial, is not the paper focus. The details aspects can be found in our past work.

An SDT initially introduced in [19] is integrated into the DOGMA framework. On one hand, it keeps the traditional tabular view of decision table. On the other hand it enriches a decision table with semantically grounded decision rules.

Let us first look at the decision table definition based on [3]:

**Definition 1.** *A decision table is a triple $<C,A,R>$, where C is the set of conditions, A the set of actions, and R the set of rules. Each condition $c_i$, $c_i \in C$ is defined as $<n_i,v_i>$ where $n_i \in N$ is a condition stub label, and $v_i \in V$ is a condition entry value. Each rule $r_j \subset R$ is defined as a function $r_j$: $V^N \rightarrow A$.*

Definition 1 describes the kernel elements of a decision table – condition (e.g. 'License type' – 'C' in Table 1), action (e.g. 'to hire' – '*' in Table 1) and decision rules (e.g. the second column in Table 1). SDT is formalized as:

**Definition 2.** *An SDT is $< \Gamma, C_l, A_l, R_l >$. $\gamma \in \Gamma$ is a context identifier pointing to the original decision table and its settings. $C_l$ is the set of condition lexons and $A_l$ is the set of action lexons. Both $C_l$ and $A_l$ are defined inside the same context identifier $\gamma$. $R_l$, which is generated through a formal ontological commitment, is a rule set that includes commitment axioms $r_c \in R_c$ and semantic-grounded decision rules $r_a \in R_a$; $\{R_a \cup R_c\} \subseteq R_l$.*

An SDT follows the layered structure of DOGMA framework: the *lexon base layer* that contains decision and action lexons, the *ontological commitment layer* that carries semantically grounded decision rules.

## 4 The Construction Method

In order to construct an SDT, we need to build its lexon base and the commitment layer based on Definition 2. Our construction method thus incorporates two main steps – establishing lexon base layer and collaborating lexons with semantically grounded decision rules.

### 4.1 Establish Decision Lexons

During the process of establishing decision lexons, we try to build the connections between decision items in a decision table and concepts in a domain ontology. An item can be mapped *directly* to a concept if the ontology already contains the conceptual definition of this item. In the PoCeHRMOM project, we observe that this kind of direct mapping is rather rare in most cases. Therefore, we need a viable mapping method.

Nowadays, there are several available annotation methods as listed in [17]. Current annotation method research investigates both manual and automated annotation algorithms. The automated annotation algorithms yield certain degree of precision. In [17], authors assert that "the fully automatic creation of semantic annotations is an unsolved problem" caused by the problem of *annotation acquisition bottle neck.* This recalls the problem we mentioned earlier in the paper: not all the concept definitions can be found in the domain ontology. Therefore, we shall investigate a semi-automatic mapping method[4] that involves domain experts.

There are two main approaches to annotating unstructured information (from decision tables and the tacit knowledge of decision makers) with structured concepts (from domain ontologies). One approach is to map the *keywords* of an item to the concepts in domain ontologies. Keyword-based mapping techniques are quite common in the literatures of information retrieval [1]. However, simple keyword-based mapping techniques don't really solve the problem of meaning ambiguities.

Another approach is triggered by the idea of *concept mapping by its extended meaning/sense*, with which the authors try to tackle disambiguation problems in the linguistic domain [16]. Similarly, an idea of *concept mapping with its extension* is evoked in the domain of knowledge management. It is initially proposed to improve human's knowledge-building and understanding by mapping the concepts amongst machines and human (experts and end users) [14]. On one hand, this kind of concept mapping facilitates representational standardization without increasing the burden on users. But on the other hand, it lacks formal syntax and the semantic of the concepts and their relationships. As Eskridge et al. discussed in [4], some formal relationships, such as the "is-a" or "child-of" relationship, need to be identified.

Fortunately, we have ontologies to define such *formal relationships*. For instance, the ontological *subsumption* relation defines the formal semantics of the "is-a" relationship. As an ontology describes the semantics of concepts precisely in a domain, it can thus solve the problem of meaning ambiguities.

Based on the discussion above, we try to achieve the *"concept mapping by its extended meaning"* in ontology engineering by designing a three-step method:

- Step 1 - Finding conceptually relevant lexons in a domain ontology (automatically), where *formal concepts* and their *relations* are defined;
- Step 2 - Building connections that reflect the links between those lexons with different contexts in order to find the *candidates of concept extension* (semi-automatically);
- Step 3 - Mapping unstructured concepts to the candidates of concept extension in the domain ontology (manually).

With regards to the question of finding relevant lexons of a concept in the existing ontology, we first look at the relations between linguistics and ontologies. A formal ontology is to describe meta-level categories used to model the world in order to refer to the entities or real objects in the world [8]. According to Guarino, ontologies collect metaphysical descriptions of concepts. For example, a *cook* in English or a *chef* in French refers to the same concept – a worker in a restaurant – in the context of restaurant staff. In this sense, we argue that ontology is language independent.

[4] Most of the current annotation methods are either focused on the linguistic indexing and authoring, or on the meta-data modeling. We try to solve the mapping problem based on its fundamental characteristics.

However, there is a strong dependence between linguistics and concepts in ontologies. In [15], Qmair asserts that there are strong relations and dependences between *concepts* and their *linguistic terms*. Changing linguistic terms may affect the intended *meaning* of their concepts. The linguistic terms should not be completely excluded or ignored by the conceptualization of ontologies [9]. In addition, lexons in the lexon bases are *lexical* representation of concepts [13]. Therefore, it is important to use linguistics to express *meanings* of concepts in ontologies.

Hence, we argue that there is a distinction between the linguistic level and the ontology level; and the mapping from linguistics to concepts bridges these two levels. Step 1 is taken at the linguistic level. Step 2 and Step 3 are taken at the semantic level.

On Step 1, we adopt a *Natural Language Process* to find linguistic *roots* of user specified decision terms and find lexons that contain the *root* (or a *synonym*[5] of the *root*). A *root* is a basic lexical unit of a word. It contains the *content semantics* of a word and cannot be reduced to smaller units [12]. We take the item "driving experience" in Table 1 as an example. The term "driving experience" contains two *linguistic roots*: "drive" and "experience". We thus get selected lexons from the ontologies, such as <γ, **drivers** license, subclass of, super class of, license>, <γ, **drive**way, is a, is, roadway>, <γ, **driver**, is a, is, person> for "drive".

On step 2, we build a *rooted-tree* for each resulting lexon. The tree is populated as it has ontological relations with other lexons within its context (i.e. Fig. 1). We use ORM [10] to visualize lexons and their constraints, for ORM has excellent conceptual modeling facilities.

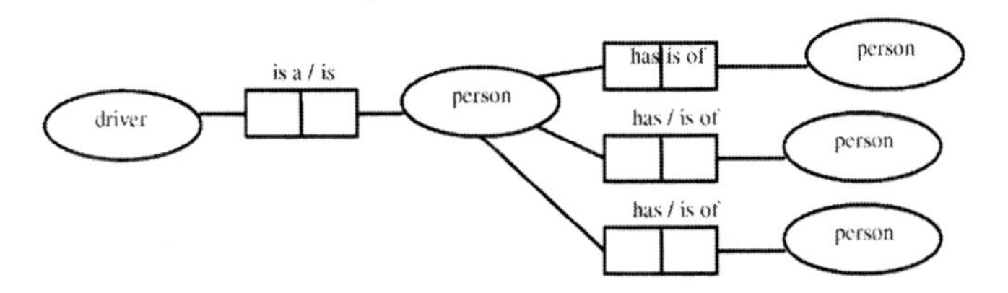


**Fig. 1.** Based on the lexon <γ, **driver**, is a, is, person>, a "driver"-rooted tree example within the context of "occupation" is built. In practice, it can be populated with more generations.

After constructing such a lexon tree, we map the decision items to a specific lexon (or lexons) with formal ontology roles to achieve the goal of constructing *formal semantics* for each mapping (Step 3). We assume that all the relevant concepts are already stored in the domain ontology. We use the *formal ontology roles* to build this mapping by using the formal ontology roles.

As discussed in [8], *mereology* (the theory of the part-whole relation) and *topology* (intended as the theory of the connection relation) are mainly used to describe formal ontology roles in general. Other formal ontology roles can be found in the literatures. We group the formal ontological roles into five categories for the annotation:

- "is-a" taxonomical roles (e.g. "subtype of" relationship);
- "part-of" merelogical roles;

[5] We use the part-of-speech tagging techniques [12] and SynSet from WordNet [5] to find synonyms of the root.

- "instance-of" roles;
- "property-of" roles;
- "equivalent" roles.

For example, we can apply the *"is-a" taxonomical* roles to describe a subtype relation between "drivers license" and "license". The "part-of" ontological roles can be applied to define a link between "person" and "leg". We use the roles in the "property-of" category to represent the "has" relation between "license holder" and "drivers license". The ontological "instance-of" roles can be used to depict the relation between "TOYOTA" and "brand". Concerning to the "equivalent" ontological roles, we take the relation between "car" and "auto" as an example.

Accordingly, we build the "is-a" *subsumption* roles (*is-a/is*) between the item "driving experience" in Table 1 and the lexon term node "professional experience" defined in the domain ontology. The "on property" relationship (*has/is of*) in the ontology role category of "property-of" is constructed between "driving experience" and "driver". Table 1 illustrates the mapping results of Table 1.

**Table 2.** Mapping result of Table 1 (trimmed)

| Decision Term | Role | Co-role | Lexon Term | Ontological relation[6] |
|---|---|---|---|---|
| previous relevant occupation | is a | is | occupation | Subtype (subtype-of) |
| previous relevant occupation | has | is of | occupation type | onProperty (property-of) |
| driving experience | is of | Has | driver | onProperty (property-of) |
| driving experience | is a | is | professional experience | Subtype (subtype-of) |
| driving experience | has | is of | value | onProperty (property-of) |
| license type | is of | Has | drivers license | onProperty (property-of) |
| number of accidents | is of | Has | accident | onProperty (property-of) |
| ... | ... | ... | ... | ... |
| to train | isA | Is | action | Instance (instance-of) |
| to fire | isA | Is | action | Instance (instance-of) |
| to reject | isA | is | action | Instance (instance-of) |

When the mapping is built, the results are stored in the domain ontology. Then, we use various means, such as ORM, Conceptual Graph and Description Logic formulas, to capture more refined meanings of the concepts in the ontologies. This topic is beyond the paper focus, we refer to our past researches for the issue of how to use ORM to model an ontology. Researches on how to establish semantic calculus to ensure the semantic equivalence will be studied in the future.

## 4.2 Construct Semantically Grounded Decision Rules

High level pseudo-natural languages are often used to elaborate basic knowledge units in the domain of knowledge engineering. As described in [11], rules in a high level pseudo-natural language are elegant, expressive, straightforward and flexible means

[6] The corresponding formal ontological role categories are indicated between parentheses.

of expressing knowledge. For example, we can illustrate a decision rule "a bus driver *must at least have* a license with type C" for Table 1 as[7]:

```
{<γ, bus driver, has, is of, drivers license>
AND <γ, drivers license, has, is of, license type>}
VALUE_RANGE (license type) = {C, D, E}.
```

The lexon <γ, bus driver, has, is of, drivers license> predefined in the domain ontology and the lexon <γ, drivers license, has, is of, license type > in Table 1 are connected with AND, which indicates the *conjunction* relationship between these two lexons. The value range of "license type" is declared by the statement: VALUE_RANGE (license type) = {C, D, E}, which means that "a license type can only be C, D or E". We thus transform the decision rule "a bus driver *must at least have* a license with type C" into "a bus driver has a drivers license AND the drivers license has a license type; the license type should be in the VALUE RANGE of C, D OR E".

This rule is further designed as a statement with the formal DOGMA commitment language (Section 3) as:

$$(P2= [\gamma, \text{bus driver, has, is of, drivers license}],\\ P3= [\gamma, \text{drivers license, has, is of, license type}])\\ : P2 \wedge P3, (\text{license type}) \in \{C,D,E\}. \quad (1)$$

By now, we have introduced a concrete construction method. A tool has been developed to assist this constructing method (Fig. 2).

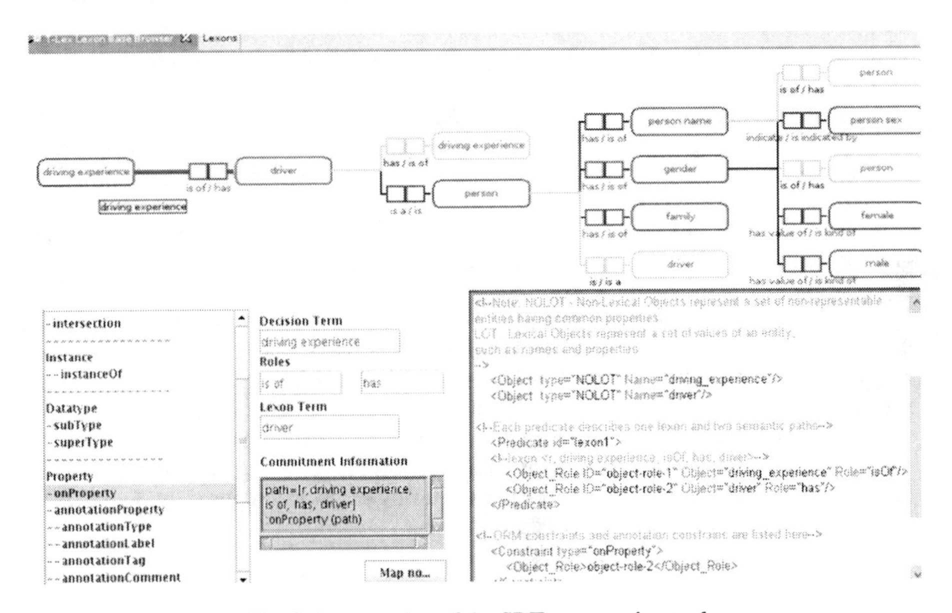


**Fig. 2.** A screenshot of the SDT construction tool

---

7 Letters in capital cases and mathematical notations are reserved as the keywords in our high level pseudo-natural language.

The decision item *"driving experience"* is mapped to the concept *"driver"* in the ontology (Fig. 2). The formal ontological role – *onProperty* - is used for the mapping. We visualize the "driving experience"-rooted ORM tree and provide semantically grounded decision rules in the Commitment Information textbox. The result is stored in an XML file for sharing. We integrate the SDT construction tool into the DOGMA Studio Workbench V1.0 developed at the VUB STAR Lab.

After constructing an SDT, we store its lexons and commitments in Prolog to reason it as we did in [19]. If we slightly modify Table 1 by adding an extra decision column that contains the information as "a candidate was a bus driver and he has a driver's license of type B". This extra column violates the decision rule "a bus driver *must at least have* a license with type C". When we check the conceptual consistency of the columns of this table with the tool in [19], we get a warning message.

## 5 Related Work

As one of the very useful and important decision objects of the decision supporting systems, decision tables have already been studied during many years. Many researchers already tried to combine semantics with decision tables in several application domains.

Goedertier et al. [6] rather focus on the rules and the semantics of the case data and activity flows in business process models, which we shall place at the application layer, i.e. as part of the use of an ontology; they do not discuss the decision table's content (conditions and decisions) and the semantics of the content. We made the semantic analysis on both the decision table itself and its content. And we expect that some of their ideas can be adopted in the future to derive a richer annotation model for our semantic decision tables.

Shiffman et al. in [18] present decision tables as a technique to reduce the complexity of using a medical knowledgebase. They demonstrate how (medical domain) semantics may be exploited ad hoc to simplify these decision tables, thereby indeed turning these tables also, in our terminology, into specific applications committed to such ontologies. In our paper we are instead interested rather in the general knowledge engineering and Knowledge Management principles underlying this process.

Colomb [2] mainly treats decision table related ontologies as "task ontologies" in a domain. They are applied in a medical application context, but his results are general, albeit that the notion of DT used in [2] is restricted compared to the decision table standards CSA Z243.1-1970. The main idea of decision objects in Colomb's paper indeed is that adoption of the correct paradigm aids expert system developers in reducing the size of the problem space. His task ontology on the other hand could be made part of (medical) decision support ontologies in our approach, i.e. SDT + medical domain knowledge + task descriptions.

## 6 Conclusion and Future Work

We introduced the *Semantic Decision Table* (SDT) to express the semantics of traditional decision tables during the decision making processes in order to tackle the problem of ambiguity and provide the conceptual reasoning facility (Section 2). An

SDT, which is annotated properly with domain ontologies, transforms a traditional decision table into semantically grounded decision rules. In this paper, we focus on the SDT construction method. A tool has been developed to assist the SDT construction and to store SDT in XML format.

We consider SDT as an instrument that supports *collaborative* decision making systems. We will focus more on the *conceptual consistency* and *validity* issues of the SDT. Moreover, typical situations of where SDT are suitable in this kind of decision making will be explored in the future.

**Acknowledgments.** It's the authors' pleasure to thank Dr. Mustafa Jarrar and Jan Demey for reviewing and discussing the paper. We shall thank all the STAR members for their comments on the paper. This research is partly supported by the IWT PoCeHRMOM project IWT-TETRA-50115.

## References

[1] Baeza-Yates, R., Ribeiro-Neto, B.: Modern information Retrieval. Addison-Wesley, London (1999)

[2] Colomb, R.M.: Representation of propositional expert systems as partial functions. Artificial Intelligence 109, 1–2 (1999)

[3] CSA, Z243.1-1970 for Decision Tables, Canadian Standards Association (1970)

[4] Eskridge, T., Hayes, P., Hoffman, R.: Formalizing the Informal: a Confluence of Concept Mapping and the Semantic Web. In: Proc. of the Second International Conference on Concept Mapping, San Jod, Costa Rica (2006)

[5] Fellbaum, C. (ed.): Wordnet, an Electronic Lexical Database. MIT Press, Cambridge (1998)

[6] Goedertier, S., Vanthienen, J.: Rule-based Business Process Modeling and Execution. In: Proceedings of the IEEE EDOC Workshop on Vocabularies Ontologies and Rules for The Enterprise (VORTE 2005). CTIT Workshop Proceeding Series, Enschede (2005) ISSN 0929-0672

[7] Gruber, T.R.: Toward Principles for the Design of Ontologies Used for Knowledge Sharing. Int. Journal of Human-Computer Studies 43, 907–928 (1995)

[8] Guarino, N.: Formal Ontology and Information Systems. In: Proceedings of FOIS'98, 98th edn., pp. 3–15. IOS press, Amsterdam (1998)

[9] Jarrar, M.: Towards the notion of gloss, and the adoption of linguistic resources in formal ontology engineering. In: Proceeding of the 15th International World Wide Web Conference, WWW2006, Edinburgh, Scotland, ACM Press, New York (2006)

[10] Halpin, T.A.: Information Modeling and Relational Databases: From Conceptual Analysis to Logical Design. Morgan Kaufman Publishers, San Francisco (2001)

[11] Hopgood, A.: Intelligent Systems for Engineers and Scientists, 2nd edn. CRC press LLC, Boca Raton (2000)

[12] Jurafsky, D., Martin, J.H.: Speech and Language Processing: An Introduction to Natural Language Processing. In: Computational Linguistics and Speech Recognition, Prentice-Hall, Englewood Cliffs (2003)

[13] Meersman, R.: Ontologies and Databases: More Than a Fleeting Resemblance. In: OES SEO 2001 RomeWorkshop, Luiss Pub. (2001)

[14] Novak, J., Gowin, D.: Learning How to Learn. Cambridge University Press, New York (1984)

[15] Qmair, Y.: Foundations of Arabic philosophy. Dar El-Machreq. Beirut (1991) ISBN 2-7214-8024-3

[16] Pustejovsky, J.: The Generative Lexicon. Journal of Computational Linguistics, pp. 409–441 (1991)

[17] Reeve, L., Han, H.: Survey of Semantic Annotation Platforms. In: Proc. Of the 20th Annual ACM Symposium on Applied Computing, Web Technologies and Application Track, pp. 1634–1638. ACM Press, New York (2005)

[18] Shiffman, R.N., Greenes, R.A.: Rule set reduction using augmented decision table and semantic subsumption techniques: application to cholesterol guidelines. In: Proceedings of Annual Symp. Computer and Application Medical Care, pp. 339–343 (1992)

[19] Tang, Y., Meersman, R.: Towards Building Semantic Decision Table with Domain Ontologies. In: Man-chung, C. Liu, J.N.K., Cheung, R., Zhou, J. (eds.) Proceedings of International Conference of information Technology and Management (ICITM2007), ISM Press (January 2007) ISBN 988-97311-5-0

# Artificial Immune Recognition System Based Classifier Ensemble on the Different Feature Subsets for Detecting the Cardiac Disorders from SPECT Images

Kemal Polat[1], Ramazan Şekerci[2], and Salih Güneş[1]

[1] Selcuk University, Dept. of Electrical & Electronics Engineering, 42075, Konya, Turkey
{kpolat,sgunes}@selcuk.edu.tr
[2] Lange Camp 16, 47139 Duisburg, Germany
rsekerci@hotmail.de

**Abstract.** Combining outputs of multiple classifiers is one of most important techniques for improving classification accuracy. In this paper, we present a new classifier ensemble based on artificial immune recognition system (AIRS) classifier and independent component analysis (ICA) for detecting the cardiac disorders from SPECT images. Firstly, the dimension of SPECT (Single Photon Emission Computed Tomography) images dataset, which has 22 binary features, was reduced to 3, 4, and 5 features using FastICA algorithm. Three different feature subsets were obtained in this way. Secondly, the obtained feature subsets were classified by AIRS classifier and then stored the outputs obtained from AIRS classifier into the result matrix. The exact result that denote whether subject has cardiac disorder or not was obtained by averaging the outputs obtained from AIRS classifier into the result matrix. While only AIRS classifier obtained 84.96% classification accuracy with 50-50% train-test split for diagnosing the cardiac disorder from SPECT images, classifier ensemble based on AIRS and ICA fusion obtained 97.74% classification accuracy on the same conditions. The accuracy of AIRS classifier utilizing the reduced feature subsets was higher than those exploiting all the original features. These results show that the proposed ensemble method is very promising in diagnosis of the cardiac disorder from SPECT images.

## 1 Introduction

In this paper, AIRS classification algorithm and new feature subsets obtained from ICA algorithm are combined to improve the classification accuracy of SPECT images dataset. The aim of this process is both to improve the classification accuracy of cardiac disorders from SPECT images dataset and to introduce a new classifier ensemble in pattern recognition field.

Aggregating outputs of multiple classifiers into a committee decision is one of the most important techniques for improving classification accuracy. A variety of schemes have been proposed for aggregating outputs of multiple classifiers. The mostly used approaches include the majority vote, averaging, weighted averaging, the

R. Wagner, N. Revell, and G. Pernul (Eds.): DEXA 2007, LNCS 4653, pp. 45–53, 2007.

Bayesian approach, the fuzzy integral, the Dempster–Shafer theory, the Borda count, aggregation through order statistics, probabilistic aggregation, the fuzzy templates, and aggregation by a neural network [1], [2]. In this work, we have used the averaging method for aggregating outputs of AIRS classifiers.

Modern medicine generates huge amounts of image data that can be analyzed and processed only with the use of specialized computer software. Since imaging techniques like SPECT, PET, and MRI can generate gigabytes of data per day. There are many advantages of computerized analysis of data over human analysis: lower price, shorter time, automatic recording of analysis results, consistency, relatively inexpensive re-use of previous solutions [3].

In literature, various studies have been considered related with diagnosing the cardiac disorders from SPECT images dataset. Ümit et al. obtained 88.24% and 93.58% classification accuracies with RBF and GRNN algorithms [4]. While Lukasz et al. achieved 84% classification accuracy using CLIP3 machine learning algorithm [3], Kurgan et al. reached to 86.1% and 90.4% classification accuracy by way of CLIP4 and CLIP4 ensemble machine learning algorithms [5].

In this study, the proposed method consists of two parts. Firstly, the dimension of SPECT images dataset, which has 22 binary features, was reduced to 3, 4, and 5 features using FastICA algorithm. In this way, three different feature subsets were obtained. Secondly, the obtained feature subsets were classified by AIRS classifier and then the outputs obtained from AIRS classifier were stored into the result matrix. The exact result that denote whether subject has cardiac disorder or not was obtained using averaging the outputs obtained from AIRS classifier into the result matrix. Whereas only AIRS classifier obtained 84.96% classification accuracy with 50-50% train-test split for diagnosing the cardiac disorder from SPECT images, classifier ensemble based on AIRS and ICA fusion obtained 97.74% classification accuracy on the same conditions.

## 2 SPECT Images Dataset

### 2.1 Acquiring of SPECT Images

SPECT imaging is used as a diagnostic tool for myocardial perfusion. The patient is injected with radioactive tracer (in our case Tl-201). Then two studies are performed, one 10-15 min. after injection during maximal stress – called stress study (stress image), and one 2-5 hours after injection – called rest study (rest image) have been performed. The studies are collected as two sets of three-dimensional images. All the images represent LV muscle perfusion that is proportional to distribution of radioactive counts within the myocardium. Cardiologists compare stress and rest studies in order to detect abnormalities in the LV perfusion [3].

There are other visualization methods used for cardiac SPECT images. One is bull's-eye method that is based on projection of 3D image of LV into 2D plane by radial projection into spherical coordinates, or into combination of spherical and cylindrical coordinates. Another family of methods is connected with 3D surface rendering of the LV; they use gated blood-pool SPECT images in order to visualize motion of the heart muscle. There are also studies that concern dynamic cardiac scenes interpretation. Cardiac motion analysis in general enables to identify pathologies related to myocardial anomalies or coronary arteries circulation deficiencies. Similar to the technique described in this paper, they use 2D LV contour images to perform quantitative and qualitative evaluation of the heart functions [3]. The SPECT images dataset is taken from UCI (University of California institute) machine learning database [6].

### 2.2 Diagnosis of Cardiac Disorders

In diagnosis of cardiac disorder, while the first stage is the diagnosing regionally for lest ventricle muscles, the second stage is to conduct the final diagnose helping from all the regional diagnosis relating the left ventricle. First of all, SPECT Image was processed by image processing algorithms in the performing of these stages. Then, some rules were made up to recognize the healthy and disorder images by cardiologists [3, 4].

The seven classes (Normal, Reversible, Partially Reversible, Artifact, Fixed, Equivocal and Reverse Redistribution) are recognized for regional diagnosis and then eight classes (Normal, Ischemia, Infarct, Infarct and Ischemia, Artifact, Equivocal, Reverse Redistribution, and the LV Dysfunction) are determined for fully diagnosis. The dataset describes diagnosing of cardiac Single Proton Emission Computed Tomography (SPECT) images. Each of the patients is classified into two categories: normal and abnormal. The database of 267 SPECT image sets (patients) was processed to extract features that summarize the original SPECT images. As a result, 44 continuous feature patterns were created for each patient. The pattern was further processed to obtain 22 binary feature patterns [3, 4, 5]. There are 55 normal (0) and 212 abnormal (1) subjects in SPECT image dataset [6].

## 3 The Proposed Method

### 3.1 Overview

The proposed classifier ensemble consists of two parts: dimensionality reduction part using ICA and classification part using AIRS classifier. The block scheme of proposed method is shown in Figure 1.

### 3.2 FastICA Algorithm: Dimensionality Reduction Process

In the FastICA algorithm, to be described below, the initial step is whitening or sphering. By a linear transformation, the measurements $x_i(k)$ and $x_j(k)$ for all i, j are made uncorrelated and unit-variance [7]. The whitening facilitates the separation of the underlying independent signals [8]. In [9], it has been shown that a well-chosen compression, during this stage, may be necessary in order to reduce the overlearning (overfitting), typical of ICA methods. The result of a poor compression choice is the productions of solutions practically zero almost everywhere, except at the point of a single spike or bump. In the studies reported in this paper, the number of important sources (both in the artifact detection and the averaged evoked response experiments)

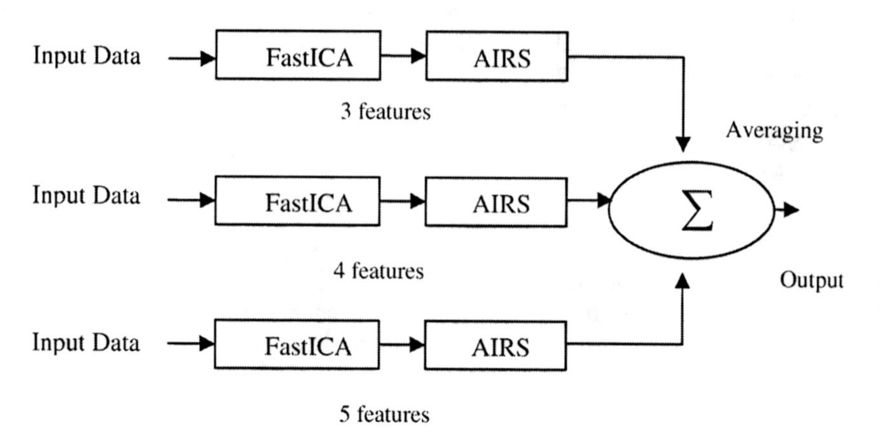


**Fig. 1.** The block scheme of proposed method

is assumed to be smaller than the total amount of sensors used, justifying such signal compression. The whitening may be accomplished by PCA projection: $v(k) = Vx(k)$ with $\{Ev(k)v(k)^T\} = I$. The whitening matrix V is given by; $\Lambda^{-1/2}U^T$ where $\Lambda = diag[\lambda_1,...,\lambda_m]$ is a diagonal matrix with the eigenvalues of the data covariance matrix $\{Ev(k)v(k)^T\}$, and U is a matrix with the corresponding eigenvectors as its columns. The transformed vectors are called white or sphere, because all directions have equal unit variance. In terms of v(k ), the model(1) becomes;

$$v(k) = VAs(k) \tag{1}$$

and we can show that matrix W=VA is orthogonal. Therefore, the solution is now sought in the form:

$$\hat{s}(k) = W^T v(k) \tag{2}$$

Uncorrelation and independence are equivalent concepts in the case of Gaussian distributed signals. PCA is therefore sufficient for finding independent components. However, Standard PCA is not suited for dealing with non-Gaussian data, where independence is a more restrictive requirement than uncorrelation. Several authors have shown [9, 10], that higher-order statistics are required to deal with the independence criterion. According to the Central Limit Theorem (CLT), the sum of independent random variables, with identical distribution functions approaches the normal distribution as m tends to infinity. We may thus replace the problem of finding the independent source signals by a suitable search for linear combinations of the mixtures that maximize a certain measure of non-Gaussianity. In FastICA, as in many other ICA algorithms, we use the fourth-order cumulant also called the kurtosis. For the th source signal, the kurtosis is defined as $kurt(s_i) = E\{s_i^{\,4}\} - 3[E\{s_i^{\,2}\}]^2$. $E\{.\}$ denotes the mathematical expectation value of the bracketed quantity. The kurtosis is negative for source signals whose amplitude has sub-Gaussian probability densities (distributions flatter than Gaussian), positive for super-Gaussian (sharper than Gaussian), and zero for Gaussian densities. Maximizing the norm of the kurtosis leads to the identification of non- Gaussian sources. Consider a linear combination $y = w^T v$ of a white random vector, with $\|w\| = 1$. Then $E\{y^2\} = 1$ and $kurt(y) = E\{y^4\} - 3$ whose gradient with respect to w is $4E\{v(w^T v)^3\}$. The FastICA is a fixed point algorithm which, maximizing the absolute value of the kurtosis, finds one of the columns of the separating matrix W (noted w) and so identifies one independent source at a time. The corresponding independent source signal can then be found using (3). Each th iteration of this algorithm is defined as;

$$w_1^* = E\{v(w_{l-1}{}^T v)^3\} - 3w_{l-1} \tag{3}$$

$$w_l = w_l^* / \|w_l^*\| \tag{4}$$

In order to estimate more than one solution and up to a maximum of, the algorithm may be run repeatedly. It is, nevertheless, necessary to" remove the information contained in the solutions already found, to estimate a different independent component each time.

### 3.3 Artificial Immune Recognition System: Classification Process

AIRS is a resource limited supervised learning algorithm inspired from immune metaphors. In this algorithm, the used immune mechanisms are resource competition, clonal selection, affinity maturation and memory cell formation. The feature vectors presented for training and test are named as Antigens while the system units are called as B cells. Similar B cells are represented with Artificial Recognition Balls (ARBs) and these ARBs compete with each other for a fixed resource number. This provides ARBs, which have higher affinities to the training Antigen to improve. The memory cells formed after the whole training Antigens were presented are used to classify test Antigens. The algorithm is composed of four main stages, which are initialization, memory cell identification and ARB generation, competition for resources and development of a candidate memory cell, and memory cell introduction. We give the details of our algorithm below.

**1. Initialization:** Create a set of cells called the memory pool (M) and the ARB pool (P) from randomly selected training data.

**2. Antigenic Presentation:** for each antigenic pattern do:

(a) Clonal Expansion: For each element of M, determine its affinity to the antigenic pattern, which resides in the same class. Select the highest affinity memory cell (mc) and clone mc in proportion to its antigenic affinity to add to the set of ARBs (P).

(b) Affinity Maturation: Mutate each ARB descendant of the highest affinity mc. Place each mutated ARB into P.

(c) Metadynamics of ARBs: Process each ARB using the resource allocation mechanism. This process will result in some ARB death, and ultimately controls the population. Calculate the average stimulation for each ARB, and check for termination condition.
(d) Clonal Expansion and Affinity Maturation: Clone and mutate the randomly selected subset of the ARBs left in P based on their stimulation level.
(e) Cycle: While the average stimulation value of each ARB class group is less than a given stimulation threshold go to step 2.c.
(f) Metadynamics of Memory Cells: Select the highest affinity ARB of the same class as the antigen from the last antigenic interaction. If the affinity of this ARB with the antigenic pattern is better than that of the previously identified best memory cell mc then add the candidate (mc-candidate) to memory set M. If the affinity of mc and mc-candidate are below the affinity threshold, remove mc from M.

**3. Classify:** Classify data items using the memory set M. Classification is performed in a k-Nearest Neighbor fashion with a vote being made among the k closest memory cells to the given data item being classified.

These steps are repeated for each training antigen. After training, test data are presented only to memory cells. k-NN algorithm is used to determine the classes in test phase. For more detailed information about AIRS, the reader is referred to [11], [12].

## 4 The Empirical Results

In this section, we used the performance evaluation methods including classification accuracy, sensitivity and specifity analysis, and ROC curves to evaluate the proposed method.

### 4.1 Results and Discussion

In this paper, we described a new classifier ensemble based on combining ICA and AIRS classifier for diagnosing the cardiac disorders from SPECT images dataset. The proposed classifier ensemble consists of two parts. In the first stage, the dimension of SPECT images dataset that has 22 binary features was reduced to 3, 4, and 5 features using FastICA algorithm and three different feature subsets were obtained. Secondly, the obtained feature subsets were classified by AIRS classifier and then the outputs obtained from AIRS classifier were stored into the result matrix. The exact result that denote whether subject has cardiac disorder or not was obtained by averaging the outputs obtained from AIRS classifier into the result matrix. While only AIRS classifier obtained 84.96% classification accuracy with 50-50% train-test split for diagnosing the cardiac disorder from SPECT images, classifier ensemble based on AIRS and ICA fusion obtained 97.74% classification accuracy on the same conditions.

The obtained classification accuracies, number of features, and sensitivity and specifity values of AIRS classifier, combining ICA (3 features) and AIRS, combining ICA (4 features) and AIRS, combining ICA (5 features) and AIRS, and proposed method based on averaging of outputs obtained from AIRS classifier are shown in Table 1.

In classification task, we also used ROC curves to compare used methods. ROC curves for AIRS classifier, combining ICA (3 features) and AIRS, combining ICA (4 features) and AIRS, combining ICA (5 features) and AIRS, and proposed method based on averaging of outputs obtained from AIRS classifier for SPECT images data-set are shown in Figure 2. The bigger area means that we have better classifier than the other one, which has smaller area. This curve figure has shown that ROC curve of proposed method is bigger than those of AIRS and the other methods.

We compared our results with previously reported methods. Table 2 gives the classification accuracies of AIRS classifier, combining ICA (3 features) and AIRS, combining ICA (4 features) and AIRS, combining ICA (5 features) and AIRS,

**Table 1.** The obtained classification accuracies, sensitivity and specifity values for AIRS, and combining ICA and AIRS, and proposed method with 50-50% train-test split

| Method Used | Number of Features | Classification Accuracy (%) | Sensitivity (%) | Specificity (%) |
|---|---|---|---|---|
| AIRS | 22 | 84.96 | 88.39 | 66.67 |
| ICA + AIRS | 3 | 95.49 | 96.29 | 92.00 |
| ICA + AIRS | 4 | 96.99 | 99.03 | 89.65 |
| ICA + AIRS | 5 | 89.47 | 95.09 | 70.96 |
| **Averaging of outputs obtained from AIRS classifier** | - | **97.74** | **99.04** | **92.85** |

**Table 2.** Proposed method's classification accuracy for diagnosing the cardiac disorders from SPECT images dataset with classification accuracies obtained by other methods in literature

| Author | Method | Classification Accuracy (%) |
|---|---|---|
| Lukasz et al. (2001) [3] | CLIP3 | 84.00 |
| Kurgan et al. (2002) [5] | CLIP4 | 86.10 |
| Kurgan et al. (2002) [5] | CLIP4 Ensemble | 90.40 |
| Ümit et al. (2004) [4] | RBF | 88.24 |
| Ümit et al. (2004) [4] | GRNN | 93.58 |
| Our Study (2007) | AIRS | 84.96 |
| Our Study (2007) | ICA + AIRS | 95.49 |
| Our Study (2007) | ICA + AIRS | 96.99 |
| Our Study (2007) | ICA + AIRS | 89.47 |
| Our Study (2007) | **Averaging of outputs obtained from AIRS classifier** | **97.74** |

proposed method based on averaging of outputs obtained from AIRS classifier, and previous methods for diagnosing the cardiac disorders from SPECT images dataset.

The accuracy of AIRS classifier utilizing the reduced feature subsets was higher than those exploiting all the original features. These results show that the proposed ensemble method is very promising in diagnosis of the cardiac disorder from SPECT images. As can be seen from Table 2, the proposed method obtained the highest classification accuracy among earlier methods and other methods used in this study.

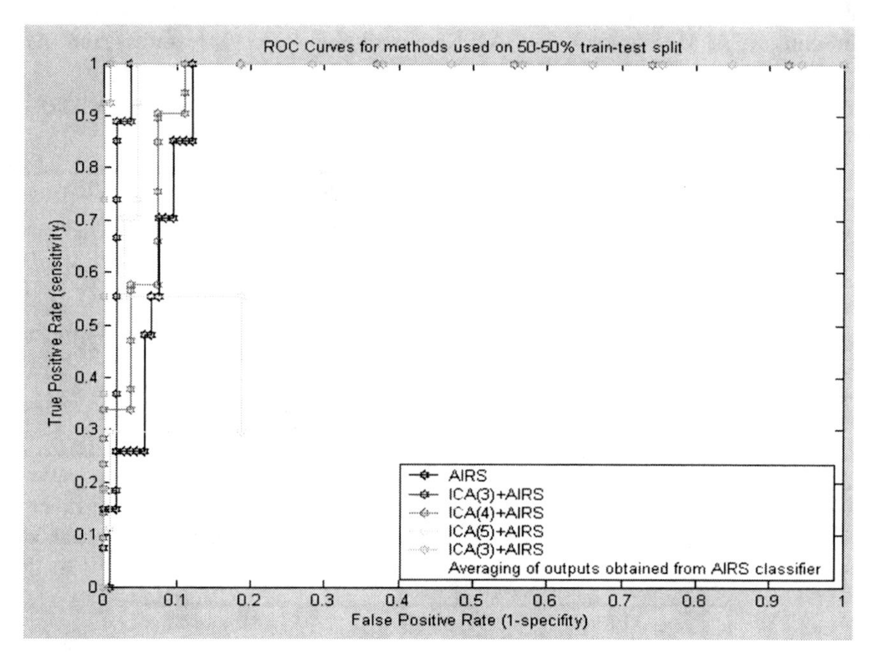


**Fig. 2.** The ROC curves of AIRS classifier, combining ICA (3 features) and AIRS, combining ICA (4 features) and AIRS, combining ICA (5 features) and AIRS, and proposed method based on averaging of outputs obtained from AIRS classifier with 50-50% train-test split for SPECT images dataset

# 5 Conclusions and Future Work

In pattern recognition applications, multiple classifier systems including medical decision support systems, classification, clustering, and image recognition are used. In this paper, a new classifier fusion based on combining artificial immune recognition system and independent component analysis for detecting the cardiac disorders from SPECT image dataset, which is a common disease among public. The results strongly suggest that the proposed classifier ensemble approach can aid in the diagnosis of cardiac disorders. It is hoped that more interesting results will follow on further exploration of data.

**Acknowledgments.** This study has been supported by Scientific Research Project of Selcuk University.

## References

1. Verikas, A., Lipnickas, A., Malmqvist, K., Bacauskiene, M., Gelzinis, A.: Soft combination of neural classifiers: A comparative study. Pattern Recognition Lett. 20, 429–444 (1999)
2. Bacauskiene, M., Verikas, A.: Selecting salient features for classification based on neural network committees. Pattern Recognition Letters 25(16), 1879–1891 (2004)
3. Kurgan, L.A., Cios, K.J., Tadeusiewicz, R., Ogiela, M., Doodenday, L.S.: Knowledge discovery approach to automated cardiac SPECT diagnosis. Artifical Intelligence in Medicine 149–169 (2001)
4. Bakırcı, Ü., Yıldırım, T.: Diagnosis of Cardiac Problems From SPECT Images by Feedforward Networks, SİU 2004. In: IEEE 12. Sinyal İşleme ve İletişim Uygulamaları Kurultayı, pp. 103–105. Kuşadası (2004)
5. Kurgan, L., Cios, K.J.: Ensemble of Classifiers to Improve Accuracy of the CLIP4 Machine Learning Algorithm. In: Accepted for the SPIE's International Symposium on Sensor Fusion: Architectures, Algorithms, and Applications VI (2002)
6. UCI Machine Learning Repository, last arrived (February 2007), http://www.ics.uci.edu/~mlearn/MLRepository.html
7. Hyvärinen, A., Oja, E.: A fast fixed-point algorithm for independent component analysis. Neural Computation 9, 1483–1492 (1997)
8. Karhunen, J., Oja, E., Wang, L., Vigário, R., Joutsensalo, J.: A class of neural networks for independent component analysis. IEEE Trans. Neural Networks 8(3), 486–504 (1997)
9. Hyvärinen, A., Särelä, J., Vigário, R.: Spikes and bumps: Artefacts generated by independent component analysis with insufficient sample size. In: presented at the Int. Workshop on Independent Component Analysis and Blind Separation of Signals (ICA'99), Aussois, France (1999)
10. Bell, A., Sejnowski, T.: An information-maximization approach to blind separation and blind deconvolution. Neural Computation 7, 1129–1159 (1995)
11. Watkins, A.B.: Exploiting Immunological Metaphors in the Development of Serial, Parallel, and Distributed Learning Algorithms. PhD dissertation, University of Kent, Canterbury (March 2005)
12. Polat, K., Güneş, S., Tosun, S.: Diagnosis of heart disease using artificial immune recognition system and fuzzy weighted pre-processing. Pattern Recognition 39(11), 2186–2193 (2006)

# A Multisource Context-Dependent Semantic Distance Between Concepts

Ahmad El Sayed, Hakim Hacid, and Djamel Zighed

University of Lyon 2
ERIC Laboratory- 5, avenue Pierre Mendès-France
69676 Bron cedex - France
{asayed, hhacid, dzighed}@eric.univ-lyon2.fr

**Abstract.** A major lack in the existing semantic similarity methods is that no one takes into account the context or the considered domain. However, two concepts similar in one context may appear completely unrelated in another context. In this paper, our first-level approach is context-dependent. We present a new method that computes semantic similarity in taxonomies by considering the context pattern of the text corpus. In addition, since taxonomies and corpora are interesting resources and each one has its strengths and weaknesses, we propose to combine similarity methods in our second-level multi-source approach. The performed experiments showed that our approach outperforms all the existing approaches.

## 1 Introduction

Comparing two objects relevantly is still one of the biggest challenges and it now concerns a wide variety of areas in computer science, artificial intelligence and cognitive science. The end-goal is that our computational models achieve a certain degree of "intelligence" that makes them comparable to human's intentions over objects. That's obviously a hard task especially that two objects sharing any attribute(s) in common may be related by some abstract 'human-made' relation.

Beyond managing synonymy and polysemy, many applications need to measure the degree of semantic similarity between two words/concepts[1]; let's mention: Information retrieval, question answering, automatic text summarization and translation, etc. A major lack in existing semantic similarity methods is that no one takes into account the context or the considered domain. However, two concepts similar in one context may appear completely unrelated in another context. A simple example for that: While *blood* and *heart* seem to be very similar in a general context, they represent two widely separated concepts in a domain-specific context like medicine.

[1] In the following, 'words' is used when dealing with text corpora and 'concepts' is used when dealing with taxonomies where each concept contains a list of words holding certain sense.

R. Wagner, N. Revell, and G. Pernul (Eds.): DEXA 2007, LNCS 4653, pp. 54–63, 2007.

Thus, our first-level approach is context-dependent. We present a new method that computes semantic similarity in taxonomies by considering the context pattern of the text corpus. In fact, taxonomies and corpora are interesting resources to exploit. We believe that each one has its strength and weakness, but using them both simultaneously can provide semantic similarities with multiple views on words from different angles. We propose to combine both methods in our second-level multisource approach to improve the expected performances.

The rest of this paper is organized as follows: Section 2 introduces quickly some semantic similarity measures. Our contribution dealing with a context-dependent similarity measure is described in Section 3. Section 4 presents the experiments made to evaluate and validate the proposed approach. We conclude and give some future works in Section 5.

## 2 Semantic Similarity in Text

### 2.1 Knowledge-Based Measures

In this work, we focus on taxonomies since we don't need a higher level of complexity. A number of successful projects in computational linguistic have led to the development of some widely used taxonomies like Wordnet[13] (generic taxonomy) and Mesh[2] (for the medical domain). Many taxonomy-based measures for semantic similarity have made their appearance; they can be grouped into edge-based measures and node-based measures.

- **Edge-based Measures.** First, calculating similarities simply relied on counting the number of edges separating two nodes by an *'is-a'* relation [15]. Since specific concepts may appear more similar than abstract ones, the depth was taken into account by calculating either the maximum depth in the taxonomy [10] or the depth of the most specific concept subsuming the two compared concepts [19]. Hirst [8] considers that two concepts are semantically similar if they are connected by a path that is not too long and that does not change direction too often.
- **Node-based Measures.** This approach came to overcome the unreliability of edge distances and based its similarities on the information associated with each node. This information can be either a node description (Feature based measures) or a numerical value augmented from a text corpus (Information content measures).

**Feature based measures.** In this category, we can cite the measure of Tversky [18] which assumes that the more common characteristics two objects have and the less non common characteristics they have, the more similar the objects are:

$$sim(c1, c2) = \frac{|C1 \bigcap C2|}{|C1 \bigcap C2| + k\,|C1/C2| + (k-1)\,|C2/C1|} \quad (1)$$

[2] http://www.nlm.nih.gov/mesh/

**Information Content measures.** The Information content (IC) approach was first proposed by Resnik[16]. It considers that the similarity between two concepts is "the extent to which they share information in common". Therefore, an IC value, based on a concept frequency in a text corpus, is assigned to each node in the taxonomy. IC represents the amount of information that a concept holds. After assigning IC values for each concept, Resnik defines the similarity between two concepts as the IC value of their Most Informative Subsumer (MIS).

Jiang[9] proposed next a model derived from the edge-based notion by adding the information content as a decision factor. Jiang assigns a link strength (LS) for each *"is-a"* link in the taxonomy which is simply the difference between the IC values of two nodes. Jiang has reached a success rate of 84.4% which led it to outperform all the other taxonomy-based measures. A similar measure to Jiang was proposed by Lin[11] which doesn't only consider the IC of the most informative subsumer but the IC of the compared concepts too.

### 2.2 Corpus-Based Measures

Semantic similarity measures can also be derived by applying statistical analysis on large corpora and by using Natural language Processing (NLP) techniques. The advantage is that corpus driven measures are self-independent; they don't need any external knowledge resources, which can overcome the coverage problem in taxonomies. Three main directions have been pursued in this category of approaches:

- **Co-occurrence-Based Similarity.** This approach study the words co-occurrence or closeness in texts with the assumption that frequent words pairs reveal the existence of some dependence between these words. The first measure was introduced to computational linguistics by Church [4] as the Mutual Information (MI). Among the works we can quote here are those of Turney [17] who showed that Pointwise Mutual Information (PMI) computed on a very large corpus (the web) and using a medium-sized co-occurrence window can be efficiently used to find synonyms. Turney applied the PMI method to TOEFL synonym match problem and obtained an impressive success rate of 72.5% which exceeds by 8% the average foreign student making the test.
- **Context-based Similarity.** This approach is based on the intuition that similar words will tend to occur in similar contexts [3]. Vector-space model is used here as a semantic measuring device. Hindle's approach [7] considers lexical relationship between a verb and the head nouns of its subject and object. Nouns are then grouped according to the extent to which they appear in similar environments. Dagan [5] propose the $L_1$ norm measure to overcome the zero-frequency problems of bigrams. Turney [17] also proposed an extension of PMI which is an application of PMI on multiple words contexts.
- **Latent Semantic Analysis (LSA).** LSA introduced by [6] came to overcome the high-dimensionality problem of the standard vector space model

especially for the context-based methods. First text is represented as a matrix rows stand for words and columns stands for contexts and each cell contains some specified weight (frequency for instance). Next, Singular Value Decomposition (SVD) is applied to the matrix in order to analyze the statistical relationships among words in a collection of text. In SVD, a rectangular matrix is decomposed into the product of three other matrices and then recomposed to a single compressed bidimensional matrix. Finally, LSA similarity is computed in a lower dimensional space, in which second-order relations among terms and texts are exploited. The similarity of two words is measured by the cosine of the angle between their corresponding compressed row vectors.

## 3 A Context-Dependent Similarity Measure

Context definition varies from one research area to another. Considering context is motivated by the fact that it can bring additional information to the reasoning process. Similarity judgments are made with respect to representations of entities, not with respect to entities themselves [12]. Thus, having a changeable representation, one can make any two items similar according to some criteria. To prevent this, a context may be used in order to focus the similarity assessment on certain features of the representation excluding irrelevant information. Barsalou[2] presents a nice example supporting the context-dependency and explaining the instability of similarity judgments.

In text, comparing two concepts doesn't make any sense if we ignore the actual context. Let's take the example of *heart* and *blood*. In a general context, these two concepts can be judged to be very similar. However, if we put ourselves in a medical context, *heart* and *blood* define two largely separated concepts. They will be even more distant if the context is more specifically related to *cardiology*. Our context-dependent approach suggest to adapt semantic similarities to the target corpus since it's the entity representing the context or the domain of interest in most text-based applications. This method is inspired by the information content theory [16] and by the Jiang[9] measure described above.

### 3.1 Problems with Information Content Measures

As we stated before, IC measures are mainly based on the concept frequency in a text corpus. According to the measure's purpose, we can show two main limitations for that approach:

- *On concept informativeness.* We believe that it's inaccurate to consider infrequent concepts as more informative than frequent ones. We argue that concept frequency is not a good decisive factor for concept informativeness. We follow Nuno's point of view [1] assuming that the taxonomic structure in WordNet is organized in an enough meaningful way to measure IC. We can simply say that the more hypernyms a concept has the more information

it expresses. Nuno have shown that at least similar results can be obtained without using a text corpus.
- *On context-dependency.* If the motivation behind measuring the IC from a text corpus is to consider the actual context, we argue that the probability of encountering a concept in a corpus is not a sufficiently adaptive measure to determine whether it's representative for a given context. Thus, IC cannot meaningfully reflect the target context.

### 3.2 A New Context-Dependency Based Measure

Our approach tends to compute semantic similarities by taking into account the target context from a given text corpus. In order to represent the context, we assign weights for taxonomy's concepts according to their Context-Dependency $CD$ to a corpus $C$. The goal is to obtain a weighted taxonomy, where 'heavier' subtrees are more context representative than 'lighter' subtrees. This will allow us to calculate semantic similarities by considering the actual context. Therefore, lower similarity values will be obtained in 'heavy' subtrees than 'light' subtrees. Thus, in our *heart/blood* example, we tend to give a high similarity for the concept couple in a general context, and a low similarity in a specific context like medicine.

As we said earlier, it's clear that a concept's frequency alone is not enough to determine its context-dependency. A concept very frequent in some few documents and absent in many others cannot be considered to be "well" representative for the corpus. Thus, the number of documents where the concept occurs is another important factor that must be considered. In addition to that, it's most likely that a concept $c_1$ -with a heterogeneous distribution among documents - is more discriminative than a concept $c_2$ with a monotone repartition which can reveal less power of discrimination over the target domain (Experimentations made assess our thesis).

Consequently, we introduce our $CD$ measure which is an adapted version of the standard $tf - idf$. Given a concept $c$ , $CD(c)$ is a function of its total frequency $freq(c)$, the number of documents containing it $d(c)$, and the variance of its frequency distribution $var(c)$ over a corpus $C$ :

$$CD(c) = \frac{log(1 + freq(c))}{log(N)} * \frac{log(1 + d(c))}{log(D)} * (1 + log(1 + var(c))) \quad (2)$$

Where $N$ denotes the total number of concepts in $C$ and $D$ is the total number of documents in $C$. The log likelihood seems adaptive to such purpose since it helps to reduce the big margins between values. This formula ensures that if a concept frequency is 0, its $CD$ will equals 0 too. It ensures also that if $c$ have an instance in $C$, its $CD$ will never be 0 even if $var(c) = 0$.

Note that the $CD$ of a concept $c$ is the sum of its individual $CD$ value with the $CD$ of all its subconcepts in the taxonomy. The weights propagation from the bottom to the top of the hierarchy is a natural way to ensure that a parent even with a low individual $CD$ will be considered as highly context-dependent if its children are well represented in the corpus(see Figure 1).

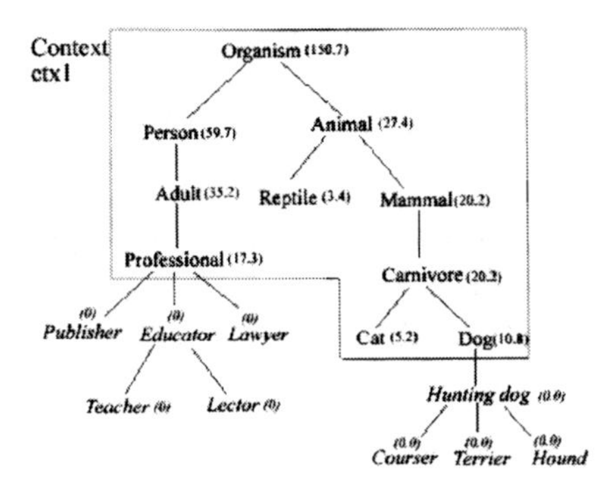


**Fig. 1.** A taxonomy extract showing $CD$ values assigned in the context *cont*1

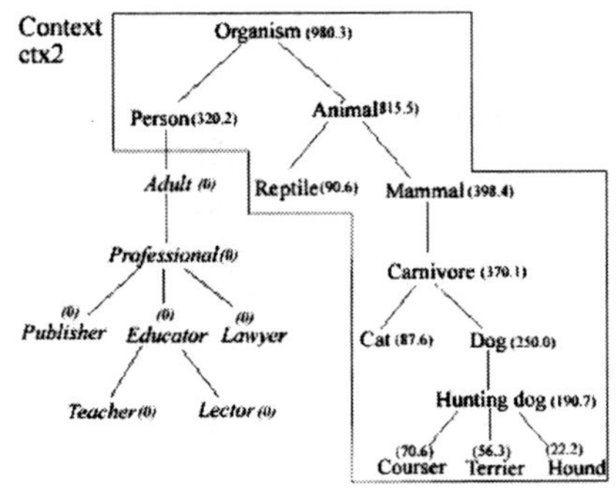


**Fig. 2.** A taxonomy extract showing $CD$ values assigned in the context *cont*2

To compare two concepts using the $CD$ values, we assign a Link Strength ($LS$) to each 'is-a' link in the taxonomy. Assume that $c_1$ subsumes $c_2$, the $LS$ between $c_1$ and $c_2$ is then calculated as follows:

$$LS(c_1, c_2) = CD(c_1) - CD(c_2)$$

Then the semantic distance[3] is calculated by summing the log likelihood of $LS$ along the shortest path separating the two concepts in the taxonomy.

$$Dist(c_1, c_2) = \sum_{c \in SPath(c_1, c_2)} log(1 + LS(c, parent(c)))$$

Where $SPath$ denotes the shortest path between $c_1$ and $c_2$.

Let us illustrate the method with an example. Consider the taxonomy extract shown in figure 1. The related context $ctx1$ taken from a corpus $C_1$ is represented by the subtree where $CD$ values are greater than 0. In $ctx1$, we notice that the corpus is likely general (talking about persons, professionals, carnivores,etc.). The obtained semantic distance between *Cat* and *Dog* in $ctx1$ is 2,2 while it's 4,5 in $ctx2$ illustrated in figure 2 where it seems to be more specialized in the animal domain. This states that *Cat* and *Dog* are closer in $ctx1$ than in $ctx2$ which respect human intuitions given the two different contexts.

### 3.3 A Corpus-Based Combination Measure

The promising rates attained by the corpus-based word similarities techniques and especially for the co-occurence-based ones has pushed us to combine them with our context-dependent measure in order to reach the best possible rates. However, two similar words can appear in the same document, paragraph, sentence, or a fixed-size window. It's true that smaller window size can help identifying relations that hold over short ranges with good precisions, larger window

[3] Our measure deals with distance which is the inverse of similarity.

size, yet too coarse-grained, allows to detect large-scale relations that could not been detected with smaller windows.

Consequently, we've choose to combine both techniques in order to view relations at different-scales. At the low scale, we use the PMI measure described above with a window size of 10 words (Table 1 Cooc). At the large scale, we calculate the Euclidian distance between words vectors where each word is represented by its tf.idf values over the documents (Table 1 Vecto).

## 4 Evaluation and Results

### 4.1 The Benchmark

In this study, Wordnet is used along with a corpus of 30,000 web pages in order to evaluate the proposed approach. The web pages are crawled from a set of news web sites (reuters.com, cnn.com, nytimes.com...).

The most intuitive way to evaluate a semantic similarity/distance is to compare machine ratings and human ratings on a same data set. A very common set of 30 word pairs is given by Miller and Charles [14]. M&C asked 38 undergraduate students to rate each pair on a scale from 0 (no similarity) to 4 (perfect synonymy). The average rating of each pair represents a good estimate on how similar the two words are. The correlation between individual ratings of human replication was 0.90 which led many researchers to take 0.90 as the upper bound ratio. For our evaluations, we've chosen the M&C subset of 28 words pairs which is the most commonly used subset for that purpose. Note that since our measure calculates distance, the M&C distance will be: $dist = 4 - sim$ where 4 represent the maximum degree of similarity.

### 4.2 Results

When comparing our distance results with the M&C human ratings, the context-dependency $CD$ method gave a correlation of 0.83 which seems to be a very promising rate (See Table 1).In view of further improvements, we evaluated multiple combination strategies.

First, at the taxonomy level, we combine our $CD$ measure with the feature-based measure (Feat) proposed by Tversky (equation 1:

- T1: $Dist = CD.Feat \Rightarrow$ Correlation = 0.83
- T2: $Dist = Feat.\sqrt{CD} \Rightarrow$ Correlation = 0.867

Second, at the corpus level, we combine the vectorial (Vecto) with the PMI measure (Cooc):

- C1: $Dist = Vecto.Cooc \Rightarrow$ Correlation = 0.649
- C2: $Dist = \alpha.vecto + \beta.coocc(\alpha = 0.3, \beta = 0.7) \Rightarrow$ Correlation = 0.564

Finally, at the overall level, we combine the taxonomy-based measures with the corpus-based measures:

- A1: $Dist = T1.Cooc \Rightarrow$ Correlation = 0.810
- A2: $Dist = T2.Cooc \Rightarrow$ Correlation = 0.833
- A3: $Dist = (1 + log(1 + CD)).(1 + log(1 + Cooc)) \Rightarrow$ Correlation = 0.884
- A4: $Dist = (1 + log(1 + T1)).(1 + log(1 + Cooc)) \Rightarrow$ Correlation = 0.890

Our method shows an interesting result whether on an individual level (CD) or on a combination level (T1-A4). We can notice that by using multiple resources (taxonomy and corpus) we could reached a correlation rate of 0.89 (Table 1 - A4) which is not too far from human correlations of 0.905. The obtained rate led our approach to outperform the existing approaches for semantic similarity (see Table 2).

**Table 1.** Similarity Results from the different strategies and their correlation to M&C Means

| Word Pair | M&C | CD | Feat | Vecto | Cooc | T1 | T2 | C1 | C2 | A1 | A2 | A3 | A4 |
|---|---|---|---|---|---|---|---|---|---|---|---|---|---|
| car-automobile | 0,08 | 1 | 0,52 | 2,801 | 0,332 | 0,52 | 0,52 | 0,93 | 1,073 | 0,173 | 0,173 | 2,178 | 1,826 |
| gem-jewel | 0,16 | 1 | 0,768 | 2,762 | 0,398 | 0,768 | 0,768 | 1,099 | 1,107 | 0,306 | 0,306 | 2,26 | 2,096 |
| journey-voyage | 0,16 | 3,783 | 0,847 | 2,765 | 0,439 | 3,203 | 1,647 | 1,214 | 1,137 | 1,407 | 0,723 | 3,499 | 3,323 |
| boy-lad | 0,24 | 1,635 | 0,81 | 2,77 | 0,389 | 1,325 | 1,036 | 1,078 | 1,103 | 0,515 | 0,403 | 2,616 | 2,449 |
| coast-shore | 0,3 | 1,426 | 0,862 | 2,762 | 0,416 | 1,229 | 1,03 | 1,149 | 1,12 | 0,512 | 0,429 | 2,543 | 2,429 |
| magician-wizard | 0,5 | 1 | 0,768 | 2,779 | 0,571 | 0,768 | 0,768 | 1,587 | 1,233 | 0,438 | 0,438 | 2,458 | 2,279 |
| midday-noon | 0,58 | 1 | 0.554 | 2,748 | 0,559 | 0.554 | 0.554 | 1,536 | 1,216 | 0,309 | 0,309 | 2,445 | 2,08 |
| furnace-stove | 0,89 | 14,182 | 0,886 | 2,79 | 0,355 | 14,182 | 3,766 | 0,99 | 1,086 | 5,028 | 1,335 | 4,849 | 4,849 |
| food-fruit | 0,92 | 8,489 | 1 | 2,793 | 0,324 | 8,489 | 2,914 | 0,905 | 1,065 | 2,751 | 0,944 | 4,163 | 4,163 |
| bird-cock | 0,95 | 3,606 | 0,858 | 2,79 | 0,513 | 3,094 | 1,629 | 1,431 | 1,196 | 1,587 | 0,836 | 3,574 | 3,407 |
| bird-crane | 1,03 | 4,286 | 0,86 | 2,79 | 0,499 | 3,687 | 1,781 | 1,392 | 1,186 | 1,839 | 0,889 | 3,744 | 3,575 |
| tool-implement | 1,05 | 2,01 | 0,85 | 2,768 | 0,392 | 1,708 | 1,205 | 1,085 | 1,105 | 0,67 | 0,472 | 2,797 | 2,657 |
| brother-monk | 1,18 | 1,473 | 0,905 | 2,765 | 0,506 | 1,333 | 1,098 | 1,399 | 1,184 | 0,674 | 0,555 | 2,685 | 2,603 |
| crane-implement | 2,32 | 8,982 | 1 | 2,754 | 1 | 8,982 | 2,997 | 2,754 | 1,526 | 8,982 | 2,997 | 5,589 | 5,589 |
| lad-brother | 2,34 | 12,745 | 0,842 | 2,762 | 0,398 | 12,643 | 3,262 | 1,099 | 1,107 | 5,028 | 1,297 | 4,833 | 4,823 |
| journey-car | 2,84 | 25,653 | 1 | 2,804 | 0,41 | 25,653 | 5,065 | 1,15 | 1,128 | 10,508 | 2,075 | 5,753 | 5,753 |
| monk-oracle | 2,9 | 13,64 | 1 | 2,776 | 0,49 | 13,64 | 3,693 | 1,36 | 1,176 | 6,683 | 1,81 | 5,153 | 5,153 |
| food-rooster | 3,11 | 13,53 | 1 | 2,798 | 0,597 | 13,53 | 3,678 | 1,67 | 1,257 | 8,077 | 2,196 | 5,397 | 5,397 |
| coast-hill | 3,13 | 7,24 | 1 | 2,762 | 0,422 | 7,24 | 2,691 | 1,166 | 1,124 | 3,054 | 1,135 | 4,203 | 4,203 |
| forest-graveyard | 3,16 | 21,004 | 0,902 | 2,77 | 1 | 18,939 | 4,133 | 2,77 | 1,531 | 18,939 | 4,133 | 6,927 | 6,76 |
| shore-woodland | 3,37 | 15,095 | 0,903 | 2,762 | 0,464 | 13,635 | 3,509 | 1,282 | 1,153 | 6,328 | 1,629 | 5,219 | 5,088 |
| monk-slave | 3,45 | 11,302 | 1 | 2,773 | 1 | 11,302 | 3,362 | 2,773 | 1,532 | 11,302 | 3,362 | 5,942 | 5,942 |
| coast-forest | 3,58 | 14,736 | 0,898 | 2,765 | 0,408 | 13,235 | 3,448 | 1,128 | 1,115 | 5,404 | 1,408 | 5,042 | 4,907 |
| lad-wizard | 3,58 | 11,853 | 1 | 2,765 | 1 | 11,853 | 3,443 | 2,765 | 1,53 | 11,853 | 3,443 | 6,017 | 6,017 |
| chord-smile | 3,87 | 15,701 | 1 | 2,762 | 1 | 15,701 | 3,963 | 2,762 | 1,529 | 15,701 | 3,963 | 6,46 | 6,46 |
| glass-magician | 3,89 | 17,276 | 1 | 2,784 | 1 | 17,276 | 4,156 | 2,784 | 1,535 | 17,276 | 4,156 | 6,613 | 6,613 |
| noon-string | 3,92 | 16,53 | 1 | 2,759 | 0,573 | 16,53 | 4,066 | 1,581 | 1,229 | 9,47 | 2,329 | 5,614 | 5,614 |
| rooster-voyage | 3,92 | 24,853 | 1 | 2,762 | 1 | 24,853 | 4,985 | 2,762 | 1,529 | 24,853 | 4,985 | 7,2 | 7,2 |
| **Correlation** | **0,905** | **0.830** | **0.619** | **0.256** | **0.649** | **0.830** | **0,867** | **0.649** | **0.564** | **0.81** | **0.833** | **0.884** | **0.890** |

Our method shows an interesting result whether on an individual or on a combination scale. A part of its interesting correlation coefficient of 0.83, our CD method has the advantage to be context-dependent, which means that our results vary from one context to another. We argue that our measure could perform better if we "place" human subjects in our corpus context. In other terms, our actual semantic distance values reflect a specific context that doesn't necessarily match with the context of the human subjects during the R&C experiments.

**Table 2.** Comparison between the principal measures and our two-level measure

| Similarity method | Type | Correlation with M&C |
|---|---|---|
| Human replication | Human | 0,901 |
| Rada | Edge-based | 0,59 |
| Hirst and St-Onge | Edge-based | 0,744 |
| Leacock and Chodorow | Edge-based | 0,816 |
| Resnik | Information Content | 0,774 |
| Jiang | Information Content | 0,848 |
| Lin | Information Content | 0,821 |
| CD | Context-Dependent | 0,830 |
| our multisource measure | Hybrid | 0,890 |

# 5 Conclusion and Future Work

We have shown the importance of considering the context when calculating semantic similarities between words/concepts. We've proposed a Context-Dependent method that takes the taxonomy as a principal knowledge resource, and a text corpus as a similarity adaptation resource for the target context. We've proposed also to combine it with other taxonomy-based and corpus-based methods. The results obtained from the experiments show the effectiveness of our approach which led it to outperform the other approaches. A much better way to evaluate the method and compare it with others is to perform context-driven human ratings, where human subjects will be asked to rank a same set of words pairs in different contexts. The machine correlation computed next according to each context will be able to show more significantly the added-value of our approach.

# References


1. And, N.S.: An intrinsic information content metric for semantic similarity in wordnet
2. Barsalou, L.: Intraconcept similarity and its application for interconcept similarity. Cambridge University Press, Cambridge (1989)
3. Christopher, H.S.: MANNING. Foundations of statistical natural language processing (1999)
4. Church, K.W., Hanks, P.: Word association norms, mutual information, and lexicography. In: Proceedings of the 27th. Annual Meeting of the Association for Computational Linguistics, Vancouver, pp. 76–83. Association for Computational Linguistics (1989)
5. Dagan, I., Lee, L., Pereira, F.C.N.: Similarity-based models of word cooccurrence probabilities. Machine Learning 34(1-3), 43–69 (1999)
6. Furnas, G.W., Deerwester, S.C., Dumais, S.T., Landauer, T.K., Harshman, R.A., Streeter, L.A., Lochbaum, K.E.: Information retrieval using a singular value decomposition model of latent semantic structure. In: Chiaramella, Y. (ed.) SIGIR, pp. 465–480. ACM Press, New York (1988)

7. Hindle, D.: Noun classification from predicate-argument structures. In: Meeting of the Association for Computational Linguistics, pp. 268–275 (1990)
8. Hirst, G., St-Onge, D.: Lexical chains as representation of context for the detection and correction malapropisms (1997)
9. Jiang, J.J., Conrath, D.W.: Semantic similarity based on corpus statistics and lexical taxonomy (1997)
10. Leacock, C., Chodorow, M., Miller, G.A.: Using corpus statistics and wordnet relations for sense identification. Computational Linguistics 24(1), 147–165 (1998)
11. Lin, D.: An information-theoretic definition of similarity. In: Proc. 15th International Conf. on Machine Learning, pp. 296–304. Morgan Kaufmann, San Francisco (1998)
12. Medin, D.: Psychological essentialism. Cambridge University Press, Cambridge (1989)
13. Miller, G.A.: Wordnet: A lexical database for english. Commun. ACM 38(11), 39–41 (1995)
14. Miller, G.A., Charles, W.: Contextual correlated of semantic similarity. Language and Cognitive Processes 6, 1–28 (1991)
15. Rada, R., Mili, H., Bicknell, E., Blettner, M.: Development and application of a metric on semantic nets. IEEE Transactions on Systems, Man, and Cybernetics 19(1), 17–30 (1989)
16. Resnik, P.: Semantic similarity in a taxonomy: An information-based measure and its application to problems of ambiguity in natural language. J. Artif. Intell. Res. (JAIR) 11, 95–130 (1999)
17. Turney, P.D.: Mining the Web for synonyms: PMI–IR versus LSA on TOEFL. In: Flach, P.A., De Raedt, L. (eds.) ECML 2001. LNCS (LNAI), vol. 2167, p. 491. Springer, Heidelberg (2001)
18. Tversky, A.: Features of similarity. Psychological Review 84, 327–352 (1977)
19. Wu, Z., Palmer, M.: Verb semantics and lexical selection. In: 32nd. Annual Meeting of the Association for Computational Linguistics, New Mexico State University, Las Cruces, New Mexico, pp. 133–138 (1994)

# Self-healing Information Systems

Barbara Pernici

Department of Electronics and Information, Politecnico di Milano
I-20133 Milano, Italy
barbara.pernici@polimi.it

**Abstract.** Information systems in highly dynamic collaborative environments are based on the composition of services and processes from different organizations and systems. In most cases, such an environment is not under the control of a single organization and therefore requires a particular attention to its management to ensure correct functionalities. Different types of faults can occur in the system, both at the functional level and in terms of reduced quality of service. Such faults may cause failures in one or more participating processes, which are hindering the correct completion of business processes towards their required goals. In the presentation, the management of cooperating processes will be discusses, following the approach being proposed within the WS-DIAMOND European project: management of failures is driven by the diagnosis of causes that are leading to the failures, and repair actions are performed on processes and services coordinating repair actions on services. Such an approach allows an autonomic behavior of information systems, reducing the need of design efforts to anticipate possible combinations of predicted exceptions. Business process management is based on the SH-BPEL (Self-healing Business Process Language) approach proposed at Politecnico di Milano to manage and repair business processes in a flexible service-oriented framework.

R. Wagner, N. Revell, and G. Pernul (Eds.): DEXA 2007, LNCS 4653, p. 64, 2007.

# A Faceted Taxonomy of Semantic Integrity Constraints for the XML Data Model*

Khaue Rezende Rodrigues and Ronaldo dos Santos Mello

Universidade Federal de Santa Catarina,
Departamento de Informática e Estatística, Caixa Postal 476,
Florianópolis, SC, Brazil 88040-900
{khaue,ronaldo}@inf.ufsc.br

**Abstract.** Some work in the literature deal with Semantic Integrity Constraints (SIC) for XML and propose taxonomies of SIC for the XML data model. However, these taxonomies are incomplete and do not mention a basis for the classification. We propose a faceted taxonomy of SIC for XML data model that tries to fulfill these limitations. Our proposal is based on previous related taxonomies for the relational and XML data models, providing a classification that can give support to expressiveness analysis of SIC specification languages for XML data, as well as XML database management systems integrity control. We demonstrate, through examples, that our taxonomy is more comprehensive than other related taxonomies available in the literature.

**Keywords:** XML, data integrity, integrity constraint taxonomy.

## 1 Introduction

XML is a W3C recommendation widely used for semistructured data representation and interchange [1]. Such increasing manipulation of XML data has originated large XML repositories. Because of this, some information systems like Semantic Data Integration Systems (SDIS) (several related work are presented in [19]) and Database Management System (DBMS) [7, 20, 27, 28] are directing their efforts to deal with the XML format. However, XML data management has some open issues or issues under work, like integrity maintenance [12]. In fact, integrity specification is a required feature of a data model [24]. Its support is relevant to several actual research areas, like XML DBMS data consistency and query processing in XML-based SDIS.

On the other hand, integrity maintenance has been widely researched in the context of structured data models, like relational and object-oriented models. For Relational Database Management Systems (RDBMS), for example, we have a theory and standards for integrity constraint specification and control, available even in the classical and early database literature [8, 10, 21, 22, 24, 29]. Although XML data be more complex than structured data, mainly because schemas for XML are very

* This work is partially supported by the DIGITEX Project of CNPq Foundation. CTInfo Process Nr.: 550.845/2005-4.

R. Wagner, N. Revell, and G. Pernul (Eds.): DEXA 2007, LNCS 4653, pp. 65–74, 2007.

irregular, in fact, XML and relational data models have many similar points, like data type binding and cardinality constraints for relationships. Hence, the main advantage of a relational-based approach for XML data integrity constraint management is the use of a consolidated theoretical foundation. Work like [11, 12, 17, 18, 25, 28] has demonstrated this research tendency through comparatives between these data models and/or adaptations of relational constraint management to XML data.

For integrity maintenance, we mean the guarantee of consistency states and state transitions for data sets. In order to provide such maintenance, we need to specify rules, or rather, Integrity Constraints (IC) [21, 24]. We may distinguish two not mutually exclusive forms of IC: *syntactic* and *semantic* IC. The syntactic IC regards to the consistency guarantee about the structural specification, and semantic IC regards to the real intention of data. Given its major importance, we focus, in this paper, on the treatment of Semantic IC (SIC) for XML data.

XML Schema [2] may be considered the main (W3C) recommendation for syntactic and even semantic specification of XML data. However, its support for SIC specification is limited [4, 11, 13, 15, 18]. Other important W3C initiatives are RDF (Resource Description Framework) [3] and OWL (Ontology Web Language) [3] schema languages, as well as SWRL (Semantic Web Rule Language) [5] rule language. Meanwhile, these proposals focus on knowledge representation and not SIC for XML instances, being therefore, out of the scope of this paper.

Several open issues or issues under work may be distinguished in the context of SIC for XML: *(i)* a standard language for SIC specification (there are many proposals [6, 9, 13, 15, 18], but there is no an agreement); *(ii)* a suitable taxonomy of SIC for XML, that can give support to the expression power evaluation of SIC specification languages [11, 13, 15, 17, 18]; *(iii)* decidability and satisfiability problems [11, 14]; and *(iv)* the use of SIC for query translation and decomposition [25]. This paper focus on item *(ii)*, motivated by the fact that the proposals in the literature are not suitable as a basis for an expressivity evaluation of SIC definition languages for the XML data model or XML DBMS integrity control (IC) modules because they are very heterogeneous, do not present a foundation to justify their categories, or are incomplete. Some of them analyze SIC over few points of view or properties, and classify them on hierarchical categories (*enumerated classification* [23]).

Therefore, we propose a well-founded, flexible and comprehensive SIC taxonomy for the XML data model that intends to overcome such limitations. We consider it *flexible* because it is specified as a less restricted and extensible classification form, named *faceted classification* [23]. It is also *comprehensive* because it is generic enough to deal with SIC components (described in Section 3) and categories presented by related work. Furthermore, we argue that it is *well-founded* because we base the taxonomy facets and categories in SIC components [21] and existing taxonomies for the relational [8, 10, 22, 24, 29] and XML data models [4, 11, 13, 14, 15, 17, 18, 26].

This paper is organized as follows. Section two briefly discusses related work in the context of the relational and XML data models. Section three describes the proposed SIC taxonomy for XML data model, and compares it against related work. Finally, section four is dedicated to the conclusion.

## 2 Related Work

Database classical and early literature presents some proposals of SIC taxonomies for the relational model [8, 10, 22, 29]. In Santos et. al. [29], it is presented a taxonomy based on the following aspects: *(i) origin*, *(ii) substance* (state transition, intention, etc); *(iii) form of specification*; and *(iv) application mode*. In Date [8], we have a taxonomy that classifies SIC as follows: *(i) domain*; *(ii) attribute*; *(iii) tuple*; *(iv) database*; *(v) state transition*; *(vi) key*; *(vii) referential integrity*; and *(viii) check point* (the moment that the IC is verified). In Codd [24], this last category is called *integrity points*. In Silberschatz et. al. [22], we have the following classification: *(i) domain*; *(ii) key*; *(iii) relationship forms*; and *(iv) referential integrity*. In Elmasri and Navathe [10], the taxonomy is: *(i) domain*; *(ii) key*; *(iii) entity*; and *(iv) referential integrity*[1].

With regard to XML data model, some work present comparatives among SIC specification languages, but they analyze ICs through specific and heterogeneous points of view [13, 15, 16, 18]. Despite of the great number of related initiatives, few work in the literature provide a taxonomy of SIC for XML data. Some of them present a so-called *complete* taxonomy [13, 15, 17, 18] while others present a limited set of categories in order to delimitate the scope of their work [4, 11, 14, 31, 26].

The work of Buneman et. al. [4] focuses on key constraints, distinguishing two categories: *(i) key* – derived from classical data models; and *(ii) relative key* – composed by a *key* and a *path expression* (regular and *XPath* expressions). Benedikt et. al. [26] proposes the following categories: *(i) type*; *(ii) key*; and *(iii) inclusion* – based on foreign key constraints. In Pavlova et. al. [17], we have: *(i) integrity* (semantic consistency of data) – composed by categories like *type*, *path* and *complex constraints*; and *(ii) data validity* – where is introduced a group called *temporal validity constraints.* Its last category represents a relevant contribution because of the consideration of temporal aspects in SIC.

In Arenas et. al. [14], the main categories are the following: *(i) absolute* and *relative IC*, distinguishing constraints that require a complete analysis of the XML document, or only a part of it; and *(ii) unary* and *multi-attribute* IC, that categorizes them according to the number of involved attributes. The main contribution of this work is to classify IC according to the amount of data that is needed to verify the constraint. (we call it *constraint range*).

In Fan and Siméon [11], the main category is called *XML basic constraints*, with a focus on constraints related to relationship and unicity, like *key*, *foreign key* and others. A positive point is that they base their categories on structured data models. However, this proposal is incomplete because it is argued that only a subset of the SIC universe is considered.

The taxonomy of Jacinto et. al. [15] organizes the SIC in a small group of specific categories and does not take into consideration aspects like constraint range and the SIC syntactic structure (we call it *constraint form*). In fact, a negative point is that a subset of constraints is grouped in a non-specific category called *other constraints.* Therefore, it suffers from the same drawback of the previous work with respect to comprehensiveness, and does not present a theoretical basis for its categories.

Hu and Tao taxonomy [13] introduces some contributions, considering constraint form (a category called *form of the constraint*), as well as the *kind of imposed*

[1] More details about these taxonomies are available in the respective references.

*limitation* (distinguished in *dynamic* and *static* constraints). Even so, it does not deal with constraint range and also does not mention a foundation for its categories.

The proposal of Lazzaretti and Mello [18] is based on the relational model and also in Date [8]. However, it does not consider constraint range and constraint form. Analogous to other work [13, 15, 17], its taxonomy is incomplete, being not suitable for the comparison of expressiveness of SIC specification languages.

## 3 The Proposed Taxonomy

Our taxonomy of SIC for XML data (Figure 1) aims at being comprehensive and flexible, fulfilling the limitations of related work (as stated in section 1). The taxonomy distinguishes several *facets* of a SIC. The concept of facet, as a component of a faceted classification[2] [23], allows classifying a SIC in different points of view, with the purpose of providing a basis for expressiveness evaluation. The proposed facets are based on constraint components and concepts for the relational data model [21, 24] also discussed in Santos [29]. Hence, the proposed categories that compose each facet are based on taxonomies and concepts for the relational data model [8, 10, 22, 29] and contributions of related work [4, 11, 13, 14, 15, 17, 18, 26].

1 **Constraint value:** Considers restrictions to XML data values.
- **State.** Restricts the value of one or more simple elements and attributes;
  - **Dynamic.** Imposes different limitations for similar XML instances;
  - **Static.** Imposes the same limitation for similar XML instances.
- **State transition.** Restricts the value transitions of one or more simple elements and attributes.
  - **Dynamic.** Imposes different state transition limitations for similar XML instances:
  - **Static.** Imposes the same state transition limitations for similar XML instances:.

2 **Constraint range**: Considers the amount of nodes in an XML document or an XML repository that are involved in the SIC specification or being constrained by it.
- **Data item.** Considers the content of an unique attribute or simple element;
- **Tuple.** Considers the content of a set of attributes and/or simple elements from a complex element;
- **Element.** Considers the content of a set of attributes and/or sub-elements of a same complex element;
- **Repository.** Considers the content of a set of attributes and/or simple elements of distinct complex elements.

3 **Constraint form:** Considers the used notation for defining the SIC.
- **Based on boolean expressions.** Evaluates a predicate and return true or false.
  - **Simple.** Considers only one element/attribute:
  - **Composed.** Considers two or more elements/attributes.
- **Based on conditional rules.** Considers embedded boolean expressions. being able or not to perform actions on the XML instances in order to guarantee data integrity.
  - **Simple.** Considers only one conditional expression;
  - **Composed.** Considers two or more nested conditional expression.

4 **Constraint check point**
- **Immediate.** Verifies the constraint in the moment that the operation on XML instances occurs:
- **Delayed.** Verifies the constraint in some moment after the operation on XML instances occurs.

5 **Constraint action**: Considers the kind of action performed by the SIC when a violation occurs.
- **Informative.** Presents an error message and rollbacks the operation;
- **Active** Executes other operations on XML instances to maintain data integrity.

**Fig. 1.** Proposed taxonomy of SIC for the XML data model

According to Santos [21], a constraint is composed by five components: *(i) constraining* (data objects used in the constraint specification); *(ii) constrained* (data objects that are constrained); *(iii) restrictive condition* (a logic expression to be

[2] The faceted classification was proposed by Ranghanathan [23] and argues that the information may be considered over several aspects or properties called *facets*.

validated that connects *constrained* and *constraining* data objects); *(iv) check point* (the moment that the constraint should be checked); and *(v) violation actions* (the actions to be performed after the violation of a constraint in order to maintain data integrity). On matching these components, we may define that data integrity maintenance related to a constraint is performed through the validation, in a specific *check point*, of *constrained objects* with regard to *restrictive conditions* that comprise *constraining objects*, performing *violation actions* in case of failure.

A faceted classification coupled to constraint components is a good choice for SIC because it allows to distinguish IC with relation to the main properties that a system that deals with SIC have to consider. We organize our SIC taxonomy according to the following facets: *(i) constraint value*, meaning the kind of limitation that the SIC is able to impose on data values – based on the *constrained* component; *(ii) constraint range*, meaning the amount of data in the constraint specification – based on the *constraining and constrained* components; *(iii) constraint form*, meaning the constraint specification syntax – based on the *restrictive condition* component; *(iv) constraint check point* – based on the *check point* component, and the *integrity points* concept [24]; and *(v) constraint action* – based on the *violation actions* component.

The next sections analyze the relevance of each facet and its categories, in order to raise the contribution of our taxonomy, as well as the limitations of related work. The XML instance in Figure 2 is used as basis for the examples ahead.

```
<professor firstName="John" middleName="Albert" lastName="Data" >
   <salary>5000.00</salary>
   <graduated>
      <university>SCFU</university>
      <degree>PhD</degree>
   </graduated>
</professor>
```

**Fig. 2.** Example of an XML instance

### 3.1 Constraint Value

This facet is related to the constraint component named *constrained*. It is useful to distinguish SIC that impose restrictions over a data state (*state* category) from more complex SIC that impose limitations over valid state transitions of data (*state transition* category). We also distinguish between *static constraints*, that always impose the same restriction on a same data class, and *dynamic constraints*, that can apply different restrictions. These categories are relevant to evaluate how expressive is a SIC specification language. XML Schema for example, does not represent *dynamic constraints*, while XCML (XML Constraint Markup Language) [13] does (Table 1). Thus, XCML is more expressive than XML Schema with respect to this facet. In your turn, XML DBMS Tamino [7] is able to define only *state* constraints, while XML DBMS eXist [20] and Timber [28] does not have even such support. Thus, these systems are expressiveless with respect to this facet.

We define the *state* category based on the *domain* category presented in proposals for the relational model [8, 10, 18, 22], as well as related work [11, 15]. The *state transition* category is mainly derived from taxonomies for the relational data model [8, 18, 29], being also discussed in the context of XML by Lazzaretti and Mello [18].

In order to compare this facet with related work, consider the following SIC example: *"the salary of a professor cannot be reduced, and must be higher than 50.00 if the professor is graduated"*. This SIC depends on a comparison of old and new values of an element (*salary*) as well as the existence of the *graduated* element to define valid states for *salary*. This SIC cannot be classified in the taxonomies of Fan and Siméon [11] and Jacinto et. al. [15] because they do not define categories for dealing with state transition and dynamic constraints. In Hu and Tao [13], this SIC may be classified as *dynamic constraint in relation to the kind of imposed limitation*, but this category does not deal with state transitions. In Lazzaretti and Mello [18], we have a category for state transition, but not one for dynamic SIC. In our taxonomy, we classify this SIC as *Dynamic State Transition*, capturing its whole intention.

### 3.2 Constraint Range

This facet is based on the constraint components named *constraining* and *constrained*. It is responsible to distinguish the elements or attributes that compose a constraint (acting either as constrained or constraining), and its relationship levels (distance between nodes on an XML hierarchical structure, also named *range level* in this paper). This facet is important to an expressiveness evaluation because it helps to define the kinds of relationship among XML elements or attributes that a SIC specification language must be able to represent. For the same reason, it is also important for an XML DBMS when it is necessary to navigate through hierarchical paths in order to check a constraint. In fact, a SIC specification language can represent different range levels for each kind of constraint present in other facets.

The *constraint range* facet considers the different hierarchical levels that can be checked by a SIC for XML data. Because of this, categories of relational data model [8, 10] like *tuple*, *database* [8] and *entity* [10], as well as XML related work [11, 14, 15, 18] were taking as basis.

Consider the following SIC: *"the name of a professor cannot exceed 30 characters and must be composed by a degree, firstName, middleName, and lastName"*. The work of Fan and Siméon [11] and Hu and Tao [13] do not have categories similar to this facet. In Jacinto et. al. [15], it is possible to classify this SIC in the category called *dependencies between two document nodes*, but this category considers only the relationship between two elements, and not the range of the overall XML document. In the same way, the work of [14, 18] deal with a small set of categories, being not detailed. In Arenas, the SIC can be classified as a generic *multi-attribute* SIC. In Lazzaretti and Mello [18], the example can be classified as a *database* SIC, but it only distinguishes if elements or attributes are in the same hierarchic level or not.

In our taxonomy, it is possible to classify the example in the *Element* category, because it involves attributes and one sub-element in a same complex element. It is more detailed than related work, presenting four range levels for a XML SIC.

### 3.3 Constraint Form

This facet is based on the constraint component named *restrictive condition* and some related work that consider such aspect [4, 11, 13]. The main contribution of this facet is to classify the kind of used notation for SIC specification. Hu and Tao [13] raises the importance of this facet for SIC specification languages, claiming that a rule-based SIC allows more natural and concise specifications of many types of IC than IC based on boolean expressions. This is specially important to the design of an SIC language and, as a consequence, of an XML DBMS IC control because it represent how the SIC will be expressed by the users, and if it is possible to specify it in a *boolean form* or a *conditional form*.

Although some related work deal with similar aspects [4, 11], only Hu and Tao [13] consider this facet, and we base it on this work. However, it does not deal with other facets considered by our proposal. A discussion about different kinds of notations and their expressiveness is out of the scope of this paper. Hu and Tao [13] present similar analysis. As a matter of general exemplification, we give the following SIC: "*"the salary of a professor must be 5000.00 if the professor has the PhD degree"*." This SIC can be classified in our taxonomy as *Based on Conditional Rules – Simple* because it could be defined as a trigger that updates a professor *salary* to 5000.00 when its *degree* becomes *PhD*.

### 3.4 Constraint Check Point

This facet is based on the similar constraint component presented in Santos [21] and Codd [24]. This facet aims at distinguishing the moment that the *restrictive condition* that composes the SIC should be verified. Considering the proposal of Date [8] for the relational model, we define two categories for this facet: *immediate* and *delayed*.

A delayed SIC could be a constraint that, in the scope of a transaction, must be verified only at the commit time. Such aspect is enforced by Codd [24]. It argues that some kinds of IC should allow that data should pass by inconsistent states until reach a new consistent state. This aspect is relevant, for example, in the evaluation of XML DBMS transaction controls, if we desire that a SIC over data could be checked in some moment after its update by the transaction. Tamino and eXists XML DBMS, for example, supports only *immediate* SIC, being not flexible for such a control.

One example in this context is the following: "*given that 80% of the professors must have a PhD degree, if an update of professor elements is executed, the SIC that guarantees such restriction must be verified only after all professor instances be updated"*. In our taxonomy, this is an example of *delayed* SIC because the constraint should not be verified immediately after/before the update of each professor element, but only when the operation ends. No related taxonomy considers such facet.

### 3.5 Constraint Action

This facet is based on the constraint component named *violation actions,* as well as work about validation mechanisms for XML specification languages [9, 13, 15, 18]. This facet considers the kind of action that should be taken when an integrity violation occurs.

Consider the following example: "*a professor with Master degree cannot have a salary lower than 2000.00"*. Supposing that a professor receives the *Master* value in its *degree* element, and its *salary* element not is incremented to *2000.00*, we may apply one of the two alternatives: *(i)* an update operation is executed to increase the element *salary* (*active constraint*); or *(ii)* a message error is generated and a rollback operation is performed, (*informative constraint*). The execution of these alternatives depends on the XML DBMS transaction and IC control capabilities, as well as the capabilities of the SIC specification language, raising the importance of this facet. The XML DBMS Tamino and eXist, for example, implements only *informative constraints*. With respect to related work, no one of the proposed taxonomies defines this facet and distinguishes this constraint component. Thus, the main contribution of this facet is to consider the actions that a SIC may hold on XML data in the context of an XML DBMS IC control, for example.

### 3.6 Taxonomy Application

In order to exemplify the application of our taxonomy in the evaluation of existing SIC definition languages and XML DBMS, Table 1 presents a comparative of the usage of our facets by them.

**Table 1.** SIC specification languages for XML and XML DBMS vs. taxonomy facets

| SIC definition languages and XML DBMS | Constraint Value | | Constraint Range* | Constraint Form | | Constraint Action | Constraint Check Point |
|---|---|---|---|---|---|---|---|
| | State | State Transaction | | Boolean Expression | Conditional Rules | | |
| 1. XML Schema [2] | Static | No | Element | Simple | No | Informative | Immediate |
| 2. Schematron [9] | Static | No | Repository | Composed | Simple | Informative | Immediate |
| 3. XCSL [15] | Dynamic | No | Repository | Composed | Simple | Informative | Immediate |
| 4. XCML [13] | Dynamic | No | Repository | Composed | Composed | Informative | Immediate |
| 5. XDC [18] | Dynamic | Dynamic | Repository | Composed | Composed | Active | Immediate |
| 6. eXists XML DBMS [20] | No | No | No | No | No | No | No |
| 7. Tamino XML DBMS [7] | Static | No | Element | Simple | No | Informative | Immediate |
| 8. Timber XML DBMS [28] | No | No | No | No | No | No | No |
| *If defined, it denotes the most comprehensive supported category in the facet. | | | | | | | |

It is important to note that the consideration of all facets is limited or heterogeneous. For example, few SIC languages are able to specify *state transactions* and/or *composed conditional rules*, which are important features of a SIC control. We also observe that the attendance of the *Constraint Action* and *Check Point* facets are not satisfactory, with most cases restricted to *informative actions* and *immediate check points*.

Furthermore, we observe that XDC and Tamino are much robust languages and XML DBMS with respect to SIC treatment, respectively. Most of the XML DBMS, in particular, suffer from the absence of a SIC control mechanism.

## 4 Conclusion

This paper proposes a SIC taxonomy for XML data that classifies SIC based on the constraint components in a faceted form. The motivation for such work is the lack of a SIC taxonomy that could give better support to an expressiveness evaluation of: *(i)* SIC specification languages in the context of XML data model; and *(ii)* XML DBMS modules that deal with SIC. As presented in Section 2, related taxonomies are heterogeneous, do not deal with the main aspects of SIC, or do not present a basis do justify their categories.

In fact, the facets of our taxonomy are generic enough to be useful for other data models. However, we consider its specific application to XML data model. For example, the *Constraint Range* facet is suitable to the hierarchical structure of XML data, being not applicable to the relational model. As pointed out before, our main intention is to contribute to the research on integrity maintenance issue for XML data.

We argue again that the contribution of this paper is the definition of a *well-founded, flexible* and *comprehensive* taxonomy that analyze a SIC with respect to the definition of constraint *components* focused on the relational model [21]. It is *flexible* because a taxonomy in a faceted form can be easily extended [23]. It is *comprehensive* because the set of proposed facets deals with all constraint components, making it generic enough. Furthermore, as demonstrated through examples in Section 3, it is more *comprehensive* than other proposals because it considers constraint components that they do not consider. Despite of that, we take care to define the only necessary categories (based on the related work [4, 11, 13, 14, 15, 17, 18, 26]) to reach the desired contributions in this subject (items *(i)* and *(ii)* described above). Thus, we claim that our taxonomy is more indicated for expressivity evaluation. In fact, a faceted taxonomy based on constraint components is a good choice because it denotes all the components that should be considered for any system that intends to deal with SIC, like a XML DBMS. Even with some adaptations in all taxonomy categories to become it more comprehensive, we called it *well-founded* because we base it on the IC literature for the relational data model [8, 10, 18, 21, 22, 24, 29] and on the contributions of related work [4, 11, 13, 14, 15, 17, 18, 26]. Finally, our taxonomy is also a contribution to the problem of lack of foundation for XML data SIC.

Future work include studies for proof of completeness of the taxonomy, and a comparative analysis of XML DBMS for evaluating their expressivity and robustness with respect to SIC specification languages and IC control. We also intend to apply our taxonomy in an integration process called *BInXS* [19]. In this context, the taxonomy will be useful as a guide for specifying SIC categories that could be extracted from heterogeneous XML sources in order to improve the semantic aspects of the data to be integrated. At the DIGITEX project, such taxonomy will be useful in the discovery of implicit SIC for XML instances to be matched.

## References

1. XML (Extensible Markup Language) (June 2007), Available at: http://www.w3.org/xml
2. XML Schema (June 2007), Available at: http://www.w3.org/xml/schema
3. Semantic Web (June 2007), Available at: http://www.w3.org/2001/sw
4. Buneman, P., et al.: Keys for XML. In: WWW 2001, Hong Kong, China, pp. 201–210 (2001)
5. SWRL (June 2007), Available at: http://www.w3.org/submission/swrl
6. Clark, J., Murata, M.: RELAX NG Specification. Technical report. In: Organization for the Advancement of Structured Information Standards (OASIS) (2001)
7. Schöning, H.: Tamino - A DBMS designed for XML. In: ICDE 2001, pp. 149–154 (2001)

8. Date, C.J.: An Introduction to Database Systems, 8th edn. Addison-Wesley, London (2003)
9. Dodds, L.: Schematron: Validating XML Using XSLT. In: XSLT UK Conference, England (2001)
10. Elmasri, R., Navathe, S.B.: Fundamentals of Database Systems, 4th edn. Addison-Wesley, London (2003)
11. Fan, W., Siméon, J.: Integrity Constraints for XML. Journal of Computer and System Sciences 66(1), 254–291 (2003)
12. Fan, W.: XML Constraints: Specification, Analysis, and Applications. In: International Workshop on Database and Expert Systems Applications, pp. 805–809 (2005)
13. Hu, J., Tao, L.: An Extensible Constraint Markup Language: Specification, Modeling, and Processing. In: XML Conference and Exhibition, U.S.A (2004)
14. Arenas, M., et al.: On Verifying Consistency of XML Specifications. In: ACM PODS, U.S.A, pp. 259–270 (2002)
15. Jacinto, M.H., et al.: XCSL: Constraint Specification Language. In: Latin American Conference on Informatics (2002)
16. Lee, D., Chu, W.W.: Comparative Analysis of Six XML Schema Languages. ACM SIGMOD Record 29, 76–87 (2000)
17. Pavlova, E., et al.: Constraints for Semistructured Data. In: Russian Conference on Digital Libraries, Russian (2000)
18. Lazzaretti, A.T., Mello, R.S.: A Domain Integrity Constraint Control for XML Documents. In: Brazilian Symposium on Databases, Brazil, pp. 115–129 (2005)
19. Mello, R.S., Heuser, C.A.: BInXS: A Process for Integration of XML Schemata. In: Pastor, Ó., Falcão e Cunha, J. (eds.) CAiSE 2005. LNCS, vol. 3520, pp. 151–166. Springer, Heidelberg (2005)
20. Meier, W.: eXist: An Open Source Native XML Database. In: Chaudhri, A.B., Jeckle, M., Rahm, E., Unland, R. (eds.) Web, Web-Services, and Database Systems. LNCS, vol. 2593, pp. 169–183. Springer, Heidelberg (2003)
21. Santos, C.S.: Caracterização Sistemática de Restrições de Integridade em Bancos de Dados. PhD Thesis, Informatics Department-PUC, Brazil, Rio de Janeiro (1980)
22. Silberschatz, A., et al.: Database System Concepts, 5th edn. McGraw-Hill, New York (2005)
23. Ranganathan, S.R.: The Colon Classification. In: Artandi, S. The Rutgers Series on Systems for the Intellectual Organization of Information, vol. IV. Graduate School of Library Science, Rutgers University, New Brunswick, NJ (1965)
24. Codd, E.F.: Data Models in Database Management. In: ACM SIGMOD 1980, U.S.A, pp. 112–114. ACM Press, New York (1980)
25. Deutsch, A., Tannen, V.: Reformulation of XML Queries and Constraints. In: Calvanese, D., Lenzerini, M., Motwani, R. (eds.) ICDT 2003. LNCS, vol. 2572, pp. 225–241. Springer, Heidelberg (2002)
26. Benedikt, M., et al.: Capturing Both Types and Constraints in Data Integration. In: ACM SIGMOD 2003, U.S.A, pp. 277–288. ACM Press, New York (2003)
27. Bourret, R.: XML consulting, writing and research (XML and Databases). Available at: Jun (2007), http://www.rpbourret.com
28. Jagadish, H.V., et al.: TIMBER: A Native XML Database. VLDB Journal 11(4), 274–291 (2002)
29. Santos, C.S., et al.: Towards Constructive Axiomatic Specifications. In: ACM SIGMOD 1980. pp. 183–185 (1980)

# Beyond Lazy XML Parsing*

Fernando Farfán, Vagelis Hristidis, and Raju Rangaswami

School of Computer and Information Sciences
Florida International University
{ffarfan,vagelis,raju}@cis.fiu.edu

**Abstract.** XML has become the standard format for data representation and exchange in domains ranging from Web to desktop applications. However, wide adoption of XML is hindered by inefficient document-parsing methods. Recent work on lazy parsing is a major step towards alleviating this problem. However, lazy parsers must still read the entire XML document in order to extract the overall document structure, due to the lack of internal navigation pointers inside XML documents. Further, these parsers must load and parse the entire virtual document tree into memory during XML query processing. These overheads significantly degrade the performance of navigation operations. We have developed a framework for efficient XML parsing based on the idea of placing internal physical pointers within the document, which allows skipping large portions of the document during parsing. The internal pointers are generated in a way that optimizes parsing for common navigation patterns. A double-Lazy Parser (2LP) is then used to parse the document that exploits the internal pointers. To create the internal pointers, we use constructs supported by the current W3C XML standard. We study our pointer generation and parsing algorithms both theoretically and experimentally, and show that they perform considerably better than existing approaches.

**Keywords:** XML, Document Object Model, Double Lazy Parsing, Deferred Expansion, XPath.

## 1 Introduction

XML has become the de facto standard format for data representation and exchange in domains ranging from the Web to desktop applications. Examples of XML-based document types include Geographic Information Systems Markup Language (GML) [8], Medical Markup Language (MML) [17], HL7 [10], and Open Document Format (ODF) [21]. This widespread use of XML requires efficient parsing techniques. The importance of efficient XML parsing methods was underscored by Nicola and John [19], showing that the parsing stage is processor and memory consuming, needing main memory as much as ten times the size of the original document.

There are two de facto XML parsing APIs, DOM [2] and SAX [22]. SAX reads the whole document and generates a sequence of events according to the nesting of the

* This project was supported in part by the National Science Foundation Grant IIS-0534530 and by the Department of Energy Grant ER25739.

R. Wagner, N. Revell, and G. Pernul (Eds.): DEXA 2007, LNCS 4653, pp. 75–86, 2007.

elements, and hence it is not possible to skip reading parts of the document as this would change the semantics of the API. On the other hand, DOM allows users to explicitly navigate in the XML document using methods like `getFirstChild()` and `getNextSibling()`. DOM is the most popular interface to traverse XML documents because of its ease of use. Unfortunately, its implementation is inefficient since entire subtrees cannot be skipped when a method like `getNextSibling()` is invoked. This also leads to frequent "Out of Memory" exceptions. In contrast to SAX, parsing a document using DOM could potentially avoid reading the whole document as the sequence of navigation methods may only request to access a small subset of the document. In this work we focus on parsing using a DOM-like interface.

Lazy XML parsing has been proposed (e.g., [25, 20]) to improve the performance of the parsing process by avoiding the loading of unnecessary elements. This approach substitutes the traditional eager evaluation with a lazy evaluation as used by functional programming languages [1]. The architecture shown in Figure 1., based on the terminology of [20], consists of two stages. First, a pre-parsing stage extracts a virtual document tree, which stores only node types, hierarchical structure information, and references to the textual representation of each node. After this structure is obtained, a progressive parsing engine refines this virtual tree on demand, expanding the original virtual nodes into complete nodes with values, attributes, etc. as they are needed.

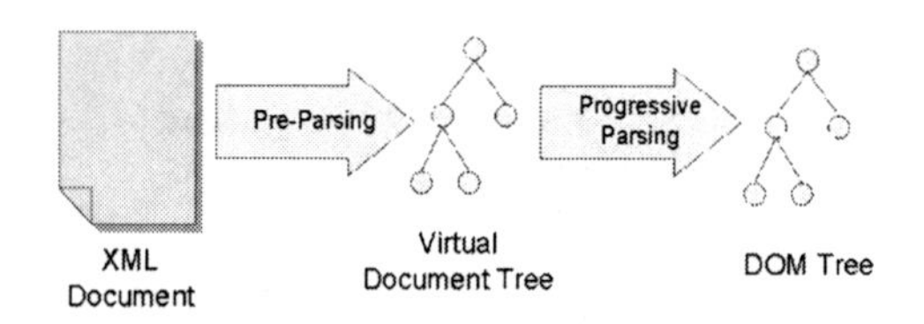


**Fig. 1.** Lazy XML Parser Architecture

Clearly, the lazy parsing technique is a significant improvement. However, it still suffers from the high initial cost of pre-parsing (Figure 1) where the whole document must be read before the lazy/progressive parsing starts. The pre-parsing stage is inevitable due to the lack of internal physical pointers (or something equivalent) within the XML document. Further, the entire virtual document tree must be loaded and processed in main memory during the progressive-parsing stage, i.e. during query processing.

**Overview of Approach:** We call our XML parsing approach double-Lazy Parsing (2LP) because both stages in Figure 1 are lazy, in contrast to previous works where only the second stage is lazy. The first stage is performed offline, when the document is partitioned into a set of smaller XML files, then interlinked using XInclude [26] pointers. The optimal partition size is computed by considering the random versus sequential access characteristics of a hard disk.

The second stage parses a partitioned document, reading a minimal set of partitions to perform the sequence of navigation commands. 2LP loads (pre-parses using the terminology of Figure 1) the partitions in a lazy manner (only when absolutely necessary). In the case of DOM, we maintain an overall DOM tree $D(T)$ which is initially the DOM tree of the root partition $P_0$ of the XML tree $T$. Then $D(T)$ is augmented with the DOM trees $D(P_i)$ of the loaded partitions $P_i$.

Our approach dramatically reduces the cost of the pre-parsing stage by only pre-parsing a typically small subset of the partitions. Furthermore, our approach leads to significantly faster progressive-parsing times than traditional lazy parsing, as we show experimentally, due to the fact that whole subtrees are skipped.

To complement lazy partition loading, our approach also performs lazy unloading of inactive partitions (described in Section 2) if the total amount of main memory used by the DOM tree exceeds a threshold. Hence, in addition to a fast pre-parsing stage, our method also allows DOM-based parsing with limited memory resources. Note that previous lazy parsing techniques can also in principle achieve this to a smaller degree; the virtual document tree must still be stored in memory at all time. However, this optimization is not used in the current implementation of the Xerces DOM parser.

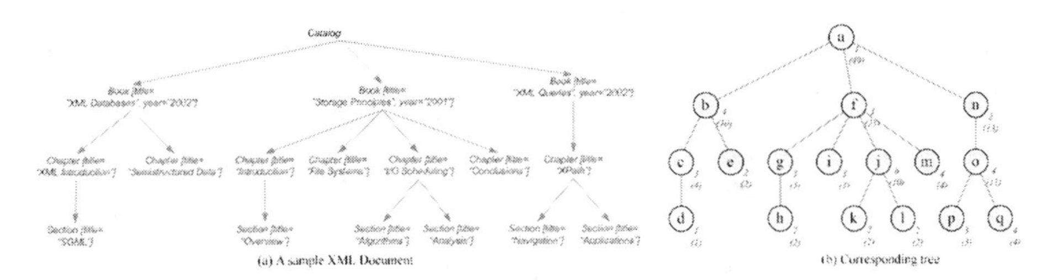

(a) A sample XML Document (b) Corresponding tree

**Fig. 2.** Sample XML Document and Corresponding Tree

A drawback of the 2LP approach is that the XML document is split into a set of smaller XML documents/files. Unfortunately, the XML standard does not support an alternative physical pointer construct (XPointer [29] is logical and not physical) due to the complication this would incur during cross-platform document exchange. We argue and demonstrate in the rest of this paper that the performance gains in XML document navigation far outweigh the drawbacks of document splitting. Further, if physical pointers are introduced for XML in the future, our work can be immediately applied.

This paper makes the following contributions: *(1)* We develop a framework to allow efficient XML parsing, which improves both the pre-parsing and progressive parsing time as well as the memory requirements of both parsing phases. *(2)* We present algorithms to perform partitioning and double-Lazy XML Parsing (2LP) for DOM-like navigation. Note that 2LP-enabled documents are backward compatible i.e., they can be parsed by current XML parsers. *(3)* We show how the theoretically optimal partition size can be computed assuming knowledge of the navigation patterns on complete XML trees and the hard disk characteristics. *(4)* We study our partitioning and parsing algorithms both theoretically and experimentally. Experiments on various XML navigation patterns, including XPath, confirm our theoretical results and show consistent and often dramatic improvement in the parsing times.

The rest of the paper is organized as follows: We describe our double-Lazy parsing techniques in Section 2. Section 3 presents techniques for partitioning the original document into smaller subtrees. Our experiments are discussed in Section 4. We present related work in Section 5. Finally, Section 6 discusses our conclusions and future work.

## 2 2LP on Partitioned XML Documents

Let T be the original XML document, and $P_0, \ldots P_n$ be the partitions to which *T* was split during the partitioning stage (elaborated in Section 3). $P_0$ is the root partition, since it contains the root element of *T*. Figure 3(a) shows an example of a partitioned XML tree. All the partitions are connected by XInclude elements, containing the URI to the partition file. The XInclude elements are represented in the figure by *b'*, *f'*, *j'*. Note that by creating a partition (e.g., $P_2$), the key result is that we facilitate skipping the subtree rooted at this partition. That is, by creating partition $P_2$ we can access directly node *a* from node *f'* to node *n*.

The XML representation of two of the partitions in Figure 3(a) is shown in Figure 3(b). Partition $P_0$ contains the document root and is then the root partition. The subtree rooted at the first *Book* element was partitioned and the *Book* element has been replaced by the XInclude pointer to the XML document of Partition $P_1$. This additional element added to the tree upon partitioning will hold the reference to the root of the partition's subtree. We explain this in detail in Section 3.

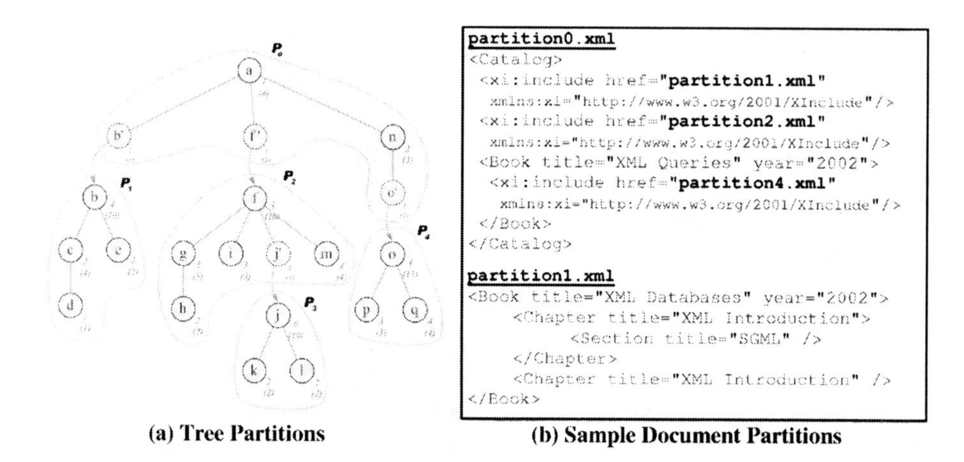


(a) Tree Partitions (b) Sample Document Partitions

**Fig. 3.** Partitioned XML Tree and Document Partitions

Figure 4 describes the process of loading (pre-parsing) a partition. After loading a partition, progressive parsing occurs as needed. The `loadPartition()` method replaces, in the working DOM tree, the XInclude pointer element e with the DOM tree of the partition that *e* points to.

To ensure the double-lazy processing of the partitions, we need to decide when it is absolutely necessary for a partition to be loaded. Intuitively, a partition must be loaded when a navigation method (e.g., `getFirstChild()`) cannot be executed without doing so, that is, the return value of the method cannot be computed otherwise.

```
procedure loadPartition(XIncludeElement e) {
1     newPartitionRoot =
          preParse(e.getAttribute("href"));
2     replace(e, newPartitionRoot);
          /*replace e by newPartitionRoot
          in the DOM tree*/
  }
```

**Fig. 4.** Load Partition Algorithm

Note that if limited memory is available, we unload inactive partitions as needed. A partition is inactive if none of its nodes appear on the path from the root of the XML document to the currently accessed XML node. Traditional replacement techniques can be used to decide which inactive partition to unload like LRU.

We now discuss the 2LP versions of the key DOM methods that may trigger the loading of a partition: `getFirstChild()`, `getNodeName()` and `getTextContent()`. Note that the `getNextSibling()` method cannot trigger a partition loading, because even if the sibling node is an XInclude pointer, we do not have to load the partition before the user asks for the details of the returned node.

```
Node getFirstChild() {
1     if this is XIncludeElement {
2       loadPartition(this);
3     }
4     return firstChild;
  }
```

**Fig. 5.** 2LP version of getFirstChild()

Figure 5 presents the logic to decide the loading of a partition for the `getFirstChild()` method. The original method only returns the `firstChild` member of the current object ("`this`"). In our modification, the loading is performed if the current node is an XInclude element, and it will replace the current object with the root element of the loaded partition. Thus, instead of returning directly the first child of the XInclude node, we return the first child of the root element of the partition.

**Example 2.1.** *Consider the partitioned XML document depicted in Figure 3 (a). Let's consider the root-to-leaf navigation pattern a→f→j→k. We start by parsing and traversing the root partition* $P_0$*. The first node-step, a, is satisfied in* $P_0$*, but to satisfy the second node-step, f, we need to follow the XInclude pointer to partition* $P_2$*. After pre-parsing* $P_2$*, we progressively parse it to reach f. We need to satisfy the last two node-steps by following the pointer to partition* $P_3$*, pre-parsing it to then progressively parse the desired nodes. In this example, we omitted the traversal of partitions* $P_1$ *and* $P_4$*.* □

**Example 2.2.** *Consider again the XML document in Figure 3 (a). Now consider the XPath query /Catalog/Book[@title="Storage Principles"]/Chapter. The acute reader can verify that this query requires loading all the partitions, even when we lazily process the document.* □

Note that in Example 2.2 we had to load partition $P_1$ just to read an attribute of its root element. To save such unnecessary partition loadings we extend the attributes of the XInclude element to contain additional information about the root element of the partition. This may save the loading of a partition when only information about its root node is required. Thus, the partition will be loaded only if the information needed by the navigation is not included in the pointer element. The data duplication to implement this idea is minimal, as shown in Section 4, since internal XML nodes are typically small.

**Table 1.** Inclusion Levels

| *Inclusion Level* | *Data to Include* | *Attribute Name* |
|---|---|---|
| NONE | None | N/A |
| TAG | Tag *(Default)* | xiPartitionTag |
| TAG_ATR | Tag + Attributes | xiPartitionAtr |
| TAG_ATR_TX | Tag + Attributes + Text | xiPartitionTxt |

Table 1 summarizes the different inclusion levels regarding the data from the partition's root element to duplicate in the corresponding XInclude element. The names of the attributes used to store this data in the XInclude element are also displayed. For the TAG_ATR level, we use a single attribute whose value has the form *field1=value1&field2=value2&...*

**Example 2.2. (continued)** *If we extend the XInclude elements depicted in Figure 3(b) according to Inclusion level TAG_ATR and execute the same XPath query, we will find the necessary information about the tag names and attribute values in the XInclude pointer elements. Thus, partitions $P_1$ and $P_4$ will not be loaded, since the attribute values added to the XInclude pointer can help us discriminate which "Chapter" elements satisfy the attribute condition without loading the partition.* □

The detailed code for the `getNodeName()` and `getTextContent()` methods, which varies according to the inclusion level, is available in [5] due to lack of space.

## 3 Partitioning the XML File

Our main goal when partitioning XML documents is to minimize the 2LP parsing time needed for navigating on the document. Other works (i. e. Natix [13, 14, 18]) have addressed the problem of partitioning the XML documents for storage purposes. Our goal here is to minimize the partitions accessed for a given request.

The key criterion to partition the original document is the number of blocks that each partition will span across the hard disk drive (i.e., the partition size). This size criterion is independent of the particular tree-structure (or schema if one exists) and the query patterns, and is shown to lead to efficient partitioning schemes (Section 4). The rationale behind this is that disk I/O performance is dictated by the average size of I/O requests when accesses are random [3]. The size criterion also allows performing a theoretical study of the optimal partition size. In the future, we plan to experiment with more complex partitioning criteria, like using different sizes for deep and shallow partitions to adapt the partition techniques to the underlying XML schema or to other physical characteristics of the document.

It must be noted that if information about the semantics and usage of the XML document is available, it can be used to further optimize the partitioning of the document. For instance, to partition a Mars document [16] we may consider the page boundaries as candidate partitioning points.

**Partitioning Algorithm:** The key idea of the algorithm is a bottom-up traversal of the XML tree, where nodes are added to a partition until the size threshold (in number of blocks) is reached. We show how the partition size is calculated in Section 4. Since we are using XInclude to simulate the physical pointers, we need to comply with the XInclude definition and hence provide partitions that are themselves well-formed XML documents. Thus, our partitions need to have exactly one root element and include a single subtree. This constraint leads to having a few very large partitions since every XML document typically has very few nodes with very high fan-out (e.g., *open_auctions* node in XMark [4]). Still, as shown in Section 4, this does not degrade the parsing performance as these partitions usually need to be fully navigated by XPath queries.

The partitioning algorithm, which is detailed in [5], recursively traverses T in a bottom-up fashion, calculates each subtree's size, and if this size exceeds the partitioning threshold, moves the entire subtree to a new XML document and a new XInclude pointer replaces its root node in the original XML file. Also, depending on the inclusion level flag, specific information of the partition's root node will be added to the newly created XInclude element. Fig. 3 shows the resulting partitioned XML tree for the XML tree of Fig. 1(b) with a threshold of ten blocks per partition. Node $b'$ is the XInclude element which points to the partition rooted at node b. The same holds for nodes $f'$, $j'$, $o'$.

**Partition Size:** To obtain an appropriate value for the partition size, we conduct the following analysis for the root-to-leaf navigation pattern. The details of the cost model and the derivations are available in [5]. Note that performing a similar analysis for general XPath patterns is infeasible due to the complexity and variety of the navigation patterns and axes. In particular, we calculate the average access time to navigate from the root to each of the leaves of the XML document. In Section 4 we show that using the theoretically obtained partition sizes leads to good results for general XPath queries as well. When the XML document is not partitioned (and hence 2LP is not applicable), the average cost of a root-to-leaf traversal is given by the following equation:

$$Cost_{root-leaf}^{noPart} = t_{rand} + N \cdot t_{transf} \tag{1}$$

where $N$ is the number of blocks in $T$, $t_{rand}$ is the random access time needed to reach the root of the tree and $t_{transf}$ is the time required to transfer one block of data for the specific disk drive. Note that the whole tree must be read (pre-parsed in Figure 1) to create the intermediate structure used to later progressively parse the document. No cost is assigned to the progressively parsing since the document has been already loaded in memory during pre-parsing. An equivalent cost model has been derived for the case where the tree has been segmented into equally sized partitions:

$$Cost_{root-leaf}^{Part} = \frac{\ln N}{\ln x}( t_{rand} + x \cdot t_{transf} ) \tag{2}$$

where $x$ is the number of nodes in a partition.

Taking the first derivative with respect to x of the right hand side and equating it to zero provides the optimal partition size.

## 4 Experiments

In this section, we evaluate our XML Partitioning and 2LP schemas. First, we evaluate the theoretical model on the partition size proposed in Section 3. Second, we measure the performance of our techniques with two navigation patterns, root-to-leaf patterns and XPath queries. The experiments were run on a 2.0GHz Pentium IV workstation with 512MB of memory running Linux. The workstation has a 20GB Maxtor D740X disk.

**Evaluation of the Theoretical Model:** We generated XML files of various sizes using the XMark generator [24]. We applied the partitioning algorithm to these documents, with several partition sizes (in blocks) to compare our theoretical model described in Section 3 against experimental results performing the same type of root-to-leaf navigation patterns detailed in [5]. Note that throughout the experiments the 2LP parser is used for partitioned documents and the Xerces for un-partitioned.

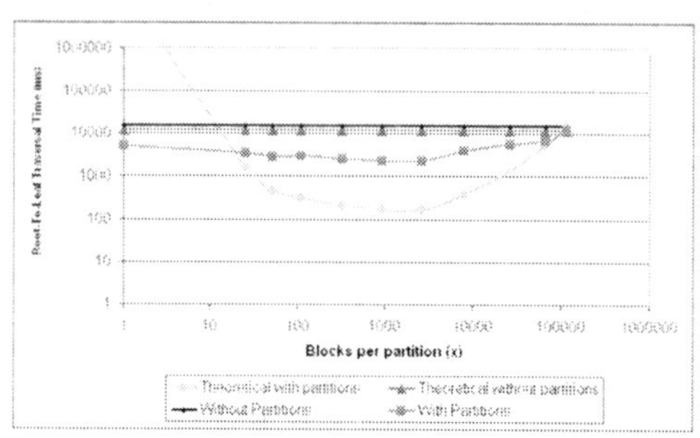


**Fig. 6.** Average Traversal Time for Partition Sizes

Fig. 6 shows the average time to traverse all the root-to-leaf paths for an XML document with XMark factor 0.5 (50MB), running on a Maxtor D740X hard drive as detailed in [5]. The theoretical curves are based on the model presented in Section 3. Notice that the scale is logarithmic and the patterns of the graphs are similar, with a slight deviation in the experimental graph. The gap between the theoretical and experimental graphs is caused because the theoretical model does not take into account the processing overhead and memory requests needed for navigating these paths, but only the I/O time involved. From the graph, we can infer the optimal size of the partition to be 2680 disk blocks, which is approximately one Megabyte. In [5] we show that the theoretical partition size is very close to the experimental one for various document sizes.

**Performance Evaluation:** We now present the evaluation of our approach using two types of navigation patterns, root-to-leaf traversals (also used in [6]) and XPath queries. The results for XPath carry to XQuery as well, since XQuery queries are typically evaluated by combining the results of the involved XPath queries. We adopt the "standard" XPath evaluation strategy described in [7]. As explained in Section 2, the comparisons assume that the XML document has not been already parsed before a query or navigation pattern, that is, we measure both the pre-parsing and progressive parsing times of Figure 1. We measure three time components in the total execution time:

Pre-Parsing: The Xerces parser uses its deferred expansion node feature by initially creating only a simple data structure that represents the document's branching and layout. This phase requires scanning the whole document to retrieve this structure. For un-partitioned documents, it means that the first time we load the file, the whole document has to be traversed and processed; for partitioned documents, every time we process a new partition, it is pre-parsed to create the logical structure in memory.

Progressive Parsing: As the navigation advances, this initial layout built in the pre-parsing phase is refined, and all the information about the nodes is added to the skeleton. This phase is performed only on the visited nodes and will have the same behavior in both un-partitioned and partitioned documents.

Inclusion: This phase is introduced by the 2LP components, and captures the time required to include and import the new partition into the working document. This component does not apply to un-partitioned documents.

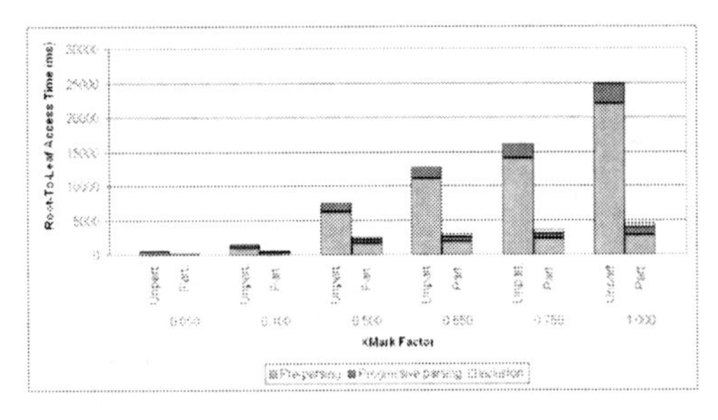

Fig. 7. Root-To-Leaf Access Cost

**Root-to-leaf traversal cost:** Figure 7 shows the average access cost in milliseconds for the root-to-leaf access patterns, comparing the performance for different XMark factors. To compute the average time, we sampled 10% of the leaves of each document, adding each tenth leaf into the sample, and performed root-to-leaf traversals for each sampled leaf. A traversal in this case results in a sequence of parent-to-first-child and sibling-to-next-sibling operations in order to reach the desired leaf. These experiments were performed with the theoretical optimal partition size and the NONE inclusion level (the inclusion level does not impact the simple root-to-leaf traversals).

Note that in addition to the pre-parsing time, 2LP offers a significant improvement of the progressive parsing time as well. This is due to the fact the partitions are equivalent to physical pointers like node to sibling, which are not available in a traditional virtual document tree. These pointers avoid the loading and progressive parsing of unnecessary subtrees.

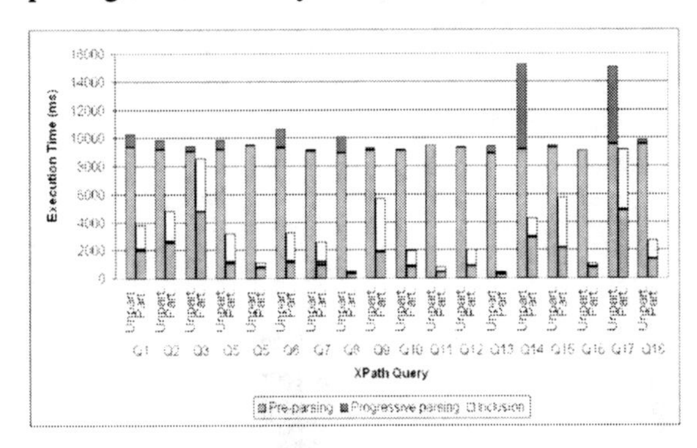


Fig. 8. Average XPath Query Performance

**XPath query cost:** Our second experiment executes a set of XPath queries over the XML data. We selected the performance queries from XPathMark[4]; these queries exploit several execution constructs of the XPath syntax and several navigation axes to illustrate the behavior of our algorithms under a large range of circumstances. The complete list of queries can be found in [5]. We have included the performance queries from XPathMark [4], that is, the ones that test the execution time and not specific XPath functional aspects. We added more queries to have a larger input set in order to obtain more reliable results.

For this set of experiments, we used several XML document sizes corresponding to various XMark factors. Once again, we use the theoretical partition size for partitioning the XML documents. We used the default inclusion level (TAG) for these experiments.

Figure 8 shows the average performance of such queries for three datasets with XMark factors 0.500, 0.750 and 1.000. We see how for un-partitioned files, the pre-parsing time is always similar, since the whole document has to be processed to load the initial layout. For partitioned files, only the required partitions are processed, leading to significant reduction in the pre-parsing phase in most of the cases. We can observe that the partitioned documents perform consistently better than the un-partitioned ones. We have some cases in which the performance of the partitioned documents is almost equal to the performance of the original files. These cases, such as $Q_3$, $Q_9$, $Q_{14}$ and $Q_{15}$, need to traverse most sections of the tree, requiring the inclusion of most partitions.

In the cases of $Q_9$, $Q_{14}$ and $Q_{17}$, we load the partition rooted at *open_auctions,* which has a size of 15MB (due to the fact that each partition must be a well-formed XML document, as explained in Section 3). Pre-parsing and progressively parsing this large partition penalizes these queries and they almost match the execution time of the un-partitioned version. However, in a typical scenario, such large partitions must be completely accessed, except for the rare case when a navigation pattern specifies a child at a particular position (e.g., 1000-th child).

The inclusion time component varies correspondingly to the size of the partitions that have to be included into the working document. We see then that the inclusion component for $Q_3$, $Q_9$, $Q_{14}$ and $Q_{15}$ is large, but again this is caused by the large size of the *open_auctions* partition required to satisfy all these four queries. For these same queries we found large segments of time consumed by the Inclusion operation. The reason is that we rely on the `Document.importNode()` method provided by DOM which traverses the whole imported XML tree and updates the owner document for every single node. Even when the tree is already in memory, this operation is CPU intensive, delaying the process of including the new partition.

**Inclusion levels:** We experimented with different inclusion levels, obtaining practically no space overhead, and observing that the TAG_ATR level is generally the best choice. We show these results in detail in [5].

# 5 Related Work

Noga et al. [20] introduce the idea of Lazy Parsing as presented in Section 1. The virtual document tree can potentially be stored on disk to avoid the pre-parsing stage; however, the entire virtual document tree has to still be read from disk. If a similar technique would be used with 2LP, only the needed portion of the virtual document tree will have to be read to answer the request. Schott and Noga apply these Lazy Parsing ideas to the XSL transformations [23]. Kenji and Hiroyuki [12] have also proposed a lazy XML parsing technique applied to XSLT stylesheets, constructing a pruned XML tree by statically identifying the nodes that will be referred during the transformation process.

There has been progress in developing XML pull parsers [27] for both SAX and DOM interfaces. Also, [28] presents a new API built just one level on top of the XML tokenizer, claiming to be the simplest and the most efficient engine for processing XML.

Van Lunteren et al. [15] propose a programmable state machine technique that provides high performance in combination with low storage requirements and fast incremental updates. A related technique has been proposed by Green et al. [9] to lazily convert an XPath query into a Deterministic Finite Automata (DFA), after which they submit the XML document to the DFA in order to solve the query. They propose a lazy construction opposed to an eager creation, since constructing the DFA with the latter technique can lead to an exponential growth in the size of the DFA. Kiselyov [11] presents techniques to use functional programming to construct better XML Parsers.

## 6 Conclusions

Lazy XML parsing is a significant improvement to the performance of XML parsing but to achieve higher levels of performance there is a need to further optimize the pre-parsing phase during which the whole document is read, as well as the progressive parsing phase during which a query is processed. In this paper, we address this problem by enabling laziness in the pre-parsing phase and allowing skipping the processing of entire (unwanted) subtrees of the document during the progressive parsing phase. To do so, we have proposed a mechanism to add physical pointers in an XML document by partitioning the original document and linking the partitions with XInclude pointers. We have also proposed 2LP, an efficient parsing algorithm for such documents, that implements pre-parsing laziness. These techniques significantly improve the performance of the XML parsing process and can play a significant role in accelerating the wide adoption of XML.

## References

1. Abramsky, S.: The Lazy Lambda Calculus. In: Turner, D. (ed.) Research Topics in Functional Programming, AddisonWesley, London (1990)
2. Document Object Model (DOM) (2006), http://www.w3.org/DOM/
3. Dimitrijevic, Z., Rangaswami, R.: Quality of Service Support for Real-time Storage Systems. In: IPSI (2003)
4. Franceschet, M.: XPathMark: An XPath Benchmark for the XMark Generated Data. In: XSym (2005)
5. Farfán, F., Hristidis, V., Rangaswami, R.: Beyond Lazy XML Parsing Extended Version (2007), http://www.cs.fiu.edu/SSS/beyondLazyExt.pdf
6. Gil, J., Itai, A.: How to pack trees. Journal of Algorithms 32(2), 108–132 (1999)
7. Gottlob, G., Koch, C., Pichler, R.: Efficient Algorithms for Processing XPath Queries. In: VLDB (2002)
8. Geography Markup Language (2006), http://opengis.net/gml/
9. Green, T.J., Miklau, G., Onizuka, M., Suciu, D.: Processing XML streams with deterministic automata. In: ICDT (2003)
10. Health Level Seven XML (2006), http://www.hl7.org/special/Committees/xml/xml.htm
11. Kiselyov, O.: A Better XML Parser Through Functional Programming. In: Krishnamurthi, S., Ramakrishnan, C.R. (eds.) PADL 2002. LNCS, vol. 2257, pp. 209–224. Springer, Heidelberg (2002)
12. Kenji, M., Hiroyuki, S.: Static Optimization of XSLT Stylesheets: Template Instantiation Optimization and Lazy XML Parsing. In: DocEng (2005)

13. Kanne, C.C., Moerkoette, G.: Efficient storage of XML data. In: ICDE 1998 (1999)
14. Kanne, C.C., Moerkoette, G.: A Linear-Time Algorithm for Optimal Tree Sibling Partitioning and its Application to XML Data Stores. In: VLDB (2006)
15. van Lunteren, J., Engbersen, T., Bostian, J., Carey, B., Larsson, C.: XML Accelerator Engine. In: First International Workshop on High Performance XML Processing (2004)
16. Mars Reference: Version 0.7. Adobe Systems Inc., http://download.macromedoa.com/pub/labs/mars/mars_reference.pdf
17. Medical Markup Language (2006), http://www.ncbi.nlm.nih.gov/entrez/query.fcgi?cmd=Retrieve& db=PubMed&list_uids=10984873&dopt=Abstract
18. Natix (2006), http://www.dataexmachina.de/
19. Nicola, M., John, J.: XML Parsing: a Threat to Database Performance. In: CIKM (2003)
20. Noga, M., Schott, S., Löwe, W.: Lazy XML Processing. In: ACM DocEng, ACM Press, New York (2002)
21. OpenDocument Specification v1.0 (2006), http://www.oasis-open.org/committees/download.php/12572/OpenDocument-v1.0-os.pdf
22. Simple API for XML (SAX) (2006), http://www.saxproject.org/
23. Schott, S., Noga, M.: Lazy XSL Transformations. In: ACM DocEng, ACM Press, New York (2003)
24. Schmidt, A., Waas, F., Kersten, M.L., Carey, M.J., Manolescu, I., Busse, R.: XMark: A Benchmark for XML Data Management. In: VLDB (2002)
25. Apache Xerces2 Java Parser: Apache XML Project (2006), http://xml.apache.org/xerces-j/
26. XML Inclusion (2006), http://www.w3.org/TR/xinclude/
27. XML Pull Parsing (2006), http://www.xmlpull.org/index.shtml
28. XML Pull Parser (2006), http://www.extreme.indiana.edu/xgws/xsoap/xpp/
29. XML Pointer Language Version 1.0 (2006), http://www.w3.org/TR/WD-xptr

# Efficient Processing of XML Twig Pattern: A Novel One-Phase Holistic Solution

Zhewei Jiang[1], Cheng Luo[1], Wen-Chi Hou[1], Qiang Zhu[2], and Dunren Che[1]

[1] Computer Science Department, Southern Illinois University Carbondale, Carbondale, IL 62901, U.S.A.
{zjiang,cluo,hou,dche}@cs.siu.edu
[2] Department of Computer and Info. Science, University of Michigan Dearborn, MI, 48128, U.S.A.
qzhu@umich.edu

**Abstract.** Modern twig query evaluation algorithms usually first generate individual path matches and then stitch them together (through a "merge" operation) to form twig matches. In this paper, we propose a one-phase holistic twig evaluation algorithm based on the TwigStack algorithm. The proposed method applies a novel stack structure to preserve the holisticity of the twig matches. Without generating intermediate path matches, our method avoids the storage of individual path matches and the path merge process. Experimental results confirm the advantages of our approach.

## 1 Introduction

XML has become a widely accepted standard for data exchange and integration over the Internet. The ability to process XML queries efficiently plays an important role in the deployment of the XML technology in the future.

The XML twig queries retrieve document elements through a joint evaluation of multiple path expressions [8]. Modern XML twig query processing approaches [1,9,2,6,4], including the Twigstack [2] and other approaches based on it, typically first decompose a twig query into a set of binary patterns or single paths and then search for matches for these individual patterns/paths. Finally, these matches are stitched together to form the answers to the twig query. The overheads incurred in the two-phase approach could be large since the cost to output and then input the individual matches and finally merge them to form twig matches can be substantial, especially when the number of matching paths is large.

To address this problem, we propose a one-phase holistic twig evaluation algorithm that outputs twig matches in their entireties without a later merge process. Instead of outputting individual path matches as soon as they are formed, our method holds the path matches until entire twig matches are formed. The new algorithm yields no intermediate results (to be output and then input), and requires no additional merge phase. Experimental results show that our algorithm compares favorable to Twigstack [2] and Twig2Stack [11].

R. Wagner, N. Revell, and G. Pernul (Eds.): DEXA 2007, LNCS 4653, pp. 87–97, 2007.

The rest of the paper is organized as follows. Section 2 gives a brief survey of twig processing. Section 3 details our one-phase holistic twig evaluation algorithm. Section 4 presents the experimental results. Section 5 concludes.

## 2 Backgrounds

Many twig query evaluation algorithms [1,2,5,9] have been proposed in the literature. Bruno et al. [2] designed the notable and efficient algorithm TwigStack. The algorithm pushes only nodes that are sure to contribute to the final results onto stacks for ancestor-descendent queries. Like most other algorithms, it is a two-phase algorithm, a path matches generation phase followed by a merge phase. Aiming to eliminate the two-phase overhead, Twig2Stack [11], a bottom-up evaluation algorithm, was proposed. Unfortunately, it may push nodes that do not contribute to the final results onto stacks, resulting in some extra work. It utilizes PathStack [2] to reduce runtime memory usage, however, at the price of increased stack manipulation complexity. TJFast [7] employs an innovative encoding scheme, called the extended Dewey code. Although this code can represent the relationships of nodes on a path elegantly, tremendous storage overhead is incurred, especially for longer paths. There are also some other algorithms [6,4] that attempt to improve the performance of Twigstack for parent-child queries for which Twigstack is suboptimal.

## 3 A One-Phase Holistic Twig Join Algorithm

In this section, we present a one-phase twig query evaluation algorithm, called the HolisticTwigStack, based on the TwigStack. The new algorithm preserves all the strengths of the TwigStack and yet has no the aforementioned two-phase overhead. In the following, we illustrate the shortcomings of the TwigStack and Twig2STack by an example.

**Example.** Consider the data tree (a) and twig query (b) in Fig1.

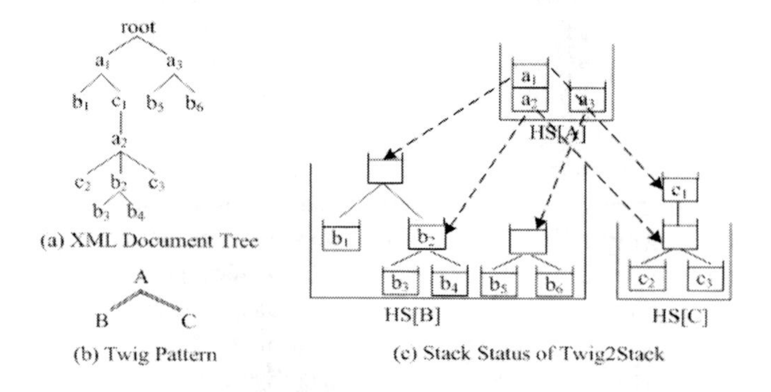


**Fig. 1.** Example of Holistic Twig Join Algorithm

TwigStack generates 12 path matches in the first phase: $a_1/b_1$, $a_1//b_2$, $a_1//b_3$, $a_1//b_4$, $a_2/b_2$, $a_2//b_3$, $a_2//b_4$ for A//B and $a_1/c_1, a_1//c_2, a_1//c_3, a_2/c_2, a_2/c_3$ for A//C. Several nodes, such as $a_1$ and $a_2$, appear multiple times in the matches, resulting in a large intermediate result (larger than the entire data tree itself). A 2-way merge is then needed to merge path matches into twig matches.

Twig2Stack is a one-phase algorithm. Although it does not have the two-phase overhead of TwigStack, it lacks the important advantage of TwigStack, i.e., not pushing any node that does not contribute to the twig matches onto a stack. In the example, Twig2Stack pushes non-contributing nodes $a_3$, $b_5$, and $b_6$, onto stacks. It also creates additional stacks, such as the one connecting $c_2$ and $c_3$, to speed up later query processing, at the cost of increased space complexity.

### 3.1 Notations

Like most twig query processing approaches [1,2,5,9], we adopt the region code scheme. Each node in the XML document tree is assigned a unique 3-ary tuple: (*leftPos*, *rightPos*, *LevelNo*), which represents the left, right positions, and level number of the node, respectively. As in all the stack-based algorithms, there is a stream $T_q$ associated with each pattern node q of the twig query. The elements in each stream are sorted by their leftPos.

We define the *Top_Branch_Node* as the branch node in the twig pattern at the highest level. The *Top_Branch_Node* and the nodes above it in the twig pattern are called *Upper_Nodes*. *Lower_Nodes* refer to nodes that are below the *Top_Branch_Node* in the twig pattern. For example, the *Top_Branch_Node* in Fig 1(b) is A. The document nodes that have the same type of *Top_Branch_Node*, *Upper_Node* and *Lower_Nodes* are referred as *Top_Branch_Element*, *Upper_El -ement* and *Lower_Element*, respectively. We also define ClosestPatternAncestor(e) as the closest ancestor of the element e in the document tree according to the query pattern. For example, in Fig 1(a),ClosestPatternAncestor($b_2$)=ClosestPatternAncestor($b_3$)= ClosestPatternAncestor($b_4$)=$a_2$. We also define a pattern sibling element of e as an element in the document tree that (i) has the same node type as e; (ii) shares the same closest pattern ancestor with e, and (iii) does not have the ancestor-descendent relationship with e. In our example, $b_3$ is a pattern sibling element of $b_4$.

### 3.2 Stack Structure

The overheads incurred in the TwigStack algorithm are caused by the"hasty" output of the individual path matches generated when a leaf element is encountered, which splits the twig matches. We also observe that individual path matches are to be merged together based on the common branch nodes to form the twig matches. In order to keep the holisticity of the twigs, we opt to delay the output of individual path matches and their merging by holding them in lists of stacks until all elements under the same *Top_Branch_Element* are processed. To hold multiple matching paths, we associate each *Upper_Node* with a single stack and *Lower_Node* with lists of stacks, as compared with one stack for each query node in the TwigStack.

Like other stack-based algorithms, an element is pushed onto a stack whose top element is an ancestor of the incoming element. In addition, we require that the incoming element must have the same closest pattern ancestor as the top element. For an element that does not satisfy the above conditions, it will be stored in a new stack. Elements that have the same closest pattern ancestor will be linked together for convenient access. Fig 2 shows the stack structure of our algorithm, named HolisticTwigStack, of the given example in Fig 1. In the structure, we have represented the closest pattern ancestor relationships by solid arrows, e.g., $a_1$ to $S_{B1}$. The stacks of the same node type that share the same closest pattern ancestor are linked by dotted arrows. Take $S_{B2}$ and $S_{B3}$ as an example. They are linked together by using a dotted arrow pointing from $b_3$ to its sibling element $b_4$. Only 9 nodes are stored in HolisticTwigStack, which is less than Twig2Stack and TwigStack.

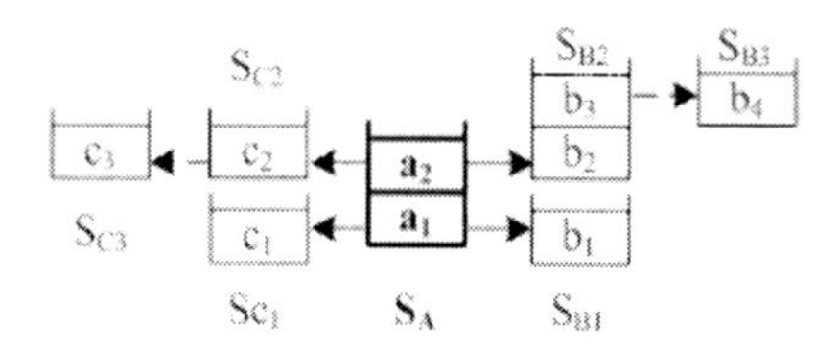


**Fig. 2.** Data Structure of HolisticTwigStack

It is observed that the commonly referenced ancestor-descendent relationships between query nodes, such as A//B and A//C, can be easily inferred from the closest pattern ancestor relationships by assuming that the ancestor elements of a stack "inherit" the relationships (i.e., the solid arrows) of their descendants (in the same stack). For example, $a_1$ could inherit the solid links of $a_3$, that is, $a_1$ implicitly also has links to $S_{B2}$ and $S_{C2}$ and thus the ancestor-descendent relationships (i.e., A//B, A//C) that $a_2$ has with other query nodes. The twig matches for the elements in the root stack can be constructed easily by traversing the links following the query pattern.

## 3.3 Algorithm

Our algorithm, named *HolisticTwigStack*, is presented in Algorithm 1. The algorithm computes the answer to a query twig pattern Q in one phase.

Elements are checked for satisfaction of structural relationships in the same way as is done in TwigStack by recursively calling the function *getNext()*, which is defined in TwigStack [2]. We attempt to withhold the elements sharing the same *Top_Branch_Element* in memory until the element with the disjoint range arrives. The reason lies in the fact that the arrival of the disjoint element actually eliminate the possibility of the subtree rooted at this *Top_Branch_Element* to participate any further match. Like in the TwigStack, an incoming element e is pushed onto a stack (line 18) only if it is surely to contribute to a twig match (line 5); otherwise, we simply advance to next element (line 20). Before pushing e onto a stack, we further check if it falls beyond the range covered by the top element in the stack of *Top_Branch_Node*, namely, $S_{TBN}$ (line 6). Note that the top element covers only a subrange of other lower elements' ranges in the same stack, so the elements in $S_{TBN}$ are visited top-down until $S_{TBN}$ is empty or the current top element can cover the incoming element e . If e is disjoint with the current top element of the $S_{TBN}$ and all the paths under it have already been formed, twig matches sharing the same *Top_Branch_Element* are output(line 8). Furthermore, the top element of $S_{TBN}$ is popped out and its ancestor element in $S_{TBN}$ (if there is any) inherit its relationship with its descendent elements in twig pattern(line 14). Otherwise, all the elements in the stacks are cleaned out (line 12). The same process also need to be conducted over the stacks of *Upper_Node* in order to clear all the nodes that are unnecessary for future matches(line 17). After having reached the end of streams (end(Q)), we output the remaining twig matches related to root elements left in *S_root* (line 22-30).

The basic idea of procedure *MoveElementToStack()* is given below. More details can be found in [10]. If the incoming element e is an *Upper_Element*, we simply push it onto the corresponding stack. If e is a *Lower_Element*, a more complicated process is involved. Assume the type of e is q and the stacks for the *Lower_Node* q have been numbered in their order of creation as $S_{q_1}, S_{q_2}, \ldots, S_{q_n}$. For each incoming element of type q , we check if it is a descendent of the top element of the last stack $S_{q_n}$ and if it has the same closest pattern ancestor as the top element of the stack. (The correctness of only checking the last stack will be given in the next subsection.) If so, we push the element onto the last stack $S_{q_n}$; otherwise, we push the element onto a new stack $S_{q_{n+1}}$. In the latter case, we shall check if it has the same closest pattern ancestor as any top element of a stack on an existing list. If so, we append $S_{q_{n+1}}$ to the end of that list; Otherwise, we directly link $S_{q_{n+1}}$ to its closest pattern ancestor.

*ShowTwigSolution()* is called to output twig matches rooted at the current root element. The twig matches can be formed by following the solid and dotted links between stacks and the ancestor-descendent relationships between the elements in the same stacks. Interested readers are referred to [10] for details of the algorithm.

### 3.4 Analysis of Algorithm

In this section, we show the correctness of the *HolisticTwigStack* algorithm. Due to the space limitation, readers are referred to [10] for details and proofs.

First, we introduce some terms and properties of TwigStack. subtreeNodes(q) include node q itself and all its descendants in the query pattern Q. An element has a minimal descendent extension if there is a solution for the sub-query rooted at q, composed entirely of the head elements for subtreeNodes(q). Here, the head element of q, denoted as $h_q$, is defined as the first element in $T_q$ that participates in a solution for the sub-query rooted at q [2].

TwigStack ensures that the element $e_q = next(T_q)$ is pushed onto the stack if and only if (i) element next($T_q$) has a descendent element $e_{q_i}$ in each of the stream $T_{q_i}$, for $q_i$=children(q), and (ii) each of the element $e_{q_i}$ recursively satisfies the first property [2].

**Algorithm 1:** HolisticTwigStack (Q)

```
1  begin
2      while not end(Q) do
3          q=getNext(Q);
4          e=the current first element in the stream of q;
5          if (q is of root type) or (at least one element in S_parent(q) covers e) then
6              while (!Empty(S_TBN) and e is disjoint with S_TBN's top element) do
7                  if (no leaf stack is empty) then
8                      ShowTwigSolution(S_root, S_root.size-1);
9                  end
10                 TopElem=s.pop();
11                 if Empty(S_TBN) then
12                     Clean all child stacks of s;
13                 else
14                     Update the links between stacks;
15                 end
16             end
17             Remove all the elements in each S_Upper_Node that are disjoint with e
               and update the links appropriately;
18             MoveElementToStack(q, e);
19         end
20         AdvanceList(q);
21     end
22     while not empty(S_root) do
23         ShowTwigSolution(S_root, S_root.size-1);
24         S_root.Pop();
25         if not Empty(S_root) then
26             Update the links between stacks;
27         else
28             Clean all child stacks;
29         end
30     end
31 end
```

**Lemma 1.** Let $e_1, e_2, \ldots, e_m$ be the sequence of elements pushed onto the stacks during the execution of the algorithm. Then, $e_1.left < e_2.left < \ldots < e_m.left$.

**Lemma 2.** The elements popped out of stack of *Upper_Node* at line 10, 17 24 and elements deleted at line 12 and line 28 from the stack lists of *Lower_Node* participate no further matches. So, the deletions are safe.

Earlier in the *MoveElementToStack()* procedure, when an element is to be pushed onto an appropriate stack, instead of examining all the stacks on the list we only check if it is a descendent of the top element of the last stack (of its type) and if it has the same closest pattern ancestor as the top element. This optimization is based on the following lemma.

**Lemma 3:** For each incoming *Lower_Element*, either it can only be a descendent of the top element of the last stack (of its type), or it is not a descendent of any top element of the stacks (of its type).

**Lemma 4:** MoveElementToStack(q) correctly places the elements onto stacks and links elements to their closest pattern ancestors.

Thus the elements in the stacks can be reached in any case as long as it participates the final match, which is guaranteed by the property of getNext(). The relevant proof can be found in [2].

**Theorem 1:** Given a twig query pattern Q and an XML document tree D, Algorithm HolisticTwigStack correctly returns all answers to Q on D.

**Theorem 2:** Given a query twig pattern Q, comprising of n nodes and only ancestor-descendent edges, over an XML document D, Algorithm *HolisticTwigStack* has the worst-case I/O and CPU time complexities linear in the sum of sizes of the n input lists and the output list. Furthermore, the worst-case space complexity of *HolisticTwigStack* is the sum of the sizes of the n input lists.

Let us make simple comparisons with TwigStack. Our algorithm may take a little more CPU time in stack manipulation, but TwigStack would require extra time and space to store and merge the individual path matches. It is important to note that our stacks store twig matches rooted at elements that are currently in the stack of *Top_Branch_Node*, while TwigStack stores all the twig matches, in the form of individual path matches. Furthermore, nodes shared by multiple paths would have to be stored repeatedly in individual paths. Therefore, our algorithm in general uses much less space than TwigStack. For example, we are given a twig pattern where the root A has k children: $A//C_1[//C_2]\ldots[//C_k]$ and the document tree has n A nodes matching the given pattern. Each node $A_i$ ($1 \leq i \leq n$) has $n_{i_1}$ $C_1$ children, $n_{i_2}$ $C_2$ children,...,$n_{i_k}$ $C_k$ children. Thus the total number of matches is $\sum_{i=1}^{n}\prod_{j=1}^{k} n_{ij}$. Both methods need the same time $O(\sum_{i=1}^{n}\prod_{j=1}^{k} n_{ij})$ to form all the combinations. But our method requires only $O(Max_{i,1\leq i\leq n}(\sum_{j=1}^{k} n_{ij})+1)$ space while TwigStack needs $O(\sum_{i=1}^{n}\sum_{j=1}^{k}(n_{ij}+1)) = O(\sum_{i=1}^{n}\sum_{j=1}^{k} n_{ij}+nk)$ to store the intermediate results. The space consumption of TwigStack can be further exacerbated when the twig pattern is deeper or more common nodes are shared by leaf nodes. In worst case, the intermediate result size of TwigStack is O(K×P)=O(J), where K is the sum of the lengths of the input lists for all leaf nodes, and P is the length of the longest root-to-leaf path in the twig pattern.

## 4 Experimental Results

In this section, we evaluate the performance of the *HolisticTwigStack* algorithm against TwigStack and Twig2Stack using both synthetic and real datasets. Twig2Stack was shown to have better perfromance than TJFast [7] in [11], and thus the latter is ommitted here.

### 4.1 Experimental Set-Up

The synthetic datasets consist of XMark, TreeBank, and other datasets generated by the random generator in [2]. The depths of the randomly generated trees vary from 5 to 10, fan-outs from 2 to 10, and the number of labels is set at 7. DBLP is the real dataset used in the experiments.

We use three types of twig queries in the experiments: Q1 represents a set of shallow but wide queries (dept=2, width= 4 to 5, in the patterns); Q2 a set of deep but narrow queries (dept= 3 to 5, width=2) and Q3 balanced queries (dept= 2 to 3 and width= 2 to 3).

We store the intermediate results of TwigStack in memory. This provision, eliminates the expensive output/input cost of the TwigStack, however, at the price of memory consumption.

### 4.2 Experimental Results

TwigStack's time is broken into two parts to reflect the cost of the two-phase algorithm with the lower and upper parts of the bars in the figures corresponding to the first and second phases, respectivley. Fig3 (a) and (b) show the performance on the random datasets. As observed, HolisticTwigStack is the fastest. TwigStack is slow for two reasons: (i) the recursive execution of ShowSolutionWithBlocking is time-consuming; and (ii) the scan and merge of the intermediate results in the second phase requires an extra amount of time. The entire execution times of TwigStack are 18%, 7% and 34% longer than ours for the three types of queries, respectively. Twig2Stack turns out to be the slowest. It requires nodes to be pushed/popped onto/out of stacks twice before forming the final results, incurring higher cost in stack manipulation. It takes 2.2, 1.9 and 1.3 times longer than HolisticTwigStack for the three types of queries, respectively. Fig3 (b) shows the space utilization. TwigStack uses, on average, 7,328, 5,208 and 9,192 bytes for Q1, Q2 and Q3, respectively, while ours uses only 1,068, 1,464 and 984 bytes for the same queries. This is because TwigStack stores all path matches in memory (due to our provision to reduce time consuming disk I/O), while ours stores only twig matches that are rooted at elements currently in the root stack, a subset of theirs, in memory. As for the Twig2Stack, although it does not store intermediate results, it may push nodes that do not contribute to the final results onto stacks, referred to as non-optimality. Recall that TwigStack and our method (based on the TwigStack), guarantees such optimality. It uses more space, from 9% to 209%, than our method, though better than TwigStack.

Figures 3(c) and(d) show the results of XMark. TwigStack consumes 8% to 18% more time than ours, as shown in (c) and 4 to 5,186 times more space than ours, as shown in (d). Please note in some cases, the space of HolisticTwigStack and Twig2Stack is too small to be shown. Usually, the larger the number of matches, the larger the space usage ratio. Our method is also much faster than Twig2Stack; it consumes only 15%, 96% and 59% of Twig2Stack's time. However, the space utilizations are almost same. This is due to the uniform and balanced tree structure of XMark and the relatively uniform distribution of query matches.

Similar results are observed on TreeBank, which are shown in (e) and (f).

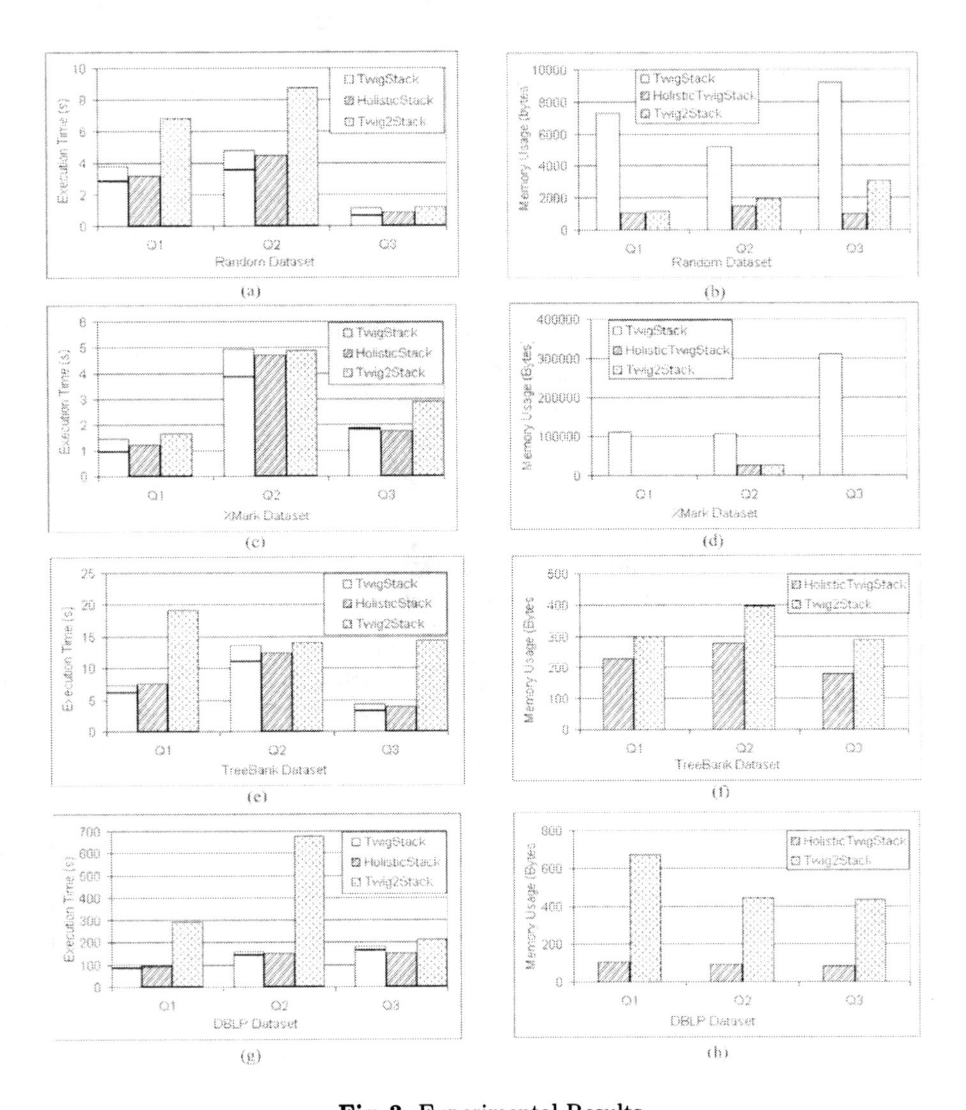


**Fig. 3.** Experimental Results

The results of DBLP are shown in (g) and (h). Although TwigStack is only slightly slower than ours, its intermediate result sizes are too large (around 100,000 times larger than the other two methods) to show in the chart. Twig2Stack uses much more time, 3.1, 4.4, and 1.4 times more, than our method, due to its complex stack manipulation and overhead for processing "non-productive" nodes in the stacks (i.e., the non-optimality). Note that Twig2Stack uses much more space than ours in this experiments than in the XMark and TreeBank as there are more "non-productive" being pushed onto stacks.

In summary, our algorithm generally runs faster and requires less memory than TwigStack and Twig2Stack. The larger the query result sizes, the better our algorithm, compared with TwigStack. The more complex the tree structures, the better our algorithm, compared with Twig2Stack.

## 5 Conclusion

In this paper, we propose an efficient one-phase holistic twig pattern matching algorithm based on the TwigStack. We lower the expensive time and space overhead incurred in the two-phase algorithms by devising a novel stack structure to hold matching paths until entire twig matches that share the same *Top_Branch_Element* are formed. Experimental results have confirmed that our method is significantly more efficient in all cases tested.

## References

1. Al-Khalifa, S., Jagadish, H.V., Koudas, N., Patel, J.M., Srivastava, D., Wu, Y.: Structural Joins: A Primitive for Efficient XML Query Pattern Matching. In: Proceedings of IEEE ICDE Conference, pp. 141–152. IEEE Computer Society Press, Los Alamitos (2002)
2. Bruno, N., Koudas, N., Srivastava, D.: Holistic Twig Joins: Optimal XML Pattern Matching. In: SIGMOD Conference, pp. 310–321 (2002)
3. Chen, S., Li, H., Tatemura, J., Hsiung, W., Agrawal, D., Candan, K.: Twig2Stack: Bottom-up Processing of Generalized-Tree-Pattern Queries over XML Documents. In: SIGMOD Conference, pp. 283–294 (2006)
4. Chen, T., Lu, J., Ling, T.W.: On Boosting Holism in XML Twig Pattern Matching Using Structural Indexing Techniques. In: ACM SIGMOD international conference, pp. 455–466 (2005)
5. Jiang, H., Wang, W., Lu, H., Yu, J.X.: Holistic Twig Joins on Indexe XML Documents. In: Proceedings of the 29th VLDB conference, pp. 310–321 (2003)
6. Lu, J., Chen, T., Ling, T.W.: Efficient Processing of XML Twig Patterns with Parent Child Edges: A Look-ahead Approach. In: Proceedings of CIKM, pp. 533–542 (2004)
7. Lu, J., Ling, T.W., Chan, C.-Y., Chen, T.: From Region Encoding To Extended Dewey: On Efficient Processing of XML Twig Pattern. In: VLDB, pp. 193–204 (2005)
8. Polyzotis, N., Garofalakis, M., Ioannidis, Y.: Selectivity Estimation for XML Twigs. In: ICDE, pp. 264–275 (2004)

9. Zhang, C., Naughton, J., DeWitt, D., Luo, Q., Lohman, G.: On Supporting Containment Queries in Relational Database Management Systems. In: SIGMOD, pp. 425–436 (2001)
10. Jiang, Z., Luo, C., Hou, W., Zhu, Q., Wang, C.-F.: An Efficient One-Phase Holistic Twig Join Algorithm for XML Data (2005), `http://www.cs.siu.edu/~zjiang`
11. Chen, S., Li, H., Tatemura, J., Hsiung, W., Agrawal, D., Candan, K.: Twig2Stack: Bottom-up Processing of Generalized-Tree-Pattern Queryies over XML Documents. In: VLDB, pp. 283–294 (2006)

# Indexing Set-Valued Attributes with a Multi-level Extendible Hashing Scheme

Sven Helmer[1], Robin Aly[2], Thomas Neumann[3], and Guido Moerkotte[4]

[1] University of London, United Kingdom
[2] University of Twente, The Netherlands
[3] Max-Planck-Institut für Informatik, Germany
[4] University of Mannheim, Germany

**Abstract.** We present an access method for set-valued attributes that is based on a multi-level extendible hashing scheme. This scheme avoids exponential directory growth for skewed data and thus generates a much smaller number of subqueries for query sets (so far fast-growing directories have prohibited hash-based index structures for set-valued retrieval). We demonstrate the advantages of our scheme over regular extendible hashing both analytically and experimentally. We also implemented a prototype and briefly summarize the results of our experimental evaluation.

## 1 Introduction

Efficiently retrieving data items with set-valued attributes is an important task in modern applications. These queries were irrelevant in the relational context since attribute values had to be atomic. However, newer data models like the object-oriented (or object-relational) models support set-valued attributes, and many interesting queries require a set comparison. An example would be to find persons who match a job offering. In this case the query set *required-skills* is a subset of the persons' set-valued attribute *skills*. Note that we assume to work on a large number of objects, but with limited set cardinality. We believe that this is the most common case found in practice. This belief is backed by our observations on real applications for object-oriented or object-relational databases (as found, for example, in product and production models [6] and molecular databases [21]).

One way to support the efficient evaluation of queries is by employing index structures. Hash-based data structures are among the most efficient access methods known, allowing retrieval in nearly constant time. Nevertheless, when applying hash-based techniques to set-valued retrieval on secondary storage we have to meet two main challenges. As it is too expensive to completely reorganize hash tables on secondary storage, dynamic hashing schemes, like linear hashing [15] and extensible hashing [4], are used. However, dynamic hashing schemes exhibit exponentially growing directory sizes on skewed data. (Even if the employed hash function works reasonably well, it cannot offset the effect of multiple copies of certain sets.) Moreover, evaluating set-valued queries on hash tables is difficult: in order to access all subsets/supersets of a query set, we have to generate all possible subsets/supersets of the query set and probe the hash table with

R. Wagner, N. Revell, and G. Pernul (Eds.): DEXA 2007, LNCS 4653, pp. 98–108, 2007.

them. Obviously, in the average case this will have an exponential running time. However, many of the generated sets are redundant, as the respective entries in the directory of the hash table point to the same (shared) buckets or are empty.

We propose a dynamic multi-leveled hashing scheme to remedy this situation. As we have shown in [10], this hashing scheme can handle skewed data much better than existing schemes. Here we focus on adapting this index structure to retrieving data items with set-valued attributes efficiently. We demonstrate that hash-based schemes are a viable approach to indexing set-valued attributes.

The remainder of this work is organized as follows. The following section describes related work and the context of our work. We give a brief introduction to superimposed coding and show how to apply signatures to set-valued retrieval in Section 3. Section 4 contains a short description of our (regular) multi-level hashing scheme, while Section 5 describes how this scheme is adapted to set-valued retrieval. In Section 6 we summarize the results of our experimental evaluation. Section 7 concludes the paper.

## 2 Related Work

Work on the evaluation of queries with set-valued predicates is few and far between. Several indexes dealing with special problems in the object-oriented [2] and the object-relational data models [18] have been invented, e.g. nested indexes [1], path indexes [1], multi indexes [16], access support relations [13], and join index hierarchies [22]. These index structures focus on evaluating path expressions efficiently.

One of the dominant techniques for indexing set-valued attributes is superimposed coding, where sets are represented by bit vector signatures. Existing techniques for organizing signatures include: sequential files [12], hierarchical organization (signature trees [3], Russian Doll Trees [7]), and partitioning (S-tree split [19], hierarchical bitmap index [17]).

At first glance, methods from text retrieval appear to be similar to set retrieval. However, text retrieval methods (like [23]) focus on partial-match retrieval, that is, retrieving supersets of the query set. Set retrieval also supports subset and exact queries, which are relevant and common for example in molecular databases (e.g. searching for characteristic parts of a large molecule).

## 3 Preliminaries

### 3.1 Querying Set-Valued Attributes

Let us assume that our database consists of a finite set $O$ of data items $o_i$ $(1 \leq i \leq n)$ having a finite set-valued attribute $A$ with a domain $D$. Let $o_i.A \subseteq D$ denote the value of the attribute $A$ for some data item $o_i$. A *query predicate* $P$ consists of a set-valued attribute $A$, a finite *query set* $Q \subseteq D$, and a *set comparison operator* $\theta \in \{=, \subseteq, \supseteq\}$. A query of the form $\{o_i \in O | Q = o_i.A\}$ is called an *equality query*, a query of the form $\{o_i \in O | Q \subseteq o_i.A\}$ is called a *subset query*, and a query of the form $\{o_i \in O | Q \supseteq o_i.A\}$ is called a *superset query*. Note that *containment queries* of the form $\{o_i \in O | x \in o_i.A\}$ with $x \in D$ are equivalent to subset queries with $Q = \{x\}$.

### 3.2 Signature-Based Retrieval

*Superimposed coding* is a method for encoding sets as bit vectors. It uses a coding function to map each set element to a bit field of length $b$ ($b$ is the *signature length*) such that exactly $k < b$ bits are set. The code for a set (also known as the set's *signature*; abbreviated as *sig*) is the *bitwise or* of the codes for the set elements [5,14].

The following properties of signatures are essential (let $s$ and $t$ be two arbitrary sets):

$$s \; \theta \; t \Longrightarrow \text{sig}(s) \; \theta \; \text{sig}(t) \;\; \text{for} \; \theta \in \{=, \subseteq, \supseteq\} \tag{1}$$

where $\text{sig}(s) \subseteq \text{sig}(t) := \text{sig}(s) \& \tilde{\ } \text{sig}(t) = 0$ and $\text{sig}(s) \supseteq \text{sig}(t) := \text{sig}(t) \& \tilde{\ } \text{sig}(s) = 0$ (& denotes *bitwise and* and ˜ denotes *bitwise complement*).

As set comparisons are very expensive, using signatures as filters is helpful. Before comparing the query set $Q$ with the set-valued attribute $o_i.A$ of a data item $o_i$, we compare their signatures $sig(Q)$ and $sig(o_i.A)$. If $sig(Q) \; \theta \; sig(o_i.A)$ holds, then we call $o_i$ a drop. If additionally $Q \; \theta \; o_i.A$ holds, then $o_i$ is a *right drop*; otherwise it is a *false drop*. We have to eliminate the false drops in a separate step. However, the number of sets we need to compare in this step is drastically reduced as only drops need to be checked.

There are three reasons for using signatures to encode sets. First, they are of fixed length and hence very convenient for index structures. Second, set comparison operators on signatures can be implemented by efficient bit operations. Third, signatures tend to be more space efficient than explicit set representation.

## 4 Multi-level Hashing

As in other dynamic hashing schemes (e.g. [4,15]), our multi-level hashing index (MLH index) is divided into two parts, a directory and buckets. In the buckets we store the full hash keys of and pointers to the indexed data items. We determine the bucket into which a data item is inserted by looking at a prefix $h_g$ of $g$ bits of a hash key $h$. Let us take a look at a non-hierarchical hashing scheme first. It has a directory with $2^g$ entries, where $g$ is called the *global depth* of the hash table. The prefix $h_g$ identifies one of these entries and we follow the link in this entry to access the corresponding bucket.

On the other hand, in our MLH index things are done differently. We also check the prefix of a hash key to find the right bucket, but the length of the prefix that we check may vary depending on the level in the directory where we finally find the correct bucket (our hashing scheme is not necessarily balanced).

### 4.1 General Description

Due to space constraints, we can only give a brief description; for details see [9,10]. We employ a multi-level extendible hash tree in which hash tables share pages according to a buddy scheme. In this buddy scheme, *z-buddies* are hash tables that reside on the same page and whose stored hash keys share a prefix of $z$ bits. Consequently, all buddy hash tables in our tree have the same global depth $z$.

Let us illustrate our index with an example. We assume that a page can hold $2^n$ entries of a hash table directory. Furthermore, we assume that the top level hash table directory (also called the root) is already filled, contains $2^n$ different entries at the moment, and that another overflow occurs (w.l.o.g. in the first bucket). In this case, we allocate a new hash table of global depth 1 (beneath the root) to distinguish the elements in the former bucket according to their $(n+1)$st bit. However, we do this not only for the overflowing bucket, but also for all 1-buddies of this bucket. The hash tables for the buddies are created in anticipation of further splits. All of these hash tables can be allocated on a single page, resulting in the structure shown in Figure 1.

In a naive hierarchical hash tree, we would have allocated just one hash table with depth $n$ for the overflowed bucket. If other buckets overflow, we allocate new recursive hash tables for them as well. The main problem with naive hash trees is waste of memory: almost all entries in these newly allocated hash tables share the same buckets, i.e. we do not need a directory with depth $n$ yet. At first glance our scheme does not seem that much different, as we also allocate a whole page. However, due to the data skew we expect splits near buckets that have already split. Even when the anticipated splits do not occur, we can eliminate unnecessary directory pages.

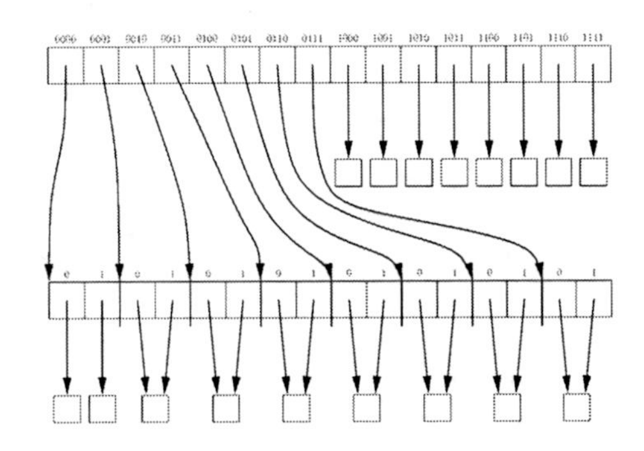

**Fig. 1.** Overflow in our multi-level hash tree

If another overflow occurs in one of the hash tables on level 2, causing it to grow, we increase the global depth of all hash tables on this page by 1, doubling their directory sizes. We now need two pages to store these tables, so we split the original page and copy the

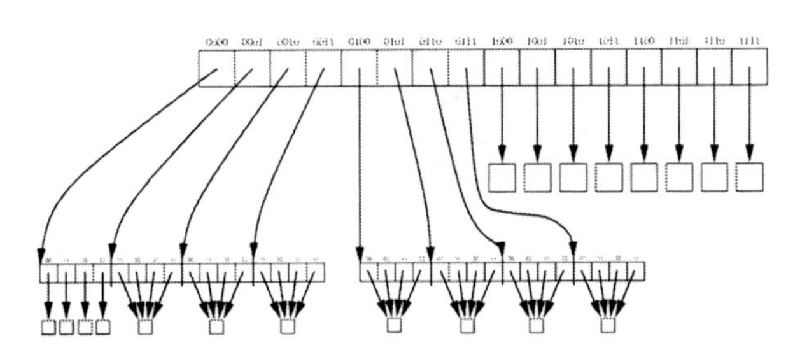

**Fig. 2.** Overflow on the second level

```
lookup(hashkey, value) {
  currentLevel = 0;

  while(true) {
    pos = relevant part of hashkey
          (for current level);
    determine offset;
    pos = pos + offset;

    nodeId = slot[pos];
    if(nodeId is a null-pointer) {
      return false;
    }
    if(node is bucket) {
      search in bucket;
      return answer;
    }
    currentLevel++;
  }
}
```

(a) Pseudocode for lookups

```
lookup(hashkey, buddies, curdepth, localwidth) {
  for all nodes n in buddies {
    subhash = hashkey[b-curdepth...b-depth-width];
    for all sub-/supersets s of subhash {
      pos = buddyoffset(n) + s;
      nodeId = slot[pos];
      if(nodeId is not a null pointer) {
        if(node is an unmarked bucket) {
          scan(node);
          add content to answer;
          mark(node);
        }
        else {
          add nodeId and localwidth to children;
        }
      }
    }
  }
  for all c in children {
    lookup(hashkey, c.nodeId,
           curdepth+localwidth, c.localwidth);
  }
}
```

(c) Pseudocode for looking up sub-/supersets

```
insert(hashkey, value) {
  currentLevel = 0;

  while(true) {
    pos = relevant part of hashkey
          (for current level);
    determine offset;
    pos = pos + offset;

    nodeId = slot[pos];
    if(nodeId is a null-pointer) {
      allocate new bucket;
      insert pointer to bucket into hash table;
      insert data item;
      return;
    }
    if(node is bucket) {
      if(node is not full) {
        insert data item;
        return;
      }
      if(local depth of bucket <
         global depth of table) {
        split bucket;
        adjust hash table;
        insert(hashkey, value);
        return;
      }
      if(global depth < bits per level) {
        split inner node;
        adjust buddies;
        insert(hashkey, value);
        return;
      }
      insert new level;
      insert(hashkey, value);
      return;
    }
    currentLevel++;
  }
}
```

(b) Pseudocode for insertions

**Fig. 3.** Pseudo-code for multi-level hashing

content that does not fit to a new page. Then we adjust the pointers in the parent directory. The left half of the pointers referencing the original page still point to this page, the right half to the new page (see Figure 2).

The space utilization of our index can be improved by eliminating pages with unnecessary hash tables. The page on the right-hand side of the second level in Figure 2 is superfluous, as the entries in the directories of all hash tables point to a single bucket, i.e. all buckets have local depth 0. In this case, the page is discarded and all buckets are connected directly to the hash table on the next higher level.

Due to our buddy scheme, we have a very regular structure that can be exploited. Indeed, we can compute the global depths of all hash tables (except the root) by looking at the pointers in the corresponding parent table. Finding $2^{n-i}$ identical pointers there means that the referenced page contains $2^{n-i}$ $i$-buddies of global depth $i$. Consequently, we can utilize the whole page for storing pointers, as no additional information has to be kept.

### 4.2 Lookups

Lookups are easily implemented (for the pseudocode see Figure 3(a)). We have to traverse inner nodes until we reach a bucket. On each level we determine the currently relevant part of the hash key. This gives us the correct slot in the current hash table. As more than one hash table can reside on a page, we may have to add an offset to access the right hash table. Due to the regular structure, this offset can be easily calculated. We just shift the last $n - i$ bits of the relevant pointer in the parent table by the size of a hash table on the shared page. If $n - i = 0$, we do not need an offset, as only one hash table resides on this page. If we reach a bucket, we search for the data item. If the bucket does not exist (no data item is present there at the moment), we hit a NULL-pointer and can abort the search.

### 4.3 Insertions

After finding the bucket where the new data item has to be inserted (using the lookup procedure), we have to distinguish several cases for inserting the new item (for the pseudocode see 3(b)). We concentrate on the most difficult case, where an overflow of the bucket occurs and the global depth of the hash table on the current level increases. The other cases can be handled in a straightforward manner.

If the hash table has already reached its maximal global depth (i.e. it resides alone on a page), we add a new level with $2^{n-1}$ hash tables of global depth 1 to the existing index structure (comparable to Figure 1). If we have not reached the maximal global depth yet (i.e. the hash table shares a page with its buddies), the global depth of all hash tables on this page is increased by 1. The hash tables on the first half of the page remain there. The hash tables on the second half of the page are moved to a newly allocated page. Then the pointers in the parent hash table are modified to reflect the changes. We optimize the space utilization at this point if we discover that the buckets of all hash tables in one of the former halves have a local depth of one (or are not present yet). In this case (compare the node in the lower right corner of Figure 2) we do not need this node yet and connect the buckets directly to the parent hash table.

## 5 Adapting ML-Hashing to Set-Valued Queries

Using a (non-hierarchical) hashing scheme in a naive way to evaluate a set-valued query is quite straightforward. All the hashing keys employed in our scheme are made up of signatures encoding sets. When processing a query we first determine the signature of the query set via superimposed coding. Depending on the type of the query (subset or superset query) we generate all supersets or all subsets of the query signature's prefix $h_g$ and initiate subqueries with all of these generated sets. When we reach a bucket, we compare the full query signature to all signatures stored there to decide whether to access a data item or not. For our multi-level hashing scheme we generate the relevant supersets and subsets of the query signature on demand on each level of the data structure. If we encounter buckets on our way down we also compare the full query signature to the signatures stored in each bucket. Figure 3(c) shows this algorithm in pseudocode (parameters for the lookup function are the hashkey, a set consisting of the root node, current depth 0, and the local width of the root table). For insertions the same code as in Figure 3(b) is used.

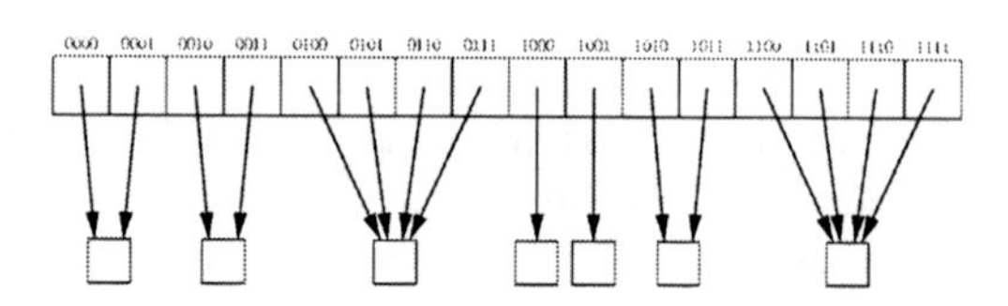

**Fig. 4.** Accessing a non-hierarchical hash table

## 5.1 Example

The following example demonstrates the difference between non-hierarchical hashing schemes and our multi-level approach. Let $A$ be a non-hierarchical hash table with a global depth of four. We wish to obtain the supersets of our query set $Q$ with signature sig($Q$) = 001011101110. The relevant prefix of sig($Q$) is 0010, and for $A$ we must now generate all eight superset prefixes, namely 0010, 0011, 0110, 1010, 0111, 1011, 1110, and 1111. Thus, for a non-hierarchical hash table, we must start eight subqueries to access three of the seven buckets (see also Figure 4).

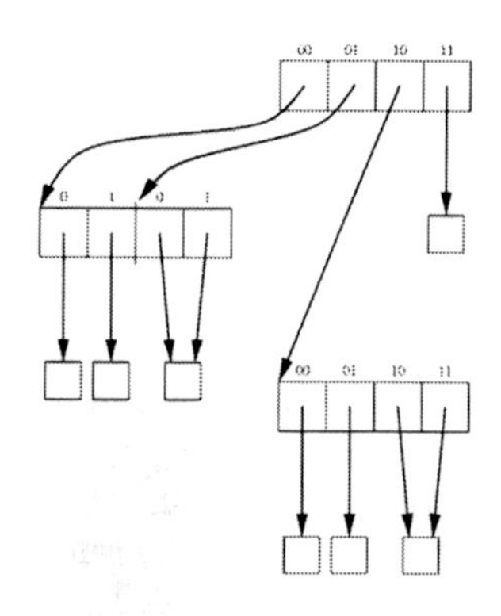

**Fig. 5.** Accessing a multi-level hash table

For our multi-level hashing approach, on the other hand, we begin by generating only the top level supersets 00, 01, 10, and 11 and then the superset 1 for the hash table on the left-hand side of the second level, followed by the supersets 10 and 11 for the hash table on the right-hand side of the second level (see also Figure 5). Thus, we need to generate only seven rather than eight supersets; at first glance, this may not seem like huge savings, but the next section will show that the savings grow when the tables are larger.

## 5.2 Comparison of ML-Hashing with Regular Extendible Hashing

If skewed data is inserted into a hash table, the directory of a non-hierarchical hashing scheme grows exponentially. This is bad news for the naive method of generating all subset or supersets, as on average we have to generate

$$\frac{2^{\lfloor \frac{g}{2} \rfloor} + 2^{\lceil \frac{g}{2} \rceil}}{2} \tag{2}$$

signatures (including the original prefix of sig($Q$)).[1] For large values of $g$ this is clearly infeasible. The worst thing is that most of these signatures are generated needlessly. Hash tables containing skewed data look a lot like the one depicted in Figure 4. In this example sixteen entries share seven buckets, which means that most of the subqueries will access the same buckets over and over again.

How do we cope with this situation? First of all, our MLH index can handle skewed data much better than other dynamic hashing schemes resulting in a much smaller directory. Summarizing the results from [9,10], in which we have substantiated our claim experimentally, we can say that the main idea is to unbalance the hierarchical directory of our hash table on purpose. We did this because obviously we are unable to change the fact that skewed data has been inserted into our hash table, meaning that we have many data items on our hands whose hash keys share long prefixes. In order to distinguish these data items we need a hash table with a large depth. However, we want to make sure that other data items are not "punished" for this. Second, when generating subsets and supersets of query signatures while evaluating set-valued queries, we do not generate them en bloc for the whole prefix. Instead, we generate the appropriate subsets and supersets for each level separately. On each level we have hash tables with a maximum depth of $n$, so we have to generate $\frac{2^{\lfloor \frac{n}{2} \rfloor}+2^{\lceil \frac{n}{2} \rceil}}{2}$ signatures on average. We have to do this for each level we look at. Let us assume that the largest prefix we distinguish in our MLH index is $g$. Then we generate

$$\frac{2^{\lfloor \frac{n}{2} \rfloor} + 2^{\lceil \frac{n}{2} \rceil}}{2} \cdot \lfloor \frac{g}{n} \rfloor + \frac{2^{\lfloor \frac{g \bmod n}{2} \rfloor} + 2^{\lceil \frac{g \bmod n}{2} \rceil}}{2} \tag{3}$$

signatures in the average case.[2]

Formula (3) does not yet consider that we can have hash tables with different depths on the same level in our directory. If the left page on the second level in Figure 2 were to split again, this would result in two pages containing two hash tables with depth three each. The other page on the second level is unaffected by this, still keeping its four hash tables with depth two. So in the worst case we have to generate signatures for each depth up to $n$ on each level (except the first; if $g < n$ use Formula (2)):

[1] Here we assume that on average half of the bits in a signature are set to 0 and half are set to 1. This is the case if the parameters $b$ and $k$ (the size of a signature and the number of set bits per hash value) have been optimized correctly.

[2] If we traverse all levels of the directory.

$$\frac{2^{\lfloor \frac{n}{2} \rfloor} + 2^{\lceil \frac{n}{2} \rceil}}{2} + \left( \sum_{i=1}^{n} \frac{2^{\lfloor \frac{i}{2} \rfloor} + 2^{\lceil \frac{i}{2} \rceil}}{2} \right) \cdot \lfloor \frac{g-n}{n} \rfloor + \sum_{i=1}^{g \bmod n} \frac{2^{\lfloor \frac{i}{2} \rfloor} + 2^{\lceil \frac{i}{2} \rceil}}{2} \qquad (4)$$

For a closed-form formula of (4) see our technical report [8]. Figure 6 compares the number of generated signatures for our hierarchical directory versus a non-hierarchical directory. As can be clearly seen, the curves for the hierarchical directories break away at some point from the exponentially growing curve for non-hierarchical directories. This happens when the top-level directory page reaches $n$, the maximum depth of the hierarchical hash tables.

In summary we can say that our MLH index is suited better for set-valued retrieval than other hash-based indexes, because it does not need exponential running time for generating the subqueries and it is able to cope better with data skew.

## 6 Summary of Experimental Evaluation

Due to space constraints, we can only give a summary of the experimental evaluation here. For a detailed description see [8].

MLH clearly shows the best behavior among all the index structures we compared it to: a sequential signature scan [12], an extensible signature hashing scheme [11], and two S-tree approaches [20] (one with a linear splitting and one with a quadratic splitting algorithm). It is best both in terms of the number of page accesses and total running time when evaluating subset queries. While for uniformly or mildly skewed data, ESH achieves a performance comparable with that of MLH, the drawbacks of ESH become apparent when the data is heavily skewed: in that case, ESH suffers due to directory growth and the exponential cost of generating subqueries. The scanning methods (SIGSCAN and SETSCAN) which have mainly been added as a reference are not able to compete with MLH either. The big surprise is the hierarchical S-tree index structure. In contrast to the results presented in [20] we show that the tree-based access methods are not suitable for indexing set-valued attributes, because they do not scale - the prevalence of all-1

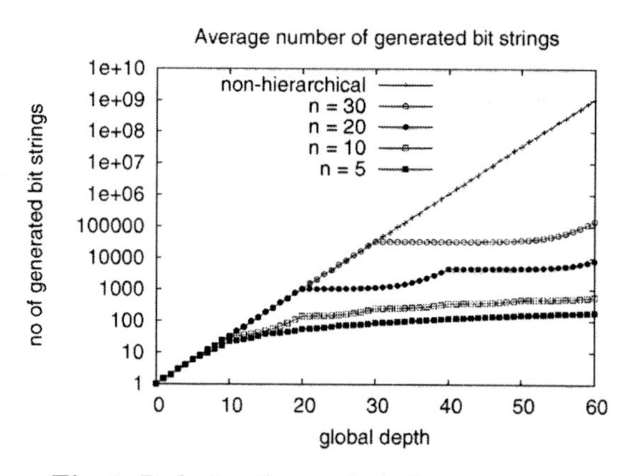


**Fig. 6.** Reducing the number of generated sets

nodes nullifies the inner nodes' filtering capacity. The superiority of hash-based schemes for equality queries does not come as a big surprise, since point queries are the strong point of hash table approaches. We have demonstrated that even in terms of index size, MLH copes extremely well with skewed data: unlike ESH the directory does not grow exponentially. Instead, the growth is linear, much as for lightly skewed or uniformly distributed data.

## 7 Conclusion

We presented the first secondary-storage, hash-based access method for indexing set-valued attributes that is able to outperform other index structures for set retrieval. Until now the fast directory growth of hash-based schemes has prevented their use for evaluating queries with subset and superset queries, as the number of subqueries that had to be submitted was exponential in the size of the directory. Our approach generates a number of subqueries linear in the global depth of the hash table. We demonstrated the competitiveness of our index structure analytically (and experimentally).

Although superimposed coding and dynamic hashing schemes have attracted some attention when they first appeared, they were not able to make their way into industrial strength database systems. One of the main reasons was their susceptibility to skewed data, which robust, data-driven index structures like $B^+$-trees were able to handle much better. Our multi-level hashing scheme represents an interesting compromise between data-driven and space-driven data structure and could renew the interest in hash-based, superimposed coding schemes.

## References

1. Bertino, E., Kim, W.: Indexing techniques for queries on nested objects. IEEE Trans. on Knowledge and Data Engineering 1(2), 196–214 (1989)
2. Cattell, R. (ed.): The Object Database Standard: ODMG 2.0. Morgan Kaufmann, San Francisco (1997)
3. Deppisch, U.: S-tree: A dynamic balanced signature index for office retrieval. In: Proc. of the 1986 ACM Conf. on Research and Development in Information Retrieval, Pisa (1986)
4. Fagin, R., Nievergelt, J., Pippenger, N., Strong, H.R.: Extendible hashing – a fast access method for dynamic files. ACM Transactions on Database Systems 4(3), 315–344 (1979)
5. Faloutsos, C., Christodoulakis, S.: Signature files: An access method for documents and its analytical performance evaluation. ACM Transactions on Office Informations Systems 2(4), 267–288 (1984)
6. Grobel, T., Kilger, C., Rude, S.: Object-oriented modelling of production organization. In: Tagungsband der 22. GI-Jahrestagung, Karlsruhe, September 1992, Springer, Heidelberg (1992)
7. Hellerstein, J.M., Pfeffer, A.: The RD-tree: An index structure for sets. Technical Report 1252, University of Wisconsin at Madison (1994)

8. Helmer, S., Aly, R., Neumann, T., Moerkotte, G.: Indexing Set-Valued Attributes with a Multi-Level Extendible Hashing Scheme. Technical Report BBKCS-07-01, Birkbeck, University of London, http://www.dcs.bbk.ac.uk/research/techreps/2007/
9. Helmer, S., Neumann, T., Moerkotte, G.: A robust scheme for multilevel extendible hashing. Technical Report 19/01, Universität Mannheim (2001), http://pi3.informatik.uni-mannheim.de
10. Helmer, S., Neumann, T., Moerkotte, G.: A robust scheme for multilevel extendible hashing. In: Yazıcı, A., Şener, C. (eds.) ISCIS 2003. LNCS, vol. 2869, pp. 220–227. Springer, Heidelberg (2003)
11. Helmer, S., Moerkotte, G.: A performance study of four index structures for set-valued attributes of low cardinality. VLDB Journal 12(3), 244–261 (2003)
12. Ishikawa, Y., Kitagawa, H., Ohbo, N.: Evaluation of signature files as set access facilities in OODBs. In: Proc. of the 1993 ACM SIGMOD, Washington, pp. 247–256. ACM Press, New York (1993)
13. Kemper, A., Moerkotte, G.: Access support relations: An indexing method for object bases. Information Systems 17(2), 117–146 (1992)
14. Knuth, D.E.: The Art of Computer Programming. In: Sorting and Searching, Addison Wesley, Reading, Massachusetts (1973)
15. Larson, P.A.: Linear hashing with partial expansions. In: Proc. of the 6th VLDB Conference, Montreal, pp. 224–232 (1980)
16. Maier, D., Stein, J.: Indexing in an object-oriented database. In: Proc. of the IEEE Workshop on Object-Oriented DBMSs, Asilomar, California (September 1986)
17. Morzy, M., Morzy, T., Nanopoulos, A., Manolopoulos, Y.: Hierarchical bitmap index: An efficient and scalable indexing technique for set-valued attributes. In: Kalinichenko, L.A., Manthey, R., Thalheim, B., Wloka, U. (eds.) ADBIS 2003. LNCS, vol. 2798, pp. 236–252. Springer, Heidelberg (2003)
18. Stonebraker, M., Moore, D.: Object-Relational DBMSs: The Next Great Wave. Morgan Kaufmann, San Francisco (1996)
19. Tousidou, E., Bozanis, P., Manolopoulos, Y.: Signature-based structures for objects with set-valued attributes. Information Systems 27(2), 93–121 (2002)
20. Tousidou, E., Nanopoulos, A., Manolopoulos, Y.: Improved methods for signature-tree construction. The Computer Journal 43(4), 301–314 (2000)
21. Will, M., Fachinger, W., Richert, J.R.: Fully automated structure elucidation - a spectroscopist's dream comes true. J. Chem. Inf. Comput. Sci. 36, 221–227 (1996)
22. Xie, Z., Han, J.: Join index hierarchies for supporting efficient navigation in object-oriented databases. In: Proc. Int. Conf. on Very Large Data Bases (VLDB), pp. 522–533 (1994)
23. Zobel, J., Moffat, A., Ramamohanarao, K.: Inverted files versus signature files for text indexing. Technical Report CITRI/TR-95-5, Collaborative Information Technology Research Institute (CITRI), Victoria, Australia (1995)

# Adaptive Tuple Differential Coding

Jean-Paul Deveaux, Andrew Rau-Chaplin, and Norbert Zeh

Faculty of Computer Science, Dalhousie University, Halifax NS Canada
jpdeveaux@starcatcher.ca, arc@cs.dal.ca, nzeh@cs.dal.ca

**Abstract.** It is desirable to employ compression techniques in Relational OLAP systems to reduce disk space requirements and increase disk I/O throughput. Tuple Differential Coding (TDC) techniques have been introduced to compress views on a tuple level by storing only the differences between consecutive ordered tuples. These techniques work well for highly regular data in which the differences between tuples are fairly constant but are less effective on real data containing either skew or outliers. In this paper we introduce Adaptive Tuple Differential Coding (ATDC), which employs optimization techniques to analyze blocks of tuples to detect large tuple differences, with the purpose of isolating them to minimize their negative effect on the compression of neighbouring tuples. Our experiments show that this new algorithm provides an increase in compression ratio of 15–30% over TDC on typical real datasets.

## 1 Introduction

Many types of information systems, particularly Relational On-Line Analytical Processing (ROLAP) systems, must store ordered multi-dimensional views on disk. Data compression is often critical to their success due to the massive size of the views involved. A properly implemented compression algorithm can save disk space and reduce the overall amount of time required to answer queries, as long as the overhead required to compress and decompress the data is less than the reduction in disk I/O time resulting from the compression.

In order to use compression in a live database environment, compression and decompression has to be fast, and the basic database functionality (insert, query, update in place) has to be retained. This rules out standard general-purpose compression techniques, as they are too computationally expensive and their strength lies in compressing large files rather than, say, individual disk blocks.

Tuple Differential Coding (TDC) techniques initially introduced by Ng and Ravishankar [6] work by storing differences between consecutive tuples and provide view compression that is often superior to traditional compression techniques both in terms of compression ratio achieved and compression and decompression time required. However, while TDC methods perform well on databases where the gaps between successive tuples are reasonably small and constant, they often deteriorate on real data containing either skew or outliers. For an overview of compression techniques as applied to information systems, see [3,5].

Given a sequence of $n$ tuples $T = [t_1, \ldots, t_n]$ to be encoded, TDC breaks it into subsequences, each of which is stored in a separate block. The subsequence

R. Wagner, N. Revell, and G. Pernul (Eds.): DEXA 2007, LNCS 4653, pp. 109–119, 2007.

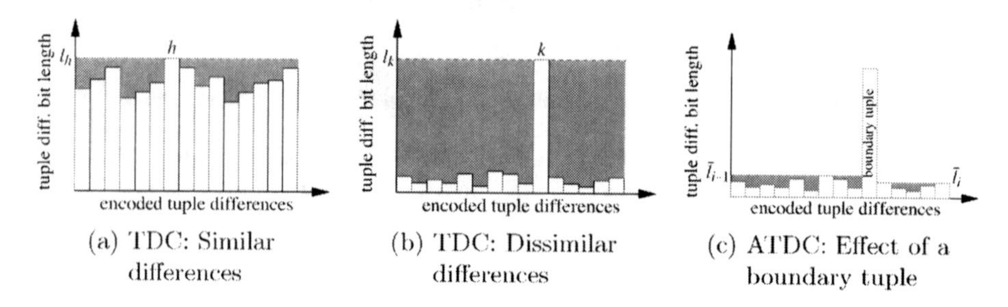


(a) TDC: Similar differences

(b) TDC: Dissimilar differences

(c) ATDC: Effect of a boundary tuple

**Fig. 1.** Encoded tuple difference values where the shaded area represents wasted space

$T[a,b] = [t_a, \ldots, t_b]$ stored in a given block is encoded as follows: For a tuple $t_j$, let $\phi(t_j)$ be its "standard" encoding discussed in Section 2.1, let $\Delta(t_j) = \phi(t_j) - \phi(t_{j-1})$, let $l_j = \lceil \log_2(\Delta(t_j) + 1) \rceil$, and let $l^*$ be the number of bits required to encode any tuple using encoding $\phi$. Then the sequence $T[a,b]$ is stored as the sequence $[\phi(t_a), \Delta(t_{a+1}), \ldots, \Delta(t_b)]$, plus a constant amount of meta-data discussed later. The $\phi(t_j)$ value is stored as a length-$l^*$ bit string. Each of the $\Delta(t_j)$ values is stored using $l_{a+1,b} = \max_{a+1 \le j \le b} l_j$ bits. This saves $(b-a)(l^* - l_{a+1,b})$ bits of space compared to storing values $\phi(t_a), \ldots, \phi(t_b)$ explicitly. The number of tuples stored in a block varies and depends on the number of bits needed to encode the tuples. In particular, tuples are added to blocks one by one. If so far, tuples $t_a, \ldots, t_j$ have been added to the current block, the next tuple $t_{j+1}$ is added to the same block if $l^* + (j+1-a)l_{a+1,j+1}$ is no more than the number of bits that fit in a block; otherwise, $t_{j+1}$ starts a new block.

Since the number of bits used to store the difference values in each block is determined by the largest difference value in the block, TDC performs best when the differences $\Delta(t_j)$ in a block are small and do not vary much; more precisely, when the $l_{a+1,b}$ value for the sequence $T[a+1,b]$ is by only a small constant factor $\alpha > 1$ greater than the average number of bits required to encode these difference values: $l_{a+1,b} \le \alpha \cdot \sum_{j=a+1}^{b} l_j/(b-a)$.

This is illustrated in Figure 1(a). In this figure, the largest difference $\Delta(t_h)$ occurs in position $h$, forcing us to encode all differences in $T[a+1,b]$ using $l_{a+1,b} = l_h$ bits. The shaded area represents the number of bits wasted by storing the differences as fixed-length bitstrings compared to encoding each value $\Delta(t_j)$ in $l_j$ bits. However, since most encoded tuple difference values in this block require close to $l_h$ bits to be encoded, padding them to length $l_h$ does not waste much space in this case. Figure 1(b), on the other hand, shows a scenario where a single difference value (at position $k$) forces us to use significantly more than $\sum_{j=a+1}^{b} l_j$ bits to encode the difference values in the block. As a result, the wasted bits represented by the grey area now account for a major portion of the space used to store the difference values in the block.

This phenomenon should not be unexpected in a dataset made up of real (as opposed to synthetic) data. Real data will contain clusters and other patterns not always found in synthetic datasets. Large tuple differences are representative of "gaps" in real data and should at least be tolerated, if not anticipated.

In this paper we introduce Adaptive Tuple Differential Coding (ATDC), which employs greedy optimization techniques to analyze blocks of tuples to detect large tuple differences, with the purpose of isolating them to minimize their negative effect on the compression of neighbouring tuples. Our experiments show that ATDC provides an increase in compression ratio of 15–30% over standard TDC on typical real datasets, depending on the distribution of the data in the domain space. Additionally, the ATDC algorithm proves to be very robust in situations where there are large gaps due to data skew or outliers. Our experiments show that a good implementation of the ATDC algorithm does not incur any performance penalty compared to reading and writing uncompressed data and compared to a variant of TDC called XTDC [5]. Given the increasing gap between processor speeds and disk access rates, we believe that the savings in I/O-time obtained using the ATDC algorithm will outweigh the investment in compression and decompression time in the near future, thus making ATDC a simple and effective tool for increasing the performance of modern information systems that store relational tables on disk.

Section 2 discusses the ATDC algorithm. Section 3 discusses our experimental results. Concluding remarks are given in Section 4.

## 2 The ATDC Algorithm

As we have already done for the TDC algorithm in the introduction, we discuss the ATDC algorithm in the context of storing a sequence $T = t_1, \ldots, t_n$ of tuples in a sequence of blocks so that each block can be decoded in isolation.

Structurally, ATDC follows the basic framework of the TDC method. The tuples in the given sequence $T = [t_1, \ldots, t_n]$ are converted into integers $\phi(t_i)$ using an encoding function $\phi$. Then the sequence $T$ is divided into subsequences $T_1, \ldots, T_k$, where $T_i = [t_{a_i}, \ldots, t_{b_i}]$, $1 = a_1 < \ldots < a_k \leq b_k = n$, and $b_i = a_{i+1} - 1$, for $1 \leq i < k$. We call the sequences $T_1, \ldots, T_k$ *chunks*. The first tuple $t_{a_i}$ in each chunk $T_i$ is stored explicitly using its encoding $\phi(t_{a_i})$, while all subsequent tuples $t_j$ in $T_i$ are represented by their difference values $\Delta(t_j)$. We call $t_{a_i}$ the *boundary tuple* of $T_i$. Each chunk $T_i$ is stored as the sequence $[n_i, \bar{l}_i, \phi(a_i), \Delta(t_{a_i+1}), \ldots, \Delta(t_{b_i})]$, where $n_i = b_i - a_i + 1$ is the number of tuples in $T_i$ and $\bar{l}_i = l_{a_i+1,b_i}$; $\phi(a_i)$ is stored as a length-$l^*$ bit string and each $\Delta(t_j)$ is stored as a length-$\bar{l}_i$ bit string.

The key difference between TDC and ATDC is the definition of chunks and their association with physical disk blocks. TDC declares a tuple $t_j$ to be the boundary tuple of a new chunk $T_i$ whenever adding $t_j$ to $T_{i-1}$ would increase the number of bits required to encode $T_{i-1}$ beyond the capacity of a disk block. Each disk block then stores one chunk.

ATDC on the other hand chooses boundary tuples to be those tuples $t_j$ whose difference values $\Delta(t_j)$ are significantly higher than those of the tuples in their vicinity, as these are exactly the tuples that negatively affect the compression ratio of TDC. (We describe the exact choice of these tuples in Section 2.2.) The space savings resulting from this strategy are visualized in Figure 1(c), which shows the same sequence as Figure 1(b), but encoded using ATDC. By storing tuple $t_k$ as a boundary tuple, we can store the tuples preceding and succeeding $t_k$ in significantly fewer bits than using TDC. The penalty we pay is that we need to store $\phi(t_k)$ explicitly, as well as values $n_i$ and $\bar{l}_i$ for the chunk $T_i$ starting with boundary tuple $t_k$. Let $c_{\mathbf{boundary}}$ be the number of bits required to store the values $n_i$, $\bar{l}_i$, and $\phi(t_k)$. It is worthwhile to make $t_k$ a boundary tuple if the space savings for storing the difference values before and after $t_k$ exceed $c_{\mathbf{boundary}} - l_k$.

As a result of choosing boundary tuples as described above, chunks may vary greatly in size, and we may end up storing more than one chunk in each disk block. The boundary tuples in all chunks in a disk block are encoded using the same number of bits, as are the tuple differences in each chunk. The number of bits used to encode tuple differences in different chunks, however, may differ.

More precisely, in order to fill disk blocks completely and in order to avoid loading the whole data set to be encoded into memory, we apply the following strategy to pack the data into blocks: Assuming that the blocks we have filled so far store tuples $t_1, \ldots, t_{a-1}$, we load the next $m = b - a + 1$ tuples $t_a, \ldots, t_b$ into memory, encode these tuples using one of the encoding functions discussed in Section 2.1 and apply one of the boundary tuple selection algorithms from Section 2.2 to this sequence. We then iterate over the sequence of tuples and pack them into the current physical block until it is full. For each tuple $t_j$, if it is marked as a boundary tuple, we add the meta-information of its chunk and $\phi(t_j)$ to the physical block. For any other tuple $t_j$, we encode $\Delta(t_j)$ using $\bar{l}_i$ bits, where $T_i$ is the chunk containing $t_j$.

We choose the number $m$ of tuples to be packed into a disk block so that we expect that their encoding requires at least the number of bits that can fit into a block. Thus, we may end up storing only a subsequence $[t_a, \ldots, t_{b'}]$, $b' \leq b$, of tuples in the current block. Tuples $t_{b'+1}, \ldots, t_b$ are reconsidered when filling the next block. On the rare occasion when encoding tuples $t_a, \ldots, t_b$ uses less than a block-full of space, we double the number of tuples we consider and restart the whole packing procedure for the current block.

The following two subsections discuss the two main factors affecting compression ratio and compression/decompression time: the choice of the encoding function $\phi$ and the selection of boundary tuples.

### 2.1 Encoding Tuples

We consider two encoding functions: a *mixed-radix* (MR) and a *bit-shift* (BS) encoding. Using the MR-encoding, a tuple $t = (a_1, \ldots, a_d)$ is encoded as

$$\phi(a_1, \ldots, a_d) = \sum_{i=1}^{d} \left( a_i \prod_{j=i+1}^{d} \mathrm{card}(j) \right),$$

where $\mathrm{card}(j)$ denotes the cardinality of dimension $j$ (ie, the values in this dimension are integers between 0 and $\mathrm{card}(j) - 1$). Using the BS-encoding, each value $a_j$ is encoded using $\lceil \log_2 \mathrm{card}(j) \rceil$ bits, and the value of $\phi(a_1, \ldots, a_d)$ is the concatenation of these bit strings.

The advantage of the MR-encoding is that it uses the minimum number of bits necessary to encode all possible tuple values in the given view. It is computationally expensive to decode, however, as computing $\phi^{-1}(e)$ incurs one division per dimension. An encoding using the BS-encoding can be expected to waste half a bit per dimension on average, but its inverse is extremely fast to compute using simple bit-shift operations. We investigate the impact of this trade-off on compression ratio and compression/decompression time in our experiments.

## 2.2 Selection of Boundary Tuples

The key step in the ATDC algorithm is the selection of boundary tuples. We describe this process as if we were to apply it to the whole sequence $T = [t_1, \ldots, t_n]$. Recall, however, that we apply it only to subsequences $[t_a, \ldots, t_b]$ expected to be long enough to fill a disk block in encoded form.

We define the set of boundary tuples incrementally, starting with $t_1$ as the only initial boundary tuple. Given the current sequence of boundary tuples, we inspect subsequences $T[a, b]$ that currently do not contain any boundary tuples, choose a candidate tuple $t_j$, $a \leq j \leq b$, in each such sequence and decide whether making $t_j$ a boundary tuple reduces the amount of space necessary to encode $T[a, b]$; if so, we add $t_j$ to the set of boundary tuples.

We present two algorithms that choose subsequences $T[a, b]$ and the candidate boundary tuples $t_j$ in these subsequences differently. Both algorithms satisfy the condition that encoding $T[a, b]$ takes $(b - a + 1) \cdot l_j$ bits if $t_j$ is not chosen as a boundary tuple, and $(j - a) \cdot l_{a,j-1} + c_{\text{boundary}} + (b - j) \cdot l_{j+1,b}$ bits if $t_j$ is chosen as a boundary tuple. Thus, choosing $t_j$ as a boundary tuple reduces the amount of space required to store the sequence $T[a, b]$ if and only if $(j - a) \cdot l_{a,j-1} + c_{\text{boundary}} + (b - j) \cdot l_{j+1,b} < l_j \cdot (b - a + 1)$. In this case, we say that tuple $t_j$ satisfies the *boundary tuple condition with respect to sequence* $T[a, b]$.

*Top-down boundary tuple selection.* The tuples that are most likely to decrease the cost of storing subsequences of $T$ when chosen as boundary tuples are those whose tuple differences are large relative to the tuple differences of their neighbours. Our first algorithm, called the *top-down boundary tuple selection algorithm* or TOP-DOWN, is based on this observation and can be described recursively: Initially, let $t_1$ be the only boundary tuple and invoke TOP-DOWN on the sequence $T[2, n]$. Given a sequence $T[a, b]$, let $i$ be the index such that $a \leq i \leq b$ and $l_i = \max_{a \leq j \leq b} l_j$. If tuple $t_i$ satisfies the boundary tuple condition w.r.t. sequence $T[a, b]$, we add $t_i$ to the set of boundary tuples and recurse on sequences $T[a, i - 1]$ and $T[i + 1, b]$. Otherwise, we choose no boundary tuples in $T[a, b]$.

When implemented naively, this algorithm takes $O(n^2)$ time. Next we describe a faster method to implement the TOP-DOWN algorithm, which runs in $O(n)$ time. The key is to store the entire tuple sequence in a *Cartesian tree* [8]. This is a binary tree in which each node has a *key* and a *priority*. The tree is a binary search tree w.r.t. the keys, that is, for every node, the keys in its left subtree are less than the key of the node itself, which in turn is less than the keys in the right subtree; and it is heap-ordered w.r.t. the priorities, that is, no node has a

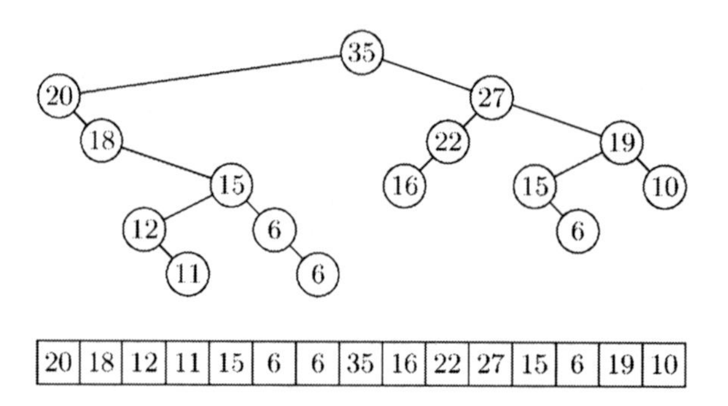


**Fig. 2.** Example of a Cartesian tree. The values in the boxes denote tuple differences.

greater priority than its parent. In the context of boundary tuple selection, the key of a tuple $t_j$ is its position $j$ in the tuple sequence; its priority is the number of bits $l_j$ required to store $\Delta(t_j)$ (see Figure 2).

A Cartesian tree for a sequence of elements sorted by their keys can be built in linear time [1]. Every node $v$ in the tree can be seen as representing the subsequence $T_v$ consisting of all tuples stored at descendant nodes of $v$. The construction algorithm is easily augmented to compute the size of $T_v$ for every node $v$ and store this size with $v$. Once the tree is given, implementing procedure Top-Down in $O(n)$ time is straightforward. In particular, testing whether a node $v$ satisfies the boundary tuple condition translates into the condition $|T_v| \cdot l_v > |T_x| \cdot l_x + c_{\text{boundary}} + |T_y| \cdot l_y$, where $x$ and $y$ are the children of $v$ in $T$.

*Bottom-up boundary tuple selection.* The problem with the top-down approach is that it is short-sighted in nature: by deciding that a tuple $t_j$ is an unsuitable choice for a boundary tuple, we do not consider any further tuples in its subtree. It may be, however, that $t_j$ by itself is not a good boundary tuple, while choosing $t_j$ together with additional tuples in its subtree leads to significant compression.

Our second tuple selection algorithm, the *bottom-up boundary tuple selection algorithm* or Bottom-Up, shown in Algorithm 1, tries to address this weakness. In order to choose the boundary tuples in a subsequence $T_v$ represented by a node $v$, it first considers $v$'s left and right subtrees in isolation and chooses boundary tuples for the subsequences represented by these trees. It then considers the subsequence $T[a, b]$ of $T_v$, where $a - 1$ and $b + 1$ are the indices of the last boundary tuple chosen in the left subtree and the first boundary tuple chosen in the right subtree, respectively. If $v$ satisfies the boundary tuple condition w.r.t. sequence $T[a, b]$, it is added to the sequence of boundary tuples.

In procedure Bottom-Up, not choosing a node $v$ as a boundary tuple does not prevent us from choosing boundary tuples in $v$'s subtree, as was the case in Top-Down. The down-side of Bottom-Up is that it always traverses the whole tree, while Top-Down can be expected to stop recursing after visiting only a small portion of the tree. Our experiments investigate the resulting trade-off between compression time and compression ratio.

**Algorithm 1.** Bottom-Up($v$): Returns a triple $(B, l_p, l_s)$, where $B$ is a list of boundary tuple indexes in $T_v$, $l_p$ is the number of bits required to encode each tuple difference up to the first boundary tuple in $B$, and $l_s$ is the number of bits required to encode each tuple difference after the last boundary tuple in $B$

```
1  if v has a left child
2      then (L, l_{p,l}, l_{s,l}) ← Bottom-Up(left(v))
3      else (L, l_{p,l}, l_{s,l}) ← (∅, 0, 0)
4  if v has a right child
5      then (R, l_{p,r}, l_{s,r}) ← Bottom-Up(right(v))
6      else (R, l_{p,r}, l_{s,r}) ← (∅, 0, 0)
7  if L = ∅
8      then a ← index of the leftmost tuple in T_v
9      else a ← (last entry in L) + 1
10 if R = ∅
11     then b ← index of the rightmost tuple in T_v
12     else b ← (first entry in R) − 1
13 if v satisfies the boundary tuple condition w.r.t. sequence T[a, b]
       ▷ This is easily checked using indices a, b, values l_{s,l} and l_{p,r}, and l_v = priority(v).
14     then return the triple (B, l_{p,l}, l_{s,r}), where B is the concatenation of L, a new list
                   node storing the index of v, and R.
15     else if L = ∅
16              then l_p ← l_v
17              else l_p ← l_{p,l}
18          if R = ∅
19              then l_s ← l_v
20              else l_s ← l_{s,r}
21          return the triple (B, l_p, l_s), where B is the concatenation of L and R.
```

## 3 Experimental Analysis

We evaluated the performance of the ATDC algorithm on a collection of different types of datasets, both in terms of the compression ratios achieved and the time required for compression and decompression. Both the Bottom-Up and Top-Down boundary tuple selection algorithms were included in the tests, as were the bit-shifting and mixed-radix encoding techniques. The results of these tests were compared with tests run using an implementation of the XTDC variant of TDC proposed in [5], as well as a Bit Compression algorithm [6].

We conducted all our experiments on an Intel P4 2.8GHz processor with 512KB L2 cache and 1GB Dual-channel DDR333 RAM on a motherboard using the Intel 875P chipset and equipped with a 3ware 7506-8 parallel ATA RAID controller and 3 Maxtor MaxLine Plus II 250GB drives (ATA/133) in RAID Level 5 configuration. The operating system was Debian Linux 2.6.8.

The compression achieved by the Bit Compression algorithm served as the baseline against which the compression ratios of the XTDC and ATDC algorithms were compared. Also, as with the tests conducted in [5], we considered only the attribute dimensions in the compression ratios: since the measure dimensions could have been of some non-categorical datatype, it could not be safely assumed that they could be compressed using the same technique that was applied to the attribute dimensions; so we decided to store them explicitly.

The timing tests we performed on the datasets consisted of measuring the *round-trip compression time* (RTT). The RTT consists of compression time and decompression time. The *compression time* represents the amount of time required to compress the tuples from their normalized form into a set of boundary tuples and tuple difference values and write them to disk. The *decompression time* refers to the amount of time required to read every boundary tuple and tuple difference value from the disk and convert them back to normalized tuple form. The results of these timing tests include the read and write times for the raw datasets that consisted of 32-bit integers, to show that even with virtually unlimited storage space, there are still speed advantages to using compression.

The uniform and skewed synthetic datasets used in our tests were created using the OLAP data generator described in [2]. Real data was extracted from the HYDRO1K database developed by the U.S. Geological Survey [7].

### 3.1 ATDC vs. GZIP Compression

Our first experiment compares the performance of ATDC to that of GZIP, a popular compression tool based on the Lempel-Ziv compression scheme.

Table 1 represents a comparison between GZIP compression and ATDC compression on a selection of the datasets used in this project. It should be noted that the compression ratios shown in this table are relative to the size of the original database file and not to the Bit-compressed version of the file as in the other experiments documented in this paper. The reason is that the GZIP algorithm did not achieve any compression whatsoever on the Bit-compressed version of the database, due to the fact that packing tuples tends to eliminate the long common sequences in the input that make GZIP effective.

As we can see, ATDC achieves compression ratios that are roughly twice as high as those achieved by GZIP on the same dataset. Not only are the ATDC compression results much better, the ATDC compression algorithm compresses data in approximately 1/4th the time required by GZIP.

### 3.2 ATDC vs. TDC Compression

Our final set of experiments compares the performance of ATDC with that of the XTDC algorithm. We consider the standard bit compression [6] to be the baseline against which we compare our compression ratios. To demonstrate that compression in general can lead to significant overall performance improvements, we also include round-trip time measurements for reading and writing the raw tuple sequence without even bit compressing it.

**Table 1.** GZIP compression vs ATDC compression

| Database | GZIP Size | GZIP Ratio | ATDC Size | ATDC Ratio |
|---|---|---|---|---|
| Uniform Synthetic | 123,749,391 | 5.818:1 | 57,230,264 | 12.581:1 |
| Skewed Synthetic (1.0) | 59,534,163 | 9.406:1 | 22,217,928 | 25.205:1 |
| HYDRO1K-Africa | 131,143,020 | 8.232:1 | 75,554,360 | 14.228:1 |
| HYDRO1K-North America | 96,134,225 | 8.271:1 | 62,448,276 | 12.732:1 |

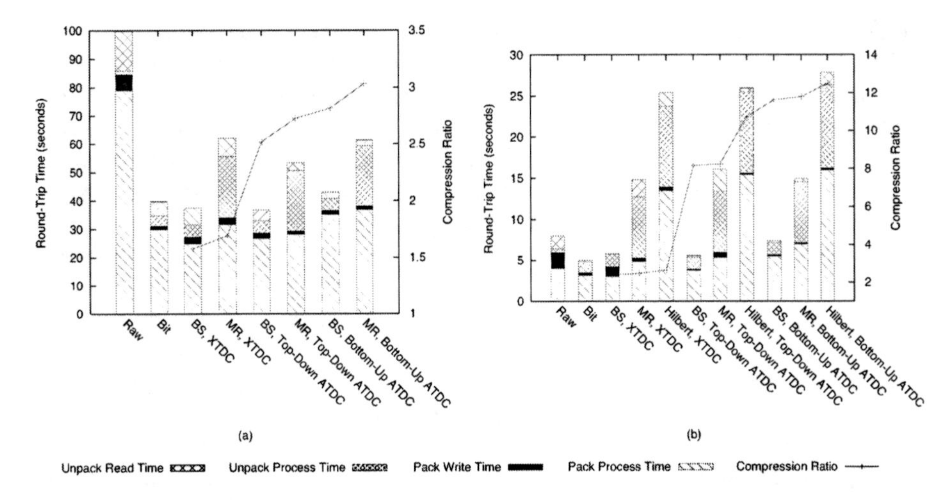


**Fig. 3.** Average round-trip times and compression ratios: (a) HYDRO1K Africa dataset, $n = 29,987,509$ tuples. (b) Synthetic data, $n = 20,000,000$ tuples. Zipf skew $z = 1.5$.

We tested these methods on a range of datasets. Due to lack of space, we report results on only two data sets here: the Africa data set from the HYDRO1K database [7] (Figure 3(a)) and a synthetic data set skewed with Zipf factor 1.5 (Figure 3(b)). The behaviour of the algorithms on other datasets is similar.

For the Africa dataset, we observe that *all* compression methods significantly decrease the round-trip time compared to the raw dataset. All our adaptive methods produce significantly higher compression ratios than the XTDC algorithm, with the bottom-up approach outperforming the top-down approach and the mixed-radix encoding leading to better compression ratios than the bit-shift approach. This aligns well with our expectations about the performance of these algorithms. Also in line with our expectations, the higher compression ratio in both cases is bought at the expense of increased round-trip times. In the case of the bottom-up algorithm, the time is lost during the compression phase; the mixed-radix encoding requires more time to decode. Note, however, that even using the bottom-up algorithm in combination with bit-shift encoding, the round-trip time is only slightly higher than using bit compression and XTDC. For the top-down algorithm and bit-shift encoding, the round-trip time is slightly lower than using bit compression and XTDC. In general, in terms of compression ratio bottom-up ATDC is superior; however, top-down ATDC may be preferred in applications where the round-trip compression time is critical.

For our skewed data set, the round-trip time improvement over the raw tuple sequence is less pronounced than on the Africa data set; the mixed-radix methods even lead to significantly higher round-trip times, due to high decompression cost. Consistent with the behaviour on the Africa dataset, the bit-shift compression methods produce round-trip times that are competitive with bit compression and XTDC but, due to the skew, now lead to significantly improved compression ratios. It is interesting to note that, on this data set, there seems to be little gain in compression ratio when using mixed-radix instead of bit-shift encoding. Since the mixed-radix encoding leads to a significantly increased decompression cost, it therefore cannot be considered useful for compressing this type of data.

Figure 3(b) also includes the compression ratios and compression times when applying our ATDC variants to a tuple sequence based on a Hilbert space encoding, which has been shown to be effective in improving query time in parallel OLAP query processing [4]. Although we used an efficient Hilbert implementation that was largely based on simple bit-shifts, the increased processing cost had a detrimental effect on round-trip times, increasing it even beyond that incurred by the mixed-radix methods. It is interesting to note, however, that the compression ratio does not deteriorate; on the contrary, the Hilbert encoded sequences lead to the highest observed compression ratios for each of the XTDC, top-down ATDC, and bottom-up ATDC algorithms. Thus, our compression method can be applied effectively in combination with Hilbert encodings in applications where the cost required for computing the Hilbert encoding is justified. In particular, our ATDC method is potentially a useful tool for reducing the amount of data exchanged between processors in parallel processing of OLAP queries.

## 4 Conclusion

We have demonstrated that the use of the ATDC algorithm has the potential of providing improved compression over existing algorithms on tuple-based statistical datasets. The algorithm was shown to be effective and robust on synthetically-generated datasets, both uniform and skewed, as well as on real-world datasets. In particular ATDC using a bit-shift encoding achieves both high compression ratios and low round-trip compression times. Furthermore, it must be emphasized that, as the gap between processor and I/O speeds grows, optimizing compression methods like ATDC, which may be encapsulated in the storage layer, will become increasingly appealing.

## References

1. Bender, M.A., Farach-Colton, M., Pemmasani, G., Skiena, S., Sumazin, P.: Lowest common ancestors in trees and directed acyclic graphs. Journal of Algorithms 57, 75–94 (2005)
2. cgmLab: OLAP data generator (2000), `http://cgmlab.cs.dal.ca/downloadarea/`
3. Chen, Z., Seshadri, P.: An algebraic compression framework for query results. In: ICDE, pp. 177–188 (2000)
4. Dehne, F., Eavis, T., Rau-Chaplin, A.: Parallel multi-dimensional ROLAP indexing. In: Proc. Int'l Symposium on Cluster Computing and the Grid, 2003, pp. 86–93 (2003)

5. Liang, B.: Compressing data cube in parallel OLAP systems. Master's thesis, Carleton University (2004)
6. Ng, W.K., Ravishankar, C.V.: Block-oriented compression techniques for large statistical databases. Knowledge and Data Engineering 9(2), 314–328 (1997)
7. US Geological Survey. HYDRO1k elevation derivative database (2003), `http://edcdaac.usgs.gov/gtopo30/hydro/index.asp`
8. Vuillemin, J.: A unifying look at data structures. Communications of the ACM 23, 229–239 (1980)

# Space-Efficient Structures for Detecting Port Scans

Ali Şaman Tosun*

Department of Computer Science
University of Texas at San Antonio
tosun@cs.utsa.edu

**Abstract.** Port scans aim to detect the services running on a computer to find vulnerabilities of a computer. Although detecting port scans using a database system is possible, it requires too much space and computational overhead and is not feasible under high load. In this paper, we propose space-efficient structures to detect parameterized versions of port scans. We investigate both exact and approximate structures for the problems. Proposed schemes are lightweight, require low space overhead, low computational overhead and can handle high load.

## 1 Introduction

Port scanning is the process of connecting to TCP and UDP ports on the target system to determine what services are running or are in a listening state. Identifying listening ports is critical to determining the services running, and consequently the vulnerabilities present. Many protocols and applications have reserved port numbers. For example, http protocol uses port 80, Microsoft SQL Server uses port 1433 and MySQL uses port 3306. Each machine has $2^{16}$ ports and by scanning through all the ports and testing whether any applications are running on these ports reveals a lot about the system. Many port scanning utilities are available. Utilities for Unix/Linux include strobe, netcat and nmap. Utilities for windows include SuperScan, WinScan and ipEye. Given the number of tools available for port scanning, detecting port scans is important. A connection between two machines can be identified using 5-tuple $(IP_{src}, Port_{src}, IP_{dest}, Port_{dest}, Time)$ where $IP_{src}$ is the IP address of the machine initiating the connection, $Port_{src}$ is the port number used for the connection by the initiating machine. $IP_{dest}$ is the IP address of the destination machine, $Port_{dest}$ is the port number for the destination machine and $Time$ is the timestamp of the connection. This 5-tuple can be used to detect portscans.

We address the following problems to detect portscans in this paper. We represent the problems with parameters that can be set by the system administrators.

- PORT-ALERT: alert when there are connection attempts to $\geq \tau$ ports of a machine

* Partially supported by Center for Infrastructure Assurance and Security at UTSA.

R. Wagner, N. Revell, and G. Pernul (Eds.): DEXA 2007, LNCS 4653, pp. 120–129, 2007.

- PORT-NODE-ALERT: alert when there are connection attempts to $\geq \tau$ ports of a machine from $\leq \kappa$ nodes.
- PORT-NODE-TIME-ALERT: alert when there are connection attempts to $\geq \tau$ ports of a machine from $\leq \kappa$ nodes in $\leq \theta$ time.

PORT-ALERT problem only counts if at least $\tau$ ports were used in connection attempts. It does not distinguish the number of machines used in connection attempts. PORT-NODE-ALERT problem also counts the number of distinct machines used in connection attempts. Although distributed port scans are possible, too many connection attempts from fewer nodes indicate a port scan. PORT-NODE-ALERT problem does not include the time dimension. Connection attempts in a short period indicate a port scan. So, we include the time threshold $\theta$ in problem PORT-NODE-TIME-ALERT.

Port scans can be detected by inserting all the relevant information into a database and by querying the database. However, this requires too much space and computational overhead and is not feasible under high load. Using the 5-tuples $(IP_{src}, Port_{src}, IP_{dest}, Port_{dest}, Time)$ approach for connections requires continuous updates to the database, continuous queries and more than 10 bytes storage per tuple.

Port scan detection can be implemented at the entry point of the network and needs to handle all the connections to all the machines on the network. To have a practical system the following properties are desirable.

- *Low space requirement:* Since too many connections are handled, space requirement per connection should be low. Attacker can send lots of data filling the data structures. So, data structures with constant space requirement or sublinear space requirement such as $O(\log n)$ or $O(\log \log n)$ are desirable.
- *Low computational overhead:* The data structure needs to be updated real-time. To handle all the connections computational overhead per connection should be low.

## 2 Related Work

In this section, we discuss related work on bitmap indexes, bloom filters, space-efficient data structures and port detection.

Bitmap indexes were introduced in [20]. Several bitmap encoding schemes have been developed, such as equality [20], range [8], interval [9], and workload and attribute distribution oriented [15]. Numerous performance evaluations and improvements have been performed over bitmaps [8,23,25,27]. While the fast bitwise operations afforded by bitmaps are perhaps their biggest advantage, a limitation of bitmaps is the index size.

Bloom Filters are used in many applications in databases and networking including query processing [17,18,19], IP traceback [21,22], per-flow measurements [10,16], web caching [12,13] and loop detection [24]. A survey of Bloom Filter (BF) applications is described in [7]. A BF computes $k$ distinct independent

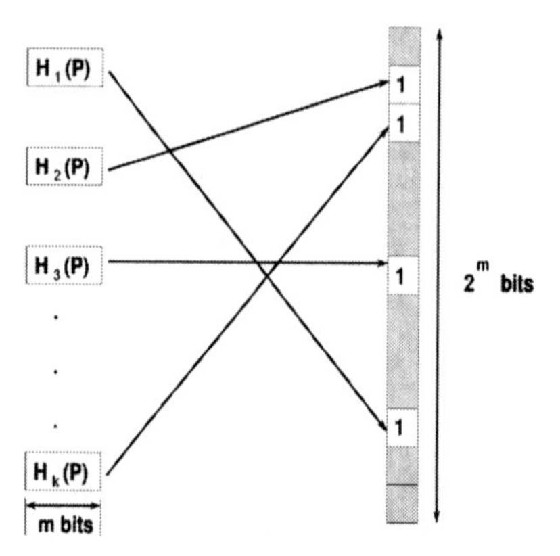


**Fig. 1.** A Bloom Filter

uniform hash functions. Each hash function returns an $m$-bit result and this result is used as an index into a $2^m$ sized bit array. The array is initially set to zeros and bits are set as data items are inserted. Insertion of a data object is accomplished by computing the $k$ hash function results and setting the corresponding bits to 1 in the BF. Retrieval can be performed by computing the $k$ digests on the data in question and checking the indicated bit positions. If any of them is zero, the data item is not a member of the data set (since member data items would set the bits). If all the checked bits are set, the data item is stored in the data set with high probability. It is possible to have all the bits set by some other insertions. This is called a *false positive*, i.e., BF returns a result indicating the data item is in the filter but actually it is not a member of the data set. On the other hand, BFs do not cause *false negatives.* It is not possible to return a result that reports a member item as a non-member, i.e., member data items are always in the filter. Operation of a BF is given in Figure 1.

Exact solution to find the number of distinct elements in $n$ items require $\Omega(n)$ space. Approximate data structures that estimate the number of distinct elements are proposed to reduce the space complexity. Counting sketches [14] and extensions [1,5] estimates the count in one pass using small amount of storage. FM sketches [14] use $O(\log n)$ space to approximate the count. Using linear hash functions sketch size can be reduced to $O(\log \log n)$[11].

## 3 System Model

Each machine on the Internet has a 32 bit IP address typically represented as $A.B.C.D$. Each letter corresponds to decimal representation of 8 bits. For example, 129.115.28.66 is the IP address of a machine in CS department. Each machine has $2^{16}$ ports. A connection between two machines includes port numbers as well. For example, a connection from machine $IP_1$ and port number

$p_1$ to machine $IP_2$ and port number $p_2$. Each process that requires communication has a port number that it can use for communication. Port numbers are used to demultiplex the packets received by the machine. To make routing easier IP addresses are assigned in blocks. For example, 256 IP addresses represented by 129.115.28.* are assigned to CS department. A block of IP addresses is called a subnet and represented as $A.B.C.D/X$. $X$ denotes the number of bits common to all the machines in the subnet. Above block can be represented as 129.115.28.0/24. In this paper, we develop a common data structure to store all the information for a subnet. We use the notation $n$ to denote the number of machines in the subnet. For the subnet 129.115.28.0/24, the value of $n$ is 256.

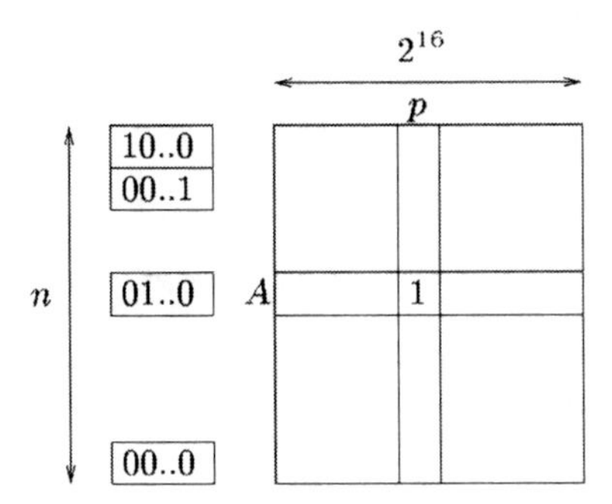


**Fig. 2.** Bitmap Approach for PORT-ALERT

## 4 Proposed Schemes

In this section, we discuss how to detect each time of alert using both exact and approximate techniques.

### 4.1 PORT-ALERT

PORT-ALERT Problem can be represented using a bitmap. Let $n$ denotes the number of machines in the subnet. A $n \times 2^{16}$ bitmap can be used to store connection attempts to the machines in the subnet. Block diagram of the approach is given in figure 2. When a connection attempt to machine $A$ on port $p$ is received, the corresponding entry in the bitmap is set to indicate this operation. In addition to the bitmap we use an array of counters to store how many ports are used. When a counter reaches the threshold $\tau$, an alert is raised. Bitmap approach is exact since it stores all the connection attempts to ports.

Bitmap approach requires $n2^{16}$ bits for the bitmap and $16n$ bits for the counters. Since there are $2^{16}$ ports, 16 bits are needed for each counter. For a 256 node subnet space requirement is about 2 megabytes which is manageable. During normal operation on the network, most of the bits will be 0. To reduce the space requirement bitmap compression schemes can be used. Several compression techniques have been developed in order to reduce bitmap size and at the

**Algorithm 1.** PORT-ALERT Insertion Algorithm for connection to (IP,p)

1: $index = GetIndex(IP)$
2: **if** $bitmap(index,p) == 0$ **then**
3: $\quad bitmap(index,p) \leftarrow 1$
4: $\quad portcount[index] \leftarrow portcount[index] + 1$
5: $\quad$ **if** $portcount[index] \geq \tau$ **then**
6: $\quad\quad ALERT$
7: $\quad$ **end if**
8: **end if**

same time maintain the advantage of fast operations [2,3,23,26]. However, since access to the bitmap is a bit at a time, the only practical approach here is to use bloom filter based bitmap compression [4].

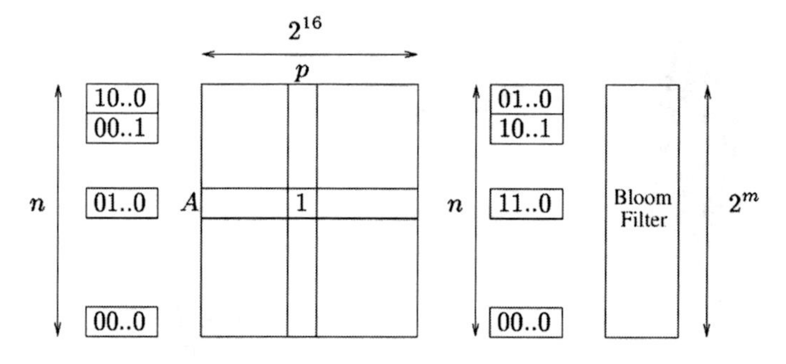


**Fig. 3.** Scheme for PORT-NODE-ALERT

## 4.2 PORT-NODE-ALERT

Our approach to PORT-NODE-ALERT problem is based on the scheme for PORT-ALERT problem. In addition, we maintain counters to count the number of distinct nodes the connection attempts come from. Exact solution to find the number of distinct elements in $m$ items require $\Omega(m)$ space. So, we use a Bloom filter to approximate the count. The Bloom filter is used to determine whether the pair of source and destination IP addresses $(IP_1, IP_2)$ was seen earlier or not. If the pair is an unseen pair, then the corresponding count is incremented. Bloom filter allows us to use a single data structure for the entire subnet instead of using separate data structures for each machine on the subnet. This simplifies the overall design. The algorithm is given in algorithm 2.

Bloom filters can have false positives, i.e., filter returns a result indicating the data item is in the filter but actually it is not. False positive rate can be controlled by the number of hash functions and the size of the filter. We next investigate how large the bloom filter should be to accommodate a given number of users with low false positive rate. We use the parameter $s$ for the number of connection attempts, $k$ for the number hashes and $m$ for the size of the filter.

**Algorithm 2.** PORT-NODE-ALERT Insertion Algorithm for connection from $IP_1$ to $(IP_2, p)$

1: $index = GetIndex(IP_2)$
2: **if** $bitmap(index, p) == 0$ **then**
3: $bitmap(index, p) \leftarrow 1$
4: $portcount[index] \leftarrow portcount[index] + 1$
5: **end if**
6: **if** $BloomFilter(IP_1||IP_2) = false$ **then**
7: $BloomInsert(IP_1||IP_2)$
8: $nodecount[index] \leftarrow nodecount[index] + 1$
9: **end if**
10: **if** $portcount[index] \geq \tau$ and $nodecount[index] \leq \kappa$ **then**
11: $ALERT$
12: **end if**

Ideally we want a data structure whose size depends on the number of entries $s$.Assume that we use a bloom filter whose size $m$ is $\alpha s$ where alpha is an integer denoting how much space is allocated as a multiple of $s$. The false positive rate of the bloom filter can be expressed as

$$(1 - (1 - \frac{1}{\alpha s})^{ks})^k \approx (1 - e^{-\frac{ks}{\alpha s}})^k = (1 - e^{-\frac{k}{\alpha}})^k \tag{1}$$

False positives result in counter values to be below what they should be. Since the connection attempt appears to be in the filter, it is never added and counter is not incremented.

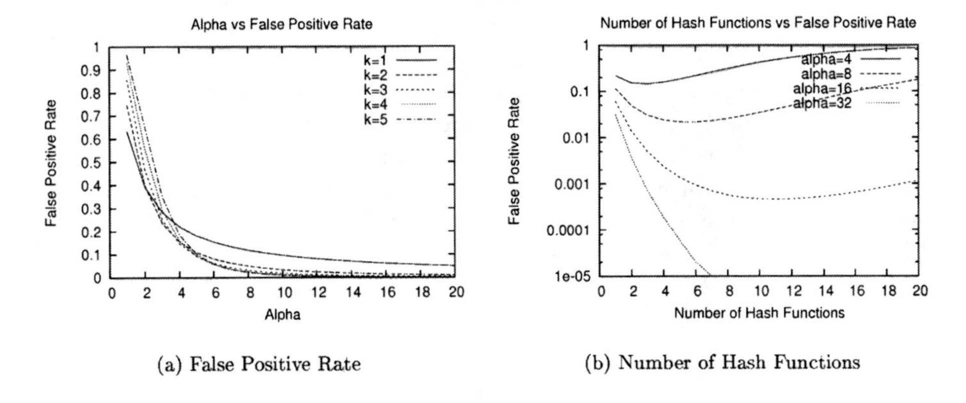


(a) False Positive Rate

(b) Number of Hash Functions

**Fig. 4.** False Positive Rate of Bloom Filter

## 4.3 PORT-NODE-TIME-ALERT

Our approach to PORT-NODE-TIME-ALERT problem is based on the scheme for PORT-NODE-ALERT problem. Since solving the query exactly requires to

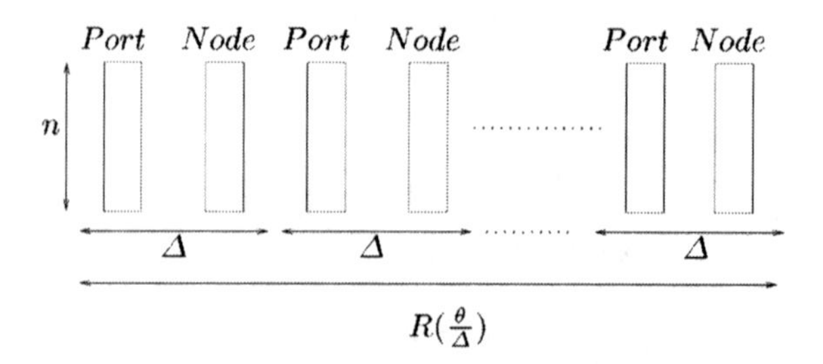


**Fig. 5.** Scheme for PORT-NODE-TIME-ALERT

much space, we use an approximate mechanism. We divide the time into intervals of size $\Delta$ and use the scheme described in PORT-NODE-ALERT to store the data for each interval. A new data structure is maintained for each interval of size $\Delta$. Given a time threshold $\theta$, the query uses the last $R(\frac{\theta}{\Delta})$ intervals to answer the query where $R()$ denotes the round function. Given a time interval $\theta$, the difference between $\theta$ and estimated time $R(\frac{\theta}{\Delta})\Delta$ is $< \frac{\Delta}{2}$. As the interval size $\Delta$ increases, the difference between $\theta$ and estimated time increases while the space requirement decreases. For older intervals only the counters need to be maintained. The bitmap and the bloom filter can be deallocated when the interval is over. Block diagram of the scheme is given in figure 5. For each old interval, $n$ counters to maintain port information and node information is needed. Number of intervals that needs to be stored is based on the time threshold $\theta$ and is given by $R(\frac{\theta}{\Delta})$. To support a large range of $\theta$ values, large number of intervals needs to be stored. Let $C_p$ denote the size of a counter to store the number of ports and let $C_n$ denotes the size of a counter to store the number of distinct machines the connection attempts come from. For each interval $n(C_p+C_n)$ space is needed to store the counters. For example, for a subnet with 256 machines and 2 byte counters space requirement per interval is 1 Kbyte which is quite small. So, interval size can be kept small (and smaller sized counters can be used).

A special set of counters to estimate the values of portcount and nodecount for a time interval of $\theta$ are maintained. These counters maintain the sum of counts for the last $R(\frac{\theta}{\Delta})$ intervals. Let $portcount_i[j]$ be the count of the number of ports for machine $j$ in interval $i$ and let $c$ be the current interval. The special counter *Sumportcount* has the following value

$$Sumportcount[j] = \sum_{k=c-R(\frac{\theta}{\Delta})+1}^{c} portcount_k[j], 0 \leq j \leq n \tag{2}$$

When an interval is over, the counters are updated to reflect the new set of intervals in use. This can be done by removing the count for the interval that falls outside the window. The new interval has counts of 0 and need not be added to the sum. This can be formally stated as

$$Sumportcount[j] = Sumportcount[j] - portcount_{c-R(\frac{\theta}{\Delta})+1}[j], 0 \leq j \leq n \quad (3)$$

A special counter *Sumnodecount* is maintained to estimate the value of nodecount for a time interval of $\theta$. The process to maintain the counters is similar to *Sumportcount*. We leave the details due to space restrictions.

For each new connection attempt, the counters *portcount* and *nodecount* for the current interval and the special counters *Sumportcount* and *Sumportcount* are updated if necessary.

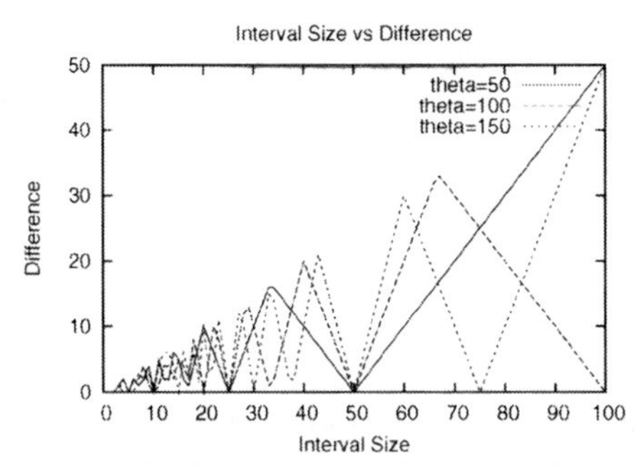


**Fig. 6.** Time Difference using Fixed-sized Intervals

Using fixed-sized intervals have some shortcomings. First, it is possible to bound the difference between $\theta$ and approximated value, however it is not possible to bound the number of connection attempts missed. The time difference for several values of $\theta$ is given in figure 6. Second, the false positive rate of bloom filter depends on the number of connection attempts in that interval. Since we don't have a bound on the number of connection attempts it is not possible to provide performance guarantees. Using variable-sized intervals solves these problems. A new interval is created when a threshold number of connections are received. Using this approach space complexity depends on the number of connections received. Using variable-sized intervals *sumportcount* can be updated as follows

$$Sumportcount[j] = \sum_{I_k \cap [t-\theta,t] \neq \emptyset} portcount_k[j], 0 \leq j \leq n \quad (4)$$

Maintenance of the variable *sumportcount* is similar to the fixed-sized interval case. Counters are updated whenever new item is added and when an interval $I_k$ falls out of the interval $[t-\theta, t]$, the counts for the interval $I_k$ are subtracted from *sumportcount*.

Counters maintained can also be used for visualization. For example, a surface plot showing *portcount* for a subset of the data is given in figure 7. A portscan that lasts a couple of intervals can be seen in the figure.

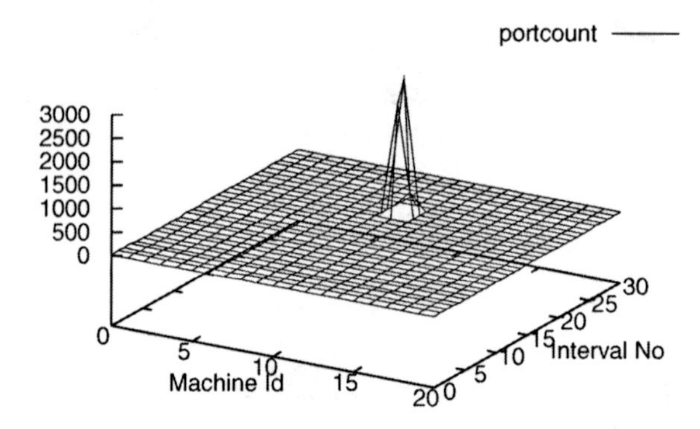


**Fig. 7.** Visualization of portcount

# 5 Conclusion

Although detecting port scans using a database system is possible, it requires too much space and computational overhead. In this paper, we propose space-efficient structures to detect parameterized versions of port scans. Parameters including number of ports scanned, number of source IP addresses involved in the scan and the time interval for the scan can be set by the system administrator. Proposed schemes are lightweight, require low space overhead, low computational overhead and can handle high load.

# References

1. Alon, N., Matias, Y., Szegedy, M.: The space complexity of approximating the frequency moments. Journal of Computer and System Sciences 58(1), 137–147 (1999)
2. Amer-Yahia, S., Johnson, T.: Optimizing queries on compressed bitmaps. In: The VLDB Journal, pp. 329–338 (2000)
3. Antoshenkov, G.: Byte-aligned bitmap compression. In: Data Compression Conference, Oracle Corp, Nashua, NH (1995)
4. Apaydin, T., Canahuate, G., Ferhatosmanoglu, H., Tosun, A.Ş.: Approximate encoding for direct access and query processing over compressed streams. In: $32^{nd}$ International Conference on Very Large Data Bases, pp. 457–846 (2006)
5. Bar-Yossef, Z., Jayram, T.S., Kumar, R., Sivakumar, D., Trevisan, L.: Counting distinct elements in a data stream. In: RANDOM (2002)
6. Bloom, B.: Space/time tradeoffs in hash coding with allowable errors. Communications of the ACM 13(7), 422–426 (1970)
7. Broder, A., Mitzenmacher, M.: Network Applications of Bloom Filters: A Survey. In: Proceedings of the 40th Annual Allerton Conference on Communication, Control, and Computing, pp. 636–646 (2002)
8. Chan, C.Y., Ioannidis, Y.E.: Bitmap index design and evaluation. In: Proceedings of the 1998 ACM SIGMOD international conference on Management of data, pp. 355–366. ACM Press, New York (1998)

9. Chan, C.Y., Ioannidis, Y.E.: An efficient bitmap encoding scheme for selection queries. SIGMOD Rec. 28(2), 215–226 (1999)
10. Feng, W.c., Kandlur, D.D., Saha, D., Shin, K.G.: Stochastic Fair Blue: A Queue Management Algorithm for Enforcing Fairness. In: Proc. of INFOCOM, vol. 3, p. 1520–1529 (April 2001)
11. Durand, M., Flajolet, P.: Loglog counting of large cardinalities. In: Di Battista, G., Zwick, U. (eds.) ESA 2003. LNCS, vol. 2832, Springer, Heidelberg (2003)
12. Fan, L., Cao, P., Almeida, J., Broder, A.: Summary Cache: A Scalable Wide-Area Web Cache Sharing Protocol. In: IEEE/ACM Transactions on Networking, Canada, ACM Press, New York (2000)
13. Fan, L., Cao, P., Almeida, J., Broder, A.: Web cache sharing. Collaborating Web caches use bloom filter to represent local set of cached files to reduce the netwrok traffic. In: IEEE/ACM Transactions on Networking, ACM Press, New York (2000)
14. Flajolet, P., Martin, G.N.: Probabilistic counting algorithms for database applications. Journal of Computer and System Sciences 31(2) (1985)
15. Koudas, N.: Space efficient bitmap indexing. In: Proceedings of the ninth international conference on Information and knowledge management, pp. 194–201. ACM Press, New York (2000)
16. Kumar, A., Xu, J.J., Wang, J., Li, L.: Algorithms: Space-code bloom filter for efficient traffic flow measurement. In: Proceedings of the 2003 ACM SIGCOMM conference on Internet measurement, October 2003, ACM Press, New York (2003)
17. Mishra, P., Eich, M.H.: Join processing in relational databases. In: ACM Computing Surveys (CSUR), March 1992, ACM Press, New York (1992)
18. Mullin, J.K.: Estimating the size of joins in distributed databases where communication cost must be maintained low. In: IEEE Transactions on Software Engineering, IEEE Computer Society Press, Los Alamitos (1990)
19. Mullin, J.K.: Optimal semijoins for distributed database systems. IEEE Transactions on Software Engineering 16, 558–560 (1990)
20. O'Neil, P.E., Quass, D.: Improved query performance with variant indexes. In: Proceedings of the 1997 ACM SIGMOD international conference on Management of data, pp. 38–49. ACM Press, New York (1997)
21. Snoeren, A.C.: Hash-based IP traceback. In: ACM SIGCOMM Computer Communication Review, ACM Press, New York (2001)
22. Snoeren, A.C., Partridge, C., Sanchez, L.A., Jones, C.E., Tchakountio, F., Schwartz, B., Kent, S.T., Strayer, W.T.: IP Traceback to record packet digests traffic forwarded by the routers. IEEE/ACM Transactions on Networking (TON) (December 2002)
23. Stockinger, K.: Bitmap indices for speeding up high-dimensional data analysis. In: Proceedings of the 13th International Conference on Database and Expert Systems Applications, pp. 881–890. Springer, Heidelberg (2002)
24. Whitaker, A., Wetherall, D.: Detecting loops in small networks. In: 5th IEEE Conference on Open Architectures and Network Programming (OPENARCH) (June 2002)
25. Wu, K., Otoo, E.J., Shoshani, A.: A performance comparison of bitmap indexes. In: Proc. Conf. on 10th International Conference on Information and Knowledge Management, pp. 559–561. ACM Press, New York (2001)
26. Wu, K., Otoo, E.J., Shoshani, A.: Compressing bitmap indexes for faster search operations. In: SSDBM, Edinburgh, Scotland, pp. 99–108 (July 2002)
27. Wu, M.C.: Query optimization for selections using bitmaps. In: Proceedings of the 1999 ACM SIGMOD international conference on Management of data, pp. 227–238. ACM Press, New York (1999)

# A Dynamic Labeling Scheme Using Vectors

Liang Xu, Zhifeng Bao, and Tok Wang Ling

School of Computing, National University of Singapore
{xuliang,baozhife,lingtw}@comp.nus.edu.sg

**Abstract.** The labeling problem of dynamic XML documents has received increasing research attention. When XML documents are subject to insertions and deletions of nodes, it is important to design a labeling scheme that efficiently facilitates updates as well as processing of XML queries. This paper proposes a novel encoding scheme, vector encoding which is orthogonal to existing labeling schemes and can completely avoid re-labeling. Extensive experiments show that our vector encoding outperforms existing labeling schemes on both label updates and query processing especially in the case of skewed updates. Besides, it has the nice property of being conceptually easy to understand through its graphical representation.

## 1 Introduction

XML[6] has become a standard to represent and exchange data on the web, and there is a lot of interest in query processing over XML data. The techniques used to facilitate XML queries can be classified into two categories: structural index approach[10] and labeling approach[12,4,11]. We focus on labeling approach which requires smaller storage space, yet efficiently determines ancestor-descendant(A-D) and parent-child(P-C) relationships between any two nodes in the XML documents.

Although most existing labeling schemes work well on querying static XML documents, their performances degrade significantly for dynamic XML documents as updating requires re-labeling[12,4,11] or label size increases very fast for skewed insertions although re-labeling can be avoided[8]. When XML documents are dynamic, it is of great interest to design a labeling scheme that can avoid re-labeling while having controllable size for skewed insertions.

The main contributions of this paper are summarized as follows:

- We propose a novel compact labeling scheme: vector encoding which can be applied to different labeling schemes and is easy to understand.
- Vector encoding completely avoids re-labeling for updates in XML doc.
- We conduct experiments to show that vector encoding performs better than existing schemes, especially in the case of skewed insertions.

R. Wagner, N. Revell, and G. Pernul (Eds.): DEXA 2007, LNCS 4653, pp. 130–140, 2007.

## 2 Related Work

Current labeling schemes include containment scheme[12], prefix scheme[4] and prime scheme[11]. Due to space constraint we only focus on containment scheme and QED encoding which are most relevant to this paper.

Based on **containment labeling scheme**[12], every node is assigned three values: "start", "end" and "level". For any two nodes $u$ and $v$, $u$ is an ancestor of $v$ iff $u.start < v.start$ and $v.end < u.end$. Node u is the parent of node $v$ iff $u$ is an ancestor of $v$ and $v.level - u.level = 1$. Although containment scheme is efficient for determining A-D and P-C relationships, the insertion of a node $n$ will lead to re-labeling of all the ancestor nodes of $n$ and all the nodes after $n$ in document order. To solve the re-labeling problem, [5] uses Float-point values for the start and end of the intervals. However, in practice, Float-point is represented physically with a fixed number of bits. As a result, at most 18 nodes can be inserted at a fixed place when consecutive integer values are used for initial labeling.

[8] proposes a novel encoding method: **QED encoding** that completely avoids re-labeling. Four numbers "0", "1", "2" and "3" are used for encoding and each number is stored with two bits, i.e. "00", "01", "10" and "11". The number "0" is reserved as the separator. An important feature of QED is that it is based on lexicographical order, i.e. "0" $\prec$ "1" $\prec$ "2" $\prec$ "3". This encoding scheme allows a QED code to be inserted between any two existing QED codes while preserving lexicographical order. For example, "113" can be inserted between "112" and "12" whereas "1122" can be inserted between "112" and "113". QED encoding method is orthogonal to existing labeling schemes. However, the label size of QED increases very fast in the event of skewed insertion. Especially in the case that new codes are inserted after a fixed code, the size of the new code increases by 2 bits per insertion.

## 3 Preliminaries

For the complete proofs of the theorems in this section, please refer to [9] which is an extended version of this paper. We ignore most of the proofs due to space constraint.

A *vector* is an object with magnitude and direction. A two dimensional vector consists of binary tuples and is represented as (x, y). In this paper, to make design simple and avoid precision problem, we only consider vector whose x and y components are positive integers. Figure 1(a) gives a graphical interpretation of a vector V that lies in the first quadrant of the X-Y plane.

Let A=$(x_1, y_1)$ and B=$(x_2, y_2)$ be two vector. *Addition* of A and B and *Multiplication* of a scalar r and a vector A are defined as:

$$A + B = (x_1 + x_2, y_1 + y_2) \quad (1)$$

$$r \cdot A = (r * x_1, r * y_1) \quad (2)$$

Figure 1(b) shows the graphical representation of vectors A, B, A+B, 2·A+B and A+2·B.

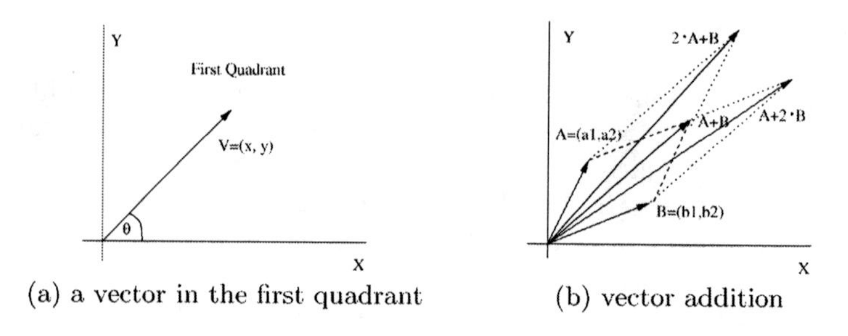


(a) a vector in the first quadrant (b) vector addition

**Fig. 1.** Graphical representation of vector and vector addition

**Definition 1.** ***(Gradient)*** *The* Gradient *of a vector V=(x, y) (denoted by G(v)) is defined as y/x.*

In Figure 1(a), vector V makes an angle $\Theta$ with the X axis. The *Gradient* of V is y/x, or equivalently, tan($\Theta$).

**Theorem 1.** *Given two vectors A=$(x_1, y_1)$ and B=$(x_2, y_2)$ in the first quadrant of the x-y plane, G(A) > G(B) iff $y_1x_2 > x_1y_2$.*

*Example 1.* G((19,5))=5/19; G((19,5))>G((16,3)) since $5 \times 16 > 19 \times 3$.

It is important to note that although *Gradient* is defined in terms of division, the comparison of the *Gradients* of two vectors can be done via multiplication.

**Theorem 2.** *Let A, B, C be three vectors in the first quadrant of the x-y plane such that C=A+B and G(A) > G(B), then G(A) > G(C) > G(B).*

*Proof.* assume that A=$(x_1, y_1)$ and B=$(x_2, y_2)$, then C=$(x_1 + x_2, y_1 + y_2)$. Since G(A) > G(B), we have from Theorem 1, $y_1x_2 > x_1y_2$. Therefore,

$$G(C) = \frac{y_1 + y_2}{x_1 + x_2} = \frac{y_1x_2 + y_2x_2}{(x_1 + x_2)x_2} > \frac{x_1y_2 + y_2x_2}{(x_1 + x_2)x_2} = \frac{y_2}{x_2} = G(B) \quad (3)$$

Similarly, we can prove that G(A) > G(C). Therefore, G(A) > G(C) > G(B).

*Example 2.* The sum of vectors (19,5) and (16,3) is (35,8), G((19,5))>G((35,8)) since $35 \times 5 > 19 \times 8$; G((35,8))>G((16,3)) since $16 \times 8 > 35 \times 3$.

**Theorem 3.** *Let A, B be two vectors in the first quadrant of the x-y plane such that G(A) > G(B), we can find infinite number of vectors whose Gradients are between G(A) and G(B).*

## 4 Vector Encoding

In this section, we introduce our vector encoding scheme which completely avoids re-labeling upon insertion. Table 1 shows different encoding schemes for numbers from 1 to 18. Details of how QED encoding is performed are in [8]. For vector encoding, we first assign vector (1,0) to the start position in the range and (0,1) to the end position. Then we work recursively by assigning the middle position of the current range a vector that equals to the sum of two vectors that correspond to the start and end position in each iteration. The formal encoding algorithm is presented in Algorithm 1.

**Theorem 4.** *Let I and J be two decimal numbers and $V_I$ and $V_J$ be their corresponding vector codes generated by Algorithm 1, we have:* $I<J$ *iff* $G(V_I)<G(V_J)$.

*Example 3.* Given that the range of integers is from 1 to 18, we assign vector (1,0) (of *Gradient* 0) to the start position in the range which is 1; and (0,1) (of *Gradient* $+\infty$) to the end position in the range which is 18, i.e. v(1)=(1,0) and v(18)=(0,1). Next we apply Algorithm 1 to recursively encode the remaining positions.

**Iteration 1** The middle position in the range [1, 18] can be found by: *middle* = $\lceil(1+18)/2\rceil = 10$. Hence v(middle)=v(10)=v(1) + v(18)=(1,0) + (0,1)=(1,1).

**Iteration 2** Now that the range [1,18] is divided into two ranges:[1, 10] and [10, 18]. The middle position of [1, 10] is $\lceil(1+10)/2\rceil = 6$; and the middle position of [10, 18] is $\lceil(10+18)/2\rceil = 14$. Therefore v(6)=(1,0)+(1,1)=(2,1) and v(14)=(1,1)+(0,1)=(1,2). This process continues until all the positions are encoded, we omit the remaining iterations here.

**Definition 2.** ***(vector order)*** *The order of vector encodings is based on the numerical ordering of the Gradients of the vectors.*

Table 1 also gives the *Gradients* of the vectors for each row (we define 1/0 to be $+\infty$). It can be seen that the numerical order of the *Gradients* indeed follow the order of the decimal numbers. It is worth noting that the *Gradients* shown in Table 1 are for illustration purpose only. From theorem 1, we can compare the *Gradients* of two vectors using multiplication instead of division, our method does not involve the calculation of *Gradients* and therefore does not suffer from the float-point precision problem in [5].

**Table 1.** Comparison of different encoding schemes

| Decimal number | QED | vector | Gradient (accurate to 0.01) | Decimal number | QED | vector | Gradient (accurate to 0.01) |
|---|---|---|---|---|---|---|---|
| 1 | 112 | (1,0) | 0 | 10 | 223 | (1,1) | 1 |
| 2 | 12 | (5,1) | 0.2 | 11 | 23 | (3,4) | 1.33 |
| 3 | 122 | (4,1) | 0.25 | 12 | 232 | (2,3) | 1.5 |
| 4 | 13 | (3,1) | 0.33 | 13 | 3 | (3,5) | 1.67 |
| 5 | 132 | (2,5) | 0.4 | 14 | 312 | (1,2) | 2 |
| 6 | 2 | (2,1) | 0.5 | 15 | 32 | (2,5) | 2.5 |
| 7 | 212 | (5,3) | 0.6 | 16 | 322 | (1,3) | 3 |
| 8 | 22 | (3,2) | 0.67 | 17 | 33 | (1,4) | 4 |
| 9 | 222 | (4,3) | 0.75 | 18 | 332 | (0,1) | $+\infty$ |

**Algorithm 1.** *VectorEncoding*

**input:** $n$ is a positive integer
**output:** return the vector codes in *vcode* for numbers from 1 to n
//*vcode* is an array of n vectors
1: $vcode[0] = (1, 0)$
2: $vcode[n-1] = (0, 1)$
3: RecEncoding($vcode, 0, n-1$)
4: return *vcode*
Procedure RecEncoding($vcode, start, end$)
1: $m = \lceil (start + end)/2 \rceil$
2: **if** $m == end$ return
3: $mV = vcode[start] + vcode[end]$
4: $vcode[m] = mV$
5: RecEncoding($vcode, start, m$)
6: RecEncoding($vcode, m, end$)

**Algorithm 2.** *LabelTheLeafNodeToBeInserted*

**input:** $n$ is the leaf node to be inserted
**output:** return *VContainment* label of $n$
//$v1$, $v2$ are two bounding vectors
//$l$ is the level of n
1: **if** $n$ has preceding sibling(s) $v1$=$cps.endV$
//$cps$ is the closest preceding sibling of $n$
2: **else** $v1$=$p.startV$ //$p$ is the parent of $n$
3: **if** $n$ has following sibling(s) $v2$=$cfs.startV$
//$cfs$ is the closest following sibling of $n$
4: **else** $v2$=$p.endV$
5: return FindNewLabel($v1$,$v2$,$l$)
Procedure FindNewLabel($v1$,$v2$,$l$)
1: **if** $GS(v1) > GS(v2)$
return ($v1$+$v2$,$v1$+2·$v2$,$l$)
2: **else** return (2·$v1$+$v2$,$v1$+$v2$,$l$)

## 4.1 Encoding Delimiters

When labels are stored for future reuse, we need to encode delimiters. 0 is reserved as delimiter in QED whereas vector codes use UTF8[7] encoding to process delimiters. In UTF8, a variable number of bytes are used to encode different integer values. A vector $V$ is stored sequentially as $V.x$, $V.y$ where $V.x$ and $V.y$ are encoded using UTF8 encoding.

## 4.2 Application of Vector Encoding Scheme

Our vector encoding scheme is orthogonal to specific labeling schemes. It can be applied to existing labeling schemes while keeping the original labeling order. In this paper we apply vector encoding scheme to containment scheme and the resulting labeling scheme is called *VContainment* scheme.

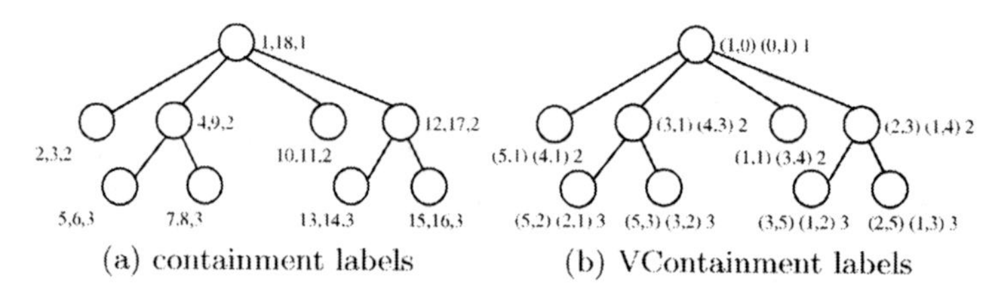


**Fig. 2.** Applying vector encoding scheme to containment scheme

*Example 4.* Figure 2 shows an example of applying vector encoding to containment scheme. The *start* and *end* value of the original containment labels are replaced by their corresponding vector codes (see Table 1 for details). The resulting *VContainment* labels are of format (*startV*, *endV*, *level*) where *startV*, *endV* are two vectors. It is easy to verify that the property of containment scheme holds. For example, Node((2,3),(1,4),2) is the parent of node((3,5),(1,2),3) as G(2,3) < G(3,5) < G(1,2) < G(1,4) and 2+1=3.

# 5 Support Updating

For dynamic XML documents, especially the ones that require frequent updates, it is important to make the update cost as low as possible. One of the most important features of vector encoding scheme is that it can completely avoid re-labeling when updates take place. This section provides elaboration on how vector encoding handles updates efficiently. We start by showing how updates can be performed in *VContainment* scheme without re-labeling, then analyze how updates can be optimized in a general context.

## 5.1 Updating in VContainment Scheme

With *VContainment* scheme, the deletion of a leaf node or an internal node has no side effect. However, handling insertions may require some consideration. First we introduce a definition which we use to measure the size of a vector.

**Definition 3.** ***(Granularity Sum)*** *The Granularity Sum of a vector* $V = (x, y)$ *(denoted by* $GS(v)$*) is defined as* $x+y$*.*

To find a vector between two vectors in vector order, we want its $GranularitySum$ to be as small as possible so that the resulting label size is small.

**Inserting Leaf Node.** Assume all *VContainment* labels are of format ($startV$, $endV$, $level$). $n$ is the node to be inserted and $p$ is the parent of $n$. $cps$ is its closest preceding sibling(if exists). $cfs$ is its closest following sibling(if exists). If $n$ is a leaf node, to maintain the correctness of *VContainment* scheme, the following inequalities should hold.

1. $G(p.startV) < G(n.startV) < G(n.endV) < G(p.endV)$
2. $G(cps.endV) < G(n.startV)$ 3. $G(n.endV) < G(cfs.startV)$

Note the second inequality is only applicable if $n$ has preceding sibling(s), and the third inequality is applicable if $n$ has following sibling(s). In any case, the $startV$ and $endV$ of $n$ will be bounded by two vectors, we call these two vectors *bounding vectors*. The new label of $n$ can be found by basically finding a pair of vectors between the two *bounding vectors*. Details on how the label of $n$ is found are presented in Algorithm 2.

**Inserting Non-leaf Node.** The case that $n$ is a non-leaf node is similar to the previous case except that another inequality needs to be enforced. Assume $fc$ is the first child of $n$ and $lc$ is the last child of $n$.

4. $G(n.startV) < G(fc.startV) < G(lc.endV) < G(n.endV)$

We ignore the details of insertion of non-leaf node here.

The core operation of Algorithm 2 is to find two vectors between $v1$ and $v2$ in terms of vector order. In Algorithm 2, the two vectors are either $v1 + v2$ and $v1 + 2 \cdot v2$ or $2 \cdot v1 + v2$ and $v1 + v2$. Based on Theorem 2, we can prove in both cases the two vectors are between $v1$ and $v2$ in vector order. Figure 1(b) shows the graphical representation of the vectors. It can be observed that if we keep inserting before or after a fixed node, the resulting label increases constantly by the *Granularity Sum* of that node. Although this method is simple and efficient, the resulting vector may not yield the minimum *Granularity Sum*. Analysis on the optimization of insertion will be presented in the next subsection. There is no re-labeling involved in the insertion, *VContainment* scheme can support efficient updates without re-labeling any existing labels.

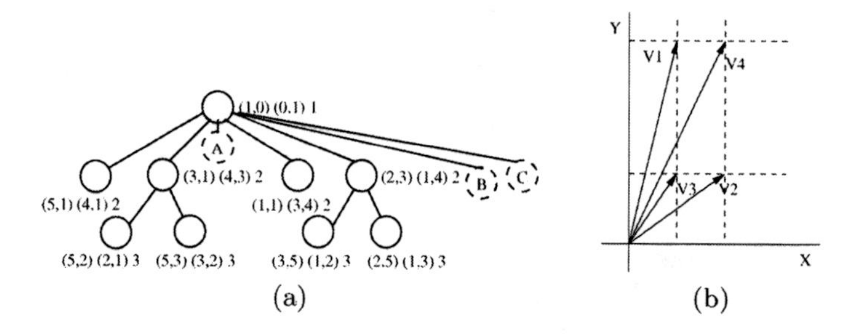


**Fig. 3.** Insertion in VContainment scheme

*Example 5.* In Figure 3(a), when inserting node A having both left sibling and right sibling, its *startV* and *endV* are bounded by *endV* of its closest left sibling and *startV* of its closest right sibling, i.e. (4,3) and (1,1). Moreover, GS(1,1)=2<7=GS(4,3). Therefore, the *startV* of A is $v1 + v2 = (5, 4)$ whereas *endV* is $v1 + 2 \cdot v2 = (6, 5)$. When inserting node B which has only left sibling, its *startV* and *endV* are bounded by the *endV* of its closest left sibling the *endV* of its parent, i.e. (1,4) and (0,1). Therefore, the *startV* of B is $v1 + v2 = (1, 5)$ whereas *endV* is $v1 + 2 \cdot v2 = (1, 6)$. Similarly, when we continue to insert C as the last child of the root, its *startV* is $v1 + v2 = (1, 7)$ and *endV* is $v1 + 2 \cdot v2 = (1, 8)$.

### 5.2 Analysis on Insertion

When vector encoding scheme is applied to different labeling schemes including containment scheme, the core operation of insertion is to find the vector between two consecutive vectors in vector order. The choice is not unique, actually there are infinitely many vectors possible(Theorem 3); however, to slow down the increase rate of labels, we would want to find the vector that has the smallest *Granularity Sum* possible. Although we can always use the sum of the two consecutive vectors, the resulting vector may not yield the minimum *Granularity Sum*. Theoretically speaking, the vector that has the smallest *Granularity Sum* can always be found through enumeration, but this can make insertion very expensive to perform. However, we have found that based on the relative positions of the two consecutive vectors, it may be possible to optimize insertion without incurring much additional computational cost. Assume the consecutive vectors are $A = (x_1, y_1), B = (x_2, y_2)$, the vector to be inserted is $C = (x, y)$, insertion may be optimized for the following cases.

**Case 1.** $x_1 = x_2$ or $y_1 = y_2$. For example, let A and B be V1 and V3 in Figure 3(b) respectively. Since $x_1 = x_2$, we can choose $x = x_1$ and y to be an integer between $y_1$ and $y_2$ when $y_1 > y_2 + 1$. When $y_1 = y_2 + 1$, we can choose C to be the sum of A and B. The case when $y_1 = y_2$ is similar.

**Case 2.** ($x1 < x2$ and $y1 > y2$) or ($x1 > x2$ and $y1 < y2$). For example , let A and B be V1 and V2 in Figure 3 (b) respectively. We can choose C to be V3=$(x_1, y_2)$ or V4=$(x_2, y_1)$ since both V3 and V4 are between V1 and V2 in vector order. V3 may be preferred since it has smaller *Granularity Sum.*

## 6 Experiments and Evaluation

We have implemented the *VContainment* scheme in JAVA and used SAX from Sun Microsystems as the XML parser. We compare our labeling scheme with QED-containment scheme as they both completely avoid re-labeling. The updating cost of previous labeling schemes[12,4,11] are much higher than QED as re-labeling is very expensive to perform[8].

XMark[3], Shakespeare's play[1], Treebank and DBLP[2] data sets have been used to compare the performance of the labeling schemes. Our experiments are performed

### 6.1 Label Generation

To compare the label generation, we choose one of the documents from Shakespeare's play and enlarge it by 10 times. XMark document is generated using scaling factor 0.005. The time needed to generate labels mostly depends on the size the XML documents and the number of nodes in the XML documents. From the results in Table 2, as the size of the data set gets larger, the generation time of QED and vector labels increase accordingly. However, generating vector labels is much faster than QED as its generation mostly consists of simple calculations.

**Table 2.** Comparison of label generation

| Data Set | File Size(MB) | No. of Nodes(K) | QED Time(Sec) | vector Time(Sec) | QED Size(MB) | vector Size(MB) |
|---|---|---|---|---|---|---|
| XMark | 0.55 | 8.5 | 0.142 | 0.018 | 0.05 | 0.06 |
| Shakespeare's play | 2.16 | 49 | 0.86 | 0.26 | 0.31 | 0.39 |
| Treebank | 82 | 2437.7 | 33.8 | 9.1 | 19.2 | 27.0 |
| DBLP | 127 | 3332.1 | 50.9 | 14.6 | 26.9 | 37.8 |

### 6.2 Uniform and Skewed Insertions

All the four data set have been used to test the performance of the two labeling schemes upon two kinds of insertions: uniform insertion and skewed insertion, and showed similar trends. Here we present the results for XMark data set.

For uniform insertions, firstly we insert one node between every two consecutive nodes. Then we gradually increase the insertions by one at a time up to six. The results are shown in Figure 4(a) and (b). The vector labels are represented using bit strings that correspond to the UTF8 representation of the labels to accommodate dynamic increase in size. But the overhead of UTF8 encoding makes the size of vector labels approximately 20 percent larger than that of QED (including initial labels). The insertion time of QED however is about 50 percent more than that of vector labels.

For skewed insertions, we keep inserting *mail* element after the last *mail* element whose parent is *mailbox*. The results (Figure 4(c) and (d)) illustrate more significant advantage of vector labels. The insertion time of QED is almost four times of that of vector labels while the size of QED labels increases much faster than vector labels upon insertions. Since for skewed insertions, the length of the new QED label increases by 1 or 2 bits per insertion, while the size of the new vector label remains unchanged unless its value exceeds the current range in which case the label size increases by 1 byte. However, such increases occur infrequently and the size of vector labels increases rather slowly upon skewed insertions comparing with QED labels.

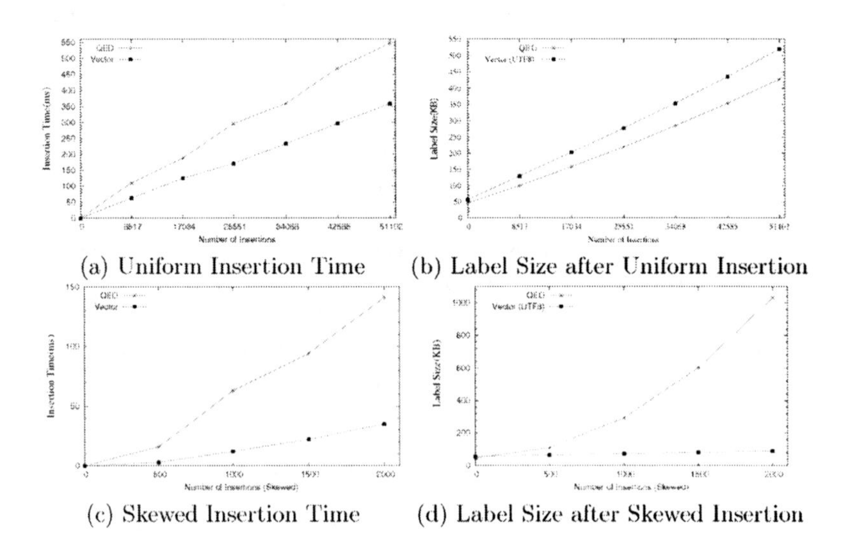


(a) Uniform Insertion Time (b) Label Size after Uniform Insertion

(c) Skewed Insertion Time (d) Label Size after Skewed Insertion

**Fig. 4.** Comparison of Uniform and Skewed Insertion for XMark data set

### 6.3 Query Time

We compare the query time using all the four data sets. Here we only present the results for XMark data set due to space constraint, the other data sets show similar results. The set of queries we used are shown in Figure 5(a). Figure 5(b), (c) and (d) show the comparison of query time on the original data and the data after uniform and skewed insertion. Notice that the time used for determining A-D and P-C relationships only constitute a fraction of the whole query time. In all these cases, query time of vector labels is faster than that of QED labels. The difference is most significant for the case of skewed insertion as when the length of QED label gets larger, lexicographical order is more expensive to compare. The comparison we show here is independent of the algorithm that is used to evaluate the queries. Basically all the algorithms involves determination of A-D and P-C relationships which is more efficient to compute using vector labels.

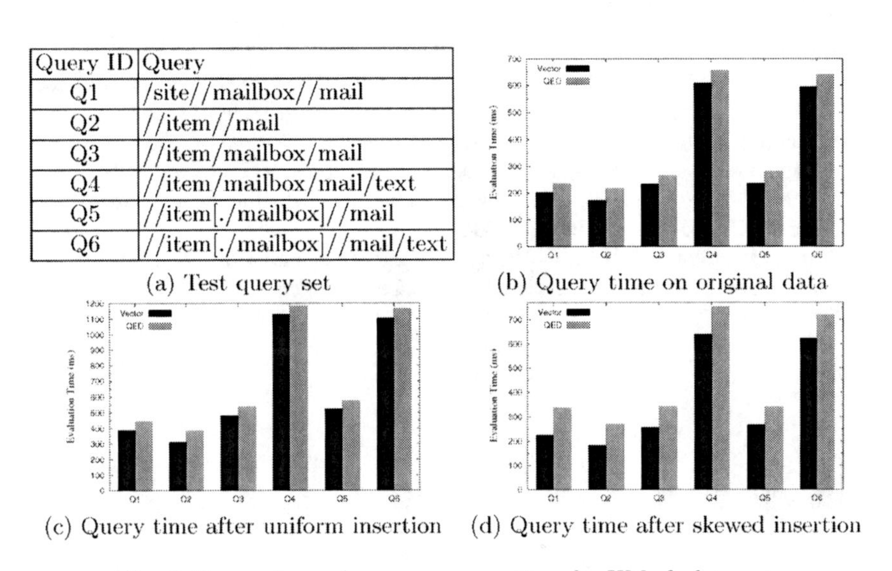

| Query ID | Query |
|---|---|
| Q1 | /site//mailbox//mail |
| Q2 | //item//mail |
| Q3 | //item/mailbox/mail |
| Q4 | //item/mailbox/mail/text |
| Q5 | //item[./mailbox]//mail |
| Q6 | //item[./mailbox]//mail/text |



(a) Test query set

(b) Query time on original data

(c) Query time after uniform insertion

(d) Query time after skewed insertion

**Fig. 5.** Comparison of query response time for XMark data set

## 7 Conclusion

In this paper, we have proposed a novel encoding scheme: vector encoding which is easy to understand and can be applied to various existing labeling schemes to completely avoid re-labeling. We have shown how it can be applied to containment scheme, and how insertions can be optimized. Finally it is experimentally proved that vector encoding outperforms existing schemes on updates and query processing especially in the case of skewed insertion.

We have focused on handling insertion in this paper. Currently we are extending our work to optimize both deletion and insertion. How to control the label size in a dynamic XML document where deletion and insertion frequently occur will be an interesting topic to explore.

## References


1. NIAGARA Experimental Data. http://www.cs.wisc.edu/niagara/data.html
2. University of Washington XML Repository. http://www.cs.washington.edu/research/xmldatasets/
3. XMark - An XML Benchmark Project. http://monetdb.cwi.nl/xml/downloads.html
4. Abiteboul, S., Alstrup, S., Kaplan, H., Milo, T., Rauhe, T.: Compact labeling scheme for ancestor queries. SIAM J. Comput. (2006)
5. Amagasa, T., Yoshikawa, M., Uemura, S.: QRS: A Robust Numbering Scheme for XML Documents. In: ICDE (2003)
6. Bray, T., Paoli, J., Sperberg-McQueen, C.M., Maler, E., Yergeau, F.: Extensible markup language (XML) 1.0, 4th edn., W3C recommendation (2006)
7. Yergeau, F.: UTF8: A Transformation Format of ISO 10646. Request for Comments (RFC) 2279 (January 2003)
8. Li, C., Ling, T.W.: QED: a novel quaternary encoding to completely avoid re-labeling in XML updates. In: CIKM (2005)
9. Liang, X., Zhifeng, B., Wang, T.L.: A Dynamic Labeling Scheme using Vectors (Extended), http://www.comp.nus.edu.sg/~xuliang/dlsv2007.pdf
10. McHugh, J., Abiteboul, S., Goldman, R., Quass, D., Widom, J.: Lore: A database management system for semistructured data. In: SIGMOD Record (1997)
11. Wu, X., Lee, M.L., Hsu, W.: A Prime Number Labeling Scheme for Dynamic Ordered XML Trees. In: ICDE (2004)
12. Zhang, C., Naughton, J.F., DeWitt, D.J., Luo, Q., Lohman, G.M.: On Supporting Containment Queries in Relational Database Management Systems. In: SIGMOD (2001)

# A New Approach to Replication of XML Data

Flávio R.C. Sousa, Heraldo J.A. Carneiro Filho, and Javam C. Machado

GRoup of computer networks, software Engineering and systems (GREat),
Federal University of Ceara, Fortaleza, Brazil
{flavio,heraldo}@great.ufc.br, javam@ufc.br
http://www.great.ufc.br

**Abstract.** XML has become a widely used standard for data exchange in several application domains. In order to manage data in this format, Native XML Databases (NXDBs) are being proposed and implemented. Even though, currently there are a number of available NXDBs, few of them provide replication mechanisms. This paper presents RepliX, a mechanism for replication of XML data based on group communication. With the purpose of validating RepliX, experiments were conducted to measure its performance.

**Keywords:** Native XML Databases, Replication, Group Communication.

## 1 Introduction

XML (*Extensible Markup Language*) [15] has become a widely used standard for data representation and exchange in several application domains. The growing usage of XML creates a need to manage data in this format. Native XML Databases (NXDBs) [5] are being proposed and implemented to target this demand. NXDBs store XML documents according to a graph logical structure, in which the nodes represent elements and attributes and the edges define the element/sub-element and element/attribute relationships. These systems implement many characteristics that are common to traditional databases, such as storage, indexing, query processing, transactions and replication.

The management of XML data is complex. This is due, mainly, to the following characteristics (i) *data model* - XML documents are represented by a graph-based data model, which increases the complexity of its structure (ii) *heterogeneity* - a XML document may have a sub-element completely absent or repeated several times. The flexibility in representing XML data makes it difficult to typify, store, and process such documents.

Regarding query processing, the XML model does not have a formal algebra yet. The W3C has developed formal semantics to the XPath and XQuery languages. This semantics allows for the identification of ambiguities in the language and aids the formal verification process. However, it is complex, making it difficult to perform decomposing operations. The current protocols of concurrency control to XML data still have limitations, and the majority of existing solutions

R. Wagner, N. Revell, and G. Pernul (Eds.): DEXA 2007, LNCS 4653, pp. 141–150, 2007.

block the whole document, offering low concurrency level and performance. Regarding the fragmentation or partitioning of XML data, the existing works try to adapt techniques of traditional systems as a solution to this problem.

Data replication techniques have been used to improve availability, performance, and scalability in relational and object-oriented database management systems [7]. The flexibility of the XML model, though, introduces some distinct challenges, thus new replication techniques ought to be developed. Current approaches to replication of XML data try to adapt existing concepts to the XML model [13] [16] [4]. However, few NXDBs provide replication mechanisms, and there is no work explicitly evaluating the performance and scalability aspects of such mechanisms. Among the several types of replication protocols, the group communication abstraction (GC) is an efficient technology to implement these protocols, since it provides reliability guarantees that simplify the application of fault-tolerance techniques [3]. Group communication primitives have been efficiently used to develop replication protocols, in both synchronous and asynchronous approaches [14].

This work describes RepliX [11], a replication mechanism for XML data. The solution we describe utilizes synchronous and asynchronous protocols, group communication primitives, and contemplates the characteristics of XML data. RepliX makes it possible to reduce response time in query processing and improve the performance of NXDBs. This paper is organized as follows. In section 2, RepliX is presented and its algorithms are discussed. Several performance tests of RepliX are presented in section 3. Section 4 summarizes related work, and section 5 presents our conclusions.

## 2 RepliX

RepliX is a mechanism for replication of XML data that takes into account the main limitations of XML data management, such as fragmentation and concurrency control. We extended the PDBREP [1] protocol in order to consider the characteristics of the XML model. For that, we modified PDBREP's synchronization method among update replicas, based on the protocol proposed by [14], to minimize conflicts during updates to XML data and to ensure that accessed data is always up-to-date. RepliX uses *one-copy serializability* [2] as its model of correctness.

The RepliX architecture can be observed in Figure 1. RepliXDriver is the component that provides access to the mechanism. This driver is a simple interface to execute transactions, encapsulating most of RepliX's functionalities and providing an abstraction layer for the developer over the whole architecture. RepliXCoordinator is composed by three sub-components: the *scheduler*, which is responsible for identifying what kind of transaction has arrived, the *router*, which decides where each transaction is going to be executed, and the *load balancer*, which balances the load across the active sites. RepliXNode is the component attached to each database. This component accesses the local database

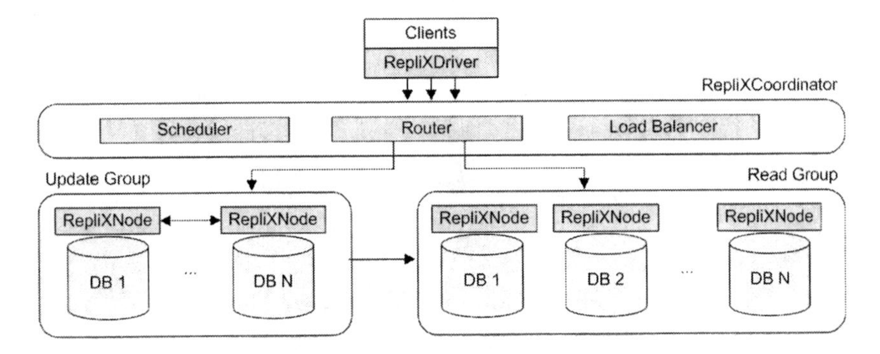


**Fig. 1.** RepliX Architecture

and executes the requested transactions, returning the results to RepliXDriver, which then returns them to the application.

### 2.1 Specifications

The system is composed by N sites, divided into two groups: a read group, which executes read-only transactions, and an update group, which executes update transactions. Transactions that contain only read operations are considered to be read transactions. If the transaction contains at least one update operation (insertion, update or deletion), it is classified as an update transaction. The set of N sites is not fixed, since the number of active sites changes as sites are being removed or added.

The strategy of partitioning the sites into groups is an important aspect of RepliX. We used this approach aiming to improve system performance and decrease the number of conflicts during update operations, since the concurrency control for XML data still has significant limitations. With this strategy, only a part of the active sites has to be modified for each update transaction.

When it receives a transaction, RepliXCoordinator, shown in Algorithm 1, analyzes its operations in order to forward it to the appropriate group. This analysis is performed checking the contents of each operation. RepliXCoordinator manages the sites of each of the two groups and uses a round-robin algorithm to choose which of them will execute the transaction, in order to distribute the load among the sites.

If a transaction is forwarded to an update group, one of its sites will receive it. This site is called primary and it is responsible for checking conflicts with other transactions that might be executing locally. It sends a multicast to the other sites in this group, called secondaries of the primary that sent the multicast. The transactions executed in the update group receive a unique identifier, which allows for its identification by RepliX.

Algorithm 2 shows the tasks that are executed by the primary. If the begin message is received, the transaction is initiated in the local database, its opera-

**Algorithm 1.** - Coordinator algorithm

```
1: procedure process_transactions
2:     loop
3:         transaction ← coordinator.receive_transaction();
4:         if transaction contains update operation then
5:             site ← coordinator.get_site(coordinator.write_site_id);
6:             coordinator.write_site_id ← coordinator.write_site_id + 1;
7:             remote_transaction ← site.begin(transaction);
8:             return remote_transaction;
9:         else
10:            site ← coordinator.get_site(coordinator.read_site_id);
11:            coordinator.read_site_id ← coordinator.read_site_id + 1;
12:            remote_transaction ← site.begin(transaction);
13:            return remote_transaction;
14:        end if
15:    end loop
16: end procedure
```

tions are executed within the initiated transaction and the results are returned to the client. When the client requests the transaction to commit, the update operations (write set) of the transaction are sent to the other sites in the update group (secondary sites) using the total order primitive of the group communication system. The primary site waits for confirmations of the certification tests executed in each secondary. Then, another multicast is sent to the secondary sites so that they can commit the transaction in the local databases. Following that, the primary site commits the transaction locally and sends the transaction write set to the sites in the read group. Later, the read sites will apply these modifications to their local databases.

**Algorithm 2.** - Primary site algorithm

```
1: procedure process_transactions
2:     loop
3:         msg ← primary_site.receive_message();
4:         if msn.type = begin then
5:             site_primary.begin;
6:             results ← primary_site.execute(msg.operations);
7:         else if msg.type = commit then
8:             primary_site.multicast(update_group, msg, write_set);
9:             primary_site.wait_for_certification_confirmation();
10:            if conflict_in_certification = true then
11:                primary_site.multicast(update_group, msg, abort);
12:            else
13:                primary_site.multicast(update_group, msg, commit);
14:            end if
15:            primary_site.commit();
16:            primary_site.multicast(read_group, msg, write_set)
17:        end if
18:    end loop
19: end procedure
```

Algorithm 3 describes the behavior of the secondary sites. When a message sent by the primary is received, the secondary sites execute a certification test to check if the updates conflict with any active transaction and, if not, a confirmation message is sent to the primary site. If the received message is a commit, a transaction is initiated in the local database and its write set operations received previously are executed and committed.

**Algorithm 3.** - Secondary site algorithm

```
1: procedure receive_transactions
2:     loop
3:         msg ← secondary_site.receive_message();
4:         if msg.type = write_set then
5:             result ← secondary_site.certification_test(msg.operations);
6:             return  result;
7:         else if msg.type = commit then
8:             secondary_site.begin();
9:             secondary_site.execute(msg.operations.write_set);
10:            secondary_site.commit();
11:        else if msg.type = abort then
12:            secondary_site.commit(msg.operations);
13:        end if
14:    end loop
15: end procedure
```

The certification test used in secondary sites is shown in Algorithm 4. This test checks for conflicts between transactions by comparing its read sets and write sets. After the certification test, the transaction is confirmed, aborting any executing local transactions (in the secondary site) that are in conflict with the transaction sent by the primary.

**Algorithm 4.** - Certification test algorithm

```
1: procedure certification_test(operations)
2:     for all transaction in secondary_site do
3:         if (operations.read_set = transaction.operations.write_set) or
4: (operations.write_set = transaction.operations.read_set) or
5: (operations.write_set = transaction.operations.write_set) then
6:             return  false;
7:         end if
8:     end for
9: end procedure
```

The modifications of the update group are serialized and sent to the read group through a multicast with a FIFO ordering property. These modifications are added to local queues and executed later on, preserving the original order of execution in the update group. The read group executes two types of transactions: propagation and refresh transactions. Propagation transactions are executed in order to commit pending updates during a site's idle time, that is, when there are no read operations or refresh transactions being executed. Refresh transactions are executed to process the transactions in the local queue of the read site.

The read site receives messages from clients or from the update group. When it receives begin messages, originated from clients, the queue is inspected, checking if there are pending modifications received from the update group that should be executed. The data is locked to allow for a refresh transaction to execute the pending updates. When the execution ends, the data of the site is unlocked. Then, a transaction is initiated in the local database, its operations are executed within the initiated transaction, and the results are returned to the client.

If the message is an update, originated from one of the sites of the update group, the received write set is placed at the end of the queue at the read site so that it will be executed later on through a refresh or propagation transaction. When the site is idle, meaning it is not executing any transactions, the queue is inspected for pending updates. The data is locked and a propagation transaction containing the modifications in the queue is executed. The data is locked to prevent the execution of new transactions before the end of the propagation transaction, thus preventing these new transactions from reading out-of-date data. At the end of the execution, the site is unlocked and it is free to execute any waiting transactions. From the users' standpoint, though, RepliX is synchronous, since they always access up-to-date data.

**Algorithm 5.** - Read site algorithm

```
1: procedure process_transactions
2:     loop
3:         msg ← read_site.receive_message();
4:         if msg.type = begin then
5:             if queue.is_empty() = false then
6:                 read_site.lock(msg.data_item);
7:                 read_site.execute_refresh_transaction(queue);
8:                 read_site.unlock(msg.data_item);
9:             end if
10:            read_site.begin()
11:            results ← read_site.execute(msg.operations);
12:            return results;
13:        else if msg.type = update then
14:            queue.add(msg.operations.write_set)
15:        end if
16:        while read_site.is_busy() = true do
17:            if queue.is_empty() = false then
18:                read_site.lock(msg.data_item);
19:                read_site.execute_propagation_transaction(queue);
20:                read_site.unlock(msg.data_item);
21:            end if
22:        end while
23:    end loop
24: end procedure
```

## 2.2 Evaluation

RepliX was mainly developed in Java, with minor parts of the source code written in C++. RepliXDriver consists of a Java library that can be used by applications to access the replicated system through RepliX's interface. RepliXCoordinator is implemented as a stand-alone service, also in Java. RepliXDriver and RepliX-Coordinator communicate using Java's RMI. Each RepliXNode is also a service written in Java that, through a database-independent abstraction layer, uses the database-specific driver to access the underlying NXDB. Our NXDB of choice was Sedna, an open-source NXDB written in C++ by the ISPRAS [6]. The communication between RepliXCoordinator and each RepliXNode uses RMI, but the nodes communicate amongst them using the Spread group communication system [12]. Some of the information needed by RepliXNode's algorithms were not accessible through Sedna's standard Java library. In order to make such data available to RepliXNode, we made small modifications in C++ to parts of the Sedna source code and extended Sedna's Java library to retrieve this data seamlessly. In order to generate the databases, we used the XMark benchmark for XML data [10] and added the update operations proposed by [11].

To evaluate RepliX, we assumed a set of sites S={$S_1$...$S_N$}. Each site $S_i$ has a database and contains a complete copy of the data, performing transaction management locally. The database uses the *two-phase locking* protocol to assure concurrency control. We also considered a set of clients C={$C_1$...$C_M$}, which generate the transactions. In order to process a transaction *t*, a client C connects to RepliX and submits a transaction t, which RepliX forwards to the site $S_i$.

The environment used in the experiments consisted of a cluster of PCs connected through an Ethernet Hub. Each PC has a 3.0 GHz processor, 1 GB of RAM, 100 Mbit/s full-duplex network interface, and runs the Windows XP operating system. The database consists of a 10 MB XML document and each transaction is composed of 10 operations. We compared the standard Sedna database against RepliX on top of Sedna. RepliX was composed of 11 sites and 3 of them were part of the update group.

The response time and throughput were measured considering different numbers of clients, each client submitting 100 transactions, 80% of those being read transactions. Figure 2 shows the average response time. The left chart increases as more clients are added and the stand-alone Sedna presents an inferior result compared to RepliX's. The reason for this behavior is that Sedna is overloaded quickly. Meanwhile, RepliX distributes the update and read transactions among the several replicas. In addition, the efficient message exchange service provided by the Spread system was key to improve the response time.

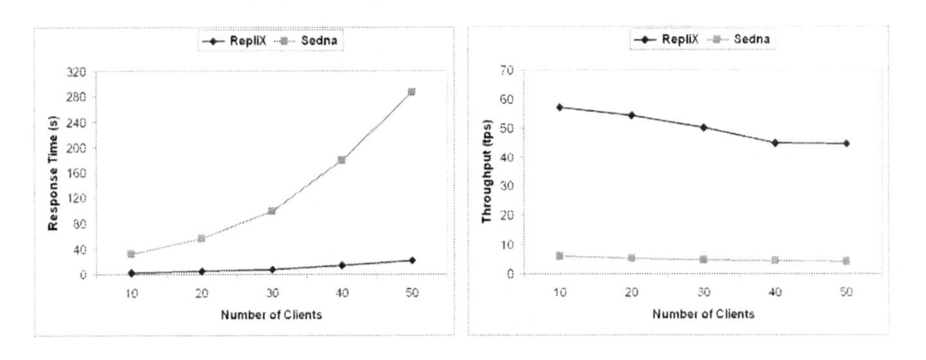


**Fig. 2.** Response Time and Throughput

Figure 2 shows the throughput. In the chart, we can observe that the throughput decreases as the number of clients is increased. The stand-alone Sedna presents low throughput values and they remain almost constant throughout the experiments. This happens because Sedna, as most centralized systems, has a limit on the amount of transactions it can process and manage. RepliX, on the other hand, is able to improve the throughput rate by distributing the transactions among the active sites. When the number of clients reaches 30, the throughput decreases considerably because every replica starts to get overloaded. With 40 and 50 clients, the throughput decreases less significantly. Generally, RepliX presented better throughput rates than the standard Sedna, given the number of clients evaluated.

The proportion of update transactions is an important parameter, and can significantly affect performance. Figure 3 presents a chart with the results of this experiment. The response time of the standard Sedna grows rapidly with the addition of more update transactions. For RepliX, the response time increases gradually until 40% of updates. One of the reasons for it is that the transactions are distributed among the update and read groups, improving processing efficiency. From 50% of updates on, the response time increases a great deal. This is due to the fact that more messages are being sent from the update group to the read group. During the execution of this experiment, we observed a small amount (less than 2%) of aborts in the update group, due to the certification test.

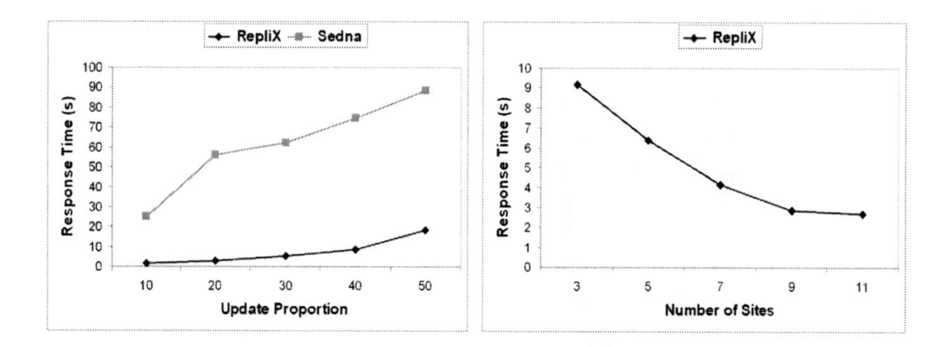


**Fig. 3.** Update Proportion and Scalability

Figure 3 shows the results of the scalability test. The addition of more sites improved RepliX's performance. With 3 sites, RepliX presented a considerable response time, but with 7 sites the improvement was notable. From 9 sites on, the response time tends to remain constant. RepliX's synchronization strategy, which updates initially the sites from the update group, does not have to update every single replica every time a database is updated, thus improving scalability.

## 3 Related Work

The eXist database system [4] provides a replication mechanism based on the JGroups system [8] to synchronize the replicas. eXist uses a primary copy protocol, propagating modifications in a synchronous fashion. When an update operation is sent to a site in the replicated group, this site sends an update to the primary copy and the secondary copies are locked. When the primary copy executes the update, it is propagated to the secondary copies, which later on are unlocked to process new requests.

The X-Hive database system [16] also presents a mechanism based on the primary copy protocol, but executes updates asynchronously. Updates are stored in a log and sent later on to the secondary copies. Since the modifications are executed asynchronously, it is possible that the applications accessing secondary copies might read out-of-date data. The Tamino database system [13] offers two replication approaches: a primary copy protocol and a Two-Phase Commit (2PC) protocol. When Tamino is configured to work with the former, replication works similarly to the X-Hive database system. With the latter execution proceeds much like in traditional databases.

Although eXist uses group communication primitives, these primitives are used only to provide a reliable message exchange service. X-Hive and Tamino apply traditional techniques to provide replication. However, these techniques are not appropriate for replication of XML data. The primary copy protocol used in the related work is not fault-tolerant, nor does it improve scalability [9]. The 2PC protocol used by Tamino shows low performance and a high number of conflicts [7] because it involves every replica in each update process. None of the related work presents results that attest the efficiency of the mechanisms and solutions proposed. Besides, the majority of them offer little portability, since they are implemented at the kernel of the database.

## 4 Conclusion

This work presented RepliX, a replication mechanism for XML data that combines synchronous and asynchronous protocols with group communication primitives that allow for efficient replication of such data. Particularly, RepliX considers characteristics of XML data, providing efficient ways to replicate data in this format, improving NXDBs' performance. RepliX was evaluated considering different aspects of replication. Analyzing the results of the experiments conducted, we observed a significant improvement in both performance and availability of the Sedna database system by using RepliX, even in scenarios with a large proportion of update transactions.

As future work, we intend to conduct a comparative study between RepliX and other replication protocols such as 2PC. Another important aspect is the development of query decomposing strategies that will allow RepliX to work with partial replication. We are planning to evaluate RepliX in different environments, such as WANs, with the intent of determining the performance variation introduced by network latency. Finally, we intend to further investigate the number of *aborts* that ocurr in the update group.

**Acknowledgments.** The authors would like to acknowledge the contribution of the Institute for System Programming - Russian Academy of Sciences in the development of this work.

## References

1. Akal, F., Türker, C., Schek, H-J., Breitbart, Y., Grabs, T., Veen, L.: Fine-grained replication and scheduling with freshness and correctness guarantees. In: VLDB '05: Proceedings of the 31st international conference on Very large data bases, pp. 565–576 (2005)
2. Bernstein, P., Newcomer, E.: Principles of transaction processing: for the systems professional. Morgan Kaufmann Publishers, San Francisco (1997)
3. Birman, K.: Reliable Distributed Systems: Technologies, Web Services, and Applications. Hardcover (2005)
4. eXist. `http://exist.sf.net`
5. Fiebig, T., Helmer, S., Kanne, C-C., Moerkotte, G., Neumann, J., Schiele, R., Westmann, T.: Anatomy of a native XML base management system. VLDB J. 11(4), 292–314 (2002)
6. Fomichev, A., Grinev, M., Kuznetsov, S., Sedna: A Native XML DBMS. In: Wiedermann, J., Tel, G., Pokorný, J., Bieliková, M., Štuller, J. (eds.) SOFSEM 2006. LNCS, vol. 3831, pp. 272–281. Springer, Heidelberg (2006)
7. Gray, J., Helland, P., O'Neil, P., Shasha, D.: The dangers of replication and a solution. In: SIGMOD '96: Proceedings of the ACM SIGMOD International Conference on Management of Data, pp. 173–182 (1996)
8. JGroups Toolkit. `http://www.jgroups.org`
9. Özsu, T., Valduriez, P.: Principles of distributed database systems, 2nd edn. Prentice-Hall, Englewood Cliffs (1999)
10. Schmidt, A., Waas, F., Kersten, M.L., Carey, M.J., Manolescu, I., Busse, R.: XMark A benchmark for XML data management. In: Proceedings of 28th International Conference on Very Large Data Bases (VLDB 2002), pp. 974–985 (2002)
11. Sousa, F.R.C.: RepliX: A Mechanism for Replication of XML Data. Master's thesis, Federal University of Ceara, Fortaleza, Brazil (2007)
12. Spread Toolkit. `http://www.spread.org`
13. Tamino XML Server. `http://www.softwareag.com/tamino`
14. Wu, S., Kemme, B.: Postgres-R(SI): Combining replica control with concurrency control based on snapshot isolation. In: ICDE '05: Proceedings of the 21st International Conference on Data Engineering, pp. 422–433 (2005)
15. Extensible Markup Language. `http://www.w3c.org/xml`
16. X-Hive: native XML database. `http://www.x-hive.com`

# An Efficient Encoding and Labeling Scheme for Dynamic XML Data

Xu Juan, Li Zhanhuai, Wang Yanlong, and Yao Rugui

School of Computer Science and Technology, Northwestern Polytechnical University, Xi'an 710072, China

xuj@mail.nwpu.edu.cn, lizhh@nwpu.edu.cn, wangyl@mail.nwpu.edu.cn, yaory@hotmail.com

**Abstract.** It is important to process the updates when nodes are inserted into or deleted from the XML tree. However, all the existing labeling schemes have high update cost. In this paper, we innovatively introduce a concept of Forbidden Code Segment (*FCS*), and then propose a novel and efficient encoding approach, called Extended Lexicographical Order encoding based on Forbidden Code Segment (*FCS-ELO* Encoding), whose codes are more compact than CDBS and QED codes. The most important characteristic is that our *FCS-ELO* labeling scheme can gracefully handle arbitrary update patterns and *completely* avoid re-labeling in XML updates, which is not at the sacrifice of query performance. We deliver the detailed theoretic analyses and experiments to show that, the proposed labeling scheme is superior to all the existing dynamic labeling schemes to process updates in terms of the incremental label size and the time for updating.

**Keywords:** Forbidden Code Segment (*FCS*), Lexicographical Order, labeling scheme, re-labeling, updates.

## 1 Introduction

XML [1] is widely used as a standard of information exchange and representation on the web. Much research has been undertaken on providing flexible indexing and query mechanisms on XML. The labeling schemes [2, 3, 4] are widely used in XML query processing, which can efficiently determine the ancestor-descendant and parent-child relationships between any two elements with small storage space.

If the XML is static, the current labeling schemes can efficiently process different queries. However, if the XML is frequently updated, large amounts of nodes need re-labeling, which is costly and becomes a performance bottleneck. Therefore, in this paper, we focus on how to efficiently update the XML documents and dramatically decrease the update cost without sacrifice of query performance. Our contributions in this paper can be summarized as follows:

- We propose a novel *FCS-ELO* encoding, which supports that *FCS-ELO* codes can be inserted between any two *consecutive FCS-ELO* codes *with the orders kept and without re-encoding* the existing codes.

R. Wagner, N. Revell, and G. Pernul (Eds.): DEXA 2007, LNCS 4653, pp. 151–161, 2007.

- *FCS-ELO* labeling scheme can *completely* avoid re-labeling in XML updates.
- We perform comprehensive experiments to demonstrate the merits of *FCS-ELO*.

The rest of the paper proceeds as follows. We first review the related works in Section 2. In Section 3, we propose a novel *FCS-ELO* encoding, and analyze its mean label size and application scopes. Section 4 presents comprehensive experiments to illustrate the performance. Finally we conclude in Section 5.

## 2 Related Works

In this section, we present several dynamic labeling schemes, which can partially or completely solve the problem of re-labeling in XML updates.

Amagasa *et al* [5] give a *Float-point Number Containment Labeling Scheme* in which every node is assigned three values: "start, end, level", and the "start"s and "end"s of the intervals are float-point values. For any two nodes *u* and *v*, *u* is an ancestor of *v* iff *u*.start < *v*.start and *v*.end < *u*.end. Node *u* is a parent of node *v* iff *u* is an ancestor of *v* and *v*.level - *u*.level = 1. For float-point number is represented in a computer with a fixed number of bits, this scheme cannot completely avoid re-labeling in XML updates [5, 8]. Furthermore, the comparison of float-point values will decrease the query performance [4, 8].

*OrdPath* [9] is an extended DeweyID [8] labeling scheme. OrdPath only uses the odd numbers at the initial labeling. When the XML is updated, it uses the even number between two odd numbers to concatenate another odd number. OrdPath can partially avoid re-labeling in XML updates because of the overflow problem [10]. Meanwhile, its query performance is worse than that of DeweyID since OrdPath needs more time to decide the prefix levels based on the even and odd numbers [11].

Changqing Li [11] proposes a novel *compact dynamic binary string* (*CDBS*) encoding. CDBS codes can be inserted between any two consecutive CDBS codes with the orders kept and without re-encoding the existing codes. The CDBS encoding is so compact that the size of CDBS is as small as the binary number encoding of consecutive integers. However, the CDBS cannot completely solve the re-labeling in frequent updates because of the overflow problem.

*QED* [10] uses four quaternary numbers "0", "1", "2" and "3", and each quaternary number is stored with 2 bits. Only "1", "2" and "3" appear in the QED code itself; "0" is used as the separator to separate different codes. Till now, the QED labeling scheme can really avoid re-labeling in XML updates. However, it is not the most compact, i.e. its size is 4/3 times larger than that of V-CDBS[11], and its update cost is larger than that of V-CDBS. When skewed frequent updating, its label size increment is 2 times over V-CDBS.

**Motivation.** CDBS can efficient process dynamic XML data, but it cannot completely avoid re-labeling; QED can avoid re-labeling, but its label size is large, and update and query performance is not as good as CDBS. With the consideration of the advantages of CDBS and QED, in this paper, we propose a novel *FCS-ELO* encoding. The *FCS-ELO* has two excellent benefits: 1) our *FCS-*

*ELO* is very compact, whose size is smaller than those of CDBS and QED; 2) *FCS-ELO* codes can be inserted between any two consecutive *FCS-ELO* codes without any re-labeling.

## 3 An Extended Lexicographical Order Encoding Based on Forbidden Code Segment

In this section, we first introduce the definitions. And then we propose a novel encoding approach. Finally we analyze the label size and its application scopes.

### 3.1 Encoding Algorithm

First of all, we introduce some correlative conceptions.

**Definition 3.1 (Lexicographical Order Code, LO Code)** *Lexicographical order code is a binary string* $C=(b_1b_2\cdots b_R)$, *where* $R$ *is the code size;* $b_R=1$ *and* $b_i\in\{0,1\}$, $1\le i\le R-1$.

**Definition 3.2 (Lexicographical order = and $\prec$, LO)** *Given two LO codes* $C_1=(b_1^1b_2^1\cdots b_T^1)$ *and* $C_2=(b_1^2b_2^2\cdots b_R^2)$,

*(1)* $C_1$ *is considered to be lexicographically equal to* $C_2$ *iff they are exactly the same, that is,* $C_1=C_2\Leftrightarrow b_i^1=b_i^2, 1\le i\le R=T$ *;*

*(2)* $C_1$ *is considered to be lexicographically smaller than* $C_2$ *(* $C_1\prec C_2$ *) iff*

*(a) the lexicographical comparison of* $C_1$ *and* $C_2$ *is bit by bit from left to right. If the current bit of* $C_1$ *is 0 and the current bit of* $C_2$ *is 1, then* $C_1\prec C_2$ *and stop the comparison, this case can be represented by* $b_i^1=b_i^2, b_k^1=0, b_k^2=1, 1\le i<k\le\min(T,R)$ *; or*

*(b)* $C_1$ *is a prefix of* $C_2$*, that is,* $b_i^1=b_i^2, 1\le i\le T<R$.

**Definition 3.3 (Code Segment, CS)** *Given a LO code* $C=(b_1b_2\cdots b_R)$, *a code segment is a segment of several consecutive bits, which can be represented by* $CS=(b_ib_{i+1}\cdots b_k)$, $1\le i\le k\le R$.

**Definition 3.4 (Forbidden Code Segment, FCS)** *Support that the code* $C=(b_1b_2\cdots b_{U-1}b_U\cdots b_R)$, *a forbidden code segment with parameter U is a code segment* $FCS_U=(e_1e_2\cdots e_K)$, *which is impossible to appear within the code segment* $(b_U\cdots b_R)$.

Just like what is described in Definition 3.1, all *LO* codes are ended with "1". In order to decrease the code size and easily detect *FCS*, we select $FCS_U=(11\cdots1)$. Here, the number of consecutive "1" is *K*. Next we introduce *FCS-ELO* code.

**Definition 3.5 (Extended Lexicographical Order Code Based on Forbidden Code Segment, FCS-ELO Code)** *Given that the flag field* $F = (f_1 f_2 \cdots f_Q)$ *and the forbidden code segment* $FCS_U = (11\cdots1)$ *with K consecutive "1", where U=2^Q-1, an FCS-ELO code* $\tilde{C}$ *can be constructed as*

*(a) if the size of LO code* $C = (b_1 b_2 \cdots b_R)$ *is smaller than U,* $\tilde{C}$ *is simply the concatenation of F and C, that is,* $\tilde{C} = F \oplus C = (f_1 f_2 \cdots f_Q b_1 b_2 \cdots b_R)$, *and* $\oplus$ *represents the concatenation operator, F records the size of C, which is equal to R;*

*(b) if the size of LO code* $C = (b_1 b_2 \cdots b_{U-1} b_U b_{U+1} \cdots b_R)$ *is equal to or larger than U,* $\tilde{C}$ *is the concatenation of F ,C, the first (K-1) bits of* $FCS_U$ *and a "0", that is,* $\tilde{C} = F \oplus C \oplus 11\cdots1(K-1) \oplus 0 = (f_1 f_2 \cdots f_Q b_1 b_2 \cdots b_{U-1} b_U b_{U+1} \cdots b_R 11\cdots10)$ , *where F is set to be U=2^Q-1;* $FCS_U$ *should never appear in the code segment* $(b_U b_{U+1} \cdots b_R)$ *and the number of consecutive "1" between* $b_R$ *and the last "0" is K-1.*

In Definition 3.5, we should give 2 comments: 1) the number of *consecutive* "1" between $b_R$ and the last "0" is *K-1* because of fully utilizing the ended "1" ($b_R$) in *C*; 2) we attach a "0" at the end for the sake of efficiently separating the "1" in the *FCS* from the "1" in the head of next code.

With respect to *FCS-ELO* codes, its code size can be defined as follows.

**Definition 3.6 (Code Size of FCS-ELO Code)** *The code size for a given FCS-ELO code* $\tilde{C} = (f_1 f_2 \cdots f_Q b_1 b_2 \cdots b_R)$ *(the case (a) in Definition 3.5) or* $\tilde{C} = (f_1 f_2 \cdots f_Q b_1 b_2 \cdots b_{U-1} b_U b_{U+1} \cdots b_R \; e_1 e_2 \cdots e_K 0)$ *(the case (b) in Definition 3.5) is the size of its corresponding LO code* $C = (b_1 b_2 \cdots b_R)$, *which is equal to* $R$.

Without special comments, *C* is the corresponding *LO* code of the *FCS-ELO* code $\tilde{C}$ in this paper.

**Theorem 3.1.** Given an *FCS-ELO* code $\tilde{C}$, where the size of the flag field is Q and $FCS_U = (11\cdots1)$ with K "1"s, we can always obtain the code size of $\tilde{C}$.

Due to space limitations, we omit the proofs of all the theorems and corollaries. The following Theorem 3.2 guarantees the existence of an *FCS-ELO* code between the given two *FCS-ELO* codes.

**Theorem 3.2.** Given two *FCS-ELO* codes $\tilde{C}_S$ and $\tilde{C}_L$, FCS and the record size, we can always find an FCS-ELO code $\tilde{C}_M$ which satisfies $C_S \prec C_M \prec C_L$.

And the algorithm to obtain inserted *FCS-ELO* code is easily concluded in Fig. 1.

**Algorithm 1**: **ObtainInsertedCode**( $\tilde{C}_S$ , $\tilde{C}_L$ )

**Input**: $\tilde{C}_S$ and $\tilde{C}_L$ are both *FCS-ELO* codes satisfying $C_S \prec C_L$ , where $FCS_U = (11\cdots1)$ with *K* "1"s and the size of the flag field is *Q*; the corresponding *LO* codes of $\tilde{C}_S$ and $\tilde{C}_L$ are $(b_1^S b_2^S \cdots b_T^S)$ and $(b_1^L b_2^L \cdots b_R^L)$

**Output**: *FCS-ELO* code $\tilde{C}_M$ satisfying $C_S \prec C_M \prec C_L$

**Description**:

1: $T$=**size**( $\tilde{C}_S$ )

2: $R$=**size**( $\tilde{C}_L$ ) // Get the code sizes of $\tilde{C}_S$ and $\tilde{C}_L$

3: **if** $T < R$ **then**

4: **if** $R+2 < 2^Q - 1$ **then** //***Case (1)***

5: $\tilde{C}_M = (R+2)_2 \oplus (b_1^L b_2^L \cdots b_{R-1}^L 01)$

// $(x)_2$ represents the binary version of $x$

6: **else if** $R+2 \geq 2^Q - 1$ **then** //***Case (2)***

7: $\tilde{C}_M = (2^Q - 1)_2 \oplus (b_1^L b_2^L \cdots b_{R-1}^L 01) \oplus 11\cdots1(K-1) \oplus 0$

// $11\cdots1(K-1)$ represents the (k-1) "1"s in *FCS*

8: **end if**

9: **else if** $T \geq R$ **then**

10: **if** $T+1 < 2^Q - 1$ **then** //***Case (3)***

11: $\tilde{C}_M = (T+1)_2 \oplus (b_1^S b_2^S \cdots b_T^S 1)$

12: **else** $T+2 \geq 2^Q - 1$ **then**

13: **if** $T \geq 2^Q - 2 + K - 1$ **then**

14: **if** $C_S$ is not ended with *K*-1 "1"s **then** //***Case (4)***

15: $\tilde{C}_M = (2^Q - 1)_2 \oplus (b_1^S b_2^S \cdots b_T^S 1) \oplus 11\cdots1(K-1) \oplus 0$

16: **else if** $C_S$ is ended with K-1 "1"s **then** //***Case (5)***

17: $\tilde{C}_M = (2^Q - 1)_2 \oplus (b_1^S b_2^S \cdots b_T^S 01) \oplus 11\cdots1(K-1) \oplus 0$

18: **end if**

19: **else if** $T < 2^Q - 2 + K - 1$ **then** //***Case (6)***

20: $\tilde{C}_M = (2^Q - 1)_2 \oplus (b_1^S b_2^S \cdots b_T^S 1) \oplus 11\cdots1(K-1) \oplus 0$

21: **end if**

22: **end if**

23: **end if**

24: **return** $\tilde{C}_M$

**Fig. 1.** Algorithm to obtain the inserted FCS-ELO code

We can easily conclude 2 corollaries from Theorem 3.2.

**Corollary 3.1.** $\tilde{C}_M$ *returned by* Algorithm *1 is an FCS-ELO code.*

**Corollary 3.2.** *Given any* two *FCS-ELO codes* $\tilde{C}_S$ *and* $\tilde{C}_L$, $C_S \prec C_L$, *we can always find P FCS-ELO codes* $\tilde{C}_{M1}, \tilde{C}_{M2}, \cdots, \tilde{C}_{MP}$ , *satisfying* $C_S \prec C_{M1} \prec C_{M2} \prec \cdots \prec C_{MP} \prec C_L$.

Theorem 3.2, Corollary 3.1 and Corollary 3.2 guarantee that we can insert nodes without re-labeling in XML updates when *FCS-ELO* codes are used to label nodes in XML tree. This is the first important property of our *FCS-ELO* codes.

Till now, we can formerly conclude the encoding algorithm listed as Fig. 2.

```
Algorithm 2: Encoding(N)
Input: A positive integer N
Output: The FCS-ELO codes for numbers 1 to N
Description:
1:suppose there is one more number 0 before the first number, and
one more number (N+1) after the last number
2:SubEncoding(codeArr, 1, N) // codeArr is an array with size (N+2)
3:discard the 0th and (N+1)th elements of codeArr
Procedure SubEncoding(codeArr, PS, PL)
/* SubEncoding is a recursive procedure; codeArr is an array, PS
and PL are the small and large position respectively */
1: PM=round((PS+PL)/2)
2: if PS+1<PL then
3:   codeArr[PM] = ObtainInsertedCode(codeArr[PS],codeArr[PL])
4:   SubEncoding(codeArr, PS, PM)
5:   SubEncoding(codeArr, PM, PL)
6: end if
```

**Fig. 2.** Encoding algorithm

### 3.2 Code Size Analysis

**CDBS.** To encode $N$ numbers, the mean size of the V-CDBS codes is [11]

$$\left(N\log_2(N+1)-N+\log_2(N+1)+N\log_2(\log_2(N+1))\right)/N \tag{1}$$

**QED.** The mean size of the $N$ QED codes is [10]

$$\left(2N\log_3(N+1)+2\log_3(N+1)+N\right)/N \tag{2}$$

**FCS-ELO.** Considering *FCS-ELO* codes with the specified *FCS* "111" and the record field size $Q$, when the encoding number $N$ is smaller than $2^Q$-1, the mean size of the FCS-ELO codes is similar to the V-CDBS codes, and can be formulated by (3).

$$\left(N\log_2(N+1)-N+\log_2(N+1)+NQ\right)/N \tag{3}$$

However, when $N \geq U = 2^Q - 1$, the mean size in this case can not be deduced easily. Suppose that the number of *FCS-ELO* codes with code size $n$ bits is $a(n)$. For difficultly deducing the close expression for $a(n)$, we just give the recursive equation.

$$\begin{aligned} b(n) &= b(n-2)+2\sum_{i=U}^{n-U-2} b(i)+2,\ n \geq U+2 \\ a(n) &= b(n)\times 2^{U-1} \end{aligned} \tag{4}$$

The initial condition is $a(U)$=$2^{U\text{-}1}$ and $a(U+1)$= $3\cdot 2^{U-2}$. Considering maximum code size $n$, 2 bits "11" in $FCS_3$, 1 bit separator "0", the record field size $Q$, and the codes with size smaller than $U$, the mean size of *FCS-ELO* codes is formulated by

$$\left(\sum_{i=1}^{U-1}(i+Q)2^{i-1}+\sum_{i=U}^{n}(i+3+Q)\times a(i)\right)\Bigg/\left(\sum_{i=1}^{U-1}2^{i-1}+\sum_{i=U}^{n}a(i)\right) \quad (5)$$

The mean code sizes for different encoding approaches are plotted in Fig. 3, where the line marked with "FCS-ELO(1)" corresponds to the case of $N < 2^Q - 1$ and the line marked with "FCS-ELO(2)" to the case of $N \geq 2^Q - 1$. From Fig. 3, we can make 2 comments: 1) the mean size of *FCS-ELO* codes is not larger than that of V-CDBS codes, and smaller than that of QED codes, which means our *FCS-ELO* codes are more compact; 2) considering the lines marked with "FCS-ELO (1)" and "FCS-ELO (2)", we can find out that the selection of $Q$ satisfying $N < 2^Q - 1$ mostly brings out much smaller mean code size. Therefore, in the initial encoding, $Q$ should be selected to be the minimum integer satisfying $N < 2^Q - 1$.From the analysis, we can conclude that *FCS-ELO* codes are rather compact, which is the second important property.

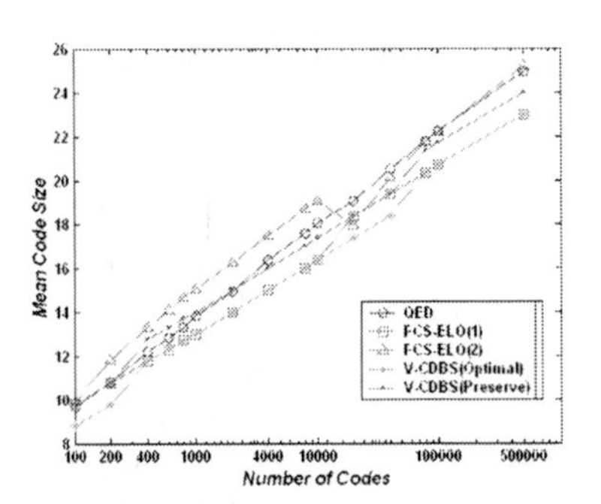


**Fig. 3.** Average code size for different encoding approaches

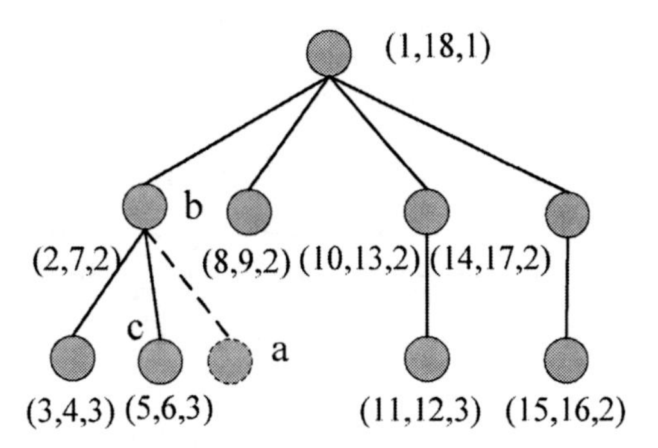


**Fig. 4.** Update

### 3.3 Application Scopes of FCS-ELO Codes

Our *FCS-ELO* can be applied to other labeling schemes [6, 8] that need to maintain the orders in updates. When we replace the "start"s and "end"s of the containment scheme [7] with *FCS-ELO* codes, an *FCS-ELO* containment labeling scheme is formed, called *FCS-ELO-Containment.* We can insert arbitrary number of nodes in any position without re-labeling, when *FCS-ELO-Containment* is used.

In *FCS-ELO* encoding, we use *FCS* as the separator to identify the different codes. Therefore, *FCS-ELO* will never encounter the overflow problem, and the *FCS-ELO* labeling scheme can completely avoid re-labeling in updates. Example 3.1 illustrates that our *FCS-ELO* labeling scheme can completely avoid re-labeling with the orders kept in XML updates. We use decimal numbers in Fig. 4 denoting the "start"s and "end"s for simplicity, but in practice, these numbers are stored using our *FCS-ELO* codes.

**Example 3.1.** To insert a node "a" in Fig. 4, we should insert 2 numbers between the "end" of node c "6" and the "end" of the node b "7". If we use the traditional labeling scheme, we cannot insert any numbers between "6" and "7", and we must re-label all the nodes in the tree. However, our *FCS-ELO* codes for "6" and "7" are "101:01001" and "100:0101". Based on the Algorithm 1, we can construct 2 inserted codes "**110:010011**" and "**111:0100111110**" as the "start" and "end" of node "a". Obviously, the corresponding *LO* codes satisfy "01001" $\prec$ "**010011**" $\prec$ "**0100111**" $\prec$ "0101". We need not re-label any existing nodes, but we can keep the containment scheme working correctly. It is similar for the insertion in the other positions.

## 4 Performance Study

CDBS has best performance in query and update processing and most compact label size among dynamic labeling schemes [11]; and QED can completely avoid re-labeling [10], so the experimental results are compared with CDBS and QED. All the schemes are implemented in Visual C++ 6.0 and the experiments are carried out on a 2.8 GHz Pentium 4 processor with 2 GB RAM running Windows XP Professional.

**Table 1.** Test datasets

| Datasets | Topics | # of files | Max/ Total of nodes |
|---|---|---|---|
| D1 | Movie | 490 | 125/26044 |
| D2 | Department | 19 | 2840/48542 |
| D3 | Actors | 482 | 1110/56995 |
| D4 | Play | 4 | 179690/392996 |
| D5 | Shakespeare's play | 37 | 6636/179689 |
| D6 | NASA | 2436 | 6022/533193 |

Table 1 shows the test datasets D1~D6, which are all real-world XML data from [14]. In addition, with respect to the preserve space for further insertion, we select ceil(log2(log2(2$H$+1))+1) and ceil(log2(log2(2$H$+1))) bits to record the real size of V-CDBS and *FCS-ELO*; $H$ is maximum total node number for each file in the dataset.

### 4.1 Performance Study on Static XML

In this section, we evaluate the performance on static XML, including label size, encoding time and query response time.

#### 4.1.1 Label Size

Fig. 5 shows the mean label sizes of different labeling schemes on datasets D1~D6. For *FCS-ELO-Containment*, it has smaller mean label size than QED-Containment [10] and CDBS-Containment [11], which means that our *FCS-ELO-Containment* is the most compact among the given 3 labeling schemes.

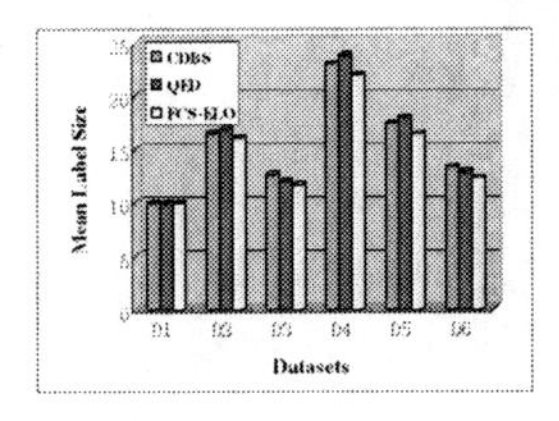


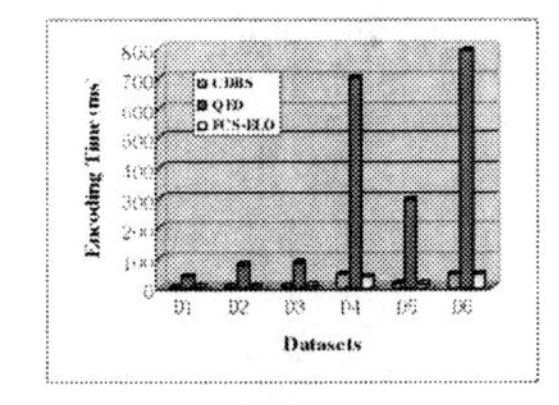


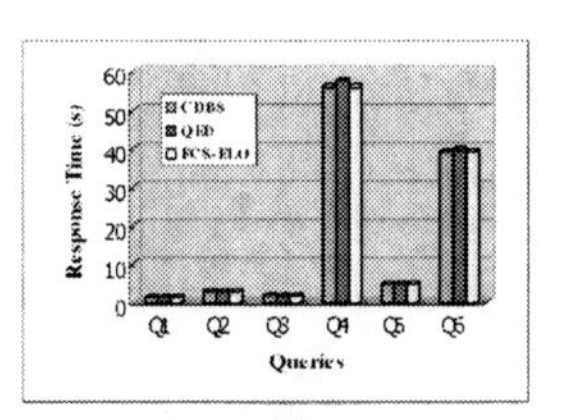


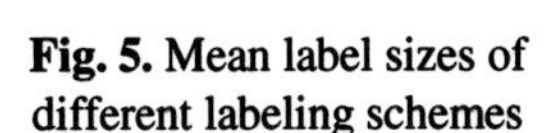
**Fig. 5.** Mean label sizes of different labeling schemes

**Fig. 6.** Encoding time of different labeling schemes

**Fig. 7.** Response time of queries Q1~Q6

### 4.1.2 Encoding Time

Fig. 6 shows the encoding times of different labeling schemes on datasets D1~D6. The encoding time of *FCS-ELO-Containment* is as longer as that of CDBS-Containment, however, only 1/15~1/10 of the encoding time of QED-Containment. Our scheme needs much fewer encoding time over QED mainly because QED-Containment encodes longer binary string and has very time-consuming division operation by "3". Therefore, our *FCS-ELO-Containment* is rather efficient in terms of encoding time.

### 4.1.3 Query Response Time

As described in [8], we scale up D5 10 times to test the queries. Table 2 shows the ordered and un-ordered queries (Q1~Q6) [11] and the number of nodes retrieved. Fig. 7 shows the response time of queries Q1~Q6. We can see our *FCS-ELO-Containment* quickly responds to the queries, whose response time is approximately similar to those of QED-Containment and CDBS-Containment. We can draw a conclusion that the superior properties of our scheme are not at the sacrifice of query performance.

**Table 2.** Test queries on the scaled dataset D5

| | Queries | # of nodes retrieved |
|---|---|---|
| Q1 | /play/act[4] | 370 |
| Q2 | /play//personae[.title]/pgroup[.//grpdescr]/persona | 2690 |
| Q3 | /play/personae/persona[12]/proceding-silling::* | 4240 |
| Q4 | //act[2]/following:speaker | 184060 |
| Q5 | //act/scene/speech | 309330 |
| Q6 | /play/*//line | 1078330 |

## 4.2 Performance Study on Frequent Updates

Next, we discuss the performance of 2 kinds of frequent updates. Section 4.2.1 evaluates the update cost of frequent insertions randomly at different places, called *Uniformly Frequent Update*. Section 4.2.2 studies the update performance of frequent insertions always at a fixed place, *called Skewed Frequent Update*, which is the worst case. We select one XML file Hamlet in D5 to test the update performance.

### 4.2.1 Uniformly Frequent Update

The Hamlet file has totally 6636 nodes. We insert 6635 nodes between every two consecutive nodes of the 6636 nodes. Based on the new file after insertion, we insert another 13270 nodes. And we repeat this kind of insertion 6 times.

Fig. 8 and 9 show the incremental label size and update time after each insertion series. When uniformly frequent update is encountered, CDBS-Containment, QED-Containment and *FCS-ELO-Containment* need to modify 1, 2 and a little more than 1 bits of the neighbor label to get the label of the inserted node respectively. Therefore, our *FCS-ELO-Containment* has cheaper update cost than QED-Containment, but has the same update cost as CDBS-Containment.

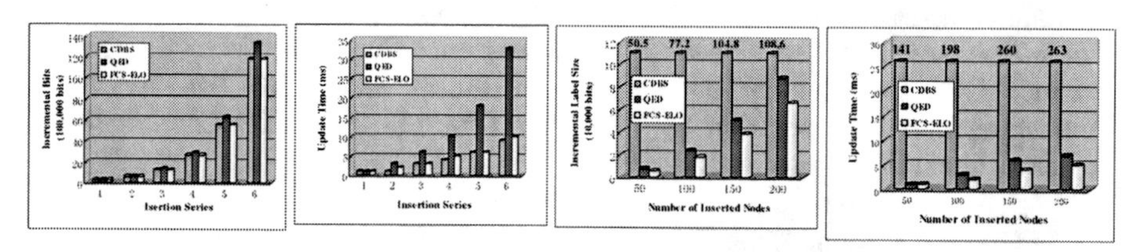


**Fig. 8.** Incremental bits for insertion series

**Fig. 9.** Update time for each insertion series

**Fig. 10.** Incremental label size for skewed frequent update

**Fig. 11.** Update time for skewed frequent update

### 4.2.2 Skewed Frequent Update

In order to test the performance of skewed frequent update, we randomly select a place in the Hamlet file, at which we consecutively insert 200 nodes. The simulated total incremental label size and update time are showed in Fig. 10 and 11, and we can see that *FSC-ELO-Containment* has the best performance while CDBS-Containment has the worst performance both in the terms of label size and update time. *FCS-ELO-Containment* and QED-Containment will never re-label for the XML updates. However, CDBS-Containment will certainly encounter re-labeling when the fixed bits number cannot represent the real codes' size in frequent insertions. In addition, our *FCS-ELO-Containment* has cheaper update cost than QED-Containment.

## 5 Conclusion

In this paper, we have proposed a novel and efficient *FCS-ELO* encoding, which can be broadly applied to different labeling schemes to *completely* avoid re-labeling in XML updates with cheaper update cost. All the theoretic analyses and experimental results demonstrate that our *FCS-ELO* labeling scheme has more compact labels and much cheaper update cost over all existing dynamic labeling schemes.

**Acknowledgement.** This paper is supported by National Science Foundation of China No. 60573096.

## References

1. Bray, T., Paoli, J., Sperberg-McQueen, C.M., Maler, E., Yergeau, F.: Extensible Markup Language (XML) 1.0, 3rd edn., W3C recommendation (2000)
2. Abiteboul, S., Kaplan, H., Milo, T.: Compact Labeling Schemes for Ancestor Queries. In: Proc. of SODA, pp. 547–556 (2001)
3. Agrawal, R., Borgida, A., Jagadish, H.V.: Efficient Management of Transitive Relationships in Large Data and Knowledge Bases. In: Proc. of SIGMOD, pp. 253–262 (1989)
4. Wu, X., Lee, M.L., Hsu, W.: A Prime Number Labeling Scheme for Dynamic Ordered XML Trees. In: Proc. of ICDE, pp. 66–78 (2004)
5. Amagasa, T., Yoshikawa, M., Uemura, S.: QRS: A Robust Numbering Scheme for XML Documents. In: Proc. of ICDE, pp. 705–707 (2003)
6. Li, Q., Moon, B.: Indexing and Querying XML Data for Regular Path Expressions. In: Proc. of VLDB, pp. 361–370 (2001)
7. Zhang, C., Naughton, J., DeWitt, D., et al.: On Supporting Containment Queries in Relational Database Management Systems. In: Proc. of ACM SIGMOD, pp. 425–436. ACM Press, New York (2001)
8. Tatarinov, S., Viglas, K.S., Beyer, J.: Shanmugasundaram, et al: Storing and Querying Ordered XML Using A Relational Database System. In: Proc. of SIGMOD, pp. 204–215 (2002)
9. O'Neil, P.E., O'Neil, E.J., Pal, S., Cseri, I., et al.: ORDPATHs: Insert-Friendly XML Node Labels. In: Proc. of SIGMOD, pp. 903–908 (2004)
10. Li, C., Ling, T.W.: QED: A Novel Quaternary Encoding to Completely Avoid Re-labeling in XML Updates. In: Proc. of CIKM, pp. 501–508 (2005)
11. Li, C., Ling, T.W., Hu, M.: Efficient Processing of Updates in Dynamic XML Data. In: Proc. of ICDE, pp. 13–22 (2006)
12. NIAGARA Experimental Data. Available at: http://www.cs.wisc.edu/niagara/data

# Distributed Semantic Caching in Grid Middleware

Laurent d'Orazio, Fabrice Jouanot, Yves Denneulin, Cyril Labbé, Claudia Roncancio, and Olivier Valentin

Laboratoire d'Informatique de Grenoble, France,
firstname.lastname@imag.fr

**Abstract.** This paper proposes a flexible caching solution to improve query evaluation in grids. It reduces both, data transfer and query computation, by adopting a distributed semantic caching approach. Our proposal introduces multi-scale cache cooperation including single site cooperation between object caches and distributed context aware cooperation between several query caches. Different cache miss resolution protocols are introduced for query evaluation and experimented in a grid data management for bioinformatics applications.

## 1 Introduction

Efficient data sharing technology is mandatory for large scale distributed systems especially for applications handling large data sets. A natural way to improve performances in this context is the use of caching solutions. Such solutions are particularly relevant in data grid management. Typical grid architectures consist of sites interconnected through high bandwidth networks, providing new caching perspectives. This article presents a distributed and semantic caching approach, as well as different cooperations of caches to supply a scalable system, applied in a grid middleware.

Generally resolving a cache miss consists in retrieving documents via servers. However these servers might become a bottleneck, due to computation or/and data transfer. That is why using other caches (called siblings) to resolve a cache miss, contacting servers in the last resort, can help reducing both load on servers and the amount of data transferred. Such a technique, referred as *distributed caching*, is well known in file systems [7] and in the Internet [5].

*Semantic caching* [13], [8] allows to exploit resources in the cache and knowledge contained in the queries themselves. As a consequence, it enables effective reasoning, delegating part of the computation process to the cache, reducing both data transfer and the load on servers. When a query is posed at a cache, it is split into two disjoint pieces: (1) a probe query, which retrieves the portion of the result available in the local cache, and (2) a remainder query, which retrieves any missing tuples in the answer from the server. If the remainder query exists then it is sent to the server for processing.

R. Wagner, N. Revell, and G. Pernul (Eds.): DEXA 2007, LNCS 4653, pp. 162–171, 2007.

In our caching solution, we aim to separate cache management from the resolution process and data transfer from the query evaluation. This two-level separation of concerns is the base of our contribution: a novel approach called *distributed semantic cache* and using *locality-based resolution*. The consequence of *locality-based resolution* is to limit the cooperation to a group of caches according to relevant neighbourhood strategy. We also propose to use *dual cache* [10] as it separates query from objects to optimize the query evaluation process. By combining both the approaches, we obtain a flexible system that aims at improving both data transfer and query processing in query evaluation. Our solution has been experimented in a grid data management middleware with bioinformatics applications.

This paper is organized as follows. Section 2 presents our proposition whereas section 3 presents a performance analysis in a middleware for data management on grids. Related work is described in section 4. Section 5 concludes this paper and gives research perspectives.

## 2 Distributed Semantic Caching

Scalability can be achieved using distributed and semantic caching. *Locality-based resolution* and *dual cache* are promising solutions for distributed and semantic caching respectively in grid environments. *Distributed semantic caching* combines both approaches, providing an architecture flexible enough to configure it according to application requirements and environment constraints. In particular, *distributed semantic caching* proposes several resolution strategies. In the following, section 2.1 presents *locality-based resolution*, section 2.2 introduces *dual cache* and section 2.3 discusses resolution potentiality.

### 2.1 Locality-Based Resolution

To make a system highly scalable, cache miss resolution has to be carefully considered. In fact, servers may become bottlenecks, making it relevant to use other caches during cache miss resolutions. However, not all caches are useful and it is important to determine the set of caches to contact. One way to do that is to regroup caches according to a given *locality*.

*Locality* may be based on different characteristics. In query evaluation middleware, we will consider *geographic* and *semantic locality*. *Geographic locality* like *proximity* [16] or *neighbourhood* [12] attempts to limit the resolution process according to distance. As a consequence, data transfers will be limited in a small area, avoiding congestion in the external network. *Semantic locality*, similar to *group of interest* [18] or *virtual community* [3] makes caches having common interests cooperate. In fact, the probability of having a cache hit increases with the ratio of interest sharing. Thus it may be relevant to contact a cache far away, since it is likely to save evaluation process.

Choosing a good *locality* depends on the application context. Traditional semantic caches are formed of regions, representing the objects answering a given query. Regions aim at reducing computation cost as well as communication for data retrieval. However, they do not distinguish both aspects, making difficult to choose an adequate *locality*. *Dual cache* is a solution to this problem.

### 2.2 Dual Cache

*Dual cache* is designed to improve query evaluation over data sources distributed across a grid. It corresponds to a semantic cache solution integrating light weight query management capabilities, captured in a `Query Manager`. The `Query Manager` provides tools to analyse query (equivalence, inclusion, etc.) and evaluating queries on entries in the cache [13], [8]. *Dual caches* are intended to be deployed on user, proxy or in special cases on server sites. They rely on the resources of these machines to improve query evaluation.

*Dual cache* attempts to maximize advantages of semantic caching which are the reduction of both data transfers and query computation. It clearly distinguishes these two goals by managing a couple composed of a query cache and an object cache as illustrated in figure 1. The query cache manages query results. Entries are identified by a query signature. The entry itself is the query answer stored as the set of identifiers of the relevant objects, `answer(Qi)=SetOf{ObjIdk}`. Objects themselves are in the related object cache, not in the query cache. When a new query is evaluated, answer retrieval implies loading the corresponding objects. So, a new query cache entry leads to object cache updates. *Dual cache* uses partially pre-calculated queries which are close to the concept of a view in database management systems. Each entry `Qi` of the query cache plays the role of a pre-calculated query and `SetOf{ObjIdk}` is its answer. As objects themselves are stored in an object cache which has no obligation to synchronize its content with the query cache, pre-calculated query or views may be fully or partially materialized

When a user query `Qj` is submitted to the *dual cache*, it may result in hits or misses in the query and the object caches. There is a query hit, if a `Qi` of the query cache can be used to answer `Qj`. Otherwise there is a query miss. In this case, `Qj` is sent to the appropriate servers in the standard way or using the current resolution protocol (see section 2.3). If there is a query hit, object misses may or may not occur depending on the current state of the object cache. Since query and object caches can initiate a resolution, servers must provide access by query and identifiers. The object cache and query cache can use their own resolution protocol and several protocols can be adopted through the grids to enhance the global caching performance.

### 2.3 Locality-Based Resolution in Dual Cache

Several resolution strategies can be used with a *dual cache*. A query is processed by a *dual cache* according to the algorithm presented in figure 3. Cache miss can be resolved using or not cooperation between query caches according to a *locality-based resolution* `lbr1`. During the process, the object cache can be contacted according to the algorithm presented in figure 4. Like query caches, object caches can communicate with each other to resolve their miss, according to a *locality-based resolution*

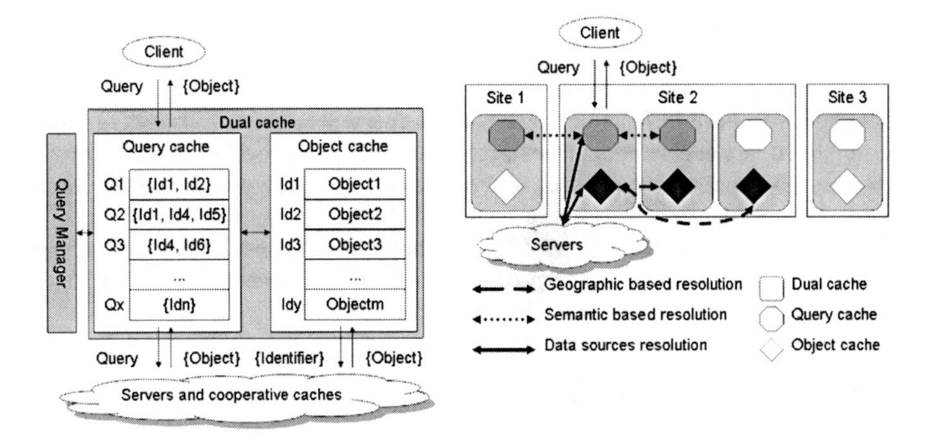


**Fig. 1.** Dual cache

**Fig. 2.** Cache miss resolution protocols in dual cache

`lbr2` that may differ from `lbr1`. Note that *locality-based resolution* is optional for both caches. As a consequence, $(n+1)^2$ caching strategies can be proposed, with n the number of *locality-based resolutions*.

```
idObjList ← QueryCache.lookup(query)
if idList ≠ null then {query hit}
    objList ← ObjCache.load(idObjList)
else {query miss}
    if servers resolution then {no locality-based resolu-
    tion 1}
        (idObjList,objList) ← servers.load(query)
    else {locality-based resolution 1}
        idObjList ← siblings.lookup(query)
        if idObjList ≠ null then {query hit on siblings}
            objList ← ObjCache.load(idObjList)
        else {query miss on siblings}
            (idObjList,objList) ← servers.load(query)
        end if
    end if
    QueryCache.add(query,idObjList)
end if
return objList
```

**Fig. 3.** Query processing by dual cache

```
for each id in idObjList do
    obj ← ObjCache.lookup(id)
    if obj ≠ null then {id hit}
        objList.add(obj)
    else {id miss}
        missIdObjList.add(id)
    end if
end for
if missIdObjList is not empty then
    if servers resolution then {no locality-based resolu-
    tion 2}
        missObjList ← servers.load(missIdObjList)
    else {locality-based resolution 2}
        missObjList ← siblings.lookup(missIdObjList)
        if missObjList = null then {object miss on sib-
        lings}
            missObjList ← servers.load(missIdObjList)
        end if
        ObjCache.add(missObjList)
    end if
end if
return objList
```

**Fig. 4.** Objects retrieval from object cache

In this paper, we will focus on *geographic locality-based resolution* for object caches and *semantic locality-based resolution* for query caches, resulting in four different strategies for a *dual cache*: *basic*, *geographic*, *semantic* and *semantic geographic*. Figure 2 gives an illustration of these strategies.

*Basic resolution:* When a *dual cache* uses a *basic resolution*, no cooperation is used. In other words, both query and object caches directly contact servers when a cache miss occurs. Such a policy is useful when servers are efficient.

*Geographic resolution:* If retrieving objects from servers is expensive, *geographic resolution* can be used, making object caches in a same location (site 2 in our example) cooperate, avoiding external data transfers.

*Semantic resolution:* If the bottleneck is on query processing, *semantic resolution*, making query caches cooperate, is relevant. Since such cooperation is strongly related on the semantics, it may concern query caches on different sites (site 1 and site 2 in our example).

*Semantic geographic resolution:* When the bottleneck is on object retrieval and query evaluations, using *semantic geographic resolution* is useful. Such a policy is a merger of *semantic* and *geographic resolutions.*

# 3 Performance Analysis

This section reports our experiences using *distributed semantic caching* in *Gedeon* [19], a middleware for data management in grids. Our main purpose is to show the impact of the different strategies proposed in section 2.3: *basic*, *geographic*, *semantic* and *semantic geographic*. In *Gedeon*, caches are read-only and modifications on servers are rare. Thus, consistency issues will not be addressed. In addition, due to the server querying capabilities, the instantiated `Query Manager` only considers query containment using query signature [6] and evaluations are reduced to selection and conjunction operators.

## 3.1 Testbed Configuration

**Experimental data set and query server.** Experiments have been done using *Swiss-Prot*[1], a biological database of protein sequences. It consists of a large ASCII file (750Mb) composed of about 210,000 sequence entries, each uniquely identified. *Gedeon* middleware provides a direct access to an entry through its identifier. It also provides query evaluation capabilities. Queries are composed of conjunctions and disjunctions of selection terms of the form `Attribute_name op value`. In the particular case of *Swiss-Prot*, `op` is often the `contain` operator and `value` is often a string. Evaluations result in a set of entries matching the query.

A *Java* and *Fractal*[2] version of *ACS* [9] has been used to instantiate *dual cache* and the different protocols proposed in section 2.3. In our experiments, the size of the object cache is 325Mb, corresponding to 50% of *Swiss-prot.* The query cache uses a size of 10 Mb, which is enough to store all the generated queries. All caches use the *LRU* replacement strategy.

**Workload generation.** Classical workloads used in benchmarks, like TPC[3], or Polygraph[4] for instance, do not consider semantically related queries, whereas we consider it as an important behavior for semantic caching. We use *Rx*, a synthetic

[1] http://expasy.org/sprot/
[2] http://fractal.objectweb.org/
[3] http://www.tpc.org/
[4] http://polygraph.ircache.net/

semantic workload [15]. Queries correspond to progressive refinements. The first query is general and the following ones are more and more precise and thus reduce the set of matching elements. In a *Rx* workload, *x* is the ratio of subsumed queries. For example, with *R50*, half of the queries will be issued by constraining former queries. In presented experiments, workload is composed of queries corresponding to a single selection term, or to conjunctions of between two to four selection terms. In order to simulate a context with semantic locality we choose for our experiments a *R40* workload.

In addition to the semantic locality, we introduce the notion of *community*. *Community* is used to group users having the same interests. The requests from the members of a community tend to focus on a particular subset of records. In the particular case of *Swiss-prot*, we have created groups of interest according to the tree of life. Each record belongs to one of four different groups : *Eukaryota*, *Archaea*, *Viruses* and *Bacteria*. Thus, for each of these groups, we defined a community of users supposed to be specifically interested in this group. In our experiments, 60% of the queries issued by any users concerns the records shared by its *community*. The last 40% requests are uniformly distributed among the other records.

**Performance metrics.** One of the most important metrics to study is the mean response time which is strongly related to the hit ratio. But the server's load and the amount of data transferred from servers to clients are also important metrics to be taken into account. As a matter of fact using a cache saves servers and network resources. As a consequence selected performance metrics involve: mean response time, hit ratio and the amount of data transfered.

### 3.2 Experiments in a Grid

The proposed solution has been deployed and evaluated on the French grid platform Grid5000[5]. Clusters at three different sites (Rennes, Nancy and Sophia-Antipolis) have been used. Nodes in these clusters are respectively: Sun Fire V20z 2x AMD Opteron 248 2.2GHz, 2GB memory and SCSI disk; HP ProLiant DL145G2 2x AMD Opteron 246 2.0GHz, 2GB memory and SATA disk; Sun Fire X4100 2x dual core AMD Opteron 275 2.2GHz, 4GB memory and SAS disk. For all clusters, the internal network is 1Gbit/s switched ethernet, whereas the inter-cluster network is a 10Gbit/s wide area network. The database has been partitioned in three equally sized files, managed by one node on each cluster. When a query is submitted, it is forwarded to the three clusters for a parallel evaluation. Results are then aggregated on the cache to build the final answer. Clients generate fifty queries according to the *R40* workload and each of them uses a local cache.

*Geographic resolution:* Figure 5 presents the impact of *geographic resolution* on performance indices. Figure 5(a) presents the amount of data transfered per query from servers to clients and figure 5(b) the ratio of queries evaluated on servers according to the number of clients/caches. Experiments have been done starting from

[5] http://www.grid5000.fr/

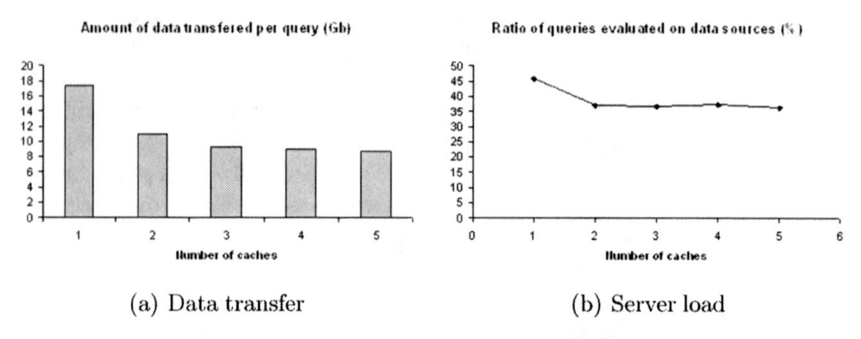


(a) Data transfer (b) Server load

**Fig. 5.** Cache cooperation based on geographic locality

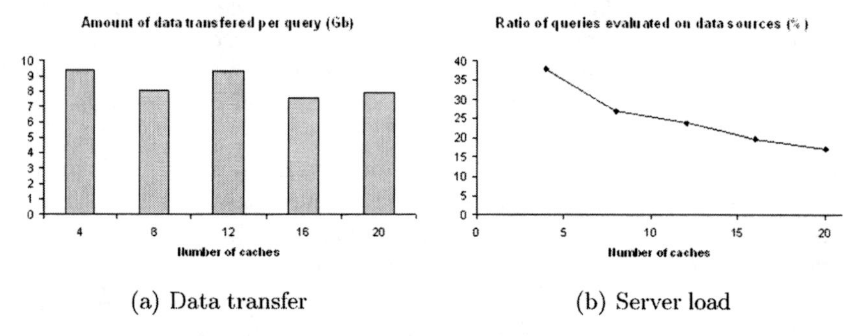


(a) Data transfer (b) Server load

**Fig. 6.** Cache cooperation based on semantic locality

one to five clients, deployed on separate nodes at Sophia-Antipolis. It can be seen that increasing the number of caches in the cluster, enables to reduce the external bandwidth consumption. However, when more than three caches are used, the volume of data stay stable, since the available resources are enough to store all relevant data. Thus adding more caches in the cooperation is useless. The number of caches is related to their size. For example, the same results will be obtained with a greater number of smaller caches. It is also important to note that *geographic resolution* has no impact on the ratio of queries evaluated on servers.

*Semantic resolution:* Figure 6 presents the study of the same performance metrics for *semantic resolution.* In this experiment, clients belong to one of the four available communities presented in 3.1. Each client uses a *dual cache* and cache cooperation is based on *semantic locality.* Each measure corresponds to an addition of one client in each community, uniformly distributed on the three clusters (growing up to seven nodes at Sophia-Antipolis and Rennes, six at Nancy). Figure 6(b) shows that when the number of caches increase, the ratio of query evaluated on servers decrease. Unlike with *geographic resolution* for objects, adding other query caches seem to be relevant. In fact, the number of possible queries is far beyond the number of objects, thus several query caches are necessary to store all of them. Note that the volume of data transfered (figure 6(a)) is not affected by *semantic resolution*.

| | Response time | Evaluation on servers | Transfered data |
|---|---|---|---|
| Basic | 44,1 s | 35 % | 9.0 Gb |
| Geographic | 43,7 s | 35 % | 8.8 Gb |
| Semantic | 28,4 s | 17 % | 7.9 Gb |
| Semantic geographic | 23,4 s | 17 % | 5.1 Gb |

**Fig. 7.** Specific performance metrics in a grid context

*Mixing semantic and geographic resolutions in a grid context:* Table 7 presents the results of the different resolution protocols proposed in 2.3 with twenty clients using a local cache. Globally, using cooperation enables to reduce mean response time. *Semantic resolution* reduces the number of evaluations on servers, as well as the amount of data transfered, since data sources are used via identifiers access avoiding to retrieve already stored objects. *Geographic resolution* alone enables a small reduction of the bandwidth consumption. In fact, most of the resolutions are done using query access. However, when it is used with *semantic resolution*, performances are greatly increased.

## 4 Related Work

This paper tackles different domains related to caching. In this section we present some of the main works related to each domain: grid, semantic and cooperative caching.

Some cache solutions for grid data management follow a mediation like approach. *ICM* [1] focuses on the problem of network latency. It proposes to store data in distributed databases replicated across the grid. User SQL queries are submitted to the cache, that decomposes them into sub-queries for local and remote domains, and composes afterwards the final results. [4] offers semantic cache functionalities by using hierarchical cache architecture. A kind of global cache federates grid node caches by using a global catalogue. A metadata catalogue helps to localize data in data sources. Even if both solutions aim at optimizing data transfer in grids, none of them has focused on the impact of minimizing computation.

Many proposals exist in semantic caching literature. They can be decomposed in two main categories, solutions separating query and objects [13] or not [8] [6]. *Dual cache* belongs to the first category. However it differs from traditional approaches by using two different caches enabling on the one hand to relaxe consistency between queries and objects and on the another hand to use different resolution protocols for each cache. In addition to these main approaches, other solutions can be found. [11] caches parts of queries that can be reused for further evaluations. This solution considers semantic aspects but does not manage probe and remainder queries. [17] proposes a cache of views in a centralized system. Caching views is quite interesting, but as object caching is not considered such a solution may be of limited use in a distributed environment. On the same principles, [14] proposed caching views if they cannot be obtained using already materialized ones.

Resolution protocols have been intensively studied in web caching. In this context many protocols have been proposed. They can be decomposed in three main categories [2] : flooding, hash-based and directory based. In this paper, a flooding resolution has been used. This choice is orthogonal to *locality-based resolution* and can be changed according to the context.

## 5 Conclusion

This paper presents a *distributed semantic caching* solution. Using the separation of concerns principle, such a solution clearly distinguishes cache management from cache miss resolution process, and computation from data transfer. It results in a mixing of *locality-based resolution* and *dual cache* leading to new opportunities. Fine configuration of the global cache strategy can be done to maximize both data transfer using cooperation between object caches and query evaluation using cooperation between query caches. Experiments have shown the relevance of such a solution in a grid context using a data management middleware. Our proposal saves computation time since it maximizes computation sharing between caches and reducing amount of data transfer by limiting external communication with a cooperation of object caches in a small area.

Future work remains. First we plan to study various application contexts for our solution, in particular warehouse oriented systems, where consistency issues can be relaxed. Such issues remain to be considered in our proposal. Our perspectives also include the study of cache solutions using different kind of querying capabilities (filtering, grouping, ordering, etc.). In addition, we want to investigate the impact of replacement strategies, specifically cooperative ones. For example, to avoid entries to be evicted, they can be placed in other caches. Finally, we are interested in proposing a self-adaptive and autonomous cache, in order to provide effective solutions in a dynamic environment.

## Acknowledgement

Thanks to N. Jayaprakash for rem the French Ministry of Research and *Institut National Polytechnique de Grenoble* for financial support. Experiments presented in this paper were carried out using the Grid'5000 experimental testbed, an initiative from the French Ministry of Research through the ACI GRID incentive action, INRIA, CNRS and RENATER and other contributing partners.

## References

1. Ahmed, M.U., Zaheer, R.A., Qadir, M.A.: Intelligent cache management for data grid. In: Proc. of the Australian WS on Grid computing and e-research, pp. 5–12 (2005)
2. Barish, G., Obraczka, K.: World wide web caching: Trends and techniques. Communications Magazine, IEEE 38(5), 178–184 (2000)
3. Brunie, L., Pierson, J.-M., Coquil, D.: Semantic collaborative web caching. In: Proc. of the 3rd Int. Conf. on Web Information Systems Engineering, pp. 30–42 (2002)
4. Cardenas, Y., Pierson, J.-M., Brunie, L.: Uniform Distributed Cache Service for Grid Computing. In: Proceedings of the International Workshop on Database and Expert Systems Applications, pp. 351–355 (2005)
5. Chankhunthod, A., Danzig, P.B., Neerdaels, C., Schwartz, M.F., Worrell, K.J.: A hierarchical internet object cache. In: USENIX Annual Technical Conf., pp. 153–164 (1996)
6. Chidlovskii, B., Borghoff, U.M.: Signature file methods for semantic query caching. In: Proc. of the 2nd European Conf. on Research and Advanced Technology for Digital Libraries, pp. 479–498 (1998)
7. Dahlin, M., Wang, R.Y., Anderson, T.E., Patterson, D.A.: Cooperative caching: Using remote client memory to improve file system performance. In: Proc. 1st Symposium on Operating Systems Design and Implementation, pp. 267–280 (1994)
8. Dar, S., Franklin, M.J., Jonsson, B.T., Srivastava, D., Tan, M.: Semantic data caching and replacement. In: Proc. of the 22nd Int. Conf. on VLDB, pp. 330–341 (1996)
9. d'Orazio, L., Jouanot, F., Labbé, C., Roncancio, C.: Building adaptable cache services. In: Proc. of the 3rd Int. WS on Middleware for Grid Computing, pp. 1–6 (2005)
10. d'Orazio, L., Valentin, O., Jouanot, F., Denneulin, Y., Labbé, C., Roncancio, C.: Services de cache et intergiciel pour grilles de données. In: 22ème journées Bases de Données Avancées (2006)
11. Finkelstein, S.: Common expression analysis in db applications. In: Proc. of the ACM SIGMOD Int. Conf. on Management of data, pp. 235–245. ACM Press, New York (1982)
12. Gadde, S., Chase, J., Rabinovich, M.: A taste of crispy squid. In: Proc. of the WS on Internet Server Performance (1998)
13. Keller, A.M., Basu, J.: A predicate-based caching scheme for client-server db architectures. The VLDB Journal 5(1), 35–47 (1996)
14. Lee, K.C.K., Leong, H.V., Si, A.: Semantic query caching in a mobile environment. SIGMOBILE Mob. Comput. Commun. Rev. 3(2), 28–36 (1999)
15. Luo, Q., Naughton, J.F., Krishnamurthy, R., Cao, P., Li, Y.: Active query caching for db web servers. In: 3rd Intl. WS on The WWW and DB, pp. 92–104 (2001)
16. Rabinovich, M., Chase, J., Gadde, S.: Not all hits are created equal: cooperative proxy caching over a wide-area network. Comput. Netw. ISDN Syst. 30(22-23), 2253–2259 (1998)
17. Roussopoulos, N.: An incremental access method for viewcache: concept, algorithms, and cost analysis. ACM Transactions on DB Systems 16(3), 535–563 (1991)
18. Tay, T.T., Feng, Y., Wijeysundera, M.N.: A distributed internet caching system. In: Local Computer Networks, pp. 624–633 (2000)
19. Valentin, O., Jouanot, F., d'Orazio, L., Denneulin, Y., Roncancio, C., Labbé, C., Blanchet, C., Sens, P., Bonnard, C.: Gedeon, un intergiciel pour grille de données. In: Conf. Française en Système d'Exploitation (2006)

# Multiversion Concurrency Control for Multidimensional Index Structures

Walter Binder[1], Samuel Spycher[2], Ion Constantinescu[2], and Boi Faltings[2]

[1] University of Lugano, CH–6900 Lugano, Switzerland
walter.binder@unisi.ch
[2] Ecole Polytechnique Fédérale de Lausanne (EPFL), CH–1015 Lausanne, Switzerland
{samuel.spycher,ion.constantinescu,boi.faltings}@epfl.ch

**Abstract.** Prevailing concurrency control mechanisms for multidimensional index structures, such as the Generalized Search Tree (GiST), are based on locking techniques. These approaches may cause significant overhead in settings where the indexed data is rarely updated and read access is highly concurrent. In this paper we present the Multiversion-GiST (MVGiST), which extends the GiST with Multiversion Concurrency Control. Beyond enabling lock-free read access, our approach provides readers a consistent view of the whole index structure, which is achieved through the creation of lightweight, read-only versions of the GiST that share unchanging nodes amongst themselves. Our evaluation confirms that for low update rates, the MVGiST significantly improves scalability w.r.t. the number of concurrent accesses when compared to a traditional, locking-based concurrency control mechanism.

## 1 Introduction

Concurrency control and manipulation of data with transactional semantics have been key issues of information systems. During the last decade complex solutions have been proposed, some generic, others more efficient and usually tailored to specific use-cases.

Based on their usage pattern we can identify two different major kinds of information systems: On the one hand databases, where essentially the frequency with which data in the system is changed is comparable to the number of data reads. And on the other hand systems such as On Line Analytical Processing (OLAP) tools [1], directories and content repositories, where fast access to multiple views of multidimensional data is more important than the rate at which data can change. These latter kind of systems usually trade space requirements for speed, and often have an extremely high read to write ratio.

On modern databases, the most successful concurrency control techniques to date make a similar space-speed trade-off through the use of Multiversion Concurrency Control (MVCC) [2]. This technique brings databases closer to OLAP-like systems by allowing read-only transactions to execute without any need for synchronization with read/write transactions.

In this paper we evaluate a MVCC strategy through which transactional semantics can be efficiently supported within multidimensional, tree-based data indexes. Our system is built on two well known concepts:

R. Wagner, N. Revell, and G. Pernul (Eds.): DEXA 2007, LNCS 4653, pp. 172–181, 2007.

Firstly, we propose an algorithm based on Multiversion Concurrency Control for concurrent access to the index structure. MVCC is a technique that manages access to shared data by replicating and versioning it where needed. By replicating data upon modification, read accesses are isolated from updates to the index. Therefore, reads have a consistent and unchanging view of the data and do not interfere with update processes through locking of individual nodes, such as in other concurrency control schemes. Our implementation of MVCC provides for a fixed snapshot of the index across multiple read operations. The reader is essentially free to request a new snapshot at any time if available, or retain the same snapshot for as many read operations as he wishes up to a specified timeout constraint.

Secondly, the index structure we chose to implement our MVCC design on is the Generalized Search Tree (GiST) [3]. The GiST is a balanced tree which contains algorithms for navigating as well as modifying the tree structure. The tree stores keys and record references in its leaf nodes, and the inner nodes contain predicates and references to their child nodes. These predicates evaluate true for any key in their child nodes. This hierarchy of predicates is essentially what is common to all tree-based index structures. The GiST itself is, however, not a fully implemented search tree, but a generic structure that provides a 'template' index structure for most of the tree-based access methods, which the user defines by extending the GiST.

The Multiversion-GiST (MVGiST) is our implementation of a concurrent index structure based on MVCC and the GiST. The MVGiST combines the flexibility and query power of the GiST with the high reader concurrency and multi-read consistency offered by MVCC. This paper provides an evaluation of the MVGiST, which demonstrates its efficiency compared to a locking-based technique. Its advantages render it interesting for applications based on multidimensional index structures that have a high read/write ratio and that depend on consistency across multiple queries. An example use-case, where we recently applied the MVGiST, is a directory indexing web service advertisements in a way that enables efficient, automated service composition [4].

The rest of this paper is structured as follows: Section 2 summarizes the features of the GiST. Section 3 presents the MVGiST, introducing the design principles and describing its structure. In Section 4 we evaluate performance and scalability of the MVGiST, comparing it with the locking-based concurrency control scheme for the GiST presented in reference [5]. Finally, Section 5 discusses related work and Section 6 concludes this paper.

## 2 Generalized Search Tree (GiST)

In the following we give an overview of the GiST; details can be found in reference [3].

A search tree is a balanced tree with (usually) high fanout. The internal nodes are used as index, and the leaf nodes contain the actual data. Every internal node has a series of keys and pointers to its child nodes. A query on the tree must supply a predicate $q$. Starting from the root node a query checks for consistency of $q$ with the keys associated with the child nodes, and moves to a child node if its key is consistent with $q$. It traverses the tree in this manner until it reaches the leaf nodes containing the data that match the query. In classical trees, predicates are constrained to specific types, such as range predicates, where keys delineate a range $[c_{min}, c_{max}]$, and a predicate is of the form $c_{min} \leq i \leq c_{max}$. But essentially a search key may be any arbitrary predicate that holds for each datum below the key.

A search tree is therefore *a hierarchy of categorizations, in which each categorization holds for the data stored under it in the hierarchy*. By exposing the key methods and the tree re-balancing methods to the user, arbitrary search trees may be constructed, which is exactly what is accomplished with the GiST. In a single piece of code, it unifies the common functionality of search trees; the user of the GiST only needs to provide the necessary extensions for the type of tree that is desired.

The GiST has a variable fanout between $kM$ and $M$, where $M$ is the maximum number of child nodes and $2/M \leq k \leq 1/2$, except for the root node, which may have fanout between 2 and $M$. Inner nodes contain $(p, ptr)$ pairs, where $p$ is a predicate that functions as a search key, and $ptr$ references another node. Leaf nodes contain the same pairs, but here $ptr$ identifies some tuple of user data.

## 3 Multiversion-GiST (MVGiST)

Proposed solutions to concurrency control in multidimensional index structures [5,6,7,8] synchronize individual operations on the tree. However, there are application domains, such as e.g. web service directories [4], where long-lasting read sessions may be required. In this paper, we use the expression 'read session' to denote a series of read operations issued from one client on an unchanging version of the tree.

### 3.1 Multiversion Concurrency Control

MVCC is a database technique that adds versioning to shared data, i.e., every write on a data item $x$ creates a new version of $x$. Since writes do not overwrite each other, this gives greater flexibility to the database system in its ordering of conflicting operations. However, executions containing writes and subsequent reads on the same data are not generally serializable, because writes may only become visible to readers after a certain delay.

MVCC has existed for many years, and there are several algorithms that exploit multiversions, which all work on the same basis. To our knowledge there has been no prior implementation of MVCC for the GiST. Reference [2] gives an in-depth view of MVCC for database systems.

### 3.2 Assumptions

The following assumptions underly the design of the MVGiST:

1. Read accesses are much more frequent than updates.
2. High concurrency for read accesses (high number of concurrent read sessions).
3. Read sessions must offer a consistent view of the tree data; they have to be isolated from concurrent updates to data and index structure.
4. Read accesses shall not be delayed.

5. Updates may become visible with a significant delay, but feedback concerning the update (success/failure) shall be returned immediately.
6. The duration of a read session can be limited (timeout).

### 3.3 MVGiST Structure

The restrictions for the MVGiST with respect to concurrent access leads us to a design in which the readers access a tree which is separate from the write tree. A periodic full replication of the write tree would however be far too costly both for memory and performance considerations. Therefore, at read tree creation, only the nodes that have been modified since the last tree replication are actually copied to the new read tree. Every write tree node contains a reference to its corresponding read node twin. The following is a generic outline of the algorithm for read tree management: After

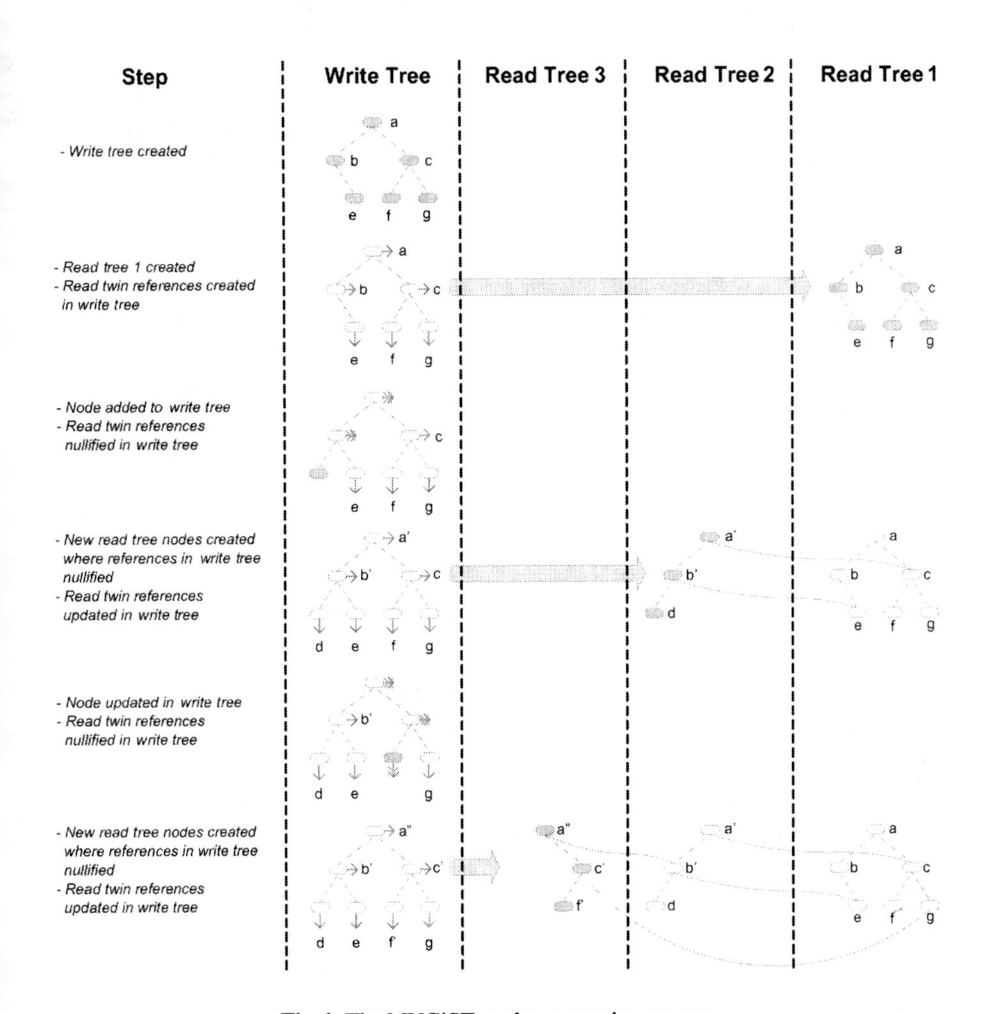


**Fig. 1.** The MVGiST read tree creation process

instantiation of the MVGiST, the write tree is populated with existing data. Then a `createReadTree` procedure is called. This procedure recurses down the tree, instantiates[1] a read node for every write node whose read twin reference is null (all nodes, when `createReadTree` is first called), and places an additional reference to it in the array of child nodes of the new read node's parent read node, thus creating a complete read tree. The new read tree root is now assigned to the current read tree root reference. All write nodes now have references to their corresponding read twins. The read tree remains constant for read access, while the write tree continues to be modified. All modified nodes in the write tree have their read twin references nullified. All references to read twins on the paths from the root to the modified nodes are nullified as well.

After a certain number of write tree modifications, the `createReadTree` procedure is called again, and only the nodes whose read twin references are null are duplicated. These new read nodes contain in their child node array references to other new read nodes as well as references to existing read nodes from the first read tree for the regions of the tree that have not been modified since the last `createReadTree` call. The new read root now becomes the current read root reference, and incoming readers receive the new read root to access the read tree. As soon as all readers have left a specific version of the read tree (either through timeout or on completion), the read nodes that are not referenced by other read tree versions will now gradually be removed.

Fig. 1 visualizes the process of read tree node creation. Bear in mind that this diagram is only schematic, and does not mean that a new read tree is created after every write node update.

An important point is that the MVGiST is far more lightweight than other concurrency control schemes as to the size of the tree nodes themselves, especially with respect to the read tree nodes. These nodes only contain the absolute minimum which is necessary for downward tree navigation. The write nodes have only one variable more than the non-concurrent GiST implementation: the reference to the read node twin. Other concurrency control schemes often maintain multiple lists and locks on a per-node basis in addition to the standard GiST node elements.

### 3.4 Supported Operations and Synchronisation Issues

There are three general operations that need to be distinguished for the MVGiST:

- Read session: one or more read queries, possible timeout.
- Write operation: batch of inserts and deletes to be completed on the write tree.
- Read tree creation: creates a new read tree, does not split write batches.

Note that for the read trees to mirror consistent states across the data tuples, certain sets of writes applied to the write tree must not be interrupted by a read tree creation. This is the case if there is some form of semantic dependency between the tuples to be written. A minimum such dependency could be update atomicity, i.e., an update consisting of a separate delete and insert operation must not be separated by the read tree creation process. A replication that splits such a set of updates would produce inconsistent read trees w.r.t. the data tuple semantics. We therefore define write operations as

[1] Instantiation of read nodes involves deep-copying of predicates.

batches of insert and delete calls, and a replication must be constrained only to begin after an entire batch has committed. Updates on the write tree and read tree creations are currently serialized, since tree modification is anticipated to be far less frequent than read access.

### 3.5 Read Tree Creation Strategies

Below we consider the freshness and memory consumption of MVGiST read trees. From these considerations, we derive two simple read tree creation strategies.

An important issue is the freshness of the most recent read tree (differences between write tree and read tree). The parameters involved are the frequency of tree modification by writers and the frequency of read tree creation. The freshness of a read tree $R$ is indirectly proportional to the number of changes in the write tree since the creation of $R$.

Memory consumption is proportional to the number of read trees in existence plus the write tree itself. The maximum duration of a read session $t$ (which can be enforced by a timeout) and the time between subsequent read tree creations $c$ control how many read trees can be active at the same time; the number of active read trees is $1 + \lceil \frac{t}{c} \rceil$. If we consider the size of the MVGiST constant (i.e., all updates consist of a delete and an insert operation), an approximate measure for the memory used by the MVGiST is $S_{write} + S_{read} + \lceil \frac{t}{c} \rceil * S_{read} * \delta_{update}$, where $S_{write}$ is the size of the write tree, $S_{read}$ is the size of the initial read tree, and $\delta_{update}$ is the average percentage of updated nodes between successive read tree creations. In order to minimize memory consumption, $c$ and $t$ may be chosen such that $t \leq c$, in which case only two read trees have to be kept in memory.

The MVGiST provides two built-in read tree creation strategies, freshness-triggered and timing-triggered (the user of the MVGiST may also implement different strategies). For both strategies, there is a dedicated thread $T_{update}$ that is responsible of processing updates and of periodically creating a new version of the read tree. $T_{update}$ gets batches of update requests from a synchronized queue $Q$.

In the case of freshness-triggered read tree creation, $T_{update}$ keeps track of the number of updates since the last read tree creation. If this number exceeds a given threshold, $T_{update}$ creates a new read tree before obtaining the next batch of update requests from $Q$.

In the case of timing-triggered read tree creation, $T_{update}$ guarantees a given minimum time span $c$ between consecutive read tree creations. The actual time span between read tree creations is $c + \epsilon$ ($\epsilon \geq 0$), and $T_{update}$ aims at minimizing $\epsilon$ (if there is a batch of updates in progress, $\epsilon > 0$). $T_{update}$ checks whether a new read tree is to be created before and after processing each batch of updates.

## 4 Evaluation

In this Section we evaluate performance and scalability of the MVGiST. All experiments operate on an initial tree with fanout of 4–8, which stores 10 000 keys. For the purposes of our evaluation, the choice of keys was unimportant, since no implementation differences of these exist between the competing access structures (we used an

R-tree implementation for our benchmark). Read operations search for keys which are known to be stored in the tree (i.e., all read operations are guaranteed to succeed). Updates consist of one delete and one insert operation, and in order to keep the tree size constant, deletion is guaranteed to succeed and insertion stores a new unique key in the tree. We analyze the throughput achieved by the MVGiST, measured as the number of read resp. update operations per second, for different workloads and different levels of concurrency (1–100 concurrent clients). A workload is a mix of read and update operations (0–100% updates).

In order to assess the strengths and drawbacks of the MVGiST, we compare the MVGiST with a traditional, locking-based concurrency control scheme for the GiST as presented by Kornacker et al. [5]. Because to the best of our knowledge, there is no implementation of this concurrency control mechanism implemented in Java (implementation language of the MVGiST), we developed our own Java reference implementation, henceforth named KCGiST (Kornacker's Concurrent GiST). The scheme presented by Kornacker et al. represents a concurrent access system complete with logging and recovery facilities. To reduce unfairness in the comparison, we restricted ourselves exclusively to the concurrency and consistency aspects of this scheme. A description of the KCGiST is given in Section 5. Please note that the KCGiST supports transactions and ensures repeatable read isolation [9], whereas the MVGiST provides a constant read-only view of the whole index tree. But since we found no concurrency control mechanisms for the GiST with functionality similar to the MVGiST, and because several other such mechanisms also employ Kornacker's system as a benchmarking reference, this was the obvious choice.

The benchmark to measure the throughput was set up to create identical workload for the two competing concurrency control schemes. Each client executes a randomly generated workload (a mix of read and update operations) with a given percentage of update operations. The workload is represented as a list which is processed sequentially by the client. Each client executes as a separate thread. In the case of the KCGiST, all read and update operations issued by the client threads directly access the same tree. A dedicated thread $T_{cleanup}$ takes care of cleaning up the logically erased entries in the tree. In the case of the MVGiST, only read operations are performed directly by the client threads, which obtain the most recent version of the read tree upon each read request. Update requests are attached to a common, synchronized queue, which is handled by the dedicated, high-priority thread $T_{update}$. Our benchmark uses the freshness-triggered read tree creation strategy explained in Section 3.5. $T_{update}$ creates a new read tree whenever it has processed 1 000 update requests.

The parameters of our measurements are the percentage of updates in the workload (0–100%) and the level of concurrency (1–100 client threads). For each setting, we execute the KCGiST resp. MVGiST benchmark 15 times and take the median of the measured throughput values. After each run, we force garbage collection. As platform we used a machine with 4 CPUs (2 dual-core Xeon 3GHz) and 4GB of RAM, running a 64-bit Windows XP installation. We employed the Sun JDK 1.5.0 with its 64-bit Hotspot Server Virtual Machine. We disabled background processes as much as possible in order to ensure consistent system conditions.

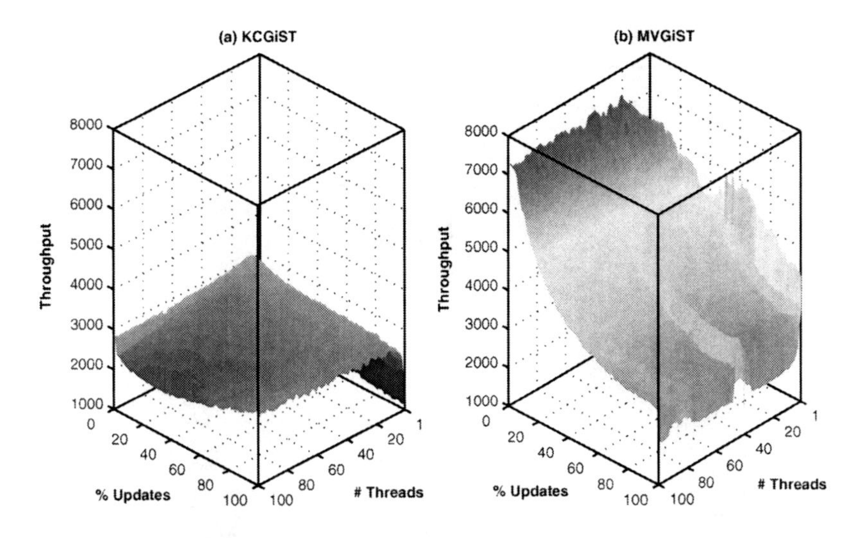


**Fig. 2.** Throughput of KCGiST (left) and MVGIST (right), depending on the level of concurrency (1–100 client threads) and the percentage of updates in the workload (0–100%). Number of leaf nodes: 10 000; fanout: 4–8; median of 15 runs. MVGiST read tree creation after each 1 000 updates.

Fig. 2 shows 3D surface plots of the measured throughput, depending on the level of concurrency (number of client threads) and the percentage of updates in the workload. For a lower percentage of updates, the MVGiST achieves about 2,5 times the throughput of the KCGiST. However, the throughput of the MVGiST significantly degrades with an increasing percentage of updates, whereas the throughput of the KCGiST remains rather stable independently of the workload. The KCGiST throughput does not suffer from a high update percentage, because node reorganisation is not frequent, as the tree is kept at a constant size. In contrast, for the MVGiST, the overhead due to read tree creation increases with a higher update percentage. In addition, updates are serialized, because only a single thread ($T_{update}$) can access the write tree.

For a lower percentage of updates, both concurrency control schemes scale well with the number of client threads. Fig. 3 depicts two slice planes of Fig. 2 for a workload with 0% resp. 20% updates. In the case of 0% updates, an increasing number of threads does not deteriorate throughput. However, a higher percentage of updates impacts scalability for both concurrency control mechanisms.

If there are only very few client threads (1–3), the throughput drops significantly, because some CPUs of our multiprocessor machine are idle. Interestingly, in a setting with very few client threads, the throughput of the MVGiST increases with the percentage of updates in the workload, reaching a peak at about 50% updates, before dropping again (see Fig. 2). The reason for this behavior is that the MVGiST uses a dedicated thread $T_{update}$ to process update requests. On a multiprocessor, $T_{update}$ may execute (accessing the write tree) in parallel with client threads processing read requests (accessing a read tree).

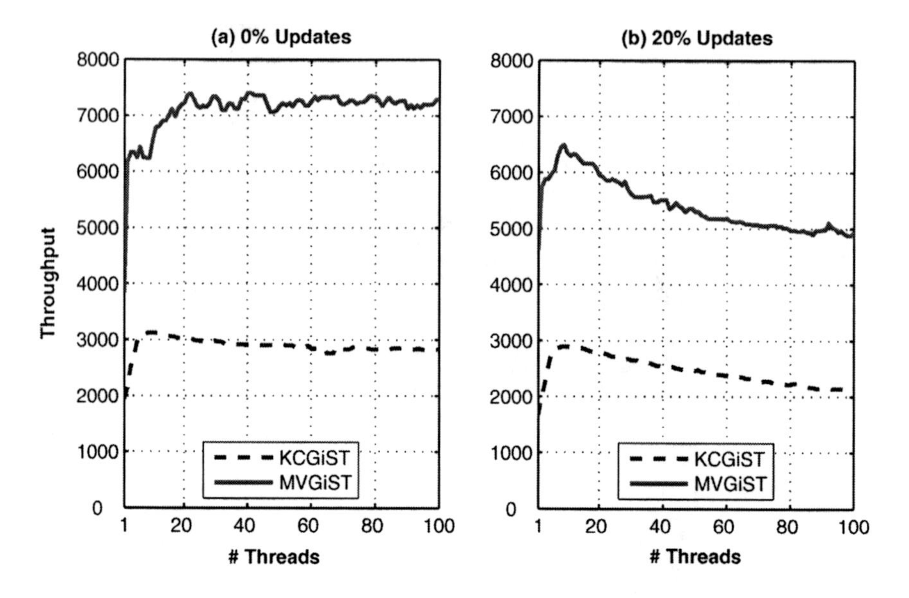


**Fig. 3.** Two slice planes of Fig. 2: Throughput depending on the level of concurrency (1–100 client threads) for a workload with 0% updates (left) resp. 20% updates (right)

## 5 Related Work

Reference [5] introduced the initial, locking-based concurrency control mechanism for the GiST. We call this scheme KCGiST. The KCGiST achieves basic concurrent protection by adding node-locks and two other components to the tree: node sequence numbers and rightlinks from every node to its split off right twin. This allows every operation traversing the tree to detect nodesplits when switching from one node to the next, and ensures concurrent access protection (S-Mode for readers, X-mode for writers) when an operation is within a node. The KCGiST implements repeatable read isolation [9]. This level of transactional isolation implies that if a search operation is run twice within the same transaction, it must return exactly the same result. The KCGiST achieves this by relying on a hybrid mechanism of two-phase locking of data records and avoidance of phantom insertions through predicate locking.

In reference [6] a node-locking based approach to concurrency on multidimensional index structures is optimized followingly: Simultaneous node-locking is avoided when updating bounding predicates by directly modifying the indexes while operations traverse down the tree. This scheme is then extended to reduce the blocking overhead during node-splits through local copying of the nodes, processing of the node-split, and then copying back the resulting changes.

An alternative concurrency control mechanism for the GiST is discussed in reference [7], which uses granular locking instead of predicate locking. In granular locking, the predicate space is divided into a set of lockable resource granules. Transactions acquire locks on granules instead of on predicates. The locking protocol guarantees that if two transactions request conflicting locks on predicates $p$ and $q$ such that $p \wedge q$ is satisfiable, then the two transactions will request conflicting locks on at least one granule in common.

In [8] the authors describe an enhanced concurrency control algorithm that reduces blocking time during split operations. They avoid lock coupling during bounding-predicate updates with a partial lock coupling technique, and have developed an update method which allows readers to access nodes where updates are being performed.

## 6 Conclusion

In this paper we promoted and evaluated the MVGiST, an index structure that is capable of dealing with large amounts of multidimensional data and that can be extended to nearly all types of tree-based access methods. The fact that the MVGiST supports consistency of data across queries makes it attractive for applications where some form of consistency beyond transactional isolation is required. Having tested the performance of the MVGiST under various load conditions, we think we have succeeded in developing an efficient concurrent access scheme for the GiST, especially for application domains with a high read/write ratio. Within these environments, for which the MVGiST was developed, the following conclusions can be made: the MVGiST can outperform a locking-based reference scheme by a factor of about 2,5. Due to the non-blocking nature of its read access, it scales very well with an increasing number of concurrent read accesses.

## References

1. Chaudhuri, S., Dayal, U.: An overview of data warehousing and olap technology. SIGMOD Rec. 26(1), 65–74 (1997)
2. Bernstein, P.A., Goodman, N.: Multiversion concurrency control – theory and algorithms. TODS 8(4), 465–483 (1983)
3. Hellerstein, J.M., Naughton, J.F., Pfeffer, A.: Generalized search trees for database systems. In: Dayal, U., Gray, P.M.D., Nishio, S. (eds.) Proc. 21st Int. Conf. Very Large Data Bases, VLDB, pp. 562–573. Morgan Kaufmann, San Francisco (1995)
4. Binder, W., Spycher, S., Constantinescu, I., Faltings, B.: An evaluation of multiversion concurrency control for web service directories. In: 2007 IEEE International Conference on Web Services (ICWS-2007), Salt Lake City (2007)
5. Kornacker, M., Mohan, C., Hellerstein, J.M.: Concurrency and recovery in generalized search trees. In: Peckman, J.M. (ed.) Proceedings, ACM SIGMOD International Conference on Management of Data: SIGMOD 1997. May 13–15, 1997, Tucson, Arizona, pp. 13–15. USA (1997)
6. Kanth, K., Serena, D., Singh, A.: Improved concurrency control techniques for multi-dimensional index structures. In: IPPS '98: Proceedings of the 12th. International Parallel Processing Symposium, pp. 580–586. IEEE Computer Society, Washington (1998)
7. Chakrabarti, K., Mehrotra, S.: Efficient concurrency control in multidimensional access methods. In: SIGMOD '99: Proceedings of the 1999 ACM SIGMOD International Conference on Management of Data, pp. 25–36. ACM Press, New York (1999)
8. Song, S.I., Kim, Y.H., Yoo, J.S.: An enhanced concurrency control scheme for multidimensional index structures. IEEE Transactions on Knowledge and Data Engineering 16(1), 97–111 (2004)
9. American National Standards Institute: ANSI X3.135-1992: Information Systems – Database Language – SQL (1992)

# Using an Object Reference Approach to Distributed Updates

Dalen Kambur, Mark Roantree, and John Murphy

Interoperable Systems Group, School of Computing, Dublin City University, Glasnevin, Dublin, Ireland

**Abstract.** With the Object-Reference (ORef) approach, the traditional object-oriented model is extended with references to act as a canonical model. Our ORef model facilitates the storage of localised behaviour and provides a precise definition for updating objects across distributed information systems. Both these characteristics are realised using an Object Pool connectivity mechanism that connects local and global object pairs.

## 1 Introduction

The context of our research is the EGTV project [1] which examines issues related to replacing specialised and mutually incompatible multimedia stores with inexpensive, general-purpose object databases that store the multimedia material as persistent objects. The formats of these objects and their operations directly depend on the database and are thus, heterogeneous. These heterogeneities must be resolved to provide universal access to the multimedia objects. We adopt the federated database approach [2] which employs a canonical model as the common interface to the participating databases. Federated client applications access canonical objects which are translations of original objects. Conversely, updates to the canonical objects must be propagated back to the original, underlying objects. Both translation of objects and propagation of updates must be provided as part of the federated services, and be invisible to client applications. In the domain of multimedia federations, a particular challenge is to provide behaviour as an integral part of the canonical model to resolve the heterogeneous behaviour of component databases. We examined a number of candidate object models, identified their deficiencies and based on these findings we designed our Object-Reference (ORef) model [3] as the canonical model.

The problem tackled in this work is the efficient translation of objects that also supports propagating updates and invoking behaviour. In the described multimedia federation database environment the resolution of this problem is crucial to connect client applications and servers. This provides the motivation for our research. The contribution of this paper is the mechanism of *Object Pool Pairs* (OPP) that provides such connectivity.

Our discussion is structured as follows: in §2 we discuss related research; in §3 we introduce the ORef modelling concepts; in §4 we describe the details of OPP connectivity; in §5 we examine implementation considerations; and finally in §6 we conclude this paper and assess the contribution of this work.

R. Wagner, N. Revell, and G. Pernul (Eds.): DEXA 2007, LNCS 4653, pp. 182–191, 2007.

## 2 Related Research

The LOQIS [4] project demonstrated the benefits of introducing references into native object-oriented database models which include the capability to define behaviour and provide a precise definition of the query results. These benefits are crucial for multimedia federated systems which, however, were not addressed.

In the IRO-DB project [5], relational and ODMG [6] databases are integrated using the ODMG model. Firstly, the local database schema is translated into a canonical External Schema with no semantic enrichment nor defining behaviour. Secondly, the External Schema is integrated into the Interoperable Layer using a CORBA-based Communication Layer [7] and restructured using the OQL [6]. Behaviour is defined in the Interoperable Layer using accessor operations that were generated for each property. Such approach results in performance penalty compared to the *direct access* to properties provided in the ORef model.

The MOOD project [8] uses the C++ object model as the native database model and extends the SQL query language with constructs for invoking behaviour. The behaviour is defined using C++ and compiled into a dynamically loadable library using a standard compiler. This approach permits the reuse of already existing behaviour definitions and for this reason it is also used in the ORef Architecture. However, the MOOD project was primarily concerned with providing the technical basis for storing behaviour rather then the issues of the underlying object model which are crucial with database federations.

In the COCOON project [9], federated heterogeneous database schemas are expressed using the COCOON model. Properties of *base objects* are fully encapsulated and updates are only possible using *methods*. A *view object* reuses the properties of the corresponding base object thus an object-preserving semantics is deployed. This project demonstrated that the object-preserving semantics is the key to the updatability of properties, hence we follow this approach. On the negative side, this project focused on COCOON databases only and unlike our approach which uses more standard object databases (O-R and ODMG).

The MultiView [10] project integrated GemStone databases using an *object-slicing* technique that provided multiple inheritance. This technique separated an object's properties into multiple *implementation objects*. Object-slicing was transparent to client applications and behaviour as properties were access using accessor operations that identified the correct implementation object. *Virtual objects* were created to represent the query result allowing the properties of source objects to be combined into new virtual objects. Object-slicing provides an identification of properties and is in our research adapted to be used with standard object-oriented programming languages.

## 3 ORef Modelling Concepts and Architecture

In this section, we give a concise description of an ORef Model introducing the concepts required for understanding the functionality of OPPs with a more complete description in [3]. Then, we position the ORef model in the ORef Architecture to provide the context for discussion on OPP functionality.

### 3.1 The ORef Model

The basic concepts of the ORef Model include: objects, types, references, relationships and behaviour. The ORef model is an object-oriented model featuring *types* that specify the *behaviour* and the *structure* for all *objects* of the type. Similar to object identifiers in traditional object models, an ORef object is assigned with an *orefOID* that uniquely identifies the object. We use the single concept of a *reference* to point to an object. This is the sole concept present in the model with such a capability and and is achieved by embedding the *orefOID* to the referenced object within the reference.

A type may be *simple*, *collection* or *user-defined*. Objects of *simple types* have a simple value. Objects of *collection types* may contain other objects using the reference mechanism. Finally, objects of *user-defined types* are complex, and are composed of multiple objects each of which takes the role of a *property* of the complex object. Property objects are independent of complex objects and may be properties of multiple objects. This independence forms the basis of the ORef *query language semantics* (QLS) which specifies that the result of a query is a set of complex objects that reuse the properties of source objects. Furthermore, this provides the basis for object updates.

An ORef relationship is an association between two user-defined types. For every relationship, each participating type contains a collection property which points to the related objects on the other side of the relationship. Updates to one side of the relationship are propagated to the related objects [3].

The behaviour of a type consists of multiple *operations* that are applied against a *target object* which itself may have multiple objects as *parameters. Mutable operations* allow target object or properties to be modified. *Immutable operations* do not allow updates to objects.

### 3.2 The ORef Architecture

The basis of the ORef Architecture is the traditional federated database architecture introduced in [2] in which heterogeneous component databases are integrated using a canonical data model. This model is also used by federated applications to access federated data. The ORef Architecture is concerned with integration of component object databases for multimedia federations using the ORef model. Component databases are based on either object-oriented ODMG or object-relational SQL:1999 [11] standard. Only the state of objects is integrated, and not their behaviour because: (1) the behaviour of ODMG databases is part of client applications from which it cannot be decoupled; and (2) the behaviour of SQL:1999 databases is *black-box* as there is no standard mechanism to determine objects that are accessed and modified by the behaviour. Once integrated into the federation, these databases may subsequently be provided with behaviour. Existing ORef federations may themselves participate in a new ORef federation. In this case, in addition to the object state, the behaviour may also be integrated as the modified objects are recorded (discussed later in 4.3). In [12] we analysed updates to objects based upon behaviour declaration.

In the ORef Architecture, both client applications and federated services that include a query processor and stored operations require an *object context* which is a container for all ORef objects. A *local object pool* (LOP) provides ORef objects by implementing three functions: (1) *materialising* ORef objects upon request, (2) *propagating updates*, and finally, (3) *disposing* ORef objects no longer used. Each LOP is a service of the ORef Architecture and it connects to a *source object pool* (SOP) that provides source objects which are manipulated using the listed functions. A SOP may either be a database, or an ORef federation and in the latter case, the SOP is known as the *Remote Object Pool* (ROP). The LOP and the SOP are tightly bound in an object pool pair (OPP) and the internal operation is described in the next section.

# 4 Object Pool Pairs Connectivity

The core functions of an object pool pair (OPP) are to materialise objects, to propagate updates to objects, and to dispose of objects. Orthogonal to these core functions, the issues of concurrent access and transactions in the ORef Architecture are discussed in [13]. These core functions are transparently invoked through ORef references. A reference may be a *local*, *neighbour* or *remote* reference. A local reference points to an object already materialised in the object context by storing the object's *orefOID*. Both neighbour and remote references point to objects outside of the object context. Such objects must be materialised first and then references are *transformed* into corresponding local references. A neighbour reference points to an object in the source object pool (SOP) which is directly connected to the object context. This reference contains the object's original source identifier *sourceOID* and the identifier of the OPP. The OPP identifier must be utilised as the original *sourceOID* is not sufficiently unique for object identification. A remote reference points to an object that cannot be reached through any directly connected SOP. This remote reference contains a chain of OPP identifiers that specify the sequence in which OPPs should be accessed to locate the source object, and its respective *sourceOID*. The OPP chain is organised as follows: (1) the first listed OPP is directly connected to the object context; (2) the second and following listed OPP are each connected to the OPP listed immediately prior; and finally, the SOP of the last listed OPP contains the target object identified using the original *sourceOID*.

A client application initiates updates where objects may be modified in the client application's object context or any other directly or indirectly connected object pool. When the client application modifies an object's properties, the object is modified within the client application's object context. However, when the client application invokes an operation, the object may either be updated in the SOP or a connected object pool depending on whether the operation updates the properties or invokes other operations. Thus, when a client application updates an object, its properties at the LOP and SOP side of one object pool pair will have different values. Propagating updates to synchronise this object may be delayed as long as objects are subsequently accessed and modified in the same object pool. This particular feature supports an efficient mechanism for preservation of the consistency of objects across OPPs (detailed in §4.3).

In the remainder of this section we first introduce object pool processing principles and then continue with an explanation of the theoretical background behind object materialisation and update propagation. *Object disposal* is invoked when all references to an object are removed and updates to the object are propagated. Further details on object disposal relate more to the implementation rather than theoretical aspects and are thus, omitted.

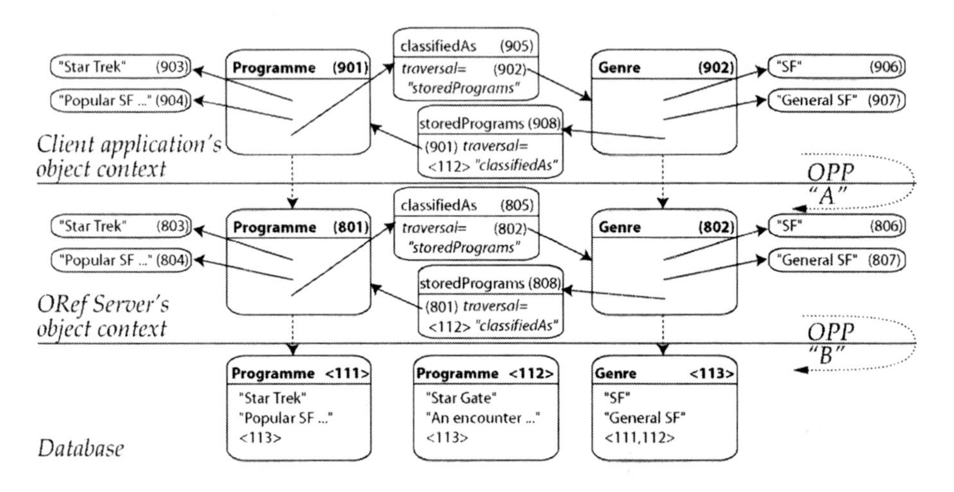


**Fig. 1.** Connected Object Pool Pairs

The scenario used in the examples in this section is illustrated in *figure 1* in which a client application displays the names of stored TV programmes and their genres. The OPP "A" is created to provide the client application with an object context connecting to the ORef Server (the top two layers). In order to access original source objects, this requires the ORef Server's object context provided by the chain OPP "B" (bottom two layers). Programmes and genres are modelled using classes `Programme` and `Genre` respectively. A relationship that connects a `Programme` to a `Genre` is named `classifiedAs`. Its traversal is named `storedPrograms` and connects a `Genre` to multiple `Programme` objects. Furthermore, a broken arrow (across the context line) illustrates the link each object preserves to its source object. In the diagram, the OPP identifiers are omitted from references for the sake of simplicity. The *orefOID* object identifiers are shown in parentheses as for example (901) where the *sourceOID* source object identifiers are shown using angle brackets as for example $\langle 111 \rangle$.

## 4.1 Processing Principles

An object pool is initially empty and is populated with materialised objects when their properties or behaviour are accessed. An object may be materialised but also updated in multiple object pools. The following requirements must be satisfied to preserve the consistency of objects across different object pools: (1) each object is materialised only once in a single object pool; and (2) properties of a single object have the same values when accessed in any object pool.

To satisfy both requirements, an object pool maintains both a *register of materialised objects* and a *set of modified objects* for each directly connected OPP. This register is optimised for fast retrieval of *orefOIDs* using *sourceOIDs* as keys, where the *orefOID* of modified objects are stored with the set. The OPP functions update both the register (of materialised objects) and the set (of modified objects) as discussed in the remainder of this section.

### 4.2 Materialisation

Only objects pointed by neighbour and remote references may be materialised as local object references are already materialised.

*A neighbour reference.* A neighbour reference consists of an OPP identifier and a *sourceOID* which references the original object on the SOP side. To materialise the object, the OPP's register is searched using the *sourceOID*, and if an object is found, the materialisation is completed by transforming the neighbour reference into a local reference constructed using the located *orefOID*.

If the object is not found, it is located using its *sourceOID* in the SOP and read into a new materialised object. The register is extended to include this object to prevent multiple materialisations. Values of the object's properties are copied from the SOP with the exception of properties that contain references. These properties must be adjusted to include the original source OPP identifier. This inclusion of source OPP identifiers involves the following conversions: (1) a local reference from the SOP into a neighbour reference in the object context; (2) a neighbour reference in the SOP into a remote reference in the object context; and finally, (3) a remote reference in the SOP into a remote reference in the object context that includes the source OPP.

In the OPP "B" in *figure 1,* when a database operation accesses the names of `Programmes` that correspond to the SF `Genre` using the relationship `storedProgrammes`, it accesses objects pointed to by a local reference (801) and a neighbour reference ⟨112⟩. While the object pointed by the local reference (801) can be readily accessed as it was materialised before, the object pointed to by the neighbour reference ⟨112⟩ must be materialised first, and then the neighbour reference transformed into a local reference.

*A remote reference.* A remote reference consists of the original *sourceOID* of an object and a chain of the OPP identifiers. A remote reference is materialised in multiple steps, each corresponding to one of the OPPs listed in the chain, starting from the last OPP listed. In each step, a neighbour reference that corresponds to the OPP is constructed, materialised and then used in subsequent steps. Firstly, the neighbour reference is constructed in the last OPP listed using its identifier and the original *sourceOID*. The object pointed to by this reference is materialised, and the neighbour reference transformed into a local reference containing the *orefOID* of the object. As demonstrated previously, materialising a neighbour reference ensures that the object is materialised only once. Secondly, this *orefOID* is used along with the identifier of the penultimate OPP to construct and materialise a neighbour reference, again ensuring that the object has been materialised only once. The same process is continued until the first OPP in the chain and this results in materialising the object, and transforming the original remote reference into a local reference that points to the materialised object. We have shown that the object is materialised only once in each OPP of the chain and therefore, we satisfy the first consistency requirement.

In the OPP "A" from *figure 1,* a client application obtained the SF `Genre` object (902) in order to retrieve the names of the SF programmes using the relationship `storedPrograms`. The `Star Trek` programme (901) was first materialised in the ORef Server's object context as object (801) from object ⟨111⟩. Then, this newly materialised object (801) is used to materialise object (901). The materialisation of the `Star Gate` programme ⟨112⟩ will take the same steps.

### 4.3 Update Propagation

An OPP may be *Synchronised, Local Current* or *Remote Current* as indicated in the state transition diagram in *figure 2* and described in the remainder of this section. The *Remote Current* state is only possible when the LOP connects to an ORef federation and not when the LOP connects to a database as no database operations may be invoked, hence no changes may occur.

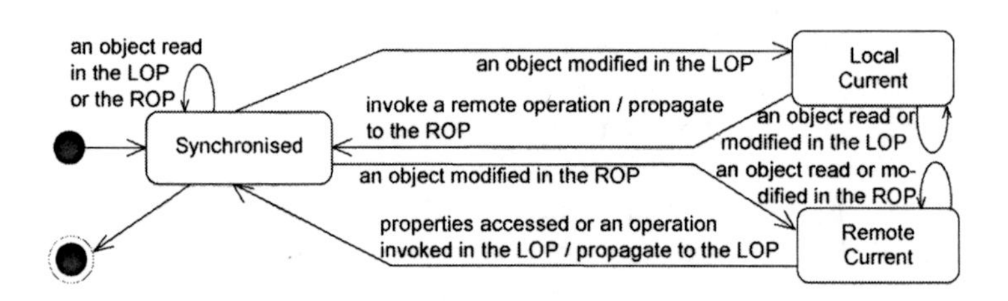


**Fig. 2.** An OPP State Transition Diagram

An OPP is initially in the *Synchronised* state and subsequent requests to materialise objects preserve this state. Reading properties or invoking immutable operations does not change the state of the OPP.

A *Synchronised* OPP changes the state to *Local Current* when an object is modified in the LOP by updating object's properties. Further changes and access to properties in the LOP preserve the *Local Current* state.

To synchronise an OPP that is in the *Local Current* state, local modifications are applied to the SOP. In this process, from the LOP's set of modified objects the *sourceOIDs* of modified objects and current values of their properties are extracted. Then, *sourceOIDs* are used to identify the original objects at the SOP side and their properties are updated to the current values which are simply copied. The collection and relationship properties are an exception as these contain references to objects in the LOP, hence they must be adjusted to point to corresponding objects in the SOP. To adjust a local reference, the *sourceOID* of the original object is used to materialise a local reference in the SOP. Neighbour and remote references are adjusted to remove the OPP identifier added in the process of materialisation as explained in §4.2. This converts a neighbour reference into a local SOP reference, and a remote reference into a remote SOP or a neighbour SOP reference.

An OPP changes the state to *Remote Current* when a mutable operation defined in the ROP is invoked. As the ROP connects to other object sources using OPPs each of which records modified objects in a set, the union of these sets is the set of objects that are modified in the ROP. Further invocations of mutable ROP operations preserve the *Remote Current* state.

An OPP in the *Remote Current* state is *Synchronised* by applying the source object modifications to the LOP. The set of modified source objects is transferred to the LOP which then re-reads all listed objects that are already materialised in the LOP using the materialisation process described in §4.2.

### 4.4 Optimisation and Efficiency

Client applications, database operations and the query processor in the ORef Architecture operate on ORef objects that are materialised in an object context provided using object pool pairs. These ORef objects are translations of source objects and they provide the capability to define behaviour and views. Our overall strategy in the OPP implementation is aimed at minimising overheads associated with maintenance processing of objects. Each object is materialised only once, on request, and only the reference that requested materialisation is transformed into a local reference. Other references pointing to the same object remain intact until they are used to access the object. Then, the register of materialised objects is consulted to retrieve *orefOID* for the requested object, if available. This approach eliminates the overhead associated with updating all references that point to the same object, as they may never be used. For example, when a persistent class is materialised, many of the relationships are not used, hence they do not need to be resolved to local references. This approach supports *prefetching object pages* which is a traditional database technique of reading a page of co-located objects instead of a single object in order to reduce network associated overheads and improve performance. The co-located objects are placed in the register of materialised objects so that they can be retrieved in subsequent materialisation requests. With regard to object updates, all copies of objects present in different object pools must be consistent, but only when they are accessed. The OPP mechanism accumulates the changes on one side which are then propagated prior to the objects on the other side of the OPP being accessed. This mechanism transfers only modified objects. However, it removes the need for constant re-synchronisation after each update and thus, eliminates the associated network and processing overhead.

## 5 Implementation

Our prototype is based upon Versant ODMG database, the omniORB CORBA engine and the Microsoft Visual C++ compiler and is functionally complete where the performance comparison is planned for our future research. The common features of the ORef model and standard object-oriented programming languages simplify our implementation. Specifically, programing language objects are identified using memory addresses. The attributes of these objects also have their own addresses which closely resembles *orefOID* semantics and allows one-to-one mapping to an *orefOID*.

References are implemented using C++ templates that embed automatic materialisation, source-updating, reference-counting and object disposal mechanisms. ORef types are mapped into C++ classes that were manually written for simple types, and generated for user-defined types using the `Definition Processor`. This is a low-level component of the ORef Architecture that also generates libraries and maintains metadata. The ORef properties are mapped to attributes, and operations to C++ methods. Classes also include internal ORef methods that transfer object state using the *CORBA Bridge* which is a communication component based on CORBA [14].

Both local (LOP) and remote (ROP) object pools are implemented as C++ classes. The LOP class provides the LOP functions and the client part of the *CORBA Bridge* including the low-level CORBA-based transfer. All these functions are implemented as methods and are transparently invoked from the class interfaces provided for ORef types. The ROP class extends the LOP class with the server part of the *CORBA Bridge* providing remote access and includes low-level functions for transferring objects, native database access protocols, and support for ORef operations. Operation definitions are compiled into dynamic libraries and loaded on request. The dynamic libraries also contain internal methods that materialise, pack and transport the state of objects, and that are used by the *CORBA Bridge* or native database access protocols.

## 6 Conclusions

This paper described the novel approach of Object Pool Pairs (OPP) for integration and updating heterogeneous object databases using the ORef Architecture and the ORef model. Our approach uses the direct addressability of both objects and properties introduced by the ORef model using *references*. References have a key role in providing stored operations for objects using a standard programming language and also provide a clear definition of object updatability. The OPP connectivity mechanism was designed to support both features while minimising the network transfer by delaying propagation using a batch mechanism.

Future research may include a *partial object materialisation* where only the accessed properties of an object are materialised. This approach may improve the performance when a client application uses only a subset of the properties and particularly when this subset does not include the multimedia content.

The ORef Architecture uses CORBA as the base communication protocol to provide a platform independent object transport. In many applications CORBA is replaced by Web Service oriented solutions which communicate using using XML documents. This however, requires objects to be transferred using a character-based XML encoding for binary data and involves a significant overhead. This is an interesting problem and will be addressed in our future research.

## References


1. Smeaton, A.F., Roantree, M.: Research in Information Management at Dublin City University. SIGMOD Record 31(4), 121–126 (2002)
2. Sheth, A., Larson, J.: Federated database systems for managing distributed, heterogeneous and autonomous databases. ACM Computing Surveys 22(3), 183–226 (1990)
3. Kambur, D., Bećarević, D., Roantree, M.: An Object Model Interface for Supporting Method Storage. In: Kalinichenko, L.A., Manthey, R., Thalheim, B., Wloka, U. (eds.) ADBIS 2003. LNCS, vol. 2798, Springer, Heidelberg (2003)
4. Subieta, K., Beeri, C., Matthes, F., Schmidt, J.: A Stack-Based Approach to Query Languages. In: Proceedings of the Second International East/West Workshop, pp. 159–180. Springer, Heidelberg (1994)
5. Busse, R., Fankhauser, P., Neuhold, E.J.: Federated schemata in ODMG. In: East/West Database Workshop, pp. 356–379. Springer, Heidelberg (1994)
6. Catell, R., Barry, D.: The Object Data Standard: ODMG 3.0. Morgan Kaufmann Publishers, San Francisco (1999)
7. Ramfos, A., Busse, R., Platis, N., Fankhauser, P.: CORBA based data integration framework. In: Proceedings of the Third International Conference on Integrated Design and Process Technology, IDPT. pp. 176–183 (1998)
8. Dogac, A., Dengi, C., Kilic, E., Ozhan, G., Ozcan, F., Nural, S., Evrendilek, C., Halici, U., Arpinar, B., Koksal, P., Kesim, N., Mancuhan, S.: A multidatabase system implementation on CORBA. In: 6th Int Workshop on Research Issues in Data Engineering: Nontraditional Database Systems, pp. 2–11 (1996)
9. Scholl, M., Laasch, C., Rich, C., Schek, H., Tresch, M.: The COCOON object model. Technical Report 211, Dept of Computer Science, ETH Zurich (1994)
10. Rundensteiner, E.A.: MultiView: A Methodology for Supporting Multiple Views in Object-Oriented Databases. In: Proceedings of the 18th International Conference on Very Large DataBases (VLDB'92), Vancouver, British Columbia, pp. 187–198. Morgan Kaufmann Publishers, San Francisco (1992)
11. Gulutzan, P., Pelzer, T.: SQL-99 Complete, Really. R&D Books (1999)
12. Kambur, D., Roantree, M.: Storage of Complex Business Rules in Object Databases. In: 5th International Conference on Enterprise Information Systems (ICEIS 2003), pp. 294–299 (2003)
13. Bećarević, D.: An Object Query Language for Multimedia Federations. PhD thesis, School of Computing, Dublin City University (2004)
14. Henning, M., Vinoski, S.: Advanced CORBA Programming with C++. Addison-Wesley, London (1999)

# Towards a Novel Desktop Search Technique

Sujeet Pradhan

Kurashiki University of Science and the Arts
Nishinoura 2640, Tsurajima
Kurashiki City, 712–8505 Japan
sujeet@cs.kusa.ac.jp

**Abstract.** The most serious challenges Personal Information Management Systems face today are the results of having to deal with a large number of heterogeneous types of data from diverse data sources, but having no means of managing and searching them in a convenient, unified fashion. We argue that simplicity and flexibility are essential attributes for the next-generation search tools to respond to these challenges. This paper lays out specific issues to realizing such a tool in the context of desktop search and ties them to existing search techniques employed by Database Management Systems and Information Retrieval — the two leading disciplines in search technology. We propose a novel technique for desktop search and show how our combined database and information retrieval approach to searching heterogeneous desktop data is going to benefit a large community of users.

## 1 Introduction

Most Personal Information Management Systems (PIMS) today face a daunting task of dealing with large collections of data from diverse sources. These data are not limited to plain, unstructured text files or structured data that can be easily fit into a conventional Database Management Systems (DBMS). For example, a personal desktop may typically contain an extremely heterogeneous collection of data including text, pictures, music, emails, XML, LaTeX and Microsoft Office documents scattered across a hierarchy of folders. What we lack today is a means of managing and searching them in a convenient, unified fashion.

Recently, this issue has gained considerable attention both from industry as well as research community. While several popular vendors such as Microsoft[15], Apple[2] and Google[7] have been offering keyword-based desktop search tools, their search range is limited to the file system managed by an underlying Operating System (OS). They seriously lack capability of retrieving a particular segment of a document's contents[5]. For example, if we wish to search for a particular section in the contents of a LaTeX file, these tools will return the name of this file (with the full contents), instead of only *that* desired section. However naive in their approach, keyword-based desktop search tools are nevertheless an important first step toward searching a mixed collection of data.

In the database research community, there has been a lot of emphasis on the need of new principles for managing a heterogeneous collection of data[9]. Recently in [5], a

R. Wagner, N. Revell, and G. Pernul (Eds.): DEXA 2007, LNCS 4653, pp. 192–201, 2007.

graph data model and a new XPath-like query language has been proposed for managing and accessing one's personal data scattered across various data sources such as desktop PC, email servers and so on. However, similar to several other database languages such as SQL and XQuery, the proposed query language is very complex. Moreover, it inherits one of the serious drawbacks of XPath or XPath-like query languages, that is, users are expected to have knowledge of the underlying structure of the data that they are going to query. This *inconvenience* is discouraging to a large section of naive (desktop) users who are already overwhelmed by a huge volume of data having no fixed schema or structure.

We argue that, similar to challenges in several other new applications[3], rather than perceiving them as mere database issues, the challenges in desktop search must be understood in a wider perspective, if these challenges are to be met effectively for the benefit of a wider audience. This paper takes an integrated 'database/information retrieval' (DB/IR) approach to searching a *desktop dataspace* which is a heterogeneous collection of data in a personal desktop. We identify not only general, but also several specific requirements and challenges in this approach. A particularly important issue that we highlight is how to achieve DB-like performance gains in this integrated DB/IR query platform.

Section 2 describes a set of major requirements and challenges for an integrated DB/IR approach to searching a desktop dataspace. Section 3 describes our unified data model capable of representing a collection of heterogenous data units and their contents. In Section 4, we describe our novel query processing technique to meet several issues that we are going to describe in Section 2. Related work is described in Section 5. Finally in Section 6, we draw conclusions and outline directions for future work.

## 2 Challenges and Issues

Here, we elaborate on two major challenges that need to be met by the next generation desktop search tools. Note that our objective is not to replace the functionalities that currently available tools offer. Rather, our goal is to extend these functionalities in order to achieve more effective search against a collection of heterogeneous data in a desktop.

### 2.1 Data Model Issue

Current desktop search tools rely heavily on the file system managed by the underlying Operating System. These tools fail to exploit the structural information mostly found in structured and semi-structured documents such as LaTeX, XML and XHTML files or even in Microsoft Office files (e.g. Word Documents and Powerpoint slides). As a result, users are unable to retrieve a particular portion or portions of a file. We believe that the partial retrieval of the contents of a file is already an essential requirement as the size of the files that we handle keep getting larger and larger in terms of their contents. Current Desktop Search Tools employ conventional Information Retrieval techniques and thus retrieve whole files as search results. For example in Fig. 1, any attempt made

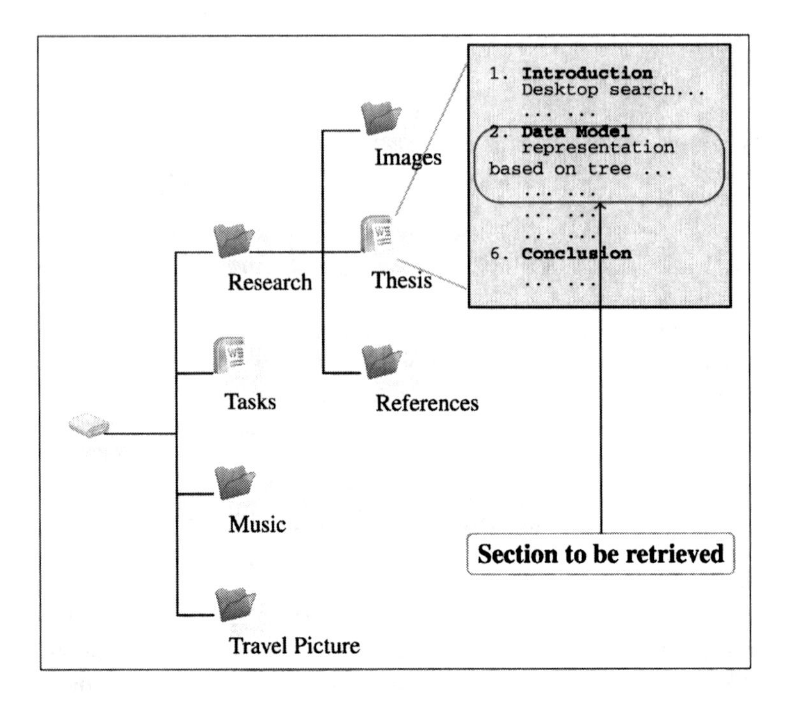


**Fig. 1.** Retrieval of a section of a file's contents

to retrieve the section of a file would result in retrieving the whole file instead. Users will then have to perform secondary search inside the file to locate exactly what he or she was looking for. Our first challenge is how to support such partial retrieval of a file in the context of keyword-based desktop search. The major issue then is to define a data model for uniform representation of a conventional file system and contents of files so that search can be carried out seamlessly across these heterogeneous collections of data.

### 2.2 Query Processing Issue

Unlike SQL or XQuery-based database queries, keyword-based search is known to be imprecise in nature. In other words, given a set of keywords, the exact definition of answers to this search has to be defined depending upon the nature of target data and the application in consideration. In the context of desktop search, this impreciseness nature of keyword search becomes even more prominent. For example, consider a query represented by the keywords *"Nepal"* and *"JAL"* against a dataspace in Fig. 2. Currently available desktop search tools will fail to provide any answers since the query keywords are split across multiple data units (two separate emails), whereas users would be happy to obtain the segment ⟨**Travel, 078, 092**⟩ (two emails containing the query keywords and the folder containing these two emails) as a single appropriate answer to this query.

Our second challenge is to how to process such search requests in order to compute appropriate answers. The main issue here would be how to do it efficiently.

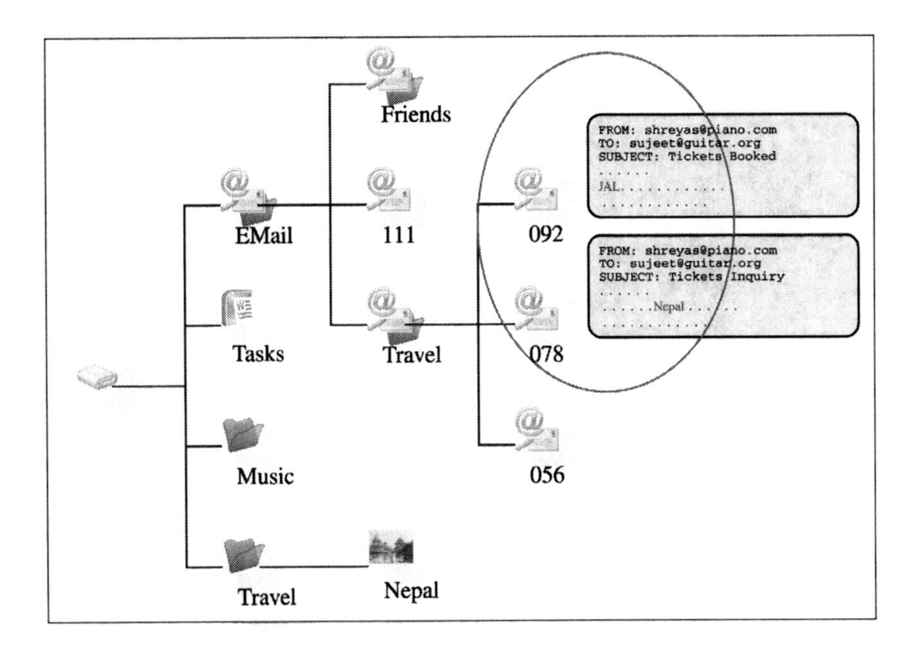


**Fig. 2.** Query keywords scattered across multiple data units

## 3 Unified Data Model

In this section, we explain our data model for unified representation of heterogeneous data in a desktop. As we stated above, the search functionality of a desktop search tool should not be limited to files and folders only. This functionality should be extended beyond and should be able to handle partial retrieval of the files' contents. Our objective is to achieve this extension using a unified data model and a unified query processing technique. That is, users would be able to perform seamless search over a large collection of heterogeneous desktop data regardless of size, type and/or structure of the underlying data. Note that in our case, an answer to a query may not necessarily be a single data unit; it may be composed of several related heterogeneous data units such as a section of a text file and an image file.

In order to achieve unified representation of several heterogeneous data units, an appropriate data structure is essential. File systems have been using hierarchical data structure right from the beginning since they naturally represent files and folders scattered across a desktop. In fact, several other structural information about various desktop dataspace can be represented by a similar hierarchical structure. For example, structured/semi-structured text files such as LaTeX, XML and XHTML have contents having natural hierarchical structure. Moreover, users generally organize Emails, HTML files in a hierarchy of appropriate folders. We can even represent the contents of a Powerpoint file by this structure. Therefore, we believe the simplest way to represent the most desktop data is to extend the hierarchical data model employed by current Operating Systems and accommodate these additional logical structural information among several data units. From here onward, we write *dataspace* to refer to all the retrievable data units managed in an integrated fashion; that is 1) files and files and folders managed by conventional files systems and 2) data units contained inside files.

It should be noted that logical modeling of a file's contents based on their structural information has also an important significance from a database point of view. Generally users organize their desktop data by keeping related files/folders under the same folder. Therefore, the current file system represents not only physical but also, to some extent, logical representation of the data. Up until now, this task needs to be done manually and could be very labor intensive. If we are to extend desktop search further to the contents of a file, naturally, manual organization of data will be even more cumbersome. The objective of our data model is also to offer database-like support for heterogeneous data management so that some kind of logical data independence can be achieved. For example, a LaTeX file may have several physical representations. Several sections of its contents can either be stored either in a single file or can be stored across multiple files (one file for each section). However, its logical representation is unique independent of its physical representation. Therefore, from the database management point of view, it is rather easier to focus on the logical representation of data. Logical data independence has played an important role in the success of Relational Database Management Systems[14]. As one can expect that the volume of data to be handled in a desktop will continue to grow, this kind of database support for desktop data management will be inevitable in the near future[5].

### 3.1 Basic Definitions

**Desktop Dataspace:** Formally, the Desktop Dataspace (or simply a *dataspace*) is defined to be a rooted ordered tree $\mathcal{D} = (\texttt{N}, \texttt{E})$ with a set of nodes N and a set of edges $\texttt{E} \subseteq \texttt{N} \times \texttt{N}$. There exists a distinguished root node from which the rest of the nodes can be reached by traversing the edges in E. Each node except the root has a unique parent node.

Each node n of the dataspace tree represents a data unit of the dataspace. This data unit is the smallest unit that is expected to be retrieved as a search result. Therefore, a node may represent not only a file or a folder, it may also represent a section of a LaTeX file or a slide of a Powerpoint file. Each node is associated with a set of attribute/value pairs for storing meta-data about the data unit. For example, a node representing a LaTeX file may have a set of attributes such as $\{< \texttt{type} = \texttt{tex} >, < \texttt{created\_date} = 2006-2-12 >, ..., < \texttt{file\_size} = 20\texttt{kb} >\}$. Similarly, a node representing an XML element may have a set of attributes such as $\{< \texttt{type} = \texttt{xmlElem} >, ..., < \texttt{element\_length} = 75 >\}$. Moreover, there is a function keywords(n) that returns the representative keywords of the corresponding data unit in n.

The tree is ordered and there is a pre-defined ordering among the sibling nodes. Although this ordering may have little significance in a conventional file system, here, it is essential in order to preserve the topology of the contents of a file (for example a LaTeX file). Therefore, the nodes are arranged in such a way that the depth-first pre-order traversal of the tree would preserve the topologies of contents of each and every file. We write $\texttt{nodes}(\mathcal{D})$ for all the nodes N.

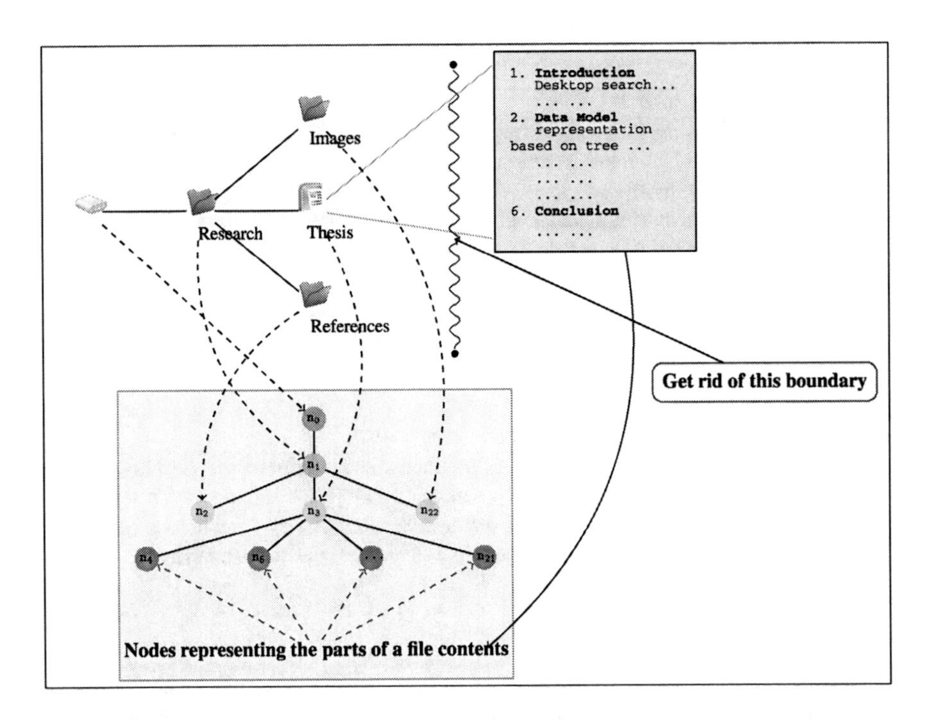


**Fig. 3.** A partial dataspace and its natural representation by our data model

**Dataspace Fragment:** Formally, a Dataspace Fragment (or simply a *fragment*) is defined to be $\mathtt{f} \subseteq \mathcal{D}$ where $\mathcal{D}$ is a Desktop Dataspace as defined above, $\mathtt{nodes(f)} \subseteq \mathtt{nodes}(\mathcal{D})$ and the subgraph induced by $\mathtt{nodes(f)}$ in $\mathcal{D}$ is also a rooted tree.

In this paper, a fragment is denoted by a subset of nodes in a dataspace tree – the tree induced by which is also a rooted ordered tree. A fragment may consist of only a single node (the smallest data unit).

Figure 3 shows how the conventional files and folders, and contents of files in a desktop dataspace can be represented by our unified data model. The set of nodes $\langle \mathtt{n1}, \mathtt{n3}, \mathtt{n4}, \mathtt{n5} \rangle$ represents a fragment of this dataspace with n1 being the root. Hereafter, unless stated otherwise, the first node of a fragment represents the root of the tree induced by it. For clarity, we refer to a single-node fragment simply as a node.

## 4 Query Processing

Note that query processing techniques used in conventional Information Retrieval are not applicable to meet our requirements. For a typical query described in Section 2.2, they either return no results or several isolated answers without considering any logical relationships between those units. In this paper, we adopt a database-like query processing technique for effectively generating the answers to those kinds of queries described in Section 2.2.

**Table 1.** Core operations of TALZBRA

| Operations | Definitions |
|---|---|
| Selection | Supposing F be a set of fragments of a given dataspace, and P be a predicate which maps a dataspace fragment into *true* or *false*, a *selection* from F by the predicate P, denoted by $\sigma_P$, is defined as a subset $F'$ of F such that $F'$ includes all and only fragments satisfying P. Formally, $\sigma_P(F) = \{f \mid f \in F, P(f) = true\}$. |
| Fragment Join | Let $f_1, f_2, f$ be fragments of the dataspace tree $\mathcal{D}$. Then, fragment join between $f_1$ and $f_2$ denoted by $f_1 \bowtie f_2$ is f iff<br>1. $f_1 \subseteq f$,<br>2. $f_2 \subseteq f$ and<br>3. $\nexists f'$ such that $f' \subseteq f \wedge f_1 \subseteq f' \wedge f_2 \subseteq f'$ |
| Pairwise Fragment Join | Let $F_1$ and $F_2$ be two sets of fragments in a dataspace tree $\mathcal{D}$, *pairwise fragment join* of $F_1$ and $F_2$, denoted by $F_1 \bowtie F_2$, is defined as a set of fragments yielded by taking *fragment join* of every combination of an element in $F_1$ and an element in $F_2$ in a pairwise manner. Formally,<br>$$F_1 \bowtie F_2 = \{f_1 \bowtie f_2 \mid f_1 \in F_1,\ f_2 \in F_2\}.$$ |
| Powerset Fragment Join | Let $F_1$ and $F_2$ be two sets of fragments in a dataspace tree $\mathcal{D}$, *powerset fragment join* between $F_1$ and $F_2$, denoted by $F_1 \bowtie^* F_2$, is defined as a set of fragments produced by applying *fragment join* operation to an arbitrary number (but not 0) of elements in $F_1$ and $F_2$. Formally,<br>$$F_1 \bowtie^* F_2 = \{\bowtie (F_1' \cup F_2') \mid F_1' \subseteq F_1, F_2' \subseteq F_2, F_1' \neq \phi, F_2' \neq \phi\}$$<br>where $\bowtie (\{f_1, f_2, \ldots, f_n\}) = f_1 \bowtie \ldots \bowtie f_n$. |

Recently we proposed an algebra for processing keyword queries against a set of schemaless XML documents[13]. We term this algebra as TALZBRA in this paper. TALZBRA consists of a set of algebraic operations and several logical optimization techniques. The core operations of TALZBRA will be the foundation for processing a typical keyword-based desktop search query. However, the optimization techniques described in [13] are not enough for our purpose and will be discussed further below. The formal definitions (slightly modified for our purpose in this paper) of these operations are given in Table 1. Readers are requested to refer to our previous work [13] for an elaboration of their algebraic properties and other details.

**Query:** A query is denoted by $Q_P\{(k_1)_{C1}, (k_2)_{C2}, ..., (k_m)_{Cm}\}$ where $k_j$ is called a query term and $C_j$ is a corresponding query keyword predicate for all $j = 1, 2, \ldots, m$ and P is a selection predicate.

We write $k \in \texttt{keywords}(n)$ to denote query term k appears in the keywords associated with the node n.

**Query Answer:** Given a query $\mathtt{Q}_{\mathsf{P}}\{(\mathtt{k}_1)_{\mathsf{C1}}, (\mathtt{k}_2)_{\mathsf{C2}}, ..., (\mathtt{k}_m)_{\mathsf{Cm}}\}$, answer $\mathtt{A}$ to this query is a set of document fragments defined to be $\{\mathtt{f} \mid (\forall(\mathtt{k})_{\mathsf{C}}) \in \mathtt{Q})\exists \mathtt{n} \in \mathtt{f} : \mathtt{n}$ is a leaf node of $\mathtt{f} \wedge \mathtt{k} \in \mathtt{keywords(n)} \wedge \mathsf{C}(\mathtt{n}) = \text{true} \wedge \mathsf{P}(\mathtt{f}) = \text{true}\}$.

Therefore, $\mathtt{Q}_{\mathsf{size}<4}\{(\mathtt{Nepal})_{\mathsf{type=email}}, (\mathtt{JAL})_{\mathsf{type=email}}\}$ represents a query asking for fragments meeting the following conditions. The fragment should consist of at least one node associated with each query term `Nepal` and `JAL`, and the attribute types of these nodes should be `email`. Moreover, the size (that is number of nodes consisting of an answer fragment) should be smaller than 4.

Intuitively, an answer to a query is a dataspace fragment consisting of several structurally-related logical data units. Each keyword in the query must appear in at least one data unit that constitutes the fragment. These data units must also satisfy the query keyword predicates. In addition, the fragment must satisfy the selection predicate(s) specified in the query. Query keyword predicates such as the one $\mathsf{type} = \mathsf{email}$ are relatively easy to process. Therefore, for the sake of clarity, we exclude them from further discussion in this paper.

A query represented by $\{\mathtt{k}_1, \mathtt{k}_2\}$ and a selection predicate $\mathsf{P}$ against a dataspace $\mathcal{D}$ can be evaluated by the following formula.

$$\mathtt{Q}_{\mathsf{P}}\{\mathtt{k}_1, \mathtt{k}_2\} = \sigma_{\mathsf{P}}(\mathtt{F}_1 \bowtie^* \mathtt{F}_2)$$

where $\mathtt{F}_1 = \sigma_{\mathsf{keyword}=k_1}(\mathtt{F})$, $\mathtt{F}_2 = \sigma_{,\mathsf{keyword}=k_2}(\mathtt{F})$ and $\mathtt{F} = \mathtt{nodes}(\mathcal{D})$.

The basic idea here is, by applying *powerset fragment join* operation, we generate every possible dataspace fragment that can be considered an appropriate candidate answer to the query. However, this is the most inefficient way of processing a query. The question then is how to do it efficiently. In the next section, we describe DB-like technique for performance gain.

### 4.1 Query Optimization

One of the main principles for algebraic manipulation in conventional database systems is to perform *selection* as early as possible[14]. Our goal here is to apply the same principle to our query mechanism so that we can eliminate as many unnecessary fragments as possible at an early stage of query processing without affecting the end result. However, in order to achieve this goal, we must ensure that *selection* can indeed be pushed down in the query evaluation tree; that is even if we perform *selection* ahead of *join*, we are still guaranteed to obtain the same desired result.

We describe a class of filters, having certain property that enables selection operation to be pushed down in the query tree. We call the filters that fall in this class as *optimal filters*.

### 4.2 Optimal Filter

Before we describe an *optimal filter*, we first describe a filter having anti-monotonic property that we proposed in [13]. Given a fragment $\mathtt{f}$, a filter $\mathsf{P}$ is anti-monotonic iff

$$\forall \mathtt{f}' \subseteq \mathtt{f} : \mathsf{P}(\mathtt{f}) = \mathit{true} \Rightarrow \mathsf{P}(\mathtt{f}') = \mathit{true}$$

Thus, if a fragment $\mathtt{f}$ satisfies a filter predicate P, then all sub-fragments of $\mathtt{f}$ also satisfies P. Therefore, a filter predicate such as $\mathsf{size}(\mathtt{f}) \leq 4$ is an anti-monotonic filter. Both conjunction and disjunction of anti-monotonic filters have also anti-monotonic property. That is, if $\mathsf{P}_1$ and $\mathsf{P}_2$ are two distinct anti-monotonic filters, then $\mathsf{P}_1 \wedge \mathsf{P}_2$ and $\mathsf{P}_1 \vee \mathsf{P}_2$ are also anti-monotonic filters. Construction of more complex anti-monotonic filters is possible and therefore we can expect a significant performance gain by developing practically useful filters having anti-monotonic property. However, such filters are limited in desktop search applications. That is why, we extend this notion of anti-monotonic filters to accommodate a larger number of practically useful filters that would enable query optimization.

We describe a class of filters called *optimal filters* having the following property. Given two distinct fragments $\mathtt{f}_1$ $\mathtt{f}_2$, a filter P is optimal iff

$$\mathsf{P}(\mathtt{f}_1 \bowtie \mathtt{f}_2) = \text{true} \Rightarrow \mathsf{P}(\mathtt{f})_1 = \text{true} \wedge \mathsf{P}(\mathtt{f})_2 = \text{true}$$

Note that the anti-monotonic filters described in [13] are essentially *optimal filters* too. However, not all *optimal filters* have an anti-monotonic property.

## 5 Related Work

There has been growing concern over the necessity of new principles for managing a heterogeneous collection of data in the database community[9]. In this context, [5] addresses the issues in managing and searching a personal dataspace consisting of heterogeneous collections of data form various data sources. Table 2 shows the comparison of our approach with the one described in [5]. The approach proposed in [5] is purely database-based ignoring a large community of naive users. Our query expressiveness may be less powerful than the one described in [5], but nonetheless, more flexible and will be more useful to a wider audience. The importance of DB/IR integrated approach for next generation applications have been emphasized in [3]. It also fully justifies the approach considered in this paper.

Much of the work regarding DB/IR approach to searching are relevant to our work. Systems to support keyword queries in relational databases are described in [1][10][12]. Similarly, keyword-based queries against XML data have been studied in [8][11][6][4][13].

## 6 Conclusions

Desktop search tools are essential for retrieving the desired data units from a huge collection of heterogeneous data in one's desktop. In this paper, we focused on two major challenges, which currently available desktop search tools have failed to address, in a typical keyword-based search system. The first challenge is the partial retrieval of a file contents and second challenge is to computation of appropriate answers in case query keywords are splits across multiple data units. In order to meet these challenges, we proposed a unified data model supported by a novel search technique. We believe that our integrated DB-IR approach which preserves the simplicity of conventional IR technique while taking advantages of database-style query processing will certainly benefit a large community of naive desktop users. Experimental evaluation of the proposed technique is our immediate future work.

**Table 2.** Comparison of our approach with iMeMex approach

| | iMeMex | Our Approach |
|---|---|---|
| Type of data that can be handled | Heterogeneous | Heterogeneous |
| Approach | Pure Database | Integrated DB-IR |
| Data Model | Unified Graph Model | Unified Rooted Ordered Tree |
| Query Processing | XPath-based Query Language | Database-like Query Processing based on TALZBRA |
| Merits | Precise Query Semantics | Easy Query Interface/Flexible |
| Demerits | Complex Query Syntax (not suitable for general users) | Imprecise Query Semantics |

## References


1. Agrawal, S., Chaudhuri, S., Das, G.: DBXplorer: A system for keyword-based search over relational databases. In: ICDE, pp. 5–16 (2002)
2. Apple Mac OS X Spotlight. http://www.apple.com/macosx/features/spotlight/
3. Chaudhuri, S., Ramakrishnan, R., Weikum, G.: Integrating DB and IR technologies: What is the sound of one hand clapping? In: CIDR, pp. 1–12 (2005)
4. Cohen, S., Mamou, J., Kanza, Y., Sagiv, Y.: XSEarch: A semantic search engine for XML. In: Proc. of 29th VLDB, pp. 45–56 (2003)
5. Dittrich, J.-P., Vaz Salles, M.A.: iDM: a unified and versatile data model for personal dataspace management. In: VLDB, pp. 367–378 (2006)
6. Florescu, D., Kossman, D., Manolescu, I.: Integrating keyword search into XML query processing. In: International World Wide Web Conference, pp. 119–135 (2000)
7. Google Desktop, http://desktop.google.com/
8. Guo, L., Shao, F., Botev, C., Shanmugasundaram, J.: XRank: ranked keyword search over XML documents. In: SIGMOD, pp. 16–27. ACM, New York (2003)
9. Halevy, A.Y., Franklin, M.J., Maier, D.: Principles of dataspace systems. In: PODS, pp. 1–9 (2006)
10. Hristidis, V., Papakonstantinou, Y.: DISCOVER: Keyword search in relational databases. In: VLDB, pp. 670–681 (2002)
11. Li, Y., Yu, C., Jagadish, H.V.: Schema-free XQuery. In: Proc. of 30th VLDB, pp. 72–83 (2004)
12. Liu, F., Yu, C.T., Meng, W., Chowdhury, A.: Effective keyword search in relational databases. In: SIGMOD Conference, pp. 563–574 (2006)
13. Pradhan, S.: An algebraic query model for effective and efficient retrieval of XML fragments. In: VLDB, pp. 295–306 (2006)
14. Ullman, J.D.: Principles of Database and Knowledge-Base Systems, vol. II. Computer Science Press (1989)
15. Windows Desktop Search. http://www.microsoft.com/windows/desktopsearch

# An Original Usage-Based Metrics for Building a Unified View of Corporate Documents

Guillaume Cabanac[1], Max Chevalier[1,2], Claude Chrisment[1], and Christine Julien[1]

[1] IRIT-SIG, UMR 5505, Toulouse 3 University, France
[2] LGC, ÉA 2043, Toulouse 3 University, France
{cabanac,chevalier,chrisment,julien}@irit.fr

**Abstract.** Nowadays, organizational members manage the huge amount of digital documents that they exploit at work. To do that, they organize documents into individual hierarchies. Actually, these documents are really parts of a company's capital as they reflect past experiences, present competences and impending expertise. Unfortunately, even if corporate documents represent high value-added material, they still mostly remain unknown from the organization as a whole. That is the reason why this paper proposes to build a unified view of corporate documents. Our approach is complementary to current content-based ones because it relies on an original metrics related to documents usage within an organization.

## 1 Introduction

In modern organizations such as companies, institutions, R&D laboratories, etc. people constantly need to search for documents in order to accomplish their tasks. This individual activity generates amounts of documents that may interest many other members that have similar needs. However, these nuggets of information are too often unknown to others because they mostly remain into each worker's personal document space. Nevertheless it would be worth gathering and exposing documents in a collective way in order to support communities of practice that tend to improve members' efficiency.

To do that, research works have proposed to visualize organization-wide documents by measuring inter-document similarity. These approaches commonly use lexical or even semantic metrics that only consider document contents. In fact, this really neglects how people use and organize documents. However, we think that we can benefit from it: observing how people use and classify documents may provide us with a clue to how much they are related regarding people's practices. Therefore, the proposed unified view is based on the way individuals organize corporate documents into their own hierarchies. This view relies on a usage-based similarity computed through a unique data structure called *multitree*. The aforementioned hierarchies consist of people's bookmarks, file systems, etc. The unified view enables people to discover new relations between documents that would have been undiscovered with approaches only based on document contents. Lastly it provides query and navigation features that enable users to find interesting corporate documents that may be of interest for their tasks.

R. Wagner, N. Revell, and G. Pernul (Eds.): DEXA 2007, LNCS 4653, pp. 202–212, 2007.

This paper is organized as follows: section 2 describes our research context, that is building a unified document view from individually organized ones. With respect to this context, we also describe and point weaknesses of lexical and semantic metrics that are commonly used for computing inter-document similarity. Section 3 formally describes the *usage-based inter-document metrics* employed for building the unified view. Lastly, we discuss strengths and weaknesses of our approach before giving some insights into future research works in section 4.

## 2 Context and Motivations

This section introduces our research context: organizational work mediated by digital documents. We focus on how they are retrieved, organized into hierarchies and actually poorly shared. Therefore, we describe ways for building a unified document view by extracting documents from every organizational members' Personal Information Spaces (PIS) [1]. A such unified representation would enable each member to view and access related documents existing within the organization. However, we underline strong weaknesses of current approaches that only consider document contents. We argue that they do not capitalize enough on individual efforts: document organization resulting from their use in daily people's tasks. This motivates our outlined approach: exploiting a usage-based rather than content-based metrics to build the unified view.

### 2.1 Organizational Context

At work, people make extensive use of information sources—including the Web—from which they retrieve documents that are relevant to accomplish their tasks. Indeed, information along with these documents is a raw material for modern organizations that earn money by analyzing, combining, enriching, etc. information and its digital media—documents—for making a profit. Following this exploitation of documents, people need to structure their PIS into individual document hierarchies [1] for various purposes. In fact, this is mostly done for finding documents later, for building a legacy and for sharing them [2]. Despite these expected objectives, we underline in the following section that documents are not fully exploited at the corporate level as a whole. That is the reason why we consider them as a quiescent capital.

### 2.2 Corporate Documents: A Quiescent Capital

People's individual activities generate amounts of documents that may be of great interest to other organizational members. However, such documents are very often kept in personal information spaces without really being shared. One argument from social psychology may be that information is perceived as power, so sharing documents would imply loosing power. People can also hesitate to share because they do not know if they will get something interesting in return. These arguments are difficult to take into account so we rather explore other trails. Other arguments explaining why information is not very shared have been pointed out by a recent field study of organizational work mediated by digital documents [3]. First, information is scattered in multiple locations all over most organizations. In fact, available information is unknown from people because there is no single access point to it. Second, organizational members are not professional searchers. They often have learned to search for information on-the-job, by themselves. Thus, without an adapted training, most people do not know where to look, how to search efficiently, etc. Third, people are inundated with too much information, mostly coming from email messages. They spend a lot of time extracting interesting information from huge amounts of noisy documents. In our opinion, a predominant reason is that sharing takes time and implies a significant cognitive overload. We investigate this latter argument, considering three main ways that workers use to *manually* spread documents. *i*) By emailing colleagues, which implies thinking about which contacts may benefit from the aforementioned documents. Choosing recipients is an highly cognitive task whereas it does not allow to spread documents outside one's circle of contacts. This approach is merely chosen for small group collaboration [4]. *ii*) By sending documents to a mailing list, whose members have subscribed for. They are often topic specific, concerning some profession for example [5]. As regards the sender, evaluating documents relevancy with respect to a mailing list can be tricky as he does not know every of its members' needs. Therefore, some people may self-censor whereas others may send unrelated material, overwhelming subscribers' mailboxes. *iii*) By publishing documents on organizational intranets. For this approach, the effort is transposed from the sender to each organizational member who has to actively search for these documents. As modern intranets can consist of millions of Web pages—5.5 millions for the IBM intranet—employees spend a large percentage of their time searching for information [6]. *Automatic* document delivery via recommender systems [7] can also be cited along with its drawbacks. Such systems exploit users' profiles and combine demographic, cognitive, collaborative, etc. strategies to diffuse documents. Their drawbacks mainly concern profile updating latency (when a user's interests evolve). Moreover, the way documents are organized regarding workers' tasks is not exploited.

Manual as well as automatic document sharing strategies are both expensive in time and cognitive load. That is why more human-centered strategies have been proposed in the Knowledge Management field, e.g. communities of practice aim to connect people so that they can share nuggets of information, exchange experience and solve problems in their area of experience [5]. All things considered, efficiently giving access to relevant documents is a difficult task to achieve, either manually or automatically. Indeed, people often loose precious time searching for or recreating already organization-wide scale encountered information [3]. As a solution we propose to gather documents in a unified view based on an original input: the way workers use documents. The interest of such a view is twofold: it improves organizational documents access through a user-friendly interaction meanwhile it presents a global vision of them. This view reflects the way documents are related to activities rather than to their contents.

### 2.3 How to Unify Corporate Documents?

Previous sections suggest that modern organizations can benefit from a unified view of corporate documents. We can at least explore two directions. *i*) Providing a shared space where people collaboratively manage a unique hierarchy of documents. *ii*) Exploiting people's individual hierarchies in order to automatically gather them in a unified view. We do not support the first alternative because [8] has noticed that people want to keep individual documents under control. Moreover, they prefer to organize hierarchically rather than to rely on a system that would not provide a global view of documents, e.g. a search engine. In addition, when sharing a common hierarchy nobody is free to organize documents the way he wants. This forces people to adhere to a "single thought." Lastly, each member would have to keep documents that he is not interested in. On the contrary, the second alternative enables people to only keep documents they really are interested in. Moreover, they can organize them the way they feel it the more efficient. That is why we discuss how to unify multiple individual document hierarchies into a unified view. In addition, visualizing corporate documents can be achieved following two ways. On the one hand, we consider Hierarchical Agglomerative Clustering (HAC) and Self-Organizing Maps (SOM) Data Mining approaches, which are content-based. On the other hand, we mention the social tagging visualization called "tag cloud," which can be considered as an early attempt to build a usage-based unified view.

**Data Mining Approaches.** In order to build a unified document view, one must be able to evaluate how documents are related to each other, i.e. how much they are similar. This can be achieved by a traditional Information Retrieval process [9] via the cosine measure associated with the Vector Space Model, for instance. Many Data Mining algorithms exploit such techniques to provide visual representations of a set of documents. For example, [10] describe algorithms for automatically organizing documents per contents based on the HAC algorithm. Another example may be the SOM algorithm [11].

**Social Tagging Approach.** A current trend called *social bookmarking* consists in sharing document pointers associated with user-contributed *tags* [12]. In order to obtain a unified view of documents thanks to these tags, a specific visualization called a *tag cloud* [13] has been proposed.

**Weaknesses of Current Approaches.** Content-based comparison of documents suffers many weaknesses that are related to synonymy, homonymy, stylistic devices such as metaphors ... This is a reason why advanced approaches such as the Latent Semantic Indexing algorithm [14] have been used. Even if partially solving lexical metrics weaknesses, such approaches cannot identify documents that are often used together, according to a specific activity. The tag cloud visualization also suffers from content-based issues as individuals can use different tags that actually represent a single concept. All things considered, current approaches *never* consider an actual human contribution resulting from a highly cognitive effort: documents organization. In order to take into account this aspect in a unified document view, we define in section 3 a specific *inter-document usage-based metrics* based on workers' individual hierarchies.

## 3 An Original Usage-Based Unified Document View

In this section, we describe an approach for building a usage-based unified view of corporate documents. We thus detail how to benefit from individuals' cognitive efforts reflected by their organized documents. Then, we formalize an *inter-document usage-based metrics* that is computed from individual hierarchies of documents. Lastly, we present how the unified view is implemented, along with concrete applications regarding information visualization and retrieval tasks.

As people find and exploit interesting documents regarding their tasks, they store them for various purposes. To do that, their favorite organization is a folder hierarchy [1] because it really reflects how documents are related regarding people's tasks. When inserting a document, the act of deciding which folder is best representative or even creating a new one from scratch is a highly cognitive task [15]. In spite of involved cognitive efforts, hierarchical organization is appreciated because it allows individuals not only to keep documents under control [8] but also to represent them as a whole. Moreover, when lexical-based metrics is static because based on contents only, usage-based metrics is dynamic as it relies on evolving hierarchies. To sum up, hierarchical document organization conveys a high value that is mostly unexploited by the current approaches presented above. That is why we describe in the following sections how to build a usage-based unified view of documents from multiple hierarchies.

### 3.1 Modeling Documents Usage: the Multitree Data Structure

In order to identify patterns of document usage, we need to represent multiple users' document hierarchies (excluding folders that users specify as "miscellaneous") into a unique data structure. Following previous research works [16], we model these hierarchies using a *multitree* that groups together users' documents along with their paths, see figure 1.

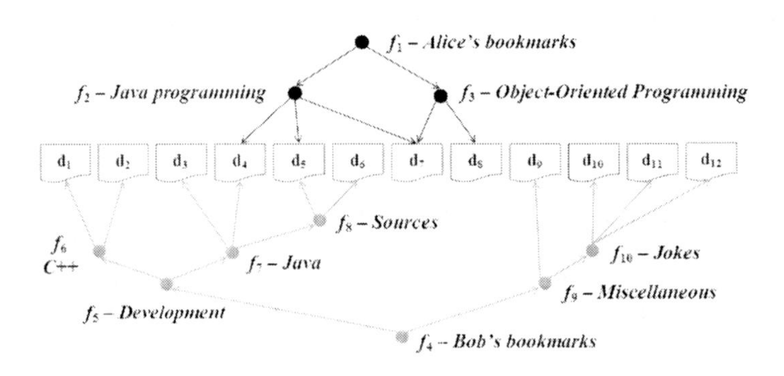


**Fig. 1.** A multitree consisting of two users' hierarchies

**Definition 1.** A multitree $\mathcal{M}= \langle D, F, U, R_D, R_F, R_U \rangle$ is a sextuplet where $D = \{d_1, \ldots, d_n\}$ is a set of documents, $F = \{f_1, \ldots, f_m\}$ is a set of folders and $U = \{u_1, \ldots, u_l\}$ is a set of users. Moreover, we define the following relations:

- $R_D$ is a binary relation on $D \times F$ that is called *document membership.* $(d_i, f_j) \in R_D$ means that the $d_i$ document is a direct child of the $f_j$ folder.
- $R_F$ is a binary relation defined on $F \times F$ that is called *folder membership.* $(f_i, f_j) \in R_F$ means that the $f_i$ folder is a direct child of the $f_j$ folder.
- $R_U$ is a binary relation defined on $U \times F$ that is called *root membership.* $(u_i, f_j) \in R_U$ means that the $u_i$ user owns the $f_j$ root of his hierarchy.

Furthermore, we define $R_F^+ : F \to F$ as the function (1) that returns the $p$ direct "parent" folder that contains a given $f$ folder. If $f$ is one of the roots of the multitree then $R_F^+(f) = \lambda$ where $\lambda$ figures the null value.

$$R_F^+(f) = p \mid \exists (f, p) \in R_F \tag{1}$$

**Definition 2.** Let $\mathcal{G}$ be the graph associated with the multitree $\mathcal{M}$. A vertex of $\mathcal{G}$ is either a node (representing a folder) or a leaf (representing a document) while $R_D \cup R_F$ are edges of $\mathcal{G}$. A *path* from a root $r$ to $d$ is a sequence denoted $/r/f_1/f_2/\ldots/f_k/d$ such that $f_1 R_F r, f_2 R_F f_1, \ldots, d R_D f_k$. The direct descendant folder $f_1 \in F$ of the $r$ root is called a *branch*, it is formally defined by the $b : F \to F$ function (2).

$$b(f) = \begin{cases} \lambda & \text{if } R_F^+(f) = \lambda \\ f & \text{if } b(R_F^+(f)) = \lambda \\ b(R_F^+(f)) & \text{else} \end{cases} \tag{2}$$

Thanks to the multitree data structure, we compute an inter-document usage-based similarity. We detail how this is achieved in the following section.

### 3.2 Computing an Inter-Document Usage-Based Similarity

In this section, we detail an inter-document usage-based similarity which considers users' patterns of organization reflected by their document hierarchies (definition 4). This depends on inter-folder similarity (definition 3).

**Definition 3.** Following research works on URL similarity [17], we provide the $\sigma_F : F^3 \to [0, 1]$ function (3) that evaluates the usage-based similarity of two folders. Concretely, the depth and number of common ancestors of two given folders are the two main criteria observed for evaluating their similarity.

$$\sigma_F(b, f_1, f_2) = 1 - \frac{s(f_1, m(f_1, f_2)) + s(f_2, m(f_1, f_2))}{s(f_1, b) + s(f_2, b) + 2} \tag{3}$$

The $s : F^2 \to \mathbb{N}_+$ function[1] (4) returns the number of "steps," i.e. edges in the path from $f_1$ to $f_2$, that are assumed to be in the same $b$ branch. To do that, we

[1] The $\ominus$ operator is the *symmetric set difference*, corresponding to the exclusive OR (XOR) in Boolean logic: $A \ominus B = (A \cup B) \setminus (A \cap B) = (A \setminus B) \cup (B \setminus A)$.

define the $a : F \rightarrow F$ function (5) that computes the set of ancestors of a given (included) $f$ folder. Moreover, the $m : F^2 \rightarrow F$ function (6) returns the "least common ancestor" of two folders $f_1$ and $f_2$, i.e. the folder that is an ancestor of both $f_1$ and $f_2$ and that has the greatest depth. This uses the $d : F \rightarrow \mathbb{N}_+$ function that gives the depth of a folder.

$$s(f_1, f_2) = |a(f_1) \ominus a(f_2)| \tag{4}$$

$$a(f) = \begin{cases} \varnothing & \text{if } f = \lambda \\ \{f\} \cup a(R_F^+(f)) & \text{else} \end{cases} \tag{5}$$

$$m(f_1, f_2) = f \mid \forall (f, f') \in (a(f_1) \cap a(f_2))^2 \quad (f \neq f') \wedge (d(f) > d(f')) \tag{6}$$

Our final aim is to evaluate inter-document usage-based similarity. Remembering that documents of the multitree come from at least one user's hierarchy, we identify common patterns in document organization. For example, if people always classify a group of documents in a same folder or in a similar way, this means that people find them similar, for any reason related to their usage [8]. Thus, we have to observe repeated patterns of document organization: the more people organize the same collection of documents in the same manner, the more these documents are considered as usage-based similar.

**Definition 4.** The $\sigma_D : D^2 \rightarrow [0, e]$ symmetric function (7) computes a usage-based similarity of two documents, provided that they are reachable from at least one branch in the multitree. Since their insertion in a same branch is the consequence of human cognitive effort, the system deduces that they have something in common according to their owner's need. In (7) the $R_D^+ : D \times F \rightarrow F$ function (8) returns the $f$ folder that contains a given $d$ document, provided that $f$ is in the $b$ branch.

$$\sigma_D(d_1, d_2) = \frac{e^{\frac{u}{|U|}}}{|B|} \sum_{b \in B} \sigma_F(b, R_D^+(d_1, b), R_D^+(d_2, b)) \tag{7}$$

$$R_D^+(d, b) = f \mid (\exists (d, f) \in R_D) \wedge (b \in a(f)) \tag{8}$$

Moreover, $B = b(d_1) \cap b(d_2)$ is a set of branches that both $d_1$ and $d_2$ have in common. Finally, $u$ is the number of users that have a branch both containing $d_1$ and $d_2$. The $e^{\frac{u}{|U|}}$ expression models the fact that the more people classify two given documents in the same branch, the more these documents are usage-based similar. The second part of (7) takes into account the average distance of folders containing the two documents.

### 3.3 Applying the Usage-Based Metrics: A Documents Unified View

As a first application we construct a unified view, i.e. a map of corporate documents thanks to the usage-based metrics proposed in the previous section. This is an original and complementary approach as opposed to classical content-based ones. The proposed unified view is a 2D similarity graph (figure 2 (a)) where vertices are documents that are interconnected by edges of variable length reflecting inter-document usage-based similarity (definition 4). To do that, we first compute a $n \times n$ document-document matrix whose values are similarities. Second, we fit the obtained matrix in a spring-embedder model [18] in order to position vertices with respect to the computed usage-based similarities.

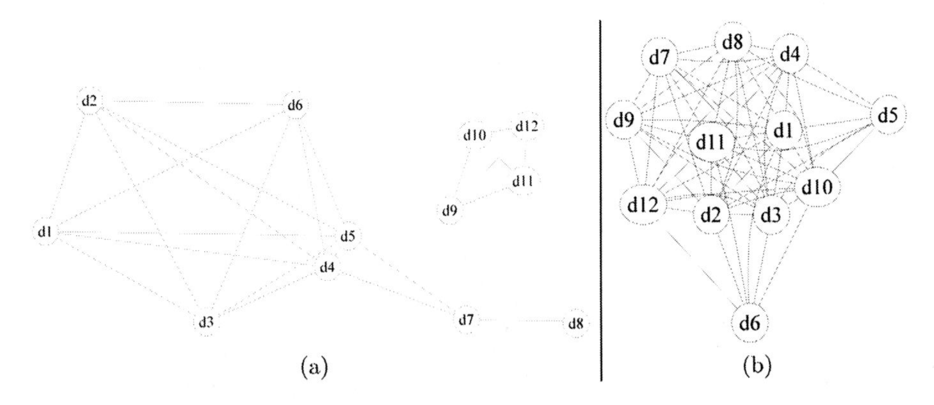


**Fig. 2.** Usage-based (a) *vs* content-based (b) similarity graphs from documents represented in the figure 1

Figure 2 compares the graph obtained by a *usage*-based (a) *vs* a classical *content*-based (b) similarity. It represents the documents from the two hierarchies of figure 1. On the one hand, one may distinguish on the *usage*-based view two separate document clusters, each one gathering usage-related documents. These clusters do not appear on the *content*-based unified view since those documents do not share any common term. This is likely to happen with large organizational documents. On the other hand, one may notice that the most *usage*-related documents $d_4$ and $d_5$ are not so close regarding *content*-based metrics. Moreover, $d_1$ and $d_{10}$ seem to be similar regarding their contents whereas they are not used in the same way: they belong to distinct clusters in the usage-based view. To sum up, the proposed usage-based unified view enables people to discover yet unidentified document relations, i.e. by using content-based metrics only. In order to help a user to understand why documents are usage-based similar in the unified view we label each vertex thanks to their most representative folder paths extracted from the branches of the multitree. Moreover, each edge connecting two given documents is labeled by common paths that are found in connected vertices. Associating this feature with manually adjustable zoom and clustering levels, one can dynamically drill-down the organizational documents unified view.

**Implementation.** In order to experiment our approach, algorithms described by definitions 3 and 4 have been implemented in the Java language. As for computing the usage-based similarity graph from the aforementioned computed matrix, we used the Graphviz[2] implementation. Concerning experimental data, we have gathered document hierarchies created by individuals through their common activities. In fact, we have asked 14 colleagues for their hierarchies of bookmarks. The corpus that we have constituted contains 4,079 documents (resp. 486 folders) with an average of 291 documents (resp. 34 folders) by user. A resulting graph can be seen on `http://www.irit.fr/~Guillaume.Cabanac/UBgraph/`. This process takes an average of 5 seconds, which makes it possible to provide users with up-to-date visualization.

**Discussion.** Summing up works presented in this paper, usage-based metrics and its implementation have the following strengths. *i*) They enable systems to compute inter-document similarity without having to be aware of their contents at all. This is a strong point since Web contents are very volatile: numerous bookmarks quickly become broken links as Web sites evolve. Moreover, usage-based similarities are more dynamic than content-based ones because they are computed thanks to evolving document hierarchies rather than fixed document contents. *ii*) Usage-based metrics exploit document organization, resulting from human-contributed cognitive efforts. To our knowledge, metrics related to organization of documents have not been defined nor studied this way before. *iii*) Usage-based metrics allow to represent documents on a cartography, i.e. a 2D map. Such a map can be used for both modalities of information retrieval: querying and navigating. *iv*) It enables users to find multilingual documents. Indeed, if documents written in different languages are stored in a same hierarchy branch, usage-based metrics identify them as similar. So, users of the retrieval feature may benefit from multilingual results returned for a monolingual query. For example, a query in English may return English documents along with French ones if they are similar enough in the sense of definition 4.

## 4 Conclusion and Future Works

Modern technologies of the Information Society we live in make it possible to access increasingly growing amounts of digital documents. People are used to organize and to classify encountered interesting documents into personal information spaces according to their needs and daily tasks.

In this paper, we argue that corporate organization-wide documents can be considered as a quiescent capital. Indeed, we show that people spend time and efforts searching for relevant information, that mostly remain unknown to other members with similar needs. Moreover, most information searched for is very often already present within the organization. That is why we propose to build a unified view in order to facilitate the access to corporate documents. This view may promote work-based communities, i.e. communities of practice that aim to increase organizational members' efficiency. First, we underline that current approaches (content-based) are not sufficient enough for reflecting the real use

[2] Graphviz is an open source graph visualization software, cf. `http://graphviz.org`.

of documents. Second, we introduce and formally describe an *inter-document usage-based metrics* which is used to build the unified view. Lastly, we discuss its implementation. In the proposed approach, representing a large amount of documents in a unified view would be impossible without clustering them first. As a perspective, we plan to explore alternative visual representations that would better handle large corpora. A second perspective aims to merge usage-based and content-based views. Indeed, we consider that coupling these two approaches may provide users with additional clues, helping them to understand the immaterial capital of their organization. Our current prototype not only provides a "proof of concept" but also an experimental framework. Thus, we plan to experiment and improve the proposed unified view along with the proposed usage-based metrics. A first step already involves the participation of corporate members of our laboratory. This will allow us to evaluate the acceptability of this unified view. Moreover, by the observation of knowledge workers' daily tasks, we will be able to quantify productivity gains. Finally, we foresee that experience reports will help us to improve the accuracy of our approach as a whole.

## References


1. Abrams, D., Baecker, R., Chignell, M.: Information Archiving with Bookmarks: Personal Web Space Construction and Organization. In: CHI '98: Proceedings of the SIGCHI conference on Human factors in computing systems, pp. 41–48. ACM Press/A-W Publishing Co., New York (1998)
2. Kaye, J.J., Vertesi, J., Avery, S., Dafoe, A., David, S., Onaga, L., Rosero, I., Pinch, T.: To Have and to Hold: Exploring the Personal Archive. In: CHI '06: Proceedings of the SIGCHI conference on Human Factors in computing systems, pp. 275–284. ACM Press, New York (2006)
3. Feldman, S.: The high cost of not finding information. KM World magazine 13(3) (2004)
4. Noël, S., Robert, J.M.: Empirical Study on Collaborative Writing: What Do Co-authors Do, Use, and Like? Comput. Supported Coop. Work 13(1), 63–89 (2004)
5. Millen, D.R., Fontaine, M.A.: Improving Individual and Organizational Performance through Communities of Practice. In: GROUP '03: Proceedings of the 2003 international ACM SIGGROUP conference on Supporting group work, pp. 205–211. ACM Press, New York (2003)
6. Dmitriev, P.A., Eiron, N., Fontoura, M., Shekita, E.: Using Annotations in Enterprise Search. In: WWW '06: Proceedings of the 15$^{th}$ international conference on World Wide Web, pp. 811–817. ACM Press, New York (2006)
7. Montaner, M., López, B., de la Rosa, J.L.: A Taxonomy of Recommender Agents on the Internet. Artif. Intell. Rev. 19(4), 285–330 (2003)
8. Jones, W., Phuwanartnurak, A.J., Gill, R., Bruce, H.: Don't Take My Folders Away!: Organizing Personal Information to Get Things Done. In: CHI '05: CHI '05 extended abstracts on Human factors in computing systems, pp. 1505–1508. ACM Press, New York (2005)
9. Salton, G., Wong, A., Yang, C.S.: A Vector Space Model for Automatic Indexing. Commun. ACM 18(11), 613–620 (1975)
10. Godoy, D., Amandi, A.: Modeling user interests by conceptual clustering. Inf. Syst. 31(4), 247–265 (2006)

11. Kohonen, T.: Self-Organizing Maps, 3rd edn. Springer, Secaucus, NJ (2001)
12. Hammond, T., Hannay, T., Lund, B., Scott, J.: Social Bookmarking Tools (I): A General Review. D-Lib Magazine 11(4) (2005)
13. Marlow, C., Naaman, M., Boyd, D., Davis, M.: HT06, Tagging Paper, Taxonomy, Flickr, Academic Article, To Read. In: HYPERTEXT '06: Proceedings of the 17$^{th}$ conference on Hypertext and hypermedia, pp. 31–40. ACM Press, New York (2006)
14. Deerwester, S.C., Dumais, S.T., Landauer, T.K., Furnas, G.W., Harshman, R.A.: Indexing by Latent Semantic Analysis. JASIS 41(6), 391–407 (1990)
15. Rucker, J., Polanco, M.J.: Siteseer: personalized navigation for the web. Commun. ACM 40(3), 73–76 (1997)
16. Chevalier, M., Chrisment, C., Julien, C.: Helping People Searching the Web: Towards an Adaptive and a Social System. In: ICWI 2004: Proceedings of the 3$^{rd}$ International Conference WWW/Internet, IADIS, pp. 405–412 (2004)
17. Jaczynski, M., Trousse, B.: WWW Assisted Browsing by Reusing Past Navigations of a Group of Users. In: Smyth, B., Cunningham, P. (eds.) EWCBR 1998. LNCS (LNAI), vol. 1488, pp. 160–171. Springer, Heidelberg (1998)
18. Eades, P.: A Heuristic for Graph Drawing. Congressus Numerantium 42, 149–160 (1984)

# Exploring Knowledge Management with a Social Semantic Desktop Architecture

Niki Papailiou[1], Dimitris Apostolou[2], Dimitris Panagiotou[1], and Gregoris Mentzas[1]

[1]Information Management Unit, Institute of Communication and Computer Systems, National Technical University of Athens, 9 Iroon Politechniou street, 15780 Athens, Greece
{nikipa, dpana, gmentzas}@mail.ntua.gr
[2]Department of Informatics, University of Piraeus, Karaoli & Dimitriou 80, 18534 Piraeus, Greece
dapost@unipi.gr

**Abstract.** The motivation of this paper is to research the individual and the team levels of knowledge management, in order to unveil prominent knowledge needs, interactions and processes, and to develop a software architecture which tackles these issues. We derive user requirements, using ethnographic methods, based on user studies obtained at TMI, an international management consultancy. We build the IKOS software architecture which follows the p2p model and relies on the use of Social Semantic Desktop for seamless management of personal information and information shared within groups. Finally, we examine the way our approach matches the requirements that we derived.

**Keywords:** Knowledge Management; Semantic Web; Semantic Desktop.

## 1 Introduction

The discipline of knowledge management addresses four levels of knowledge management: individual, team, organizational and inter-organizational [14]. However, the main focus of research as well as of commercial projects up to now has been the organizational level, which has been analysed from several points of view, mainly the strategic, process, technology and organisational ones [5], [15]. In parallel, various attempts have been made to delve into the personal and team levels.

At the personal level, the phrase "personal information management" was first used in the 1980s [11] in the midst of popular excitement over the potential of the personal computer to greatly enhance the human ability to process and manage information. Today, personal information management is meant to support activities such as acquisition, organization, maintenance and retrieval of information captured, used and applied by individuals [22].

At the team level, research focused on knowledge networking, building on the assertion that useful knowledge does not only exist in individuals, but is continuously produced and revised in social processes [10]. Individual knowledge workers participate in a variety of knowledge networks, which in turn constitute a specific

R. Wagner, N. Revell, and G. Pernul (Eds.): DEXA 2007, LNCS 4653, pp. 213–222, 2007.

type of social networks. The term "social network" is often used to refer to "a specific set of linkages among a defined set of actors, with the additional property that the characteristics of these linkages as a whole may be used to interpret the social behavior of the actors involved" [21]. Consequently, the term "network" designates a social relationship between actors. Knowledge network are social networks, which are assembled in order to create, revise and transfer knowledge, for the purpose of creating value [21].

Hence, a typical knowledge worker is both handling personal information management tasks on daily basis and also participating in a variety of knowledge networks in the context of his/her work.

A large number of tools has recently emerged supporting personal (e.g. [20], [18], [4]) as well as social information and knowledge management (e.g. community tools, wikis, social tagging). A knowledge worker is typically using a variety of such tools, often switching between different tools when moving from one assignment to another. This has created the need for novel means to enable users to seamlessly manage both their personal information, tasks, contacts, etc. and the information and knowledge shared between them and their knowledge networks. Semantic technologies play an important role in the development of such tools [3], [12].

Our motivation in this paper is to research the individual and the team levels of knowledge management, in order to unveil prominent knowledge needs, interactions and processes, and to develop a software architecture which tackles these issues. We focus on a recent research direction related to the emergence of the Social Semantic Desktop [7], [19]. The Social Semantic Desktop *(see http://www.semanticdesktop.org/)* aims to provide a comprehensive semantics-based software framework for extending the personal desktop into a collaboration environment that supports both personal information management and the sharing and exchange across social relations.

In this paper we present IKOS, a software architecture which is based on the Social Semantic Desktop framework. Our architecture aims to support the integration of personal and collaborative information and knowledge management in the context of knowledge intensive firms; specifically we examine the case of professional business service firms (e.g. consulting, legal, advertising). After studying the daily work of employees within a typical consulting firm, we develop in Section 3 the IKOS architecture. IKOS uses and extends a number of components developed within the Nepomuk project. Nepomuk is an EC-funded Integrated Project developing a set of Social Semantic Desktop services *(see http://nepomuk.semanticdesktop.org)*. In Section 4 we present how the user requirements extracted from the use cases are fulfilled by the proposed architecture. The final section presents our conclusions and areas for further work.

## 2 User Studies

In order to extract user requirements for knowledge workers working in professional business service firms, we analyzed use cases obtained at TMI *(see http://www.tmiworld.com/)*, an international management consultancy. TMI is operating through a network of local partners in 40 countries. It is organized according to a licensor-licensee model. The TMI licensor organization is responsible for the development of TMI strategy and its dissemination to local partners, whereas licensees need to incorporate strategic issues to the local level. This business cooperation framework adopts a strong partnership approach and strengthens the collaboration between different partners.

In order to develop user requirements, we conducted user research at TMI using ethnographic methods such as contextual observations and interviews. We shadowed people as they did their work and then we interviewed them. We asked the informants about the processes and procedures they used to get their work done and how they collaborated with others. From our study we extracted conclusions about typical processes and we created personas [2], [9]. Although personas are fictitious, they are based on the knowledge of real users and therefore identify users' behaviour patterns, motivation, expectations, goals, skills, attitudes and environment. Using these typical processes and personas we developed two use cases representing the knowledge creation and sharing work processes within TMI. The first use case depicts customer-specific product sales processes (i.e. training programs or consulting processes), while the second describes the development of a management-supported knowledge network within TMI. Tables 1 and 2 give a brief overview of the use cases.

**Table 1.** Use Case 1: Development of customized products

| Personas | Use case steps |
|---|---|
| • Nasim is the manager responsible for booking sales meetings with prospective clients.<br>• Alistair is the sales manager – he prepares sales meetings and he plans follow-up meetings<br>• Karen is the product developer and the trainer – she creates the new customer specific products, performs the course and writes a report<br>• Josephine is the project assistant – she prepares the material for the courses and books training locations for big courses | 1. *Book clients*: Nasim searches for companies that could be interested in TMI's offering. He makes a list with these companies, he prepares a call sheet and he gives calls in order to book meetings with prospective clients<br>2. *Prepare a presentation*: Alistair prepares a sales presentation. He searches for similar past TMI projects, related emails and knowledgeable TMI managers and trainers.<br>3. *Prepare the sales meeting*: Alistair prepares the sales meeting, i.e. he designs a solution and writes a proposal. He reviews the existing TMI offerings (proposals and products), examines customer related information and searches for TMI experts.<br>4. *Develop the customized product*: Karen finds out all standard products developed by TMI (i.e. not client specific solutions), examines them and chooses the most suitable. Karen defines the structure and duration of the customised product and develops the new customised product.<br>5. *Prepare the course material*: Josephine collects all information connected to the project, examines the related emails, books the material needed for the course and sends it directly to the training location.<br>6. Course implementation: Karen performs the course. |

**Table 2.** Use case 2: Development of a knowledge network

| Personas | Use case steps |
|---|---|
| • John is a senior manager and community sponsor - he establishes the mission of the community, its expected outcomes, and insures its exposure in the organization. He also helps to secure needed resources.<br>• Alistair, the sales manager, is the community leader – he is a Subject Matter Expert.<br>• Nasim, the manager responsible to book meetings, is the community facilitator - responsible for clarifying communications, drawing out the reticent, ensuring that dissenting points of view are heard and understood, posing questions to further discussion and keeping discussions on topic.<br>• Karen and Emma, product developers, are core team members – they embody deep knowledge of their practice or domain, participate actively in the community, learn and share their learning. | 1. *Decide to create a knowledge network*: John recognizes the need for development of common consulting practices within TMI and decides to create a new knowledge network consisting of TMI workers from various offices worldwide. He defines together with Alistair and Nasim the specific goals, activities and value of the group to the company.<br>2. *Find members*: Alistair provides overall guidance and leadership and identifies the strategic fit of the group. Nasim finds TMI employees who are suitable to get core team members of the network. Karen and Emma are meet as peers seeking unity, mutual understanding and common context. Often, Nasim connects core team members and he organizes sub-communities according to the interest/expertise of the network members.<br>3. *Share knowledge*: Nasim helps by searching, retrieving and transferring information to requests for the group knowledge and content. Team members work with one other and collaborative activities take place.<br>4. *Produce results*: John and Alistair work with core team members to set group boundaries, mission, norms, and values and make policy changes as needed. Nasim and the core team members organize TMI's knowledge in order to identify knowledge gaps.<br>5. *Deliver results to the company*: Core team members prepare deliverables and transfer their results to the company. Alistair diagnoses and maintains network health and Nasim measures and evaluates the network's contributions to the company.<br>6. *The knowledge network expires*: After the accomplishment of the community goal, the network expires. |

## 3 The IKOS Architecture

IKOS constitutes a combination of a personal information management system and a group support system. Figure 1 illustrates the architecture of IKOS, which follows the peer-to-peer model and relies on the use of Social Semantic Desktop services to support the requirements of knowledge workers for seamless management of personal information and information shared within ad-hoc groups.

IKOS' client application comprises *Search* and *Workspaces*, the two graphical interfaces the users can use to interact with IKOS. Through the *Search* interface users can retrieve content that matches exactly the query keywords as well as semantically similar content both from their personal desktops and from the public space of other users' desktops. Users are also able to navigate through available content by exploiting relationships between different content items.

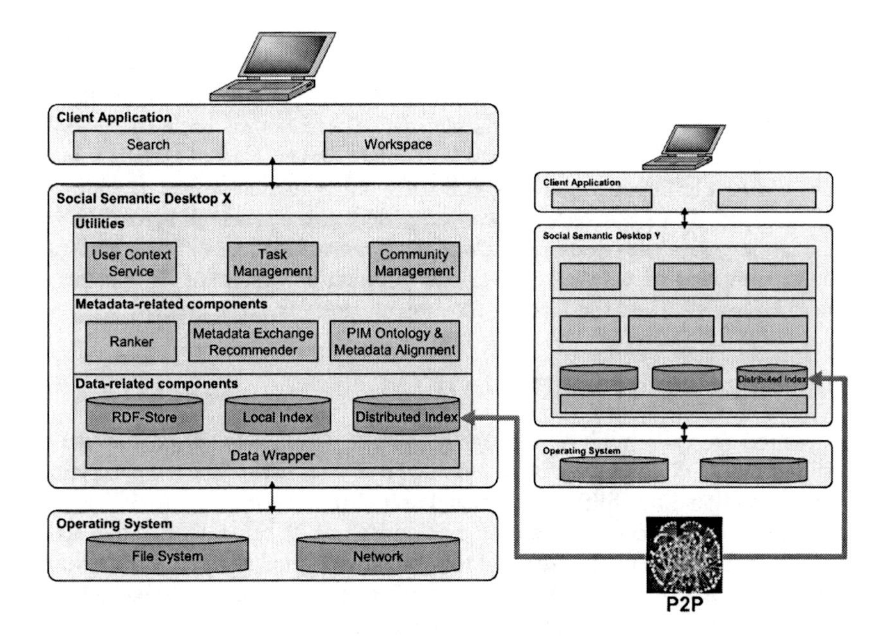


**Fig. 1.** The IKOS Architecture

*Workspaces* give users the opportunity to create spaces supporting the accomplishment of their tasks. Workspaces can support either the personal work of each user or the collaborative work of groups. In the case that more than one person are involved in the workspace, a list of the participants and information about them are provided by the system. In each workspace, related resources (email messages, office files, pdfs, web bookmarks) from the public space of the desktop peers can be stored. Moreover, users involved in a workspace have the possibility to discuss in chat rooms. Workspaces also include task management support for the users (assigning tasks, deadlines, etc.). Finally, workspaces provide team and personal calendar support.

IKOS' business logic is provided by the following Social Semantic Desktop components that make their functionalities available as services. The *User Context Service component* supports the observation of and reasoning about a user's current work context. It elicits knowledge about the current goals of the user that is useful for tuning content structuring, annotation and retrieval. The *Task Management* component provides functionalities such as personal task modelling, personal work scheduling, personal work trigger and control, task delegation, task model preservation, task model abstraction, task model reuse and task model retrieval. The *Community Management* component allows community identification and analysis by exploiting p2p network data and transactions. It provides functionalities such as community detection and labelling, community structure analysis and detection of trends and threads. The *Ranker* component utilises ranking algorithms based on shared metadata and generic user ranking information that are used to enhance the retrieval of shared resources. The *Metadata Exchange Recommender* provides a mechanism for metadata exchange between users. It uses data gathered from the community to discover relationships between items in the knowledge base, in order to interact and find relevant material in the community of users. The *Personal Information Management Ontology (PIMO) & Metadata Alignment* component hosts the personal ontologies and implements metadata alignment methods and. The *RDF-Store* component is used to store all crawled content and associated metadata in RDF. The *Local Index* component allows full-text and semantics-based search in the personal desktop. The *Distributed Index* component extends search across the public spaces of other users' desktops. The *Data Wrapper* component extracts and queries full-text content and metadata from various information systems (file systems, web sites, mail boxes, etc.) and file formats (documents, images, etc.).

## 4 Matching Requirements with IKOS

In this section we examine how the user requirements extracted from the use cases are fulfilled by IKOS. First we outline a typical usage scenario. Then, we explain how the system functionalities fulfil the extracted user requirements.

In a typical usage scenario Karen (see also Figure 2) can select among existing information resources (emails, files, folders, bookmarks, images., etc.) from her local desktop for addition to IKOS. The crawler component extracts full-text index from the selected resources while the PIMO component explicitly extracts or implicitly infers metadata for the selected resources, according to the current Karen's context. For example, existing file folders and email folders can be automatically associated with PIMO concepts. These resources are represented in the RDF syntax, identified by their URIs and integrated automatically into the local repository. For manual annotation of resources with metadata, the interface should filter to show suggested metadata by analyzing the resource text. Moreover, metadata are modelled in an ontology that represents Karen's conceptualization of her domain. This allows a more fine-grained of desktop resource classification than the one provided by most operating systems that only allow one file to exist in exactly one folder.

At any point, Karen can intervene and append her PIMO structures by adding deleting, or renaming concepts and relations between the concepts and by annotating any concept using free text and also data properties. PIMO concepts can be used for tagging resources. The PIMO is also used to generate a graphical knowledge map the user can use to browse through it. A key aspect in our framework is the ability to import existing ontologies, such as corporate ontologies, by copying them. These ontologies can then be used as templates for users to build their PIMOs upon.

Of course the most important aspect for Karen is to be able get answers to specific queries. Queries can be entered by clicking in the graphical knowledge map or manually as text. The crawler's inference engine tries to answer the query by routing it to both the local index and the distributed index. When the distribute index is accessed, alignment between local and distributed metadata should be performed in order to be able to retrieve content that is modelled according to different but similar ontological structures. A corporate ontology of the domain has a key role in the metadata alignment process by acting as a public identifier that can be used to identify the concepts that two different PIMOs include. We assume that the two PIMOs developed by two users that share similar working contexts and that are probably based on the same corporate ontology are overlapping enough in order to be aligned.

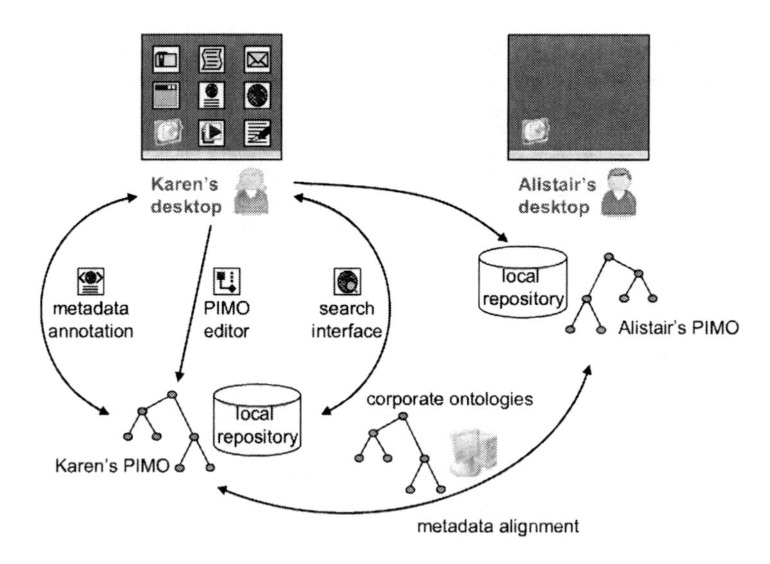


**Fig. 2.** Typical Usage Scenario of IKOS

In the following we examine how the IKOS architecture matches the user requirements described in Section 2.

*Need for easy search:* Knowledge workers in TMI often need to search and retrieve information in order to accomplish their tasks. Therefore, users need easy information search. When Alistair prepares the presentation for the sales meeting (Use Case 1 / Step 2) or writes a proposal (Use Case 1 / Step 3), he searches for relevant information. Also, Karen (Use Case 1 / Step 4), Josephine (Use Case 1 / Step 5) and Nasim (Use Case 2 / Step 2) search for information. IKOS currently provides full-text and semantics-based search while in the future it will also provide community detection support for locating experts in certain sectors, domains, fields of interest or related projects.

*Ability to query for content residing not only in corporate systems but also in colleagues' computers:* Often, knowledge workers in TMI search for information that another employee within TMI has retrieved in the past. The repeat of the queries means time cost for the employees and therefore its avoidance is required. Karen searches for customer related information to create the customised product (Use Case 1 / Step 4), while Alistair has retrieved the same information for the sales meeting (Use Case 1 / Step 2). IKOS gives the possibility of peer-2-peer sharing of desktop files. It allows searching across the public spaces of peers and downloading retrieved content.

*Avoidance of separate queries to find different types of items:* When knowledge workers create a new document, they search separately about different types of items, e.g. documents, emails and experts. However, separate queries are costly for employees. When Alistair prepares the presentation for the sales meeting, he must search separately about relevant TMI products, emails, customer related information and TMI experts (Use Case 1 / Step 2). The same happens to Alistair (Use Case 1 / Step 3) and Karen (Use Case 1 / Step 4). IKOS provides means to associate thematically items, such as email messages, office files, pdfs, web bookmarks, calendar item and events, in order to build a network of interrelated information resources, and it supports seamless searching across thematically connected information resources.

*Need to use parts from different documents:* In order to create a new document, knowledge workers in TMI need to use different parts from various information resources. Alistair (Use Case 1 / Step 3) and Karen (Use Case 1 / Step 4) use parts from different TMI products to prepare their documents/presentations. Within IKOS, we plan to develop a component that will permit annotating parts of documents of popular applications formats. Furthermore, these annotations will be used to provide recommendations to users, given the subject and scope of the new document they are working on (e.g. which fragments or related documents to use).

*Need for task management support:* TMI knowledge workers require task management support, since they have to perform standard tasks in order to accomplish the customer specific product sales processes. In order to write a proposal for the sales meeting, Alistair has to follow standard steps, i.e. review TMI's offering, search and examine customer related information and consult other TMI experts (Use Case 1 / Step 3). The same happens by Nasim (Use Case 1 / Step 1), Alistair (Use Case 1 / Step 2), Karen (Use Case 1 / Step 4) and Josephine (Use Case 1 / Step 5). The task management component of the Social Semantic Desktop will provide personal task management functionalities such as task monitoring, task logging, and task resources and relations storage. Moreover, it will provide support for collaborative tasks with functionalities such as task delegation and task pattern management that will allow re-use of prominent task patterns.

*Need for tools supporting community management:* For each knowledge network, TMI managers have to analyse network activity, control network progress and evaluate members' contribution. Alistair and Nasim have to define group activities (Use Case 2 / Step 1). Costas and Alistair have to set group boundaries, mission, norms and values and to make policy changes as needed (Use Case 2 / Step 4). Later, Alistair has to diagnoses and maintains network health, and Nasim has to measure and evaluate core team members' contribution (Use Case 2 / Step 5). IKOS will in the future support community identification. For example, given a user in the community and a particular keyword, it will locate users with similar interests. Moreover, given a community of users, it will identify emerging trends such as tags or resources becoming popular within the community.

*Need for a knowledge repository supporting the knowledge network:* Knowledge network members require a repository to store resources that are relevant to their network. Existing resources must be located (Use Case 2 / Step 3), organised (Use Case 2 / Step 4) and examined (Use Case 2 / Step 5). Members can store relevant resources within a dedicated workspace created to support the group. Hence, the workspace can be used as knowledge repository for the knowledge network. Further, knowledge network members can search and retrieve network resources (office documents, pdfs, emails, web links, calendar events).

*Need for tools supporting communication among knowledge network members:* Knowledge network members have to communicate and collaborate with each other. Core network members have to meet as peers (Use Case 2 / Step 2) and to participate in collaborative activities (Use Case 2 / Step 3). IKOS provides team and personal calendar support and it helps members to organize their activities related to the group and be reminded about group events and deadlines. Future extensions include the integration of synchronous communication tools such as chat tools and development of an asynchronous communication tool that communication channels used by network members (e.g. email, blogging).

## 5 Conclusions and Future Work

Knowledge workers need tools that seamlessly organize and manage both personal as well as collective information and knowledge. In this paper we presented early work on IKOS, a software architecture based on the Social Semantic Desktop framework. IKOS aims to provide a solution that integrates personal and social knowledge management support using semantic technologies.

Although we have not yet conducted a formal evaluation of IKOS, user feedback from testing a prototype has been positive and confirms that IKOS is in-line with our primary goal to seamlessly support personal and social knowledge-intensive processes of knowledge workers. Users perceived improvements in integration of content creation and processing within their everyday working habits as well as in reuse of desktop resources and mental models in a community context and vice-versa. In turn, these improvements facilitate wider and easier involvement of knowledge workers in project teams, easier reach to colleagues and experts for advice and information, and persistency of knowledge shared.

Our future work includes: (a) development of a Wiki-based collaborative environment to make information creation and sharing within workspaces more flexible; (b) workspace enhancement with synchronous and asynchronous communication support, (c) integration of the Nepomuk Ontological Network Miner component for semi-automatic metadata annotation and term disambiguation; (d) integration of community detection and analysis support tools, (e) development of an annotating component that will allow adding metadata to parts of documents, (f) integration of the Nepomuk task management component, Finally, we plan a thorough usability testing and impact analysis of IKOS within TMI.

**Acknowledgments.** This work has been partially funded by the European Commision Information Society Technologies programme (NEPOMUK FP6-027705).

## References

1. Applehans, W., Globe, A., Laugero, G.: Managing Knowledge: A Practical Web-Based Approach. Addison-Wesley, Reading (1999)
2. Calabria, T.: An introduction to personas and how to create them (Retrieved February 2, 2006) from Step Two Designs PTY LTD (2004), Web site: http:www.steptwo.com.au/papers/kmc_personas/index.html
3. Caldwell, F., Linden, A.: PKN and Social Networks Change Knowledge Managemen, (Retrieved March 10, 2006) from Gartner (2004), Web site: www.gartner.com
4. Cheyer, A., Park, J., Giuli, R.: Integrate. Relate. Infer. Share. In: Gil, Y., Motta, E., Benjamins, V.R., Musen, M.A. (eds.) ISWC 2005. LNCS, vol. 3729, Springer, Heidelberg (2005)

5. Davenport, Prusak: Working Knowledge, Harvard Business School Press, Boston, Massachusetts (1998)
6. Dawson, R.: Developing Knowledge-Based Client Relationships, The Future of Professional Services, Butterworth-Heinemann (2000)
7. Decker, S., Frank M.: The Social Semantic Desktop, Technical Report, September 10, 2006 (2004)
8. Ehrig, M., Tempich, C., Broekstra, J., van Harmelen, F., Sabou, M., Siebes, R., Staab, S., Stuckenschmidt, H.: SWAP - ontology-based knowledge management with peer-to-peer technology. In: Sure, Y., Schnurr, H.P. (eds.) Proceedings of the 1st National "Workshop Ontologie-basiertes Wissensmanagement (WOW2003) (2003)
9. Goodwin, K.: Perfecting your personas, (Retrieved April 16, 2007) from cooper (2007) Web site: http:www.cooper.com/insights/journal_of_design/articles/perfecting_your_ personas_1.html
10. Heller-Schuh, B., Kasztler, A.: Analyzing Knowledge Networks in Organizations. In: Proceedings of I-KNOW 2005, Graz, Austria, June 29-July 1, 2005 (2005)
11. Lansdale, M.: The psychology of personal information management. Applied Ergonomics 19(1), 55–66 (1988)
12. Linden, A.: Semantic Web Drives Data Management, Automation and Knowledge Discovery, (Retrieved March 14, 2006) (2005) from: www.gartner. com
13. Maier, R., Haedrich, T.: Centralized Versus Peer-to-Peer Knowledge Management Systems. Knowledge and Process Management Systems 13(1), 61–77 (2006)
14. Mentzas, G., Apostolou, D., Abecker, A., Young, R.: Knowledge Asset Management. Springer, Heidelberg (2002)
15. Nonaka, Takeuchi.: The Knowledge-Creating Company: How Japanese Companies Create the Dynamics of Innovation. Oxford Univ. Press, Oxford (1995)
16. OVUM.: Knowledge Management: Building the Collaborative Enterprise (1999)
17. Parameswaran, M., Susarla, A., Whinston, A.B.: P2P Networking: An Information-Sharing Alternative. IEEE Computer Society Press, Los Alamitos (2001)
18. Quan, D., Huynh, D., Karger, D.: Haystack: A Platform for Authoring End User Semantic Web Applications. In: International Semantic Web Conference, Springer, Heidelberg (2003)
19. Sauermann, L., Bernardi, A., Dengel, A.: Overview and Outlook on the Semantic Desktop. In: Gil, Y., Motta, E., Benjamins, V.R., Musen, M.A. (eds.) ISWC 2005. LNCS, vol. 3729, Springer, Heidelberg (2005)
20. Sauermann, L.S., Schwarz.: Introducing the Gnowsis Semantic Desktop. In: Sauermann, L.S (ed.) Proceedings of the International Semantic Web Conference, Springer, Heidelberg (2004)
21. Seufert, A., von Grogh, G., Bach, A.: Towards knowledge networking. Journal of Knowledge Management 3(3), 180–190 (1999)
22. Teevan, J., Jones, W., Bederson, B.B.: Personal Information Management. Communications of the ACM, 40–43 (2006)
23. Tsui, E.: Technologies for Personal and Peer-to-Peer Knowledge Management (last accessed 1 June 2007), www.csc.com/abo utus/lef/mds67_off/ uploads/P2P_ KM.pdf

# Classifying and Ranking: The First Step Towards Mining Inside Vertical Search Engines

Hang Guo[1], Jun Zhang[2], and Lizhu Zhou[1]

[1] Computer Science & Technology Department
100084, Tsinghua University, Beijing, China
guohang@mails.tsinghua.edu.cn, dcszlz@mail.tsinghua.edu.cn
[2] IBM China Software Develop Lab
100084, Beijing, China
zhjun@cn.ibm.com

**Abstract.** Vertical Search Engines (VSEs), which usually work on specific domains, are designed to answer complex queries of professional users. VSEs usually have large repositories of structured instances. Traditional instance ranking methods do not consider the categories that instances belong to. However, users of different interests usually care only the ranking list in their own communities. In this paper we design a ranking algorithm –ZRank, to rank the classified instances according to their importances in specific categories. To test our idea, we develop a scientific paper search engine–CPaper. By employing instance classifying and ranking algorithms, we discover some helpful facts to users of different interests.

## 1 Introduction

With the booming of Internet resources, Internet is becoming the most important source of information. The most popular way to retrieve information is to use search engines, such as Google, Yahoo, etc. However, most of them support only simple keywords queries. If queries are too complicated, the search engines become less helpful. For example, a user may want to know *"where to buy a book about database lower than 10$ in New York"*. Traditional search engines do not support such a complex query because their data are not structured. Vertical Search Engines (VSEs), which usually work on specific domains, are designed to solve the problem. Their data are structured instances (objects) such as products, papers, persons, organizations, and so on. Many technologies (i.e., instance identification [14], automatic generated wrappers [11], manual wrappers rules [1], etc.) have made it possible to extract structured instances from web pages. Typical vertical search engines include CiteSeer[1], LawCrawler[2], Google Book[3], Libra [12], etc.. Users can retrieve instances by structured query languages or well-designed interfaces.

[1] http://citeseer.ist.psu.edu/
[2] http://web.findlaw.com
[3] http://books.google.com/

R. Wagner, N. Revell, and G. Pernul (Eds.): DEXA 2007, LNCS 4653, pp. 223–232, 2007.

As in relational databases, we can discover helpful facts in the vertical search engines by employing data mining techniques. Previous work [12] has discovered meaningful facts in vertical search engines by analyzing the links between instances. There are many mining algorithms to process the instances. In this paper we mainly discuss instance ranking techniques.

The most popular web page ranking algorithms are PageRank [2] and HITS [7], which are developed by Google and IBM respectively. Similar to web pages, structured instances are also linked to each other. The idea of PageRank and HITS are followed by ObjectRank [2], XRank [8] and PopRank [12]. ObjectRank uses heuristic authority transfer rates to rank objects according to given keywords. XRank proposes an unified link analysis framework called "link fusion" to consider both the inter- and intra- type link structure among multi-type interrelated objects. PopRank tries to learn the transfer rate according to the partial ranking given by the domain experts. However, in many vertical search engines such as the blog search engine, news search engine and product search engine, it is quite hard, if possible, to get partial ranking lists from experts.

An important problem of the current instance ranking techniques is that they do not consider the category of instances. Practically users of different interests usually care only the ranking list in their own communities. Important instances in one topic could be meaningless in another. For instances, both *IEEE Trans Software Engineering* and *IEEE Trans Database* are top journals on computer science. For those software engineering researchers, the former are more important. However, database researchers prefer the latter. The problem goes worse when there are many categories in the VSE, such as the news search engine. An important news for some users may be meaningless for others. Therefore we need an adaptive ranking algorithm to satisfied the needs of different users. In this paper we propose an instance ranking algorithm named ZRank to rank instances by their importances in specific categories. ZRank has following features:

- The ranks of instances are decided by their link structure. Similar to PageRank, if a instance is linked by many important instances, its rank is high.
- Instances of different concepts (classes) have different transfer rates. The transfer rates are heuristic priors.
- The ranks of instances vary with categories. In a given category c, if instance $i$ is associated with c, $i$'s neighbors will receive more "authority" from $i$ in each iteration. That means $i$'s neighbors are probably important instances in c.

We have developed a vertical search engine demo, called CPaper, to test our ideas. CPaper is a scientific paper search engine on computer science. Users can retrieve three types of instances: papers, authors and conference(journals). All the instances are automatically categorized before being ranked.

By employing the instance classifying and ranking algorithms, we have discovered helpful facts with CPaper. For instance, users can find how the interests of a conference i.e., VLDB or a researcher move in the last 20 years. CPaper also shows the most important papers and journals in each topic. These features make our systems more helpful than traditional VSEs like CiteSeer.

The remainder of the paper is organized as follows. Section 2 introduces the framework of CPaper. Our classification method is presented in Section 3. Section 4 introduces ZRank. Section 5 presents the classifying and ranking results in CPaper. Finally the paper is concluded in Section 6.

## 2 CPaper

Fig. 1 illustrates the framework of CPaper. Most instances are extracted from pages of DBLP[4]. Concept Schema in Fig 1 refers to the concepts and their relations in the given domain. It is used in the Instance Collection and Instances Ranking components. Fig 2 shows the schema used in CPaper. Each instance is directly linked to others when they are extracted. For example, pages in DBLP list researchers and their publications. We collect these pages, generate instances and link them according to their relations. Each instance act as a node in a huge graph. We use Berkeley Database System[5] to store the graph. Users access the search engine repository through well-designed interfaces. In CPaper users can choose to submit complicated queries other than a set of keywords. Details of the extracting process are introduced in [4].

The dashed lines are the instance mining components. They can help users get interesting facts. These facts include the research topics of a researcher, the ranks of papers, researchers, and journals. The task of the Instances Classification component is to automatically classify all the instances stored in repository.

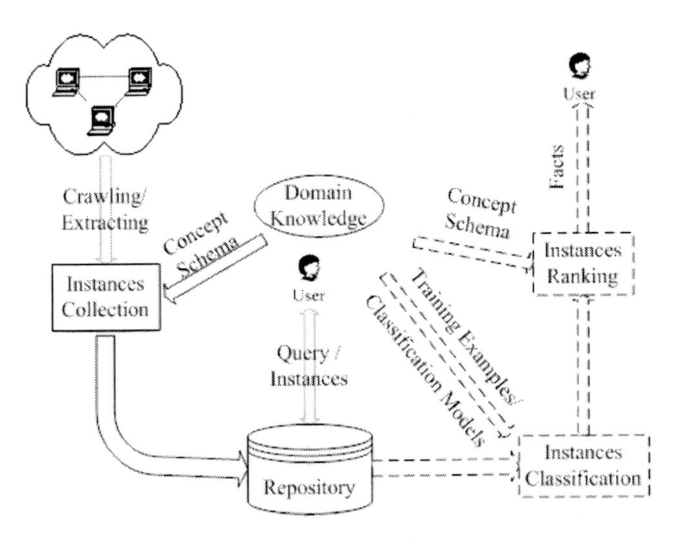


**Fig. 1.** CPaper Framework

[4] http://www.informatik.uni-trier.de/ ley/db/
[5] http://www.oracle.com/database/berkeley-db.html

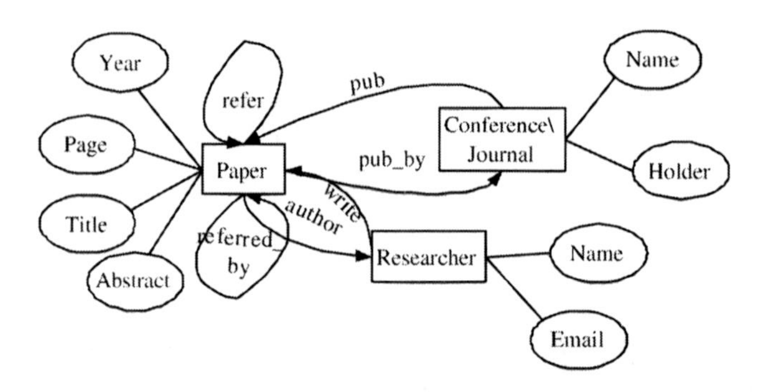


**Fig. 2.** Concept Schema in CPaper

Then the Instance Ranking component rank the instances according to their classification results. Users can rank instances by categories.

## 3 Instances Classification

The classification process is shown in Fig 3. It is not possible to classify the instances altogether. In CPaper, we classify the paper instances first. Researchers are classified according to the papers they have written. Similarly, journals are classified according to the papers they have published.

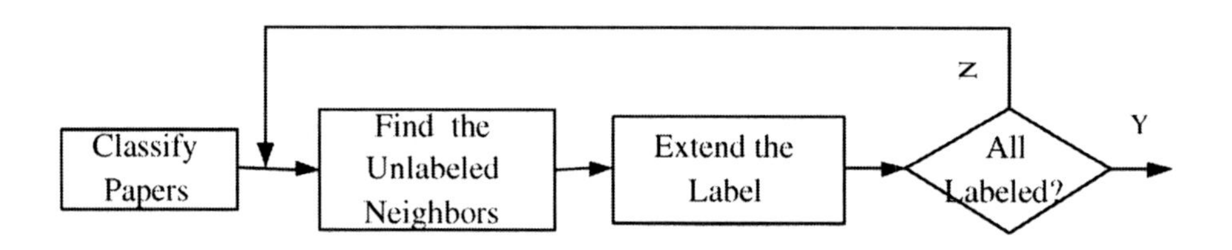


**Fig. 3.** Classification Process

Substantial efforts have been made in the literature for plain text classification [13]. Some of the well-established automatic document categorization techniques are NaiveBayes [10], SVM [6], and so on. Some vertical search engines, like CiteSeer, have employed automatic classifying techniques. Traditional text classification models can be used in paper classification. However, many instances have short title (headline) area. The importances of these titles have been well exploited in IR [5,9]. Therefore title words should play more important roles than other words in classification. In our previous work [3], we have improved the performance of the classifiers by putting higher weight on the important words. Before classifying, we multiply the weight of title words by a prior parameter $\theta$. We set higher $\theta$ in papers because their titles are long and well-written. For informal documents like newsgroup articles, $\theta$ is relatively lower. Normally it is set to 1.5. We choose MNB [10] classifier to classify papers.

When papers are classified, we start to classify other instances. The classification predication of an author is the average of the predications of his papers. And so are the journals. For example, suppose we have two categories $C_1$ and $C_2$, Tom has published two papers, namely paper $P_1$ and paper $P_2$. $p(C_1|P_1) = 0.1$, $p(C_2|P_1) = 0.9$. $p(C_1|P_2) = 0.8$, $p(C_2|P_2) = 0.2$. Then Tom is classified into $C_2$. And $p(C_1|Tom) = 0.45$, $p(C_2|Tom) = 0.55$.

## 4 Instances Ranking

### 4.1 PageRank

Our instance ranking method follows the idea of PageRank, which is used by Google. The idea is: if a page is important, it is linked by other important pages. Suppose $R^{k+1}$ is the rank matrix after $k+1$ iteration, there is [6]:

$$R^{k+1} = d * T \times R^k \tag{1}$$

where T denotes the transfer matrix of web pages, which means how much "authority" in a page is "transferred" to other pages in each iteration. If the out links of page $i$ is denoted as *out(i)*, there is:

$$T[i,j] = \begin{cases} 0 & : 0 \quad \text{there is no hyperlink from page i to page j;} \\ 1/out(i) & : 0 \quad \text{otherwise} \end{cases}$$

d is the damping factor. It is usually set to 0.85.

Similar to webpages, the instances of different concepts are linked. Important instances must be refered to other important instances. We take the CPaper as an example. According to Fig. 2, there is:

- Good papers usually refer to good papers.
- Good journals usually publish good papers.
- Good papers are usually written by good authors.

Based on this observation, we can explore the use of PageRank on instances.

### 4.2 ZRank

Our instance ranking algorithm is called *ZRank*. The difference between ZRank and PageRank lies in the transfer matrix, denoted as $\tilde{T}_c(I * I)$. Here $I$ is the number of instances in the domain and c indicates a category. $\tilde{T}_c(I * I)$ has following features:

[6] We use a simplified version of PageRank since the users' activities are not considered in this paper.

- In the traditional PageRank algorithm, all pages are of the same class. But in vertical search engines, instances are of different concepts. The relations between different concepts are usually weighted. For example, paper A is not written by well-know authors but it is referred by good papers. Paper B is written by a famous author but it is only referred by ordinary papers. In our system A is ranked higher than B. In other words $\tilde{T}[i,j]$ is affected by the relations between concepts.
- As mentioned in Section 1, ZRank considers the label of instances. If a paper is highly associated with a category, the ranks of its authors should be increased in the category. Therefore it should "transfer" more "authority" to its authors than other papers. On the contrary, a paper not related to this category "transfers" very little "authority" to its authors, even if it is an important paper. The transfer matrixes in different categories should be different. As a result, an important instance in a category may be nothing in another.

Following this idea, our transfer matrix in category c is calculated as:

$$\tilde{T_c} = L \bullet (M \times \dot{T} \times M^T) \bullet B_c \tag{2}$$

here $\bullet$ is a matrix operation that $M_1 \bullet M_2[i,j] = M_1[i,j] * M_2[i,j]$.

$L(I * I)$ is the linkage matrix of instances. If the out links of instance $i$ is denoted as *out(i)*, there is,

$$L[i,j] = \begin{cases} 0 & : \text{ there is no links from instance i to instance j;} \\ 1/out(i) & : \text{ otherwise} \end{cases}$$

Suppose C is the number of concepts in this domain, $M(I * C)$ is the membership matrix that,

$$M[i,j] = \begin{cases} 1 & : \text{ instance } i \text{ is the instance of concept } j \\ 0 & : \text{ otherwise} \end{cases}$$

$\dot{T}(C * C)$ is the transfer matrix of concepts. $\dot{T}$ is a decided by experts in this domain. We take CPaper as an example. $\dot{T}_{CPaper}$ is shown in Fig. 4.

$B_c(I * I)$ shows the relationship between instances and category c.

$$B_c[i,j] = \begin{cases} 3 & : \ p(c|i) > 0.3 \\ 10 * p(c|i)) & : \ 0.3 \geq p(c|i) \geq 0.1 \\ 1 & : \ \text{otherwise} \end{cases}$$

Here $p(c|i)$ is the classification predication calculated by the Instance Classification component. It represents the possibility that instance $i$ is associated with category c. The ranking algorithm is shown in Fig 5. When $|R_{k+1} - R_k| < \epsilon$, the ranking algorithm is finished. Here $\epsilon$ is the threshold. In CPaper the algorithm finishes in at most 20 iterations for each category. In ZRank, the dumping factor d is set to 0.45.

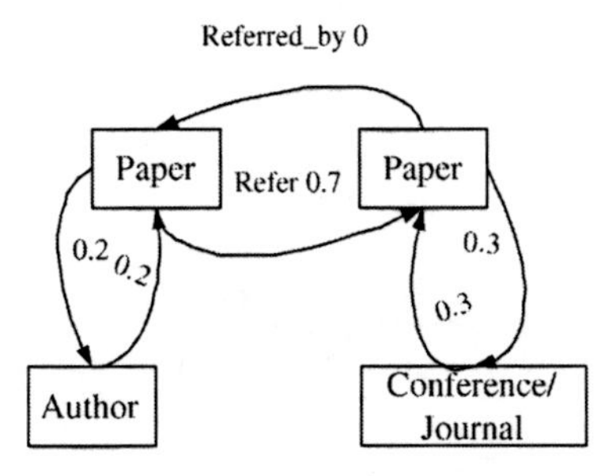


**Fig. 4.** $\dot{T}_{CPaper}$

---

**Input:** $\dot{T}$, L, M, $\epsilon$, $B_c$,d
**Output:** Rank Matrix $R_c$

**(a) Initialization**
1 $R_0 = \{1, 1, \ldots, 1\}$
2 $\tilde{T}_c = L \bullet (M \times \dot{T} \times \bar{M}) \bullet B_c$

**(b) Iteration**
3 **do**
4 $R_{k+1} = d * \tilde{T} \times R_k$
5 **while** $(|R_{k+1} - R_k| > \epsilon$ **and** $k < 20)$

---

**Fig. 5.** Pseudo Codes for ZRank

# 5 Application

The classifying and ranking algorithm are used in CPaper. There are 338,190 papers, 201,843 researchers and 2,885 conferences/journals in CPaper. They are classified into 14 categories and 98 subcategories. We take CPaper as an example to show the impact of the classifying and ranking algorithms.

## 5.1 Impact of Classifying

Traditional classifiers can hardly classify instances like conferences and authors. Our system presents such instances as mixtures of different topics.

**Analyse on A Well-known Conference.** Fig 6 shows what kind of papers are accepted by VLDB in 1999. The information is important for researchers who want to submit papers to that conference.

**Analyse on A Researcher.** Fig 7 shows Jiawei Han's[7] interests on DW.

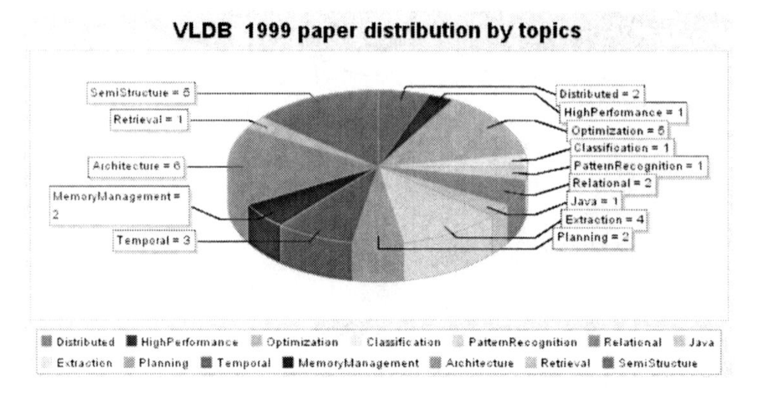


**Fig. 6.** Papers Published in VLDB 1999

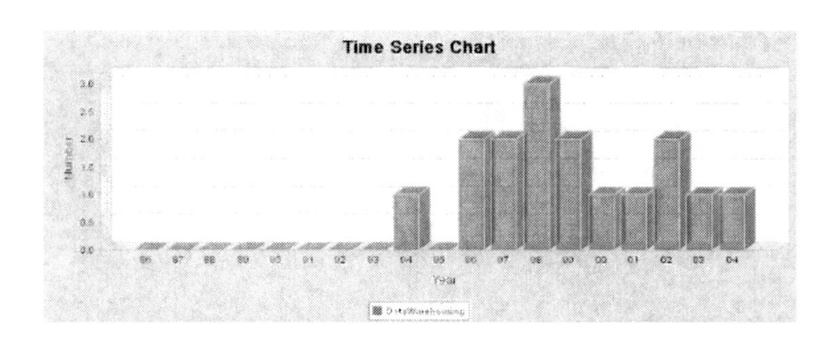


**Fig. 7.** Papers on Data Warehousing Published by Jiawei Han

**Table 1.** Top-10 Papers on Database

| | Paper Title |
|---|---|
| 1 | A Relational Model of Data for Large Shared Data Banks. |
| 2 | System R : Relational Approach to Database Management. |
| 3 | The Notions of Consistency and Predicate Locks in a Database System. |
| 4 | The Entity-Relationship Model - Toward a Unified View of Data. |
| 5 | SEQUEL : A Structured English Query Language. |
| 6 | Access Path Selection in a Relational Database Management System. |
| 7 | Database Abstractions : Aggregation and Generalization. |
| 8 | The Functional Data Model and the Data Language DAPLEX. |
| 9 | Implementation of Integrity Constraints and Views by Query Modification. |
| 10 | On the Semantics of the Relational Data Model. |

[7] http://www.cs.uiuc.edu/ hanj

### 5.2 Impact of Ranking

Table 1 shows the top papers on database. Table 2 shows the ranking list of journals in software engineering and database. The ranks of most journals are different in the two list. *IEEE TRANS SOFTWARE ENG* is the top journal in software engineering while *ACM TRANS DATABASE SYST* is the top journal in database.

**Table 2.** Top-10 Journals on Database and Software Engineering

| | Software Engineering* | Database |
|---|---|---|
| 1 | IEEE TRANS SOFTWARE ENG | ACM TRANS DATABASE SYST |
| 2 | AUTOMATED SE | COMMUN ACM |
| 3 | J ACM | IEEE TRANS KNOWL DATA ENG |
| 4 | ACM COMPUT SURV | IEEE TRANS SOFTWARE ENG |
| 5 | COMMUN ACM | SIGMOD RECORD |
| 6 | J SOFTWARE ENG KNOWL ENG | J ACM |
| 7 | INF SYST | INF SYST |
| 8 | IEEE TRANS KNOWL DATA ENG | ACM COMPUT SURV |
| 9 | SIGMOD RECORD | IEEE DATA ENG BULL |
| 10 | IEEE DATA ENG BULL | INF PROCESS LETT |

* VLDB and J. VLDB are not distinguished in CPaper, therefore we do not take J. VLDB into consideration.

## 6 Conclusion

Vertical Search Engines become a hot topic recently. Traditional instance ranking algorithms does not consider the categories of instances. To overcome this problem we propose ZRank, a structured instance ranking algorithm, to rank the classified instances. We develop a paper search engine–CPaper to test the algorithm. By employing instance classifying and ranking algorithms, we discovered some helpful facts to users of different communities.

Besides classifying and ranking, we are trying to employ more mining algorithms like association rule mining and social network mining on VSEs.

## Acknowledgement

this work is supported by National Nature Science Foundation of China under Grant No.60520130299.

## References

1. Arocena, G.O., Mendelzon, A.O.: Weboql: Restructuring documents, databases, and webs. In: Proc of ICDE (1998)
2. Balmin, A., Hristidis, V., Papakonstantinou, Y.: ObjectRank: Authority-based keyword search in databases. In: Proc. of VLDB (2004)

3. Guo, H., Zhou, L.: Segmented document classification: Problem and solution. In: Bressan, S., Küng, J., Wagner, R. (eds.) DEXA 2006. LNCS, vol. 4080, pp. 41–48. Springer, Heidelberg (2006)
4. Guo, Q., et al.: A highly adaptable web extractor based on graph data model. In: Proc. of 6th Asia Pacific Web Conference (April 2004)
5. Jin, R., Hauptmann, A.G., Zhai, C.X.: Title language model for information retrieval. In: Proc. of SIGIR (2002)
6. Joachims, T.: Text categorization with support vector machines: learning with many relevant features. In: Proc. of 10th European Conference on Machine Learning, Chemnitz (1998)
7. kleinberg, J.: Authoritative sources in a hyperlinked environment. Journal of the ACM (1999)
8. Botev, C., Guo, L., Shao, F., Shanmugasundaram, J.: Xrank: Ranked keyword search over xml documents. In: Proc. of SIGMOD (2003)
9. Lam-Adesina, A.M., Jones, G.J.F.: Applying summarization techniques for term selection in relevance feedback. In: Proc. of 24th SIGIR (2001)
10. McCallum, A., Nigam, K.: A comparison of event models for naive bayes text classification. In: Proc. of AAAI workshop on Learning for Text Categorization, pp. 41–48. American Association for AI (July 1998)
11. Meng, X., Hu, D., Li, C.: Sg-wrap: A schema-guided wrapper generator. In: Proc of ICDE (2002)
12. Nie, Z., Zhang, Y., Wen, J., Ma, W.: Object-level ranking: bringing order to web objects. In: Proc. of WWW, pp. 567–574. ACM Press, New York (2005)
13. Sebastiani, F.: Machine learning in automated text categorization. ACM Computing Surveys 34 (2002)
14. Tejada, S., Knoblock, C., Minton, S.: Learning domain-independent string transformation weights for high accuracy object identification. In: Proc of KDD (2002)

# Progressive High-Dimensional Similarity Join

Wee Hyong Tok, Stéphane Bressan, and Mong-Li Lee

School of Computing
National University of Singapore
{tokwh,steph,leeml}@comp.nus.edu.sg

**Abstract.** The Rate-Based Progressive Join (RPJ) is a non-blocking relational equijoin algorithm. It is an equijoin that can deliver results progressively. In this paper, we first present a naive extension, called neRPJ, to the progressive computation of the similarity join of high-dimensional data. We argue that this naive extension is not suitable. We therefore propose an adequate solution in the form of a Result-Rate Progressive Join (RRPJ) for high-dimensional distance similarity joins. Using both synthetic and real-life datasets, we empirically show that RRPJ is effective and efficient, and outperforms the naive extension.

## 1 Introduction

Conventional distance similarity join algorithms batch process datasets that reside on local storage. The algorithms are blocking. They are unsuitable for progressively computing the similarity join of streams of high-dimensional data as they cannot produce results progressively, i.e. as soon as data is available.

Key issues in the design of a progressive algorithm are the management of main memory and the flushing policy. Indeed while main memory is limited, data is potentially incoming in very large quantities. One must make sure to keep in memory data most likely to participate in the production of results. An effective and efficient solution for relational equijoins, the Rate-based Progressive Join (RPJ), was proposed in [1]. We proposed a more efficient algorithm in [2]. Our algorithm is a Result Rate-based Progressive relational equijoin (RRPJ). It uses statistics on production of results to determine the partitions to be flushed to disk whenever memory is full. One of the advantages of using a result rate-based method is that it is independent of the model of data. In [3], we showed that the principle could be applied to the processing of spatial joins. In this paper, we consider high-dimensional distance similarity joins. We propose an effective and efficient algorithm for the progressive computation of the distance similarity join of streaming high-dimensional data, using limited main memory.

The main contributions of the paper are:

- a novel progressive high-dimensional similarity join algorithm. The algorithm uses a result-rate-based flushing strategy. It is an extension of our work on progressive relational equijoin [2] for relational data.

R. Wagner, N. Revell, and G. Pernul (Eds.): DEXA 2007, LNCS 4653, pp. 233–242, 2007.

- an extensive performance analysis. We compare our proposed algorithm with a naive extension of RPJ [1](that we call neRPJ). We make use of both synthetic and real-life datasets.

The rest of the paper is organized as follows: In section 2, we discuss related work. In section 3, we present the framework for progressive similarity join on high dimensional data, and propose two flushing strategies. We conduct an extensive performance analysis in section 4. We conclude in section 5.

## 2 Related Work

We present conventional distance similarity join algorithms for high-dimensional data in subsection 2.1. We present generic progressive join algorithms and similarity join algorithms and discuss their limitations in subsection 2.2.

### 2.1 Non-progressive Distance Similarity Joins for High-Dimensional Data

Many efficient distance similarity joins [4,5,6,7,8] have been proposed for high-dimensional data. To facilitate efficient join processing, similarity join algorithms often relies on spatial indices. R-trees (and variants) [9], X-tree [10] or the $\epsilon$-kdb tree [4] are commonly used. The Multidimensional Spatial Join (MSJ) [5,11] sorts the data based on their Hilbert values, and uses a multi-way merge to obtain the result. The Epsilon Grid Order (EGO) [7] orders the data objects based on the position of the grid-cells. The main limitation of conventional distance similarity join algorithms is that they are designed mainly for datasets that reside locally. Hence, they are not able to deliver results progressively.

### 2.2 Progressive Joins and Progressive Similarity Joins

Most progressive joins are relational equijoin of the XJoin [12] family. There is currently no known efficient algorithm of the XJoin family for the progressive computation of similarity joins of high-dimensional data. One of the most recent proposal is the Rate-based progressive Join (RPJ) presented in [1]. It is a hash-based relational equijoin for data streams. The authors recognize the need for an effective flushing strategy for the data that resides in main memory. They propose a probabilistic model of the data distribution of the data streams. Partitions residing in main memory are flushed based on the model. In [2], we discuss several limitations of RPJ. We also observe that the result output statistics can be used directly and propose the Result Rate-based Progressive Join (RRPJ) for relational data.

The authors of [1] remark that extending RPJ to the processing of other joins than relational equijoins, and in particular similarity joins of high-dimensional data is a challenge. Indeed, the probability model for other than relational data is complex and hard to determine analytically.

In [13], a generic sort-merge framework for handling progressive joins, called the Progressive Merge Join (PMJ) was proposed. Though PMJ can be applied for various model of data, it is not able to deliver initial results quickly. In [14], the Hash-Merge Join (HMJ) was proposed. HMJ relies on in-memory hash partitions and a probe-insert paradigm similar to XJoin in order to deliver initial results quickly. Similar to PMJ, HMJ uses a sort-merge paradigm for joining disk-based data.

Since the Result Rate-based Progressive Join (RRPJ) that we have proposed for relational data relies on output statistics, it can be effectively extended and adapted to other than relational data. We propose such an extension in this paper for distance similarity join for high-dimensional data (which is symmetric in nature).

## 3 Progressive Similarity Join on High-Dimensional Data

In this section, we discuss how we can design progressive similarity join algorithms for high-dimensional data. Our goal is to deliver initial results quickly and subsequent results with a high throughput (i.e. progressively). We present the problem definition in subsection 3.1. We present a framework for progressive similarity join in Section 3.2. In subsection 3.3, we propose two methods for determining the high-dimensional data to be flushed to disk whenever memory is full, namely: a naive extension of RPJ (neRPJ) and the novel Result-Rate Based Progressive Join (RRPJ).

### 3.1 Problem Definition

We consider two d-dimensional bounded data streams R and S. We refer to data from R and S as $R_i$ and $S_j$ respectively ($0 \leq \mathrm{i} \leq |R|$, $0 \leq \mathrm{j} \leq |S|$), where $|R|$ and $|S|$ are the total number of data objects in R and S respectively. Each data point consists of $d$ values. Given a data point $R_i$, the values are ($r_{i1}$, $r_{i2}$, ..., $r_{id}$), where $r_{ix}$ denotes the x-th value ( $1 \leq \mathrm{x} \leq \mathrm{d}$). Similarly, for a data point $S_j$, the values are ($s_{j1}$, $s_{j2}$, ..., $s_{jd}$).

The results of a similarity join between R and S, *SimJoin(R,S)*, consists of all object pairs ($R_i$, $S_j$), where $D_d(R_i, S_j) \leq \epsilon$, Here, we consider without loss of generality $D_d$ to be the Euclidean distance, where $D_d(R_i, S_j) = (\sum_{x=1}^{d} |(r_{ix} - s_{jx})^2|)^{\frac{1}{2}}$. $\epsilon$ is a user-defined threshold, which determines the maximum dis-similarity between $R_i$ and $S_j$. Notice that the similarity join is symmetrical.

Main memory is limited. Whenever it is full, some of the in-memory data needs to be flushed to disk. Our objective is to identify the data less likely to produce results and to sacrifice them in order to maximize throughput. We first look for a partitioning of main memory that helps in the production of results as well as in the flushing. An effective partitioning of in-memory tuples, allows limiting the search to matching partitions when an incoming tuple is inserted into a given partition. It also allows flushing entire partitions rather than individual tuples. In the case of high-dimensional similarity joins, a meaningful partition is a *d*-dimensional grid. Partitions are cells of the grid.

### 3.2 Grid-Based Similarity Join

We use the probe-and-insert approach as described in [15] and [2].

**Probing (Algorithm 1).** Whenever a new tuple, $t_d$, arrives (from one of the data streams), it is used to probe the in-memory tuples from the other data stream. In order to efficiently identify the tuples to be probed, a *d*-dimensional grid is used to partition the data space. The scanning for potential result tuples is restricted to the cell in which $t_d$ falls into and to its neighboring grid cells (those within $\epsilon$ distance of the border of the grid cell).

In Algorithm 1, Line 1 finds the grid cell in which $t_d$ falls into; whereas Line 2 identifies the cells that are within the $\epsilon$-distance of the grid cell, $g$. Once the cells are identified, Line 3-6 then checks whether each tuple, $t$, in the grid can be joined by checking the Euclidean distance between $t_d$ and $t$.

We keep track of the number of results produced by each grid cell using a counter, *numResults* (Line 8). Once the probing of the grid cell $c$ is completed, we update the statistics for the grid cell (Line 9).

**Algorithm 1:** Probing

**Data** : $t_d$ - Newly Arrived d-dimension tuple used to probe the other grid
**Result** : R, Results of the Similarity Join
**begin**

```
1    g = findGridCell(t_d) ;
2    n = findGridCellInNeighbourhood(g,ε) ;
3    for ( GridCell c in (g ∪ n)) do
4        numResults = 0 ;
5        for ( Tuple t in c) do
6            if ( D_d(t_d, t) ≤ ε ) then
7                R = R ∪ (t_d,t) ;
8                numResults++ ;
9        Update statistics for c ;
10   Return R;
end
```

**Insertion and Flushing (Algorithm 2).** We then identify the grid cell in which the new tuple should be inserted (Line 1). If there is space, $t_d$ is inserted into its own grid. If memory is full, we invoke *FlushDataToDisk*() which flushes data to disk to make space for newly arrived tuples. We then insert $t_d$ into the grid cell $g$ (Line 4).

---

**Algorithm 2:** Insertion and Flushing

---

```
Data : t_d - Newly Arrived d-Dimension tuple to be inserted
begin
1   g = findGridCell(t_d) ;
2   if ( memory is full() ) then
3       FlushDataToDisk();
4   Insert t_d into g ;
end
```

---

For each $i$th cell of the grid ( with $1 \leq i \leq n$, where n is the total number of grid cells), we maintain a count. The cells to be flushed are determined based on this value. The two flushing strategies that we propose differ in the way the value is computed (described in Section 3.3) and the partitions to be flushed are selected. Partitions are flushed until *NumFlush* (user-defined) tuples have been flushed.

### 3.3 Flushing Strategies

**Naive Extension to RPJ (neRPJ).** We present a simple extension to RPJ, called Naive Extension to RPJ (neRPJ) for high-dimensional data. The neRPJ algorithms maintains the *neRPJ* value, that is the number of data in a cell divided by the total number of data. The opposite cell (that is the matching cell in the other streams partition) to the grid cell with the smallest *neRPJ* value is flushed.

In the relational case, the mapping of a cell in one stream to the opposite stream is 1-to-1. When we probe for result tuples, we probe only a single cell from the opposite data stream. However, when dealing with high-dimensional data, besides probing the corresponding cell from the opposite data stream, we need to probe the neighboring cells (those within $\epsilon$ distance) as well. When neRPJ flushes an opposite cell; it might have inadvertently flushed a cell that could produce results at a later time.

**Result Rate-based Flushing (RRPJ).** The $Th_i$ value is an estimate of the productivity of the i-th cell (with $1 \leq i \leq n$, where $n$ is the total number of cells used to store the data). In the equation below, $R_i$ is the total number of results produced by the i-th cell and $N_i$ is the total number of tuples in the i-th cell.

$$Th_i = \frac{Ri}{N_i} \tag{1}$$

The RRPJ algorithm maintains the $Th_i$ value (Equation 1). In RRPJ, the grid cells with the smallest values are flushed.

# 4 Performance Analysis

We implemented all the algorithms in C++, and conduct the experiments on a Pentium 4 2.4 Ghz PC (1GB RAM). Unless otherwise stated, the parameters presented in Table 1 are used for the experiments. Similar to [2], we refer to the proposed result rate-based method for high-dimensional data as the Result-Rated Based Progressive Join (RRPJ). In addition, we also included a Random method as a baseline. Whenever memory is full, the Random method randomly selects a grid cell to be flushed to disk.

**Table 1.** Experiment Parameters and Values

| Parameter | Values |
|---|---|
| Disk Page Size | 4096 bytes |
| Number of cells Per Dimension | 4 |
| Memory Size, M | 1000 pages |
| Number of points per disk page | 85 |
| Number of MBRs flushed to disk | 10% of M |
| Dataset Size (for 2 streams) | 500K data points |
| Similarity Distance Join Threshold, $\epsilon$ | 0.1, 0.2, 0.3 |

In this section, we compare the performance of the algorithms (RRPJ, neRPJ and *Random*). We measure the number of result tuples generated (y-axis) vs percentage of data that have arrived (x-axis). In all experiments, we assume that there are two finite d-dimensional datasets. Each dataset is characterized by the data distribution and the order of arrival of the data. In Section 4.1, we use a uniform and skewed datasets. For the skewed dataset, we also consider various correlations between the data distributions - Harmony and Reverse [1]. In addition, we compare the performance of the algorithms in two extreme cases. In the first case (Section 4.2), we use a 'checkered' dataset. In the second case (Section 4.3), we consider the case where the data in some of the grid cells are non-uniformly distributed. In Section 4.4, we validate the effectiveness of the proposed algorithm for real-life data using the COREL [16] dataset.

## 4.1 Uniform and Skewed Dataset

The goal of these experiments is to compare the performance of the algorithms using uniform and skewed datasets. In addition, we also vary the order of arrival of the data. When the data from the datasets are uniformly distributed, each tuple is equally probable to contribute to a result. Hence, all flushing strategies are equally efficient and as good as random. This is illustrated in figure 1.

In the next experiment, we consider skewed dataset. We simulate clustered data by dividing the space into a $d$-dimensional grid, and by varying the cardinality of the grid cells based on a Zipfian distribution. We set the skewed factor for the Zipfian distribution, $\sigma$ to be 1.0. Thus, some grid cells have more data than

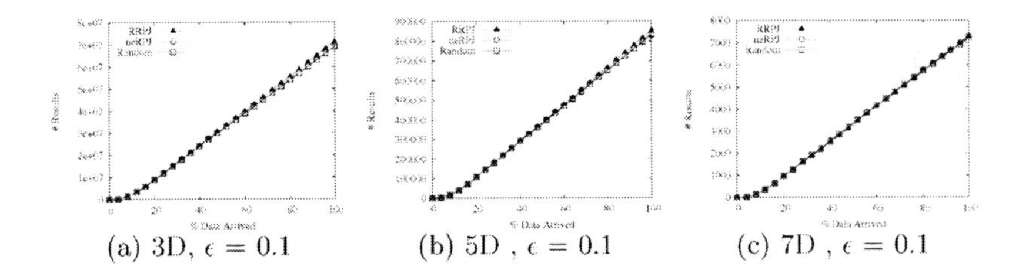

(a) 3D, $\epsilon = 0.1$ (b) 5D , $\epsilon = 0.1$ (c) 7D , $\epsilon = 0.1$

**Fig. 1.** Varying Dimension: Uniform Dataset

others. In addition, we also investigated the impact of the correlation between the two data streams on the datasets (we use two schemes used in [1] called *HARMONY* and *REVERSE*). In the *HARMONY* scheme, corresponding clusters on each stream have the same density of data. In the *REVERSE* dataset, corresponding clusters have reverse densities (according to the grid numbering). In addition, we use a third scheme in which data is reverse and arrive in a random order.

RRPJ outperforms the other methods in all cases. It is more the case with a *REVERSE* randomized dataset (figure 4a-c) than with a *REVERSE* figure (3a-c), than again with a *HARMONY* dataset (figure 2a-c). In other words, RRPJ is capable of adapting to the irregularities of the datasets distribution and arrival.

### 4.2 Checkered Data

We now consider the extreme case in which data is generated by alternating the cells in which the data falls into on each stream. In one dataset, only the even cells contain data; and in the other dataset, only the odd cells contain data. Thus, the data in the two data streams are somehow 'disjoint'. We refer to the dataset as the checkered dataset.

From figure 5, we can see that RRPJ outperforms neRPJ. This is because whenever memory is full, neRPJ first determines the cells with the lowest neRPJ values and flushes the cells in the other data stream. However, this might not be the optimal decision, since the cell that is flushed could be a cell that could contribute to a large number of results. Recall that in a high-dimensional similarity join, we do not just scan the corresponding cell, but also its immediate neighborhood. Since RRPJ determines the results for each cell, and flushes cells with the lowest $Th_i$ values (and not the cell from the other data stream), it is able to differentiate between cells that contribute to large number of results from cells that do not.

### 4.3 Non-uniform Data Within Cells

The worst-case scenario for RPJ is when the local uniformity assumption for cells does not hold. We construct such a data set by having cells where the majority of the data in one cell do not entirely 'join' with the data in the other

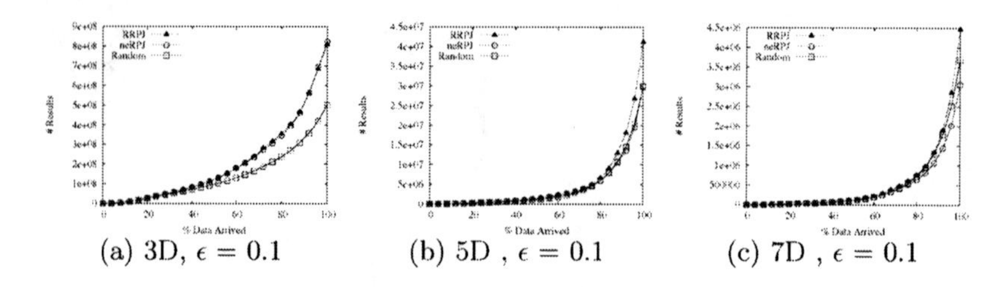


(a) 3D, $\epsilon = 0.1$ (b) 5D , $\epsilon = 0.1$ (c) 7D , $\epsilon = 0.1$

**Fig. 2.** Varying Dimension: Skewed Dataset - Harmony

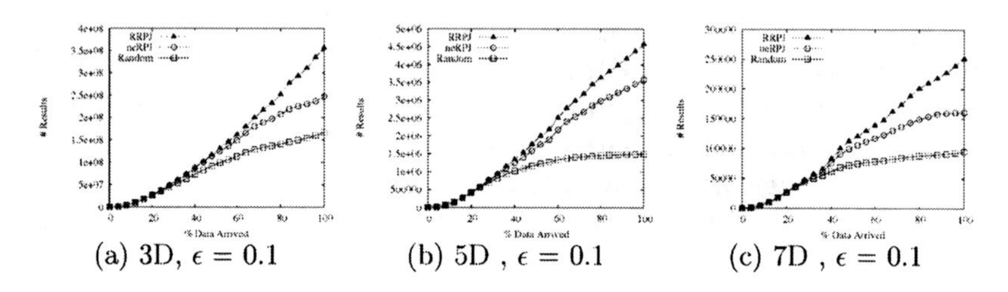


(a) 3D, $\epsilon = 0.1$ (b) 5D , $\epsilon = 0.1$ (c) 7D , $\epsilon = 0.1$

**Fig. 3.** Varying Dimension: Skewed Dataset - Reverse

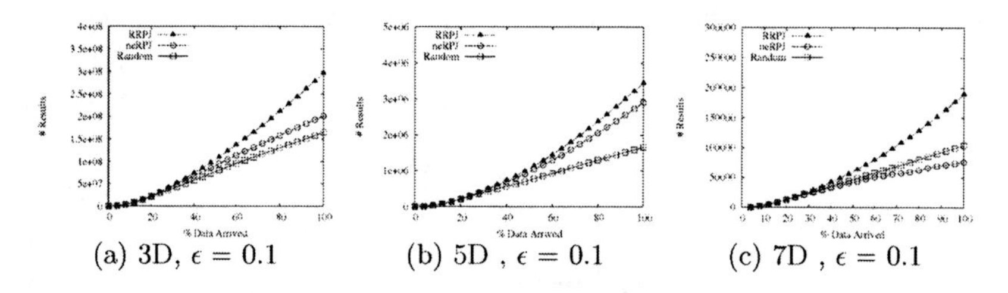


(a) 3D, $\epsilon = 0.1$ (b) 5D , $\epsilon = 0.1$ (c) 7D , $\epsilon = 0.1$

**Fig. 4.** Varying Dimension: Skewed Dataset - Reverse (Randomize arrival)

cell. We refer to this as non-uniformity within cells. We restrict the range of values for some of the dimensions, which we refer to as *non-uniform dimensions.* For each non-uniform dimension, we limit the random values generated to be in the range [0,0.5] for one dataset, and [0.6,1.0] for the corresponding data set. Given a d-dimensional dataset, we set d/2 of the dimensions to be non-uniform dimensions, and the remaining to be uniform dimensions. The results are presented in figure 6, where we observed that RRPJ performs much better than neRPJ. This is because neRPJ relies on a local uniformity assumption for the data within cells, which does not entirely hold in this worst-case scenario. In contrast, RRPJ do not suffer from this problem because it tracks the statistics on the result output of cells. In figure 6(c), we make use of $\epsilon = 0.3$ in order to produce readable figure, but verified that the result for various $\epsilon$ values are consistent.

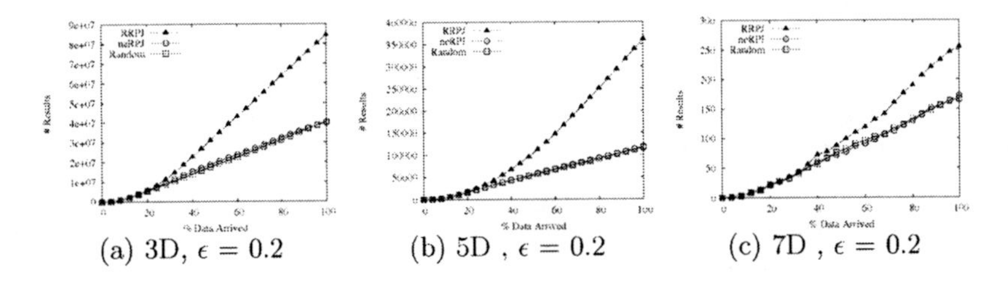


(a) 3D, $\epsilon = 0.2$ (b) 5D , $\epsilon = 0.2$ (c) 7D , $\epsilon = 0.2$

**Fig. 5.** Varying Dimension: Checkered Dataset

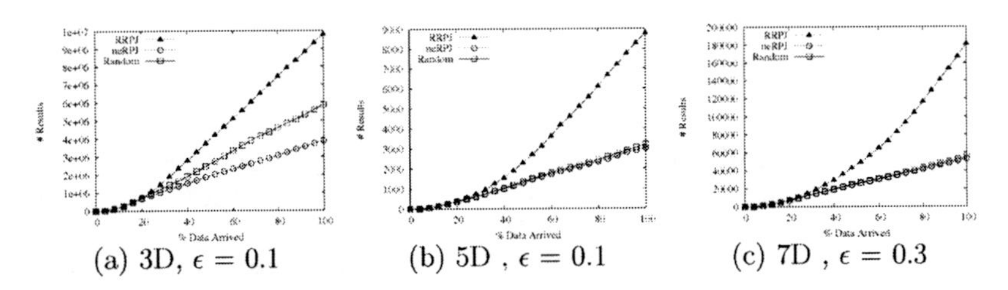


(a) 3D, $\epsilon = 0.1$ (b) 5D , $\epsilon = 0.1$ (c) 7D , $\epsilon = 0.3$

**Fig. 6.** Varying Dimension: Non-Uniform Data Within Cells

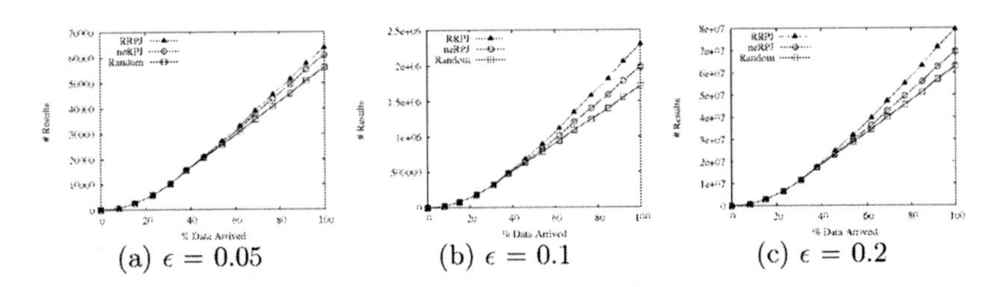


(a) $\epsilon = 0.05$ (b) $\epsilon = 0.1$ (c) $\epsilon = 0.2$

**Fig. 7.** Varying $\epsilon$: COREL Dataset, 9D

### 4.4 Real-Life Datasets

Finally, we validate the effectiveness of the proposed RRPJ algorithm for real-life datasets. In this experiment, we use the Corel (Color Moment) dataset [16]. The Corel dataset consists of 9 dimensional features for 68,040 images. We created two data streams by randomizing the order of the data for both datasets. We then perform a self-join on the data. From figure 7, we can observe that RRPJ outperforms neRPJ and Random in all cases for varying $\epsilon$. This further reinforces the advantages from using a result-rate based approach.

## 5 Conclusion

In this paper, we have proposed a novel progressive high-dimensional similarity join algorithm. The algorithm uses a result-rate based flushing strategy. It is an extension of our previous work on progressive relational equijoin [2] to the case of high-dimensional data. We have conducted an extensive performance analysis, comparing our proposed algorithm with a naive extension of RPJ [1] (a state-of-the-art progressive relational join), called neRPJ, to high-dimensional data . Using both synthetic and real-life datasets, we have shown that our proposed method, RRPJ, outperforms neRPJ by a large margin and is therefore both effective and efficient. In contrast to conventional similarity join algorithms, RRPJ can deliver results progressively and maintain a high result throughput. We are currently exploring the application of the RRPJ flushing strategy to the design of join algorithms for progressive query processing of multiple XML data streams.

## References

1. Tao, Y., Yiu, M.L., Papadias, D., Hadjieleftheriou, M., Mamoulis, N.: RPJ: Producing fast join results on streams through rate-based optimization. In: SIGMOD, pp. 371–382 (2005)
2. Tok, W.H., Bressan, S., Lee, M.-L.: RRPJ: Result-rate based progressive relational join. In: DASFAA, pp. 43–54 (2007)
3. Tok, W.H., Bressan, S., Lee, M.-L.: Progressive spatial joins. In: SSDBM, pp. 353–358 (2006)
4. Shim, K., Srikant, R., Agrawal, R.: High-dimensional similarity joins. In: ICDE, pp. 301–311 (1997)
5. Koudas, N., Sevcik, K.C.: High dimensional similarity joins: Algorithms and performance evaluation. IEEE Transactions on Knowledge and Data Engineering 12(1), 3–18 (2000)
6. Böhm, C., Braunmüller, B., Breunig, M.M., Kriegel, H.-P.: High performance clustering based on the similarity join. In: CIKM, pp. 298–305 (2000)
7. Böhm, C., Braunmüller, B., Krebs, F., Kriegel, H.-P.: Epsilon grid order: An algorithm for the similarity join on massive high-dimensional data. In: SIGMOD, pp. 379–388 (2001)
8. Kalashnikov, D.V., Prabhakar, S.: Fast similarity join for multi-dimensional data. Inf. Syst. 32(1), 160–177 (2007)
9. Guttman, A.: R-trees: A dynamic index structure for spatial searching. In: SIGMOD, pp. 47–57 (1984)
10. Berchtold, S., Keim, D.A., Kriegel, H.-P.: The x-tree: An index structure for high-dimensional data. In: VLDB, pp. 28–39 (1996)
11. Koudas, N., Sevcik, K.C.: High dimensional similarity joins: Algorithms and performance evaluation. In: ICDE, pp. 466–475 (1998)
12. Urhan, T., Franklin, M.J.: XJoin: Getting fast answers from slow and bursty networks. Technical Report CS-TR-3994, University of Maryland (1999)
13. Dittrich, J.-P., Seeger, B., Taylor, D.S., Widmayer, P.: Progressive merge join: A generic and non-blocking sort-based join algorithm. In: VLDB, pp. 299–310 (2002)
14. Mokbel, M.F., Lu, M., Aref, W.G.: Hash-merge join: A non-blocking join algorithm for producing fast and early join results. In: ICDE, pp. 251–263 (2004)
15. Wilschut, A.N., Apers, P.M.G.: Dataflow query execution in a parallel main-memory environment. In: PDIS, pp. 68–77 (1991)
16. Corel image features dataset (1999), `http://kdd.ics.uci.edu/`

# Decomposing DAGs into Disjoint Chains

Yangjun Chen

Department of Applied Computer Science
University of Winnipeg
Winnipeg, Manitoba, Canada R3B 2E9
y.chen@uwinnipeg.ca

**Abstract.** In this paper, we propose an efficient algorithm to decompose a directed acyclic graph (DAG) into chains, which has a lot of applications in computer science and engineering. Especially, it can be used to store transitive closures of directed graphs in an economical way. For a DAG *G* with *n* nodes, our algorithm needs $O(n^2 + bn\sqrt{b})$ time to find a minimized set of disjoint chains, where *b* is *G*'s width, defined to be the largest node subset *U* of *G* such that for every pair of nodes $u, v \in U$, there does not exist a path from *u* to *v* or from *v* to *u*. Accordingly, the transitive closure of *G* can be stored in $O(bn)$ space, and the reachability can be checked in $O(\log b)$ time. The method can also be extended to handle cyclic directed graphs.

## 1 Introduction

Let *G* be a DAG. A chain cover of *G* is a set *S* of disjoint chains such that it covers all the nodes of *G*, and for any two nodes *a* and *b* on a chain $p \in S$, if *a* is above *b* then there is a path from *a* to *b* in *G*. In this paper, we discuss an efficient algorithm to find a minimized *S* for *G*.

As an application of this problem, consider the transitive closure compression. Let $G(V, E)$ be a directed graph (digraph for short). Digraph $G^* = (V, E^*)$ is the reflexive, transitive closure of *G* if $(v, u) \in E^*$ iff there is a path from *v* to *u* in *G*. (See Fig. 1(a) and (b) for illustration.) Obviously, if a transitive closure (*TC* for short) is physically stored, the checking of the ancestor-descendant relationship (whether a node is reachable from another node through a path) can be done in a constant time. However, the materialization of a whole transitive closure is very space-consuming. Therefore, it is desired to find a way to compress a transitive closure, but without sacrificing too much the query time.

The idea of using disjoint chains to compress a transitive closure was first suggested by Jagadish [6], who proposed an $O(n^3)$ algorithm to decompose a DAG into a minimized set of disjoint chains by reducing the problem to a network flow problem.

Once *G* is decomposed, its transitive closure can be represented as follow.

(1) Number each chain and number each node on a chain.
(2) The *j*th node on the *i*th chain will be assigned a pair $(i, j)$ as its index.

R. Wagner, N. Revell, and G. Pernul (Eds.): DEXA 2007, LNCS 4653, pp. 243–253, 2007.

In addition, each node $v$ on the $i$th chain will be associated with a pair sequence of length $k$ - 1: $(1, j_1) \ldots (i-1, j_{i-1}) (i+1, j_{i+1}) \ldots (k, j_k)$ such that for any index $(x, y)$ $(x \neq i)$ if $y \leq j_x$ it is a descendant of $v$, where $k$ is the number of the disjoint chains. Of course, any node below $v$ on the $i$th chain is a descendant of $v$. In this way, the space overhead is decreased to $O(kn)$ (see Fig. 1(c) for illustration).

More importantly, it is very convenient to handle such a data structure in relational databases. We need only to establish a relational schema of the following form:

Node(Node_id, *label*, *label_sequence*, ...),

where *label* and *label_sequence* are utilized to accommodate the label pair and the label pair sequence associated with the nodes of $G$, respectively. Then, to retrieve the descendants of node $v$, we issue two queries as below.

```
Q1: SELECT  label              Q2:  SELECT  *
    FROM    Node                    FROM    Node
    WHERE   Node_id = v             WHERE   φ(label, label_sequence, y).
```

Let the label pair obtained by evaluating the first query $Q_1$ be $y$. Then, by evaluating the function $\phi$(label, label_sequence, $y$) in the second query $Q_2$, we will get all those nodes, whose labels are subsumed by $y$ or whose label sequences contain a label subsumed by $y$. A label pair $(x, y)$ is said to be subsumed by another pair $(x', y')$ if $x = x'$ and $y \leq y'$.

Since each label sequence is sorted according to the first element of each pair in the sequence, we need only $O(\log_2 k)$ time to check whether a node $u$ is a descendant of $v$.

As demonstrated in [13], $k$ is equal to $b$, the width of $G$, defined to be the largest node subset $U$ of $G$ such that for every pair of nodes $u, v \in U$, there does not exist a path from $u$ to $v$ or from $v$ to $u$.

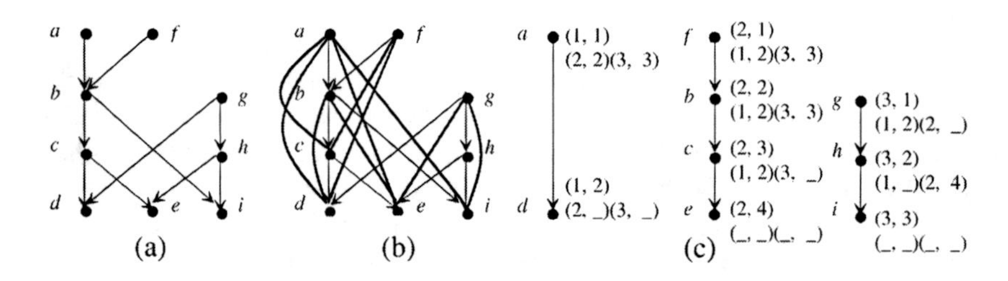


**Fig. 1.** DAG, transitive closure and graph encoding

Finally, we note that this technique can also be employed for cyclic graphs since we can collapse each *strongly connected component* (SCC) to a single node while maintaining all the reachability information (see 3.3 of [6]).

There are some other efforts to compress transitive closures. The method discussed in [4] is based on the so-called tree encoding. This method requires $O(\beta e)$ time to generate a compressed transitive closure, where $\beta$ is the number of the leaf nodes of a spanning tree covering all the nodes of $G$. $\beta$ is generally much larger than $b$. The time and space complexities of this method are bounded by O(β$n$) and O(logβ), respectively. In [12], a quite different method is proposed, using the so-called *PQ*-Encoding, by which each node is associated with an interval and a bit-vector that is of the length equal to the number of slices. A slice is just a subset of the nodes and the size of a slice is bounded by a constant. So the number of the slices must be on O($n$). Therefore, this method needs O($n^2$) space in the worst case, not suitable for a database environment.

In this paper, we propose a new algorithm for decomposing DAGs into a minimized set of disjoint chains. Its time complexity is bounded by $O(n^2 + bn\sqrt{b})$, which enables us to generate a compressed transitive closure efficiently.

The remainder of the paper is organized as follows. In Section 2, we give some basic concepts and techniques related to our algorithm. Section 3 is devoted to the description of our algorithm to decompose a DAG into chains. In Section 4, we prove the correctness of the algorithm and analyze its computational complexities. Finally, a short conclusion is set forth in Section 5.

## 2 Graph Stratification and Bipartite Graphs

Our method for finding a minimized set of chains is based on a DAG stratification strategy and an algorithm for finding a maximum matching in a bipartite graph. Therefore, the relevant concepts and techniques should be first reviewed and discussed.

### 2.1 Stratification of DAGs

Let $G(V, E)$ be a DAG. We decompose $V$ into subsets $V_1, V_2, ..., V_h$ such that $V = V_1 \cup V_2 \cup ... \cup V_h$ and each node in $V_i$ has its children appearing only in $V_{i-1}, ..., V_1$ ($i = 2, ..., h$), where $h$ is the height of $G$, i.e., the length of the longest path in $G$. For each node $v$ in $V_i$, we say, its level is $i$, denoted $l(v) = i$. We also use $C_j(v)$ ($j < i$) to represent a set of links with each pointing to one of $v$'s children, which appears in $V_j$. Therefore, for each $v$ in $V_i$, there exist $i_1, ..., i_k$ ($i_l < i$, $l = 1, ..., k$) such that the set of its children equals $C_{i_1}(v) \cup ... \cup C_{i_k}(v)$.

Such a DAG decomposition can be done in O($e$) time, by using the following algorithm, in which we use $G_1/G_2$ to stand for a graph obtained by deleting the edges of $G_2$ from $G_1$; and $G_1 \cup G_2$ for a graph obtained by adding the edges of $G_1$ and $G_2$ together. In addition, $(v, u)$ represents an edge from $v$ to $u$; and $d(v)$ represents $v$'s outdegree.

**Algorithm.** *graph-stratification*($G$)
**begin**
1. $V_1$ := all the nodes with no outgoing edges;
2. **for** $i = 1$ to $h - 1$ **do**
3. { $W$ := all the nodes that have at least one child in $V_i$;
4. **for** each node $v$ in $W$ **do**
5. {let $v_1, ..., v_k$ be $v$'s children appearing in $V_i$;
6. $C_i(v) := \{v_1, ..., v_k\}$;
7. **if** $d(v) > k$ **then** remove $v$ from $W$;};
8. $G := G/\{(v, v_1), ..., (v, v_k)\}$;

9. $d(v) := d(v) - k$;
10. $V_{i+1} := W$;
11. }
**end**

In the above algorithm, we first determine $V_1$, which contains all those nodes having no outgoing edges (see line 1). In the subsequent computation, we determine $V_2$, ..., $V_h$. In order to determine $V_i$ ($i > 1$), we will first find all those nodes that have at least one child in $V_{i-1}$ (see line 3), which are stored in a temporary variable $W$. For each node $v$ in $W$, we will then check whether it also has some children not appearing in $V_{i-1}$, which can be done in a constant time as demonstrated below. During the process, the graph $G$ is reduced step by step, and so does $d(v)$ for each $v$ (see lines 8 and 9). First, we notice that after the $j$th iteration of the out-most **for**-loop, $V_1$, ..., $V_{j+1}$ are determined. Denote $G_j(V, E_j)$ the changed graph after the $j$th iteration of the out-most **for**-loop. Then, any node $v$ in $G_j$, except those in $V_1 \cup ... \cup V_{j+1}$, does not have children appearing in $V_1 \cup ... \cup V_j$. Denote $d_j(v)$ the outdegree of $v$ in $G_j$. Thus, in order to check whether $v$ in $G_{i-1}$ has some children not appearing in $V_i$, we need only to check whether $d_{i-1}(v)$ is strictly larger than $k$, the number of the child nodes of $v$ appearing in $V_i$ (see line 7).

During the process, each edge is accessed only once. So the time complexity of the algorithm in bounded by $O(e)$.

As an example, consider the graph shown in Fig. 1(a) once again. Applying the above algorithm to this graph, we will generate a stratification of the nodes as shown in Fig. 2.

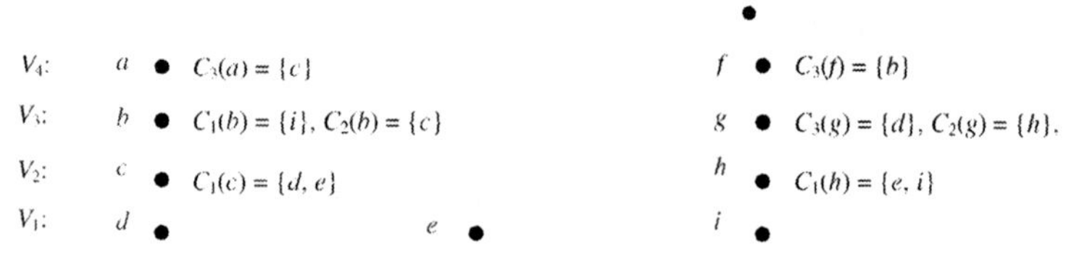

$V_4$: $a$ • $C_3(a) = \{c\}$ $f$ • $C_3(f) = \{b\}$
$V_3$: $b$ • $C_1(b) = \{i\}$, $C_2(b) = \{c\}$ $g$ • $C_3(g) = \{d\}$, $C_2(g) = \{h\}$,
$V_2$: $c$ • $C_1(c) = \{d, e\}$ $h$ • $C_1(h) = \{e, i\}$
$V_1$: $d$ • $e$ • $i$ •

**Fig. 2.** Illustration for DAG stratification

In Fig. 2, the nodes of the DAG shown in Fig. 1(a) are divided into four levels: $V_1 = \{d, e, i\}$, $V_2 = \{c, h\}$, $V_3 = \{b, g\}$, and $V_4 = \{a, f\}$. Associated with each node at each level is a set of links pointing to its children at different levels below.

## 2.2 Concepts of Bipartite Graphs

Now we restate two concepts from the graph theory, which will be used in the subsequent discussion.

**Definition 1.** (*bipartite graph* [2]) An undirected graph $G(V, E)$ is bipartite if the node set $V$ can be partitioned into two sets $T$ and $S$ in such a way that no two nodes from the same set are adjacent. We also denote such a graph as $G(T, S; E)$.
For any node $v \in G$, *neighbour*($v$) represents a set containing all the nodes connected to $v$.

**Definition 2.** (*matching*) Let $G(V, E)$ be a bipartite graph. A subset of edges $E' \subseteq E$ is called a *matching* if no two edges have a common end node. A matching with the largest possible number of edges is called a *maximum matching*, denoted as $M_G$.

Let $M$ be a matching of a bipartite graph $G(T, S; E)$. A node $v$ is said to be *covered* by $M$, if some edge of $M$ is incident with $v$. We will also call an uncovered node *free*. A path or cycle is *alternating*, relative to $M$, if its edges are alternately in $E/M$ and $M$. A path is an *augmenting path* if it is an alternating path with free origin and terminus. In addition, we will use $free_M(T)$ and $free_M(S)$ to represent all the free nodes in $T$ and $S$, respectively.

Much research on finding a maximum matching in a bipartite graph has be conducted. The best algorithm for this task is due to Hopcroft and Karp [5] and runs in $O(e \cdot \sqrt{n})$ time, where $n = |V|$ and $e = |E|$. The algorithm proposed by Alt, Blum, Melhorn and Paul [1] needs $O(n^{1.5}\sqrt{e/(\log n)})$ time. In the case of large $e$, the latter is better than the former.

## 3 Algorithm Description

The main idea of our algorithm is to construct a series of bipartite graphs for $G(V, E)$ and then find a maximum matching for each bipartite graph by using Hopcroft-Karp's algorithm. All these matchings make up a set of disjoint chains and the size of this set is minimal.

During the process, some new nodes, called *virtual nodes*, may be introduced into $V_i$ ($i = 2, ..., h$; $V = V_1 \cup V_2 \cup ... V_h$), to facilitate the computation. However, such virtual nodes will be eventually resolved to obtain the final result.

In the following, we first show how a virtual node is constructed. Then, the algorithm will be described.

We start our discussion with the following specification:

$M_i$ - the found maximum matching of $G(V_{i+1}, V_i; \boldsymbol{C}_i)$, where $\boldsymbol{C}_i = C_i(v_1) \cup ... C_i(v_k)$ and $v_l \in V_{i+1}$ ($l = 1, ..., k$).

$M_i'$ - the found maximum matching of $G(V_{i+1}, V_i'; \boldsymbol{C}_i')$, where $V_i' = V_i \cup$ {all the virtual nodes added into $V_i$} $\cup$ {all the free nodes in $C_{i-1}(v) \cup ... C_1(v)$ for $v \in V_{i+1}$}. $\boldsymbol{C}_i' = \boldsymbol{C}_i \cup$ {all the new edges incident to the virtual nodes in $V_i'$} $\cup$ {all the edges incident to the free nodes in $C_{i-1}(v) \cup ... C_1(v)$ for $v \in V_{i+1}$}.

In addition, for a graph $G$, we will use $V(G)$ to represent all its nodes and $E(G)$ all its edges.

**Definition 3.** (*virtual nodes*) Let $G(V, E)$ be a DAG, divided into $V_1, ..., V_h$ (*i.e.*, $V = V_1 \cup ... \cup V_h$). The virtual nodes added into $V_i$ ($i = 1, ..., h - 1$) are the new nodes constructed as below.

1. If $i = 1$, no virtual nodes are created for $V_1$. $V_1' := V_1$; $\boldsymbol{C}_1' := \boldsymbol{C}_1$; $M_1' := M_1$.
2. If $i > 1$, let $M_{i-1}'$ be a maximum matching of $G(V_i, V_{i-1}'; \boldsymbol{C}_{i-1}')$. Let $v$ be a free node in $free_{M_{i-1}'}(V_{i-1}')$. Let $u_1, ..., u_k$ be the covered nodes appearing in $V_{i-1}'$ such that each

$u_g$ ($g$ = 1, ..., $k$) shares a covered parent node $w_g$ (i.e., $(w_g, u_g) \in M_{i-1}$') with $v$. Let $v_{g1}$, ..., $v_{gj_g}$ be all the parents of $u_g$ in $V_i$. Construct a virtual node $v$' (to be added into $V_i$), labeled with $v[(w_1, u_1, \{v_{11}, ..., v_{1j_1}\}), ..., (w_k, u_k, \{v_{k1}, ..., v_{kj_k}\})]$. Let $o_1$, ..., $o_l$ be the nodes in $V_{i+1}$, which have at least one child node appearing in $\{v_{11}, ..., v_{1j_1}\} \cup ... \cup \{v_{k1}, ..., v_{kj_k}\}$. Establish the edges ($o_1$, $v$'), ..., ($o_l$, $v$'). Establish a virtual edge from $v$' to $v$ (so $v$ is not considered as a free node any more). Denote $V_i$' = $V_i \cup$ {all the virtual nodes added into $V_i$} $\cup$ {all the free nodes in $C_{i-1}(v) \cup ... C_1(v)$ for $v \in V_{i+1}$}. Denote $C_i$' = $C_i \cup$ {all the new edges} $\cup$ {all the edges incident to the free nodes in $C_{i-1}(v) \cup ... C_1(v)$ for $v \in V_{i+1}$}.

The following example helps for illustration.

**Example 1.** Let's have a look at the graph shown in Fig. 1(a) once again. The bipartite graph made up of $V_2$ and $V_1$, $G(V_2, V_1; C_1)$, is shown in Fig. 3(a) and a possible maximum matching $M_1$ of it is shown in Fig. 3(b).

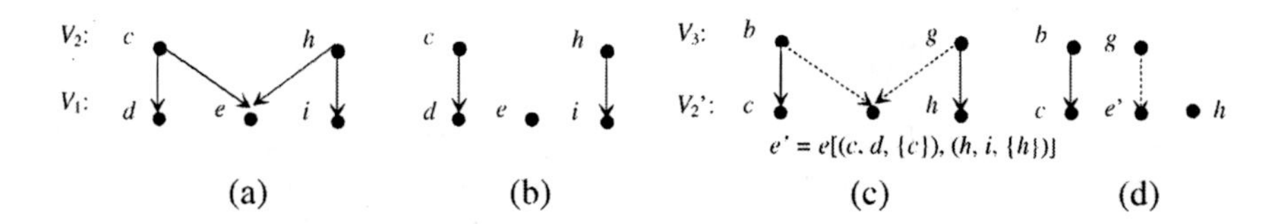


**Fig. 3.** Bipartite graphs and maximum matchings

Relative to $M_1$, we have a free node $e$.

For this free node, we will construct a virtual node $e$', labeled with $e[(c, d, \{c\}), (h, i, \{h\})]$, as shown in Fig. 3(a). In addition, two edges ($b$, $e$') and ($g$, $e$') are established according to Definition 3.'

The graph shown in Fig. 3(c) is the second bipartite graph, $G(V_3, V_2'; C_2')$. Assume that the maximum matching $M_2$' found for this bipartite graph is a graph shown Fig. 3(d).

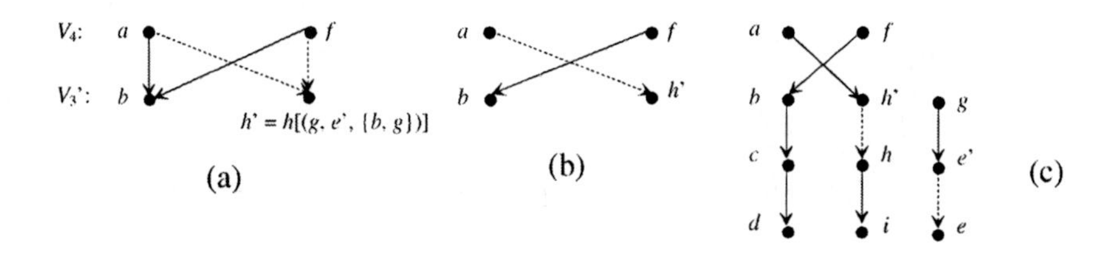


**Fig. 4.** Illustration for virtual node construction

Relative to $M_2$', $h$ is a free node, for which a virtual node $h$' = $h[(g, e', \{b, g]$ will be constructed as illustrated in Fig. 4(a). This shows the third bipartite graph, $G(V_4, V_3'; C_3')$, which has a unique bipartite graph $M_3$' shown in Fig. 4(b).

Now we consider $M_1 \cup M_2' \cup M_3'$. This is a set of three chains as illustrated in Fig. 4(c). In order to get the final result, all the virtual nodes appearing in those chains have to be resolved.

Therefore, we will have a two-phase process. In the first phase, we generate virtual nodes and chains. In the second phase, we resolve all the virtual nodes.

**Algorithm.** *chain-generation*($G$'s stratification) (*phase 1*)
input: $G$'s stratification.
output: a set of chains
**begin**
1. find $M_1$ of $G(V_2, V_1; \boldsymbol{C}_1)$; $M_1' := M_1$; $V_1' := V_1$; $\boldsymbol{C}_1' := \boldsymbol{C}_1$; $i := 2$;
2. **for** $i = 2$ **to** $h - 1$ **do**
3. { construct virtual nodes for $V_i$ according to $M_{i-1}'$;
4. let $U$ be the set of the virtual nodes added into $V_i$;
5. let $W$ be the newly generated edges incident to the virtual nodes in $V_i$;
6. $V_i' := V_i \cup U$; $\boldsymbol{C}_i' := \boldsymbol{C}_i \cup W$;
7. find a maximum matching $M_i'$ of $G(V_{i+1}, V_i'; \boldsymbol{C}_i')$;
8. }
**end**

The algorithm works in two steps: an initial step (line 1) and an iteration step (lines 2 - 8). In the initial step, we determine a $M_1$ of $G(V_2, V_1; \boldsymbol{C}_1)$. In the iteration step, we repeatedly generate virtual nodes for $V_i$ and then find a $M_i'$ of $G(V_{i+1}, V_i'; \boldsymbol{C}_i')$. The result is $M_1 \cup M_2' \cup ... \cup M_{h-1}'$.

After the chains for a DAG are generated, we will resolve all the virtual nodes involved in those chains. To do this, we search each chain top-down. Whenever we meet a virtual node $v'$ along an edge $(u, v')$, we do the following:

1. Let $v[(w_1, u_1, \{v_{11}, ..., v_{1j_1}\}), ..., (w_k, u_k, \{v_{k1}, ..., v_{kj_k}\})]$ be the label of $v'$.
2. If there exists an $i$ such that $u$ is an ancestor of some node in $\{v_{i1}, ..., v_{ij_i}\}$, do the following operations:

   (i) Replace $(w_i, u_i)$ with $(w_i, v)$.
   (ii) Remove $(u, v')$ and $v'$.
   (iii) Add $(u, u_i)$.

See the following example for a better understanding.

**Example 2.** Searching the chains shown in Fig. 4(c), we will first meet $h'$, whose label is $h[(g, e', \{b, g\})]$. Since $b$ is a descendant of $a$, we will (i) replace $(g, e')$ with $(g, h)$, (ii) remove $(a, h')$ and $h'$, and (iii) add $(a, e')$ (see Fig. 5(a) for illustration).

Next we will meet $e'$, whose label is $e[(c, d, \{c\}), (h, i, \{h\})]$. Since $c$ is a descendant of $a$, we will (i) replace $(c, d)$ with $(c, e)$, (ii) remove $(a, e')$ and $e'$, and (iii) add $(a, d)$. The result is shown in Fig. 5(b).

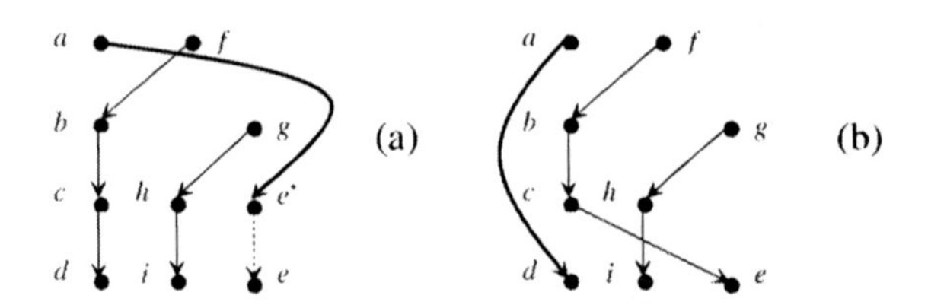


**Fig. 5.** Illustration for virtual node resolution

By this example, we should pay attention to the following properties:

(1) When we resolve $h' = h[(g, e', \{b, g\})]$, we will check whether $b$ or $g$ is a descendant of $a$. For this purpose, we need only to check $a$'s child nodes since $(a, h')$ is an edge on the chain.

(2) After $h'$ is resolved, $e' = e[(c, d, \{c\}), (h, i, \{h\})]$ becomes connected to $a$ and will be resolved next. Similarly, we will check whether $c$ or $d$ is a descendant of $a$. But we only need to check whether they are $b$'s child nodes instead of searching $G$ from $a$ again. It is because $e'$ appears in $h[(g, e', \{b, g\})]$ (showing that $e'$ is a virtual child node of $b$) and $b$ is the only child node of $a$.

In this way, we search the graph $G$ only once for resolving the virtual nodes along a chain.

The following is the formal description of this process.

**Algorithm.** *virtual-resolution*($C$) (*Phase 2*)
input: $C$ - a chain set obtained by executing the algorithm *chain-generation*.
output: a set of chains containing no virtual nodes
**begin**
1. $C' := \Phi$;
2. **while** $C$ not empty **do**
3. {choose a chain $l$ from $C$ such that the first virtual node on $l$ appears at the highest level (the tie is resolved arbitrarily);
4. let $l'$ be the chain containing no virtual node after resolving virtual nodes along $l$;
5. $C' := C' \cup \{l'\}$; $C := C/\{l'\}$;
6. }

**end**

## 4 Correctness and Computational Complexities

In this section, we prove the correctness of the algorithm and analyze its computational complexities.

**Proposition 1.** The size of the chains generated by Algorithm *chain-generation*( ) is minimum.

*Proof.* Let $S = \{l_1, ..., l_g\}$ be the set of the chains generated by *chain-generation*( ). For any chain $l_i$ and any two nodes $v$ and $u$ on $l_i$, if $v$ is above $u$, there must be a path from $v$ to $u$. By the virtual node resolution, this property is not changed. Let $S' = \{l_1', ..., l_g'\}$

be the chain set after the virtual node resolution. Then, for any $v$' and $u$' on $l_i$', if $v$' is above $u$', we have a path from $v$' to $u$'.

Now we show that $g$ is minimum.

First, we notice that the number of the chains produced by the algorithm *chain-generation* is equal to

$$N_h = |V_1| + |\,free_{M_1}(V_2)\,| + |\,free_{M_2'}(V_3)\,| + \ldots + |\,free_{M_{(h-1)}'}(V_h)\,|.$$

We will prove by induction on $h$ that $N_h$ is minimum.

Initial step. When $h = 1, 2$, the proof is trivial.

Induction step. Assume that for any DAG of height $k$, $N_k$ is minimum. Now we consider the case when $h = k + 1$:

$$N_{k+1} = |V_1| + |\,free_{M_1}(V_2)\,| + |\,free_{M_2'}(V_3)\,| + \ldots + |\,free_{M_k'}(V_{k+1})\,|.$$

If $|\,free_{M_1}(V_1)\,| = 0$, no virtual node will be added into $V_2$. Therefore, $V_2 = V_2'$. In this case,

$$N_{k+1} = |V_2| + |\,free_{M_2}(V_3)\,| + |\,free_{M_3'}(V_4)\,| + \ldots + |\,free_{M_k'}(V_{k+1})\,|.$$

In terms of the induction hypothesis, it is minimum.

If $|\,free_{M_1}(V_1)\,| > 0$, some virtual nodes are added into $V_2$ and the corresponding edges are added into $\mathbf{C}_2$. Removing $V_1$, we get the stratification of another graph $G$' (with all the leaf nodes being in $V_2$'), which is of height $k$ and has the same minimal decomposition as $G$. For $G$', the number of the chains produced by the algorithm *chain-generation* is equal to

$$N_k' = |V_2'| + |\,free_{M_2'}(V_3)\,| + |\,free_{M_3'}(V_4)\,| + \ldots + |\,free_{M_k'}(V_{k+1})\,|.$$

Let $V_2' = W_1$, $V_3 = W_2$, ..., $V_{k+1} = W_k$. We have

$$N_k' = |W_1| + |\,free_{L_1}(W_2)\,| + |\,free_{L_2'}(W_3)\,| + \ldots + |\,free_{L_{(k-1)}'}(W_k)\,|,$$

where $L_1 = M_2$' and $L_i' = M_{(i+1)}$' ($i = 2, \ldots, k - 1$).

In terms of the induction hypothesis, $N_k$' is minimum. So $N_{k+1} = N_k$' is minimum. This completes the proof.

According to [6], the number of the chains in a minimized set is equal to $b$.

In the following, we analyze the computational complexities of the algorithm.

**Lemma 1.** The time complexity of the algorithm *chain-generation* is bounded by $O(n^2 + bn\sqrt{b})$.

*Proof.* The cost for generating a virtual node $v$' for node $v$ can be divided into two parts: $cost_1$ and $cost_2$. $cost_1$ is the time spent on establishing new edges, which is bounded by $O(n^2)$ since for each actual node at most $h$ virtual nodes will be constructed and the number of the new edges incident with all these virtual nodes is bounded by $O(n)$.

$cost_2$ is the time for the edge inheritance from node $v$. It is bounded by a constant.

The time for finding a maximum matching of $G(V_{i+1}, V_i'; \mathbf{C}_i)$ is bounded by

$$O(\sqrt{|V_{i+1}| + |V_i'|} \cdot |C_i'|). \quad \text{(see [5])}$$

Therefore, the total cost of this process is

$$O(n^2) + O(\sum_{i=1}^{h-1} \sqrt{|V_{i+1}| + |V_i'|} \cdot |C_i'|) \leq O(n^2 + (\sqrt{b} \sum_{i=1}^{h-1} b \cdot |V_i|) = O(n^2 + bn\sqrt{b}).$$

**Lemma 2.** The time complexity of the algorithm *virtual-resolution* is bounded by $O(n^2)$.

*Proof.* During the process, the virtual nodes will be resolved level by level. At each level, only $O(|C_i'|)$ edges will be visited. Therefore, the total number of the visited edges is bounded by

$$O(\sum_{i=1}^{h-1} |C_i'|)) = O(n^2).$$

From Lemma 1 and Lemma 2, we have the following proposition.

**Proposition 2.** The time complexity for the whole process to decompose a DAG into a minimized set of chains is bounded by $O(n^2 + bn\sqrt{b})$.

The space complexity of the process is bounded by $O(e + bn)$ since during the execution of the algorithm *chain-generation* at most $bn$ new edges are added.

## 5 Conclusion

In this paper, a new algorithm for finding a chain decomposition of a DAG is proposed, which is useful for compressing transitive closures. The algorithm needs $O(n^2 + bn\sqrt{b})$ time and $O(e + bn)$ space, where $n$ and $e$ are the number of the nodes and the edges of the DAG, respectively; and $b$ is the DAG's width. The main idea of the algorithm is a DAG stratification that generates a series of bipartite graphs. Then, by using Hopcropt-Karp's algorithm for finding a maximum matching for each bipartite graph, a set of disjoint chains with virtual nodes involved can be produced in an efficient way. Finally, by resolving the virtual nodes in the chains, we will get the final result.

## References


[1] Alt, H., Blum, N., Mehlhorn, K., Paul, M.: Computing a maximum cardinality matching in a bipartite graph in time $O(n^{1.5}\sqrt{e/(\log n)})$ *Information Processing Letters*, 37 (1991), 237-240

[2] Asratian, A.S., Denley, T., Haggkvist, R.: Bipartite Graphs and their Applications, Cambridge University (1998)

[3] Banerjee, J., Kim, W., Kim, S., Garza, J.F.: Clustering a DAG for CAD Databases. IEEE Trans. on Knowledge and Data Engineering 14(11), 1684–1699 (1988)

[4] Chen, Y., Cooke, D.: On the Transitive Closure Representation and Adjustable Compression. In: SAC'06, April 23-27, Dijon, France, pp. 450–455 (2006)

[5] Hopcroft, J.E., Karp, R.M.: An $n^{2.5}$ algorithm for maximum matching in bipartite graphs. SIAM J. Comput. 2, 225–231 (1973)

[6] Jagadish, H.V.: A Compression Technique to Materialize Transitive Closure. ACM Trans. Database Systems 15(4), 558–598 (1990)

[7] Keller, T., Graefe, G., Maier, D.: Efficient Assembly of Complex Objects. In: Proc. ACM SIGMOD conf. Denver, Colo., pp. 148–157 (1991)

[8] Kuno, H.A., Rundensteiner, E.A.: Incremental Maintenance of Materialized Object-Oriented Views in MultiView: Strategies and Performance Evaluation. IEEE Transactions on Knowledge and Data Engineering 10(5), 768–792 (1998)

[9] Teuhola, J.: Path Signatures: A Way to Speed up Recursion in Relational Databases. IEEE Trans. on Knowledge and Data Engineering 8(3), 446–454 (1996)

[10] Wang, H., Meng, X.: On the Sequencing of Tree Structures for XML Indexing. In: Proc. Conf. Data Engineering, Tokyo, Japan, pp. 372–385 (April 2005)

[11] Zhang, C., Naughton, J., DeWitt, D., Luo, Q., Lohman, G.: On Supporting Containment Queries in Relational Database Management Systems. In: Proc. of ACM SIGMOD Intl. Conf. on Management of Data, California (2001)

[12] Zibin, Y., Gil, J.: Efficient Subtyping Tests with PQ-Encoding. In: Proc. of the 2001 ACM SIGPLAN conf. on Object-Oriented Programming Systems, Languages and Application, Florida, pp. 96–107 (October 14-18, 2001)

[13] Dilworth, R.P.: A decomposition theorem for partially ordered sets. Ann. Math. 51, 161–166 (1950)

# Evaluating Top-k Skyline Queries over Relational Databases

Carmen Brando, Marlene Goncalves, and Vanessa González

Universidad Simón Bolívar, Departamento de Computación y TI, Apartado 89000
Caracas 1080-A, Venezuela
{carmen,mgoncalves,vgonzalez}@ldc.usb.ve

**Abstract.** Two main languages have been defined to allow users to express their preference criteria: Top-k and Skyline. Top-k ranks the top $k$ tuples in terms of a user-defined score function while Skyline identifies non-dominated tuples, i.e. such tuples that does not exists a better one in all user criteria. A third language, Top-k Skyline, integrates them. One of the drawbacks of relational engines is that they do not understand the notion of preferences. However, some solutions for Skyline and Top-k queries have been integrated into relational engines. The solutions implemented outside the core query engine have lost the advantages of true integration with other basic database query types. To the best of our knowledge, none of the existing engines supports Top-k Skyline queries. In this work, we propose two evaluation algorithms for Top-k Skyline which were implemented in PostgreSQL, and we report initial experimental results that show their properties.

## 1 Introduction

Currently, many applications, such as decision support or customer information systems, may take advantage of preference queries in order to find best answers. Thus, in the case of a decision support system, a possible query could be to determine the best customers who have made many purchases and have little or no complaints. In this example, two user criteria have been defined: many purchases and few complaints. In a SQL query, these criteria may be specified using an ORDER BY clause where the number of purchases is sorted ascendingly, and the number of complaints is sorted descendingly. Additionally, the relational engine must sort the answers and the user must discard all irrelevant data. Thus, the number of answers to be analyzed by the user might be enormous and declarative query languages like SQL might not be the most suitable ones to evaluate user preferences.

Hence, SQL was extended to consider user criteria and for this purpose, two preference languages were defined: Top-k and Skyline. The first, produces the top $k$ tuples based on a user-defined score function that induces a total order. The second, identifies non-dominated tuples based on a multicriteria function defined by the user. A tuple dominates another one if it is as good or better than the other in all criteria and better in at least one criterion. The multicriteria function,

R. Wagner, N. Revell, and G. Pernul (Eds.): DEXA 2007, LNCS 4653, pp. 254–263, 2007.

that seeks to maximize or minimize criteria, induces a partially ordered set and, in consequence, there are several optimal answers.

Top-k and Skyline are two different ways to solve a preference query. On one hand, Top-k requires that a score or weight function be defined and this might not be natural for all users. On the other hand, Skyline retrieves tuples where all criteria are equally important and a score function cannot be defined. However, the number of Skyline answers may be smaller than required by the user, who needs at least $k$. Given that Skyline does not discriminate its answers in the same way as Top-k does and that Top-k does not allow to solve queries where a score function can not be defined a priori, Top-k Skyline was defined as a unified language to integrate them. Top-k Skyline allows to get exactly the top $k$ from a partially ordered set stratified into subsets of non-dominated tuples. The idea is to partition the set into subsets (strata) consisting of non-dominated tuples and to produce the Top-k of these partitions.

Many solutions have been given for Skyline and Top-k, being the best those that integrate them into a relational engine because these avoid performance degradation. Performance improvement is due to the query evaluation being done inside the database system, instead of retrieving tuple by tuple in a midleware application and then, computing the Skyline or Top-k. Regarding that these solutions have demonstrated to be successful, we propose to integrate Top-k Skyline queries into a relational engine. To the best of our knowledge, none of the existing engines supports Top-k Skyline queries. In this work, we propose two evaluation algorithms for Top-k Skyline which were implemented in PostgreSQL, and we report initial experimental results that show their properties.

Finally, this paper comprises five sections. In Section 2, we introduce a motivating example. In Section 3, we briefly describe Skyline and Top-k as Related Work. In Section 4, we present the definition for Top-k Skyline. In section 5, we describe our solutions for Top-k Skyline queries. In Section 6, we report our initial experimental results. And, finally, in Section 7, the concluding remarks and future work are pointed out.

## 2 Motivating Example

Lets consider that a government program runs a scholarship contest. For this purpose the applicants must submit their resumes, providing information on their academic and professional performance and also an application form with data concerning their annual income. The summarized information is organized in the relational table:

*Candidate(name, degree, GPA, publications, income),*

where *degree* refers to the last academic degree achieved, *GPA* to the corresponding GPA in a scale from 1 to 5, *publication* represents the number of scientific publications writen, and *income* relates to the annual income in euros (€). Table 1 shows some tuples for this relation.

The resulting candidates with best academic merits must include the ones with the highest degree, the greater number of publications and the best GPA. Each

**Table 1.** Candidate relation

| name | degree | GPA | publications | income |
|---|---|---|---|---|
| Sue Smith | MsC | 3.6 | 3 | 25,000 |
| Hansel Bauer | PhD | 4.0 | 8 | 30,000 |
| Jamal Jones | MsC | 4.2 | 2 | 25,000 |
| Kim Tsu | PhD | 4.3 | 3 | 27,500 |
| María Martínez | BEng | 4.5 | 1 | 33,000 |

of these criteria are equally important, making score functions inappropriate for this selection. All the applicants that satisfy these criteria should make finalists. Thus, a candidate can only be chosen to win a scholarship if and only if there is no other candidate with higher degree, number of publications and GPA. Thus, the finalists are Hansel Bauer, Kim Tsu and María Martínez. Sue Smith and Jamal Jones, are disregarded given they have lower degrees, GPAs and number of publications than any other.

Additionally, the government program might have to choose a limited number of winners, say $k = 2$. Under these new circumstances, it is necessary to define a new criterion to discriminate between the finalists in order to only yield $k$. Therefore, the awarded students will be the ones that better qualify given a score function defined by the panel of judges, e.g. a score function defined on their income to aid the candidates with the lowest ones. From the finalists, the judges select the new scholarship winners: Hansel Bauer and Kim Tsu, because María Martínez has higher income than the other two.

We have now intuitively evaluated a preference query. On one side, *Skyline* determined the finalists without restricting the result to exactly $k$ tuples. On the other hand, *Top-k* returned the $k$ winners from the finalists with the lowest income. If we were only to apply the score function to the candidates, without the previous *Skyline*, results could differ, because Top-k might favor some candidates that could be dominated by others and these are considered false results under Skyline criteria. Note that if we apply only Skyline the results could be incomplete when the number of Skyline answers is lower than $k$. Neither Skyline nor Top-k can solve this kind of problem, it is then necessary a solution such as Top-k Skyline for the problem of selecting student scholarship winners. With Top-k Skyline, the answers are guaranteed to be sound and complete for this case.

## 3 Related Work

Several algorithms have been proposed in order to identify the Skyline in relational database systems: Divide and Conquer extension, Block-Nested-Loops (BNL), Sort-Filter-Skyline (SFS) and LESS (Linear Elimination Sort for Skyline). All of them [9] scan the entire table and compute the Skyline. Also, progressive Skyline algorithms are introduced in [19], [14], [18], meanwhile algorithms for distributed systems are presented in [2], [3], [15], [12].

On the other hand, Top-k solutions try to avoid probing a user-defined score function on all of the tuples and to stop as early as possible. One of the first approaches was done by Carey and Kossman [5], [6], they proposed a new SQL operator called STOP AFTER $k$ that indicates how many objects are required. Then, some algorithms were defined that focus on the problem of minimizing the number of probes [8], [17], [1], [4], [7], [16], [13].

Recently, solutions for the combination of Skyline and Top-k have been defined [3], [10], [15]. In general, these solutions calculate the first stratum or Skyline with some sort of post-processing. None of these solutions identify the $k$ best answers considering situations like the one described in the previous section.

## 4 Top-k Skyline

Given a set $T = \{t_1, \ldots, t_n\}$ of database tuples, where each tuple $t_i$ is characterized by $p$ attributes $(A_1, \ldots, A_p)$; $r$ score functions $s_1, \ldots, s_r$ defined over some of those attributes, where $s_i : O \to [0, 1]$; a combined score function $f$ defined over combinations of the score functions $s_1, \ldots, s_r$ that induces a total order of the tuples in T; and a multicriteria function $m$ defined also over some of the score functions $s_1, \ldots, s_r$, which induces a partial order of the tuples in T. We define Strata according to multicriteria function $m$ through the recursion presented in Definition 1. For simplicity, we suppose that the scores related to the multicriteria function are maximized.

*Definition 1a (Base Case: First Stratum $R_1$ or Skyline)*

$$R_1 = \left\{ \begin{array}{r} t_i \in T / \neg \exists t_j \in T : (s_1(t_i) \leq s_1(t_j) \wedge \cdots \wedge s_r(t_i) \leq s_r(t_j) \\ \wedge \exists q \in \{1, ..., r\} : s_q(t_i) < s_q(t_j)) \end{array} \right\}$$

*Definition 1b (Inductive Case: Stratum $R_i$)*

$$R_i = \left\{ \begin{array}{l} t_l \in T / t_l \notin R_{i-1} \wedge \neg \exists t_u \in (T - \cup_{j=1}^{i-1} R_j) : \\ \qquad (s_1(t_l) \leq s_1(t_u) \wedge \cdots \wedge s_r(t_l) \leq s_r(t_u) \\ \qquad \wedge \exists q \in \{1, ..., r\} : s_q(t_l) < s_q(t_u)) \end{array} \right\}$$

These two definitions establish that the tuples comprising each stratum will only be dominated by others in strata prior to their own. An efficient solution to the Top-k Skyline problem should avoid building all strata. In fact, it should only create the necessary ones. A Stratum $R_i$ of $R =< R_1, \ldots, R_n >$ is necessary if and only if exist strata $R_1, \ldots, R_i, \ldots, R_v$, where $v \leq n$ and $v$ is the minimum number of strata that satisfy $| \cup_{i=1}^{v} R_i| \geq k$. On the other hand, it should only include the tuples from the last necessary stratum with the highest score values until there are $k$ tuples in the answer. The answer for a relational Top-k Skyline query includes all the tuples in the strata $R_1, \ldots, R_{v-1}$ (Previous Necessary Strata), plus those in $R_v$ (Last Necessary Stratum) with the highest scores in $f$ required to complete $k$ tuples. Thus, we define Previous Necessary Strata and Last Necessary Stratum in Definition 2.

*Definition 2.1 (Previous Necessary Strata $R_{ps}$)*

$$R_{ps} = \{\cup_{i=1}^{v-1} R_i / | \cup_{i=1}^{v} R_i| \geq k > | \cup_{i=1}^{v-1} R_i|\}$$

*Definition 2.2 (Last Necessary Stratum $R_{lt}$)*

$$R_{lt} = \{R_v / | \cup_{i=1}^{v} R_i| \geq k > | \cup_{i=1}^{v-1} R_i|\}$$

Finally, the conditions to be satisfied by the answers of a relational Top-k Skyline query are given in Definition 3.

*Definition 3 (Relational Top-k Skyline TKS)*

$$TKS = \{ t_i \in T / t_i \in R_{ps} \vee (t_i \in R_{lt} \wedge \neg \exists^{k-|R_{ps}|} t_j \in R_{lt} : (f(t_j) > f(t_i))) \}$$

## 5 Our Proposed Solutions for Top-k Skyline

Two kinds of Skyline algorithms have been proposed in relational databases. The first kind scans the entire input and the second does not necessarily scans all the tuples, because it is index-based. In this work, we do not regard index-based algorithms, primarily considering that efficient access to data does not affect performance of a Skyline query as much as the multicriteria function evaluations do; secondly, although the Skyline can be precomputed through indexes, the Skyline of a set of attributes can not be calculated from the skylines of subsets of attributes, and viceversa; thirdly, the preceding option may be invalid when a new tuple is inserted into the database such that it dominates some tuples from the index and, in consequence, the entire index must be recalculated; finally, Skyline might be calculated from a set of materialized tuples -instead of base tables- and the index can only be applied to base tables.

Block-Nested-Loops (BNL), Sort-Filter-Skyline (SFS) and LESS (Linear Elimination Sort for Skyline) are three relevant algorithms for Skyline computation in relational databases because of their performance [9]. The BNL algorithm scans the entire table while it maintains a window of non-dominated tuples, which could be replaced by any other tuple that is seen later on. In Algorithm 1, we present a BNL extension for Top-k Skyline queries in a relational setting. The Extended Block-Nested-Loops (EBNL) algorithm receives a relation $R$, a multicriteria function $m$ and a combined score function $f$ as input and produces the Top-k Skyline tuples in terms of $m$ and $f$. The iteration corresponding to steps 5 through 30, calculates the necessary strata. Meanwhile, the one from 7 to 27 corresponds to BNL, with the difference that here, dominated tuples are not discarded, instead, they are stored into a temporary file $R1$, so they can be used to determine the next stratum, if necessary. If the first stratum contains $k$ tuples (condition verified on step 5) the algorithm stops, else the input set is replaced with the temporary file $R_1$ (step 29) and hence, the necessary strata are built one at a time, without partitioning all the data, until there are $k$ or more Top-k Skyline tuples. In steps 7 through 27, BNL algorithm is executed without discarding dominated tuples. When a tuple $p$ is read from the input set (steps 9-21), $p$ is compared against all the tuples in the window (steps 10-19) and: if $p$ is dominated by any tuple in the window, then $p$ is inserted into a temporal file $R_1$; else, if $p$ dominates any tuples in the window, these dominated tuples are removed from the window and inserted into a temporary file $R_1$, and $p$ is inserted into the window; and if $p$ is non-dominated, then it is inserted into the window. If, in any of the mentioned situations where $p$ must enter the window, there is not enough room in it, $p$ is inserted into another temporary file $R_2$ in order to be processed in the next iteration of the algorithm. Finally, if the temporary file $R_2$ is not empty, the input set is replaced with it to resume computing the current stratum (steps 22-24).

**Algorithm 1.** Extended Block-Nested-Loops Algorithm

```
 1: INPUT: R: relation; m: multicriteria function; f: combined score function.
 2: OUTPUT: Top-k Skyline tuples.
 3: Initialize i ← 1; count ← 0;
 4: Create a window w of incomparable tuples in main memory;
 5: while count < k and exist tuples in R do
 6:    Initialize P_i ← ∅; continue ← true
 7:    while (continue) do
 8:       Get the first tuple t from R;
 9:       while exist tuples in R do
10:          if some tuple t_1 from w dominates t then
11:             t is inserted into the temporal table R_1;
12:          else if t dominates some tuples from w then
13:             insert t into w;
14:             delete dominated tuples from w and insert them into R_1;
15:          else if no tuple t_1 from w dominates t and there is enough room in w then
16:             t is inserted into the window w;
17:          else if no t_1 from w dominates t and there is not enough room in w then
18:             t is inserted into a temporal table R_2;
19:          end if
20:          Get the next tuple t from R;
21:       end while
22:       if exist tuples in R_2 then
23:          R ← R_2;
24:       else
25:          continue ← false;
26:       end if
27:    end while
28:    Evaluate f for all tuples in w; copy tuples from w to P_i;
29:    count ← count + size(P_i); i ← i + 1; R ← R_1;
30: end while
31: return Top-k Skyline tuples;
```

Similarly, we extended SFS for Top-k Skyline computing in relational database contexts. SFS could be regarded as a BNL variant, since it only requires a previous sorting step based on a topological order compatible with the Skyline criteria and does not need window tuple replacement like BNL does (steps 12-14 of Algorithm 1), because of the mentioned initial topological sort; for further details see [9]. Steps 10 through 19 of Algorithm 1 are replaced by steps 1 through 7 of Algorithm 2. Additionally, the first statement that SFS must execute is a sorting on the input, based on the topological order compatible with the multicriteria function.

**Algorithm 2.** A portion of Extended Sort-Filter-Skyline Algorithm

1: **if** some tuple $t_1$ from $w$ dominates $t$ **then**
2: $t$ is inserted into the temporal table $R_1$;
3: **else if** no tuple $t_1$ from $w$ dominates $t$ and there is enough room in $w$ **then**
4: $t$ is inserted into the window $w$;
5: **else if** no $t_1$ from $w$ dominates $t$ and there is not enough room in $w$ **then**
6: $t$ is inserted into a temporal table $R_2$;
7: **end if**

Finally, Algorithm LESS initially sorts tuples as SFS does, but presents two improvements over it: in the first ordering phase, it uses an elimination-filter window to discard dominated tuples quickly and it combines the last phase of the sort algorithm with the Skyline filter phase of SFS to eliminate remaining dominated tuples. For our Top-k Skyline problem, LESS is not easily extensible for the stratification of data because the Skyline order would only be profited by the first stratum, while the tuples that are inserted into a temporary file to be used in the determination of the following strata become randomly ordered, and multiple sorting phases would be necessary to discard the dominated tuples of each stratum.

## 6 Experimental Study

Our experimental study was performed on **PostgreSQL 8.1.4**. We have extended PostgreSQL to include a logical operator for Top-K Skyline. This operator has two physical implementations, namely EBNL and ESFS, and is evaluated after all the relational operators, i.e. scan, join, sort, etc. This is due to the semantics of Top-k Skyline queries. Since our PostgreSQL extension does not optimize these queries, only minor changes to the parser, rewriter and optimizer were necessary in order to integrate this new operator.

This study consisted of experiments running over a relational table with 25,000 tuples. The table contains an identifier and six real number columns that represent the scores. Values of numeric columns vary from 0 to 1. The attribute values were generated following a uniform data distribution.

The algorithms were executed on a Sun Fire V240 with 2 UltraSPARC IIIi processors of 1503MHz, 2 GB of memory and an Ultra160 SCSI disk of 146 GB, running SunOS 5.10.

Ten queries were generated randomly, characterized by the following properties: (a) there is only one table in the FROM clause; (b) the attributes in the multicriteria function and combined score function are chosen randomly from the attributes of the table using a uniform distribution; (c) the optimization type for each attribute of the multicriteria function is selected randomly considering MIN and MAX directives; (d) the number of attributes of the multicriteria function is two, four and six; and (e) $k$ corresponds to 3% of the data size.

In this experimental study, the number of multicriteria function evaluations and the time taken by each algorithm were measured. Figure 1 reports the average of multicriteria function evaluations (x$10^{-6}$) and average time (in x$10^{-4}$ miliseconds) used by each algorithm. Average time is the average of ten queries ran against the table with data generated according to a uniform distribution.

We have observed that EBNL requires more multicriteria function evaluations than ESFS. This could be attributed to the benefits gained through sorting as a first step in ESFS. Besides, EBNL is more efficient on smaller Skylines and the strata of the experimental queries were not small. Also, EBNL behavior deteriorates as the dimensions grow, because strata size increases. On the other hand, ESFS remains stable and it is not affected by strata size.

Finally, EBNL requires more time than ESFS does. This is because of the number of multicriteria function evaluations. Additionaly, ESFS time is not affected by the sorting phase because it reduces the number of multicriteria function evaluations.

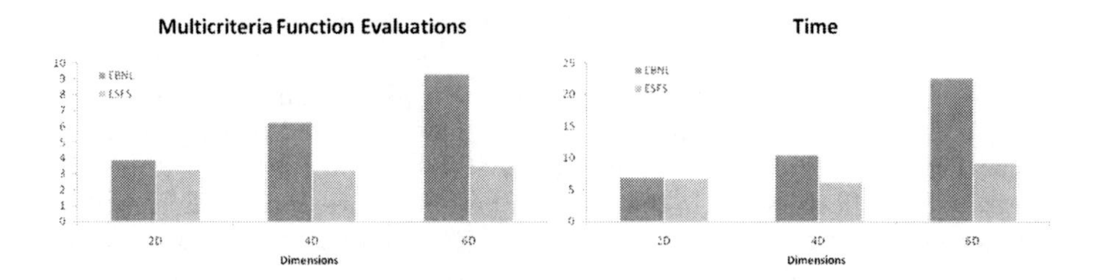


**Fig. 1.** Uniform data

Similarly, we ran experiments over a database that contains real web data. We created a table with information from Zagat Survey Guides [20]. This study consisted of experiments running over a relational table with, approximately, 16,876 tuples. The table contained an identifier and five real number columns that represent the scores. Values of numeric columns vary from 0 to 30. Any given column may have duplicate values.

Five queries were generated randomly, characterized by the following properties: (a) there is only one table in the FROM clause; (b) the attributes in the multicriteria function and combined score function are chosen randomly from the attributes of the table using a uniform distribution; (c) the optimization type for each attribute of the multicriteria function is selected randomly considering

MIN and MAX directives; (d) the number of attributes of the multicriteria function is four; (e) $k$ corresponds to 1%, 3% and 5% of the data size.

For this study, the number of multicriteria function evaluations (x$10^{-6}$) and the time (x$10^{-3}$ miliseconds) taken by each algorithm were measured. Figure 2 reports the results of the experiments for the real data. Similarly to uniform data, the number of multicriteria function evaluations is a little lower and requires less time for the ESFS algorithm.

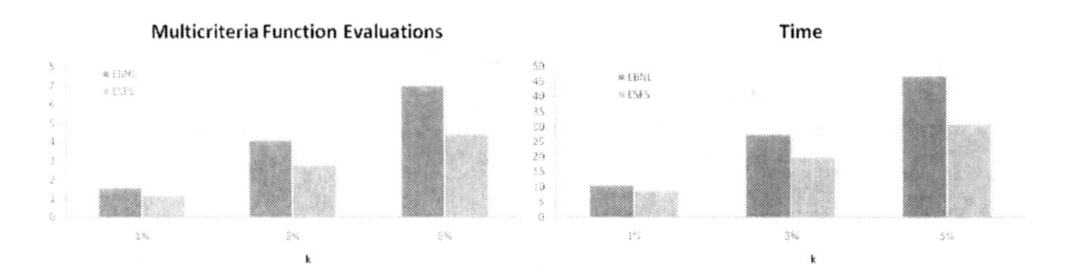


**Fig. 2.** Real data

## 7 Conclusions and Future Work

Top-k Skyline queries determine exactly the Top-k considering a score function from a partially ordered set stratified into subsets of non-dominated tuples in terms of a multicriteria function. Existing algorithms cannot solve Top-k Skyline problems. On one hand, Skyline algorithms do not restrict the result to exactly $k$ tuples. On the other hand, Top-k algorithms might return results that would be false under Skyline criteria. Therefore, we have extended two algorithms for evaluating Top-k Skyline queries in relational databases. Both algorithms were integrated into PostgreSQL in the quest for better performance. Both algorithms are complete and build less strata than a naive solution. Initial experimental results show that ESFS performs less multicriteria function evaluations and requires less evaluation time than EBNL.

PostgreSQL does not optimize Top-k Skyline queries, so it does not benefit from full integration with the relational engine and fails to take advantage of the interesting orders that could be present in the data. In the future, we plan to extend PostgreSQL optimizer so that it considers Top-k Skyline queries when calculating costs and planning the execution.

Also, EBNL and ESFS could be more efficient if they would not require one table scan per stratum, this should be discussed in the future. Finally, an asymptotic complexity analysis should be performed to compare the proposed algorithms with the straightforward implementation of Top-k skyline.

## Acknowledgments

This work was supported by FONACIT under Project G-2005000278.

## References


1. Balke, W-T., Güntzer, U., Kiebling, W.: Towards Efficient Multi-Feature Queries in Heterogeneous Environments. In: Proceedings of the IEEE International Conference on Information Technology: Coding and Computing (ITCC), pp. 622–628 (April 2001)
2. Balke, W-T., Güntzer, U., Zheng, J.: Efficient Distributed Skylining for Web Information Systems. In: Bertino, E., Christodoulakis, S., Plexousakis, D., Christophides, V., Koubarakis, M., Böhm, K., Ferrari, E. (eds.) EDBT 2004. LNCS, vol. 2992, pp. 256–273. Springer, Heidelberg (2004)
3. Balke, W-T., Güntzer, U.: Multi-Objective Query Processing for Database Systems. In: Proceedings of the International Conference on Very Large Databases (VLDB), pp. 936–947 (September 2004)
4. Bruno, N., Gravano, L., Marian, A.: Evaluating Top-k Queries over Web-Accessible Databases. In: Proceedings of International Conference on Data Engineering (ICDE), vol. 29(4), pp. 319–362 (2002)
5. Carey, M., Kossman, D.: On saying Enough Already! in SQL. In: Proceedings of the ACM SIGMOD Conference on Management of Data, pp. 219–230 (May 1997)
6. Carey, M., Kossman, D.: Reducing the Braking Distance of a SQL Query Engine. In: Proceedings of VLDB, pp. 158–169 (August 1998)
7. Chang, K., Hwang, S-W.: Optimizing Access Cost for Top-k Queries over Web Sources: A Unified Cost-Based Approach. Technical Report UIUCDS-R-2003-2324, University of Illinois at Urbana-Champaign (March 2003)
8. Fagin, R.: Combining Fuzzy Information from Multiple Systems. Journal of Computer and System Sciences (JCSS) 58(1), 216–226 (1996)
9. Godfrey, P., Shipley, R., Gryz, J.: Maximal Vector Computation in Large Data Sets. In: Proceedings of VLDB, pp. 229–240 (2005)
10. Goncalves, M., Vidal, M.E.: Preferred Skyline: A Hybrid Approach Between SQLf and Skyline. In: Andersen, K.V., Debenham, J., Wagner, R. (eds.) DEXA 2005. LNCS, vol. 3588, pp. 375–384. Springer, Heidelberg (2005)
11. Goncalves, M., Vidal, M.E.: Top-k Skyline: A Unified Approach. In: Proceedings of OTM (On the Move) 2005 PhD Symposium, pp. 790–799 (2005)
12. Huang, Z., Jensen, C.S., Lu, H., Ooi, B.C.: Skyline Queries Against Mobile Lightweight Devices in MANETs. In: Proceedings of ICDE, pp. 66–77 (2006)
13. Ilyas, I.F., Aref, W.G., Elmagarmid, A.K.: Supporting Top-k Join Queries in Relational Databases. In: Proceedings of VLDB, pp. 754–765 (2003)
14. Kossman, D., Ransak, F., Rost, S.: Shooting Stars in the Sky: An Online Algorithm for Skyline Queries. In: Proceedings of VLDB, pp. 275–286 (2002)
15. Lo, E., Yip, K., Lin, K-I., Cheung, D.: Progressive Skylining over Web-Accessible Databases. Journal of Data and Knowledge Engineering 57(2), 122–147 (2006)
16. Natsev, A., Chang, Y-CH., Smith, J.R., Li, CH.-S., Vitter, J.S.: Supporting Incremental Join Queries on Ranked Inputs. In: Proceedings of VLDB, pp. 281–290 (2001)
17. Nepal, S., Ramakrishnan, M.V.: Query Processing Issues in Image (Multimedia) Databases. In: Proceedings of ICDE, pp. 22–29 (1999)
18. Papadias, D., Tao, Y., Fu, G., Seeger, B.: Progressive Skyline Computation in Database Systems. ACM Transactions Database Systems 30(1), 41–82 (2005)
19. Tan, K-L., Eng, P-K., Ooi, B.C.: Efficient Progressive Skyline Computation. In: Proceedings of VLDB, pp. 301–310 (2001)
20. Zagat Survey Guides: available at `http://www.zagat.com`

# A P2P Technique for Continuous k-Nearest-Neighbor Query in Road Networks

Fuyu Liu, Kien A. Hua, and Tai T. Do

School of EECS, University of Central Florida, Orlando, FL, USA
{fliu,kienhua,tdo}@cs.ucf.edu

**Abstract.** Due to the high frequency in location updates and the expensive cost of continuous query processing, server computation capacity and wireless communication bandwidth are the two limiting factors for large-scale deployment of moving object database systems. Many techniques have been proposed to address the server bottleneck including one using distributed servers. To address both of the scalability factors, P2P computing has been considered. These schemes enable moving objects to participate as a peer in query processing to substantially reduce the demand on server computation, and wireless communications associated with location updates. Most of these techniques, however, assume an open-space environment. In this paper, we investigate a P2P computing technique for continuous kNN queries in a network environment. Since network distance is different from Euclidean distance, techniques designed specifically for an open space cannot be easily adapted for our environment. We present the details of the proposed technique, and discuss our simulation study. The performance results indicate that this technique can significantly reduce server workload and wireless communication costs.

## 1 Introduction

With the advances in wireless communication technology and advanced positioning systems, a variety of location based services become available to the public. Among them, one important service is to continuously provide *k-nearest-neighbor* (*k*NN) search for a moving object. Early research effort has focused on moving query over static points of interest. Recently, interest has been shifted to monitoring moving queries over moving objects, e.g., "Give me the five nearest BMW cars while I am driving on Colonial Drive." This new type of query, demanding constant updates from moving objects to keep the query results accurate, raises a great challenge.

A simple mobile query processing system consists of a centralized server and a large number of moving objects. There are two scalability issues for such systems: (1) query processing cost, and (2) location update cost. Addressing the first issue has been the focus of the majority of the existing work [2, 4, 5, 6, 7, 12, 13, 16, 17]. These researches focus on query processing techniques and do not worry about the communication cost associated with location updates. To address the second issue, namely update cost, using distributed servers has been proposed [14] to leverage the

R. Wagner, N. Revell, and G. Pernul (Eds.): DEXA 2007, LNCS 4653, pp. 264–276, 2007.

aggregate bandwidth of the servers. Another interesting idea for reducing location updates is to use *safe regions* [8, 15] or *thresholds* [19], where an object moving within a *safe region* or a *threshold* does not need to update its location. To address both issues, i.e., expensive query processing cost and intensive location updates, *Peer-to-peer* (P2P) techniques were investigated in [1, 3, 9, 20]. In these schemes, each moving object participates in query processing as a peer by monitoring nearby queries and updating their result if the object's new location affects the query results. The benefits of this strategy are twofold. First, server computation is no longer a bottleneck (i.e., first scalability issue); and second, moving objects need to update their location much less frequently (i.e., second scalability issue).

Most P2P solutions [1, 3, 20] assume an open space environment, where the distance between two objects is the straight line distance between them. In real-life scenarios, many moving objects (e.g., cars) are restricted to move on a network (e.g., road network). Since the distance between two objects in a network is defined as the shortest network distance between them, techniques developed specifically for an open space environment cannot be easily extended to a road network. A more recent P2P technique has been proposed in [9] for dynamic range queries over a network.

In this paper, we focus on *k*NN queries over road networks. Unlike range queries, there are no fixed ranges for *k*NN queries and as objects move around, the ranges constantly change. Therefore, new definitions and new techniques must be developed to address the challenge. We solve the problem by proposing an efficient P2P solution. In our approach, each moving object monitors queries in the neighboring road segments, and will update a query result maintained on a server if the object becomes one of the *k*NN or is no longer one of the *k*NN of the affected query. Besides saving server computation costs, this scheme reduces communications as much less messages are communicated between objects and server compared with a centralized solution.

The contributions of this paper are summarized as follows:

- We introduce a novel way to define the *range* for *k*NN queries in road networks.
- We propose a P2P solution to process *k*NN queries over road networks which has less server computation cost and communication cost.
- We provide simulation study to show the benefits of using the proposed P2P solution.

The remainder of this paper is organized as follows. Related work is discussed in Section 2. Section 3 covers formal definitions and background information. The proposed solution is introduced in Section 4. In Section 5, we present the simulation study. Finally, Section 6 concludes the paper.

## 2 Related Work

Mouratidis et al. [12] studied the *k*NN monitoring query problem in road networks, where query and data objects all move around. However, their techniques only focused on reducing server workload without worrying about the communication cost and the update cost. As we pointed out in the introduction section, these costs will undermine the scalability of the system. Recently, Wu et al. [20] proposed a distributed solution to answer moving *k*NN queries; nevertheless, the proposed solution is only applicable to open space environments.

To the best of our knowledge, the work most related to ours is the research presented by Jensen et al. in [11], in which an algorithm was given for continuous *k*NN queries. This algorithm takes a client-server approach with the server keeps the location information of all the clients. For a given new query, the server performs a *k*NN search to identify a *Nearest Neighbor Candidate* set (NNC set) and a *distance limit*. This information is sent to the query object, which subsequently needs to repeatedly estimate distances between the clients in the NNC set and the query object to maintain the query result. When the number of clients in the NNC set with a distance to the query object greater than the *distance limit* exceeds a predefined certain threshold, the query object needs to contact the server to refresh the NNC set. A drawback of this approach is the potentially low accuracy in the *k*NN approximation because the criterion employed to refresh the NNC set does not consider the clients outside the NNC set, which could become the query's *k*NNs even when the criterion is still satisfied.

In summary, although there have been a tremendous amount of work in *k*NN query processing, there is no existing P2P solution for such queries in a road network environment, which allows all objects to participate in query processing in order to reduce both server computation and communication costs.

## 3 Preliminaries

In this section, we first define the underlying spatial network, and then give definitions for moving objects, *k*NN queries and monitoring regions.

**Definition 1. (Network)** A network is modeled as an undirected graph $G = (N, E)$, where $N$ is a set of nodes, and $E$ is a set of edges. An edge is expressed as $<n_i, n_j>$, where $n_i$ and $n_j$ represent the *start* node and the *end* node. To avoid ambiguity, we use a numbering system such that $i$ is always less than $j$. The distance between two nodes $n_i$ and $n_j$ is denoted by $d(n_i, n_j)$, which is the shortest network distance from $n_i$ to $n_j$.

Please note that for simplicity, a road network is modeled as an undirected graph where edges are considered to be bidirectional, but our techniques can be easily extended to networks with unidirectional edges. Also, in this paper, road segment and edge are used interchangeably whenever there is no confusion.

**Definition 2. (Edge Distance)** Based on the types of nodes connecting two edges together, we classify the distance between two edges into the following four types: *SS*, *SE*, *ES*, and *EE*. We call the resultant distance associated with a specific type as *Edge Distance*. For example, the distance type is *SS* if both nodes are *start* (*S*) nodes; the distance type is *SE* if one node is *start* (*S*) node while another is *end* (*E*) node. Formally, given $e_i = <n_{is}, n_{ie}>$ and $e_j = <n_{js}, n_{je}>$, $d_{xy}(e_i, e_j) = d(n_{ix}, n_{jy})$, where $x, y \in \{S, E\}$. To make the definition complete, we add an extra distance type called *SAME* (*SM*) to cover the case when the two edges are identical. Formally, if $i = j$, $d_{SM}(e_i, e_j) = 0$, otherwise, $d_{SM}(e_i, e_j) = \infty$. As a result, the shortest distance between any two edges $e_i$ and $e_j$ can be expressed as: $d(e_i, e_j) = min_{type \in \{SM, SS, SE, ES, EE\}}\{ d_{type}(e_i, e_j)\}$.

**Definition 3. (Moving Object)** A moving object is represented by a moving point in the road network. At any one time, an object $o$ can be described as $<e, pos, direction, speed, reportTime>$, where $e$ is the edge that $o$ is moving on and $pos$ is the distance from $o$ to the *start* node of $e$. The value of *direction* is set to 1 if $o$ is moving from the *start* node of $e$ to the *end* node of $e$; otherwise, it is set to -1. *reportTime* records the time when the *pos* is reported.

Distance between any two objects $o_i$ and $o_j$, denoted as $d(o_i, o_j)$, is the shortest network distance from $o_i$ to $o_j$. It can be calculated as below.

**Property 1.** Assume the positions of objects $o_i$ and $o_j$ are denoted as $<e_i, pos_i>$ and $<e_j, pos_j>$, where $e_i = <n_{is}, n_{ie}>$, $e_j= <n_{js}, n_{je}>$, and the lengths of $e_i$ and $e_j$ are $e_i.length$ and $e_j.length$, respectively. The distance between $o_i$ and $o_j$ can be calculated as the minimum of the following five items:

$d(o_i, o_j) = min\{d_{SM}(e_i, e_j) + | pos_i - pos_j|, d_{SS}(e_i, e_j)+ pos_i + pos_j , d_{SE}(e_i, e_j)+ pos_i + e_j.length - pos_j , d_{ES}(e_i, e_j)+ e_i.length - pos_i + pos_j , d_{EE}(e_i, e_j)+ e_i.length - pos_i + e_j.length - pos_j \}$

**Property 2.** For a moving object, with *pos*, *direction*, *speed*, and *reportTime* all known, and provided that the moving object still moves on the same edge, the new position of the moving object at current time *currentTime* can be calculated as $(currentTime - reportTime) \times speed \times direction + pos$.

**Definition 4. (k-Nearest-Neighbor Query)** A $k$NN query $q$ can be denoted as $<o, k>$, where $o$ is the object issuing the query (or the focus of the issued query), and $k$ is the number of nearest neighbors interested in. Denote the set of all other moving objects (i.e. excluding $o$) as $O$, a $k$NN query $q$ returns a subset $O' \subseteq O$ of $k$ objects, such that for any object $o_i$ in $O'$ and any object $o_j$ in $(O - O')$, $d(o_i, o) \leq d(o_j, o)$.

For a given $k$NN query $<o, k>$, we call the object $o$ as the *query object*, all objects in the set $(O - O')$ as the *data objects*. Among all objects in the query results $O'$, we name the object that has the largest distance to $o$ as the *kNN object*, and all other objects in the set $O'$ as the *(k-i)NN objects*, with $i = 1, \ldots, k$-1.

**Definition 5. (Range of $k$NN Query)** Given a $k$NN query $q = <o, k>$, with object $o$ moving on edge $e_o$. Assume the *kNN object* for this query is object $o_{NN}$, which is moving on edge $e_{NN}$, then the *range* of the $k$NN query is defined as $e_o.length$ if $e_o$ and $e_{NN}$ are identical, otherwise, the *range* is defined as $e_o.length + e_{NN}.length + d(e_o, e_{NN})$. Please note that the *range* is the allowed maximum distance between the *query object* and the *kNN object* given that both objects move on their own edges.

As shown in the following Definition 6, this *range* concept is utilized to prune out objects that certainly can not become query result.

**Definition 6. (Monitoring Region)** A monitoring region of a $k$NN query is a set of edges that can be reached by the query's *range* while the *query object* and the query's *kNN object* both move within their own current edges. Formally, for a query $q = <o, k>$ where $o$ moves on edge $e$, if the query's range is $q.range$, then its monitoring region $r = \{e_i \mid e_i \in E, d(e, e_i) < q.range \}$. If an edge is included in a query's monitoring region, we say that this edge *intersects* with the query's monitoring region.

The monitoring region can be computed with a depth-first search by expanding from the *start* and the *end* node of edge *e*. The detailed algorithm is omitted. The interested reader is referred to [9]. The output of the algorithm, denoted by *mrOutput*, has the following format: $mrOutput = \{<e_i, type, distance> \mid e_i \in E, type \in \{SM, SS, SE, ES, EE\}, distance = d_{type}(e_i, e) < q.range\}$. For an object moving on edge $e_{i'}$ in that monitoring region, it stores locally a subset of the above *mrOutput* as $\{<e_i, type, distance> \mid <e_i, type, distance> \in mrOutput, e_i = e_{i'}\}$, to facilitate computing its distance to the *query object*. As a result, moving objects do not need to store the whole road network and perform the computation-intensive shortest-path algorithm. This is considered as one nice feature of our proposed technique.

To illustrate the above definitions, we give an example below. A partial road network is drawn in Figure 1, where nodes are denoted as $n_1$, $n_2$, etc. Each edge's length is indicated by the number close to that edge. Notations like $n_1n_2$, $n_1n_3$, are used to represent edges. Assume that there is one object *A* (represented by a star) moving on edge $n_1n_4$, and we are interested in its 2-NNs, which have been determined to be *B* and *C* (represented by triangles). Based on Definition 4, *A* is the *query object*, *B* is the *(k-i)NN object*, *C* is the *kNN object*, and all other objects (represented by circles) are *data objects*. Since *C* is moving on edge $n_2n_3$, and the shortest distance between edge $n_1n_4$ and $n_2n_3$ is 1 (through edge $n_3n_4$), based on Definition 5, the range of this query is computed as the sum of the lengths of edge $n_1n_4$ and edge $n_2n_3$, then added by 1, which gives (3 + 2 + 1) = 6. The monitoring region is then computed by expanding from both nodes ($n_1$ and $n_4$) of edge $n_1n_4$. The results are shown in the figure by the thick edges. All objects moving in the monitoring region will monitor this query.

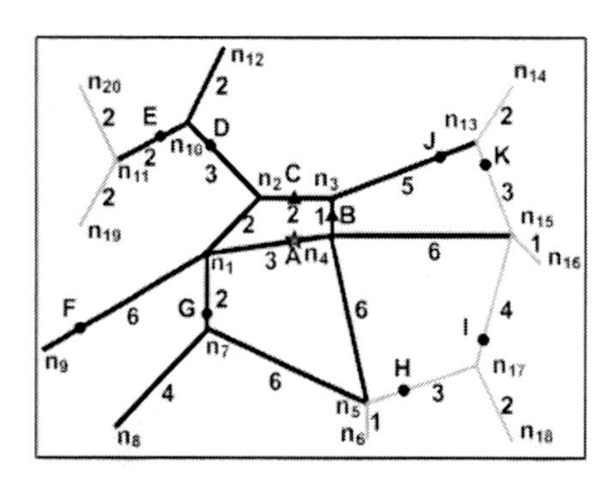


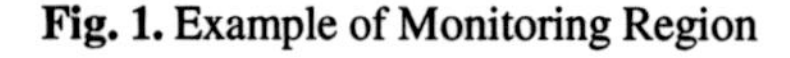
**Fig. 1.** Example of Monitoring Region

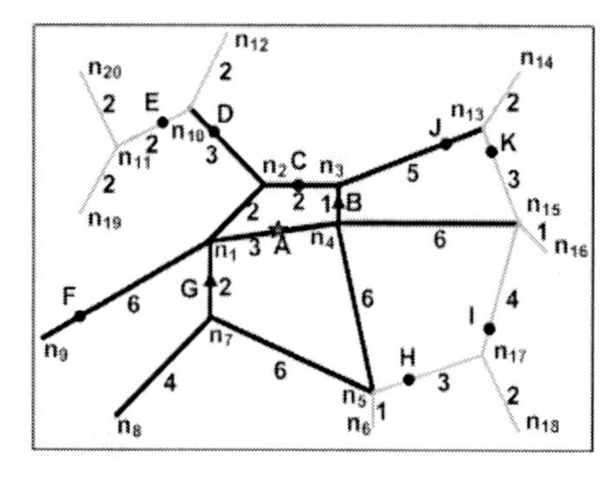


**Fig. 2.** Example of Message Processing

To deal with long road segments, such as highways, we set a maximum for the allowed segment length. Any segment that is longer than this maximum will be divided into multiple shorter pieces. We add a virtual node at each position where the original segment is divided, and the resultant shorter segments become virtual edges. In our system, we do not differentiate virtual node (edge) from real node (edge).

# 4 Proposed Solution

## 4.1 Assumptions and System Overview

We have the following two assumptions: (1) every moving object is equipped with some positioning devices. (2) every moving object has some computing power for data processing.

The proposed system adopts a server-client architecture. On the server side, all information about *query objects, kNN objects, (k-i)NN objects,* and queries are stored. The server determines the monitoring region for each query and sends the query to objects moving in that monitoring region. The moving objects save the received information in their local storage space. Periodically, based on the saved information, a moving object needs to calculate its distance to the *query object*, and compares that distance with the distance from the *kNN object* to the *query object*. For a *data object*, if it moves closer to the *query object* than the *kNN object*, it sends a message to the server to trigger an update. Similarly, for a *(k-i)NN object*, if it moves further away from the *query object* than the *kNN object*, it also notifies the server.

## 4.2 Server Data Structure

A number of excellent disk-based storage structures have been proposed for road networks [10, 18]. Any of these techniques can be easily adapted for our network database to achieve good access locality and therefore low I/O cost.

There are mainly three tables used: (1) a *query-object-table* to store *query objects* in the form of *<oid, eid, pos, direction, speed, reportTime>*, (2) a *query-table* to store monitoring queries in the form of *<qid, oid, k, kNN object, (k-i)NN objects, mrOutput>*, where the *mrOutput* is the output from the algorithm for monitoring region calculation, as mentioned in Section 3, and (3) a *segment-query-table*, where for each edge, the *qid*s of all queries whose monitoring regions *intersect* with that edge are stored. An entry in this table has the form of *<eid,* {*qid*}>. To facilitate the initialization step (to be discussed in Section 4.4), we also keep track of how many objects currently moving on each edge.

## 4.3 Moving Object Data Structure

A moving object stores all queries whose monitoring regions intersect with the edge where it is moving. We use a table for that need. For a moving object moving on edge *e*, Each entry in the table has the following format: *<qid, oid, eLength,* {*<e, type, dist>*}, *nn_oid, nn_eLength,* {*<e', nn_type, nn_dist>*}>, where *qid* is the query id, *oid* is the corresponding *query object*'s id, *eLength* is the length of the edge that the *query object* is on, and {*<e, type, dist>*} stores a subset of *mrOutput* (the attribute inside the *query-table* on the server), where each tuple specifies the edge distance type and the actual edge distance from the moving object's segment to the *query object*'s segment. With *eLength* and the set {*<e, type, dist>*}, the moving object can calculate its distance to the *query object*. Similarly, *nn_oid* denotes the *kNN object*'s object id, *nn_eLength* is the length of the edge where the *kNN object* is on, and {*<e', nn_type, nn_dist>*} stores a set of tuples which help to calculate the distance from the *kNN* *object* to the *query object*. Please note that *e'* is the edge where the *kNN object* is moving on. In order to estimate the locations of the *query object* and the *kNN object* at different time units other than at the saved *reportTime*, we also store the information about the *query object* and the corresponding *kNN object* on moving objects.

### 4.4 Initialization

For every new moving object that enters the system, it needs to report its location, heading, and speed to the server. The server determines and sends the moving object a set of queries that should be monitored. If the new moving object is a *query object*, the server calculates the first set of *k* nearest neighbors in the following four steps:

(1) Since the server knows how many objects are moving on each edge, by comparing with the requested number *k*, the server can decide the set of edges to send a probe message. The probe message has the format of *<qid, oid, pos, eLength,* {*<e, type, dist>*}>, where *pos* is the position of the *query object* on its edge, and other parameters have the same meanings as those discussed in Section 4.3.

(2) After the probe message is received by all moving objects moving on the identified edges, based on Property 1, moving objects can calculate their distances to the *query object* and send the distances back to the server.

(3) The server compares all the returned distances and picks the *k* smallest ones. The moving objects with the *k* smallest distances are the initial *k* nearest neighbors. Among the *k* identified objects, the one with the largest distance is the *kNN object*.

(4) With the *kNN object* known, the server calculates the query's range using Definition 5, computes the query's monitoring region with Definition 6, and sends a message, which contains the information of the *query object* and the *kNN object*, to all objects in the monitoring region.

### 4.5 Message Processing

For a given query, there are four different types of objects, namely, *query object, data object, kNN object*, and *(k-i)NN object*. Please note that since the system as many *k*NN queries, for a moving object, it can assume multiple roles. Below we list different types of messages sent out from moving objects, and discuss how the server responds to these messages.

#### 4.5.1 Switch Segment Message

Every moving object needs to monitor its own location on the segment it is moving on. If its position on that segment is less than zero or greater than the segment's length, it knows that it has moved to a new segment. At this time, the moving object reports to the server and requests for the new segment's length and a new set of queries. For each query, the server sends the *query object* and the *kNN object*'s information with the relevant edge distances, and the lengths of the edges where the *query object* and the *kNN object* are moving on, respectively. With the received information, later on, the moving object can estimate the position of the *query object* and the *kNN object*. Using saved edge distances, the moving object can calculate its distance to the *query object* and the distance from the *kNN object* to the *query object*.

However, if the moving object itself is also a *query object* (or a *kNN object*) for some query, the monitoring region for that query needs to be updated. The server performs the following three tasks: (1). update the *query-object-table* (or the *query-table*) with the object's new location. (2). compute a new monitoring region for the query and update the *query-table*. (3). send out messages to notify moving objects in the old monitoring region to stop monitoring the query, and notify moving objects in the new monitoring region to add this query for monitoring. If the moving object is a *(k-i)NN object* for some query, although there is no need for monitoring region re-computation, the server needs to update the *query-table* accordingly.

### 4.5.2 Enter Query Message

For a *data object*, it periodically checks its distance to the monitored *query object*, and compares with the distance from the *kNN object* to the *query object*. If it is getting closer to the *query object* than the *kNN object* is, a message is sent to the server to indicate that it is currently part of the query result.

The server first estimates the current positions of the saved *kNN object* and *(k-i)NN objects* to decide which object should be replaced by the new-coming object. Then the server updates the *query-object-table*. If the replaced object is the *kNN object*, the server also calculates a new monitoring region based on the new *kNN object*, and notifies all affected objects.

### 4.5.3 Exit Query Message

For a *(k-i)NN object*, since it could move further away from the *query object* and become the *kNN object*, it needs to periodically monitor its distance to the *query object* and compare with that of the *kNN object*. Once the distance is larger than the distance from the *kNN object* to the *query object*, the *(k-i)NN object* needs to report to the server.

After the server receives this type of message, it replaces the current *kNN object* with the one sending out the message, re-calculates the monitoring region, and notifies all affected objects.

### 4.5.4 Other Messages

Other than the three types of messages described above, there are some other scenarios when a moving object needs to contact server. When a *query object* (or a *kNN object*) changes its speed, it sends the update to the server, and the server updates the *query-object-table* (or the *query-table*) and forwards the update to relevant moving objects. Similarly, when a *(k-i)NN object* changes its speed, it also notifies the server, and the server just updates the *query-table* (i.e. No need to send the update to moving objects).

### 4.5.5 Optimization

For the sake of clarity, we have assumed that the server can receive only one message per time unit. In reality, the server bandwidth is more plentiful and many messages should be able to arrive at the server per time unit. To reduce server computation and communication cost, for all the messages received during a given time unit requiring monitoring region re-computation (such as "Exit Query Message" and "Enter Query Message"), immediately after the message is received, the server only updates the query result to keep the result accurate. And the server waits until the end of that time unit to re-compute the monitoring region and sends out message to notify moving objects to update their monitoring queries.

### 4.6 An Example

In this section, we use the same example as the one used in Section 3 to show how a *data object* keeps monitoring its distance to the *query object* and the distance from the *kNN object* to the *query object*.

For example, at time *t*, as shown in Fig. 1, the *query object A* is at position 2.5 on edge $n_1n_4$, the *kNN object C* is at position 1 on edge $n_2n_3$, and a *data object G* is at position 1.5 on edge $n_1n_7$. Since *G* is inside the query's monitoring region, it has *A* and *C*'s information saved locally. Besides, it stores the edge distance {<$n_1n_7$, *SS*, 0>} and the length of edge $n_1n_4$, to determine the distance from itself to *A*. To calculate the distance from *C* to *A*, it also has the edge distances {<$n_2n_3$, *SS*, 2>, <$n_2n_3$, *SE*, 3>, <$n_2n_3$, *ES*, 4>, <$n_2n_3$, *EE*, 1>} and the length of edge $n_2n_3$ saved.

At time (*t*+1), as shown in Fig. 2, *data object G* moves to position 1 on edge $n_1n_7$. It estimates the new position of *A* on edge $n_1n_4$ using Property 2 and gets 1.5. Similarly, it estimates the new position of *C* on edge $n_2n_3$ as 1.1. Then with Property 1, it computes its distance to *A* as 2.5 (calculated as: 1 + 1.5 + 0), while the distance from *C* to *A* is 3.4 (through the edge distance <$n_2n_3$, *EE*, 1>, calculated as: (2 - 1.1) + 1.5 + 1). Since its distance to the *query object* is less than the distance from the *kNN object* to the query object, it sends an enter query message to the server. The server replaces *C* with *G* as the new *kNN object*, determines the new query range, and re-computes the monitoring region. In this example, the new query range is 5 (sum of the lengths of edge $n_1n_4$ and edge $n_1n_7$), and the new monitoring region is drawn as thick edges in Figure 2. As we can see, data object *E* on edge $n_{10}n_{11}$ is no longer in the monitoring region.

## 5 Performance Study

We implemented a simulator to measure the performance of our proposed technique. For a system designed to process monitoring queries, the server could easily become a bottleneck. Whether or not a system can reduce server computation and communication cost is very critical, as a result, we choose the following performance metrics.

- **Server workload.** This cost is measured as the total number of edges accessed in order to answer queries. This is a good measure because server workload consists of I/O time and CPU time, while I/O time is more dominant.
- **Communication cost.** We measure this cost by counting the messages sent out from both the server and the client to reflect the bandwidth consumption.

For server workload, we compare our technique with one popular centralized solution: *query index* [8], which was originally designed for an open space environment. To make the comparison fair, the *query index* scheme is adapted for a road network environment. In the adapted scheme, queries are indexed by a segment-query table (similar to the table used in our technique), where for each segment, all queries whose *query object* can reach that segment within the distance from the *query object* to the *kNN object* are saved. Every time when a moving object sends its updated location to the server, based on the segment where the moving object is on, the server retrieves all queries for that segment from the segment-query table. Then the server computes the distances from the moving object to *query objects* to determine if the moving object belongs to any query result. Also, every time when a *query object*'s location is updated or its *kNN object*'s location is updated, the server updates that segment-query table.

We also compare communication cost to a centralized approach, which we name it as *Query Blind Optimal* (QBO) method. In the QBO method, moving objects only need to contact the server when they switch segments or change speeds. When an object moves to a new segment, the server sends back the new segment's length to help the object to determine when it moves out of that segment. At each time unit, the server estimates all moving objects' locations and answers all *k*NN queries. This method is optimal on communication cost if we assume that moving objects do not have any knowledge about queries, which is why we call it *Query Blind Optimal* method. Besides this QBO method, we also have a naïve method which serves as a basis for comparison. In this naïve method, all moving objects report to the server when their locations change, as a result, there is no need for the server to send messages back to the clients.

### 5.1 Simulation Setup

Our simulation is based on a terrain of $50 \times 50$ square miles. We generate a synthetic road network by first placing nodes randomly on the terrain, and then connect nodes together randomly to form edges. There are 2000 nodes and 4000 edges in our setup, with the longest edge as 3 miles. Moving objects are placed randomly on edges with initial speeds and directions. Among all the moving objects, some are specified as *query objects* with a pre-defined number ($k$) of interested nearest neighbors. The speeds are in the range of [0.1, 1] mile/min, following a Zipf distribution with a deviation of 0.7. When an object moves close to a road intersection, it moves to a randomly picked segment. At each time unit, there are a certain percentage of objects changing their speeds. The threshold for changing speed is set as 0.1 mile/min. The time step parameter for the simulation is one minute. We run simulation for 10 times and compute the average as the final output. Each simulation lasts for 200 time units. The simulation was run on a Pentium 4 2.6GHz desktop pc with 2GB memory.

In the experiments, we vary different parameters, as listed in Table 1, to study the scalability of the proposed system. If not otherwise specified, the experiment takes the default values.

**Table 1.** Simulation Parameters

| Parameter Name | Value Range | Default Value |
|---|---|---|
| Number of Moving Objects | [50000, 100000] | 100000 |
| Number of Queries | [10, 1000] | 200 |
| Number of Nearest Neighbors ($k$) | [1, 20] | 5 |
| Percentage of Objects Changing Speed per Time Unit | [2, 50] | 10 |

### 5.2 Simulation Results

Figure 3 shows the impact of number of queries on server workload and communication cost. Please note that in Fig 3.a, the vertical axis is in logarithmic scale. The plot shows that both the proposed technique and the *Query Index* method incur more server workload with the increases in the number of concurrent queries. A comparison indicates that the proposed approach is about 50 times better than the *Query Index* method. This huge savings can be attributed to the computations carried out on moving objects, which greatly reduce server workload.

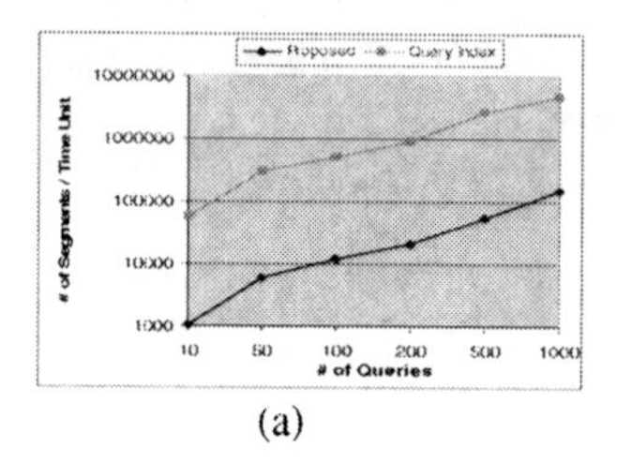


(a)

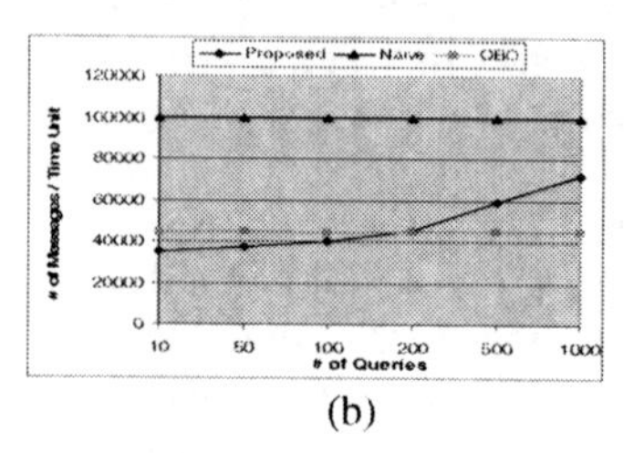


(b)

**Fig. 3.** Effect of number of queries on (a) server workload and (b) communication cost

In Fig 3.b, we compare the communication cost of the proposed technique with those of a Naïve approach and the QBO method. We observe that the curve of the naïve method is flat and it displays the highest communication costs. This is expected since every object updates its location at every time step. The curve of the QBO method is also flat because the communication cost is primarily introduced by objects when they move to new segments or change their speeds; and the occurrences of such activities are independent of the number of concurrent queries. The communication cost of the proposed technique increases as the number of queries increases. This can be explained as follows. Since the P2P strategy needs to update the query results maintained on the remote server, the objects have more query updates to perform with the increases in the number of concurrent queries resulting in a higher communication cost. Nevertheless, the proposed technique performs very well (i.e., comparable to the QBO) for numbers of queries as high as 200. Its performance worsens when the numbers of concurrent queries is greater than 200. Under this circumstance, we note that distributed servers can be used to accommodate the increase in the communication costs. In such an environment, our P2P technique would require a smaller number of distributed servers since it is able to reduce server workload and the demand on server bandwidth.

In Figure 4, we vary the other three parameters to study their effects on communication cost. In Fig 4.a, the number of moving objects is varied from 50000 to 100000. As we can see, for all the three studied methods, the number of messages increases as the number of moving objects increases. Fig 4.b studies the effect of increasing the percentage of objects changing speed per time unit from 2% to 50%. The result shows that for the naïve method, the curve is a flat line as in Fig 3.b. Both our technique and the QBO technique incur more communication cost when there are more objects changing speeds at every time step. Compared to the QBO technique,

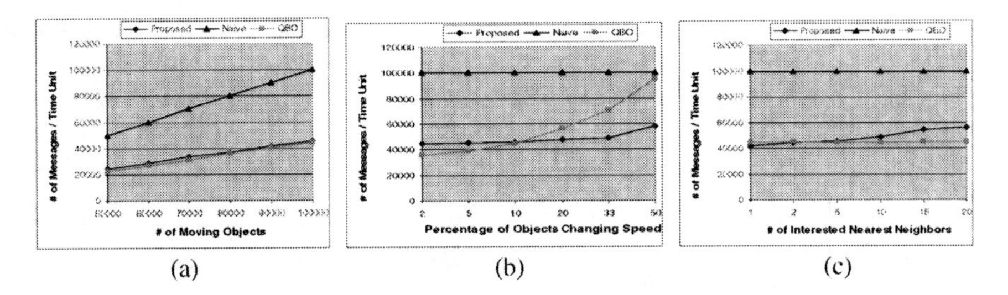


**Fig. 4.** Effect of (a) number of moving objects, (b) percentage of objects changing speed per time unit, (c) number of interested nearest neighbors, on communication cost

our technique has a much less steeper curve because among all objects that change speeds, only *query objects*, *kNN objects*, and *(k-i)NN objects*, which combined account for a small fraction of the total number of objects, need to contact the server, however, in the QBO technique, all objects changing speeds have to report to the server. We also try to vary the number of interested nearest neighbors (*k*) and the result is shown in Fig 4.c. From the plot, we observe that both the naïve method and the QBO technique are not affected by the number of requested nearest neighbors. For our technique, more messages are needed if there are more nearest neighbors to be found. This is expected since a bigger *k* will make more objects into *(k-i)NN objects*, and quite probably, larger monitoring regions are demanded. Consequently, higher communication cost is necessary.

## 6 Conclusions

In this paper, we introduced a P2P technique for continuous *k*NN queries in a network environment. To the best of our knowledge, this is the first P2P solution that fully leverages the computation power of all peers to address the *k*NN problem. This scheme utilizes mobile computing power to reduce server workload and the number of location updates necessary. We presented the detailed design and gave simulation results to show the performance advantages of the proposed technique. When compared to an adapted *Query Index* method, our approach incurs about 50 times less server load. In terms of communication cost, the proposed technique performs comparable to a *Query Blind Optimal* scheme when there are as many as 200 concurrent queries. As the number of concurrent queries increases, the moving objects need to communicate more frequently to update the query results maintained on the server.

## References

1. Cai, Y., Hua, K.A., Cao, G.: Processing Range- Monitoring Queries on Heterogeneous Mobile Objects. In: MDM (2004)
2. Mouratidis, K., Hadjieleftheriou, M., Papadias, D.: Conceptual Partitioning: An Efficient Method for Continuous Nearest Neighbor Monitoring. In: SIGMOD (2005)

3. Gedik, B., Liu, L.: MobiEyes: Distributed Processing of Continuously Moving Queries on Moving Objects in a Mobile System. In: Bertino, E., Christodoulakis, S., Plexousakis, D., Christophides, V., Koubarakis, M., Böhm, K., Ferrari, E. (eds.) EDBT 2004. LNCS, vol. 2992, Springer, Heidelberg (2004)
4. Hu, H., Lee, D.L., Xu, J.: Fast Nearest Neighbor Search on Road Networks. In: Ioannidis, Y., Scholl, M.H., Schmidt, J.W., Matthes, F., Hatzopoulos, M., Boehm, K., Kemper, A., Grust, T., Boehm, C. (eds.) EDBT 2006. LNCS, vol. 3896, Springer, Heidelberg (2006)
5. Kolahdouzan, M.R., Shahabi, C.: Voronoi-Based K Nearest Neighbor Search for Spatial Network Databases. In: VLDB, pp. 840–851 (2004)
6. Hu, H., Lee, D.L., Lee, V.C.S.: Distance Indexing on Road Networks. In: VLDB (2006)
7. Xiong, X., Mokbel, M., Aref, W.: SINA: Scalable Incremental Processing of Continuous Queries in Spatio-temporal Databases. In: SIGMOD (2004)
8. Prabhakar, S., Xia, Y., Kalashnikov, D.V., Aref, W.G., Hambrusch, S.E.: Query indexing and velocity constrained indexing: Scalable techniques for continuous queries on moving objects. IEEE Trans. on Computers 51(10) (2002)
9. Liu, F., Do, T.T., Hua, K.A.: Dynamic Range Query in Spatial Network Environments. In: Bressan, S., Küng, J., Wagner, R. (eds.) DEXA 2006. LNCS, vol. 4080, Springer, Heidelberg (2006)
10. Shekhar, S., Liu, D.R.: CCAM: A Connectivity-Clustered Access Method for Networks and Network Computations. IEEE TKDE 9(1) (1997)
11. Jensen, C.S., Kolar, J., Pedersen, T.B., Timko, I.: Nearest Neighbor Queries in Road Networks. In: Proc. ACMGIS, pp. 1–8 (2003)
12. Mouratidis, K., Yiu, M.L., Papadias, D., Mamoulis, N.: Continuous Nearest Neighbor Monitoring in Road Networks. In: VLDB, pp. 43–54 (2006)
13. Cho, H., Chung, C.: An Efficient and Scalable Approach to CNN Queries in a Road Network. In VLDB, pp. 865–876 (2005)
14. Wang, H., Zimmermann, R., Ku, W.S.: Distributed Continuous Range Query Processing on Moving Objects. In: Bressan, S., Küng, J., Wagner, R. (eds.) DEXA 2006. LNCS, vol. 4080, pp. 655–665. Springer, Heidelberg (2006)
15. Hu, H., Xu, J., Lee, D.L.: A Generic Framework for Monitoring Continuous Spatial Queries over Moving Objects. In: SIGMOD (2005)
16. Yu, X., Pu, K.Q., Koudas, N.: Monitoring k-nearest neighbor queries over moving objects. In: IEEE ICDE, IEEE Computer Society Press, Los Alamitos (2005)
17. Xiong, X., Mokbel, M., Aref, W.: SEA-CNN:Scalable Processing of Continuous K-Nearest Neighbor Queries in Spatio-temporal Databases. In: IEEE ICDE, IEEE Computer Society Press, Los Alamitos (2005)
18. Papadias, D., Zhang, J., Mamoulis, N., Tao, Y.: Query Processing in Spatial Network Databases. In: VLDB (2003)
19. Mouratidis, K., Papadias, D., Bakiras, S., Tao, Y.: A Threshold-Based Algorithm for Continuous Monitoring of k Nearest Neighbors. IEEE TKDE 17(11), 1451–1464 (2005)
20. Wu, W., Guo, W., Tan, K.L.: Distributed Processing of Moving K-Nearest-Neighbor Query on Moving Objects. In: IEEE ICDE, IEEE Computer Society Press, Los Alamitos (2007)

# Information Life Cycle, Information Value and Data Management

Rudolf Bayer

Institut für Informatik
Technische Universität München
Boltzmannstrasse 3
D-85748 Garching

**Abstract.** Data volumes grow extremely fast. Storage capacity and cost can easily cope with that growth, but data access is critical. However, most information has a surprisingly short life cycle. This has a deep impact on data management systems and leads to a new storage architecture consisting of the two subsystems FileCache and FileStore with many advantageous properties w.r. to reliability, availability and cost. A multidimensional database of metadata is used to organize the data. This architecture is also suitable for complete multimedia biographies.

**Keywords:** Information Life Cycle Management (ILM), Information Value, Data Management, FileCache, FileStore, UB-Tree, Hierarchical Storage Management (HSM).

## 1 Preface

The topic of his paper has been largely ignored by academic computer science, but it is very important for commercial end even private IT. Part of this paper is speculative, based on only a few measurements, personal observations and plausible hypotheses. On this basis a new architecture for Information Life Cycle (ILM) and Hierarchical Storage Management (HSM) is proposed.

## 2 Some Basic Facts on Data and Storage

The tremendous growth of data volumes and the life cycle of information have a fundamental impact on data management. In industry data volumes are growing rapidly, a factor of 1.7 per year is widely assumed. Fileservers of 10 TB serving 1.000 users are common, with personal shares of well below 10 GB. In private environments data volumes might be growing even much faster due to significant shifts in document types: automatic recording of voluminous digital photos, videos and sound tracks is even more common in the private sector than in industry. The project MyLifeBits [1] assumes a growth of 1 GB/month/person, which is well below my own private observations and estimates.

R. Wagner, N. Revell, and G. Pernul (Eds.): DEXA 2007, LNCS 4653, pp. 277–286, 2007.

Is the subjective value of information growing at the same rate as our data volumes?

Data are stored primarily on hard disks, whose price per GB has fallen dramatically over several years and is presently at 0.5 €/GB or 500 €/TB. Jim Gray [2] estimates that in a few years the price will be 500 €/PB.

The bottom line is that capacity and prices of storage are moving faster than we can capture, and that storage is essentially free. This raises an interesting question about reclamation of storage: To reclaim 1 GB of storage on your HD you have to inspect and to manually delete at least 1000 files, pictures or emails, a tremendous waste of time and mental effort, and then you have recovered a storage capacity costing just 50 cents. This is obviously a bad idea and the reason that hardly anybody does it privately. However, in industry the capacity of personal shares on fileservers is often limited, e.g. to less than 5 GB, and forces employees to clean up their drives/shares periodically, a very poor management decision.

Storage of data is one aspect, a second aspect is access to data. The ratio of access time - i.e. to scan the complete content of a HD - to capacity has been worsening steadily over many years. At a raw disk speed of 50 MB/s the effective transfer rate as seen by an application today is more like 5 MB/s. If data access is brute force scanning, like for grepping or backing up files, it takes less than one second for 1 MB, 3 minutes for one GB, 2 days for 1 TB and 5.5 years for a PB. Therefore access time is critical and often longer than the available time windows, e.g. for backup of large fileservers over the weekend, and even parallelism can bring only limited relief.

Simply storing data is obviously of no use, if the data cannot be accessed in reasonable time. Therefore, storage is not the issue, what we really need is something quite different from simple storage: we need memory. For this paper I simply define memory as the combination of storage and access:

***Memory = Storage + Access***

As we have just seen, storage is available almost for free. Therefore we concentrate on access. There are only a few possibilities to achieve good access:

- The state of the art is to organize data in the hierarchical file directory of the operating system and to memorize where you put your data: This seems to be an impossible task for large data collections. It is identical to the problem, which has frustrated librarians for centuries, they never know, into which category and shelf they should put a book, since by definition interesting books elude a unique, narrow, linear classification. Organizing books or datafiles is an inherently multidimensional problem, which cannot be solved by forcing them into the linear organization principles of book shelves or file directories. Since it is so hard to remember where exactly you put your data, you probably try to clean up and reorganize your file directories in order to avoid chaos and lengthy search. This requires considerable discipline and time. Most people give up again after a short "good try".
- The next possibility of access is to use a full text index like typical search engines. However, fulltext search has the disadvantage of yielding result sets whose cardinality is unpredictable and difficult to control. It depends on the

search terms that are used in the query. To be on the safe side, people use simple search terms and accept the order in which the search engines present the results. This simply means that the sorting of the results, e.g. by the page rank algorithm of Google, determines in the end what the world is reading, certainly not an ideal situation.

- Another alternative is to consider your data objects as a large data warehouse and to use a database of metadata in order to find and access your data. Since the classification problem is inherently multidimensional, this approach seems natural.

## 3 Usage of Data

There is little discussion in academia of how data are actually used by people. An interesting question is what the bandwidth of a white collar person is. My measurements of the archive system of the Leibniz Computer Centre in Munich indicate, that on average scientists touch at most 10 files per day with a volume of less than 10 MB. Similar measurements in a large corporation indicate, that data have a very short life cycle of only a few days from creation through modification to the last access. Some measurements are shown for several directories on fileservers in the following table:

- **User directories**
  - 71% 2 days
  - 84 % 3 days
  - 58 % 4 days
  - 50 % 1 day
- **Project directories**
  - 91 % 7 days
  - 100% 1 day
  - 91 % 1 day
- **Group directories**
  - 76 % 1 day
  - 39 % 1 month
  - 84 % 1 day
  - 85 % 1 month

I caution that these are isolated measurements, but they are similar to some other measurements and might well be typical. My hypothesis is that data have indeed a very short life cycle, comparable to a daily newspaper. But nevertheless they are often stored for many years on premium storage and backed up anew every weekend as complete disk images, although most of these data are unchanged since the previous backup.

This short life cycle coincides with the observation, that the value of information is not independent of time, but often decays rapidly. Let me give some examples:

- If you are late to catch a train, the only interesting information of highest value to you personally is the track from which the train leaves. As soon as you are sitting in the right train, this information is of no further interest and its value drops to zero instantly.

- As soon as an order of goods has been shipped, the date at which the order was received is usually of little interest and therefore value, whereas the shipping date is now valuable in order to estimate the projected delivery date.
- During the presentation for this paper, the corresponding powerpoint document is of highest value to me, after the presentation the value drops dramatically.

It is surprising that these quite obvious observations have not influenced the way our computer systems organize and manage data.

## 4 The FileCache and FileStore Architecture

In this chapter I propose a storage architecture, called FileCache and FileStore, which takes into account the hypothesis, that data have a short life cycle and that the value of most information decays rapidly. The following picture shows the typical architecture of fileservers today: Many clients, e.g. 1.000, are supported by a large central fileserver, which itself is regularly backed up via the Backup-System B, e.g. every night or every weekend, to guard against the loss of data. There are many backup systems B on the market, details of B are of no further interest in this paper. Old, stale data are often stored on such fileservers for a long time and backed up again and again, even if nothing has changed.

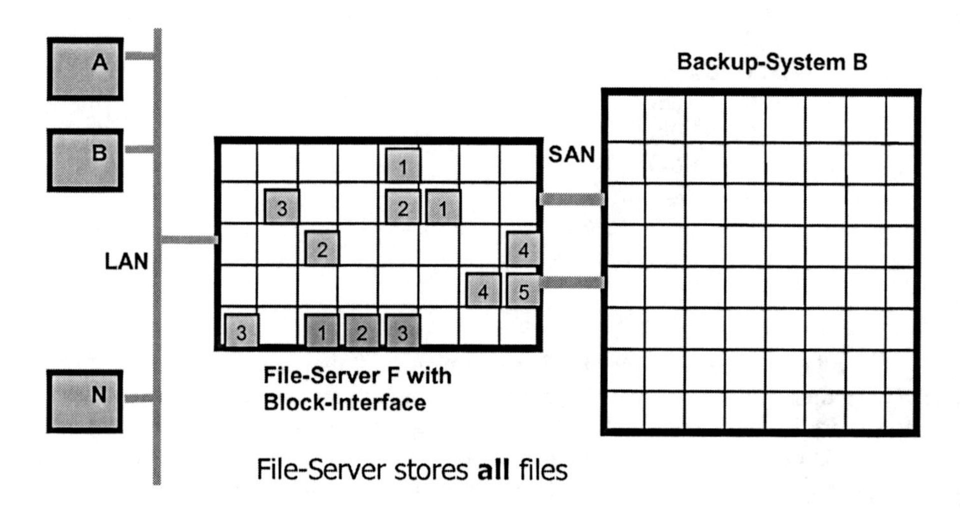


The little squares in the File-Server S indicate a block oriented interface, where the blocks of a certain file have the same colour and are usually, but not always, scattered over the fragmented disk. To retrieve a file requires many random accesses to the disk.

The new architecture arises from the simple idea to split the fileserver into two largely independent subsystems, the FileCache and the FileStore. This is quite similar to the split between the classical mainstore and various processor caches. Notice some significant differences between the FileCache and the FileStore: The FileCache is much smaller than the FileStore and the FileStore is not fragmented, since the files are allocated sequentially.

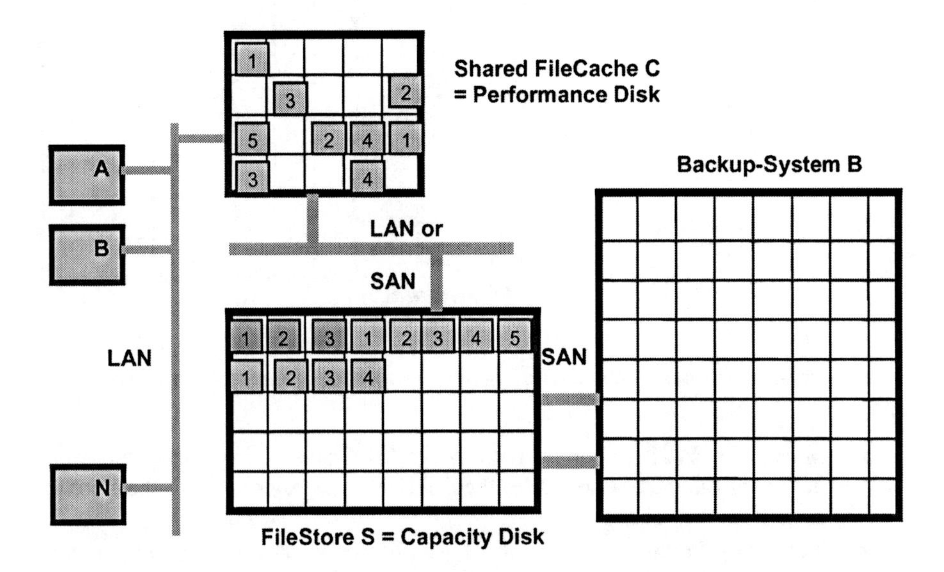


Like in multiprocessor systems an arbitrary number of FileCaches can be served by a central FileStore as in the following picture. In addition, some clients might have their own local caches, yielding the typical multilevel cache as we know it from multiprocessor systems.

The interaction protocol between FileCaches and FileStore is very similar to conventional caches:

- All files are principally stored on the FileStore, which may also maintain an arbitrary number of versions of data.
- All active files are in the FileCache and are copied to the FileStore when they change, using a write through policy. This policy can use various strategies, like observing file changes in real time or scanning the file directories in regular, short intervals. Such intervals might be from a few seconds to an hour depending on the nature of the application.
- Files are purged from the FileCache according to flexible rules, e.g. after a certain time since the last access or if the FileCache has been filled beyond a certain threshold. Purging of files from the FileCache is fast and cheap, since they have already been copied to the FileStore. Thus purging simply means deleting from the FileCache and adjusting the entry in the meta-database.

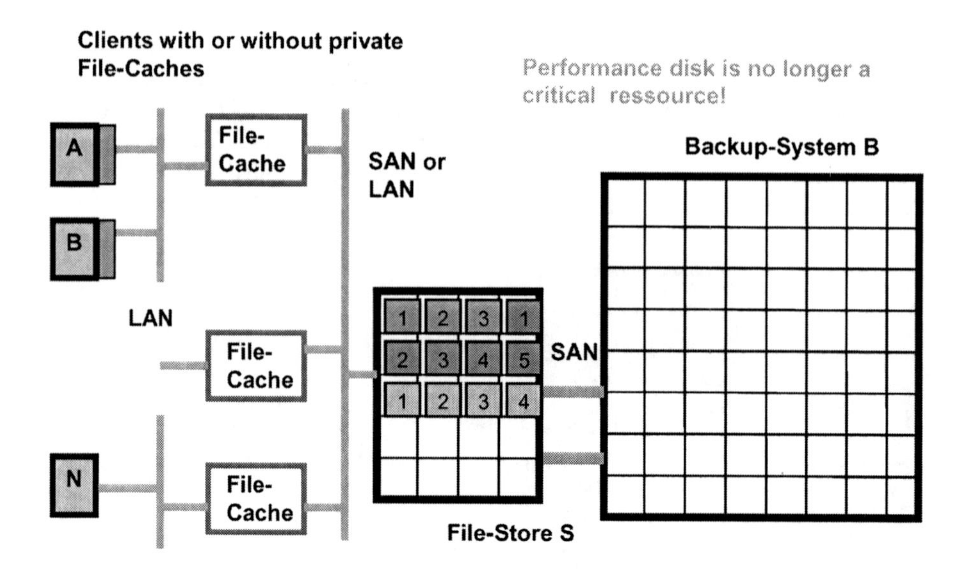


This architecture has many advantageous properties compared to classical file servers:

- **Mirroring** of all important data in the FileCache and in the FileStore: by definition, active data reside in the FileCache and are copied to the FileStore as soon as they change
- **True FileCache:** The backup of the fileserver is replaced by the write through technique of classical cache-management algorithms
- **Backup:** The classical backup via archive systems is only needed for the FileStore, it runs as a background service to provide continuous backup. Due to compression and sequential placement of files on the FileStore the backup should be faster by at least a factor of 10 for a full backup. With incremental backup the time critical backup windows disappear completely, since the usual workload happens on the FileCaches and the FileStore itself has a very low workload
- **Failure Modes:** FileCache and FileStore have independent failure modes
- **Recovery of FileCache:** If the FileCache fails it is reloaded from the FileStore on demand in analogy to reloading the processor cache after a process switch. This amounts to instant, nearly lossless recovery of data after a failure of a FileCache.
- **Recovery of FileStore:** If the FileStore fails, the files in the FileCache are not affected. Only minimal impact on normal processing is noticeable, namely only if an old file is needed which is no longer in the FileCache and happens to reside on a failed volume of the FileStore. This should be a very rare event. Recovery of the FileStore for the failed volume runs as a background service, while mirroring of the FileCaches continues as normal, just to other functioning volumes of the FileStore

- **Storage Capacity:** Probably less than 10% of the total data volume is needed as capacity for FileCaches. With compression only about 50% of the total data volume is needed as capacity on the FileStore
- **Storage Classes:** High performance fibrechannel-disks should be used only for large central FileCaches, cheap SATA-disk for the FileStore and local caches
- **Cost:** The total cost of such an architecture should be much lower than for conventional file-servers
- **Availability** of data is obviously extremely high, comparable to a PLATIN system with complete real time mirroring technique
- **No lost work:** The loss of data and work is limited by the write through interval of the FileCaches, which should be at most a few minutes

The extremely short life cycle of data hypothesized before suggests that very small FileCaches of 10 % of the stored data volumes should suffice. Cache management algorithms like LRU replacement can probably be taken from main memory caching and adapted easily to file caching. An interesting modification of conventional algorithms might be replacement strategies depending on the type of files which is related to the length of life cycles, e.g. ZIP and PDF files seem to have particularly short life cycles and could be replaced after a very short time.

## 5 FileCache Architecture for Databases

The following approach might make the FileCache architecture suitable for databases: The key idea is to distinguish between life and stale data in a database. Therefore, we simply split a relation R into two disjoint tables R1 and R2. R1 shall contain live data and R2 stale data, therefore R = R1 + R2. A separating predicate, e.g.

$$\rho = (\text{create\_date} > \text{last\_archive\_date})$$

distinguishes between the tuples in R1 or R2 resp.

$R1 := \rho\,(R)$
$R2 := \text{not}\ \rho\,(R)$

R is defined as the relational view R1 + R2

An archiving transaction runs as a cron job to move tuples from R1 to R2.

The classification of data into live and stale data can easily be generalized to an arbitrary number of stages. Just consider the standard order and delivery example. Here the order table could be split into:

**R1 = orders received,**
**R2 = orders in production,**
**R3 = orders shipped,**
**R4 = orders under warranty,**
**R5 = orders in archive**
**View R = R1 + R2 + R3 + R4 + R5**

A concrete example using just two tables would need the following declarations:

```
declare table R1 (order_received datetime, ...)
declare table R2 (order_received datetime, ...)
create view R as
    select * from R1 union select * from R2
declare table Archive_Date (last_move datetime, ...)
declare @move_date datetime
```

The following transaction moves stale data older than three weeks from R1 to R2:

```
begin tran
        select @move_date = DATEADD (DAY, -21, GETDATE())
        insert into R2
            select * from R1 where order_received < @move_date
        delete from R1 where order_received < @move_date
        delete Archive_Date
        insert into Archive_Date values (@move_date )
commit tran
```

The user and application programmer only sees relation R and should not be bothered with this complex internal splitup of R. He might formulate a query like

query q(R) = ***select * from R where*** $\gamma(R)$

Since the internal splitup of R is known to the query optimizer via the system tables, it can rewrite this query as:

***if*** $(\gamma$ ***and*** $\rho)(R)$ ***= empty then*** $\gamma(R2)$
***else if*** $(\gamma$ ***and not*** $\rho)(R)$ ***= empty then*** $\gamma(R1)$
***else*** $\gamma(R)$

## 6 Integration with Information Lifecycle Management

In industry the placement of files, emails, etc in hierarchical storage systems, in particular in long term archiving systems, is usually controlled via complex rules dictated by legal requirements like the Oxley-Sabanes act in the US. These rules can be used by the database for the metadata - which is part of the FileStore subsystem – to decide whether to delete data or to move them to an archive instead. In addition this database could easily retain the metadata also for the archived data and guarantee fast search and retrieval even of archived data.

This raises the question which metadata should be collected and stored in the meta-database. The collection of metadata must be fully automatic to be effective. Even semantic annotations to the metadata must be fully automatic and must not be done by people, the latter would be by far too unreliable. In addition a fulltext index could be maintained as an option. It seems that at least the following metadata can and should be collected automatically:

- Domain and user
- Author
- Legal owner
- Directory path
- Filename
- Version number
- File extension
- Time of creation
- Time of last update
- Time of last access
- GPS position where the file was created, this is interesting for pictures and videos
- URI of original file
- Signature
- etc.

Now it is clear that our data collection has outgrown ordinary file directories and is in reality a multidimensional data warehouse of data objects. To organize the metadata for such a data warehouse multidimensional indexes like UB-trees [3], [6] are required and very efficient.

## 7 The Quest for Eternity and Perfect Personal Memories

Man seems to have an innate quest for eternity and spends considerable effort in time and money to record and preserve his achievements and experiences as long as possible. In the past only rulers and the very rich could afford to create monuments in order to be remembered long after their days on earth. Egyptian pharaos and Roman emperors forced a significant part of their people to create glorious monuments [4], the very rich like the Medici spent large sums to preserve their memory in splendid works of art [5].

Today, everybody can preserve as much of her life as she wants in a multimedia biography of arbitrary detail [1]. The project MyLifeBits arrives at estimates for the datavolumes and the number of objects that are assembled over a lifetime: 1 TB of data and about 1 million data objects. The cost of 1 TB of cheap disk storage is 500 € today (2007), i.e. as much as a digital camera which most people can and do afford. The recording equipment – digital camcorders, cameras, notebooks - is available and widespread. To put a million objects into a file directory does not seem feasible, but search engines and databases for metadata solve the problem. The metadata for a complete personal memory are the same as mentioned in the previous chapter and can be recorded automatically. The database to store, organize and index these metadata to solve the access part of the memory problem is at most 1 GB and fits easily on a memory stick.

This means that we have the affordable technology today to record personal multimedia memories of arbitrary detail and quality. Will people do it? Will they spend a significant part of their lives to watch the other parts on multimedia devices?

The previous examples in this chapter and the way people use digital media today suggest that many will.

Even if people do not actively record their lives, others do it already today. The moment you use your mobile phone your position is recorded by your mobile phone provider with sufficient accuracy to reconstruct your complete lifetrack on this earth. Presently these data are stored only for a limited time because of privacy acts, but we all experienced how quickly laws are changed under exceptional circumstances by the government on behalf of law enforcement and with the justification of increased security in the fight against terrorism. The sheer amount of data is no longer prohibitive, recording 60 positions per day, i.e. one position every 10 minutes you are awake during your whole life just requires 50-100 MB of data and fits on your memory stick in addition to your meta-database mentioned above.

What is the impact of such perspectives on our personal life? The ability to forget certain details of the past might be essential for our survival and for our interactions as social human beings. With a complete and perfect personal memory we no longer forget, and nobody knows what the consequences will be.

We computer scientists have already created the technology to record and organize complete personal life memories. We will probably have little control on how this technology will be used in the future. During a recent discussion with colleagues there was unanimous agreement, that this technology will be used by many. The least we can and must do is to think about and to discuss the consequences, and to caution those who are starting to use it.

## References

1. Bell, G.: http://research.microsoft.com/barc/mediapresence/MyLifeBits.aspx
2. Gray, J.: http://www.research.microsoft.com/~Gray/talks/Gray%20IIST%20 Personal%20 Petabyte%20Enterprise%20Exabyte.ppt
3. Bayer, R.: The universal B-Tree for multidimensional Indexing: General Concepts. In: World-Wide Computing and Its Applications'97 (WWCA'97), Tsukuba, Japan (March 10-11, 1997)
4. Trajan's Column: http://cheiron.humanities.mcmaster.ca/~trajan/
5. Gozzoli, B.: Procession of the Magi, http://www.wga.hu/frames-e.html?/html/g/gozzoli/ 3magi/ index.html
6. Markl, V., Ramsak, F., Pieringer, R., Fenk, R., Elhardt, K., Bayer, R.: The TransBase HyperCube RDBMS: Multidimensional Indexing of Relational Tables. In: Proc. of 17th ICDE, Heidelberg, Germany (2001), http://wwwbayer.in.tum.de/cgi-webcon/webcon/ lehrstuhldb/details/Veroeffentlichungen/num/4/1

# Vague Queries on Peer-to-Peer XML Databases

Bettina Fazzinga, Sergio Flesca, and Andrea Pugliese

DEIS, University of Calabria
Via P. Bucci, 87036 Rende (CS) Italy
{bfazzinga, flesca, apugliese}@deis.unical.it

**Abstract.** We propose a system, named *VXPeer*, for querying peer-to-peer XML databases. VXPeer ensures high autonomy to participating peers as it does not rely on a global schema or semantic mappings between local schemas. The basic intuition is that of "vaguely" evaluating queries, i.e., computing partial answers that satisfy new queries obtained by transformation of the original ones, then combining these answers, possibly on the basis of limited knowledge about the local schemas used by peers (e.g., key constraints). A specific query language, named *VXPeerQL*, allows the user to declare constraints on the query transformations applicable. The system retrieves partial answers, using an intelligent routing strategy, then attempts at combining those referring to the same real-world object.

## 1 Introduction

Peer-to-Peer systems (P2P) are massively distributed systems aimed at sharing large amounts of resources. In the P2P paradigm, no *apriori* distinction exists between clients and servers: on the basis of its particular needs, each node in a P2P network can act as a client or server for other nodes. Several kinds of P2P infrastructures have been proposed in the recent past. In *unstructured* P2P systems, no restriction is imposed on the resources shared by peers and on their placement in the network topology [11]. In *hybrid* P2P systems, some distinguished peers (*super-peers*) act as resource information indices, that maintain meta-information about the resources made available by the different peers, and are possibly organized in P2P networks themselves [6,17,18]. Finally, *structured* systems support efficient addressing and lookup capabilities by strictly constraining both the distribution of resources and the network topology [21,23].

As regards data management in P2P networks, classical systems used simple $(name, file)$ pairs as their logical data model, and keyword-based querying schemes. Recently, the need of supporting more complex models, such as relational and semistructured, has become a major issue. In particular, XML data management in P2P networks has received significant attention [15]. Many recent proposals exploit semantic schema mapping techniques (possibly obtained through some form of automation) to obtain proper views over XML data coming from different sources. These approaches are very powerful in expressing complex XML queries on P2P databases. However, they suffer from some shortcomings. First, human effort is often needed in order to define the mappings, which limits the flexibility of the network. Moreover, as semantic correspondences are typically combined transitively among different peers, data provided by a peer can be correctly interpreted only if the peer is reachable through a "chain" of

R. Wagner, N. Revell, and G. Pernul (Eds.): DEXA 2007, LNCS 4653, pp. 287–297, 2007.

semantic correspondences during query evaluation; therefore, the typical volatility of peers becomes a critical aspect. Finally, since every peer expresses queries on its own local schema, it can only retrieve information mentioned in that schema.

In this paper we propose a hybrid P2P system, named *VXPeer*, that is oriented at XML data retrieval. The system guarantees high peer autonomy and enables the retrieval of meaningful information over the network using XPath-based queries. In our approach, peers are allowed to store and/or export data employing different schemas and terms, without the need to define semantic mappings with respect to their neighbors. Moreover, in each different peer, a same object can be represented with respect to different properties, by means of what we call *partial* data. Our system attempts at retrieving the objects that satisfy a given query, even if their description is spread across many peers. In addition, the system intelligently narrows the search of peers that are able to contribute to query results, by exploiting compact synopses of their XML data.

As a motivating example, consider the scenario described in Fig. 1. Three different peers store data about the same book, but each peer employs a different schema and focuses on different properties of the book (Figs. 1(a), 1(b), and 1(c)). In this scenario, a user interested in finding information about books written by the author "Silberschatz" and having a price lower than 70, may issue an XPath query $q$ of the form `//book[author='Silberschatz'][price<70]`. Fig. 1(d) shows a *tree pattern* query [16] corresponding to $q$, where the dashed box denotes the output node. None of the XML fragments shown in Figs. 1(a), 1(b), and 1(c) is an exact answer to $q$; therefore, the evaluation of $q$ over the three fragments yields no result. However, the peers contain enough information to correctly characterize the book of interest and therefore answer $q$.

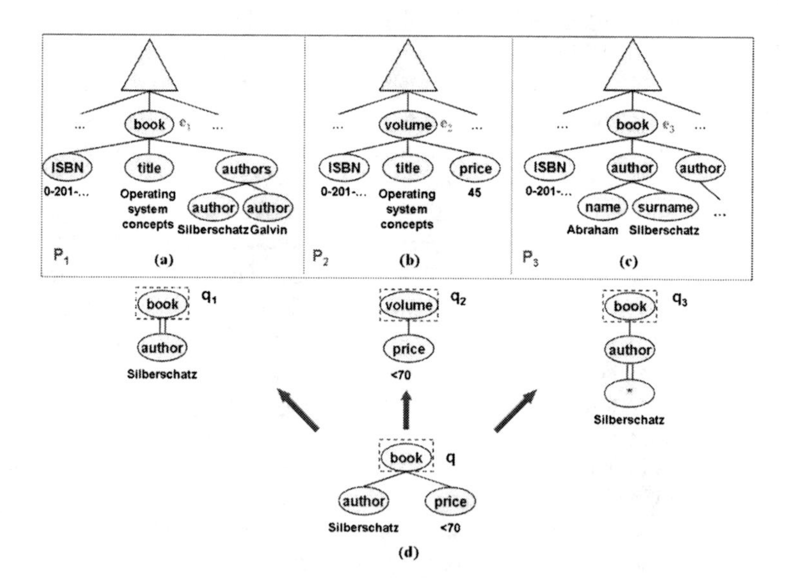


**Fig. 1.** Motivating example

## 2 Vague Queries on P2P XML Databases

In this section we define a query language, named *VXPeerQL*, whose flexibility enables users to find the information they are interested in, even when such information is disseminated in different peers of the network. The language is essentially a specialization of XPath. For the sake of conciseness, we do not consider attributes in the definition of the language.

As observed in Section 1, different peers may provide information about the same object, and in general each peer may describe a subset of the object's characteristics. In the example of Fig. 1, indeed, every peer provides an XML element containing information about a same book, but none of these elements satisfies query $q$. However, the book *is* an answer to $q$, as it has been written by "Silberschatz" and its price is 45. Thus, an XML P2P system should return an object describing the book as an answer to query $q$.

A basic query evaluation strategy on P2P XML databases would be that of collecting all available XML elements, merging those associated with the same object, and finally check whether the merged elements satisfy a given query. Obviously, it is preferable to retrieve only those elements which satisfy some of the conditions expressed by the query, thus avoiding the transmission of useless data. This can be done by obtaining transformed versions of the query where some of the conditions are relaxed, in order to match elements which are candidate to be answers to the query (*partial answers*). We call this process *vague query evaluation*.

For instance, in the scenario of Fig. 1, we can transform $q$ into three queries $q_1, q_2$, and $q_3$, to be evaluated over the three peers. In particular, $q_1$ is obtained by relaxing the parent-child relationship between the `book` and `author` elements to an ancestor-descendant one, and removing the subtree rooted at the `price` element. $q_2$ is obtained by renaming the root element from `book` to `volume`, and removing the subtree rooted at the `author` element. Finally, $q_3$ is obtained by removing the subtree rooted at the `price` element and adding a $*$-labeled descendant to the `author` element. Hence, the answer to $q$ can be computed by first issuing $q_1, q_2$, and $q_3$ on $P_1, P_2$, and $P_3$, respectively, then merging the retrieved elements to provide the set $\{e_1, e_2, e_3\}$ as the final answer to $q$.

VXPeerQL is based on XPath, and in addition it supports several basic transformations applicable to XPath expressions. Specifically, the *basic transformations* applicable to an XPath step of the form axis::$l[f]$ are the following:

- *node renaming*: $l$ is replaced with a different label $l'$;
- *node deletion*: the step is replaced with descendant-or-self::$*[f]$;
- *edge relaxation*: the step is replaced with descendant::$l[f]$.

Moreover, transformations are applicable to XPath predicates, which appear in the leaf nodes of the corresponding tree pattern query. Let $f$ be an XPath comparison predicate of the form `text() ` *op* ` 'text'`, where *op* is a comparison operator; the language supports the following transformations:

- $*$-*node insertion*: $f$ is replaced by descendant::$*[f]$;
- *relaxation of equality predicate*: if the comparison operator used in $f$ is $=$, $f$ is replaced with `contains(text(),'text')`;
- *predicate deletion*: $f$ is deleted.

The application of a basic transformation to an XPath expression yields a transformed version of the original expression and implies a transformation cost. The cost implied by the application of a node renaming is weighted with a value representing the semantic dissimilarity between the specified label and the labels found in the data. The dissimilarity value between labels is computed by a semantic distance function provided by an external ontology. Given an XPath expression $t$ and a transformed version $t'$ of $t$, the cost associated with $t'$ is the sum of the transformation costs of the basic transformations applied to $t$ for obtaining $t'$. In order to avoid answers that are too different from those requested by the original expression, only answers satisfying relaxed expressions whose overall transformation cost is under a user-specified *cost threshold* are considered.

VXPeerQL provides two additional ways of characterizing the answers to be collected. First, *prohibited* transformations can be specified. Second, some transformations can be *marked* and the maximum number ($max_m$) of marked transformations applicable can be specified. In the remainder, we will focus on the three basic query types:

- *exact* queries, where no transformation is allowed;
- *count-based* queries, where some transformations are marked and $max_m$ is specified;
- *cost-based* queries, where no transformation is marked.

As an example of VXPeerQL query, consider the one shown in Fig. 2. Here, all basic transformations are allowed except the relaxation of the child axis between nodes `book` and `price`. Moreover, the deletions of nodes `price` and `author` are marked. Hence, in this case, specifying $max_m = 1$ would mean that query answers obtained by removing both `author` and `price` nodes must not be considered part of the result.

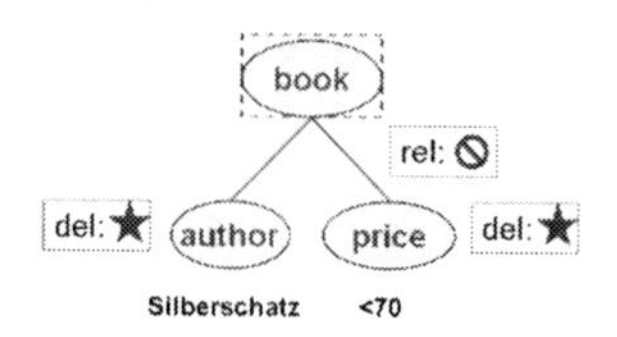


**Fig. 2.** VXPeerQL query

The relaxation of tree pattern queries has been proposed in [2,22], where the cost associated with the transformations indicates the "semantic" distance between the transformed query and the original one. We adopt a similar approach to query relaxation. In addition, as we aim at combining partial results coming from different peers, we distinguish two levels at which to apply relaxation: during the retrieval of partial answer (*local* relaxation) and when building final answers (*global* relaxation). Every query is thus evaluated with respect to local and global cost thresholds and number of allowed transformations. Local values are in general assumed to be higher than global ones, as query evaluation should be less restrictive when gathering candidate elements than when computing final answers: deeper modifications should be allowed to the original query, but when partial answers are merged, the resulting elements should better satisfy the original query.

## 3 The VXPeer System

Our proposed system implements the simple architecture shown in Fig. 3. In particular, the left-hand side of the figure depicts the modules implemented by peers, and its right-hand side depicts the modules implemented by super-peers. Each peer is connected to a single super-peer.

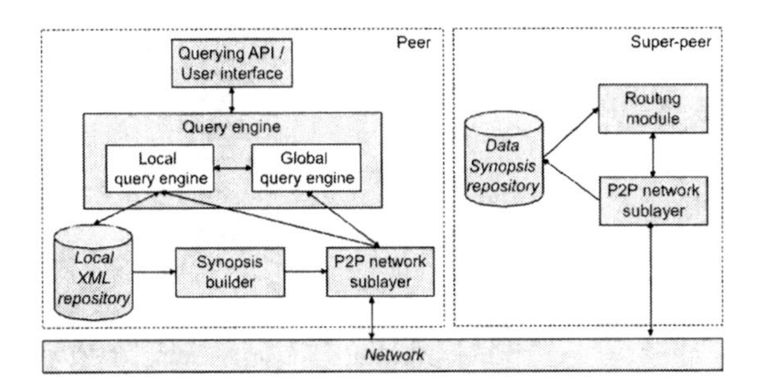


**Fig. 3.** VXPeer architecture

Besides the underlying database management subsystem, the architecture of peers comprises four main modules: the *P2P network sublayer*, the *Synopsis builder*, the *Querying API/User interface*, and the *Query engine*. The P2P network sublayer manages the interactions with the underlying network. The synopsis builder computes concise representation of the stored XML data (whose structure will be detailed in the following), and sends them to the super-peer of reference, through the P2P network sublayer. The querying API/user interface module manages the interactions with users. It provides an API for submitting queries in their textual form and collecting results. A user interface allows the user to (*i*) specify queries in both graphical and textual form; (*ii*) obtain a graphical representation of the results as they are received (as it will be clearer in the following, the systems aims at firstly contacting the peers that are likely to provide results); (*iii*) decide, on the basis of his/her degree of satisfaction, when to stop the process. The query engine implements the query evaluation algorithm and the logic for combining partial answers coming from different sources. These functionalities are managed separately by two submodules:

- The *Local query engine* applies the vague query evaluation process over the local XML database, producing partial answers. Such answers may bring along part of the local schema, for extracting proper information (e.g., key constraints) which is subsequently used to evaluate the degree of dissimilarity among different XML elements. Moreover, each XML element in the answers is paired with the query transformation cost introduced to match the actual data. The results of the local query evaluation process are returned to the global query engine if the query was submitted to the local peer, otherwise they are sent back through the P2P network sublayer. The local query engine also connects to an external ontology (not shown in the figure) that provides the semantic distance function between two element names.

- The *Global query engine* is employed when a query is issued locally. It forwards the query to the super-peer of reference and collects answers through the P2P network sublayer, then completes the global query evaluation process by joining the partial results obtained and returning them to the user through the querying API. More details on the overall vague query evaluation process are given in Section 3.1.

The architecture of super-peers comprises three main modules: the *Synopsis repository*, the *P2P network sublayer*, and the *Routing module*. The P2P network sublayer receives data synopses from peers and stores them into the repository. Moreover, it receives vague queries from peers and passes them to the routing module. The routing module works in co-operation with the other super-peers. It gathers data synopses from its local repository and from the repositories of other super-peers, then it applies an efficient routing strategy that, by exploiting the information in the synopses, is capable of (*i*) reducing the number of query issued on non-relevant peers, i.e., peers whose local schema ensures that the local query evaluation would not provide results; (*ii*) giving priority to peers that will possibly provide more results. This routing strategy is based on selectivity estimation techniques that make use of *XSketch* data synopses [12], and will be described in Section 3.2.

### 3.1 Evaluating Vague Queries

We now briefly describe the overall process of vague evaluation of a VXPeerQL query over a set of peers. The process is composed of 4 main steps:

***Local evaluation.*** The query is evaluated over each peer by possibly applying suitable transformations. At each step of the evaluation, the evaluation algorithm tries to apply all possible transformations to a query step that make it match the available data while not violating the local thresholds.

***Joining.*** The partial answers (XML elements) yielded by the local evaluation which are likely to refer to the same object are joined. This process is based on a function measuring the dissimilarity *of the objects two elements describe* by looking at their keys. XML elements whose dissimilarity value is under a certain *join threshold* are grouped in sets (named *vague XML elements*). The output of this step is a set of vague XML elements, each representing a query answer and having an "overall" transformation cost.

***Selection.*** Vague elements whose associated transformation cost is under the global threshold are selected.

***Pruning.*** As the joining step may produce vague elements that are subsets of others and thus needless in the final result, redundant vague elements are pruned from the result. Note that, as shown in [9], if the dissimilarity function has certain properties, it is feasible to prune intermediate results during query evaluation, thus notably increasing the efficiency of the query evaluation process.

A more detailed description of the vague query evaluation process can be found in [9].

### 3.2 Routing VXPeer Queries

In this section we describe the routing strategy implemented by the super-peers in VXPeer. This routing strategy is based on the use of the *XSketch* synopses proposed in [12].

The XSketch synopsis associated with an XML document is a graph whose nodes represent sets of elements in the document that have the same name. Each node in the synopsis is annotated with the cardinality and the shared element name of the corresponding set. An edge between two nodes $n_1, n_2$ represents a parent-child relationship between an element in $n_1$ and an element in $n_2$. Moreover, the edge from $n_1$ to $n_2$ is labeled with `F` iff every element in $n_1$ has at least one child in $n_2$; the edge is instead labeled with `B` iff for every element in $n_2$, its parent is in $n_1$. For instance, in the document represented by the synopsis in Fig. 4, (*i*) there are 2 `book` and 4 `title` elements; (*ii*) each `book` and `paper` has a `title`; (*iii*) each `book` has an `isbn`, and `isbns` are children of `books` only; (*iv*) the 4 `authors` elements are children of both `books` and `papers`, but all of the 11 `author` elements are children of `authors` elements.

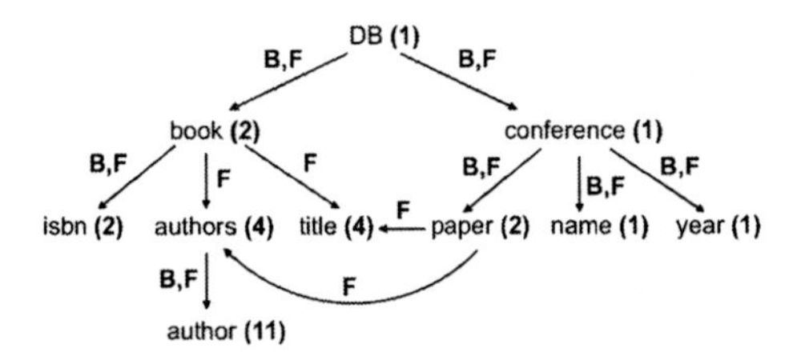


**Fig. 4.** An example XSketch synopsis

An XSketch synopsis can be exploited to estimate the selectivity of an XPath expression, that is the number of XML elements that are selected by the expression. In general, the selectivity estimation of an XPath expression $q$ using a synopsis $\mathcal{S}$ is performed by first computing the whole set of embeddings of $q$ in $\mathcal{S}$, then summing up the selectivity associated with each embedding. In particular, our algorithm uses the algorithm proposed in [12] to compute the selectivity of an XPath query $q$ w.r.t. a node $n$ of the synopsis, denoted as $sel(q, n)$.

Selectivity estimation is used by the VXPeer query routing module to compute an overall *score* given to a synopsis with respect to a VXPeerQL query. This score is then employed to drive routing decisions, i.e., more priority is given to the peers whose synopses exhibit higher scores. The score given to a synopses w.r.t. a query is computed by first removing textual filters from the query, as this information is not represented by the synopsis. Then, for each node in the synopsis, the selectivity of the transformed versions of the query w.r.t. the node is computed. Since a transformed query does not represent all the original query conditions, we weigh the selectivity associated with a node in the synopsis with the "relative" cost of the transformed query which selects the node. Given a VXPeerQL query $q$ and a transformed query $rq$, the relative transformation cost of $rq$ is given by the cost for obtaining $rq$ from $q$ ($cost(rq, q)$) divided by the maximum transformation cost for every possible transformed query obtainable from $q$ ($maxcost(q)$). The score given to a synopsis $\mathcal{S}$ w.r.t. a query $q$ is defined as follows:

$$score(q, \mathcal{S}) = \sum_{n \in \mathcal{S}} max_{rq \in \mathcal{R}(q,n)} \left( sel(rq, n) * \frac{cost(rq, q)}{maxcost(q)} \right)$$

where $\mathcal{R}(q, n)$ is the set of transformed queries obtainable from $q$, without violating cost thresholds, that return node $n$ when evaluated on $\mathcal{S}$. Note that the formula correctly rules out non-output nodes as no transformed query exists for them under the cost thresholds.

## 4 Experimental Results

In this section we show some preliminary experimental results assessing the effectiveness of our proposed system. The experiments have been performed by issuing queries against a network of 100 peers providing clinical and diagnostic data. The peers adopted different schemas (partially shared with the other peers) and differently-structured keys comprising social security, fiscal, and personal data. 10 of the 100 peers acted as super-peers and were part of a fully-connected network; each of them was connected to 9 peers. The total data size was 100MB. The 4 queries issued (Q1, Q2, Q3, and Q4) returned elements describing patients, with decreasing selectivity: Q1 imposed conditions on a specific disease and year of hospitalization; Q2 and Q3 looked for patients who suffered from two and three specific diseases, respectively; Q4 also imposed conditions on the surgeries undergone by the patients. We employed three different global cost thresholds, corresponding to the 50% (*high*), 30% (*medium*), and 10% (*low*) of the maximum cost of the transformed versions of the queries. The local cost threshold has been set equal to the global one increased by a 25%. Moreover, we made the conservative choices of setting the global maximum number of marked transformations applicable equal to 1 and the local one equal to the 50% of the number of marked transformations in each query. The experiments have been run on a LAN with 100 Pentium IV machines each equipped with 512MB RAM. The queries have been executed using a two-minutes timeout.

Fig. 5(left) shows the number of correct answers returned. Here, a vague element is assumed to be an incorrect answer if either it contains an element describing an object that is not an answer to the query, or if it contains two elements describing different

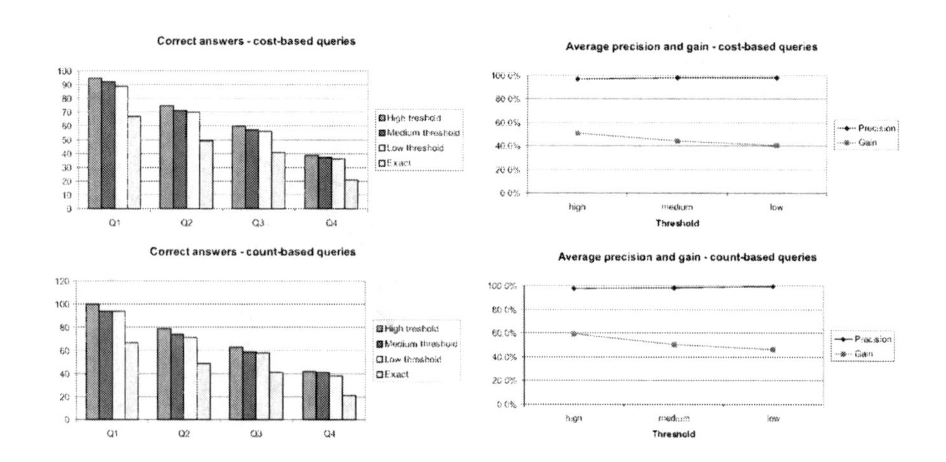


**Fig. 5.** Experimental results

objects. The figure compares the number of correct answers obtained by adopting cost-based and count-based evaluation with the baseline of exact evaluation. Fig. 5(right) reports the average *precision* obtained, defined as the ratio between the number of correct answers and the total number of answers. We do not report the *recall*, that is the ratio between the number of correct answers and total number of objects satisfying the query, as it is impractical to assess the total number of satisfying objects. We instead report a parameter estimating the recall increment obtained by relaxing the query; this parameter, called *gain*, is defined as $ans/exAns - 1$ where $ans$ is the number of correct answers to the query, and $exAns$ is the number of correct answers to the exact version of the query. The experimental results show that in all cases relaxed queries allow the retrieval of more answers than exact queries (the gain was 45.5% on average for cost-based queries and 52.2% for count-based queries) with a generally high precision (97.8% on average for cost-based queries and 98.3% for count-based queries).

We also evaluated our proposed scoring function by looking at how the number of partial answers returned by peers is related to the score given to their synopses. Fig. 6 reports the percentage of partial answers retrieved as the evaluation proceeds; the X-axis reports the percentage of peers already contacted (we recall that peers are contacted in decreasing score order). We averaged the values over the 4 cost-based queries with medium threshold. The results obtained show that the routing policy gives proper priority to the peers that are more likely to contribute to the query results. Specifically, in the case depicted in the figure, more than 80% of the total number of answers are returned to the user after having accessed just 40% of the contributing peers.

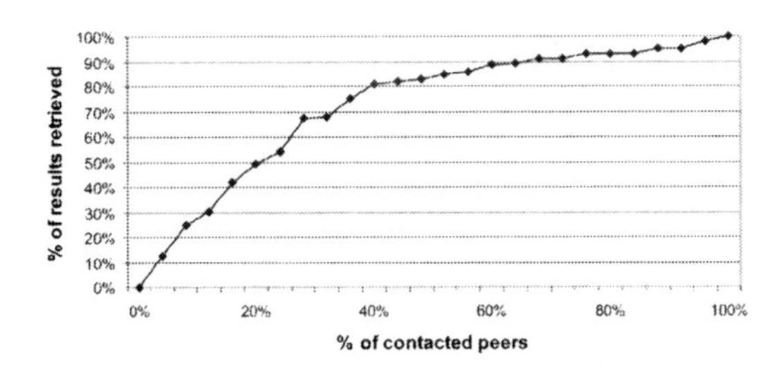


**Fig. 6.** Effect of the routing policy

## 5 Related Work and Conclusions

Significant research efforts have recently been made to make P2P infrastructures capable of supporting complex data models and expressive querying languages [1,5,13,19]. Several solutions have also been proposed to the problem of XML data sharing and querying in the context of P2P networks [7,8,14,20,24,25]. For instance, in the HepToX [7] system, schema correspondences are established using visual annotations and queries against a local peer schema are translated according to the schemas of its neighbors using schema correspondences. In DBGlobe [20], data are wrapped by services and distributed indexes are used to discover peers that offer an appropriate service. Information exchanged between peers is expressed in a language supporting calls to services embedded in XML data. In PIAZZA [24] each peer stores the XMLSchema of its data and mappings between this schema and those of its neighbors. Query processing relies on the propagation of the reformulated query to the neighbors, according to the mappings. Techniques for pruning paths in the reformulation process and for minimizing the reformulated query are also applied.

Approaches to the problem of vague XML querying have also been proposed, targeted at query answering over single XML documents [2,3,4,10,22,25]. None of the proposed approaches tackles the problem of integrating partial XML data coming form different sources. In particular, [22] proposes a query language that allows node insertion, deletion, and renaming, and employs a bottom-up query evaluation algorithm based on a list algebra. Costs are associated with labels and possible node renamings are chosen in advance and completely specified by the user, independently of the available data. In [2], allowed transformations are node renaming, leaf deletion, subtree promotion, and edge relaxation. The evaluation algorithm works with complex join plans that embed all possible transformations. The proposed querying mechanism only supports node renamings w.r.t. fixed name hierarchies that must be provided.

In this paper we proposed a system for querying P2P XML databases. The system ensures high autonomy to participating peers, as it does not rely on a global schema or semantic mappings. Nonetheless, it enables the retrieval of meaningful results by combining partial answers coming from different peers, through the use of (limited) knowledge about their local schemas. We also proposed an intelligent routing technique that proved useful in improving the efficiency of the distributed query evaluation process.

## References

1. Aberer, K., Cudré-Mauroux, P., Hauswirth, M., Van Pelt, T.: GridVine: Building internet-scale semantic overlay networks. International Semantic Web Conference, pp. 107–121 (2004)
2. Amer-Yahia, S., Cho, S., Srivastava, D.: Tree pattern relaxation. In: Jensen, C.S., Jeffery, K.G., Pokorný, J., Šaltenis, S., Bertino, E., Böhm, K., Jarke, M. (eds.) EDBT 2002. LNCS, vol. 2287, pp. 496–513. Springer, Heidelberg (2002)
3. Amer-Yahia, S., Koudas, N., Marian, A., Srivastava, D., Toman, D.: Structure and content scoring for XML. In: VLDB (2005)
4. Amer-Yahia, S., Lakshmanan, L.V.S., Pandit, S.: FleXPath: Flexible structure and full-text querying for XML. In: SIGMOD Conference, pp. 83–94 (2004)
5. Arenas, M., Kantere, V., Kementsietsidis, A., Kiringa, I., Miller, R.J., Mylopoulos, J.: The hyperion project: From data integration to data coordination. In: SIGMOD Record (2003)
6. http://www.bittorrent.com
7. Bonifati, A., Chang, E.Q., Ho, T., Lakshmanan, L.V.S., Pottinger, R.: HePToX: Marrying XML and heterogeneity in your P2P databases. In: VLDB, pp. 1267–1270 (2005)
8. Comito, C., Patarin, S., Talia, D.: PARIS: A peer-to-peer architecture for large-scale semantic data integration. In: DBISP2P (2005)
9. Fazzinga, B., Flesca, S., Pugliese, A.: Vague queries on heterogeneous XML data sources. Technical Report (2006), http://si.deis.unical.it/ apugliese/vague.pdf
10. Fuhr, N., Großjohann, K.: XIRQL: An XML query language based on information retrieval concepts. ACM Trans. on Information Systems (2004)

11. http://www.gnutella.com
12. Polyzotis, N., Garofalakis, M.: XSKETCH Synopses for XML Data Graphs. ACM Transactions on Database Systems 31(3), 1014–1063 (2006)
13. Huebsch, R., Hellerstein, J.M., Lanham, N., Loo, B.T., Shenker, S., Stoica, I.: Querying the internet with PIER. In: VLDB, pp. 321–332 (2003)
14. Kokkinidis, G., Christophides, V.: Semantic query routing and processing in P2P database systems: The ICS-FORTH SQPeer middleware. In: Lindner, W., Mesiti, M., Türker, C., Tzitzikas, Y., Vakali, A.I. (eds.) EDBT 2004. LNCS, vol. 3268, pp. 486–495. Springer, Heidelberg (2004)
15. Koloniari, G., Pitoura, E.: Peer-to-peer management of XML data: Issues and research challenges. In: SIGMOD Record (2005)
16. Miklau, G., Suciu, D.: Containment and equivalence for a fragment of XPath. J. ACM 51(1), 2–45 (2004)
17. http://www.napster.com
18. Nejdl, W., Wolf, B., Qu, C., Decker, S., Sintek, M., Naeve, A., Nilsson, M., Palmér, M., Risch, T.: EDUTELLA: A P2P networking infrastructure based on RDF. In: WWW, pp. 604–615 (2002)
19. Ng, W.S., Ooi, B.C., Tan, K.L., Zhou, A.: PeerDB: A P2P-based system for distributed data sharing. In: ICDE 2003, pp. 633–644 (2003)
20. Pitoura, E., Abiteboul, S., Pfoser, D., Samaras, G., Vazirgiannis, M.: DBGlobe: A service-oriented P2P system for global computing. In: SIGMOD Record (2003)
21. Ratnasamy, S., Francis, P., Handley, M., Karp, R.M., Shenker, S.: A scalable content-addressable network. In: SIGCOMM 2001 (2001)
22. Schlieder, T.: Schema-driven evaluation of approximate tree-pattern queries. In: Jensen, C.S., Jeffery, K.G., Pokorný, J., Šaltenis, S., Bertino, E., Böhm, K., Jarke, M. (eds.) EDBT 2002. LNCS, vol. 2287, pp. 514–532. Springer, Heidelberg (2002)
23. Stoica, I., Morris, R., Karger, D.R., Kaashoek, M.F., Balakrishnan, H.: Chord: A scalable peer-to-peer lookup service for internet applications. In: SIGCOMM 2001, pp. 149–160 (2001)
24. Tatarinov, I., Halevy, A.Y.: Efficient query reformulation in peer-data management systems. In: SIGMOD Conference (2004)
25. Theobald, A., Weikum, G.: Adding Relevance to XML. In: WebDB (Informal Proceedings), pp. 35–40 (2000)

# Proximity Search of XML Data Using Ontology and XPath Edit Similarity

Toshiyuki Amagasa[1,2], Lianzi Wen[1], and Hiroyuki Kitagawa[1,2]

[1] Graduate School of Systems and Information Engineering,
Department of Computer Science
[2] Center for Computational Sciences,
University of Tsukuba
1–1–1 Tennodai, Tsukuba 305–8573, Japan
{amagasa, kitagawa}@cs.tsukuba.ac.jp, moon@kde.cs.tsukuba.ac.jp

**Abstract.** XML data is explosively increasing, and a large amount of XML data, in which similar contents are described using different tag names and structures, have been emerging as a consequence. In such a situation, one cannot write a query against such XML data unless he/she knows the structure of the data. In this research, we propose a scheme to cope with this problem. Specifically, we expand XPath queries by replacing tag names with similar ones with the help of ontologies. In addition, we try to realize (structural) proximity matching of path expressions using edit similarity, which is a similarity measure based on edit distance. We also discuss application of SSJoin, which is an operator to support similarity joins in relational database systems, for speeding up the proposed scheme. We finally show the effectiveness of the proposed method by a series of experimentations.

## 1 Introduction

For the past several years, XML (Extensible Markup Language) [1] has become a ubiquitous format for electronic data representation and exchange. XML is a meta-language for (semi)structured data; any tree-structure can be represented in terms of nested elements in plain text. For this brevity, growing number of applications in B2B, B2C, Web services, e-Sciences, and e-Governments use XML as their data formats.

Because of the explosive diffusion of XML, huge amount of information resources written in XML is available in the Internet. For this reason, it is essential that we are able to locate portions of XML data of interest. XPath (XML Path Language) [2] is a language for locating arbitrary portions of XML data in terms of path expressions. Due to its importance, XPath has been used in many applications and related standards, such as XSLT [3] and XQuery [4], as a (simple) query language or a sub-language.

A remarkable observation here is that there are many cases where similar contents are described with different XML schemata[1]. Let us take bibliographic

[1] We use the term "XML schema" to denote schemata for XML data in a general sense, and discriminate it from W3C XML Schema.

R. Wagner, N. Revell, and G. Pernul (Eds.): DEXA 2007, LNCS 4653, pp. 298–307, 2007.

databases for example. As is well-known for researchers, many academic societies, such as ACM and IEEE-CS, publish their bibliographic data in digital libraries, and some of them are also provided in the form of XML data. Although such data have similar contents, their (document) structures are different with each other in many cases. Such situations give rise to several problems. One is diversity of tag names among XML data, that is, similar contents may be described in different tag names. For example, information about papers may be described using `paper` tag in an XML data. However, in another data, `article` tag may be used. Another problem is structural difference among XML data. For example, in an XML data, authors of a paper may be represented with a path like `/papers/paper/authors/author`, while `/bib/paper/author` may be used in another data. In such situations, an XPath query, which is tailored for an XML data, cannot be applied to other XML data unless the underlying XML schema is unified.

To cope with this problem, we propose a scheme that allows us to apply an XPath query designed for known XML data to those XML data whose structure is not known. Specifically, given an XPath query, we attempt to expand it with the help of ontologies, by rewriting tag names with different ones which are expected to describe similar contents. Another technique proposed in this paper is proximity matching of path expressions using *edit similarity*. The *edit similarity* is a similarity measure between a pair of sequences, which can be computed from the edit distance between them. In addition, in order to compute edit similarity efficiently, we employ SSJoin [5], which is a primitive operator for similarity joins in relational databases. This enables us to match large amount of different but similar path expressions efficiently.

The rest of this paper is organized as follows. Section 2 describes some related works. In Section 3, our proposed scheme is presented and Section 4 is to discuss the proposed system based on the scheme, The effectiveness is evaluated in Section 5, and Section 6 concludes this paper.

## 2 Related Work

In the context of relational databases, similarity joins have been well studied in order for proximity matching of structured data [6].

Meanwhile, XML has been dominated as a de facto standard of data representation, and many researchers have devoted on the issues of proximity search and similarity join on XML data. Liang et al. [7] proposed a scheme for similarity join over two distinct XML data. In their approach, for each pair of extracted XML subtrees, similarity of text contents and path expressions are computed, and they are joined if the similarity is higher than a given threshold. Although they address the problem of proximity matching over XML path expressions, their approach is quite simple in the sense that they just count the number of shared tag names over two distinct path expressions.

In order to deal with the heterogeneity of XML data, a number of researchers addressed the problem of proximity searching of XML data. For example,

Amer-Yahia et al. [8] proposed a scheme for proximity search on XML data using the technique called *tree pattern relaxation*. A user query is given in terms of a tree pattern consisting of nodes and weighted edges. A tree pattern can be relaxed in order to control the number of matching query results. For a given query, the ranking of candidate results is done by computing scores of each candidate in consideration of tree pattern weights.

It is well known that *tree edit distance* (TED) [9] has been used to measure structural similarity over XML fragments. It is defined as the minimum number of operations over nodes, such as insertion, deletion, and replacement, necessary to transform a tree to another. However, TED is costly to compute, and it is therefore difficult to apply them to massive data, consequently.

In this paper, we attempt to investigate a scheme for proximity matching of path expressions rather than subtrees, with the aim of introducing more detailed measurements based on ontologies and edit similarity. In particular, most of the related works are parameterized by lexical choices, whereas our proposed scheme uses ontological information to overcome the limitations. Furthermore, we make the best use of the power of RDBMSs to speed up the processing.

## 3 The Proposed Scheme

For the purpose of proximity matching between a pair of path expressions, we attempt to use ontologies for expanding tag names and edit similarity for evaluating structural similarity between the paths.

### 3.1 Tag Name Expansion Using Ontologies

As discussed in the introduction, it is often the case that different tag names are used to describe similar facts and/or contents for real XML data. For example, `article` may be used in a bibliographic XML data, while `paper` may be used in another data.

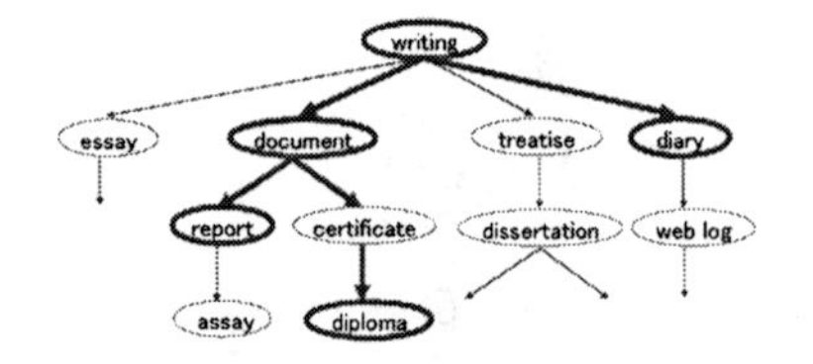


**Fig. 1.** An example of ontology

To cope with this problem, we employ ontologies that are controlled vocabularies for specific problem domains. An ontology is usually represented as a graph structure[2], where a concept (or a term) corresponds to a node in the graph, and edges among the nodes represent relationships among the concepts. It turns out that for a given concept (or a term), similar ones can be obtained by finding neighboring nodes. In particular, we use the concept of LCAs (Least Common Ancestors) to decide whether given pair of terms are similar or not. For a pair of distinct nodes in a DAG, the LCA is the least ancestor node shared by the nodes. A pair of nodes is hence expected

[2] We assume that an ontology is represented as a DAG (Directed Acyclic Graph).

to represent similar concepts, if the maximum distance between the LCA and the nodes being considered is smaller than a given threshold,

Taking an ontology in Figure 1 for example, the LCA of `report` and `diploma` is `document`, and the LCA of `diploma` and `diary` is `writing`. Suppose that the threshold is 2, we decide that `report` and `diploma` are similar, but `diploma` and `diary` are not.

**Types of ontologies.** When considering our objective, it appears that there are several choices on the types of ontologies.

**General purpose thesauri.** Ontologies in this category collect general terms like "title" and "author". A most well-known example may be the WordNet [10]. Such ontologies are suitable for those XML data which contain general terms as their tag names, while they are not applicable in such cases that XML data include domain specific terms, composite words, acronyms, or symbols.

**Domain specific ontologies.** Besides general purpose thesauri, there have been many ontologies dedicated for their specific domains. Gene Ontology [11] is an example. They are useful when tags are named using domain specific terminologies.

**Tailored ontologies.** When sets of tag names or the schemas of the XML data, which are going to be queried, are known in advance, it is likely that users can construct a dedicated (relatively small) ontology consisting of the known vocabulary. This approach has the advantage that users can design an ontology taking account of the requirements of the application. However, it is costly to construct, and maintain as well, when the vocabulary or the schema is huge.

In fact, we can choose any of the above approaches or their combinations. So, it is important to decide the choice by taking into account the characteristics of the XML data. In the following discussion, we use the WordNet as the ontology. Notice that the discussion can be applied other kind of ontologies without loss of generality.

**WordNet.** WordNet [10] is an English thesaurus. The basic component for describing semantics of words is *synset*, which is a set of synonyms. Semantic information among terms, such as *hypernym*, *hyponym*, *holonym*, and *meronym*, is given as relationships among synsets. As a consequence, the entire data structure can be represented as a graph, where synsets are nodes and their relationships are edges.

In order to make WordNet processible in computers, there have been several projects that attempt to convert WordNet into machine readable formats. In this work, we employ RDF/OWL Representation of WordNet [12] in that WordNet is represented using RDF (Resource Description Framework) [13]. It is consisting of several RDF files, and we use `wordnet-senselabels.rdf`, in which correspondence between synsets and words is described, and and `wordnet-hyponym.rdf`, in which relationship among synsets is described.

**Tag name retrieval using SPARQL.** Since an ontology is represented in RDF, for the purpose of retrieving similar tag names using (the above mentioned) distance from LCA, we can make use of query languages for RDF. Taking a closer look at such a query, it is actually a subgraph matching problem. SPARQL (SPARQL Query Language for RDF) [14], which is still a W3C working draft, but is expected to be a standardized query language for RDF, has such a functionality. We do not go into the detail of the language due to the page limitation, but such a query can be processed in the following way:

1. Find synsets $S_w$ containing the tag name $w$ from `wordnet-senselabels.rdf`.
2. Find synsets $S_{wi}$ s.t. the distance between its LCA, identified with $S_w$, is less than a given threshold from `wordnet-hyponym.rdf`
3. Collect all the words contained in $S_{wi}$ using `wordnet-senselabels.rdf`.

### 3.2 Proximity Matching of Path Expressions Based on Edit Similarity

Edit distance is a similarity measure between a couple of strings. It is defined as the minimum number of point mutations required to change one string into another, where a point mutation is change, insertion, or deletion of a letter. There are several variations depending on the way how point mutations are defined and weighted. For example, humming distance only permits insertion and deletion. In this work, we use Levenshtein distance, where change, insertion, and deletion are permitted and are equally weighted, due to its brevity.

From the definition, for a given couple of strings, the edit distance between them is affected by the lengths, which is an undesirable when using it as a similarity measure. In order to cancel those effects, we can make use of *edit similarity* [5], which can be obtained from edit distance by the following formula:

$$ES(\sigma_1, \sigma_2) = 1.0 - \frac{ED(\sigma_1, \sigma_2)}{max(|\sigma_1|, |\sigma_2|)}$$

where $\sigma_1$ and $\sigma_2$ denote strings being compared, and $ED$ and $ES$ denote edit distance and edit similarity, respectively.

When comparing a given couple of path expressions, what we all have to do is to regard each location step (tag name) as an alphabet. Suppose that we attempt to compute the edit similarity between $p_1$ = `/articles/article/author` and $p_2$ = `/article/authors/author`. The edit distance can be computed as follows:

| | | | | |
|---|---|---|---|---|
| $p_1$ | /articles | /article | | /author |
| $p_2$ | | /article | /authors | /author |
| Cost | 1 | 0 | 1 | 0 |

From the result that $ED(p_1, p_2) = 2$, we have $ES(p_1, p_2) = 1.0 - \frac{2}{3} = 0.33\ldots$. Notice that it is quite time consuming if we compute edit similarities for all possible combinations over the sets of path expressions. To cope with the problem, we make use of a technique to accelerate the processing.

### 3.3 Using Set Similarity Join (SSJoin) for Improving Performance

*Set similarity join (SSJoin)* [5] is a novel operator for RDBMSs to accelerate processing proximity matching over strings, originally developed for the data cleaning problem. In this work we attempt to use SSJoin for efficient edit similarity computation over large number of path expressions.

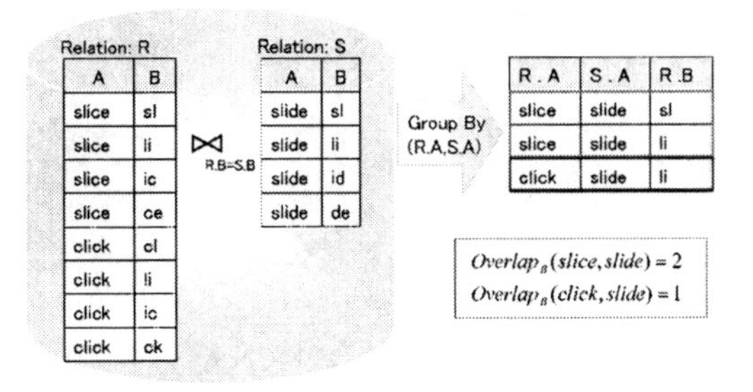


**Fig. 2.** An overview of SSJoin

Figure 2 illustrates how similarities are computed in SSJoin. Every string (path expressions in this work) stored in a database is decomposed into $q$-grams and stored in another relational table $R$. $q$ can be arbitrary integers, but we use 2 in this example. For a given set of query strings, we also convert them into bigrams and store in another table $S$. Then, we equi-join the tables, and count the cardinalities for each combination of strings, which is denoted as $Overlap_B(R.A, S.A)$.

An important notice here is that we can obtain rough estimation of edit distance by the following formula:

$$ED(R.A, S.A) \geq \frac{max(|R.A|, |S.A|) - q + 1}{q} - \frac{Overlap_B(R.A, S.A)}{q}$$

Due to the fact that the edit distance (and edit similarity as well) is guaranteed to be larger than the real value, we can use it to filter out unnecessary candidates. Finally, we refine the resulting candidates by computing real edit similarities in order to get rid of false negatives.

As we can see, the total performance could be significantly improved, because the number of candidate results is reduced by SSJoin, and it ends up with reducing the number of costly edit similarity computation. When applying SSJoin to our scheme, all that we need to do is to regard each tag name as an alphabet.

## 4 System Overview

Figure 3 depicts an overview of our proposed system. The system resides outside of an XML database, and acts as its subsystem for query expansion. We extract all occurrences of distinct path expressions from XML data stored in the XML database, compute $q$-grams of the path expressions, and store them in an underlying relational database (`pathngram`) beforehand.

Given a user's query ($p = /p_1/p_2/\ldots/p_n$) and a threshold $\alpha$, we first extract all tag names from each location step ($p_i, \ldots, p_n$), and try to find similar tag names for them ($P_1, \ldots, P_n$) using ontologies. We implement this step using SPARQL as mentioned above. Having retrieved similar tag names, we obtain candidate path expressions by computing Cartesian product over them, i.e.,

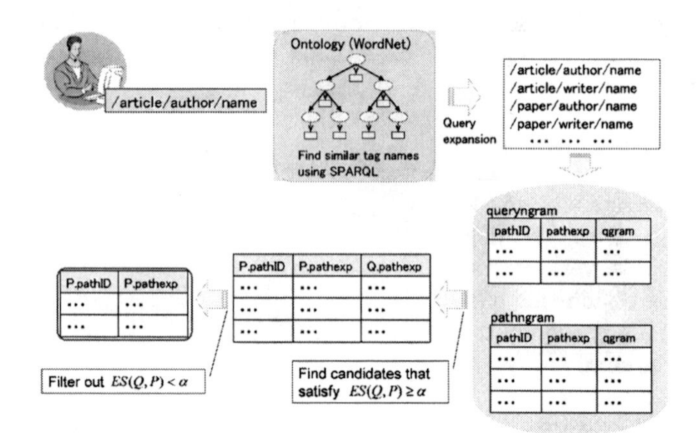


**Fig. 3.** System overview

$P = P_1 \times P_2 \times \ldots \times P_n$. Finally, we omit those path expressions that do not really exist in the XML database.

The next step is to decompose the user's query into $q$-grams, store them in another table (**queryngram**), and perform SSJoin over the tables, **pathngram** and **queryngram**, for filtering out unnecessary candidates whose similarities are less than threshold ($\alpha$) from $P$. For the remaining candidates $P$, we calculate real edit similarities, and get rid of false negatives.

*Coping with descendant axis.* The readers can easily suspect that we need a special care about such queries that include descendant axis (`//`). A workaround for the problem is to rewrite (or expand) the original query to queries without descendant axis. For example, suppose that we are going to process **//title**. What we need to do is to find such path expressions stored in the database that end with `title`, such as:

- **/article/prolog/title**,
- **/article/body/section/title**, and
- **/article/body/section/subsec/title**.

As a consequence, we can query the database with these expanded path expressions instead of the original query (**//title**).

For other kinds of axes, we can apply a scheme to rewrite queries including backward axes to those that consist of forward axes [15].

## 5 Experimental Evaluation

We have conducted several experimentations to evaluate the effectiveness of our proposed scheme.

### 5.1 Experimental Setup

We used 4-way Intel(R) Xeon(TM) 3.0GHz with 6GB memory running Red Hat Enterprise Linux 4.0. The program was implemented using J2SE 1.5, and we used PostgreSQL 8.1.0 as the underlying RDBMS.

**Table 1.** Dataset statistics

| Data | Size (KB) | # paths |
|---|---|---|
| SIGMOD Record | 464 | 12 |
| DBLP | 357,284 | 164 |
| Wikipedia | 266,108 | 10 |
| XBench | 10,608 | 38 |

We used XML versions of SIGMOD Record[3] and DBLP bibliography[4], Wikipedia's abstract[5], and synthetic data generated by XBench [16]. Table 1 shows a statistics of the dataset.

In this experiment, we firstly checked the feasibility of the proposed scheme from the viewpoint of accuracy. Specifically, we created a set of benchmark queries, whose target is SIGMOD Record data, and tried to find corresponding path expressions from other dataset. We then tested the effectiveness of the method.

### 5.2 Experimental Results

**Accuracy.** In our preliminary experiment, we had a problem that irrelevant path expressions were ranked higher, i.e., both `/dblp/book/author` (1st) and `/dblp/book/series` (2nd) were ranked higher for the query `/SigmodRecord/issues/issue/articles/article/authors/author`. This is due to the nature of edit distance that any alphabet (tag name for this case) can be replaced by another one with a fixed cost. To cope with this problem, we added an additional refinement step in that candidate paths were filtered out if their tag names at the bottom are irrelevant to that of the query. In the above case, `/dblp/book/series` may be filtered out, if the maximum distance between `series` and `author` and their LCA is larger than a given threshold.

Tables 2 (a)–(f) show the top 5 query results for the benchmark queries. It appears that the proposed scheme is successful in retrieving similar path expressions. The percentages of correct answers for the queries were: (a) 100%, (b) 83.3%, (c) 80%, (d) 75%, (e) 71.4%, (f) 80%, and 81.3% on average.

**Efficiency.** We investigated the efficiency of the proposed scheme, in particular to see the benefit from SSJoin. We conducted two cases where the maximum distance from LCA is 1 and 2. For each case, we tested SSJoin with 1-gram and 2-gram, and without SSJoin as the baseline.

Figure 4 illustrates the elapse time for the benchmark queries. We can observe that the larger the maximum LCA distance the more time it takes to process, because the number of candidate path expressions increases significantly as the LCA distance grows. However, it seems that SSJoin is quite successful in reducing the computational cost. Note that some 2-gram results are missing. This is due to the fact that path expressions (queries and the ones that extracted from

[3] `http://www.acm.org/sigmod/record/xml/`

[4] `http://www.informatik.uni-trier.de/~ley/db/`

[5] `http://en.wikipedia.org/wiki/Wikipedia:Database_download`

**Table 2.** Query results (top 5)

$p_1$ /SigmodRecord/issues/issue/articles/article/authors/author

| pathID | pathexp | ES |
|---|---|---|
| 184 | /dblp/article/author | 0.285 |
| 7 | /article/prolog/authors/author | 0.285 |
| 96 | /dblp/book/author | 0.285 |
| 69 | /dblp/mastersthesis/author | 0.142 |
| 62 | /dblp/phdthesis/author | 0.142 |

$p_2$ /SigmodRecord/issues/issue

| pathID | pathexp | ES |
|---|---|---|
| 103 | /dblp/book/series | 0.333 |
| 67 | /dblp/phdthesis/series | 0.333 |
| 40 | /dblp/proceedings/series | 0.333 |
| 73 | /dblp/article/journal | 0.333 |
| 195 | /dblp/proceedings/journal | 0.333 |

$p_3$ /SigmodRecord/issues/issue/number

| pathID | pathexp | ES |
|---|---|---|
| 125 | /dblp/article/number | 0.25 |
| 186 | /dblp/phdthesis/number | 0.25 |
| 158 | /dblp/proceedings/number | 0.25 |
| 202 | /dblp/inproceedings/number | 0.25 |
| 187 | /dblp/book/month | 0.25 |

$p_4$ /SigmodRecord/issues/issue/volume

| pathID | pathexp | ES |
|---|---|---|
| 89 | /dblp/book/volume | 0.25 |
| 176 | /dblp/article/volume | 0.25 |
| 196 | /dblp/proceedings/volume | 0.25 |
| 103 | /dblp/book/series | 0.25 |
| 67 | /dblp/phdthesis/series | 0.25 |

$p_5$ /SigmodRecord/issues/issue/articles/article/title

| pathID | pathexp | ES |
|---|---|---|
| 131 | /dblp/article/title | 0.333 |
| 120 | /dblp/mastersthesis/title | 0.333 |
| 71 | /dblp/book/title | 0.333 |
| 5 | /article/prolog/title | 0.166 |
| 219 | /feed/doc/title | 0.166 |

$p_6$ /SigmodRecord/issues/issue/articles/article

| pathID | pathexp | ES |
|---|---|---|
| 55 | /dblp/article | 0.2 |
| 195 | /dblp/proceedings/journal | 0.2 |
| 73 | /dblp/article/journal | 0.2 |
| 33 | /article/epilog | 0.2 |
| 38 | /article/prolog/genre | 0.2 |

documents) are too short to extract enough 2-grams for applying SSJoin. So, we are thinking about the possibility to adaptively combine 1-gram and 2-gram (or more) depending on the length of queries in the future work.

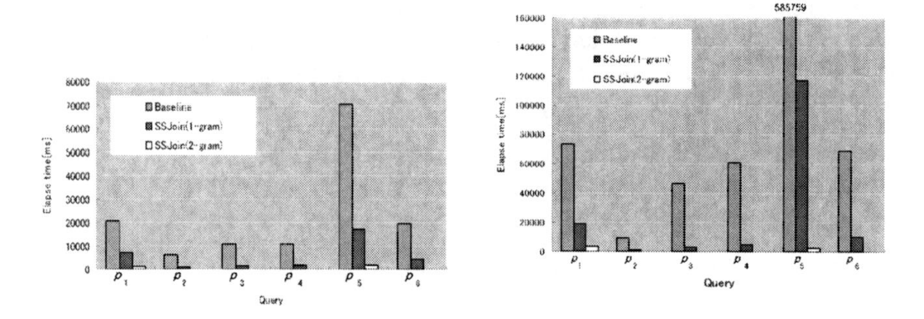


**Fig. 4.** Elapse time: LCA dist = 1 (left), LCA dist = 2 (right)

## 6 Conclusions

In this paper we proposed a novel scheme for applying a path query to those XML data whose structures are not known. The core idea is to apply ontologies for finding similar tag names, and to introduce edit similarity for performing proximity matching among slightly different path expressions. The experimental results suggest that the proposed scheme seems to be reasonable in both accuracy and effectiveness.

In the future, we plan to introduce weighted edit similarity in that the edit cost for similar tag names is less than that of different tag names. We also plan to apply the proposed scheme to the problem of similarity join over XML data.

## Acknowledgments

This study has been supported by Grant-in-Aid for Scientific Research of JSPS (#18650018 and #19700083) and of MEXT (#19024006).

## References

1. W3C: Extensible Markup Language (XML) 1.0, 3rd edn., Recommendation (April 2004), http://www.w3.org/TR/xml/
2. W3C: XML Path Language (XPath) Version 1.0. Recommendation (November 1999), http://www.w3.org/TR/xpath.html
3. W3C: XSL Transformations (XSLT) Version 1.0. Recommendation (November 1999), http://www.w3.org/TR/xslt
4. W3C: XQuery 1.0: An XML Query Language. Recommendation (January 2007), http://www.w3.org/TR/xquery/
5. Chaudhuri, S., Ganti, V., Kaushik, R.: A primitive operator for similarity joins in data cleaning. In: Proc. ICDE 2006, p. 5 (2006)
6. Cohen, W.W.: Data integration using similarity joins and a word-based information representation language. ACM Transactions on Information Systems (TOIS) 18(3), 288–321 (2000)
7. Liang, W., Yokota, H.: A path-sequence based discrimination for subtree matching in approximate XML joins. In: Proc. The 2nd Int'l Special Workshop on Databases for Next-Generation Researchers (SWOD), p. 116 (2006)
8. Amer-Yahia, S., Cho, S., Srivastava, D.: Tree pattern relaxation. In: Jensen, C.S., Jeffery, K.G., Pokorný, J., Šaltenis, S., Bertino, E., Böhm, K., Jarke, M. (eds.) EDBT 2002. LNCS, vol. 2287, pp. 496–513. Springer, Heidelberg (2002)
9. Zhang, K., Shasha, D.: 11. In: Tree pattern matching. Pattern Matching Algorithms, Oxford University Press, Oxford (1997)
10. WordNet a lexical database for the English language, http://wordnet.princeton.edu/
11. The Gene Ontology project, http://www.geneontology.org/
12. RDF/OWL Representation of WordNet (2006), http://www.w3.org/,/03/wn/wn20/
13. W3C: Resource Description Framework (RDF): Concepts and Abstract Syntax (February 2004) Recommendation (2004), http://www.w3.org/TR/,/REC-rdf-concepts-20040210/
14. W3C: SPARQL Query Language for RDF, Working Draft (October 2006), http://www.w3.org/TR/rdf-sparql-query/
15. Olteanu, D., Meuss, H., Furche, T., Bry, F.: XPath: Looking Forward. In: Chaudhri, A.B., Unland, R., Djeraba, C., Lindner, W. (eds.) EDBT 2002. LNCS, vol. 2490, pp. 109–127. Springer, Heidelberg (2002)
16. XBench – A Family of Benchmarks for XML DBMSs, http://se.uwaterloo.ca/~ddbms/projects/xbench/

# Cooperative Data Management for XML Data

Katja Hose and Kai-Uwe Sattler

Department of Computer Science and Automation,
TU Ilmenau, Germany
{katja.hose, kus}@tu-ilmenau.de

**Abstract.** Emerging non-standard applications like the production of high-quality spatial sound pose new challenges to data management. Beside the need for a flexible transactional management of complex hierarchical scene descriptions a main requirement is the support of cooperative processes allowing a group of authors to edit a scene together in a distributed environment. Based on previous work on cooperative and non-standard transactions we present in this paper a transaction model and protocol for XML databases addressing this issues.

## 1 Introduction

In recent years, XML has been widely established as a data exchange format but also as a native data format for (semi-)structured data. XML data management is particularly well suited in application domains where a fixed structure of data is too restrictive and where hierarchical structures have to be represented. In this way, XML data management can be seen as a successor of object-oriented database technology. An example of an emerging application domain from which we derive the motivation for our work presented here is media production, e.g. the production of movies, sounds, and graphics. The authoring process in this domain is typically characterized by

(1) the incremental construction of scenes which are often represented as graphs with varying structures.
For example, sound production in high quality spatial sound systems, like the IOSONO system[1], is based on an object-oriented modelling of scenes. The scene is rendered at runtime to compute the signals for a large number of loudspeakers installed in the listening room. In such a system, a scene consists of several audio objects with properties as well as spatial and temporal relationships describing position, movement, start time and duration of the sound [1]. Scenes are organized in several layers such as dialog, foley, effects, atmosphere and music.
(2) the duration and cooperativeness of the construction process.
In bigger projects we already encountered situations where up to 60 sound designers in up to 8 different studios – located all over the world – are working simultaneously on one sound production project. Each designer is

[1] www.iosono-sound.com

R. Wagner, N. Revell, and G. Pernul (Eds.): DEXA 2007, LNCS 4653, pp. 308–318, 2007.

responsible for a part of a scene, e.g. modelling special effects, while others model the actors' dialogs. Both groups must know the current state of the other group's work because for example the volume of a speaker has to be set with respect to the background noise.

Basically, (1) can be addressed by representing a scene as an XML graph stored in a database. Today, commercial DBMS provide support for XML data management, either by treating XML documents as CLOB objects or by "shredding" the document into a set of tables. In both cases, the transactional support is restricted to the native (relational) structure and is not adopted to the special characteristics of (hierarchical) XML data.

The alternative of using a native XML DBMS promises a better handling of XML data, e.g. wrt. querying, but regarding transactional support these systems are more or less still in their infancies.

Orthogonal to the scheme of storing the scene data we can look at the issue of supporting cooperative work (2) in different scenarios:

**Single User:** The simplest and in fact non-cooperative case is the situation where a single author (sound designer) is working on the sound production. Though, basic database functionality like persistence and recovery is required, advanced techniques for distributed and cooperative operation are not needed.

**Workgroup:** When multiple designers work on the same scene together in a workgroup, their updates have to be synchronized. In this scenario a central repository is needed but can be extended by local caches or databases for better response times.

**Workspace:** In a large-scale scenario multiple studios (maybe situated on different continents) work on the same scene data. Here, permanent connections to a central repository (as in the case of the workgroup scenario) may not be assumed. Therefore, the different users work on their own workspaces requiring a decoupled synchronization technique similar to version control systems like CVS.

In our work we focus on the workgroup scenario because it is the most interesting one for the intended application of sound production. In any case, the notion of transaction is required and because of the characteristics of work we need long running, nested transactions. Furthermore, in order to allow a cooperative work on the same scene (e.g. on different layers of a scene) modifications should be visible to other authors before the end of the global transaction.

Based on these observations and the above mentioned challenges of data management in media authoring processes, we present in this paper an approach using an open-nested transaction model for XML data supporting cooperative processes. The remainder of the paper is structured as follows. In Section 2 we give a brief survey on related work. Based on this, we introduce in Section 3 our cooperative transaction model. Next, we describe the protocol implementing this model by a combination of locking and notification in Section 4. Finally, we conclude the paper and point out to future work.

## 2 Related Work

For many years transactions have been used to enable concurrent access to a database where each transaction adheres to the ACID properties. A very common means to enforce serializability are lock protocols. They have advantages like low complexity and disadvantages like limiting concurrency. Thus, many variants have been developed; 2PL [2] and hierarchical locks [3] to mention only a few.

In order to overcome the disadvantage of limited concurrency advanced transaction models have been developed by relaxing some of the ACID properties. One principle is to divide a transaction into several smaller ones. This concept is for example used by nested transactions [4,5] and multi-level transactions [6,7]. With the advent of CAD systems with typically long transactions chained transactions and sagas [8] have been proposed. Whereas both of these concepts require the definition of subtransactions in advance, split/join transactions [9, 10] can be used to determine subtransactions at runtime.

As XML became more important DBMS were extended by XML modules. At first, it was sufficient to support XML as input and output format. But since this limits querying abilities, first concepts for native XML databases have been developed. In general, classic lock protocols are too restrictive in terms of concurrency for use in XML DBMS. One possibility is to use path based locks as proposed by several groups [11, 12, 13, 14]. Natix [15] is a native XML DBMS that recognizes that transaction management needs a non-traditional approach. However, the authors focus more on recovery and isolation, and use a lock manager supporting multi granularity locking and strict two-phase locking.

Another possibility to realize concurrent access are protocols based on the taDOM model, e.g., [16, 17]. They use multi granularity locking, apply the concept of intensional locks on the path from the root node to the context node, and provide lock conversion. taDOM models attributes and text content as special nodes. By this, attributes or text can be locked without locking the original XML node as a whole. Lock granularity and lock escalation can be adapted according to the users needs.

As indicated in the introduction we need to allow multiple users to work on the same XML document simultaneously. The approaches on XML databases mentioned above do net (yet) pay attention to the requirements imposed by our scenario where we have deal with long transactions adhering to an open nested transaction model.

## 3 Transaction and Cooperation Model

The basic architecture of our solution is a client-server-model where the server holds the latest version of the data and coordinates the clients. An arbitrary number of clients can be connected to the server. Both client and server have a DBMS to manage their data. The server maintains the global copy ensuring data consistency. The client uses its local DBMS to manage its local copy of the portion of the data that it downloaded from the server. Since we are working with XML data, we chose Berkeley DB XML[2] for both client and server. For the local DBMS on client side, we can use any standard transaction model since only one user is working on the local copy. Using the DBMS, however, the client may also benefit from database functionality like persistence and recovery.

In the following we at first identify a transaction model that fulfills the requirements stated in the introduction. Then, we describe how notifications can be used to make non-committed updates visible to other users.

### 3.1 Transaction Model

As motivated in the introduction transactions may endure a rather long time period. In order to support concurrent access to the common global database we need a transaction model that efficiently supports long transactions but still allows multiple users to work concurrently on the same data. This suggests the use of a nested transaction model [4, 5] where a global transaction is divided at runtime into several subtransactions. In our model a subtransaction is not vital for the global transaction. Thus, when a subtransaction is aborted not all changes have to be undone but only those that have been made by the aborted subtransaction. The global transaction goes on until the global commit or the global abort.

**Table 1.** Primitives at Client and Server

| | Client | Server |
|---|---|---|
| user-level primitives | query<br>update<br>savepoint<br>revert<br>begin, abort, commit | checkout<br>checkin<br>subcommit, publish, subbegin<br>subabort, subbegin<br>begin, abort, commit |
| system-level primitives | lock, unlock<br>subscribe, unsubscribe | lock, unlock<br>subscribe, unsubscribe |

Having started a transaction the user issues commands on the client side. The client communicates with the server to initiate corresponding actions. Table 1 lists the main primitives that are supported by client and server instances. They can be divided into two groups: user-level primitives and system-level primitives. The former are issued more or less by the user himself, the latter by the system transparently to the user.

Figure 1 illustrates a sample sequence of primitives that are issued in a typical transaction. When the user starts to work a global transaction is started (*begin*). Since we are using a nested transaction model, the first subtransaction starts at the same time (*subbegin*). The *query* primitive allows the user to read elements of the XML document that are identified using XPath. Hence, this primitive corresponds to the classic read operation. Its counterpart, i.e. a write operation,

[2] http://www.sleepycat.com/products/bdbxml.html

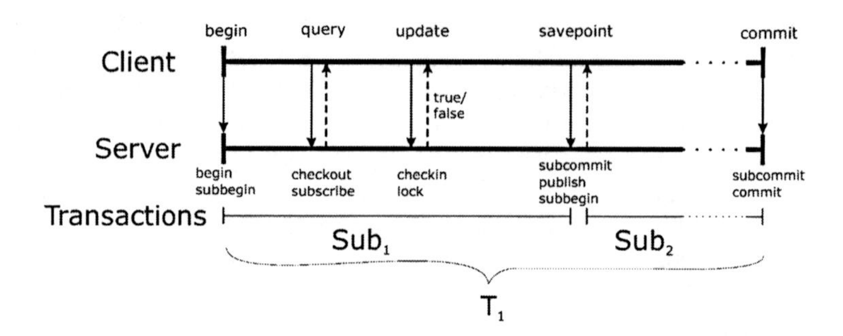


**Fig. 1.** Basic Sequence of Actions for Transactions

is represented by the *update* primitive. The *query* primitive results in a *checkout* at the server retrieving the latest version of the read data. The *update* primitive of the client is realized by the *checkin* primitive at the server. This, however, requires another step: first the server is trying to *lock* the data that is to be updated. If the lock has been acquired the data is updated, if not the attempt to update is refused.

When the user decides for a *savepoint* the currently running subtransaction commits and the next subtransaction is implicitly started. All changes that have been made by the first subtransaction are now visible to other users – but may not yet be changed by them. In case a subtransaction is aborted, all changes that have been made since the beginning of the subtransaction are undone without any effect on the global transaction. Finally, when the global transaction commits, the current subtransaction is committed and all locks are released.

### 3.2 Cooperation Model

As mentioned above other users might need to see the changes of committed subtransactions before the global transaction submits. In order to achieve such a cooperative transaction model we use the concept of notifications. For this purpose the server maintains a list of *Listeners*. Each entry is defined by two pieces of information: the identifier of the corresponding client and an XPath expression that indicates a subtree of the XML document. Whenever a subtransaction of another client commits, all clients that have registered for the affected data are notified using the *publish* primitive (see Figure 1). When the client reads data in a transactional context using the *query* primitive the client is implicitly registered at the server for updates concerning the read data.

Figure 2 illustrates this concept in a situation where two transactions are working on the same version of the data. Both transactions start with checking out version $V1$ from the server, they implicitly register for updates concerning the retrieved data. Both transactions ($T1$ and $T2$) are now starting to work concurrently. When subtransaction $Sub1.1$ of $T1$ commits, the changes are propagated to the client that $T2$ is running on. The client updates the affected portion of local data. Afterwards, the data of both clients corresponds to subversion $V1.1$.

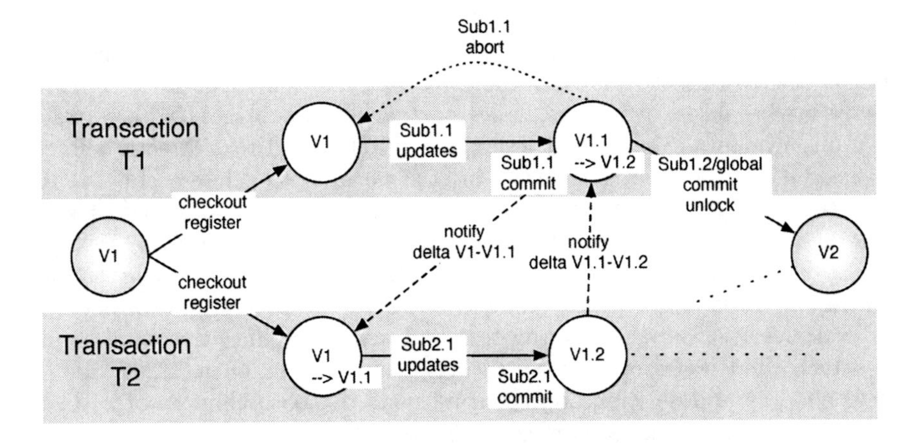


**Fig. 2.** Example for Versioning Using Listeners

By the use of our locking protocol that we use for synchronization (Section 4) it is not possible that both transactions changed the same data records.

In case another transaction $T3$ would now checkout the current version from the server it would retrieve the basic version $V1$ and all updates of already committed subtransactions so that the local version of $T3$'s data would correspond to subversion $V1.1$.

After subtransaction $Sub1.1$ of $T1$ has committed assume that now subtransaction $Sub1.1$ of $T2$ commits. Then, the client that $T1$ runs on retrieves the updates made by $Sub2.1$ of $T2$. Both clients now have the same subversion $V1.2$. Finally, the global transaction $T1$ commits. The server now retrieves the changes, releases the locks still held by $T1$, and converts the global version to $V2$. After having notified the client that $T2$ runs on, the corresponding client updates its local data, and now works on version $V2$. In case $T2$ now aborts, then the changes made by the already committed subtransactions of $T2$ are undone and the registered clients are notified. In case $T2$ submits the the server creates version $V3$ and notifies the registered clients.

## 4 Transaction Protocol

In this section we at first identify operations that need to be supported and their demands on a suitable locking protocol. Then, we present such a protocol that we use for synchronization. Finally, we show how to increase concurrency by using a special internal representation.

### 4.1 Operations and Lock Compatibility

The set of relevant operations that we need to support consists of: (i) edit, (ii) delete, (iii) add, and (iv) move. XML documents consist of nodes organized in a hierarchical structure. Each node may contain text and attributes and might have attached child node. To change the content of a node's text or attribute, the *edit* operation is used. Obviously, this requires an exclusive lock so that no other user is able to edit attributes and text at the same time. The *delete* operation can be used to delete attributes, nodes, text, or whole subtrees that are attached to a given node $n$. Since the deletion of attributes and text directly affects $n$, we need $n$ to be locked exclusively. In case we want to delete a node $n_1$ (or a subtree rooted with $n_1$) that is attached to $n$ as a child, we only need a shared lock on $n$ and an exclusive lock on $n_1$ (and on all subtree nodes).

The *add* operation can be used to attach attributes, text, child nodes, or subtrees to node $n$. Since text and attributes are an integral part of $n$ we need an exclusive lock on $n$ for adding text and attributes. Intuitively, all we need to attach child nodes or subtrees to $n$ is a shared lock on $n$. Thus, adding a child node $n_1$ and deleting another child node $n_2$ (both children of $n$) by two concurrent transactions is possible at the same time.

The *move* operation can be used to move attributes, nodes, text, and subtrees from node $n$ to node $m$. We can treat this as a deletion followed by an insertion so that we can use the two operations introduced above. Table 2 shows the lock matrix that results from the considerations discussed so far.

**Table 2.** Operations and required locks when attributes and text *cannot* be locked apart from the corresponding node – XL = Exclusive Lock, SL = Shared Lock

| | edit | delete | add |
|---|---|---|---|
| attribute | XL on node | XL on node | XL on node |
| text | XL on node | XL on node | XL on node |
| node | – | XL on node, SL on parent | SL on parent |
| subtree | – | XL (root and descendants), SL on parent | SL on parent |

This is still rather restrictive. Under the assumption that we can lock attributes and text independently from the corresponding XML nodes, we can achieve a higher level of concurrency: e.g. concurrent operations on attributes and text are now compatible with each other. Furthermore, these operations are compatible with concurrent deletion and addition of child nodes. Table 3 shows the resulting lock matrix.

Since we have to deal with only two kinds of locks (exclusive and shared), we can apply the tree protocol that has been developed for hierarchical databases.

**Table 3.** Operations and required locks when attributes and text *can* be locked apart from the corresponding node – XL = Exclusive Lock, SL = Shared Lock

| | edit | delete | add |
|---|---|---|---|
| attribute | XL on attr. | XL on attr., SL on corr. node | SL on corr. node |
| text | XL on text | XL on text, SL on corr. node | SL on corr. node |
| node | – | XL on node, SL on parent | SL on parent |
| subtree | – | XL (root and descendants), SL on parent | SL on parent |

Thus, the remainder of this section first presents the tree protocol. Then, we point out how to manage locks on attributes and text without locking the whole node.

### 4.2 Tree Protocol

In contrast to most lock protocols the tree protocol [3] does not imply two phase locking (2PL). It has been designed for use in hierarchically organized data structures and thus can be used for XML data as well. The basic variant of this protocol only knows one kind of locks: exclusive locks. The advanced variant – that we consider – also knows non-exclusive locks. Any transaction $T_i$ that adheres to the following rules satisfies the advanced tree protocol and leads to a serializable schedule:

1. At first, $T_i$ locks any node of the hierarchy – provided that any existing locks are compatible
2. $T_i$ locks each node at most once – there is no lock conversion
3. locks may be released at any time – in contrast to 2PL protocols
4. $T_i$ may lock node $u$ if and only if it is currently holding a lock on a predecessor (father) of $u$
5. $LS(T_i) \subseteq \mu(L(T_i))$, where $LS(T_i)$ is the set of shared locks held by $T_i$, $L(T_i)$ the set of all locks (shared or exclusive) held by $T_i$, and $\mu(W) = \{v \in W | \textit{there exists at most one } w \in W \textit{ such that } v \textit{ and } w \textit{ are neighbors}\}$

Note that if $T_i$ satisfies conditions $1-4$ it fulfills the basic tree protocol [18]. The fifth condition assures that deadlocks cannot occur when allowing non-exclusive locks.

With respect to Section 4.1 using the tree protocol means that locks are acquired based on the tree protocol and may be released before the commit. However, in our implementation those locks that are required by the supported operations (Table 3) need to be held until the global transaction commits. Only then can be guaranteed that a rollback is possible without side-effects on other transactions.

### 4.3 XML Representation

In order to improve concurrency on XML nodes especially with respect to editing attributes and text, we apply the taDOM concept [19] and adopted it to our needs.

**Representing XML Documents for Fine Grained Locking.** taDOM [19] has originally been designed for supporting a fine grained lock granularity for documents that are accessed by the DOM API. In contrast to our application and access methods the DOM API knows functions like *getAttributeNode()*, *getValue()* etc. In order to support these functions efficiently, the authors split up XML nodes into several parts, e.g. each XML node is represented by one element node $n_e$, one text node $n_t$ (as child of $n_e$) and one string node $n_s$ (as child of $n_t$). If the original XML node contained attributes then an additional attribute root node $n_{AR}$ is inserted as child of the $n_e$. For each attribute, $n_{AR}$ has one attribute child node $n_{a_i}$ where each $n_{a_i}$ has a string node that contains the attribute value.

Since we do not aim to support such APIs we do not need such a large number of nodes. Thus, we reduce the overhead by simply splitting up an XML element node into one element node $n_e$, one child for the contained text $n_t$, one attribute root node $n_{AR}$ with one attribute child node $n_{a_i}$ for each attribute – $n_{a_i}$ contains all information about the attribute. In short, we keep text nodes and string nodes together in one node. The separation of attributes from the original node has two advantages: first, locking attributes apart from their nodes is possible and second, locking all attributes at once promises to be low effort.

Figure 3 illustrates these concepts with an example. It shows a small extract of a sample XML file and its corresponding representation where we distinguish between *element nodes*, *text nodes*, *attribute root nodes*, and *attribute nodes.* Inner nodes of XML documents are represented by element nodes.

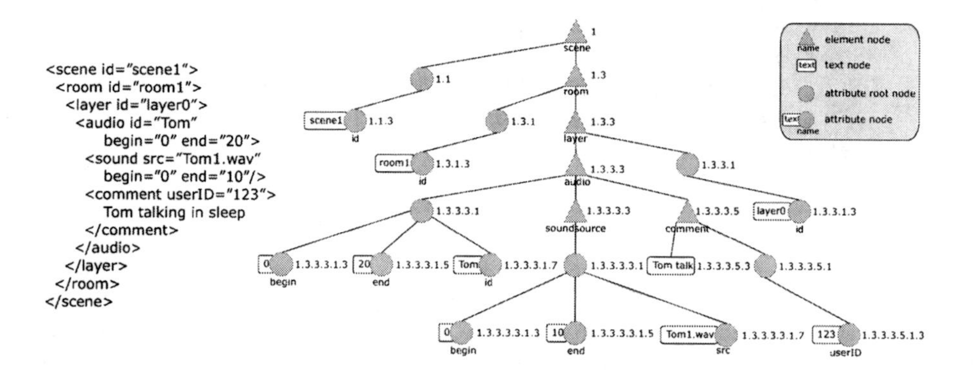


**Fig. 3.** Internal XML Representation

**DeweyIDs.** By using DeweyIDs [20] we assign a unique identifier to each node and thus enable an efficient management of read and write sets. As we have borrowed the idea of separating attributes from elements and using an extra attribute root node to group them, we also adopt most of the adaptations of DeweyIDs that were made to support taDOM. DeweyIDs are based on the decimal classification and serially number all nodes in the same level with odd numbers. With the exception of the document root node, the number 1 is only assigned to attribute root nodes. This makes finding and identifying attributes easy. The ID of a node is defined as the conjunction of the parent-ID and the assigned number separated by a point. Thus, the prefix of each ID reveals its level and unambiguously identifies the parent node and all ancestor nodes. Since initially only odd numbers are used new nodes may be added at any position. For instance, between DeweyIDs 1.3 and 1.5 we may add nodes with the following

IDs: 1.4.3, 1.4.5, etc. Figure 3 gives an example for the DeweyID numbering scheme that we use in our implementation.

## 5 Conclusion

In this paper we have addressed the problem of concurrent access and modification to XML documents. The application scenario demands that users may see changes of other users whose global transactions have not yet committed. To solve this problem we proposed an open nested transaction model that uses the advanced tree protocol for synchronization. Notifications take care of propagating recent updates to registered clients, so that they are always up-to-date. Future work will consider to further increase concurrency. One possibility to achieve this is releasing locks already with the commit of a subtransaction so that other transactions may update the same elements before the global transaction submits. Introducing compensating transactions might be a solution to the problem. However, it remains a task for future work to define such compensating transactions and maybe determine their actions without extensive interaction with the user.

## References

1. Heimrich, T., Reichelt, K., Rusch, H., Sattler, K., Schröder, T.: Modelling and streaming spatiotemporal audio data. In: OTM Workshops, pp. 7–8 (2005)
2. Bernstein, P.A., Hadzilacos, V., Goodman, N.: Concurrency Control and Recovery in Database Systems. Addison-Wesley Publishing Company, London (1987)
3. Silberschatz, A., Kedem, Z.: Consistency in hierarchical database systems. Journal of the ACM 27, 72–80 (1980)
4. Härder, T., Rothermel, K.: Concepts for transaction recovery in nested transactions. In: SIGMOD '87, pp. 239–248 (1987)
5. Moss, J.E.: Nested transactions: an approach to reliable distributed computing. Massachusetts Institute of Technology, Cambridge (1985)
6. Weikum, G.: Principles and realization strategies of multilevel transaction management. ACM Transactions on Database Systems 16, 132–180 (1991)
7. Weikum, G., Schek, H.J.: Concepts and applications of multilevel transactions and open nested transactions. Database transaction models for advanced applications, 515–553 (1992)
8. Garcia-Molina, H., Salem, K.: Sagas. In: SIGMOD '87, pp. 249–259 (1987)
9. Kaiser, G., Pu, C.: Dynamic restructuring of transactions. In: Database Transaction Models for Advanced Applications, pp. 265–295 (1992)
10. Pu, C., Kaiser, G., Hutchinson, N.: Split-transactions for open-ended activities. In: VLDB '88, pp. 26–37 (1988)
11. Choi, E., Kanai, T.: XPath-based Concurrency Control for XML Data. In: DEWS 2003 (2003)
12. Dekeyser, S., Hidders, J.: Path Locks for XML Document Collaboration. In: WISE '02, pp. 105–114 (2002)
13. Dekeyser, S., Hidders, J.: Conflict scheduling of transactions on XML documents. In: ADC '04, pp. 93–101 (2004)

14. Jea, K., Chen, S., Wang, S.: Concurrency Control in XML Document Databases: XPath Locking Protocol. In: ICPADS '02 , p. 551 (2002)
15. Fiebig, T., Helmer, S., Kanne, C., Moerkotte, G., Neumann, J., Schiele, R., Westmann, T.: Natix: A Technology Overview. In: Web, Web-Services, and Database Systems, pp. 12–33 (2003)
16. Haustein, M., Härder, T.: Adjustable Transaction Isolation in XML Database Management Systems. In: Proc. 2nd Int. XML Database Symposium, pp. 173–188 (2004)
17. Haustein, M., Härder, T., Luttenberger, K.: Contest of XML Lock Protocols. In: VLDB 2006, pp. 1069–1080 (2006)
18. Kedem, Z., Silberschatz, A.: Controlling Concurrency Using Locking Protocols (Preliminary Report). In: FOCS, pp. 274–285 (1979)
19. Haustein, M.P., Härder, T.: taDOM: A Tailored Synchronization Concept with Tunable Lock Granularity for the DOM API. In: Kalinichenko, L.A., Manthey, R., Thalheim, B., Wloka, U. (eds.) ADBIS 2003. LNCS, vol. 2798, pp. 88–102. Springer, Heidelberg (2003)
20. O'Neil, P., O'Neil, E., Pal, S., Cseri, I., Schaller, G., Westbury, N.: ORDPATHs: insert-friendly XML node labels. In: SIGMOD '04, pp. 903–908 (2004)

# C-ARIES: A Multi-threaded Version of the ARIES Recovery Algorithm

Jayson Speer and Markus Kirchberg

Information Science Research Centre, Department of Information Systems,
Massey University, Private Bag 11 222, Palmerston North 5301, New Zealand
M.Kirchberg@massey.ac.nz

**Abstract.** The ARIES recovery algorithm has had a significant impact on current thinking on transaction processing, logging and recovery. In this paper, we present the C-ARIES algorithm, which extends the original algorithm with the capability to perform transaction aborts and crash recovery in a highly concurrent manner. Concurrency is achieved by performing transaction aborts and the Redo and Undo recovery phases on a page-by-page basis. An additional benefit of C-ARIES is that the database system can commence normal processing at the end of the Analysis phase, rather than waiting for the recovery process to complete.

## 1 Introduction

Introduced by Mohan et al. [1], the ARIES (Algorithm for Recovery and Isolation Exploiting Semantics) algorithm has had a significant impact on current thinking on database transaction logging and recovery. It has been incorporated into IBM's DB2 Universal Database, Lotus Notes and a number of other systems [2].

ARIES, like many other recovery algorithms, is based on the WAL (Write Ahead Logging) protocol that ensures recoverability of databases in the presence of a crash. However, ARIES' Repeating History paradigm sets it apart from other WAL based protocols. The repeating history paradigm involves returning the database to the exact state it was in before the crash occurred and allows ARIES to support properties such as fine granularity locking, operation logging and partial rollbacks.

### 1.1 Contribution

We propose an adaptation of the original ARIES recovery algorithm. The proposed C-ARIES algorithms extends the original algorithm with the capability to perform transaction aborts during normal processing and crash recovery in a highly concurrent manner. Concurrency is achieved by performing transaction aborts and the Redo and Undo crash recovery phases on a page-by-page basis. An additional benefit of our approach is that the database system can be returned to normal processing at the end of the Analysis phase, rather than waiting for the recovery process to complete.

R. Wagner, N. Revell, and G. Pernul (Eds.): DEXA 2007, LNCS 4653, pp. 319–328, 2007.

### 1.2 Outline

This paper is organised as follows: Section 2 introduces the original ARIES recovery algorithm. Sections 3 to 5 present the proposed concurrent ARIES (i.e. C-ARIES) adaptation. Finally, Section 6 concludes this paper.

## 2 ARIES

This section provides a brief overview of the original ARIES algorithm [1,3].

ARIES, like many other algorithms, is based on the WAL protocol that ensures recoverability of a database in the presence of a crash. All updates to all pages are logged. ARIES uses a log sequence number (LSN) stored on each page to correlate the state of the page with logged updates of that page. By examining the LSN of a page (called the PageLSN) it can be easily determined which logged updates are reflected in the page. Being able to determine the state of a page w.r.t. logged updates is critical whilst repeating history, since it is essential that any update be applied to a page once and only once. Failure to respect this requirement will in most cases result in a violation of data consistency.

Updates performed during forward processing of transactions are described by Update Log Records (ULRs). However, logging is not restricted to forward processing. ARIES also logs, using Compensation Log Records (CLRs), updates (i.e. compensations of updates of aborted / incomplete transactions) performed during partial or total rollbacks of transactions. By appropriate chaining of CLR records to log records written during forward processing, a bounded amount of logging is ensured during rollbacks, even in the face of repeated failures during crash recovery. This chaining is achieved by 1) assigning LSNs in ascending sequence; and 2) adding a pointer (called the PrevLSN) to the most recent preceding log record written by the same transaction to each log record.

When the undo of a log record causes a CLR record to be written, a pointer (called the UndoNextLSN) to the predecessor of the log record being undone is added to the CLR record. The UndoNextLSN keeps track of the progress of a rollback. It tells the system from where to continue the rollback of the transaction, if a system failure were to interrupt the completion of the rollback.

Periodically during normal processing, ARIES takes fuzzy checkpoints in order to avoid quiescing the database while checkpoint data is written to disk. Checkpoints are taken to make crash recovery more efficient.

When performing crash recovery, ARIES makes three passes (i.e. Analysis, Redo and Undo) over the log. During Analysis, ARIES scans the log from the most recent checkpoint to the end of the log. It determines 1) the starting point of the Redo phase by keeping track of dirty pages; and 2) the list of transactions to be rolled back in the Undo phase by monitoring the state of transactions. During Redo, ARIES repeats history. It is ensured that updates of all transactions have been executed once and only once. Thus, the database is returned to the state it was in immediately before the crash. Finally, Undo rolls back all updates of transactions that have been identified as active at the time the crash occurred.

# 3 C-ARIES: The Multi-threaded ARIES Algorithm

In the next three sections, we introduce a multi-threaded version of the ARIES recovery algorithm (referred to as *C-ARIES*).

In this section, basic concepts such as modifications to data structure, logging and checkpointing are outlined. Based on these concepts, Section 4 discusses crash recovery processing in greater detail. Subsequently, Section 5 considers necessary modifications to transaction rollback during normal processing.

C-ARIES preserves the desirable properties of the original ARIES algorithm. However, enhancements were made to the Redo and Undo phases of the crash recovery process, whereby these phases are now performed on a page-by-page basis. This results in a much higher degree of concurrency since operations that would normally be performed serially can now be performed concurrently. This page-by-page technique also provides the basis for the improved method to transaction aborts during normal processing.

It was important not to impose any unnecessary costs on the algorithm in terms of logging, since this is purely overhead on the system that offers no benefit until recovery is required [4]. The extra logging required for C-ARIES is very small with a single extra field being added to some log record types and fields removed from others.

## 3.1 Logging

ARIES and also C-ARIES require that LSNs be monotonically increasing. This, however, is not a burden but rather a benefit. It allows a direct correspondence between a log record's physical and logical address to be maintained.

However, in order to adopt a page-by-page approach to crash recovery and transaction abort, a number of modifications must be made to the way the ARIES algorithm performs logging, these are as follows:

**Modification of the CLR.** In C-ARIES, extensive changes are made to the CLR, both in terms of the information it contains and the way in which it is used:

1. The `UndoneLSN` field replaces the `NextUndoLSN` field. Whereas the `UndoNextLSN` records the LSN of the next operation to be undone, the `UndoneLSN` records the LSN of the operation that was undone.
2. The `PrevLSN` field is no longer required for the CLR record.
3. CLR records are now used to record undo operations during normal processing only, the newly defined SCR records (refer below) is used to record undo operations during crash recovery.

The rationale behind these modifications can be understood by observing the differences in how C-ARIES and the original ARIES algorithm perform undo operations and how this affects the information required by C-ARIES. In ARIES, operations are undone one at a time in the reverse order to which they were performed by transactions. However, in order to increase concurrency, C-ARIES can perform multiple undo operations concurrently, where updates to individual pages are undone independently of each other.

**Definition of the SCR.** The Special Compensation log Record (SCR) is almost identical to the modified version of the CLR, the only differences being:

1. The record type field (SCR rather than CLR).
2. When SCR records are written.

During normal rollback processing, operations are undone in the reverse order to which they were performed by individual transactions. However, during crash recovery rollback, operations are undone in reverse order that they were performed on individual pages. Having separate log records for compensation during recovery and normal rollback allows us to exploit this fact.

**PageLastLSN Pointers.** The `PageLastLSN` pointer is added to all ULR, SCR and CLR records. It records the LSN of the record that last modified an object on this page. Recording these `PageLastLSN` pointers provides an easy method of tracing all modifications made to a particular set of objects (stored on a page).

### 3.2 Fuzzy Checkpoint

As with ARIES, a fuzzy checkpoint is performed in order to avoid quiescing the database while checkpoint data is written to disk. The following information is stored during the checkpoint: Active transaction table and `DirtyLSN` value.

For each active transaction, the transaction table stores the following data:

- `TransId`, i.e. an identifier of the active transaction.
- `FirstLSN`, i.e. the LSN of the first log record written for the transaction.
- `Status`, i.e. either *Active* or *Commit*.

Given the set of pages that were dirty at the time of the checkpoint, the `DirtyLSN` value points to the record that represents the oldest update to any such page that has not yet been forced to disk.

## 4 C-ARIES: Crash Recovery

With C-ARIES, recovery remains split into three phases: Analysis, Redo and Undo. However, recovery takes place on a page-by-page basis, where updates are reapplied (Redo phase) and removed from (Undo phase) pages independently from one another. The Redo phase reapplies changes to each page in the exact order that they were logged and the Undo phase undoes changes to each page in the exact reverse order that they were performed. Since the state of each page is accurately recorded (by use of the `PageLSN`), the consistency of the database will be maintained during such a process.

### 4.1 Data Structures

Data collected during the Analysis pass is stored in the following data structures:

**Transaction Status Table.** The *transaction status (TransStatus)* table determines the final status of all transactions that were active at some time after the last checkpoint. This information is used to determine whether changes made to the database should be kept or discarded. The TransStatus table holds:

- `TransId`, i.e. the identifier of the active transaction.
- `Status`, i.e. the status of the transaction, which determines whether or not it must be rolled back. Possible stares are: *Active*, *End* and *Commit*.

Any transaction with status '*Active*' is declared a *Loser Transaction (LT)*, whilst all other transactions are declared *Winner Transactions (WTs)*.

**Page Link List.** The *page link (PLink)* list provides a linked list of records for each modified page. This list is used in the Redo phase to navigate forwards through the log. For each page that has CLR, SCR or ULR records, such a PLink list, which is an ordered list of all LSNs that modified that page, is created.

**Page Start List.** The *Page Start List* determines, for each page, from where to commence recovery. It holds the following field: `PageId`, i.e. the page identifier.

During the forward scan of the log, the first time the algorithm encounters a log record for a page $P_j$, it creates a Page Start List entry for page $P_j$. The Page Start List captures all pages that are to be visited during the Redo and Undo phases of recovery. Thus, we can lock those pages exclusively on behalf of the recovery algorithm and make available all other pages for normal processing[1].

**Page End List.** The Page End List is an optional data structure intended to optimise the Undo phase of recovery by accurately determining where this phase should stop processing. The Page End List has the following fields:

- `PageId`, i.e. the identifier of the page.
- `TransId`, i.e. the identifier of the transaction that has modified the page.
- `EndLSN`, i.e. the LSN of the last record to process this page during Undo.

Given a set of records, the rule for creating and updating an entry for some page $P_j$ in the Page End List proceeds as follows:

1. The first time a log record written for page $P_j$ by transaction $T_k$ is encountered, the following entry should be inserted into the Page End List: ( `PageId`, `TransId`, `EndLSN` ) = ( $P_j$, $T_k$, LSN from record ).
   Each page should have only one entry in the Page End List for each transaction that has written a log record for it.

[1] Note, at this stage there was no access to persistent data other than that of the log.

2. Once the scanning of all records back to `ScanLSN` is completed and the set of all winner and loser transactions is known, the following actions are performed:
   (a) Delete all entries where `TransId` is in TW.
   (b) For each page, retain only the entry with the lowest `EndLSN`.

Finally, there will be at most a single Page End List entry for any page. The `EndLSN` value indicates where the Undo phase will terminate for that page.

**Undone List.** The *Undone List* stores a list of all operations that have been previously undone. It has the following fields:

- `PageId`, i.e. the identifier of the page.
- `UndoneLSN`, i.e. the LSN of the record that has been undone.

During the scan of the log, whenever the algorithm encounters a CLR record, it adds an entry to the Undone List.

### 4.2 Analysis Phase

During the Analysis phase, the algorithm collects all data that is required to restore the database to a consistent state. This involves performing a forward scan through the log, collecting the data required for the Redo and Undo phases of recovery. The Analysis phase of the recovery process is comprised of three steps, being: Initialisation, Data Collection and Completion.

**Step 1: Initialisation.** Initialisation involves reading the most recent checkpoint in order to construct an initial TransStatus table and determine the start point (`ScanLSN`) for the forward scan of the log.

*TransStatus Table.* For each transaction stored in the Active Transaction table, a corresponding entry is created in the TransStatus table.

*Start Point.* The log scan starts from the lowest LSN of either the `DirtyLSN`, or the lowest `FirstLSN` of any active transaction in the TransStatus table.

**Step 2: Data Collection.** During the forward scan of the log, data for all data structures as discussed in Section 4.1 is collected. The type of record encountered during the log scan determines the data that is collected and into which data structure(s) it is stored. The records from which the Analysis phase collects data are Commit Log Record, End Log Record, ULR, CLR, and SCR.

*Commit Log Record.* Each time a commit log record is encountered, an entry is inserted into the TransStatus table for the transaction with status set to *Commit*. Any existing entries for this transaction are replaced.

*End Log Record.* Each time an end log record is encountered, an entry is inserted into the TransStatus table for the transaction with status set to *End*.

*Update Log Record (ULR).* Upon encountering an ULR:

- If no entry exists in the TransStatus table for this transaction, then an entry is created for this transaction with status set to *Active*.
- Add an entry to the PLink list.
- Create a Page Start List entry and a Page End List entry as required.

*Compensation Log Record (CLR).* Upon encountering a CLR, the same step as for ULRs are performed and an entry is added to the Undone List.

*Special Compensation Log Record (SCR).* Same as for an ULR record.

**Step 3: Completion.** Once the forward scan of the log is completed, the recovery algorithm acquires an exclusive lock on all pages identified in the Page Start List. Subsequently, the DBS can commence normal processing. Only those pages that are locked for recovery will remain unavailable.

Now, the Redo phase may be entered. Once all loser transactions are known, a page can potentially enter the Undo phase – Page End Lists are not required but rather help the algorithm to determine where to terminate. In the absence of Page End Lists, Undo terminates as soon as the `ScanLSN` record is reached.

### 4.3 Redo Phase

The Redo phase is responsible for returning each page in the database to the state it was in immediately before the crash. For each page in the Page Start list, the redo algorithm will spawn a thread that 'repeats history' for that page. Given a page $P_j$, history is repeated by performing the following tasks:

1. Start by considering the oldest log record for page $P_j$ that was written after `PageLSN`. This requires reading the page into main memory.
2. Using the PLink lists for page $P_j$, move forward through the log until no more records for this page exist.
3. Each time a redoable record is encountered, reapply the described changes. Redoable records are: SCR, CLR and ULR records.

Once the thread has processed the last record for this page, the recovery algorithm may enter the Undo phase. The recovery algorithm may enter the Undo phase for different pages at different times, for example page $P_1$ might enter the Undo phase while page $P_2$ is still in the Redo phase.

Once the recovery algorithm has completed the Redo phase for all pages, an End Log Record can be written for all transactions whose status is *Commit* in the TransStatus table. For expediency, this can be deferred until after the Undo phase is complete if so desired.

### 4.4 Undo Phase

The Undo phase is responsible for undoing the effects of all updates that were performed by loser transactions. The thread that was spawned for the Redo phase will now begin working backwards through the log undoing all updates to the page that were made by loser transactions:

1. Work backwards through the log using the `PageLastLSN` pointers processing each log record until all updates by loser transactions have been undone.
2. Each time an SCR or ULR record is encountered, take the following actions:
   - (SCR): Jump to the record immediately preceding the record pointed to by the `UndoneLSN` field. The `UndoneLSN` field indicates that during a previous invocation of the recovery algorithm, the updates recorded by the record at `UndoneLSN` have already been undone.
   - (ULR): If the update was not written by a loser transaction or has previously been undone, then no action is taken. Otherwise:
     (a) Write an SCR record that describes the undo action to be performed with the `UndoneLSN` field set equal to the LSN of the ULR record whose updates have been undone.
     (b) Execute the undo action described in the SCR record written.

Once the thread has completed processing all records back to `EndLSN`, the page can be unlocked and, thus, made available for normal processing again. The advantage of allowing each page to be unlocked individually is that the database can return to normal processing as quickly as possible. Once the recovery algorithm has completed the Undo phase for all pages, an End Log Record can be written for all transactions whose status is *Active* in the TransStatus table.

### 4.5 Crashes During Crash Recovery

By preserving ARIES' paradigm of repeating history, it can be guaranteed that multiple crashes during crash recovery will not affect the outcome of the recovery process. The Redo phase ensures that each update lost during the crash is applied exactly once by using the `PageLSN` value to determine which logged updates have already been applied to the page. Since all compensation operations are logged during the Undo phase, the Redo phase and the nature of the Undo phase ensure that compensation operations are also performed exactly once. The `UndoneLSN` plays a similar role in C-ARIES as the `UndoNextLSN` does in ARIES.

## 5 C-ARIES: Rollback During Normal Processing

Having defined the algorithm for rollback of transactions during crash recovery, it is now necessary to do the same for normal processing. There are two main classes of schedules that must be considered when defining a rollback algorithm, these are: Schedules with cascading aborts and schedules without.

The case where cascading aborts do not exist is trivial, where rolling back a transaction simply involves following the `PrevLSN` pointers for the transaction backwards undoing each operation as it is encountered. Since cascading aborts do not exist in these schedules, no consideration need be given to conflicts between the aborting transaction and any other transactions.

The case where cascading aborts do exist is a great deal more complex, since rolling back a transaction may necessitate the rollback of other transactions. Each time an operation is undone, it is necessary to consider which transactions, if any, must be rolled back in order to avoid database inconsistencies.

In ARIES, rollback of transaction $T_i$ involves undoing each operation in reverse order by following the PrevLSN pointers from one ULR record to the next. Whenever an undo operation for transaction $T_i$ conflicts with an operation in some other transaction $T_j$, a cascading abort of transaction $T_j$ must be initiated. Transaction $T_i$ must then suspend rollback and wait for transaction $T_j$ to rollback beyond the conflicting operation before it can recommence rollback.

Clearly this is not the most efficient method, since the rollback of the entire transaction is suspended due to a single operation being in conflict. A more desirable method is to suspend rollback only of those operations that are in conflict and to continue rollback of all other operations. It is also desirable to trigger the cascading abort of all transactions in conflict as early as possible. By taking advantage of multi-threading, it is possible to roll back a transaction on a page-by-page basis. This allows a transaction in rollback to simultaneously:

- Trigger multiple cascading aborts,
- Suspend rollback of updates to pages whilst waiting for other transactions to roll back, and
- Continue rolling back updates that do not have any conflicts.

Partial rollback of transactions is achieved by establishing save points [5] during processing, then at some later point requesting the rollback of the transaction to the most recent save point. This can be contrasted with total rollback that removes all updates performed by the transaction.

### 5.1 Sketch of Algorithms

Rollback of a transaction is achieved by the use of a single 'Master Thread' that is responsible for coordinating the rollback process and multiple 'Slave Threads' that are responsible for the rollback of updates made to individual pages.

The *master thread* is responsible for coordinating the rollback of a transaction by performing the following actions:

- Triggering the cascading abort of transactions as required.
- Undoing all update operations that are not in conflict with update operations from other active transactions. By consulting the lock table in the usual way, the concurrency control manager is able to detect these conflicts.
- Spawning a new slave thread whenever a conflict detected requires the undo of updates to a page be delayed while other transaction(s) roll back.

The algorithm terminates once the master thread has reached the save point and has received a done message from all slave threads spawned.

The *slave thread* is responsible for undoing all updates made by the transaction to a single page. This thread must not undo any operation until any conflicting operation(s) have been undone. Once the slave thread has completed rolling back all changes to the page, it sends a message to the master thread.

**Optimisation.** In rollback, it is possible for both the master thread and the slave threads to reduce the frequency with which they check for conflicts between the current operation and operations belonging to other transactions. Given a ULR record written for page $P_j$ by transaction $T_i$, it is only necessary to check for conflicts if the last ULR record written for page $P_j$ was not written by transaction $T_i$. The `PageLastLSN` pointers are ideally suited for determining whether or not the last ULR record written for page $P_j$ was written by transaction $T_i$.

## 6 Conclusion

In this paper, we presented the C-ARIES algorithm, which extends the original ARIES recovery algorithm with the capability to perform transaction aborts and crash recovery in a highly concurrent manner. Concurrency is achieved by performing transaction aborts and the Redo and Undo crash recovery phases on a page-by-page basis. Additional enhancements are included that decrease the time taken for the DBS to recover from a crash and reduce the time that the database remains unavailable for normal processing.

It should also be mentioned that the proposed algorithm offers support for recovery from isolated hardware failures (e.g. a single-page restore after a torn write) is provided. Moreover, the proposed algorithm readily permits to exploit a common and very effective optimisation, namely logging of disk writes.

## References

1. Mohan, C., Haderle, D.J., Lindsay, B.G., Pirahesh, H., Schwarz, P.M.: ARIES: a transaction recovery method supporting fine-granularity locking and partial rollbacks using write-ahead logging. ACM Transactions on Database Systems (TODS) 17, 94–162 (1992)
2. Mohan, C.: ARIES family of locking and recovery algorithms (2004), On the Internet at `http://www.almaden.ibm.com/u/mohan/ARIES_Impact.html`
3. Mohan, C.: Repeating history beyond ARIES. In: Atkinson, M.P., Orlowska, M.E., Valduriez, P., Zdonik, S.B., Brodie, M.L. (eds.) Proceedings of 25th International Conference on Very Large Data Bases, pp. 1–17. Morgan Kaufmann, San Francisco (1999)
4. Mohan, C., Treiber, K., Obermarck, R.: Algorithms for the management of remote backup data bases for disaster recovery. In: Proceedings of the 9th International Conference on Data Engineering, pp. 511–518. IEEE Computer Society Press, Washington (1993)
5. Gray, J., McJones, P., Blasgen, M., Lindsay, B., Lorie, R., Price, T., Putzolu, F., Traiger, I.: The recovery manager of the System R database manager. ACM Computing Surveys (CSUR) 13, 223–242 (1981)

# Optimizing Ranked Retrieval

Thomas Neumann

Max-Planck-Institut Informatik
Saarbrücken, Germany
neumann@mpi-inf.mpg.de

**Abstract.** Ranked retrieval plays an important role in explorative querying, where the user is interested in the top $k$ results of complex ad-hoc queries. In such a scenario, response times are very important, but at the same time, tuning techniques, such as materialized views, are hard to use. Therefore it would be highly desirable to exploit the top-$k$ property of the query to speed up the computation, reducing intermediate results and thus execution time. We present a novel approach to optimize ad-hoc top-$k$ queries, propagating the top-$k$ nature down the execution plan. Our experimental results support our claim that integrating top-$k$ processing into algebraic optimization greatly reduces the query execution times and provides strong evidence that the resulting execution plans are robust against statistical misestimations.

## 1 Introduction

Ranked queries occur during explorative querying, where users want to get an impression of the available data. As the data set is typically huge, a user is often only interested in the top-$k$ most relevant/important entries, where $k$ is usually small. A typical example query, aiming to find out the ten orders with the most revenue placed in 2006 by customers from France, is shown below:

```
SELECT   *
FROM     Customers C, Orders O
WHERE    C.id=O.cid AND C.country="France" AND O.year=2006
ORDER BY O.revenue DESC
LIMIT    10
```

The natural way to evaluate this query is to first optimize and execute the query without the *LIMIT* part, including sorting the (potentially huge) intermediate result, and eventually return the top ten entries. However, this involves an expensive join, producing and sorting thousands of tuples that are discarded later. This is not only wasteful, but irritating to the user, who usually issues explorative queries in an ad-hoc fashion and waits for the results.

A more promising approach would be to somehow push the *LIMIT* operator down, such that only the largest orders are joined in the first place, reducing the join effort and the number of produced tuples. While this approach seems similar to the standard optimization technique of pushing selections down, it involves

R. Wagner, N. Revell, and G. Pernul (Eds.): DEXA 2007, LNCS 4653, pp. 329–338, 2007.

one particular difficulty: How many tuples from *Orders* have to be joined such that ten output tuples can be produced? More general, how does $k$ change when pushing a *top k* operator (the equivalent of the *LIMIT* statement) down another operator? This question is difficult to answer, as it highly depends on the data distribution (in this case, on the customers from France). In fact, we will show that it typically cannot be answered without looking at the individual data items, which is infeasible to do during query optimization. Optimization techniques for ranked queries therefore need to use an approach more sophisticated than simply pushing the *top k* operators down.

Existing work, for example the seminal RankSQL paper [1], usually concentrates on the efficient calculation of the (potentially aggregated) ranking functions, *revenue* in our simple example. We assume that this problem is already solved, and augment the query optimizer to more aggressively use these ranking functions. Our novel optimization approach generalizes the rewriting approach based on equivalences for usage in ranked queries. We present an execution model that efficiently copes with estimation errors that might happen during the optimization, always guaranteeing the correct result as it would have been computed by a classical database system.

The remainder of this paper is structured as follows: After reviewing related work in Section 2, we present an algebraic formulation of our approach in Section 3. We discuss physical operators and their optimization in Sections 4 and 5. Our approach is evaluated in Section 6, with final conclusion in Section 7.

## 2 Related Work

Ranked retrieval is very common in IR applications, where data is retrieved and aggregated according to some measure of quality (score). An overview and instance optimal algorithms for this kind of problem is presented in the seminal paper by Fagin [2]. It calculates thresholds to decide which tuple might make it into the final *top k* and updates the thresholds while processing tuples. Many variations of this approach have been proposed, some try to exploit statistical information to improve threshold estimations (see e.g. [3]), some try to minimize communication effort etc. Another approach transforms the top-$k$ query into range queries by guessing the required ranges [4]. However, all these algorithms only support "simple" queries, i.e., IR-style queries without complex operators.

A more database-oriented approach for ranked queries is the RankSQL proposal [1,5]. It integrates the decision about rank computation and usage into the query optimization step and extends the relation algebra into a rank-aware algebra for rank computations. This is orthogonal to the problem we try to solve here, we simply assume that the ranking attribute is known. Related, they suggest an efficient joins strategy when the ranking attribute is derived from multiple relations (more details in [6]). Another database-oriented approach is presented in [7]. It does not consider query optimization but proposes an execution strategy related to our approach. The main idea is to partition the input data on the domain of the ranking attribute. If the output is ranked on the attribute $a$, a condition $a \geq \alpha$ is introduced at the base relation for $a$. If this condition removes too many tuples from the result, the query is restarted with a new condition $\alpha > a \geq \beta$ to produce the missing tuples and so on. The main difficulty of this approach is predicting $\alpha$ and $\beta$. The paper only briefly suggests using histograms, but it is unclear how to handle complex queries.

# 3 Ranked Queries

## 3.1 An Operator for Ranked Queries

In order to optimize ranked queries using standard techniques, we have to define an algebraic operator that expresses the limitation to the top-most entries. Note that we are only interested in a well-defined operator, i.e., an operator that produces a deterministic result, as otherwise optimization becomes difficult. This implies that we require a well defined tuples order (i.e. a *ORDER BY* clause) when optimizing *LIMIT* queries, that is queries of the form:

```
SELECT ... ORDER BY a DESC LIMIT k
```

We concentrate on descending single-attribute *ORDER BY* conditions here, multi-attribute *ORDER BY* conditions can be translated into this form by using a suitable encoding (see [1] for more details on multi-attribute ranking).

While queries of this form provide a well-defined meaning of ranking, they still bear the problem of potentially producing non-deterministic results in the case of ties for the ranking attribute. As there is obviously no natural ordering for ties, we instead extend the *LIMIT* to include all tuples with the same ranking value as the $k$-th result tuple. This leads to the following operator definition:

$$top_a k(R) := \{t | t \in R \wedge |\{t' | t' \in R \wedge t'.a > t.a\}| < k\}$$

The operator $top_a k$ produces the $k$ tuples with the largest value of $a$ (more than $k$ if there are ties for the $k$-th place).

## 3.2 Equivalences for Ranked Queries

Besides the integration into the query processing algebra, the efficient execution of ranked queries also requires a proper integration into the query optimizer. Query optimizers rely upon algebraic equivalences to find plan alternatives to eventually create better (faster) query execution plans.

Regarding the interaction of *top k* operators with themselves, we observe that the operators determine the maximum number of tuples that pass through them, effectively reducing the $k$ for later operators:

$$top_a k_1(top_a k_2(R)) \equiv top_a \min(k_1, k_2)(R)$$

When pushing the *top k* operator below some other operators, we have to make sure that the operators never eliminate a tuple from the final *top k* result.

Furthermore, we can usually only push a *top k* operator down by duplicating it, as the operator itself might create new ties, e.g. a many-to-many join with a selectivity > 1. This observation leads to the following push down equivalences for binary operators:

$$\begin{aligned} top_a k(R_1 \cup R_2) &\equiv top_a k(top_a k(R_1) \cup top_a k(R_2)) & \\ top_a k(R_1 \times R_2) &\equiv top_a k(top_a k(R_1) \times R_2) & \text{if } a \in \mathcal{A}(R_1) \\ top_a k(R_1 \bowtie R_2) &\equiv top_a k(top_a k(R_1) \bowtie R_2) & \text{if } a \in \mathcal{A}(R_1) \end{aligned}$$

Unfortunately these equivalences are limited in application. Finding more general equivalences is not an easy task, as we will discuss in the next subsection.

### 3.3 Equivalences Are Not Enough

To efficiently optimize *top k* operators, we want to push them down commonly used operators, for example selections. This can be quite difficult, as can be seen by the following example. Consider the query *top* $k(\sigma(R))$. We would like to transform it into an equivalent query $\sigma(top\ k'(R))$, pushing the *top k* below the selection. The question is how to choose $k'$. Consider the example illustrated in Figure 1 with $top_a 2$, $\sigma_{b=1}$. For the relation $R_1$, $top_a 2(\sigma_{b=1}(R))$ is equivalent to $\sigma_{b=1}(top_a 3(R))$, while for $R_2$ it is not. The difference for $R_2$ is caused by the tuple [3,2] that "kicks out" the desired tuple [2,1] in the intermediate result computed by $top_a 3(R)$. This illustrates that the *top k* push-down is highly data-dependent, being sensitive even to minor variations.

| a | b |
|---|---|
| 5 | 1 |
| 4 | 2 |
| 2 | 1 |
| 1 | 2 |

$R_1$

| a | b |
|---|---|
| 5 | 1 |
| 2 | 1 |

$top_a 2(\sigma_{b=1}(R_1))$

| a | b |
|---|---|
| 5 | 1 |
| 2 | 1 |

$\sigma_{b=1}(top_a 3(R_1))$

| a | b |
|---|---|
| 5 | 1 |
| 4 | 2 |
| 3 | 2 |
| 2 | 1 |

$R_2$

| a | b |
|---|---|
| 5 | 1 |
| 2 | 1 |

$top_a 2(\sigma_{b=1}(R_2))$

| a | b |
|---|---|
| 5 | 1 |

$\sigma_{b=1}(top_a 3(R_2))$

**Fig. 1.** *top k* push-down with different data instances

This has unpleasant consequences: first, it is nearly impossible to calculate the new $k'$ after a push down without looking at the data tuples, which is not feasible during optimization. Second, a mistake in the estimation of the new $k'$ can lead to an incorrect result, which is unacceptable. Therefore, we cannot expect to find hard equivalences for pushing *top k* operators down in general.

Instead, we have to use a weaker concept. The key idea is to estimate how $k$ changes, and compensate misestimations at runtime. For this purpose we introduce a *soft* operator for *top k*, coined *softtop k*. This *softtop k* operator initially behaves like a normal *top k* operator, but produces more tuples if needed. In our above example, when rephrasing the query to $top_a 2(\sigma_{b_1}(softtop_a 3(R))$, the new operator guarantees the correct result as follow: for $R_1$, the plan directly produces the correct output; for $R_2$, conceptually, the $softtop_a 3$ operator will initially only let 3 tuples pass, but will produce the missing tuple as $top_a 2$ requests it. We can consider a *softtop k* operator as a prediction that we only need $k$ tuples from its input. Pipeline breakers that are followed by a *softtop k* operator will try to produce only $k$ tuples initially, leaving some of their input unprocessed. We will look at the operator model in more detail in Section 4.

Predicting the required number of tuples is not without dangers: while a misprediction in the plan will still produce the correct result, it can cause the query optimizer to underestimate the costs of a plan, resulting in an inefficient execution plan. The problem can be mitigated by using statistics and learning optimizers [8,9], but a misprediction can still occur. The operators described in Section 4 handle a misprediction gracefully if it is within reasonable bounds. If the runtime system detects that the prediction is off by orders of magnitude, it is preferable to re-optimize the full query [9,10].

### 3.4 Adding Soft *top k* Operators

The *softtop k* operator can be introduced by starting from a regular *top k* operator (where the required cardinality is known exactly) and pushing the resulting *softtop k* operator down, changing $k$ as the expected cardinality changes:

$$
\begin{aligned}
top_a k(R) &\equiv top_a k(softtop_a k(R)) \\
softtop_a k(\sigma(R)) &\equiv \sigma(softtop_a \tfrac{k}{s}(R)) \qquad \text{where } s = \tfrac{|\sigma(R)|}{|R|} \\
softtop_a k(R_1 \bowtie R_2) &\equiv (softtop_a \tfrac{k}{s}(R_1)) \bowtie R_2 \text{ where } s = \tfrac{|R_1 \bowtie R_2|}{|R_1|}, a \in \mathcal{A}(R_1)
\end{aligned}
$$

When pushing the operator down, an adjusted $k'$ is derived by dividing the former $k$ by $s$, which is the selectivity used during query optimization. This estimation model makes some simplified assumptions: first, it assumes that the *top k* attribute $a$ is independent from the selection/join predicate. Second, the factor $s$ is usually not known exactly, but can itself only be estimated from statistics. However these are the standard problems of query optimization, where cardinality estimates and, thus, cost functions are always inaccurate to some degree. Many techniques have been proposed to increase the accuracy of estimations (e.g. [11]) and are applicable here as well. With our new *softtop k* operator, misestimation can lead to a wrong cost estimation, but the produced result will still be correct.

The equivalences above illustrate how a *softtop k* operator can be pushed down. However the *softtop k* operators represent information also useful for other operators, therefore they are not really pushed down, but instead replicated:

$$softtop_a k(\sigma(R)) \equiv softtop_a k(\sigma(softtop_a \frac{k}{s}(R)))$$

By using replication instead of push downs, the plan is annotated with expected cardinality assumptions, which helps the query optimizer and the pipeline-breaking operators.

## 4 Operators

The logical $top_a k$ operator can be implemented as a physical operator in two ways. Either, it reads the whole input and keeps the top $k$ entries in memory, discarding the rest. After the whole scan, it produces the $k$ entries. Alternatively, if the input is already sorted on $a$, it reads and produces the first $k$ tuples and then stops. Here, the second strategy is often preferable: The first alternative always has to read its complete input, even though it avoids sorting. The second alternative requires sorted input, but the sort can often be done early, avoiding sorting (and reading) the whole intermediate result.

The logical $softtop_a k$ operator could similarly be implemented in two ways. It could either initially behave like a regular $top_a k$ operator, i.e. read the whole input, keep the top $k$ entries, and eventually discard the rest. But a $softtop\ k$ operator must be able to produce more than $k$ tuples if requested (e.g. $k$ was misestimated), requiring a restart in this implementation. The implementation that we will assume throughout this paper expects an input that is sorted on attribute $a$, which makes the $softtop_a k$ operator trivial: it just passes its input to its consumer, producing as many tuples as needed. The parameter $k$ only affects the cost model and the behavior of pipeline breakers.

The main purpose of the $softtop\ k$ operators is to give runtime hints to the pipeline breaking operators following them. In particular, a pipeline breaker should try to read only $k$ tuples if possible. This principle can be illustrated by looking at a hybrid hash join. For simplicity and without loss of generality, we assume that the left-hand side of the join is a $softtop\ k$ operator and the right hand side is an arbitrary execution plan. The hash operator first reads the right-hand side fully, partitions the data into partitions on disk, or, if possible, in memory. After this step, it reads $k$ tuples from the left-hand side, partitions them, and joins the partitions with the right-hand side. If the estimation of $k$ was correct, this should be enough to produce the required number of output tuples. If not, i.e. the hash operator is requested to produce more tuples, it can request another $k$ tuples from the $softtop\ k$ operator in the attempt to produce more output tuples. Conceptually, this approach converts pipeline breakers into "pipelining" operators for the $softtop\ k$ side, in the sense that they only process chunks of $k$ tuples instead of the whole input.

## 5 Integration into the Query Optimizer

We now look at the problem of integrating $softtop_a k$ operators into the query optimizer. Considering $softtop_a k$ operators clearly increases the search space, as they can be added at any point where $a$ is available without affecting the query result. In fact, this increase seems to be rather tremendous, as the operator has a parameter $k$, and different $k$ might lead to different optimal plans. For a transformative approach this is not a problem, as the operator is pushed down and therefore "knows" the required $k$. But for the standard technique of building plans bottom-up, it is unclear which operators follow and therefore which $k$ is needed. They key insight to tackle this problem is that for any given partial plan, only one value of $k$ makes sense: the minimal $k$ such that all output tuples are still produced. Of course we do not know the value and cannot calculate it easily, but we can estimate it — and the estimations should be unaffected by operators that are still missing: the (unknown) operators that are above the partial plan will obviously not remove any of the tuples that will make it into the final results, therefore they do not affect $k$.

The estimation is carried out in two phases: The base estimation (concerning a single relation) is done before the plan generation, while the estimation for more complex plans is calculated bottom up during the search. Before the plan generation, we construct an arbitrary execution plan and use the transformative techniques from Section 3.4 to push the $softtop_a k$ down to the applicable relation containing $a$. If $a$ is a compound, we just take any single attribute from $a$ for the purpose of the estimation. The $k$ at the relation now tells us how many of the top tuples of this relation we expect in the final result. In the case of compound values for $a$, we could not compute the top tuples yet, but this does not affect the expected required cardinality. We use the $k$ as base estimation for the parameter $k$, and propagate it up: During the plan generation, we now update the estimation for $k$ by reversing the transformations from Section 3.4. This gives as an estimation for $k$ at any plan that contains the base relation with $a$. Note that this estimation for $k$ does not increase the search space, just like cardinality it only requires an additional attribute for each partial plan.

# 6 Evaluation

In this section we study the benefits of our approach and show its robustness towards misestimations. We first compare it to previously published work. Due to the limited scope of the other work, we then perform experiments of our own on more general queries, and finally study the effect of misestimations on the query runtime. All experiments were done on Athlon64 3500+ with 1GB of memory running Linux 2.6.17. To avoid the complex area of buffer space management, all operators were given 1 MB each to store intermediate results. This is somewhat unfair towards the optimizing approaches (which need less memory in some operators), but makes the comparison more reproducible. As database back end we used a custom textbook database system.

## 6.1 Comparison with Existing Approaches

Most of the previously published related work addresses different aspects of ranked retrieval, therefore comparisons cannot be done easily. One exception is [7], that uses partitioning of the input data to speed up ranked retrieval. Unfortunately, most examples in [7] are very simple from an optimizers point of view (they consider only a single relation), which does not allow for meaningful comparisons. Their only example with more than one relation was Query 4, that joined two relations (see Figure 2a).

```
SELECT    *
FROM      Emp e, Dept d
WHERE     e.age > 20 AND
          d.budget > 1000 AND
          e.works_in=d.dno
ORDER BY  e.salary DESC
LIMIT     10;
```

(a)

```
SELECT    *
FROM      Emp e, Dept d, Tasks t
WHERE     e.age > 20 AND
          d.budget > 1000 AND
          e.works_in=d.dno AND
          t.done_by=e.eno AND
          t.duration > 5
ORDER BY  e.salary DESC
LIMIT     10;
```

(b)

**Fig. 2.** SQL queries used in the experiments

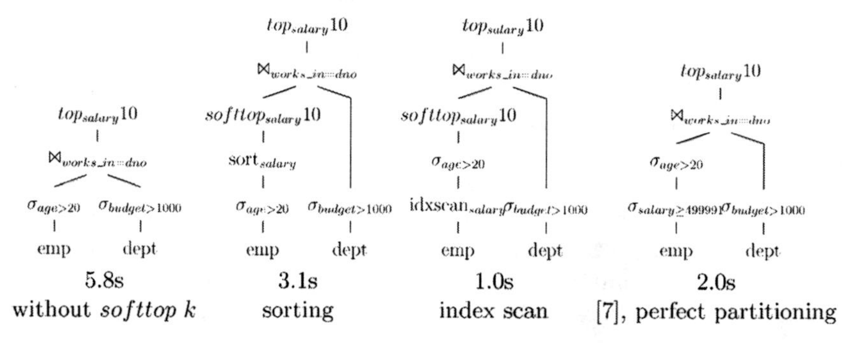


**Fig. 3.** Query Plans and Execution Time for Query from Figure 2 (a)

We create a randomized data set as described in [7], consisting of 500.000 tuples in *Emp* and 100.000 in *Dept*, for a total database size of about 65 MB. We then considered four different execution strategies using blockwise nested loop joins, the last one corresponding to the plan in the paper: First, a normal execution plan with a final *top* 10 operator, then a plan using *softtop k* operators and sorting, a plan using an index scan instead of sorting and finally the partitioning approach from [7]. The paper itself did not make it clear how the partitioning should be done. We therefore use perfect partitioning on the salary ($\sigma_{salary \geq 499991}$ produces just the required tuples from *Emp*), although this requires prior knowledge and could not be done in practice. The corresponding execution plans and the query executions times are shown in Figure 3.

Adding *softtop k* operator clearly improved the runtime compared to a standard *top k* execution. When the plan requires sorting, the gain is not that large, as more than 70% of all data has to be sorted here. When the sorting can be avoided, the plan is much faster, about a factor of 5 compared to standard *top k*. The partitioning scheme from [7] is in the middle between of these two alternatives, even though we use a perfect partitioning scheme. As described in the paper, a misestimation would require restarting the query, i.e. executing it at least twice. Therefore even the sort based *softtop k* approach seems more attractive here, as it is only somewhat slower than the partitioning scheme but very robust concerning misestimations, as we will see below.

### 6.2 Larger Queries

The above *top k* query taken from [7] consists of a join of two relations, where the *top k* can be pushed down to the larger of the two relations. This is scenario is very favorable for *top k* processing, as we can expect large gains by reducing the largest partition. To get a more difficult query, we extended the data set by adding a relation *tasks(done_by,duration,description)* containing tasks with descriptions. Each employee handles 0-10 tasks, where each task has a duration between 1-20. The *tasks* relation contains about 2.5 Mio. tuples, which makes it the largest relation in the combined query (shown in Figure 2 b). The join with the *tasks* relation is expensive, as can be seen on the left-hand side of Figure 4).

| | without sorting index | | |
|---|---|---|---|
| | *softtop k* | | scan |
| runtime [s] | 4.7s | 3.5s | 2.5s |

hash joins

| | without sorting index | | |
|---|---|---|---|
| | *softtop k* | | scan |
| runtime [s] | 4.7s | 2.4s | 0.2s |

index nested loop joins

**Fig. 4.** Results for Query from Figure 2 (b)

The *softtop k* optimization can only reduce the runtime up to a certain degree, as the costs for the *tasks* relation cannot be avoided. The query optimizer can cause much larger gains, if the necessary building blocks are available. The query above can benefit greatly from indices on *dno* and *done_by*: The query optimizer can then use index nested loop joins, which significantly decreases the runtime, as shown on the right-hand side of Figure 4. When using an index scan, the database only has to lookup a few of tuples and can produce the answer almost instantaneous. This demonstrates that *softtop k* operators have to be integrated into the query optimizer: index nested loop joins are expensive, they were only selected because the query optimizer expected few lookups.

### 6.3 Sensitivity to Misestimations

A critical aspect of *top k* optimization is the effect of misestimations. Overestimations are usually fine (except some wasted work), but underestimations can lead to too few tuples and thus requires some kind of query restarting, potentially doubling the execution time. Our approach of adding *softtop k* operators handles underestimations more gracefully, as additional tuples can be produced as needed. This can be seen in Figure 5: The figure shows results for the previous query with a varying degree of misestimation, both for hash joins and for

| estimation error | -90% | -50% | -25% | 0% | +25% | +50% | +90% |
|---|---|---|---|---|---|---|---|
| no *softtop k*, hash joins [s] | 4.7 | 4.7 | 4.7 | 4.7 | 4.7 | 4.7 | 4.7 |
| *softtop k*, hash joins [s] | 3.9 | 3.8 | 3.7 | 3.7 | 3.7 | 3.7 | 3.7 |
| *softtop k*, bnl joins [s] | 5.4 | 4.6 | 4.6 | 3.8 | 3.8 | 3.8 | 3.8 |

**Fig. 5.** Effect of Misestimations, *top* 1000

blockwise nested loop joins. Both approaches are unaffected by overestimations here, as the additional effort is minor compared to the rest of the query. For underestimations, hash joins have to partition and join another block of $k$ tuples, but this does not affect the runtime much. Blockwise nested loops are more sensitive, as additional passes over the data are required for underestimations. Note that index nested loop join results from above are unaffected by misestimations, as tuples are produces on demand. The only danger there is that the query optimizer might decide to use a different join technique due to overestimated costs. Overall the *softtop k* approach handles misestimations reasonably well, in particular hash joins are a robust choice against misestimations.

## 7 Conclusion

We proposed two algebraic operators, *top k* and *softtop k*, to efficiently process and optimize ranked queries. Both offer new equivalences when optimizing ranked queries, and can be integrated into a dynamic programming approach. Our experimental results show that query processing can greatly benefit from these operators, and that the approach is robust against misestimations. Future work should include more precise statistics to estimate $k$ after a push-down, in particular taking correlations into account.

## References

1. Li, C., Chang, K.C.C., Ilyas, I.F., Song, S.: Ranksql: Query algebra and optimization for relational top-k queries. In: SIGMOD (2005)
2. Fagin, R., Lotem, A., Naor, M.: Optimal aggregation algorithms for middleware. In: PODS (2001)
3. Theobald, M., Weikum, G., Schenkel, R.: Top-k query evaluation with probabilistic guarantees. In: VLDB (2004)
4. Chaudhuri, S., Gravano, L., Marian, A.: Optimizing top-k selection queries over multimedia repositories. TKDE 16(8), 992–1009 (2004)
5. Ilyas, I.F., Shah, R., Aref, W.G., Vitter, J.S., Elmagarmid, A.K.: Rank-aware query optimization. In: SIGMOD, pp. 203–214 (2004)
6. Ilyas, I.F., Aref, W.G., Elmagarmid, A.K.: Supporting top-k join queries in relational databases. In: VLDB, pp. 754–765 (2003)
7. Carey, M.J., Kossmann, D.: Reducing the braking distance of an sql query engine. In: VLDB (1998)
8. Markl, V., Megiddo, N., Kutsch, M., Tran, T.M., Haas, P.J., Srivastava, U.: Consistently estimating the selectivity of conjuncts of predicates. In: VLDB (2005)
9. Stillger, M., Lohman, G.M., Markl, V., Kandil, M.: Leo - db2's learning optimizer. In: VLDB, pp. 19–28 (2001)
10. Avnur, R., Hellerstein, J.M.: Eddies: Continuously adaptive query processing. In: SIGMOD (2000)
11. Ilyas, I.F., Markl, V., Haas, P.J., Brown, P., Aboulnaga, A.: Cords: Automatic discovery of correlations and soft functional dependencies. In: SIGMOD (2004)

# Similarity Search over Incomplete Symbolic Sequences

Jie Gu and Xiaoming Jin

Software School of Tsinghua University
guj05@mails.tsinghua.edu.cn,
xmjin@tsinghua.edu.cn

**Abstract.** Reliable measure of similarity between symbolic sequences is an important problem in the fields of database and data mining. A lot of distance functions have been developed for symbolic sequence data in the past years. However, most of them are focused on the distance between complete symbolic sequences while the distance measurement for incomplete symbolic sequences remains unexplored. In this paper, we propose a method to process similarity search over incomplete symbolic sequences. Without any knowledge about the positions and values of the missing elements, it is impossible to get the exact distance between a query sequence and an incomplete sequence. Instead of calculating this exact distance, we map a pair of symbolic sequences to a real-valued interval, i.e, we propose a lower bound and an upper bound of the underlying exact distance between a query sequence and an incomplete sequence. In this case, similarity search can be conducted with guaranteed performance in terms of either recall or precision. The proposed method is also extended to handle with real-valued sequence data. The experimental results on both synthetic and real-world data show that our method is both efficient and effective.

## 1 Introduction

It is a common requirement to determine the similarity between two symbolic sequences for a large number of applications such as information retrieval and pattern recognition[1][2]. Due to its importance, similarity measurement of symbolic sequences has received considerable attention in the past years. The core of similarity measurement lies in the similarity function, which maps a pair of sequences to a real number. Distance functions are analogous, except that the higher the distance, the lower the similarity. We will use the two terms interchangeably, depending on which interpretation is more suitable. While most of the previous work has been focused on the distance function for complete symbolic sequences where all elements of the sequence are available, incomplete symbolic sequences have been ignored largely. Symbolic sequence is defined to be incomplete if several number of the its elements are missing. In real-world applications, the elements of symbolic sequence can be missing for various reasons, including: the data were not recorded by the observer, the data collection

R. Wagner, N. Revell, and G. Pernul (Eds.): DEXA 2007, LNCS 4653, pp. 339–348, 2007.

equipment was not functioning properly, or the data were deemed to be bad by quality control procedures. Then there is a natural requirement for similarity measurement for incomplete symbolic sequences.

In this paper, we mainly tackle the problem of similarity search of incomplete symbolic sequences. This problem is challenging since the both the positions and values of missing elements in a symbolic sequence are unknown in most cases. Then it is impossible for us to determine the exact similarity, but this does not mean no useful similarity information can be extracted from incomplete symbolic sequences.

In our method, the Hamming distance is employed as the basic similarity model since it is most popular and has been long used in similarity measurement for symbolic sequences[4]. But it should be noted that our technique can be applied to any other distance metrics without any substantial modification. We propose a novel method for incomplete symbolic sequence which can map a pair of symbolic sequences(*one is the complete query sequence, the other is an incomplete sequence in the data set*) to a real-valued interval, whose two end points are the lower bound and the upper bound of the underlying exact Hamming distance. It is unnecessary to know anything about the missing elements to get the lower(upper) bound in our approach. With the proposed lower(upper) bound, similarity search can be conducted over incomplete sequences data set with guaranteed performance in terms of either precision or recall(*more details in Section 3.3*). Since the brute-force approach is prohibitive in time complexity, we mainly propose a dynamic programming technique which can reduce the computation cost from $O(n * C_m^n)$ to $O(m * (m - n))(n < m)$, where $m$ and $n$ are the lengths of the two symbolic sequences being compared. We also propose several filtering techniques to further speed up the similarity search process.

The remainder of this paper is organized as follows. We formulate in section 2 the motivation of our proposed similarity function. In section 3, the algorithm of our function is presented. We extend our method to real-valued sequence in section 4. In section 5, the experimental result is reported.

## 2 Problem Description

### 2.1 Distance Function for Complete Symbolic Sequence

A symbolic sequence $X$ comprises an ordered list of symbols, i.e., $X$=($X_1$, $X_2$, ..., $X_n$), where $n$ is the length of $X$. Each element in $X$, denoted as $X_i (1 \leq i \leq n)$, is a symbol from a global symbol set $\mathcal{D}$ which is called a dictionary. The content of $\mathcal{D}$ varies among different applications, e.g., $\mathcal{D} = \{A, C, G, T\}$ in DNA database.

Similarity search over symbolic sequences is an operation that retrieves all the sequences similar to a query sequence from a large sequence data set. Similarity measurement is the core subroutine in this process, where a certain distance function is adopted(high similarity low distance). Before move on to the topic of similarity of incomplete symbolic sequences, we first have a brief review of the distance function and similarity search for complete symbolic sequences. Below we give a variant of the Hamming distance:

**Definition 1.** ***Distance Function for Complete Symbolic Sequences*** *$X$=($X_1, X_2, ..., X_n$) and $Y$=($Y_1, Y_2, ..., Y_n$) are two symbolic sequences with n elements. The distance between X and Y can be defined as follows:*

$$D(X,Y) = \frac{\sum_{i=1}^{n} \delta(X_i, Y_i)}{n}, \delta(X_i, Y_i) = \begin{cases} 0 & if\ X_i == Y_i \\ 1 & otherwise \end{cases}$$

Take two symbolic sequence $X$=($ADCABADC$) and $Y$=($ABCABADA$) for example, $D(X,Y)$=0.25 since ($X_2$=$D$)$\neq$($Y_2$=$B$) and ($X_8$=$C$)$\neq$($Y_8$=$A$).

Based on the definition of distance described above, the similarity search over complete sequences can be formulated as follows:

**Definition 2.** ***$\epsilon$-Approximation Sequence Search*** *Given a set of sequences S and a query sequence Q together with a threshold $\epsilon(0 \leq \epsilon \leq 1)$, the $\epsilon$-Approximation Sequence Search is to retrieve all the sequences from S that satisfy the following inequality:*

$$D(Q, S_i) \leq \epsilon, S_i \in S$$

## 2.2 Distance Function for Incomplete Symbolic Sequence

A symbolic sequence is said to be incomplete if several elements in it are missing. The most popular technique dealing with missing elements is imputation which produces an artificial value to replace a missing element. For an incomplete symbolic sequence, a new artificial complete sequence will be obtained after imputation. For example, given an incomplete symbolic sequence $X$=$(A, D, ?, B)$ where '?' represents a missing element whose value is unknown, we can impute the missing element with 'C' and get $X'$=$(A, D, C, B)$. $X'$ is referred as the complete form of an incomplete symbolic sequence $X$.

The challenge in our problem is that we have no priori knowledge such as positions and values about the missing elements, which makes it impossible to get a deterministic complete form of an incomplete sequence. In the above example, if the position of the missing element '?' is unknown, there may be 4 possible forms of $X'$ since '?' may appear in all the 4 positions of $X$. It should be noted that it is only the case when we impute the missing element with a constant value, if the missing element can be replaced by various values, the number of possible complete forms will expand. Given the huge amounts of possible complete forms, it is impossible to measure the exact distance between a query symbolic sequence and an incomplete symbolic sequence. However, we argue that a lower bound and an upper bound for this unaccessible exact distance can be derived. The problem can be formulated as:

*Given a query sequence $Q$=$(Q_1, Q_2, ..., Q_m)$, for an incomplete symbolic sequence $S$=$(S_1, S_2, ..., S_n)$ whose length of complete form should be $m(n < m)$, calculate two distances $D_{LB}(Q,S)$ and $D_{UB}(Q,S)$ such that*

$$D_{LB}(Q,S) \leq D(Q,S) \leq D_{UB}(Q,S)$$

*where $D(Q,S)$ is the exact distance between $Q$ and $S$.*

The proposed problem can be divided into two subproblems:

* What values should be used to replace the missing elements?
* How to determine the positions for missing elements in order to get the lower(upper) bound?

# 3 Proposed Approach

## 3.1 Imputation Method

We first fix the positions of missing elements and show how to solve the first subproblem. It is straightforward to find that the values to replace the missing elements in an incomplete sequence should be determined by the query sequence.

**Lemma 1.** ***Query Sensitive Imputation:*** *Given a query sequence $Q$ and an incomplete sequence where the positions of missing elements are known, the lower bound of $D(Q,S)$ can be derived by imputing the missing elements in $S$ with the elements in corresponding positions of the query sequence $Q$. The upper bound can be obtained by imputing the missing elements with values that are different from the elements in corresponding positions of the query sequence $Q$.*

*Proof.* Consider the distance function proposed in **Definition 2**, the distance between $Q$ and $S$ can be expressed as:

$$D(Q,S) = \frac{\sum_{i=1}^{n} \delta(Q_i^1, S_i^1) + \sum_{i=1}^{m-n} \delta(Q_i^2, S_i^2)}{m}$$

where $S^1$ represents the elements in $S$ that are not missing and $S^2$ is the missing part. $Q^1$ and $Q^2$ are the corresponding parts in $Q$. By the proposed query sensitive imputation, $\sum_{i=1}^{m-n} \delta(Q_i^2, S_i^2)$ can be reduced to 0. Then

$$D_{LB}(Q,S) = \frac{\sum_{i=1}^{n} \delta(Q_i^1, S_i^1)}{m} \leq D(Q,S)$$

Similarly, the upper bound can be obtained by maximizing $\sum_{i=1}^{m-n} \delta(Q_i^2, S_i^2)$ to $m - n$, then

$$D_{UB}(Q,S) = \frac{\sum_{i=1}^{n} \delta(Q_i^1, S_i^1) + m - n}{m} \geq D(Q,S)$$

Now we provide a running example to illustrate this method in more detail. Suppose the query sequence is $Q$=$(B, A, C, D, B)$ and the incomplete sequence is $S$=$(B, ?, E, D, ?)$ with two elements missing, if we impute the two missing elements with 'A' and 'B' respectively, the complete form of $S$ will be $S'$=$(B, A, E, D, B)$, then $D_{LB}(Q, S)$=$D(Q, S')$=0.2. If the two missing elements are imputed with values other than 'A' and 'B', say, 'B' and 'E', we then get $S'$=$(B, B, E, D, E)$. Consequently, $D_{UB}(Q, S)$=$D(Q, S')$=0.6. The exact distance $D(Q, S)$ must satisfy $0.2 \leq D(Q, S) \leq 0.6$.

### 3.2 Determine Positions of Missing Elements

In last subsection, it is shown that when the positions of missing elements are fixed, by query sensitive imputation, we can get the lower bound and upper bound of the exact distance between the query sequence and the incomplete sequence. Now the first subproblem has been solved and the original problem proposed in **Section.3** has been reduced to the problem below:

> *Given a query sequence $Q$=$(Q_1, Q_2, ..., Q_m)$, for an incomplete symbolic sequence $S$=$(S_1, S_2, ..., S_n)$ whose length of complete form should be $m(n < m)$, find the $m$-$n$ positions such that the lower(upper) bound obtained by applying query sensitive imputation at the $m$-$n$ positions is minimal(maximal).*

For all the $m$ positions $(1, 2, 3, ..., m)$, we refer to any arbitrary $n$ positions as an $n$-position combination. A simple three-phase approach to the above problem can be described as:

- Enumerate all possible $(m$-$n)$-position combinations.
- Apply query sensitive imputation in every $(m$-$n)$-position combination and calculate the corresponding lower bound and upper bound
- Select out the minimal lower bound and the maximal upper bound among all the $(m$-$n)$-position combinations.

The seemingly straightforward method is prohibitive since enumerating all the possible $(m$-$n)$-position combinations is an extremely time-consuming task.

**Lemma 2.** *Given an incomplete symbolic sequence $S$ with $n$ elements missing, if the complete form of $S$ is of length $m(n < m)$, the number of possible $n$-position combinations for the missing elements is $C_m^n$.*

Now we propose a more efficient method to tackle the explosion of position combinations. Given a query sequence $Q$ of length $m$ and an incomplete sequence $S$ of length $n(n < m)$, a $m \times n$ matrix $M$ is proposed to represent the pairwise distance of elements in $Q$ and $S$. $M[i][j]$=0 if $Q_i$=$S_j$, $M[i][j]$=1 otherwise. A dynamic programming approach can be applied to the matrix to find the position combinations that corresponding to the minimal lower bound and maximal upper bound.

**Fig.1** shows the running process of our algorithm.The query sequence is $Q = (A, C, D, C, D, A, B)$, it should be noted the query sequence is given by the

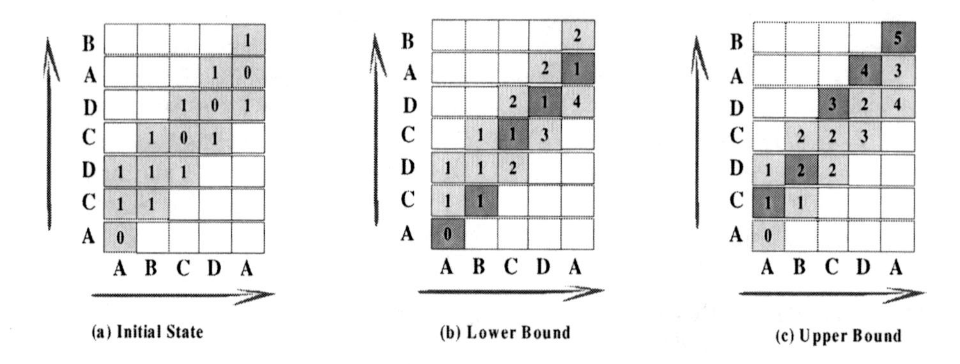


**Fig. 1.** Example of the Calculating Algorithm

user and is always complete. The incomplete sequence is $S = (A, B, C, D, A)$, whose complete form should have the same length with $Q$. **Fig.1(a)** is the initial state of the matrix. The minimal element in the rightmost column of the matrix in **Fig.2(b)** is $M[6][5]=1$, then the lower bound distance is $\frac{1}{7} \approx 0.14$, where 7 is the length of $Q$. This means that, if we impute the missing elements of $Q$ in its $3_{rd}$ and $7_{th}$ position with the corresponding element in $Q$, the distance between $Q$ and the complete form $S'$ of $S$ is minimal. In this case, $S'=(A, B, \mathbf{D}, C, D, A, \mathbf{B})$, where the symbol in bold is imputed by the proposed query sensitive imputation. The maximal element in the rightmost column of the matrix in **Fig.2(c)** is $M[7][5]=5$, then the upper bound distance is $\frac{5}{7} \approx 0.71$, meaning that if $S$ is imputed to be $S'=(\mathbf{A}, A, B, \mathbf{C}, C, D, A)$, the distance between $Q$ and $S'$ is maximal. Now we have successfully mapped the two sequences $Q$ and $S$ to the real-valued interval $[0.14, 0.71]$, where 0.14 and 0.71 are the lower bound and the upper bound of the underlying exact distance between $Q$ and the incomplete sequence $S$.

### 3.3 Usage in Similarity Search

Now we discuss how to employ the proposed method in similarity search over symbolic sequences. Given a database of sequences, a query sequence $Q$, a threshold $\epsilon$, the goal is to find sequences in the database within distance $\epsilon$ from $Q$. For each pair $(Q, S)$:

- if $\epsilon \leq D_{LB}(Q, S)$, add $S$ to the pruned set
- if $\epsilon \geq D_{UB}(Q, S)$, add $S$ to the true result set
- if $D_{LB}(Q, S) < \epsilon < D_{UB}(Q, S)$, add $S$ to the candidate set

After this process, the sequences in the true result set must be the qualified data. If all the sequences in the candidate set are excluded from the final result set, we have a precision as high as 100% but with a relatively low recall. The recall can be improved to be 100% if all the sequences in the candidate set is included in the final result data set, but the precision will be degraded in this case. Thus the candidate set can be used depending on the accuracy requirement of different applications.

## 4 Extension to Real-Valued Sequence

The algorithm proposed in previous sections is designed for sequences whose elements are discrete symbols. In this section, we show that our method can also be extended to handle with real-valued sequences. We also exploit the Hamming distance as the basic similarity model. When dealing with real-valued sequences, its form is as follows:$D(X,Y) = \sum_{i=1}^{n} |X_i - Y_i|$. Given a query sequence $Q$ and an incomplete sequence $X$, there is no upper bound of their exact distance when they are both real-valued. For example, for $Q$=$(2,9,3,9,0,7,2)$ and $X$=$(7,5,7,?,1,2,8)$, the missing element '?' in $X$ is totally unknown and can be infinitely large. Thus $D(Q,X)$ can be infinitely large and no upper bound exists. However, the lower bound for $D(Q,X)$ does exist and can be calculated with the same algorithm presented. Then we can process similarity search over incomplete real-valued sequences database with a recall of %100, i.e., all qualified data objects will be retrieved. Moreover, there is an efficient method to speed up the similarity search process with this lower bound.

Suppose that $Q$=$(Q_1,Q_2,...,Q_m)$ is given as an real-valued query sequence, $X$=$(X_1,X_2,...,X_n)$ is an incomplete real-valued sequence, $\epsilon$ is a given threshold, we want to determine whether $D_{LB}(Q,X) < \epsilon$. We define the data structure $Q^L$=$(Q_1^L,Q_2^L,...,Q_n^L)$ and $Q^U$=$(Q_1^U,Q_2^U,...,Q_n^U)$, where

$$Q_i^L = \min\{Q_i,..,Q_{i+m-n}\}, Q_i^U = \max\{Q_i,..,Q_{i+m-n}\}$$

A lower bound of $D_{LB}(Q,X)$ can be derived with the proposed $Q^L$ and $Q^U$,

$$D'_{LB}(Q,X) = \sum_{i=1}^{n} g(X_i, W_i^L, W_i^U), where$$

$$g(X_i, Q_i^L, Q_i^U) = \begin{cases} |X_i - Q_i^L| & \text{if } X_i \le Q_i^L \\ |X_i - Q_i^U| & \text{if } X_i \ge Q_i^U \\ 0 & \text{otherwise} \end{cases}$$

It is obvious that $D'_{LB}(Q,X) < D_{LB}(Q,X)$, then it is possible for us to avoid calculating $D_{LB}(Q,X)$ explicitly since if $D'_{LB}(Q,X) > \epsilon$, it can be concluded safely that $D_{LB}(Q,X) > \epsilon$.

## 5 Experimental Evaluation

In this section, we present a group of experiments to show the efficiency and effectiveness of the proposed method. For this purpose, we use both real-world and synthetic data sets. All experiments were conducted on a PC with a Pentium 4 processor running at 1.7 GHz and 512M of main memory. The programming language was Java with JDK 1.6.

### 5.1 Data Set

All the sequences in the following data sets are complete, we made each of them to be incomplete by eliminating several elements randomly. By this way, we can compare the query result of our approach with the ground truth.

**NSF Abstract Title Data Set:** This data set consists of strings extracted from the NSF Research Awards Abstracts.[1] We selected those strings of length 15 from this data set and used them in our experiments.

**Synthetic Data Set:** This is a synthetic data set with 5,000 symbolic sequences generated from dictionary $\mathcal{D}=\{A, G, C, T\}$ randomly. Each of the sequences is of length 20.

### 5.2 Precision and Recall

In this subsection, the effect of our method in similarity search is evaluated. As mentioned in **subsection 3.3**, the precision and recall of the query result can be adjusted according to the accuracy requirement. In **Fig.2(a,b)**, we show the precision of the query result when all the sequences in the candidate set are considered to be qualified(the recall is 100% in this case). As the ratio of missing elements increases, the precision decreases since more false candidate are included in the final result. But the overall performance of the proposed method is nice, since the precision is considerately high when the ration of missing elements remains at a low level. In **Fig.2(c,d)**, the recall of the query result, when all candidate set are excluded from the final result, is described. Analogous to the precision, the recall also degrades as more elements in the sequence are missing. From the figure, we can see that the lower bound and upper bound should be employed in different situations. When the query threshold is small , it is rationale to use the lower bound in the query process since a considerably high precision is achievable. If the query threshold is large, the upper bound is a better choice due to the higher recall.

### 5.3 Tightness of Lower Bound and Upper Bound

In this subsection we examine the tightness of the proposed lower bound and upper bound in an intuitive way. As descried in **Figure.3**, we show the relationship of the three distance values. We first chose 50 pairs of symbolic sequences randomly from the synthetic data set, and calculate their hamming distance as the true distance. Then for each pair, we make one of the sequences incomplete by eliminating some of its elements. Then we calculate the lower bound and upper bound of the true distance according to the incomplete distance and another complete distance. This figure shows that both the lower bound and the upper bound is tight to the true distance. When the ration of missing elements decreases, the two bounds get more tighter.

---

[1] http://kdd.ics.uci.edu/databases/nsfabs/nsfawards.html

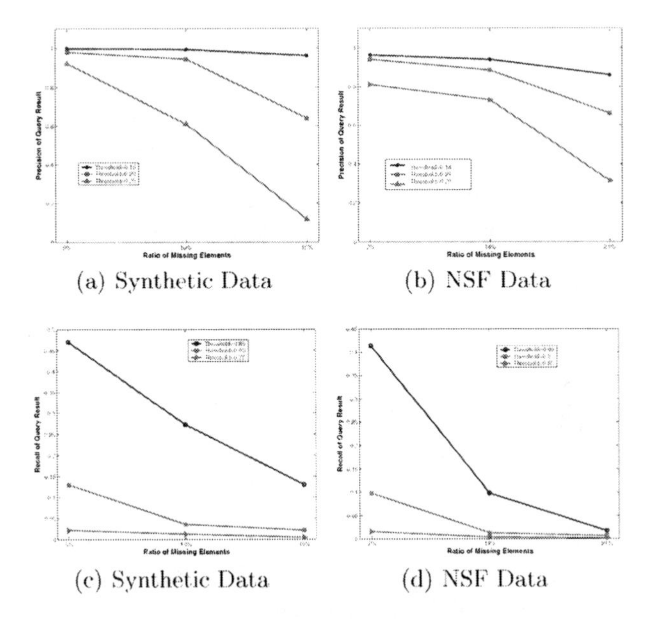


(a) Synthetic Data (b) NSF Data

(c) Synthetic Data (d) NSF Data

**Fig. 2.** Precision&Recall

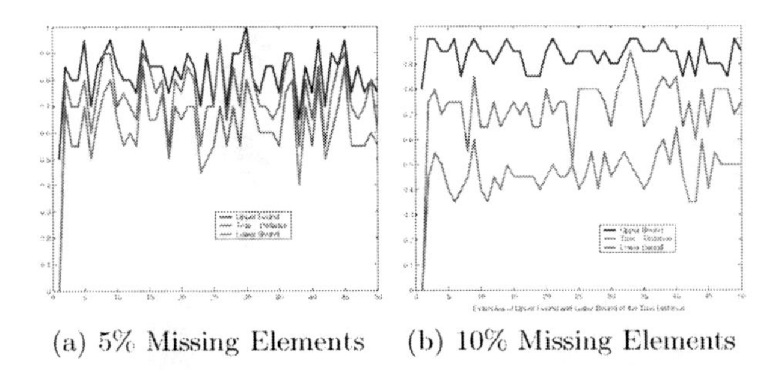

(a) 5% Missing Elements (b) 10% Missing Elements

**Fig. 3.** Example of Lower(Upper) Bound

## 6 Conclusion

In this paper, we propose a strategy for similarity search over incomplete symbolic sequence data set. Though the exact distance between a query sequence and an incomplete sequence cannot be reached, our approach can determine the lower bound and upper bound of this exact distance. Thus similarity search can be conducted with a guaranteed performance. We also extend this method to real-valued sequences together with a pruning technique to further speed the query process.

## Acknowledgement

The work was supported by the NSFC 60403021.

## References

1. Agrawal, R., Faloutsos, C., Swami, A.N.: Efficient Similarity Serach in Sequence Database. In: Proc. Conference of Foundations of Data Organization and Algorithms (1993)
2. Agrawal, R., Lin, K.I., Sawhney, H.S., Shim, K.: Fast Similarity Search in the Presence of Noise, Scaling, and Translation in Time-Series Databases. VLDB Journal (1995)
3. Park, S., Chu, W.W., Yoon, J., Won, J.: Similarity Search of Time-warped Subsequences via a Suffix Tree. Information Systems 28(7), 867–883 (2003)
4. Chakrabarti, K., Garofalakis, M.N., Rastogi, R., Shim, K.: Approximate Query Processing Using Wavelets. The VLDB Journal (2000)
5. Kahveei, T., Singh, A.: Variable Length Queries for Time Series Data. In: Proc. of The ICDE (2001)
6. Yi, B.-K., Jagadish, H.V., Faloutsos, C.: Efficient Retrieval of Similar Time Sequence Under Time Warping. In: Proceedings of the 14th International Conference on Data Engineering (ICDE'98) (1998)
7. Zdonik, S., Cetintemel, U., Cherniack, M., Convey, C.: Monitoring Streams:a New Class of Data Management Applications. In: Proc. VLDB (2002)
8. Hsul, C.: Efficient Searches for Similar Subsequences of Different Lengths in Sequence Databases. In: Proceedings of the 16th International Conference on Data Engineering

# Random Multiclass Classification: Generalizing Random Forests to Random MNL and Random NB

Anita Prinzie and Dirk Van den Poel

Department of Marketing, Ghent University, Tweekerkenstraat 2, 9000 Ghent, Belgium
{Anita.Prinzie, Dirk.VandenPoel}@UGent.be

**Abstract.** Random Forests (RF) is a successful classifier exhibiting performance comparable to Adaboost, but is more robust. The exploitation of two sources of randomness, random inputs (bagging) and random features, make RF accurate classifiers in several domains. We hypothesize that methods other than classification or regression trees could also benefit from injecting randomness. This paper generalizes the RF framework to other multiclass classification algorithms like the well-established MultiNomial Logit (MNL) and Naive Bayes (NB). We propose Random MNL (RMNL) as a new bagged classifier combining a forest of MNLs estimated with randomly selected features. Analogously, we introduce Random Naive Bayes (RNB). We benchmark the predictive performance of RF, RMNL and RNB against state-of-the-art SVM classifiers. RF, RMNL and RNB outperform SVM. Moreover, generalizing RF seems promising as reflected by the improved predictive performance of RMNL.

## 1 Introduction

Random Forests (RF), introduced by Breiman [Breiman] to augment the robustness of classification and regression trees, have been applied successfully in many domains. RF is a bagged classifier building a 'forest' of decision trees splitting at each node on the best feature out of a random subset of the feature space. Whereas bagging enhances the robustness of the original base classifier, random feature selection improves the accuracy in domains characterized by many input variables, with each one containing only a small amount of information. It seems logical, that methods other than decision trees, might also benefit from the exploitation of these two sources of randomness, i.e. random inputs (bagging) and random feature selection.

Firstly, most algorithms suffer the curse of dimensionality and therefore they will benefit from random feature selection. Given the susceptibility of many methods to Hughes phenomenon (on increasing the number of features as input to the classifier over a given threshold, the accuracy decreases) and the tendency towards huge input spaces, most algorithms benefit from *any* type of feature selection. However, as the dimensionality of the feature space grows, a complete search is infeasible. Hence, random feature selection might be an acceptable solution. Secondly, building an ensemble, e.g. bagging, typically results in significant improvements in accuracy. Therefore, this paper investigates the performance improvement of classification

R. Wagner, N. Revell, and G. Pernul (Eds.): DEXA 2007, LNCS 4653, pp. 349–358, 2007.

algorithms by injecting randomness by adopting a randomized ensemble approach. We generalize the RF framework to two classification algorithms, MultiNomial Logit, a Random Utility (RU) model explaining unordered multiple choices using a random utility function, and naive Bayes, a probabilistic classifier simplifying Bayes' Theorem. We propose Random MNL (RMNL) as a new bagged classifier combining a forest of MNLs estimated with randomly selected features. Analogously, we introduce Random Naïve Bayes (RNB). The performance of RF, RMNL and RNB is benchmarked against state-of-the art SVMs [Vapnik]. We illustrate our RMNL and RNB on a multiclass classification problem; a cross-sell case. The results are promising as generalizing RF to MNL substantially improves predictive performance.

## 2 Methodology

### 2.1 Random Forests (RF)

Random Forests (RF) [Breiman] is a highly accurate machine-learning algorithm far more robust than decision trees and capable of modeling huge feature spaces. RF is a bagged classifier combining a collection of $T$ classification or regression trees (i.e. forest of trees), here $T$ classification trees. Each tree $t$ is grown on a different bootstrap sample $S_t$ containing randomly drawn instances with replacement from the original training sample. Besides bagging RF also employs random feature selection. At each node of the decision tree $t$, $m$ features are selected at random out of the $M$ features and the best split selected out of these $m$. Each decision tree is grown using CART methodology to the largest extent possible. An instance is classified into the class having the most votes of over all $T$ trees in the forest, i.e. Majority Voting.

Breiman [Breiman] estimates the importance of each feature on *out-of-bag* (oob) data, cf. in each bootstrap sample about 1/e instances are left out. Randomly permute the feature $m$ in the oob data and put the data down the corresponding tree. Subtract the number of votes for the correct class in the feature-$m$-permuted data from the number of correct votes in the untouched data and average over all trees $T$ in the forest. This is the *raw importance score* for feature $m$ from which the z-score is derived by dividing the raw score by its standard error.

The exploitation of randomness make RF accurate classifiers in several domains. While bagging increases stability of the original decision trees, the random feature selection enhances the 'noise' robustness, yielding error rates that compare even favorably to Adaboost [Freund and Shapire].

### 2.2 Random MultiNomial Logit (RMNL)

Within multinomial-discrete choice modeling [Ben-Akiva], RU models define a random utility function $U_{ik}$ for each individual $i$ for choice $k$ belonging to choice set $D_K$ with $K > 2$ (cf. multiclass). This random utility is decomposed into a deterministic and stochastic component (1):

$$U_{ik} = \beta' x_{ik} + \varepsilon_{ik} \tag{1}$$

where x is a matrix of observed attributes which might be choice (e.g. price of product) or individual specific (e.g. age of customer), $\beta$' is a vector of unobserved marginal utilities (parameters) and $\varepsilon_{ik}$ is an unobserved random error term (i.e. disturbance term or stochastic component). Different assumptions on the error term of the random utility function $U_{ik}$ give rise to different classes of models. In this paper, we apply the MultiNomial Logit (MNL, independent and i.i.d. disturbances). To date, the MultiNomial Logit (MNL) model is the most popular RU model due to its closed-form choice-probability solution [Baltas]. The probability of choosing an alternative $k$ among $K_i$ choices for individual $i$ can be written as in (2). The classifier utilizes the maximum a posteriori (MAP) decision rule to predict the class for individual $i$. MNL exhibits great robustness but is susceptible to multicollinearity.

$$P_i(k) = \frac{\exp(x'_{ik}\,\beta)}{\sum_{k \in K} \exp(x'_{ik}\,\beta)} \tag{2}$$

We will estimate a MNL model incorporating all features. This model might serve as a benchmark for the Random MNL.

Just like the instable decision trees (cf. RF), even a robust classifier like MNL could benefit from injecting randomness by random input selection and random feature selection. Where decision trees performance improves mainly because bagging enhances stability, we hypothesize that MNLs performance will increase because random feature selection reduces the estimation bias due to multicollinearity. We prefer to combine this random feature selection with bagging as it can still improve the stability of an even robust base classifier like MNL. Therefore, inspired by RF, we propose Random MNL (RMNL) as a new bagged classifier combining a forest of R MNLs estimated with $m$ randomly selected features on the $r$-th bootstrap sample. Firstly, just like RF builds $T$ classification trees on bootstrap samples $S_t$, in RMNL each MNL $r$ is estimated on a different bootstrap sample $S_r$ containing randomly drawn instances with replacement from the original training sample. Secondly, this bagging is used in tandem with random feature selection. To classify an observation put the input vector 'down' the $R$ MNLs in the 'forest'. Each MNL votes for its predicted class. Finally, unlike RF, we assess the predictive value of the bagged predictor using the adjusted Majority Vote (aMV) as each $r$th MNL delivers continuous outputs, i.e. posterior probabilities.

We utilize the *out-of-bag* (oob) to assess the feature importances [Breiman].

### 2.3 Random Naïve Bayes (RNB)

Naive Bayes (NB) is a probabilistic classifier simplifying Bayes' Theorem by *naively* assuming class conditional independence. Although this assumption leads to biased posterior probabilities ((3), Z is a scaling factor), the ordered probabilities of NB result in a classification performance comparable to that of classification trees and neural networks [Langley].

$$p(C|F_i,\ldots,F_m) = \frac{1}{Z}\, p(C) \prod_{i=1}^{n} p(F_i|C) \tag{3}$$

Notwithstanding NB's popularity due to its simplicity combined with high accuracy and speed, its conditional independence assumption rarely holds. There are mainly two approaches [Zhang, Jiang and Su] to alleviate this naivity: 1) Selecting attribute subsets in which attributes are conditionally independent (cf. selective NB; [Langley and Sage]), or 2) Extending the structure of NB to represent attribute dependencies [AODE, Web et al. 2005]. We adopt the first approach and hypothesize that NB's performance might improve by random feature selection. Analogous to AODE, we build an *ensemble*, but unlike AODE, we combine zero-dependence classifiers. To decrease the variance of the ensemble, we build a bagged NB classifier. Hence generalizing RF to NB, Random Naive Bayes (RNB) is a bagged classifier combining a 'forest' of *B* NBs. Each *b*th NB is estimated on a bootstrap sample $S_b$ with *m* randomly selected features. To classify an observation put the input vector 'down' the *B* NBs in the 'forest'. Each NB votes for its predicted class. Finally, unlike RF, we assess the predicted class of the ensemble by adjusted Majority Vote (aMV) as each *b*th NB delivers continuous posterior probabilities.

We estimate the importance of each feature on oob data [Breiman].

The predictive performance of RF, MNL, RMNL, NB, RNB and a multi-class one-against-one SVM [Vapnik] with RBF-kernel function is evaluated on a separate test set. Given the objective to classify cases correctly in all classes *K* and the small class imbalance, a weighted PCC (each class-specific $PCC_k$ is weighted with the relative class frequency $f_k$,) [Prinzie] is more appropriate than a PCC [Morrison, Barandela et al.]. Secondly, we benchmark the model's performance to the proportional chance criterion $Cr_{pro}$ rather than the maximum chance criterion $Cr_{max}$ [Morrison]. A final evaluation criterion is the Area Under the receiver Operating Curve (AUC) [Fawcett]. A multiclass AUC results from averaging *K* binary AUCs (one-against-all).

# 3 A CRM Cross-Sell Application

The methodological framework is applied on scanner data of a major home-appliances retailer to analyze customers' cross-buying patterns in order to support cross-sell actions. The objective is to predict in what product category the customer will acquire his next durable. We partition the home-appliance product space into nine product categories Hence, $Y \in \{1, 2, \ldots, 9\}$, K=9. Y has prior distribution $f_1 = 9.73\%$, $f_2 = 10.45$, $f_3 = 20.49$, $f_4 = 12.64$, $f_5 = 11.70$, $f_6 = 9.74$, $f_7 = 8.67$, $f_8 = 8.13$ and $f_9 = 8.45$. We randomly assigned 37,276 ($N_1$) customers to the estimation sample and 37,110 ($N_2$) customers to the test sample. For each customer we constructed a number of predictors X building a general customer profile (e.g. purchase profile, brand loyalty, socio-demographical information) as well as capturing sequential patterns in customer's purchase behavior (the order of acquisition of durables - ORDER, and the duration between purchase events - DURATION) [Prinzie].

# 4 Results

## 4.1 Random Forests (RF)

We estimated RF with 500 trees (default), balanced (higher weights for smaller classes, [Breiman]), on a range of *m* values starting from the square root of M (default); m=441$^{1/2}$. We engage in a grid search with main step size 1/3 of the default setting. Table 1 reports some of the results and shows the sensitivity of RF to *m*. On the estimation data, a balanced RF with 500 trees, *m*=336 delivers the best performance: wPCCe=21.04%, PCCe=21.67% and AUCe=0.6097.

**Table 1.** Estimation performance of RF

| m | wPCCe | PCCe | AUCe |
|---|---|---|---|
| 21 | 19.74 | 20.38 | 0.6007 |
| 42 | 20.20 | 20.88 | 0.6057 |
| 63 | 20.56 | 21.23 | 0.6071 |
| 84 | 20.63 | 21.25 | 0.6061 |
| 168 | 20.98 | 21.59 | 0.6089 |
| 231 | 20.86 | 21.56 | 0.6090 |
| 294 | 21.01 | 21.69 | 0.6114 |
| 336 | 21.04 | 21.67 | 0.6067 |

## 4.2 MultiNomial Logit (MNL) and Random MNL (RMNL)

**MNL.** We estimated a MNL model with M non-choice specific parameters (89 corresponding to RF's 441). A stable solution was not obtained. Alternatively, adopting a wrapper approach, we firstly selected the best features within three types of covariates (general, purchase order and duration). We subsequently compared four MNL models: 1) General, 2) General and Order, 3) General and Duration and 4) General, Order and Duration. The third MNL model delivered the highest performance (wPCCe= 19.75, PCCe=22.00 with $Cr_{pro}$=12.28% and AUCe=0.5973).

**RMNL with R=100 (RMNL_100).** Initially, we combine 100 MNLs (R=100) estimated on 100 bootstrap samples with *m* ($m \leq M$) randomly selected features. We take the square root of M; m=89$^{\wedge 1/2}$ as default parameter setting and, subsequently, engage in a grid search with main step size 1/3 with *m* in [3, 84]. Unfortunately, MNL models with more than 48 dimensions failed to estimate (cf. multicollinearity). Table 2, R_MNL (R=100) gives an overview of the results. The highest performance is observed for *m*=48 (wPCCe=21.25, PCCe=26.87, AUCe=0.6491, $Cr_{pro}$=12.28%).

**RMNL combining MNLs with 10% highest wPCC (RMNL_10).** Combining only the MNLs for a given *m* with the 10% highest wPCCe might improve the accuracy [Dietterich]. We refrain from evaluating a) combining the 10% with highest PCCe or AUCe, as the wPCCe is the main criterion and b) the sensitivity to the number of

**Table 2.** Estimation performance of RMNL

| m | R_MNL (R=100) | | | 10_R_MNL (R=10) | | |
|---|---|---|---|---|---|---|
| | wPCCe | PCCe | AUCe | wPCCe | PCCe | AUCe |
| 3 | 11.53 | 21.41 | 0.6163 | 19.30 | 23.93 | 0.6232 |
| 9 | 15.60 | 23.62 | 0.6270 | 19.69 | 24.98 | 0.6315 |
| 15 | 18.36 | 24.98 | 0.6328 | 20.56 | 26.33 | 0.6403 |
| 21 | 19.33 | 25.56 | 0.6390 | 21.09 | 26.78 | 0.6436 |
| 27 | 19.74 | 25.90 | 0.6423 | 21.14 | 26.63 | 0.6435 |
| 33 | 20.37 | 26.53 | 0.6458 | 21.59 | 27.13 | 0.6468 |
| 42 | 20.91 | 26.69 | 0.6480 | 21.82 | 27.31 | 0.6477 |
| 48 | 21.25 | 26.87 | 0.6491 | 22.01 | 27.33 | 0.6489 |

classifiers combined. Table 2, column 10_R_MNL reports that, analogous to RMNL_100 (R=100), the highest predictive performance is attained for *m*=48.

### 4.3 Naive Bayes (NB) and Random Naive Bayes (RNB)

**NB.** To assess the value of RNB, we benchmark it with a Laplace estimation of NB with all features M (441). We preprocessed numeric attributes by Fayad's and Irani's supervised discretization method [Fayad and Irani]. NB's wPCCe is comparable to MNLs (wPCCe= 19.74, PCCe=21.69, AUCe=0.5982).

**RNB with B=100 (RNB_100).** We investigate the value of injecting randomness for NB as an attempt to mitigate its class conditional independence assumption. We build an ensemble of 100 NBs (B=100) with *m* ($m \leq M$) randomly selected features. Similar to RF and RMNL, we engage in a grid search for *m* starting from the square root of M ($441^{1/2}$) with step size 1/3. Table 3, first column, reports some of the results. The highest performance is measured for RNB_100 with 42 (*m*) randomly selected features: wPCCe=19.83, PCCe=20.00, AUCe=0.6100.

**Table 3.** Estimation performance for RNB

| m | RNB_100 | | | RNB_10 | | |
|---|---|---|---|---|---|---|
| | wPCCe | PCCe | AUCe | wPCCe | PCCe | AUCe |
| 7 | 17.46 | 23.46 | 0.6141 | 18.96 | 22.29 | 0.6071 |
| 14 | 19.41 | 22.74 | 0.6134 | 19.39 | 22.26 | 0.6078 |
| 28 | 19.74 | 22.31 | 0.6115 | 19.87 | 22.99 | 0.6081 |
| 42 | 19.83 | 22.15 | 0.6100 | 20.19 | 22.39 | 0.6122 |
| 56 | 19.78 | 22.05 | 0.6083 | 20.13 | 22.28 | 0.6115 |
| 77 | 19.78 | 21.97 | 0.6066 | 20.21 | 22.31 | 0.6096 |
| 133 | 19.73 | 21.78 | 0.5508 | 20.03 | 21.98 | 0.6048 |
| 294 | 19.78 | 21.75 | 0.6011 | 19.85 | 21.82 | 0.6008 |

**RNB combining NBs with 10% highest wPCC (RNB_10).** Analogous to RMNL, we address whether combining only the 10% best classifiers (based on wPCCe) of the ensemble might improve its accuracy [Dietterich 1997]. The results (Table 3, second column), show a maximum improved wPCCe (+ 0.74 pctp) for *m*=77, with corresponding PCCe (+ 0.62 pctp). Benchmarked against NB, an increase of almost 1 pctp is observed for RNB_10s AUCe with smaller comparable improvements for wPCCe and PCCe. Hence, generalizing RF to NB slightly enhances the accuracy of NB, but not as substantially as for generalizing RF to MNL.

## 4.4 Support Vector Machines (SVM)

We benchmark the performance RF, MNL, RMNL, NB and RNB against that of state-of-the SVM with RBF kernel [LIBSVM] determined by parameters (C,γ). Numerical attributes are scaled from [P1, P99] to the range [-1,+1]. Each SVM (C,γ) is estimated on the scaled training data omitting instances having at least one attribute's value outside the range [-1,+1]. The optimal (C,γ) is determined as the SVM with the highest cross-validation accuracy over 5 folds via parallel grid search. In a first step, using a coarse grid with $C=2^{-5}, 2^{-3}, \ldots, 2^{13}$ and $\gamma=2^{-15}, 2^{-13}, \ldots, 2^{3}$ (100 SVMs) we identified $(2^{7},2^{-11})$ as a "better" region on the grid. In a second step, we conducted a best-region-only grid search around $(2^{7},2^{-11})$ with $C=\{2^{6}, 2^{6.585}, 2^{7}, 2^{8}, 2^{8.585}\}$ and $\gamma=\{2^{-9.415}, 2^{-10}, 2^{-11}, 2^{-11.415}, 2^{-12}\}$ preserving the same difference between subsequent C/γ values. Taking the wPCCe as main criterion, the best predictive performance is measured for $(2^{7}, 2^{-11.415})$: wPCCe=18.66%, PCCe=25.01% and AUCe=0.6111.

## 4.5 Predictive Model Evaluation on Test Data

We assess the robustness of the results on the estimation sample by applying the best RF (*m*=336, balanced), the best MNL (general and order), the best RMNL (*m*=48, RMNL_10), the original NB, the best RNB (*m*=77, RNB_10) and the best SVM $(2^{7},2^{-11.415})$ on a separate test sample ($N_2$=37,110). For SVM we first estimate the SVM $(2^{7},2^{-11.415})$ on the full scaled training data (not cross validation data) excluding instances with at least one attribute with value outside [-1,+1] and applied this SVM on the full scaled test sample including instances with values outside [-1,+1].

Table 4 and Fig. 1 clearly corroborate the estimation findings. The arrows in Fig. 1 show the improvement in accuracy from MNL or NB as compared to RMNL and RNB. Both MNL and NB benefit from injecting randomness, but generalizing RF is most profitable for MNL. Clearly, random feature selection addresses the multicollinearity problem of MNL thereby sincerely enhancing predictive performance. The smaller advantage of injecting randomness for NB might stem from diminished (as compared to full NB), but still considerable dependence between the attributes in the subset of *m* randomly selected features. Overall, the highest test performance is measured for RMNL_10 (*m*=77). Note that only RMNL_10 achieves to combine a high PCC (26.41%) with a high wPCC (21.06%). For this best test model (RMNL_10), we determine whether its *k* AUCs are statistically different from those of RF, MNL, NB, RNB_10 and SVM. Per product category, we employ the non-parametric test by DeLong et al. [DeLong] to determine whether the areas under

**Table 4.** Predictive test performance

| | wPCCt | PCCt | AUCt |
|---|---|---|---|
| RF | 20,66 | 21,39 | 0.6090 |
| MNL | 19,75 | 21,84 | 0.5626 |
| RMNL | 21,06 | 26,41 | 0.6322 |
| NB | 19,27 | 21,05 | 0.5899 |
| RNB_10 | 19,61 | 21,56 | 0.5983 |
| SVM | 18,92 | 25,24 | 0.6188 |

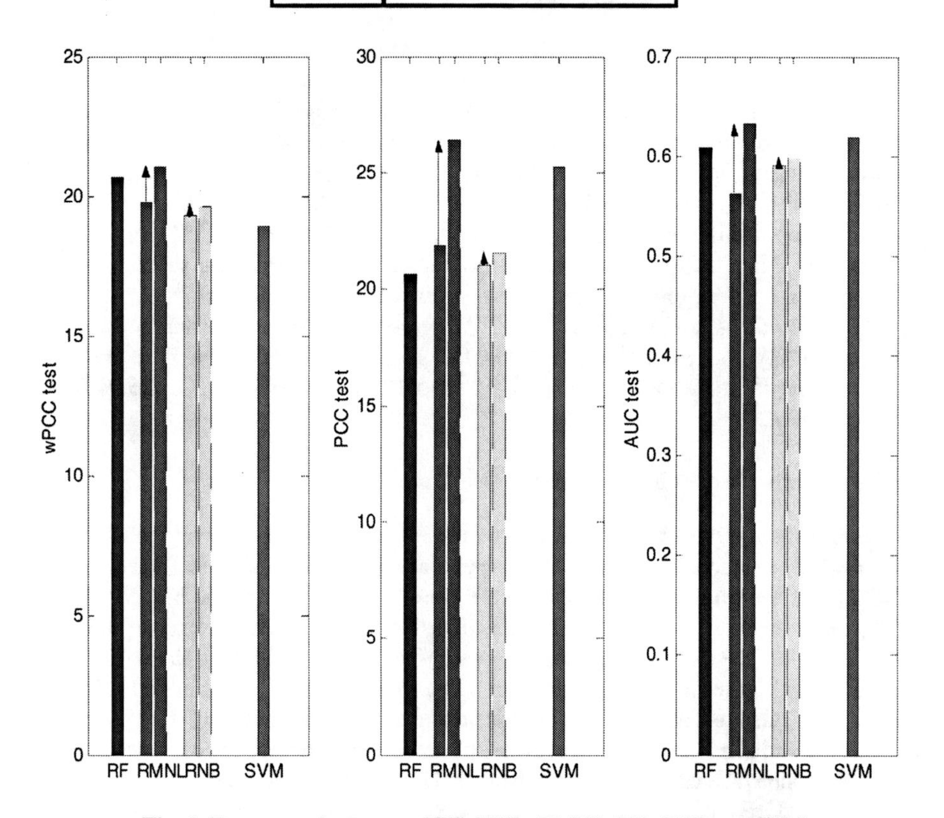


**Fig. 1.** Test set performance of RF, MNL, RMNL, NB, RNB and SVM

the ROC curves (AUCs) within a product category are significantly different. All AUCs on the test set are statistically significant at α=0.05 except for 10_R_MNL_SVM, k=4 and k=5. Finally, also the *k* AUCs for NB and RNB_10 are significantly statistically different at α=0.05.

### 4.6 Feature Importance

From a CRM cross-sell action perspective, it is vital to gain insight in which features drive cross-buying propensities. Therefore, we assess the importance of the seleceted features in the RF, RMNL_10 and RNB_10 models.

Table 5 lists the top-10 features for RMNL_10 (best overall model) together with their z-score calculated on oob data as well as their appropriate rank in RF and RNB_10. A serious loss in predictive accuracy occurs when dropping features on the number of (different) appliances acquired per product category, the gender of the customer, the order of acquisition of home appliances and the time until a first acquisition or between repeated acquisitions in a product category.

**Table 5.** Top-10 features

| Rank | Description | z | RF | RNB |
|---|---|---|---|---|
| 1 | monetary, depth and width | 29.27 | | 18 |
| 2 | monetary, depth and width | 24.91 | | 9 |
| 3 | socio-demo | 19.70 | 14 | 41 |
| 4 | order | 16.01 | | 78 |
| 5 | duration | 9.48 | 4 | 63 |
| 6 | order | 9.21 | | 81 |
| 7 | order | 7.69 | 51 | 15 |
| 8 | order | 4.86 | 7 | 34 |
| 9 | socio-demo | 4.84 | 35 | |
| 10 | brand-loyalty | 4.74 | 16 | 11 |

## 5 Conclusion

RU models are robust multiclass models dominating marketing choice modelling due to their micro-economic theoretical underpinnings. Unfortunately, these RU models are un-suited to model choice settings with many features as they suffer heavily from multicollinearity. The latter might prevent convergence and seriously distorts parameter estimate interpretation. As to date, RU models lack any feature selection, we propose Random MNL, employing bagging in tandem with random feature selection, as alternative.

## Acknowledgments

Our thanks go to Ghent University for funding computer infrastructure (Grantno. 011B5901). Dr Anita Prinzie is a Postdoctoral Fellow of the Research Foundation, Flanders (FWO Vlaanderen).

## References

1. Baltas, G., Doyle, P.: Random utility models in marketing: a survey. Journal of Business Research 51(2), 115–125 (2001)
2. Barandela, R., Sánchez, J.S., Garcia, V., Rangel, E.: Strategies for learning in class imbalance problems. Pattern Recognition 36(3), 849–851 (2003)

3. Ben-Akiva, M., Lerman, S.R.: Discrete Choice Analysis: Theory and Application to Travel Demand. The MIT Press, Cambridge (1985)
4. Breiman, L.: Random Forests. Machine Learning 45(1), 5–32 (2001)
5. Chang, C.C., Lin, C.J.: LIBSVM: A library for support vector machines (2001), Software available at http://www.csie.ntu.edu.tw/~cjlin/libsvm
6. DeLong, E.R., DeLong, D.M., Clarke-Pearson, D.L.: Comparing the areas under two or more correlated receiver operating characteristic curves: a nonparametric approach. Biometrics 44, 837–845 (1988)
7. Dietterich, T.G.: Machine-Learning Research – Four current directions. AI Magazine 18(4), 97–136 (1997)
8. Fawcett, T.: ROC Graphs: Notes and Practical Considerations for Researchers. Technical Report HPL-2003-4, HP Laboratories (2003)
9. Fayyad, U.M., Irani, K.B.: Multi-interval discretization of continuous-valued attributes for classification learning. In: Proceedings of the 13th International Joint Conference on Artificial Intelligence, pp. 1022–1027. Morgan Kaufmann, San Francisco (1993)
10. Freund, Y., Shapire, R.: Experiments with a new boosting algorithm. In: Machine Learning: Proc. of the Thirteenth International Conference, pp. 148–156 (1996)
11. Langley, P., Iba, W., Thomas, K.: An analysis of Baysian classifiers. In: Proceedings of the Tenth National Conference on Artificial Inteligence, pp. 223–228. AAAI Press, Stanford (1992)
12. Louviere, J., Street, D.J., Burgess, L.: A 20+ retrospective on choice experiments. In: Wind, Y., Green, P.E. (eds.) Marketing Research and Modeling: Progress and Prospectives, Academic Publishers, New York (2003)
13. Morrison, D.G.: On the interpretation of discriminant analysis. Journal of Marketing Research 6, 156–163 (1969)
14. Prinzie, A., Van den Poel, D.: Predicting home-appliance acquisition sequences: Markov/Markov for Discrimination and survival analysis for modelling sequential information in NPTB models. Decision Support Systems (accepted 2007), http://dx.doi.org/10.1016/j.dss.2007.02.008
15. Vapnik, V.N.: Statistical Learning Theory. John Wiley & Sons, New York (1998)
16. Zhang, H., Jiang, L., Su, J.: Hidden Naive Bayes. In: Proceedings of the Twentieh National Conference on Artificial Inteligence, AAAI Press, Stanford (2005)

# Related Terms Clustering for Enhancing the Comprehensibility of Web Search Results

Michiko Yasukawa and Hidetoshi Yokoo

Department of Computer Science, Gunma University
1-5-1 Tenjin-cho, Kiryu, Gunma, 376-8515 Japan
{michi,yokoo}@cs.gunma-u.ac.jp

**Abstract.** Search results clustering is useful for clarifying vague queries and in managing the sheer volume of web pages. But these clusters are often incomprehensible to users. In this paper, we propose a new method for producing intuitive clusters that greatly aid in finding desired web search results. By using terms that are both frequently used in queries and found together on web pages to build clusters our method combines the better features of both "computer-oriented clustering" and "human-oriented clustering". Our evaluation experiments show that this method provides the user with appropriate clusters and clear labels.

## 1 Introduction

Recently, web search engines have become indispensable for everyday life. People all over the world use them to find various information about businesses, education, hobbies, etc. Search engines (e.g. Google[1]) have become so advanced that they can search a vast number of web pages for exact matches of strings (terms and phrases) in less than a second. However, search engine users can have difficulties in finding information that fits their needs. The following are problems often encountered during web searches.

- **User query ambiguity:** If users are unable to be specific and concise in expressing their information needs, their sessions with search engines take more time than needed. Search engines require concise, correctly-spelled terms, which precisely express the users information needs for best results. It is difficult for ordinary users to provide such terms.
- **Unmanageable number of search results:** Another problem for search engine users is the sheer volume of search results. A search engine may return thousands of URLs, titles and summaries (snippets). It is no longer possible to check the entirety of results manually. It would be helpful to users if the results could be understood with a simple look.

The information retrieval techniques related to these two problems are the following.

[1] Google, http://www.google.com/

R. Wagner, N. Revell, and G. Pernul (Eds.): DEXA 2007, LNCS 4653, pp. 359–368, 2007.

Query Logs of a Search Engine

| timestamp | search term | access from |
|---|---|---|
| 2006.12.31 01:23 | **salsa** cd music | xx.xx.121.22 |
| 2007.01.01 02:01 | **salsa** dance | xx.xx.122.32 |
| 2007.01.02 03:10 | belly dance | xx.xx.21.113 |
| 2007.01.02 03:21 | **salsa** sauce recipe | xx.xx.102.31 |
| ... | **salsa** dance step | ... |
| ... | ... | ... |
| ... | **salsa** learn dance step | ... |
| ... | ballroom dance | ... |
| 2007.01.31.23:21 | homemade **salsa** recipe | ... |
| 2007.02.01 01:22 | **salsa** step learn dvd | ... |

A collection of Related Terms

| count | related term |
|---|---|
| 1325 | dance |
| 945 | sauce |
| 887 | step |
| 683 | learn |
| 624 | recipe |
| 155 | homemade |
| ... | ... |

terms related to “salsa” during a certain period

**Fig. 1.** Example of Query Logs and Related Terms

- **Query recommendation** ([1], [2]): Pairs or triads of terms are recommended to users to help their queries become more specific and relevant. These terms have been obtained from query logs of the search engine, and selected as those that are related to frequently-asked questions or to the most popular queries. The recommended queries are given as a simple list, which is neither organized nor grouped. Therefore, the user must use trial and error to find the correct query.
- **Search results clustering** ([3], [4], [5]): Clustering works well for clarifying vague queries. It can also be useful for showing users the major topics of the returned results. The main disadvantage of clustering is difficulty in building meaningful and understandable clusters and in giving proper labels to clusters.

Despite its disadvantage, search results clustering is a promising solution for users to enhance their search activities and experiences. The main purpose of this research is to design a method that makes intuitive clusters from search results. We have taken a hint from the Query Recommendation approach to develop a new method, which we have designed to solve the aforementioned problems. The clusters of our method are built by grouping already-collected terms that are related to the query issued by the user. Similar or associative related terms are grouped into a cluster, and web pages corresponding to each cluster are shown to the user with an understandable label.

## 2 Preliminaries and Definitions

### 2.1 Query Logs and Related Terms

Unless query recommendation or search results clustering is available, search engine users submit query terms accompanied by the related terms in order to obtain more relevant results. For example, when $user^a$ submits a query $Q^a = \{\underline{salsa}, dance, lesson\}$, he/she may recognize that the search term $q_1^a = \{salsa\}$ is accompanied by the related terms $w_1^a = \{dance\}$ and $w_2^a = \{lesson\}$.

On the other hand, $user^b$ may submit a query $Q^b = \{salsa, \underline{dance}, lesson\}$, in which the search term is $q_1^b = \{dance\}$ and the related terms are $w_1^b = \{salsa\}$ and $w_2^b = \{lesson\}$, while $user^c$ may issue a query $Q^c = \{\underline{salsa}, \underline{dance}, lesson\}$, in which the search terms are $q_1^c = \{salsa\}$ and $q_2^c = \{dance\}$, the related term is $w_1^c = \{lesson\}$. For simplicity, we consider only $user^a$-type cases where one query consists of one search term and one or more related terms.

A collection of related terms can be obtained from query logs of a search engine (Fig. 1). Let $V^k(q) = \{w_1, w_2, \cdots, w_k\}$ be a collection of $k$ related terms to a search term $q$. Let *qid_term*(*qid*, *term*) be a *relation* between the identifier *qid* of a query and a term *term* included in the query. Here, the collection of related terms $V(salsa)$ can be derived by the following SQL statement[2].

```
select count(*), term from qid_term
where term<>'salsa' and
qid in (select qid from qid_term where term='salsa')
group by term order by count desc;
```

Query recommendation that suggests $l$ related terms is possible by presenting the enumeration of related terms, such as $V^l(salsa) = \{dance, sauce, step, recipe\}$ for $l = 4$. However, it is not helpful for users. The terms should be organized or grouped.

## 2.2 Search Results and Document-Term Matrix

Suppose that a search engine returns $m$ resutls $R^m(q)$ for query $q$. Here, let *results_coll*(*rid*, *URL*, *title*, *summary*) be a *relation* between the identifier *rid* for a result and *URL*, *title*, *summary*. A tuple within the *results_coll* corresponds to a document, i.e., a web page. Consequently, search results $R^m(q)$ retrieved by a search term $q$ have links to $m$ documents $D^m(q) = \{d_1, d_2, \cdots, d_m\}$. The documents collection $D^m(q)$ corresponds to a collection of $n$ terms, $T^n(q) = \{t_1, t_2, \cdots, t_n\}$. Let *did_term*(*did*, *term*) be a *relation* between the identifier *did* for a document and a term *term* included in the document. The collection of terms $T^n(q)$ can be derived by the following SQL statement.

```
select count(*), term from did_term
group by term order by count desc;
```

A relationship between the documents in $D^m(q)$ and the terms in $T^n(q)$ can be described in a document-term matrix. The element $f_{ij}$ of the matrix is the *term frequency*, which represents the number of occurrences of the term $t_j$ in the document $d_i$. Note that the terms are sorted by the *collection frequency* $cf_j$, which represents the total number of occurrences of the term $t_j$ in $D^m(q)$. An example of search results for "Salsa" and the document-term matrix are shown in Fig. 2.

[2] We do not use SQL in actual implementation. See Section 3.3.

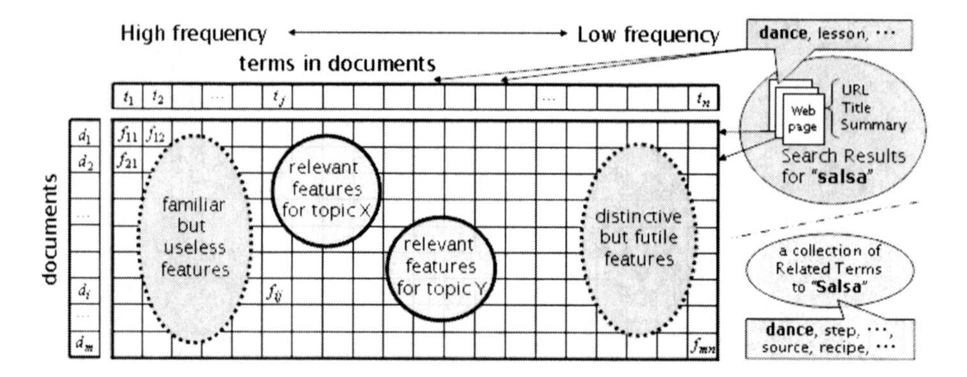


**Fig. 2.** Example of Search Results and Document-Term Matrix

### 2.3 Naive Clustering Methods and Their Problems

With the document-term matrix, we may have the following algorithms.

1. Documents clustering
   The most naive method for the clustering is *documents clustering.* Documents are grouped into disjoint clusters by the similarities in their terms frequencies. Usually, the clusters are labeled by the feature-terms. This method is simple and easy, but the clusters and labels become meaningless when the documents are intermingled with miscellaneous topics and concepts. The two dotted oval parts in the matrix (Fig. 2) degrade the quality and intelligibility of clusters.
2. Terms clustering
   Another naive method for the clustering is *terms clustering.* Terms are grouped in disjoint clusters by the similarity of documents. And then, the feature documents are assigned to the terms clusters.
   In general, important terms are the most frequent terms in the documents. Therefore, the clustering of frequent terms could be meaningful. Actually, we found that it is much more descriptive than the naive documents clustering. However, the clusters are quite familiar or descriptive but they do not correspond to the users interests well. The left dotted oval part in the matrix (Fig. 2) shows such features. The clustering independent from users interests cannot be useful when there is a discrepancy between the tendency of entire documents and typical users' interests.

## 3 The Method: Related Terms Clustering

### 3.1 Vocabulary Selection

In our method, the collection of related terms is exploited to build meaningful and useful clusters. The collection contains user-friendly terms which may reflect potential purposes or situations of many users. Therefore, the collection can be the important set of objects for the clustering. If the product set $E(q) = V^k(q) \cap T^n(q)$ has more terms than needed for the clustering, a subset $E^l(q)$, which consists of $l$ related terms is obtained by the following algorithms.

**Basic Method: Order by Related Terms.** In this algorithm, related terms are selected by the priority of the number of times occurred in query logs. Let *vid_term*(*vid*, *term*) be a *relation* between the identifier *vid* of a related term and the related term *term*. Similarly, let *tid_term*(*tid*, *term*) be a *relation* between the identifier *tid* of a term and the term *term*. The $l$ related terms can be derived by the following SQL statement.

```
select vid,term from vid_term
where term in (select term from tid_term)
order by vid limit l;
```

This algorithm was evaluated in [15] and mostly effective. However, some related terms are too specialized for ordinary users to understand. For example, suppose that two unknown strange strings are the name of a band that plays salsa music and the name of a restaurant that is famous for salsa source, respectively. The cluster including such terms cannot be understood at a glance. Some other related terms are more easy to understand but they do not produce meaningful clusters when they have few good features in the matrix.

**Advanced Method: Order by Frequent Related Terms.** In this algorithm, related terms are selected by the priority of the *collection frequency*. A term $t_j$, which has high collection frequency $cf_j$, is one of the important terms. However, if its *document frequency* is very high or low, it tends to form a useless or futile cluster. Here, the document frequency $df_j$ is the number of documents that term $t_j$ occurs in. Therefore, the document frequency as well as collection frequency in the matrix should be considered. Let *tid_df*(*tid*, *df*) be a *relation* between the identifier *tid* of a related term and a document frequency *df* of the related term. To limit the extent of the document frequency, let us define two thresholds: the maximum threshold *df_max* and the minimum threshold *df_min*. The $l$ related terms can be derived by the following SQL statement.

```
select tid,term from tid_term
where term in (select term from vid_term) and
tid in (select tid from tid_df where df <= df_max and df >= df_min
order by tid limit l;
```

The related terms that have moderately high frequency in the matrix are expected to be good descriptors both for documents and for users.

### 3.2 Building the Cluster

How to group related terms has an impact on the reasonability and the intelligibility of the clusters. Even if the selected terms are understandable, unsuccessful grouping makes them meaningless and useless. The Hierarchical Bayesian Clustering (HBC) algorithm is proposed in [8]. The algorithm has been shown to be effective to automatic thesaurus construction based on verb-noun relations in [9]. Although users' query is not necessarily a grammatical expression, such the algorithm is expected to be effective for the related terms clustering because the queries as $\{salsa, order, cd\}$ and $\{salsa, learn, dance, step\}$ are some kind of interrogative sentences presented to the search engine. The algorithm based on HBC for building $L$ clusters of related term is the following.

```
(1) Initialize
  (1-a) Create term clusters, each of which has a single term.
  (1-b) Calculate the similarity P(c|x).
(2) Repeat the process
  (2-a) Merge the pair of clusters that have the maximum and the
        second maximum sums of P(c|x).
(3) Evaluate the condition
  (3-a) If the number of the clusters is more than L, repeat (2-a).
  (3-b) If the number of the clusters is L, end.
```

Here, $P(c|x)$ is the probability of the term $x \in D^m(q)$, which is included in the cluster $c$, and defined as follows[8].

$$P(c|x) = P(c) \sum_{i=1}^{m} \frac{P(d_i|c)P(d_i|x)}{P(d_i)}$$

### 3.3 Web Application Prototype

We have developed a prototype of clustering meta search engine. The prototype is implemented by C (for clustering) and PHP (for web interface and pre-process). The process of the system is the following. (1) Data Collection: To retrieve search results and web pages (cache pages), we utilize the Search API of Yahoo!JAPAN[3] and MSN[4]. The keyword tool of Overture[5] is utilized to acquire a collection of related terms. (2) Preprocess: A piece of text data is extracted from each web page, stop words are removed, and the remaining terms are stemmed by the Porter Stemming Algorithm[13]. (3) Clustering: The document-term matrix is created, the clusters of related terms are built, and stemmed terms are reversed to the normal terms, documents are assigned to the clusters and the clusters are shown to the user. The sample of clustering is shown in Fig. 3.

## 4 Evaluation

In this section, we evaluate the proposed method with the test data and the prototype system. The test data was built from actual data. User study is also conducted for the evaluation from the user's viewpoint.

[3] Yahoo!JAPAN Developer Network, http://developer.yahoo.co.jp/
[4] MSN Search SDK, http://msdn.microsoft.com/msn/
[5] Keyword Selecter Tool, http://inventory.overture.com/

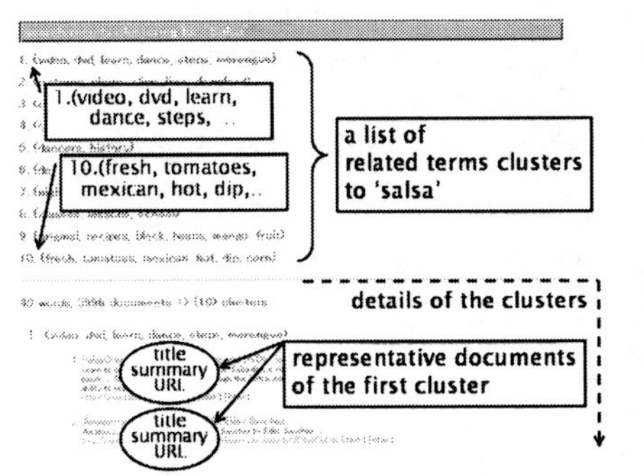


Fig. 3. Example of Web Application

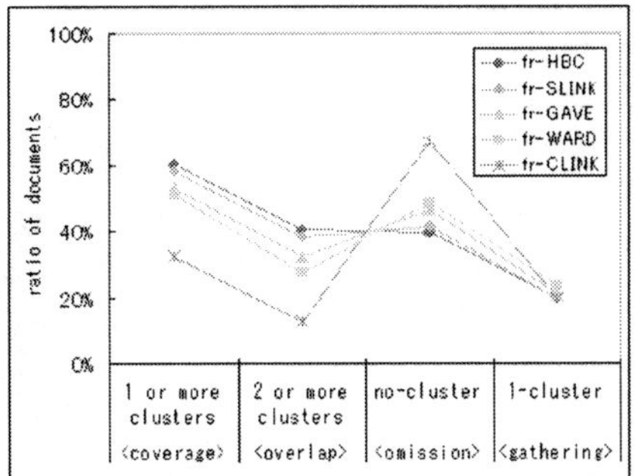


Fig. 4. Comparison of Similarity Measure

## 4.1 Comparison of Clustering Quality

**Test Data.** The topic data of NTCIR-3 WEB[14] were utilized as test queries. At first, we collected the related terms to these queries. Among the 105 queries, 81 queries have one or more related terms. Next, we retrieve search results for the queries. Some queries have few search results and/or no related term. Except for those scarcity queries, we collected the search results and related terms of the 40 queries to setup the test data. In the test data, each query has 1900 or more cached web pages, and 80 or more related terms. Based on the studies of [12] and [15], we have adjusted each method to build 10 clusters and label them with 40 terms. The minimum and maximum document frequencies were set 5% and 50% of the number of documents in the collection respectively.

**Quality Measure of Clusters.** When a single document have multiple topics, it may belong to multiple, say $N$, clusters. In that sense, clusters are not necessarily disjoint. Usually, a cluster may (a) cover a lot of interesting documents, (b) overlap with other clusters, (c) omit uninteresting documents, or (d) separate dissimilar documents and group similar documents.

In order to examine these characteristics of the clusters, we consider the number of documents that are assigned to $N$ clusters. For various $N$, we represent its ratio to the total number of documents by

(a) coverage: one or more clusters ($N \geq 1$),
(b) overlap: two or more clusters ($N \geq 2$),
(c) omission: no cluster ($N = 0$),
(d) gathering: a single cluster ($N = 1$).

Note that the sum of (b) and (c) is equal to (a) for each query. Since the sum of (b), (c) and (d) is 100% for each query, there is a trade-off between (b), (c) and (d).

**Comparison with General Distance Measures.** In order to investigate the proposed method, which is based on HBC, other general distance measures for agglomerative clustering[11] are introduced to the evaluation experiments. They are SLINK (single linkage method), CLINK (complete linkage method), GAVE (group average method) and WARD (Ward's method).

The distance measure functions $D(C_1, C_2)$ of them are defined as follows.

(1) SLINK: $D(C_1, C_2) = min_{x_1 \in C_1, x_2 \in C_2} D(x_1, x_2)$
(2) CLINK: $D(C_1, C_2) = max_{x_1 \in C_1, x_2 \in C_2} D(x_1, x_2)$
(3) GAVE: $D(C_1, C_2) = \frac{1}{n_1 n_2} \sum_{x_1 \in C_1} \sum_{x_2 \in C_2} D(x_1, x_2)$
(4) WARD: $D(C_1, C_2) = E(C_1 \cup C_2) - E(C_1) - E(C_2)$
$E(C_1) = \sum_{x \in C_1} (D(x, c_i))^2$

In the proposed method (described in Section 3.1 Advanced Method), the HBC algorithm (described in Section 3.2) is compared with the general distance measures. Its results are shown in Fig. 4. CLINK has remarkably a higher omission ratio and a lower gathering ratio than others. This means that it cannot build informative clusters. Clusters should be informative as well as comprehensible. Except for CLINK, the other distance measures yield no significant differences. Although SLINK is close to HBC, it is slightly inferior to HBC. In summary, HBC is preferable in that it has higher coverage, higher overlap and lower omission.

**Comparison with Naive Methods.** The proposed methods, *Related Terms Clustering* (proposed-r) and *Frequent Related Terms Clustering* (proposed-fr) described in Section 3.2 are compared with the naive methods, *Documents Clustering* (naive-doc) and *Terms Clustering* (naive-term) described in Section 2.3. The behavior of the algorithm depends on the commonality of the frequent terms and related terms. Here, the commonality ratio is of the top 40 frequent terms and related terms. Except for the outliers, the typical ratio is between 30% and 60%. The comparisons of the methods are divided into three groups according to the ratio. When the ratio is 40% or more, *frequent related terms* closely resembles *related terms.* In Fig. 5, the ratio is 30% or more and less then 40%. It shows that the proposed-fr is between the proposed-r and the two naive methods. The proposed-fr is more explanatory than the proposed-r. However, the proposed-r is less complicated than the proposed-fr. Which method can produce more comprehensible clusters for users? It is evaluated in the following.

### 4.2 Comparison of Clustering Comprehensibility

We compared the proposed-r with the naive methods in [15]. The user study has shown that the clusters built by the proposed-r are more understandable for users since the proposed-r is more human-oriented than the naive methods.

Here, let us compare the proposed-fr with the proposed-r. We conducted a new experiment where 12 university students were shown the clusters produced by the two methods on 6 queries. The students glanced through a set of clusters, and evaluated the set subjectively on a 4-point scale: (1) not understandable, (2) not so understandable, (3) rather understandable, (4) understandable. The

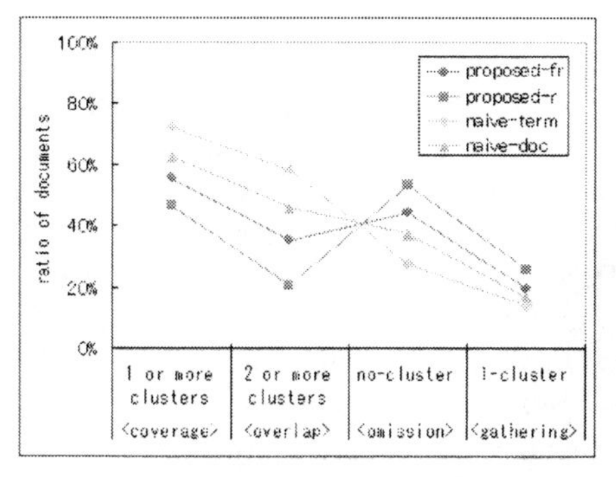


**Fig. 5.** Comparison of Methods (30-40%)

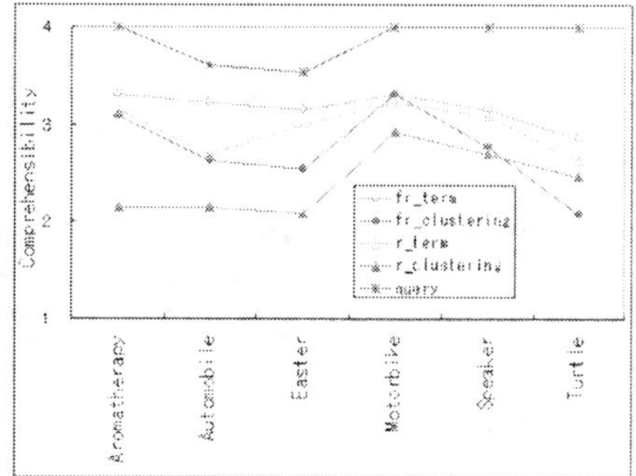


**Fig. 6.** Comprehensibility by Queries

average comprehensibility is shown in Fig. 6. The proposed-fr produces more comprehensible terms and clusters. As for the query "Turtle," the clusters were not comprehensible because its frequent related terms did not have good features. In such case, the documents collection should be expanded to include more related terms.

# 5 Discussion

Clustering of query logs for query recommendation (e.g. [1], [2]) and mining of query logs for improving efficiency of search engine (e.g. [6], [7]) were proposed. These researches utilize not only query logs but also users' clickthrough data, which is also recorded by the search engine. On the other hand, our algorithm utilizes query log, but does not utilize clickthrough data. Methods for automatic thesaurus construction (e.g. [9], [10]) were proposed, and general or universal thesauri were built. Different from them, main subject of our research is building dynamic, just-in-time thesaurus-like clusters, which are useful for a user to understand a certain search results. Methods for search results clustering (e.g. [3], [4], [5]) were proposed. Different from these methods, the related terms clusters are labeled from the user's viewpoint with our proposed method since the labels of clusters consist of extracted terms from a collection of related terms to the query. The collection is acquired from many users' query logs of the search engine and reflect users' information needs. The algorithm was evaluated both in the quality of labels and in the user study and it was found to be effective for building understandable clusters. The documents clustering and the terms clustering are more "computer-oriented" and effective to discover knowledge (i.e. data mining). On the other hand, the related terms clustering is rather "human-oriented" and effective to share knowledge (i.e. social filtering). Visualized and interactive user interfaces (e.g. clusty[6]) are thought to be useful. Development of an advanced user interface is the future work.

[6] Clusty the clustering search engine, http://clusty.com/

## 6 Conclusion

Our method enables easier viewing of search results by the creation of comprehensive clusters. We first form related terms using words that often appear together on web pages and are frequently used together in search engine queries. These related terms are the most reliable way to conduct concise searches. The HBC, an algorithm based on a stochastic model, is used to separate and collect related terms to form understandable clusters. This method can also provide clusters that contain synonyms and/or terms related to the original query. This allows the results to be displayed in a succinct and concise manner. In the future, we plan to investigate how to better present our clusters in an easy-to-navigate fashion. We also intend to develop an intuitive visual interface.

## References

1. Baeza-Yates, R.A., Hurtado, C.A., Mendoza, M.: Query Recommendation Using Query Logs in Search Engines. In: Lindner, W., Mesiti, M., Türker, C., Tzitzikas, Y., Vakali, A.I. (eds.) EDBT 2004. LNCS, vol. 3268, pp. 588–596. Springer, Heidelberg (2004)
2. Zhang, Z., Nasraoui, O.: Mining search engine query logs for query recommendation. WWW2006, pp. 1039–1040 (2006)
3. Zamir, O., Etzioni, O.: Web Document Clustering: A Feasibility Demonstration. SIGIR 1998, pp. 46–54 (1998)
4. Kohonen, T.: Self-Organizing Maps of Massive Document Collections. IJCNN 2000 2, 3–12 (2000)
5. Osinski, S.: Improving Quality of Search Results Clustering with Approximate Matrix Factorisations. In: ECIR 2006, pp. 167–178 (2006)
6. Beeferman, D., Berger, A.L.: Agglomerative clustering of a search engine query log. KDD 2000, 407–416 (2000)
7. Wen, J.R., Nie, J.Y., Zhang, H.J.: Query clustering using user logs. ACM Trans. Inf. Syst. 20(1), 59–81 (2002)
8. Iwayama, M., Tokunaga, T.: Hierarchical Bayesian Clustering for Automatic Text Classification. IJCAI 1995, 1322–1327 (1995)
9. Tokunaga, T., Iwayama, M., Tanaka, H.: Automatic Thesaurus Construction based on Grammatical Relations. IJCAI, 1308-1313 (1995)
10. Schutze, H., Pedersen, J.: A Cooccurrence-Based Thesaurus and Two Applications to Information Retrieval. Inf. Process. Manage. 33(3), 307–318 (1997)
11. Manning, C.D., Schutze, H.: Foundations of Statistical Natural Language Processing. Mit Pr (1999)
12. Kaki, M.: Optimizing the number of search result categories. CHI 2005, 1517–1520 (2005)
13. PorterStemmer, `http://tartarus.org/~martin/PorterStemmer/index.html`
14. Eguchi, K., Oyama, K., Ishida, E., Kando, N., Kuriyama, K.: NTCIR-3 WEB: An Evaluation Workshop for Web Retrieval. NII Journal 6, 31–56 (2003)
15. Yasukawa, M., Yokoo, H.: Web Search Based on Clustering of Related Terms Acquired from Search Log. IEICE Trans. Inf. Syst (Japanese) J90-D(2), 269–280 (2007)

# Event Specification and Processing for Advanced Applications: Generalization and Formalization*

Raman Adaikkalavan[1] and Sharma Chakravarthy[2]

[1] CIS Department, Indiana University South Bend
raman@cs.iusb.edu
[2] CSE Department, The University of Texas At Arlington
sharma@cse.uta.edu

**Abstract.** Event processing is being used extensively in diverse application domains. Simple and composite events play a critical role in event processing systems and were identified based on application domains. They were formally defined using detection-based (*point-based*) and occurrence-based (*interval-based*) semantics over various consumption modes. Even though both the semantics are required they are insufficient for handling emerging applications such as *information security, stream and sensor data processing* systems. Generalizing the event specification and detection is inevitable for supporting these new applications that were not foreseen by extant systems. First, we motivate the need for generalization using applications from diverse domains. Second, we generalize and formalize primitive and composite events. Finally, we briefly discuss how generalized events can be detected using event registrar graphs.

## 1 Introduction

Active (or Event-Condition-Action) rules used for situation monitoring are considered as one of the most general formats for expressing rules. There has been a lot of work done in the area of event detection and specification in the form of ECA rules. As the event component was the least understood (conditions correspond to queries, and actions correspond to transactions) part of the ECA rule, there was a large body of work on the language for event specification. A number of event processing systems using ECA rules have been proposed and implemented in the literature [1,2,3,4,5,6,7,8,9]. Recently, there have also been some work on semantic events [10].

The above mentioned works provide well-defined point-based and interval-based event semantics. Although both event semantics can be used for event processing, they are inadequate for supporting many newer application domains. The main shortcoming of both the semantics is that they are solely based on timestamp of the event occurrence. In this paper we generalize and formalize events and briefly discuss the generalized event detection. We discuss the existing temporal semantics in Section 2. We motivate the need for event generalization using various real time applications in Section 3. Event generalization is discussed in Section 4 and generalized event detection is discussed in Section 5. Section 6 has conclusions.

* This work was supported, in part, by NSF grants IIS-0326505, EIA-0216500, MRI-0421282, IIS-0534611.

R. Wagner, N. Revell, and G. Pernul (Eds.): DEXA 2007, LNCS 4653, pp. 369–379, 2007.

## 2 Background: Events and Temporal Semantics

An event is "any *occurrence of interest* in an application, system, or environment", which can be either primitive (e.g., depositing cash) or composite (e.g., depositing cash, followed by withdrawal of cash). Primitive events occur at a point in time (i.e., time of depositing) and composite events occur over an interval (i.e., starts at the time cash is deposited and ends when cash is withdrawn). With current event processing systems, primitive events are detected at a point in time, whereas the composite events can be detected either at the *end* of the interval (i.e., detection- or point-based semantics, where start of the interval is not considered) [1,2,3,4] or can be detected *over* the interval (i.e., occurrence- or interval-based semantics) [5,6,7,8,9]. To the best of our knowledge, all the active databases operate using *temporal* semantics. In this paper, we use Snoop [3] and SnoopIB [9] that was developed as the event specification component of the ECA rule formalism used as a part of the Sentinel active *object-oriented* database [11]. Main motivation to use Snoop and SnoopIB is that they support expressive event specification using *both* point- and interval-based semantics in various event consumption modes.

**Definition 1 (Point-based Event).** *A point-based event $E[t]$ is a function from the time domain ($T$) onto the boolean values. "$E : T \rightarrow \{True, False\}$". It is given by*

$$E[t] = \begin{cases} T(rue) & \textit{if event E occurs at time point } [t] \\ F(alse) & \textit{otherwise.} \end{cases}$$

Similarly, *an interval-based event $E[t_s, t_e]$ is a function from the interval domain ($I$) onto the boolean values. "$E : I \rightarrow \{True, False\}$".* $t_s$ and $t_e$ represent the start and end time of the event, respectively.

### 2.1 Primitive Events

Primitive events are the basic building blocks in an event processing system and are derived from the application domains. Below we define them formally[1].

**Definition 2 (Point-based Primitive Event).** *An event* E *occurs atomically at a point $[t]$ on the time line. It is detected at $[t]$ and is expressed as $\mathcal{D}(E[t])$. It is defined as "$\mathcal{D}(E[t]) \triangleq \exists t\ (E[t])$;"*

**Definition 3 (Interval-based Primitive Event).** *An event* E *occurs atomically at a point $[t]$ on the time line. It is detected over an interval $[t, t']$, where $[t]$ is the start time and $[t']$ is the end time and ($t = t'$). It is expressed as $\mathcal{O}(E[t, t'])$ and is defined as "$\mathcal{O}(E[t, t']) \triangleq \exists t = t'\ (E[t, t'])$;"*

Based on the object-oriented paradigm, Snoop and SnoopIB allow two primitive event definition types using function signatures (instance- and class-level), as shown below.

$E_1 = \mathcal{F}(formal\ parameters)$; $\quad E_2 = \mathcal{X} \rightarrow \mathcal{F}(formal\ parameters)$;

When a function $\mathcal{F}$ is invoked, events $E_1$ and $E_2$ should be raised. However, event $E_2$ is raised only when object $\mathcal{X}$ has invoked the function i.e., condition on the object.

[1] In this paper, we use $\mathcal{D}$ to represent detection- or point-based and $\mathcal{O}$ to represent occurrence- or interval-based semantics.

In addition to function signatures, events can also be based on time, transaction, etc. On the other hand, ODE [12] allows primitive events to be associated with *masks*. For instance, an event "$buyStock(double\ price)$ && $price > 500$" is detected when the function "buyStock" is invoked and $price > 500$. Masks (predicates) can only be based on the formal parameters of the function or based on the state of the database object.

## 2.2 Composite Events

Composite events are defined by composing more than one primitive or composite event using *event operators* [3,9]. Using *temporal composition conditions*, an event operator defines how a composite event needs to be composed and detected.

**Definition 4 (Point-based Composite Event).** *A composite event $\mathcal{D}(E)$ occurs over an interval $[t_s, t_e]$ and is detected at a time point $[t_e]$, where $t_s$ is the start time of* initiating *event, and $t_e$ is the end time of* detecting *event. "$\mathcal{D}(Eop\ (E_1, \ldots E_n), [t_e]);$"*

**Definition 5 (Interval-based Composite Event).** *A composite event $\mathcal{O}(E)$ occurs and is detected over an interval $[t_s, t_e]$. It is defined as "$\mathcal{O}(Eop\ (E_1, \ldots E_n), [t_s, t_e]);$"*

*Eop* represents an n-ary event operator (e.g., binary). $(E_1, \ldots E_n)$ represents the constituent events. When the event operator's temporal composition conditions are satisfied, a composite event is detected. In a composite event, *Initiator* starts the detection, *detector* detects and raises the event, and *terminator* ends the detection. Based on the operator semantics the same event can act as initiator/detector/terminator. Formalization of all other operators can be found in [3,9]. Below we explain two operators.

**SEQUENCE** $(E_1 \gg E_2)$**:** A binary event operator that captures the sequential occurrence of constituent events is raised when event $E_1$ occurs before event $E_2$. It is *detected* when $E_2$ occurs. It is formally defined in point and interval semantics as:

$$\mathcal{D}(E_1 \gg E_2, [t_2]) \triangleq \exists t_1, t_2(\mathcal{D}(E_1, [t_1]) \wedge \mathcal{D}(E_2, [t_2]) \wedge (t_1 < t_2));$$
$$\mathcal{O}(E_1 \gg E_2, [t_s, t_e]) \triangleq \exists t_s, t_e, t, t'(\mathcal{O}(E_1, [t_s, t]) \wedge \mathcal{O}(E_2, [t', t_e]) \wedge (t_s \leq t < t' \leq t_e));$$

In the above, events $E_1$ and $E_2$ can be primitive or composite. In the first definition, event $E_1$ is detected at $[t_1]$ and $E_2$ at $[t_2]$. Sequence event is detected if $t_1 < t_2$. With interval-based, sequence is detected if end time of $E_1$ is less than start time of $E_2$.

**NOT** $O(\neg(E_2)(E_1 \gg E_3), [t_1, t_2])$**:** Non-occurrence of an event $E_2$ in between two other events $E_1$ and $E_3$ triggers the NOT event.

The above temporal semantics is based on the unrestricted event consumption mode where no constituent event is dropped after participating in an event detection. In order to avoid the unnecessary event detection, *event consumption modes* [9] such as Recent, Continuous, Chronicle, and Cumulative were defined based on the application domains. Similar to primitive events, composite events can also be associated with "Mask" in ODE. But, "Any mask predicate applied to a composite event, unlike the mask predicates of logical (primitive) events can only be evaluated in terms of the 'current' state of the database. [12]" Thus masks associated with composite events are very restricted as they are just constant integer expressions.

# 3 Temporal Semantics Limitations

We discuss two different application domains and show that event processing based on temporal semantics *alone* is not sufficient. Even though there are other domains, we are restricting to only two due to space constraints [13,14]. We will also discuss only the need for the composite event generalization as it subsumes primitive events.

## 3.1 Stream Data Processing

Many stream applications (e.g., smart homes) seem to not only need computations on data streams, but these computations also generate interesting events and several such events have to be composed, detected and monitored for taking necessary actions [13,15,16]. In an automobile *accident detection and notification* system, based on the linear road benchmark, each expressway is modeled as a linear road, and is further divided into equal-length segments. Each registered vehicle reports its location periodically (say, every 30 seconds) using a sensor. Based on this *location stream* data, we can detect a car accident in a near-real time manner. Detecting an accident (or traffic jam) has at least three requirements. (1) IMMOBILITY: whether a car is immobile for four (or n) consecutive time units, (2) SPEED REDUCTION: whether there is at least one car that has reduced its speed by 30% or more, and (3) SAME SEGMENT: determining whether the car identified in (2) is in the same segment and it follows the car identified in (1).

Assume that *location stream* sends inputs to two continuous queries; CQ1 checks every car for immobility, and CQ2 checks for speed reduction. Primitive events *Eimm* and *Edec* are defined on CQ1 and CQ2. An accident is modeled using the *sequence* event operator as event *Eacc*, as an accident is detected when *Eimm* happens before *Edec*. In addition to the sequence condition, both the cars should be from the *same segment*. CQ1 and CQ2 generate primitive events and raise them along with their attributes, and notify events *Eimm* and *Edec*, respectively. Attributes of both events are:

Eimm: ***(timestamp, carId**, speed, expWay, lane, dir, **segmentId)***
Edec: ***(timestamp, carId**, speed, expWay, lane, dir, **segmentId**, decreaseInSpeed)*

*Eimm* occurs at 10:00 a.m. ***(10:00 a.m, 1**, 0 mph, I123, 3, NW, **104)***, and *Edec* occur at 10:03 a.m. and 10:04 a.m. ***(10:03 a.m., 2**, 40 mph, I123, 1, NW, **109**, 45%)*, and ***(10:04 a.m., 5**, 20mph, I123, 4, NW, **104**, 40%)*.

With current event processing, tuples with *carId 1* (10:00 am) and *carId 2* (10:03 am) triggers the event *Eacc*. Similarly tuples with *carId 1* and *carId 5* trigger the event *Eacc*. Thus, the important condition that both the cars should be from the same segment can only be checked in the condition part of the rule after the event *Eacc* is detected. As a result, this introduces a high overhead on the rule processing system (and the event processing system) to deal with large number of unnecessary events as the condition can evaluate to false in most cases. The above example can be modeled using instance level events of Snoop(IB) (but not in other event processing languages) or masks. However, all the instances need to be pre-defined (or known previously) which may be impossible in a system where the data streams' attribute values are dynamic. In addition, instance level events introduce a high overhead for computation.

As mentioned earlier, event consumption modes play a critical role in event processing systems. In our example, using modes will reduce duplicate accident detections, which is not possible with relational operators. When modes are used, same segment condition *cannot* be checked in the rule, since it will lead to incorrect event detection [13,14]. Hence, the two problems discussed above clearly shows there is an inevitable need for enhancing event detection semantics in addition to or in lieu of temporal semantics. Please refer [13,14] for detailed discussion.

### 3.2 Information Security

Access control is considered as one of the pillars of information security. In role-based access control (RBAC) [17], users and objects are *assigned* to one or more roles. Thus, users should be *active* in the role that has the required permissions, before access is granted. On the other hand, composite constraints can be placed on the role activations, so that users are allowed to activate a role only when those constraints are satisfied. Below we explain two examples, for more examples please refer to [14].

Consider a prerequisite role constraint, that requires *any* user to be active in a role (e.g., A) for activating another role (e.g., B). This constraint can be modeled using a *sequence* event. The first constituent event captures the activation of role A. The second constituent event captures the activation request for role B. In this case, any user (or object) can activate (or invoke activation function) the role. Consider a scenario when user Tom is activating role A at 10:00 a.m. and user Bob is requesting an activation for role B at 10:05 a.m. As the current semantics is based on time this will lead to an event detection, allowing Bob to activate the role. In other words, Bob should have activated role A before role B. Thus, there needs to be a condition that requires both the users (or userId's) to be same, similar to the *same segment* condition in the previous example.

Capturing violations are critical in the security domain. Currently, only the events that follow a particular order are considered for event detection (i.e., complete events). For example, with a Sequence operator there are two constituent events where the first event should occur ahead of the second event. *What happens when a second event occurs without the first event's occurrence?* With extant event systems, the second event is dropped and it is a limitation, since this occurrence of the second event without the first event can be a violation of a security policy.

Consider the security policy; "*Allow a user to enter the pregnant ward in a hospital from a virus ward only when the user has visited the hygienize ward.*" When the access is granted without hygienizing it can allow the user to spread the virus. This policy can be modeled with the NOT operator $(O(\neg(E_{HS})(E_{VW} \gg E_{PW}), [t_1, t_2]))$; user entering virus ward is modeled as *initiator* event $E_{VW}$, entering hygienize ward is $E_{HS}$ and entering pregnant ward is *detector/terminator* $E_{PW}$. With NOT operator, when the user enters virus ward, does not enter hygienize ward, and enters pregnant ward, then the NOT event is detected. On the other hand, what happens when the user enters the pregnant ward directly, or from the hygienize ward without going to the virus ward (i.e., no initiator)? With the current semantics, NOT event is not detected and events are dropped when; 1) all $E_{VW}$, $E_{HS}$ and $E_{PW}$ occur, 2) only $E_{HS}$ and $E_{PW}$ occur, or 3) when $E_{PW}$ occurs alone. In our example, the user should be allowed to enter the pregnant ward when all three events occur. Thus, *incomplete* (partial) and *failed* events have to be captured in addition to *complete* events.

# 4 Event Generalization

Event generalization should accommodate the existing temporal semantics and should allow users to use point- or interval-based with or without the generalization.

Each event has a well-defined set of *attributes* based on the *implicit* and *explicit* parameters, that provide the necessary information about that event. *Implicit* parameters are optional and contain system and user defined attributes, such as: event name, and time of occurrence ($t_{occ}$). These parameters are defined internally based on the application domain and are collected at the time when the event is raised. *Explicit* parameters are collected from the event itself and values for these parameters are assigned when it is raised. When a function is defined as a primitive event, then explicit parameters are just the formal parameters of that function. In this paper we represent optional implicit parameters as "$[A_{i1}, A_{i2}, \ldots, A_{il}]$" and available explicit parameters as "$(A_{x1}, A_{x2}, \ldots, A_{xm})$".

Currently, event operators just utilize the timestamp for composition conditions. Computations using the parameters are carried out in the condition and action procedures of the ECA rule, which is not sufficient, as discussed previously. Thus generalizing events by allowing composition conditions to be based on both implicit and explicit parameters seems natural as they provide all the necessary information about an event. In order to generalize events, two types of expressions are constructed based on the parameters and are associated with the events. Implicit expressions ($\mathcal{I}_{expr}$) are based on the implicit parameters and explicit expressions ($\mathcal{E}_{expr}$) are based on the explicit parameters. Both these expressions can be based on logical operators, relational operators, etc. such as $<, >, <=, >=, !=, =, \in, \subset$.

## 4.1 Generalized Primitive Event

**Definition 6 (Generalized Primitive Event).** *A generalized primitive event $G(E)$ can be a point- or interval-based primitive event with conditional expressions based on $\mathcal{I}_{expr}$ and $\mathcal{E}_{expr}$. They are formally defined as*

$$G_P(E[t]) \triangleq \exists t\ (G_P((E, [t]) \wedge (\mathcal{I}_{expr} \wedge \mathcal{E}_{expr})));$$
$$G_O(E[t, t']) \triangleq \exists t = t'\ (G_O((E, [t, t']) \wedge (\mathcal{I}_{expr} \wedge \mathcal{E}_{expr})));$$

In both the generalized definitions, primitive event is detected if it occurs, and both the conditional expressions return TRUE. Below, generalized primitive event for Snoop and SnoopIB is defined. *eName* corresponds to the name of the event, $\mathcal{F}$ corresponds to the name of the function on which the event is defined. Only one implicit parameter, "ObjInstance" is shown. Both the expressions evaluate to either TRUE or FALSE.

$$Event\ eName = (\mathcal{F}(objInstance, A_{x1},\ A_{x2},\ \ldots, A_{xr}), (\mathcal{I}_{expr} \wedge \mathcal{E}_{expr}));$$

Four possible types of primitive events based on $\mathcal{I}_{expr}$ and $\mathcal{E}_{expr}$ are:

$$\begin{aligned}
(\mathcal{I}_{expr} \in \emptyset) \wedge (\mathcal{E}_{expr} \in \emptyset) \rightarrow E_1 &= \mathcal{F}(A_{x1},\ A_{x2},\ \ldots, A_{xr});\\
(\mathcal{I}_{expr} \notin \emptyset) \wedge (\mathcal{E}_{expr} \in \emptyset) \rightarrow E_2 &= \mathcal{X} \rightarrow \mathcal{F}(A_{x1},\ A_{x2},\ \ldots, A_{xr});\\
(\mathcal{I}_{expr} \in \emptyset) \wedge (\mathcal{E}_{expr} \notin \emptyset) \rightarrow E_3 &= (\mathcal{F}(A_{x1},\ A_{x2},\ \ldots, A_{xr}) : \mathcal{E}_{expr});\\
(\mathcal{I}_{expr} \notin \emptyset) \wedge (\mathcal{E}_{expr} \notin \emptyset) \rightarrow E_4 &= (\mathcal{X} \rightarrow \mathcal{F}(A_{x1},\ A_{x2},\ \ldots, A_{xr}) : \mathcal{E}_{expr});
\end{aligned}$$

Events $E_1$ and $E_2$ are class-level and instance-level events, same as the existing event specifications shown in Section 2. Event $E_1$ is raised when function $\mathcal{F}$ is invoked by any object, and event $E_2$ is raised when it is invoked by an object $\mathcal{X}$ (i.e., $\mathcal{I}_{expr} = \mathcal{X}$ is TRUE). Event $E_3$ is detected when the method is invoked by any object but must satisfy $\mathcal{E}_{expr}$. Event $E_4$ is raised when both $\mathcal{I}_{expr}$ and $\mathcal{E}_{expr}$ return TRUE. Even though some of the conditions can be specified by existing systems via masks, not all possible conditions can be specified. In all, our generalization is simple and powerful. As shown below, events $EP3$, $EP4$ and $EP5$ have $\mathcal{I}_{expr}$ and events $EP2$ and $EP5$ have $\mathcal{E}_{expr}$.

$EP1 = setPrice(price);$ $\quad EP2 = (setPrice(price), (price > 100));$
$EP3 = (setPrice(price), (t_{occ} > 18.00));$
$EP4 = (setPrice(price), (stockId = GOOG));$
$EP5 = (setPrice(price), ((stockId = GOOG) \wedge (price > 500)));$

Although the effect of generalization can be achieved by moving both $\mathcal{I}_{expr}$ and $\mathcal{E}_{expr}$ to the `rules` it will be inefficient - due to unnecessary rule processing as all events are raised, but filtered in the rules. Another critical issue is that when the expressions are moved to the rules, those events cannot be used as part of composite events.

### 4.2 Generalized Composite Event

Similar to the primitive event definition generalization, we have generalized the current event operator definitions, so that composition condition can be based on $\mathcal{I}_{expr}$ and $\mathcal{E}_{expr}$ and not just based on time.

**Definition 7 (Composite Event).** *A generalized composite event $G(E)$ can be a point- or interval-based composite event with conditional expressions based on $\mathcal{I}_{expr}$ and $\mathcal{E}_{expr}$. They are formally defined as*

$$G_D(Eop\ (E_1, \ldots E_n), (\mathcal{I}_{expr}\ \wedge\ \mathcal{E}_{expr}), [t_e]);$$
$$G_O(Eop\ (E_1, \ldots E_n), (\mathcal{I}_{expr}\ \wedge\ \mathcal{E}_{expr}), [t_s, t_e]);$$

- $G_D$ and $G_O$ correspond to generalized point-based and interval-based events.
- *Eop* represents an n-ary event operator. Some of the event operators are; And, Or, Sequence, Not, Plus, Periodic, Aperiodic, Periodic*, and Aperiodic*.
- $(E_1, \ldots E_n)$ are the constituent events. For example, $E_1$ can be primitive/composite.
- $[t_s]$ is the start time of the initiator and $[t_e]$ is the end time of the detector. Point-based events have $[t_e]$ as the time of occurrence and interval-based events have $[t_s, t_e]$ as the time of occurrence.
- Implicit parameter expression $\mathcal{I}_{expr}$ subsumes existing point- and interval-based semantics. For instance, a binary event operator with events $E_1$ and $E_2$ can have $\mathcal{I}_{expr} = t_{occ}(E_1)\ \theta\ t_{occ}(E_2)$, where $t_{occ}$ represents the timestamp of event occurrence, and $\theta$ can be any operator $<, >, \leq, \geq, =, \neq, \in, \ldots$
- $\mathcal{E}_{expr}$ has conditions based on the explicit parameters. For instance, a binary event operator with the events $E_1$ and $E_2$ can have $\mathcal{E}_{expr} = E_1(A_{xi})\ \theta\ E_2(A_{xj})$, where attributes $E_1(A_{xi})$ and $E_2(A_{xj})$ have values from the same domain.
- composite event is detected *iff* all the above mentioned expressions return TRUE. We assume that all these conditional expressions must not be empty at the same time, otherwise it will detect the composite event always.

As opposed to the alternate primitive event generalization, the effect of composite event generalization *cannot* be achieved by moving both $\mathcal{I}_{expr}$ and $\mathcal{E}_{expr}$ to rules when consumption modes are involved. This is because composite events will be detected incorrectly if $\mathcal{I}_{expr}$ or $\mathcal{E}_{expr}$ are moved to rules as explained in Section 3.

***Temporal Semantics using Generalized Event Formalization:*** As an example, below we formalize the generalized Sequence operator ($E_1 \gg E_2$) and show how the generalization can incorporate the existing temporal semantics using $\mathcal{I}_{expr}$.

$$G_D(E_1 \gg E_2, [t_2]) \triangleq \exists t_1, t_2 (G_D(E_1, [t_1]) \wedge G_D(E_2, [t_2]) \wedge (\mathcal{I}_{expr} \ \wedge \ \mathcal{E}_{expr}));$$
$$\mathcal{I}_{expr} = (t_1 < t_2);$$
$$G_O(E_1 \gg E_2, [t_s, t_e]) \triangleq \exists t_s, t_e, t, t' (G_O(E_1, [t_s, t]) \wedge G_O(E_2, [t', t_e]) \wedge (\mathcal{I}_{expr} \ \wedge \ \mathcal{E}_{expr}));$$
$$\mathcal{I}_{expr} = (t_s \le t < t' \le t_e);$$

***Complete, Incomplete and Failed Events:*** As composite events combine more than one event, they are detected only when the event `completes` in the current event detection systems. We categorize these events as *complete* events. Actions corresponding to an event can be performed *iff* that event is complete. Even though event completion is necessary in many situations it is not required in all the domains. Consider the prerequisite role constraint explained in Section 3.2. When user `Bob` tries to activate role `B` without activating role `A`, activation request should be declined and notified. This requires an additional capability of current event detection semantics paradigm to infer that a constituent event (always the detector) of a composite event has been detected, but not other events to complete the detection of the composite event. To identify such occurrences the *If-Then-Else* mechanism is proposed (because the *If-Then* mechanism only detects complete events). This allows for additional actions to be taken when the detector occurs and the event is not completed because of the non-occurrence of other constituent events. We term these events as `incomplete` events.

As explained in Section 3.2, in addition to complete and incomplete events there can be other type of events. Consider a NOT event operator. When the first and third event occur without the middle event it is a complete event. When the second and third event occur without the first event or when the third event occurs without the first event then they are considered as *incomplete* events. On the other hand, when all the first, second and third event occur, a NOT event is not detected and is categorized as *failed* event. In general, "A *complete* event E occurs when, i) initiator occurs, ii) all the constituent events occur, and iii) detector occurs." "An *incomplete* (partial) event E occurs when i) no initiator occurs, ii) other constituent events can occur, and iii) detector occurs." "A *failed* event E occurs when i) initiator occurs, ii) other constituent events occur, and iii) detector occurs, but the event fails because some constituent event that should not occur has occurred."

***Generalized Event Operators - Stream Processing:*** Below we show how the automobile accident notification example defined in Section 3.1 can be modeled using the generalized event definition. For brevity, we will explain it using the point-based semantics. Consider the attributes of events $Eimm$ and $Edec$ from Section 3.1.

$$G_D(Eimm \gg Edec, [t_2]) \triangleq \exists t_1, t_2 (G_D(Eimm, [t_1]) \wedge G_D(Edec, [t_2]) \wedge (\mathcal{I}_{expr} \ \wedge \ \mathcal{E}_{expr}));$$
$$\mathcal{I}_{expr} = (t_1 < t_2); \qquad \mathcal{E}_{expr} = (Eimm.segmentId.equals(Edec.segmentId));$$

In the above event, $\mathcal{I}_{expr}$ requires event $Edec$ to follow $Eimm$ and $\mathcal{E}_{expr}$ requires both of them to be from the same segment. Below we consider the same event occurrences from Section 3.1 and show what events are detected. In the event occurrences, *carId 1* (10:00 a.m.) is combined with *carId 5* (10:04 a.m.) as both of them are from the same segment 104, as opposed to the current event semantics where *carId 1* and *carId 2* are combined to form the composite event.

***Generalized Event Operators - Information Security:*** Similar to the above, role activation problem discussed in Section 3.2 can be addressed. On the other hand, the example from Section 3.2 where a user is controlled from entering the pregnant ward based on the history is modeled using a NOT operator. We define a NOT event

$$\begin{aligned}\texttt{Event } Esec &= (NOT(E_{VW}, E_{HS}, E_{PW}) \wedge (\mathcal{I}_{expr} \ \wedge \ \mathcal{E}_{expr}));\\ \mathcal{I}_{expr} &= (Point/Interval\ Semantics);\\ \mathcal{E}_{expr} &= (E_{VW}.userId = E_{HS}.userId = E_{PW}.userId);\end{aligned}$$

With the NOT operator, first event $E_{VW}$ is the initiator, last event $E_{PW}$ is the detector and middle event $E_{HS}$ is the non-occurrence event. The predicate relates all the events with the same user for controlling each user simultaneously. This condition can be modified to include specific roles, users, and so forth.

**Case 1:** When the user enters virus ward, $Esec$ is initiated. When the user tries to enter pregnancy ward $E_{PW}$ is raised and $Esec$ is detected as $E_{HS}$ has not occurred. As both the initiator and detector have happened according to the semantics, a *complete* event is raised. Thus, this triggers the complete rule that in turn denies the user to enter the pregnancy ward. **Case 2:** Assume the user proceeds from the virus ward to hygiene stop and then to pregnancy ward. In this case the middle event occurs, thus making $Esec$ a *failed* event. In other words, the non-occurrence has failed. The failed rule associated with the event is triggered and the user is allowed to enter the pregnancy ward. **Case 3:** When the users proceeds from hygiene stop to pregnancy ward or directly to the pregnancy ward, an *incomplete* event is triggered. As the initiator did not occur this is an incomplete event and the incomplete rule is triggered. As the user has not entered the virus ward before, this rule allows the user to enter the pregnancy ward.

## 5 Event Registrar Graphs (ERG)

We have implemented the generalized event detection in the Sentinel active database [11]. ERGs (or extended event detection graphs) record event occurrences as and when they occur and keep track of the constituent event occurrences over the time interval they occur. ERGs are acyclic graphs, where each event pattern is a *connected tree*. In addition, event sub-patterns that appear in more than one event pattern are shared. ERG shown in Figure 1 has two leaf nodes and each of them represent a simple or primitive event. Similarly the

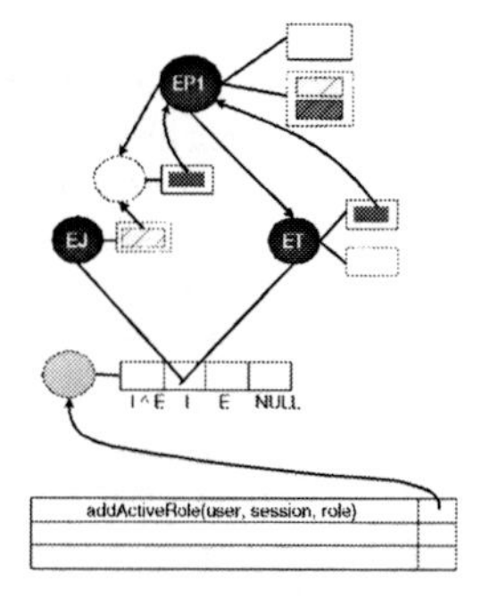


**Fig. 1.** Event Registrar Graph

internal node represents the complex event. The ERG as a whole represents a composite event. In Figure 1, the complex event is a binary event operator (e.g., AND), thus having two child events. Extant event detection graphs (EDGs) [3] are extended to handle primitive and composite event generalization, to support implicit and explicit event expression computations. Please refer [14] for more detailed explanations of ERGs and limitations of EDGs.

## 6 Conclusions

We discussed the need for generalizing event processing with some critical examples from two different domains. We generalized and formalized the traditional primitive and composite events. We explained how existing temporal semantics can be supported using the generalization. Even though the generalization seems to be simple on the face of it, it is powerful and inevitable; 1) as it reduces the high overhead, 2) detects events correctly, and 3) extends the situations that can be monitored, thus catering a larger class of applications. In addition, generalization also introduced some significant challenges while detecting complete, incomplete and failed events with implicit and explicit expressions. We also briefly discussed event detection using event registrar graphs.

## References


1. Gehani, N.H., Jagadish, H.V., Shmueli, O.: Composite Event Specification in Active Databases: Model & Implementation. In: Proc. of VLDB, pp. 327–338 (1992)
2. Gatziu, S., Dittrich, K.R.: Events in an Object-Oriented Database System. In: Proceedings of Rules in Database Systems (September 1993)
3. Chakravarthy, S., Mishra, D.: Snoop: An Expressive Event Specification Language for Active Databases. DKE 14(10), 1–26 (1994)
4. Zimmer, D.: On the semantics of complex events in active database management systems. In: Proc. of the ICDE, p. 392. IEEE Computer Society, Washington (1999)
5. Roncancio, C.: Toward Duration-Based, Constrained and Dynamic Event Types. In: Andler, S.F., Hansson, J. (eds.) ARTDB 1997. LNCS, vol. 1553, pp. 176–193. Springer, Heidelberg (1999)
6. Galton, A., Augusto, J.: Two Approaches to Event Definition. In: Hameurlain, A., Cicchetti, R., Traunmüller, R. (eds.) DEXA 2002. LNCS, vol. 2453, pp. 547–556. Springer, Heidelberg (2002)
7. Mellin, J., Adler, S.F.: A formalized schema for event composition. In: Proc. of Conf on Real-Time Computing Systems and Applications, Tokyo, Japan, pp. 201–210 (March 2002)
8. Carlson, J., Lisper, B.: An Interval-based Algebra for Restricted Event Detection. In: Larsen, K.G., Niebert, P. (eds.) FORMATS 2003. LNCS, vol. 2791, pp. 121–133. Springer, Heidelberg (2004)
9. Adaikkalavan, R., Chakravarthy, S.: SnoopIB: Interval-Based Event Specification and Detection for Active Databases. DKE 59(1), 139–165 (2006)
10. Nagargadde, A., Varadarajan, S., Ramamritham, K.: Semantic Characterization of Real World Events. In: Zhou, L.-z., Ooi, B.-C., Meng, X. (eds.) DASFAA 2005. LNCS, vol. 3453, pp. 675–687. Springer, Heidelberg (2005)
11. Chakravarthy, S., Anwar, E., Maugis, L., Mishra, D.: Design of Sentinel: An Object-Oriented DBMS with Event-Based Rules. IST 36(9), 559–568 (1994)

12. Gehani, N.H., Jagadish, H.V., Shmueli, O.: Event Specification in an Object-Oriented Database. In: Proc. of SIGMOD, San Diego, pp. 81–90 (June 1992)
13. Jiang, Q., Adaikkalavan, R., Chakravarthy, S.: MavEStream: Synergistic Integration of Stream and Event Processing. In: Schwentick, T., Suciu, D. (eds.) ICDT 2007. LNCS, vol. 4353, Springer, Heidelberg (2006)
14. Adaikkalavan, R.: Generalization and Enforcement of Role-Based Access Control Using a Novel Event-based Approach. Ph.D. dissertation, UTA, [Online]. Available (2006), `http://itlab.uta.edu/ITLABWEB/Students/sharma/theses/Ada06PHD.pdf`
15. Garg, V., Adaikkalavan, R., Chakravarthy, S.: Extensions to Stream Processing Architecture for Supporting Event Processing. In: Bressan, S., Küng, J., Wagner, R. (eds.) DEXA 2006. LNCS, vol. 4080, pp. 945–955. Springer, Heidelberg (2006)
16. Wu, E., Diao, Y., Rizvi, S.: High-performance complex event processing over streams. In: SIGMOD Conference 2006, pp. 407–418 (2006)
17. RBAC Standard: ANSI INCITS 359-2004 (2004)

# An Evaluation of a Cluster-Based Architecture for Peer-to-Peer Information Retrieval

Iraklis A. Klampanos and Joemon M. Jose

Department of Computing Science
University of Glasgow
United Kingdom

**Abstract.** In this paper we provide a full-scale evaluation of a cluster-based architecture for P2P IR, focusing on retrieval effectiveness. We observe that there is a significant difference in performance between the architecture we examine and a centralised index. After inspecting our experimental methodology and our results, we provide evidence that suggests that this discrepancy is due to the information clustering algorithms employed throughout. The construction errors of the resource descriptions as well as the failure of the clustering mechanisms to discover the structure of the smallest of peer-collections lead to erroneous query routing. We proceed further to show experimentally how content replication and relevance-feedback mechanisms can help to alleviate the problem.

## 1 Introduction

Information retrieval (IR) over peer-to-peer (P2P) networks is a challenging problem that is frequently referred to in the IR literature ([1,2,3,4,5], etc.). A number of architectures have been proposed that address various instantiations of this problem. It is clear that different applications of P2P networks will pose different challenges for IR. Popular applications of P2P IR include digital libraries, open information-sharing and others ([1,2] etc.). Information clustering is often used by various studies as an architectural component or as a tool for achieving realistic evaluation environments. However, the application of clustering in P2P IR may lead to errors in the cluster centroids. These errors are caused by the inadequate information that describes the constituent objects. However, the effects of this problem have not been studied within the context of P2P IR and so we do not know the extent of the problem, let alone which solutions could be applied in order to amend it. These are the issues that this paper contributes insight and solutions for.

In this paper we provide a wide-scale experimental evaluation of a cluster-based P2P IR architecture [2], using a set of testbeds that were devised for this purpose [6]. Through clustering, this architecture attempts to organise the shared content into semantically-related peer-groups. The testbeds employed are totally independent of the experimental evaluation process itself. As our initial effectiveness results are poor, we provide insight into what may be causing this behaviour and we propose solutions that we justify experimentally.

In the next section we present the cluster-based architecture our study is based on as well as the experimental testbeds we use for our experiments. In Section 3 we present

R. Wagner, N. Revell, and G. Pernul (Eds.): DEXA 2007, LNCS 4653, pp. 380–391, 2007.

an initial evaluation that is targeted on retrieval performance. In Section 4 we narrow down our evaluation on a near-optimal (for retrieval purposes) subset of the original testbeds. We use this smaller collection in order to focus on various individual aspects of the architecture, isolate potential problems and suggest potential solutions. Finally, in Section 5 we present our conclusions and provide pointers for future work.

## 2 Related Work

### 2.1 A P2P IR Architecture

We base our evaluation on an architecture [2] that employs clustering at two levels: first, in order to derive usable resource description vectors from the participating information providers and, subsequently, in order to generate content-aware peer-groups (CAGs). The ultimate goal is to form groups of peers that share similar content. The main hypothesis behind this organisation is that it can, potentially, increase the retrieval effectiveness through selective query routing, *i.e.* bypassing irrelevant information sources. Content-based network organisation also increases efficiency since it avoids uninformed query-routing strategies, such as query flooding.

Another property of this architecture is that it is hybrid (*i.e.* there exist super-peers with additional administrative responsibilities) and service-oriented (please refer to [2] for the exact services that are identified). For our evaluation purposes peers are either hubs, *i.e.* peers that are responsible for managing connections and routing messages, or information providers, *i.e.* peers that share documents with the rest of the network.

### 2.2 Testbeds for Evaluating P2P IR

The evaluation of P2P IR systems is an intimidating task due to the potential size of the network and the total volume of the shared information. An additional challenge is posed by factors having to do with the distribution of documents among the peer-collections, the concentration of relevant documents in the evaluation testbeds etc. Different potential applications of P2P IR technologies exhibit different such properties. Since these factors, generally, affect retrieval performance, they have to be taken into account during evaluation.

We performed our evaluation using the testbeds proposed in [6]. These testbeds are based on TREC's WT10g collection and are designed to address a number of P2P IR applications through different document distributions and concentrations of relevant documents. The individual testbeds used are the following:

**ASISWOR.** This testbed is designed to reflect the properties of open information-sharing environments. It exhibits a steep power-law distribution of documents. In this testbed, each web-domain of WT10g corresponds to a peer-collection.

**UWOR.** This is a testbed designed to address P2P IR in environments where the documents are uniformly distributed across the participating information providers. Such environments may include strict DRM environments, networks of devices with restricted resources etc.

**DLWOR.** This testbed aims to reflect a digital-library setting. The number of collections are less than in the ASISWOR, making the individual collections larger in average.

**DLLC.** This is a testbed originally proposed and made available by Lu and Callan in [1] and it also addresses the problem of P2P IR in digital libraries.

## 3 Initial Evaluation

### 3.1 Methodology and Parameters

The evaluation we present in this paper is simulation-driven. For simulating this architecture one has to take under consideration a number of parameters that affect its behaviour. These parameters, having to do with content representation and network topology, are presented in the following sections.

**Content Descriptions.** In the proposed architecture, content descriptions are used at two stages: by information providers that advertise their content to hubs and by hubs that organise the network performing some kind of clustering. In this study, content descriptions are either term-frequency (TF) or binary vectors [1].

**Network Topology.** The topology of the network depends primarily on how the hubs group the information providers. For the evaluation of this architecture we implemented two different approaches to this organisation. The first is to cluster the clusters of the information providers using single-pass clustering in its simplest form (*Simple* topology). The second alternative is to use a fixed number of CAGs, as attractors for the information providers. For the experiments that follow, we used the largest relevant document for each topic of WT10g as CAG attractors (*Fixed* topology).

### 3.2 IR-Related Results

For our evaluation we assessed the underlying document collection (WT10g) against the standard 100 TREC topics as a centralised index. Even though these results are not directly comparable to the results from the P2P architecture, they provide a point of reference for discussion. The results of the centralised index run are presented in Table 1.

**Table 1.** IR effectiveness for WT10g as a centralised index

| *Topics* | *Relevant* | *Retrieved* | *Rel. Retrieved* | *P@10* |
|---|---|---|---|---|
| 100 | 5,980 | 97,048 | 3,817 | 0.2960 |

***Simple* Single-Pass Topology.** For the Simple topology, a simple single-pass clustering algorithm [8] was used to cluster peer-centroids into CAGs. For this, we did not cap the number of CAGs to be created. The results for IR effectiveness can be seen in Table 2.

[1] Even though binary vectors are thought to lead to worse IR effectiveness, it has been reported [7] that there is no evidence to suggest that are inferior to TF vectors for clustering.

**Table 2.** IR effectiveness across non-replication testbeds for the Simple topology

| *Testbed* | *Threshold* | *CAGs* | *Topics* | *Relevant* | *Retrieved* | *Rel. Retrieved* | *P@10* |
|---|---|---|---|---|---|---|---|
| | 0.05 | 57 | 46 | 3,562 | 7,525 | 83 | 0.0196 |
| ASISWOR | 0.1 | 145 | 22 | 1,050 | 3,184 | 27 | 0.0000 |
| | 0.2 | 559 | 16 | 954 | 1,979 | 41 | 0.0125 |
| | 0.05 | 70 | 30 | 2,248 | 5,900 | 23 | 0.0233 |
| UWOR | 0.1 | 203 | 28 | 2,064 | 5,050 | 49 | 0.0250 |
| | 0.2 | 523 | 10 | 437 | 1,400 | 14 | 0.0600 |
| | 0.05 | 44 | 35 | 2,051 | 5,300 | 36 | 0.0057 |
| DLWOR | 0.1 | 126 | 17 | 952 | 2,700 | 9 | 0.0059 |
| | 0.2 | 471 | 16 | 1,112 | 1,850 | 48 | 0.0063 |
| | 0.05 | 17 | 20 | 1,226 | 3,076 | 12 | 0.0100 |
| DLLC | 0.1 | 64 | 14 | 745 | 1,776 | 26 | 0.0286 |
| | 0.2 | 272 | 9 | 606 | 887 | 15 | 0.0111 |

**Table 3.** IR effectiveness across non-replication testbeds for the Fixed topology

| *Testbed* | *Threshold* | *Topics* | *Relevant* | *Retrieved* | *Rel. Retrieved* | *P@10* |
|---|---|---|---|---|---|---|
| | 0.05 | 61 | 3,987 | 10,912 | 158 | 0.0393 |
| ASISWOR | 0.1 | 37 | 2,328 | 6,829 | 56 | 0.0189 |
| | 0.2 | 15 | 773 | 2,277 | 19 | 0.0067 |
| | 0.05 | 55 | 3,725 | 10,100 | 101 | 0.0164 |
| UWOR | 0.1 | 37 | 2,320 | 6,500 | 28 | 0.0108 |
| | 0.2 | 14 | 761 | 2,300 | 13 | 0.0000 |
| | 0.05 | 59 | 3,892 | 10,950 | 182 | 0.0492 |
| DLWOR | 0.1 | 37 | 2,328 | 6,900 | 61 | 0.0054 |
| | 0.2 | 13 | 759 | 2,000 | 26 | 0.0231 |
| | 0.05 | 56 | 3,800 | 9,150 | 152 | 0.0286 |
| DLLC | 0.1 | 34 | 2,272 | 5,700 | 52 | 0.0206 |
| | 0.2 | 13 | 621 | 1,600 | 23 | 0.0615 |

The column entitled *Threshold* corresponds to the threshold that was used for the document clustering as well as for the query routing that took place after the topology was created. The column *CAGs* shows the number of CAGs that were created with the given threshold. The column *Topics* shows the number of topics that were successfully routed to the network for matching. This number depends on the routing threshold. The initiating hub only routes a query to a CAG if its similarity to the CAG's centroid is higher than this threshold. The column *Relevant* shows the number of relevant documents for the number of topics that responses were given for. This comes from the relevance assessments provided by TREC for WT10g. The column *Retrieved* shows the number of documents that were retrieved in total, while *Rel. Retrieved* shows the number of relevant documents that were retrieved. Last, *P@10* is the precision achieved for the first 10 results in the result list, averaged over all the topics that got evaluated.

From this table we can see that there is a significant difference in retrieval effectiveness when compared to the results we obtained for the centralised index of Table 1. Even though these results may seem rather poor, one has to keep in mind a number of factors that are known to affect retrieval. First, WT10g is a web collection and therefore its documents cannot be expected to be of the same quality as the ones in other collections of documents such as collections of journal articles. Another important factor is the lengths of the documents. In the web, most documents are very small. This affects matching and, more importantly for this architecture, clustering. Very small documents (like very large documents) are harder to relate to other documents and classify automatically. Therefore, these results are not as surprising as they may seem at first, especially since no measures have been taken to counteract the aforementioned issues.

***Fixed* Topology.** For this topology we created a fixed number of CAGs based on the 100 TREC topics and their relevance assessments. We took the largest relevant documents for all the topics and used them as attractors for the rest of the documents. This gave us a topology of 94 CAGs – 2 topics have no relevant documents while 4 more did not attract any other documents apart from themselves. The retrieval effectiveness results we obtained are shown in Table 3. In this table, *Threshold* corresponds to the routing thresholds only, since we did not threshold similarity during the CAGs creation. The rest of the columns have the same meaning as their counterparts in Table 2, explained in the previous Section.

It can be seen in Table 3 that the effectiveness for this topology is very low and comparable to that exhibited by the Simple topology presented in the previous Section. This may seem unexpected as a result. Indeed, we included this alternative topology expecting to achieve significantly higher retrieval effectiveness, especially since the attractor documents were based on the topics that we would eventually evaluate against. This is a strong hint that there is a more important factor involved that impedes effectiveness. We believe that this factor has solely to do with the formation of cluster centroids and we will be analysing it further in Section 4.

## 4 Evaluating on an Optimal Testbed

In this section we re-assess the architecture using a small and near-optimal testbed based on the ASISWOR testbed of Section 2.2. We used ASISWOR as a base for our near-optimal testbed because it addresses openly available information-sharing environments and, as such, it is arguably the most generally applicable environment for the given architecture. We choose a smaller and more manageable testbed in order to better analyse and understand the P2P IR architecture and, therefore, to discover its pathological sources in a better controlled environment.

### 4.1 Characteristics and Conditions

**Testbed Characteristics.** The minimal ASISWOR testbed is near-optimal for the IR-based evaluation we will be presenting because it has a very high concentration of relevant documents. It was derived by randomly removing non-relevant documents from peer-collections also randomly picked. It consists of 4834 documents in total, spanning 1316 peers. 2267 of these documents are the relevant documents of the 100 standard

TREC topics while the rest were left intentionally in order to preserve some minimal distortion.

The, relatively to the total number of documents, large number of peer-collections ensured some skewness in the document distribution. This skewness is an important property that makes the ASISWOR testbed realistic and so even partially retaining it in the minimal testbed is important. The maximum number of documents a peer-collection has is 137, while 71% of the collections have 1 or 2 documents.

*Minimal ASISWOR as a Centralised Collection.* Similarly to the previous section, we provide the testbed's IR behaviour as a centralised corpus. These results are shown in Table 4.

**Table 4.** The retrieval effectiveness of minimal ASISWOR as a centralised collection

| *#Topics* | *Relevant* | *Retrieved* | *Rel_Retrieved* | *P@10* |
|---|---|---|---|---|
| 100 | 5980 | 47710 | 4596 | 0.6900 |

## 4.2 Evaluation Results

The overall retrieval effectiveness results are presented in Table 5. These results show that the IR effectiveness of the architecture is still at very significant odds compared to its centralised counterpart (Table 4). The sources of this discrepancy include the following:

1. The testbed does not encapsulate any structure to be found by the clustering mechanisms of the architecture.
2. The clustering mechanisms fail to discover the structure in the testbed.
3. The routing fails to locate enough relevant sources for the query to get forwarded to.

However, for this study instead of discussing these issues further, we will take them for granted, as a property of a realistic environment for P2P IR. Instead, we will pursue potential solutions that might help us to counter them[2].

**Table 5.** Results on retrieval effectiveness

| | *#Topics* | *#CAGs* | *Relevant* | *Retrieved* | *Rel_Retrieved* | *P@10* |
|---|---|---|---|---|---|---|
| *S-P* – 0.0 | 87 | 1 | 5475 | 3550 | 740 | 0.2678 |
| *S-P* – 0.05 | 19 | 10 | 1976 | 747 | 123 | 0.1737 |
| *S-P* – 0.1 | 16 | 30 | 1573 | 78 | 54 | 0.2562 |
| *S-P* – 0.2 | 16 | 139 | 1335 | 62 | 40 | 0.2125 |
| *FIXED* | 89 | 89 | 5530 | 16162 | 551 | 0.0596 |

[2] Additional experimental evidence, not presented herein, suggests that the fundamental assumptions made by both the architecture and the minimal-ASISWOR testbed hold. Hence they were omitted from this paper.

### 4.3 Compensating for Distortion

In Section 3 we showed experimentally that a two-level clustering, especially on small collections, can potentially limit the retrieval effectiveness of our P2P IR architecture. However, in our treatment, we neglected to look into a feature present in other architectures and indeed a very important feature for P2P networks in general, namely replication [4,9,10,1]. In this section we will look into whether replication can improve retrieval effectiveness. We will also look into term-weight adjustment, as this can result from relevance-feedback. Without arguing for a particular relevance-feedback implementation, we will show that weight-adjusted resource descriptions can increase the retrieval effectiveness in cluster-based P2P environments.

**Replication.** In order to assess the effect of replication on the P2P IR architecture, we implemented a replication strategy based on hypothetical popularities for the standard TREC topics. In our implementation, popularity is represented by a real number within the range $[0, 1)$ with 0 representing a topic that is not popular at all. The relevant documents to the topics are replicated to a number of peers according to their corresponding topic's popularity value, *i.e.* a document whose topic is popular has more chances to reside to another peer-collection etc. In order to calculate these popularities we used an inverse power law. Where, according to power-law, $y = \alpha x^k$, in our case, a popularity score $s_t$, for a topic $t$, is given by $s = \alpha / r^k$, where $\alpha$ is a constant that determines the popularity score for the most popular topic, $r$ is the rank of the topic with 1 being the most popular and $k$ is the exponent that determines the skewness of the output values. Once all topics have been assigned a popularity score, our algorithm iterates over all relevant documents and peer-collections and replicates documents randomly, according to their topic's score. This technique allows us to introduce realistic replication, scaled-down to the number of topics that we experiment on. For our experiments we took $\alpha = 0.9$ and $k = 2$. The $\alpha$ value ensures that no document gets replicated to all the peer-collections, while the $k$ value ensures that the trend of the popularities is not too steep so as to get meaningful replication for at least some of the topics.

For our experiments we created seven minimal testbeds with different arrangements of replicated content. This was done because of the element of randomness involved in the replication process described in the previous paragraph. The IR effectiveness results can be seen in Table 6. Comparing this table to Table 5 we notice two important differences: first, the effectiveness in the testbeds with replication is higher than in the testbed without. In particular, after the introduction of replication, for the testbeds used, we get an average P@10 of 0.4071, while in the testbed without replication, for the same threshold, *P@10* is 0.2562. On the other hand we notice that the number of topics that get to be answered (column *#Topics*) in the testbeds with the replication is much smaller (average of 1.86) than its corresponding figure for the testbed without replication (16). These two artifacts show that there is a significant improvement in effectiveness when replication is introduced, but only for the popular topics. In fact, the rest of the topics do not even get to be answered, *i.e.* their similarity to any CAG description falls below the threshold. This behaviour can be explained by looking into the cosine similarity measure that is used. When more similar documents, about a particular topic, are included in a cluster centroid (or a resource description for our purposes),

**Table 6.** Results on retrieval effectiveness on testbeds with replication. These results were obtained for a threshold of 0.1. The size of the original minimal testbed (before introducing replication) is 4834 documents.

| | *Size* | *#Topics* | *Relevant* | *Retrieved* | *Rel_Retrieved* | *P@10* |
|---|---|---|---|---|---|---|
| *Testbed* 0 | 68,469 | 3 | 316 | 600 | 169 | 0.3667 |
| *Testbed* 1 | 81,081 | 3 | 316 | 600 | 186 | 0.3333 |
| *Testbed* 2 | 65,199 | 2 | 310 | 400 | 177 | 0.5000 |
| *Testbed* 3 | 75,454 | 1 | 269 | 200 | 13 | 0.4000 |
| *Testbed* 4 | 101,168 | 1 | 269 | 200 | 101 | 0.5000 |
| *Testbed* 5 | 23,729 | 1 | 269 | 200 | 112 | 0.6000 |
| *Testbed* 6 | 153,885 | 2 | 59 | 350 | 20 | 0.1500 |
| ***Average*** | 81,283.57 | 1.86 | 258.29 | 364.29 | 111.14 | 0.4071 |

the similarity between this centroid and any topic other than the heavily replicated ones decreases. In this particular case, this decrease pushes the similarity below the lowest acceptable threshold, hence the small number of topics that get answered. Even though this seems to be a drawback, we believe it is to be expected in a large and widely available P2P information-sharing environment, where the potential number of topics are in the millions, not just one hundred. We believe that in such an environment the system could work sufficiently well for the majority of the users.

**Relevance Feedback.** For experimenting, we use the relevance assessments in order to alter the CAG centroids instead of the queries (the standard relevance-feedback application). Our goal is to counter-balance the noise that is introduced by the two-level clustering, by filtering, not augmenting, the document vectors. We assume that a relevance-feedback mechanism exists, which allows the aforementioned modification of resource-description vectors. For an original term weight $t_i$ of a CAG centroid, $t_i$ becomes $t_i + 0.5$ if $t_i$ is relevant (*i.e.* being a term of a document that is relevant to any of the TREC topics); otherwise $t_i$ becomes $t_i - 0.5$. In other words, the terms that describe relevant documents, collectively, to any of the topics, get promoted by 50% of their original weight while the rest get demoted by the same percentage. For these experiments we only adjusted the CAG centroids. Alternatively we could have also adjusted the cluster centroids that form the resource descriptions of the information-providers. This would lead to the re-clustering of these peers into new CAGs and possibly to better performance. However, while the creation of the CAG descriptions is a responsibility of the network, the creation of the cluster descriptions is a responsibility of the participating information providers. We did not want to directly adjust the cluster descriptions of the information providers since the architecture assumes that they are autonomous and trusted.

Evaluating this adjustment on the minimal ASISWOR testbed gives the results in Table 7.In Table 8 we summarise the difference in performance between the two different flavours of the architecture. From this table, apart from the aforementioned difference in P@10, we also note that the architecture with the hypothetical relevance-feedback mechanism manages to address more topics than the basic one. Beside this difference between the architectures one can observe that more topics are addressed for the higher

**Table 7.** Results on retrieval effectiveness with relevance-feedback term-weighting on the resource descriptions

| | *#Topics* | *#CAGs* | *Relevant* | *Retrieved* | *Rel_Retrieved* | *P@10* |
|---|---|---|---|---|---|---|
| *S-P* − **0.05** | 39 | 10 | 3543 | 1540 | 404 | 0.2667 |
| *S-P* − **0.1** | 32 | 30 | 3329 | 997 | 293 | 0.2844 |
| *S-P* − **0.2** | 49 | 138 | 4294 | 823 | 314 | 0.2510 |

**Table 8.** Comparison of IR effectiveness between the basic and relevance-feedback architecture

| *Threshold* | 0.05 | | 0.1 | | 0.2 | |
|---|---|---|---|---|---|---|
| | *Basic* | *RelFbk* | *Basic* | *RelFbk* | *Basic* | *RelFbk* |
| *#Topics* | 19 | 39 | 16 | 32 | 16 | 49 |
| *P@10* | 0.1737 | 0.2667 | 0.2562 | 0.2844 | 0.2125 | 0.2510 |

threshold of 0.2. In this case, this is a desirable fact, since the effective routing of more topics does not hinder the overall performance (as measured by P@10) of the system.

**Applying Weight-Adjustment along with Replication.** Having observed how the retrieval effectiveness increases when using relevance-feedback-based weight adjustment and replication separately, in this section we look into the effectiveness when both these mechanisms are applied. For the experiments presented below we adjusted the weights of the CAGs in the small testbeds we used in Section 4.3. The results in effectiveness are summarised in Table 9.

**Table 9.** Results on retrieval effectiveness on testbeds with replication and weight adjustment based on relevance-feedback. These results were obtained for a threshold of 0.1.

| | *#Topics* | *Relevant* | *Retrieved* | *Rel_Retrieved* | *P@10* |
|---|---|---|---|---|---|
| *Testbed* 0 | 10 | 1064 | 2000 | 253 | 0.3000 |
| *Testbed* 1 | 14 | 1302 | 2650 | 527 | 0.2786 |
| *Testbed* 2 | 19 | 1564 | 3500 | 476 | 0.3263 |
| *Testbed* 3 | 21 | 1940 | 3900 | 598 | 0.3190 |
| *Testbed* 4 | 18 | 1385 | 3600 | 616 | 0.2889 |
| *Testbed* 5 | 19 | 1430 | 3650 | 663 | 0.2842 |
| *Testbed* 6 | 21 | 1557 | 4200 | 680 | 0.2333 |
| *Average* | 17.43 | 1463.14 | 3357.14 | 544.71 | 0.2900 |

Comparing this to the results of Table 6, showing the retrieval effectiveness when only replication has been used, we notice that the introduction of relevance-feedback (the use of better aligned vectors to the topics) helps routing more topics than when we just used replication. In actual numbers, the average number of topics effectively routed when only replication was used is 1.86, while when both relevance-feedback and replication is used, for the same replication testbeds, the corresponding figure is 17.43. On the other hand the overall effectiveness, as measured by P@10, falls by about 11%. Because the gain in the number of topics that get routed is disproportionate to the loss of retrieval effectiveness we conclude that the use of both replication and relevance-feedback would probably benefit most P2P IR applications; however, this would still depend on the application requirements, with some applications preferring more precise results over wider query penetration.

**Comparison.** In Table 10 we present a comparison in retrieval performance across all variations of the minimal testbed. From these results we conclude that the term weighting adjustment, that could be accomplished by relevance feedback, is the most effective means to overcome the loss of information caused by clustering. We derive this conclusion after observing that, even though retrieval effectiveness does not fall – it actually increases – more topics are routed to the network. Even though not experimentally verified we anticipate that if network clusters were to change according to the new term-weights, retrieval effectiveness and query penetration would increase even more.

Replication significantly improves the effectiveness for the few topics that are popular, even though it impedes penetration. From our experiments it appears that replication and weight-adjustment complement each-other and so they could yield meaningful results if used together. However, since we only use the standard 100 TREC topics, the popular topics, used for replication, end up being very few and so we will not be expanding on it any further.

**Table 10.** Comparison in effectiveness across all variations of the minimal ASISWOR testbed

| | *Basic* | *Relevance-Feedback* | *Replication* | *Both* |
|---|---|---|---|---|
| *#Topics* | 16 | **32** | 1.86 | 17.43 |
| *P@10* | 0.2562 | 0.2844 | **0.4071** | 0.2900 |

## 4.4 Adjusting Term-Weights on Large Testbeds

So far, we have demonstrated two main points. First, that the use of clustering for large-scale P2P IR, at least on testbeds that have similar properties to ours, proves to be ineffective due to the loss of information inherent to the creation of cluster centroids. The second point is that two effective ways to amend this problem is by either introducing (or by using existing) replication and/or introducing some relevance-feedback mechanism that would help overcome the noise in the network resource descriptions. The second point has still to be demonstrated in a larger evaluation environment than the small ASISWOR-based testbed that we have used so far in this chapter.

In this section we present the retrieval effectiveness achieved in the original testbeds when weight-adjustment is used. These results can be seen in Table 11 and they demonstrate that even though the overall effectiveness across all testbeds does not rise beyond approximately 5% (for the case of DLWOR), an important difference in favour of the use of term-weighted resource description emerges, namely that the query penetration almost doubles across all the testbeds. This effect becomes more significant as the number of the topics that get routed rises alongside the retrieval performance.

The results of Table 11 confirm the results derived from earlier experiments using the minimal ASISWOR testbed of Section 4.3. The use of weighted vectors as resource descriptions, as opposed to using the original term-frequency vectors, appear to increase retrieval performance while it greatly enhances the query penetration of the network.

**Table 11.** Retrieval effectiveness in the original P2P IR testbeds in both the basic and the weight-adjustment configurations. The routing threshold is set to 0.1 while the adjusted vectors were only used for routing and not for peer-clustering.

| *Testbed* | *Configuration* | *#Topics* | *Relevant* | *Retrieved* | *Rel. Ret* | *P@10* |
|---|---|---|---|---|---|---|
| *ASISWOR* | Basic | 22 | 1050 | 3184 | 27 | 0.0000 |
| | RelFbk | 40 | 3497 | 7358 | 72 | 0.0175 |
| UWOR | Basic | 28 | 2064 | 5050 | 49 | 0.0250 |
| | RelFbk | 36 | 2907 | 6750 | 91 | 0.0361 |
| DLWOR | Basic | 17 | 952 | 2700 | 9 | 0.0059 |
| | RelFbk | **44** | 3595 | 7800 | 160 | **0.0591** |

While the effectiveness in both the minimal testbed we used in this section as well as in the large testbeds of Section 3 is by far worse than in a centralised index alternative, the findings of this Chapter are important for future studies and systems as they provide solid experimental evidence suggesting that relevance-feedback is a promising and natural evolution of current P2P IR technologies. Especially the automatic topological adaptation of a P2P network based on feedback seems to be promising as far as retrieval effectiveness is concerned.

## 5 Conclusions and Future Work

In this paper we presented a full-scale evaluation of a cluster-based P2P IR architecture, focusing on retrieval effectiveness. The architecture [2] we considered uses a two-level clustering in order to organise the shared content of the participating peers, taking no assumptions on the actual document distributions or other properties of the overall shared content. For our experiments we used a set of testbeds [6], which are based on TREC's WT10g. The use of a number of testbeds offers a more holistic view on the behaviour of the architecture we evaluate.

Our findings are the following: Employing a two-level clustering for P2P IR, especially in an open information-sharing environment, seems to amplify issues having to do with clustering itself, therefore resulting in poor retrieval performance. In particular, the noise in the resource descriptions created through clustering impedes standard IR practices such as query-routing based on cosine similarity. Building on this conclusion, we proposed replication and relevance-feedback as potential solutions for this problem and showed experimentally, using a small and manageable testbed, that both mechanisms can improve retrieval effectiveness for cluster-based P2P IR. Finally, we replicated our findings on the large testbeds we used originally. The results show that the performance of the architecture gets improved, mainly, through the significant increase of its query penetration rate.

Obvious pointers for future work include distributed relevance-feedback algorithms, devising replication strategies targeted at IR performance as well as studying the adaptability of the network given real-time changes of resource descriptions from an efficiency viewpoint.

## References

1. Lu, J., Callan, J.: Content-based retrieval in hybrid peer-to-peer networks. In: Proceedings of the twelfth international conference on Information and knowledge management, pp. 199–206. ACM Press, New York (2003)
2. Klampanos, I.A., Jose, J.M.: An architecture for information retrieval over semi-collaborating peer-to-peer networks. In: Proceedings of the 2004 ACM Symposium on Applied Computing, vol. 2, pp. 1078–1083. Nicosia, Cyprus (2004)
3. Tang, C., Xu, Z., Mahalingam, M.: Peersearch: Efficient information retrieval in peer-peer networks. Technical Report HPL-2002-198, Hewlett-Packard Labs (2002)
4. Lv, Q., Cao, P., Cohen, E., Li, K., Shenker, S.: Search and replication in unstructured peer-to-peer networks. In: ICS, New York (2002)
5. Nottelmann, H., Fuhr, N.: Comparing different architectures for query routing in peer-to-peer networks. In: Proceedings of the 28th European Conference on Information Retrieval Research (ECIR 2006) (2006)
6. Klampanos, I.A., Poznański, V., Jose, J.M., Dickman, P.: A suite of testbeds for the realistic evaluation of peer-to-peer information retrieval systems. In: Losada, D.E., Fernández-Luna, J.M. (eds.) ECIR 2005. LNCS, vol. 3408, pp. 38–51. Springer, Heidelberg (2005)
7. Tombros, A.: The Effectiveness of Hierarchic Query-Based Clustering of Documents for Information Retrieval. PhD thesis, Department of Computing Science, University of Glasgow (2002)
8. van Rijsbergen, C.J.: Information Retrieval, 2nd edn. Butterworths, London (1979)
9. Plaxton, C.G., Rajaraman, R., Richa, A.W.: Accessing nearby copies of replicated objects in a distributed environment. In: Proceedings of the ninth annual ACM symposium on Parallel algorithms and architectures, pp. 311–320. ACM Press, New York (1997)
10. Cuenca-Acuna, F.M., Martin, R.P., Nguyen, T.D.: Planetp: Using gossiping and random replication to support reliable peer-to-peer content search and retrieval. Technical Report DCS-TR-494, Department of Computer Science, Rutgers University (2002)

# A Conceptual Framework for Automatic Text-Based Indexing and Retrieval in Digital Video Collections

Mohammed Belkhatir[1] and Mbarek Charhad[2]

[1] Monash University, School of Information Technology
[2] FSM, Computer Science Department, Tunisia

**Abstract.** The growing need for 'intelligent' video retrieval systems leads to new architectures combining multiple characterizations of the video content that rely on expressive frameworks while providing fully-automated indexing and retrieval processes. As a matter of fact, addressing the problem of combining modalities for video indexing and retrieval is of huge importance and the only solution for achieving significant retrieval performance. This paper presents a multi-facetted conceptual framework integrating multiple characterizations of the visual and audio contents for automatic video retrieval. It relies on an expressive representation formalism handling high-level video descriptions and a full-text query framework in an attempt to operate video indexing and retrieval beyond trivial low-level processes, keyword-annotation frameworks and state-of-the art architectures loosely-coupling visual and audio descriptions.

## 1 Introduction and Related Work

The size, heterogeneity of content, and temporal characteristics of the video data bring about many interesting challenges to the video indexing and retrieval community. Among these challenges is the modeling task for efficient content-based indexing and user access capabilities such as querying, retrieval and browsing. In the literature, a video document is modeled based on its visual content (such as color, motion, shape, and intensity) [1,7], audio content [9,12], and semantic content in the form of text annotations [8,11,19]. Because machine understanding of the video data is still an unsolved research problem, text annotations are usually used to describe the video content according to the annotator's expertise and the purpose of this content. However, such descriptions are biased, incomplete and often inaccurate since subjected to the annotator's point-of-view. Also, manual annotation is a costly task which cannot keep pace with the ever-growing size of video collections.

Dealing with video indexing and search also entails taking into account an important number of modalities due to its multimedia aspect. For instance, a given concept (person, object ...) can be present under several forms in a video: it can indeed be seen, heard, talked of and a combination of these three characterizations can also occur. Being able to model aspects related to visual and audio contents is crucial in order to assist a human user in the tasks of querying and browsing. Also, since users are more skilled in defining their information needs using language-based descriptors [2,3,11], this modeling task is to consider a symbolic representation of the visual and

R. Wagner, N. Revell, and G. Pernul (Eds.): DEXA 2007, LNCS 4653, pp. 392–403, 2007.

audio features as textual descriptors. Indeed, a user would naturally formulate his need to being provided with videos where X is shown through the query "videos with X", or with videos where Y is speaking about X with the full-text query "videos where Y is speaking about X". The processing of the first query would consider the visual representation of the video while the processing of the second would entail searching in the audio track for a segment in which Y is the speaker and X is mentioned in the transcription. We could also imagine providing the possibility to combine both visual and audio descriptions within a unique full-text query instead of two distinct ones. This particularly brings about an exciting challenge since it is the main focus of the TRECVID evaluation campaign where a set of multimedia topics (i.e. textual transcriptions of information needs involving multiple characterizations of the visual and audio contents) are proposed to be solved. For this, the traditional keyword-based approaches in state-of-the-art video architectures would appear clearly not satisfactory since they fail to take into account aspects related to conceptual and relational descriptions of the video content. Indeed, we believe that in order to process a query such as "videos with Bill Clinton speaking and a US flag behind him" (proposed in the framework of the TRECVID topic search track), a system is to characterize visual concepts such as *Bill Clinton* and *US flag* as well as the relations between them (here the spatial relation *Behind*). Also it is to model the fact that the audio characterization of the concept *Bill Clinton* is linked to the audio relation *Speaking*.

In order to process these queries involving non-trivial information needs, we propose in this paper to integrate audio and visual descriptions within a unified full-text framework by considering:

-The specification of a rich video model featuring all characterizations of the visual and audio contents. It is based on **visual** and **audio objects**, structures abstracting visual entities and audio flow related to a video document.

-The integration of a knowledge representation formalism (conceptual graphs) in order to instantiate the video model within a video indexing and retrieval framework and therefore specify indexing, querying and matching processes.

-The specification of fully-automated processes to build and manipulate the conceptual index and query descriptions. Indeed, the strength of our approach relies in the specification of high-level descriptions of the video content while being able to process video corpus of relatively important size.

## 2 A Strongly-Integrated Model and Its Representation Formalism

A video document has a specific organization in terms of scenes, each characterizing a specific event. A video scene is itself composed by a number of shots, each of which is an unbroken image sequence captured continuously by the same camera. A shot could legitimately be compared to a word, as they are both the basic entities structuring respectively a video sequence and a text fragment. We will therefore consider a video shot as our elementary unit of description.

We propose the outline of a video model combining visual and audio/speech characterizations, each considered as a particular description **layer** of the video, to build the most exhaustive video specification.

The visual layer considers the representation of a video shot through its extracted key-frames: images providing a compact representation of the video shots. They can serve as pointers to the given portion of the video content for indexing and retrieval. This layer therefore groups all aspects of the image (keyframe) content as well as its general context and is itself seen as a bi-facetted object with the two principal facets being the visual semantics and signal facets. At the core of the visual layer is the notion of **visual object (VO),** abstract structure representing a visual entity within a keyframe. Their specification is an attempt to operate visual indexing and retrieval operations beyond simple low-level processes or object-based techniques [16] since VOs convey the visual semantics and signal information.

- The **visual semantics facet** describes the keyframe semantic content and is based on labeling VOs with a visual semantic concept. E.g., in fig. 1, the second VO (Vo2) is tagged by the semantic concept *Flag*. Its description will be dealt with in section 3.1.

- The **signal facet** describes the keyframe signal content in terms of symbolic perceptual features and consists in characterizing VOs with signal concepts. It is itself composed of three subfacets. The **color subfacet** features the keyframe signal content in terms of symbolic color words. E.g., the second VO (Vo2) is associated with color words *Blue*, *Red* and *White*. The **texture subfacet** describes the signal content in terms of symbolic texture features. E.g. Vo2 is associated with texture words *lined/striped* and *uniform*. The **spatial subfacet** specifies the spatial relationships that indicate the relative positions of VOs within a keyframe. For instance, in fig. 1, Vo2 is *covered* by (behind) Vo1. The signal facet is detailed and formalized in section 3.2.

The audio layer describes the information related to video shots through audio segment flows. It is represented by two facets:

- The **audio object facet** characterizes audio objects (AOs), abstraction of audio elements extracted from the audio flow.

- The **audio semantics facet** provides the semantic description related to each AO. It is based on concepts such as person identity, organization, location… and consists in specifying the speaker identity in each shot as well as the characterization of the audio content being spoken of. For example, in fig. 1, it represents the fact that the audio object Ao1 (corresponding to the concept *B. Clinton*) is speaking and that the audio object Ao2 is about the concept *Iraq*.

In order to instantiate this model as a video retrieval framework, we need a representation formalism capable of representing VOs and AOs as well as the visual and audio information they convey. Moreover, this representation formalism should make it easy to visualize the information related to a video. It should therefore combine expressivity and a user-friendly representation. As a matter of fact, a graph-based representation and particularly conceptual graphs (CGs) [17] are an efficient solution to describe a video and characterize its components. The asset of this knowledge representation formalism is its flexible adaptation to the symbolic approach of multimedia image and video retrieval [2,3,11]. It allows indeed to uniformly represent components of our architecture and to develop expressive and efficient index and query frameworks.

Formally, a CG is a finite, bipartite, connex and oriented graph. It features 2 types of nodes: the first one between brackets in our CG alphanumerical representation (i.e. as coded in our framework) is tagged by a concept (graphically represented by a rectangle in fig. 4) however the second between parentheses is tagged by a conceptual relation (graphically represented by a circle in fig. 4). E.g., the CG [DEXA_07]←(Name)←[Conference]→(Location)→[Regensburg] is interpreted as: the DEXA 2007 conference is held in Regensburg.

Concepts and conceptual relations are organized within a lattice structure partially ordered by the IS-A (≤) relation. E.g., Person ≤ Man denotes that the concept *Man* is a specialization of the concept *Person*, and will therefore appear in the offspring of the latter within the lattice organizing these concepts. Within the scope of the model, CGs are used to represent the video shot content in the visual and audio layers.

The indexing module provides a representation of a video shot document in the corpus with respect to the bi-layered video model. It is a CG called video shot document index graph. In fig. 1, a video shot belonging to the corpus is characterized at the conceptual level by a bi-layered representation, each layer consisting itself of several facets.

Also, as far as the retrieval module is concerned, a user full-text query is translated into a video shot conceptual representation: the video shot query graph corresponding

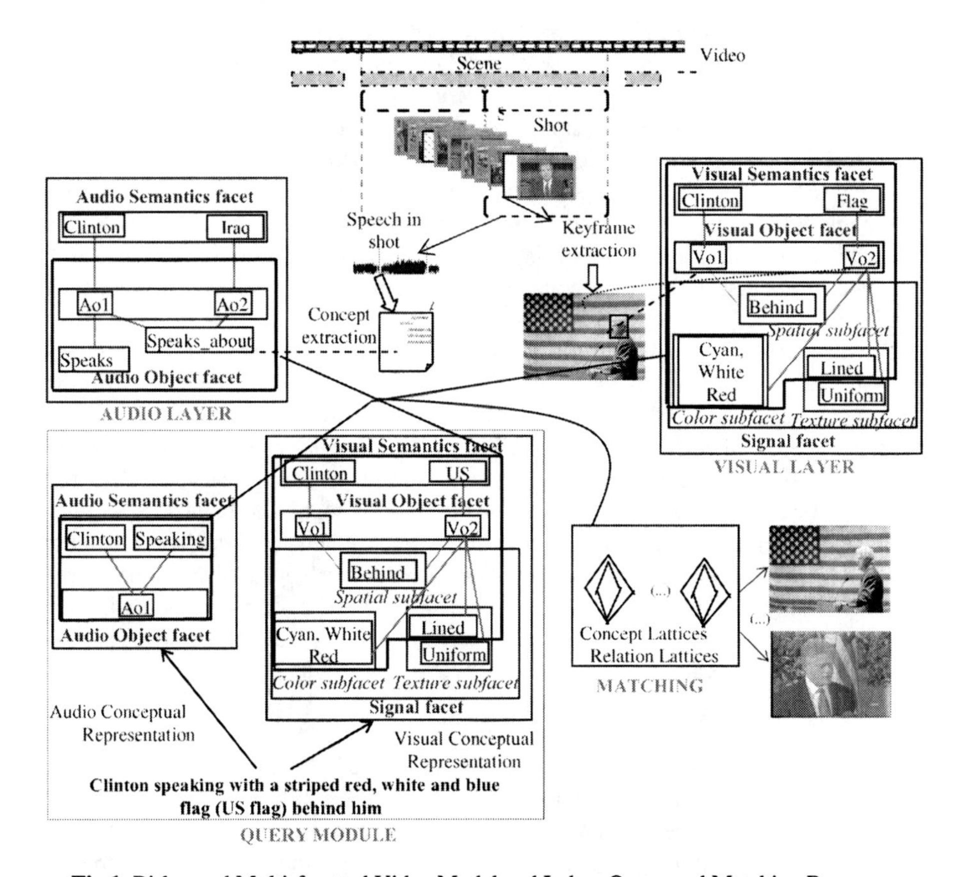


**Fig 1.** Bi-layered Multi-facetted Video Model and Index, Query and Matching Processes

to the bi-layered multi-facetted shot description. In fig. 1, the query "Find shots of Bill Clinton speaking with a US flag behind him" is translated into a multi-facetted conceptual representation for each of the visual and audio layers. Let us note that the visual semantic concept US flag which is not characterized in our semantic extraction framework (cf. 3.1) is translated into a visual conceptual representation combining both semantics (the concept *flag*) and signal features (the texture concept *striped* and the color concepts *red*, *white* and *blue*).

The video shot query graph is then compared to all conceptual representations of video shot documents in the corpus. Lattices organizing visual and audio concepts are processed and a relevance value, estimating the degree of similarity between video shot query and index graphs is computed in order to rank all video shot documents relevant to a query.

# 3 The Visual Layer

## 3.1 The Visual Semantics Facet

**Automatic extraction of Visual Concepts.** Visual concepts are learnt and then automatically extracted given a visual ontology. From the specification of twenty categories or picture scenes describing the image content at a global level (such as outdoor scene, city scene…), web-based image search engines (google, altavista) are queried by textual keywords corresponding to these picture scenes and 100 images are gathered for each query. These images are used to establish a list of visual concepts characterizing objects that can be encountered in these scenes. This list (in particular concepts related to individuals' names) is enriched with concepts provided by VideoAnnex [14] and a total of 96 visual concepts to be learnt and automatically extracted are specified.

A 3-layer feed-forward neural network with dynamic node creation capabilities is used to learn these concepts from 1000 labeled key-frame patches cropped from the training and annotation corpus of the TRECVID 2003 search task. Color and texture features are computed for each training region as an input vector for the neural network. Once the network has learnt the visual vocabulary, the approach subjects a tiled keyframe to be indexed to multi-scale, view-based recognition against these visual semantic concepts. A keyframe to be processed is scanned with windows of several scales. Each one represents a visual token characterized by a feature vector constructed with respect to the feature vectors of visual concepts previously highlighted. Recognition results are then reconciled across multiple resolutions and aggregated according to configurable spatial tessellation. (cf. [13] for further details).

**Model of the visual semantics facet.** VOs are represented by *Vo* concepts and visual semantic concepts are organized within a multi-layered lattice ordered by a specific/generic partial order (we propose a part of the lattice in fig. 2(a)). An instance of the visual semantics facet is represented by a set of canonical CGs, each one containing a *Vo* type linked through the conceptual relation *is_a* to a visual semantic concept.

The basic graph controlling the generation of all visual semantic facet graphs is: [Vo]→(is_a)→ [VSC]. E.g., the representation of the visual semantics facet for our example image in fig.1 is [Vo1]→(is_a)→[Clinton] and [Vo2]→(is_a)→[flag], translated as the first VO represents Clinton and the second VO a flag.

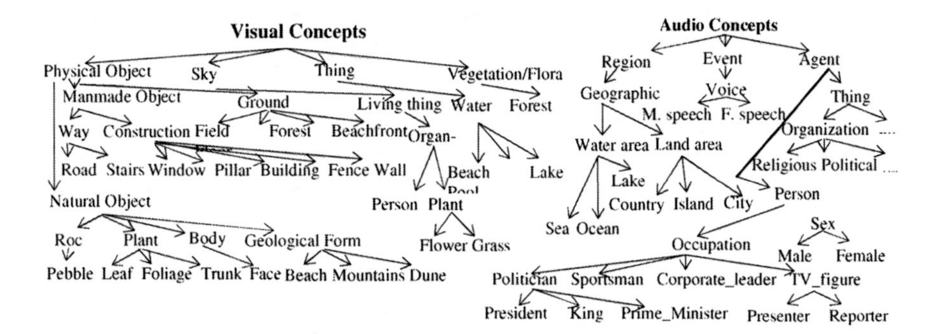


**Fig. 2.** (a) Lattice of Visual Concepts, (b) Lattice of Audio Concepts

## 3.2 The Signal Facet

The integration of signal information within the visual layer is not straightforward and involves highlighting a mapping between low-level signal features and their equivalent symbolic representation. For each of the color, texture and spatial subfacets, we propose the processes for mapping low-level to symbolic features, then present their conceptual characterizations and finally provide the algorithms for the automatic generation of the conceptual structures.

**The Color Subfacet.** The first step is to specify symbolic colors which correspond to low-level extracted features, therefore specifying a correspondence process between color names and color stimuli. Our symbolic representation of color information is guided by the research carried out in color naming and categorization [4] stressing a step of correspondence between color names and their stimuli. We will consider the existence of a formal system of color categorization and naming which specifies a set of color words. Within the scope of this paper, 11 color words ($C_1$=red, $C_2$=white, $C_3$=blue, $C_4$=grey, $C_5$=cyan, $C_6$=green, $C_7$=yellow, C8=purple, $C_9$=black, $C_{10}$=skin, $C_{11}$=orange) spotlighted in [10] are described in the HVC perceptually uniform space by a union of brightness, tonality and saturation intervals.

Each VO is indexed by a color index concept (*CIC*) featuring its color distribution by a conjunction of color words and their corresponding integer pixel percentages. The second VO (Vo2) corresponding to the semantic concept *flag* in fig. 1 is characterized by the CIC <r:40,w:45,b:15,g:0…>, interpreted as Vo2 having 40% of red, 45% of white **and** 15% of blue. CICs are elements of partially-ordered lattices, organized with respect to the query operator processed: either a Boolean or a quantification operator (at most, at least, mostly, few) explicited in [2]. **Index color graphs** link a *Vo* type through the conceptual relation *has_color* to a CIC: [Vo]→(has_color)→[CIC].

The algorithm summarizing the automatic generation of all conceptual structures of the color subfacet is as follows:

For each IO:
1. Compute the RGB value of each of its pixels
2. Map it to tonality, brightness & saturation values in the HVC perceptive space
3. Determine the associated color word considering the HVC perceptive color word partition [10]
4. Store for each color word the percentage of associated pixels
5. Generate the associated CIC and the alphanumerical color CG: [Vo]→(has_color)→[CIC]

E.g., the representation of the color subfacet for our example image in fig.1 is [Vo2]→(has_color)→[<r:40,w:45,b:15,g:0…>], translated as the second VO (Vo2) is associated with the CIC <r:40,w:45,b:15,g:0…> (i.e. 40% of red, 45% of white & 15% of blue).

**The Texture Subfacet.** Although several works have proposed the identification of low-level features and the development of algorithms and techniques for texture computation, few attempts have been made to propose an ontology for texture symbolic characterization and naming. In [5], a texture lexicon consisting of 11 high-level texture categories is proposed as a basis for symbolic texture classification. In each of these categories, several texture words which best describe the nature of the characterized texture are proposed. We consider the following texture words as the representation of each of these categories: $tw_1$=**bumpy**, $tw_2$=**cracked**, $tw_3$=**disordered**, $tw_4$=**interlaced**, $tw_5$=**lined**, $tw_6$=**marbled**, $tw_7$=**netlike**, $tw_8$=**smeared**, $tw_9$=**spotted**, $tw_{10}$=**uniform** and $tw_{11}$=**whirly**. These 11 high-level texture words, foundation of our framework for texture symbolic characterization are mapped to automatically-extracted 49-dimension vectors of Gabor energies (simulating the action of the visual cortex, where an object is decomposed into several primitives by the filtering of cortical neurons sensitive to several frequencies and orientations of the stimuli) through support vector machines [18].

Each VO is indexed by a texture concept (TC). A TC is supported by a vector structure **t** with eleven elements corresponding to texture words $tw_i$. Values t[i], $i \in [1,11]$ are booleans stressing that the texture distribution of the considered VO is characterized by the texture word $tw_i$. E.g., the second VO (Vo2) corresponding to the semantic concept *flag* in fig. 1 is characterized by the TC <B:0…D:0…L:1…U:1…>, translated as Vo2 being characterized by the texture words **striped** (lined) and **uniform**. TCs are elements of partially-ordered lattices which are organized respectively to the type of the query processed [3]. The basic graphs controlling the generation of all texture subfacet graphs links a Vo type through the conceptual relation *has_tex* to a texture concept: [Vo] →(has_tex)→[TC]

The algorithm generating all conceptual structures of the texture subfacet is as follows:

For each VO
1. Compute its associated 49-dimensions vector of Gabor energies
2. Map it to the linked texture words through a support vector machine architecture
3. Compute the posterior recognition probabilities of association [3]
4. Generate the associated TC & the texture CG: [Vo]→(has_tex)→[TC]

E.g., the representation of the texture subfacet for our example image in fig.1 is [Vo2]→(has_tex)→[<B:0…D:0…L:1…U:1…>], translated as the second VO (Vo2, representing the semantic concept *flag*) is associated with the TC <B:0…L:1…U:1…> (i.e. striped/lined and uniform).

**The Spatial Subfacet.** In order to model spatial data, we first consider a subset of the topological relations explicited in the RCC-8 theory [6]; 5 relations which are exhaustive and relevant for image querying are chosen. Considering 2 VOs (Vo1 and Vo2), these relations are ($s_1$=**C**,Vo1,Vo2): 'Vo1 partially covers (in front of) Vo2', ($s_2$=**C_B**,Vo1,Vo2): 'Vo1 is covered by (behind) Vo2', ($s_3$=**P**,Vo1,Vo2): 'Vo1 is a part of Vo2', ($s_4$=**T**,Vo1,Vo2): 'Vo1 touches Vo2 (externally connected)' and ($s_5$=**D**,Vo1,Vo2): 'Vo1 is disconnected from Vo2'. Directional relations Right ($s_6$=**R**), Left ($s_7$=**L**), Above ($s_8$=**A**), Below ($s_9$=**B**) are invariant to basic geometrical transformations (translation, scaling). Two relations specified in the metric space are based on the distances between VOs. They are the Near ($s_{10}$=**N**) and Far ($s_{11}$=**F**) relations.

Each pair of VOs are related through a spatial concept (SpC), compact structure summarizing spatial relationships between these VOs. A SpC is supported by a vector structure **sp** with eleven elements corresponding to the previously explicited spatial relations. Values sp[i], $i \in [1,11]$ are booleans stressing that the spatial relation $s_i$ links the two considered VOs. E.g., Vo1 and Vo2 are related through the SpC <C:1,C_B:0…N:1,F:0>, translated as Vo1 covering and being near to Vo2 (Vo2 being therefore behind Vo1). SpCs are elements of a partially-ordered lattice. The basic graph controlling the generation of all spatial subfacet graphs links two Vo types through the conceptual relations *agent_1* and *agent_2* to a SpC: [Vo1]←(agent_1)←[SpC]→(agent_2)→[Vo2].

The algorithm generating all conceptual structures of the spatial subfacet is as follows:

Given a pair of VOs, Vo1 and Vo2
1.Associate a topological relation to the results of interior and boundary sets of Vo1 and Vo2
2.Compare the centers of gravity of both VOs to determine the directional relations linking them
3. Compute d(Vo1_g,Vo2_g). To determine the near/far relations between Vo1 & Vo2, compare it to the measure of the spread of the distribution of centers of gravity of VOs.
4.Generate the associated SpC and the alphanumerical spatial CG: [Vo1]←(agent_1)←[SpC]→(agent_2)→[Vo2]

E.g., the representation of the spatial subfacet for our example image in fig. 1 is [Vo1]←(agent_1)←[<C:1,C_B:0…N:1,F:0>]→(agent_2)→[Vo2], translated as Vo1 covering and being near to Vo2.

**Conceptual Specification of the Visual Layer.** The conceptual representation of the visual layer is a CG obtained through the combination of CGs over the color, texture and spatial subfacets. The CGs: [Visual_Layer]→[has_VO]→[VO] are automatically added (for each visual entity characterized by a visual semantic concept as explicited in section 3.1) in order to unify conceptual representations of the three subfacets. For our example video shot of fig.1, the CG representation of the visual layer is the subgraph of fig.4 whose graphical elements have dashed contours.

# 4 The Audio Layer

## 4.1 Speaker Identification and Automatic Concept Detection

We base our automatic audio processing on the tasks of speaker identification and automatic concept detection. For this, we analyze automatic speech recognition (ASR) [9] transcripts. The ASR process is based on the processing of the audio flow in order to transcribe speech. The obtained output is a structured text with temporal descriptions.

**Speaker Identification.** We first aim at analyzing the transcription content for speaker identification in each segment using three classes of linguistic patterns. The first category is for detecting the identity of the speaker who is speaking in the current segment. For example, when the speaker introduces himself: "... this is C.N.N. news I' m [Name]... ". The second category is used to detect the identity of the speaker who has just spoken in the previous segment. The third category is for detecting the identity of the person who will speak (speaker of the next segment).

Table 1 summarizes some of the patterns that we use in our approach gathered by category.

**Table 1.** List of some linguistic patterns

| Previous segment | Curr.ent segment | Next segment |
|---|---|---|
| thank you...[name] | I'm [name] | tonight with [name] |
| thanks ... [name] | [name] CNN | ABC's [name] |
| [name] reporting | [name] ABC | [name] reports |
| ....... | ....... | ......... |

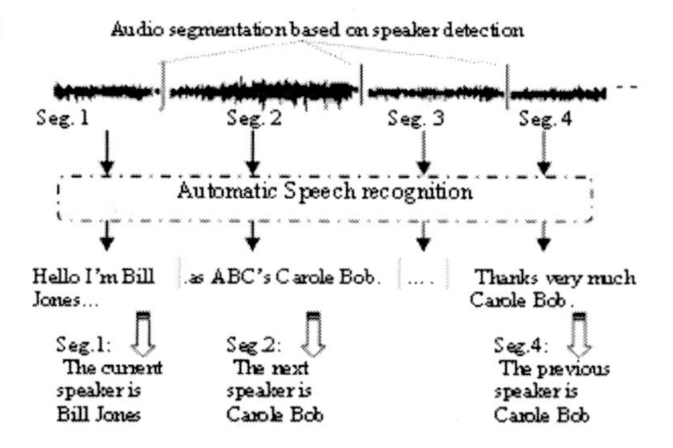


**Fig. 3.** Overview of the identity detection approach

The detection process consists in parsing each segment and identifying passages containing one of these patterns. We then apply a tool for identity recognition. For this, we use a named-entity extraction process based on two lists of concepts. The first contains a set of first names (~12400) and the second contains common words except family names. We compare neighboring words of each detected pattern with the content of the 2 lists. If neighboring words are for example elements of the first list and not present in the second list, we can estimate that they deal with a person's identity. We then infer the corresponding identity by comparing its localization with respect to a linguistic pattern.

Our approach is summarized in fig. 3 which displays the complete process for automatic speaker identification. We tested our approach on the TRECVID 2004 benchmark collection and obtained a success rate of 82 % for automatic speaker's identity recognition.

**Automatic Concept Extraction.** The other challenging task related to audio flow analysis is the specification of what is being spoken of in each audio segment. We have specified four concept classes in the extraction process: *person*, *place*, *organization* and *acronyms*. To extract this information, we parse transcriptions files to detect symbolic information by comparing their items to elements in the four concept classes (e.g. a person's identity, the name of a city, an organization etc...). We then extract, from each document, the concepts which correspond to each class. This process is based on the projection of each document on a specific domain ontology (we provide a part of its lattice representation in fig. 2(b)). We exploit linguistic patterns to specify concepts such as the titles Mr., Mrs. appearing before a person's identity or propositions like "in", "at", "from"... before a localization concept (*place*). Here is the algorithm summarizing the concept extraction process:

For all audio segments:
1. Extract AOs by projecting the transcription of the audio segment on specific ontologies
2. Collect AOs belonging to the concept category *person*
3. Collect AOs belonging to the concept category *place*
4. Collect AOs belonging to the concept category *organization*
5. If AOs belong to the concept category *person*, specify AOs corresponding to speakers using linguistic patterns.

### 4.2 A Conceptual Model for the Audio Layer

Extracted concepts are related through specific audio relations to AOs. Considering for example two AOs (Ao1 and Ao2), these relations are **spk**(Ao1) where Ao1 belongs to the concept class *person* translated as Ao1 speaks and **spk_abt**(Ao1,Ao2) translated as Ao1 is speaking about Ao2. For the audio layer modeling we use all automatically extracted concepts, they are labeled audio semantic concepts (**ASC**).

The audio layer is represented by a set of canonical CGs: [Ao1]→(spk_abt)→[Ao2], [Ao1]→(spk) and [Ao]→(is_a)→[ASC] linking AOs to ASCs.

The visual layer of our example video shot in fig.1 is given by the the subgraph of fig.4 whose graphical elements are highlighted (grey background).

## 5 Unification of CGs over Visual and Audio Layers

The conceptual representation of a video shot is a CG obtained through the combination (**join** operation [17]) of CGs over all the facets of both visual and audio layers. The CGs: [Video]→[composed_of]→[Visual_layer] and [Video]→[composed_of]→[Audio_layer] are automatically added in order to unify conceptual representations of the audio and visual layers. For the example video shot of fig.1, the unified visual/audio conceptual representation is given in fig. 4.

As far as querying is concerned, our conceptual architecture is based on a unified full-text framework allowing a user to query over both the visual and audio layers. This obviously optimizes user interaction since the user is in 'charge' of the query process by making his information needs explicit to the system. The retrieval process using CGs relies on the fact that a query is also expressed under the form of a CG.

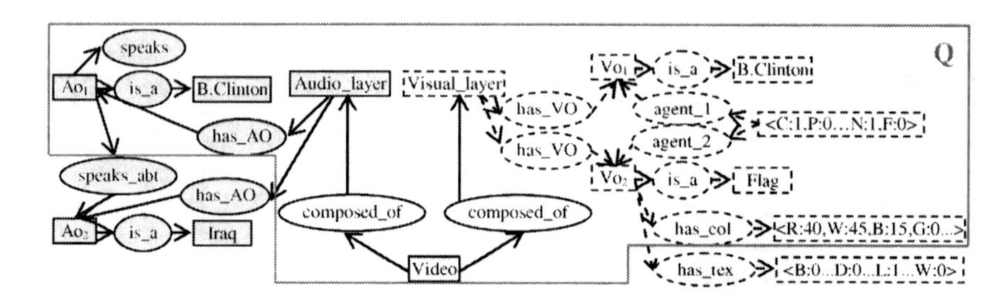


**Fig. 4.** Unified visual/audio index representation of the example video shot

The representation of a user query in our model is, like image index representations, obtained through the combination (join operation) of CGs over all the facets of both visual and audio layers. Without going into details, a trivial grammar composed of a list of the previously introduced visual and audio concepts, as well as the specified visual and audio relations allows to translate a query string into an alphanumerical CG structure. For instance, the string: "Bill Clinton speaking with a striped blue, white and red flag (US flag) behind him" is translated into the subgraph **Q** within the framed structure in fig. 4.

The matching framework is based on an extension of VanRijsbergen's logical model proposed in [15]. We define the relevance of a video shot VS with respect to a query Q as a combination of the exhaustivity and specificity measures: **Relevance(VS,Q) = F[P(VS→Q), P'(Q→VS)]**

Exhaustivity quantifies to which extent the video shot satisfies the query. It is given by the value of P(VS→Q), P being the exhaustivity function. Specificity measures the importance of the query themes within the considered video shot, it is given by the value of P'(Q→VS), P' being the specificity function.

## 6 Conclusion

We proposed the specification of a video retrieval framework unifying visual and audio characterizations within a strongly-integrated architecture to achieve greater retrieval accuracy. We introduced visual and audio objects, abstract structures representing visual and audio entities in order to operate indexing and retrieval operations at a higher abstraction level than state-of-the-art frameworks. We specified the visual and audio layers, their related facets and/or subfacets as well as their conceptual representations. Also, we presented the automated processes handling the indexing and retrieval operations of video documents with respect to both visual and audio layers. We are currently in the process of evaluating our framework on the topic retrieval task of the TRECVID evaluation campaign.

## References

1. Amato, G., Mainetto, G., Savino, P.: An Approach to a Content-Based Retrieval of Multimedia Data. Multimedia Tools and Applications 7, 9–36 (1998)
2. Belkhatir, M., Mulhem, P., Chiaramella, Y.: Integrating Perceptual Signal Features within a Multi-facetted Conceptual Model for Automatic Image Retrieval. In: McDonald, S., Tait, J. (eds.) ECIR 2004. LNCS, vol. 2997, pp. 267–282. Springer, Heidelberg (2004)

3. Belkhatir, M.: Combining semantics and texture characterizations for precision-oriented automatic image retrieval. In: Losada, D.E., Fernández-Luna, J.M. (eds.) ECIR 2005. LNCS, vol. 3408, pp. 457–474. Springer, Heidelberg (2005)
4. Berlin, B., Kay, P.: Basic Color Terms: Their universality and Evolution. UC Press (1991)
5. Bhushan, N., et al.: The Texture Lexicon: Understanding the Categorization of Visual Texture Terms and Their Relationship to Texture Images. Cognitive Science 21(2), 219–246 (1997)
6. Cohn, A., et al.: Qualitative Spatial Representation and Reasoning with the Region Connection Calculus. Geoinformatica 1, 1–44 (1997)
7. Fablet, R., Bouthémy, P.: Statistical motion-based video indexing and retrieval. In: Conf. on Content-Based Multimedia Information Access, pp. 602–619 (2000)
8. Fan, J., et al.: ClassView: hierarchical video shot classification, indexing, and accessing. IEEE Transactions on Multimedia 6(1), 70–86 (2004)
9. Gauvain, J.L., Lamel, L., Adda, G.: The LIMSI Broadcast News transcription system. Speech Communication 37, 89–108 (2002)
10. Gong, Y., Chuan, H., Xiaoyi, G.: Image Indexing and Retrieval Based on Color Histograms. Multimedia Tools and Applications II, 133–156 (1996)
11. Kokkoras, F.A., et al.: Smart VideoText: a video data model based on conceptual graphs. Multimedia Syst. 8(4), 328–338 (2002)
12. Kwon, S., Narayanan, S.: Speaker Change Detection Using a New Weighted Distance Measure. In: ICSLP, pp. 16–20 (2002)
13. Lim, J.H.: Explicit query formulation with visual keywords. ACM Multimedia, 407–412 (2000)
14. Lin, C.Y., Tseng, B.L., Smith, J.R.: VideoAnnEx: IBM MPEG-7 Annotation Tool for Multimedia Indexing and Concept Learning. IEEE ICME (2003)
15. Nie, J.Y.: An outline of a General Model for Information Retrieval Systems. ACM SIGIR, 495–506 (1988)
16. Smeulders, A.W.M., et al.: Content-based image retrieval at the end of the early years. IEEE PAMI 22(12), 1349–1380 (2000)
17. Sowa, J.F.: Conceptual structures: information processing in mind and machine. Addison-Wesley publishing company, London (1984)
18. Vapnik, V.: Statistical Learning Theory. Wiley, Chichester (1998)
19. Zhu, X., et al.: InsightVideo: toward hierarchical video content organization for efficient browsing, summarization and retrieval. IEEE Trans. on Multimedia 7(4), 648–666 (2005)

# Dimensionality Reduction in High-Dimensional Space for Multimedia Information Retrieval

Seungdo Jeong[1], Sang-Wook Kim[2], and Byung-Uk Choi[2]

[1] Department of Electrical and Computer Engineering, Hanyang University
[2] College of Information and Communications, Hanyang University
17 Haengdang-dong, Sungdong-gu, Seoul, 133-791 Korea
{sdjeong, wook, buchoi}@hanyang.ac.kr

**Abstract.** This paper proposes a novel method for dimensionality reduction based on a function approximating the Euclidean distance, which makes use of the norm and angle components of a vector. First, we identify the causes of errors in *angle estimation* for approximating the Euclidean distance, and discuss basic solutions to reduce those errors. Then, we propose a new method for dimensionality reduction that composes a set of subvectors from a feature vector and maintains only the norm and the estimated angle for every subvector. The selection of a good reference vector is important for accurate estimation of the angle component. We present criteria for being a good reference vector, and propose a method that chooses a good reference vector by using the Levenberg-Marquardt algorithm. Also, we define a novel distance function, and formally prove that the distance function consistently lower-bounds the Euclidean distance. This implies that our approach does not incur any false dismissals in reducing the dimensionality. Finally, we verify the superiority of the proposed approach via performance evaluation with extensive experiments.

## 1 Introduction

*Multimedia information retrieval* is to search for information satisfying a query condition from multimedia databases. In most previous studies, a multimedia object is represented as a *feature vector*, which quantifies its contents or features in a form of a vector. In order to express the original object sufficiently, feature vectors normally become several tens to a few hundreds dimensional [4,10,11]. We call feature vectors stored in a database *data vectors*. Also, we call feature vectors used in querying *query vectors*.

A majority of previous studies used the *Euclidean distance* as a measure for evaluating the similarity between two vectors [2,4]. Diverse indexing methods have been proposed. However, the performance of these indexing methods degrades dramatically with the increasing dimensionality of feature vectors. This has been known as *dimensionality curse* [3,11]. One of its solutions is *dimensionality reduction* that transforms feature vectors in high-dimensional space to those in low-dimensional space [1,7,5]. For simplicity, we call feature vectors of reduced low-dimensional space *low-dimensional feature vectors*.

R. Wagner, N. Revell, and G. Pernul (Eds.): DEXA 2007, LNCS 4653, pp. 404–413, 2007.

Multimedia information retrieval using dimensionality reduction consists of two steps: a *filtering step* and a *post-processing step*. We call this the *two-step searching method*. A desirable two-step searching method not only guarantees no *false dismissals* but also minimizes *false alarms* [10]. In previous researches, various mathematical transformations have been used for dimensionality reduction. The principal component analysis(PCA), the discrete cosine transform(DCT), and the discrete Fourier transform(DFT) are typical ones [8]. The methods using the PCA and the DCT have been known to outperform that using the DFT [7].

This paper deals with effective dimensionality reduction for high-dimensional applications. To guarantee no false dismissals, the distance between any pair of feature vectors in low-dimensional space should lower-bound the distance between the original feature vectors. The function for computing the Euclidean distance includes the inner product of two vectors. The *Cauchy-Schwartz inequality* defines the relationship between the norm multiplication and the inner product of two vectors [8]. The distance function substituting the multiplication of norms for their inner product is a lower-bounding function to the Euclidean distance [5]. However, this function suffers from large errors because it uses only norms. To solve this problem, Jeong et al. proposed a novel distance function that lower-bounds the Euclidean distance [6]. This function takes the *angle component* between two vectors into account as well as norms, and thus, can reduce effectively the error of the approximated distance.

We propose a novel method for dimensionality reduction based on the function approximating the Euclidean distance, which makes use of the norm and angle components of a vector. For this, we identify the causes of the errors in angle estimation for approximating the Euclidean distance, and then propose a new method for dimensionality reduction that composes a set of subvectors from a feature vector and maintains only the norm and the estimated angle for every subvector. We verify the superiority of the proposed approach via performance evaluation with extensive experiments.

## 2 Related Work

### 2.1 Lower-Bounding Function Using Cauchy-Schwartz Inequality

$<X,Y>$ denotes the inner product of two vectors $X$ and $Y$ in $n$-dimensional space. Equation (1) is the Cauchy-Schwartz inequality. It defines the upper-bound of their inner product.

$$<X,Y> \leq \|X\|\|Y\| \qquad where\ \|X\|^2 = \textstyle\sum_{i=1}^{n} x_i^2 \tag{1}$$

The Euclidean distance between two vectors is defined as equation (2).

$$D(X,Y) = \sqrt{\textstyle\sum_{i=1}^{n}(x_i - y_i)^2} = \sqrt{\|X\|^2 + \|Y\|^2 - 2<X,Y>} \tag{2}$$

We define a function approximating the Euclidean distance using the Cauchy-Schwartz inequality, $D_{cs}(X,Y)$, as in equation (3).

$$D_{cs}(X,Y) = \sqrt{\|X\|^2 + \|Y\|^2 - 2\|X\|\|Y\|} \tag{3}$$

From the definition of the Cauchy-Schwartz inequality, the relationship between two distance functions $D(X,Y)$ and $D_{cs}(X,Y)$ is represented as equation (4). Thus, $D_{cs}(X,Y)$ is a function lower-bounding the Euclidean distance $D(X,Y)$.

$$D(X,Y) \geq D_{cs}(X,Y) \quad (4)$$

Therefore, the retrieval using $D_{cs}(X,Y)$ does not incur any false dismissals. Besides, the computation is very fast because only norms instead of all the dimension values are considered.

## 2.2 Lower-Bounding Function Considering the Angle Component

$D_{cs}(X,Y)$ does not take the angle component of two vectors into account. This causes large approximation errors. To solve this problem, Jeong et al. proposed a lower-bounding function considering the angle component [6]. We can compute the Euclidean distance by using norms and angles as shown in equation (5).

$$D(X,Y) = \sqrt{\|X\|^2 + \|Y\|^2 - 2 <X,Y>} = \sqrt{\|X\|^2 + \|Y\|^2 - 2\|X\|\|Y\| cos\theta} \quad (5)$$

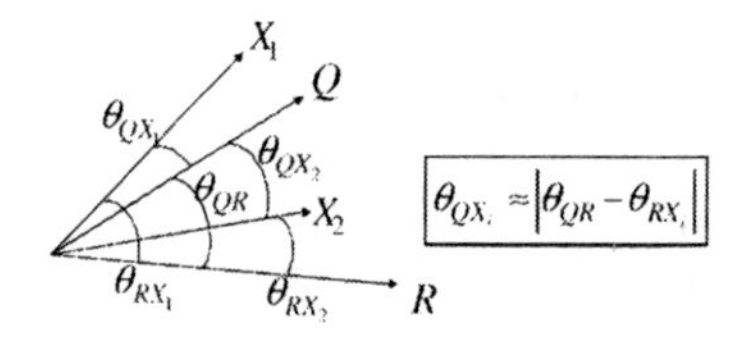


**Fig. 1.** Angle approximation using a reference vector

Jeong et al. introduced the notion of the *reference vector* $R$ to estimate the angle component between a query vector $Q$ and every data vector $X_i$ as shown in Fig. 1 [6]. If we know the angle $\theta_{RX_i}$ between $R$ and $X_i$, we can approximate the angle $\theta_{QX_i}$ between $Q$ and $X_i$ by a simple calculation given in equation (6). The angles between $R$ and every $X_i$ are computed and stored at the time of database construction. The function $D_A(Q, X_i)$, which approximates the distance to be compared with a tolerance in the filtering step, is given in equation (7).

$$\tilde{\theta}_{QX_i} = |\theta_{QR} - \theta_{RX_i}| \quad (6)$$

$$D_A(Q, X_i) = \sqrt{\|X_i\|^2 + \|Q\|^2 - 2\|X_i\|\|Q\| cos\tilde{\theta}_{QX_i}} \quad (7)$$

Although $R$, $Q$, and $X_i$ do not exist on the same plane, $\theta_{QX_i}$ is also approximated by the same equation. In this case, the approximated angle is always less than or equal to the actual angle. Therefore, any false dismissals are not introduced by this method.

**Theorem 1.** *For any pair of two vectors $X$ and $Y$, the following equation holds.*

$$D(X,Y) \geq D_A(X,Y)$$

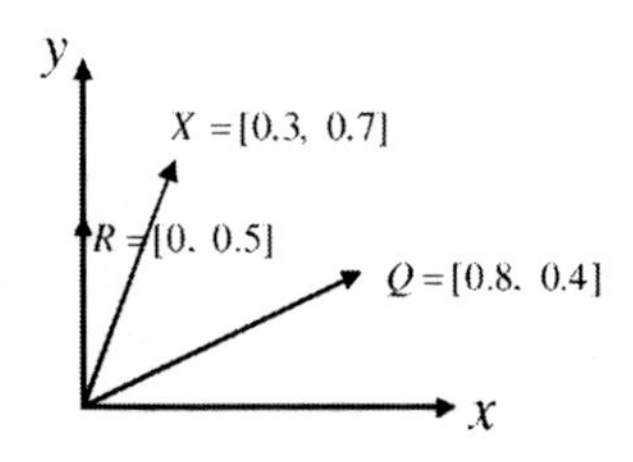


Fig. 2. An example of two-dimensional vectors

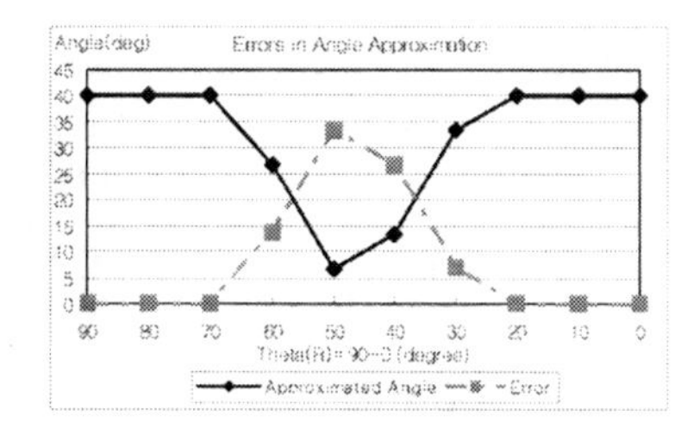


Fig. 3. Errors in angle approximation with varying angles

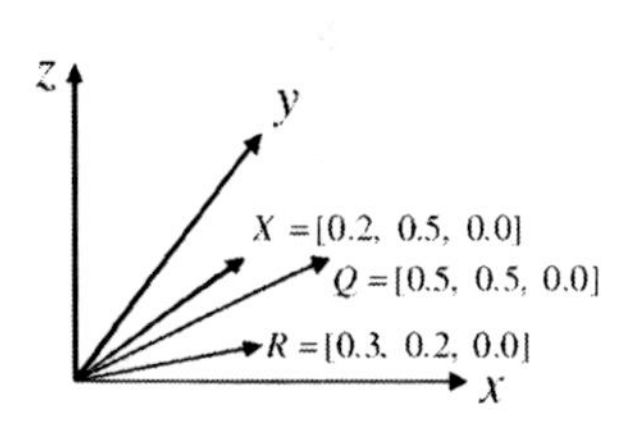


Fig. 4. An example of three-dimensional vectors

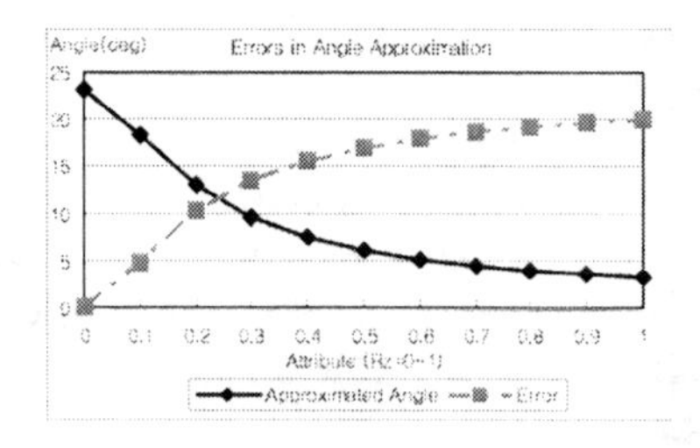


Fig. 5. Errors in angle approximation for varying values for the $z$ coordinate of a reference vector

*Proof.* Refer to [6].

$D_A(X, Y)$ has the norm and angle referred to the reference vector as arguments. Thus, this is a kind of dimensionality reduction that maps a data vector in high-dimensional space to that in *two*-dimensional space whose axes are the norm and angle.

# 3 Proposed Method

## 3.1 Error Analysis

In the angle approximation, an error of the approximated angle occurs depending on the locations among a data vector $X$, a query vector $Q$, and a reference vector $R$. Fig. 2 shows an example of three vectors $X$, $Q$, and $R$ in two-dimensional space. Fig. 3 presents the approximated angles and their errors. Here, we fix the norm of $R$ as 0.5 and vary the angle of $R$ from 90 to 0. We note that approximation errors occur if $R$ is located between $X$ and $Q$ as shown in Fig. 3.

Fig. 4 shows an example of three vectors $X$, $Q$, and $R$ in three dimensional space. The value for the $z$ coordinate of all these three vectors is 0.0. This means that these three vectors are on the *same plane*. Because three vectors are on the same plane and $R$ is not located between $X$ and $Q$, the approximated angle between $X$ and $Q$ is exactly same as the actual angle.

Fig. 5 shows the difference between the approximated and actual angles. Here, we increase the value for the $z$ coordinate of $R$ from 0 to 1. As the value for the $z$ coordinate of $R$ increases, the errors of the approximated angles also get larger. In other words, as the three vectors get closer to the same plane, the errors become smaller. To reduce such errors, we employ two kinds of reference vectors: (1) One is very close to a plane composed of $Q$ and every $X$ and (2) The other is most close to an axis.

### 3.2 Dimensionality Reduction Using Dimension Grouping

To use $D_A(X,Y)$ that lower-bounds the Euclidean distance, we store only *norms* of original vectors and *angle components* between those and the reference vector. In low-dimensional space, the probability for arbitrary three vectors to be close to the same plane is relatively high. That is, the lower the dimensionality is, the less the approximation errors are. To reflect this characteristic, we propose a *dimension grouping method* that represents high-dimensional space as a set of low-dimensional space. Because arbitrary three vectors are likely to close to one plane in low-dimensional space, we can successfully reduce the errors of the approximated angle.

We call the number of groups for dimensionality reduction the *reduced dimensionality* $k$. We also choose $k$ as the same as the number of dimensions in low-dimensional space. A vector $X'$ which is composed of some attributes among $n$ attributes of the $n$-dimensional data vector $X$ is defined as a *subvector*. Let $X_i'$ be the $i$-th subvector of $X$ and $X_i' = [x_{i1}, x_{i2}, \ldots, x_{il_i}]$ be its numerical representation. $k$ subvectors satisfy two constraints shown in equation (8), where $A(X) = \{\forall x_i | x_i \in X\}$, that is a set composed of respective attributes of $X$.

$$\{A(X_i') \cap A(X_j') | \forall i (\leq k), j (\leq k), i \neq j\} = \emptyset, \quad A(X) = \bigcup_{i=1}^{k} A(X_i') \tag{8}$$

$\sum_{i=1}^{k} l_i = n$, where $l_i$ is the number of attributes of the $i$-th subvetor. The reference vector $R$ can be also represented by $k$ subvectors $R_i'$. we can compute their norms and angles using equation (9) and equation (10), respectively, where $X_i' = [x_{i1}, x_{i2}, \ldots, x_{il_i}]$, $R_i' = [r_{i1}, r_{i2}, \ldots, r_{il_i}]$.

$$\|X_i'\| = \sqrt{\sum_{j=1}^{l_i} x_{ij}^2} \tag{9}$$

$$\theta_{X_i'} = cos^{-1}\left(\sum_{j=1}^{l_i} x_{ij} r_{ij} / \|X_i'\| \|R_i'\|\right) \tag{10}$$

If we divide $X$ into $k$ subvectors, a *reduced data vector* $X_{GA}$ is represented as equation (11). Also, we define the distance $D_{GA}(X,Y)$ between two vectors $X_{GA}$ and $Y_{GA}$ in reduced dimensional space as equation (12).

$$X_{GA} = [\|X_1'\|, \theta_{X_1'}, \|X_2'\|, \theta_{X_2'}, \ldots, \|X_k'\|, \theta_{X_k'}] \tag{11}$$

$$D_{GA}(X,Y) = \sqrt{\sum_{i=1}^{k}(\|X_i'\|^2 + \|Y_i'\|^2 - 2\|X_i'\|\|Y_i'\|cos\tilde{\theta}_{X_i'Y_i'})}$$
$$where \quad \tilde{\theta}_{X_i'Y_i'} = |\theta_{X_i'} - \theta_{Y_i'}| \tag{12}$$

### 3.3 Selection of a Reference Vector

In order to reduce approximation errors, it is necessary to select $R$ that is close to the plane formed by $Q$ and every $X_i$. In addition, to reduce the errors caused by the position of $R$, $R$ should be as close as possible to one of the axes that form data space.

We assume that the distribution of $Q$ follows that of $X_i$ [9]. Thus, to locate $R$, $X_i$, and $Q$ as close as possible to the same plane, it is necessary to select $R$ whose distance to the plane, which is approximated by all $X_i$, should be minimized. For this, we propose a method to select $R$ using the *Levenberg-Marquardt(L-M)* algorithm [8].

$V = [v_1, v_2, \ldots, v_n]$ denotes a normal vector of an $n$-dimensional plane. Numerical representation of this plane is the same as equation (13), where we represent an axis of $n$-dimensional space as $A = [a_1, a_2, \ldots, a_n]$. For this plane and $X_i$, a constant $T_i$ satisfying equation (14) exists. If the magnitude of $V$ is equal to 1, $T_i$ is the distance between $X_i$ and the plane. Rearranging equation (14) for $T_i$, we can obtain equation (15).

$$v_1 a_1 + v_2 a_2 + \ldots + v_n a_n = 0 \tag{13}$$

$$v_1(x_{i1} + T_i v_1) + v_2(x_{i2} + T_i v_2) + \ldots + v_n(x_{in} + T_i v_n) = 0 \tag{14}$$

$$T_i = -\frac{v_1 x_{i1} + v_2 x_{i2} + \ldots + v_n x_{in}}{v_1^2 + v_2^2 + \ldots + v_n^2} \tag{15}$$

A sum of squared distances from $N$ data vectors to the plane is the same as equation (16). Thus, if we could obtain $V$ that minimizes equation (16), it is the plane whose distance to all $X_i$ is minimum.

$$\sum_{i=1}^{T} T_i^2 = \sum_{i=1}^{T}(v_1 x_{i1} + v_2 x_{i2} + \ldots + v_n x_{in})^2 \tag{16}$$

In this paper, we compute $V$ using the Levenberg-Marquardt algorithm to minimize equation (16). Next, we obtain projection of every $X_i$ to this plane, and select one of a pair of projection as $R$. Here, the selected pair of projection has the largest angle. $R$ selected by our algorithm is very close to the same plane as well as an axis. Thus, we can reduce errors significantly.

### 3.4 Database Construction and Query Processing

First, we select $R$ as described in Section 3.3. Next, we compute a $k$-dimensional data vector for every $X_i$ using equations (9) and (10). In query processing, we use low-dimensional data vectors at the filtering step for obtaining candidates and original data vectors at the post-processing step for discarding false alarms.

### 3.5 Discussions

In this section, we prove that our dimensionality reduction method guarantees no false dismissals.

**Theorem 2.** *For any pair of two vectors* $X$ *and* $Y$*, the following equation holds.*

$$D(X,Y) \geq D_{GA}(X,Y)$$

*Proof.* For the subvectors $X_i' = [x_{i1}, x_{i2}, \ldots, x_{il_i}]$ and $Y_i' = [y_{i1}, y_{i2}, \ldots, y_{il_i}]$, $D(X_i', Y_i')$ can be obtained by equation (17).

$$D(X_i', Y_i') = \sqrt{\|X_i'\|^2 + \|Y_i'\|^2 \ - 2 < X_i', Y_i' >} \tag{17}$$

In the same way, $D_A(X_i', Y_i')$ can be also computed by equation (18).

$$D_A(X_i', Y_i') = \sqrt{\|X_i'\|^2 + \|Y_i'\|^2 \ - 2\|X_i'\|\|Y_i'\|cos\tilde{\theta}_{X_i'Y_i'}} \tag{18}$$

Equation (19) shows the relationship of $D(X,Y)$ and $D(X_i', Y_i')$. Also, by using equations (12) and (18), we can show the relationship between $D_{GA}(X,Y)$ and $D_A(X_i', Y_i')$ as in equation (20).

$$D(X,Y)^2 = \sum_{i=1}^{k} D(X_i', Y_i')^2 \tag{19}$$

$$D_{GA}(X,Y)^2 = \sum_{i=1}^{k} D_A(X_i', Y_i')^2 \tag{20}$$

From equations (19) and (20) with Theorem 1, we know that $D(X,Y) \geq D_{GA}(X,Y)$ always holds. Therefore, *the distance function using the dimension grouping lower-bounds the Euclidean distance.* □

## 4 Performance Evaluation

We used synthetic and real-life data sets for experiments. The synthetic data sets are composed of 20,000 to 100,000 data vectors of 25 to 200 dimensions. The real-life Corel image data set consists of 68,040 images [12], each of which is described with a 32-dimensional feature vector. We compared the performance of the proposed method with those of the previous ones using the PCA and the DCT by counting the number of candidates obtained after filtering. The hardware platform for the experiments was the PC equipped with 2.8G Pentium CPU and 512MB RAM. The software platform was the MS Windows 2000 and Visual C++ 6.0.

Fig. 6 shows the results with various kinds of reference vectors. "small(+)" denotes the result with the reference vector having very small values for all the attributes to be located close to an axis. "minus one" does that with the reference vector having -1 for all the attributes. "random1" and "random2"

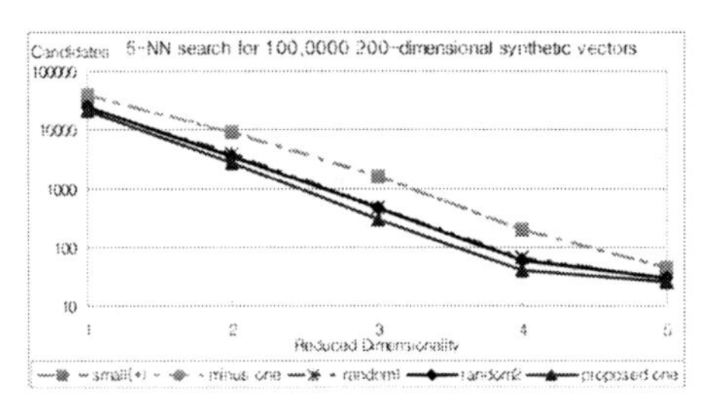


Fig. 6. Results with various reference vectors(log scale)

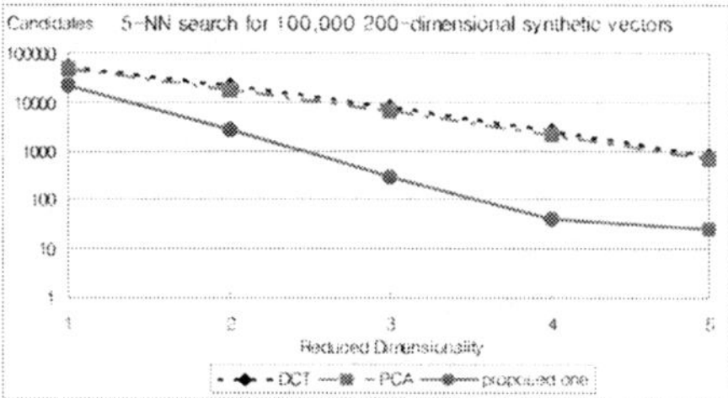


Fig. 7. Results with various reduced dimensionality(log scale)

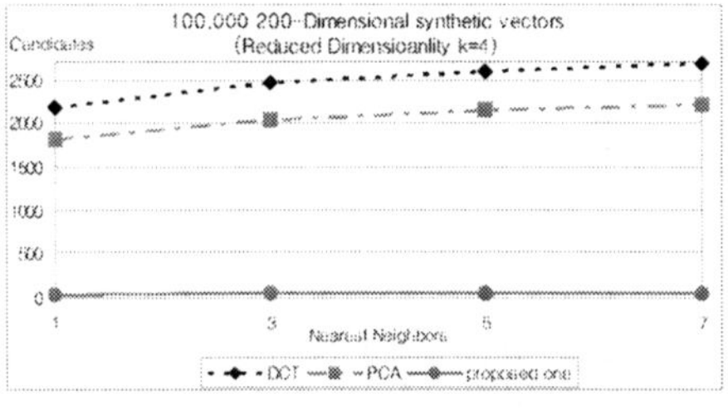


Fig. 8. Results with various tolerances

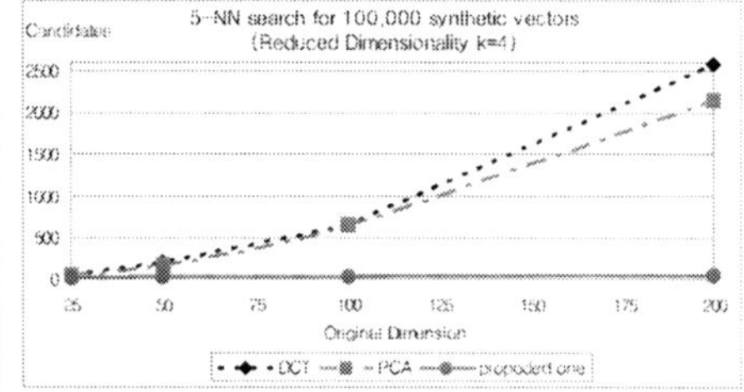


Fig. 9. Results with various dimensionality

denote the results with randomly selected reference vectors from data vectors. Finally, "proposed one" denotes the result with the reference vector selected by our selection method in Section 3.3. Our selection method outperforms the others as shown in the result because it considers both causes of errors mentioned in Section 3.1. In the following experiments, we used the reference vector selected by the this method.

Fig. 7 shows the results with varying the reduced dimensionality $k$. In all the methods, the number of candidates decreases dramatically as $k$ increases. The proposed method 2 to 60 times outperforms the method using the DCT, and 2 to 50 times outperforms the one using the PCA.

Fig. 8 shows the results with various tolerances. The number of candidates obtained in all three methods slightly increases as a tolerance increases. The proposed method produces candidates 60 and 50 times less than the methods using the DCT and the PCA, respectively.

Fig. 9 shows the results with various dimensionality. In cases of the previous methods using the PCA and the DCT, the number of candidates increases rapidly as the original dimensionality increases. However, it is kept nearly constant in the proposed method even with increasing dimensionality. In high-dimensionality such as 200, our method 60 times and 50 times outperforms the methods using the DCT and the PCA, respectively.

Fig. 10 shows the results with varying numbers of data vectors. The number of candidates in the previous methods using the PCA and the DCT increases rapidly as the number of data vectors increases. This is due to the statistical

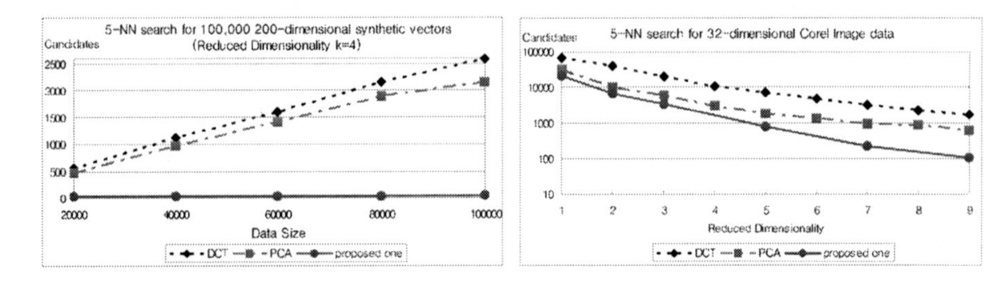


**Fig. 10.** Results with various numbers of data vectors

**Fig. 11.** Results with varying $k$ for the Corel image data set(log scale)

property that data vectors are spread over entire high-dimensional space [11]. Otherwise, the proposed method shows almost the same number of candidates in all cases. Our method 18 to 60 times outperforms the method using the DCT and 15 to 50 times outperforms the method using the PCA.

Fig. 11 shows the results using the real-life Corel image data set with varying reduced dimensionality. In case of $k = 1$, the method using the DCT retrieves every data vector as a candidate. The method using the PCA retrieves 48% of data vectors and our method does only 31% of data vectors as candidates. In other words, our method filters 69% of data vectors out, which should not be included in the final answer set by using only one reduced dimension. In case of $k = 9$, the method using the DCT selects 2.47% of data vectors and the method using the PCA does 0.89% of data vectors as candidates. Our method chooses only 0.15% of data vectors as candidates, and produces candidates about 6 times and 16 times less than the methods using the DCT and the PCA, respectively.

## 5 Conclusions

This paper has proposesd a novel method for dimensionality reduction based on a function approximating the Euclidean distance, which makes use of the norm and angle components of a vector. First, we have identified the causes of errors in angle estimation, and have discussed basic solutions to reduce those errors. Next, we have proposed a new method for dimensionality reduction that composes a set of subvectors from a feature vector and maintains only the norm and the estimated angle for every subvector. Also, we have defined a novel distance function that consistently lower-bounds the Euclidean distance. This implies that our method does not incur any false dismissals caused by reducing the dimensionality. Finally, we have performed a variety of experiments. The results show that our method improves the performance remarkably in comparison with previous ones.

## Acknowledgement

This research was supported by the MIC(Ministry of Information and Communication) of Korea under the ITRC(Information Technology Research Center) support program supervised by the IITA(Institute of Information Technology Assessment) (IITA-2005-C1090-0502-0009).

## References

1. Aggarwal, C.C.: On the Effects of Dimensionality Reduction on High Dimensional Similarity Search. In: Proc. of Int'l. Symp. on Principles of Database Systems, pp. 256–266 (2001)
2. Agrawal, R., Faloutsos, C., Swami, A.: Efficient Similarity Search in Sequence Database. In: Proc. of Int'l. Conf. on Foundations of Data Organization and Algorithms, pp. 69–84 (1993)
3. Beyer, K.S., Goldstein, J., Ramakrishnan, R., Shaft, U.: When Is Nearest Neighbor Meaningful? In: Proc. of Int'l. Conf. on Database Theory, pp. 217–235 (1999)
4. Bohm, C., Berchtold, S., Keim, D.A.: Searching in High-Dimensional Spaces-Index Structures for Improving the Performance of Multimedia Databases. ACM Computing Surveys 33(3), 322–373 (2001)
5. Egecioglu, O., Ferhatosmanoglu, H., Ogras, U.: Dimensionality Reduction and Similarity Computation by Inner Product Approximations. IEEE Trans. on Knowledge and Data Engineering, 714–726 (2004)
6. Jeong, S., Kim, S.-W., Kim, K., Choi, B.-U.: An Effective Method for Approximating the Euclidean Distance in High-Dimensional Space. In: Proc. of Int'l. Conf. on Databases and Expert Systems Applications, pp. 863–872 (2006)
7. Kanth, K.V.R., Agrawal, D., Singh, A.: Dimensionality Reduction for Similarity Searching in Dynamic Databases. In: Proc. of Int'l. Conf. on Management of Data. ACM SIGMOD, pp. 166–176. ACM Press, New York (1998)
8. Moon, T.K., Stirling, W.C.: Mathematical Methods and Algorithms for Signal Processing. Prentice-Hall, Englewood Cliffs (2000)
9. Pagel, B.-U., Six, H-W., Winter, M.: Window Query-Optimal Clustering of Spatial Objects. In: Proc. of Int'l. Conf. on Very Large Data Bases. VLDB., pp. 506–515 (1997)
10. Seidl, T., Kriegel, H.-P.: Optimal Multi-Step k-Nearest Neighbor Search. In: Proc. of Int'l. Conf. on Management of Data. ACM SIGMOD, pp. 154–165. ACM Press, New York (1998)
11. Weber, R., Schek, H.J., Blott, S.: A Quantitative Analysis and Performance Study for Similarity-Search Methods in High-Dimensional Spaces. In: Proc. of Int'l. Conf. on Very Large Data Bases. VLDB., pp. 194–205 (1998)
12. `http://kdd.ics.uci.edu/databases/CorelFeatures/CorelFeatures.html`

# Integrating a Stream Processing Engine and Databases for Persistent Streaming Data Management

Yousuke Watanabe[1], Shinichi Yamada[2], Hiroyuki Kitagawa[2,3], and Toshiyuki Amagasa[2,3]

[1] Japan Science and Technology Agency
[2] Graduate School of Systems and Information Engineering, University of Tsukuba
[3] Center of Computational Sciences, University of Tsukuba

**Abstract.** Because of increased stream data, managing stream data has become quite important. This paper describes our data stream management system, which employs an architecture combining a stream processing engine and DBMS. Based on the architecture, the system processes both continuous queries and traditional one-shot queries. Our proposed query language supports not only filtering, join, and projection over data streams, but also continuous persistence requirements for stream data. Users can also specify continuous queries that integrate streaming data and historical data stored in DBMS. Another contribution of this paper is feasibility validation of queries. Processing queries on streams with frequent inputs may cause the system to overflow its capacity. Specifically, the maximum writing rate to DBMS is a significant bottleneck when we try to store stream data into DBMS. Our system detects infeasible queries in advance.

## 1 Introduction

Advancements of device technologies and network infrastructures enable us to access data streams, which provide up-to-date information that changes over time. From a huge number of streams, we can get many kinds of information such as weather forecasts, news, stock prices, and sensor data from the real world. Requirements for stream data management are increasing and diverse. The requirements we consider in this paper are classified into four types.

1. *Continuous queries [3] on streaming data*: This type of requirement lets a system monitor streams over the long term. An example of the requirements is "When a new data unit satisfying the conditions arrives from the stream, deliver it to me."
2. *Persistence requirements for stream data*: Archiving stream data is important for applications that ought to keep logs. An example is "Continuously store stream data onto disk for later browsing",
3. *One-shot queries on historical data stored in disks*: This type of query corresponds to traditional SQL queries used to retrieve data from DBMS.

R. Wagner, N. Revell, and G. Pernul (Eds.): DEXA 2007, LNCS 4653, pp. 414–423, 2007.

4. *Integration of streaming data and historical data*: An example is failure detection using historical data. "Monitor similarity between two time-series data sequences: streaming data and historical data received in the past."

Stream processing engines [1,2,6,8] are well-designed systems to handle stream data. They mainly focus on memory processing, and execute continuous queries over stream data. Our objective, however, is to implement a data stream management system that achieves the four types of requirements listed above. We need a system that supports not only continuous query processing but also sophisticated disk management. From the standpoint of disk management, stream processing engines are not better than traditional DBMS. Thus, combining a stream processing engine and DBMS is a reasonable approach to implement a data stream management system.

This paper explains Harmonica, our data stream management system, which is based on an architecture combining a stream processing engine [12] and databases. It provides an SQL-like query language that supports continuous queries, one-shot queries, persistence requirements and integration requirements for streams and databases. Our proposed language is designed as an extension to the previous continuous query languages and SQL.

Another contribution of this paper is feasibility validation for queries including persistence requirements. Suppose we have a persistence requirement for a stream whose arrival rate is quite frequent, the output rate of the stream processing engine may exceed the maximum writing rate of the database. Since such overflows often bring data loss, it is a critical problem for some applications. Thus, we have to validate queries before starting query executions. We propose a rate-based query validation scheme to detect queries that cause the system to overflow capacity.

The remaining part of this paper consists of the following parts: Section 2 introduces an example scenario. Section 3 describes our system. Our query language is explained in Section 4. Section 5 proposes our feasibility validation scheme. Related work is summarized in Section 6. Finally, Section 7 presents conclusions and introduces future research issues.

## 2 Example Scenario

We assume an application system for an industrial plant managing gas turbines (Figure 1). Behavior of individual turbines is monitored by many types of sensors, temperature (ttxd11), revolutions per minutes (rpm), and amount of fuel (fsr). Each sensor periodically produces data each second. In this example, we need both real-time monitoring over stream data (e.g., online failure detection) and archiving stream data into persistent storage (e.g., trouble analysis).

## 3 System Architecture

Here is the architecture of Harmonica (Figure 2). The system consists of the following modules:

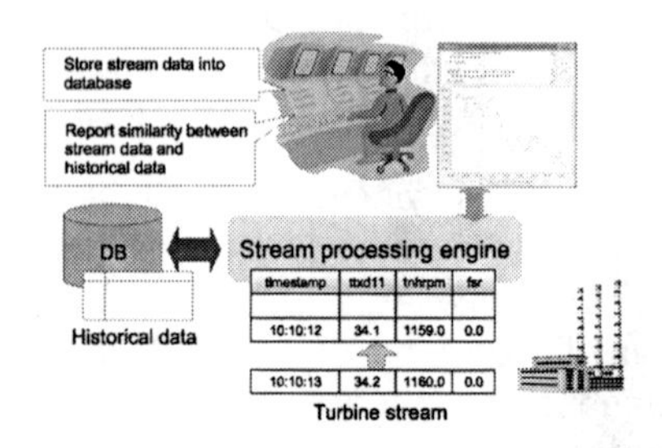


Fig. 1. Industrial plant monitoring

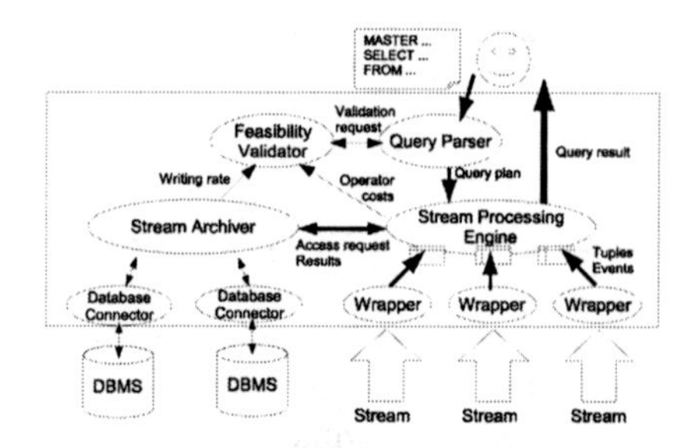


Fig. 2. System architecture

- **Query parser:** This module analyzes queries given by users, and constructs query plans. Our query language is explained in Section 4.
- **Feasibility validator:** This module decides whether or not a given query plan is feasible. We will present the validation scheme in Section 5. A query that has passed the validation process is sent to the stream processing engine. It collects the information needed to validate plans from the stream processing engine and stream archiver.
- **Stream processing engine:** StreamSpinner [12] is a stream processing engine that we implemented earlier. A characteristic of our engine is event-driven operator evaluation. When the engine receives an event-notification from a wrapper, it evaluates operators associated with the event. If a query plan contains operations to access DBMS, the stream processing engine makes requests of the stream archiver. The engine delivers the query result to the user.
- **Wrapper:** A wrapper takes charge of a stream. When it receives data from the stream, it transforms the data into a tuple. It notifies the stream processing engine of an event to trigger operator evaluation.
- **Stream archiver:** The stream archiver module manages databases. It maintains schema information on all databases, and transfers access requests to corresponding databases.
- **Database connector:** A database connector holds a database connection. The connectors to MySQL, PostgreSQL, and SQL Server are available.

We have implemented Harmonica in Java. Figure 3 is a screen-shot of our system.

# 4 Harmonica Query Language

## 4.1 Syntax

The following introduces a query language used in the system. We employ a relational model as the common data model in the system. Data streams are modeled as unbounded relations. Each delivery unit from a stream is regarded as a tuple included in the relation. A relation consists of multiple normal attributes and one special attribute to hold an arrival timestamp for each tuple.

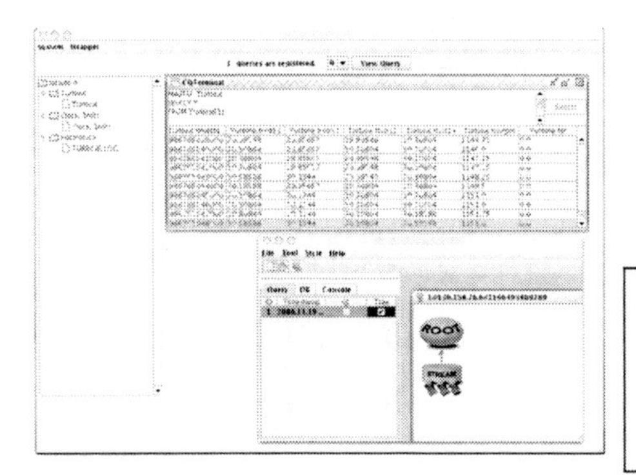

**Fig. 3.** A screen-shot of Harmonica

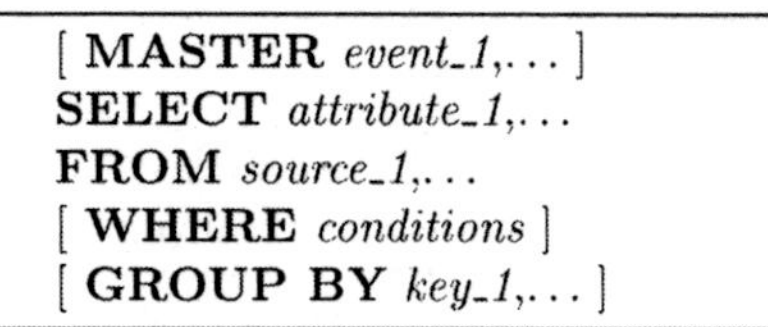

[ **MASTER** *event_1,...* ]
**SELECT** *attribute_1,...*
**FROM** *source_1,...*
[ **WHERE** *conditions* ]
[ **GROUP BY** *key_1,...* ]

**Fig. 4.** Syntax of query

The syntax of our query language is presented in Figure 4. A **MASTER** clause specifies events to trigger query processing. When new data arrives from the streams written in the MASTER clause, the query is evaluated by the stream processing engine. If a MASTER clause is omitted, the query is regarded as a traditional one-shot query. **SELECT–FROM–WHERE–GROUP BY** clauses are the same as with SQL except for the time-based window specification in the FROM clause. A window specification is used to specify sliding-window. It consists of an interval and an origin of the moving window. For example, "$Turbine[1min, now]$" in a FROM clause indicates tuples delivered from the Turbine stream within the most recent minute. A window slides as time progresses, and tuples within the range are evaluated at each event occurrence.

Although **INSERT INTO**, **CREATE TABLE** and **DROP TABLE** are omitted in Figure 4, users can give requirements using these clauses.

## 4.2 Examples

Figure 5 is continuous query filtering stream data. It means "Deliver data when its temperature exceeds 100 degrees." The MASTER clause indicates that the query is triggered by an arrival of Turbine data. Width of the temporal window is 1 millisecond, and the origin is the arrival time of the Turbine data. ttxd11 is an attribute including a value from a temperature sensor.

Figure 6 is a persistence requirement. It says "When new data arrives from the Turbine, store it into the Turbine_log table if the temperature value exceeds 100." Turbine_log is the name of a table in DBMS. rpm is an attribute including a value of revolutions per minute.

When a MASTER clause is not given in a query, the requirement is for a one-shot query. Figure 7 is an example of a one-shot query. "Retrieve an average of temperature data that arrived on June 16 from Turbine_log table." Turbine_log is a table stored in DBMS. "1day" gives the interval of the query's window, and "June 16, 2007 00:00 000" gives the origin of the window. "avg" is an aggregate function to calculate the average.

```
MASTER Turbine
SELECT ttxd11
FROM Turbine [1msec, now]
WHERE ttxd11 > 100
```

**Fig. 5.** Filtering query

```
MASTER Turbine
INSERT INTO Turbine_log
SELECT ttxd11, rpm
FROM Turbine [1msec, now]
WHERE ttxd11 > 100
```

**Fig. 6.** Persistence requirement

```
SELECT avg(ttxd11)
FROM Turbine_log [1day, "June 16, 2007 00:00 000"]
```

**Fig. 7.** One-shot query

Following is an example of a query integrating stream and DBMS. Figure 8 means "Report similarity between recent temperature values from Turbine stream and historical temperature values in turbine_log." Clock_1minute is a timer stream provided by the system. "array" is a function to convert a set of attribute values into an array. "sim" is a function to compute similarity between two arrays.

### 4.3 Query Plan

Query parser constructs a query plan from a given query. A query plan is a tree of operators. Figure 9 shows the query plan constructed from the query in Figure 8. When a data arrival from Clock_1minute is notified to the stream processing engine, the query plan is designated to be evaluated. The operators included in the plan are evaluated from bottom to top of the tree. The plan contains seven operators, including projection, cartesian-product, grouping and function evaluation.

## 5 Feasibility Validation

This section proposes a feasibility validation scheme for a query plan. Our scheme is an extension of Ayad's method [4]. However, their original method does not consider persistence requirements and integration queries for streams and databases.

Here, we assume that the average input rate of each stream is almost stable. Although bursty data arrivals may happen in data streams, our objective is to detect (in advance) queries that cause stationary overflow. We think that a system with sufficient queues can treat bursty arrivals.

### 5.1 Definition

Except for one-shot queries, a query is triggered by events specified by the MASTER clause in the query. To keep pace with data arrivals from streams, the system must finish evaluations of all operators corresponding to an event before

```
MASTER Clock_1minute
SELECT sim(S1.V, S2.V)
FROM
( SELECT array(ttxd11) AS V
  FROM Turbine [1min, now] ) AS S1,
( SELECT array(ttxd11) AS V
  FROM Turbine_log [1min, "June 16, 2007 00:00 000"] ) AS S2
```

**Fig. 8.** Integration of stream data and historical data

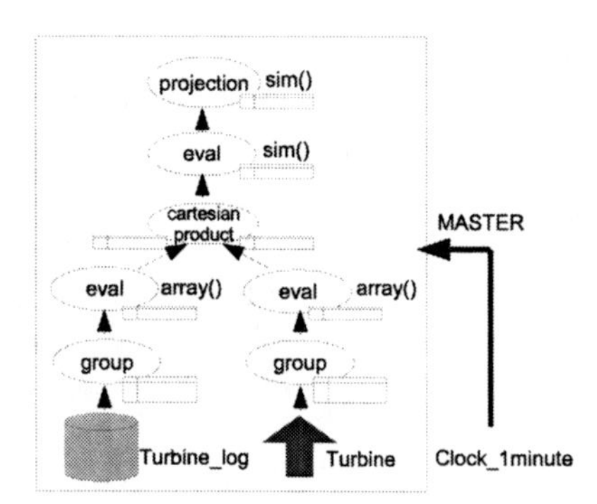


**Fig. 9.** An example of query plan

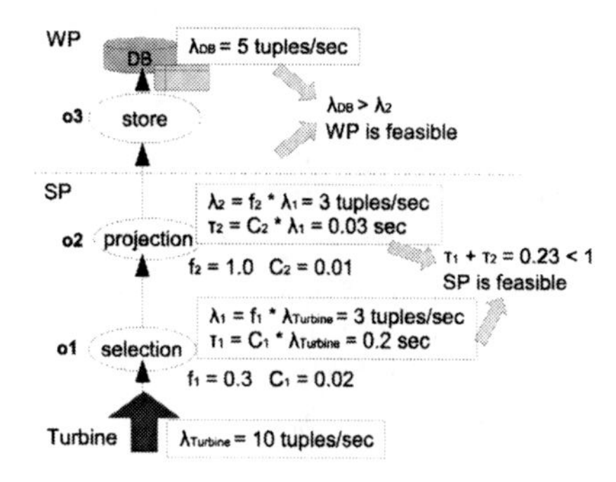


**Fig. 10.** Validation example

the next event occurs. If evaluation tasks are continuously stacked in the system, the system will eventually crash. We use "unit time" to specify the average time interval between one event and the next.

As described in Section 3, our system consists of a stream processing engine and DBMS. We can divide query plan $q$ into two parts: **stream processing part** $SP_q$, which is mainly processed by the stream processing engine, and **writing part** $WP_q$, which is processed mainly by DBMS. $SP_q$ is a subtree of the query plan constructed by removing the store operator from $q$. $WP_q$ is a store operator when $q$ is a persistence requirement. Otherwise, $WP_q$ is empty. Before defining feasibility of a whole query plan, we define feasibility of $SP_q$ and $WP_q$.

**Definition 1.** *Let $o_k$ $(1 \leq k \leq n)$ be an operator included in $SP_q$ and $\tau_k$ be the average time needed by operator $o_k$ to evaluate input tuples per unit time. A stream processing part $SP_q$ is feasible if the following condition is satisfied.*

$$\sum_{k=1}^{n} \tau_k < unit\_time \tag{1}$$

Formula 1 means that the sum of evaluation time of all operators should not exceed unit time.

**Definition 2.** *Let $\lambda_{SP}$ be an output rate produced by stream processing engine with a query plan $q$, and $\lambda_{DB}$ be the maximum writing rate to DBMS. A writing part $WP_q$ is feasible if the following condition is satisfied.*

$$\lambda_{SP} < \lambda_{DB} \tag{2}$$

Formula 2 means a number of output tuples produced by the stream processing engine should not exceed the number of maximum writing rate to DBMS.

**Definition 3.** *A query plan $q$ is feasible if and only if both the stream processing part $SP_q$ and the writing part $WP_q$ are feasible.*

### 5.2 Validation Algorithm

According to the above definitions, our algorithm consists of three steps.

1. A query plan $q$ is constructed from a query, and it is separated into the stream processing part $SP_q$ and the writing part $WP_q$ .
2. We obtain $\tau_k$, which is the average time needed by each operator $o_k$ in $SP_q$. How to estimate $\tau_k$ is explained in Section 5.3. After that, Formula 1 is validated. If $SP_q$ is feasible and $WP_q$ is not empty, we go to the next step.
3. Formula 2 is validated. An output rate of stream processing engine $\lambda_{SP}$ is obtained by the previous step. The system estimates $\lambda_{DB}$ using the method described in Section 5.4.

### 5.3 Estimating Operator Costs

Validator traverses a query plan from bottom to top. It estimates the cost of each operator based on Ayad's cost model [4]. For each operator $o_k$, we derive an output rate $\lambda_k$, a time to process tuples arriving in unit time $\tau_k$, and the number of output tuples included in the window $W_k$. $W_k$ is required to compute costs for window-join and cartesian-product. The stream processing engine provides information needed by this estimation: input rates of streams $\lambda_i$, a time to process one tuple in an operator $C_o$. We first explain how to estimate costs of unary operators such as selection, projection, grouping, and function evaluation. We then present an estimation for binary window-join and cartesian-product.

**Unary Operators.** Let $\lambda_i$ be an input rate to the operator $o_k$, $W_i$ be a number of input tuples included in the window and $f_k$ be $o_k$'s selectivity (=#outputs/#inputs). The output rate $\lambda_k$ is derived by the input rate: $\lambda_k = f_k \lambda_i$. And, $\tau_k$, the time to process tuples arriving in unit time, is also obtained from the input rate $\lambda_i$: $\tau_k = C_k \lambda_i$. Finally, we can obtain $W_k$, the number of output tuples included in the window, from the number of input tuples in the window: $W_k = f_k W_i$. In selecting and grouping operators, getting accurate selectivity values is sometimes difficult. In such cases, we use $f_k = 1$ pessimistically. For projection and function evaluation operators, we always use $f_k = 1$.

**Binary Operators.** Suppose a binary-join operator is connected to two input streams L and R. Let $\lambda_L$ and $\lambda_R$ be input rates of L and R respectively, and $W_L$, $W_R$ be numbers of tuples included in the windows of L and R respectively. A new tuple from L is compared to tuples held in the R's window, thus the number of output tuples produced by input tuples from L equals $f_k \lambda_L W_R$. In the same way, the number of output tuples produced by input tuples from R is $f_k \lambda_R W_L$. Therefore, the output rate of operator $o_k$ is obtained by the following formula.

$$\lambda_k = f_k(\lambda_L W_R + \lambda_R W_L)$$

The time to process tuples arriving in unit time is calculated as follows.

$$\tau_k = \tau_L + \tau_R = C_k(\lambda_L + \lambda_R)$$

Finally, the number of output tuples included in the window becomes as follows.

$$W_k = f_k W_L W_R$$

In the estimation for cartesian-product operator, we use $f_k = 1$.

### 5.4 Estimating the Maximum Writing Rate to DBMS

To validate Formula 2, we need $\lambda_{SP}$, which is the output rate produced by stream processing, and $\lambda_{DB}$, which is the maximum writing rate to DBMS. $\lambda_{SP}$ is obtained by the estimation in Section 5.3, because $\lambda_{SP}$ equals the output rate of the top operator in $SP_q$.

The maximum writing rate to DBMS depends on environment, such as machine power and implementations of DBMS. In addition, the rate may change according to the data size of one tuple. Generally, the writing rate for a large tuple is lower than that for a small tuple. However, measuring actual writing rates for all sizes of data is impractical. Our approach is measuring rates only for different $N$ sample sizes. With N samples, we can estimate a writing rate for other data sizes by applying linear approximation. $\lambda_{DB}$ for query plan $q$ is estimated by the following formula.

$$\lambda_{DB} \simeq rate\_estimate(tuple\_size(S_q)) \tag{3}$$

$S_q$ is a schema corresponding to the output produced by query plan $q$. *tuple_size* is a function to compute average data size based on the schema, and *rate_ estimate* is a function to estimate writing rate by applying linear approximation.

### 5.5 Validation Example

The following illustrates the validation process for the query in Figure 6. At the first step, the system constructs a query plan (Figure 10). The plan consists of selection operator $o1$, projection operator $o2$ and store operators $o3$.

Next, the system validates whether or not $SP$ is feasible. In this example, we suppose $\lambda_{Turbine}$ and $\lambda_{DB}$ are 10 and 5 (tuples/second) respectively. We also suppose $f_1$ and $f_2$ equal 0.3 and 1.0, $C_1$ and $C_2$ equal 0.02 and 0.01 respectively. Based on the estimation method in Section 5.3, we can obtain $\lambda_1 = f_1 \lambda_{Turbine} = 0.3*10 = 3$ (tuples/second). And, $\tau_1 = C_1 \lambda_{Turbine} = 0.02*10 = 0.2$ (second). For

Fig. 11. Feasibility validator

| input rate | decision | result |
|---|---|---|
| : | : | : |
| 33 tuples/s | feasible | succeed (hit) |
| 34 tuples/s | feasible | overflow (miss) |
| 35 tuples/s | not feasible | overflow (hit) |
| : | : | : |

Fig. 12. Experiment results for the query in Fig. 6

| qid | operators | estimated border | result |
|---|---|---|---|
| 1 | store | 34 tuples/s | 33 tuples/s |
| 2 | selection, store | 68 tuples/s | 66 tuples/s |
| 3 | projection, store | 35 tuples/s | 34 tuples/s |
| 4 | selection, projection, store | 70 tuples/s | 68 tuples/s |
| 5 | cartesian product, projection, store | (left) 3 tuples/s (right) 5 tuples/s | 3 tuples/s 5 tuples/s |

Fig. 13. Summary of experiments

the projection operator $o2$, $\lambda_2 = f_2 * \lambda_1 = 1.0 * 3 = 3$ tuples/second. $\tau_2 = C_2\lambda_1 = 0.01 * 3 = 0.03$ second. Since Formula 1 becomes $\tau_1 + \tau_2 = 0.2 + 0.03 = 0.23 < 1$, $SP$ is feasible.

Finally, feasibility of $WP$ is checked. Because $o2$ is located at the top of $SP$, $\lambda_{SP}$ equals $\lambda_2$. $\lambda_{SP}$ is smaller than $\lambda_{DB}$, therefore $WP$ is feasible. We can get the result that the query plan in Figure 10 is feasible. Figure 11 is a screen-shot of feasibility validator in our system.

### 5.6 Experiment

We investigated accuracy of validation results. Our environment consists of a Pentium D 3GHz, 2GB memory, Windows Vista Business, MySQL 5.0 and JDK 6. Since we chose the default parameters for MySQL, it was not tuned.

In this experiment, our system first validates the query in Figure 6 with several input rates. We then tried to execute the query in the system. The result is presented in Figure 12. Although there is a miss near the border, overall accuracy seems good. If we want to improve accuracy, we have to estimate $\lambda_{SP}$ and $\lambda_{DB}$ more precisely. We show the summary of our experiments validating 5 queries (Figure 13). Figure 13 indicates that our method can work for several types of queries.

## 6 Related Work

There is much research on stream processing engines. Aurora [1], Borealis [2], TelegraphCQ [6], NiagaraCQ [7], STREAM [8], CAPE[11] and so on. The research focuses mainly on continuous query processing in main memory. Harmonica combines the stream processing engine and DBMS, because we need the system to support not only in-memory continuous query processing but also sophisticated disk management. CQL [3] is a continuous query language for stream processing engines. The window specification in CQL does not contain a window origin. And, CQL cannot specify any event-specification. Our query language supports both facilities. In addition, we can explicitly write persistence requirements to DBMS. Load shedding [5,9] is a technique to reduce inputs when the load become quite high. These methods do not consider persistence requirements.

## 7 Conclusion

This paper described our data stream management system which integrates a stream processing engine and DBMSs. We also presented a feasibility validation scheme for persistence requirements. There are some future research issues. The first is recommendation of feasible candidate queries when an original query is detected to be infeasible. In our current scheme, once an infeasible query is found, we have to manually rewrite the requirement to pass the validation process. The second is treating streams whose input rates may dynamically change over time.

## Acknowledgement

This research was supported in part by CREST, Japan Science and Technology Agency, and Grant-in-Aid for Scientific Research (A) from Ministry of Education, Culture, Sports, Science and Technology.

## References

1. Abadi, D.J., et al.: Aurora: a New Model and Architecture for Data Stream Management. VLDB Journal 12(2), 120–139 (2003)
2. Abadi, D.J., et al.: The Design of the Borealis Stream Processing Engine. In: Proc. CIDR, pp. 277–289 (2005)
3. Arasu, A., et al.: The CQL Continuous Query Language: Semantic Foundations and Query Execution. VLDB Journal 15(2) (2006)
4. Ayad, A.M., et al.: Static Optimization of Conjunctive Queries with Sliding Windows Over Infinite Streams. In: Proc. ACM SIGMOD, pp. 419–430 (2004)
5. Babcock, B., et al.: Load Shedding for Aggregation Queries over Data Streams. In: Proc. ICDE, pp. 350–361 (2004)
6. Chandrasekaran, S., et al.: TelegraphCQ: Continuous Dataflow Processing for an Uncertain World. In: Proc. CIDR (2003)
7. Chen, J., et al.: NiagaraCQ: A Scalable Continuous Query System for Internet Databases. In: Proc. ACM SIGMOD, pp. 379–390 (2000)
8. Motwani, R., et al.: Query Processing, Resource Management, and Approximation in a Data Stream Management System. In: Proc. CIDR (2003)
9. Tatbul, N., et al.: Load Shedding in a Data Stream Manager. In: Proc. VLDB, pp. 309–320 (2003)
10. Viglas, S.D., et al.: Rate-based Query Optimization for Streaming Information Sources. In: Proc. ACM SIGMOD, pp.37–48 (2002)
11. Wang, S., et al.: State-Slice: New Paradigm of Multi-query Optimization of Window-based Stream Queries. In: Proc. VLDB, pp. 619–630 (2006)
12. StreamSpinner. `http://www.streamspinner.org/`

# Data Management for Mobile Ajax Web 2.0 Applications

Stefan Böttcher and Rita Steinmetz

University of Paderborn (Germany)
Computer Science
Fürstenallee 11
D-33102 Paderborn
stb@uni-paderborn.de , rst@uni-paderborn.de

**Abstract.** Whenever Ajax applications on mobile devices have to retrieve large XML data fragments from a remote server, a reduction of the exchanged data volume may be crucial to manage limited bandwidth and limited energy of the mobile device. We propose to use an XML compression technique that compresses an XML document to a binary directed acyclic graph (DAG) and to use DAG-based DOM evaluation on the client side. Our experiments show that the data transfer for applications like amazon or eBay can be reduced to 70% of the original data transfer needed.

## 1 Introduction

Whenever web applications shall involve small mobile devices, limited resources like energy and bandwidth require minimizing data exchange. While Ajax [7] has the advantage that new web pages can be shown on the mobile client device without a full exchange of the data that is contained in the web page, Ajax still uses a DOM model, which, in the case of XML data exchange, requires exchanging at least fragments of XML. As XML tends to be verbose, a promising optimization is to use compressed XML instead of exchanging and storing XML as a DOM tree.

The contributions of this paper are the following.

- We summarize the requirements to an Ajax engine working on compressed XML (in Section 2).
- We show how XML data can be converted into a binary directed acyclic graph (binary DAG) that can be transported from client to server in order to reduce the amount of data exchanged (in sections 3.1 and 3.2)
- We outline how to implement the navigation operations of the DOM model on top of the binary DAG, such that the client can navigate on the cached binary DAG which contains the compressed XML data (in Section 3.3).
- We describe how the binary DAG representing the DOM model of the client can be updated to modify the data stored in the client side Ajax engine (in Section 3.4)
- Finally, we present an experimental evaluation that shows that, in comparison to the standard XML exchange and the standard DOM parser in the Ajax engine, our approach significantly reduces both, the amount of data exchange and the amount of main memory storage needed on the small mobile client devices (in Section 4).

R. Wagner, N. Revell, and G. Pernul (Eds.): DEXA 2007, LNCS 4653, pp. 424–433, 2007.

### 1.1 Related Works

Our work applies XML compression techniques including navigation and manipulation of compressed XML data to Ajax-based Web 2.0 programming.

Ajax [7] is a programming technique for interactive web applications that combines an XML data representation in a DOM tree on the client with XMLHttpRequest as data exchange protocol and JavaScript as the client side programming language. The potential of Ajax for building rich web applications and the current state of the art concerning Ajax are summarized in [13].

There exist several algorithms for XML compression, some of which are DAG-based and others or not DAG-based. However, not all of them can be used to decrease communication costs and main memory consumption for Ajax applications, as they do neither support efficient search operations nor updates without prior decompression, and therefore do not allow to build a DOM model on top of them. An example for such a compression approach generating a non-searchable compressed data structure is XMILL[9].

Among the not DAG-based compression techniques, there exist several approaches to the compression of XML data that are efficiently searchable or queryable. The approaches XGrind [15], XPRESS [10] and XQueC [2] compress the tag information using dictionaries and Huffman-encoding and replaces the end tags by either a '/'-symbol or by parentheses.

XQC [11], DTD subtraction [3] and the approach presented in [14] omit all information from the XML document that is redundant, as this information can be inferred from the given DTD.

We however follow the DAG-based approaches to XML compression, like XQZip [7] which uses a DAG for compressing the structure of an XML document, and LZCS [1] and [4] which use a DAG for compressing the whole XML document, as DAGs offer the following advantages. A DAG does not only allow efficiently compressing XML documents. According to [5], the DAG also allows a more efficient search and query evaluation as could be provided on the original XML data, even for large repositories of data.

Our approach of write operations on the DAG-based DOM implementation follows the technique of DAG updates described e.g. in [5]

In contrast to XQZip [7], LZCS [1], and [5], which regard the unranked XML tree and the unranked DAG, we use a binary DAG to implement a DOM model, which allows us to simplify both, path search and modifications on the DOM model.

In comparison to all other approaches, our approach uses a binary DAG-based implementation of a DOM tree to optimize data transfer of Ajax-based Web 2.0 applications, which is an application field of increasing interest.

## 2 Requirements to an Ajax Engine Working on Compressed XML

In order to be able to present web pages to the client, Ajax needs all the operations that a DOM interface offers. Therefore, as with uncompressed XML, our client working on compressed XML shall offer all the operations that are offered by a DOM client, i.e., traversing all the axes and inserting, deleting or modifying DOM nodes or sub-trees.

As we want to exchange compressed XML and want to traverse compressed XML on the client side, this includes navigation and partial modification on compressed structures. As we use a binary DAG as the compressed XML data structure, which shares common XML structures instead of storing them multiple times, we have to support navigation and all modification operations on this binary DAG. Navigation on the shared data structures of the DAG has to know about the context in which a shared part of a DAG is used whenever a navigation operation has to leave this shared part.

Modification of an XML structure that previously has been compressed to a DAG sharing common sub-trees has to consider that modifying a common sub-tree S would change all XML fragments represented by S. Therefore, modification usually requires a re-organization of the shared structures of the DAG, and the goal is to minimize the changes of the DAG.

Finally, the client should allow for updates at arbitrarily selected points of the DOM representation of the XML data which has to be supported by the DAG implementing the DOM representation. This includes direct access to any node of the DOM representation of the XML data or the DAG representation implementing this DOM tree.

The server has to compress the XML tree structure that shall be transferred to the client in such a lossless way, that arbitrary DOM operations can still be executed on the compressed data structure. JavaScript commands that are embedded in the XML document shall be still executable when the compressed binary DAG for of the XML document is transferred to the client and is used to execute operations on the DOM model of the client.

## 3 The Solution: A DAG-Based DOM Model

Our solution consists of some preparation steps on the server followed by evaluation steps on the client. We first transform the SAX event stream of the given XML file into a stream of binary SAX-events reporting on first-child, next-sibling, and parent axes steps. As a second step on the server, we use a hash table in order to store and to reuse binary DAG sub-trees instead of re-sending binary DAG sub-trees. Then a stream of binary DAG events is sent from the server to the client. After this preparation steps, all the DOM operations can be implemented on the binary DAG on the client side.

### 3.1 Server-Side Binarization of the XML Document

In general, XML documents can be regarded as unranked trees, i.e., a parent node may have arbitrary many child nodes. For the simplification of the following steps, we have decided to transform the given XML document into a binary tree using the first-child and next-sibling axes as the 'left' and the 'right' pointers in the binary tree.

In order to transform a given SAX event stream into a stream of binary SAX events, we regard pairs of SAX events. The generation of binary SAX events from given SAX events is summarized in Figure 1.

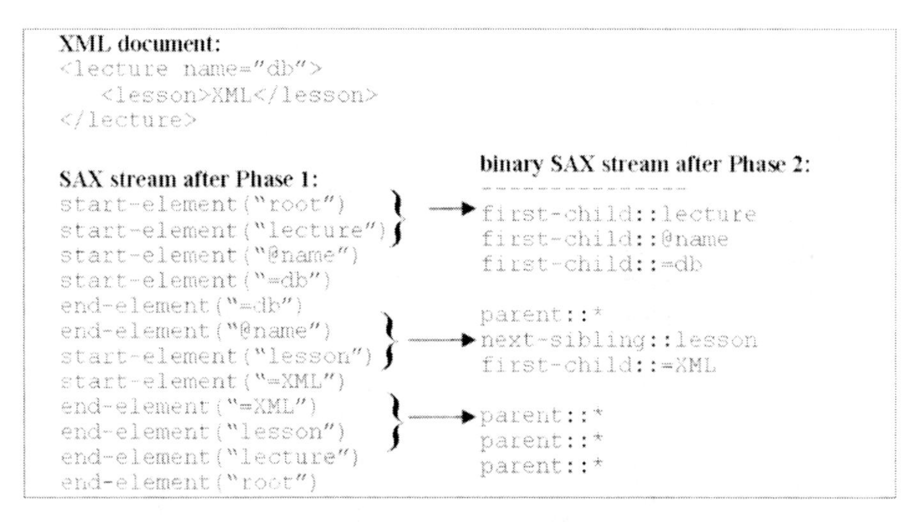


**Fig. 1.** Example XML document with the resulting SAX and binary SAX streams

Each start-element event followed by another start-element event in the SAX input stream represents a first-child-axis location step in the binary SAX event stream.

Each end-element event followed by a start-element event in the SAX input stream represents a next-sibling-axis location step in the binary SAX event stream.

Each end-element event followed by another end-element event in the SAX input stream represents a parent-axis location step in the binary SAX event stream.

The transformation of the SAX event stream into the binary SAX event stream is done in two phases.

**Phase 1:** In the first phase, we reduce the events of the SAX stream to two kinds of events: start-element(...) and end-element(...). For this purpose, we transform the SAX event character(T) generated for a text value T found in the XML document into a SAX event sequence

start-element(=T), end-element(=T).

Similar, we transform each attribute/value pair A=V found in the XML document into a SAX event sequence

start-element(@A), start-element(=V), end-element(=V), end-element(@A) .

The symbols '@' and '=' uniquely identify attributes and text nodes respectively, therefore, they are not allowed as an initial character for element-names.

The SAX event start-document is transformed into a SAX event start-element("root"), and the SAX event end-document is transformed into a SAX event end-element("root"), assuming that "root" does not exist as an element-name within the XML-document. At the end of Phase 1, the transformed SAX event stream contains only two kinds of events: start-element(...) and end-element(...).

**Phase 2:** In the second phase, we analyze pairs of start- and end-element events to transform them into the binary SAX events first-child::a, next-sibling::a and parent::*:

1. first-child: Whenever a SAX event start-element(x) is followed by a second SAX event start-element(a), 'a' is the first child of 'x'. Therefore, our binary SAX encoder generates the binary SAX event first-child::a.

2. next-sibling: Whenever a SAX event end-element(x) is followed by a SAX event start-element(a), 'a' is the next sibling of 'x'. Therefore, our binary SAX encoder generates the binary SAX event next-sibling::a.

3. parent: Furthermore, whenever a SAX event end-element(x) is followed by a second SAX event end-element(y), 'y' is the parent of 'x'. Therefore, our binary SAX encoder generates the event parent::*.

4. Whenever a start-element(x) is followed by an end-element(x), no binary SAX event is created.

Altogether, Phase 1 and Phase 2 together transform a SAX stream into a binary SAX stream of first-child::a, next-sibling::a, and parent::* events.

### 3.2 Transferring a Binary DAG

As a binary DAG is in general much more compact than a binary XML tree, we have decided to transfer a sequence of binary DAG nodes instead of a sequence of binary XML tree nodes.

This however requires transforming the binary SAX event input stream of a given XML document into a stream of binary DAG events. Each binary DAG event represents that the DAG parser has received a new input node of the binary DAG together with its pointers to the first child and the next-sibling DAG nodes. Note that, as the DAG is calculated bottom-up, we require the first-child and the next-sibling DAG nodes of an actual DAG node N to be sent before N is sent.

The construction of the binary DAG event stream from the binary SAX event stream is implemented by using a hash table for the DAG. Each new DAG node is stored in the hash table, whereas the hash table is also used to check whether or not a DAG node is new. Whenever the DAG contains already an identical node, i.e., a node with the same node name, the same first-child and the same next-sibling, this node is not stored a second time in the DAG. Instead a pointer to the already stored DAG node is used to reference the already stored node or tree or DAG. Note that the binary DAG event is a generalization of a binary SAX event, i.e., it represents that a DAG node has been read. The DAG now can be transferred from the server to the client, which is in our case done by transferring a sequence of binary DAG events.

### 3.3 DOM Read Operations Implemented on a Binary DAG

Each implementation of the DOM model requires supporting the concept of a current context node and supporting navigation starting from the current context node.

Each current context node in the DOM can be represented or identified by a unique sequence of edges starting from the root in the binary DAG, called the *DAG edge sequence representing the current context node*. Each edge of the DAG edge sequence representing the current context node is either a 'left' edge connecting a DAG node to its first-child or a 'right' edge connecting a DAG node to its next-sibling node.

When we start at the root of the binary DAG and follow the DAG edge sequence representing the current context node, we traverse a sequence of DAG nodes, which is called the *DAG path of the current context node*. The final DAG node of the DAG path of the current context node is called the *DAG node of the current context node*. The DAG node of the current context node always has the same node name as the current context node of the DOM.

In our implementation, we use a stack to store the DAG edge sequence representing the current context node together with pointers to the referenced nodes of the DAG path of the current context node, i.e., the stack contains pairs of entries

`(DAG node pointer, edge),`

where the `DAG node pointer` references a node on the DAG path of the current context node and `egde` is either 'left' or 'right'. We use this stack for all navigation operations and for insert, delete, and modification operations on the DAG as follows. Each navigation operation in the DOM is implemented by one or more operations modifying the stack, where some operations, e.g., computing a list of child nodes is composed of more basic operations, e.g., the first-child and the next-sibling axes.

**first-child, next-sibling**

Whenever the application requires moving from any current context node `C` of the DOM tree along the first-child-axis or the next-sibling-axis to a node `C2` of the DOM tree, we implement this in the DAG as follows. We simply push a pair (Dnp,edge) on the stack, where edge is 'left' if we have to use the first-child axis and edge is 'right' if we have to use the next-sibling axis in the DOM model, and Dnp is a pointer to the DAG node of `C2`. `C2` becomes the new current context node of the DOM. Note that the pointer Dnp found in the top element of the stack points to the DAG node of the new current context node (`C2`), i.e., the DAG edge sequence is correctly implemented by our stack.

**following-sibling**

Whenever the application requires returning a list of following-sibling nodes of the current context node in the DOM, we perform as many next-sibling steps as possible, and we return the collected pointers to the resulting following-sibling nodes.

**child**

Whenever the application requires returning a list of child nodes of the current context node in the DOM, we perform one first-child step and zero or one following-sibling step, and we return the collected pointers to the found child nodes.

**previous-sibling, parent-of-first-child**

When the DOM model requires moving along the previous-sibling axis or the parent-of-first-child axis to a DOM node `D`, we simply pop the top-most element from the stack. As a result, the new stack contains the DAG edge path representing the `D`, and the top-most stack element points to the DAG node of `D`.

**preceding-sibling**

Whenever the application requires returning a list of preceding-sibling nodes of the current context node in the DOM, we perform as many previous-sibling steps as possible, and we return the collected pointers to the resulting preceding-sibling nodes.

**parent**
Whenever the application requires returning the parent node of the current context node in the DOM, we perform as many previous-sibling steps as possible followed by one parent-of-first-child step, and we return the pointer to the found parent node.

**descendant-or-self**
Whenever the application requires returning a list of descendant-or-self nodes of the current context node in the DOM, we perform zero or more child steps, and we return the collected pointers to the resulting descendant-or-self nodes.

**ancestor-or-self**
Whenever the application requires returning a list of ancestor-or-self nodes of the current context node in the DOM, we perform zero or more parent steps, and we return the collected pointers to the resulting ancestor-or-self nodes.

### 3.4 DOM Write Operations Implemented on a Binary DAG

Write operations on a DOM tree implemented by our binary DAG have to consider that shared DAG structures have to be updated only once, i.e., only for one of the sub-trees represented by a DAG entry.

In order to explain the implementation of write operations, we need the concept of the top-most join node on the root path of the DAG. The *root path* is the path from the current context node to the root of the DAG. A *join node* J is a DAG node that has at least two different nodes pointing to J. The *top-most join node on the root path of the DAG* is that join node that is closest to the root on the root path of the DAG.

Whenever a current context node Nc has to be modified by either changing the label, or changing (i.e., inserting, deleting, or modifying) the 'left' or the 'right' sub-tree of Nc, we have to distinguish two cases for the implementation of this DOM operation on the binary DAG:

Case 1: There is no Join node on the root path of Nc. Then, we simply modify Pc.

Case 2: There is at least one join node on the root path of Nc. Then, we copy the DAG nodes <N0,N1,…,Nc> on the root path of the DAG to Nc, where N0 is the top-most join node on this root path into a path <N0',N1',…,Nc'>. All nodes Ni' get the same labels as the corresponding nodes Ni. All first-child or next-sibling pointers of the nodes Ni that do not point to the next node Ni+1 are copied to the corresponding nodes Ni'. Note that each of these copied pointers is either null or points to a node that was already previously in the DAG and now is a join node. Therefore, modifications on the DOM tree need only one sub-path of the root path of the DAG to be copied which is usually only a small extension of a given DAG. Thereafter, the current root path of the DAG has to be adjusted in such a way that the predecessor of N0 now points to N0' instead of N0. Finally, the modification can be done on the node Nc', which does not affect Nc, i.e., it does not affect the other parts of the DAG.

## 4 Evaluation and Results

We have implemented an XML compression system which compresses HTML and Ajax to binary DAGs representing the data structure and embedding the JavaScript commands. Our implementation has been tested on a Pentium 4 with 2.4 GHz Windows XP system with 1 GB of RAM running Java 1.5.

We have tested our technique with the following web application data sets:

- ebay: A set of article listings (printers) of the online auction website eBay.
- amazon: A set of book listings and book article pages of the bookstore amazon
- Zuggest: A web service that displays a table containing the first 10 amazon results as soon as a search word fragment is entered.
- AutoComp: An Ajax based search service including auto-completion for the online encyclopedia Wikipedia offering the 6 best choices in a combo-box as soon as a search word fragment is entered.

We have performed two test series. In the first test series, we have measured the communication costs, i.e., the size of the HTML file to be sent from the server to the client and compared this with the size of the DAG. In a second series, we have measured the main memory consumption that is needed to load the received data into the client's main memory, i.e., to load the DOM tree or the DAG.

Figure 2 summarizes the improvements that we have achieved by using a DAG instead of exchanging and using plain HTML, i.e., it compares the communication costs of the DAG with the communication costs of plain HTML and shows the relationship |DAG|/|HTML|. Furthermore, Figure 2 summarizes the relative improvements in memory consumption when using a DAG based DOM implementation in comparison to a pure DOM implementation representing the HTML files as XML tree, i.e.,

|MemoryUsageOf(DAG_based_DOM)|/|MemeoryUsageOf(HTML_DOM)| .

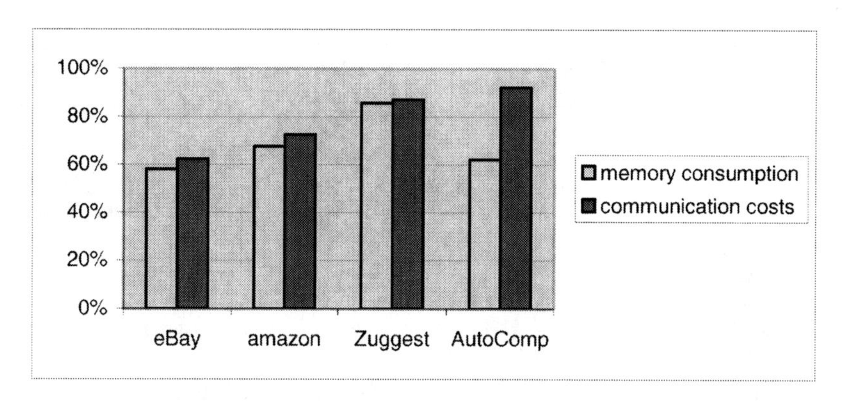


**Fig. 2.** Memory Consumption ( MemoryUsageOf ( DAG_based_DOM ) / MemeoryUsageOf ( HTML_DOM ) ) and Communication Costs ( sizeOf ( DAG ) / sizeOf ( HTML ) ) of DAG compared to plain HTML

As can be seen from Figure 2, the DAG representation of the application requires only between 62% and 92% of the size required by the plain HTML or XML representation of the data. Furthermore, the DAG-based DOM implementation requires only between 58% and 86% of the size required by the HTML or XML based DOM implementation. Therefore, we conclude that the DAG is much more suitable for energy and bandwidth saving data exchange, and the DAG-based DOM implementation is much more suitable for small mobile devices than the standard DOM representation of the XML file.

## 5 Summary and Conclusions

Whenever Web 2.0 applications based on Ajax technology have to exchange large XML data fragments with small mobile devices, the amount of data transfer can be reduced by using data compression. Our approach transforms XML data on the server side into a binary DAG of XML data that includes JavaScript commands, transfers this binary DAG to the client, and implements a DOM interface on top of the compressed binary DAG representation of the compressed data. We do not only support all navigation operations on the DOM model, but also support all insert, update, and delete modifications on the compressed binary DAG representation. Finally, our experimental evaluation has shown that, in comparison to the standard XML exchange and the standard DOM parser in the Ajax engine, our approach significantly reduces both, the amount of data exchange and the amount of main memory storage needed on the small mobile client device. Therefore, we consider it to be an interesting contribution to optimize the data exchange of Ajax-based Web 2.0 applications that involve small mobile devices.

## References

[1] Adiego, J., Navarro, G., de la Fuente, P.: Lempel-Ziv compression of structured text. In: Proceedings of the 2004 IEEE Data Compression Conference (DCC 2004), pp. 112–121 (2004)

[2] Arion, A., Bonifati, A., Costa, G., D'Aguanno, S., Manolescu, I., Pugliese, A.: XQueC: Pushing queries to compressed XML data. In: Proc. VLDB, pp. 1065–1068 (2003)

[3] Böttcher, S., Klein, N., Steinmetz, R.: XML Index Compression by DTD Subtraction. In: 9th International Conference on Enterprise Information Systems. ICEIS (to appear)

[4] Buneman, P., Choi, B., Fan, W., Hutchison, R., Mann, R., Viglas, S.: Vectorizing and Querying Large XML Repositories. In: ICDE 2005, pp. 261–272 (2005)

[5] Buneman, P., Grohe, M., Koch, C.: Path Queries on Compressed XML. In: VLDB 2003, pp. 141–152 (2003)

[6] Busatto, G., Lohrey, M., Maneth, S.: Efficient Memory Representation of XML Dokuments. In: Bierman, G., Koch, C. (eds.) DBPL 2005. LNCS, vol. 3774, pp. 199–216. Springer, Heidelberg (2005)

[7] Cheng, J., Ng, W.: XQzip: Querying Compressed XML Using Structural Indexing. In: Bertino, E., Christodoulakis, S., Plexousakis, D., Christophides, V., Koubarakis, M., Böhm, K., Ferrari, E. (eds.) EDBT 2004. LNCS, vol. 2992, pp. 219–236. Springer, Heidelberg (2004)

[8] Garrett, J.: Ajax: A New Approach to Web Applications. Adaptive path (2005), http://www.adaptivepath.com/publications/essays/archives/000385.php

[9] Liefke, H., Suciu, D.: XMill: An Efficient Compressor for XML Data. In: Proc. of ACM SIGMOD (May 2000)

[10] Min, J.K., Park, M.J., Chung, C.W.: XPRESS: A Queriable Compression for XML Data. In: Proceedings of SIGMOD (2003)
[11] Ng, W., Lam, W.-Y., Wood, P.T., Levene, M.: XCQ: A Queriable XML Compression System. In: Knowledge and Information Systems, Springer, Heidelberg (to appear, 2006)
[12] Olteanu, D., Meuss, H., Furche, T., Bry, F.: XPath: Looking Forward. In: Chaudhri, A.B., Unland, R., Djeraba, C., Lindner, W. (eds.) EDBT 2002. LNCS, vol. 2490, pp. 109–127. Springer, Heidelberg (2002)
[13] Paulson, L.D.: Building Rich Web Applications with Ajax. IEEE Computer 38 (2005)
[14] Sundaresan, N., Moussa, R.: Algorithms and programming models for efficient representation of XML for Internet applications. WWW 2001 (2001)
[15] Tolani, P.M., Hartisa, J.R.: XGRIND: A query-friendly XML compressor. In: Proc. ICDE 2002, pp. 225–234. IEEE Computer Society Press, Los Alamitos (2002)

# Data Management in RFID Applications

Dan Lin[1], Hicham G. Elmongui[1,*], Elisa Bertino[1], and Beng Chin Ooi[2]

[1] Department of Computer Science, Purdue University, USA
{lindan, elmongui, bertino}@cs.purdue.edu
[2] Department of Computer Science, National University of Singapore, Singapore
ooibc@comp.nus.edu.sg

**Abstract.** Nowadays, RFID applications have attracted a great deal of interest due to their increasing adoptions in supply chain management, logistics and security. They have posed many new challenges to existing underlying database technologies, such as the requirements of supporting big volume data, preserving data transition path and handling new types of queries. In this paper, we propose an efficient method to manage RFID data. We explore and take advantage of the containment relationships in the relational tables in order to support special queries in the RFID applications. The experimental evaluation conducted on an existing RDBMS demonstrates the efficiency of our method.

## 1 Introduction

Radio frequency identification (RFID) [6] has been around for decades, and recently, there has been greater push from governments for its adoption for more efficient manufacturing, logistics and supply-chain management, and as a measure for security enforcement and weeding out counterfeiting. Take the supply-chain management for example (Figure 1), RFID enables accurate and real-time tracking of inventory by companies throughout an entire supply chain. Specifically, data stored in RFID are captured remotely via radio waves. Information from goods tagged with RFIDs can then be read simultaneously using fixed or mobile readers rather than requiring the scanning of individual bar code. Such a better supply chain visibility with the use of RFID also means that loss of inventory will be minimized during shipment. Businesses are suggested to use RFID for better inventory control since it may reduce excess inventories and free up capital for other activities.

Unfortunately, traditional database cannot efficiently support these new applications. Tracking each individual item causes data input to increase tremendously, and volume of data is enormous. As an example, Venture Development Corporation [4] has predicted that when tags are used at the item level, Walmart supermarket will generate around 7 terabytes of data every day. Though some compression techniques have been proposed (e.g. [8]), none of them fully explore the speciality of the RFID data while supporting online tracking.

For a better understanding of the characteristics of RFID data, consider the following example of the supply-chain management. Suppose there are several warehouses

* Also affiliated with Alexandria University, Alexandria, Egypt.

R. Wagner, N. Revell, and G. Pernul (Eds.): DEXA 2007, LNCS 4653, pp. 434–444, 2007.

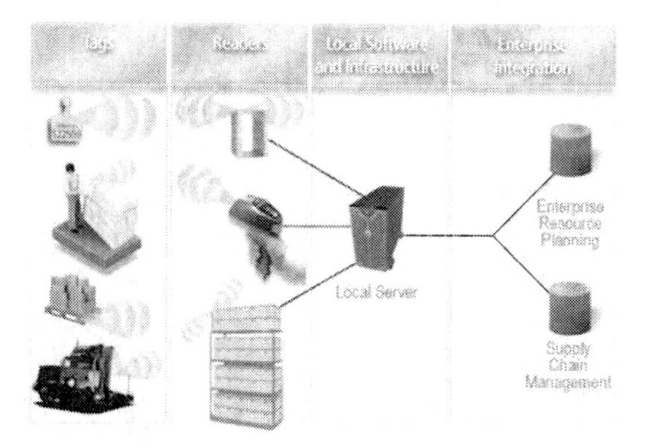


**Fig. 1.** Supply-Chain Management

and stores. Products like T-shirts, milk packages are tagged with RFIDs and shipped respectively from warehouses to stores by trucks. During the shipment, products may be reallocated or reorganized at some intermediate warehouses. All such information is recorded in a central database system when products pass through a warehouse or a store. In this scenario, suppose a type of T-shirts at a store is sold out and a customer wants to know when his order can be completed. To answer such a query, the retailer needs to check current status of the shipment. If he knows from the database that this type of T-shirts is now at a warehouse close to his store and will soon be sent to his store, he can then estimate the arrival time for the customer. Next, let us examine a more interesting but complicated situation. A retailer finds that a box of milk packages in his store is contaminated. He thus asks a query on the path of the shipment: "which place did the box of milk packages stay before arriving at my store?" If it is deemed to be contaminated in a truck, an alerting query may be issued to avoid more losses: "where is the truck now and what goods are in it?" This requires the system to quickly identify suspected trucks (which are possibly heading to other stores), and stop them to prevent possible contamination that may happen in other stores. The scenario would have been more disastrous if the movement of goods or living things causes infectious diseases to spread (for example, the breakout of SARS in Asia in 2003).

In this paper, we tackle the above problems specifically. We summarize our contributions as follows.

- We have explored the path and containment relationships in the RFID data and developed an ER-model based on it. To the best of our knowledge, it is the first time to clearly identify such inherent data connections in RFID applications so that they can be taken into account during the system design.
- We have proposed a real-time tracking system for applications in supply-chain management, manufacturing, logistics and delivery services. Both incremental updates and online queries are supported.
- We have conducted an extensive experimental study. The results demonstrate the efficiency of our system compared with the traditional method.

The rest of the paper is organized as follows. Section 2 reviews related work. Section 3 presents our proposed ER-model and discusses queries in the RFID applications.

Section 4 proposes our approaches for the RFID data management. Section 5 reports the experimental results. Finally, Section ?? gives the conclusion.

## 2 Related Work

RFID technology has posed many new challenges to database management systems [10,12]. Some IT companies are providing RFID platforms [1,2,3,5,6], through which RFID data are acquired, filtered and normalized, and then dispatched to applications. Thus high level RFID data modelling and management is up to applications. However little research has been observed in this area.

Chawathe et al. [7] presented an overview of RFID data management from a high-level perspective and introduced the idea of an online warehouse but without providing details at the level of data structure or algorithms. Later, Wang et al. [11] proposed a model for RFID data management. This model shares many common principles with the traditional models and hence is still inefficient in representing the specialty of RFID data. Hu et al. [9] proposed a bitmap data type to compactly represent a collection of identifiers, which can significantly reduce the storage overhead. However, the bitmap technique may not work well when the data in the same cluster are not continuous. As also reported by the authors, this approach might not be a good candidate for some applications like postal mail dispatch, because unlike the retail sector, the items in these applications do not lend themselves well to grouping based on a common property, thus precluding the use of bitmap for these cases.

Most recently, Gonzalez et al. [8] have proposed a new warehousing model that preserves object transitions while providing significant compression and path-dependent aggregates. The warehouse is constructed after all data have been collected. Specifically, each object is registered in the database only once at the end of its movement, which is different from traditional method that records each object at each station during its movement. This approach can largely reduce information volume. However, it may not be able to answer online queries on current status of objects, and hence it is not applicable for real-time tracking problems.

## 3 RFID Data Modeling

In this section, we will first introduce a new ER-model for the RFID data management, and then address the query types. Finally, we discuss a running example to present an overview of functions that are achieved by our approach.

### 3.1 ER-Model and Query Types

In RFID applications, it is often the case that items tagged with RFIDs move and stay together during their movements or are regrouped at some locations [8]. Consequently, queries on path and containment relationship naturally arise. In order to efficiently support these queries, we propose an ER-model that captures such internal relationships among RFID data.

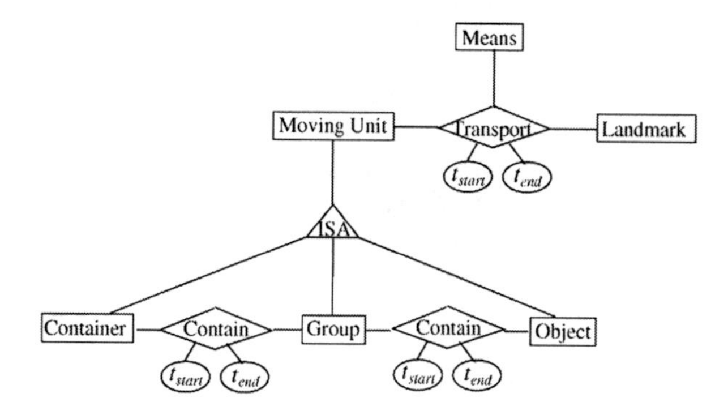


**Fig. 2.** The ER-Model

In our ER-model, there are three main entities: *landmark, means* and *moving units.* Landmarks can be warehouses, delivery centers, super markets, etc. Means can be trucks, ships or airplanes. Moving units can be moving objects (goods item), groups of objects, or containers. Figure 2 shows the relationships among the entities, where moving units are transported to some landmarks by some means from time $t_{start}$ to time $t_{end}$. There exists a hierarchy of containment relationship. That is *Object* is contained in *Group* and *Group* is contained in *Container*. Another implicit containment relationship is that of the containers and locations. Note that there may be multiple levels in the hierarchy, though our example uses only three levels.

Queries on RFID data can be categorized from different aspects. According to the query time, there are three types of queries: current queries, predictive queries and historical queries. According to the query condition, queries can be classified into two categories: ID-based queries and location-based queries. In the ID-based queries, retrieval is based on given ID information. In the location-based queries, retrieval is based on given location information. According to the information being queried, we identify two types of queries: containment-relationship queries and path-preserving queries. The *containment-relationship queries* find all objects contained in a given object at a higher level. The *path-preserving queries* retrieve path information of one or more objects under specified constraints. Queries in the last categorization mostly reflect RFID data characteristic, and hence we will address their processing in details.

### 3.2 An Illustrative Example

For illustration purpose, we adopt a simple example from the supply-chain management scenario, which will be used throughout the paper. As shown in Figure 3, there are two locations $L_1, L_2$, three containers $C_1, C_2, C_3$, and three groups $G_1, G_2, G_3$. Each group contains one object: $G_1$ contains $O_1$, $G_2$ contains $O_2$ and $G_3$ contains $O_3$. During time 0 to 5, container $C_1$ stayed at location $L_1$ and contained two groups $G_1$ and $G_2$. $C_1$ was then shipped from $L_1$ to $L_2$. After $C_1$ arrived at $L_2$, its containment was changed, where group $G_2$ was moved to container $C_2$. At time 50, a new container $C_3$ arrived at location $L_1$. Note that this example is only a part of the whole scenario. In the following discussion, we represent different entities by using their IDs. The detailed information of these entities can be stored in a separate information table, which will not affect the efficiency of the proposed method.

Regarding this example, we will examine three representative queries. The first one (denoted as $Q_1$) is "what objects are (were) in group $G$ (container $C$) at time $t$?", which is a containment-relationship query. Second, $Q_2$ is "where has object $O$ (or group $G$, container $C$) been to?". Third, $Q_3$ is "what objects (groups, containers) were shipped from $L_1$ to $L_2$ via $L_3$ and $L_4$ ($L_3$ and $L_4$ are intermediate warehouses) during time $t_1$ to $t_2$?". The last two are both path-preserving queries.

## 4 RFID Data Management

Handling a large amount of RFID data as well as providing efficient query services poses new challenges to existing database techniques. To make this point clear, we first study a straightforward method – Time-Line approach, and discuss its limitations. After that, we propose a more efficient approach – Multi-Table approach.

### 4.1 Time-Line Approach

The Time-Line approach is a naive method that stores all information in one table according to the insertion time. The format of each row in the table is $\langle Ts, Te, LID, CID, GID, OID, Means\rangle$, where $Ts$ is the arrival time, $Te$ is the leaving time, $LID$, $CID$, $GID$ and $OID$ correspond to the IDs of the location, container, group and object respectively, and $Means$ is the way the moving units being transported. Figure 4 shows how the data in the previous example is stored by using this Time-Line approach. Once there is an update on a field of the table, a new row is inserted. Here, an update could be a location update (e.g. a container reaches a new station), or a containment update (e.g. reallocation of goods in a container, or an object being delivered).

The aforementioned three queries can all be answered by combination of projection, selection and join operations. For example, $Q_2$ (to find where has object $O$ been to) can be answered as: SELECT $*$ FROM Table WHERE OID = '$O$'.

The main disadvantage of this approach is the data redundancy. Specifically, if the containment of a container (or a group) does not change frequently during the transportation, the Time-Line approach will store a lot of redundant information caused by

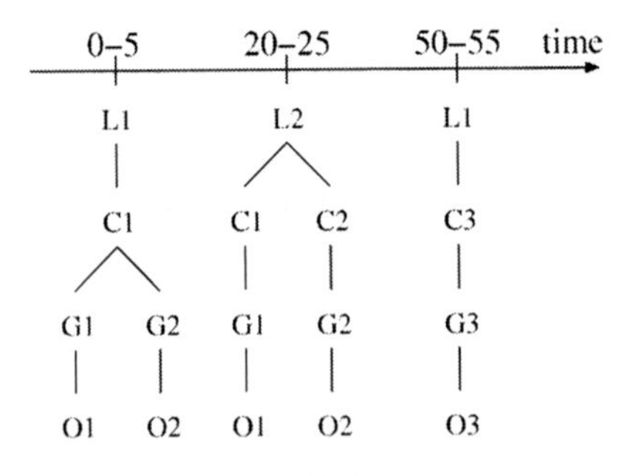


**Fig. 3.** An Example

| Ts | Te | LID | CID | GID | OID | Means |
|---|---|---|---|---|---|---|
| 0 | 5 | L1 | C1 | G1 | O1 | Truck1 |
| 0 | 5 | L1 | C1 | G2 | O2 | Truck1 |
| 20 | 25 | L2 | C1 | G1 | O1 | Truck1 |
| 20 | 25 | L2 | C2 | G2 | O2 | Truck2 |
| 50 | 55 | L1 | C3 | G3 | O3 | Truck3 |
| ⋮ | ⋮ | ⋮ | ⋮ | ⋮ | ⋮ | ⋮ |

**Fig. 4.** Time-Line Approach

the containment relationships. As shown in the example (Figure 3), $O_1$ stayed in the same container $C_1$ and the group $G_1$ when being transported from $L_1$ to $L_2$. The containment information of $O_1$ is unchanged but repeatedly stored in two records (1st and 3rd records in Figure 4). Such redundant information will unnecessarily increase the table size and result in poor performance.

### 4.2 Multi-table Approach

To alleviate the data redundancy problem and take advantage of the specialty of RFID data, we develop a Multi-Table Approach based on our proposed ER-model. Our approach adopts the following assumptions. Each object only belongs to one group, which means we do not reallocate objects to other groups. This is due to the consideration of the scenario like a box of milk packages, where a single milk package (object) usually stays at the same box (group) during its transportation. Unlike objects which are at the lowest level of the containment relationship hierarchy, groups can be reallocated to other containers, containers can be reallocated to other trucks, and so on. Moreover, groups and objects have their final destinations while containers and trucks can be reused.

Location–Container Table

| Ts | Te | LID | CID | Means |
|---|---|---|---|---|
| 0 | 5 | L1 | C1 | Truck1 |
| 20 | 25 | L2 | C1 | Truck1 |
| 20 | 25 | L2 | C2 | Truck2 |
| 50 | 55 | L1 | C3 | Truck3 |
| ⋮ | ⋮ | ⋮ | ⋮ | ⋮ |

Container–Group Table

| Ts | Te | CID | GID |
|---|---|---|---|
| 0 | 25 | C1 | G1 |
| 0 | 5 | C1 | G2 |
| 20 | 25 | C2 | G2 |
| 50 | 55 | C3 | G3 |
| ⋮ | ⋮ | ⋮ | ⋮ |

Group–Object Table

| Ts | Te | GID | OID |
|---|---|---|---|
| 0 | 25 | G1 | O1 |
| 0 | 25 | G2 | O2 |
| 50 | 55 | G3 | O3 |
| ⋮ | ⋮ | ⋮ | ⋮ |

**Fig. 5.** Three Main Relational Tables of Multiple-Table Approach

| GID | LID–Time List |
|---|---|
| G1 | <L1, 5>, <L2, 25>, ... |
| G2 | <L2, 25>, ... |
| G3 | <L1, 55>, ... |
| ... | ... |

**Fig. 6.** An Example of Group_Path Table

In the Multi-Table approach, there are two types of relational tables: the *containment table* and the *path table*. The containment table stores the information of containment relationship and the path table captures the path information of moving units.

Figure 5 gives an overview of the containment tables in our system. There are Location-Container (L-C for short) table, Container-Group (C-G) table and Group-Object (G-O) table. Each row of these tables consists of at least four fields. $[Ts, Te]$ is the time interval during which one moving unit (e.g. $GID$) stays at the same place (e.g. $CID$). In the L-C table, there is one more field – *Means*, which indicates the transportation means of the containers. Each table has corresponding history tables. Records are moved to history tables periodically (details will be covered shortly).

Figure 6 shows the structure of the path table, i.e., the Group_Path table. This table is a query-driven table, which is created during the query processing. It stores part of query results in order to facilitate new queries. Each row of this table contains two fields: a group ID $GID$ and an LID-Time list. The LID-Time list records a sequence of $\langle location, Te \rangle$ pairs, which indicates the location that the group has visited and the corresponding departure time.

In the rest of this section, we first present how to update information in the containment and path tables. Then we present the query algorithms.

**Construction.** Consider the scenario at a station, where several containers arrive at time $Ts$. First, there will be an arrival scan that reports the container IDs to the system. During their stay, their containments will be scanned and checked. If there is any change of the containments, i.e., rearrangement of goods, the system will receive new inventories for the corresponding containers. Finally, when containers leave, a departure scan is carried out and reports the departure time $Te$ to the system. From the above scenario, we identify three types of events: (i) Arrival event; (ii) Containment arrangement event; (iii) Departure event. The algorithm for each event is presented as follows.

The arrival event provides the location information of containers, and hence only the L-C Table is modified at this stage. Specifically, for each container, we will insert a new record $\langle Ts, \_, LID, CID, \_ \rangle$ to the L-C Table. The two fields $Te$ and $Means$ will be filled later when more information is received.

The containment arrangement event includes two sub-events corresponding to containers and groups respectively. We first elaborate the management of containment change in containers. If there is a reallocation in container $C_1$, $C_2$, ..., $C_n$, in the C-G table, set the $Te$ of groups that move out of the above containers to be the reallocation time, and insert a set of new records of groups that move into these containers. The event of containment arrangement of groups is triggered by object arrival or delivery. If objects $O_1$, $O_2$, ..., $O_m$ are new objects to the system, insert records like $\langle GID, O_1, Ts, \_ \rangle$ to the G-O table. If objects $O$ has been delivered, move its record from G-O table to history G-O table and set the $Te$ to be the delivery time.

Finally, we handle the departure event. The operation is simple. We only need to update the departure time $Te$ of each departure container as well as its transportation means (e.g. truck ID) in the L-C table.

Apart from the event handling, there is one more step for system optimization, which is the construction of history tables. Every certain time interval $T_{int}$, we will check L-C and C-G tables to move records with $Te$ older than current time to the history tables. Each history table has a global time interval that indicates the earliest and latest timestamps of its records. As time elapses, there may exist a set of history tables. Here, $T_{int}$ is an application dependent parameter which controls the table size. It can be set according to the speed of information grow. For example, if updates are frequent, a small value of $T_{int}$ may benefit the query retrieval.

**Query Processing.** We proceed to present algorithms for three representative queries (in Section 3.1). Note that other queries are special cases of the techniques used for these three representative queries. To speed up the search in each table, we have a clustered index on one type of ID and an unclustered index on the other.

For $Q_1$ (containment-relationship query) on location $L_1$, the search starts from the L-C table, where we obtain a list of containers at location $L_1$. Then we search the C-G table to find the groups of these containers. Finally, we retrieve the G-O table to get the objects at location $L_1$.

For $Q_2$ (path-preserving query) on object $O$, there are two main steps. The first step is to find the group that object $O$ belongs to. According to the object status (delivered or not), we can find its group ID in G-O table or history G-O tables. The second step is to find the locations that this group $G$ has visited within the life time of object $O$. Here, we may take advantage of the Group_Path table. If there exists a record with respect to the group $G$ in the Group_Path table, we further check whether this record contains sufficient information of object $O$, i.e., whether the location list contains a location with $Te$ larger than the object delivery time (or the latest update time). If yes, we report locations in the list till the one with $Te$ larger than the query time. If we can not find a corresponding record of group $G$ in the Group_Path table or the table does not contain full path of object $O$, we have to retrieve C-G table to obtain a set of containers that group $G$ ever belonged to, and then retrieve L-C table to find the locations of the containers. Finally, we need to append the query results to the Group_Path table.

The last query $Q_3$ is more complicated than previous ones since it requires to retrieve both containment and path information. The algorithm consists of following three steps. First, we find all containers at location $L_1$ during time $t_1$ to $t_2$ by searching the L-C table. Second, we find all groups of these containers and store them in a group list. The Third step is to check the Group_Path table to see if the path of each group in the group list contains a sequence of locations $\langle L_1, L_3, L_4, L_2\rangle$. If yes, we report the objects in the qualified groups with lifetime cover the query time interval. Otherwise, there could be two situations. One is that the path of the group as recorded in the Group_Path table is different from the query path, which can be safely pruned. The other situation is that the path of the group is not completed or there is not a record of this group. For this case, we need to find the locations of the group by retrieving all the containers that it ever belonged to, and retrieving all the locations of these containers. Then we check if the path of the group matches the query path. Finally, we append the group path information to the Group_Path table for the use of future queries.

## 5 Performance Study

We implemented the proposed algorithms as stored procedures in MS SQL Server 2005. For all experiments, we use a Xenon 2.0GHz CPU with 1GB of RAM. We created

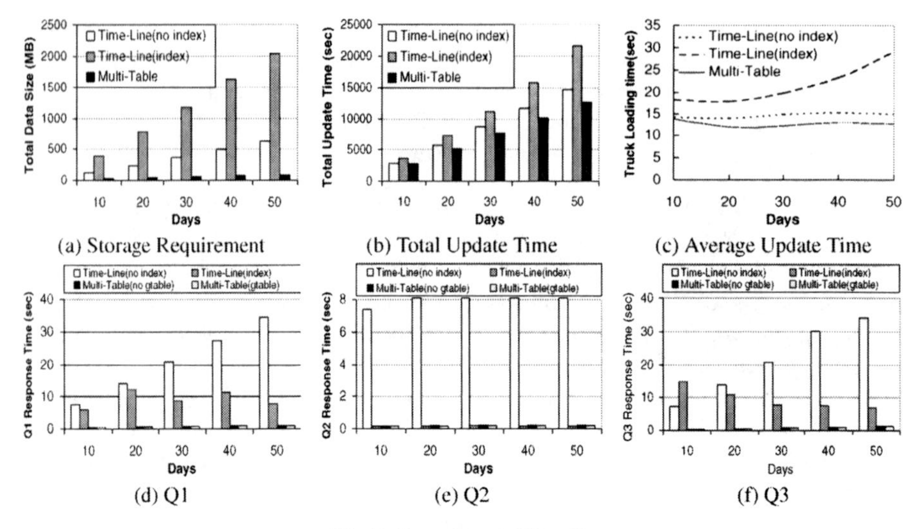


(a) Storage Requirement (b) Total Update Time (c) Average Update Time

(d) Q1 (e) Q2 (f) Q3

**Fig. 7.** Experimental Results

an application that simulates the movement of 18-wheelers between warehouses and stores. The simulated scenario is for 20 trucks and 80 warehouses. Each 18-wheeler contains 8 containers; each container holds up to 8 boxes; each box contains 12 objects. All containers, boxes, and objects are tagged with RFIDs. Upon arrival to a warehouse, the 18-wheeler is filled to completion. Upon arrival to a store, the probability that a container contains boxes for delivery is set to $(1-p)$. The probability that a box in such a container is to be delivered is also $(1-p)$. Thus objects are delivered to stores according to a geometric distribution with average numbers of hops $1/q$, where $q = 1-(1-p)^2$. We set default value of this average to 6 hops. The parameters of this simulation come from real samples of 18-wheelers. The trip from a warehouse to a store is uniform with mean equal to a day and with a standard deviation of 20 minutes.

We implemented two variants of Time-Line approaches distinguished by having index assistance or not, denoted as "Time-Line(no index)" and "Time-Line(index)" respectively. We also implemented two version of Multi-Table approaches distinguished by using the Group_Path table or not, denoted as "Multi-Table(no gtable)" and "Multi-Table(gtable)" respectively. It is worth noting that the size of the Group_Path table is ignorable compared to the total data size and the table is not involved in the data update process. Therefore we do not distinguish the two variants in the experiments regarding storage space and update performance.

**Storage Requirement.** The total data size that needs to be stored for an application is an important concern in database system design since a small data size can save cost for companies and may also benefit the system performance. To evaluate the storage efficiency, we examine the total data size stored by each approach every 10 days. Figure 7(a) shows the results, in which the Multi-Table approach requires the least storage space than the Time-Line approaches. The main reason is that the Time-Line approach maintains more redundant information. For example, if a container contains 100 items, each time the container reaches a station with the same items inside, the Time-Line approach needs to create 100 new records for all the items, whereas the Multi-Table approach only needs to create one new record corresponding to the container itself. In addition, Time-Line(index) needs more space than Time-Line(no index) to store the indexes.

**Update Performance.** Figure 7(b) plots the total update time every 10 days for each approach. It is not surprising that the insertion time of all approaches increases as time elapses due to the increased table sizes. Among all, the update time of Multi-Table approach is the shortest because it has the smallest table size (as shown in Figure 7(a)). The Time-Line(index) is the slowest approach with respect to the insertion performance. This is because Time-Line(index) needs to maintain its indexes for each update.

Figure 7(c) shows the average update time of each truck. The result again shows that the Multi-Table approach is the best. Moreover, we also observe that both the Multi-Table approach and Time-Line(no index) achieve steady performance, while the Time-Line(index) requires more time to maintain its indexes with the growth of the data size.

**Query Performance.** In the following experiments, we will evaluate three representative queries. Figure 7(d) and (f) show the average response time of $Q_1$ and $Q_3$ respectively. We can observe that Multi-Table approaches achieve the best performance, which possibly due to small data sizes that reduce the data retrieval and table join time.

Figure 7(e) shows the performance of query $Q_2$. We can see that the Time-Line(no index) is extremely slow (more than 100 times slower), and the other three approaches yield the similar performance. The slowness of the Time-Line(no index) is mainly because without any index support, it has to execute "brute-force" join operations. The Time-Line(index) is sometimes a little bit better than the Multi-Table approaches, but we should note that the Time-Line(index) requires much more space and longer update time. Another interesting observation is that the Group_Path table can reduce the query cost and its benefit increases as time passes (this effect is a little hard to be seen from the figure due to the large value of Time-Line approach).

## 6 Conclusion

In this paper, we study the important features of RFID applications, such as the hierarchy of containment relationships and path preserving in query operations. We propose an expressive ER-model. Based on the ER-model, we develop a simple yet efficient real-time tracking system for RFID data managements. Our extensive experimental results prove the significant performance improvement achieved by our system compared with a naive method.

## References


1. Developing auto-id solutions using sun java system rfid software. `http:// java.sun.com/ developer/ technical-Articles/ Ecommerce/rfid/ sjsrfid/ RFID.html`
2. Microsoft's rfid 'momentum' includes middleware platform, apps. `http://www.eweek.com/article2/0,1759,1766050,00.asp`

3. Oracle sensor edge server.
http://www.oracle.com/technology/products/sensor_edge_server
4. Venture development corporation (vdc): http://www.vdc-corp.com
5. Websphere rfid premises server.http://www-306.ibm.com/software/pervasive/ws_rfid_premises_server
6. Bornhovd, C., Lin, T., Haller, S., Schaper, J.: Integrating automatic data acquisition with business processes - experiences with sap's auto-id infrastructure. In: Proc. VLDB, pp. 1182–1188 (2004)
7. Chawathe, S., Krishnamurthy, V., Ramachandran, S., Sarma, S.: Managing rfid data. In: Proc. VLDB, pp. 1189–1195 (2004)
8. Gonzalez, H., Han, J., Li, X., Klabjan, D.: Warehousing and analyzing massive rfid data sets. In: Proc. ICDE, p. 83 (2006)
9. Hu, Y., Sundara, S., Chorma, T., Srinivasan, J.: Supporting rfid-based item tracking applications in oracle dbms using a bitmap datatype. In: Proc. VLDB, pp. 1140–1151 (2005)
10. Lampe, M., Flrkemeier, C.: The smart box application model. In: Proc. Int. Conf. of Pervasive Computing (2004)
11. Wang, F., Liu, P.: Temporal management of rfid data. In: Proc. VLDB, pp. 1128–1139 (2005)
12. Want, R.: The magic of rfid. ACM Queue 2(7), 40–48 (2004)

# When Mobile Objects' Energy Is Not So Tight: A New Perspective on Scalability Issues of Continuous Spatial Query Systems

Tai T. Do, Fuyu Liu, and Kien A. Hua

School of Electrical Engineering and Computer Science,
University of Central Florida, Orlando, FL 32816-2362
{tdo, fliu, kienhua}@cs.ucf.edu

**Abstract.** The two dominant costs in continuous spatial query systems are the wireless communication cost for location update, and the evaluation cost for query processing. Existing works address both of these scalability factors by employing the distributed computation strategy, in which some part of query processing is carried out by mobile objects. In this paper, we make one important assumption about mobile objects' energy; that is for many applications, mobile objects' energy is not limited, as opposed to the battery-powered objects assumed in existing works. Under this new assumption, we re-examine the scalability issues for continuous spatial query systems. Our examination points out that the major bottleneck of these systems is now the wide-area wireless uplink bandwidth, which has not been addressed adequately in the past. We attack the problem by leveraging the local-area wireless communication between mobile objects, leading to our proposal of a hybrid communication architecture to be used in these continuous spatial query systems. The hybrid communication architecture unifies the two communication paradigms, wide-area and local-area wireless networks. We then propose a proof of concept system, called P2MRQ (Peer-to-peer technique for Moving Range Queries), to answer continuous moving range queries over moving objects. While MobiEyes [1], an existing continuous range query system, only utilizes distributed computation, our P2MRQ is able to leverage both distributed computation and local-area wireless communication. Our performance study shows that the required wide-area wireless uplink bandwidth from P2MRQ is consistently less than that of MobiEyes; for all considered cases, P2MRQ requires at most 50% of the wide-area wireless uplink bandwidth as MobiEyes does.

## 1 Introduction

A location-based service allows the users query for information based on their own locations and/or other users' locations. The explosive growth rate of the number of location-aware mobile wireless devices, ranging from navigational systems in vehicles to handheld devices and cell phones, means future location-based services need to employ scalable architectures to support a large and growing number of users and more complex queries. The research community [2,3,1,4,5,6]

R. Wagner, N. Revell, and G. Pernul (Eds.): DEXA 2007, LNCS 4653, pp. 445–458, 2007.

has spent considerable efforts in finding scalable solutions to support continuous spatial queries, an important class of these location-based services.

Two dominant costs that dictate the scalability of continuous spatial query systems are the wireless communication cost for location update, and the evaluation cost for query processing. Many papers, for instance [2, 4], have focused on reducing the evaluation cost. To address the wireless communication cost, the authors in [6] use distributed servers to leverage the aggregate bandwidth of the servers. To address both scalability issues, the distributed computation strategy has been used [3,1,5], in which some part of query processing is directed to mobile devices. These mobile devices monitor their own locations, and they report their location updates to the server only when the location changes likely affect some query results. Leveraging the mobile device's computational capacity interestingly mitigates both the wireless communication cost and the evaluation cost. For an environment with battery-powered mobile devices, wireless communication cost is translated into both wireless bandwidth and the devices' battery power. In [3, 1, 5], wireless bandwidth consumption is reduced because location update messages are no longer sent periodically. Since mobile devices trade off high energy operations in wireless communication with low energy operations in local query processing and monitoring, mobile devices' enery is saved. In addition, allowing mobile objects to monitor the query regions directly relieves the server from the overwhelming workload of query evaluation.

We observe that existing distributed approach only tries to leverage the computational capabilities of mobile devices. This calculated decision can be traced back to the limitation of mobile devices' energy. The concern over mobile devices' energy is a valid one in a range of applications, in which mobile devices' batteries are not easily recharged during operation. For instance, tracking sensors are attached to wild animals for animal tracking, or monitoring badges are assigned to children during their field trip [3]. Since it is well-known that sending a wireless message consumes substantially more energy than running a simple procedure [7], it is a good design choice to utilize only the computational capabilities of battery-powered mobile devices. Nonetheless, there are also many applications in which the energy issue for mobile devices is not so critical, because the power source for these devices is not so limited. Such mobile devices are often found in vehicles like cars or airplanes, where mobile wireless devices share the power source with the vehicles. In this paper, we divide mobile devices into two categories based on the availability of their energy source, *energy-limited* mobile devices (i.e. battery-powered devices) and *energy-abundant* mobile devices (i.e. carrying vehicle-powered devices). Our interest is to re-examine the scalability issues for continuous spatial query systems assuming that there are only energy-abundant mobile devices. Under the new assumption, two scalability issues are still the wireless communication cost and the evaluation cost but with a subtlety, in which wireless communication cost now only means wireless bandwidth. While existing distributed computation approaches [3, 1, 5] have done a respectable job of addressing both scalability issues as discussed in the previous paragraph, the wireless bandwidth bottleneck has not been handled adequately. In these past works, communication is carried out through a wide-area wireless network (i.e. cellular network), presumably a third generation (3G) wireless data network. In 3G wireless data networks, 1xEV-DO (Evolution-Data Only) networking technology, also known as HDR (High Data Rate), is an integral part of the CDMA2000 family of 3G standards [8]. The HDR downlink channel has a data rate of 2.4 Mbps, while the HDR uplink data rate is only 153.6Kbps [9,8,10]. Since location updates from mobile devices to the server account for most of the communication in continuous spatial query systems, the limited HDR uplink data rate would easily become the communication bottleneck in existing distributed systems.

We leverage the local-area wireless network, also known as ad hoc network, to address the wide-area wireless uplink problem. First, using the ad hoc network, mobile devices send their location updates to some designated mobile devices. Then after processing these location updates, the designated mobile devices finally send the data to the server through the wide-area network. So instead of having lots of wide-area wireless connections as done in past works, we use lots of local-area wireless connections together with a few wide-area wireless connections to transmit location updates from mobile devices to the server. While the improvement for the wide-area uplink bottleneck problem could be quite obvious, the price we pay is the involvement of the local-area wireless network. This ad hoc communication consumes the local-area wireless bandwidth as well as the mobile devices' energy. Since the ad hoc network is currently unused in the existing continuous spatial query systems, in our opinion usage of the underutilized local-area wireless bandwidth should not pose any problem. Moreover, we already assume energy-abundant mobile devices in this paper; hence, energy consumed by ad hoc communication is acceptable to us too.

To the best of our knowledge, we are the first to investigate the scalability issues of continuous spatial query systems under the new, yet practical assumption of energy-abundant mobile devices. The new assumption allows us to look further into and analyze the wireless communication bottleneck found in the existing continuous spatial query systems. The rest of our paper provides technical details to solidify our findings and analysis. Section 2 discusses in details the hybrid communication architecture, a unified framework to allow the co-existence of both the wide-area wireless network and the local-area wireless network. To demonstrate the benefit and the integration process of the hybrid communication architecture into the framework of continuous spatial query systems, we propose a proof-of-concept system, called P2MRQ in Sections 3 and 4. We compare P2MRQ with MobiEyes in our performance study in Section 5. Finally, we conclude the paper in Section 6.

## 2 Communication System Architecture

### 2.1 The Hybrid Wireless Network Architecture

Hybrid wireless network architectures have been considered in the past. Wei's dissertation presents a survey on more than a dozen of recent proposed architectures [10]. Our focus in this paper is not to introduce a new or better hybrid architecture.

We only wish to have an architecture that is appropriate to allow the local-area wireless network to assist the wide-area wireless network as discussed in section 1. We start with the UCAN (Unified Cellular and Ad-hoc Network) architecture [8] since it is closest to our need. Fig. 1 shows the general architecture employed in UCAN. In UCAN, each mobile device has two radio interfaces, one for the local-area wireless network and one for the wide-area wireless network. The local-area wireless network represents communication among mobile devices using the IEEE 802.11b protocol in its ad-hoc mode. The wide-area wireless network represents connection between the base station and mobile devices within the cell coverage using the 1xEV-DO (Evolution-Data Only) protocol, also known as HDR (High Data Rate). The goal of UCAN is to improve the cell's aggregate throughput, while maintaining fairness [8]. The key idea that allows UCAN to achieve both of these contradicting goals is the *opportunistic* use of the IEEE 802.11 interfaces to improve the wide-area cell throughput. When receiving data from the base station, if a destination client experiences low HDR downlink channel rate, instead of transmitting directly to the destination, the base station transmits the data frames to another client (proxy client) with a better channel rate. These frames are further relayed, possibly through multiple clients, to the destination, using the high-bandwidth IEEE 802.11b links. The paper addresses three main technical challenges 1) proxy discovery, routing and maintenance, 2) scheduling algorithm at the base station, and 3) an incentive mechanism to encourage mobile clients to participate in message relaying. Out of these three issues, the first one is the most relevant to our paper. Due to space limitation, we omit altogether the detailed description of the solution of the first challenge as well those of the other two challenges, and refer interested readers to the original paper [8].

## 2.2 Adapting UCAN for Continuous Spatial Query Systems

Even though we adopt the UCAN framework in our paper, the targeted applications in the two papers are very different. UCAN supports applications that provide wireless Internet access to mobile users, while we support continuous

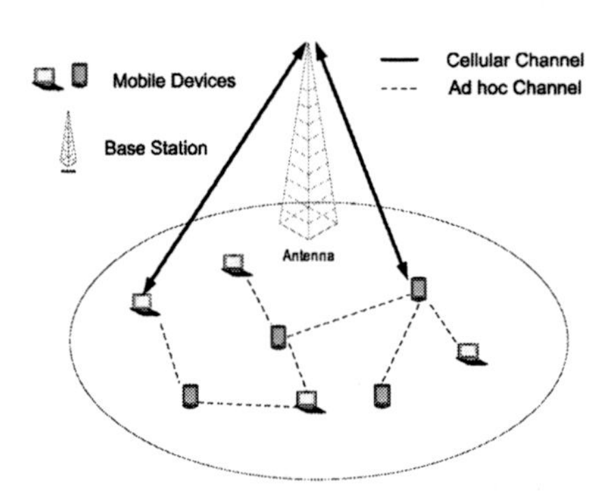


**Fig. 1.** The Unified Cellular and Ad-hoc Network Architecture

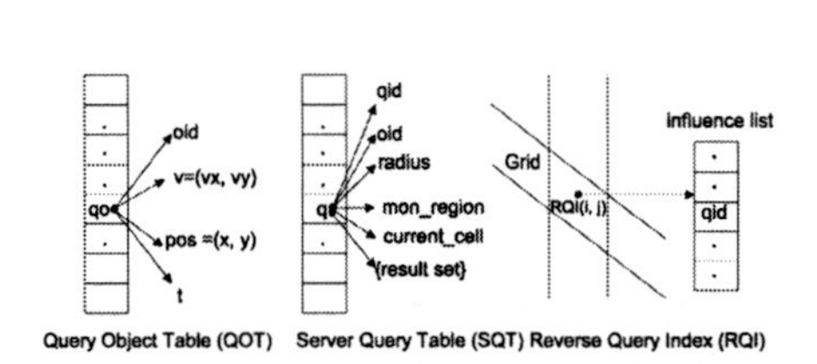


**Fig. 2.** Server-side Data Structure

spatial query systems. The data flows in UCAN are from the Internet through the base station to mobile clients. On the other hand, in our system, data are location updates generated by mobile clients themselves; hence our data flows are from mobile clients to the base station. Consequently, UCAN improves the aggregate *downlink* channel utilization, while we want to improve the aggregate *uplink* channel utilization for the wide-area wireless network.

When a mobile object (henceforth called data object) in our systems wants to send a location update message to the base station, instead of sending the message directly to the base station using the HDR uplink channel, the data object sends the message to a mobile object (henceforth called query object) using the IEEE 802.11b interface. The query object receives location update messages from a number of data objects, processes the messages and periodically sends the data to the base station through the HDR uplink channel.

Since a data object is aware of the identifier and location of the query object, which will be clear later in section 4, the data object does not have to go through the discovery phase to find the query object. The routing of messages from the data object to the query object is done using the position-based routing strategy [11]. In this paper, we choose one of the Location Aided Routing (LAR) protocols [12] for its simplicity and efficiency. Using the LAR1 protocol [12], a rectangular region is formed in which the data object and the query object are the two opposite corners of the rectangle. A mobile object only relays the message for the data object, if the mobile object's position is inside the rectangular region. As a result, the LAR1 protocol routes a message from the data object to the query object without flooding the entire local-area wireless network.

# 3 Models and Notations of the P2MRQ System

## 3.1 System Assumptions

This section summarizes the underlying assumptions that we make in this paper. 1) Each mobile device is equipped with two wireless interfaces. These two wireless interfaces, an IEEE 802.11x interface and a 3G interface, are used to enable the mobile device to operate in the ad hoc and the cellular modes, respectively. 2) Mobile devices are able to locate their positions using positioning devices such as GPS. 3) Mobile devices are able to determine their velocity vector. The velocity vector of a mobile device can easily be determined given its location and an internal timer. 4) Mobile devices have computational capabilities to carry out computational tasks. 5) Mobile devices have relatively unlimited energy source. These mobile devices share the energy source with the vehicles that carry them. As long as the carrying vehicles can be refueled, the mobile devices do not need to concern much about energy consumption. 6) Mobile devices have synchronized clocks. This assumption can be made if the mobile objects are equipped with GPS.

## 3.2 Basic Notations

- *Mobile Object, Mobile Device, Peer, and Client*: are used interchangeably in this paper.

- *Query Object*: A mobile object that is the moving center of a moving range query is called a query object.
- *Data Object*: Every mobile object is a data object, in the sense that the result set of a query contains the identifiers of mobile objects, which are currently in the query range.
- *Rectangle Shaped Region*: is defined by its lower left corner point $(lx, ly)$ and upper right corner point $(ux, up)$. Specifically, a rectangle

$$R(lx, ly, ux, uy) = \{(x, y) : lx \leq x \leq ux \wedge ly \leq x \leq uy\}$$

- *Geographical Area of Interest*: is a big rectangle $R(LX, LY, UX, UY)$, in which $LX$, $LY$, $UX$, $UY$ are system parameters to be set at the system initialization time.
- *Grid and Grid cells*: The geographical area of interest is mapped onto a grid $G$ of cells, where each cell is an $a \times a$ square area, and $a$ is a system parameter that defines the size of the grid cell. $G_{i,j}$ denotes an $a \times a$ square area representing the grid cell that is located on the $i$th row and $j$ column of the grid $G$.
- *Position to Grid Cell Mapping*: Given $pos = (x, y)$ as the current position of a mobile object $o$, the following function $f(pos)$ determines the current grid cell of $o$, i.e. the grid cell currently containing $o$:

$$f(pos) = G_{\lceil |\frac{pos.x-(UX-LX)}{a}| \rceil, \lceil |\frac{pos.y-(UY-LY)}{a}| \rceil}$$

### 3.3 Moving Query Model

Let $Q$ be the set of moving queries. A query $q$ in $Q$ is represented by a triplet: $(qid, o, radius)$, where $qid$ is the unique query identifier, $o$ is the query object, and $radius$ is the range or search radius around the query object. We consider the result of a query as the set of object identifiers of the mobile objects that are located within the area covered by the spatial region of the query. P2MRQ utilizes peers' computational capability in the distributed query processing. The following concepts are the keys in that distributed process.

- *Monitoring Region of a Moving Query*: Consider a query $q = (qid, o, radius)$, and $G_{i,j}$ is the current grid cell of the query object, i.e. $G_{i,j} = f(o.pos)$. Let $(lx, ly)$ and $(ux, uy)$ be the lower left corner and upper right corner of the grid cell $G_{i,j}$. The *Minimum Bounding Rectangle* that covers all possible areas that the spatial region of the query $q$ may move into when the query object moves within its current grid cell can be defined as $R(G_{i,j}.lx - radius, G_{i,j}.ly - radius, G_{i,j}.ux + radius, G_{i,j}.uy + radius)$. The grid region defined by the union of all grid cells that intersect with the Minimum Bounding Rectangle of the query forms the *monitoring region* of the query. The monitoring region of $q$ covers all the mobile objects that are subject to be included in the result set of $q$ when the query object stays in its current grid cell.

– *Nearby Queries of an Object*: Given a mobile object $o$, we refer to all queries whose monitoring regions intersect with the current grid cell of the mobile object $o$ the nearby queries of the object $o$.

# 4 Data Structures and Algorithms of the P2MRQ System

In P2MRQ, the server acts as the mediator coordinating mobile objects. A new query is always posed to the server. The server then computes the monitoring region of the query, based on the current reported location of the query object and the search radius. The server also determines the list of mobile objects, whose list of nearby queries now includes the new query. Using the wide-area wireless downlink channel, the server communicates updates to affected mobile objects. On the mobile objects' side, a mobile object monitors its own location, velocity vector and its nearby queries. In general, mobile objects only need to send their location updates to the server when their location changes affect one or more query results. In P2MRQ system, mobile objects use the local-area wireless communication to send their location updates to query objects, which in turn use the wide-area wireless uplink channel to send location updates in batches to the server. The subsequent sections describe in details the data structures and algorithms used in P2MRQ.

## 4.1 Data Structures

**Server Side Data Structure.** Fig. 2 shows the three main data structures used at the server. The *Query Object Table, QOT,* stores the list of query objects and their associated parameters including velocity vectors and current positions. The *Server Query Table, SQT,* keeps the list of moving range queries. Each query is associated with the query object identifier, *oid*, and the defined range, *radius*. Given the current position, *pos*, and the radius *radius* of the query object, the current cell, *current_cell*, of the query object can be inferred using the position to grid cell mapping in section 3.2, and the monitoring region, *mon_region*, of the query object can be determined as shown in section 3.3. The *Reverse Query Index, RQI,* is introduced to speed up the search of nearby queries for a data object, given that the current grid cell of the data object is known.

**Mobile Object Side Data Strcuture.** Every mobile object $o$ has a *Nearby Query Table, NQT,* and a boolean variable *hasQuery*. The schema of the *NQT* is as follows *(qid, oid, pos, v, up_time, radius, mon_region, isTarget)*, where *qid* is the identifier of the nearby query whose monitoring region *mon_region* intersects with the object's current grid cell, *oid* is the identifier of the query object associated with the nearby query, *pos* and $v$ are position and velocity vector reported by the query object from the most recent time, denoted as *up_time*. *radius* is the range of the query. The boolean variable *isTarget* indicates whether $o$ in the result set of the query based on the distance calculation from $o$ to the query object, carried out by $o$. The variable *hasQuery* is true if $o$ is also a query object. If $o$ is also a query object, $o$ maintains an additional data structure, *Object*

*Location Table, OLT= (oid, qid, report_time, isTarget). oid* is the identifier of a data object. *qid* is the query identifier of a query, whose query object is $o$. *report_time* stores the most recent time the query object receives location update from the data object. The boolean variable $isTarget$ is set to true if the data object is in the query set, based on the location update at time *report_time*. The idea is $OLT$ temporarily stores the location updates of the data objects which just recently move in or out of one of the queries associated with the query object $o$. This data structure is used solely to support our location update strategy using the hybrid communication architecture.

### 4.2 Installing Queries

Installing a moving range query into the system consists of two phases. First, the query is installed at the server side, and the server state is updated to reflect the installation of the query. Second, the query is installed at the set of mobile objects that are located inside the monitoring region of the query.

**Updating the Server State.** When the server receives a new query, with the query object's identifier *oid* and the search radius $radius$, it performs the following actions. 1) The server first checks whether the query object with identifier *oid* is already contained in the $QOT$. 2) If the query object of the query is not present in the $QOT$, the server-side installation manager needs to contact the query object of this new query and request the position and velocity information. Then the server can directly insert the entry $(oid, pos, v, t)$ into $QOT$, where $t$ is the timestamp when the object with identifier *oid* has recorded its *pos* and $v$ information. 3) The server then assigns a unique identifier *qid* to the query and calculates the current grid cell *curr_cell* of the query object and the monitoring region *mon_region* of the query. A new moving query entry $(qid, oid, radius, curr\_cell, mon\_region)$ will be created and added into the $SQT$. The server also updates the $RQI$ by adding this query with identifier *qid* to $RQI(i, j)$ if $G_{i,j} \bigcap mon\_region(qid) \neq \phi$. At this point the query is installed on the server side.

**Installing Queries on the Mobile Objects.** After installing queries on the server side, the server needs to complete the installation by triggering query installation on the mobile object side. This job is done by performing two tasks. First, the server sends an installation notification to the query object with identifier *oid*, which upon receiving the notification sets its *hasQuery* variable to true. This makes sure that the mobile object knows that it is now a query object and is supposed to report velocity changes to the server. The second task is for the server to forward this query to all objects that reside in the query's monitoring region, so that they can install the query and monitor their position changes to determine if they become the target objects of this query. This broadcast message contains information similar to the schema of the Nearby Query Table $NQT$ as in section 4.1. When an object receives the broadcast message, it checks whether its current grid cell is covered by the query's monitoring region. If so, the object installs the query into its neighbor query table $NQT$. Otherwise, the object discards the message.

### 4.3 Handling Query Objects that Change Their Velocity Vectors

Once a query is installed in the P2MRQ system, the query object needs to report to the server any significant change to its location information, including significant velocity changes or changes that move the query object out of its current grid cell. We describe the mechanisms for handling velocity changes in this section and the mechanisms for handling objects, including both query objects and data objects, that change their current grid cells in the section 4.5.

A velocity vector change from a query object, once identified as significant, will need to be relayed to the objects that reside in the query's monitoring region through the server acting as a mediator. A significant change means either change in speed or direction of the velocity vector exceeds a predefined threshold. When the query object reports a velocity change, it sends its new velocity vector, its position and the timestamp at which this information is recorded, to the server. The server first updates the $QOT$ with the information received from the query object. Then for each query associated with the query object, the server communicates the newly received information to objects located in the monitoring region of the query by using a wide-area wireless broadcast message.

### 4.4 Handling Data Objects That Change Their Spatial Relationship with Nearby Queries

A data object periodically processes all queries registered in its $NQT$. For each query, it predicts the position of the query object of the query using the velocity, time, and position information available in the $NQT$ entry of the query. Then it compares its current position and the predicated position of the query's query object to determine whether itself is covered by the query's spatial region or not. When the result is different from the last result computed in the previous time step, the object notifies the query object associated with the affected query of the change through the local-area wireless network. The message the data object sends to the query object has a similar format as the schema of $OLT$.

Upon receipt of the location update message from a data object, the query object updates its $OLT$. Using the $isTarget$ variable in the $OLT$, for each query bounded to it, the query object periodically computes two sets, namely $IN$ and $OUT$. The $IN$ and $OUT$ sets contain the new inclusion and exclusion, respectively, of data objects to the query's result set since the previous time step. The query object then sends a result update message, containing the $IN$ and $OUT$ sets, to the server through the wide-area wireless uplink channel. Upon receipt of the result update message from a query object, the server just needs to update the $SQL$ accordingly, i.e. adding (removing) data objects in the $IN$ ($OUT$) set to (from) the result set of the specified query.

One issue with the use of the hybrid communication architecture to handle mobile objects' location updates is the delay, coming from both the communication protocol and the periodic result update of the query objects. We offer the following explanations. In P2MRQ, the communication delay of a location update consists of delays on both the local-area wireless network and the wide-area wireless network. Since the bandwidth of the IEEE 802.11x protocol (11 Mbps for 802.11b) used in the local-area wireless network is much higher than the uplink bandwidth of the HDR protocol (153.6Kbps) used in the wide-area wireless network, the delay from the local-area wireless communication can be considered negligible; hence, location update delay from communication in P2MRQ is not different from that of the previous continuous spatial query systems. The delay associated with the periodic result update can be controlled by how frequently the query object sends the result update to the server. Our simulation study in section 5 shows that even with a high update frequency, the benefit of utilizing the local-area wireless network is still significant.

### 4.5 Handling Mobile Objects That Change Their Grid Cells

In P2MRQ, when a mobile object changes its current grid cell, the nearby query set of the object is also changed. In case the object is also a query object, the change also has an impact on the set of objects which are monitoring the queries bounded to this query object.

When an object changes its current grid cell, it notifies the server of this change by sending its object identifier, its previous grid cell and its new current grid cell to the server. The object also removes those queries whose monitoring regions no longer cover its new current grid cell from it nearby query table $NQT$. Upon receipt of the notification, the server uses the $QOT$ to check whether the object is a query object. If the object is not a query object, the server only needs to find out from the $RQI$ what new queries should be installed on this object and then perform the query installation on this mobile object. If the object is also a query object, the server needs to take the following actions. First, the server updates the $SQT$ with the new values of *cur_cell* and *mon_region* for all queries bounded to the query object. Second, for each affected query, the server updates the $RQI$ by removing the query off the influence list of grid cells overlapping with the old monitoring region, and inserting the query to the influence list of grid cells overlapping with the new monitoring region. Finally, the server needs to forward this change of monitoring regions to mobile objects so that they can update their nearby query tables accordingly. For each affected query, the server sends a broadcast message including both the old and new monitoring regions of the query, denoted as $mon\_region_{old}$ and $mon\_region_{new}$. Upon receiving the broadcast message, mobile objects take one of the following actions depending on the spatial relationship of their current grid cells with $mon\_region_{old}$ and $mon\_region_{new}$. 1) if its current grid cell belongs to $(mon\_region_{old} \bigcap mon\_region_{new})$, the object does not need to do anything, 2) if its current grid cell belongs to $mon\_region_{old} - (mon\_region_{old} \bigcap mon\_region_{new})$, the object removes the query off its nearby query table, 3) if its current grid cell belongs to $mon\_region_{new} - (mon\_region_{old} \bigcap mon\_region_{new})$, the object inserts the query into its nearby query table.

# 5 Performance Evaluation

## 5.1 Simulation Study

We implement a simulator to measure the performance of our proposed technique P2MRQ against the existing MobiEyes technique [1] and a naive technique. In the naive method, mobile objects periodically report their locations to the servers; hence, there is no need for the server to send messages back to the objects. We introduce the naive technique as a way to gauge how much better P2MRQ and MobiEyes improve over a do-nothing approach. We use the following two performance metrics in our simulation study: 1) *Wide-area Wireless Uplink Bandwidth*: The total number of messages sent by mobile objects to the server. 2) *Wide-area Wireless Bandwidth*: This metric includes both the uplink bandwidth and the downlink bandwidth. It is defined as the sum of the number of messages sent by mobile objects to the server and the number of messages sent by the server to mobile objects. We also study the sensitivity of the system with two variables, namely the number of queries and the number of objects. We measure the bandwidth in terms of number of messages, which we assume have the same size.

Our simulation is set up as follows. The area of interest in our simulation is a square shaped region of 10,000 square miles. The whole region is mapped into grid cells, where each grid cell has a size $a$ of 5 miles. Mobile objects are randomly placed in the region. The initial speeds of the mobile objects follow a Zipf distribution with a deviation of 0.7, and the values are between 0 and 1 mile per time unit. The initial directions of the velocity vectors are set randomly. At each time unit, one tenth of the mobile objects will change their velocities. Query objects are selected randomly from the mobile objects. In our simulation, the number of mobile objects varies from 2000 to 10000, and the number of queries is in the range of 10 to 1000. Query's range is selected randomly from the list 1, 2, 3, 4, 5 (miles). In P2MRQ, the query objects send the result update messages every one simulation time unit. Likewise, in the naive technique, the mobile objects also send their location updates to the server every one simulation time unit. For each simulation setting, we run the simulation for 10 times with different seed numbers and take the average as the final output. Each simulation run lasts for 200 time units.

Fig. 3 and Fig. 5 show the impact of number of mobile objects on the wide-area wireless bandwidth, when the number of queries is fixed at 1000. Fig. 5 is similar to Fig. 3, except it decomposes the wide-area wireless bandwidth into uplink and downlink bandwidth. We do not show the naive technique in Fig. 5, because in this technique the uplink messages accounts for all of the messages exchanged between the objects and the server. P2MRQ and MobiEyes require similar wide-area downlink bandwidth, but P2MRQ requires much less wide-area uplink bandwidth than MobiEyes does. For all cases, the required uplink bandwidth from P2MRQ is at most 50% of that from MobiEyes. In addition, P2MRQ is able to save more uplink bandwidth when there are more mobile objects, an indication that P2MRQ is scalable with respect to the number of

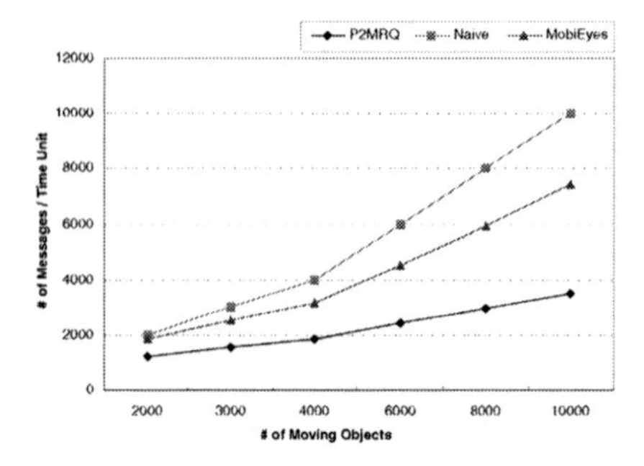


Fig. 3. Wide-area Wireless Bandwidth

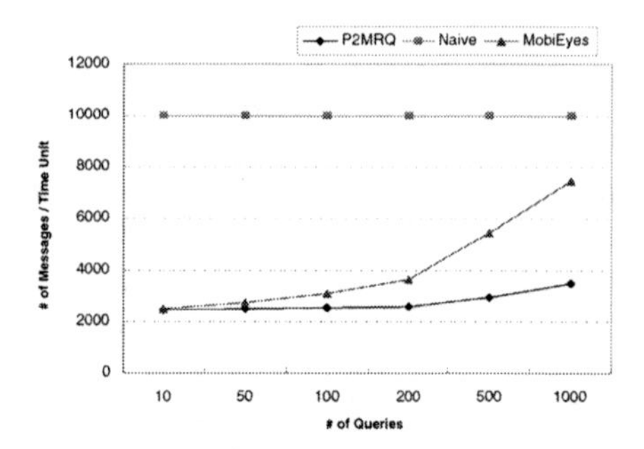


Fig. 4. Wide-area Wireless Bandwidth

mobile objects; when the number of objects is 10000, P2MRQ requires only 30% of the uplink bandwidth comparing to MobiEyes.

Similarly, Fig. 4 and Fig. 6 show the impact of number of queries on the wide-area wireless bandwidth, when the number of objects is fixed at 10000. Again, P2MRQ and MobiEyes require similar wide-area downlink bandwidth, but P2MRQ needs much less wide-area uplink bandwidth than MobiEyes does. When the number of queries increases, P2MRQ tends to save even more wide-area wireless uplink bandwidth comparing to MobiEyes.

The improved performance of P2MRQ over MobiEyes in terms of the required wide-area wireless uplink bandwidth can be easily attributed to the utilization of the local-area wireless network. Additionally, while we do not compare the evaluation cost of the two systems, we can infer with high probability that the two systems have similar the evaluation cost. First, they both employ the distributed computation approach. Second, since in these two systems computation at the server is usually followed by communication with the mobile clients, similarity in the required wide-area downlink bandwidth should indicate similarity in the evaluation cost at the server.

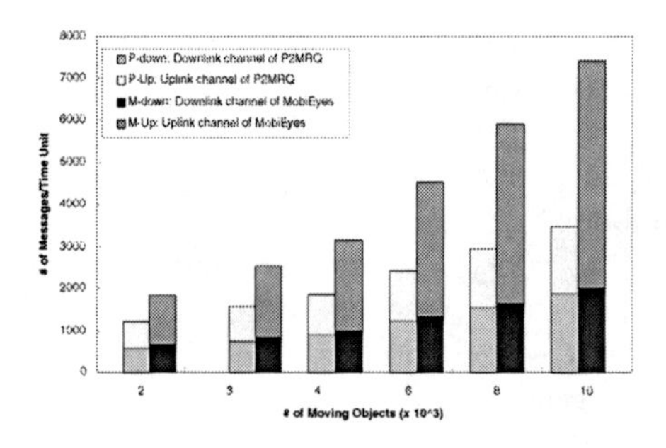


Fig. 5. Wide-area Wireless Bandwidth Breakdown

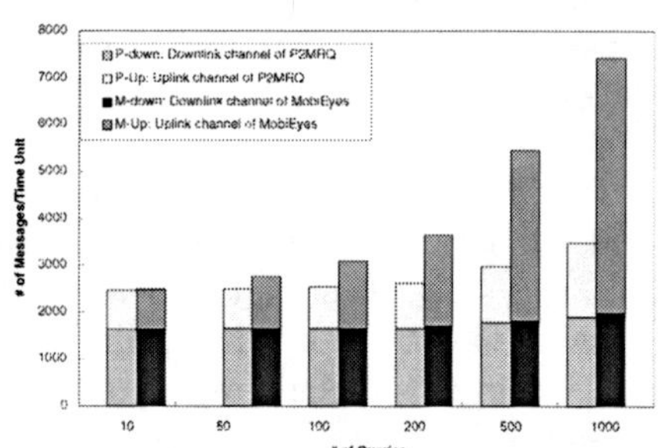


Fig. 6. Wide-area Wireless Bandwidth Breakdown

## 6 Conclusion

In this paper, we have re-examined the scalability issues in continuous spatial query systems under the new assumption on the availability of mobile devices' energy. With the assumption of energy-abundant mobile devices, we have identified the wide-area wireless uplink bandwidth as these systems' communication bottleneck, which has not received adequate attention from previous works. We propose to leverage the local-area wireless network to alleviate the burden on the wide-area uplink channel. We then propose a hybrid communication architecture, augmented from the UCAN system [8], to allow the co-existence of the wide-area wireless network and the local-area wireless network. To demonstrate the significance of the communication bottleneck problem, as well as the application of the hybrid communication architecture, we propose a proof-of-concept system, called P2MRQ, to answer continuous range queries over mobile objects. P2MRQ is able to utilize both the distributed computation approach and the local-area wireless network to successfully address both scalability issues of the system. Comparing to MobiEyes [1], P2MRQ requires only at most 50% of the wide-area uplink bandwidth as MobiEyes does.

We hope that our observations in this paper may lead to new interests in location-based services in moving databases, especially continuous spatial query services. While we do not claim that the technical details in our P2MRQ are readily applicable to other continuous spatial query systems, such as continuous k-Nearest Neighbor queries, the principle design of using the local-area wireless network should still be very helpful. As for our future works, we wish to follow a number of directions. One immediate task is to provide a formal analytical study of the P2MRQ and MobiEyes systems, using geometrical probability. The other direction is to study how demanding the P2MRQ system is on the local-area wireless network, something we skip in the performance study of this paper since for the time being we assume that the local ad hoc wireless network is still vastly underutilized.

## References

1. Gedik, B., Liu, L.: Mobieyes: Distributed processing of continuously moving queries on moving objects in a mobile system. In: Bertino, E., Christodoulakis, S., Plexousakis, D., Christophides, V., Koubarakis, M., Böhm, K., Ferrari, E. (eds.) EDBT 2004. LNCS, vol. 2992, pp. 67–87. Springer, Heidelberg (2004)
2. Prabhakar, S., Xia, Y., Kalashnikov, D.V., Aref, W.G., Hambrusch, S.E.: Query indexing and velocity constrained indexing: Scalable techniques for continuous queries on moving objects. IEEE Trans. Computers 51, 1124–1140 (2002)
3. Cai, Y., Hua, K.A.: An adaptive query management technique for efficient real-time monitoring of spatial regions in mobile database systems. In: IPCCC, pp. 259–266 (2002)
4. Mokbel, M.F., Xiong, X., Aref, W.G.: Sina: Scalable incremental processing of continuous queries in spatio-temporal databases. In: SIGMOD Conference, pp. 623–634 (2004)

5. Liu, F., Do, T.T., Hua, K.A.: Dynamic range query in spatial network environments. In: Bressan, S., Küng, J., Wagner, R. (eds.) DEXA 2006. LNCS, vol. 4080, pp. 254–265. Springer, Heidelberg (2006)
6. Wang, H., Zimmermann, R., Ku, W.S.: Distributed continuous range query processing on moving objects. In: Bressan, S., Küng, J., Wagner, R. (eds.) DEXA 2006. LNCS, vol. 4080, pp. 655–665. Springer, Heidelberg (2006)
7. Stemm, M., Katz, R.H.: Measuring and reducing energy consumption of network interfaces in hand-held devices. IEICE Trans. on Communications E80-B, 1125–1131 (1997)
8. Luo, H., Ramjee, R., Sinha, P., Li, L.E., Lu, S.: Cellular and hybrid networks: Ucan: a unified cellular and ad-hoc network architecture. In: ACM MobiCom., ACM Press, New York (2003)
9. Inc, Q.: 1xev: 1x evolution is-856 tia/eia standard. White Paper (2001)
10. Wei, H.: Integrating mobile ad hoc networks with cellular networks. PhD thesis, Department of Electrical Engineering, Columbia University (2004)
11. Giordano, S., Stojmenovic, I., Blazevic, L.: Position-Based Routing Algorithms for AdHoc Networks: A Taxonomy. In: Ad hoc Wireless Networks, Kluwer Academic Publishers, Dordrecht (2003)
12. Ko, Y.B., Vaidya, N.H.: Location-aided routing (lar) in mobile ad hoc networks. In: MOBICOM, pp. 66–75 (1998)

# Sequence Alignment as a Database Technology Challenge

Hans Philippi

Dept. of Computing and Information Sciences
Utrecht University
hansp@cs.uu.nl
http://www.cs.uu.nl/people/hansp

**Abstract.** Sequence alignment is an important task for molecular biologists. Because alignment basically deals with approximate string matching on large biological sequence collections, it is both data intensive and computationally complex. There exist several tools for the variety of problems related to sequence alignment. Our first observation is that the term 'sequence database' is used in general for textually formatted string collections. A second observation is that the search tools are specifically dedicated to a single problem. They have limited capabilities to serve as a solution for related problems that require minor adaptations. Our aim is to show the possibilities and advantages of a DBMS-based approach toward sequence alignment. For this purpose, we will adopt techniques from single sequence alignment to speed up multiple sequence alignment. We will show how the problem of matching a protein string family against a large protein string database can be tackled with q-gram indexing techniques based on relational database technology. The use of Monet, a main-memory DBMS, allows us to realize a flexible environment for developing searching heuristics that outperform classical dynamic programming, while keeping up satisfying sensitivity figures.

## 1 Introduction

There is no doubt about the importance of sequence alignment for molecular biologists. Homology searching comes down to matching a specific string, the query, to a large collection of already known strings, the database. The database can either contain nucleotide strings, based on the ACGT-alphabet or amino acid strings, based on a twenty letter alphabet. Evolutionary changes force us to deal with inexact matching. So essentially we are talking about approximate string matching on large string collections.

Traditionally, sequence databases have a pure textual format. The actual string contents are mixed with identifiers and annotation. Moreover, dedicated tools like Blast ([3], [4], [1]) and HMMER ([9]) are used for searching. In other words, if a DBMS is used at all, it is only used as a storage engine. So, the challenge for the database community is to show that the query facilities of a DBMS can simplify searching and make it more flexible.

R. Wagner, N. Revell, and G. Pernul (Eds.): DEXA 2007, LNCS 4653, pp. 459–468, 2007.

Single sequence alignment, i.e. the matching of one query string to a large string collection, has been exhaustively investigated ([5]). The exact solution is provided by a dynamic programming algorithm (Smith-Waterman). The need for quick response led to the development of Blast, a heuristic alignment tool based on q-gram indexing.

The q-gram indexing techniques on which the Blast heuristics are based, can be easily translated to a relational database environment. The indexing is realized at the logical level: the q-gram set is added as a table. The filtering process based on q-gram indexing can be concisely expressed in relational algebra. Variations in the filtering heuristics can be investigated by minor changes in the query expression.

To illustrate the versatility of the DBMS approach, we will focus on multiple sequence alignment. The notion of multiple sequence alignment deals with matching a collection of related protein strings (a so called family) to a database. This collection can be represented by a Hidden Markov Model (HMM). Models representing a protein string collection are known as profile HMM's. Matching a profile HMM to a protein string database is generally solved using Viterbi-like dynamic programming principles. The HMMER-package by Sean Eddy ([9]) is a freely available, open source implementation of these techniques. A typical matching operation using HMMER with a medium size database will take several minutes on commodity hardware. The answer is exact, in the sense that it finds all matches within some similarity distance.

In this paper, we will describe a generalized, Blast-like, heuristic search method based on q-gram indexing. Adapting these ideas to the context of multiple alignment turns out to be surprisingly straightforward, due to the flexibility that our DBMS provides. This way, our database supports both single string queries and family queries. The implementation of our methods on the Monet main-memory DBMS enables us to reduce the reponse time compared to HMMER significantly, while keeping up satisfactory sensitivity figures.

## 2 Preliminaries

In this section, we will introduce the concepts needed to discuss the domain of protein sequence alignment.

### 2.1 Strings

Our basic objects of interest are strings and q-grams. We will define them here.

- A *string* is a mapping from an integer interval $[k..n]$ to the set of characters. We have the notion of substring. Our alphabet is limited to the twenty amino acid symbols.
- A *q-gram* is a string of length $q$, a fixed number that typically is 3 for protein databases. We will use the value $q = 3$ in our examples. The term *word* is a synonym for q-gram and more common in the Blast community.

- A *position specific q-gram* is a combination of a q-gram and the position in the string where it refers to. Example: in the string ACDEG, with starting position 1, we identify the position specific q-grams $(1, ACD), (2, CDE)$, $(3, DEG)$.
- Basically, our *database* is a set of strings which all have $k = 1$.
- A *hitlist* is a set of position specific q-grams.

## 2.2 A Relational View on Q-Gram Indexing

In general, sequence databases are files in a character based format, like the Fasta format. In Fasta collections, the strings are listed interleaved with their annotation. We represent a string collection by two tables: `Strings` and `Annots`.

```
Strings(id, string)
Annots(id, annot)
```

Strings are internally identified with a system generated `id`. It maintains the connection between the protein strings, the annotation strings and the q-grams. So for each string in `Strings`, we have a describing tuple in `Annots`.

The `Q-grams` table contains a set of position specific q-grams. It serves as the index for matching hits between query data and the string database. Note that, on the physical level, we do not make use of either traditional indexing support or specialized string indexing techniques, such as suffix trees. We rely on the power of our main-memory DBMS to process the queries efficiently.

```
Q-grams(id,j,qg)
```

The position of the q-grams in the strings is denoted by a $j$ in the database strings and an $i$ in the query. The annotation table joins in (literally) at a very late stage.

BLASTP works as follows. A query string is, like the database strings, decomposed into q-grams. Suppose we have the query string denoted by `qs` and a database string by `s`.

```
qs =  CWYWRWYY
 s = RRWYWAWYYRR
```

In Table 1, we see a q-gram decomposition of the query string. We see a few q-gram matches between `qs` and `s`. $(2, WYW)$ in `qs` matches $(3, WYW)$ in `s`; $(6, WYY)$ in `qs` matches $(7, WYY)$ in `s`.

Because the distance between the matching q-grams is equal in the two strings, we say that they are 'on the same diagonal'. Technically, the notion of diagonal is represented by the difference of the q-gram positions: 6-2 = 7-3. The essence of the BLASTP filtering approach comes down to looking for two non-overlapping q-gram hits on the same diagonal within a certain distance (default 40).

To improve sensitivity for BLASTP, we also have the notion of 'similar' q-grams. The q-grams $(1, CWY)$ in `qs` and $(2, RWY)$ in `s` will generally be identified as similar, due to the notion of evolutionary distance (see [2] for further

**Table 1.** Example database and query

| **Strings** | |
|---|---|
| *id* | *string* |
| 1 | RRWYWAWYYRR |
| 2 | RRRWYWAWYWRR |
| 3 | RRWYWAAWYYRR |

| **Annots** | |
|---|---|
| *id* | *annot* |
| 1 | comments on string 1 .. |
| 2 | comments on string 2 .. |
| 3 | comments on string 3 .. |

| **Qgrams** | | |
|---|---|---|
| *id* | *j* | *qg* |
| 1 | 1 | RRW |
| 1 | 2 | RWY |
| 1 | 3 | WYW |
| ... | ... | ... |
| 3 | 10 | YRR |

| **Query** | |
|---|---|
| *i* | *qg* |
| 1 | CWY |
| 2 | WYW |
| ... | ... |
| 6 | WYY |

details). This means that we should extend the basic q-gram set of a query string with similar q-grams.

Our query string defines a hitlist containing, among others, $(1, CWY)$, $(1, RWY)$, $(2, WYW)$ and $(6, WYY)$. It depends on the parameter settings which similar q-grams will show up in our hitlist.

## 2.3 Profile HMM Matching

We now direct our attention toward the problem of matching a family of related sequences to a string database. A profile HMM ([6]) is a probabilistic model that represents a collection of related protein strings, often called a family. Figure 1 shows the basic HMM-architecture as used in the HMMER package ([9]). By

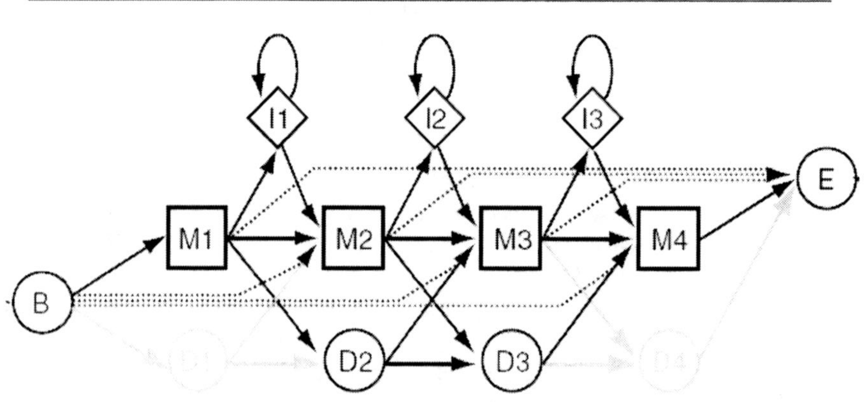


**Fig. 1.** HMM-architecture, as used in the HMMER package

matching a string to a HMM, we get a quantitative expression for the level of 'relatedness' between this string and the family that is represented by this model. Algorithms exist to calculate the optimal matching. They are based on dynamic programming techniques ([2], [5]). Note that these algorithms are exact: they find all matches with at least a specified minimal similarity.

The optimal match of a string to an HMM is represented by a path through the model. The begin and end-state are straightforward. The M-states represent a match. For each possible character, it specifies the 'emission probability', i.e. the probability of finding this character on this position. The I-states and D-states represent gaps in the match. By adding the rewards for matching and the penalties for mismatching, we get a final value expressing the relatedness.

To get a feeling for the principle, let us take a look at this small fraction of a four string family. The '-' represents a gap.

```
AFVEFEDP
GFVEFEDY
AFV-FEDP
AFVRF-DK
```

Because of these strings have length eight, we need an HMM with eight matching states. Matching state $M1$ (corresponding to the first column in the family) emits character `A` with probability 3/4 and `G` with probability 1/4. State $M2$ emits `F` with probability 1: there is 'consensus', just as in columns 3, 5, 6 and 7. State $M4$ emits character `E` with probability 2/3 and `R` with probability 1/3. State $M8$ emits `P` with probability 1/2 and emits `Y` and `K` with probability 1/4. Note that, if we would match the last string to this HMM, the optimal path for traversing the model would go from state $M5$ to state $M7$ through state $D6$, resulting in a gap in position 6.

In the model, these probabilities are transformed to the log-odds of the emission probabilities, according to the random amino acid distribution model. This log-odds conversion enables us to transform multiplication of probabilities into additions. States $D1$ and $D4$ are optional, depending on the choice to match globally (i.e. matching the whole model) or locally (i.e. matching the model partially).

A profile HMM defines a hitlist in a straightforward way. For each position we inspect the corresponding matching state with its characters and corresponding log-odds values. The most extended hitlist generated from our example family would be
$(1, AFV), (1, GFV), (2, FVE), (2, FVR), (3, VEF), (3, VRF), ..., (6, EDP)$.

We will use limited HMM-hitlists for filtering purposes, thereby focussing on position-character combinations with high probability. This choice is made at the character level, taking into consideration only the characters that have position specific emmission log-odds values meeting a threshold value $T$.

Summarizing, we see that the q-gram indexing principles can be extended from single protein query strings to profile HMM's. Both can serve as a query, because from a technical point of view, the matching object is in both cases a hitlist.

## 3 Filtering

As we have seen, the classical BLASTP approach uses the two-hit-diagonal filtering principle. Thresholds to limit the hitlist can, to some extend, be set by the user, influencing sensitivity and selectivity. In the case of profile HMM matching, we have added a tuning parameter $n$, making the number of required hits on the diagonal a user defined variable. It is clear that by increasing $n$, we increase selectivity and decrease sensitivity. Note that the principle of $n$-hit diagonal filtering corresponds to the *framecount* notion of CAFE ([11]).

We will describe how $n$-hit diagonal filtering can be expressed as a relational query. We choose to formulate the constituent expressions in an extended version of the relational algebra (RA). See [10] for details.

Note that the query, essentially a hitlist, might represent both a single string and a profile HMM. A hit represents an exact match between a q-gram in the query and a q-gram in the database. Recall that in the hitlist, at the same position, more than one (similar) q-gram may occur.

$$Hits := \pi_{id,diag \leftarrow (j-i)}(Qgrams \bowtie Query) \tag{1}$$

According to the $n$-hit diagonal filtering method, candidates are defined by $n$ hits in the same string and on the same diagonal. We first apply a self join on *Hits* to find hit pairs on the same diagonal. We need a copy of the hits table to express this self join. Hits2 should be interpreted as an alias of (or view on) the *Hits* table, not as a physical copy.

$$Hits2 := \pi_{id2,i2,j2,diag2}(Hits) \tag{2}$$

$$Pairs := \pi_{id,diag,j,j2}(Hits \bowtie_\theta Hits2) \tag{3}$$

where $\theta$ denotes the join condition:

$$\theta : (id = id2, diag = diag2, |j - j2| \leq A)$$

Here $A$ denotes the range, i.e. the maximal distance between two hits. Note that, en passant, we have by now expressed BLASTP filtering. Enforcing the $n$-diagonal is expressed by the grouping operator $\Gamma_{grp;function}$ of the relational algebra.

$$Filter := \sigma_{cnt \geq (n-1)}(\Gamma_{id,diag,j;cnt()}(Pairs)) \tag{4}$$

Note that we select on $n-1$ values within the group because $j$ itself is already one hit. The relevant strings can now be selected.

$$Candidates := Strings \ltimes Filter \tag{5}$$

This approach can be refined by taking in account the actual hit positions in the database and query string. That allows us to define a substring as a candidate for matching in stead of the complete string. Defining the boundaries however is quite tricky, because the HMM allows gapping. We will stick to the approach of pure string filtering as defined by the last step.

## 4 The Monet Approach

Our DBMS of choice is Monet (version 4.10.2), developed at the CWI in Amsterdam ([7], ([8]). To realize a full-fledged Monet based profile search tool, we should incorporate the **hmmsearch** method into our system to execute the expansion phase. For our project, this approach was to laborious, although all the required information was in the database. To get an impression whether our filtering approach would be fruitful, we simply wrote the filtered strings into a file in Fasta format and let the HMMER **hmmsearch** tool run on this selection in stead of the full database. As long as the selectivity was reasonable, the overhead of generating the output after the filtering step could be ignored.

There were two reasons to choose Monet.

- *main-memory approach*
  Monet is a main-memory DBMS (MMDBMS). The data and, possibly, indexes are supposed to be resident in main-memory. Where the classical DBMS focuses on minimizing IO, Monet gains performance by optimizing for main-memory access and by applying cache-conscious techniques. The data easily fits in main memory and it results in a very short running time for the initial join.
- *layered levels of access and extensibility*
  Monet provides the developer with a extensible relational algebra. It offers the possibility to write specific algebraic operators in C, using an API to access the binary tables, and add them to the Mil collection of standard algebraic operators. This makes sense in the case of very performance critical operations. For performance reasons, we decided to write a Mil-extension to execute the combined self-join-grouping step.

The memory requirements of our approach can be quantified easily. A protein collection in Fasta format (i.e. protein character strings and annotation mixed) has a total size of $B$ bytes. All the protein character strings together contain $N$ characters, where $N$ is practically equal to the number of q-grams. The number of protein strings is $L$. For the selection of the Swissprot database we used, the values are approximately $L = 100,000$, $N = 38,000,000$ and $B = 48,000,000$.

The tables `Strings` and `Annots` can be mapped easily to Monet, requiring $8L + B$ bytes. The `Qgrams` table is split into three columns according to Monet's binary data model. The column `id` can be represented in a minimal sense with a virtual identifier. For the column `qg`, we used two-byte integers. The total space requirement for the `Qgrams` table now is $14N$ bytes, which is about 0,5 GB for the 48 MB Swissprot selection, easily fitting in main memory.

### 4.1 HMMER

The HMMER package ([9]) by Sean Eddy provides us with several tools to build and use HMM's. Our main tool of interest is **hmmsearch**, that matches a profile HMM to a string database. **Hmmsearch** was run with an expectation value parameter $E < 0.1$. This value expresses the probability that a match is found purely by chance.

The package contains a tutorial with two prepared HMM's, that we used gratefully. The first one is the 'RNA recognition motif (rrm)' HMM. It has a size of 77 matching states. The second one is the 'globins50' alignment. It is about twice as long as the 'RNA recognition motif' HMM.

We used version 2.3.2 of HMMER.

## 5 Experiments and Discussion

Our results of the experiments on the Swissprot selection were compared to the results of running `hmmsearch` from the HMMER package.

The choice of the threshold $T$ for the emission log-odds is a bit of an art. It should be positive to make sense, but values that are too high destroy the sensitivity. We varied around $T = 1$, which turned out to be a good choice.

Selectivity was measured by simply counting the number of bytes of the resulting reduced string set after filtering, compared to the original string collection. It is expressed as a percentage, where a low value indicates a strong selectivity. Because `hmmsearch` behaves quite linear in the size of the string database, the selectivity percentage gives a good indication of the response time behaviour on the filtered string selection.

Sensitivity was measured by checking the presence of high scoring domains. Apart from complete HMM alignments, `hmmsearch` gives a list of local high scoring segments of the database. For each of these local matches, we checked if there was overlap with the candidates we found. Sensitivity is also expressed as a percentage, where a high value is good. We give figures for the complete set of matches found by HMMER (sens100) and for the top $k$ lists, where we only compare the best $k$% of the HMMER results. We did this for $k = 60, 40, 20$. The range parameter was fixed on 40 (the Blast default).

The most interesting tuning parameter is $n$, defining the number of diagonal hits within the range. We mention only the interesting values of $n$, keeping selectivity close to or less than 10%.

The tests were done on a dual processor Xeon 3.2 GHz machine with 4GB of main memory and 2MB of cache, running under Linux.

In general, we observe that, with adequate tuning, we are able to combine high top $k$ sensitivities with selectivities around or less than 10%. Note that the HMM-hitlist generation requires only a fraction of a second and gives the hitlist size, so the user has the possibility to tune the parameters before running the query. Keep in in mind that output lists will be inspected by biologists manually. Therefore, we claim that referring to the 'upper half' sensitivities is justified, analogous to the practice with web search tools.

The measurements also suggest that the log-odds threshold $T$ is less interesting as a tuning parameter. Fixing $T = 1$ and varying $n$ turns out to be a better tuning principle.

**Table 2.** Test results

| HMM = rrm; runtime hmmsearch = 92 sec<br>T=1<br>Filtering time: 7.2 sec | | | | | |
|---|---|---|---|---|---|
| $n$ | sens100 | sens60 | sens40 | sens20 | selectivity |
| 7 | 64% | 85.4% | 96.7% | 99% | 1.04% |
| 6 | 74.8% | 93.1% | 98.9% | 100% | 3.9% |
| 5 | 85.8% | 98.9% | 100% | 100% | 16.6% |
| **HMM = globins50; runtime hmmsearch = 166 sec<br>T=1<br>Filtering time: 3.3 sec** | | | | | |
| $n$ | sens100 | sens60 | sens40 | sens20 | selectivity |
| 7 | 75.7% | 100% | 100% | 100% | 0.11% |
| 6 | 81.4% | 100% | 100% | 100% | 0.2% |
| 5 | 82.6% | 100% | 100% | 100% | 1.2% |
| 4 | 87.4% | 100% | 100% | 100% | 9.2% |
| **HMM = rrm; runtime hmmsearch = 92 sec<br>T=1.2<br>Filtering time: 3.4 sec** | | | | | |
| $n$ | sens100 | sens60 | sens40 | sens20 | selectivity |
| 6 | 58.4% | 78.8% | 87.3% | 92.3% | 0.7% |
| 5 | 74.8% | 91.6% | 95.1% | 96.7% | 3.9% |
| 4 | 86.0% | 97.4% | 99.5% | 100% | 19.7% |
| **HMM = rrm; runtime hmmsearch = 92 sec<br>T=0.8<br>Filtering time: 10.2 sec** | | | | | |
| $n$ | sens100 | sens60 | sens40 | sens20 | selectivity |
| 8 | 64.3% | 89.1% | 96.2% | 98.9% | 0.7% |
| 7 | 73.5% | 93.8% | 97.8% | 100% | 2.6% |
| 6 | 84.0% | 97.8% | 100% | 100% | 9.2% |

# 6 Conclusions

Our first goal was to investigate whether main-memory database technology can be succesfully applied to biological sequence alignment. The paper shows that the q-gram indexing techniques of Blast, designed for single sequence matching, could be extended to the HMM matching problem with limited effort, due to the support of the DBMS query facilities. The filtering and its tuning possibilities are fully realized with the possibilities offered by Monet, in a rather concise way. In particular, we extended the algebra with a new operator to calculate the candidates efficiently. We were able to reach filtering times that were significantly smaller than the running times of `hmmsearch`, resulting in filtered string sets with the desired sensitivity and selectivity figures.

A secondary goal was to investigate whether Blast-like q-gram indexing techniques could be applied to profile HMM-matching. Especially the behaviour on the 'top $k$' result lists is very satisfying when we restrict ourselves to the upper

40 or 60 percent. With appropriate tuning of the querying parameters, we can combine top $k$ sensitivity figures close to 100% with selectivities of less than 10%.

## Acknowledgements

The author thanks the Monet development team at the CWI, Amsterdam, for hospitality and support during the development of the prototype.

## References

1. Korf, I., Yandell, M., Bedell, J.: Blast, O'Reilly (2003)
2. Durbin, R., Eddy, S.R., Krogh, A., Mitchison, G.: Biological Sequence Analysis. Cambridge University Press, Cambridge (1998)
3. Altschul, S.F., Gish, W., Miller, W., Myers, E.W., Lipman, D.J.: Basic local alignment search tool. Journal of Molecular Biology 215, 403–410 (1990)
4. Altschul, S.F., Madden, T.L., et al.: Gapped BLAST and PSI-BLAST: a new generation of protein database search programs. Nucleic Acids Research 25(17), 3389–3402 (1997)
5. Aluru, S. (ed.): Handbook of Computational Molecular Biology, Chapman & Hall/CRC (2005)
6. Krogh, A.: An introduction to Hidden Markov Models for biological sequences. In: Salzberg, S.L., Searls, D.B., Kasif, S. (eds.) Computational Methods in Molecular Biology, pp. 45–63. Elsevier, Amsterdam (1998)
7. Boncz, P.A., Kersten, M.L.: MIL Primitives for Querying a Fragmented World. The VLDB Journal 8, 101–119 (1999)
8. Boncz, P.A.: Monet: A Next-Generation DBMS Kernel For Query-Intensive Applications, PhD thesis, UVA, Amsterdam, The Netherlands (May 2002)
9. `http://hmmer.janelia.org`
10. Garcia Molina, H., Ullman, J.D., Widom, J.D.: Database System Implementation. Prentice-Hall, Englewood Cliffs (2000)
11. Williams, H.E., Zobel, J.: Indexing and Retrieval for Genomic Databases. IEEE Transactions on Knowledge and Data Engineering 14, 63–78 (2002)

# Fuzzy Dominance Skyline Queries

Marlene Goncalves and Leonid Tineo

Universidad Simón Bolívar, Departamento de Computación, Apartado 89000,
Caracas 1080-A, Venezuela
{mgoncalves,leonid}@usb.ve

**Abstract.** Skyline is an important and recent proposal for expressing user preferences. While no one best row exists, Skyline discards rows which are worse on all criteria than some other and retrieves non-dominated or the best ones that match user preferences. Nevertheless, some dominated rows could be interesting to user requirement, but they will be rejected by Skyline. Dominated rows could be discriminated (or ranked) by means of user preferences, but Skyline only discards dominated ones and it does not discriminate them. SQLf is a proposal for preferences queries based on fuzzy logic that allows to discriminate rows and includes user-defined terms, such as fuzzy comparison operators. In this work, we propose to flexibilize Skyline queries using fuzzy comparison operators in order to retrieve interesting dominated rows. We also introduce an evaluation mechanism for these queries and our initial experimental study shows that this mechanism has a reasonable performance.

**Keywords:** Skyline, SQLf, Flexible Querying, Fuzzy Conditions.

## 1 Introduction

Preference queries have received special attention of many database researchers. Thus, several SQL extensions have been introduced for expressing user preferences in a query and some of them are: Skyline [7] and SQLf [4].

Skyline [7] selects the best rows or all non-dominated based on a crisp multicriteria comparison. A row dominates another one if it is as good or better than the other in all multiple criteria and better in at least one criterion. In Skyline, multicriteria must be satisfied simultaneously in order to obtain the best rows.

It is known that the problem of identifying the skyline is $O(n^2)$ where "n" the number of rows in the data set [9]; it seems to be of high processing cost. In consequence, some efficient algorithms have been defined in order to evaluate such queries in a relational system [7]. The Sort-Filter-Skyline (SFS) [9] is one of the most relevant algorithms at present time for improving performance of Skyline computation in relational databases. SFS algorithm begins sorting the table, after it passes a cursor over the sorted rows and finally it discards dominated rows.

On the other hand, Skyline has another problem: dominance rigidity. There is no distinction between rows that are dominated by fare and those that are near to dominant rows. Additionally, Skyline answers are not discriminated due to the fact

R. Wagner, N. Revell, and G. Pernul (Eds.): DEXA 2007, LNCS 4653, pp. 469–478, 2007.

that Skyline is based on crisp comparison. Skyline rigidity could be solved by means of fuzzy logic. In the past, this approach has demonstrated to be the most general [2]. Thus, a SQL extension based on fuzzy sets was proposed: SQLf [4]. An interesting feature of SQLf is user-defined fuzzy comparators [4] and they might be used to flexibilize Skyline comparisons. A first effort to integrate Skyline and SQLf was made in [10], but it has not solved the Skyline rigidity.

Also, some variations of Skyline queries were presented in [12]. Ranked Skyline queries retrieve the top-k rows from the Skyline in terms of a monotonic score function. Constrained Skyline queries select Skyline rows restricted according to some constraints. Enumerating and K-dominating queries are kinds of Skyline queries that enumerate and return the number of dominated rows for each row in the Skyline. Although these extensions are more expressive, they do not solve the Skyline rigidity.

In this work, we propose a new Fuzzy Dominance Skyline operator in order to solve Skyline rigidity. It relaxes Skyline by means of fuzzy dominance comparisons among rows. Thus, a Fuzzy Dominance Skyline query selects and discriminates rows that are actually not dominated, but also rows that are near to be not dominated.

However, fuzzy queries presuppose high processing costs [3]. Some evaluation mechanisms have been proposed for SQLf queries [1][2][5]. The most relevant is that based on the Derivation Principle [2][5] because it has shown to keep low the extra cost for fuzzy conditions computation. For this reason, we propose a Derivation Principle based evaluation mechanism for the new Fuzzy Dominance Skyline queries and we present a performance study to show the feasibility of such queries.

This paper is organized as follows: Section 2 presents the Skyline semantics and its rigidity problem. The new Fuzzy Dominance Skyline operator is formally defined in the section 3. Section 4 describes implementation issues and evaluation mechanisms of Fuzzy Dominance Skyline queries. A performance analysis for these queries is shown in section 5. Finally, conclusions and further works are addressed in section 6.

## 2 Skyline Operator

Skyline [7] is a SQL extension that incorporates a new clause where user preferences are specified. For simplicity, we assume one-relation queries without the **WHERE** clause in order to understand Skyline semantics. Thus, Skyline query structure is:

```
SELECT * FROM rel SKYLINE OF a1 dir1, an dirn
```

Here, $a_1,\ldots,a_n$, named dimensions, are the attributes that user preferences range over and their domains are integers, floats or dates. The directives $dir_1,\ldots,dir_n$, may be either **MIN** or **MAX** or **DIFF**. The **MIN** and **MAX** directives specify whether the user prefers lowest or highest values, respectively. The **DIFF** directive defines the interest in retaining best choices with respect to every distinct value of that attribute. Thus, a multicriteria function is defined in the **SKYLINE OF** clause. Formally, the result set of a Skyline query is presented in (1), where $\propto_i$ is the comparator corresponding to the directive $dir_i$ ("<" **for MIN**, ">" for **MAX** and "=" for **DIFF**). This semantics is based on crisp comparison.

$$\{\ r\ \ /\ r \in rel \wedge \neg\exists_{r' \in rel}(r'.a_1 \propto_1 r.a_1) \wedge \ldots \wedge (r'.a_n \propto_n r.a_n)\} \tag{1}$$

As an example, consider that a user wants to determine the best hotels for booking. A hotel is described by an identifier (*idHotel*), the average of the room price (*avgPrice*), the quality of the service (*qService*) and the quality of the place (*qPlace*). Quality measures are numeric values between 0 and 80. Thus, a hotel can be retrieved if and only if there is no other one that has better service and better place quality. Graphically, points in Fig. 1 represent hotels in a database and Skyline was drawn as black round points.

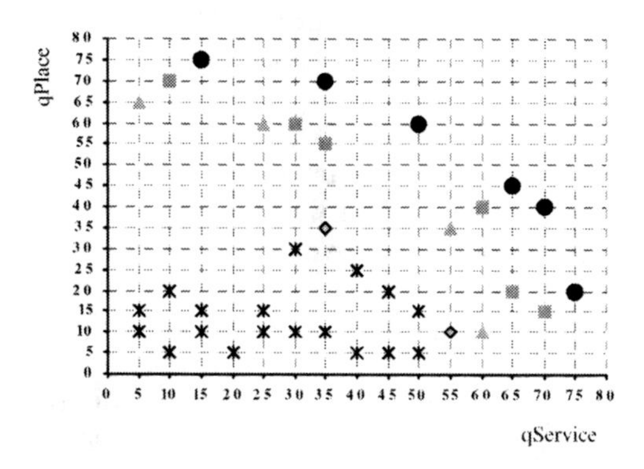


**Fig. 1.** Skyline of Best Service and Best Place Quality Hotels

Skyline, as Fig. 1 shows, left out rows that are in the frontier. That is: points that are close to the non-dominated ones. Instead of just a curve of "perfect" answers, we would like to obtain this curve but also those points that are near to it. Rows in the frontier might be interesting to the user, but they are rejected by the Skyline query. On the other hand, if we give all these points, how can the user establish difference between them? Moreover, how can the system aid user in discriminating desired answers? Fig. 1 might be rather interpreted with legend in Fig. 3 (next section).

# 3 Fuzzy Dominance Skyline

Zadeh [16] introduced fuzzy sets in 1965. In regular sets, there is a discontinuity between the set members and neighbor elements in the universe. Fuzzy sets provide the transition between the complete membership and the complete exclusion, using a membership function ranked in the real interval [0,1]. Any universe element is provided of a degree that represents its membership to the fuzzy set.

Some Skyline rows might be unsatisfactory to the user need. For example, consider a query for retrieving services and providers where selected service providers are unavailable for much time or simply they are disagreeable to the user. Many alternative solutions in the frontier are left out by Skyline. The solution to Skyline rigidity could be fuzzy sets. Thus, we can think that a row dominates other when it is much better than the other one. In this way, the querying system might provide an answer more suitable to user preferences.

In order to provide a mechanism for expressing fuzzy dominance, we propose to adopt SQLf[4] user defined fuzzy comparators that are specified with the syntax [13]:

```
CREATE COMPARATOR symbol ON domain AS expression IN fuzzy_set
```

This statement defines a linguistic comparator identified by *symbol.* A linguistic comparator is interpreted as a fuzzy binary relation. The *expression* may be **(X-Y)**, **(X/Y)** or **(X,Y)**. This *expression* specifies how to compare: based on the difference, the quotient, or by pairs. The *expressions* **(X-Y)** and **(X/Y)** may be used only in case of numeric *domain.* The **(X,Y)** *expression* may be used in any domain, even if it has not a natural order.

The $v_1$ *symbol* $v_2$ fuzzy condition's satisfaction degree is denoted as $\mu_{symbol}(v_1,v_2)$ and is given by the membership degree to the *fuzzy_set* for *expression* evaluated in **X**=$v_1$ and **Y**=$v_2$. The *fuzzy_set* may be defined by a trapezium, in case of **(X-Y)** or **(X/Y)** *expression* or by extension in case of **(X,Y)** *expression*.

A trapezium is specified by four values $(A,B,C,D)$. It defines a fuzzy set $F$ with membership function $\mu_F$ where $(V\leq A)\rightarrow(\mu_F(V)=0)$, $(A<V<B)\rightarrow(\mu_F(V)=(V-A)/(B-A))$, $(B\leq V\leq C)\rightarrow(\mu_F(V)=1)$, $(C<V<D)\rightarrow(\mu_F(V)=(D-V)/(D-C))$, $(D\leq V)\rightarrow(\mu_F(V)=0)$.

A fuzzy binary relation specified by extension is of the form $\{\mu_1/(A_1,B_1), \mu_2/(A_2,B_2), \ldots, \mu_n/(A_n,B_n)\}$. It defines a fuzzy set $F$ with membership function $\mu_F$ where $\mu_F((A_i,B_i))=\mu_i$.

Let's see an example of user defined fuzzy comparator:

```
CREATE COMPARATOR >> ON 0..80 AS (X-Y) IN (0,20,INFINIT,INFINIT)
```

With this statement, the user defines a >> fuzzy comparator according to user preference. Here, the comparison is based on the difference. The underlying fuzzy set is defined by the trapezium shape membership function of Fig. 2. For example: $\mu_{>>}(75,10)=1$, $\mu_{>>}(65,55)=.5$, $\mu_{>>}(75,70)=.25$ and $\mu_{>>}(70,75)=0$.

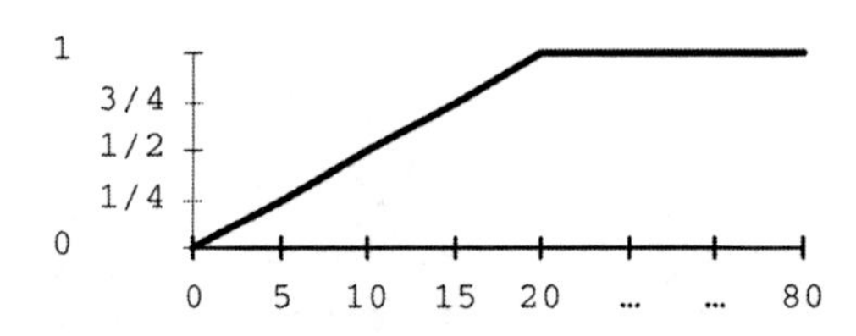


**Fig. 2.** Membership function of the fuzzy set in >> comparator definition

We propose a way to change crisp Skyline dominance with more general fuzzy comparison. Thus, the Fuzzy Dominance Skyline query structure is like:

```
SELECT * FROM rel SKYLINE OF a1 cmp1, ... , an cmpn
```

The new directive $cmp_i$ allows using any comparator: crisp and fuzzy. Directives **MIN**, **MAX** and **DIFF** and their aliases (**<**, **>** and **=**, respectively) may be used. The formal interpretation of this new querying structure becomes from the extension of the classic Skyline adding fuzzy set semantics [10].

The satisfaction degree of a conjunction is given by the minimum between the satisfaction degrees of the conditions in the conjunction. The ***inf*** operator over sets is intended to be the generalization of the minimum. The satisfaction degree of a fuzzy sentence with an existential quantifier becomes given by the highest value for the fuzzy condition under the quantification (***sup*** operator). Finally, the satisfaction degree for the negation of a fuzzy condition is the complement to one. For all these fuzzy logic operators, we have adopted the same interpretation that SQLf uses [4]. Moreover, this interpretation is commonly used for database querying [8].

The query result is a fuzzy relation. Each row $r$ in the result is pervaded of a satisfaction degree $\mu_r$ obtained from the fuzzy criteria. It is formally represented with the notation $\mu_r/r$ in a fuzzy relation specification. Rows with satisfaction degree equal to 0 are completely excluded of the result. Therefore, the Fuzzy Dominance Skyline query produces as result set of (2).

$$\left\{ \mu_r/r \;\middle/\; r \in \mathrm{rel} \wedge \left( \mu_r = 1 - \sup_{r' \in rel} \left( \inf_{i \in \{1..n\}} \mu_{cmp_i}(r'.\mathrm{a}_i, r.\mathrm{a}_i) \right) \right) \wedge (\mu_r > 0) \right\} \quad (2)$$

In a final user interface, selected rows would be automatically shown in decreasing order of their satisfaction degrees to the fuzzy criteria, as SQLf does.

Let's consider the example from Section 2. We ask for hotels with best service quality and best place quality, but based on the fuzzy comparator >> (as defined above), instead the classic **MAX** directive. Our Fuzzy Dominance Skyline query is:

```
SELECT * FROM hotel SKYLINE OF qService >>, qPlace >>
```

It retrieves more hotels than classic Skyline operator does. Moreover, the answer is discriminated. Using Fig. 3 given legend, we can see in Fig. 1 the result of this query,

● Hotels with degree 1.0 ■ Hotels with degree .75 ▲ Hotels with degree .50
◇ Hotels with degree .25 ✕ Non-retrieved Hotels

**Fig. 3.** Legend for Fuzzy Dominance Skyline of Best Service and Place Quality Hotels

We must remark that this new query feature allows user to define its own dominance comparison. It is done by means of user defined fuzzy comparators. Fuzzy Dominance Skyline may be based on different comparators in different dimensions. Thus, this new operator is more expressive than classic Skyline Operator is. The representation of user preference is richer. Moreover, Fuzzy Dominance Skyline may be defined using attributes whose domains might not have a natural order. This is made with the use of comparators interpreted as fuzzy binary relation defined by extension.

## 4 Implementation Issue

As ever, the problem of a new flexible querying feature is to provide reasonable time processing mechanisms. The naive strategy for Fuzzy Dominance Skyline consists in a nesting procedure computing all rows satisfaction degrees as follows:

| | |
|---|---|
| 1 | PROCEDURE NPS |
| 2 | Result := ∅ |
| 3 | FOR r1 IN rel LOOP |
| 4 | S := 0 |
| 5 | FOR r2 IN rel LOOP |
| 6 | m := $\inf_{i \in \{1 \ldots n\}} \mu_{cmp_i}$ (r2.a<sub>i</sub>,r1.a<sub>i</sub>) |
| 7 | IF m>s THEN s := m |
| 8 | END LOOP |
| 9 | IF s<1 THEN result := result ∪ {(1-s)/r1} |
| 10 | END LOOP |
| 11 | RETURN result |
| 12 | END NPS |

In the context of SQLf, the Derivation Principle has been proposed for keep low the added cost of fuzzy query processing[3][6]. This principle consists in deriving a classic SQL query intended for selecting rows whose satisfaction degree is greater or equal to a user given threshold *t*. If the user does not specify a threshold, rows with non-zero satisfaction degree must be selected. A variety of this kind of transformations is given in [11].

Formally, according to [6] and [15], given a fuzzy query in SLQf $\Psi$, the derived query $DQ(\Psi)$ is a regular SQL query $\Phi$ such that $support(result(\Psi)) \subseteq result(\Phi)$. Where: $result(\Psi)$ is the fuzzy set of rows retrieved by $\Psi$; remark that each row r in *result*($\Psi$) is provided of its satisfaction degree to the query $\mu_\Psi(r)$; $support(result(\Psi))$ is the regular set of elements in $result(\Psi)$ with $\mu_\Psi(r)>0$; and, finally, $result(\Phi)$ is the answer set of the regular query $\Phi$.

The basis of the principle is to derive boolean conditions from fuzzy ones. Given a fuzzy condition $fc$, a desired satisfaction degree λ, and a relational comparator ∝ in $(>,<,=,\geq$ or $\leq)$, $DC(fc,\propto,\lambda)$ is a booelan condition $bc$ such that $bc(v) \Leftrightarrow (\mu_{fc}(v) \propto \lambda)$. thus given a user defined fuzzy comparator *cmp*, $DC(v2\ cmp\ v1,=,1)$ is a crisp condition equivalent to $(\mu_{cmp}(v1,v2)=1)$ and $DC(v2\ cmp\ v1,<,1)$ is a boolean condition derived from $(\mu_{cmp}(v1,v2)<1)$ by equivalence. It is obvious the existence of such boolean conditions due to the definition of fuzzy comparators. There are distribution rules for the DC operator in order to be applied to complex fuzzy conditions. A complete list of derived conditions may be seen in [305].

We propose the following derived query for Fuzzy Dominance Skyline queries:

```
DQ(SELECT * FROM rel SKYLINE OF a1 cmp1, ... , an cmpn)=

SELECT * FROM rel AS r1 WHERE NOT EXISTS

    (SELECT * FROM rel AS r2 WHERE
         DC(r2.a1 cmp1 r1.a1,=,1) AND ... AND DC(r2.an cmpn r1.an,=,1))
```

It may proved that this is in fact a query that retrieves the same rows that the fuzzy one. The demonstration is not easy. It involves the formal semantics of Fuzzy Dominance Skyline queries given here in Section 3 above. In also involves the formal semantics of SQL nested queries. The proof must be done by equivalence based logic derivation. We do not present here the proof due to space limitation. Interested readers are referred to [13][14][15] where Derivation Principle is applied to others querying structures with corresponding formal proofs.

Previous derived query must be used in the Derivation Principle based processing method. As we can see in PROCEDURE NPS, the processing of Fuzzy Dominance Skyline queries requires an inner set processing. We introduce the notation *DIBQ(Ψ)* for the derived query for obtaining relevant rows to inner loop:

*DIBQ(*SELECT * FROM *rel* SKYLINE OF $a_1$ $cmp_1$, ... , $a_n$ $cmp_n$)=

```
SELECT * FROM rel AS r2 WHERE
     DC(r2.a1 cmp1 r1.a1,<,1) OR ... OR DC(r2.an cmpn r1.an,<,1)
```

Derivation Principle based method for Fuzzy Dominance Skyline queries, named PROCEDURE DPS, is obtained from PROCEDURE NPS with these three changes:

- In line 3, change rel with
  DQ(SELECT * FROM *rel* SKYLINE OF $a_1$ $cmp_1$, ... , $a_n$ $cmp_n$)
- In line 5, change rel with
  DIBQ(SELECT * FROM *rel* SKYLINE OF $a_1$ $cmp_1$, ... , $a_n$ $cmp_n$)
- In line 9, omit the condition check IF s<1 THEN.

## 5 Performance Analysis

We have proposed here a Fuzzy Dominance Skyline Operator that is based on user defined fuzzy comparators. It allows a richer expression of user preference regarding classic Skyline Operator. We have also proposed an evaluation mechanism for this new querying feature. But we still must to show if the processing cost of Fuzzy Dominance Skyline Operator is reasonable for a database system. Therefore, we make an experimental study of performance using formal model statistic. The idea of the performance analysis method is to explain the influence of several considered factors in the observed values from experiments. The importance of a factor is measured by the proportion of the total variation in the response that is explained by the factor.

In the experimentation: The queries were addressed to a database relation populated with 1000 random generated rows. Each row had ten integer values in the range from 1 to 30 and one string value. Each row was randomly generated following a uniform distribution. We adopted simple table structures with an index for each attribute. Queries randomly generated. Each query has its classic (crisp) and fuzzy version. We would like to compare the behavior of fuzzy query respects similar classic one. In case of Classic Skyline queries, we use the directive **MIN**, in case of Fuzzy Dominance Skyline queries, we use the directive << defined by:

```
CREATE COMPARATOR << ON 1..30 AS (X/Y) IN (0,0,0.33,1)
```

In this performance study, we just observe the *Total Spent Time* in query processing. The experiments were performed with three replicas. Considered factor where: — *The Dimension* (quantity of attributes in the criteria preference expression) with three considered levels: 2D (two dimensions), 3D (three dimensions) and 5D (five dimensions). For each one of these levels, we have used a random generated query. — *The Method*, with three considered levels: SKL (Classic Skyline with SFS algorithm), NPS (Fuzzy Dominance Skyline with Naive Processing Strategy) and DPS (Fuzzy Dominance Skyline with Derivation Principle based Strategy).

We have chosen a full factorial design for our experimental study. That is, we considered all the factors with all their levels. We think that all factors and their interactions have significant influence in the performance. This kind of design allows study the influence of each factor and all their interactions.

Experiments were run in a flexible querying system prototype on top of Oracle 9i RDBMS, running on a Red Hat Linux 8.0 operated 895MHz Pentium III computer with 512MB RAM and 20GB HDD. Query translator was coded in SWI Prolog. We have loaded the experimental results in R statistical software. The summary of the analysis of variance with the full factorial model is presented in Table 1.

**Table 1.** ANOVA of Full Factorial Model

```
                   Df  Sum Sq Mean Sq F value    Pr(>F)
Dimension           2   52242   26121  1010.6 < 2.2e-16 ***
Method              2 2217356 1108678 42895.7 < 2.2e-16 ***
Dimension:Method    4  119122   29780  1152.2 < 2.2e-16 ***
Residuals          18     465      26
---
Signif. Codes:  0 `***' 0.001 `**' 0.01 `*' 0.05 `.' 0.1 ` ' 1
```

The stochastic F-test in Table 1 shows that the model fits very well to the data. Therefore, we may take conclusions about the performance using this experiment, because it is stochastically valid. We can see also in Table 1 that the interaction between factors is very significant to explain the results (*** marks).

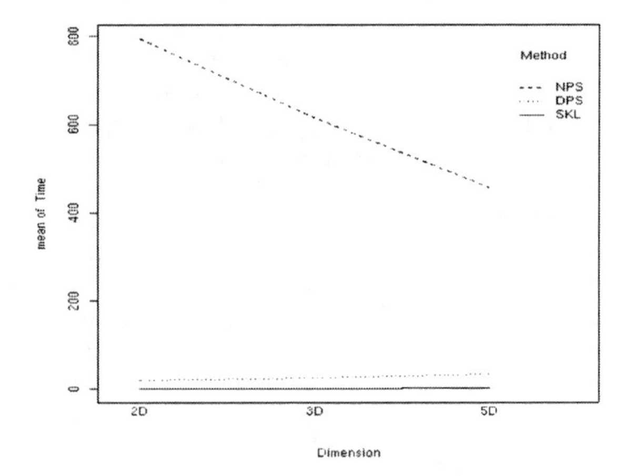


**Fig. 4.** Dimension-Method Interaction Plot

Let's analyze the interaction between factors, shown in Fig. 4. We are interested here in two comparisons: DPS vs. SKL methods, which show the over cost of Fuzzy Dominance Skyline queries against Skyline. On the other hand, DPS vs. NPS methods, that shows the advantage of using DPS for Fuzzy Dominance Skyline queries, respects the NPS.

We observe that DPS and SKL are similar in tendency: In both cases, times increase a little while Dimension does. We may explain it due to the non-dominated points increasing behavior regarding the dimension. While the criteria use more dimensions, it is less probable for a row to be dominated. Higher times for DPS respects SKL are reasonable. They seem to be in the same order. The difference obeys to the fact that with DPS we obtain a richer answer: with elements in frontier and discrimination. This shows that it is reasonable to implement a querying system with Fuzzy Dominance Skyline operator using DPS without an unacceptable impact in performance. On the other hand, times for NPS are too high respects to DPS. This is due to some extra computation and access to undesired rows. We observe that times decrease for the NPS while the Dimension increases. This is an amazing result that would be explored in future studies. Despite this behavior, NPS times are too high. It shows in fact that DPS improves processing of these kinds of queries.

## 6 Conclusions and Future Works

We have proposed a new querying operator: Fuzzy Dominance Skyline. It is an extension of Skyline based on SQLf fuzzy comparators. A more general new directive $cmp_i$ allows using any comparator both crisp and fuzzy in Skyline dominance relation. It solves the Skyline problem of rigidity, retrieving frontier rows and giving discriminated answers. Fuzzy Dominance Skyline may be defined using attributes whose domains might not have a natural order.

In order to process these queries, we have presented a Naïve Processing Mechanism and a Derivation Principle based evaluation method. The fuzzy query is evaluated over the result of a derived Boolean query. Derivation Principle takes advantage of relationship between fuzzy and classic sets with the concept of support.

An experimental performance study has been made analyzing new operator behavior respects classic Skyline queries. In the study: fuzzy queries were processed with both presented methods; and classic Skyline was processed with the SFS algorithm. The strategy based on the Derivation Principle demonstrated to keep low the extra-added cost of fuzzy query processing. Therefore, it is reasonable to implement a querying system with this new feature. It would be helpful to do more experimental studies involving non-considered factors, such as databases of different volume, queries with multiple tables.

Here, we have only dealt with the relaxation of the comparisons involved in the Skyline. In a future work we will deal with Skyline extension considering the use of fuzzy quantifiers combined with fuzzy comparators in order to provide more flexible dominance criteria. It would also be matter of further work to explore the possibility of defining specific algorithms for obtaining better performance in these new kinds of Fuzzy Dominance Skyline queries.

**Acknowledgments.** This work was supported in part by the Venezuela Foundation for Science, Innovation and Technology FONACIT Grant G-2005000278. The main reason for doing this work and all things that we do in this life is to acknowledge that person in our lives that helps us anytime: Jesus Christ the Lord of Lords. "Whatever you do, work at it with all your heart, as working for the Lord, not for men" (Colossians 3:23).

## References

1. Bosc, P., Farquhar, K., Pivert, O.: Integrating Fuzzy Queries into an Existing Database Management System: An Example. International Journal of Intelligent Systems 9, 475–492 (1994)
2. Bosc, P., Pivert, O.: Some Approaches for Relational Databases Flexible Querying. International Journal of Intelligent Systems 1(34), 323–354 (1992)
3. Bosc, P., Pivert, O.: On the efficiency of the alpha-cut distribution method to evaluate simple fuzzy relational queries. Advances in Fuzzy Systems-Applications and Theory, 251–260 (1995)
4. Bosc, P., Pivert, O.: SQLf: A Relational Database Language for Fuzzy Querying. IEEE Transactions on Fuzzy Systems 3(1) (February 1995)
5. Bosc, P., Pivert, O.: SQLf Query Functionality on Top of a Regular Relational Database Management System. Knowledge Management in Fuzzy Databases, 171–190 (2000)
6. Bosc, P., Liétard, L., Pivert, O.: Evaluation of Flexible Queries: The Quantified Statement Case. In: Proceedings of IPMU, Madrid, España, pp. 1115–1122 (2000)
7. Börzsönyi, S., Kossmann, D., Stocker, K.: The Skyline operator. In: Proceedings of International Conference on Data Engineering (ICDE), pp. 421–430 (2001)
8. Cox, E.: Relational Database Queries using Fuzzy Logic. Artificial Intelligent Expert, 23–29 (January 1995)
9. Godfrey, P., Shipley, R., Gryz, J.: Maximal Vector Computation in Large Data Sets. In: Proceedings of the Conference on Very Large Databases (VLDB), pp. 229–240 (2005)
10. Goncalves, M., Vidal, M.E.: Preferred Skyline: A Hybrid Approach between SQLf and Skyline. In: Andersen, K.V., Debenham, J., Wagner, R. (eds.) DEXA 2005. LNCS, vol. 3588, pp. 375–384. Springer, Heidelberg (2005)
11. Ma, Z.M., Yan, L.: Generalization of Strategies for Fuzzy Query Translation in Classical Relational Databases. Information and Software Technology 49(2), 172–180 (2007)
12. Papadias, D., Tao, Y., Fu, G., Seeger, B.: An Optimal and Progressive Algorithm for Skyline Queries. In: Proceedings of ACM SIGMOD, pp. 467–478. ACM Press, New York (2003)
13. Tineo, L.: Interrogaciones Flexibles a Bases de Datos Relacionales., Trabajo de Ascenso, Universidad Simón Bolívar, Caracas, Venezuela (1998)
14. Tineo, L.: Extending the power of RDBMS for Allowing Fuzzy Quantified Queries. In: Ibrahim, M., Küng, J., Revell, N. (eds.) DEXA 2000. LNCS, vol. 1873, pp. 407–416. Springer, Heidelberg (2000)
15. Tineo, L.: Una Contribución a la Interrogación Flexible de Bases de Datos: Evaluación de Consultas Cuantificadas Difusas, Tesis Doctoral, Universidad Simón Bolívar (2006)
16. Zadeh, L.A.: Fuzzy sets. Information and Control 8, 338–353 (1965)

# Pruning Search Space of Physical Database Design

Ladjel Bellatreche[1], Kamel Boukhalfa[1], and Mukesh Mohania[2]

[1] Poitiers University - LISI/ENSMA France
{bellatreche,boukhalk}@ensma.fr
[2] I.B.M. India Research Lab - India
mkmukesh@in.ibm.com

**Abstract.** Very large databases and data warehouses require many optimization structures to speed up their queries. These structures can be classified into two main categories: (1) redundant structures like mono attribute indexes, multi-attribute indexes (bitmap join indexes), materialized views, etc. and (2) no redundant structures, like horizontal partitioning and vertical partitioning. The problem of selecting any of these structures is a very crucial decision for the performance of the data warehouse. In this work, we focus on horizontal partitioning and bitmap join indexes. We first show the similarity between horizontal partitioning and bitmap join indexes. Secondly, we propose a new approach of selecting simultaneously these structures in order to reduce the query processing cost. It consists in using the horizontal partitioning schema obtained by a genetic algorithm to prune the search space of the problem of bitmap join index selection. Thirdly, we propose a greedy algorithm to select bitmap join indexes under a storage bound. Finally, we conduct several experimental studies using an adaptation of APB-1 benchmark in order to validate our proposed algorithms.

**Keywords:** Physical design, data partitioning, Bitmap join index.

## 1 Introduction

Very large databases and data warehouses store large amounts of data usually accessed by complex queries with many join operations. To speed up these queries, many optimization structures were proposed. We can divide them into two main categories: *redundant structures* like materialized views [8], advanced indexing schemes (bitmap, bitmap join indexes, etc.) [10] and *non-redundant structures* like horizontal, vertical partitioning [14] and parallel processing [15]. The main drawbacks of redundant structures are: extra storage cost and maintenance overhead. Horizontal partitioning (HP) is an important aspect of physical database design. In context of relational data warehouses, it allows tables, indexes and materialised views to be partitioned into *disjoint* sets of rows that are physically stored and accessed separately [14]. It has a significant impact on performance of queries and manageability of data warehouses. Many studies

R. Wagner, N. Revell, and G. Pernul (Eds.): DEXA 2007, LNCS 4653, pp. 479–488, 2007.

have recommended the combination of redundant and non-redundant structures to get a better performance for a given workload [14][12][16][3].

In [2], we showed that the best way to partition a relational data warehouse is to decompose the fact table based on the fragmentation schemas of dimension tables. Concretely, we (1) partition some/all dimension tables using their simple selection predicates[1], and then (2) partition the facts table using the fragmentation schemas of the fragmented dimension tables (this fragmentation is called *derived horizontal fragmentation*). This fragmentation procedure takes into consideration the star join queries requirements [2].

The number of horizontal fragments (denoted by $N$) of the fact table generated by this partitioning procedure is given by: $N = \prod_{i=1}^{g} m_i$ , where $m_i$ and $g$ are the number of fragments of the dimension table $D_i$ and the number of dimension tables participating in the fragmentation process, respectively. This number may be very large [2].

Bitmap index is probably the most important result obtained in the data warehouse physical optimization field [6]. The bitmap index is more suitable for low cardinality attributes since its size strictly depends on the *number of distinct values of the column on which it is built.* Bitmap join indexes (BJIs) are proposed to speed up join operations. In its simplest form, it can be defined as a bitmap index on a table $R$ based on a single column of another table $S$, where $S$ commonly joins with $R$ in a specific way.

Most of previous work in physical database design considered the problems of HP selection and BJI in isolation. However, both BJI and HP are fundamentally similar - both are structures that speed up query execution, pre-compute join operations and defined on selection attributes of dimension tables. Furthermore, BJIs and HP can interact with one another, i.e., the presence of an index can make a partitioned schema more attractive and vice versa. To illustrate this similarity, we consider an example in next section.

### 1.1 Similarity Between HP and BJIs: A Motivating Example

To show the similarity between HP and BJIs, we consider the following scenario that serves as a running example in this paper. Suppose we have a data warehouse represented by three dimension tables (TIME, CUSTOMER and PRODUCT) and one fact table (SALES). The population of this schema is given in Figure 1. The following query is executed on this schema.

```
SELECT Count(*)
FROM CUSTOMER C, PRODUCT P, TIME T, SALES S
WHEERE C.City='Poitiers'
AND P.Range='Beauty'
AND T.Month='June'
AND P.PID=S.PID AND C.CID=S.CID AND T.TID=S.TID
```

[1] A simple predicate p is defined by: p : $A_i$ $\theta$ $Value$, where $A_i$ is an attribute, $\theta \in \{=, <, >, \leq, \geq\}$, and Value $\in Dom(A_i)$.

[2] A star join query is a query defined on a star schema. It imposes restrictions on the dimension values that are used for selecting specific facts; these facts are further grouped and aggregated according to the user demands.

This query has three selection predicates defined on dimension table attributes *City, Range* and *Month* and three join operations. It can be executed using only HP (this strategy is called HPFIRST) or only BJI (BJIFIRST).

| C_RID | CID | Name | City |
|---|---|---|---|
| 6 | 616 | Gilles | Poitiers |
| 5 | 515 | Yves | Paris |
| 4 | 414 | Patrick | Nantes |
| 3 | 313 | Didier | Nantes |
| 2 | 212 | Eric | Poitiers |
| 1 | 111 | Pascal | Poitiers |

**Customer**

| P_RID | PID | Name | Range |
|---|---|---|---|
| 6 | 106 | Sonoflore | Beauty |
| 5 | 105 | Clarins | Beauty |
| 4 | 104 | WebCam | Multimedia |
| 3 | 103 | Barbie | Toys |
| 2 | 102 | manure | Gardening |
| 1 | 101 | SlimForm | Fitness |

**Product**

| T_RID | TID | Month | Year |
|---|---|---|---|
| 6 | 11 | January | 2003 |
| 5 | 22 | February | 2003 |
| 4 | 33 | March | 2003 |
| 3 | 44 | April | 2003 |
| 2 | 55 | May | 2003 |
| 1 | 66 | June | 2003 |

**Time**

| S_RID | CID | PID | TID | Amount |
|---|---|---|---|---|
| 1 | 616 | 106 | 11 | 25 |
| 2 | 616 | 106 | 66 | 28 |
| 3 | 616 | 104 | 33 | 50 |
| 4 | 545 | 104 | 11 | 10 |
| 5 | 414 | 105 | 66 | 14 |
| 6 | 212 | 106 | 55 | 14 |
| 7 | 111 | 101 | 44 | 20 |
| 8 | 111 | 101 | 33 | 27 |
| 9 | 212 | 101 | 11 | 100 |
| 10 | 313 | 102 | 11 | 200 |
| 11 | 414 | 102 | 11 | 102 |
| 12 | 414 | 102 | 55 | 203 |
| 13 | 515 | 102 | 66 | 100 |
| 14 | 515 | 103 | 55 | 17 |
| 15 | 212 | 103 | 44 | 45 |
| 16 | 111 | 105 | 66 | 44 |
| 17 | 212 | 104 | 66 | 40 |
| 18 | 515 | 104 | 22 | 20 |
| 19 | 616 | 104 | 22 | 20 |
| 20 | 616 | 104 | 55 | 20 |
| 21 | 212 | 105 | 11 | 10 |
| 22 | 212 | 105 | 44 | 10 |
| 23 | 212 | 105 | 55 | 18 |
| 24 | 212 | 106 | 11 | 18 |
| 25 | 313 | 105 | 66 | 19 |
| 26 | 313 | 105 | 22 | 17 |
| 27 | 313 | 106 | 11 | 15 |

**Sales**

**Fig. 1.** A sample of a data warehouse population

*HPFIRST:* The DBA can derived partitioned the fact table SALES using fragmentation schemas of dimension tables: CUSTOMER, TIME and PRODUCT based on *City, Month, Range*, respectively. Consequently, the fact table is fragmented in 90 fragments [3], where each fact fragment is defined as follows:

$$Sales\_i = SALES \ltimes CUSTOMER_i \ltimes TIME_i \ltimes PRODUCT_i,$$

$\ltimes$ represents semi-join operation. Figure 2c shows the fact fragment SALES_BJP corresponding to sales of *beauty products* realized by customers living at *Poitiers* city during month of *June.* To execute the above query, the optimizer shall rewrite it. Therefore, it loads only the fact fragment SALES_BJP. The above query is then rewritten as follows: *SELECT Count(*) FROM SALES_BJP*

*BJIFIRST:* The DBA selects an BJI on three dimension attributes: *City, Month* and *Range* as follows:

```
CREATE BITMAP INDEX sales_cust_city_prod_range_time_month_bjix
ON SALES(CUSTOMER.City, PRODUCT.Range, TIME.Month)
FROM SALES S, CUSTOMER C, TIME T, PRODUCT P
WHERE S.CID= C.CID AND S.PID=P.PID AND S.TID=T.TID
```

Figure 2a shows the generated BJI. To execute the above query, the optimizer just accesses the bitmaps corresponding to the columns representing *June, Beauty* and *Poitiers* and performs the AND operation. This example shows the

[3] We suppose we have 3 cities, 6 months and 5 ranges of product.

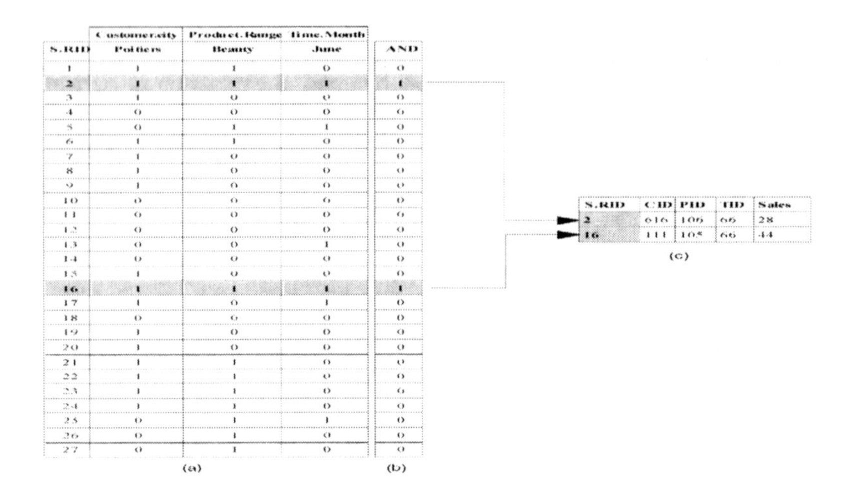


**Fig. 2.** (a) The bitmap join index (b) The result of AND operation, (c) The fact fragment

similarity between HP and BJIs. Both of them save three join operations. Contrary to HP, BJI needs to be stored and maintained. Based on this similarity we propose a new approach for selecting simultaneously HP and BJIs by pruning search space of BJIs. To the best of our knowledge, our proposed work is the first article that addresses this issue.

This paper is divided in five sections: Section 2 reviews the HP and BJI selection problem and their formalizations. Section 3 proposes our approach for selecting HP and BJIs that uses pruning rules of search space of BJI selection problem and a greedy algorithm to solve it. Section 4 gives the experimental results using an adaptation of the APB-1 benchmark. Section 5 concludes the paper by summarizing the main results and suggesting future work.

## 2 Selecting HP and BJIs

In this section, we propose our formulations of the two problems:

### 2.1 Horizontal Partitioning Selection Problem

A formulation of the HP problem is the following:
Given a data warehouse with a set of dimension tables $D = \{D_1, D_2, ..., D_d\}$ and a fact table $F$, a workload $Q$ of queries $Q = \{Q_1, Q_2, ..., Q_m\}$, where each query $Q_i$ $(1 \leq i \leq m)$ has an access frequency. The HP selection problem consists in fragmenting the fact table into a set of fact fragments $\{F_1, F_2, ..., F_N\}$ such that: (1) the sum of the query processing cost when executed on top of the partitioned star schema is minimized and (2) $N \leq W$, where $W$ is a threshold, fixed by the database administrator (DBA), representing the maximal number of

fact fragments that she can maintain. We have chosen this constraint because, in a DBMS supporting HP, each global query having conjunction of selection predicates $G$ must be rewritten using horizontal fragments. Each one is defined by a conjunction of predicates $P$ [4]. Three scenarios are possible to rewrite the query on the fragments [7]: (a) *no rewriting*: where $P \wedge G$ is *unsatisfiable*, implying that the fragment will not contribute to the answer to the query, (b) *perfect rewriting*: where $P \rightarrow G$, implying that the whole fragment will be in the answer to the query and (c) *partial rewriting*: neither $P$ *contradicts* $G$ nor $P$ *implies* $G$, implying that there may exist tuples in the fragment that will contribute to the answer to the query. The problem of rewriting global queries on horizontal fragments has been showed to be NP-Complete, if the number of selection predicates of $P$ is equal or higher than 3 [7].

To solve the HP selection problem, we have proposed a genetic algorithm [2], that we use in the present work.

### 2.2 BJIs Selection Problem

Formally, a BJI selection problem has the same formulation as HP, except the definition of its constraint. The aim of BJI selection problem is to find a set of BJIs minimizing the query processing cost and storage requirements of BJIs do not exceed $S$ (representing storage bound).

The index selection problem has been studied by many people [5], [9], [4], but there is not enough work in dealing with the BJI selection problem [1]. The main difficulty of BJI selection is the large number of dimension attributes that can participate on selection process (see Motivating example). To reduce this number, Aouiche et al. [1] proposed the use of data mining techniques (like CLOSE algorithm [13] that generates frequent itemsets). The basic idea is that if one attribute or a group of attributes are *frequently* present in the queries then it is interesting to consider them in the process of selection of BJIs. We claim that this approach does not give better results since it *uses only the frequency of appearance of attributes* to generate frequent itemsets. To ensure a better performance, there are other parameters, like size of tables, selectivity factors, cardinality of dimension attributes etc. that shall be taken into account.

## 3 Pruning Search Space of BJIs Selection Using HP

Actually, most commercial database systems support HP and BJIs. Therefore, to speed up a set of queries, the DBA have several choices: (1) using only HPFIRST, (2) using only BJIFIRST and (3) HP⟶BJI, where she combines the two structures. In the proposed work, we mainly focus on the third scenario, because HP preserves the schema of base tables [5], therefore, the obtained fragments can further be indexed.

[4] This process is called, the localization of the fragments, in the context of the distributed databases [11].

[5] A horizontal fragment has the same schema as its original base table.

In order to reduce the complexity of BJI selection problem, we propose the following approach that prunes its search space.

1. Partition the database schema using any algorithm. In this work, we use a genetic algorithm [2] according a set of queries $Q = \{Q_1, Q_2, \cdots, Q_m\}$. Our genetic algorithm generates a set of fragmentation attributes $FASET$ (dimension attributes used in definition of fragments). In our motivating example, *Month, City* and *Range* represent fragmentation attributes. Among $m$ queries, we can identify those that get benefit from HP, denoted by $Q' = \{Q'_1, Q'_2, \cdots, Q'_l\}$. This identification is done using a rate defined for each query $Q_j$ as follows:
$$rate(Q_j) = \frac{C[Q_j, FS]}{C[Q_j, \phi]} \quad (1)$$
where $C[Q_j, FS]$ and $C[Q_j, \phi]$ represent the cost of executing the query $Q_j$ on un-partitioned database and partitioned schema $FS$, respectively. The DBA has the right to set up this rate using a threshold $\lambda$: if $rate(Q_j) \leq \lambda$ then $Q_j$ is a profitable query, otherwise it is a no profitable query.
2. Among $FASET$, pick attribute(s) with a low cardinality in order to built BJIs. The set of these attributes is denoted by $BJISET$.
3. The selection of BJIs is then done using $BJISET$ and no profitable queries $Q' = \{Q'_1, Q'_2, \cdots, Q'_l\}$ using a greedy algorithm that we describe in the following section. Note that the selected BJIs shall reduce the cost of executing no profitable queries *obtained by HPFIRST*.

The architecture of our approach is summarized in Figure 3.

*Example 1.* To understand our approach let us consider the scenario based on our Motivating example, where HPFIRST generates 90 fact fragments. Assume that the DBA wants to fragment the database in only 18 fragments ($W = 18$). In this case, our genetic algorithm generates a fragmentation schema defined on dimension attributes *City* and *Month* ($FASET = \{Month, City\}$). Consequently, the attribute *Range* of PRODUCT is not taken in the fragmentation process ($BJISET = \{Beauty\}$). To speed up the whole queries, DBA may define a BJI on *Range*.

### 3.1 Greedy Algorithm for Generating BJIs

Our greedy algorithm is based on a mathematical cost model which is an adaptation of Aouiche's model [1]. It starts with configuration having a BJI defined on an attribute (of $BJISET$) with smallest cardinality (let say, $I_{min}$), and iteratively improves the initial configuration until no further reduction in total query processing cost and no violation of the storage bound. Details of the algorithm are given in Algorithm 1.

## 4 Experimental Studies

To evaluate our approach, we conduct several experimental studies. We adapt the dataset from the APB1 benchmark, by adding new attributes on

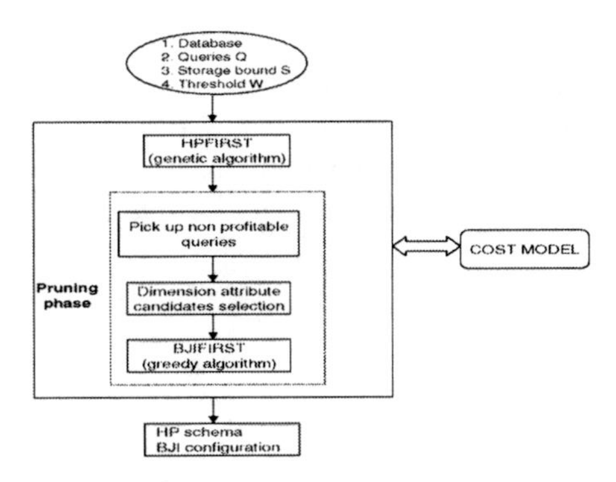


**Fig. 3.** Architecture of the proposed solution

---

**Algorithm 1.** Greedy Algorithm for BJIs Selection

---

**Inputs:** Set of non profitable queries: $Q' = \{Q'_1, Q'_2, \cdots, Q'_p\}$, $BJIASET$, $S$.
$BJI_j$: Bitmap join index defined on attribute $A_j$. $Size(BJI_j)$: storage cost of $BJI_j$
$C[Q', HPFIRST]$: cost of executing $Q'$ using HPFIRST
**Output:** $Config_{finale}$: set of selected BJIs.
**begin**
  $Config_{finale} = BJI_{min}$;
  $S := S - Size(BJI_{min})$;
  $BJISET := BJISET - A_{min}$;  *$A_{min}$ is the attribute used to defined $BJI_{min}$*
  WHILE ($Size(Config_{finale}) \leq S$) DO
    FOR each $A_j \in BJISET$ DO
      IF ($C[Q', (Config_{finale} \cup BJI_j))] < C[Q', HPFIRST]$)
      AND (($Size(Config_{finale} \cup BJI_j) \leq S$)) THEN
        $Config_{finale} := Config_{finale} \cup BJI_j$;
        $Size(Config_{finale}) := Size(Config_{finale}) + Size(BJI_j)$;
        $BJISET := BJISET - A_j$;
**end**

---

dimension tables in order to have more dimension attributes. The star schema of this benchmark has one fact table Actvars and four dimension tables: Actvars(24 786 000 tuples), Prodlevel (9 000 tuples), Custlevel (900 tuples), Timelevel (24 tuples) and Chanlevel (9 tuples). We have considered 74 queries with 17 selection attributes. These queries may contain from 1 to 6 selection attributes. Our algorithms have been implemented using Visual C++ performed under a Intel Centrino with a memory of 1 Go.

In the first experiments, we compare HPFIRST and BJIFIRST strategies. The parameters of our genetic algorithm are: *number of chromosomes* = 40, *number of generations* = 100, *crossover rate* = 75, *mutation rate* = 30 and *threshold* $W$ = 100. Figure 4 shows performance in terms of cost of evaluating the set of queries (74), by varying the storage capacity from 20 to 400 Mo. We observe that HP outperforms the BJIs when the storage bound is between 20 and 200. But when we assign more storage, BJIs outperforms better, especially for *COUNT() queries*. HP can also performs well when we relax its maintenance constraint representing the threshold $W$ (see Figure 5). In this case, we increase the probability that all selection attributes will participate in the final fragmentation schema.

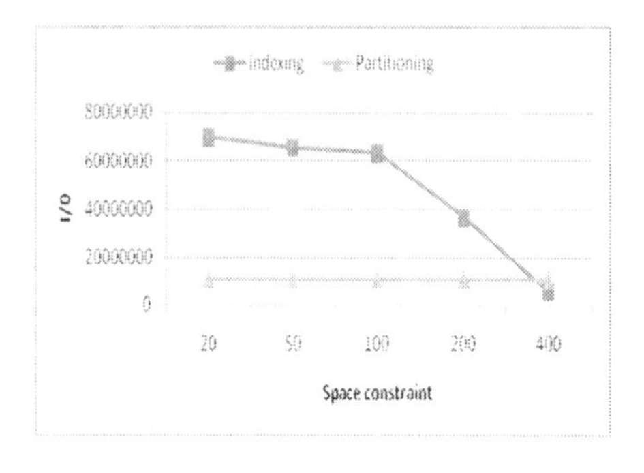


**Fig. 4.** HPFIRST vs. BJIFIRST($S$)

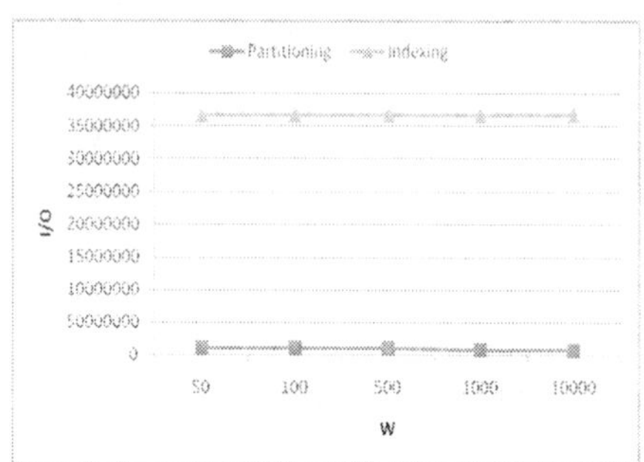


**Fig. 5.** BJIFIRST vs. HPFIRST($W$)

To evaluate our approach against other approaches, we conduct the following experiment. We use genetic algorithm with a fragmentation maintenance constraint $W = 100$ and $\lambda = 0.6$ (used to identify the profitable queries from HP). After execution of this algorithm, among 74 queries, we identify 19 no profitable queries (Q') and 8 ($HPSET = 8$) fragmentation attributes. Consequently, 9 ($BJISET = 9$) will be used to select the finale configuration of BJIs. Our greedy algorithm is then executed using $Q'$, $BJISET$ and a space bound equal 20 Mo. Three BJIs have been selected. The performance of our approach is shown in Figure 6. It reduces the cost obtained by HPFIRST by 12% and without adding extra space cost. We conduct also experiment to compare HPFIRST (with $W = 100$ and $\lambda = 0.6$.) with our approach, by varying storage space from 0 (where none BJI is selected) to 70 (all BJIs candidates are selected) in our greedy algorithm. Figure 7 shows the comparison results.

To evaluate the importance of $\lambda$ on performance of our approach, we compare it by varying $\lambda$ from 0 (all queries are non profitable from HP) to 1 (all queries are profitable from HP). Figure 8 shows that when this parameter is equal 0, our approach reduces the cost by 18% by assigning 18 Mo to the selected BJIs and when it reaches 1, its performance is equal to HPFIRST (no index). This parameter is important in physical database design. DBA can emphasize HP, by only using HPFIRST strategy (by setting up for example $\lambda$ to 1), BJIs by using only BJIFIRST strategy ($\lambda$ is set up to 0), or combining the two strategies. Thus, this approach can cater to the existing database design strategies.

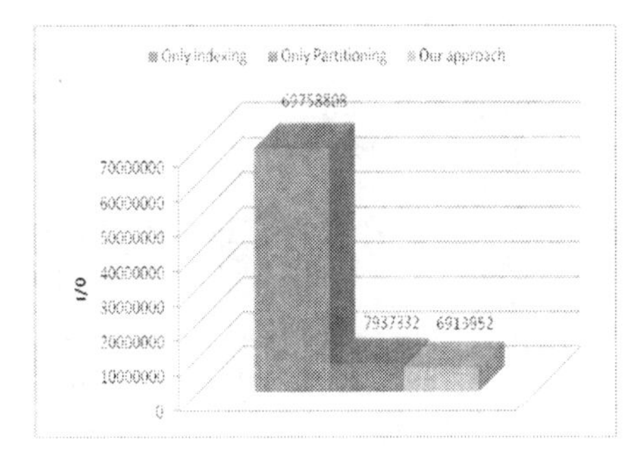


Fig. 6. Quality of our approach

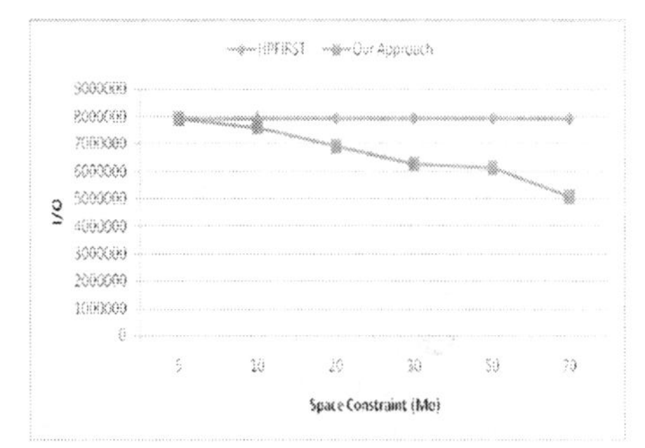


Fig. 7. Our approach vs. storage

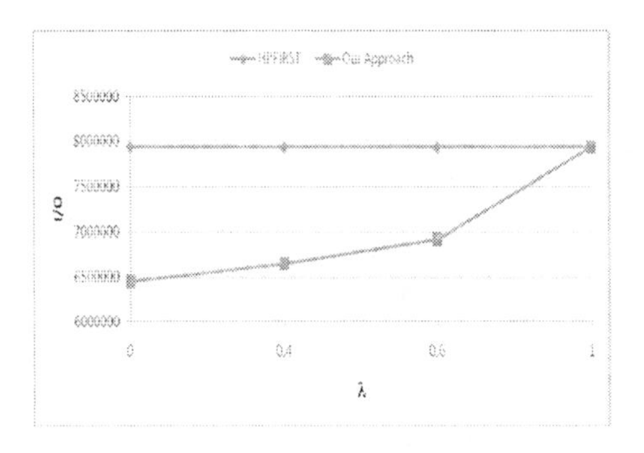


Fig. 8. Effect of lambda on performance

## 5 Conclusion

Databases have grown very large in size and accessed with complex queries having several join operations. To optimize these queries, several structures are available and supported by commercial DBMSs that we can classify into two categories: redundant and non-redundant structures. In this paper, we concentrate on one redundant structure (bitmap join indexes) and non-redundant (horizontal partitioning). By an example, we showed the strong similarity between them. We have proposed a new approach to select simultaneously a HP schema and BJIs to optimize a set of queries. It starts by selecting a HP schema using a genetic algorithm, and then selects BJIs by using greedy algorithm that considers only queries that do not get profit from HP and eliminating fragmentation attributes. This approach prunes the search space of BJI selection problem. We have conducted several experimental studies that showed the cost savings with modest storage requirements.

In the future, we plan to develop and compare a number of heuristics for pruning the exhaustive search space in order to improve the quality of the selected solution. Another issue that we should consider applying the same approach for selecting materialized views.

## References

1. Aouiche, K., Boussaid, O., Bentayeb, F.: Automatic Selection of Bitmap Join Indexes in Data Warehouses. In: Tjoa, A.M., Trujillo, J. (eds.) DaWaK 2005. LNCS, vol. 3589, Springer, Heidelberg (2005)
2. Bellatreche, L., Boukhalfa, K.: An evolutionary approach to schema partitioning selection in a data warehouse environment. In: Tjoa, A.M., Trujillo, J. (eds.) DaWaK 2005. LNCS, vol. 3589, pp. 115–125. Springer, Heidelberg (2005)
3. Bellatreche, L., Schneider, M., Lorinquer, H., Mohania, M.: Bringing together partitioning, materialized views and indexes to optimize performance of relational data warehouses. In: Kambayashi, Y., Mohania, M.K., Wöß, W. (eds.) DaWaK 2004. LNCS, vol. 3181, pp. 15–25. Springer, Heidelberg (2004)
4. Chaudhuri, S.: Index selection for databases: A hardness study and a principled heuristic solution. IEEE Transactions on Knowledge and Data Engineering 16(11), 1313–1323 (2004)
5. Chaudhuri, S., Narasayya, V.: An efficient cost-driven index selection tool for microsoft sql server. In: Proceedings of the International Conference on Very Large Databases, pp. 146–155 (August 1997)
6. Golfarelli, M., Rizzi, S.: A methodological framework for data warehouse design. In: DOLAP, pp. 3–9 (November 1998)
7. Guo, S., Wei, S., Weiss, M.A.: On satisfiability, equivalence, and implication problems involving conjunctive queries in database systems. IEEE Transactions on Knowledge and Data Engineering 8(4), 604–612 (1996)
8. Gupta, H.: Selection of views to materialize in a data warehouse. In: Afrati, F.N., Kolaitis, P.G. (eds.) ICDT 1997. LNCS, vol. 1186, pp. 98–112. Springer, Heidelberg (1996)
9. Labio, W., Quass, D., Adelberg, B.: Physical database design for data warehouses. In: Proceedings of the International Conference on Data Engineering (ICDE) (1997)
10. Oneil, P.: Multi-table joins through bitmapped join indioces. In: SIGMOD, vol. 24(03) (1995)
11. Özsu, M.T., Valduriez, P.: Principles of Distributed Database Systems, 2nd edn. Prentice-Hall, Englewood Cliffs (1999)
12. Papadomanolakis, S., Ailamaki, A.: Autopart: Automating schema design for large scientific databases using data partitioning. In: Proceedings of the 16th International Conference on Scientific and Statistical Database Management (SSDBM 2004), pp. 383–392 (June 2004)
13. Pasquier, N., Bastide, Y., Taouil, R., Lakhal, L.: Discovering frequent closed itemsets. In: Beeri, C., Bruneman, P. (eds.) ICDT 1999. LNCS, vol. 1540, pp. 398–416. Springer, Heidelberg (1998)
14. Sanjay, A., Narasayya, V.R., Yang, B.: Integrating vertical and horizontal partitioning into automated physical database design. In: Proceedings of the ACM SIGMOD International Conference on Management of Data, pp. 359–370 (June 2004)
15. Stöhr, T., Märtens, H., Rahm, E.: Multi-dimensional database allocation for parallel data warehouses. In: Proceedings of the International Conference on Very Large Databases, pp. 273–284 (2000)
16. Zilio, D.C., Rao, J., Lightstone, S., Lohman, G.M, Storm, A., Garcia-Arellano, C., Fadden, S.: Db2 design advisor: Integrated automatic physical database design. In: Proceedings of the International Conference on Very Large Databases, pp. 1087–1097 (August 2004)

# A Two-Phased Visual Query Interface for Relational Databases

Sami El-Mahgary [1] and Eljas Soisalon-Soininen[2]

[1] Computing Centre, Helsinki University of Technology, Box 1100,
FI-02015 TKK, Finland.
Sami.Mahgary@tkk.fi
[2] Laboratory of Software Technology, Helsinki University of Technology,
Box 5400, FI-02015 TKK, Finland.
Eljas.Soisalon-Soininen@tkk.fi

**Abstract.** Developing an easy-to-use, visual query tool for non-expert users to perform their own ad-hoc queries from relational databases is an active research area. The challenge lies in designing a visual query application that is both versatile and user-friendly; often expressive power comes at the expense of non-intuitive user interfaces. This work is based on a novel two-phased approach, so that users first specify the subset of data that is of interest (using a so-called principal concept), whereby the query generates an intermediary dataset. In the second phase, users query this dataset to further eliminate unwanted rows through Boolean constraints and to perform any additional operations, i.e. grouping or columns renaming. The user interface has been kept simple and the presented ideas have been successfully implemented in a prototype known as OVI-2, for use with the university student database at Helsinki University of Technology.

**Keywords:** Relational databases, visual query interfaces, query languages.

## 1 Introduction

Applications that provide a visual query interface for a database are commonly referred to as visual query systems (VQS). Developing an easy-to-use VQS for users to perform their own ad-hoc queries from relational databases has been an active research area in the last decades; a compilation in the late 1990s already counted dozens of different approaches [1,2]. While many authors use the terms VQS and query languages rather interchangeably, few approaches actually incorporate a full, 'self-standing' query language with its well-defined grammar such as [3]. Most VQS, including this work, focus on developing a user-friendly interface for ad-hoc querying that maps queries to an existing commercial query language such as SQL or OQL (Object Query Language) [4,5,6].

The goal in this project was to develop a user-friendly visual query interface for querying a relational student database consisting of a large number of relations. The requirement was that the tool be versatile, yet easy enough to use without requiring

R. Wagner, N. Revell, and G. Pernul (Eds.): DEXA 2007, LNCS 4653, pp. 489–498,2007.

any specific knowledge of the underlying database schema. The developed query interface is named OVI-2 (Oodi's Visual Interface) with the '2' emphasising the fact that there are two distinct stages to generating the query, and Oodi being the name of the student database.

One reason why developing a user-friendly and yet powerful VQS is not an easy task is that typically, versatility and expressive power come at the expense of user-friendliness. Moreover, some query operations such as universal or existential quantification do not lend themselves to a clear visual representation [7, 8]. The Visual Query Language developed by [3] is an example of a VQS built on top of a powerful language supporting universal quantification. However, a successful use of such an interface for complex queries requires expertise, as the user may need to specify database variables in the query.

There are today simply too many different VQS implementations so that it is not within the scope of this work to provide a history of the developments in the field. Based on [1,5] query interfaces may be classified into four categories, depending on how the query formulation is visualised [1]. These are (1) tabular, (2) diagrammatic, (3) iconic and (4) hybrid. To this list could be added the (5) graph-based approach (the nodes being the entities and the arcs the existing relationships), which portrays the underlying schema as a graph, such as the Gql implementation [9].

Tabular approaches are part of the first generation which produced the well-known language QBE [10]. At their simplest, tabular systems alleviate the user's task by providing a tabular representation of the underlying schema, so that users need not input the names of relations or attributes.

The diagrammatic (2) approach represents the underlying database schema with graphic elements such as rectangles (typically denoting entities), circles and lines. An icon-based system (3) relies heavily on icons and uses a specific icon for each important concept in the schema. Because finding a suitable, unique icon for each concept is not always possible, icons are often used together with a tabular or diagrammatic approach to enhance the visual appeal of the system. A hybrid interface (4) uses any combination of the previous three approaches (1-3). Although OVI-2 relies on interactive forms, it uses icons and concepts and is best characterised as a hybrid approach.

The remainder of this paper is organised as follows: Chapter 2 discusses the basics of OVI-2, and Chapter 3 presents it using a real example. Chapter 4 briefly examines implementation issues while Chapter 5 shows some users' comments. Chapter 6 compares OVI-2 to previous work and finally, Chapter 7 presents the concluding remarks.

## 2 Basic Idea Behind OVI-2

The ideal VQS should be versatile and allow even complex queries to be defined in a user-friendly fashion. An SQL-query may be considered as complex when it involves sub-queries or makes use of Boolean conditions on the group operator (i.e. the Having-clause in SQL) [6]. Special care should be taken to ensure that the query formulation can be done with as little ambiguity as possible, so that the user can be assured that the query defined is indeed what was originally intended.

Besides allowing for the selection of attributes, a user-friendly VQS should provide support for some kind of high-level data abstraction so that users can easily interact with the database schema. This can be achieved by using *concepts* which are often visualised as icons or simple boxes. Simply put, a concept is a higher-level entity that simplifies query formulation. A concept such as 'Exam' may be comprised of several attributes (e.g. exam date, exam grade). The requirement in [5] that attributes of a concept must belong to the same relation is not enforced here.

OVI-2 queries are thus performed in two separate stages, where the first stage is concerned with defining the basic result set. Users first select a single concept of primary interest, known as the *principal concept* [5]. A primary concept helps to hide some of the complexities of the database schema by showing only those concepts that are relevant to the user's query. Selecting the principal concept in OVI-2 (currently either 'Students' or 'Teachers') displays respectively either a 'Student' or a 'Teacher' form. Users then fill-in a few simple details to characterise the selected principal concept. This in effect defines the *query topic*, i.e. that subset of the schema that is shown to the user [6].

Once the user has defined the query topic, s/he can run the query against the database to obtain a simple tabular report. The user can then freely browse these results and decide (i) to either go back and refine the query topic, or (ii) proceed to the second query stage.

At the onset of the second stage, the user has at his/her disposal all the required data re-arranged into new, intuitive temporary relations. No further access to the original database server is needed in the second stage. In this respect, the presented approach differs from the traditional incremental query building techniques in the literature [6]. The user can also define simple *Boolean conjunctions/disjunctions* to further eliminate unwanted rows, and apply additional operations such as grouping. In the second stage, the query is run against the temporary relations obtained from Stage I and the final results are presented in a tabular or spreadsheet format.

**Table 1.** The basic steps for querying with OVI-2

| Step | Example of What Step Might Involve |
|---|---|
| (1) Select primary concept (Stage I) | Choosing 'Students' as Primary Concept. |
| (2) Define query topic: restrict the set using sub-concepts and simple restrictions (Stage I). Each sub-concept represents a unique set of students. | Selecting Undergraduates who started in 2005 or later and have not completed a given set of courses. |
| (3) Run query and browse results set. Proceed to Stage II to further restrict result set or return to Stage I to loosen restriction on the sub-concepts. | Browsing results to ascertain that all required data is included. |
| (4) Place Boolean conditions to eliminate unwanted rows and pick attributes to include in final result set (Stage II) | Restricting query to undergraduates who have taken over 20 courses and have registered with the university. |

## 3 A Working Example of OVI-2

This chapter presents a real-life example based on the needs of the users of the student database used at Helsinki University of Technology. Users would like to get a list of all students who have not yet completed their Swedish Language Proficiency Test (SLPT), which is a mandatory part of the graduation requirements. This need creates a challenging query as the SLPT can be completed in a number of ways; namely, by taking a single advanced Swedish course, or two upper-level Swedish courses. The following is a step-by-step overview on defining the example query using OVI-2.

*Example: What are the names of those undergraduates students who started their studies in the year 2005 or later but have not yet passed the Swedish Language Proficiency Test ?*

### 3.1 Stage I: Specifying the Sub-concepts

In the very first step, the user simply determines the query's principal concept. Currently in OVI-2, when the user starts the application s/he chooses the principal concept to be either 'Students' or 'Teachers'. In this example query, the user is interested in finding out the set of students that meet a given criteria (e.g. have completed the SLPT requirement) so the query results are directly related to students, and thus 'Students' is the query's principal concept.

Once the user has decided on the principal concept, s/he is presented with a main form for defining the query topic. Since 'Students' is the principal concept, the user now indicates what type of students s/he is interested in retrieving, e.g. the sub-concepts of interest. The 'Students' concept is divided into five sub-concepts: (1) Undergraduates, (2) Graduates, (3) Doctoral students, (4) those who have completed their doctoral studies and finally (5) Non-Degree students. One or several sub-concepts can be selected through checkmarks.

As the user is now only interested in undergraduates, a single checkmark is placed (Fig. 1). The acceptance year for undergraduates is then restricted to be at least 2005 in the form as shown in Fig. 1. Having defined the set of students to be included in the query, the user may then specify any requirements for courses taken/not taken for the given set of students (Sect. 3.2)

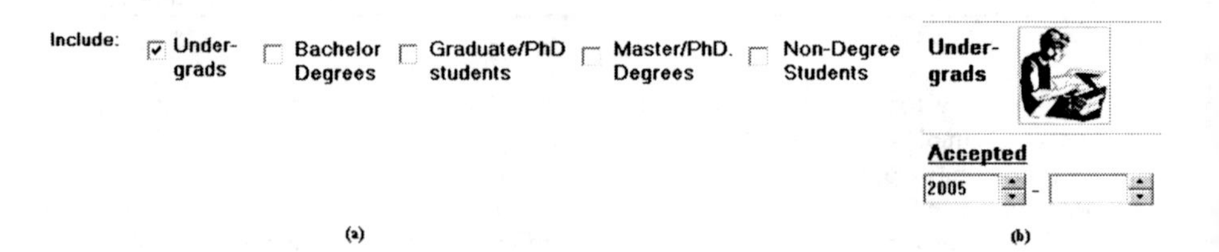


**Fig. 1.** Specifying the set of students using one to five sub-concepts derived from the principal concept 'Students' in (a). On the right, restricting undergraduates to those accepted in 2005 or later (b).

### 3.2 Stage I : Existential Quantification

With OVI-2, the courses taken/not taken are specified through a separate Course form, part of which is shown in Figure 2. In this case we're interested in students who have not passed the SLPT, which in fact is a test for negation (of existential quantification). Because SLPT can be completed in more than one way, OVI-2 makes use of a special User-Defined Course List (UDCL), where users have beforehand created their own course lists (please see Appendix for details).

Using such a list greatly simplifies the visual interface, as the user now needs only to indicate s/he is using a UDCL (Fig. 2a) and then select SLPT from a pop-up window showing all previously defined UDCLs (Fig. 2c). Finally, selecting the option 'Not yet taken!' (Fig. 2b) will eliminate all students in question who have passed the Swedish Language Proficiency Test, effectively returning those who have not yet passed it.

**Fig. 2.** Selecting a user-defined list (a) will cause the previously defined course lists to appear in a pop-up window for the user to pick from (c). The selection in Fig. 2b indicates that the specified course(s) must not have been taken by the set of students in question.

### 3.3 Stage II : Selecting the Attributes and Eliminating Unwanted Rows

At the last stage, the user can further refine the query by issuing Boolean conditions through a point-and-click interface. Attributes to be included in the final result set are selected by placing checkmarks next to them. Users can also perform various operations such as attributes renaming, along with grouping and sorting.

As an example, the result set obtained in Stage I (undergraduates who started in 2005 or later and had not taken their SLPT) could now be restricted to only those students who have completed at least 20 courses and who have not been marked as unregistered during the last two semesters, Autumn 2006 and Spring 2007. This is visualised in Fig. 3.

## 4 Implementation Issues

OVI-2 is basically a user interface (developed with Delphi, an object-oriented Pascal language) for querying a complex database schema (Oracle server). Additionally, the system maintains a data repository for storing 'lookup data' that does not change on a daily-basis. This includes basic information on Teachers, Courses, Departments and Student Majors, which are typically updated into this repository every month or so.

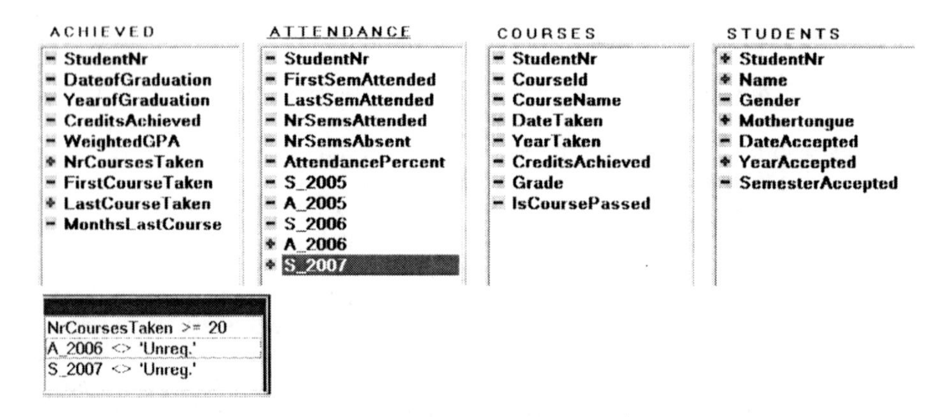


**Fig. 3.** The eight attributes (marked with a plus-sign) to be included have been selected (the data has been re-organised into four temporary relations, 'Achieved', 'Attendance', 'Courses' and 'Student'). The condition box at the bottom further restricts the set of students to those who have taken at least 20 courses and have registered during the last two semesters, Autumn 2006 and Spring 2007.

The OVI-2 implementation generates its own temporary relations (without resorting to data warehousing) based on data extracted from the database server in the first stage. As a special twist, lookup data is read from a data repository that can be quickly rebuilt from scratch through a simple mouse-click to reflect the latest state. OVI-2 also supports user settings, of which the most important is the section for defining the user-defined course list. The use of such a UDCL greatly reduces query complexity and the possibility for errors when searching for students who have completed/not completed a given set of courses.

## 5 User Comments

At the time of writing, OVI-2 had undergone limited testing by actual users (mostly University Study Counsellors), but some feedback was already available. In particular, users appreciated the following:

1. No special language needs to be learnt, nor are there possibly time consuming dragging operations to be mastered.
2. The two-phase approach allows for quick data browsing without the need to specify the attributes to be retrieved.
3. The Boolean conditions that are interactively defined in Stage II effectively filter out unnecessary rows. Determining for instance students who have been absent a given semester is easily defined using the condition box approach of Fig 3.

Regarding improvements, the OVI-2 interface could be enhanced for negating existential qualification. For instance, when looking for students who have not taken French I and French II, the interface should make it clear whether the query will return students who have not yet completed both French I and French II, or whether the query will return only those have taken neither French I nor French II (a smaller subset). Both negations are supported by OVI-2, but users may get confused as to the meaning of each negation if the difference is not clearly visualised.

## 6 Comparison with Other Approaches

The following briefly discusses four VQSs that are all aimed at mostly non-expert users and that like OVI-2, support concepts in some way.

VISIONARY uses a viewpoint (similar to a query topic) to display the underlying structure of the database in a user-friendly way, limiting the view to the part that interests the user [5]. Once the primary concept has been selected, the system displays concepts as icons and their associations as arrows with descriptive text. In some cases, the default associations are not what the user is after, so VISIONARY has capabilities for allowing the user to edit the viewpoint. That is, the user may (i) edit/disable associations between concepts, and if required, (ii) drop associations or drag them to the new concepts.

A similar interesting approach is ConQuer-II, which is also built around the theme of concepts [11]. As with VISIONARY, the user builds the query topic through various dragging operations. The attributes of the concepts are initially hidden to reduce complexity, but can be included or dropped from the query through a simple mouse-click. ConQuer-II's expressive power is well founded as it is built on top of an Object-Role Modelling approach. It does however, rely less on icons and visual aids than VISIONARY for instance.

QueloDB presents the user with a visual query frame for building the query topic in progressive steps. As with OVI-2, the user must first select a principal concept (the so-called starting class) which in the database used by the authors is one of Professors, LectureSeries, or Students [12]. The user then gradually builds the query by selecting sub-concepts that are semantically correct with respect to the selected principal concept (or a previously selected sub-concept). This allows for a rather intuitive interface. However, as the authors of QueloDB note, the expressive power is a bit limited, and although Boolean constraints are supported, negation is supported only for literals and not at the existential quantification (e.g. tuple) level.

Kaleidoquery presents some very interesting ideas on implementing a VQS. Kaleidoquery was developed for an object-oriented database, and concepts (referred to as classes) are represented visually through an icon and the name of the concept [4]. The equivalent of a principal concept (the so-called 'initial data type') is shown as the lowest icon on the visualisation screen from which the query is considered to flow to (an upward arrow represents the flow) other concepts in the query. As the query grows in complexity however, it requires more and more visualisation space and the workspace may thus become cluttered and the query's meaning harder to grasp.

Although OVI-2 and VISIONARY both share similarities (support of a principal concept and aggregation for instance), OVI-2 takes a different approach for the query topic through its two-phased technique. For one, OVI-2 is based on the assumption that attribute selection (e.g. projection) is a simple, albeit time-consuming task that should be performed only when the user has ascertained that the obtained dataset does indeed satisfy his/her requirements. This is why in the first query stage, OVI-2 automatically returns a set of fixed, pre-defined attributes which the user can then expand in the second stage to freely pick those that are actually needed for the final result set. This two-phased approach has the added benefit that it re-organises the data into more intuitive relations and generates new calculated attributes to which the user can refer to when generating the Boolean conditions in the second phase. This enhances the expressive power of OVI-2, as not all complex queries can be dealt with in a single query. Finally, deciding on the primary concept is somewhat simpler with OVI-2 than with VISIONARY, since a primary concept is simply defined as the concept to which the attributes returned by the query are related to.

Allowing the user to build the query topic through dragging and editing certainly adds to the flexibility of the query. But it may also increase the possibility for errors while designing the query. This is especially true if the system is geared towards users with little knowledge of the underlying database structure. This is why OVI-2, unlike Kaleidoquery for instance, eliminates altogether the step of building the query topic incrementally through graphical primitives, and relies instead on an interactive query form.

Moreover, defining the query topic in OVI-2 does not involve changing the associations between the concepts as with VISIONARY. Instead, the user simply defines the query topic by filling in a few items on an interactive query-form. These items are basically high-level operations which are then translated into join operations and existential quantification behind the scenes. With OVI-2, the user is thus happily unaware of the complexity of the query and can focus on those parts of the query-form relevant to the query at hand.

## 7 Conclusions

With regards to the non-expert user, this work has presented a two-stage approach for formulating user-friendly queries. In particular, the OVI-2 approach has the following advantages:

1. The interface prevents the formulation of semantically incorrect queries and queries that don't make real sense in the first place.
2. Because the query is formulated in two distinct stages, it allows for flexible queries by re-arranging the data into a more query-friendly structure.

With a visual and especially diagrammatic query language, there is always the possibility that users build a query that is semantically correct, but is not what was intended. To reduce this possibility, a VQS should not require users to specify complex operations of which the meaning may not be always clearly understood. Based on observing non-expert users, restriction operations (e.g. Boolean conditions) present little or no problem. However, requiring users to join relations or understand sub-queries is beyond the scope of most users; this kind of terminology should be reserved to database professionals. This does not of course imply that complex queries or negation at the tuple level cannot be supported by the interface, only that they be well integrated into the system so that users are not required to deal with these notions 'as such'.

While this work focused on a student database, the ideas presented should be applicable elsewhere too. For instance, if the students database analogy is extended to a typical order database schema involving Products, Customers and Orders, finding those customers who have not ordered a set of given items is quite akin to finding out students who have not completed a set of given courses.

This work has presented some of the key features found in OVI-2. Generally speaking, a VQS could probably be simplified by incorporating only the query topics that are actually needed for each principal concept selected by the user, without thus resorting to a full implementation of relational algebra operations. Nevertheless, designing a VQS that is intuitive and yet powerful enough for sophisticated queries is quite challenging and it may not be possible to accommodate all different types of database users in a single VQS implementation.

## Acknowledgments

The following persons (in alphabetical order) have supported and given valuable comments regarding OVI-2; their help is gratefully acknowledged: Mr. Petri Autio, Prof. Kalevi Ekman, Ms. Pirjo Häkkinen, Ms. Päivi Kauppinen, Ms. Katriina Korhonen, Ms. Päivi Koivunen, Mr. Harri Långstedt, Prof. Lauri Malmi, Mr. Juhani Markula, Ms. Kirsti Olamo, Ms. Vuokko Rantanen, Ms. Rita Rekonen, Ms. Pia Rydestedt, Ms. Pirjo Solin, Mr. Timo Tuhkanen and Mr. Jan von Pfaler.

## References

1. Catarci, T., Constabile, M., Levialdi, S., Batini, C.: Visual query systems for databases: A survey. J of Visual Lang. Comput. 8, 215–260 (1997)
2. Ower, V.: Development of a conceptual query language: Adopting the user-centred methodology. Comp. J. 46, 602–624 (2003)
3. Mohan, L., Kashyap, R.: A Visual Query Language for Graphical Interaction with Schema-Intensive Databases. IEEE T. Knowl. Data En. 5, 843–858 (1993)
4. Murray, N., Paton, N., Goble, C., Bryce, J.: Kaleidoquery: a flow-based visual language and its evaluation. J. of Visual Lang. Comput. 11, 151–189 (2000)
5. Benzi, F., Maio, D., Rizzi, S.: VISIONARY: a Viewpoint-based Visual Language for Querying Relational Databases. J. of Visual Lang. Comput. 10, 117–145 (1999)
6. Zhang, G., Chu, W., Meng, F., Kong, G.: Query Formulation from High-Level Concepts for Relational Databases. In: Proc. of User Interfaces to Data Intensive Systems, Los Angeles, pp. 64–74 (1999)
7. Bélières, B., Trépied, C.: New metaphors for a visual query language. In: Proc. Of International Workshop on Database and Expert Systems Applications, DEXA '96 University of Zurich, pp. 229–236 (1996)
8. Whang, K.-Y., Malhotra, A., Sockut, G.H., Burns, L., Choi, K.-S.: Two-Dimensional specification of universal quantification in a grapahical database query language. IEEE Transact. on Softw. Eng. 18, 216–224 (1992)
9. Papantonakis, A., King, P.J.H.: Gql a declarative graphical query language based on the functional mode. In: Proc of the Workshop on Advanced Visual Interfaces, Bari Italy, pp. 113–122 (1994)

10. Zloof, M.M.: Query-By-Example, a data base language. IBM Syst. J. 16, 324–343 (1977)
11. Bloesch, A., Halpin, T.: Conceptual queries using ConQuer-II. In: Embley, D.W. (ed.) ER 1997. LNCS, vol. 1331, pp. 113–126. Springer, Heidelberg (1997)
12. Bresciani, P., Nori, M., Pedot, N.: QueloDB: a Knowledge Based Visual Query System. In: Arabinia, H. (ed.) Proceedings of the 16th International Conference of Artificial Intelligence-IC-AI '2000, Las Vegas, pp. 1319–1325. Springer, Heidelberg (2000)

## Appendix: User-Defined Course Lists

User-Defined Course Lists are part of the settings in OVI-2 for defining beforehand a set of courses that is then treated by OVI-2 as a single entity. Users assign each user-defined list a unique descriptive name and may define for instance the Swedish Language Profiency Test as shown in Table 2. Any reference to the user-defined list SLTP will be translated by OVI-2 to the Boolean conditions of the courses in question as shown in Table 2.

**Table 2.** The user has defined that SLT can be complete by passing either one of the five courses (A1-A5) in the left-hand column or by taking any single combination of one of the three pairs of courses (B1-B3) shown on the right-hand column.

| Single Courses Satisfying Requir. for SLTP | | | | Course Pairs Satisfying Requir. for SLTP |
|---|---|---|---|---|
| (A1) | Lang-98.344 | (A2) | Lang-98.519 | (B1) Lang-98.5001 and Lang-98.5002 |
| (A3) | Lang-98.5166 | (A4) | Lang-98.5170 | (B2) Lang-98.013 and Lang-98.012 |
| (A5) | Lang-98.5172. | | | (B3) Lang-98.340 and Lang-98.349. |

# Wavelet Synopsis: Setting Unselected Coefficients to Zero Is Not Optimal

Chong Sun, Yan Sheng Lu, Chong Zhou, and Jun Liu

DB& Software Engnieering Lab, Department of Computer Science,Huazhong universtiy of Sci.&Tech.430074 Wuhan, China
SunChong217@hotmail.com

**Abstract.** Histogram and Wavelet synopses provide useful tools in query optimization and approximate query answering. Traditional wavelet synopsis construction algorithms treat the construction algorithms as the wavelet coefficients selection problem which is called Coefficient Thresholding. However, all these algorithms just focus on the selection of best wavelet coefficients but deal with the unselected ones naively (just setting them to zero). A key problem is whether it can achieve the optimum of error when the unselected ones are set to one single value: zero. In this paper, we consider a novel Wavelet-based Synopsis construction for the known L2 error measure which can handle the unselected wavelet coefficients effectively. We provide a comprehensive theoretical analysis and demonstrate the effectiveness of these algorithms in providing more optimal error significantly through synthetic data sets.

**Keywords:** query optimization; data reduction; wavelet synopsis; error measure.

## 1 Introduction

The reduction of massive data sets into a more manageable size is required in many database applications. Wavelet decomposition [6] provides a very effective data reduction tool, with applications in data mining [7], selectivity estimation [8], and approximate and aggregate query processing of massive relational tables [9]. In simple terms, a wavelet synopsis is extracted by applying the wavelet decomposition on an input collection (considered as a sequence of values) and then summarizing it by retaining only a select subset of the produced wavelet coefficients. The original data can be approximately acquired by wavelet reconstruction based on this compact synopsis. The wavelet reconstruction just simply treats the wavelet coefficients not in synopsis to be zero. Given an $N$ wavelet coefficients collection *Wall[n]* and space budget $B$, Wavelet Thresholding is used to determine which $B$ wavelet coefficients should be retained in the synopsis for some error measure. In other words, it is to determine which $N$-$B$ wavelet coefficients should be discarded. Traditional wavelet synopsis construction algorithms [1], [2], [3], [4], [5] view the discarded coefficients as zero, but that cannot always make the synopsis achieve the optimal error. Suppose

R. Wagner, N. Revell, and G. Pernul (Eds.): DEXA 2007, LNCS 4653, pp. 499–508,2007.

we are given the one-dimensional data vector $A$ containing the $N = 8$ data values $A$ = [2, 2, 0, 2, 3, 5, 4, 4]. First, by wavelet decomposition on $A$, we get the wavelet coefficients collection $W_{all} = [2.75, -1.25, 0.5, 0, 0, -1, -1, 0]$ .Second, we determine to retain four (e.g., $B$=4) coefficients (e.g., 2.75,-1.25,-1,-1) as wavelet synopsis of $A$ by wavelet coefficients selection for L2 error measure and view all the unselected ones (e.g., 0.5, 0, 0, 0) as zero. However, setting unselected coefficients to zero makes the L2 error value of the synopsis to 1, while if we set unselected coefficients to 1/6, the error value is changed to 2/3. Therefore, we propose two questions for the conventional method.

**Question1.** For a certain error measure, is it really optimal when setting all the coefficients not in the synopsis to zero?

**Question2.** Why just set one single value to the unselected coefficients, but not two or more?

**Contribution.** To the best of our knowledge, we are the first ones to propose these problems and give proofed answers to them. Out of the consideration of the these problem, we study the construction method for the known L2 error measure as well as propose a novel synopsis method called $\vec{X}_k$*wavlet* which handles the unselected coefficients effectively and achieves the error closer to the optimum. We also provide a comprehensive theoretical analysis and demonstrate the effectiveness of these algorithms through synthetic data sets.

**Organization.** The remainder of the paper is organized as follows. Some backgrounds and related work are provided in section 2. The answer to our first question is offered in section 3 by introducing an example and our observations with proof as well, while the second question and how to determine the multi values are answered in section 4 and also the construction methods of $\vec{X}_k$*wavlet* are given in this section. Experimental results on various synthetic data sets are outlined in section5. Finally, in section 6 we present our conclusions and the future work.

## 2 Backgrounds and Related Work

### 2.1 Haar Wavelet and Error Tree

Haar function is the common wavelet decomposition function used for data reduction. References [6], [8] introduce the wavelet decomposition procedure based on haar function in detail.

Reference [8] introduces a hierarchical structure called error tree which illustrates the key properties of the Haar wavelet decomposition. The error tree is built on top of the original data set. Each leaf node $A[i]$ ($i = 0...N$) of the error tree is associated with a data value in the original vector $A$, while each internal node $S[i]$ ($i = 0...N$) is associated with a wavelet coefficient value in $W_{all}$ . Figure 1(a) shows the Haar wavelet decomposition in the form of an error tree for the example above. Given a node $u$ in an error tree $T$, let *path(u)* denote the set of all proper ancestors of $u$ in $T$ (i.e., the nodes on the path from u to the root of $T$, including the root but not u) with coefficients. A key property of the Haar wavelet decomposition is that the reconstruction of any data value *A[i]* depends only on the values of the coefficients on *path(A[i])*; more specifically, we have

$$A_{[i]} = \sum_{S_{[j]} \in path(A_{[i]})} sign_{i,j} \bullet S_{[j]}, \; sign_{i,j} = \begin{cases} +1 & A[i] \text{ in the left child subtree of } S[j] \text{ or } j = 0 \\ -1 & \text{otherwise} \end{cases} \quad (1)$$

Thus, reconstructing any data value involves summing at most *(logN + 1)* coefficients. In our example, as demonstrated in Figure 1, *A [4] = - (-1) - (0.5) + (-1.25) + (2.75) = 2*.

## 2.2 Wavelet-Based Data Reduction: Coefficient Thresholding

Given an *N* wavelet coefficients collection *Wall[n]* and space budget *B*, the goal of Wavelet Thresholding is to determine which *B* wavelet coefficients should be retained in the synopsis, so that some overall error measure in the approximation is minimized. In other words, it determines which *N-B* coefficients should be eliminated.

Intuitively, wavelet coefficients carry different weights with respect to their importance in rebuilding the original data values. In order to equalize the importance of all wavelet coefficients, we need to normalize the final entries of $W_{all}$ appropriately. A common normalization scheme is $S_i^* = S_i / \sqrt{2^{level(S_i)}}$ in [8], while $level(S_i)$ denotes the level of error tree at which the coefficient $S_i$ appears. Retaining the *B* largest Haar-wavelet coefficients in absolute normalized value is provably optimal with respect to minimizing the overall root-mean-squared error (i.e., L2-norm average error) in the data compression. More formally, let $R_{[i]}$ denote the approximately data value reconstructed (from wavelet synopsis) for cell *i*, and L2 error measure can be expressed as

$$Opt = \min \sum_{1 \le i \le n} (A_{[i]} - R_{[i]})^2 \text{ (For the given amount of space B).} \quad (2)$$

## 2.3 Definition and Problem Formulation

**Definition.** $\vec{X}_k wavelet$ Synopsis is a 3-item tuple defined as $\vec{X}_k wavelet ::= (Ws, \vec{X}_k, \rho_k)$. $Ws$ and $\overline{Ws}$ denote the two collections of selected and unselected wavelet coefficients respectively. $\vec{X}_k$ means we will allocate *k* different values (e.g., $x_1 ... x_k$ stored in vector $\vec{X}_k$) to the entries in $\overline{Ws}$. The symbol $\rho_k$ records the allocation plan from $\overline{Ws}$ to $\vec{X}_k$. In fact, $\rho_k$ is a mapping relation between them, we will discuss the problem of storing the mapping relation with low space cost in later section. Given a $\vec{X}_k wavelet$ synopsis *t*, *L2norm*(*t*) denotes the value of L2 error produced by $t$, if $L2norm(t)$ achieves the optimal error we consider $t$ as the optimal synopsis denoted by $(\vec{X}_k wavelet)^{opt}$.

**Problem Formulation.** The wavelet synopsis created in the traditional method can be represented in our formal definition as $\vec{0}_1 wavelet = \{\text{Ws}, \overline{X_1} = <0>, \phi\}$. It means we just allocate one single value to the unselected coefficients and the value is zero, such that recording the mapping relation between $\overline{Ws}$ and $\vec{X}_k$ is unnecessary and $\rho_k$ is denoted by $\phi$. From the definition, these two questions can be formalized.

**Question1.** Is it correct that $\vec{0}_1 wavelet$ is $(\vec{X}_1 wavelet)^{opt}$ ?

**Question2.** Is it correct that $L2norm((\vec{X}_k wavelet)^{opt})$ is no less than $L2norm((\vec{X}_{k+1} wavelet)^{opt})$ ?

## 3 $(\vec{X}_1 wavelet)^{opt}$ Construction

### 3.1 A Counterexample

Returning to the same example in section 1, we check whether the answer to question 1 is 'yes'. Replacing the unselected coefficients in the error tree $T$ with symbol x and doing the wavelet reconstruction based on the modified error tree $XT$ produce the approximated data collection in which every entry is an expression containing symbol $x$. Further more, evaluating the L2 error of this collection can produce the L2 error function of x, which is a quadratic function (e.g., $Ax^2 + Bx + C$). Figure 1(b) illustrates this procedure.

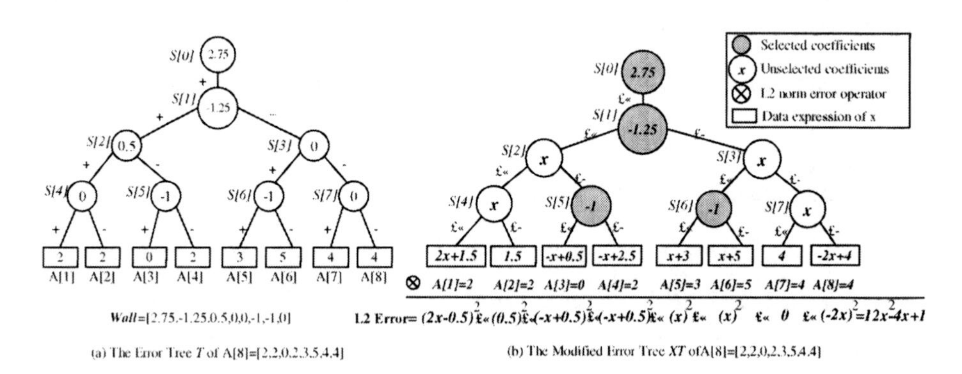


**Fig. 1.** An error tree and a modified error tree

In this example, the L2 error function is $12x^2 - 4x + 1$. Obviously, when $x = 1/6$, this function can achieve the minimum 2/3, in other words, $\vec{0}_1 wavelet \neq (\vec{X}_1 wavelet)^{opt}$ and the L2 optimum is 2/3 not 1 (respecting to $x = 0$).

### 3.2 Theorem and Proof

**Theorem 1.** Setting all the unselected coefficients to zero cannot produce the optimal wavelet synopsis for the L2 error measure. In other words, $\overrightarrow{0_1}wavelet$ is not $(\overrightarrow{X_1}wavelet)^{opt}$.

**Proof.** Given a modified error tree $XT$ and its L2 error function as $f(x) = Ax^2 + Bx + C$, from the definition of $XT$ and the calculation procedure of L2 error based on it, we can know $A$ is no less than zero. If $A$ were zero, then no entry in the leaf nodes of $XT$ would contain symbol $x$, so $A$ is bigger than zero. With the method of evaluating the minimum of quadratic function with one variable, we can get the following equations $\begin{cases} df/dx = 2Ax + B = 0 \\ d^2f/dx^2 = 2A > 0 \end{cases}$. When $x = -B/2A$ function $f$ gets the minimum $(4AC - B^2)/4A$.

### 3.3 Construction Algorithm of $(\vec{X}_1wavelet)^{opt}$

Fig.2 shows our construction algorithm. Note that, some functions (e.g., waverec, L2norm, diff) in our algorithm need symbolic computation, but most programming languages just support numerical computation. For this reason, we choose matlab to implement our algorithm.

```
Procedure SingleXOpt( A , B )
Input: Array A = [A1,...,An] of N original data items, Space budget B (number of retained coefficients).
Output: Struct X1Wavelet {Ws, Xopt} is the optimal wavelet synopsis, X1Wavelet.Ws is the collection of selected coefficients, X1Wavelet.Xopt is the value setting to all unselected coefficients
 begin
1.   Wall := wavedec( A ) // Haar wavelet decomposition, all coefficients retain in Wall.
2.   X1Wavelet.Ws := wavethreshold( Wall,B ) // Coefficients Thresholding.
3.   Xmark( Wall , X1Wavelet.Ws ) // Mark all the unselect ones with symbol x.
4.   // Haar wavelet reconstruction, XA is an N-item array each item in XA is an
5.   // expression of symbol x.
6.   XA := waverec( Wall )
7.   FX := L2norm( XA, A ) // Compute the L2 error expression FX.
8.   DiffFX := diff( FX, 'x' ) // Compute the differential expression of FX.
9.   X1Wavelet.Xopt := solve( DiffFX, 'x') // Compute the equation of x
10.  return (X1Wavelet)
 end.
```

**Fig. 2.** Construction Algorithm of $(\vec{X}_1wavelet)^{opt}$

# 4 Construction Algorithms of $\vec{X}_k wavelet$

A *B*-item wavelet synopsis occupies *2B+1* space. The total space is divided into two parts: *2B* space is allocated to retain indices and values for each selected coefficient and *1* space is used to record the default value of all the unselected coefficients. When the space allocated to the unselected coefficients is fixed, the L2 error of synopsis decreases with the increase of the number of selected coefficients. Meanwhile, when the number of selected coefficients is fixed, the L2 error of synopsis also decreases with the increase of the number of default values to which the unselected coefficients are mapped. Accordingly, the space allocation plan of traditional algorithms may not be the optimal and we can check all plans to find the best one.

Section 4.1 shows that when the selected coefficients collection is fixed, the more numbers we allocate to make the unselected coefficients map to, the smaller error of the synopsis we can get. In section 4.2, the space-efficient method is introduced to deals with the unselected coefficients by multi default values.

## 4.1 Theorems and Proof

**Theorem 2.** $L2norm((\vec{X}_k wavelet)^{opt})$ is no less than $L2norm((\vec{X}_{k+1} wavelet)^{opt})$
Proof. (Constructive proof method)
We can view the $(\vec{X}_k wavelet)^{opt}$ as some $\vec{X}_{k+1} wavelet$, so $L2norm((\vec{X}_k wavelet)^{opt})$ and $L2norm(\vec{X}_{k+1} wavelet)$ are equivalent. Because the error value of any $\vec{X}_{k+1} wavelet$ is 0no less than $L2norm((\vec{X}_{k+1} wavelet)^{opt})$, we can get that $L2norm((\vec{X}_k wavelet)^{opt})$ is no less than $L2norm((\vec{X}_{k+1} wavelet)^{opt})$. The following example shows the details.

Suppose $\overline{Ws} = \{s_4, s_5, s_7, s_8\}$ and the $(\vec{X}_2 wavelet)^{opt}$ (depicted in figure 3 (a)) is given. We can viewed $(\vec{X}_2 wavelet)^{opt}$ as a $\vec{X}_3 wavelet$ (depicted in figure 3 (b)). Although the two wavelet synopses are different, the unselected coefficients collection they produced are the same (both of which are $\{s_4=4, s_5=6, s_7=4, s_8=6\}$) so that $L2norm$ $((\vec{X}_2 wavelet)^{opt})$ equals $L2norm$ $(\vec{X}_3 wavelet)$. Because $L2norm$ $(\vec{X}_3 wavelet)$ is no less than $L2norm$ $((\vec{X}_3 wavelet)^{opt})$, we can get that $L2norm$ $((\vec{X}_2 wavelet)^{opt})$ is no less than $L2norm$ $((\vec{X}_3 wavelet)^{opt})$.

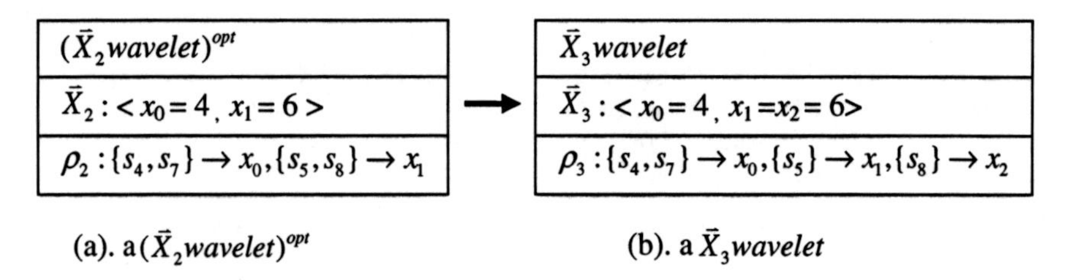


(a). a $(\vec{X}_2 wavelet)^{opt}$ (b). a $\vec{X}_3 wavelet$

**Fig. 3.** an example of viewing a $(\vec{X}_2 wavelet)^{opt}$ as a $\vec{X}_3 wavelet$

### 4.2 Near Optimal Construction of $(\vec{X}_k wavelet)^{opt}$

Dealing with unselected coefficients with multi default values, we need to do two things: one is to map all unselected coefficients to the multi default values; the other is to determine the multi default values.

The second one is easy. Given the mapping relation $\rho_k$, the L2norm error of wavelet synopsis is a k-variable quadratic function $f(x_1, \ldots, x_k)$ and the multi default values ($\vec{X}_k$) are determined by solving the system of equations:

$$\begin{cases} \partial f(x_1,\ldots,x_k)/\partial x_1 = 0 \\ \quad\vdots \\ \partial f(x_1,\ldots,x_k)/\partial x_k = 0 \end{cases} \tag{3}$$

The key problem is how to create the mapping relation and record it in a low space. Our method is described as follows.

Similar to Equi-Width histogram, we partition the sorted $\overline{Ws}$ into $k$ adjacent Equi-width intervals and map all coefficients in the $i^{th}$ interval to the $i^{th}$ component of $\vec{X}_k$, so that we can map $k$ intervals to $k$ components of $\vec{X}_k$. What's more, it just needs 1 space to keep the width of interval from which the mapping relation can be derived. Using this method, it just needs k+1 space (in the wavelet synopsis) to handle the unselected coefficients collection with k default values.

However, producing the $\rho_k$ in this way cannot always make the theorem 2 satisfied. For example, a '*k*=2' optimal solution can be viewed as a 'k = 4' feasible solution but not as a 'k =3' one, so it is hard to tell whether the 'k=3' solution is less than the 'k=2' solution for the L2 error. Figure 4 illustrates this problem. Although there is such a problem, our method is still useful to create near optimal solution and to decrease the L2 error, for the theorem 3.

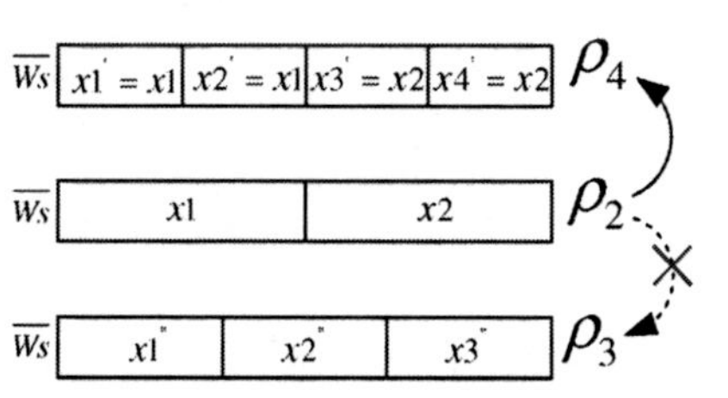


**Fig. 4.** "*k*=2" solution can only create "*k*=4" solution

**Theorem 3.** The Equi-width partition can guarantee that $L2norm(\vec{X}_k wavelet), k = C^i$ is larger than $L2norm(\vec{X}_k wavelet), k = C^{i+1}$, while $C$ is any natural number lager than 1.

The proof is similar to the theorem 2. We can also view a k=Ci wavelet synopsis as a k=Ci+1 synopsis. Such that $L2norm(\vec{X}_k wavelet), k = C^i$ is larger than $L2norm(\vec{X}_k wavelet), k = C^{i+1}$.

Figure 5 demonstrates the construction algorithm of near optimal synopsis $(\vec{X}_k wavelet)^{near-opt}$.

```
Procedure KX_NearOpt( A , B, K )
Input: Array A = [A_1,...A_n] of n original data items, Space budget B (number of
retained coefficients), Space budget K (number of the variables which unselected
ones map to).
Output: Struct XkWavelet {Ws, KXopt, P} is the optimal wavelet synopsis,
X1Wavelet.Ws is the vector of selected coefficients, X1Wavelet.KXopt is a k-item
vector which record values set to all unselected coefficients, X1Wavelet.P is a k-item
array used to record the end point of each partitioned(from the unselected coefficients
collection )interval.
begin
1.   // Haar wavelet decomposition, all coefficients are retained in Wall.
2.   Wall := wavedec(A)
3.   XkWavelet.Ws := wavethreshold( Wall,B ) // Coefficients Thresholding.
4.   // partition the unselected coefficients into K adjacent intervals.
5.   XkWavelet.P := partition(Wall , Ws, K )
6.   // Mark the K adjacent intervals with symbols x_1...x_k.
7.   KXmark( Wall , Ws, P)
8.   // Haar wavelet reconstruction, XA is an n-item array each item in XA is a
9.   // symbol expression.
10.  XA := waverec( Wall )
11.  // Compute the L2 error expression FX.
12.  FX := L2norm( XA,A )
13.  // Compute the partial differential expressions of FX.
14.  for i = 1 to k do
15.    DiffFX[i] := diff( FX, xi)
16.  end
17.  XkWavelet.Xopt[] :=solve( DiffFX[])// Compute the system of equations
18.  return X1Wavelet
end.
```

**Fig. 5.** Equi-width construction Algorithm of $(\vec{X}_k wavelet)^{near-opt}$

# 5 Experimental Evaluations

In this section, we present the results of an empirical study we have conducted through the algorithmic techniques developed in this article for synopses optimized for *L2* error measure. The primary objective of our study is to verify theorems proposed above and to demonstrate the improvement that the new synopsis can achieve comparing to the traditional ones. Our experimental study has made the following comparisons for the *L2* error measure:

- The comparison between $(\overrightarrow{X_1} wavelet)^{opt}$ and the traditional wavelet synopsis;
- The comparison between $(\overrightarrow{X_1} wavelet)^{opt}$ and $\vec{X}_k wavelet^{Near-opt}$ constructed by equi-width scheme.

We ran our techniques against several different one-dimensional synthetic data distributions, generated as follows. First, a Zipf data generator was used to produce Zipf frequencies for various levels of skew (controlled by the *z* parameter of the Zipf), numbers of distinct values *N*, and max frequency values. We varied the *z* parameter from 0.5 (low skew) to 2.0 (high skew), the frequency dataset size $N$ between 128 and 1,024, and the max count between 1k and 16k. Next, a random permutation step was applied on the generated data set. We used a Zipf distribution generator coded by Kenneth J. Christensen (University of South Florida) which can be found at [10].

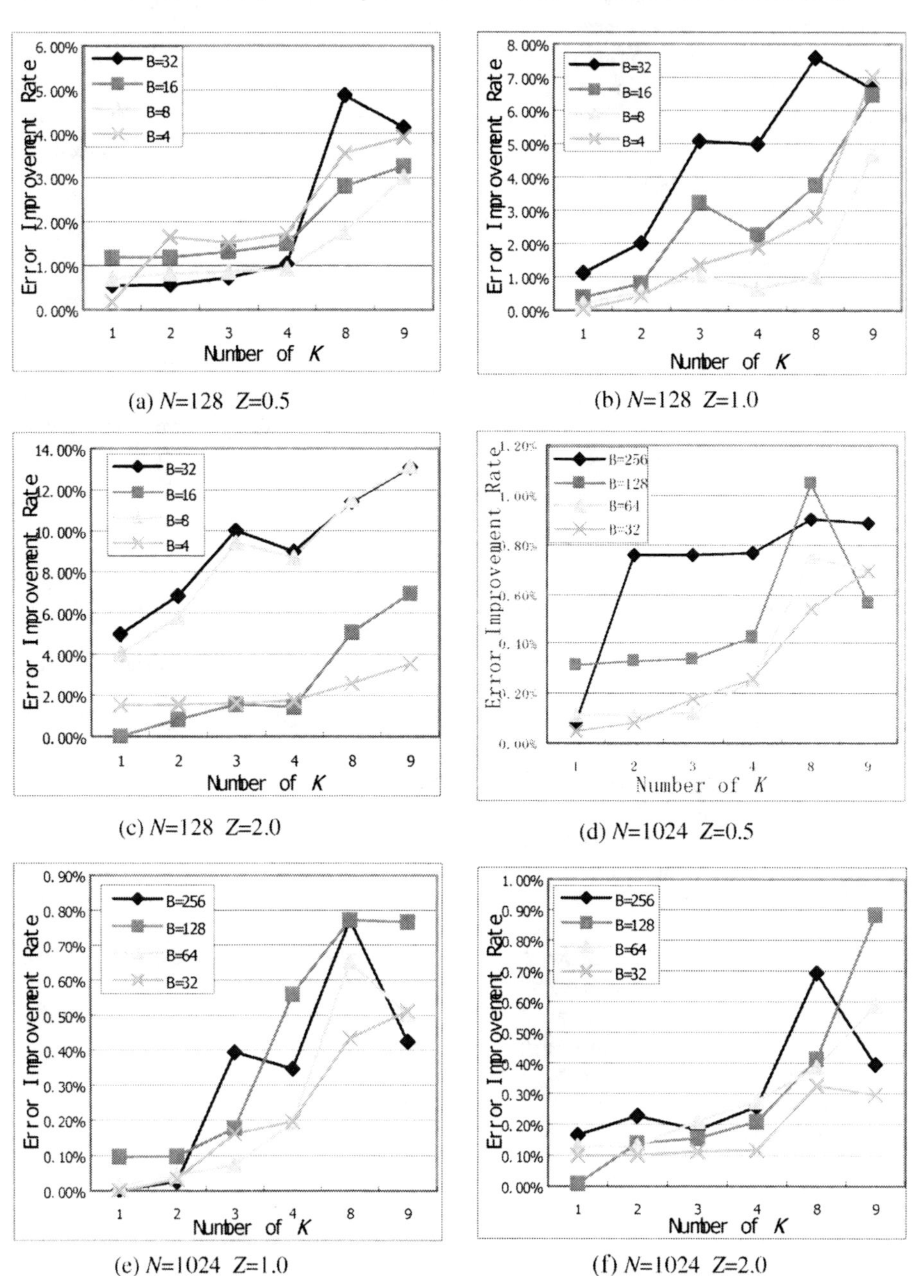


(a) $N$=128 $Z$=0.5 (b) $N$=128 $Z$=1.0

(c) $N$=128 $Z$=2.0 (d) $N$=1024 $Z$=0.5

(e) $N$=1024 $Z$=1.0 (f) $N$=1024 $Z$=2.0

**Fig. 6.** Experiment Results

Each picture of fig.6 illustrates the experiment results of one data distribution. We vary the $K$ parameter and $B$ parameter on every data distribution. The $Y$ axis shows the error improve rate of our wavelet synopsis (corresponding to traditional synopsis) and $X$ axis shows the number of $K$. As fig.6 shows, all of our synopses are more optimal than traditional ones and our theorems are supported by the experiment results. Most of error improve rate functions increase with $k$, but there are some exceptions, for example: in fig.6(a) , when $B$=32, the error improve rate of $k$=9 is less than that of $k$=8 ,but the error improve rate increases in the order of $k$=1,2,4,8 and in the order of $k$=1,3,9. This situation is caused by the equi-partition method and Theorem 3 gives the explanation of it.

## 6 Conclusions and Future Work

In this paper, we point out that processing the unselected wavelet coefficients of wavelet synopses effectively is also very important, and provide effective solutions at the same time. For now, our solutions are off-line. We will try to create the (approximate) algorithms for the time series model and for the dynamic maintenance of our synopsis in the future.

## References


1. Garofalakis, M.N., Kumar, A.: Wavelet synopses for general error metrics. ACM Trans. Database Syst. 30(4), 888–928 (2005)
2. Garofalakis, M.N., Gibbons, P.B.: Probabilistic wavelet synopses. ACM Trans. Database Syst. 29, 43–90 (2004)
3. Garofalakis, M.N., Gibbons, P.B.: Wavelet synopses with error guarantees. In: SIGMOD Conference, pp. 476–487 (2002)
4. Karras, P., Mamoulis, N.: One-Pass Wavelet Synopses for Maximum-Error Metrics. VLDB, pp. 421–432 (2005)
5. Muthukrishnan, S.: Subquadratic Algorithms for Workload-Aware Haar Wavelet Synopses. In: Ramanujam, R., Sen, S. (eds.) FSTTCS 2005. LNCS, vol. 3821, pp. 285–296. Springer, Heidelberg (2005)
6. Blatter, C.: Wavelets: A Primer. A K Peters (1998)
7. Li, T., Li, Q., Zhu, S., Ogihara, M.: A survey on wavelet applications in data mining. SIGKDD Explorations Newsletter 4(2), 49–68 (2002)
8. Matias, Y., Vitter, J.S., Wang, M.: Wavelet-based histograms for selectivity estimation. In: Proc. of SIGMOD Conf., pp. 448–459 (1998)
9. Chakrabarti, K., Garofalakis, M., Rastogi, R., Shim, K.: Approximate query processing using wavelets. VLDB Journal 10(2-3), 199–223 (2001)
10. http://www.cs.unc.edu/~vivek/home/stenopedia/zipf/genzipf.c

# A Logic Framework to Support Database Refactoring

Shi-Kuo Chang[1], Vincenzo Deufemia[2], Giuseppe Polese[2], and Mario Vacca[2]

[1] University of Pittsburgh, Department of Computer Science
6101 Sennott Building, Pittsburgh, PA, USA, 15260
chang@cs.pitt.edu

[2] Università di Salerno, Dipartimento di Matematica e Informatica
Via Ponte don Melillo, 84084 Fisciano (SA), Italy
{deufemia,gpolese,mvacca}@unisa.it

**Abstract.** We propose a formal framework for database refactoring, analyzing both the changes to the database schema, and their impact on queries. The framework defines a logic model of changes, and views the database refactoring process as an agent based one. The agent tries to discover and resolve inconsistencies, and it is modeled as a problem solver capable to perform changes triggered upon the detection of database schema anomalies. The framework can be considered a first step towards the automation of the database refactoring process.

## 1 Introduction

Waterfall methodologies have their weakness in their incapability to cope with changes, which makes maintenance considerably an expensive process. For this reason, incremental and iterative methodologies were introduced [12]. They view system development as a step by step process, with the introduction of new functionalities to meet user needs. The main problem arising in both paradigms is the complexity in facing the effects of changes. Therefore, an increased automated support in this task would result in a reduction of efforts and costs, especially in incremental methodologies, because it would make them more systematic.

Changes are often necessary to reflect the continuous evolution of the real world, which causes frequent changes in functional requirements. This entails frequent modifications to the software, yielding a gradual decay of its overall quality. For this reason, many researchers in this field have developed software refactoring techniques [15]. Software refactoring is intended as the restructuring of an existing body of code, aiming to alter its internal structure without changing its external behavior [5]. It consists of a series of small behavior preserving transformations, which altogether can produce a significant software structural change. Moreover, system modifications resulting in changes to the database structure are also relatively frequent [21]. These changes are particularly critical, since they affect not only the data, but also the application programs relying on them [1,10].

R. Wagner, N. Revell, and G. Pernul (Eds.): DEXA 2007, LNCS 4653, pp. 509–518, 2007.

Several disciplines have faced the problem of managing the effects of database schema changes. In particular, *schema modification* has faced the problem of changing the schema of a populated database. In addition to this, *schema evolution* pursues the same goal, but it tries to avoid loss of data. Alternatively, *schema versioning* performs modifications of the schema, but it keeps old versions to preserve existing queries and application programs running on it. Although schema versioning faces the problem of query and application programs preservation, it considerably increases the complexity and the overhead of the underlying DBMS. Finally, *database refactoring* aims to modify the database schema, and to change the corresponding application programs accordingly. In other words, database refactoring is the process of slowly growing a database, modifying the schema by small steps, and propagating changes to the queries [1].

So far research on database refactoring has led to the definition of several methodologies [1]. However, no significant contribution has been provided towards the automation of this process. This is mainly due to the lack of formal approaches, like those developed for schema versioning and schema evolution [3,6,11,17,18]. Nevertheless, these approaches use models that do not consider queries, hence they do not analyze the impact of schema changes on queries and application programs. In this paper we propose a formal framework for database refactoring, analyzing both the changes to the database schema and their impact on queries and application programs. The framework defines a logic model of changes, and views the database refactoring process as an agent based one. Here, the goal of agents is to discover and resolve inconsistencies. The agent is modeled as a problem solver capable to perform changes which are triggered upon the detection of database schema anomalies.

The use of elementary operators can already be found in many other approaches (see, for example, [3,6,18]), but their application relies on designer decisions or it is strongly coupled with the model features. By triggering such operators upon the detection of anomalies, our approach can potentially reduce the designer effort, providing the basis to automate the database refactoring process.

The paper is organized as follows. In section 2 we discuss related works, while section 3 introduces the approach we propose to automate database refactoring. In the sections 4 and 5 a more detailed discussion about the proposal is provided. Finally, conclusions and future works are provided in Section 6.

## 2 Related Work

Database refactoring is a relatively new research topic [1], and no formal approaches have been proposed for dealing with it. On the contrary, many theoretical models exist for schema evolution and schema versioning [3,6,11,17,18].

Nowadays, database researchers agree on the fact that schema evolution and versioning introduce two main problems: the *semantics of changes*, and the *change propagation*. The former requires determining the effects of changes on the schema, whereas the second analyzes the consequences of changes on data.

So far two kinds of theoretical models have been proposed: the *invariant and rule model* [3,17], and the *axiomatic or formal model* [6,18].

The *invariant and rule model* is based on the ORION object-oriented data model [3,17]. It is structured into three components: a set of properties of the schema (*invariants*), a set of schema changes, and a set of *rules*. The invariants state the properties of the schema (for example, the classes are arranged in a lattice structure), whereas rules help detecting the most meaningful way of preserving the invariants when the schema changes. This model of schema evolution yields two important issues: completeness and soundness of the schema evolution taxonomy. Both of them have been proved only for a subset of the schema change operations.

The *axiomatic model* has three basic components: *terms*, *axioms*, and *changes* [18]. The basic concept underlying this model is the *type* (analogous to the concept of class in ORION), which is in turn characterized by the *terms*. Examples of terms are the lattice of types and the set of type properties. The *axioms* state the properties of the terms, like the properties of the lattice of types. Changes on the schema are performed by means of three basic change operations: *add*, *drop*, and *modify*. The problem of the semantics of changes is solved by re-computing the entire lattice using the axioms. The model satisfies the properties of soundness and completeness.

An approach based on the axiomatic model is provided in [6], and it models schema versioning from a logical and computational point of view. In particular, it proposes a semantic and formal framework based on Description Logic [2]. The basic elements of the model are: classes and their attributes, schema (a set of class definitions), and elementary schema change operators. A basic concept underlying the framework is the *legal database instance*, which, informally represents a database instance satisfying all the constraints. This notion allows broadening the number of consistencies that are considered as reasoning problems, according to the style of Description Logic. Finally, all the consistency problems considered have been proved as decidable.

The approaches based on the two models mentioned above present two main limitations. The former regards the explosion of rules when facing more general schema changes, whereas the second regards the fact that they are all suited to the object-oriented data model. Although a taxonomy of change operations for the relational model has been proposed [20], it does not represent a complete model.

The refactoring of relational databases entails facing two important problems, which cannot be managed through the two models above: the variability of schema properties, and the propagation of changes into queries. In this paper we face both these problems.

## 3 Database Refactoring Through Epistemic Logic

Epistemic logic is the logic of knowledge [7,16]. It deals with the reasoning mechanisms of knowledge and with the process of *belief revision*, i.e., the evolution of a base of beliefs. In epistemic logic there are three kinds of belief changes: *expansion*, *revision*, and *contraction*. The first change refers to the addition of a belief to a base, the second is related to the addition of an inconsistent belief to a base that causes the deletion of other beliefs, and finally, the third takes into account the retraction of a belief. Epistemic logic deals with both the formulation of *postulates* for belief revision and the *constructions* of the revision process.

Database refactoring can be seen as a *revision process*. In fact, an example of schema change (together with its queries) is the *addition* of a functional dependency, which might cause the split of a table (revision) in order to keep the schema in a certain normal form. It is easy to notice that changes in a database schema depend on the properties holding in it. For instance, the addition of an attribute might only entail the modification of the table in which it is added, but it might also require more complex changes. In fact, the new attribute might alter the degree of normalization of the table if it depends only from a portion of the primary key, or it might require the introduction of new referential integrity constraints in case it coincides with the primary key of another table in the schema. Therefore, the process of database refactoring is not simply a composition of elementary changes, but it implies more sophisticated reasoning tasks, like detec! ting inconsistencies.

If we look at the schema as a knowledge base, the refactoring becomes a process of changes in the knowledge, and hence it can be interpreted as an epistemic process, which can be naturally modeled through Epistemic Logic. Within this view, it becomes natural to see refactoring as an agent managed process aiming to operate on the schema in order to perform the required changes, and trying to preserve original properties in terms of knowledge and queries.

We abide by the Thagard conception [22], which views concepts like data structures. Since a data structure can be modeled as a signature with axioms [13], we will see a database schema as a kind of data structure, and will focus on those changes involving elements in the signature (for instance, the addition or the deletion of an attribute or a functional dependency). Therefore, we need to precisely define both the knowledge on which the agent operates, and the behavior of the agent. In order to do this, we need to define

- the features of the schema;
- the allowed change requirements;
- the reasoning mechanisms of the agent.

When a change requirement arises, the agent has to decide the actions to perform. For example, when the agent receives a request of adding a new attribute, it might decide to also add one or more new functional dependencies involving the attribute. Thus, the agent is a kind of problem solver.

# 4 A Formalization of the Database Refactoring Problem Using Predicate Logic

In this section we formalize the problem of database refactoring using predicate logic. To this end, in the following we introduce the notations that will be used throughout the paper.

Let $\Sigma$ be the set of all the attribute symbols, $D$ the set of types, $N$ the set of names, and $V$ a set of variables, $A = \{(n,t)|n \in N, t \in D\}$ the set of attributes, $R = \{(n, a_1 \times \ldots \times a_m)|\ n \in N, a_1, a_2, \ldots, a_m \in \Sigma\}$ the set of relations, and $\Phi = \{(n, a_1 \times \ldots \times a_k \rightarrow b)|a_1, a_2, \ldots, a_k, b \in \Sigma, n \in N\}$ the set of functional dependencies. In order to express the schema properties, we will use the following functions and predicates: $table(R)$ to state if $R$ is a relation, $attr(R)$ returning the set of attributes of table $R$. Queries are non-recursive, function-free, and Datalog formatted [4], i.e., a query is formed by a head and a body. The head is a couple $(name, X)$ with $X \in V^n$; the body is a conjunction of predicates on $X$. The functions $body(Q)$ and $var(Q) \subseteq V$ return the body of a query $Q$, and the set of its variables, respectively. Variables are labeled with the attribute to which they refer. For instance, $x_a$ indicates that $x$ is a variable referring to attribute $a$. Moreover, $FD(f)$ is a predicate that is true when $f$ is a functional dependency, $LHS(f)$ (resp. $RHS(f)$) is a function returning the set of attributes on the left (resp. right) hand side of $f$, and finally, $table(f)$ returns the table to which $f$ refers to.

**Definition 1.** *A database system $K$ is a quintuple $K = (A, T, F, Q, P)$, where $A \subseteq \Sigma$, $T \subseteq R$, $F \subseteq \Phi$, $Q$ is a set of queries, and $P$ is a set of properties (propositions) involving elements of $A$, $T$, and $F$.*

*Example 1.* Let us consider a database system storing data about employees of a company, and having a query for retrieving all employees of the Computer Science personnel department can be represented by $K = (A, T, F, Q, P)$ where

$$
\begin{aligned}
A = &\{Employee\_ID, Name, Department\_ID, Salary, Address\} \\
T = &\{R(Employee\_ID, Name, Department\_ID, Salary, Address)\} \\
F = &\{f_1 : Employee\_ID \rightarrow Name; f_2 : Employee\_ID \rightarrow Department\_ID; \\
&\quad f_3 : Employee\_ID \rightarrow Salary, f_4 : Employee\_ID \rightarrow Address\} \\
Q = &\{q(x, y, w, z) \equiv R(x, y, \text{“}CS\text{”}, w, z)\} \\
P = &\{1) primary\_key(R, Employee\_ID) \\
&\quad 2) \forall r \in T\ \exists k \subseteq Attr(r) \text{ such that } primary\_key(r, k) \\
&\quad 3)\ key\_dep(r, k) \equiv \forall a \in (attr(r) - k) \\
&\quad (\exists f \in F \text{ such that } (LHS(f) = k \wedge RHS(f) = \{a\}) \wedge \\
&\quad (\neg \exists f \text{ such that } (LHS(f) \neq k \wedge RHS(f) = \{a\})) \\
&\quad 4) \forall r \in T\ (primary\_key(r, k) \rightarrow key\_dep(r, k))\}
\end{aligned}
$$

The properties in $P$ state that every relation has a primary key, and the attributes fully depend on the primary key only.

**Definition 2.** *A database system $K = (A, T, F, Q, P)$ is said to evolve towards a database system $K' = (A', T', F', Q', P')$ iff there are four functions*

$\varepsilon_{Attr} : \mathcal{P}(\Sigma) \to \mathcal{P}(\Sigma)$
$\varepsilon_{Table} : \mathcal{P}(R) \to \mathcal{P}(R)$
$\varepsilon_{Constr} : \mathcal{P}(\Phi) \to \mathcal{P}(\Phi)$
$\varepsilon_{P} : \mathcal{P}(Prop) \to \mathcal{P}(Prop)$

*where $\mathcal{P}$ is the power set operator and Prop is the set of all propositions on A, T, F;*

*and a substitution $\theta = (n_1 \leftarrow expr_1, \ldots, n_k \leftarrow expr_k)$ where $n_j$ are names and $expr_j$ are expressions constituted by either single names or their conjunctions, such that*

$A' = \varepsilon_{Attr}(A)$
$T' = \varepsilon_{Table}(T)$
$F' = \varepsilon_{Constr}(F)$
$P' = \varepsilon_{P}(P)$
$Q' = \{q' | \ q' = q\theta \text{ with } q \in Q\}$ *i.e., $q'$ is obtained by applying $\theta$ to $q$.*

For sake of brevity, when no confusion occurs, we use symbol $\varepsilon$, named "evolution", to refer to the four functions together with the substitution. We also write $D' = \varepsilon(D)$.

The semantics of the database systems modeled through logic frameworks is usually specified by interpretation functions (e.g., [2]). An interpretation $I$ is a couple $(\Delta^I, \cdot^I)$ where $\Delta^I$ is a domain and $\cdot^I$ is an interpretation function providing set theoretic interpretations. For instance, the interpretation of a relation $R$ having two attributes is $R^I \subseteq \Delta^I \times \Delta^I$.

Given a database system $K$, a *database instance* on $K$, denoted by $\Delta(D)$, is an interpretation in $K$. The interpretation of a query $q \in Q$ , denoted with $q^I$, is the set of all tuples in the database satisfying $q$. Two queries $q$ and $q'$ are equivalent in a database instance if and only if they produce the same answers. The following definition introduces the concept of query equivalence under the projection operator, which will be used for defining the concept of refactored systems.

**Definition 3.** *A query $q$ is equivalent to a query $q'$ under the projection operator $\pi$, denoted by $q \equiv_\pi q'$, if and only if*

$$\pi_{(var(q) \cap var(q'))}(q^I) = \pi_{(var(q) \cap var(q'))}(q'^I)$$

*where $I$ is an interpretation function.*

Now we are ready to introduce a formal definition of *refactoring*.

**Definition 4.** *A database system $K = (A, T, F, Q, P)$ is said refactored in $K' = (A', T', F', Q', P')$ if and only if*

*i.* $\forall q \in Q \ \exists q' \in Q'$ *such that* $q \equiv_\pi q'$
*ii.* *if* $\forall \Delta(K) \ \Delta(K) \models P$ *then* $\forall \Delta(K') \ \Delta(K') \models P'$

Refactoring functions are particular kinds of evolution functions preserving the results of queries and the properties of the database system. For instance, if $D$ is in third normal form, then also $D'$ must be in the same normal form. Notice that schema evolution is a special case of refactoring. In fact, if $Q = Q'$ and $P = P'$ refactoring reduces to schema evolution.

*Example 2.* Let us consider the database system $D$ introduced in example 1 and the following evolution functions:

$\varepsilon_{Attr}(A) = A$
$\varepsilon_{Table}(T) = (T - \{R\}) \cup \{R_1, R_2\}$
$\varepsilon_{Constr}(F) = F \cup \{\ f_5:\ Department\_ID \rightarrow Address\}$
$\theta = (R \leftarrow R_1 \wedge R_2)$

where $R_1$ and $R_2$ have attributes $(Employee\_ID, Name, Salary, Department\_ID)$ and $(Department\_ID, Address)$, respectively.

By applying $\varepsilon$ on $D$ we obtain the database system $D' = (A', T', F', Q', P')$ with

$A' = \{Employee\_ID, Name, Department\_ID, Salary, Address\}$
$T' = \{R_1(Employee\_ID, Name, Salary, Department\_ID), R_2(Department\_ID, Address)\}$
$F' = \{f_1 : Employee\_ID \rightarrow Name; f_2 : Employee\_ID \rightarrow Department\_ID; f_3 : Employee\_ID \rightarrow Salary, f_4 : Employee\_ID \rightarrow Address, f_5 :\ Department\_ID \rightarrow Address\}$
$Q' = \{q(x, y, w, z) \equiv R_1(x, y, w, \text{“}CS\text{”}) \wedge R_2(\text{“}CS\text{”}, z)\}$
$P' = (P - \{primary\_key(R, Employee\_ID)\} \cup \{primary\_key(R_1, Employee\_ID), primary\_key(R_2, Department\_ID)\}$

## 5 The Process of Database Refactoring

As the refactoring is an agent based process, in order to realize the required changes, the agent has to operate on the schema in a way that preserves the properties of the knowledge and of the queries. Two kinds of approaches can be used to accomplish this task: axiom based and constructive. The former is based on a set of postulates, known in the literature as *postulates for belief revision* [7,14]. The constructive approaches use *propositions* and *programs* for handling changes in the knowledge [8].

In the proposed refactoring process we use the constructive approach and build the evolution operator $\varepsilon$ by using *propositions*, *questions*, and *change operations*. A question is denoted with $?p$, where $p$ is a proposition.

An example of change operation is the splitting of a table $t$ after the introduction of a new functional dependency $f$, which could be described in the following way:

$split\_table(t, t', t'', f) \leftarrow$
$(A' = A \wedge$
$T' = (T - \{t\}) \cup \{t', t''\} \wedge$

$$F' = F \wedge$$
$$Q' = \{q' |\ var(q') = var(q),\ body(q') = \rho(body(q), t, t' \wedge t'')\} \wedge$$
$$attr(t') = attr(t) - RHS(f) \wedge$$
$$attr(t'') = LHS(f) \cup RHS(f))$$

The database refactoring process is based on the following predicates: *Consistent(change-operation)*, *Hold(p)*, and *Resolve(change-operation, p)*. The former is true when the set of properties $P'$ obtained by the application of the *change-operation* is consistent. The second is true when proposition $p$ holds. Finally, the third is true when proposition $p$ holds after the application of *change-operation.*

These logical operations can be expressed using the $K$ operator of epistemic logic [9]. The $K$ operator is applied to a proposition $p$ using the expression $Kp$, whose meaning can be informally expressed by "it is known that $p$". As a consequence, *Hold(p)* can be expressed as $Kp$, *Consistent(change-operation)* as $K(\forall p \in \epsilon(P)).p$, whereas *Resolve(change-operation, p)* as $Kp$ applied after the *change-operation.*

The agent uses the previous predicates to submit questions or to answer questions according to rules like the following:

$$\frac{\neg Consistent(change - operation)}{?\exists \omega Resolve(\omega, p)}$$

$$\frac{\neg Consistent(change - operation)}{?\exists x\ \neg Hold(x)}$$

$$\frac{\neg Hold(\neg \exists x.P(x))}{?Resolve(add(x), \exists x P(x))}$$

$$\frac{\neg Hold(\exists x.P(x))}{?Resolve(drop(x), \exists x P(x))}$$

$$\frac{?Resolve(\omega, p)}{?Consistent(change - operation)}$$

$$\frac{\neg Resolve(\omega, p)}{?\exists \omega'((\omega' \neq \omega) \wedge Resolve(\omega', p))}$$

For instance, if an inconsistency on a proposition $p$ arises, the first rule suggests the agent to ask the question "Does there exist a change operation resolving the inconsistency?".

In general, the *reasoning process* of the agent has a question as starting point, and a change operation as ending point. The process of answering a question like the previous one is a problem solving process, since it involves the choice of a change operation. This is made through heuristics, as it usually happens in the problem solving domain [19].

*Example 3.* Let us consider the database system of example 1. When the agent receives a request of adding a functional dependency

$$f_5 : \mathit{Department\_ID} \rightarrow \mathit{Address}$$

it processes the following questions (answers are visualized in bold):

$?Consistent(add(f_5))$ **NO**
$?\exists x\ \neg Hold(x)$ **YES** $x = (LHS(f_5) \neq \mathit{Employee\_ID} \wedge RHS(f_5) = \{\mathit{Address}\})$
$?Resolve(drop(f_5), P)$ **NO**
$?Resolve(split\_table(R, R', R''), f_5)$ **YES**
$?Consistent(split\_table(R, R', R''))$ **NO**

## 6 Conclusions and Future Works

We have presented a formal framework for database refactoring based on epistemic logic. The framework defines a logic model of changes, and uses an agent to discover and resolve inconsistencies, and to analyze the impact of changes on queries.

In the future we would like to investigate several important issues. Firstly, it is necessary to study the system of rules and their properties. We also need the agent to be capable of making decisions. Thus, we should make the agent more autonomous and should equip it with problem solving heuristics. Moreover, we need the agent to be more communicative, in order to base its decisions also on user suggestions. For example, adding a functional dependency is a serious decision, and it would be desirable having the agent ask for user support. We would also like to investigate the possibility to exploit the second generation of epistemic logic that is based on the erotetic logic [9].

Finally, we would like to investigate the possibility of using visual language based tools capable of supporting the database refactoring process directly on the database conceptual or logic schema by means of special gesture operators.

## References

1. Ambler, S.W., Sadalage, P.J.: Refactoring databases: Evolutionary database design. Addison-Wesley, London (2006)
2. Baader, F., Calvanese, D., McGuinness, D.L., Nardi, D., Patel-Schneider, P.F.: The description logic handbook: Theory, implementation, and applications. Cambridge University Press, Cambridge (2003)
3. Banerjee, J., Kim, W., Kim, H.J., Korth, H.F.: Semantics and implementation of schema evolution in object-oriented databases. In: Proceedings of the 1987 ACM SIGMOD International Conference on Management of Data, pp. 311–322. ACM Press, New York (1987)

4. Bonner, A.J.: Hypothetical Datalog: Negation and linear recursion. Rutgers University (1989)
5. Du Bois, B., Van Gorp, P., Amsel, A., Van Eetvelde, N., Stenten, H., Demeyer, S., Mens, T.: A discussion of refactoring in research and practice. Technical report, n. 2004-03, University of Antwerp, Belgium (2004)
6. Franconi, E., Grandi, F., Mandreoli, F.: A general framework for evolving schemata support. In: Proceedings of SEBD 2000, pp. 371–384 (2000)
7. Gärdenfors, P.: Belief revision: An introduction. In: Belief Revision, pp. 1–20. Cambridge University Press, Cambridge (1992)
8. Gerbrandy, J.: Dynamic epistemic logic. In: Moss, L.S., Ginzburg, J., de Rijke, M. (eds.) Logic, Language and Computation, vol. 2, pp. 67–84. CSLI Publications, Stanford (1999)
9. Hintikka, J.: A second generation epistemic logic and its general significance. In: Hendricks, et al. (eds.) Knowledge Contributors, Synthese Library no. 322, Kluwer Academic Publishers, Dordrecht (2003)
10. Karahasanovic, A.: Identifying impacts of database schema changes on application. In: Proceedings of the 8th Doctoral Consortium at the CAiSE*01, pp. 93–104 (2001)
11. Lakshmanan Laks, V.S., Sadri, F., Subramanian, I.N.: On the logical foundations of schema integration and evolution in heterogeneous database systems. In: Ceri, S., Tsur, S., Tanaka, K. (eds.) DOOD 1993. LNCS, vol. 760, pp. 81–100. Springer, Heidelberg (1993)
12. Larman, C., Basili, V.R.: Iterative and incremental development: A brief history. IEEE Computer 36(6), 47–56 (2003)
13. Luo, Z.: Program specification and data refinement in type theory. Mathematical Structures in Computer Science 3(3), 333–363 (1993)
14. Maghsoudi, S., Watson, I.: Epistemic logic and planning. In: Negoita, M.G., Howlett, R.J., Jain, L.C. (eds.) KES 2004. LNCS (LNAI), vol. 3213, pp. 36–45. Springer, Heidelberg (2004)
15. Mens, T., Tourwé, T.: A survey of software refactoring. IEEE Transactions on Software Engineering 30(2), 126–139 (2004)
16. Meyer, J.J., Van Der Hoek, W.: Epistemic logic for AI and computer science. Cambridge University Press, Cambridge (1995)
17. Nguyen, G., Rieu, D.: Schema evolution in object-oriented database systems. Rapports de Recherche 947 (1988)
18. Peters, R.J., Özsu, M.T.: An axiomatic model of dynamic schema evolution in objectbase systems. ACM Transactions on Database Systems 22(1), 75–114 (1997)
19. Polya, G.: How to Solve It. Princeton University Press, Princeton (1957)
20. Roddick, J.F., Craske, N.G., Richards, T.J.: A taxonomy for schema versioning based on the relational and entity relationship models. In: Elmasri, R.A., Kouramajian, V., Thalheim, B. (eds.) ER 1993. LNCS, vol. 823, pp. 137–148. Springer, Heidelberg (1994)
21. Roddick, J.F.: A survey of schema versioning issues for database systems. Information and Software Technology 37(7), 383–393 (1995)
22. Thagard, P.: Conceptual revolutions. Princeton University Press, Princeton (1992)

# An Iterative Process for Adaptive Meta- and Instance Modeling

Melanie Himsl[1], Daniel Jabornig[1], Werner Leithner[1], Peter Regner[1], Thomas Wiesinger[1], Josef Küng[1], and Dirk Draheim[2]

[1] FAW-Institute
Johannes Kepler University of Linz, Austria
{mhimsl,djabornig,wleithner,pregner,twiesinger, jkueng}@faw.uni-linz.ac.at
[2] Software Competence Center Hagenberg
draheim@acm.org

**Abstract.** In this paper we propose a practice for organizing modeling activity. We see substantial, successful modeling efforts in enterprises, e.g., in our logistics, manufacturing, banking and insurance projects, even without model-driven engineering metaphor. The focus of our discussion is the working domain expert. The working domain expert desires tool support, service support, and adaptivity of the modeling approach. We discuss these topics in the proven framework of the IT Infrastructure Library.

**Keywords:** Metamodeling, Iterative Modeling Process, Adaptive Modeling, Model Evolution, IT-Infrastructure Library (ITIL).

## 1 Introduction

In this paper we investigate modeling as a tailorable common activity in companies, we investigate modeling languages as tailorable modeling tools. An adaptive metamodeling tool that has been designed with respect to the needs of the working domain expert serves as the basis of this discussion. We present our thoughts in terms of the IT Infrastructure Library.

Adaptivity of modeling languages is a major driving issue in the community of Model-Driven Architecture [1], [2] (MDA), which is the current automatic programming [3], [4] metaphor. Modeling is pervasive in modern enterprises; however, it is so without automatic programming metaphor. Of course, modeling, and visual modeling in particular, is used in Software development projects. With respect to Software development, there are different opinions about the role and the importance of modeling. For example, the Rational Unified Process (RUP) [5] is based on modeling – it is model-driven. On the other hand, in agile processes like Extreme Programming (XP) [6] modeling is deemphasized. Despite that we see severe modeling efforts in companies, in both vertical and horizontal projects, not only Software development projects but projects [7] in general. For example, we currently see huge business process re-

R. Wagner, N. Revell, and G. Pernul (Eds.): DEXA 2007, LNCS 4653, pp.519–528,2007.

documentation projects in major enterprises. In day-to-day projects we see modeling activities in business reengineering, logistics, supply chain management, industrial manufacturing and so on. Even if models are not intended as blueprints in Software development projects they add value. Why? They foster communication between stakeholders, because they enforce a certain standardization of the respective domain language. Therefore, they speed up requirement elicitation and than serve as a long-time documentation of system analysis efforts.

Modeling is here to stay. Research in model-driven engineering is important. In this paper we have a different focus on modeling than model-driven engineering. We have a look at the working domain [8] expert. Often, it is necessary to adapt the modeling method and, in particular, to adapt the used modeling language to the current needs of the domain. It may become necessary to introduce new modeling elements, to deprecate an existing model element, to add properties to an existing modeling element, to detail the semantics or to change the appearance of a model element. However, changes to a modeling apparatus must be done in a disciplined manner. Ideally, all stakeholders in the project should agree upon changes. At least, there should be an authority in the project, a senior modeler so to speak, who coordinates and eventually allows or rejects changes.

With respect to a disciplined approach to adaptive modeling, it is a good practice to use a modeling tool. A modeling tool not only eases modeling, it also helps in streamlining the modeling efforts. All stakeholders have the same view onto modeling efforts as the set of modeling elements supported by the used tool. Now, if adaptivity is an issue, the modeling tool should allow for changes to the supported modeling languages. If it is not possible to change the modeling language at all, the only work-around is the introduction of names for specific modeling elements and specific properties along with naming conventions. However, the loose introduction of specific names weakens the aforementioned streamlining effect of the modeling tool. Therefore, the modeling tool should ideally support metamodeling. Metamodeling capabilities introduce the flexibility needed for modeling language adaptation. At first sight, metamodeling capabilities do not come at the price of weakening the streamlining effect of the tool, because changes to the modeling language are now done systematically in the framework of the defined metamodeling features. However, if every stakeholder in the project gets access to the metamodeling features, the described risks again exist. A concrete risk is, for example, that a modeler is tempted to introduce modeling elements from his or her own modeling idiolect. Another risk is that modelers start modeling on the metamodeling level, i.e., introducing concrete elements of the concrete modeling problem as modeling elements into the method. All this can lead to an uncoordinated rank growth of the metamodel. It remains a task to define guidelines for modeling and metamodeling activities.

The introduction of metamodeling not only has the risk of uncontrolled changes to the modeling apparatus. A converse kind of phenomenon can also be observed. Some stakeholders are challenged with metamodeling concepts. They just want to do their modeling job good and do not want to learn about metamodeling and metamodeling tool features. They want to communicate their problems with the modeling apparatus and get the problems fixed by a tool smith. This means, the working modeler typically perceives the modeling tool as a Software service. For the working modeler a change request to the modeling language is not different from a change request to any other feature of the modeling tool. This means that for the working modeler the adaptivity of the modeling language just belongs to the changeability of the tool in the sense of Software product quality – see ISO standard 9126 [9].

With this paper we want to give guidance for projects that show modeling by domain experts. Most importantly, we incorporate metamodeling into the considerations in order to achieve the desired flexibility. On the other hand, we want stay in control of the metamodeling activity. In general, the domain expert has the characteristics of an end user [10] and therefore he or she wants to be supported by the modeling tool as a Software service. For these reasons, we consequentially define a Software service support process for the modeling scenario backed by metamodeling capabilities. We do this in the terminology of ITIL for the reason of maximal reuse of concepts, because ITIL is the most comprehensive, most widely used collection of best practices in managing Software services and the group of ITIL users is steadily growing.

## 2 Addressed Problem Areas

Because of the different domains modeling is used for, it is important to represent the domain specific knowledge in the modeling techniques. The level of abstraction and the genericness of standard techniques often constrict a complete modeling of a specific issue. This leads to an incorrect description of the reality or to an inappropriate abstraction level. For example the area of business process management includes a wide range of activities. They reach from quality management to the introduction of standard software. Each activity needs a custom process modeling technique which should be able to define the knowledge of a specific branch, company or assignment.

To grant an accurate representation of the modeling domain, a customized modeling technique is necessary. In practice it is often impossible to adapt the given modeling techniques. However if adaptations would be possible either the users are generally prevented from making adaptations or they must have special skills and knowledge like knowing specific programming or scripting languages.

Furthermore, to find an optimal solution for a domain specific modeling technique it is necessary to traverse several building cycles to get an ideal description of the subject. The reason therefore is that the requirements often can't be completely defined before the technique definition and they first arise during the modeling of the domain. The requirements can also vary if changes in the modeling domain occur. Changed requirements need an adaptation of the modeling technique and probably an adaptation of all models which have already been defined with that technique.

## 3 A Concept for an Iterative Meta- and Instance Modeling Process

At the "Institute of Applied Knowledge Processing" (FAW) of the Johannes Kepler University Linz a concept was developed for graphical meta- and instance modeling. The goal behind the concept is to enable users to create or adapt modeling techniques to their specific research area. The creation and/or adaptation should be done on a meta-layer and the usage of the method takes place on an instance-layer.

On the meta-layer the user has the possibility to define the elements the modeling method should consist of. Also attributes and visual properties can be defined for elements and connections between those elements can be attached. Based on a created metamodel any number of instance models can be produced. The metamodeling concept makes the user feel free to define a modeling method which fits exactly to his research area.

The metamodeling is done graphically to keep the users away from learning programming- or scripting languages for defining a metamodel and to facilitate the definition of a modeling method. So the user can build metamodels intuitively by connecting objects and connections in a graphical editor like on the instance-layer. That implies that the representation of the elements is the same on the instance and the meta-layer: objects are displayed as figures and connections as edges.

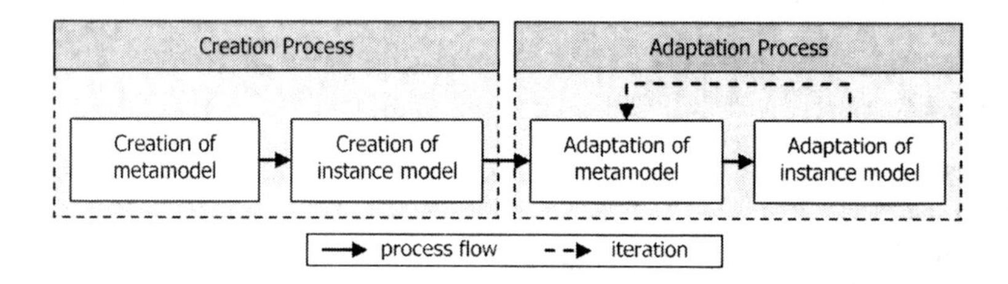


**Fig. 1.** The iterative modeling process

Another goal and basic requirement for the concept is the support of an iterative modeling process. That means adaptations of the metamodel can be made although instances already exist. This process will be displayed in figure 1 and includes the following four steps:

1. Creation of a metamodel
2. Creation of a instance model
3. Adaptation of the metamodel
4. Adaptation of the instance model

The first step in the iterative modeling process is the creation of a metamodel. The metamodel defines model elements, which are available for modeling at the instance-layer. In the second step an instance model will be created. Each instance model has exactly those characteristics defined in the metamodel. These first two steps describe together the creation process of an instance model.

If it is necessary to extend the instance model (for example with additional elements or attributes) or to add certain restrictions e.g. due to domain evolution, the underlying metamodel has to be changed. This adaptation of the metamodel is the third step of the iterative modeling process. As the instance model is an instance of the changed metamodel, it follows that the instance model has to be adapted. This step is performed with an integrated transformation engine. The adaptation of the metamodel and the following adaptation of the instance model belong to the adaptation process. The number of iterations in the adaptation process is not limited.

# 4 A Concept for Iterative Meta- and Instance Modeling Within an Organizational Process

Metamodeling and particularly the previously described iterative modeling process can lead to several unintended situations when they are implemented in an inappropriate way within an organizational structure. The main reason therefore is that a metamodel has to define a common language for a specific domain and should be constant until an adaptation is inescapable. Areas where such adaptations are often mandatory were described previously and can lead to inconsistencies in the company-wide library of defined metamodels.

Due to that fact, one major criterion for a successful organizational implementation of metamodeling is the iterative metamodeling process, which describes the resulting adaptation of instances after metamodel changes. This process prevents that several different versions of a specific metamodel are used in an organization. In other words, there always will be only one valid version of a metamodel for all its instances, and that leads to consistency as well as transparency.

As a metamodel defines a common language that is used at least company-wide or possibly across depended companies, any changes on a metamodel, especially if there exists a huge number of instances, has to be planned carefully and coordinated with all stakeholders concerned. For that reason it is obligatory to implement standardized and predefined change- and authorization-processes. Those processes should be integrated in a main modeling support process that defines activities, roles and permissions for all occurring incidents. Then the main process has to be implemented in the organizational structure. Figure 2 visualizes this "Modeling Support Process" (MSP) on an abstract level by defining its main sub-processes and results.

Prerequisite for the MSP is the prevention from any redundancies. That can be achieved at best by the integration of a *central repository* for all defined metamodels. The repository not only contains metamodels but also any changes made to them. By that it becomes obvious how a metamodel evolved. With additional information about why a metamodel was changed, who changed it and the primary incidents that caused the change, an ex-post view is practicable and often recommended to evaluate the improvement.

The origin of a change in a metamodel can be any request expressed by a stakeholder, who is somehow affected by the metamodel. Mostly a request will occur because the stakeholder uses the metamodel to create instances and wants an improvement of the modeling technique. But not always every request is a reason for a change in the metamodel. Some requests will also be a result of disinformation about the correct usage, possibilities and domain of a given metamodel. Within the MSP there has to be a central point where such a request will be accepted and handled by following a well defined process. At this point the request has to be rated and it has to be detected if there are requirements for a change or if the request can be responded with information about the correct usage.

If the incoming request can not be handled directly there may be a problem with the metamodel or room for improvement and that has to be considered in detail. Basis for that process are records of known problems and workarounds that can be also part of the central repository. If there are no recorded workarounds nor there can be workarounds created, the main decision that has to be made is if the request desires a change in the metamodel. As basis for decision the importance of the request and of course the number of similar requests will be crucial. If, as result of this problem-rating process, the change of the metamodel is decided, a "Request for Change" (RFC) has to be created.

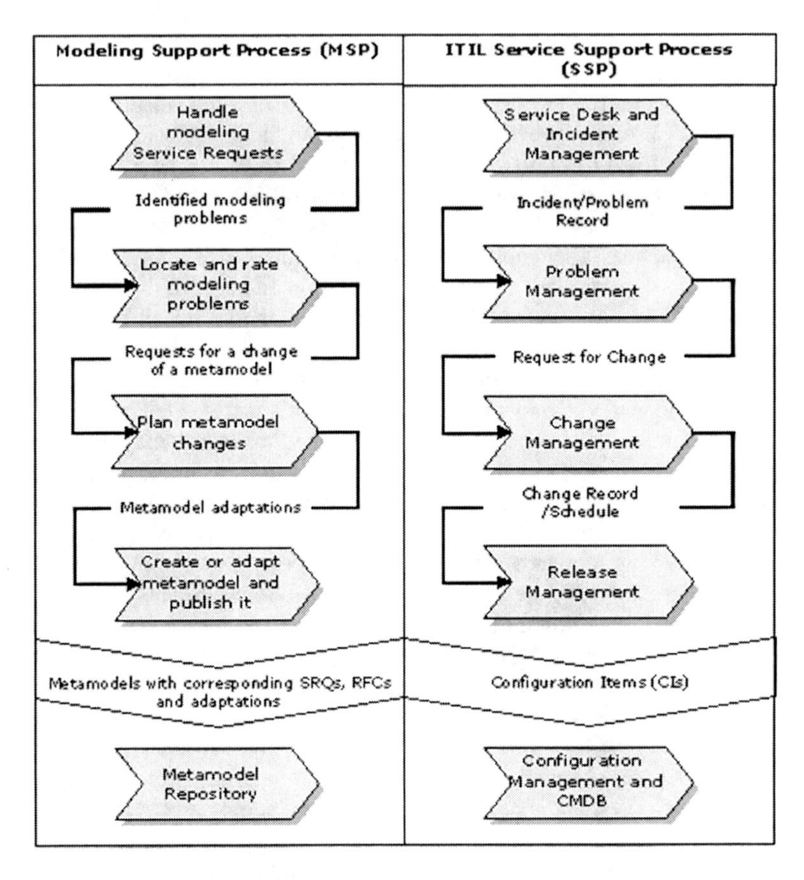


**Fig. 2.** The *Modeling Support Process (MSP)* describes the recommended activities for the organizational integration of an adaptive and iterative meta- and instance modeling approach. The left swimlane shows the MSP sub-processes, the process flow and the main results of each sub-process. Back-links between each sub-process exist but are omitted for abstraction. The MSP is an instance of the well-established *Service Support Process (SSP)* defined by the *IT Infrastructure Library (ITIL)*. The right swimlane shows the five SSP sub-processes by mapping them to the corresponding MSP sub-processes.

Any change of the metamodel affects all its instances and with that any stakeholder that is dependent on one of these instances. Those consequences make it obvious and essential that any change must be planned in detail and that all effects have to be considered in advance. This has to be done in a well-defined process that has a comprehensive change plan and schedule as result.

After the detailed planning process the results have to be realized through an adaptation of the concrete metamodel. In most cases it makes sense to summarize planned changes of a metamodel to one major release. Beside that, a new or adapted metamodel requires an update of all documentations, descriptions and training information. All stakeholders have to be informed about the new metamodel release and additional data of the performed changes has to be provided to them.

All processes described in this chapter are intended to be wide scalable. In the minimum case a single person within an organizational unit is responsible for all sub-processes of the MSP. Main requirement for the scalability is a detailed definition of roles and activities assigned to them.

### 4.1 MSP Meets ITIL Service Support

In practice modeling is an activity that is used in several different areas within a company, dependent on the type of company and the domain that has to be visualized. But at least the service of modeling and modeling tools provided to users are a key domain of every IT-Management. During the years the way how IT-Services are provided to users has changed. Now it is a requirement that services provided are strongly focused on bringing value to users and supporting business processes to enhance business performance. Due to that the focus is now on the implementation of a *service-oriented IT-Management.*

When talking about service-oriented IT-Management the IT-Infrastructure-Library (ITIL) developed by the British Central Computer and Telecommunications Agency (CCTA) is the de-facto standard and becomes more and more popular not at least because of the strong support of some major developers and users. ITIL is a best-practice framework for IT-Service-Management and consists of five process areas where the "Service Delivery" [11] and the "Service Support" [12] processes describe the administrative tasks of delivering IT-Services to users including the definition of "Service Level Agreements" (SLAs) and the operative tasks of supporting IT-Service users.

Within a service-oriented IT-Management approach like ITIL, modeling, as part of the IT-Management service-portfolio, will be provided as an IT-Service to users. *"Service Delivery"* is responsible for defining SLAs for the modeling service and has to plan the availability, capacity, continuity and financial requirements that need to be maintained as a result of the SLAs.

The *"Service Support"* as defined in ITIL has to deal with incidents that occur by "Service Requests" from users and by monitoring the IT-Service. As the figure 2 shows, the complete previously defined "Modeling Support Process" (MSP) can be mapped to the "Service Support Process" (SSP) defined by ITIL. Moreover the specified process can be seen as an instance of the ITIL-"Service Support"-Process for the specific domain *"Modeling"* and fully integrated as part of an overall IT-Service Support Process within a service-oriented IT-Management.

## 5 A Platform for Iterative Meta- and Instance Modeling

The iterative meta- and instance modeling process has been implemented in a prototypic modeling platform by the use of open-source technologies. As a main feature an intuitive and flexible metamodel definition language was integrated to allow a simplified graphical definition of metamodels and to support the iterative modeling process.

The tool defines a meta-layer for textual or graphical definition of metamodels that are persisted in an integrated or central metamodel repository. To enable the creation of metamodel instances an instance-layer has been implemented. Both the meta- and instance-layer are integrated as modules within the platform and can be optionally removed to create either only a metamodeling- or instance modeling tool. Beside that the access to each module is role dependent and can be restricted by an administration module that manages roles, users and user groups.

The role specific access to modules is especially important to integrate into the Modeling Support Process (MSP) that was specified previously in this work. Only with the definition of privileged roles it is possible to access the meta-layer. This and a central metamodel repository prevent from the decentralized definition or adaptation of metamodels by unauthorized users. To integrate completely into the MSP the platform will be extended to send Service Requests (SRQs) and to handle replies to that requests.

## 6 Related Work

In general, best practices in project management are [7] actually polymorphic collections of techniques – both with respect to activities and work products. The meta activity of designing or tailoring a concrete project management guideline is not in issue in general. Not so for the domain of Software development. For example, the Rational Unified Process considers adaptivity from the outset – the RUP classifies itself as a tailorable process. Actually, in the RUP the tailoring of the process is addressed as part of the project, i.e., the process instance, so that dynamic changes of the Software process are considered. However, the tailoring of the process can be done only in terms of the predefined process entities and work product items of the RUP and the UML. In particular, the work product items are not subject to design. The latter is also a true for OMG's Software Process Engineering Metamodel Specification (SPEM) [13]. This is pointed out in [14] and a co-design for process and work products is proposed. The results of this strand of work [14, 15] are now available as ISO standard 24744 [16] – the Software Engineering Metamodel for Development Methodologies (SEMDM). In SEMDM, special emphasis is put on the semantic foundations of metamodeling [17]. However, if the tailoring of the work products during the project lifecycle is an issue, the appropriate framework is given by a meta-process model like the generic spiral model [18], where the next phase planning segment is a natural place for such endeavor. Our interest in this paper is different. We do not focus on the process of metamodeling nor on the process incorporating metamodeling, but an orthogonal service support process for modeling supported by metamodeling. We believe that this is an interesting viewpoint, because it allows for improving important commonalities of the different modeling efforts in companies. Furthermore, we are interested in enterprise modeling efforts in general, not only in modeling efforts in Software development projects.

## 7 Further Work

One problem addressed by the concepts in our paper is integrated metamodeling and model migration, i.e., the challenge to keep existing models in synch with the evolving metamodel. As an instance of this problem we have discussed the deletion of a metamodel entity. A non-trivial example is the refinement of a metamodel entity into several new entities. Currently, we elaborate sophisticated support mechanisms like guidelines, a versioning concept, and a customizable tool feature for automatic model migration. This strand of work is analog to the ubiquitous problem of integrated model evolution and data migration in model-driven engineering of multi-tier architectures [19].

## 8 Conclusion

*We observed:*

- We need awareness for the working domain expert as a modeler, independent from model-driven engineering approaches.
- The domain expert has the characteristics of an end-user. He or she is no Software development expert.
- The domain expert wants to use the modeling tool as Software service.
- Adaptivity of the modeling approach is a key success factor in projects.
- Metamodeling offers the appropriate support for adaptivity.
- Adaptivity and metamodeling capabilities bear the risk of uncontrolled rank growth of the metamodel.

*We contributed:*

- We have developed a combined modeling and metamodeling tool that is oriented towards the needs of the working domain expert.
- We have unified the above observations in the proven framework of the IT Infrastructure Library.
- We have proposed concrete guidance for the organization of modeling and metamodeling efforts in enterprises. In particular, we elaborated a service support process for the enabling metamodeling tool.

## References

1. Atkinson, C., Kühne, T.: The Role of Metamodeling in MDA. In: Proceedings of WISME@UML2002 – International Workshop in Software Model Engineering (2002)
2. Soley, R.: Model Driven Architecture, white paper formal/02-04-03, draft 3.2, Object Management Group (November 2003)

3. Czarnecki, K., Eisenecker, U.W.: Generative Programming – Methods, Tools, and Applications. Addison-Wesley, London (2000)
4. Parnas, D.L.: Software Aspects of Strategic Defense Systems. In: Software Engineering Notes, ACM Sigsoft, vol. 10(5), ACM Press, New York (1985)
5. Jacobson, I., Booch, G., Rumbaugh, J.: The Unified Software Development Process. Addison-Wesley, London (1999)
6. Beck, K.: Extreme Programming Explained – Embrace Change. Addison-Wesley, London (2000)
7. Duncan, W.R.: A Guide to the Project Management Body of Knowledge. Project Management Institute (1996)
8. Bjorner, D.: On Domains and Domain – Engineering Prerequisites for Trustworthy Software – A Necessity for Believable Project Management. Domain Engineering and Digital Rights Group (April 2006)
9. ISO Technical Committee JTC 1/SC7. ISO/IEC 9126-1:2001. Software Engineering – Product quality – Part 1: Quality model. International Organization of Standardization (2006)
10. Jones, C.: C. End-User Programming. In: IEEE Computer, vol. 28(9), pp. 68–70. IEEE Press, New York (1995)
11. Office of Government and Commerce: Service Delivery (IT Infrastructure Library). Stationery Office (2001)
12. Central Computing and Telecommunications Agency: Service Support (IT Infrastructure Library). The Stationery Office (2002)
13. Object Management Group: Software Process Engineering Metamodel Specification. OMG document formal/2002-11-14, OMG (2002)
14. Gonzalez-Perez, C., Henderson-Sellers, B.: Templates and Resources in Software Development Methodologies. Journal of Object-Technology 4(4) (May 2005)
15. Gonzalez-Perez, C., Henderson-Sellers, B.: An Ontology for Software Development Methodologies and Endeavours. In: Ontologies in Software Engineering and Software Technology, pp. 123–151. Springer, Heidelberg
16. ISO Technical Committee JTC 1/SC7. ISO/IEC 24744:2007. Software Engineering – Metamodel for Development Methodologies. International Organization of Standardization (2007)
17. Gonzalez-Perez, C., Henderson-Sellers, B.: On the East of Extending a Powertype-based Methodology Metamodel. In: Proceedings of WoMM2006 – the 2nd Workshop on Metamodelling and Ontologies. LNI, vol. 96, pp. 11–25 (2006)
18. Boehm, B.W.: A Spiral Model of Software Development and Enhancement. IEEE Computer 21(5), 61–72 (1988)
19. Bordbar, B., Draheim, D., Horn, M., Schulz, I., Weber, G.: Integrated Model-Based Software Development, Data Access and Data Migration. In: Briand, L.C., Williams, C. (eds.) MoDELS 2005. LNCS, vol. 3713, Springer, Heidelberg (2005)

# Compiling Declarative Specifications of Parsing Algorithms*

Carlos Gómez-Rodríguez, Jesús Vilares, and Miguel A. Alonso

Departamento de Computación, Universidade da Coruña (Spain)
{cgomezr, jvilares, alonso}@udc.es
Campus de Elviña, s/n - 15071 A Coruña (Spain)
Tel: +34 981 16 70 00 - Fax: +34 981 16 71 60

**Abstract.** The parsing schemata formalism allows us to describe parsing algorithms in a simple, declarative way by capturing their fundamental semantics while abstracting low-level detail. In this work, we present a compilation technique allowing the automatic transformation of parsing schemata to efficient executable implementations of their corresponding algorithms. Our technique is general enough to be able to handle all kinds of schemata for context-free grammars, tree adjoining grammars and other grammatical formalisms, providing an extensibility mechanism which allows the user to define custom notational elements.

## 1 Introduction

The process of parsing, by which we obtain the structure of a sentence as a result of the application of grammatical rules, is a highly relevant step in the automatic analysis of natural language sentences. Parsing schemata, described in [14], provide a formal, simple and uniform way to describe, analyze and compare different parsing algorithms. The notion of a parsing schema comes from considering parsing as a deduction process which generates intermediate results called *items*. Each item contains a piece of information about the sentence's structure, and a successful parsing process will produce at least one *final item* containing a full parse tree for the sentence or guaranteeing its existence. An initial set of items is directly obtained from the input sentence, and the parsing process consists of the application of inference rules, called *deductive steps*, of the form $\frac{\eta_1 \dots \eta_m}{\xi} \Phi$ that allow us to infer the item specified by its consequent $\xi$ from those in its antecedents $\eta_1 \dots \eta_m$. *Side conditions* ($\Phi$) specify the valid values for the variables appearing in the antecedents and consequent, and may refer to grammar rules or specify other constraints that must be verified in order to infer the consequent.

* Partially supported by Ministerio de Educación y Ciencia (MEC) and FEDER (TIN2004-07246-C03-01, TIN2004-07246-C03-02), Xunta de Galicia (PGIDIT05PXIC30501PN, PGIDIT05PXIC10501PN, Rede Galega de Procesamento da Linguaxe e Recuperación de Información) and Programa de Becas FPU (MEC).

R. Wagner, N. Revell, and G. Pernul (Eds.): DEXA 2007, LNCS 4653, pp. 529–538, 2007.

A schema specifies the steps that must be executed and the intermediate results that must be obtained in order to parse a given string, but it makes no claim about the order in which to execute the steps or the data structures to use for storing the results. Their abstraction of low-level details makes parsing schemata very useful, allowing us to define parsers in a simple and straightforward way. Comparing parsers, or considering aspects such as their correctness and completeness or their computational complexity, also becomes easier if we think in terms of schemata. However, when we want to actually test a parser and check its results, we need to implement it in a programming language, so we have to abandon the high level of abstraction and worry about implementation details that were irrelevant at the schema level. The technique presented in this paper automates this task, by compiling parsing schemata to Java language implementations of their corresponding parsers.

## 2 From Declarative Descriptions to Program Code

Our compilation process proceeds according to the following principles:

- A class is generated for each deductive step in the schema.
- The generated implementation will create an instance of this class for each possible set of values satisfying the side conditions that refer to production rules.
- The classes representing deductive steps have an `apply` method which tries to apply the deductive step to a given item. If the step is in fact applicable to the item, the method returns the new items obtained from the inference. In order to achieve this functionality, the method works as follows: first, it checks if the given item matches any of the step's antecedents. For every successful match found, the method searches for combinations of previously-generated items in order to satisfy the rest of the antecedents. Each combination of items satisfying all antecedents corresponds to an instantiation of the step variables which is used to generate an item from the consequent.
- The execution of deductive steps in the generated code is coordinated by a *deductive parsing engine*. This is a schema-independent algorithm, and therefore its implementation is the same for any schema:

```
steps = {deductive step instances};
items = {initial items};
agenda = [initial items];
For each deductive step with an empty antecedent (s) in steps {
 result = s.apply([]);
 items.add(result);
 agenda.enqueue(result);
 steps.remove(s);
}
While agenda not empty {
 curItem = agenda.removeFirst();
 For each deductive step applicable to curItem (p) in steps {
  result = p.apply(curItem);
  items.add(result);
  agenda.enqueue(result);
 }
}
return items;
```

The algorithm works with the set of all items that have been generated (either as initial hypotheses or as a result of the application of deductive steps) and an *agenda*, implemented as a queue, which contains the items we have not yet tried to trigger new deductions with. When the agenda is emptied, all possible items will have been generated, and the presence or absence of final items in the item set at this point indicates whether or not the input sentence belongs to the language defined by the grammar. The correctness and completeness of this algorithm can easily be proved by induction. The parse forest can be recovered easily from the item set, as in [1].

## 2.1 Indexing

The implementation described above will only be efficient if we can efficiently access items and deductive steps. In particular, implementation of the operations checking if a given item exists in the item set (implicitly used by the `items.add` operation in the pseudocode above) and searching the item set for all items satisfying a certain specification (used by the `apply` method of deductive steps) affects the resulting parser's computational complexity. An inefficient implementation of any of these operations will give as result a parser with a computational complexity above the expected theoretical bounds for the corresponding algorithms. In order to maintain the theoretical complexity, we must provide constant-time access to items. In this case, each single deduction takes place in constant time, and the worst-case complexity is bounded by the maximum possible number of step executions: all complexity in the generated implementation is inherent to the schema.

In order to achieve this, we generate indexing code allowing efficient access to the item set. Two distinct kinds of indexes are generated for each schema, corresponding to the operations mentioned before: *existence indexes* are used to check whether an item exists in the item set, and *search indexes* allow us to search for items conforming to a given specification. Apart from items, deductive steps are also indexed in *deductive step indexes.* These indexes are used to restrict the set of "applicable deductive steps" for a given item, discarding those known not to match it. Deductive step indexes usually have no influence on computational complexity with respect to input string size, but they do have an influence on complexity with respect to the size of the grammar, since the number of deductive step instances depends on grammar size when production rules are used as side conditions.

Our indexing mechanism is explained in detail in [7]. As an example of how the adequate indexes can be determined by a static analysis of the schema prior to compilation, we analyze the case where we have a deductive step of the form

$$\frac{[a, d, e, g] \qquad [b, d, f, g]}{(consequent)} \; c\ e\ f\ g$$

where each lowercase letter represents the set of elements (be them grammar symbols, string positions or other entities) appearing at particular positions in the step, so that $a$ stands for the set of elements appearing only in the first antecedent item, $e$ represents those appearing in the first antecedent and side condition, $g$ those appearing in both antecedents and side condition, and the rest of the letters represent the other possible combinations as can be seen in the step. In this example, we consider only two antecedents for the sake of simplicity, but the technique is general and can be applied to deductive steps with an arbitrary number of antecedents.

In this case, the following indexes are generated:

1. One deductive step index for each antecedent, using as keys the elements appearing both in the side condition *and* in that particular antecedent: therefore, two indexes are generated using the values $(e, g)$ and $(f, g)$. These indexes are used to restrict the set of deductive step instances applicable to items. As each instance corresponds to a particular instantiation of the side conditions, in this case each step instance will have different values for $c$, $e$, $f$ and $g$. When the deductive engine asks for the set of steps applicable to a given item $[w, x, y, z]$, the deductive step handler will use the values of $(y, z)$ as keys in order to return only instances with matching values of $(e, g)$ or $(f, g)$. Instances of the steps where these values do not match can be safely discarded, as we know that our item will not match any of both antecedents.
2. One search index for each antecedent, using as keys the elements appearing in that antecedent which are also present in the side condition *or* in the other antecedent. Therefore, a search index is generated by using $(d, e, g)$ as keys in order to recover items of the form $[a, d, e, g]$ when $d$, $e$ and $g$ are known and $a$ can take any value; and another index using the keys $(d, f, g)$ is generated and used to recover items of the form $[b, d, f, g]$ when $d$, $f$ and $g$ are known. The first index allows us to efficiently search for items matching the first antecedent when we have already found a match for the second, while the second one can be used to search for items matching the second antecedent when we have started our deduction by matching the first one.
3. One existence index using as keys all the elements appearing in the consequent, since all of them are instantiated to concrete values when the step successfully generates a consequent item. This index is used to check whether the generated item already exists in the item set before adding it.

As this index generation process must be applied to all deductive steps in the schema, the number of indexes needed to guarantee constant-time access to items increases linearly with the number of steps. However, in practice we do not usually need to generate all of these indexes, since many of them are repeated or redundant. For example, if we suppose that the sets $e$ and $f$ in our last example contain the same number and type of elements, and elements are ordered in the same way in both antecedents, the two search indexes generated would in fact be the same, and our compiler would detect this fact and generate only one. In practical cases, the items used by different steps of a parsing schema usually have the same structure, so most indexes can be shared among several deductive steps and the amount of indexes generated is small.

All the generated indexing code is placed into two classes (the *item handler* and the *deductive step handler*) whose function is to provide efficient access to items and deductive steps, responding to queries issued by the deductive parsing engine.

## 2.2 Elements in Schemata

The variety of elements that may be present in parsing schemata poses an interesting difficulty if we want our technique to be general enough to cope with all sorts of schemata. The schemata notation is open, and any mathematical object could potentially appear as part of the definition of a schema.

As it is obviously impossible to provide a system that will recognize any kind of element that we could potentially include in a schema, but neither do we want our compiler to be limited to certain types of elements, we have defined an extensibility mechanism which allows us to define new elements that can be handled by the system in an easy way. For this purpose, we will classify all notational elements into four basic types, according to the treatment they should receive during code generation. Any new kind of element added to the system should be classified into one of these types:

- *Simple Elements:* Atomic, unstructured elements, which can be instantiated or not in a given moment. When simple elements are instantiated, they take a single value from a set of possible values, which can be bounded or not. Values can be converted to indexing keys. Examples of simple elements are grammar symbols, integers, string positions, probabilities...
- *Expression Elements*: These elements denote expressions which take simple elements or other expressions as arguments. For example, $i+1$ is an expression element representing the sum of two string position arguments, and $tree[A, B]$ is an expression over nonterminal symbols. Feature structures and logic terms are also represented by this kind of elements. When all simple elements in an expression are instantiated to concrete values, the expression will be treated as a simple element whose value is obtained by applying the operation it defines (for example, summation). For the code generator to be able to do this, a Java expression must be provided as part of the expression element type definition, so that, for example, sums of string positions appearing in schemata can be converted to Java integer sums in the generated implementation. Unification of feature-structures has been implemented in this way.
- *Composite Elements*: Composite elements represent sequences of elements whose length must be finite and known. Composite elements are used to structure items. For instance, the Earley item $[A \rightarrow \alpha.B\beta, i, j]$ is represented as a composite element with three components: the first one is in turn a composite element, representing a grammar rule, while the remaining two are simple elements which denote string positions.
- *Sequence Elements*: These elements denote sequences of elements of any kind whose length is finite, but only becomes known when the sequence is

**Table 1.** Information about the grammars used in the experiments: total number of symbols, nonterminals, terminals, production rules, distribution of rule lengths, and average rule length

| Grammar | $\|N \cup \Sigma\|$ | $\|N\|$ | $\|\Sigma\|$ | $\|P\|$ | Epsilon | Unary | Binary | Other | Rule length |
|---|---|---|---|---|---|---|---|---|---|
| Susanne | 1,921 | 1,524 | 397 | 17,633 | 0% | 5.26% | 22.98% | 71.76% | 3.54 |
| Alvey | 498 | 266 | 232 | 1,485 | 0% | 10.64% | 50.17% | 39.19% | 2.4 |
| Deltra | 310 | 282 | 28 | 704 | 15.48% | 41.05% | 18.18% | 25.28% | 1.74 |

instantiated to a concrete value. The strings appearing in Earley items are examples of sequence elements, being able to represent symbol strings of any length. The code generator must take this fact into account when generating matching code for these elements.

In order to add a new kind of element to the schema compiler, the user will have to define it as a subclass of one of these four basic types, and implement that type's interface by following some simple guidelines. In addition to this, the user must provide one or more regular expressions in order to specify the format of the strings representing the new kind of element in schemata definition files. These expressions can be included in a global configuration file or directly in the schema files that will use the element. The schema parser will use the regular expressions to identify our new type of element in schema files. When one of these elements is found in a schema, the compiler will dynamically load the corresponding class and instantiate it by using Java's reflection mechanisms, thus avoiding the need to recompile the system in order to add new element classes. This makes our technique highly extensible, and easily allows us to work with schemata containing all kinds of non-predefined items.

## 3 Experimental Results

We have used our technique to generate implementations of three popular parsing algorithms for context-free grammars: CYK [9,15], Earley [3] and Left-Corner [10]. The schemata we have used describe recognizers, and therefore their generated implementation only checks sentences for grammaticality by launching the deductive engine and testing for the presence of final items in the item set. However, these schemata can easily be modified to produce a parse forest as output [1]. If we want to use a probabilistic grammar in order to modify the schema so that it produces the most probable parse tree, this requires slight modifications of the deductive engine, since it should only choose the item with the highest probability when several items are available to match an antecedent.

The three algorithms have been tested with sentences from three different natural language grammars: the English grammar from the Susanne corpus [11], the Alvey grammar [2] (which is also an English-language grammar) and the Deltra grammar [12], which generates a fragment of Dutch. The Alvey and Deltra grammars were converted to plain context-free grammars by removing their arguments and feature structures. The test sentences were randomly generated by starting with the axiom and randomly selecting nonterminals and rules to perform expansions, until valid sentences consisting only of terminals were produced. Note that, as we are interested in measuring and comparing the performance of the parsers, not the coverage of the grammars; randomly-generated sentences are a good input in this case: by generating several sentences of a given length, parsing them and averaging the resulting runtimes, we get a good idea of the performance of the parsers for sentences of that length. Table 1 summarizes some facts about the three grammars, where by "Rule Length" we mean the average length of the right-hand side of a grammar's rules.

For Earley's algorithm, we have used the schema described in [14]. For the CYK algorithm, grammars were converted to Chomsky normal form (CNF), since this is a precondition of the algorithm. In the case of the Deltra grammar, which is the only one of our test grammars containing epsilon rules, we have used a weak variant of CNF allowing epsilon rules. For the Left-Corner parser, the schema used is the *sLC* variant described in [14].

Performance results[1] for all these algorithms and grammars are shown in table 2. The following conclusions can be drawn from the measurements:

- The empirical computational complexity of the three algorithms is below their theoretical worst-case complexity of $O(n^3)$, where $n$ denotes the length of the input string. In the case of the Susanne grammar, the measurements we obtain are close to being linear with respect to string size. In the other two grammars, the measurements grow faster with string size, but are still far below the cubic worst-case bound.
- CYK is the fastest algorithm in all cases, and it generates less items than the other ones. This may come as a surprise at first, as CYK is generally considered slower than Earley-type algorithms, particularly than Left-Corner. However, these considerations are based on time complexity relative to string size, and do not take into account complexity relative to grammar size. In this aspect, CYK is better than Earley-type algorithms, providing linear — $O(|P|)$ — worst-case complexity with respect to grammar size, while Earley is $O(|P|^2)$. Therefore, the fact that CYK outperforms the other algorithms in our tests is not so surprising, as the grammars we have used have a large number of productions[2]. The greatest difference between CYK and the other two algorithms in terms of the amount of items generated appears with the Susanne grammar, which has the largest number of productions. It is also worth noting that the relative difference in terms of items generated tends

[1] The machine used for these tests was a standard laptop: Intel 1500 MHz Pentium M processor, 512 MB RAM, Sun Java Hotspot virtual machine (version 1.4.2_01-b06) and Windows XP.

[2] It is possible to reduce the computational complexity of Earley's parser by applying some transformations to the schema. Even in this case, CYK performs better than Earley's algorithm due to the smaller number of items generated: $O(|N \cup \Sigma|n^2)$ for CYK vs. $O(|G|n^2)$ for Earley, where $|G|$ denotes the size of the grammar measured as the number of productions plus the summation of the lengths of all productions.

**Table 2.** Performance measurements for generated parsers

| Grammar | String length | Time Elapsed (s) | | | Items Generated | | |
|---|---|---|---|---|---|---|---|
| | | CYK | Earley | LC | CYK | Earley | LC |
| Susanne | 2 | 0.000 | 1.450 | 0.030 | 28 | 14,670 | 330 |
| | 4 | 0.004 | 1.488 | 0.060 | 59 | 20,945 | 617 |
| | 8 | 0.018 | 4.127 | 0.453 | 341 | 51,536 | 2,962 |
| | 16 | 0.050 | 13.162 | 0.615 | 1,439 | 137,128 | 7,641 |
| | 32 | 0.072 | 17.913 | 0.927 | 1,938 | 217,467 | 9,628 |
| | 64 | 0.172 | 35.026 | 2.304 | 4,513 | 394,862 | 23,393 |
| | 128 | 0.557 | 95.397 | 4.679 | 17,164 | 892,941 | 52,803 |
| Alvey | 2 | 0.000 | 0.042 | 0.002 | 61 | 1,660 | 273 |
| | 4 | 0.002 | 0.112 | 0.016 | 251 | 3,063 | 455 |
| | 8 | 0.010 | 0.363 | 0.052 | 915 | 7,983 | 1,636 |
| | 16 | 0.098 | 1.502 | 0.420 | 4,766 | 18,639 | 6,233 |
| | 32 | 0.789 | 9.690 | 3.998 | 33,335 | 66,716 | 39,099 |
| | 64 | 5.025 | 44.174 | 21.773 | 133,884 | 233,766 | 170,588 |
| | 128 | 28.533 | 146.562 | 75.819 | 531,536 | 596,108 | 495,966 |
| Deltra | 2 | 0.000 | 0.084 | 0.158 | 1,290 | 1,847 | 1,161 |
| | 4 | 0.012 | 0.208 | 0.359 | 2,783 | 3,957 | 2,566 |
| | 8 | 0.052 | 0.583 | 0.839 | 6,645 | 9,137 | 6,072 |
| | 16 | 0.204 | 2.498 | 2.572 | 20,791 | 28,369 | 22,354 |
| | 32 | 0.718 | 6.834 | 6.095 | 57,689 | 68,890 | 55,658 |
| | 64 | 2.838 | 31.958 | 29.853 | 207,745 | 282,393 | 261,649 |
| | 128 | 14.532 | 157.172 | 143.730 | 878,964 | 1,154,710 | 1,110,629 |

to decrease when string length increases, at least for Alvey and Deltra, suggesting that CYK could generate more items than the other algorithms for larger values of $n$.

- Left-Corner is notably faster than Earley in all cases, except for some short sentences when using the Deltra grammar. The Left-Corner parser always generates fewer items than the Earley parser, since it avoids unnecessary predictions by using information about left-corner relationships. The Susanne grammar seems to be very well suited for Left-Corner parsing, since the number of items generated decreases by an order of magnitude with respect to Earley. On the other hand, the Deltra grammar's left-corner relationships seem to contribute less useful information than the others', since the difference between Left-Corner and Earley in terms of items generated is small when using this grammar. In some of the cases, Left-Corner's runtimes are a bit slower than Earley's because this small difference in items is not enough to compensate for the extra time required to process each item due to the extra steps in the schema, which make Left-Corner's matching and indexing code more complex than Earley's.
- The parsing of the sentences generated using the Alvey and Deltra grammars tends to require more time, and the generation of more items, than that of

the Susanne sentences. This happens in spite of the fact that the Susanne grammar has more rules. The probable reason is that the Alvey and Deltra grammars have more ambiguity, since they are designed to be used with their arguments and feature structures, and information has been lost when these features were removed from them. On the other hand, the Susanne grammar is designed as a plain context-free grammar and therefore its symbols contain more information.
- Execution times for the Alvey grammar quickly grow for sentence lengths above 16. This is because sentences generated for these lengths tend to be repetitions of a single terminal symbol, and are highly ambiguous.

## 4 Conclusions

In this paper, we have presented a compilation technique which allows us to automatically transform a parsing schema into an implementation of the algorithm it describes, keeping the theoretical computational complexity of the algorithm. This makes our work different from the parsing machine described by Shieber *et al.* in [13], a Prolog implementation of a deductive parsing engine which can also be used to implement parsing schemata; however, its input notation is less declarative, since schemata have to be programmed in Prolog, and it does not support automatic indexing, so the resulting parsers are inefficient unless the user programs indexing code by hand, abandoning the high abstraction level. Another alternative for implementing parsing schemata is the Dyna language [4], which can be used to implement some kinds of dynamic programs; but it has a complex notation, clearly less declarative than ours, which is specifically designed for denoting schemata: in our approach, the user only has to write the schema without worrying about implementation details. In addition, we provide an extensibility mechanism that allows the user to add new kinds of elements to schemata apart from the predefined ones.

Compilation of parsing schemata has been shown very useful for the design, analysis and prototyping of parsing algorithms, as it has allowed us to test them (even variants with "tricks" that improve practical performance in some cases) and check their results and performance without having to implement them in a programming language. As we have seen by comparing the performance of CYK, Earley and Left-Corner parsers for several grammars, not all algorithms are equally suitable for all grammars. In this work we provide a quick way to evaluate several parsing algorithms in order to find the best one for a particular application.

Our compilation technique is not limited to working with context-free grammars, since all grammars in the Chomsky hierarchy can be handled in the same way as context-free grammars, and other formalisms can be added by defining element classes for their rules using the extensibility mechanism. In this way, we have used our compiler to generate implementations for some of the most popular parsers for tree adjoining grammars (TAG) [8]. A detailed explanation of the performance results obtained by applying our compilation technique to TAG parsers can be found at [5,6].

Currently, we are applying our compilation technique to generate robust, error-correcting parsers for context-free grammars and tree adjoining grammars.

## References

1. Billot, S., Lang, B.: The structure of shared forest in ambiguous parsing. In: Proc. of the 27th Annual Meeting of the Association for Computational Linguistics, Vancouver, British Columbia, Canada, pp. 143–151. ACL (June 1989)
2. Carroll, J.A.: Practical unification-based parsing of natural language. Technical Report no. 314, University of Cambridge, Computer Laboratory, England. PhD Thesis (1993)
3. Earley, J.: An efficient context-free parsing algorithm. Communications of the ACM 13(2), 94–102 (1970)
4. Eisner, J., Goldlust, E., Smith, N.A.: Dyna: A declarative language for implementing dynamic programs. In: Proceedings of ACL 2004 (Companion Volume), Barcelona, pp. 218–221 (July 2004)
5. Gómez-Rodríguez, C., Alonso, M.A., Vilares, M.: On theoretical and practical complexity of TAG parsers. In: Monachesi, P. Penn, G., Satta, G., Wintner, S. (eds.) FG 2006: The 11th conference on Formal Grammar. Malaga, Spain, July 29-30, 2006, ch. 5, pp. 61–75, Center for the Study of Language and Information, Stanford (2006)
6. Gómez-Rodríguez, C., Alonso, M.A., Vilares, M.: Generating XTAG parsers from algebraic specifications. In: Proceedings of the 8th International Workshop on Tree Adjoining Grammar and Related Formalisms. Sydney, July 2006, pp. 103–108, Association for Computational Linguistics, East Stroudsburg (2006)
7. Gómez-Rodríguez, C., Alonso, M.A., Vilares, M.: Generation of indexes for compiling efficient parsers from formal specifications. In: Moreno-Díaz, R., Pichler, F., Quesada-Arencibia, A. (eds.) Computer Aided Systems Theory. LNCS, Springer, Heidelberg (2007)
8. Joshi, A.K., Schabes, Y.: Tree-adjoining grammars. In: Rozenberg, G., Salomaa, A. (eds.) Handbook of Formal Languages, Beyond Words, ch. 2, vol. 3, pp. 69–123. Springer, Heidelberg (1997)
9. Kasami, T.: An efficient recognition and syntax algorithm for context-free languages. Scientific Report AFCRL-65-758, Air Force Cambridge Research Lab., Bedford, Massachussetts (1965)
10. Rosenkrantz, D.J., Lewis II, P.M.: Deterministic Left Corner parsing. In: Conference Record of 1970 Eleventh Annual Meeting on Switching and Automata Theory, Santa Monica, pp. 139–152. IEEE Computer Society Press, Los Alamitos (1970)
11. Sampson, G.: The Susanne corpus, Release 3 (1994)
12. Schoorl, J.J., Belder, S.: Computational linguistics at Delft: A status report, Report WTM/TT 90–09 (1990)
13. Shieber, S.M., Schabes, Y., Pereira, F.C.N.: Principles and implementation of deductive parsing. Journal of Logic Programming 24(1-2), 3–36 (1995)
14. Sikkel, K.: Parsing Schemata — A Framework for Specification and Analysis of Parsing Algorithms. Springer, Heidelberg (1997)
15. Younger, D.H.: Recognition and parsing of context-free languages in time $n^3$. Information and Control 10(2), 189–208 (1967)

# Efficient Fragmentation of Large XML Documents

Angela Bonifati[1] and Alfredo Cuzzocrea[2]

[1] ICAR Inst., National Research Council, Italy
bonifati@icar.cnr.it
[2] DEIS Dept., University of Calabria, Italy
cuzzocrea@si.deis.unical.it

**Abstract.** Fragmentation techniques for XML data are gaining momentum within both distributed and centralized XML query engines and pose novel and unrecognized challenges to the community. Albeit not novel, and clearly inspired by the classical *divide et impera* principle, fragmentation for XML trees has been proved successful in boosting the querying performance, and in cutting down the memory requirements. However, fragmentation considered so far has been driven by semantics, i.e. built around query predicates. In this paper, we propose a novel fragmentation technique that founds on structural constraints of XML documents (size, tree-width, and tree-depth) and on special-purpose structure histograms able to meaningfully summarize XML documents. This allows us to predict bounding intervals of structural properties of output (XML) fragments for efficient query processing of distributed XML data. An experimental evaluation of our study confirms the effectiveness of our fragmentation methodology on some representative XML data sets.

## 1 Introduction

An imminent development of XML processing is undoubtly making it as fast and efficient as possible. Query engines for XML are being designed and implemented, with the specific goal of employing indexes to improve their performance [10]. Others [23] employ statistics to cost the most frequently asked queries, or use classical algebraic techniques [16] to optimize query plans.

On the other hand, XML query processors suffer from main-memory limitations that prevent them from processing large XML documents. While content-based predicates can be used to project down parts of documents, an XML query engine which is parsimonious in resources, may still enable a further resizing of the obtained projection/query results. This may also happen in many resource-critical contexts, such as a distributed database, or a stream processor. The advantages of XML fragmentation are already being proved in an XML query engine [4,5] or in a distributed setting [3]. Fragmentation of XML documents as proposed by the previous works has been based on semantics, whereas in this paper we work out a novel kind of fragmentation, which is orthogonal to the first and is only guided by the structural properties of an XML document.

Given an XML document, modeled w.l.g. as a tree, there exist several ways of splitting it into subtrees, which may be semantically driven or structurally driven. Usually, query processors decides to apply projections and selections beforehand in order

R. Wagner, N. Revell, and G. Pernul (Eds.): DEXA 2007, LNCS 4653, pp. 539–550, 2007.

to reduce the amount of data to be manipulated in memory during query evaluation. Notwithstanding the effectiveness of pushing algebraic operators within the query plan, it may happen that the size of intermediate results are still too large to fit in memory. If for instance we consider a 100MB XMark document, and a query $Q_1$ asking for open auctions sold by people owning a credit card (`creditcard` being an optional element) and for closed auctions sold by any people, the result query plan would look like the one shown in Fig. 1 (a). The two branches of the join operator(s) would in such a case be of size 29.6MB(6.1MB) and 17.09MB(11.8MB), respectively. Along with the size, the intermediate results can be also resized w.r.t. tree-width and tree-depth constraints that may affect the query processing time as well. These can be prohibitively large for the join branches above, i.e. 70 (38) and 9 (2) respectively for the left-hand-side join operator. If structure-driven fragmentation is employed, the subtrees output by the selections and projections can be resized to smaller pieces according to the structural constraints and can be then processed per piece. Fig. 1(b) pictures the result of fragmentation on the two operands of the join(s).

A suitable application of structure-driven fragmentation is streaming XML processing (e.g. [8]). Stream query processors are mainly memory-based, thus motivating the use of smaller fragments given for instance by top-down navigation of the original document. Our fragmentation could be employed to obtain smaller XML messages input to the stream, carefully designed to not exceed specified memory requirements at query runtime. Thus, using the three structural constraints altogether allows us to obtain approximately uniform fragments, e.g. to be used in a uniform stream. Finally, distributed and parallel query processing may leverage the fragmentation of the original documents, in order to improve their performance. This issue will be further discussed in our experimental study on *XP2P* [3], a P2P-based infrastructure we have developed.

Coming back to our problem, we can state it as follows. Let $t$ be an XML tree, $w$ a tree-width constraint, $d$ a tree-depth constraint and $s$ a tree-size constraint, we split $t$ into valid fragments $f$ of $t$ such that $size(f) <= s$, $tw(f) <= w$ and $td(f) <= d$ and $\nexists f' \neq f$ such that $f \cap f' \neq \emptyset$, $size$, $tw$ and $td$ being functions returning the size of $f$, the maximum width of $f$, and the maximum depth of $f$, and $f'$ being a valid fragment of $t$, respectively. Specifically, we consider performance issues of an arbitrary XML processor for what concerns (*i*) the aspect of fragmenting a given XML document, and (*ii*) the aspect of querying the fragmented representation of a given XML document. To this end, we propose an innovative approach for efficiently supporting XML document fragmentation via *structural constraints*, according to which a given XML document is fragmented by imposing "range-shaped" constraints to size, tree-width and tree-depth of output fragments. We name the resulting fragmentation technique as *structure-driven fragmentation of XML documents*.

Although a set of heuristics performing this kind of fragmentation can be easily devised, a key problem is determining the values of structural constraints input to the above heuristics, given that the search space is prohibitive at large. To alleviate the problem, we introduce special-purpose *structure histograms* that report the constraint values for the fragments of a given document. We then present a *prediction algorithm* that probes those histograms to output the expected number of fragments, when fixed input values of the constraints are used. This number is obtained in dependence on

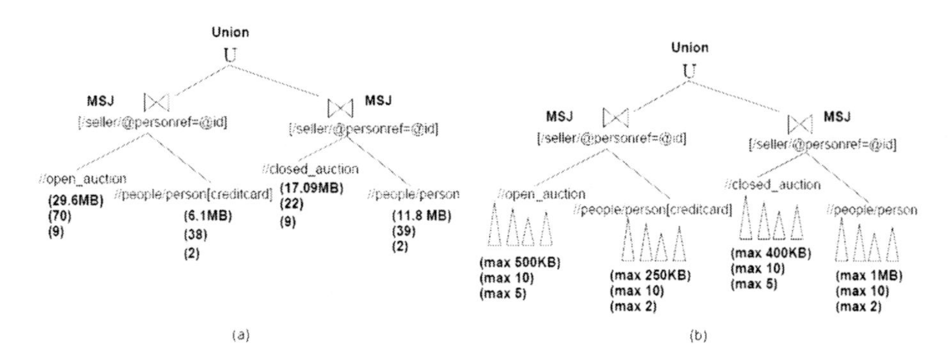


**Fig. 1.** Query plan of query $Q_1$ without (a) and with (b) the fragmentation operator applied

structural properties of the input document, thus constituting a value that "summarizes" these properties. Furthermore, we study how to relax the fixed constraints by means of classical distributions. The overall approach we propose is codified within a novel set of heuristics, called *SimpleX*, which, to the best of our knowledge, is the first proposal addressing the XML data fragmentation problem via structural constraints. Finally, we also provide an experimental evaluation of *SimpleX* that clearly shows the effectiveness of our fragmentation methodology in a relevant real-life scenario drawn by a P2P setting and against some representative XML data sets.

The rest of the paper is organized as follows: Section 2 shows the *SimpleX* heuristics for structure-driven fragmentation; Section 3 describes the structure histograms and their use in-support-of the prediction task; Section 4 presents a variety of experiments that probe the effectiveness of our techniques; Section 5 discusses the related work; finally, Section 6 states conclusions and further research.

## 2 *SimpleX*: Simple Top-Down Heuristics for Shredding an XML Document

The fragmentation problem stated above is a problem with linear cost function and integer constraints, which is intrinsically exponential. To effectively explore the search space, we have designed a set of simple top-down heuristics for document fragmentation, *SimpleX*. They all have in common the fact that they start at the root of the document and proceed in a top-down fashion. At each step the current subtree width, depth and size are checked against the constraints $w$, $d$, $s$. If the constraints are satisfied, the subtree becomes a valid fragment and is pruned from the document to constitute a separate valid XML document. A new node containing as PC-data the path expression of the obtained fragment will then replace the given subtree in the original document. If instead the constraints are not satisfied, the algorithm inspects the next subtree in the XML tree according to the criteria assessed by the heuristic.

A first criterion to select the next subtree is for instance given by the order of visit, i.e. depth-first or breadth-first. We call these variants *in-depth* and *in-width*. Fig. 2

**Table 1.** Sizes of subtrees of Fig. 2

| *Node* | *Size* (KB) | *Node* | *Size* (KB) | *Node* | *Size* (KB) |
|---|---|---|---|---|---|
| site | 145 | people | 100 | catgraph | 45 |
| person1 | 20 | person2 | 50 | person3 | 30 |
| edge1 | 15 | edge2 | 10 | edge3 | 20 |

represents an XML tree compliant to the XMark DTD, whose subtree sizes [1] are reported in Table 1 as absolute numbers (dots in Fig. 2 represent PC-data elements whose sizes appear in Table 1).

Applying for instance the in-depth heuristics with constrained size $s = 100$ and depth $d = 2$ and unconstrained width $w$, the XML tree gets fragmented as in Fig. 2 (a), whereas with $s = 100$, $d = 2$ and $w = 1$, it gets fragmented as shown in Fig. 2 (b). The application of the other heuristics on the sample tree is omitted for conciseness.

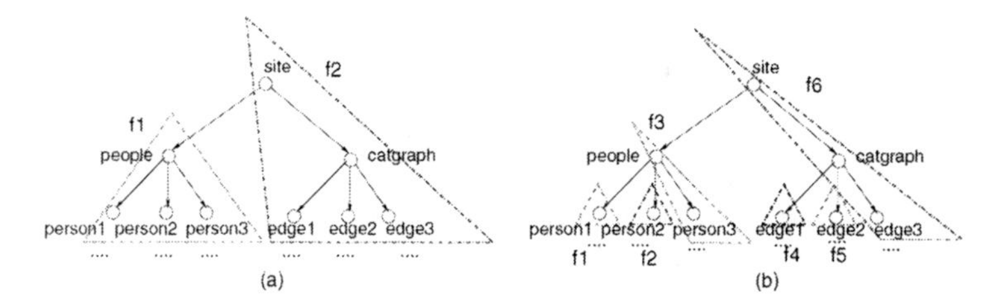


**Fig. 2.** A sample XMark tree fragmented with one of the *SimpleX* heuristics and two (three) constraints in (a) (in (b))

*SimpleX* [2] is one possible set of simple heuristics among the various ones that can be applied for shredding a document (e.g. bottom-up or random-access). In principle, there is no better heuristics than any other, as it actually depends on the structure of the document. Our aim in this paper is not finding the best heuristic, but instead to show how to tune the fragmentation constraints for *SimpleX* heuristics, if summary data structures are employed. In fact, note that the constraints of the problem statement introduced in Section 1 may turn to be incompatible if randomly specified, thus possibly leading to empty solutions.

Being the search space prohibitively large, a key problem is determining the values of structural constraints input to the above heuristics. To alleviate the problem, we have designed the structure histograms, which let determine correct combinations of the constraint values, without actually doing the fragmentation beforehand. The histograms have been implemented within an analysis module that uses algorithms to predict an interval for the number of fragments produced by the heuristics. This is particularly

[1] The subtree depth and width can be easily inferred from Fig. 2 and are omitted for space reasons.

[2] In the remainder, and in the experiments, we will simply indicate the set of heuristics as *SimpleX*. We mean that we apply all the heuristics in the set and pick the results of the most efficient one at each run.

interesting for large XML data sets and query results, as it offers a visual summarization tool that can be inspected at any time for prediction. We introduce the structure histograms next.

## 3 Structure Histograms

Given an XML document $X$, the structure histograms present $X$ as *summarized* by counting the fragments (i.e., the sub-trees) in $X$ such that these fragments hold the following structural properties: (*i*) the fragment size $s$, (*ii*) the fragment tree-depth $d$, and (*iii*) the fragment tree-width $w$.

Formally, let $X$ be an XML document, let $p$ be a structural property defined on $X$, let $D_p = [p_{min}, p_{max}]$ be the value domain of $p$, a *class* $\Delta p$ is defined on $D_p$ as follows: $\Delta p = [p'_{min}, p'_{max}]$, such that $p_{min} \leq p'_{min} \leq p'_{max} \leq p_{max}$. Then, a structure histogram built on $X$, denoted by $H_S(X, p, \Delta p)$, grouping $p$ by an aggregation function $f$ = COUNT (thus, reporting the *frequency* of the fragments) over $\Delta p$-wise steps, is a tuple $\langle D_p, H_p \rangle$, such that each bucket $b(\Delta p)$ in the co-domain $H_p$ counts the fragments in $X$ having a value of the (structural) property $p$ ranging $\Delta p = [p'_{min}, p'_{max}]$. We call $H_S$ a one-dimensional histogram computed over $p$. Moreover, to support parametric summarization of XML data and thus improve the fragmentation prediction, we introduce the *parametric structure histogram* $H_S^{\mathcal{P}}(X, p, \Delta p)$, which is an extension of the previous histogram, where $\mathcal{P}$ is a fixed structural property w.r.t. which the histogram over $p$ is computed. Specifically, $H_S^{\mathcal{P}}$ is a two-dimensional histogram computed over $\langle \mathcal{P}, p \rangle$.

In our fragmentation framework, we make use of the following structure histograms summarizing a given XML document $X$: (*i*) the *Tree-Size Structure Histogram* $H_S$ $(X, s, \Delta s)$, which summarizes $X$ w.r.t. the size $s$ (i.e., $p = s$); (*ii*) the *Tree-Depth Structure Histogram* $H_D(X, d, \Delta d)$, which summarizes $X$ w.r.t. the tree-depth $d$ (i.e., $p = d$); (*iii*) the *Tree-Width Structure Histogram* $H_W(X, w, \Delta w)$, which summarizes $X$ w.r.t. the tree-width $w$ (i.e., $p = w$); (*iv*) the *Max-Tree-Size Parametric Structure Histogram* $H_D^{\mathcal{S}}(X, s, \Delta d)$, which, fixed the size $s$ by computing the max value (i.e., $\mathcal{P} = \text{MAX}(s)$), summarizes $X$ w.r.t. the tree-depth $d$ (i.e., $p = d$).

More precisely, given an input XML document $X$, and a structural property $p$, we build the output structure histogram $H_S(X, p, \Delta p)$ by setting the input parameters $D_p$ and $\Delta p$ as follows (it should be noted that the input parameter $p$ is directly set by the target user/application): (*i*) $D_p = [0, MaxValue]$, such that $MaxValue$ is the maximum value of the structural property $p$ among all the fragments in $X$, (*ii*) $\Delta p = \mathcal{N} \cdot |D_p|$, such that $0 \leq \mathcal{N} \leq 1$ is an empirically set parameter, and $|D_p|$ is the cardinality of $D_p$. Examples of such structure histograms for the subtree in Fig. 2 are sketched in Table 2. Note that building the histograms for an arbitrary XML tree is necessarily exponential in the worst case, but our heuristics can significantly trim the number of inspected fragments.

A user (or application) willing to partition a document who knows how many fragments he/she wants to obtain, may want to know the values of constraints that let exactly obtain that number of fragments. In other words, he/she would like to properly tune the constraints values. Moreover, constraints as specified by the user may not be compatible among each other or the final results may be biased to the data set

**Table 2.** $H_D$, $H_S$ (partial) and $H_W$ for the sample XMark tree of Fig. 2

$H_D$

| $D$ | $f$ |
|---|---|
| 2 | 1 |
| 1 | 2 |
| 0 | 6 |

$H_S$

| $S$ | $f$ |
|---|---|
| 145 | 1 |
| 100 | 1 |
| 45 | 1 |
| 20 | 1 |
| ... | ... |

$H_W$

| $W$ | $f$ |
|---|---|
| 2 | 1 |
| 3 | 2 |
| 0 | 6 |

inherent structure. In order to automatize the task of deciding the constraint values, we have devised algorithm *predictInterval* that predicts the range of frequencies by inspecting the structure histograms. The algorithm pseudocode is shown in Fig. 3. For space reasons, we limit ourselves to discuss the algorithm on the XMark sample of Fig. 2 and show that it lets predict the range of frequencies quite sharply. Earlier, we have pointed out that if we disregard the width $w$ in (a), we obtain a rather different fragmentation w.r.t. (b), where $w$ has a non-null value. We start by looking at the histogram $H_D$ reported in Table 2 and we remark that for a value of depth $d = 1$, we would obtain two fragments. If we look at the histogram $H_W$, this in turn tells that there are two nodes with width $w = 3$, and these nodes cannot be part of the same fragment if $w$ is chosen to be 1. In such a case we would generate 6 fragments out of those nodes. Thus, only by looking at $H_D$ and $H_W$, we learn that the number of fragments shall be in the range $[2, 6]$. If we further add the third constraint $s$, the upper bound of the above range may raise or not, depending on whether the fragments so far obtained satisfy or not the value of $s$. This leads to choose a value of $s$ from histogram $H_S$, that pessimistically corresponds to the size of the largest subtree located at depth $d = 1$ (e.g. subtrees rooted in nodes `people` and `catgraph` in Fig. 2), information that we learn from an $H_D^S$ histogram. In this particular example, we can choose for instance a size $s$ equal to 100, thus obtaining the fragmentation shown in Fig. 2 (b) quite straightforwardly.

As we have seen, choosing correct values for the input constraints of the fragmentation algorithm is a non trivial task. An incorrect value for such constraints would lead to too many fragments or too few of them, or even to an empty solution in some cases. Indeed, there may exist random values of $w$, $d$ and $s$, which turn out to be incompatible among each other. In order to alleviate this problem, we let the constraints vary along classical distributions (such as Uniform, Gauss, Zipf), thus relaxing constraints with such distributions. Thus, along with choosing fixed bounds for $s, w, d$, we assign ranges to them according to those distributions. This is further motivated by the fact that an XML document contains "unbreakable" pieces of text (such as PC-data, entities etc.) that needs to be taken into account in the choice of the constraint values (especially for the size constraint). By empirically comparing the output of *SimpleX* against the baseline case given by a constant distribution, we will show below that non-constant distributions have in general a better behavior.

```
   algorithm predictInterval(H_D: tree-depth structure histogram,
        H_S: tree-size structure histogram,
        H_W: tree-width structure histogram,
        s_0, d_0, w_0: size, depth, width constraints): return [f_min, f_max]
 1   Let d_0 be the chosen depth in H_D // alternatively, w_0 in H_W
 2      such that H_D(d_0, Δd_0, f_0)
 3   Pick the max width w_max in H_W //alternatively, d_max in H_D
 4   Let f_min = f_0, f_max = sum(f_0,w_max − w_0)
 5   Let s_max the max size at depth d_0 in H_D^S
 6   Let f_s_max the corresponding frequency in H_S
 7      If s_0 >= s_max
 8         return [f_min, f_max]
 9      else {
10         f_max += f_s_max
11         for each w_i in the interval w_max − w_0 in H_W
12            f_max += f_w_i * (w_i − w_0)
13      }
14   return [f_min, f_max]
```

**Fig. 3.** Algorithm *predictInterval*

# 4 Experimental Assessment

We have conducted an experimental study aimed at showing the effectiveness of our structure-driven fragmentation methodology. The experiments are divided into three classes. First, we build the structure histograms for representative XML data sets, and show their use to decide the final values of constraints. Secondly, we define and measure the accuracy error of fragmentation using the *SimpleX* set of heuristics, when fixed constraints are employed against the cases (baseline) in which the constraints vary with classical distributions (e.g. Uniform, Gauss, Zipf). Finally, we demonstrate the utility of fragmentation in a distributed setting, such as a DHT-based P2P network. In such a case, we measure the impact of fragmentation on network performance against the expected ideal behavior.

All the experiments have been performed on a machine with a 1.2 GHz processor, 512 MB RAM, and running Linux Suse 9.1. We uniquely identify each fragment with its absolute root-to-leaf path expression. Each fragment stores with extra sub nodes the path expression of subfragments and separately the path expression of its parent fragment. We have presented this data model in [3]. Notice that any data model (such as [5] for instance) other than ours can be adopted here to represent the fragments. Finally, the data sets and queries employed in the study are summarized in Table 3.

Fig. 4 shows the structure histograms for the Nasa data set. In order to improve readability, we separately report the complete histograms in Fig. 4 and some of the frequency values in Table 4. Note that such histograms have a size of the order of $KB$, thus being reasonably small. For instance, considering a triple $d_0 = 5$, $w_0 = 1200$ and $s_0 = 230KB$, and applying the Algorithm in Fig. 3, we obtained a prediction range equal to $[1200, 2500]$. Similar results for the other data sets of Table 3 and other values of the constraints are omitted for space reasons. Moreover, notice that the histograms

**Table 3.** XML documents and queries used

| *Document d* (MB) | *# elems.* | *maxDepth* | *maxWidth* | *provenance* |
|---|---|---|---|---|
| XMARK (113) | 3,332,129 | 11 | 25,500 | [20] |
| XMARK (30) | 501,705 | 11 | 7649 | [20] |
| NASA (24) | 476,645 | 7 | 2434 | [19] |
| FourReligiousWorks-Bom (1.5) | 7,656 | 5 | 79 | [9] |

| *Query* | *Description* |
|---|---|
| $QD_i$ | `FOR $p IN XPathExpr RETURN $p` |

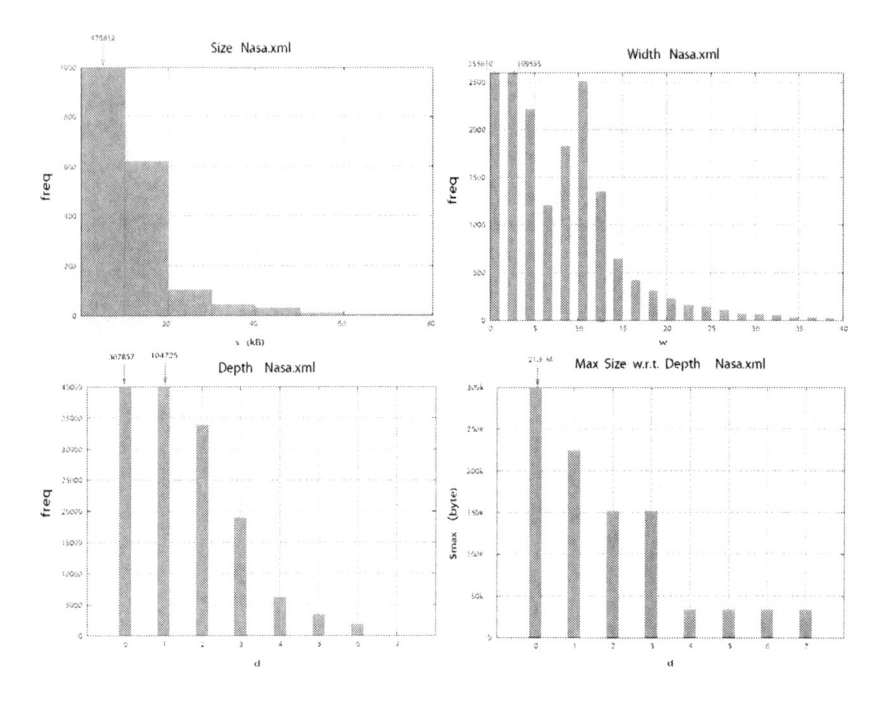


**Fig. 4.** Tree-Size ($H_S$), Tree-Width ($H_W$), Tree-Depth ($H_D$) and, finally, Max-Tree-Size Parametric ($H_D^S$) structure histograms for the Nasa data set

also allow a user to quickly discard incompatible values of constraints as they summarize only valid constraints values.

We have put at work the proposed heuristics on the data sets of Table 3. We have considered various values of parameters $s$, $w$ and $d$ and run the heuristics with fixed values of these parameters and with their variations as given by classical distributions (e.g. Uniform, Gauss, Zipf). We define the accuracy error $e_c$ of fragmentation w.r.t. constraint $c$ as follows. Let $c_a$ be the average value of the size (tree-depth and tree-width, resp.) obtained with *SimpleX* heuristics, $c_0$ the fixed value of constraint input to algorithm *predictInterval* (see Fig. 3), $c_{min}$ and $c_{max}$ the interval of the constraint value as obtained via the particular distribution, then the accuracy error of fragmentation is

**Table 4.** Window frames of histograms in Figure 4 depicting some of the frequency values

| $[s_1 - s_2]$ (KB) | *freq* |
|---|---|
| [50 - 100] | 20 |
| [100 - 150] | 8 |
| [150 - 200] | 6 |
| [200 - 250] | 1 |

| $w$ | *freq* |
|---|---|
| 674 | 1 |
| 730 | 1 |
| 1188 | 1 |
| 2434 | 1 |

| $d$ | *freq* |
|---|---|
| 4 | 6100 |
| 5 | 3430 |
| 6 | 1854 |
| 7 | 1 |

**Table 5.** Application of *SimpleX* to NASA with/without distribution

| *Distribution* | # *fragments* | *Avg.* # *nodes* | $e_s$ | $e_d$ | $e_w$ |
|---|---|---|---|---|---|
| None | 1879 | 253 | 0.97 | 0.6 | 0.97 |
| Uniform | 61 | 7813 | 0.02 | 0.57 | 0.89 |
| Gauss | 57 | 8362 | 0.09 | 0.42 | 0.88 |
| Zipf | 71 | 6713 | 0.1 | 0.57 | 0.9 |

given by the formula $\frac{|c_0 - c_a|}{c_0}$ for fixed constraints, and by the formula $\frac{|\frac{c_{min}+c_{max}}{2} - c_a|}{|\frac{c_{min}+c_{max}}{2}|}$ for non-fixed constraints.

As an example, Table 5 shows the obtained results with the NASA data set and value of constraints: $s$ = 230KB, $w$ = 1200 and $d$ = 5. The lower is the accuracy error, the better is the matching of the heuristics with the fragmentation constraints. It can be noticed that the case when the constraints are strict upper bounds leads to fairly more fragments than the cases when distributions are applied. On average, fragmentation via distributions obtains lower accuracy errors than the case when distributions are not used. Results with other data sets and other values of constraints showed the same trend.

As we already discussed, there exist several applications of our fragmentation strategy, which justify its effectiveness. Here, we present some experiments that have been performed on *XP2P*, our DHT-based P2P simulator [3]. For each experiment, we have scattered a certain number of fragments in the network obtained with our structure-driven fragmentation. We then measured the network scalability when both varying the number of peers and the number of queries.

Fig. 5 (a) shows the nr. of hops versus the number of peers when XMark(30) has been divided into 1000 fragments with the constraint values predicted by our analysis tool. Here we have considered exactly as many queries of kind $QD_i$ (see Table 3) as the number of fragments, each query being propagated to the successor peers as dictated by the current peer list of successors (i.e. at logarithmic distance). It can be noticed that the case where XML structure-driven fragmentation is used closely tracks the original Chord [3] logarithmic curve. Finally, Fig. 5 (b) shows the nr. of hops when varying the nr. of fragments for XMark(30) data set within a network of 500 peers. The fragmentation slightly increases the number of hops, if compared with the constant curve that represents no fragmentation.

[3] The original Chord simulator only stores on each peer an identifier of resources. In *XP2P*, we have extended it to store XML fragments.

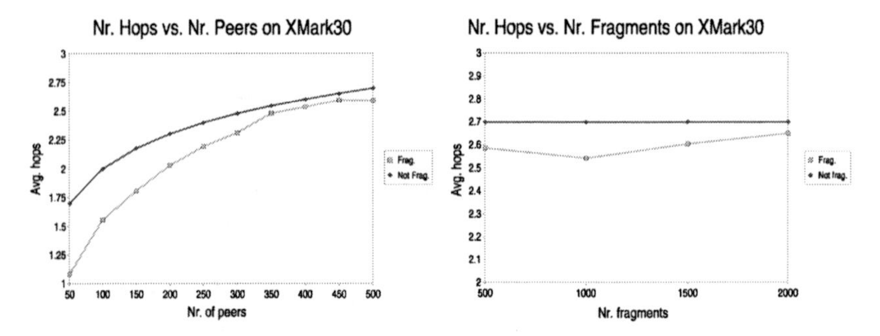


**Fig. 5.** Nr. of hops w.r.t. nr. of peers with 1000 fragments (a); Nr. of hops w.r.t. nr. of fragments with 500 peers (b). In both cases, the number of queries of kind $QD_i$ equals the number of fragments.

## 5 Related Work

The advantages of fragmenting relational databases are well established [17] as both horizontal fragmentation [7], which splits a given relation into disjoint sub-relations, and vertical fragmentation [13], which projects a given relation onto a subset of its attributes. More recently, fragmentation techniques have been adapted to object-oriented [1], semi-structured [14], and native XML [4] databases.

[15] proposes an innovative approach for supporting the distribution of XML databases via horizontal fragmentation. It employs a query-oriented cost model taking into account the most frequently asked queries, and uses heuristics to optimize the fragmentation based on the efficiency of such queries. Being query-driven, this approach does not consider the structural properties of XML data, as we do in our proposal. *XFrag* [4] is a framework for processing XQuery-formatted queries on XML fragments in order to reduce memory processing. This work focuses on how to employ fragments to make query optimizations, thus strengthening our proposal. Several query optimization techniques have been presented for XML data, among which [16] and [11]. While the former relies on algebraic projections, the latter is based on tree-automata. These techniques can be combined with ours to let the processor evaluate queries in parallel on multiple fragments.

[6] and [18] propose summary data structures for XML twigs and paths that let derive an estimation of queries selectivity by using statistical methods, such as histograms or wavelets. Notwithstanding the importance of the above data structures for query optimization, our histograms are instead aimed at predicting the number of fragments of an XML document when applying *SimpleX* heuristics. The latter prediction could not be inferred by looking at the above data structures. [22] proposes the *position histograms*, which allow the estimation of both simple and complex pattern query answers. Furthermore, when XML schema information is available, they employ the so-called *coverage histograms* that extend the former and allow the target XML database to be better summarized. Differently from ours, those histograms help estimating the sizes of child and descendant steps in path expressions. Histograms are used for a rather different purpose in our framework, as stated above.

Finally, *XRel* [21] is a path-based approach to store XML documents into RDBMS and retrieve them afterwards. While the path identification of fragments is similar to ours, the focus of the paper is on building an extension of relational databases for XML data.

## 6 Conclusions and Future Work

We have presented a fragmentation strategy for XML documents that is driven by structural constraints. To the best of our knowledge, this is the first work addressing such a problem. We further offer the user or the application a prediction of the "outcome" of the fragmentation, by means of the so-called structure histograms. By means of distributions, we are able to vary the constraint values thus improving the fragmentation performance.

We are currently developing new classes of heuristics. The first one uses additional data structures in combination with histograms in order to make the prediction more precise. The other one considers schemas of XML documents (when available) during the prediction. Moreover, another research direction we are considering consists in providing full support to *XML join queries* via devising ad-hoc heuristics that focus on the fragment size, which is a critical parameter affecting the computational cost due to evaluating such queries. Finally, we plan to embed our fragmentation tool and its analysis module in an existing XQuery engine.

## References


1. Bellatreche, L., Karlapalem, K., Simonet, A.: Algorithms and Support for Horizontal Class Partitioning in Object-Oriented Databases. Distributed and Parallel Databases 8 (2000)
2. Bohannon, P., Freire, J., Roy, P., Simeon, J.: From XML Schema to Relations: A Cost-based Approach to XML Storage. In: Proc. of ICDE (2002)
3. Bonifati, A., Cuzzocrea, A.: Storing and Retrieving XPath Fragments in Structured P2P Networks. Data & Knowledge Engineeering 59 (2006)
4. Bose, S., Fegaras, L.: XFrag: A Query Processing Framework for Fragmented XML Data. In: Proc. of WebDB (2005)
5. Bremer, J.M., Gertz, M.: On distributing xml repositories. In: Proc. of WebDB (2003)
6. Chen, Z., Jagadish, H.V., Korn, F., Koudas, N., Muthukrishnan, S., Ng Raymond, T., Srivastava, D.: Counting Twig Matches in a Tree. In: Proc. of ICDE (2001)
7. Ezeife, C., Barker, K.: A Comprehensive Approach to Horizontal Class Fragmentation in a Distributed Object based System. Distributed and Parallel Databases 3 (1995)
8. Florescu, D., Hillery, C., Kossman, D., et al.: The BEA/XQRL Streaming XQuery Processor. In: Proc. of VLDB (2003)
9. Ibiblio.org web site (2004), Available at `http://www.ibiblio.org/xml/books/biblegold/examples/baseball/`
10. Jagadish, H.V., Al-Khalifa, S., Chapman, A., Lakshmanan, L.V., Nierman, A., Paparizos, S., Patel, J., Srivastava, D., Wiwatwattana, N., Wu, Y., Yu., C.: Timber: a Native XML Database. VLDB Journal 11 (2002)
11. Koch, C.: Efficient Processing of Expressive Node-Selecting Queries on XML Data in Secondary Storage: A Tree Automata-based Approach. In: Proc. of VLDB (2003)

12. Krishnamurthy, R., Chakaravarthy, V.T., Naughton, J.F.: On the Difficulty of Finding Optimal Relational Decompositions for XML Workloads: A Complexity Theoretic Perspective. In: Proc. of ICDT (2003)
13. Lin, X., Orlowska, M., Zhang, Y.: A Graph-based Cluster Approach for Vertical Partitioning in Databases Systems. Data & Knowledge Engineeering, 11 (1993)
14. Ma, H., Schewe, K.D.: Fragmentation of XML Documents. In: Proc. of SBBD (2003)
15. Ma, H., Schewe, K.D.: Heuristic Horizontal XML Fragmentation. In: Proc. of CAiSE (2005)
16. Marian, A., Simeon, J.: Projecting XML Documents. In: Proc. of VLDB (2003)
17. Ozsu, M., Valduriez, P.: Principles of Distributed Database Systems. Alan Apt (1999)
18. Polyzotis, N., Garofalakis, M.N.: Statistical synopses for graph-structured XML databases. In: Proc. of SIGMOD (2002)
19. University of Washington's XML repository (2004), Available at `http://www.cs.washington.edu/research/xml/datasets`
20. Xmark: An XML Benchmark Project (2002), Available at `http://monetdb.cwi.nl/xml/`
21. Yoshikawa, M., Amagasa, T., Shimura, T., Uemura, S.: XRel: A Path-based Approach to Storage and Retrieval of XML Documents Using Relational Databases. ACM Transactions on Internet Technology 1 (2001)
22. Wu, Y., Patel, J., Jagadish, H.: Using Histograms to Estimate Answer Sizes for XML Queries. Information Systems 28 (2003)
23. Zhang, N., Haas, P., Josifovski, V., Lohman, G., Zhang, C.: Statistical Learning Techniques for Costing XML Queries. In: Proc. of VLDB (2005)

# Locating and Ranking XML Documents Based on Content and Structure Synopses

Weimin He, Leonidas Fegaras, and David Levine

University of Texas at Arlington, CSE
Arlington, TX 76019-0015
{weiminhe,fegaras,levine}@cse.uta.edu

**Abstract.** We present a new framework for indexing, locating and ranking XML documents based on content and structural synopses extracted from the documents. Instead of indexing each single element or term in a document, we extract a structural summary and a small number of data synopses from the document, which are indexed in an efficient way suitable for query evaluation. Our query language is XPath extended with full-text search. The result of query evaluation is a ranked list of document locations that best match the query. We propose a novel aggregated ranking scheme, which is integrated into the query evaluation to score the documents based on those data synopses. Our experimental evaluation shows that our indexing scheme outperforms the standard XML indexing scheme based on inverted lists and our ranking scheme is effective in terms of precision and recall.

## 1 Introduction

With the proliferation of XML as the data format for a wide variety of web data repositories, extensive work has been motivated on designing powerful query languages, developing efficient indexing and query evaluation algorithms, and proposing effective ranking schemes over XML data [1,2,4,5,6]. Khalifa *et al* [1] propose a bulk algebra called TIX, which integrates simple IR scoring schemes into a traditional pipelined query evaluator for an XML database. In [2], the authors propose XML scoring methods that account for both structure and content while considering query relaxations. Carmel *et al* [4] present an extension of the vector space model for searching XML collections via XML fragments and ranking results by relevance. XSEarch [6] is a semantic search engine that extends simple keyword search by incorporating keyword context information into the query, i.e., each query term is a keyword-label pair instead of a single keyword.

However, most of existing proposals combine structure indexes and inverted lists extracted from XML documents to fully evaluate a full-text query against these indexes and return the actual XML fragments as query answers. In general, these approaches perform costly containment joins among long inverted lists in order to evaluate a full-text XML query, which may not be suitable for online interactive and data-intensive web applications. In this paper, we present

R. Wagner, N. Revell, and G. Pernul (Eds.): DEXA 2007, LNCS 4653, pp. 551–561, 2007.

a novel framework for indexing, locating and ranking XML documents based on condensed summaries extracted from the structural and textual content of the documents. In our framework, a user can publish concise summaries of its local XML documents onto the server. As an XML document is published, the document itself remains at the publisher's site, only essential meta-data are extracted from the document and published onto the server. The server will be responsible for indexing these meta-data and answering queries from any client based on these meta-data. The result of the query evaluation is a ranked list of document locations that best match the query. A document location includes the important meta information about the document, such as the IP address of the document owner, the document schema and the document description. Upon receiving these meta information, the user may choose some interesting document locations from the ranked answer list, requesting the document owner to evaluate the original query over the actual document and return the XML fragments as query answers. We believe that our framework can serve as an infrastructure for a wide range of web applications, such as online interactive XML data exploring systems and XML search engines in a peer-to-peer environment.

In summary, our main contributions are the following:

- We present an effective framework for searching XML documents based on content and structure synopses.
- We propose an efficient XML meta-data indexing scheme that is suitable for a full-text XPath evaluation.
- We introduce an effective aggregated ranking scheme to score an XML document based on our proposed data synopses.
- We experimentally validate the efficiency of our indexing scheme and demonstrate the effectiveness of our ranking scheme.

## 2 Query Specification and Document Indexing

Our query language is XPath extended with a full-text search predicate $e \sim S$, where $e$ is an XPath expression. This predicate returns true if at least one element from the sequence returned by $e$ matches the *search specification*, $S$. A search specification is a simple IR-style boolean keyword search that takes the form

$$\text{“term”} \ \mid \ S_1 \ \underline{\text{and}} \ S_2 \ \mid \ S_1 \ \underline{\text{or}} \ S_2 \ \mid \ ( \ S \ )$$

where $S$, $S_1$, and $S_2$ are search specifications. A term is an indexed term that must be present in the text of an element returned by the expression e. As a running example used throughout the paper, the following query Q:

```
//proceedings//article[abstract ~ "XML"]
   [body//paragraph ~ "index" and "rank"]/title
```

searches for the titles of all articles whose abstract contains the keyword "XML" and whose body contains the terms “index” and “rank” in their paragraph.

As an XML document is published, the following meta-data are indexed: *Structural Summary(SS)*, *Content Synopses(CS)*, and *Positional Filters(PF)*.

**Structural Summary.** A structural summary [7] is a tree that captures all unique paths to the data in an XML document. An example is shown in Figure 1(a). Each node in an SS has a tagname and a unique id. Note that one SS node may be associated with many elements in the actual document.

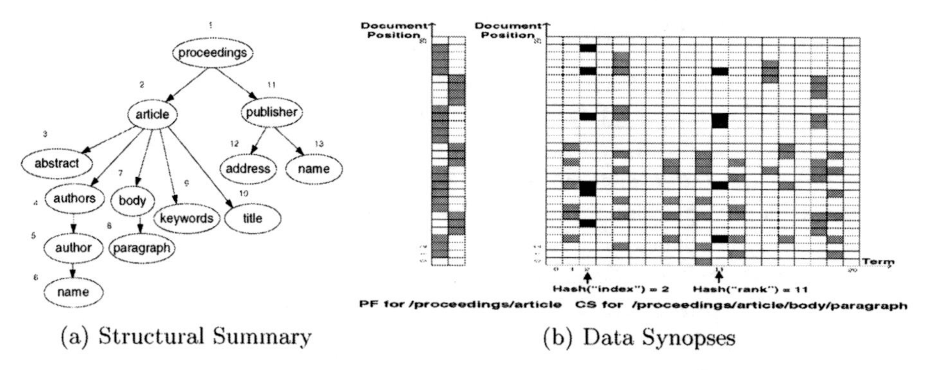


(a) Structural Summary (b) Data Synopses

**Fig. 1.** Structural Summary & Data Synopses Examples

**Content Synopses.** A node in an SS is called a text node if the element associated with that node contains text data in an XML document. If an XML document does not contain intermixed data, a text node is a leaf node in SS. To capture the textual content of a document, for each text node $k$ in the structural summary $S$ of document $D$, we construct a *content synopsis* $H_p^D$ to summarize the textual data associated with $k$ in $D$. The label path $p$ from the root of S to $k$ is used as the index key in DB. $H_p^D$ is a bit matrix of size $L \times W$, where $W$ is the number of term buckets and $L$ is the document positional ranges of the elements that directly contain terms associated with node $k$. The positional information is represented by the document order of the begin/end tags of the elements. More specifically, for each term $t$ contained directly in an element associated with $k$, whose begin/end position is $b/e$, we set all matrix values $H_p^D[i, \text{hash}(t) \bmod W]$ to one, for all $\lfloor b \times L/|D| \rfloor \leq i \leq \lfloor e \times L/|D| \rfloor$, where 'hash' is a string hashing function and $|D|$ is the document size. That is, the $[0, |D|]$ range of tag positions in the document is compressed into the range $[0, L]$. For example, the content synopsis for SS node **paragraph** is illustrated on the right in Figure 1(b). As we can see, after the term "rank" is hashed to the term bucket 11, we obtain a bit vector that has 4 one-bit ranges(emphasized by black color). Each one-bit range represents a **paragraph** element that directly contains "rank" in the document. The start/end of the range is the document order of begin/end tag of the **paragraph** element. We can evaluate the search predicate body//paragraph $\sim$ "index" and "rank" in the query Q by bitwise *anding* the vectors $H_8$["index"] and $H_8$["rank"], which are the $2nd$ and $11th$ columns respectively in the CS(emphasized by black color) in Figure 1(b). If all bits in the resulting bit vector are zeros, the corresponding document does not satisfy the search predicate because it does not have both the term "index" and "rank" in the same **paragraph** element. Note that the subscript number 8 in $H_8$["index"] is the id number of node **paragraph** in SS.

**Positional Filters.** Although the positional information in $CS$ enforces the constraint that the terms in a single search predicate must be in the same element associated with the predicate, it can not ensure that different elements associated with different search predicates are contained in the same element in a document. For example, given the relevant bit vectors $H_8$["index"], $H_8$["rank"], and $H_3$["XML"] only, we can not enforce the containment constraint in Q that the article whose abstract contains "XML" must be the same article whose paragraph contains "index" and "rank". To address this problem, for each non-text node $n$ in the structural summary, we construct another type of data synopsis, called *Positional Filter*, denoted by $F_p^D$. $F_p^D$ is a bit matrix of size $L \times M$, where $L$ is the document positional ranges of the elements associated with node $n$ that is reachable by the label path $p$, and $M$ is the number of bit vectors in $F_p^D$. The positional filter for SS node **article** is demonstrated on the left in Figure 1(b). The 7 one-bit ranges indicate there are 7 **article** elements in the document.

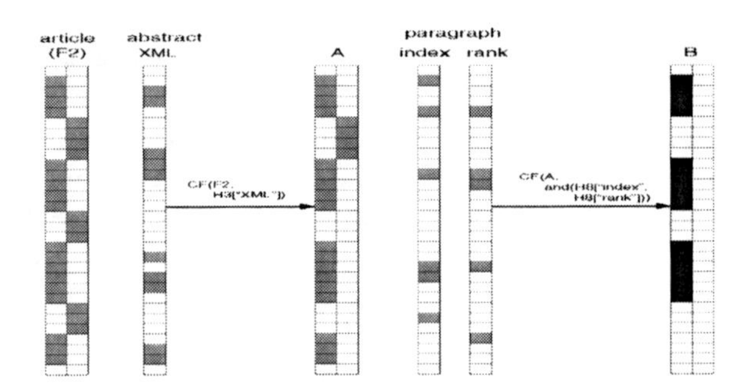


**Fig. 2.** Containment Filtering Illustration

**Containment Filtering.** We can enforce the element containment constraints in the query using an operation called *Containment Filtering*. Let $F$ be a positional filter of size $L \times M$ and $V$ be a bit vector extracted from a content synopsis whose size is $L \times W$. The *Containment Filtering* copies a continuous range of one-bits from $F$ to the resulting positional filter $F'$ if there is at least one position within this range in which the corresponding bit in $V$ is one. Figure 2 shows how to employ containment filtering to determine whether a document is likely to satisfy the query Q. First, we do a containment filtering between the initial positional filter $F_2$ and the bit vector $H_3$["XML"]. In the resulting positional filter $A$, only 5 one-bit ranges out of 7 in $F2$ are left. Counting from bottom to top, the $2nd$ and $4th$ one-bit ranges in $F2$ are discarded in $A$ because there is no any one-bit range in $H_3$["XML"] that intersects with the $2nd$ or $4th$ range, which means that the $2nd$ or $4th$ **article** element does not contain an **abstract** element that contains the term "XML". Similarly, we can do containment filtering between $A$ and the resulting bit vector derived from the bitwise *anding* between the bit vectors $H_8$["index"] and $H_8$["rank"]. The 3 one-bit ranges left in $B$ indicate 3 **article** elements in the document satisfy all the search predicates in Q, thus the document is considered to satisfy the query.

# 3 Query Processing

We briefly describe the query processing in our framework in this section. In our framework, the first step in evaluating an XPath query is deriving a query footprint from the query. A query footprint($QF$) captures the essential structural components and all the *entry points* associated with the search predicates. For example, the query footprint of Q is:

`//proceedings//article:1[abstract:2][body//paragraph:3]/title`

The numbers 1, 2, and 3 are the numbers of entry points in $QF$ that indicate the places where data synopses are needed for query evaluation. Our query footprint derivation algorithm is omitted due to the space limitation.

In the server local indexes, each node $k$ in a structural summary $S$ is encoded by the triple $(b, e, l)$, where $b/e$ is the begin/end numbering of $k$ and $l$ is the level of $k$ in $S$. We leverage the iterator model in relational databases to form a pipeline of iterators derived from the query footprint to retrieve all matching structural summaries. Meanwhile,we also derive the full label paths from the structural summary that match the entry points in $QF$. In the example query Q, the label paths are `/proceedings/article`, `/proceedings/article/abstract`, and `/proceedings/article/body/paragraph`. Using these label paths as keys, the corresponding data synopses are retrieved from DB and qualified document locations are filtered out using containment filtering and returned to the client.

# 4 Aggregated Ranking

Since a query footprint may match a large number of structural summaries and there may be a large number of documents that match each SS, it is desirable to rank all the qualified documents using a scoring function. We first extend $tf * idf$ ranking to score a document and then enhance it with a positional weight derived from containment filtering. Finally, we combine term proximity with the enhanced scoring to further improve the quality of ranking results.

## 4.1 Extended TF*IDF Scoring

We consider a path-term pair as the unit for content scoring. The formal definitions of *TF* and *IDF* scores of a path-term pair are given below.

**Definition 1.** ***TF Score of a Path-Term Pair.*** *Let D be an XML document associated with the pair* $(p, t)$*, where p is a full text label path from its structural summary and t is a term. Let* paths(D) *be the set of tuples* $(t_x, p_x, b_x, e_x, i_x)$ *for all terms* $t_x$ *in D, where* $b_x/e_x$ *is the begin/end position of the element that* *directly contains* $t_x$, $p_x$ *is a full label path that reaches* $t_x$, *and* $i_x$ *is the document position of* $t_x$ *in* $D$. *The* TF *score of* $(p,t)$ *relevant to* $D$ *is defined as:*

$$TF^D(p,t) = |\{i_x|(t_x,p_x,b_x,e_x,i_x) \in paths(D) \wedge \ p = p_x \ \wedge \ t = t_x\}| \tag{1}$$

Basically, $TF^D(p,t)$ counts the number of $(p,t)$ pairs in the document $D$. If $t$ occurs $n$ times in the same element reachable by $p$, $(p,t)$ will be counted $n$ times.

**Definition 2.** ***IDF Score of a Path-Term Pair.*** *Let* $N$ *be the total number of documents in the corpus. Let* $D_j$, $1 \leq j \leq N$, *be an XML document associated with the pair* $(p,t)$, *where* $p$ *is a full text label path derived from SS matching and* $t$ *is the term in the query. The* IDF *score of* $(p,t)$ *is defined as:*

$$IDF(p,t) = log\left(\frac{N_p}{N(p,t)}\right) \tag{2}$$

*where* $N_p$ *and* $N(p,t)$ *are calculated as follows:*

$$N_p = \sum_{j=1}^{N} |\{j|(t_x,p_x,b_x,e_x,i_x) \in paths(D_j) \ \wedge \ p = p_x\}| \tag{3}$$

$$N(p,t) = \sum_{j=1}^{N} |\{(t_x,p_x)|(t_x,p_x,b_x,e_x,i_x) \in paths(D_j) \wedge \ p = p_x \ \wedge \ t = t_x\}| \tag{4}$$

Basically, $N_p$ counts the total number of documents that contain path $p$ and $N(p,t)$ counts the total number of documents that contain $(p,t)$ in the corpus.

### 4.2 Enhanced Scoring with Positional Weight

A path-term pair $(p,t)$ corresponds to a positional bit vector $V$ in the content synopsis associated with $p$. A one-bit range in $V$ represents an element that contains $t$ and is reachable by $p$. For instance, in Figure 2, the bit vector $H_8$["index"] corresponds to the pair (/proceedings/article/body/paragraph, "index") and it contains 5 one-bit ranges. The number of one-bit ranges in the vector reflects the *TF* score of the pair (/proceedings/article/body/paragraph, "index"). However, after the bitwise *anding* operation, only 3 one-bit ranges are left in the resulting bit vector, which indicates that among those 5 **paragraph** elements, only 3 of them contain both "index" and "rank". Similarly, after the containment filtering between the positional filter of **article**($F_2$) and $H_3$["XML"], only 5 **article** elements are left out of 7 in the resulting positional filter($A$). To make the weight calculation of a path-term pair more accurate, we introduce the *positional weight*, which is the percentage of qualified path-term pairs found during the containment filtering or bitwise *anding* operation.

**Definition 3.** ***Positional Weight of a Path-Term Pair.*** *Let* $D$ *be an XML document,* $PF_0^D(p,t)$ *be the PF associated with* $(p,t)$, *and* $PF^D(p,t)$ *be the result from containment filtering or bitwise anding operation. In addition, let* $N_{PF_0^D(p,t)}$

*be the number of one-bit ranges in* $PF_0^D(p,t)$ *and* $N_{PF^D(p,t)}$ *be the number of one-bit ranges in* $PF^D(p,t)$*. The positional weight of* $(p,t)$ *in* $D$ *is defined as:*

$$PW^D(p,t) = \frac{N_{PF^D(p,t)}}{N_{PF_0^D(p,t)}} \tag{5}$$

Combining the *TF* score, *IDF* score, and the positional weight, the definition of the weight of $(p,t)$ in $D$ is determined by the following equation:

$$W^D(p,t) = PW^D(p,t) \times TF^D(p,t) \times IDF(p,t) \tag{6}$$

Finally, we give the definition of the enhanced content score of $D$ relevant to $Q$.

**Definition 4.** ***Enhanced Content Score of a Document.*** *Let* $Q$ *be the query and* $D$ *be an XML document. Let* $W_i^Q(p_i^Q, t_i^Q)$ *be the weight of the path-term pair* $(p_i^Q, t_i^Q)$ *in* $Q$ *and* $W_i^D(p_i^D, t_i^D)$ *be the weight of the corresponding path-term pair* $(p_i^D, t_i^D)$ *in the document* $D$*. The enhanced content score of* $D$ *relevant to* $Q$ *is defined as*

$$ECS(D,Q) = \frac{\sum_{i=1}^{n} W_i^Q(p_i^Q, t_i^Q) \times W_i^D(p_i^D, t_i^D)}{\sqrt{\sum_{i=1}^{n} W_i^Q(p_i^Q, t_i^Q)^2} \times \sqrt{\sum_{i=1}^{n} W_i^D(p_i^D, t_i^D)^2}} \tag{7}$$

*where* $n$ *is the number of path-term pairs.*

## 4.3 Combined Scoring with Term Proximity

We incorporate term proximity into the scoring to further improve the ranking scheme. We first use the size of the *lowest common ancestor* (LCA) of the full label paths derived from structural summary matching to measure the depth term proximity.

**Definition 5.** ***Depth Term Proximity.*** *Let* $Q$ *be the query and* $D$ *be an XML document. Let* $(p_i^D, t_i^D)$*,* $1 \le i \le n$*, be a matching path-term pair in* $D$*. Let* $r_{lca}$ *be the root of the tree rooted at* $LCA$ *of all paths* $p_i^D$ *and* $DIST(p_i^D, r_{lca})$ *be the number of steps between the leaf node of* $p_i^D$ *and* $r_{lca}$*. The* depth term proximity *of* $D$ *is defined as*

$$DTP(D,Q) = \frac{1}{\sum_{i=1}^{n} DIST(p_i^D, r_{lca})} \tag{8}$$

At the end of containment filtering, a non-zero positional filter $PF$ is derived for each qualified document. Each one-bit range in $PF$ represents an element that is associated with the $PF$ entry in the query footprint. The smaller the size of this element, the closer the search terms in the document. In addition, more one-bit ranges in $PF$ indicate the document contains more qualified elements, so the document should be ranked higher. Thus, we use the average length of one-bit range and the number of one-bit ranges in the $PF$ to measure the width term proximity.

**Definition 6.** ***Width Term Proximity.*** *Let $Q$ be the query and $D$ be an XML document. Let $PF$ be the final positional filter after the containment filtering. Let $N^{PF}_{obr}(D,Q)$ be the total number of one-bit ranges in $PF$ and $L^{PF}_{avg}(D,Q)$ be the average length of one-bit ranges in $PF$. The* width term proximity *of $D$ is defined as*

$$WTP(D,Q) = \frac{N^{PF}_{obr}(D,Q)}{L^{PF}_{avg}(D,Q)} \tag{9}$$

The final score of the document $D$ is determined by the following equation:

$$S(D,Q) = ECS(D,Q)^{\alpha} \times DTP(D,Q)^{\beta} \times WTP(D,Q)^{\gamma} \tag{10}$$

where $\alpha$, $\beta$, and $\gamma$ are experimental parameters.

## 5 Experimental Results

We have implemented our framework using Java (J2SE 6.0) and Berkeley DB Java Edition 3.2.13 [3] was employed as the storage manager. Our experiments were carried out on a WindowsXP machine with 2.8GHz CPU and 512M memory. The datasets we used were synthetically generated from the XMark and XBench [10] benchmarks. The main characteristics of our datasets and data synopses size are summarized in Table 1. For each dataset, our query workload is 10 full-text XPath queries exhibiting different sizes, query structures and search predicates. From Table 1, we can see that our data synopses are small enough to make our system scalable.

**Table 1.** Data Set Characteristics and Data Synopses Size

| Data Set | Data Size (MB) | Files | Avg. File Size (KB) | Avg. SS Size (KB) | Avg. CS Size (KB) | Avg. PF Size (KB) |
|---|---|---|---|---|---|---|
| XMark | 55.8 | 11500 | 5 | 0.413 | 0.305 | 0.016 |
| XBench | 1050 | 2666 | 394 | 0.427 | 2.012 | 0.174 |

**Efficiency of Indexing Scheme.** To demonstrate the efficiency of our Data Synopses Indexing(DSI) scheme, we compared it with traditional Inverted List Indexing(ILI) scheme. Note that we didn't compare our indexing scheme with XCluster [8] because the proposed XML synopses in [8] are mainly used for selectivity estimation, rather than for locating and ranking XML documents. Figure 3 shows that for XMark dataset, ILI consumes over 2 times index build time, disk space and query response time than DSI. For XBench dataset, DSI is much more efficient than ILI because XBench data is text-centric data generated from Text-Centric Multiple Document(TC/MD) class, which is more suitable for the index comparison experiments. In fact, DSI consumes less than 8% index build time than ILI and the index size of DSI is about 3% of that of ILI. The query response time of DSI is over 40 times faster than ILI.

We first measured our ranking scheme based on content similarity, then we incorporated the positional weight and term proximity into the ranking function

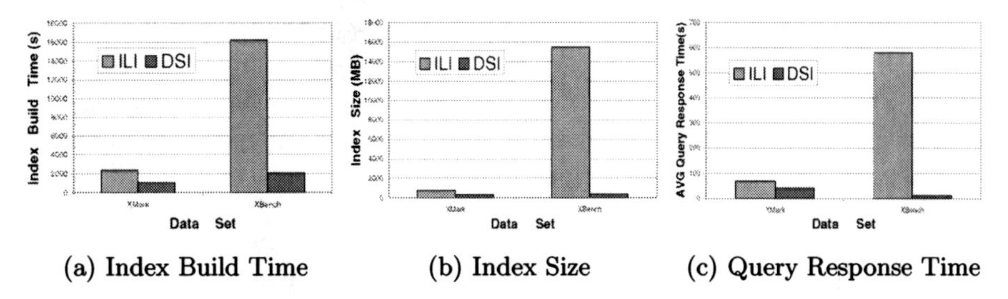


(a) Index Build Time (b) Index Size (c) Query Response Time

**Fig. 3.** Comparison between ILI and DSI

to demonstrate the improvement of ranking scheme. To construct the accurate relevant set for each query, we exploited Qizx/open [9] to evaluate the query over each dataset to obtain the strict relevant set. In the following figures, the **width factor** is the ratio between the width of a content synopsis and the number of terms the associated SS node contains. The **height factor** is the ratio between the height of a data synopsis and the number of begin/end tags in the document.

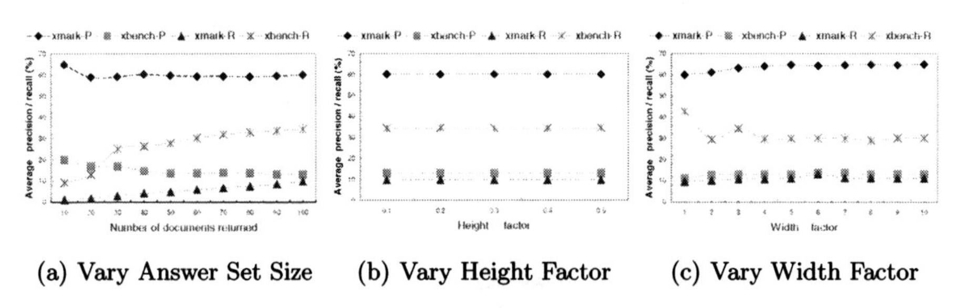


(a) Vary Answer Set Size (b) Vary Height Factor (c) Vary Width Factor

**Fig. 4.** Effectiveness of Content Scoring

**Effectiveness of Content Scoring.** We first fixed the size of data synopses to measure the average precision and recall of a query. The results in Figure 4(a) show that as the number of returned documents increases, the average precision(recall) of a query over each dataset drops(increases) smoothly, which indicates that our scoring function can effectively rank the relevant documents on the top of the ranked list so that the precision does not drop too much. We then varied the height factor to see its impact on precision and recall. Figure 4(b) shows that as the height factor varies, the precision and recall almost remain at the same value for each dataset. This is expected because when the height of data synopses is reduced, a query may get more false positives, but our ranking function can effectively rank the most relevant documents close to the top, while moving false positives near the bottom of the answer set so that the precision and recall almost do not change. Finally, we varied the width factor to see its impact on precision and recall. As we can see from Figure 4(c), the precision and recall change a little more than those in Figure 4(b). This result implies that if we want to decrease the size of data synopses to reduce the data storage overhead but still keep high precision, the height factor should be adjusted first.

**Effectiveness of Aggregated Scoring.** Since most existing ranking schemes [2,6] focus on ranking XML elements in original XML documents, rather than ranking XML documents based on data synopses, it is inappropriate to make direct comparisons with those ranking schemes. Instead, we fixed the number of returned documents to 50 and compared the three ranking schemes over our data synopses: Content Similarity Scoring(CS), Enhanced Scoring with Positional Weight(CS-PW) and Combined Scoring with Term Proximity(CS-PW-TP). Figure 5 shows that for both datasets, the average precision(recall) of CS-PW-TP is higher than that of CS-PW, which is in turn higher than the precision(recall) of CS. Note that for XBench(XMark) dataset, the average precision(recall) is a little lower because its relevant set is smaller(larger).

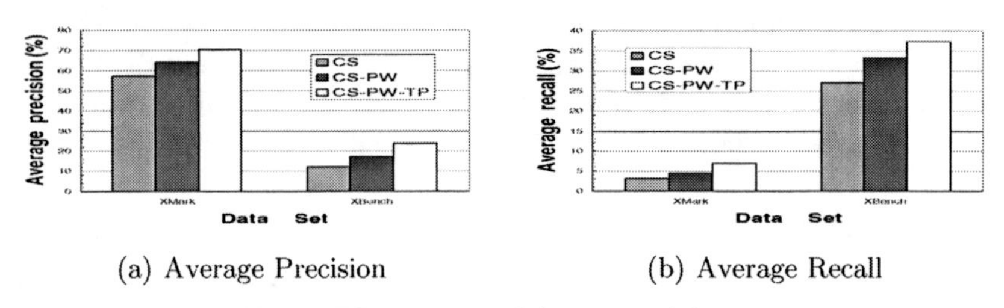


(a) Average Precision (b) Average Recall

**Fig. 5.** Effectiveness of Aggregated Scoring

# 6 Conclusion

We presented a framework for indexing, querying and ranking XML documents based on extracted content and structural synopses. We introduced an aggregated ranking scheme that scores an XML document based on content similarity, positional weight and term proximity. Our experiments show that our data synopses indexing scheme outperforms the standard XML indexing scheme based on inverted lists and our ranking scheme is effective in terms of precision and recall.

**Acknowledgments.** This work is supported in part by NSF under grant IIS-0307460.

# References

1. Al-Khalifa, S., Yu, C., Jagadish, H.V.: Querying Structured Text in an XML Database. In: Proc. of ACM SIGMOD, San Diego, pp. 4–15 (2003)
2. Amer-Yahia, S., Koudas, N., Marian, A., Srivastava, D., Toman, D.: Structure and Content Scoring for XML. In: Proc. of VLDB, Trondheim, pp. 361–372 (2005)
3. Berkeley, D.B.: `http://www.sleepycat.com/`
4. Carmel, D., Maarek, Y.S., Mandelbrod, M., Mass, Y., Soffer, A.: Searching XML Documents via XML Fragments. In: Proc. of ACM SIGIR, Toronto, pp. 151–158 (2003)

5. Clarke, C.: Controlling Overlap in Content-Oriented XML Retrieval. In: Proc. of ACM SIGIR, Salvador, pp. 314–321 (2005)
6. Cohen, S., Mamou, J., Kanza, Y., Sagiv, Y.: XSEarch: A Semantic Search Engine for XML. In: Proc. of VLDB, Berlin, Germany, pp. 45–56 (2003)
7. Kaushik, R., Bohannon, P., Naughton, J.F., Shenoy, P.: Updates for Structure Indexes. In: Proc. of VLDB, Hongkong, China, pp. 239–250 (2002)
8. Polyzotis, N., Garofalakis, M.: XCluster Synopses for Structured XML Content. In: Proc. of 22nd Int. Conference on Data Engineering, Atlanta, vol. 62 (2006)
9. Qizx/open. `http://www.axyana.com/qizxopen/`
10. XBench. `http://se.uwaterloo.ca/~ddbms/projects/xbench/`

# MQTree Based Query Rewriting over Multiple XML Views*

Jun Gao, Tengjiao Wang, and Dongqing Yang

Department of Computer Science and Technology,
Peking University, Beijing, China, 100871
{gaojun,tjwang,dqyang}@pku.edu.cn

**Abstract.** Using XML views to answer the XML query is an important query optimization strategy especially in the distributed environment. Although many methods have been proposed to handle the single XML view rewriting, they will lead to the redundant computation cost due to the shared paths among different XML views. This paper handles the query rewriting over multiple views by organizing the multiple XML views into a tree called MQTree, in which the shared sub paths among the multiple views have been merged in a top down fashion. In addition, this paper designs a MQTree based query rewriting method. The candidate query rewriting plans are generated over MQTree directly. In order to reduce the validation cost of the candidate query rewriting plans, the preliminary validation is made at the granularity of the path query {//,/,*} over the MQTree first, which prunes the candidate views further and provides the intermediate results for the plans validation at the granularity of the whole tree. The final experiments show the efficiency and effectiveness of our method.

## 1 Introduction

The query rewriting problem is a fundamental problem in many database research areas, including data integration, query optimization, semantic cache, etc. The problem can be formulated informally as follows: given a query $q$ and a view $v$ defined over a database $D$, find a query $q'$ which runs over $v$, the result $q'(v)$ of $q'$ over $v$ is contained or the same as $q(D)$ of $q$ over $D$ [8].

With the XML emerging as information representation and exchange standard in the Internet, XML is used in the distributed computing environment increasingly. How to rewrite the XML query using XML views also receives high attention. Since the XML query supports much more features than the relational query, the XML query rewriting faces more challenges. The current study [7] shows that the XML query containment, one of the fundamental steps in the XML query rewriting, is a Co_NP problem even when only commonly used features including {//,/,*,[]} of XPath are considered.

* Supported by Project 2006AA01Z230 under the National High-tech Research and Development of China, Project 60503037 under National Natural Science Foundation of China (NSFC), Project 4062018 under Beijing Natural Science Foundation (BNSF).

R. Wagner, N. Revell, and G. Pernul (Eds.): DEXA 2007, LNCS 4653, pp. 562–571, 2007.

The existing methods handle the problem from different perspectives. The XPath {//,/,*,[]} query rewriting is studied [1] to generate a sound but incomplete equivalent plan in a polynomial time if the index is regarded as a special case of the XML materialized view. The XPath query rewriting problem with more limited features [11] is studied to produce a sound and complete equivalent rewriting plan. The maximally contained query rewriting plan is discussed in [9] on XPath {//,/,[]} with and without the consideration of the schema.

There are maybe multiple XML views due to the possible different query patterns in real applications of the distributed computation. The paper [2] has discussed how to select XML queries to cache from a set of queries workload. In order to exploit all possible data from each cached XML view to answer the new XML query, the naive idea is to employ existing methods to generate rewriting plans for each view separately and union the results from all plans. However, the shared common paths among the multiple XPath views indicate the redundant cost in the query rewriting. Due to the high cost in the XML query rewriting plans generation and validation, the redundant computation cost should not be ignored.

This paper studies the problem on how to generate maximally contained XPath {//,/,*,[]} rewriting plans over multiple views. More specifically, our contributions can be summarized as follows:

- We propose an XML view management framework to organize the multiple XML views as a whole. We merge the shared paths of the XML view tree patterns in a top down fashion to generate a MQTree, which indicates that the redundant computation cost caused by the shared paths among the multiple views can be reduced.
- We propose a MQTree based query rewriting method. The candidate query rewriting plans are generated over the MQTree directly. In order to reduce the validation cost of the candidate query rewriting plans defined with XPath query {//,/,*,[]}, we make the preliminary validation at the granularity of the path query {//,/,*} over the MQTree first. We implement the containment mapping at the granularity of the whole tree with the intermediate results in the preliminary validation.
- We implement extensive experiments to illustrate the efficiency and effectiveness of our method.

The rest of the paper is organized as follows: We introduce background knowledge and define the problem formally in section 2. We propose a MQTree based query rewriting method, including the generation and validation method for candidate query rewriting plans in section 3. The extensive experimental evaluations are presented in section 4. Related works are discussed in section 5. The paper is summarized and the future work is discussed in section 6.

## 2 Background Knowledge

This section reviews some background knowledge including the XML query and the query rewriting problem, and defines the problem in this paper.

### 2.1 XPath Query

XPath is a basic mechanism to select nodes in the XML tree [3]. The features of XPath discussed in this paper include {/} for parent/child relationship, {//} for ancestor/descendant relationship, {*} for wildcards, {[]} for the predicates. The XPath expression can be represented by an XPath tree pattern [7], which can be defined as:

**Definition 1 (XPath tree pattern).** *An XPath query can be represented by a tree pattern* $Q = (N, E, r)$, *where* $N$ *represents a set of element nodes,* $E$ *represents a set of relation in term of* $e = (n_1, n_2)$, *where* $n_1 \in N$, $n_2 \in N$. *For each* $e \in E$, $type(e)$ = *"AD" (short for ancestor/descendant) or "PC" (short for parent/child).* $r$ *denotes the root node of* $Q$. *The target element node of XPath query is called* ***the return node****. The path from the root node to the return node is called* ***the main path****. The XPath* $q$ *without {[]} can be regarded as a special case of the XPath tree pattern, which is called* ***the path query****. We call a path query* $p$ ***the root-leaf path*** *when* $p$ *starts at the root node and ends at the leaf node.*

Two important notations, query equivalence and query containment, are used to determine the validation of the XML query rewriting plan [7]. Given two XPath expressions $p_1$ and $p_2$, we call $p_1$ **is contained in** $p_2$ if $p_1(D) \subseteq p_2(D)$ for any XML document $D$; we call $p_1$ **is equivalent to** $p_2$ if $p_1$ is contained in $p_2$ and $p_2$ is contained in $p_1$. The containment between two XPaths indicates the existence of a containment nodes mapping between two XPath tree patterns.

**Definition 2 (The containment nodes mapping between XPath pattern).** *Given two XPath tree pattern {//,/,*,[]}* $p_1$ *and* $p_2$, $p_2$ *is contained in* $p_1$, *the containment nodes mapping* $M$ *from* $p_1$ *to* $p_2$ *should meet the following requirements: 1) the root node of* $p_1$ *is mapped to the root node of* $p_2$*; 2) for each sub path* $p_{11} = (n_{11}, n_{12})$ *in* $p_1$, *the mapped sub path* $M(p_{11}) = (M(n_{11}), M(n_{12}))$ *in* $p_2$ *contained in* $p_{11}$*; 3) the main path of* $p_1$ *is mapped to the main path of* $p_2$.

### 2.2 XML Query Rewriting over Single XML View

**Definition 3 (XML query rewriting).** *Given an XPath query tree pattern* $q$ *over an XML document* $T$, *an XML view* $v$ *over the same XML* $T$, *suppose a query* $q_1$ *can be evaluated over* $v$, *where the results* $q_1(v)$ *of* $q_1$ *evaluated over* $v$ *is a sub set of results* $q(T)$ *of* $q$ *evaluated over* $T$, $q_1$ *is called the query* $q$ *rewriting plan over view* $v$.

Given an XML query $q$ and an XML view $v$, the query rewriting plan $q_r$ should meet the requirement illustrated in Fig 1: if we construct **a composite query** $q_c$ by merging the root node $q_r$ and the return node of $v$, we can establish a containment mapping $M$ from $q$ to $q_c$. We call node $n$ in the main path of $q$ **the pivot node** if $n$ is mapped to the return node of $v$ under $M$. We also notice that the main path of $v$ is contained in the prefix path from the root node to the pivot node in $q$.

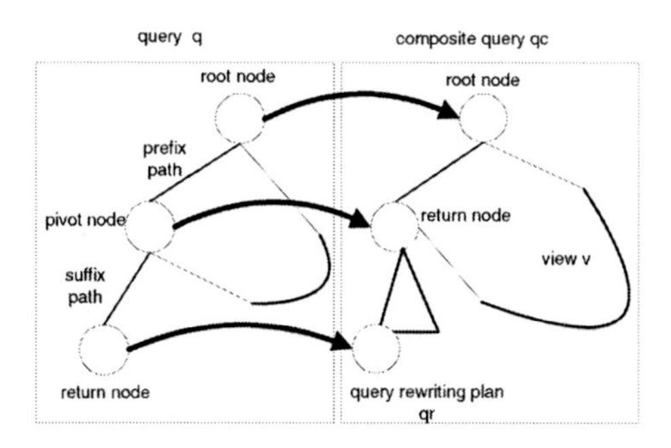


**Fig. 1.** Containment Mapping from a query to a view

In order to generate the maximally contained query rewriting plans, the suffix path rooted with the pivot node in the main path of the query can be a part of the query rewriting plan $q_r$. As for the root-leaf query $s$ which is not the main path of $q$, it may also have the pivot node $n$, which indicates that the suffix path rooted with $n$ may also a part of the query rewriting plan $q_r$.

### 2.3 Problem Definition

With the above definitions, the problem in this paper can be defined as follow:

*Given an XML Database $T$, a set of predefined materialized views $V$ defined with the XPath {//,/,*,[]} over $T$, how to generate the maximally contained query rewriting plans for an XPath{//,/,*,[]} query $q$ over views set $V$ efficiently?*

## 3 MQTree Based Query Rewriting over the Multiple XML Views

Since the cost of the query rewriting plan generation over the single XML view is expensive, the shared paths among the multiple XML views should not be ignored. In order to improve the efficiency of the query rewriting over the multiple XML views, we organize the XML view tree patterns into a MQTree(short for **M**ultiple **Q**ueries tree) via merging the shared paths in a top down fashion, and design a query rewriting method over the MQTree directly.

**Definition 4 (The MQTree for XPath tree patterns).** *Given an XPath tree patterns set $Q = \{q_1, \ldots q_n\}$ defined over the same XML document, each node in $q_i (1 \leq i \leq n)$ assigned with a unique $id[q_i]$ for $q_i$ and $isMain$ to indicate the node in the main path, the MQTree $T = (V, E, r)$ is constructed in the following way: the root node $r$ of $T$ is initialized with the $IDS[r] = \{id[q_0]\} \cup \ldots \{id[q_n]\}$ to indicate $r$ is the mapped node for the root node of $q_i (1 \leq i \leq n)$; if there is one path starting from the root node to node $n$ in MQTree which is equivalent to the path from the root node to node $m$ in $q_i$, $IDS[n] = IDS[n] \cup \{id[q_i]\}$, $n$ is the mapped node for $m$; otherwise $m$ is added under node $k$ in $T$ which is the mapped node for the parent node of $n$ in $q_i$, and $IDS[m] = \{id[q_i]\}$.*

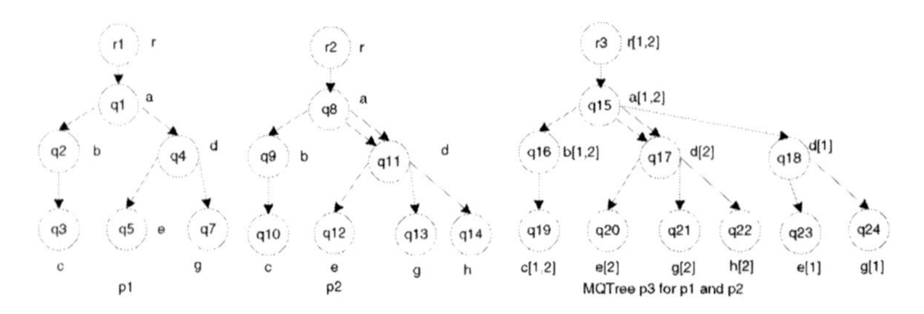


**Fig. 2.** An Example of MQTree

Fig. 2 illustrates a MQTree $p_3$ from two XPath tree patterns $p_1$ and $p_2$. The single(double) line represents $PC(AD)$ relation between nodes. We notice that the root node of MQTree is merged from the root node of $p_1$ and $p_2$. Node $q_3$ in $p_1$ and node $q_{10}$ in $p_2$ can be merged into $q_{19}$ in MQTree since the path expression from $r_1$ to $q_3$ in $p_1$ is equivalent to the expression from $r_2$ to $q_{10}$ in $p_2$, while node $q_4$ in $p_1$ can not be merged with node $q_{11}$ in $p_2$ due to the different sub path expressions.

### 3.1 The Candidate Query Rewriting Plans Generation over the MQTree

The candidate query rewriting plan generation can be reduced to the problem on how to locate pivot nodes in the main path of the query against the main path of views. We discover pivot nodes in a top down fashion with the nodes mapping encoded in the form of the nodes pair $(n_q, n_v)$, where $n_q$ is a node in the main path of a query $q$, $n_v$ is a node in the main path of a view $v$ indicated by $isMain$ in the MQTree. The whole process can be described as follows:

The root node of a MQTree $t$ and the root node of a query $q$ is combined into an initial nodes pair and added it into a queue $Q$. We repeat the following process until $Q$ is empty. We remove the first nodes pair $(n_q, n_v)$ from $Q$, and add all possible next nodes pairs, $(n_{q1}, n_{v1})$ or $(n_q, n_{v1})$, into $Q$ according to $type(n_q, n_{q1})$, $type(n_v, n_{v1})$ and the elements annotated on $n_{q1}$ and $n_{v1}$, where $n_{q1}$ is a child node of $n_q$ and $n_{v1}$ is a child node of $n_v$. The rules to add new nodes pair should also guarantee a validated nodes mapping. For example, in the case $type(n_q, n_{q1}) = AD$ and $type(n_v, n_{v1}) = AD$, we add a nodes pair $(n_q, n_{v1})$ into $Q$, and add a node pair $(n_{q1}, n_{v1})$ into $Q$ if $n_{q1}$ and $n_{v1}$ are annotated with the same element or $n_{q1}$ is annotated with the wildcards. If there exists one nodes pair $(n_q, n_v)$ added in $Q$, where $n_v$ is the return node for view $v$, $n_q$ can be regarded as a pivot node for the main path of query $q$. The final candidate rewriting plan takes the form of $(q, n_s, v)$, where $q$ is a XML query, $n_s$ is a pivot node in the main path of query $q$, and $v$ is an XML view in the MQTree.

### 3.2 The Validation of Candidate Rewriting Plan over MQTree

With the candidate plan in the form of $(q, n_s, v)$, we know there is a containment mapping $M$ from the path from the root node to $n_s$ in $q$ to the main path of the view. In the following, we need to determine whether $M$ can be extended to other parts of the XML tree.

The establishing of a containment mapping between two XPath tree pattern {//,/,*,[]} is a Co_NP problem as discussed in [5,7]. The existing studies also reveal that it takes polynomial time to implement the path query containment. We can exploit such a feature to validate the candidate query rewriting plans preliminarily over the MQTree first.

**The Preliminary Validation for the Candidate Query Rewriting Plan.** The containment mapping $M$ from a query $q$ to the composite query $q_c$ indicates the following properties: for each root-leaf path query $p$ in $q$, we can 1) locate the mapped path $M(p)$ in $v$ which is contained in $p$, or 2) the main path of $v$ is contained in a prefix path of $p$.

In other words, if property 1 or property 2 are not satisfied either, we know that we can not establish the containment mapping. Since property 1 and property 2 can be tested in a polynomial time, we can remove the views which do not satisfy the properties efficiently. Such a validation method is called **the preliminary validation.**

We also notice that we still need to validate the candidate plan at the granularity of the whole XML tree even after the preliminary validation is passed. In order to reduce the cost in the next step, we record the mapped nodes in the preliminary validation step on key nodes in the query tree pattern.

**Definition 5 (The key nodes and the XPath segment in the XPath).** *The key nodes set $K$ in XPath tree pattern $q$ includes all branching nodes, leaf nodes and the root node. The sub path $s = (n_1, n_2)$ of $q$ is called the XPath segment if the start node $n_1$ and the end node $n_2$ are in $K$, and there is no other key nodes in the middle of $s$.*

For each root-leaf path $p$ in query $q$, the preliminary test runs on path query $v$ in the MQTree, where the $IDS[n]$ of node $n$ in $v$ is the superset of candidate views set $C$ in the candidate plan generation. We detect whether $p$ satisfies property 1 or property 2 with the similar method as the location of pivot nodes in the section 3.1. In the case that the properties are satisfied, we record the mapped view nodes set $map[n, p]$ for the key node $n$ in $p$; otherwise, we know the preliminary validation fails directly. If two path queries $p_1$ and $p_2$ in $q$ share the same branching node $n$, the mapped nodes set $map[n]$ for $n$ can be regarded as the intersection of $map[n, p_1]$ and $map[n, p_2]$. We repeat the process until all root-leaf paths have been handled. The final results in the preliminary test include a candidate views set $C_1$ each of which passes the preliminary validation, and the mapped nodes set annotated on the each key nodes in $q$.

**The Construction of the Containment Mapping for the Whole Tree.** The preliminary validation alone does not guarantee the soundness of the query rewriting plan. We have to validate plans at the granularity of the whole XML tree and produce the final XML query rewriting plan. Let $C$ denotes the candidate views which pass the preliminary test, we validate whether the view $v \in C$

can be used to answer query $q$. As for each key node $n$ in $q$, we select one mapped node $m$ from the mapped nodes set $map[n]$. The key nodes and their mapped nodes can formulate a containment mapping $M$ from $q$ to $v$. The containment mapping $M$ is valid if only the mapped segment $M(s)$ in $v$ is contained in the each segment $s$ in $q$, where $s$ is not in the sub tree rooted with the pivot node in the main path of $q$.

Another thing should be noticed that when a root-leaf path $p$ other than the main path in $q$ is mapped to the main path of view $v$ under a valid mapping $M$, we know that $p$ contains a pivot node $n$. We need to merge the suffix path $p_1$ rooted with $n$ into the final query rewriting plan.

Although the validation of each candidate containment nodes mapping can be implemented in a polynomial time, the number of possible nodes mappings may be exponential. Suppose the number of the mapped nodes $n_i[m]$ for the key node node $n_i(1 \leq i \leq n)$ in $q$, we have to validate $\prod_{i=1}^{n} n_i[m]$ different containment mappings at the worst case. However, since the preliminary validation can prune much validation space, and the intersection of the mapped nodes for the key nodes from the different root-leaf path can reduce the number of the mapped nodes, the validation can be implemented efficiently for the XPath in the real applications.

## 4 Performance Study

### 4.1 Experimental Setup

The algorithms we proposed have been implemented using JDK 1.41 in a Window XP environment on a Dell Optiplex GX260 with P4 2.0GHz CPU and 512MB RAM. We select one XPath generator developed at Berkeley [12], which can generate XPaths set with the assigned probabilities for {//, /, *,[]} respectively.

Since the increase of probability for {//,*,[]} in XPath set can lead to the increase of the size of the MQTree and the cost of related operations, we generate XPaths sets with various probability of different features. Due to the space constraints, we report the experimental results on two representative XPath query sets with different size, XPaths set $P_{33}$ generated with $prob(//) = prob(*) = prob([]) = 0.3$, XPaths set $P_{11}$ generated with $prob(//) = prob(*) = prob([]) = 0.1$.

We implement the MQTree based XPath rewriting method, and compare the cost with the query rewriting over each XML view separately. In addition, we also illustrate the effectiveness of the optimization strategies proposed in MQTree, for example, the preliminary validation at the granularity of the path query.

Since the construction of the MQTree can be implemented in a linear time in the size of the XPath tree nodes, we focus on the cost of the rewriting plans generation. Given a set of generated XPath set $Q$ with the size $n$, we implement the query rewriting in the following way: for each query $q \in Q$, we assume that the queries set $(Q - \{q\})$ are materialized views $V$, and try to answer $q$ using views $V$. Therefore, we implement $n - 1$ times query rewriting for each XPath query.

### 4.2 Performance Study

Fig. 3(a) and Fig. 3(b) compare the whole rewriting cost between the method based on MQTree (denoted as MQTree) and the method to handle each view separately(denoted as Separate). We notice that the MQTree based method takes 1/5 to 1/10 cost of separate method since the shared path can only be handled once in the MQTree based method. The efficiency of MQTree increases with the increase size of the XPaths set since more XPath nodes can be merged into MQTree. In addition, the time cost of MQTree takes high cost to handle $P_{33}$ set than to handle $P_{11}$ set. This is due to that the increase of uncertain factor leads to the decrease of the merging probability for the query tree and the increase of size of the MQTree.

The validation of the XPath query rewriting plans occupies a large portion of the total cost. We propose a two-phase candidate plans validation method in this paper. In Fig. 3(c) and Fig. 3(d), we compare the time cost of MQTree with or without the optimization strategy of the preliminary validation. We notice that the preliminary validation can improve the efficiency of the XPath rewriting especially on the XPaths set $P_{33}$, since the more predicates {[]} in the XPath indicate there exist more path queries in the XPath tree pattern and the preliminary validation can prune more validation space on $P_{33}$.

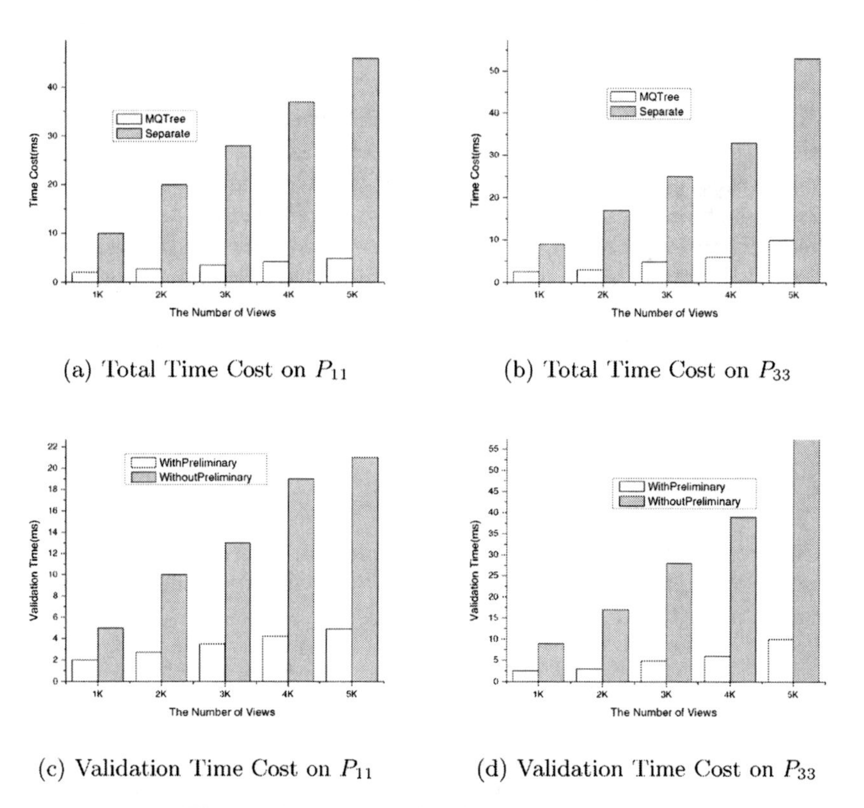


(a) Total Time Cost on $P_{11}$

(b) Total Time Cost on $P_{33}$

(c) Validation Time Cost on $P_{11}$

(d) Validation Time Cost on $P_{33}$

**Fig. 3.** Time cost in the XML query rewriting

## 5 Related Work

Answering query using views has been extensively studied on the relational model [8]. Two fundamental algorithms, bucket algorithm and inverse rule algorithms, have been proposed in [10,6]. Although much work has been done on the relational model, it is not a trivial work to extend the existing techniques to the nested data model, due to the semantic mismatch between two models and different expressive power of two query languages.

Some attempts have been made on the semi-structured data model or the graph data model. The method supports [13] the nested query expression and the results reconstruction. However, the work does not discuss the query with features similar to "//" or "*" in XPath. Regular path query rewriting discussed in [4] on the graph model supports regular expressions, which has the similar expression as XPath {//,/,*}, while the regular expression query does not support the result reconstruction and "[]".

The most recent progress in XPath rewriting is made at the server side [1] or at the client side [11]. The paper [1] studies the query rewriting problem if the index is regarded as a special kind of the XML view. It extends the query containment test [7] and proposes an incomplete but efficient XPath {//,/,[],*} rewriting algorithm via the dynamic programming method. The paper [11] studies the query rewriting problem at the client side. It proposes a sound and complete method to handle XML query rewriting with the more restricted features. The work [9] studies the maximally contained query rewriting for XPath{//,/,[]} and discusses the impact of the schema to reduce the number of the sub plans. Although our work also studies the maximally contained XML query rewriting, our method considers the XPath features includes {//,/,[],*} and discusses how to handle multiple XML views efficiently.

The XML query processing over the XML data stream handles millions XML query at the same time [12]. The redundancy shared computation paths are merged together to reduce the evaluation cost for the whole set. Our work shares the similar idea on the merging XPath. However, the operations on the merged structure in the data stream are totally different from those in XML query rewriting.

## 6 Conclusion

In this paper, we propose a MQTree based method to generate query rewriting plans over multiple views efficiently. The future work will study the impact of schema on the MQTree.

## References

1. Beyer, K., Cochrane, R., Pira-hesh, H., Balmin, A., Ozcan, F.: A framework for using materialized xpath views in xml query processing. In: Proc. VLDB, pp. 60–71 (2004)

2. Suciu, D., Mandhani, B.: Query caching and view selection for xml databases. In: Proc. VLDB, pp. 469–480 (2005)
3. Clark, J.: Xml path language(xpath). W3C (1999), http://www.w3.org/TR/XPath
4. Lenzerini, M., Vardi, M., Calvanese, D., Giacomo, G.D.: Rewriting of regular expressions and regular path queries. In: Proc. PODS, pp. 194–204 (1999)
5. Schwentick, T., Neven, F.: Xpath containment in the presence of disjunction, dtds, and variables. In: Proc. ICDT, pp. 315–329 (2003)
6. Mendelzon, A.O., Grahne, G.: Tableau techniques for querying information sources through global schemas. In: Proc. ICDT, pp. 332–347 (1999)
7. Suciu, D., Miklau, G.: Containment and equivalence for an xpath fragment. In: Proc. PODS, pp. 65–76 (2002)
8. Halevy, A.: Answering queries using views: A survey. VLDB Journal 10(4), 270–294 (2001)
9. Zhao, Z., Lakshmanan, L., Wang, H.: Answering tree pattern queries using views. In: Proc. VLDB, pp. 571–582 (2006)
10. Duschka, O.M., Genesereth, M.R., Keller, A.M.: Infomaster: An information integration system. In: Proc. SIGMOD, pp. 539–542 (1997)
11. zsoyoglu, Z.M., Xu, W.: Rewriting xpath queries using materialized views. In: Proc. VLDB, pp. 121–132 (2005)
12. Franklin, M., Zhang, H., Diao, Y.l., Altinel, M., Fischer, P.: Path sharing and predicate evaluation for high-performance xml filtering. In: TODS (2003)
13. Vassalos, V., Papakonstantinou, Y.: Query rewriting for semi-structured data. In: Proc. SIGMOD, pp. 455–466 (1999)

# Convex Cube: Towards a Unified Structure for Multidimensional Databases

Alain Casali, Sébastien Nedjar, Rosine Cicchetti, and Lotfi Lakhal

Laboratoire d'Informatique Fondamentale de Marseille (LIF),
CNRS UMR 6166, Université de la Méditerranée
Case 901, 163 Avenue de Luminy, 13288 Marseille Cedex 9, France
lastname@lif.univ-mrs.fr

**Abstract.** In various approaches, data cubes are pre-computed in order to efficiently answer OLAP queries. Such cubes are also successfully used for multidimensional analysis of data streams. The notion of data cube has been explored in various ways: iceberg cubes, range cubes, differential cubes or emerging cubes. In this paper, we introduce the concept of convex cube which captures all the tuples satisfying a monotone and/or antimonotone constraint combination. It can be represented in a very compact way in order to optimize both computation time and required storage space. The convex cube is not an additional structure appended to the list of cube variants but we propose it as a unifying structure that we use to characterize, in a simple, sound and homogeneous way, the other quoted types of cubes.

## 1 Introduction and Motivations

Pre-computing all the possible aggregates at various levels of granularity makes it possible to handle data cubes and efficiently answer OLAP queries [Han and Kamber, 2006]. The data cube is thus a key concept for data warehouse management. More recently, cube computation has been successfully used for multidimensional analysis of data streams [Han et al., 2005]. In this kind of applications, huge amounts of data, at a very thin granularity, are generated in continuous flows. Such flows must be scanned only once because of a threefold reason: the reading cost, the very quick changes in data and the needs of rapid reactions for users faced with data changes. In such a context, computing cubes is an interesting way to investigate the described issue.

Research work has proposed different variations around the concept of data cubes. For instance, iceberg cubes are partial cubes inspired from frequent patterns. They capture only sufficiently significant trends by enforcing minimal threshold constraints over measures [Beyer and Ramakrishnan, 1999]. Range cubes can be seen as extending the previous ones because measures are constrained in order to belong to a given range [Casali et al., 2003]. Users are then provided with trends fitting in a particular "window". New trends appearing (or established trends disappearing) when a data warehouse is refreshed or along a data stream capture are exhibited by differential cubes [Casali, 2004]. The

R. Wagner, N. Revell, and G. Pernul (Eds.): DEXA 2007, LNCS 4653, pp. 572–581, 2007.

latter can be perceived as the result of a set difference between two cubes: for instance one stored in the data warehouse and the other computed from the refreshment data or also one computed from a data stream at a precise instant and the other later. Depending on the order of the two operands, appearing or disappearing trends are exhibited. Finally, emerging cube [Nedjar et al., 2007] captures trends which are not significant at a moment but which grow significant later. In a dual way, it can exhibit relevant trends which become irrelevant. In addition with the appearing or disappearing trends of the differential cube, the emergent cube provides the decision maker with trend reversals. Such a knowledge is strongly required in multidimensional analysis of data stream [Han et al., 2005] and OLAP.

Frequently these different types of cubes, by starting with the original data cube itself [Gray et al., 1997], have not been grasped as concepts but rather as the result of queries or more efficient algorithms.

In this paper, we propose a novel and unifying structure which offers a soundly founded framework for characterizing the various quoted cubes. More precisely, our contributions are the following:

(*i*) we state the foundations of a new structure called convex cube which is based on the search space of the cube lattice [Casali et al., 2003]. The convex cube takes into account combinations of monotone and antimonotone constraints. We show that such a structure is a convex space and thus can be represented, in a very compact way, by its borders;

(*ii*) by taking benefit of the convex cube structure, we propose formal and homogeneous definitions for the data cube, iceberg cube, range cube, differential cube and emmerging cube.

The article is organized as follows. Section 2 presents the background of our proposal by briefly describing the multidimensional search space that we use: the cube lattice. In section 3, we detail the structure of convex cube. Its use for characterizing the various types of cubes is proposed in the next section.

## 2 Cube Lattice Framework

In this section, we recall the concepts of cube lattice [Casali et al., 2003] which is used to formalize the new structure proposed in this paper.

Throughout the paper, we make the following assumptions and use the introduced notations. Let $r$ be a relation over the schema $\mathcal{R}$. Attributes of $\mathcal{R}$ are divided in two sets (*i*) $\mathcal{D}$ the set of dimensions, also called categorical or nominal attributes, which correspond to analysis criteria and (*ii*) $\mathcal{M}$ the set of measures.

The multidimensional space of the categorical database relation $r$ groups all the valid combinations built up by considering the value sets of attributes in $\mathcal{D}$, which are enriched with the symbolic value ALL. The latter, introduced in [Gray et al., 1997] when defining the operator Cube-By, is a generalization of all the possible values for any dimension.

The multidimensional space of $r$ is noted and defined as follows: $Space(r) = \{\times_{A \in \mathcal{D}}(Dim(A) \cup \text{ALL})\} \cup \{(\emptyset, \ldots, \emptyset)\}$ where $\times$ symbolizes the Cartesian product, $Dim(A)$ is the projection of $r$ on the attribute $A$ and tuple $(\emptyset, \ldots, \emptyset)$ stands for the combination of empty values. Any combination belonging to the multidimensional space is a tuple and represents a multidimensional pattern.

The multidimensional space of $r$ is structured by the generalization/specialization order between tuples, denoted by $\preceq_g$. This order has been originally introduced by T. Mitchell [Mitchell, 1997] in the context of machine learning. In a datawarehouse context, this order has the same semantic as the operator ROLLUP/DRILLDOWN [Gray et al., 1997] and is used, in the quotient cube [Lakshmanan et al., 2002], to compare tuples (cells).

Let $u$, $v$ be two tuples of the multidimensional space of $r$:

$$u \preceq_g v \Leftrightarrow \begin{cases} \forall A \in \mathcal{D} \text{ such that } u[A] \neq \text{ALL}, u[A] = v[A] \\ \text{or } v = (\emptyset, \ldots, \emptyset) \end{cases}$$

If $u \preceq_g v$, we say that $u$ is more general than $v$ in $Space(r)$. In other words, $u$ captures a similar information than $v$ but at a rougher granularity level.

*Example 1.* Let us consider the relation DOCUMENT (*cf* Table 1) yielding the quantities sold by Type, City and Publisher. In the multidimensional space of our relation example, we have: (Novel, ALL, ALL) $\preceq_g$ (Novel, Marseilles, Hachette), *i.e.* the tuple (Novel, ALL, ALL) is more general than (Novel, Marseilles, Hachette) and (Novel, Marseilles, Hachette) is more specific than (Novel, ALL, ALL).

**Table 1.** Relation example DOCUMENT

| RowId | Type | City | Publisher | Qty |
|---|---|---|---|---|
| 1 | Novel | Marseilles | Hachette | 2 |
| 2 | Novel | Marseilles | Collins | 2 |
| 3 | Essay | Paris | Collins | 1 |
| 4 | Textbook | Paris | Collins | 6 |
| 5 | Essay | Marseilles | Collins | 1 |

The two basic operators provided for tuple construction are: Sum (denoted by +) and Product (noted •). The Sum of two tuples yields the most specific tuple which generalizes the two operands. The Product of two tuples yields the most general tuple which specializes the two operands. If it exists, for these two tuples, a dimension $A$ having distinct and real world values (i.e. existing in the original relation), then the only tuple specializing them is the tuple $(\emptyset, \ldots, \emptyset)$ (apart from it, the tuple sets which can be used to retrieve them are disjoined).

By providing the multidimensional space of $r$ with the generalization order between tuples and using the above-defined operators Sum and Product, we define an algebraic structure which is called cube lattice. Such a structure provides a sound foundation for several multidimensional data mining issues.

**Theorem 1.** *Let r be a categorical database relation over $\mathcal{D} \cup \mathcal{M}$. The ordered set $CL(r) = \langle Space(r), \preceq_g \rangle$ is a complete, graded, atomistic and coatomistic lattice, called cube lattice in which Meet ($\bigwedge$) and Join ($\bigvee$) elements are given by:*

1. $\forall\ T \subseteq CL(r),\ \ \bigwedge T = +_{t \in T}\ t$
2. $\forall\ T \subseteq CL(r),\ \ \bigvee T = \bullet_{t \in T}\ t$

# 3 Convex Cubes

In this section, we study the cube lattice structure faced with conjunctions of monotone and antimonotone constraints according to the generalization order. We show that this structure is a convex space which is called convex cube. We propose condensed representations (with borders) of the convex cube with a twofold objective: defining the solution space in a compact way and deciding whether a tuple $t$ belongs or not to this space.

We take into account the monotone and antimonotone constraints the most used in database mining [Han and Kamber, 2006]. They are applied on:

- measures of interest like pattern frequency, confidence, correlation. In these cases, only the dimensions of $\mathcal{R}$ are necessary;
- aggregates computed from measures of $\mathcal{M}$ and using statistic additive functions (COUNT, SUM, MIN, MAX).

We recall the definitions of convex space notion, monotone and/or antimonotone constraints according to the generalization order $\preceq_g$.

**Definition 1.** *Convex Space - Let $(\mathcal{P}, \leq)$ be a partial ordered set, $\mathcal{C} \subseteq \mathcal{P}$ is a convex space [Vel, 1993] if and only if $\forall x, y, z \in \mathcal{P}$ such that $x \leq y \leq z$ and $x, z \in \mathcal{C}$ then $y \in \mathcal{C}$. Thus $\mathcal{C}$ is bordered by two sets: (i) an "Upper set", noted $U$, defined by $U = \max_{\leq}(\mathcal{C})$, and (ii) a "Lower set", noted $L$ and defined by $L = \min_{\leq}(\mathcal{C})$.*

**Definition 2.** *Monotone/antimonotone constraints*

1. *A constraint Const is monotone according to the generalization order if and only if: $\forall\ t, u \in CL(r) : [t \preceq_g u \text{ and } Const(t)] \Rightarrow Const(u)$.*
2. *A constraint Const is antimonotone according to the generalization order if and only if: $\forall\ t, u \in CL(r) : [t \preceq_g u \text{ and } Const(u)] \Rightarrow Const(t)$.*

*Notations:* We note *cmc* (*camc* respectively) a conjunction of monotone constraints (antimonotone respectively) and *chc* an hybrid conjunction of constraints. By resuming the symbols $U$ and $L$ according to the considered case, the introduced borders are indexed by the type of the constraint in question. For instance, $U_{camc}$ symbolizes the set of the most specific tuples satisfying the conjunction of antimonotone constraints.

*Example 2.* - In the multidimensional space of our relation example DOCUMENT (*cf.* Table 1), we would like to know all the tuples for which the measure value is greater than or equal to 3. The constraint " SUM(*Quantity*) $\geq 3$ " is antimonotone. If the amont of sales by Type, City and Publisher is greater than 3, then the quantity satisfies this constraint at a more aggregated granularity level *e.g.* by Type and Publisher (all the cities merged) or by City (all the types and publishers together). In a similar way, if we aim to know all the tuples for which the quantity is lower than 6, the underlying constraint " SUM(*Quantity*) $\leq 6$ " is monotone.

**Theorem 2.** *The cube lattice with monotone and/or antimonotone constraints is a convex space which is called convex cube, $CC(r)_{const} = \{t \in CL(r)$ such that $const(t)\}$. Its upper set $U_{const}$ and lower set $L_{const}$ are:*

1. *if* $const = cmc$, $\begin{cases} L_{cmc} = \min_{\preceq_g}(CC(r)_{cmc}) \\ U_{cmc} = (\emptyset, \ldots, \emptyset) \end{cases}$

2. *if* $const = camc$, $\begin{cases} L_{camc} = (ALL, \ldots, ALL) \\ U_{camc} = \max_{\preceq_g}(CC(r)_{camc}) \end{cases}$

3. *if* $const = chc$, $\begin{cases} L_{chc} = \min_{\preceq_g}(CC(r)_{chc}) \\ U_{chc} = \max_{\preceq_g}(CC(r)_{chc}) \end{cases}$

The upper set $U_{const}$ represents the most specific tuples satisfying the constraint conjunction and the lower set $L_{const}$ the most general tuples respecting such a conjunction. Thus $U_{const}$ and $L_{const}$ result in condensed representations of the convex cube faced with a conjunction of monotone and/or antimonotone constraints.

The following corollary provides a characterization of the convex cube borders with an hybrid conjunction $chc = camc \wedge cmc$ by knowing only ($i$) either the maximal border for the antimonotone constraints ($U_{camc}$) and the monotone ones $cmc$, ($ii$) or the minimal border for the monotone constraints ($L_{cmc}$) and the antimonotone ones $camc$

**Corollary 1**

1. *Given $U_{camc}$ and $cmc$, the convex cube borders $CC(r)_{chc}$ are:*

$$\begin{cases} L_{chc} = min_{\preceq_g}(\{t \in CL(r) \mid \exists t' \in U_{camc} : t \preceq_g t' \text{ and } cmc(t)\}) \\ U_{chc} = \{t \in U_{camc} \mid \exists t' \in L_{chc} : t' \preceq_g t\} \end{cases}$$

2. *Given $L_{cmc}$ and $camc$, a condensed representation of $CC(r)_{chc}$ is:*

$$\begin{cases} U_{chc} = max_{\preceq_g}(\{t \in CL(r) \mid \exists t' \in L_{cmc} : t' \preceq_g t \text{ and } camc(t)\}) \\ L_{chc} = \{t \in L_{cmc} \mid \exists t' \in U_{chc} : t \preceq_g t'\}. \end{cases}$$

*Example 3.* Table 2 gives the borders $U_{camc}$, $U_{chc}$, $L_{cmc}$ and $L_{chc}$ of the convex cube of our relation example with the hybrid constraint "$3 \leq$ SUM($Qty$) $\leq 6$".

**Table 2.** Borders of the convex cube for $3 \leq \text{SUM}(\textit{Quantity}) \leq 6$

| | |
|---|---|
| $U_{camc}$ | (Novel, Marseilles, ALL)<br>(ALL,Marseilles, Collins)<br>(Textbook, Paris, Collins) |
| $U_{chc}$ | (Novel, Marseilles, ALL)<br>(ALL, Marseilles, Collins)<br>(Textbook, Paris, Collins) |
| $L_{cmc}$ | (Novel, ALL, ALL)<br>(Essay, ALL, ALL)<br>(Textbook, ALL, ALL)<br>(ALL, Marseilles, ALL)<br>(ALL, ALL, Hachette) |
| $L_{chc}$ | (Novel, ALL, ALL)<br>(Textbook, ALL, ALL)<br>(ALL, Marseilles, ALL) |

The characterization of the convex cube as a convex space makes it possible to know whether a tuple satisfies or not the constraint conjunction by only knowing borders of the convex cube. Actually if a conjunction of antimonotone constraints holds for a tuple of *Space*($r$) then any tuple generalizing it also respects the constraints. Dually if a tuple fulfils a monotone constraint conjunction, then all the tuples specializing it also satisfy the constraints.

*Example 4.* From the borders of the convex cube for our relation DOCUMENT (*cf.* Table 2) the following queries can be easily answered:

1. Is the sold quantity in Marseilles between 3 and 6?
2. Is the number of textbooks sold in Paris between 3 and 6?
3. Is the number of documents sold in Paris and published by Collins between 3 and 6?

The answer to the first question is yes because the tuple (ALL, Marseilles, ALL), giving the sales in the city of Marseilles, all the products and dates merged, belongs to the border $L_{chc}$. We answer the second query in a similar way because the tuple (Textbook, Paris, ALL) generalizes the tuple (Textbook, ALL, ALL) belonging to $L_{chc}$ and specializes the tuple (Textbook, Paris, Collins) included in the border $U_{chc}$. In contrast, the answer to the third question is no because the tuple (ALL, Paris, Collins) (all the purchases made in Paris and puiblished by Collins) does not specialize any tuple of $L_{chc}$ even if it generalizes the tuple (Textbook, Paris, Collins) of $U_{chc}$.

## 4 Formalization of Existing Cubes

In this section, we review different variants of data cubes and, by using the convex cube structure, we propose a characterization both simple and well founded.

### 4.1 Datacubes

Originally proposed by J. Gray *et al.* [Gray et al., 1997], The data cube according to a set of dimensions is presented as the result of all the GROUP BY that it is possible to express using a combination of dimensions. The result of any GROUP BY is called a cuboid, and the set of all the cuboids is structured within a relation noted $Datacube(r)$. The schema of this relation remains similar to the one of $r$, i.e. $\mathcal{D} \cup \mathcal{M}$ and the very same schema is used for all the cuboids (in order to perform their union) by enforcing a simple idea: any dimension which is not envolved in the computation of a cuboid (i.e. not mentioned in the GROUP BY) is provided with the value ALL. For any attribute set $X \subseteq \mathcal{D}$, a cuboid of the data cube, noted $Cuboid(X, f(\{\mathcal{M}|*\}))$, is yielded by the following SQL query:

```
SELECT [ALL,] X, f({M|*})
   FROM r
   GROUP BY X;
```

Thus a data cube can be achieved by the two SQL queries:

1. by using the operator CUBE BY (or GROUP BY CUBE according to the DBMS):

```
SELECT D, f({M|*})
   FROM r
   CUBE BY D;
```

2. by performing the union of the cuboids:

$$Datacube(r, f(\{\mathcal{M}|*\})) = \bigcup_{X \subseteq \mathcal{D}} Cuboid(X, f(\{\mathcal{M}|*\}))$$

A tuple $t$ belongs to the data cube of $r$ if and only if it exists a tuple $t'$ in $r$ which specializes $t$; else $t$ cannot be built up. As a consequence, whatever the aggregative function is, the tuples of the data cube projected over the dimensions remain invariant, only the values computed by the aggregative function vary.

**Proposition 1.** *Let $r$ be a relation projected over $\mathcal{D}$, the set of tuples (i.e. without the measure values) of the data cube of $r$ is a convex cube for the constraint "COUNT(*) $\geq 1$": $Datacube(r) = \{t \in CL(r) \mid t[Count(*)] \geq 1\}$*

Since the constraint "COUNT(*) $\geq 1$" is an antimonotone constraint (according to $\preceq_g$), a data cube is a convex cube. By applying theorem 2, we infer that any data cube can be represented by two borders: the relation $r$ which is the upper set and the tuple (ALL, ..., ALL) which is the lower set. Then we can easily assess the belonging of any tuple $t$ to the data cube of $r$: we have just to find a tuple $t' \in r$ specializing $t$.

*Example 5.* With our relation example DOCUMENT (*cf.* Table 1), the tuple (Novel, Marseilles, ALL) belongs to the data cube because it generalizes the tuple (Novel, Marseilles, Hachette) of the original relation.

In this secion, we have shown that we can characterize the datacube as a convex cube. In the same way, in the following section we use the genericity of our structure to capture various type of cubes.

### 4.2 Others Cubes

Most of the existing cubes can achieved by SQL queries or with by using by using our structure. In a first time, we present cubes, the most used in practice. Then we summarize characterizations of these in table 3.

(*i*) Inspired from frequent patterns, Beyer *et al.* introduce the Iceberg cubes [Beyer and Ramakrishnan, 1999] which are presented as tuple subset of the data cube satisfying for the measure values a minimal threshold constraint. The proposal is motivated by the following objective: the decision makers are interested in general tendencies, the relevant trends are trends sufficiently distinctive. Thus it is not necessary to compute and materialize the whole cube (the search space is pruned). This results in a significant gain for both execution time and required storage space.

(*ii*) The tuples of a range cube have measure values which fit in a given range. Such cubes place emphasis on middle tendencies, not too general and not too specific.

(*iii*) Differential cubes [Casali, 2004] result from the set difference between the data cubes of two relations $r_1$ and $r_2$. They capture tuples relevant in a relation and not existing in the other. In contrast with the previous ones, such cubes perform comparisons between two data sets. For instance in a distributed application, these data sets are issued from two different sites and their differential cube highlights trends which are common here and unknown there. For OLAP applications as well as data stream analysis, trend comparisons along time are strongly required in order to exhibit trends which are significant at a moment and then disappear or on the contrary non-existent trends which latter appear in a clear-cut way. If we consider that the original relation $r_1$ is stored in a data warehouse and $r_2$ is made of refreshment data, the differential cube shows what is new or dead.

(*iv*) Emerging cubes [Nedjar et al., 2007] capture trends which are not relevant for the users (because under a threshold) but which grow significant or on the contrary general trends which soften but not necessarily disappear. Emergent cubes enlarge results of differential cubes and refine cube comparisons. They are of particular interest for data stream analysis because they exhibit trend reversals. For instance, in a web application where continuous flows of received data describe in a detailed way the user navigation, knowing the craze for (in contrast the disinterest in) such or such URL is specially important for the administrator in order to allow at best available ressources according to real and fluctuating needs.

**Table 3.** Formalization of existing cubes

| Type of Datacube | SQL Query | Constraints | Characterization |
| --- | --- | --- | --- |
| Iceberg cubes | SELECT D, f({M\|*})<br>FROM r<br>CUBE BY D<br>HAVING<br>f({M\|*})>=MinThreshold; | $f(\mathcal{M}\|*) \geq MinThreshold$ | $CubeIcerberg(r) = \{t \in CL(r) \mid t[f(\{\mathcal{M}\|*\})] \geq MinThreshold\}$. |
| Range cubes | SELECT D, f({M\|*})<br>FROM r<br>CUBE BY D<br>HAVING f({M\|*})<br>BETWEEN MinThreshold<br>AND MaxThreshold; | $f(\mathcal{M}\|*) \geq MinThreshold$<br>$f(\mathcal{M}\|*) \leq MaxThreshold$ | $RangeCube(r) = \{t \in CL(r) \mid t[f(\{\mathcal{M}\|*\})] \geq MinThreshold$ and $t[f(\{\mathcal{M}\|*\})]\} \leq MaxThreshold\}$. |
| Differential cubes | SELECT D, f({M\|*})<br>FROM r2<br>CUBE BY D<br>HAVING<br>f({M\|*})>=MinThreshold<br>Minus<br>SELECT D, f({M\|*})<br>FROM r1<br>CUBE BY D; | $f(\mathcal{M}\|*) \geq MinThreshold$<br>*not belonging to the datacube of* $r_1$ | $DiffCube(r_2, r_1) = \{t \in CL(r) \mid t[f(\{\mathcal{M}\|*\})] \geq MinThreshold$ and $\nexists t' \in r_1 \mid t \preceq_g t'\}$. |
| Emerging cube | SELECT D, f({M\|*})<br>FROM r2<br>CUBE BY D<br>HAVING<br>f({M\|*})>=MinThreshold2<br>AND D NOT IN<br>(SELECT D<br>FROM r1<br>CUBE BY D<br>HAVING<br>f({M\|*})>MinThreshold1) | $f(\mathcal{M}\|*, r_2) \geq MinThreshold_2$<br>$f(\mathcal{M}\|*, r_1) < MinThreshold_1$ | $EmergingCube(r_2, r_1) = \{t \in CL(r_1 \cup r_2) \mid f(\mathcal{M}\|*, r_2) \geq MinThreshold_2$ and $f(\mathcal{M}\|*, r_1) < MinThreshold_1\}$. |
| Convex Cubes | SELECT D, f({M\|*})<br>FROM r<br>CUBE BY D<br>HAVING<br>cmc(t) AND camc(t) ; | montone constraint *cmc*()<br>anti-monotone constraint *camc*() | $ConvexCube(r) = \{t \in CL(r) \mid cmc(t)$ and $camc(t)\}$. |

## 5 Conclusion

In this paper, we review different declensions of the concept of data cube. With their compact representation and their efficient computation, such cubes are good candidates for multidimensional analysis of data stream. Actually, users of such dynamic applications are strongly interested by trend evolutions over time for reacting at these changes in real time.

We define a unifying structure, the convex cube, which is a formal and generic framework making it possible to characterize in a simple and sound way different variants of data cubes. The latter have frequently been seen as the result of queries or algorithms rather than concepts. We pay particular attention to confront these two visions. It results from this work an homogeneous characterization of the examined types of cubes, a didactic classification facilitating the user choice for the cube variant the most suitable, but above all a compact representation, soundly founded for the generic structure of the convex cube and applied to its specific declensions which are the iceberg, range, differential and emerging cubes.

For future work, we plan to study different ways to summarize emergent cubes by using cube closures and to extract emergent efficiently cuboids for a finer analysis in OLAP databases.

## References

Beyer, K., Ramakrishnan, R.: Bottom-Up Computation of Sparse and Iceberg CUBEs. In: Proceedings of the International Conference on Management of Data, SIGMOD, pp. 359–370 (1999)

Casali, A.: Mining borders of the difference of two datacubes. In: Proceedings of the 6th International Conference on Data Warehousing and Knowledge Discovery, DaWaK, pp. 391–400 (2004)

Casali, A., Cicchetti, R., Lakhal, L.: Cube lattices: a framework for multidimensional data mining. In: Proceedings of the 3rd SIAM International Conference on Data Mining, SDM, pp. 304–308 (2003)

Gray, J., Chaudhuri, S., Bosworth, A., Layman, A., Reichart, D., Venkatrao, M., Pellow, F., Pirahesh, H.: Data cube: A relational aggregation operator generalizing group-by, cross-tab, and sub-totals. Data Mining and Knowledge Discovery 1(1), 29–53 (1997)

Han, J., Kamber, M.: Data Mining: Concepts and Techniques. Morgan Kaufmann, San Francisco (2006)

Han, J., Chen, Y., Dong, G., Pei, J., Wah, B.W., Wang, J., Cai, Y.D.: Stream cube: An architecture for multi-dimensional analysis of data streams. Distributed and Parallel Databases 18(2), 173–197 (2005)

Lakshmanan, L., Pei, J., Han, J.: Quotient cube: How to summarize the semantics of a data cube. In: Proceedings of the 28th International Conference on Very Large Databases, VLDB, pp. 778–789 (2002)

Mitchell, T.M.: Machine learning. Computer Science. MacGraw-Hill, New York (1997)

Nedjar, S., Casali, A., Cicchetti, R., Lakhal, L.: Emerging cubes for trends analysis in OLAP databases. In: Proceedings of the 9th International Conference on Data Warehousing and Knowledge Discovery, DaWaK (2007)

Vel, M.: Theory of Convex Structures. North-Holland, Amsterdam (1993)

# Dependency Management for the Preservation of Digital Information

Yannis Tzitzikas

Computer Science Department, University of Crete, Greece, and
Institute of Computer Science, FORTH-ICS, Greece
tzitzik@ics.forth.gr

**Abstract.** The notion of dependency is ubiquitous. This paper approaches this notion from the perspective of digital information preservation. At first, an abstract notion of *module* and *dependency* is introduced. Subsequently, and for building preservation information systems, the notion of *profile* is proposed as a gnomon for deciding *representation information adequacy* (during input) and *intelligibility* (during output). Subsequently some general dependency management services for identifying and filling gaps during input and output are described and analyzed (also described as protocols that could be used in the communication between a preservation information system and information consumers and providers).

## 1 Introduction

The preservation of digital information is an important requirement of the modern society. Digital information has to be preserved not only against hardware and software technology changes, but also against changes in the knowledge of the community. According to the OAIS reference model [2], metadata are distinguished to various broad categories. One very important (for preservation purposes) category of metadata is named *Representation Information* (RI) which aims at enabling the conversion of a collection of bits to something useful. In brief, the RI of a digital object should comprise information about the Structure, the Semantics and the needed Algorithms for interpreting and managing a digital object. Figure 1 shows one corresponding part of the information model of OAIS.

In order to abstract from the various domain-specific and time-varying details, in this paper we model the RI requirements as *dependencies*. This view is very general and can capture a plethora of cases. Subsequently, we identify a set of core services for managing dependencies. These services aim at identifying the knowledge gaps (missing RI), and at computing and proposing ways to fill these gaps. These services can be used during both importing and exporting information (to and from a preservation information system). As different users (consumers or providers), or communities of users, have different characteristics (in terms of RI), we introduce the notion of DC (Designated Community) profile. Subsequently, we describe protocols (interaction schemes) that could be

R. Wagner, N. Revell, and G. Pernul (Eds.): DEXA 2007, LNCS 4653, pp. 582–592, 2007.

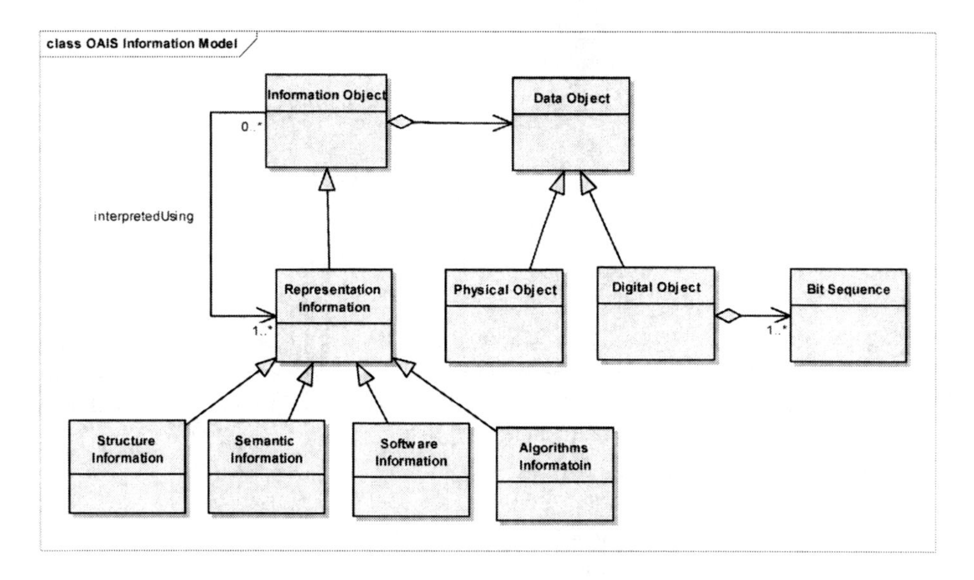


**Fig. 1.** The information model of OAIS

used in the communication between a preservation information system and information consumers and providers. This work can be exploited for building advanced preservation information systems and registries. Motivation for this work is the ongoing EU project CASPAR (FP6-2005-IST-033572) whose objective is to build a pioneering framework to support the end-to-end preservation lifecycle for scientific, artistic and cultural information.

The paper is organized as follows. Section 2 formalizes the notion of dependency and knowledge gap, while Section 3 describes interaction schemes for identifying and filling these gaps. Finally, Section 4 concludes the paper and identifies issues for further research.

## 2 Formalizing Dependencies

Let $Obj = \{o_1, \ldots o_n\}$ be set of all objects of the domain, e.g. the set of all data objects of an archive. Let $\mathcal{T}$ be the set of all *modules* (or components) that are needed for understanding/executing/managing the objects in $Obj$. We adopt a very general interpretation of the term module. It can be a software or hardware module. In addition, it could be a knowledge model expressed either formally or informally, explicitly or tacitly. For instance, it could be an English-To-Greek dictionary that is useful for a Greek-speaking person to understand a piece of text written in English. It could also be a ontology A (which could be expressed in RDF/S) that is useful for understanding the contents of a metadata file (expressed in RDF), or for understanding another ontology B (e.g. if B uses or specializes elements defined in A).

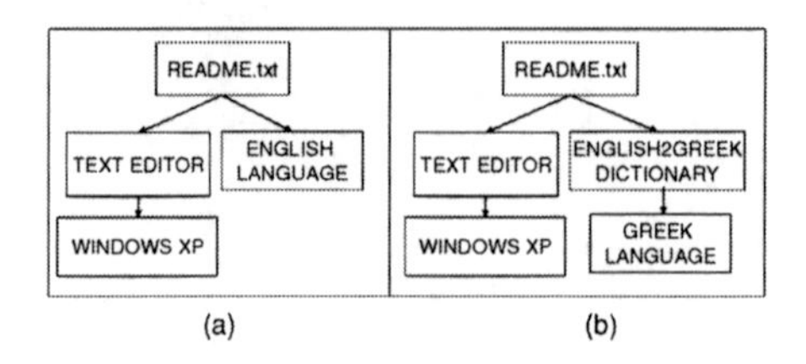


**Fig. 2.** Dependencies for a text file

There is dependency relation between modules in the sense that a module may require the availability of one or more other modules in order to function. We can model this as a graph $\Gamma = (\mathcal{T}, <)$. A relationship $t < t'$ means that $t'$ depends on $t$, e.g. it may mean that $t'$ cannot function without the existence of $t$. Below we describe some small examples (based on the needs of the CASPAR project). Figure 2(a) shows the dependencies of a text file written in English. However, a Greek-speaking consumer may define a dependency graph like the one illustrated in Figure 2(b).

FITS[1] is a standard data format that is used in astronomy. To understand such a file one needs to understand the FITS standard which is in turn described in a PDF document. To understand the keywords contained in a FITS file one needs to be able to understand the FITS dictionary (that explains the usage of keywords). Figure 3(a) illustrates these dependencies, while Figure 3(b) shows the dependencies of a digital object representing an interactive multimedia performance. Finally, an example of dependencies between formal knowledge expressed in the form of RDF Schemas is shown in Figure 4 (where fat arrows are used to denote dependencies between namespaces).

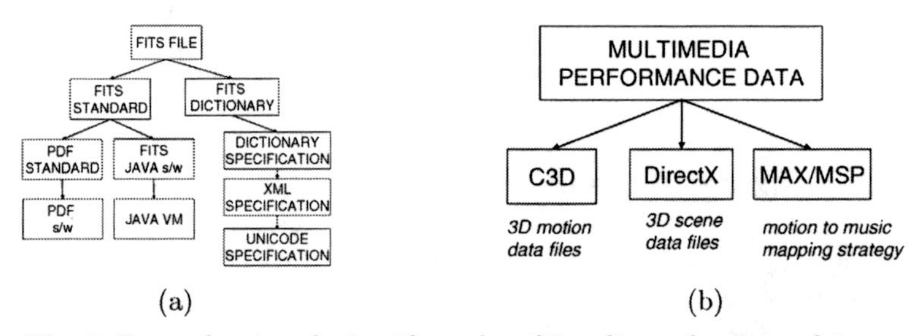


**Fig. 3.** Dependencies of scientific and multimedia performance data

A general remark is that there is no standard method for defining what a module is. For instance, we may have modules of various levels of abstraction. One module in one dependency graph could correspond to a large number of interconnected and interdependent finer modules in another dependency graph.

[1] http://fits.gsfc.nasa.gov/

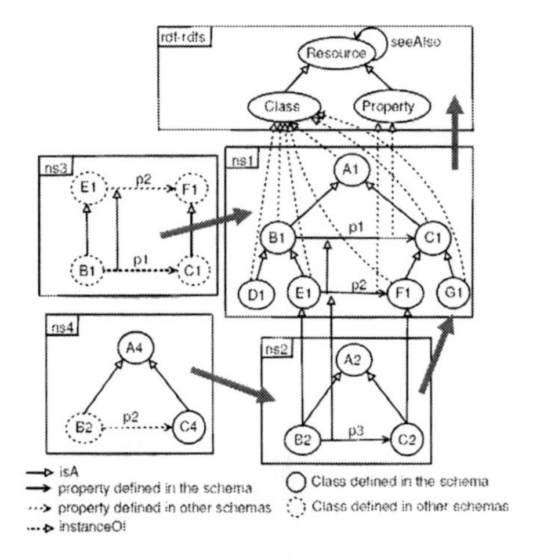


**Fig. 4.** Dependencies between RDF Schemas

For instance, the `WINDOWS XP` module in Figure 2 is actually the aggregation of several interconnected modules[2]. Hereafter we shall make the working assumption that the dependency graph is acyclic, i.e. it is a DAG. Equivalent modules (e.g. all editors that can read and edit ASCII texts) can be captured by assuming that each element of $\mathcal{T}$ is not atomic but it is a set of equivalent modules (this is like having disjunctive dependencies).

The intelligibility of a digital object, i.e. of an element in $Obj$, may require the existence of one or more modules in $\mathcal{T}$. We can model this by a binary relation $R$ ($R \subseteq Obj \times \mathcal{T}$). To keep notations simple we abuse notation and we will also use $<$ to denote $R$. So we can view all dependencies (among modules and between object and modules) as one graph $\Gamma = (T \cup Obj, <)$. For example, if the management of an object $o$ requires two modules $t_1, t_2$ where $t_2$ requires a module $t_3$ we can write $t_1 < o$, $t_2 < o$, $t_3 < t_2$. Table 1 introduces some notations that will be used in the sequel. The minimal elements of $\mathcal{T}$, i.e. the set $min(\mathcal{T})$, comprises the primitive modules which are assumed to be always available (e.g. an Operating System, a programming language, or the English vocabulary). However, probably nothing in this world is self-existent so the notion of primitive modules is actually a convention.

Regarding OAIS, we could say that the `interpretedUsing` relation of Figure 1 defines a plain dependency graph with the only difference that the nodes of this graph may be further specialized, i.e. classified under the indicative categories that are shown (e.g. `Algorithm, Semantics, Structure`, etc). In any case, the resulting object graphs would contain a dependency graph like the one we have introduced so far.

[2] Hierarchical clustered graphs could be probably used for modeling and formalizing the dependencies among modules of different granularity, but this goes beyond the scope of this paper.

**Table 1.** Notations

| Notation | Definition |
|---|---|
| $\mathcal{T}$ | the set of all modules and objects |
| $t$ | an element of $\mathcal{T}$ |
| $S$ | a subset of $\mathcal{T}$ |
| $t < t'$ | $t'$ depends on $t$ (in other words, $t'$ requires $t$) |
| $<^*$ | the transitive closure of $<$ |
| $min(S)$ | the minimal elements of $S$ w.r.t. $<^*$ |
| $max(S)$ | the maximal elements of $S$ w.r.t. $<^*$ |
| $Nr(t)$ | $\{ t' \mid t' <^* t\}$, i.e. all modules that $t$ requires |
| $Nr(S)$ | $\cup\{ Nr(t) \mid t \in S\}$ |

## 2.1 Intelligibility of Data Objects

Given an object or module $t$, we can define the required for understandability (or intelligibility) modules of $t$ as follows: $req(t) = Nr(t)$. If $S \subseteq \mathcal{T}$, then we can define $req(S) = \cup\{ req(t) \mid t \in S\}$. Let $u$ be an actor (e.g. user, or information consumer) and let $T_u$ be the modules available to him (e.g. software/hardware modules available at his computer or knowledge available at his/her mind), where $T_u \subseteq \mathcal{T}$. Now suppose that $u$ is given a set of objects $A$ ($A \subseteq Obj$). The set $A$ could be the answer of a query $q$ posed to an information system, or the result of browsing an information space, or the result of any other method (e.g. $u$ may have received the set $A$ by email). The prerequisites for understanding the set $A$ is $req(A)$. For example, consider the case illustrated in Figure 5 where $\mathcal{T} = \{t_1, \ldots, t_8\}$, $A = \{o_x, o_y\}$, and $T_u = \{t_3, t_6\}$. Since $T_u$ contains $t_6$ and none of its narrower modules $t_7$ and $t_8$, we can understand that $t_6$ is a primitive module for $u$. So we can safely make the assumption that $u$ knows $t_7$ and $t_8$. We can call this *unique module assumption (uma)*, meaning that each module is uniquely identified by its name and that its required modules are always the same. Here we have $req(o_x) = \mathcal{T}$ and $req(o_y) = \{t_3, t_6, t_7, t_8\}$. Also note that $max(req(o_x)) = t_1$ and $max(req(o_y)) = t_3$.

We can easily see that $u$ can understand an object $o$ if $max(req(o)) \subseteq T_u$. In the current example $u$ can understand $o_y$ because $max(req(o_y)) = t_3 \in T_u$, however $u$ cannot understand $o_x$ because $max(req(o_x)) = t_1 \notin T_u$.

## 2.2 Intelligibility Gaps

Consider the case of an object $o_x$ that is not understandable by $u$. In this case we can say that we have an intelligibility gap. To fill the gap, we need to find the missing modules. The set of missing modules that $u$ needs in order to understand an object $o$ are given by the formula: $Missing(o, u) = req(o) - Nr(T_u)$. In our example, $Missing(o_x, u) = req(o_x) - Nr(T_u) = \mathcal{T} - \{t_3, t_6, t_7, t_8\} = \{t_1, t_2, t_4, t_5\}$. Clearly, if $A \subseteq Obj$, then we can define $Missing(A, u) = \cup\{ Missing(o, u) \mid o \in A\}$.

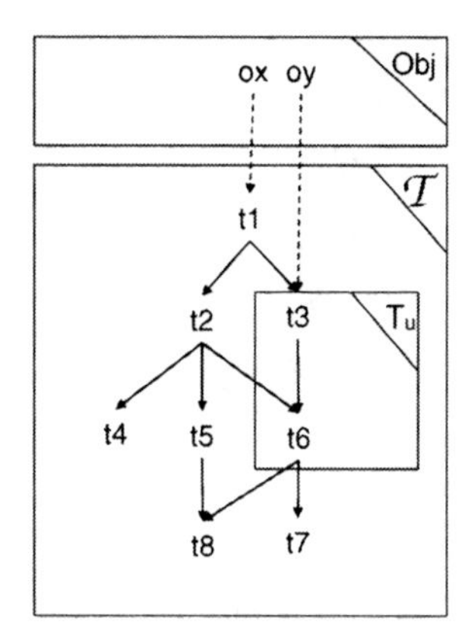


**Fig. 5.** Example of a dependency graph (between objects and modules)

Note that without the unique module assumption (uma), we could not make the assumption that $u$ knows $t_7$ and $t_8$. In that case we would have to define $Missing(o, u) = req(o) - T_u$. In our running example we would have $Missing(o_x, u) = req(o_x) - T_u = \mathcal{T} - \{t_3, t_6\} = \{t_1, t_2, t_4, t_5, t_7, t_8\}$. The relationships between two dependency graphs are specified formally below.

Consider an information provider $p$ and an information consumer $u$, each having a dependency graph $\Gamma_p$ and $\Gamma_u$ respectively.

**Definition 1.** Let $\Gamma_u = (T_u, <_u)$ and $\Gamma_p = (T_p, <_p)$ be two dependency graphs. We say that $\Gamma_u$ is *subgraph* of $\Gamma_p$, and we write $\Gamma_u \subseteq \Gamma_p$, if (a) $T_u \subseteq T_p$, and either (b1) $<_u \subseteq <_p$, or (b2) $<_u = <_{p|T_u}$.

Note that (b2) is more strict than (b1). Specifically, and regarding the relationships between the elements of $T_u$, (b1) ensures that $p$ has at least the relationships that $u$ has, while (b2) ensures that $p$ has exactly the relationships that $u$ has. For instance, $\Gamma_u$ of Figure 5 satisfies (b2). If $\Gamma_u$ did not contain the relationship $t_6 < t_3$ then it would satisfy only (b1). If $\Gamma$ did not contain the relationship $t_6 < t_3$ then it would not satisfy neither (b1) nor (b2). Note that: (b1) implies that $Nr_p(t) \supseteq Nr_u(t)$, $\forall t \in T_u$, while (b2) implies that $Nr_p(t) = Nr_u(t)$, $\forall t \in T_u$.

## 3 (Intelligibility-Aware) Interaction Schemes

Consider an information provider $p$ and an information consumer $u$. Here we describe various interaction methods that could be used for identifying and filling intelligibility gaps.

### 3.1 For Consuming (Delivering) Information

Without loss of generality we can assume a *query-and-answer* interaction scheme where $u$ sends to $p$ a query $q$ and $p$ returns a set of objects $A$. Below we describe some interaction schemes that enrich the query-and-answer interaction scheme with intelligibility-related concerns.

Note that given an object $o$ and a user $u$, for computing $Missing(o,u) = req(o) - Nr(T_u)$ one needs to be able to compute $req(o)$ and $Nr(T_u)$. If $req(o)$ or $Nr(T_u)$ are very large in size then this could cause inefficiencies (especially in a distributed setting). For this reason below we describe a number of options.

**Interaction Schemes with Fixed Number of Messages**

(A) $u$ submits a query, $p$ returns the answer with all modules that are required.
(1) $u \rightarrow p$: query($q$)
(2) $p \rightarrow u$: return($A, req(A)$)
Note that $req(A)$ does not necessarily return the modules themselves. It may return references to these modules which one could use in order to find the actual modules (e.g. for downloading and installing them). The user can identify the missing modules (i.e. those elements in $req(A)$ which are unknown to him) and proceed accordingly. However in practice $req(A)$ could be very large in size.

(B) $u$ submits her query and profile, $p$ returns answers accompanied by the missing modules.
(1) $u \rightarrow p$: query($q, T_u$)
(2) $p \rightarrow u$: return($A, Missing(A,u)$)
Note that if $T_u$ is smaller than $req(A)$ then this scheme is more efficient than (A). We can further improve the above scheme, specifically we can reduce the data that have to be exchanged, if $\Gamma_u \subseteq \Gamma_p$. In particular, in that case step (1) can be replaced by:
(1') $u \rightarrow p$: query($q, max(T_u)$)

(C) $u$ registers her profile once, $p$ returns answers accompanied by the missing modules. This scheme avoids sending the profile with each query. Instead, $u$ registers a (DC) profile $T_u$ once, which is then exploited in the subsequent query-and-answer interactions. Again the provider sends back the answer and the missing modules.
(1) $u \rightarrow p$: register($u, T_u$)
(2) $u \rightarrow p$: query($q$)
(3) $p \rightarrow u$: return($A, Missing(A,u)$)
We can further improve the above scheme, specifically we can reduce the data that have to be exchanged for the registration, if $\Gamma_u \subseteq \Gamma_p$. In particular, in that case, step (1) can be replaced by:
(1') $u \rightarrow p$: register($u, max(T_u)$)

**Progressive Interaction Schemes (with variable number of messages).** In some cases it might be useful (or efficient) to provide gradual/progressive methods for identifying and filling intelligibility gaps. Two such schemes are described bellow.

(Ai) This is a progressive version of scheme (A). Instead of sending $req(A)$, the provider at first sends only the maximal elements.

(1) $u \rightarrow p$: query($q$)
(2) $p \rightarrow u$: return($A, max(req(A))$)

The user can identify the missing modules (i.e. those elements in $max(req(A))$ which are unknown to her) and proceed accordingly. Note that $u$ could also ask again $p$ about the required modules of the elements of $max(req(A))$ and so on, i.e. the dialog could be continued as shown next. Below we use *recmsg* to denote the previously received message.

(3) $u$: repeat
(4) $u$: $M := recmsg - T_u$ // i.e. $M := max(req(A)) - T_u$
(5) $u$: If $M \neq \emptyset$ then
(6) $u \rightarrow p$: $getDirectReqsOf(M)$
(7) $p \rightarrow u$: return($max(req(recmsg))$)
(8) $u$: until $M = \emptyset$

For instance, in our running example the formula $max(req(o_x)) - T_u$ returns the highest missing module, i.e. $t_1$. The entire sequence of $M$'s is shown below:

$M_1$: $t_1$ ( $= max(req(o_x)) - T_u$)
$M_2$: $t_2$ ( $= max(req(t_1)) - T_u$)
$M_3$: $\{t_4, t_5\}$ ( $= max(req(t_2)) - T_u$)
$M_4$: $\{t_8\}$ ( $= max(req(t_5)) - T_u$)
$M_5$: $\emptyset$

Note that $t_8$ could be already known to $u$ as it is narrower than $t_6 \in T_u$.

(D) $u$ submits only the query, $p$ returns only the answer.
Here $u$ sends to $p$ only $q$. If $u$ cannot understand the result, she can send to $p$ what she did not understand. With the assumption that each object has links to its direct required modules, $u$ can identify the direct missing modules and send these to $p$ and continue in this way until reaching to her primitive elements or getting all elements of $req(A)$.

(1) $u \rightarrow p$: query($q$)
(2) $p \rightarrow u$: return($A$)
(3) $u$: repeat
(4) $u$: $M := computeDirectReqsOf(recmsg) - T_u$
(5) $u$: If $M \neq \emptyset$ then
(6) $u \rightarrow p$: $getDirectReqsOf(M)$
(7) $p \rightarrow u$: return($max(req(recmsg))$)
(8) $u$: until $M = \emptyset$

### 3.2 For Providing (Ingesting) Information

A preservation system could follow a policy of the form: the dependencies of the stored objects should be known and stored. This means that the submission of information, e.g. the submission of an object or module $t$, to the system should be accompanied by adequate representation information. In other words $req(t)$

should be known. However as there is not any objective method for deciding whether $req(t)$ is complete or not (may nothing is complete in the strict sense) we can again use the notion of profile in order to decide whether the submitted RI is complete or not (with respect to a specific profile or with respect to all profiles known by the preservation system).

As one can imagine, the provision (ingestion) of information has many similarities with the consumption (delivery) of information. We could capture the ingestion of information by changing the previously described interaction schemes. Specifically we could ignore the query submission step and consider that the user $u$ is the preservation system who wants to ingest the set of objects $A$ that $p$ sends to $u$. Fore reasons of space their detailed description is omitted.

### 3.3 Complex Objects and Other Technicalities

Let us for example consider the case of Web pages. Consider a digital file named *a.html*. The extension of the filename gives us a hint about the type of the digital object, so we may write $type(a.html) = HTML$, and as $a.html > HTML$, we may generalize and consider that for every $o \in Obj$, it holds $o > type(o)$, if $type(o)$ is known. However, an html page is a text that may contain pointers to other types of data (images, sounds, etc). In order to obtain this content, we need a HTML parser. So we could say that $computeDirectReqsOf(a.html)$ needs the availability of an HTML parser[3]. Consequently, $computeDirectReqsOf(o)$ could be as follows: $computeDirectReqsOf(o) = type(o) \cup type(o).parse(o).getContents()$. To compute all required modules of an object we have to continue analogously.

## 4 Concluding Remarks

Dependencies are ubiquitous and dependency management is an important requirement that is subject of research in several (old and new emerged) areas, from software engineering [6,7,8,1] to ontology engineering [3,5]. In software engineering the various build tools (e.g. make, gnumake, nmake, jam, ant) are definitely related, as well as the problems of installability, deinstallability and maintainability. Recall that the art of large-scale design is to minimize dependencies (recall Model Driven Architecture). However we could say that the preservation of the intelligibility of digital objects requires a generalization (or abstraction) able to capture also non software modules (e.g. explicit or implicit domain knowledge). The agenda of ontology engineering includes similar in spirit problems, e.g. the problem of how to reflect a change of an ontology to the dependent ontologies (i.e. to those that reuse or extend parts of it), which may be stored in different sites, as well as the schema evolution problem, i.e. the problem of reflecting schema changes to the underlying instances.

[3] As another example, for a .java named file we need to parse the file in order to extract all import statements, while for a .rdf named file, we need to parse it in order to extract the namespaces it uses.

A modern preservation system should be generic, i.e. able to preserve heterogeneous digital objects which may have different interpretation of the notion of dependency. The dependency relations should be specializable and configurable (e.g. it should be possible to associate different semantics to them). Focus should be given on finding, recording and curating the dependencies. For example, the `makefile` of an application program is not complete for preservation purposes. The preservation system should also describe the environment in which the application program (and the make file) will run. Recall the four worlds of an information system (Subject World, System World, Usage World, Development World) as identified by Mylopoulos [4]. Finally, the provision of notification services for risks of loosing information (e.g. obsolescence detection services) is important.

The contribution of this paper lies in specifying a generic view by adopting an abstract notion of module and dependency and by introducing the notion of DC profile. Subsequently it specified a number of core services around these notions, allowing to check and control whether the ingestion of information is complete and for computing the minimum extra information required to be delivered to ensure the intelligibility of a digital object by the consumer. Based on these services a number of interaction schemes for identifying and filling the intelligibility gaps were presented. A proof-of-concept prototype based on Semantic Web technologies has already been built. The benefits of adopting Semantic Web languages, for the problem at hand, is that although the core dependency management services need to know only a very small core ontology (defining the abstract notion of module and dependency), it is possible to refine (specialize) the dependency relation.

Issues for further research include (a) extending the framework with converters (for tackling migration/emulation), (b) studying the effects of changes in the dependency graphs (and what kind of notification services are required), and (c) studying composite modules and dependencies of different granularity.

**Acknowledgements.** This work was partially supported by the EU project CASPAR (FP6-2005-IST-033572). Many thanks to David Giaretta and the rest "CASPARtners".

## References

1. Franch, X., Maiden, N.A.M.: Modeling Component Dependencies to Inform their Selection. In: 2nd Intern. Conf. on COTS-Based Software Systems (2003)
2. International Organization For Standardization: OAIS: Open Archival Information System – Reference Model (2003), Ref. No ISO 14721:2003
3. Jarrar, M., Meersman, R.: Formal Ontology Engineering in the DOGMA Approach. In: International Conference on Ontologies, Databases and Applications of Semantics (ODBase), pp. 1238–1254 (2002)
4. Mylopoulos, J., Borgida, A., Jarke, M., Koubarakis, M.: Telos: Representing Knowledge about Information Systems. ACM Transactions on Information Systems, 8(4) (October 1990)

5. Sunagawa, E., Kozaki, K., Kitamura, Y., Mizoguchi, R.: An Environment for Distributed Ontology Development Based on Dependency Management. In: Fensel, D., Sycara, K.P., Mylopoulos, J. (eds.) ISWC 2003. LNCS, vol. 2870, pp. 453–468. Springer, Heidelberg (2003)
6. Vieira, M., Dias, M., Richardson, D.J.: Describing Dependencies in Component Access Points. In: Proceedings of The 23rd International Conference on Software Engineering (ICSE'01), Toronto, Canada, pp. 115–118 (2001)
7. Vieira, M., Richardson, D.: Analyzing dependencies in large component-based systems. ASE, 00:241 (2002)
8. Walter, M., Trinitis, C., Karl, W.: OpenSESAME: An Intuitive Dependability Modeling Environment Supporting Inter-component Dependencies. In: Procs. of 2001 Pacific Rim International Symposium on Dependable Computing, pp. 76–83 (2001)

# Constraints Checking in UML Class Diagrams: SQL vs OCL

D. Berrabah and F. Boufarès

LIPN, Paris13 University,
99 avenue J.B.Clément, 93430 Villetaneuse - France
{db,boufares}@lipn.univ-paris13.fr

**Abstract.** Numerous CASE tools are used for applications analysis and design. These tools often do not take into account all the information (structures and constraints) given in a conceptual level. So, the elements obtained at the physical level do not completely coincide with the conceptual elements. Consequently, some semantics are lost. Our goal, in this paper, is to give rules to translate some constraints not taken into account in the processes used to translate the conceptual schema. In object databases, these constraints are expressed in OCL while they are expressed, in relational databases, using active mechanisms. Consequently, these constraints are checked during databases updates.

**Keywords:** conceptual schema, constraint translation, SQL, OCL.

## 1 Introduction

The Unified Modeling Language UML [16] has become the most commonly used and most powerful formalism for applications analysis and design. Numerous CASE tools which are available on the market, such as Power AMC and Rational Rose [22 and 17], produce a very flexible environment for UML diagrams modeling. These tools do not answer all the designer's needs: many constraints are neither well defined nor translated or checked. In other word, in database (DB) design methodologies [8 and 23], the processes used to translate a UML class diagram (database conceptual schema (DBCS)) into a target database schema (TDBS) left aside the constraint satisfaction problem. The elements obtained during these processes do not completely coincide with the conceptual elements, thus bringing about some semantic losses [3]. This problem often arises when most of the constraints established in the DBCS and that reflect the universe of discourse are not translated correctly. Moreover, it should be noted that if the conceptual diagram is not valid (i.e. it contains conflicts) then this translation is useless. For instance, an XOR constraint defined on two associations can be in contradiction with a minimum multiplicity constraint equals to 1 and in addition it remains unchecked. Many kinds of constraints are well known nowadays, but little is done for possible conflicts between them. We think that adding a new step in DB design process is necessary. In this step, possible conflicts which may be generated by constraints must be detected, localized and corrected [2, 3, 5 and 4]. If the

R. Wagner, N. Revell, and G. Pernul (Eds.): DEXA 2007, LNCS 4653, pp. 593–602, 2007.

DBCS is valid, it can be translated in a specific language. Many transformation processes which concern the formal and/or non-formal translations were presented in literature. Among these translations, structure properties in a UML class diagram can be expressed through UML basic structures and the Object Constraint Language (OCL) [9]. It is also possible to translate UML diagrams into formal specification Z or B [11, 12, 20 and 21]. Up today, using case tools, only some multiplicity constraints are considered. Within SQL [8] some constraints must be translated using procedures and triggers. Unfortunately, there are no rules to generate these procedures and triggers. Consequently, they are not generated using CASE tools. Our major aim is to study both categories. First, we deal with the expression of the highest number of constraints in a target language and second we study their coherence.

In this paper, we give a comparison study on participation constraints (PC) defined on binary associations and those defined on generalization/specialization associations. The general idea is to define and express these constraints using a constraints specification language. We use event-condition-action (ECA) rules and OCL [16 and 25] to translate these constraints. This study aims to give the designer the possibility of a total coherence control of constraints in order to deal with the creation and maintenance of databases.

The structure of this paper is as follows. Section 2 synthesizes the basic principles of constraints and their role in preserving the semantics of the universe of discourse. Section 3 shows how to introduce and express participation constraints on binary associations and generalization/specialization associations with a constraint specification language. Our approach is based on OCL and trigger-based SQL scripts which are represented as ECA rules. Finally, the paper ends with a conclusion and perspectives.

## 2 Constraints

A constraint is a condition or a semantic restriction expressed in a linguistic instruction form using a textual language (Fig.1). Generally, a constraint is linked to one or more elements of the DBCS (property, class, association …). It represents semantic information associated with these elements. Constraints can have the same name but not the same signification. For instance, the participation constraints defined on classical binary associations are different from those defined on inheritance links. Some constraints are inherent to the DBCS whereas others require an explicit definition. The graphic elements offered by CASE tools do not allow expressing the most of the constraints such as dependency and participation constraints. In addition, not all the expressed constraints are translated. For instance, multiplicity constraints in a..b form ($a>1$ and $b< \infty$) are not translated in the TDBS.

Sometimes, constraints are checked using declarative constraints. Unfortunately, this solution does not always hold. Consequently, more powerful systems based on OCL or triggers have to be used. OCL is a formal specification language used to express constraints in UML or other object oriented languages. The kind of constraints which can be expressed using OCL includes invariants on the structures of the DBCS. Thus those expressions must be true for all instances of those structures.

Triggers constitute a good mean to implement referential actions. On the common database management systems (DBMS), it is necessary to use triggers to perform actions other than those defined by default. Though triggers are found in the majority of DBMS, unfortunately the execution models of the triggers vary from one DBMS to another. There are some principle components which are valid for all the systems and generally these do not change. Trigger is expressed by ECA rules [6, 7 and 10]. It is activated during DB transition state.

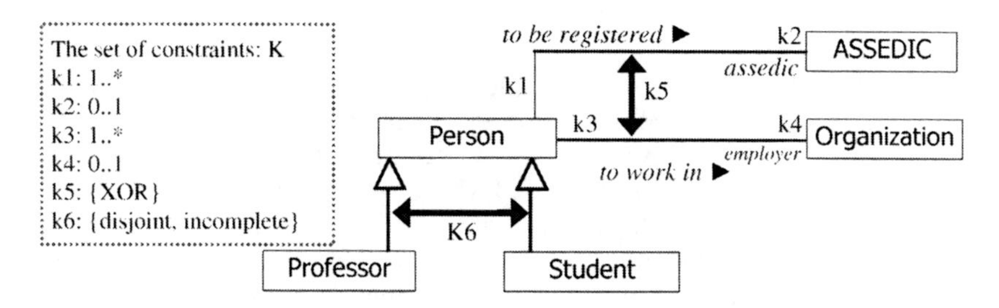


**Fig. 1.** Human resources management schema

The example of the Fig.1 represents the management of human resources. The set K of constraints defined on this DBCS is composed by two subsets of constraints, multiplicity ({ki, i=1,4}) and participation constraints ({ki, i=5,6}). The last one contains an XOR and a disjoint PCs. The XOR PC (k5) ensures that each person must be registered in ASSEDIC[1] or work in an organization but not both. The disjoint PC (k6) guarantees that a person may be either a professor or a student but not both. She/he may be neither. We have shown in [3] that the DBCS is valid if and only if the set of all its constraints is coherent. Otherwise the translation of the DBCS to a TDBS has not to be done. For instance, if k2 is equal to 1 then the constraint XOR has no signification.

## 3 Transformation Rules of Participation Constraints

Participation constraints generally refer to conditions on links between objects of a class and objects of two or more others classes. In this section, we study two kinds of participation constraints: those defined on classical binary associations and those defined on generalization/specialization links (inheritance links). In the latter, every object can belong to the generalization as well as to its specialization. Each special object has a link with exactly one general object but the reverse does not necessarily hold. Thus, the special objects indirectly have features of the more general objects.

Generally, Inheritance links are considered as a one-to-one association type. Unfortunately, this representation brings about semantic loss. Consequently, it will be strengthened with an OCL or a SQL additional constraint (Fig.2). This constraint

[1] **ASSEDIC** : "**ASS**ociation pour l'**E**mploi **D**ans l'**I**ndustrie et le **C**ommerce" means Organization for Employment in Industry and Trade.

must ensure that each special object has a link with exactly one general object but the reverse does not necessarily hold.

The additional constraint is expressed in OCL as follow,

```
Context Specialization inv:
Self.general→notEmpty()
Self.general→size()=1

Context generalization inv:
Self.special→forAll(s1,s2| s1<>s2 implies s1.general<>s2.general
```

In relational model, the translation of both kinds of links shown below strongly depends on the multiplicity constraints. Indeed, structures and multiplicity constraints translations are done simultaneously. However, the translation of the other constraints is done later (but must be done). Many structures translations rules were proposed in the literature. We retain the basic ones and which we need in this paper. So, any class is transformed into a table with a primary key; association with a maximum multiplicity equal to 1 is represented in the form of a foreign key; association which does not contain a maximum multiplicity equal to 1 is represented by a table.

The translation of DBCS constraints into a TDBS must be done in a specific language. The following subsections show how PC translations are given in OCL and SQL languages. The aim of this study is first, to remove ambiguities on definitions of these two categories of constraints; secondary, to show how to express them in OCL and SQL and then check them; and finally, to give their common points.

### 3.1 Checking PCs on Generalization/Specialization Associations

PCs defined on generalization/specialization links are mainly divided into disjoint and complete constraints. A disjoint constraint specifies whether two objects of different specializations may be related to the same object of the generalization. A complete constraint specifies whether objects of the specifications are related to all generalization-objects (Fig.2).

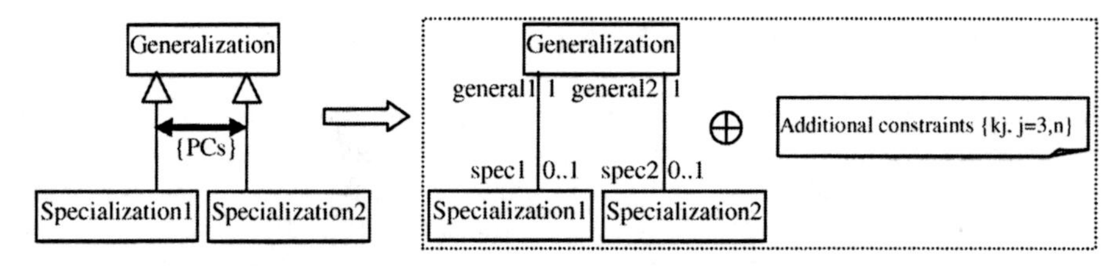


**Fig. 2.** PCs on generalization/specialization relationship

- **A disjoint constraint**

A PC may be disjoint if each general object has a link to at most one special object. In the reverse case, this constraint is said to be overlapping. To check the disjoint constraint, we must ensure that each general object is a member of at most one specialization. This is expressed in object model by adding an OCL constraint using Empty and notEmpty properties as follow.

```
Context g: generalization
inv: g.spec1→notEmpty implies g.spec2→Empty

Context g: generalization
inv: g.spec2→notEmpty implies g.spec1→Empty
```

In relational model, this constraint must be checked during the maintenance of data. To do so, an ECA rule that events are insertion or update of an object in specialization is generated. The action of this rule is rejecting the operation if that object already exists in the other specialization. Note that ECA rules are easily translated within SQL.

```
ECA
event: insert or update on Specialization1
condition: new value of object to be insert or update exists in
           Specialization2
action: reject operation
```

- **A complete constraint**

A PC defined on a generalization/specialization association is complete if each general object has a link to at least one special object, otherwise it is said to be incomplete. This can be maintained by adding OCL constraints which state that if a general object is not related to any object in one specialization then it must be related to an object in the other.

```
Context g: generalization
inv: g.spec1→Empty implies g.spec2→notEmpty

Context g: generalization
inv: g.spec2→Empty implies g.spec1→notEmpty
```

In relational model, three ECA rules are needed. The first one is to check the PC during insertion into the generalization and the others during deletion or update in specializations. These two later ECA have the same definition.

```
ECA1
event: insert on Generalization
condition: none
action: insert obligatory object in Specialization1 or Specialization2

ECA2
event: delete or update on Specialization1
condition: old value of object to be deleted or updated do not exists in
           Specialization2
action: reject operation
```

- **A disjoint and complete constraint**

A PC defined on a generalization/specialization association may be disjoint and complete at the same time. In this case each general object must be related to at least one special object and at most to one object in both specializations. This case joins together the two preceding cases, from where it is possible to use additional constraints used to express disjoint and complete PCs. That assertion can also be expressed in OCL using the logic operator xor.

```
Context g: Generalization inv:
Self.allInstances→forAll(g| g.spec1→notEmpty xor g.spec2→notEmpty)
```

When the TDBS is represented in relational model, the ECA rules generated are as follow.

```
ECA1
event: insert on Generalization
condition: none
action: insert obligatory object in one in only Specialization

ECA2
event: delete on Specialization1
condition: none
action: delete object from generalization or insert it in
          Specialization2

ECA3
event: update on Specialization1
condition: none
action: delete new object from Specialization2 if it exists there or
            reject operation
         delete old object from Generalization or insert it in
            Specialization2
```

## 3.2 Checking PCs on Binary Associations

PCs on binary associations are also called interrelation constraints. They frequently relate to the coexistence of occurrences of class objects in one or several associations (Fig.3). They can also be defined by the number of times an object of a class can participate to a connected association set. More detailed definitions of this constraint category are given by [14, 18, 2 and 15].The introduction of these constraints into a DBCS must be taken into account in order to preserve the semantics of the real world. Consequently, they must be translated in the TDBS.

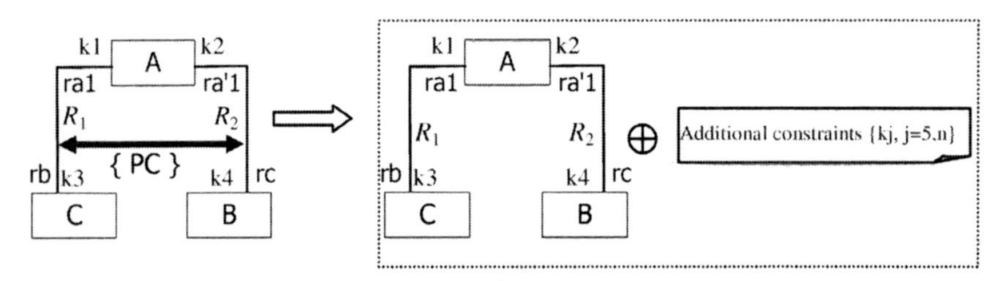


**Fig. 3.** Translation of participation constraints on binary associations

PCs control the link of objects of the same class to objects of different classes via associations. We talk about inclusion PC (for instance $R_1$ is included in $R_2$) when an A-object participating in $R_1$ must participate in $R_2$. The simultaneity PC states that an A-object must participate in both associations, simultaneously or not. We also find the exclusion PC which ensures that an A-object participates either in $R_1$ or $R_2$ or in neither of them. The totality PC means that all A-objects must participate in $R_1$ or $R_2$ or both at the same time. Finally, joined together exclusion and totality PCs define so called XOR PC which state that all A-objects must participate but only in one association. It seems that PCs on binary associations are similar to those defined on generalization/specialization associations. Though there is a great resemblance in the definition, the semantics is not the same. PCs on binary associations must also be expressed and checked. This task strongly depends on multiplicity constraints in relational model since the multiplicity constraints decide whether the participation of A-objects may appear in $R_1$ or $R_2$.

For instance, XOR constraint defined on two associations expresses that any object of a class connected to these two associations must take part in only one of the associations. It has obviously the same definition as that of the {disjoint, compete} constraint. Thus, OCL additional constraints used to express the later can be used to express the XOR constraint. The following OCL constraint can also be used.

```
Context A inv:
Self.allInstances→forAll(a |
   B.allInstances → forAll (b |
      C.allInstances → forAll (c |
         (b.ra1 → excludes(a) or c.ra'1 → excludes(a))
         and
         (b.ra1 → includes(a) or c.ra'1 → includes(a)) )))
```

```
Context A inv:
Self.allInstances→forAll(a |
   B.allInstances → forAll (b |
      C.allInstances → forAll (c |
         (b.ra1 → excludes(a) and c.ra'1 → includes(a)) or
         (c.ra'1 → excludes(a) and b.ra1 → includes(a)) )))
```

This constraint can also be expressed using the xor operator and Empty property. This is done starting from an A-object and navigating the R1 and R2 associations to refer the sets of B-objects and C-objects linked to the A-object via these associations. Consequently, one and only one of these sets must be empty and the other must not.

```
Context A inv:
Self.allInstances→forAll(a | a.rb→notEmpty xor a.rc→notEmpty)
```

In relational model, as said above, expressing XOR constraint strongly depends on the multiplicity constraint. Thus, following ECA rules are defined in general forms considering that $R_1$ and $R_2$ are translation tables of associations $R_1$ and $R_2$ respectively. In practice case, associations $R_1$ and $R_2$ are translated by foreign keys in the table A, the table B or the table C or also by creating new tables according to the multiplicity constraints.

```
ECA1
event: insert on A
condition: none
action: insert obligatory object in one of R1 or R2

ECA2
event: insert on R1
condition: if object in R2
action: reject operation

ECA3
event: delete or update on R1
condition1: if (deleting or updating) and old object is unique in R1
action1: insert this object in R2 or reject operation
condition2: if new object exist in R2
action2: delete this object from R2 or reject operation
```

Two other ECA rules are generated on $R_2$ which have the same definition as ECA2 and ECA3.

**Example**

In Fig.1, the XOR participation constraint states that a person must work in an organization or she/he must be registered at an ASSEDIC, but she/he does not have the right to do both. This constraint can be expressed, as mentioned above, using the definition of {disjoint, complete} PC defined on generalization/specialization association as follow:

```
Context p: Person inv:
Self.allInstances→forAll(p| p.assedic→notEmpty xor
                            p.employer→notEmpty)
```

In relational model, since $R_1$ and $R_2$ belong to the one-to-many association type, the participation of all A-objects in these associations appear in the table A because $R_1$ and $R_2$ are translated by the primary keys of the tables B and C respectively in A as foreign keys. Consequently, checking the XOR constraint consists in checking the values of the foreign keys which refer the tables B and C. To do so, a trigger must be generated. This one ensures that the values of these foreign keys are different and one of them is null. The trigger statement is as follow.

```
Create trigger Insert_Update_Person
Before insert or update on Person
For each row
DECLARE Const_viol EXCEPTION;
Begin
  If NEW.FK_ASS<>NEW.FK_ORG
      and (NEW.FK_ASS is null XOR NEW.FK_ORG is null)
  Then true else RAISE Const_viol;
  End If;
EXECPTION
 When Const_viol then OUTPUT.PUT_LINE('XOR constraint violated');
End Insert_Update_Course;
```

### 3.3 Recapitulative

Some times OCL and SQL must be used at the same time. For instance, in the DBCS of Fig.1, if a person is a student then she/he can not be registered in ASSEDIC. This condition must be expressed in textual language. Consequently, OCL is needed to express it. In addition, if this DBCS is translated into a relational TDBS then OCL and SQL are combined. Another solution can be done checking automatically OCL constraints using a DBMS-based approach [13].

Whatever the language used, using the one-to-one association type to represent generalization/specialization links, PCs defined on the latter can be treated in the same way that those defined on classical binary associations. The following table (Tab.1) gives for participation constraints defined on generalization/specialization associations their equivalent one on binary associations and equivalent cases to check them.

**Table 1.** Comparison of PCs defined on generalization/specialization and binary associations

| PCs defined on generalization /specialization associations | The same definition as that of: | Is checked as: |
|---|---|---|
| 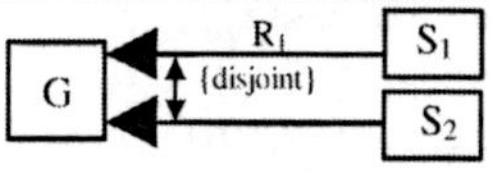 G, $R_1$, {disjoint}, $S_1$, $S_2$ | Exclusion | 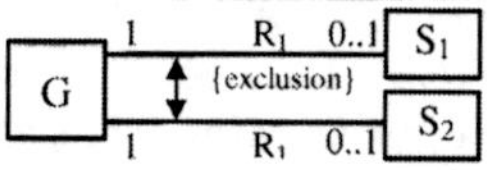 G, 1, $R_1$, 0..1, $S_1$, {exclusion}, 1, $R_1$, 0..1, $S_2$ |
| 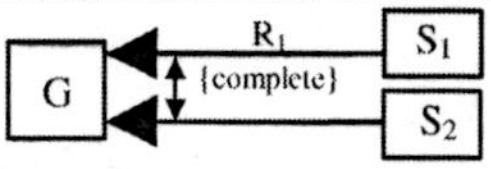 G, $R_1$, {complete}, $S_1$, $S_2$ | Totality | 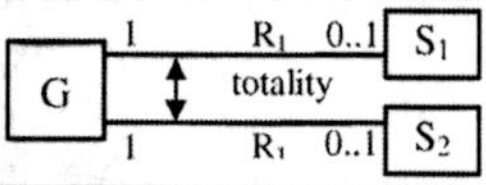 G, 1, $R_1$, 0..1, $S_1$, totality, 1, $R_1$, 0..1, $S_2$ |
| 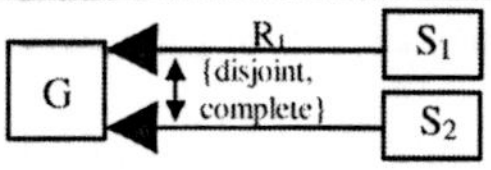 G, $R_1$, {disjoint, complete}, $S_1$, $S_2$ | XOR | 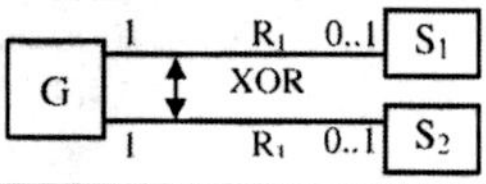 G, 1, $R_1$, 0..1, $S_1$, XOR, 1, $R_1$, 0..1, $S_2$ |

## 4 Conclusion

In this paper, we reported a systematic study of the use of participation constraints to specify assertions defined on the behavior of class object participations. Sometimes, it is necessary to use these constraints in a DBCS to satisfy customer requirements. Our aim is to remove any ambiguity from the definition of PCs. Though PCs have the same definition on binary associations as well as on generalization/specialization ones, their semantics is not the same. We translate the two categories of PCs using OCL and trigger-based SQL additional constraints to cover object and relational models. Thus, we provide a general framework for transforming PCs.

We are completing our prototype of data modeling by integrating the approach developed in this paper. This prototype, first checks the coherence of constraints defined in the DBCS [3, 4] then translates all constraints in a specific language [2]. It is also very useful for information system integration and building Datawarehoses.

## References


1. Al-Jumaily, H.T., Cuadra, D., Martinez, P.: Plugging Active Mechanisms to Control Dynamic Aspects Derived from the Multiplicity Constraint in UML. In: The workshop of 7th International Conference on the Unified Modeling Language, Portugal (2004)
2. Berrabah, D., Boufarès, F., Ducateau, C.F.: Analysing UML Graphic Constraint, How to cope with OCL. In: 3rd International Conference on Computer Science and its Applications, California (2005)
3. Berrabah, D., Boufares, F., Ducateau, C.F., Gargouri, F.: Les conflits entre les contraintes dans les schémas conceptuels de Bases de Données: UML – EER. Journal of Information Sciences for Decision Making, Special Issue of the 8th MCSEAI'04 19, 234 (2005)
4. Berrabah, D.: Etude de la cohérence globale des contraintes dans les bases de données. Ph. D. Thesis report, Laboratory CRIP5, Paris 5 University (December 2006)
5. Boufarès, F.: Un outil intelligent pour l'analyse des schémas EA. Interne Report. Informatics Laboratory of Paris Nord, University of Paris 13 France (2001)

6. Ceri, S., Widom, J.: Deriving production rules for constraint maintenance. In: Proc. of the 16th International Conference on Very Large Data Bases, pp. 566–577. Brisbane, Australia (1990)
7. Cochrane, R.J., Pirahesh, H., Mattos, N.M.: Integrating triggers and declarative constraints in SQL database systems. In: Proceedings of the 22nd International Conference on Very Large Data Bases, Mumbai, India, pp. 567–578 (1996)
8. Eisenberg, A., Melton, J., Kulkarni, K., Michels, J., Zemke, F.: SQL: 2003 has been published. ACM SIGMOD Record 33(1) (March 2004)
9. Gogolla, M., Richters, M.: Expressing UML Class Diagrams Properties with OCL. In: Object Modeling with the OCL, pp. 85–114. Springer, Heidelberg (2002)
10. Horowitz, B.: Intermediate states as a source of non deterministic behavior in triggers. In: 4th International Workshop on Research Issues in Data Engineering: Active Database Systems, Houston TX, pp. 148–155 (February 1994)
11. Laleau, A., Mammar, A.: Overview of method and its support tool for generating B from UML notations. In: Proceeding of 15th international conference on Automated Software Engineering, Grenoble, France (2000)
12. Ledru, Y., Dupuy, S.: Expressing dynamic properties of static diagrams. In: Z. Conference of Approches Formelles dans l'Assistance au Développement de Logiciels, Rennes, France (2003)
13. Marder, U., Ritter, N., Steiert, H.-P.: A DBMS-based Approach for Automatic Checking of OCL Constraints. In: OOPSLA'99-Workshop Rigorous Modeling and Analysis with the UML: Challenges and Limitations. Denver, Co. (1999)
14. Matheron, J.P.: Approfondir Merise. Tome1. Edition Eyrolles (1991)
15. Nanci, D., Espinasse, B.: Ingénierie des systèmes d'information: Merise deuxième génération. 4th edn. Edition-Vuibert (2001)
16. OMG, editor: UML 2.0., http://omg.org
17. Rational: http://www-306.ibm.com/ software/ rational/ sw-bycategory/ subcategory/ SW710.html
18. Rochfeld, A., Negros, P.: Relationship of relationships and other inter-relationship links in ER model. Data and Knowledge Engineering 9, 205–221 (1993)
19. Rumbaugh, J., Jacobson, I., Booch, G.: UML 2.0 Guide de Référence, Edition Campus Press (2004)
20. Shroff, M., France, R.B.: Towards a Formalization of UML Class Structures. In: Z. 21st IEEE Annual international computer Software and Applications Conference, pp. 646–651 (1997)
21. Soon-Kyeong, K., Carrington, D.: A formal mapping between UML models and Object-Z specifications. In: Bowen, J.P., Dunne, S., Galloway, A., King, S. (eds.) B 2000, ZUM 2000, and ZB 2000. LNCS, vol. 1878, pp. 2–21. Springer, Heidelberg (2000)
22. Sybase: http://www.sybase.com/products/information management/powerdesigner
23. Toby, J.T.: Database Modeling & Design, 3rd edn. Data Management Systems. Morgan Kaufmann, San Francisco (1999)
24. Truongm, N.T., Souquières, J.: Validation des propriétés d'un scénario UML/OCL à partir de sa dérivation en B. Conference: Approches Formelles dans l'Assistance au Développement de Logiciels, Besançon, France, pp. 99–114 (2004)
25. Warmer, J., Kleppe, A.: The Object Constraint Language: Getting Your Models Ready for MDA. 2nd edn. Paperback-Edition (2003)

# XML-to-SQL Query Mapping in the Presence of Multi-valued Schema Mappings and Recursive XML Schemas

Mustafa Atay, Artem Chebotko, Shiyong Lu, and Farshad Fotouhi

Department of Computer Science
Wayne State University
Detroit, Michigan 48202 USA
{matay, artem, shiyong, fotouhi}@wayne.edu

**Abstract.** Several query mapping algorithms have been proposed to translate XML queries into SQL queries for a schema-based relational XML storage. However, existing query mapping algorithms only support single-valued mapping schemes, in which each XML element type is mapped to exactly one relation, and do not support multi-valued mapping schemes, in which each XML element type can be mapped to multiple relations. In this paper, we propose a generic query mapping algorithm, *ID-XMLToSQL*, for a schema-based relational XML storage. To the best of our knowledge, our algorithm provides the first generic solution to the XML-to-Relational query mapping problem that is applicable to both single-valued and multi-valued mapping schemes. Moreover, our algorithm also provides an elegant solution to the query mapping problem in the presence of recursive XML schemas and recursive queries. While existing algorithms need special recursion operators, our algorithm only requires the traditional relational operators and thus, can work with all relational databases.

## 1 Introduction

Numerous researchers propose to use relational databases for storing and querying XML documents in order to get benefits of this mature technology. This approach requires algorithms to map XML schemas, documents and queries, into their relational equivalents.

An XML-to-SQL query mapping algorithm for a schema-based relational XML storage should respect the underlying XML-to-Relational schema mapping scheme. The XML-to-Relational schema mapping schemes in the literature can be classified into the following two categories:

- *Single-valued Schema Mappings.* In a single-valued schema mapping, an XML element or attribute type is mapped into exactly one single relation in the target relational schema. Thus, it shows the characteristics of a function. The *Shared* schema mapping approach introduced in [1] and *ODTDMap* approach introduced in [2] fall into this category.

R. Wagner, N. Revell, and G. Pernul (Eds.): DEXA 2007, LNCS 4653, pp. 603–616, 2007.

- *Multi-valued Schema Mappings.* In a multi-valued schema mapping, an XML element or attribute type can be mapped into more than one relation in the target relational schema. The multi-valued schema mappings do not show the characteristics of a function and thus they are harder to deal with. The *Basic* and *Hybrid* schema mapping approaches proposed in [1] fall into this category.

Although there are several query mapping algorithms for single-valued schema mapping schemes, there is no published query mapping algorithm which supports multi-valued schema mapping schemes to our best knowledge. Therefore, we propose a generic query mapping algorithm which supports both multi-valued and single-valued schema mapping schemes in this paper.

Our generic algorithm also provides an elegant solution to the XML-to-Relational query mapping problem in the presence of recursive XML schemas and recursive queries. This problem is identified as an important practical problem in the literature [3,4,5]. Recursive XML schemas are common in practice as pointed out by [6] in which 35 DTDs found to be recursive out of 60 real-world DTDs. On the other hand, recursive XML queries, which include descendant axis '//', are also common in practice.

The challenge of XML-to-SQL query mapping is that, when there is recursion both in an XML query and in its underlying XML schema, there might be infinitely many paths corresponding to the given recursive XML query. There are two elegant algorithms [4,5] in the literature which address this issue. These algorithms solve the recursion within the relational engine by using special SQL operators which are not supported by some RDBMSs. On the other hand, we solve the recursion at XML schema level without using special SQL operators.

The main contributions of this paper include the following:

1. We propose a generic query mapping algorithm, *ID-XMLToSQL*, for a schema-based relational XML storage scheme. To the best of our knowledge, our algorithm provides the first generic solution to the XML-to-Relational query mapping problem that is applicable to all relational XML storage mapping schemes proposed in the literature, including both single-valued and multi-valued schema mapping schemes.
2. We propose to convert a cyclic XML schema graph to a directed acyclic graph by unfolding the cycles in the XML schema graph to facilitate the recursive query mapping process. Thus, we can find out a finite number of matching paths on the generated acyclic graph for an arbitrary XML query including the recursive ones. Therefore, our proposed query mapping technique can be implemented on any RDBMS as it does not require using special SQL operators to capture the recursion while the existing algorithms need special recursion operators.

*Organization:* The rest of the paper is organized as follows. Section 2 gives a summary of related work. We give a motivation on generic query mapping in Section 3. Section 4 includes all necessary preliminaries for our generic query mapping algorithm. The outline of our proposed query mapping algorithm *ID-XMLToSQL* is given in Section 5. We demonstrate a performance study of the algorithm *ID-XMLToSQL* in Section 6. Finally, Section 7 concludes the paper and points out some potential future work.

## 2 Related Work

In order to query XML data stored in a relational database, one should map the XML queries into relational queries based on the underlying XML-to-Relational schema mapping scheme. Hence, we can split the XML-to-Relational query mapping algorithms into the following two categories based on the underlying schema mapping schemes:

- *Schema-less Query Mapping.* There has been a lot of work on schema-less query mapping [7,8,9,10,11,12,13,14]. In this approach, XML schema is considered to be missing or not used and a generic relational schema is generated for all XML documents. Then, a given XML query is mapped to its relational equivalent using the generic relational schema.
- *Schema-based Query Mapping.* There have been several works on schema-based query mapping [1,15,4,5,16,17,18] where an XML schema is provided and used to generate a good relational schema. The generated relational schemas vary according to the input XML schemas. Therefore, an XML-to-Relational query mapping algorithm should know and respect the underlying XML-to-Relational schema mapping to generate correct and efficient relational queries.

The problem of mapping recursive XML queries in the presence of recursive schemas studied in schema-less query mapping space [8,10]. However, their query mapping algorithms are not applicable to the schema-based query mapping space. Recently, two elegant approaches proposed in [4,5] to map recursive XML queries to their relational equivalents in the presence of recursive XML schemas.

The query mapping algorithm of [4] first derives a query graph for an input path query from the XML schema graph. Then, it partitions the query graph into strongly-connected components and generates an SQL query for each component. If a component is recursive, then, the recursion in this component is captured in the corresponding SQL query by using the *with* construct of SQL'99.

The query mapping algorithm of [5] first rewrites a given XPath query into a regular XPath expression which is capable of capturing recursion both in a DTD and in an XPath query. Furthermore, they provide an algorithm for translating regular XPath expressions to relational queries using least fixpoint (LFP) operator. The LFP operator is used to capture the recursion in the queries.

However, these recursive query mapping algorithms are not generic enough to be used with multi-valued mappings such as Basic and Hybrid introduced in [1]. Moreover, they require the usage of special SQL operators such as *with* construct of SQL'99 or LFP operator which are not supported by some RDBMSs. Our proposed *ID-XMLToSQL* algorithm overcomes these limitations.

## 3 Motivation

A generic query mapping algorithm for a schema-based relational XML storage is supposed to work with a general class of XML-to-Relational schema mappings which can be classified into two main categories as *Single-valued Schema Mappings* and *Multi-valued Schema Mappings.*

Surprisingly, there is no published XML-to-Relational query mapping algorithm in the schema-based XML storage space which is generic enough to work with the multi-valued XML-to-Relational schema mappings. The recursive query translation algorithm of [4] handles a general class of single-valued XML-to-Relational mappings. The main query translation procedure SQL() in [4] uses the function Annot() to find out the relation/column corresponding to an XML element. Neither Annot() nor SQL() support the multi-valued XML-to-Relational schema mapping. Thus, [4] is not generic enough to handle all types of mappings proposed in the literature. While the *RegToSQL* algorithm proposed in [5] supports a broad class of XPath queries, it still lacks the support for multi-valued schema mappings.

A single-valued mapping is a function which returns only one relation for an input XML element/attribute type. The target relation to retrieve an XML element or attribute can easily be determined from a single-valued mapping. Thus, single-valued mappings are relatively easier to handle during the query mapping phase.

A multi-valued mapping is not a function since it can return multiple relations for an input XML element/attribute type. This situation may cause ambiguity while a query mapping algorithm is trying to locate the target relation for an XML element type to retrieve its data. Hence, a query mapping algorithm based on a multi-valued mapping should be intelligent enough to resolve this possible ambiguity and find out the correct relation(s) to access. Thus, it is more challenging to map XML queries to relational queries under multi-valued mapping schemes than under single-valued mapping schemes.

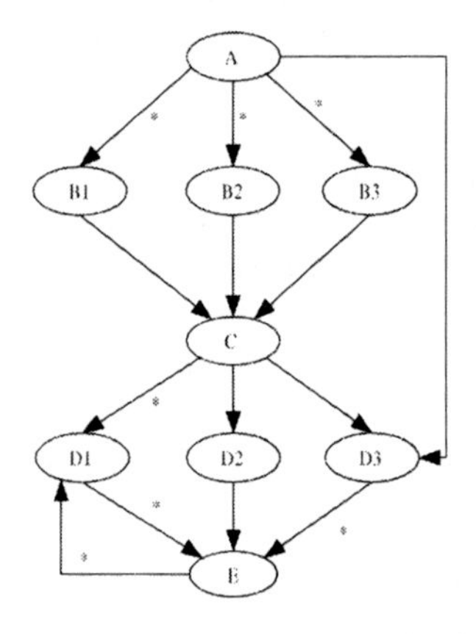


**Fig. 1.** A Sample XML Schema Graph

**Table 1.** Single-valued and Multi-valued Schema Mapping Examples

| Single-valued $\sigma$-mapping (Shared) | |
|---|---|
| **Element** | **Relation** |
| A | A |
| B1 | B1 |
| B2 | B2 |
| B3 | B3 |
| C | C |
| D1 | D1 |
| D2 | C |
| D3 | D3 |
| E | E |

(A)

| Multi-valued $\sigma$-mapping (Hybrid) | |
|---|---|
| **Element** | **Relation** |
| A | A |
| B1 | B1 |
| B2 | B2 |
| B3 | B3 |
| C | B1, B2, B3 |
| D1 | D1 |
| D2 | B1, B2, B3 |
| D3 | A, B1, B2, B3 |
| E | E |

(B)

We use a data structure to store XML-to-Relational schema mapping information. We call this data structure as $\sigma$-mapping and formally define it in Section 4.1. The $\sigma$-mappings based on *Shared* and *Hybrid* approaches for the XML schema shown in Figure 1 are given in Table 1.A and Table 1.B, respectively. We assume the XML attribute types are mapped to the same relation with their parent element types.

*Example 1.* If the XPath expression /A/B1/C/D3 is given against the XML schema graph shown in Figure 1, following will be its SQL equivalent based on a typical query mapping algorithm which generates a SQL query by joining all the relations along a path:

Select T4.ID
From $\sigma(A)$ T1, $\sigma(B1)$ T2, $\sigma(C)$ T3, $\sigma(D3)$ T4
Where T1.ID=T2.parentID And T2.ID=T3.parentID And T3.ID=T4.parentID

While it is trivial to find out the matching relations in this SQL query based on the single-valued $\sigma$-mapping given in Table 1.A, it is not straightforward to find out them in case of the multi-valued $\sigma$-mapping shown in Table 1.B. For instance, it is not clear which relation should be returned for $\sigma$(C) out of the set {B1,B2,B3} and for $\sigma$(D3) out of the set {A,B1,B2,B3}.

We propose the notion of *path-based $\sigma$-mapping* ($\sigma_p$-mapping) in Section 4.2 to resolve the ambiguity due to the multi-valued schema mapping schemes by the help of input path structure and the existing mapping information.

# 4 Preliminaries

## 4.1 Schema-Based Query Mapping

In schema-based relational XML storage, query mapping typically takes an XML query, an XML schema, the XML-to-Relational schema mapping information, which is called $\sigma$-mapping, and a database as input, produces a relational query, runs it against the database where the XML document is stored, and returns the query results as output. In the following, we formalize the notions of $\sigma$-mapping and query mapping:

**Definition 1 ($\sigma$-Mapping).** *Given an XML schema $S$ with element-type set $E$ and attribute-type set $A$, and a database schema $R$, a $\sigma$-mapping is a mapping $\sigma : (E \cup A) \rightarrow R$, such that given an attribute/element type $e$, $\sigma(e)$ is the set of relations in which the instances of $e$ will be stored.*

**Definition 2 (Query Mapping).** *A* query mapping *$QM$ is a function that assigns to each tuple $(Q, S, X, R, B, \sigma)$ a relational query $Q'$, where $Q$ is an XML query, $S$ is an XML schema, $X$ is an XML document conforming to $S$, $R$ is a database schema, $B$ is a database of $R$, $\sigma$ is a mapping from $S$ to $R$, and $Q'$ is a set of relational queries equivalent to $Q$ such that $Q'(B) \equiv Q(X)$.*

### 4.2 $\sigma_p$-Mapping

We propose to define a path-based $\sigma$-mapping ($\sigma_p$-mapping) to resolve the mapping ambiguity that arises in the presence of multi-valued schema mappings. The $\sigma_p$-mapping uses the information obtained from the path structure and $\sigma$-mapping to find a single relation for each element in the input path. Once $\sigma_p$-mapping of a particular path expression is computed, then the equivalent relational query can be constructed without any ambiguity concern.

**Lemma 1.** *Any edge in an XML schema graph $G$ is identified either as a normal-edge or a *-edge.*

*Proof.* If an element can occur at most once under its parent, then it is connected to its parent by an edge labeled by ',' or '?' in XML schema graph $G$. All the edges in $G$ labeled by ',' and '?' operators constitute normal-edges. If an element can occur more than once under its parent, then this element is connected to its parent by an edge labeled by '*' or '+' in $G$. All the edges in $G$ labeled by '*' and '+' operators constitute *-edges. Since there is no occurrence operator other than {',', '?', '*', '+'} in $G$, any edge in an XML schema graph is either a normal-edge or a *-edge.

In the following, we formalize the notions of *simple path expression* and $\sigma_p$-mapping:

**Definition 3 (Simple Path Expression).** *A simple path expression $p$ can be denoted as $/n_1/n_2/.../n_k$ where each $n_i$ is the node type of step $i$ and the axis of each step is child axis '/' which denotes parent-child relationship. The node type $n_1$ represents the root element of the XML document and $k$ represents the number of steps in $p$.*

**Definition 4 ($\sigma_p$-Mapping).** *Given an input simple path $p = /e_1/e_2/.../e_n$, $\sigma$-mapping $\sigma$, and an XML schema graph $G$, $\sigma_p(e_i)$ is defined as follows where $i = 1, 2, ..., n$:*

$$\sigma_p(e_i) = \begin{cases} \sigma(e_i), & \text{if } |\sigma(e_i)|=1 \\ e_i, & \text{if } |\sigma(e_i)| > 1 \text{ and } (e_{i-1}, e_i) \text{ is a *-edge in } G \\ \sigma_p(e_{i-1}), & \text{if } |\sigma(e_i)| > 1 \text{ and } (e_{i-1}, e_i) \text{ is a normal-edge in } G \end{cases}$$

*Example 2.* If the XPath expression p=/A/B1/C/D3 is given based on the XML schema graph shown in Figure 1, the below $\sigma_p$-mapping is produced by computing the $\sigma_p$ based on the multi-valued schema mapping shown in Table 1.B:

| $\sigma_p$ | |
|---|---|
| Element | Relation |
| A | A |
| B1 | B1 |
| C | B1 |
| D3 | B1 |

**Theorem 1 (Correctness).** *Given an input simple path expression* $p = /e_1/e_2/.../e_n$*,* $\sigma_p(e_i)$ *returns the correct and single target relation for every element* $e_i$ *in* $p$*, where* $i = 1, 2, ..., n$*.*

*Proof (Sketch).* First, $\sigma_p(e_i)$ returns the same relation as $\sigma(e_i)$ if the input element $e_i$ is mapped to a single relation.

Second, if the input element $e_i$ is mapped to multiple relations, then the type of the edge between $e_i$ and its parent $e_{i-1}$ is checked from the XML schema graph. If the edge is a *-edge, then the $\sigma_p(e_i)$ returns the relation $e_i$ since $e_i$ is mapped to a separate relation as it occurs multiple times under its parent.

Third, if the input element $e_i$ is mapped to multiple relations and the type of the edge between $e_i$ and its parent $e_{i-1}$ is a normal-edge, then the $\sigma_p(e_{i-1})$ is called to determine the target relation for $e_i$ since it is mapped to the same relation as its parent $e_{i-1}$. Recursive call to $\sigma_p(e_{i-1})$ stops whenever a single relation is returned. If all the edges from $e_1$ to $e_{i-1}$ are normal-edges, then the recursion is going to stop at $\sigma_p(e_1)$ since $e_1$ is the root element and it is always mapped to the single relation $e_1$.

All the edges in an XML schema graph fall into either normal-edge or *-edge categories as it follows from Lemma 1. As a result, $\sigma_p(e_i)$ returns the correct and the single relation corresponding to element $e_i$.

Besides multi-valued mappings, the $\sigma_p$-mapping can deal with single-valued schema mappings where it returns the same values as $\sigma$-mapping. Therefore, $\sigma_p$-mapping is sufficient to develop a generic XML-to-Relational query mapping algorithm in the presence of multi-valued schema mappings as well as single-valued schema mappings.

### 4.3 Unfolded XML Schema Graph

The challenge with translating recursive XML queries over recursive XML schemas is to identify the infinite number of matching paths in the XML schema graph. However, if we unfold the recursive XML schema based on the maximum levels of depths for each cycle in the schema graph, we can find out a finite number of matching paths for an arbitrary XML query including the recursive ones. This observation leads us to an elegant and efficient solution for the problem of translating recursive XML queries in the presence of recursive XML schemas.

We propose to convert a cyclic XML schema graph to a directed acyclic graph by unfolding the cycles in the original schema. This new schema is called *unfolded*

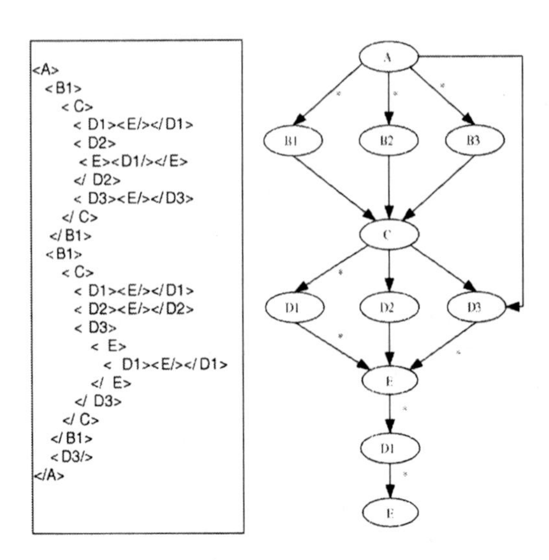
<A>
<B1>
< C>
< D1><E/></D1>
< D2>
< E><D1/></E>
</ D2>
< D3><E/></D3>
</ C>
</B1>
<B1>
< C>
< D1><E/></D1>
< D2><E/></D2>
< D3>
< E>
< D1><E/></D1>
</ E>
</ D3>
</ C>
</B1>
<D3/>
</A>
A
B1
B2
B3
C
D1
D2
D3
E
D1
E

**Fig. 2.** A Sample XML Document and its Unfolded XML Schema Graph (UXG)

*XML schema graph (UXG).* A UXG of a sample XML document, which conforms to the XML schema graph given in Figure 1, is shown in Figure 2. The formal definition of UXG is given in Definition 5.

**Definition 5 (Unfolded XML Schema Graph (UXG)).** *Given an XML schema $S$, unfolded schema of $S$ is a directed acyclic graph $UXG = (V, E, d_1, ...d_k)$, where $V$ is the set of vertices, $E$ is the set of edges, each $d_i$ is the maximum level of depth for each cycle $c_i$ in $S$ and $k$ denotes the number of cycles in $S$. Each cycle $c_i$ in $S$ is unfolded to depth $d_i$ in $UXG$ in top-down topological order. The vertices represent element types in $S$, and the edges represent their parent-child relationships. Each vertex is labeled with the name of the corresponding element type. An edge is labeled by '*' if it is incident to a vertex which can appear more than once under its parent in the corresponding XML documents, otherwise no label is used.*

A recursive XML schema $S$ can be converted into a non-recursive one in the form of a UXG $G$ by unfolding the recursion in $S$ with a finite number of occurrences of recursion that is decided from the XML documents $X$ stored in the database, such that $X$ conforms to $S$ and $G$ at the same time. In other words, $S$ and $G$ are equivalent to each other with respect to $X$.

We can create a UXG by using one of the following two approaches:

- *Static approach.* The maximum depth of each cycle in the XML schema graph is determined by the help of a domain expert and a fixed UXG is generated during the schema mapping phase. This fixed UXG is used during the query mapping regardless of the structure of underlying XML documents.
- *Dynamic approach.* The maximum depth of each cycle in the XML schema graph is initialized to 1 and a default UXG is generated during schema

mapping phase. When a new XML document is loaded to the database during the data mapping phase, the maximum depth of each cycle in the current document is found and UXG is modified if any current depth value is greater than the existing one.

Static UXG approach does not have any computation overhead during the data mapping phase. However, it may return unnecessary matching paths for a given recursive XML query. On the other hand, dynamic UXG approach associates some computational cost during the data mapping phase to maintain the UXG for minimizing the total number of matching paths for the input recursive XML queries.

The UXG graph is constructed either during the schema mapping phase or the data mapping phase. We assume bulk data is loaded to the database system first, then it is queried next in a batch-processing fashion. Therefore, the construction of UXG does not introduce additional overhead to XML-to-Relational query mapping performance since it is precomputed before query mapping phase.

## 5 ID-Based Generic Query Mapping

All the schema-based approaches proposed for XML-to-Relational query mapping in the literature have used ID-based techniques as in [4,5]. In ID-based techniques, each element is associated with a unique ID and the tree structure of the XML document is preserved by maintaining a foreign key to the parent which we call *parentID*. Each child axis '/' is translated into an equijoin between child and parent elements over their parentID and ID columns in ID-based techniques.

We propose a generic ID-based XML-to-Relational query mapping algorithm, *ID-XMLToSQL*, in this section. An outline of *ID-XMLToSQL* is given in Figure 3. The *ID-XMLToSQL* algorithm first identifies all the matching simple paths $p_i$ and $\sigma_p$-mappings $\sigma_{p_i}$ corresponding to those paths when a path expression $P$ and a UXG $G_u$ is given. Then it calls the SQL generation procedure SPathToSQL() for each simple path $p_i$ along with its mapping $\sigma_{p_i}$, and then, gets the union of the output SQL queries. We formalize the notion of a path expression as follow:

**Definition 6 (Path Expression).** *A path expression $P$ can be denoted as $a_1n_1a_2n_2...a_kn_k$ where each $n_i$ is a node type and each $a_i$ is either child axis '/' or descendant axis '//'. Each $a_in_i$ constitutes a navigation step of $P$ and $k$ is the number of steps in $P$.*

A naive XML-to-SQL query mapping procedure follows a blindfold approach. It takes an input simple path expression and generates a relational query by joining the relations corresponding to each step in the simple path expression. A sample SQL query generated using naive query mapping approach is given in Example 1.

When consecutive elements in a simple path expression are mapped to the same relation, then the naive approach unnecessarily joins the same relation

```
00 Algorithm ID-XMLToSQL
01 Input: Path Expression P, UXG G_u
02 Output: SQL query sql
03 Begin
04   Let p_i, i=1,2,...,n, be the set of all matching simple paths of P in G_u
05   Let σ_{p_i} be σ_p-mapping for the simple path p_i, i=1,2,...,n
06   sql=∅
07   sql = ∪_{i=1}^{n} SPathToSQL(p_i,σ_{p_i})
08   Return sql
09 End

00 Procedure SPathToSQL(Simple Path Expression p, σ_p-Mapping σ_p)
01 Begin
02 Use σ_p to cluster p = /e_1/e_2/.../e_m according to Definition 7
03 FromClause="From"
04 WhereClause="Where"
05 For i=1 to m do /* Construct From Clause */
06   If e_i is the first element of a cluster then
07     FromClause += "$σ_p(e_i)"
08   End If
09 End For
10 For i=2 to m do /* Construct Where Clause */
11   If e_i is the first element of a cluster then
12     WhereClause += "$σ_p(e_{i-1}).(e_{i-1}.ID) = σ_p(e_i).(e_i.parentID)"
13   End If
14   If e_i is neither first nor last element of a cluster then
15     WhereClause += "$σ_p(e_i).(e_i.ID) is not null"
16   End If
17 End For
18 sql="Select $σ_p(e_m).(e_m.ID)" + FromClause + WhereClause
19 Return sql
20 End
```

**Fig. 3.** ID-based Query Mapping Algorithm ID-XMLToSQL

with itself multiple times. For the simple path expression and its $\sigma_p$-mapping given in Example 2, corresponding SQL query will include two unnecessary self joins since the elements of last three steps in the path are mapped to the same relation.

An intelligent XML-to-SQL query mapping algorithm should be able to recognize the elements mapped to the same relations and avoid the unnecessary self-join operations. We deal with this issue in SPathToSQL() procedure. The outline of SPathToSQL() procedure is shown in Figure 3. The SPathToSQL() procedure identifies the clusters in a path expression which are the groups of elements in consecutive navigation steps mapped into the same relation. The SPathToSQL() procedure recognizes each cluster in a simple path expression and only joins the relation corresponding to the last element of a cluster to the relation corresponding to the first element of its successor cluster. Thus, it avoids the self-join problem of a blindfold query mapping approach. The notion of a cluster is formalized as follows:

**Definition 7 (Cluster).** *Given a simple path expression p and a mapping $\sigma_p$ over p, the elements of consecutive steps in p which are mapped to the same relation constitute a cluster. Hence, p can be denoted as a sequence of clusters* *such that* $p = c_1 c_2 ... c_k$ *where each* $c_i$ *is a cluster and* $k$ *is the number of clusters in* $p$.

The SPathToSQL() procedure given in Figure 3 first constructs the From clause at lines 05-09. It introduces one relation per cluster to the From clause since all the elements in a cluster are mapped to the same relation. The Where clause is constructed at lines 10-17. A transition from one cluster to another in the input path is handled at lines 11-13. A predicate of the form $\sigma_p(e_{i-1}).(e_{i-1}.ID) = \sigma_p(e_i).(e_i.parentID)$ joining the last element of the previous cluster to the first element of current cluster is added to the Where clause. As a result, the relations representing all the neighboring cluster are joined. The SPathToSQL() procedure adds an existential predicate of the form $\sigma_p(e_i).(e_i.ID)$ *is not null* for the intermediate elements of a cluster to the Where clause (lines 14-16) as it skips the intermediate elements in a cluster. Thus, it ensures that the middle elements of a cluster co-exist with the elements at each end of the cluster in the underlying XML document. The output SQL query is constructed and returned at lines 18-19.

The existential predicate *not null* is not introduced for the elements at each end of a cluster since they are already included within the join conditions of the output SQL query. Although the last element in a path expression may not be used in a join condition, we do not need to check the existence of the last element as it is used in the Select clause. We do not need to check the existence of the first element of a simple path expression, which is the root element, as all the simple paths start from the root element.

*Example 3.* If the path expression /A/D3//E is given against the UXG shown in Figure 2 and input to *ID-XMLToSQL* algorithm, *ID-XMLToSQL* calls SPathToSQL() procedure with the following simple paths identified from the UXG: (i) /A/D3/E and (ii) /A/D3/E/D1/E and, their $\sigma_p$-mappings: (i) {(A,A), (D3,A), (E,E)} and (ii) {(A,A), (D3,A), (E,E), (D1,D1), (E,E)}, respectively. Below is the generated output SQL query by our *ID-XMLToSQL* algorithm:

```
Select E.ID
From A, E
Where A.D3.ID=E.parentID
UNION ALL
Select E.ID
From A, E T1, D1, E T2
Where A.D3.ID=T1.parentID And T1.ID=D1.parentID and D1.ID=T2.parentID
```

**Theorem 2 (Time Complexity).** *The time complexity of the procedure SPathToSQL is* $O(n)$ *where* $n$ *is the number of steps in an input simple path expression* $p$.

*Proof (Sketch).* The statement at line 02 navigates $p$ once to cluster it and can be evaluated in O(n). The loop at lines 05-09 navigates $p$ once to construct the From clause and is evaluated in O(n). The loop at lines 10-16 navigates $p$ once to construct the Where clause and is executed in O(n). Thus, the time complexity of SPathToSQL() is O(n).

## 6 Experimental Study

We compare the performance of our *ID-XMLToSQL* algorithm and the recursive query translation algorithm SQLGen of [4] in this section. We used a Pentium IV computer with 2.4 GHz processor and 1 GB main memory for the experiments. The experiments were run using the Java software development kit. We minimized the usage of system resources during the experiments to get more realistic results. We ran the programs 6 times and got the average value, excluding the first run, to have more accurate results.

We used *auction.xml* document of XMark benchmark [19] as our data set to compare the performance of our proposed *ID-XMLToSQL* algorithm and SQLGen algorithm of [4]. The DTD of XMark includes several cycles, and thus, it is an appropriate XML schema for our experiments.The number of elements in the test XML document is 73,740.

We selected nine queries with particular features for the test suit. Our test query suit is shown in Table 2. All the queries in our test suit are recursive queries as they contain descendant axis '//'. All the queries return the elements which are included in a cycle in the XML schema. While the queries Q1, Q8 and Q9 include clusters of two or more elements, the queries Q2, Q3, Q5, Q7, Q8 and Q9 include shared elements which have more than one parents in the XML schema.

We implemented only a single-valued schema mapping scheme to run the two query mapping algorithms *ID-XMLToSQL* and SQLGen as SQLGen does not support multi-valued schema mapping schemes. We used a commercial relational DBMS which allows the usage of advanced SQL'99 *with* clause as it is centric to the algorithm of SQLGen. We measured the response time for each test query by running the queries generated by two algorithms separately. The experimental results are shown in Figure 4. We used logarithmic scale to increase the readability of the chart.

As can be seen from the chart, our *ID-XMLToSQL* algorithm outperformed the SQLGen algorithm in all the test queries. The main reasons for the performance difference between *ID-XMLToSQL* and SQLGen include the followings:

**Table 2.** Query Suit for Testing

| Query | Query Definition |
|---|---|
| Q1 | /site/categories/category/description//parlist |
| Q2 | //text |
| Q3 | //parlist |
| Q4 | //asia//listitem |
| Q5 | //item//listitem |
| Q6 | //asia//parlist |
| Q7 | //item/parlist |
| Q8 | /site/regions/asia/item//parlist |
| Q9 | /site/regions/asia/item//listitem |

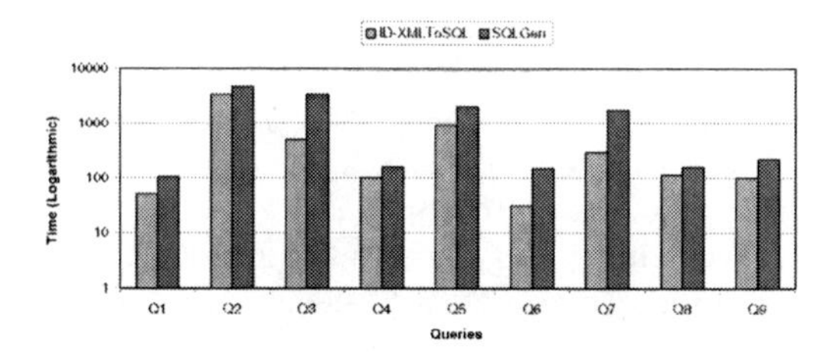


**Fig. 4.** Experimental Results for Query Mapping

- *ID-XMLToSQL* resolves the recursion at the XML schema level using pre-computed unfolded XML schema graph unlike SQLGen which resolves it inside the relational engine using recursive SQL query.
- The queries generated by SQLGen are typically more complex and larger than the ones generated by our *ID-XMLToSQL*.
- *ID-XMLToSQL* uses the notion of clustering and avoids unnecessary self-joins.

# 7 Conclusions and Future Work

In this paper, we proposed the generic XML-to-SQL query mapping algorithm ID-XMLToSQL which can be used with multi-valued schema mappings as well as with single-valued schema mappings. ID-XMLToSQL uses our proposed path-based $\sigma_p$-mapping technique to find the target relation for a given element of a path query in the presence of multi-valued schema mappings.

We proposed to convert a cyclic XML schema graph to an acyclic one by unfolding the cycles in the graph to a maximum level of depth. Thus, we are able to map the recursive XML queries over the unfolded XML schema graph to SQL queries without using special operators to capture the recursion. Therefore, our proposed query mapping algorithm can be used on any RDBMS as it uses standard SQL features unlike other recursive query mapping algorithms in the literature.

We compared the performance of our ID-XMLToSQL algorithm to SQLGen algorithm of [4] and observed that ID-XMLToSQL outperformed SQLGen for all the queries in our test suit. We consider augmenting our proposed ID-based generic query mapping algorithm with interval-based and path-based mapping schemes as a potential future work.

# Acknowledgment

The authors would like to thank Rajasekar Krishnamurthy for providing the source code of SQLGen algorithm and his cooperation, and Dapeng Liu for involving in the implementation of our ID-XMLToSQL algorithm.

## References


1. Shanmugasundaram, J., Tufte, K., Zhang, C., He, G., DeWitt, D.J., Naughton, J.F.: Relational databases for querying XML documents: Limitations and opportunities. In: VLDB, pp. 302–314 (1999)
2. Atay, M., Chebotko, A., Liu, D., Lu, S., Fotouhi, F.: Efficient schema-based XML-to-Relational data mapping. Information Systems Journal 32(3), 458–476 (2007)
3. Krishnamurthy, R., Kaushik, R., Naughton, J.F.: XML-to-SQL query translation literature: The state of the art and open problems. In: XML Database Symposium (2003)
4. Krishnamurthy, R., Chakaravarthy, V.T., Kaushik, R., Naughton, J.F.: Recursive XML schemas, recursive XML queries, and relational storage: XML-to-SQL query translation. In: Proc. of the 20th International Conference on Data Engineering, Boston, pp. 42–53 (March 2004)
5. Fan, W., Yu, J.X., Lu, H., Lu, J., Rastogi, R.: Query translation from XPath to SQL in the presence of recursive DTDs. In: Proc. of the 31sh VLDB Conference, Trondheim, Norway (2005)
6. Choi, B.: What are real DTDs like. In: WebDB Workshop (2002)
7. Deutsch, A., Fernandez, M.F., Suciu, D.: Storing semistructured data with STORED. In: SIGMOD Conference, pp. 431–442 (1999)
8. Florescu, D., Kossmann, D.: Storing and querying XML data using an RDBMS. IEEE Data Engineering Bulletin 22(3), 27–34 (1999)
9. Schmidt, A., Kersten, M., Windhouwer, M., Waas, F.: Efficient relational storage and retrieval of XML documents. In: WebDB (2000)
10. Yoshikawa, M., Amagasa, T., Shimura, T., Uemura, S.: XRel: A path-based approach to storage and retrieval of XML documents using relational databases. ACM Transactions on InternetTechnology (TOIT) 1(1), 110–141 (2001)
11. Tatarinov, I., Viglas, S., Beyer, K.S., Shanmugasundaram, J., Shekita, E.J., Zhang, C.: Storing and querying ordered XML using a relational database system. In: SIGMOD Conference, pp. 204–215 (2002)
12. Dehaan, D., Toman, D., Conses, M.P., Ozsu, T.: A comprehensive XQuery to SQL translation using dynamic interval encoding. In: SIGMOD Conference (2003)
13. Teubner, J., Grust, T., Keulen, M.V.: Staircase join: Teach a relational DBMS to watch its (axis) steps. In: VLDB Conference (2003)
14. Krishnamurthy, R., Kaushik, R., Naughton, J.F.: Efficient XML-to-Relational query translation: Where to add intelligence? In: Proc. of the 30th VLDB Conference, Toronto, Canada (2004)
15. Runapongsa, K., Patel, J.M.: Storing and querying XML data in object-relational dbmss. In: EDBT Workshops (2002)
16. Cheng, J., Xu, J.: DB2 extender for XML. IBM (2000), `http://www-4.ibm.com/software/data/db2/extenders/xmlext/`
17. Oracle: XML Database Developer's guide - Oracle XML DB Release 2 (2002), `http://otn.oracle.com/tech/xml/xmldb/content.html`
18. Microsoft: SQLXML and XML Mapping Technologies (2004), `http://msdn.microsoft.com/sqlxml/default.asp`
19. Schmidt, A., Waas, F., Kersten, M.L., Carey, M.J., Manolescu, I., Busse, R.: XMark: a benchmark for XML data management. In: VLDB, pp. 974–985 (2002)

# Efficient Evaluation of Nearest Common Ancestor in XML Twig Queries Using Tree-Unaware RDBMS

Klarinda G. Widjanarko, Erwin Leonardi, and Sourav S. Bhowmick

School of Computer Engineering, Nanyang Technological University, Singapore
Singapore-MIT Alliance, Nanyang Technological University, Singapore
{klarinda,lerwin,assourav}@ntu.edu.sg

**Abstract.** Finding all occurrences of a twig pattern in a database is a core operation in XML query processing. Recent study showed that *tree-aware* relational framework significantly outperform *tree-unaware* approaches in evaluating structural relationships in XML twig queries. In this paper, we present an efficient strategy to evaluate a specific class of structural relationship called NCA-*twiglet* in a tree-unaware relational environment. Informally, NCA-twiglet is a subtree in a twig pattern where all nodes have the same nearest common ancestor (the root of NCA-twiglet). We focus on NCA-twiglets having parent-child relationships. Our scheme is build on top of our SUCXENT++ system. We show that by exploiting the encoding scheme of SUCXENT++ we can reduce useless structural comparisons in order to evaluate NCA-twiglets. Through a comprehensive experiment, we show that our approach is not only more scalable but also performs better than a representative tree-unaware approach on all benchmark queries with the highest observed gain factors being 352.

## 1 Introduction

Finding all occurrences of a twig pattern in a database is a core operation in XML query processing, both in relational implementations of XML databases [3,6,7,8,12,13,14,19, 20], and in native XML databases [1,4,5,10,11]. Consequently, in the past few years, many algorithms have been proposed to match twig patterns. These approaches (i) first develop a labeling scheme to capture the structural information of XML documents, and then (ii) perform twig pattern matching based on the labels alone without traversing the original XML documents.

For the first sub-problem of designing appropriate labeling scheme, various methods have been proposed that are based on tree-traversal order [1, 8, 9], region encoding [4, 20], path expressions [10, 14] or prime numbers [17]. By applying these labeling schemes, one can determine the structural relationship between two elements in XML documents from their labels alone. The goal of second sub-problem of matching twig patterns is to devise efficient techniques for structural relationship matching. In general, structural relationship in a twig query may be categorized in two different classes: (a) NCA-*twiglet*, and (b) *path expression*. Given a query twig pattern $Q = (V, E)$, the *nearest common ancestor* (denoted as NCA) of two nodes $x \in V$, $y \in V$ is the common ancestor of $x$ and $y$ whose distance to $x$ (and to $y$) is smaller than the distance to $x$ of any other common ancestor of $x$ and $y$. The twig substructure rooted at such NCA

R. Wagner, N. Revell, and G. Pernul (Eds.): DEXA 2007, LNCS 4653, pp. 617–628, 2007.

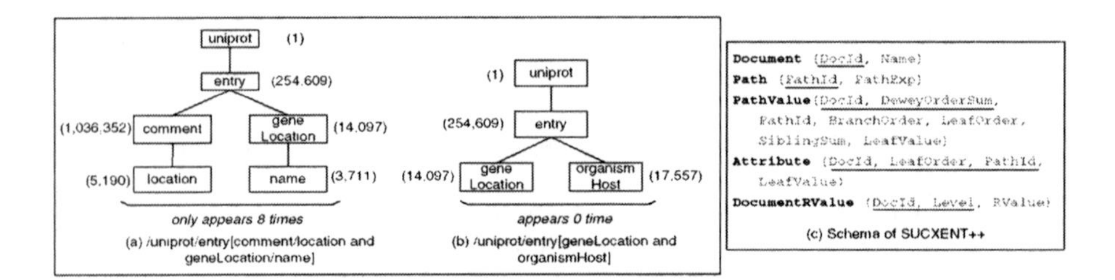


**Fig. 1.** Example of twig queries and SUCXENT++ schema

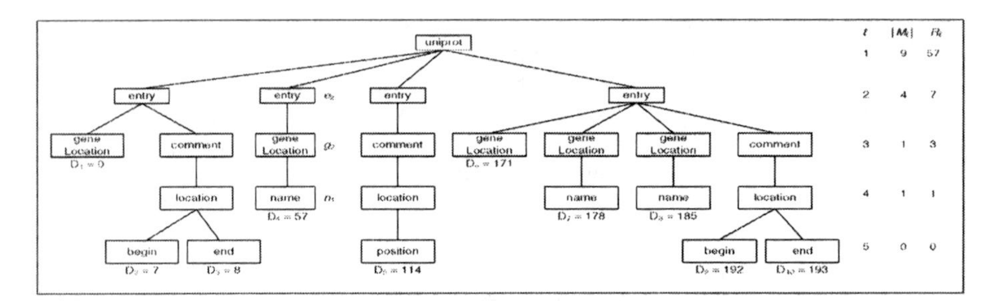


**Fig. 2.** Example of XML data

node is called NCA-*twiglet*. For example, consider the twig query in Figure 1(a). The twig structure rooted at `entry` node is an example of NCA-twiglet as it is the NCA of `location` and `name` nodes. On the other hand, *path expression* is a linear structural constraint. For example, `/uniprot/entry` is a path expression in Figure 1(a). In this paper, we focus on *efficient evaluation of* NCA-*twiglets* in a *relational implementation* of XML databases.

In literature, evaluation strategies of twig pattern matching can be broadly classified into the following three types: (a) *binary-structure matching*, (b) *holistic twig pattern matching*, and (c) *string matching*. In the *binary-structure matching* approach, the twig pattern is first decomposed into a set of binary (parent-child and ancestor-descendant) relationships between pairs of nodes. Then, the twig pattern can be matched by matching each of the binary structural relationships against the XML database, and "stitching" together these basic matches [1, 7, 9, 13, 20]. In the *holistic twig pattern matching* approach, the twig query is decomposed into its corresponding path components and each decomposed path component is matched against the XML database. Next, the results of each of the query's path expressions are joined to form the result to the original twig query [4, 10]. Lastly, approaches like ViST [15] and PRIX [11] are based on *string matching* method and transform both XML data and queries into sequences and answer XML queries through subsequence matching.

A key challenge in NCA-twiglets evaluation (as well as twig pattern matching in general) is to develop techniques that can reduce generation of large intermediate results. For instance, the binary-structure matching approaches may introduce very large intermediate results. Consider the sample document fragment from UNIPROTKB/ SWISS-PROT and the NCA-twiglet in Figures 2 and 1(a), respectively. The path match ($e2$, $g2$,

$n1$) for path `entry/geneLocation/name` does not lead to any final result since there is no `comment/location` path under $e2$. Note that this problem is exacerbated for queries that are *high-selective* but each path in the query is *low-selective*. Note that we use "high-selective" or "very selective" to characterize a twig query with few results and "low-selective" to characterize a query with many results. For example, the query in Figure 1(a) is very selective as it returns only 8 results. However, all the paths are low-selective. The number associated with each node in the queries in Figure 1 represents the number of occurrences of the path from the root node to the specific node in the XML database. Similarly, the query in Figure 1(b) is a high-selective query as it does not return any results although all the paths are low-selective. To solve this problem, the *holistic twig pattern matching* has been developed in order to minimize the intermediate results. In this approach, only those root-to-leaf path matches that will be in the final twig results are enumerated. However, when the twig query contains parent-child (PC) relationships, these solutions may still generate large numbers of useless matches [5]. Hence, in this paper we focus our attention on NCA-twiglets containing *PC relationship* and are components of *high-selective queries having low-selective paths*.

## 2 Framework and Contributions

The problem of efficiently finding NCAs in a general tree has been studied extensively over the last three decades [2]. Most of these approaches work using some mapping of the tree to a completely balanced tree, thereby exploiting the fact that for completely binary trees the problem is easier. Different algorithms differ by the way they do the mapping. However, these techniques cannot be directly used in the XML context for the following reasons. (i) Although the labels of the nodes used in some of the NCA algorithms can compute the label of NCA in constant time [2], they are not generic enough to efficiently support evaluation of various XPATH axes. Hence, the XML community has resorted to devising novel labeling schemes to support efficient twig matching. (ii) Due to the nature of XML data, the mapping of an XML tree to a completely binary tree may not be an efficient technique for processing different types of XPATH axes. Consequently, the research community has proposed various techniques on native and relational frameworks to evaluate twig queries.

### 2.1 Relational Approaches for Twig Query Processing and Our Contributions

While a variety of approaches have been proposed in the literature to process twig queries in native XML storage [4, 5, 10, 11], finding ways to evaluate such queries in relational environment has gained significant momentum in recent years. Specifically, there has been a host of work [3, 4, 6, 8, 9, 20] on enabling relational databases to be *tree-aware* by invading the database kernel to implement XML support. On the other side of the spectrum, some completely jettison the approach of internal modification of the RDBMS for twig query processing and resort to alternative *tree-unaware* approach [7, 12, 13] where the database kernel is not modified in order to process XML queries.

While the state-of-the-art tree-aware approaches are certainly innovative and powerful, we have found that these strategies are not directly applicable to relational databases.

The RDBMS systems need to augment their suite of query processing strategies by incorporating special purpose external index systems, algorithms and storage schemes to perform efficient XML query processing. Therefore, the integration of external modules into commercial relational databases could be complex and inefficient. On the other hand, there are considerable benefits in tree-unaware approaches with respect to portability as they do not invade the database kernel. Consequently, they can easily be incorporated in an off-the-shelf RDBMS. However, one of the key stumbling block for the acceptance of tree-unaware approaches has been query performance. In fact, recent results reveal that the tree-aware approaches appear scalable and, in particular, perform orders of magnitude faster than several tree-unaware approaches [3, 8]. In this paper, we explore the challenging problem of *efficient evaluation of* NCA-*twiglets* in a *tree-unaware* relational framework.

In summary, the main contributions of this paper are as follows. (a) Based on a novel labeling scheme, in Section 3, we present an efficient algorithm for determining nearest common ancestor (NCA) of two elements in an XML document. Our strategy accesses much fewer elements compared to existing state-of-the-art tree-unaware approaches in order to evaluate NCA-twiglets. Importantly, our proposed algorithm is capable of working with any off-the-shelf RDBMS without any internal modification. (b) Through an extensive experimental study in Section 4, we show that our approach significantly outperforms a state-of-the-art tree-unaware scheme (GLOBAL-ORDER [14]) for evaluating benchmark NCA-twiglets.

## 2.2 Overview of SUCXENT++ Approach

Our approach for NCA-twiglet evaluation is based on the SUCXENT++ system [12]. It is a tree-unaware approach and is designed primarily for query-mostly workloads. Here, we briefly review the storage scheme of SUCXENT++ which we shall be using in our subsequent discussion. The SUCXENT++ schema is shown in Figure 1(c). Document stores the document identifier DocId and the name Name of a given input XML document $T$. We associate each distinct (root-to-leaf) path appearing in $T$, namely PathExp, with an identifier PathId and store this information in Path table. For each leaf element $n$ in $T$, we shall create a tuple in the PathValue table.

SUCXENT++ uses a novel labeling scheme that *does not* require labeling of internal elements in the XML tree. For each leaf element it stores four additional attributes namely LeafOrder, BranchOrder, DeweyOrderSum and SiblingSum. Also, it encodes each level of the XML tree with an attribute called RValue. We now elaborate on the semantics of these attributes. Given two leaf elements $n_1$ and $n_2$, $n_1$.LeafOrder $<$ $n_2$.LeafOrder *iff* $n_1$ precedes $n_2$. LeafOrder of the first leaf element in $T$ is 1 and $n_2$.LeafOrder = $n_1$.LeafOrder+1 *iff* $n_1$ is a leaf element immediately preceding $n_2$. Given two leaf elements $n_1$ and $n_2$ where $n_1$.LeafOrder+1 = $n_2$.LeafOrder, $n_2$.BranchOrder is the level of the NCA of $n_1$ and $n_2$. The data value of $n$ is stored in $n$.LeafValue.

To discuss DeweyOrderSum, SiblingSum and RValue, we introduce some auxiliary definitions. Consider a sequence of leaf elements $C$: $\langle n_1, n_2, n_3, \ldots, n_r \rangle$ in $T$. Then, $C$ is a *k-consecutive leaf elements* of $T$ *iff* (a) $n_i$.BranchOrder $\geq k$ for all $i \in [1,r]$; (b) If $n_1$.LeafOrder $> 1$, then $n_0$.BranchOrder $< k$ where $n_0$.LeafOrder+1 = $n_1$.LeafOrder; and (c) If $n_r$ is not the last leaf element in $T$, then $n_{r+1}$.BranchOrder $< k$ where

$n_r$.LeafOrder+1 = $n_{r+1}$.LeafOrder. A sequence $C$ is called a *maximal k-consecutive leaf elements* of $T$, denoted as $M_k$, if there does not exist a $k$-consecutive leaf elements $C'$ and $|C|<|C'|$.

Let $L_{max}$ be the largest level of $T$. The RValue of level $\ell$, denoted as $R_\ell$, is defined as follows: (i) If $\ell = L_{max}-1$ then $R_\ell = 1$; (ii) If $0 < \ell < L_{max}-1$ then $R_\ell = 2R_{\ell+1} \times |M_{\ell+1}| + 1$. For example, consider the XML tree shown in Figure 2. Here $L_{max} = 5$. The values of $|M_1|$, $|M_2|$, $|M_3|$, and $|M_4|$ are 9, 4, 1, and 1, respectively. Then, $R_4 = 1$, $R_3 = 3$, $R_2 = 2 \times 3 \times |M_3| + 1 = 7$, and $R_1 = 2 \times 7 \times |M_2| + 1 = 57$. Note that due to facilitate evaluation of XPATH queries, the RValue attribute in DocumentRValue stores $\frac{R_\ell - 1}{2} + 1$ instead of $R_\ell$.

DeweyOrderSum is used to encode a element's order information together with its ancestors' order information using a single value. Consider a leaf element $n$ at level $\ell$ in $T$. $\mathsf{Ord}(n, k) = i$ *iff* $a$ is either an ancestor of $n$ or $n$ itself; $k$ is the level of $a$; and $a$ is the $i$-th child of its parent. DeweyOrderSum of $n$, $n$.DeweyOrderSum, is defined as $\sum_{j=2}^{\ell} \Phi(j)$ where $\Phi(j)$=[$\mathsf{Ord}(n, j)$-1]$\times R_{j-1}$. For example, consider the rightmost `name` element in Figure 2 which has a Dewey path "1.4.3.1". DeweyOrderSum of this element is: $n$.DeweyOrderSum $= (Ord(n, 2) - 1) \times R_1 + (Ord(n, 3) - 1) \times R_2 + (Ord(n, 4) - 1) \times R_3 = 3 \times 57 + 2 \times 7 + 0 \times 3 = 185$. Note that DeweyOrderSum is not sufficient to compute position-based predicates with `QName` name tests, *e.g.*, `entry`[2]. Hence, the SiblingSum attribute is introduced to the PathValue table. We do not elaborate further on SiblingSum as it is beyond the scope of the paper.

To evaluate non-leaf elements, we define the *representative leaf element* of a non-leaf element $n$ to be its first descendant leaf element. Note that the BranchOrder attribute records the level of the NCA of two consecutive leaf elements. Let $C$ be the sequence of descendant leaf elements of $n$ and $n_1$ be the first element in $C$. We know that the NCA of any two consecutive elements in $C$ is also a descendant of element $n$. This implies (a) except $n_1$, BranchOrder of a element in $C$ is at least the level of element $n$ and (b) the NCA of $n_1$ and its immediately preceding leaf element is not a descendant of element $n$. Therefore, BranchOrder of $n_1$ is always smaller than the level of $n$. The reader may refer to [12] for details on how these attributes are used to efficiently evaluate *ordered* XPATH axes.

## 3 Evaluation of NCA-Twiglets

In this section, we present the evaluation strategy of NCA-twiglets in SUCXENT++. We begin by formally introducing the notion of NCA-twiglet.

### 3.1 Data Model and NCA-Twiglet

We model XML documents as ordered trees. In our model we ignore comments, processing instructions and namespaces. We also ignore attributes for determining NCA as an attribute is not a child of an element. Queries in XML query languages make use of twig patterns to match relevant portions of data in an XML database. The twig pattern node may be an element tag, a text value or a wildcard "*". We distinguish between query and data nodes by using the term "node" to refer to a query node and the term "element" to refer to a data element in a document. In this paper, we focus only on parent-child relationships between the nodes in the twig pattern. Recall that existing holistic twig pattern matching approaches achieve optimality for ancestor-descendant relationships but may generate large numbers of useless matches when the twig query contains parent-child relations [5]. We now formally define NCA-*twiglet*.

**Definition 1 (NCA-Twiglet).** *Given a query twig pattern $Q = \langle V, E \rangle$, a* **NCA-*Twiglet*** *$N = \langle V_n, E_n, \Re \rangle$ in $Q$, denoted as $N \prec Q$, is a subtree in $Q$ rooted at node $\Re \in V$ such that (a) $V_n \subset V$ is a set of nodes whose nearest common ancestor is $\Re$, and (b) $E_n \subseteq E$.*

A NCA-twiglet consists of a collection of *rooted path* patterns, where a *rooted path* pattern (RP) is a root-to-leaf path in the NCA-twiglet. The level of the root $\Re$ is called NCA-*level*. For example, the NCA-twiglet in Figure 1(a) consists of the rooted paths `entry/comment/location` and `entry/geneLocation/name`. Note that each of the above RPs has a parent-child relationship between the nodes. The path from $Root(Q)$ to $\Re$ is called the *reachability path* of $N$. For instance, `/uniprot/entry` is the reachability path.

Given a NCA-twiglet $N \prec Q$ and an XML document $D$, a match of $N$ in $D$ is identified by a mapping from the nodes in $N$ to the elements in $D$, such that: (a) the query node predicates are satisfied by the corresponding database elements, wherein wildcard "*" can match any single tag; (b) the parent-child relationship between query nodes are satisfied by the corresponding database elements; and (c) the reachability path of $N$ is satisfied by the database elements. Next, we present our approach to match $N$ in $D$.

### 3.2 NCA-Twiglet Matching

Recall that in SUCXENT++ each root-to-leaf path of an XML document is encoded with the attributes LeafOrder, BranchOrder, DeweyOrderSum, and SiblingSum. Additionally each level of the XML tree is associated with a RValue. Hence, given the NCA-twiglet $N \prec Q$ and document $D$, our goal is to use these attributes to efficiently determine those root-to-leaf paths that satisfy $N$. We achieve this by using the following lemma and theorem.

**Lemma 1.** *Let $n_1$ and $n_2$ be two leaf elements in an* XML *document. If* $|n_1.\text{DeweyOrderSum} - n_2.\text{DeweyOrderSum}| < \frac{R_\ell - 1}{2} + 1$ *then the level of the nearest common ancestor is greater than $\ell$.* □

**Theorem 1.** *Let $n_1$ and $n_2$ be two leaf elements in an* XML *document. If* $\frac{R_{\ell+1} - 1}{2} + 1 \leq |n_1.\text{DeweyOrderSum} - n_2.\text{DeweyOrderSum}| < \frac{R_\ell - 1}{2} + 1$ *then the level of the nearest common ancestor of $n_1$ and $n_2$ is $\ell + 1$.* □

Due to space constraints, we do not present the proof here. The reader may refer to [16] for formal proof. We now illustrate with an example the above lemma and theorem in the context of a twig query. Consider the query in Figure 1(a) and the fragment of the PathValue table in Figure 3 (Step 1). Note that for clarity, we only

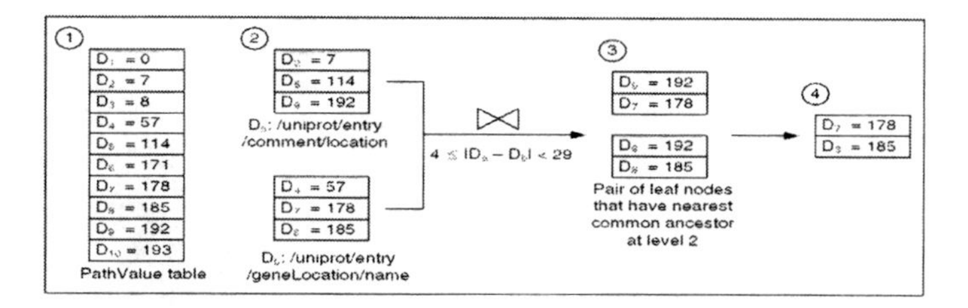


**Fig. 3.** An example of NCA-twiglet evaluation

```
evaluateNCATwiglet ( queryTwig )

01 i = 1
02 for every rootedPath in the queryTwig {
03   from_sql.add("PathValue as Vi")
04   where_sql.add("Vi.pathid in rootedPathi.getPathId()")
05   where_sql.add("Vi.branchOrder < rootedPathi.level()")
06   if (i > 1) {
07     where_sql.add("Vi.DeweyOrderSum BETWEEN
          Vi-1.DeweyOrderSum -
             RValue(rootedPathi.NCAlevel() - 1) + 1 AND
          Vi-1.DeweyOrderSum +
             RValue(rootedPathi.NCAlevel() - 1) - 1")
08   }
09   i++
10 }
11 select_sql.add("DISTINCT Vi-1.docId, Vi-1.DeweyOrderSum")
12 return select_sql + from_sql + where_sql
```

(a) *evaluateNCATwiglet* algorithm

```
XPath: /uniprot/entry[comment/location and
       geneLocation/name]

01 SELECT DISTINCT V2.DocId, V2.DeweyOrderSum
02 FROM PathValue V1, PathValue V2
03 WHERE V1.pathid in (2,3,4)
04 AND V1.branchOrder < 4
05 AND V2.docId = V1.docId
06 AND V2.pathid in (5)
07 AND V2.branchOrder < 4
08 AND V2.DeweyOrderSum BETWEEN
     V1.DeweyOrderSum - CAST(29 as BIGINT) + 1 AND
     V1.DeweyOrderSum + CAST(29 as BIGINT) - 1
```

(b) An example of Translated SQL query

**Fig. 4.** *evaluateNCATwiglet* algorithm

show the DeweyOrderSums of the root-to-leaf paths in the PathValue table. Let $D_a$ be DeweyOrderSum of the representative leaf elements satisfying `/uniprot/entry/comment/location` (second, fifth, and ninth leaf elements) and $D_b$ be DeweyOrderSum of the representative leaf elements satisfying `/uniprot/entry/geneLocation/name` (fourth, seventh, and eighth leaf elements). This is illustrated in step 2 of Figure 3. From the query we know that $D_a$ and $D_b$ have NCA at level 2 (`/uniprot/entry` level). Hence, based on Theorem 1 we can find pairs of (`location,name`) elements which have NCA at level 2. $D_a$ and $D_b$ fall on the following range: $(R_2-1)/2+1 \leq |D_a - D_b| < (R_1-1)/2+1 \Rightarrow 4 \leq |D_a - D_b| < 29$ which return the (seventh, ninth) and (eighth, ninth) leaf elements pairs (Step 3 of Figure 3). We can easily return the `entry` subtree by applying Lemma 1 on either one of the elements in the pair (Step 4). Note that since from the XPATH we know that $D_a$ and $D_b$ can not have NCA at level greater than 2, we only need to use Lemma 1 for matching NCA-twiglets. Observe that the above approach can reduce unnecessary comparison as we do not need to find the grandparent of `location` and `name` elements. We can determine the NCA directly by using the DeweyOrderSum and RValue attributes.

### 3.3 Query Translation Algorithm

Given a query twig (XPATH), the `evaluateNCATwiglet` procedure (Figure 4(a)) outputs SQL statement. A SQL statement consists of three clauses: *select_sql*, *from_sql* and *where_sql*. We assume that a clause has an `add()` method which encapsulates some simple string manipulations and simple SUCXENT++ joins for constructing valid

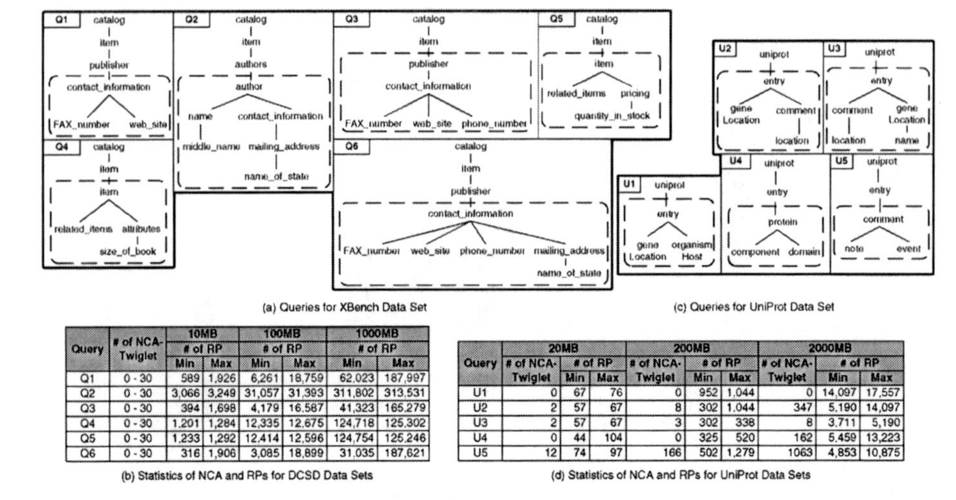


(a) Queries for XBench Data Set

(c) Queries for UniProt Data Set

| Query | # of NCA-Twiglet | 10MB # of RP Min | 10MB # of RP Max | 100MB # of RP Min | 100MB # of RP Max | 1000MB # of RP Min | 1000MB # of RP Max |
|---|---|---|---|---|---|---|---|
| Q1 | 0 - 30 | 589 | 1,926 | 6,261 | 18,759 | 62,023 | 187,997 |
| Q2 | 0 - 30 | 3,066 | 3,249 | 31,057 | 31,393 | 311,802 | 313,531 |
| Q3 | 0 - 30 | 394 | 1,698 | 4,179 | 16,587 | 41,323 | 165,279 |
| Q4 | 0 - 30 | 1,201 | 1,284 | 12,335 | 12,675 | 124,718 | 125,302 |
| Q5 | 0 - 30 | 1,233 | 1,292 | 12,414 | 12,596 | 124,754 | 125,246 |
| Q6 | 0 - 30 | 316 | 1,906 | 3,085 | 18,899 | 31,035 | 187,621 |

(b) Statistics of NCA and RPs for DCSD Data Sets

| Query | 20MB # of NCA-Twiglet | 20MB # of RP Min | 20MB # of RP Max | 200MB # of NCA-Twiglet | 200MB # of RP Min | 200MB # of RP Max | 2000MB # of NCA-Twiglet | 2000MB # of RP Min | 2000MB # of RP Max |
|---|---|---|---|---|---|---|---|---|---|
| U1 | 0 | 67 | 76 | 0 | 952 | 1,044 | 0 | 14,097 | 17,557 |
| U2 | 2 | 57 | 67 | 8 | 302 | 1,044 | 347 | 5,190 | 14,097 |
| U3 | 2 | 57 | 67 | 3 | 302 | 338 | 8 | 3,711 | 5,190 |
| U4 | 0 | 44 | 104 | 0 | 325 | 520 | 162 | 5,459 | 13,223 |
| U5 | 12 | 74 | 97 | 166 | 502 | 1,279 | 1063 | 4,853 | 10,875 |

(d) Statistics of NCA and RPs for UniProt Data Sets

**Fig. 5.** Query and data sets

SQL statements. In addition to preprocessing PathId, for a single XML document, we also preprocess RValue to reduce the number of joins.

The procedure firstly breaks the query twig into its subsequent rooted path (Line 02). Then for every rooted path, it gets the representative leaf nodes of the rooted path by using PathId and BranchOrder (Lines 04-05). After that, for the second rooted path onwards, it uses Lemma 1 to get the pair of leaf elements that have NCA at the NCA-level (Line 07). After processing the set of rooted paths, we return the DocId and DeweyOrderSum of the rightmost rooted path (Line 11) since only either one of the pairs is needed to construct the whole subtree. Finally, we collect the final SQL statement (Line 12). For example, consider the query in Figure 1(a). The output SQL statement can be seen in Figure 4(b). Lines 03-04 and Lines 06-07 are used to get the representative leaf elements of the respective rooted path. Line 08 is used to get the pair of leaf elements that have NCA at the NCA-level.

## 4 Performance Study

In this section, we present the performance results of our proposed approach and compare it with a state-of-the-art tree-unaware approach. Since there are several tree-unaware schemes proposed by the community, our selection choice was primarily influenced by the following two criteria. First, the storage scheme of representative approach should not be dependent on the availability of DTD/XML schema. Second, the selected approach must have good query performance for a variety of XPATH axes (ordered as well as unordered) for *query-mostly* workloads. Hence, we chose the GLOBAL-ORDER storage scheme as described in [14]. Prototypes for SUCXENT++ (denoted as SX), and GLOBAL-ORDER (denoted as GO) were implemented with JDK 1.5. The experiments were conducted on an Intel Pentium 4 3GHz machine running on Windows XP with 1GB of RAM. The RDBMS used was Microsoft SQL Server 2005 Developer Edition.

**Data and Query Sets:** In our experiments, we used XBench DCSD [18] as synthetic dataset and UNIPROT (downloaded from www.ebi.ac.uk/uniprot/database/download.html) as real dataset. We vary the size of XML documents from 10MB to 1GB for XBench and from 20MB to 2GB for UNIPROT. Recall that we wish to explore twig queries that are high-selective although the paths are low-selective. Hence, we modified XBench dataset so that we can control the number of subtrees (denoted as $K$) that matches the NCA-twiglet and the number of occurrences of the rooted paths. We set $K \in \{0, 10, 20, 30\}$ for XBench dataset. Note that we did not modify the UNIPROT dataset. Figures 5(a) and 5(c) depict the benchmark queries on XBench and UNIPROT, respectively. We vary the number of rooted paths in the queries from 2 to 4. The number of occurrences of subtrees that satisfies a NCA-twiglet and the minimum and maximum numbers of occurrences of rooted paths in the datasets are shown in Figures 5(b) and 5(d) for XBench and UNIPROT queries, respectively.

**Test Methodology:** Appropriate indexes were constructed for all approaches through a careful analysis on the benchmark queries. Particularly, for SUCXENT++ we create the following indexes on PathValue table: (a) unique clustered index on PathId and DeweyOrderSum, and (b) non-unique, non-clustered Index on PathId and BranchOrder. Furthermore, since our dataset consists of a single XML document, we removed the DocId column from the tables in SX and GO. Prior to our experiments, we ensure that statistics had been collected. The bufferpool of the RDBMS was cleared before each run. Each query was executed 6 times and the results from the first run were always discarded.

Since GO and SX have different storage approaches, the structure of the returned results are also different. Recall from Section 3.2, the goal of our study is to *identify* subtrees that matches the NCA-twiglet. Hence, we return results in the *select* mode [14]. That is, we do not reconstruct the entire matched subtree. Particularly, for the GO approach, we return the identifier of the root of the subtree (without its descendants) that matches the NCA-twiglet. Whereas for SX, we return the DeweyOrderSum of the root-to-leaf path of the matching subtree. This path must satisfy the rightmost rooted path of the NCA-twiglet. For example, for the query in Figure 1(a), we return the identifiers of the `entry` elements in GO and the DeweyOrderSums of the root-to-leaf paths containing the rightmost rooted path `entry/geneLocation/name` elements in SX. Lastly, for SX we enforce a "left-to-right" join order on the translated SQL query using query hints. The performance benefits of such enforcement is discussed in [12].

**NCA-twiglet evaluation times:** Our experimental goal is to measure the evaluation time for determining those subtrees that match a NCA-twiglet with a specific reachability path in the twig queries in Figure 5. Figures 6(a) and 6(b) depict the NCA-twiglet evaluation times of SUCXENT++ and GLOBAL-ORDER, respectively. Figure 6(c) depicts the evaluation time for UNIPROT data set. We observe that SX significantly outperforms GO for all queries with the highest observed factor being 352 (Query *U5* on 2GB dataset). Particularly, SX is orders of magnitude faster for high-selective queries. Observe that for XBench dataset, when $K = 0$, SX is up to 332 times faster (Query *Q6* on 1GB dataset)

| ID | 10MB | | | | 100MB | | | | 1000MB | | | |
|---|---|---|---|---|---|---|---|---|---|---|---|---|
| | K=0 | K=10 | K=20 | K=30 | K=0 | K=10 | K=20 | K=30 | K=0 | K=10 | K=20 | K=30 |
| Q1 | 25.60 | 27.00 | 28.60 | 27.40 | 142.80 | 157.40 | 151.80 | 158.20 | 1,586.80 | 1,564.80 | 1,574.20 | 1,596.60 |
| Q2 | 61.00 | 62.40 | 62.60 | 69.80 | 430.40 | 418.20 | 475.20 | 427.00 | 3,846.80 | 3,631.20 | 3,994.80 | 3,619.60 |
| Q3 | 68.80 | 85.80 | 85.80 | 84.20 | 100.00 | 182.60 | 185.60 | 189.00 | 990.80 | 1,664.80 | 1,620.60 | 1,640.40 |
| Q4 | 26.00 | 28.00 | 27.60 | 27.40 | 205.20 | 168.20 | 194.40 | 180.00 | 1,980.40 | 1,995.40 | 1,950.60 | 1,985.20 |
| Q5 | 63.80 | 75.40 | 62.20 | 75.60 | 180.20 | 186.80 | 192.80 | 189.20 | 1,994.40 | 1,947.60 | 1,963.20 | 1,952.60 |
| Q6 | 92.00 | 100.40 | 112.40 | 114.00 | 83.60 | 257.00 | 239.20 | 237.60 | 617.20 | 2,161.20 | 2,159.00 | 2,159.00 |

(a) SUCXENT++ (XBench, in msec)

| ID | 10MB | | | | 100MB | | | | 1000MB | | | |
|---|---|---|---|---|---|---|---|---|---|---|---|---|
| | K=0 | K=10 | K=20 | K=30 | K=0 | K=10 | K=20 | K=30 | K=0 | K=10 | K=20 | K=30 |
| Q1 | 609.60 | 603.00 | 936.40 | 717.60 | 8,478.80 | 8,316.60 | 8,862.20 | 6,066.80 | 114,717.60 | 83,035.80 | 80,979.60 | 84,053.80 |
| Q2 | 599.00 | 467.00 | 443.20 | 448.80 | 6,080.00 | 5,996.40 | 5,451.40 | 6,640.00 | 349,974.00 | 236,906.20 | 226,509.20 | 225,286.20 |
| Q3 | 698.20 | 736.00 | 957.20 | 675.00 | 5,510.60 | 5,565.20 | 5,458.40 | 5,464.60 | 80,954.80 | 78,541.20 | 76,998.20 | 80,571.40 |
| Q4 | 494.80 | 474.00 | 473.80 | 763.80 | 3,522.80 | 4,250.40 | 4,727.20 | 4,598.80 | 107,910.20 | 108,039.40 | 71,082.40 | 111,399.20 |
| Q5 | 553.60 | 492.60 | 570.20 | 655.60 | 4,915.40 | 4,959.40 | 4,010.60 | 4,161.60 | 70,275.20 | 111,901.60 | 71,232.80 | 75,587.40 |
| Q6 | 762.40 | 1,205.00 | 792.60 | 970.60 | 7,516.40 | 6,925.40 | 7,907.20 | 7,948.60 | 204,835.40 | 249,687.40 | 259,323.20 | 238,127.40 |

(b) Global Order (XBench, in msec)

| ID | SUCXENT++ | | | Global Order | | |
|---|---|---|---|---|---|---|
| | 20MB | 200MB | 2000MB | 20MB | 200MB | 2000MB |
| U1 | 9.00 | 23.40 | 163.60 | 201.00 | 1,424.60 | 17,281.60 |
| U2 | 9.00 | 21.00 | 156.80 | 363.80 | 2,512.60 | 23,839.20 |
| U3 | 5.00 | 12.20 | 86.00 | 266.00 | 2,675.20 | 23,705.20 |
| U4 | 5.40 | 14.40 | 123.60 | 14.00 | 618.80 | 6,870.80 |
| U5 | 5.00 | 22.20 | 123.60 | 438.80 | 2,800.00 | 43,502.20 |

(c) UniProt (in msec)

**Fig. 6.** Performance results

and on average 56 times faster than GO. This is significant in an environment where users would like to issue exploratory ad hoc queries. In this case, the user would like to know quickly if the query returns any results. If the result set is empty then he/she can further refine his/her query accordingly.

SX is significantly faster than GO because of the following reasons. Firstly, SX uses an efficient strategy based on Theorem 1 to reduce useless comparisons. Furthermore, the number of join operations in GO is more than SX. For example, for *Q6*, GO and SX join six tables and four tables, respectively. Secondly, GO stores every element of an XML document whereas SX stores only the root-to-leaf paths. Consequently, the number of tuples in the Edge table is much more than that in the PathValue table.

## 5 Related Work

We first compare our proposed approach with existing tree-unaware techniques [7, 12, 13, 14, 19]. Note that we do not compare our work with tree-aware relational schemes [1,3,6,8,9,20] as these techniques modify the database internals. Our approach differs from these tree-unaware techniques in the following ways. First, we use a novel and powerful numbering scheme that only encodes the leaf elements and the levels of the XML tree. In contrast, most of the tree-unaware approaches encode both internal and leaf elements. Second, the translated SQL of SUCXENT++ does not suffer from large number of joins. Third, all previous tree-unaware approaches, reported query performance on XML documents with small/medium sizes – smaller than 500 MB. We investigate query performance on large synthetic and real datasets (up to 2GB). This gives more insights on the scalability of the state-of-the-art tree-unaware approaches for twig query processing.

In our previous work [12], we focused on efficiently evaluating ordered path expressions rather than tree-structured queries. In this paper, we investigate how the encoding scheme in [12] can be used for efficiently processing NCA-twiglet, a specific class of structural relationship in a twig pattern query.

## 6 Conclusions

The key challenge in XML twig pattern evaluation is to efficiently match the structural relationships of the query nodes against the XML database. In general, structural relationship in a twig query may be categorized in two different classes: path expression and NCA-twiglet. A path expression enforces linear structural constraint whereas NCA-twiglet specifies tree-structured relationship. In this paper, we present an efficient strategy to evaluate NCA-twiglets having parent-child relationship in a tree-unaware relational environment. Our scheme is build on top of SUCXENT++ [12]. We show that by exploiting the encoding scheme of SUCXENT++ we can reduce useless structural comparisons in order to evaluate NCA-twiglets. Our results showed that our proposed approach outperforms GLOBAL-ORDER [14], a representative *tree-unaware* approach for all benchmark queries. Importantly, unlike tree-aware approaches, our scheme does not require invasion of the database kernel to improve query performance and can easily be built on top of any off-the-shelf RDBMS.

## References

1. Al-Khalifa, S., Jagadish, H.V., Patel, J.M., et al.: Structural Joins: A Primitive for Efficient XML Query Pattern Matching. In: ICDE (2002)
2. Alstrup, S., Gavoille, C., Kaplan, H., Rauhe, T.: Nearest Common Ancestors: A Survey and a new Distributed Algorithm. In: SPAA (2002)
3. Boncz, P., Grust, T., van Keulen, M., Manegold, S., Rittinger, J., Teubner, J.: MonetDB/XQuery: A Fast XQuery Processor Powered by a Relational Engine. In: SIGMOD (2006)
4. Bruno, N., Koudas, N., Srivastava, D.: Holistic Twig Joins: Optimal XML Pattern Matching. In: SIGMOD (2002)
5. Chien, S., Li, H-G., Tatemura, J., et al.: Twig$^2$Stack: Bottom-up Processing of Generalized-Tree-Pattern Queries over XML Documents. In: VLDB (2006)
6. DeHaan, D., Toman, D., Consens, M.P., Ozsu, M.T.: A Comprehensive XQuery to SQL Translation Using Dynamic Interval Coding. In: SIGMOD (2003)
7. Florescu, D., Kossman, D.: Storing and Querying XML Data using an RDBMS. IEEE Data Engg. Bulletin 22(3) (1999)
8. Grust, T., Teubner, J., Keulen, M.V.: Accelerating XPath Evaluation in Any RDBMS. In: ACM TODS, vol. 29(1) (2004)
9. Li, Q., Moon, B.: Indexing and Querying XML Data for Regular Path Expressions. In: VLDB (2001)
10. Lu, J., Ling, T.W., Chen, T.Y., Chen, T.: From Region Encoding to Extended Dewey: On Efficient Processing of XML Twig Pattern Matching. In: VLDB (2005)
11. Rao, P., Moon, B.: PRIX: Indexing and Querying XML Using Prufer Sequences. In: ICDE (2004)
12. Seah, B.-S, Widjanarko, K.G., Bhowmick, S.S., Choi, B., Leonardi, E.: Efficient Support for Ordered XPath Processing in Tree-Unaware Commercial Relational Databases. In: DASFAA (2007)

13. Shanmugasundaram, J., Tufte, K., et al.: Relational Databases for Querying XML Documents: Limitations and Opportunities. In: VLDB (1999)
14. Tatarinov, I., Viglas, S., Beyer, K., et al.: Storing and Querying Ordered XML Using a Relational Database System. In: SIGMOD (2002)
15. Wang, H., Park, S., Fan, W., Yu, P.S.: ViST: A Dynamic Index Method for Querying XML Data by Tree Structures. In: SIGMOD (2003)
16. Widjanarko, K.J., Leonardi, E., Bhowmick, S.S.: Efficient Evaluation of Nearest Common Ancestor in XML Twig Queries Using Tree-Unaware RDBMS. Technical Report (2007), Available at
`http://www.cais.ntu.edu.sg/~assourav/TechReports/nca-TR.pdf`
17. Wu, X., Lee, M., Hsu, W.: A Prime Number Labeling Scheme for Dynamic Ordered XML Trees. In: ICDE (2004)
18. Yao, B., Tamer Özsu, M., Khandelwal, N.: XBench: Benchmark and Performance Testing of XML DBMSs. In: ICDE (2004)
19. Yoshikawa, M., Amagasa, T., Shimura, T., Uemura, S.: XRel: a path-based approach to storage and retrieval of xml documents using relational databases. ACM TOIT 1(1), 110–141 (2001)
20. Zhang, C., Naughton, J., Dewitt, D., Luo, Q., Lohmann, G.: On Supporting Containment Queries in Relational Database Systems. In: SIGMOD (2001)

# Exclusive and Complete Clustering of Streams

Vasudha Bhatnagar and Sharanjit Kaur

Department of Computer Science, University of Delhi, Delhi, India
{vbhatnagar,skaur}@cs.du.ac.in

**Abstract.** Clustering for evolving data stream demands that the algorithm should be capable of adapting the discovered clustering model to the changes in data characteristics.

In this paper we propose an algorithm for exclusive and complete clustering of data streams. We explain the concept of completeness of a stream clustering algorithm and show that the proposed algorithm guarantees detection of cluster if one exists. The algorithm has an on-line component with constant order time complexity and hence delivers predictable performance for stream processing. The algorithm is capable of detecting outliers and change in data distribution. Clustering is done by growing dense regions in the data space, honouring recency constraint. The algorithm delivers complete description of clusters facilitating semantic interpretation.

## 1 Introduction

Clustering of data streams aids summarization of data characteristics and finds important applications in both scientific and commercial domains. The result of clustering on streaming data is an approximation of data characteristics according to some predefined properties like centroid, radius etc. [6]. The approximation arises because of single scan constraint on data streams, unlike traditional clustering methods where multiple scans over data lead to exclusive clustering [7]. Traditionally clustering of data streams is carried out using window-based models, where recent points are given due importance in clustering [3].

Sub-cluster maintenance is the most common approach for clustering evolving data streams. The issue of likely (poor) approximation during clustering of data streams using this approach has been addressed in [6]. In this paper we address three other important issues related to clustering of data streams.

1. Complete clustering: Pyramidal time window [1], sliding window [5] and damped window [2,8] models are some of the commonly used approaches for clustering evolving data streams. The effect of these models is to discount historical data in a continuous manner. These approaches suffer from the deficiency that if the speed of the stream suddenly changes or the interval between two consecutive clusterings is large, some of the historical data may be discounted before being used in the construction of the model. Though, the parameters like fading factor are tuned on the basis of anticipated speed

R. Wagner, N. Revell, and G. Pernul (Eds.): DEXA 2007, LNCS 4653, pp. 629–638, 2007.

of the stream, these parameters do not change dynamically. This may happen if a small outlier cluster appears for a period of time that is much smaller compared to the time interval between two clusterings. Thus it is not guaranteed that all the clusters that appeared after the last clustering will always be detected leading to incomplete clustering.
Figure 1 shows a small cluster marked X, appearing right after time $T_1$, when clustering was performed. The other prominent clusters also form during this time. The next clustering at time $T_2$ may not report cluster X if the weight of the micro-cluster corresponding to X falls below a prespecified threshold because of passing of relatively longer interval of time.
We propose the notion of complete clustering where it is guaranteed that a point recieved between two consecutive clusterings will be accounted for. I.e. every data point is reported as a member of a cluster, noise or an outlier at least once.

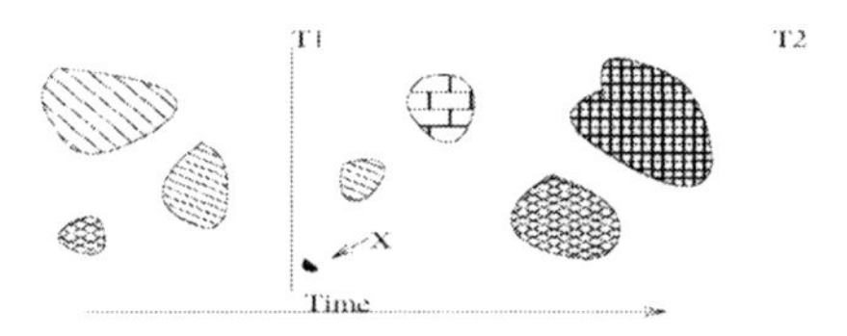


**Fig. 1.** Cluster X fades out before the next clustering at $T_2$ resulting in incomplete clustering

2. Predictable performance of on-line component: The on-line component of clustering algorithm for streams must be efficient so that there is no data loss. In micro-cluster based approaches, the complexity of the stream processing component is O($dn$) where $d$ is the number of dimensions and $n$ is the number of micro-clusters. Since the number of micro-clusters may keep on changing with time [1,4], the performance of the on-line component may be unpredictable.
3. Human-Centric description of clusters: Clusters discovered in the streaming data are commonly described by the cluster features [1,4]. In case of arbitrary shaped clusters this description is not of much help for the end-user. In short, the clustering approximations delivered by the stream clustering algorithms are not amenable to easy interpretation.

### 1.1 Our Approach

To achieve **Ex**clusive and **C**omplete **C**lustering, we propose ExCC algorithm for streams which has an on-line component with predictable performance (Section 2.1). The second component of the algorithm performs clustering on demand (Section 3) and may be run either on-line or offline depending on requirement.

The clustering approximation is improved by pruning the outdated regions in data space just before clustering. This ensures *completeness* and a cluster, how so ever small (X in Fig. 1), that showed up in the stream after the last clustering is detected and reported. The salient features of the algorithm are:

1. Maintaining detailed data distribution along with the speed of streaming data for clustering of streams (Section 2.1, 2.2).
2. On-line component takes constant time for stream processing (Section 2.1)
3. Outliers are detected on-the-fly and noise is reported separately (Section 2.4)
4. Exclusive clustering with user centric description like boundary of cluster, signature of seed etc. (Section 3)

### 1.2 Related Works

The single scan for data streams has motivated several good algorithms like CluStream [1], HPStream [2] and DenStream [4]. These algorithms are based on the philosophy of micro-cluster maintenance and use either density or distance criteria to discover clusters on-demand. GCHDS is a grid based algorithm for high dimensional data streams [8] and discovers clusters by applying connected component analysis on the grid. These algorithms are not robust with respect to outliers in the stream and may result into incomplete clustering.

The micro-cluster maintenance approach for clustering of data streams has been dissected and critically analyzed in [6]. It has been shown with the help of carefully crafted counter-examples that this approach may lead to violation of basic requirements of clustering. I.e., it may happen that although a point exists in one cluster but it maybe close to the center of another cluster.

## 2 ExCC Algorithm

The proposed ExCC algorithm implements a multidimensional grid to maintain the detailed distribution of data along all dimensions. The actual speed of the different data distributions is also maintained in the grid. Each cell in the grid represents a region in data space. The grid is pruned before clustering so as to discard obsolete data regions. Clustering is performed by combining the significant regions in the data space, taking both recency and density into account.

Freak data points are reported on-the-fly to the user as outliers and noisy points are reported separately. We explain these features and the methodology in following subsections.

### 2.1 Grid Structure

Grid structure is suitable for stream clustering because the processing time of a data point in a grid, which is crucial in stream mining algorithms, is a known constant [7,9]. A point is placed in the grid only on the basis of the values of its dimensional components, thereby avoiding the need of distance computation function. At any time instance, the grid structure $G$ contains the detailed data distribution of *recent* data points coming in the stream $S$ and maintains only those nodes which have at least one point.

Given the dimension set $\Delta = \{d_1, \ldots, d_d\}$ , let $l_i$ and $h_i$ respectively be the lowest and the highest data values along dimension $d_i$, as known to the domain expert. The range $r_i = [l_i, h_i]$ of $d_i$ is divided into $k$ (even) number of equi-width intervals $(l_i^1, h_i^1), ]l_i^2, h_i^2), \ldots, ]l_i^k, h_i^k)$, such that $l_i^1 = l_i, h_i^k = h_i$. $k$ is a user defined parameter and determines the granularity of the grid. Keeping number of intervals same for all dimensions simplifies notation, even though the proposed framework does not have any limitation in this respect. In the rest of the paper, we use $k$ as the number of intervals for all dimensions.

Initially the grid $G$ is empty. It gets populated as incoming data points are inserted. A data point is stored using $d$ internal nodes at $d$ levels in $G$, each corresponding to $d$ dimensions. A leaf node (cell) in the grid corresponds to the region in the data space determined by the intersection of one interval from each dimension. Number of data points in a cell indicates the density of the region. The cell also stores the rate at which the data points are getting added to the region. Though the actual memory requirement of the grid will depend on the data distribution in the stream, memory bound of $O(k^d)$ may be used as a broad guideline to decide the granularity of the grid.

**Lemma 1.** *The insertion time for a data point in the grid is O(d), where d is the dimension of a data point.*

*Proof.* Insertion of a data point involves two tasks viz. i) finding the interval for each dimension value ii) inserting the data point in grid. For $d$ dimensions, task (i) can be completed in O($d$) time. Since the depth of the grid is exactly $d$, task (ii) takes O($d$) time. Hence total insertion time per point is O($d$).

The constant order time complexity guarantees predictability of the performance of the on-line component of the algorithm.

### 2.2 Speed of the Stream

Intuitively, in a fast stream, data decays relatively faster than in a slow stream. To be fair enough to all data regions, we maintain the speed of the incoming data at each region in the data space. Varying data speed for a cell in the grid is averaged out and is used to compute its recency.

**Definition 1.** *Let $C$ be an arbitrary cell in the grid with count $n$, by virtue of accumulating points $p_1, \ldots, p_n$ at times $t_1, \ldots, t_n$ respectively. The average inter-arrival time $aat^C$ is computed as follows.*

$$aat^C = \begin{cases} 1 & n = 1 \\ \frac{1}{n} \sum_{j=1}^{n-1} (t_{j+1} - t_j) & n > 1 \end{cases} \quad (1)$$

Based on the average inter-arrival time, the speed of the stream at cell $C$ is defined as $Sp^C = \frac{1}{aat^C}$. The speed of the stream (speed at root node) is denoted by $Sp^R$. Please note that the notion of speed is conceptual and in computations we make use of $aat$'s.

### 2.3 Grid Pruning

Unbounded nature of the stream necessitates design of a mechanism to keep a check on the size of the grid. We prune the grid just before clustering and remove the non-*recent* cells. Non-recent cells denote those regions in data space which have not seen significant addition of data points since the last clustering or even longer. These regions indicate that currently the stream is not supportive of the particular data distribution and hence are candidates for pruning.

If $t_{curr}$ is the current time, $t_l^C$ is the time when the last point was added to the cell and *Numpoints* is the number of points actually arrived in stream since the last clustering, the cell is pruned iff the following criterion holds.

$$\frac{t_{curr} - t_l^C}{Numpoints} \geq aat^C \tag{2}$$

The computation of recency is optimistic in the sense that as soon as a data point is added to a cell, its aat is reduced in anticipation that more data points will be added to it in near future and hence, retained. This permits regions like shown in Figure 1 to be reported to the user, even though the trend was observed for a short duration and gives complete clustering.

**Lemma 2.** *The pruning function does not prune any region (cell in the grid) that was added in the grid since last clustering, irrespective of the speed of the stream and length of the period between two clusterings.*

*Proof.* Let two consecutive clusterings be done at times $t_0$ and $t_1$. During this time, say $n$ points came in the stream and only one point was received in a newly created cell $C$ immediately after $t_0$. We show that $C$ is never pruned at $t_1$.

For sake of simplicity, let $t_0 = 0$ and $t_1 = t$. $C$ will be pruned iff $\frac{t-0}{n} \geq 1$ (refer Eq. 2). But the LHS is $aat^R$, which is always $\leq 1$ irrespective of $t$. Thus $C$ will never be pruned.

### 2.4 Handling Data Drift and Outliers

Since the data space $\mathcal{D} = [l_1, h_1] \times \ldots \times [l_d, h_d]$ is based on the experience of the human experts in the domain, it is necessary to address the situation when a data point falls outside $\mathcal{D}$.

A point $p_x(v_{x1}, \ldots, v_{xd}) \notin \mathcal{D}$ is actually an aberration and is considered anomalous. However if there is a drift in the underlying data generation process, the algorithm should be able to detect it and redefine $\mathcal{D}$.

An anomalous point $p_x$ is recognized by assessing its distance outside the boundary of $\mathcal{D}$, which also indicates its *extent of outlying*. Given $r_i$ as range of dimension $d_i$, let $\delta_i$ be the distance of $p_x$ from the boundary of $\mathcal{D}$ along $d_i$.

$$\delta_i = \begin{cases} 0 & \text{if } l_i \leq v_{xi} \leq h_i \\ \frac{1}{r_i} * min(|v_{xi} - l_i|, |v_{xi} - h_i|) & \text{otherwise} \end{cases} \tag{3}$$

$p_x$ is an outlier iff $\exists i$, such that $\delta_i \geq 1$. A outlier is identified on-the-fly by the algorithm in constant time. In other cases where $0 < \delta_i < 1$, we wait-and-watch before declaring the data point as an outlier.

Arrival of an anomalous point once in a while is an aberration in the data generating process. In case this happens more often than expected, a change in data distribution is signified. The algorithm uses average arrival rate of an anomalous point as the indicator of the frequency of their arrival in $S$.

The algorithm parks an anomalous point in a *Hold Queue* (HQ) instead of immediately reporting it to the user. The objective is to gather evidence whether the point is actually an outlier, or there is a indication of shift in data distribution. The *Hold Queue* is processed periodically to determine the nature of the data points it contains.

Let $aat_i^j$ denote the arrival rate of data points in the $j^{th}$ interval of dimension $d_i$. Under the assumption that an out-of-range data point is a spill over from either first or last interval of dimension $d_i$, $aato_i$ is compared with $aat_i^1$ and $aat_i^k$ leading to the following three cases:

1. In case $aato_i$ is one, the singleton is clearly an outlier and is reported.
2. If $aato_i \ll aat_i^1$ or $aat_i^k$, a temporary data drift is concluded and reported. It can be construed as some change in the distribution along this dimension, which has probably now reversed. Malfunctioning of a sensor in a sensor network, corrected subsequently, could lead to a situation like this.
3. If $aato_i \approx aat_i^1$ or $aat_i^k$ then a definite change in the data distribution along the dimension in question can be concluded, motivating investigation of the data generating process.

Data drift (Case 3) is handled by expanding the grid along the dimension $d_i$. Grid expansion is carried out by merging the adjacent intervals in all the nodes at the level corresponding to $d_i$, and inserting relevant points from the HQ in the grid. Reconstruction of the grid redefines $\mathcal{D}$ and the time required for this expansion depends on the location of the dimension.

## 3 Clusters Generation

Clusters are generated from the grid by identifying regions that are adjacent to the dense region in the data space. Each dense region that is not adjacent to other dense region denotes a cluster. This strategy of growing clusters together with Lemma 2 ensures completeness and exclusiveness of the clustering scheme.

After pruning, a pool of eligible cells is available for clustering. The algorithm chooses the dense cell (*seed*) from unused cells on the basis of weight[1].

**Definition 2.** *The weight $W^C$ of the cell $C$, with average inter arrival time $aat^C$ and number of points $n^C$ is computed as*

$$W^C = \frac{n^C}{aat^C} \tag{4}$$

[1] An experienced user can optionally use the weight to further prune the pool of cells to be clustered by specifying an appropriate threshold.

The cell with highest weight is selected as the *seed* for cluster formation.

### 3.1 Connectivity and Clustering

As mentioned earlier, in order to discover clusters from grid, cells *connected* to the *seed* need to be identified. Definition 3 formalizes the intuitive notion of *connectedness* of two cells.

**Definition 3.** *Let $C_1(k_1, \ldots, k_d)$ and $C_2(k'_1, \ldots, k'_d)$ be two cells with respective dimensional intervals shown in brackets. $C_1$ and $C_2$ are connected iff $\forall\, i$ either $k_i = k'_i$ or $|k_i - k'_i| = 1$.*

Clusters are generated iteratively using a greedy approach starting with the *seed*. As soon as a cluster is found, the member cells of the discovered cluster are removed from the pool, and search begins for another cluster with a new seed. Thus in one iteration all clusters are found including the smallest one.

This strategy gives rise to exclusive clusters of arbitrary shapes and varied sizes and is guaranteed not to miss a cluster if one exists. The worst case complexity of the procedure is $O(NumCell^{NumClust})$, where $NumCell$ is the number of cells in the pool and $NumClust$ is the number of clusters discovered.

The flip side of the completeness property of this algorithm is that the number of reported clusters may be very large. We propose an optional step in which, all clusters which overlap on at least half of the dimensional space and their corresponding seeds are adjacent, are merged.

### 3.2 Cluster Description

For each discovered cluster, signature of the seed, its boundary and density are reported to the user. Seed signature along with the boundary of the cluster gives an idea of the shape of the cluster. Analysis of the spread of each dimension in the cluster gives an idea of the compactness along the dimension. Thus a dimension along which all cells in the cluster have same interval, is the most compact dimension (i.e. 100 % compact). To the best of our knowledge, this is the most explicit reported description of a cluster by a stream clustering algorithm.

### 3.3 Noise Detection

The data points that do not belong to any of the discovered clusters as per the specified quality criterion represent the noise in the data. The quality criterion filters out non-significant regions and reports the same separately. Since the noise is a subjective notion, the quality criterion may vary depending on the application. For example, density of the cluster, % of dimensions with desired compactness etc... could be used for noise detection.

## 4 Experimental Study

We implemented the algorithm in ANSI C with no optimizations, and compiled using g++ compiler (3.3.2-2) and executed on Intel Centrino processor with 256

MB RAM, running Linux (kernel 2.4.22-1). All the experiments reported here are done on a simulated stream with uniform speed. The data sets used are:

i) Intrusion Detection data set (KDD cup 99) available at [10], consists of 494,021 records, each having 42 attributes (34 continuous and 8 categorical). Each record corresponds to either normal class or an attack class. We performed experiments with 23 classes using 34 continuous attributes.

ii) Forest Cover type data set available at [11] has 581,012 observations each with 54 attributes out of which 10 quantitative attributes are used in the reported experiment.

### 4.1 Performance of On-Line Component of ExCC

We performed experiments with both data sets to support Lemma 1. We simulate a stream with $aat \approx 0$ so as to get the accurate stream processing time. Figure 2 shows processing time of a point for Forest data set. The time remains nearly constant for fixed $k$ and marginally increases with increase in the number of intervals ($k$). Figure 3 shows total processing time for KDD Cup data using window of size 50K . The time shows a linear trend with little variation because of changes in data distribution leading to creation of new cells.

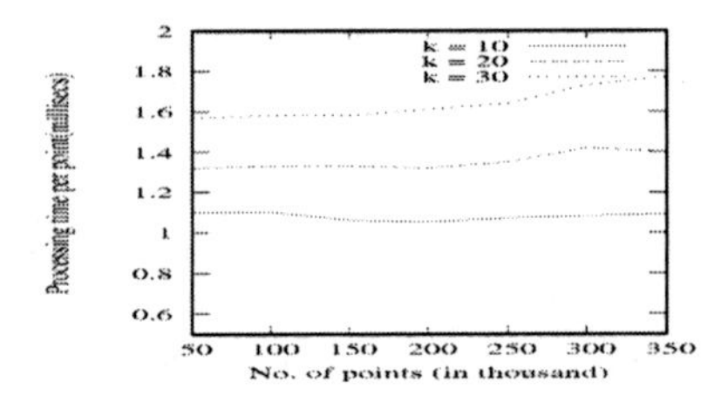


**Fig. 2.** Per point processing time (Forest Data)

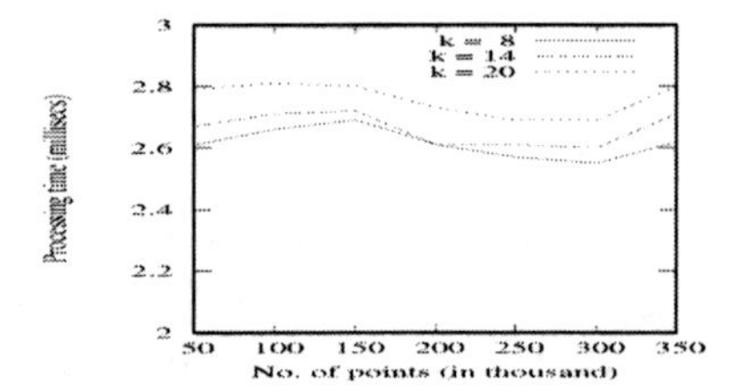


**Fig. 3.** Total processing time (Intrusion Data)

### 4.2 Clustering Quality Evaluation

We investigated the effect of granularity of the grid on purity of the clustering scheme using 150,000 records of intrusion data. Figure 4 shows cluster purity for different values of $k$ in a single window. Cluster purity increases initially with increasing $k$, and then dips before stabilizing. We also compared the clustering quality of *ExCC* algorithm with *CluStream* and *HPStream* using 34 continuous dimensions of intrusion data. We recreate the experiment reported in [2] to make the comparison fair and fix $k = 14$ in ExCC. Figure 5 shows that cluster purity is better for *ExCC* than that for *CluStream*. Marginal lowering of cluster purity of *ExCC* compared to *HPStream* is explained by the latter's capability to perform projection.

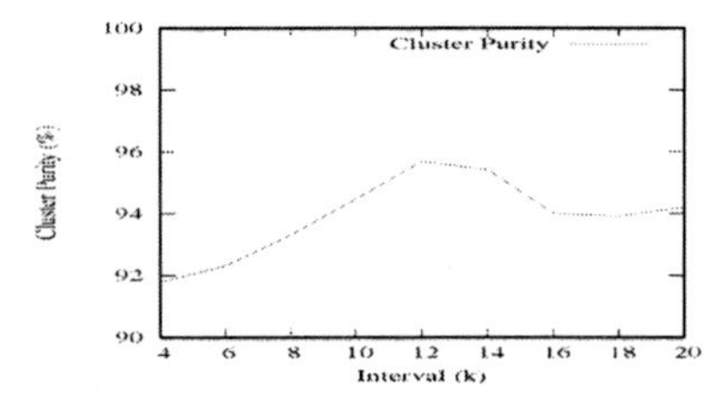


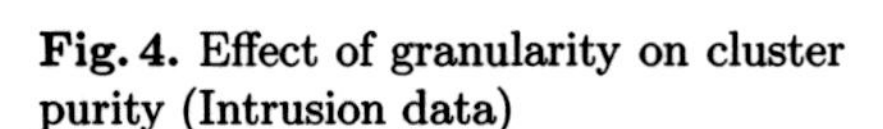

**Fig. 4.** Effect of granularity on cluster purity (Intrusion data)

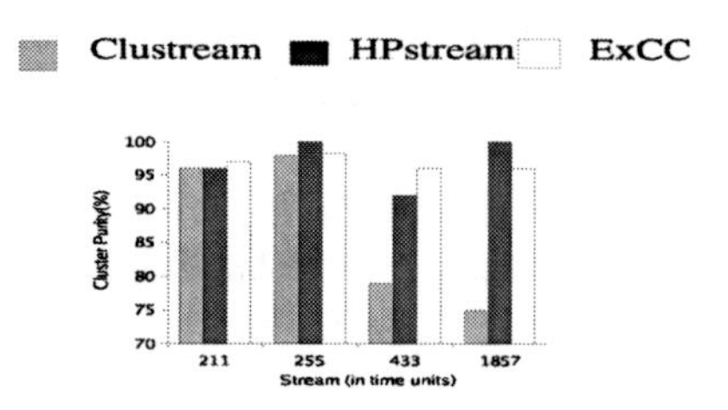


**Fig. 5.** Comparison of clustering quality for Intrusion data

## 4.3 Testing for Complete Clustering

We used Intrusion Data to demonstrate *completeness* of clustering. We picked up two small classes LAND and WAREZMASTER containing 21 and 20 records respectively. We placed records of class WAREZMASTER in the beginning of file and records of LAND after 150000 records in same file to simulate two small clusters (attacks). The data was streamed with $aat \in U[0,1]$ and clustering was performed after 150,000 and 400,000 records. The completeness property guarantees that attack WAREZMASTER occurring in the beginning of the first window and attack LAND occurring in the beginning of the second window (larger) are not missed.

In the first time window, the algorithm detected a 100% pure cluster of WAREZMASTER class containing 15 records. The remaining 5 records got merged with the cluster with NORMAL as a dominant class. In the second window 100 % pure cluster for LAND class was detected containing 14 records. The remaining seven records get merged with other clusters.

Table 1 shows the sample output of ExCC algorithm for the first window. We have highlighted only the targeted cluster and two other prominent clusters. It is interesting to see that the *smurf* cluster is a small and high density data region

**Table 1.** Cluster Description using 150000 records of KDD CUP '99

| Attack Class | Cells | Points | Dimensions Compactness | Cluster Description | Purity (%) |
|---|---|---|---|---|---|
| Smurf | 17 | 30398 | 0-24,29-33 (100%)<br>27 (85.71%)<br>25,26,28 (71.43%)<br>15,16 (50%) | $0 \leq d_0 \leq 4166.36$<br>$0 \leq d_1 \leq 4.95e+7$<br>$0 \leq d_2 \leq 368248$<br>$0 \leq d_3 \leq 0.2143$... | 100 |
| Neptune | 229 | 41047 | 0-14,16-19,28-33 (100%)<br>20,27 (92.86%) 22,25,<br>26 (71.43%) 23 (64.29%)<br>18 (50%) 15,21 (42.86%) | $0 \leq d_0 \leq 4166.36$<br>$0 \leq d_1 \leq 4.95e+7$<br>$0 \leq d_2 \leq 368248$<br>$0 \leq d_3 \leq 0.2143$... | 100 |
| Warez Master | 1 | 15 | 0-33 (100%) | $0 \leq d_0 \leq 4166.36$<br>$0 \leq d_1 \leq 4.95e+7$<br>$4.78e+6 \leq d_2 \leq 5.15e+6$... | 100 |

while *neptune* class is spread over much larger region. The output also indicates the compactness of the dimensions.

**Acknowledgment.** We are thankful to Naveen Kumar and Manoj Aggarwal for their help in understanding of KDD cup data and related experiments.

## 5 Conclusion

Completeness and exclusiveness of stream clustering algorithms is an important characteristics. We developed a novel algorithm which is guaranteed not to miss a cluster if one exists. The data points are mapped in data space and clustering is obtained by growing dense regions. The algorithm provides a better approximation of clusters in streaming data by ensuring that each point lies in the most appropriate cluster. The discovered clusters are reported to the user in terms of boundary in the data space, facilitating semantic interpretation. Experiments related to clustering purity, timings and completeness gave promising result.

## References

1. Aggarwal, C.C., Han, J., Yu, P.S: A Framework for Clustering Evolving Data Streams. In: VLDB conference, pp. 81–92 (2003)
2. Aggarwal, C.C., Han, J., Yu, P.S: Framework for Projected Clustering of High Dimensional Data Streams. In: VLDB conference. Canada, pp. 852–863 (2004)
3. Barbára, D.: Requirements of Clustering Data Streams. SIGKDD 3, 23–27 (2002)
4. Cao, F., Ester, M., Qian, W., Zhou, A.: Density-Based Clustering over an Evolving Data Stream with Noise. In: SIAM, pp. 326–337 (2006)
5. Dong, G., Han, J., Lakshmanan, L.V.S., et al.: Online Mining of Changes from Data Streams: Research Problems and Preliminary Results. ACM SIGMOD (2003)
6. Orlowska, M.E., Sun, X., Li, X.: Can Exclusive Clustering on Streaming Data be Achieved? SIGKDD 8, 102–108 (2006)
7. Maimon, O., et al.: Data Mining and Knowledge Discovery Handbook. Springer, Heidelberg (2004)
8. Lu, Y., Sun, Y., Xu, G., Liu, G.: A Grid-Based Clustering Algorithm for High-dimensional Data Streams. ADMA. China (2005)
9. Agrawal, R., et al.: Automatic Subspace Clustering of High Dimensional data for Data Mining application. In: ACM SIGMOD (1998)
10. KDD CUP 99 Intrusion Data: `http://kdd.ics.uci.edu//databases/kddcup99/kddcup99.html`
11. University of California at Irvine: UCI Machine Learning Repository, `http://www.ics.uci.edu/~mlearn/MLSummary`

# Clustering Quality Evaluation Based on Fuzzy FCA

Minyar Sassi[1], Amel Grissa Touzi[1], and Habib Ounelli[2]

[1] Ecole Nationale d'Ingénieurs de Tunis
Bp. 37, Le Belvédère 1002 Tunis, Tunisia
{minyar.sassi,amel.touzi}@enit.rnu.tn
[2] Faculté des Sciences de Tunis
Campus Universitaire -1060 Tunis, Tunisia
habib.ounelli@fst.rnu.tn

**Abstract.** Because clustering is an unsupervised procedure, clustering results need be judged by external criteria called validity indices. These indices play an important role in determining the number of clusters in a given dataset. A general approach for determining this number is to select the optimal value of a certain cluster validity index. Most existing indices give good results for data sets with well separated clusters, but usually fail for complex data sets, for example, data sets with overlapping clusters. In this paper, we propose a new approach for clustering quality evaluation while combining fuzzy logic with Formal Concept Analysis based on concept lattice. We define a formal quality index including the separation degree and the overlapping rate.

**Keywords:** Clustering Quality, Overlapping Rate, Separation Degree, Validity Index, Formal Concept Analyis, Fuzzy Concept Lattice.

## 1 Introduction

Fuzzy clustering allows objects of a data set to belong to several clusters simultaneously, with different degrees of membership. The data set is thus partitioned into a number of fuzzy partitions (clusters) [1].

Despite being a very effective technique, difficulties arise when evaluating the quality of clusters.

So, evaluating the quality of the clustering results is an important issue in cluster analysis. Because clustering is an unsupervised procedure, clustering results need be judged by an external criterion.

For low dimensional data sets (1-, 2- or 3-dimensional), humans can also evaluate the clustering results by visual observation. For high dimensional data sets (more then 3-dimentional), there is no objective criterion for evaluating the clustering results; they are assessed using a cluster validity index.

Depending on the type of clustering approach (crisp or fuzzy), there are various validity indices designed for evaluating the clustering results [2]. The general principle of these indices consists on minimizing the compactness within a cluster and maximizing the separation between clusters.

R. Wagner, N. Revell, and G. Pernul (Eds.): DEXA 2007, LNCS 4653, pp. 639–649, 2007.

These measures play an important role in determining the number of clusters. It is expected that the optimal value of the cluster validity index should be obtained at the true number of clusters. A general approach for determining the number of clusters is to select the optimal value of a certain cluster validity index. Whether a cluster validity index yields the true number of clusters is a criterion for the validity index.

Most existing indices give good results for data sets with well separated clusters, but usually fail for complex data sets, for example, data sets with overlapping clusters. One of the main reasons for this problem is that many fuzzy clustering methods fail to distinguish between partially overlapped clusters [3].

Because they disregard lack of considering the theoretical characterization of the overlapping phenomenon, they often yield questionable results for cases involving overlapping clusters [4].

To cure this problem, we propose to use conceptual scaling theory [5] based on an extension of Formal Concept Analysis (FCA) [6] which permits us to:

- Visualizing the clusters results will help us in interpreting and distinguishing overlapping clusters, and hence,
- Evaluating the quality of clusters while calculating a separation degree and an overlapping rate for a given clustering.

The rest of the paper is organized as follows. Section 2 discusses the backgrounds in FCA based on concepts lattices and Conceptual scaling. Section 3 presents our quality evaluation process. Section 4 concludes the paper and gives some future works.

## 2 Backgrounds

FCA provides a conceptual framework for structuring, analyzing and visualizing data, in order to make them more understandable [6]. In FCA, application domains are organized and structured according to concept lattices. In this section, we discuss about concept lattices and conceptual scaling.

### 2.1 Concept Lattices

The reason for the introduction of FCA was to relate the mathematically oriented theory of lattices and orders to practical problems [6,7].

In 1979, Wille [6] recognized that this description could be formalized by the introduction of 'formal concepts' of a given data table, which consists of a set $G$ of object, a set $M$ of attributes and a binary relation $I \subseteq G \times M$. Then the triple $K = (G, M, I)$ is called a formal context, representing just a set of statements of the form 'object $g$ has attribute $m$', written '$g\ I\ m$'.

The basic definition of a 'formal concept' of $K$ is based on two well-known operations: For any subset $X \subseteq G$ we are interested in the set $X\uparrow$ of all common attributes of $X$, defined formally by $X\uparrow := \{m \in M \mid \forall g \in X \quad g\ I\ m\}$ and dually for any $Y \subseteq M$ we are interested in the set $Y\downarrow$ of all common objects of $Y$, defined formally by $Y\downarrow := \{g \in G \mid \forall m \in Y \quad g\ I\ m\}$. A formal concept of a formal context $K$ is a pair $(A, B)$ where $A \subseteq G$, $B \subseteq M$ and $A\uparrow = B$ and $B\downarrow = A$. $A$ is called the extent, $B$ the intent of $(A, B)$.

The set of all formal concepts of $K$ is denoted by $B(K)$. The conceptual hierarchy among concepts is defined by set inclusion: For $(A_1,B_1)$, $(A_2,B_2)\in B(K)$ let $(A_1,B_1)\le(A_2,B_2)\Leftrightarrow A_1\subseteq A_2$ (which is equivalent to $B_2\subseteq B_1$.

An important role is played by the object concepts $\gamma(g):=(\{g\}\uparrow\downarrow,\{g\}\uparrow)$ for $g\in G$ and dually the attribute concepts $\mu(m):=(\{m\}\downarrow,\{m\}\downarrow\uparrow)$ for $m\in M$.

The pair $(B(K),\le)$ is an ordered set, i.e., $\le$ is reflexive, anti-symmetric, and transitive on $B(K)$.

## 2.2 Conceptual Scaling

An arbitrary ternary relation on a set $G$ of 'objects' is a special case of a ternary relation among three sets of objects. In formal descriptions of measurements by data tables the following three sets play a fundamental role: $A$ set $G$ of 'objects', a set $M$ of 'measurements' and a set $W$ of values which are related by a ternary relation whose elements $(g,m,w)$ are interpreted as 'object $g$ has at measurement $m$ the value $w$'. That leads to the following definition of a many-valued context $(G,M,W,I)$ as a quadruple of four sets, where the elements of $G$ are called 'objects', the elements of $M$ 'many-valued attributes', the elements of $W$ 'values', and $I$ is a ternary relation, $I\subseteq G\times M\times W$, such that for any $g\in G$, $m\in M$ there is at most one value w satisfying $(g,m,w)\in I$. Therefore, a many-valued attribute $m$ can be understood as a (partial) function, and we write $m(g)=w$ *iff* $(g,m,w)\in I$. A many-valued attribute m is called complete *iff* for any $g\in G$ there is (exactly one) $w\in W$ such that $m(g)=w$. $(G,M,W,I)$ is called complete if each $m\in M$ is complete [7].

The central process in conceptual scaling theory is the construction of a formal context $S_m=(W_m,M_m,I_m)$ for each $m\in M$ such that $W_m\supseteq_m G:=\{m(g)|g\in G\}$. Such formal contexts, called conceptual scales, represent a contextual language about the set of values of $m$. Usually one chooses $W_m$ as the set of all 'possible' values of $m$ with respect to some purpose. Each attribute $n\in M_m$ is called a scale attribute. The set $n\downarrow=\{w\ |wI_m n\}$ is the extent of the attribute concept of $n$ in the $S_m$ scale. Hence, the choice of a scale induces a selection of subsets of $W_m$. The set of all intersections of these subsets constitutes just the closure system of all extents of the concept lattice of $S_m$.

The granularity of the language about the possible values of m induces in a natural way a granularity on the set $G$ of objects of the given many-valued context, since each object g is mapped via m onto its value $m(g)$ and $m(g)$ is mapped via the object concept mapping $\gamma_m$ of $S_m$ onto $\gamma_m(m(g)): g\rightarrow m(g)\rightarrow\gamma_m(m(g))$.

Hence the set of all object concepts of $S_m$ plays the role of a frame within which each object of $G$ can be embedded. For two attributes $m$, $m'\in M$ each object $g$ is mapped onto the corresponding pair: $g\rightarrow(m(g),m'(g))\rightarrow(\gamma_m(m(g)),\gamma_{m'}(m'(g)))\in B(S_m)\times B(S_{m'})$.

The standard scaling procedure, called plain scaling, constructs from a scaled many-valued context $((G,M,W,I),(S_m \mid m\in M))$, consisting of a many-valued context $(G,M,W,I)$ and a scale family $(S_m \mid m\in M)$ the derived context, denoted by $K := (G,\{(m,n) \mid m\in M, n\in M_m\},J)$, where $gJ(m,n)$ *iff* $m(g)I_m n$ $(g\in G, m\in M, n\in M_m)$.

The concept lattice $B(K)$ can be (supremum-) embedded into the direct product of the concept lattices of the scales [8]. That leads to a very useful visualization of multidimensional data in so-called nested line diagrams [9]).

## 3 The Quality Evaluation Process

As we have mention in section 1, evaluating the quality of clusters is an important issue in cluster analysis. It often based on a clustering validity index. The general principle of these indices consists on minimizing the compactness within a cluster and maximizing the separation between clusters. Most existing criteria give good results for data sets with well separated clusters, but usually fail for complex data sets, for example, data sets with overlapping clusters.

In this paper, we use conceptual representation of clustering results which permits us to formally calculate the compactness and the separation degrees which permits us to evaluate the quality of clusters. However, there are many situations in which uncertainty information also occurs. For example, it is sometimes difficult to judge whether an object belongs totally to an attribute or not. Traditional conceptual representation is hardly able to represent such vague information. To tackle this problem, we propose to combine fuzzy logic [10] with FCA as Fuzzy FCA (FFCA). Once this structure is built, we calculate a certain similarity distance based on membership degrees. This distance permits us to evaluate the compactness and the separation of the clustering result.

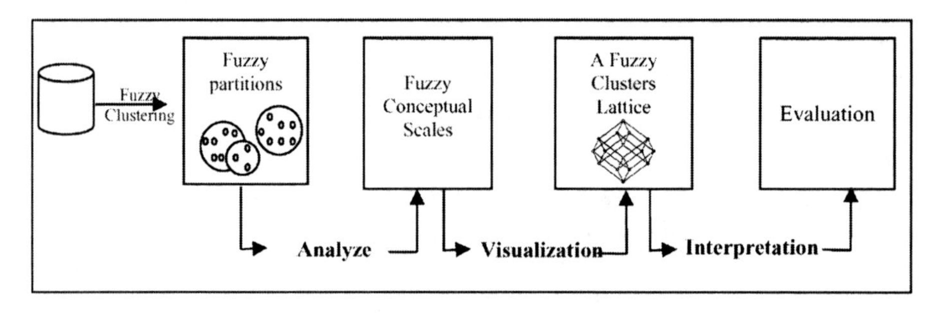


**Fig. 1.** The Quality Evaluation Process

The principle of our quality evaluation process determines three steps. The first step consists of analysing the fuzzy clusters for a given dataset based on fuzzy conceptual scaling. The second step consists of visualizing the results based on fuzzy

Formal Concept Analysis. This allows deducing overlapping between clusters. The third step consists of evaluating the quality of clustering results which includes the separation between clusters and compactness within a cluster. Fig. 1 shows the proposed approach.

### 3.1 Analyze

Fuzzy clustering methods allow objects to belong to several clusters simultaneously, with different degrees of membership. A data, set $X$ is thus partitioned into $C$ fuzzy partitions (clusters). In many applications training data relates individual objects to attributes that take on several values. For the generation of fuzzy formal context, we propose to relate objects with the clusters of each attribute that take on several values. These values represent the membership degrees of each object in each cluster. Fuzzy formal context incorporate fuzzy clustering, to represent vague information.

**Definition 1.** A fuzzy conceptual scale for a set $Y \subseteq M$ is a (single-valued) fuzzy formal context $S_Y := (G_Y, M_Y, I_Y = \varphi(G_Y \times M_Y))$ with $G_Y \subseteq \times_{m \in Y} W_m$.
The idea is to allow objects $G$ to belong to several clusters simultaneously. We replace the attribute values in $W_m$ with different degrees of membership. Each relation $(g,m) \in I_Y$ has a membership value $\mu(g,m)$ in $[0,1]$. The sum of the values of each fuzzy conceptual scale is equal to 1.

**Definition 2.** Given a fuzzy conceptual scale $S_Y := (G_Y, M_Y, I_Y = \varphi(G_Y \times M_Y))$, we define $\alpha - Cut(S_i) = (C(S_i))^{-1}$ where $C(S_i)$ is the number of clusters of scale $S_i$.

*Example:* Table. 1 present the results of fuzzy clustering applied to price and surface scales. For price scale, fuzzy clustering generate three clusters (C1,C2 and C3) for surface attribute, two clusters (C4,C5). Table 1 shows the fuzzy conceptual scales for price and surface attributes with $\alpha - Cut$. In this example, $\alpha - Cut(price) = 0.3$ and $\alpha - Cut(surafce) = 0.5$.

**Table 1.** Fuzzy Conceptual Scales with $\alpha - Cut$ for price and surface attributes

| | Price | | | Surface | |
|---|---|---|---|---|---|
| | C1 | C2 | C3 | C4 | C5 |
| A1 | - | 0.5 | 0.4 | 0.5 | 0.5 |
| A2 | 0.3 | 0.6 | - | - | 0.6 |
| A3 | 0.7 | - | - | 0.7 | - |
| A4 | - | 0.4 | 0.5 | - | 0.8 |
| A5 | - | 0.4 | 0.4 | 0.6 | - |
| A6 | 0.5 | 0.3 | - | 0.5 | 0.5 |

### 3.2 Visualization

Traditional FCA is hardly able to represent fuzzy properties from uncertainly data. To tackle this problem, we use a new technique that incorporates fuzzy logic into FCA as Fuzzy Formal Concept Analysis (FFCA), in which uncertainty information is directly represented by a real number of membership value in the range of [0,1]. So we give some defined the so called Fuzzy Formal Context, the Fuzzy Formal Concept Analysis and the similarity concept.

**Definition 3.** Given a fuzzy formal context $K=(G,M,I)$ and an $\alpha-Cut$, we define $X^*=\{m\in M|\forall g\in X:\mu(g,m)\geq\alpha-Cu\}$ for $X\subseteq G$ and $Y^*=\{g\in G|\forall m\in Y:\mu(g,m)\geq\alpha-Cu\}$ for $Y\subseteq M$. A fuzzy formal concept (or fuzzy concept) of a fuzzy formal context $(G,M,I)$ with an $\alpha-Cut$ is a pair $(X_f=\varphi(X),Y)$ where $X\subseteq G$, $Y\subseteq M$, $X^*=Y$ and $Y^*=X$. Each object $g\in\varphi(X)$ has a membership $\mu_g$ defined as $\mu_g=\min_{m\in Y}\mu(g,m)$. Where $\mu(g,m)$ is the membership value between object $g$ and attribute $m$, which is defined in $I$. Note that if $Y=\{\ \}$ then $\mu_g=1$ for every $g$.

Generally, we can consider the attributes of a formal concept as the description of the concept. Thus, the relationships between the object and the concept should be the intersection of the relationships between the objects and the attributes of the concept. Since each relationship between the object and an attribute is represented as a set of membership values in fuzzy formal context, then the intersection of these membership values should be the minimum of these membership values, according to fuzzy theory [8].

**Definition 4.** Let $(A_1,B_1)$ and $(A_2,B_2)$ be two fuzzy concepts of a fuzzy formal context $K=(G,M,I=\varphi(G\times M))$.

$(\varphi(A_1),B_1)$ is a the subconcept of $(\varphi(A_2),B_2)$ denoted as $(\varphi(A_1),B_1)\leq(\varphi(A_2),B_2)$ if and only if $\varphi(A_1)\subseteq\varphi(A_2)$ $(\Leftrightarrow B_2\subseteq B_1)$.

Equivalently, $(A_2,B_2)$ is the superconcept of $(A_1,B_1)$.

**Definition 5.** A fuzzy concept lattice of a fuzzy formal context $K$ with an $\alpha-Cut$ is a set $C$ of all fuzzy concepts of $K$ with the partial order $\leq$ with the $\alpha-Cut$ value. We noted as $\Im(C)$.

**Definition 6.** The similarity of a fuzzy formal concept $C_1=(\varphi(A_1),B_1)$ and its subconcept $C_2=(\varphi(A_2),B_2)$ is defined as:

$$S(C_1C_2)=\frac{|\varphi(A_1)\cap\varphi(A_2)|}{|\varphi(A_1)\cup\varphi(A_2)|} \tag{1}$$

*Exemple:* The corresponding fuzzy concept lattices of fuzzy context presented in table 1 are given by the following fuzzy lattices. These are illustrated in fig. 2.

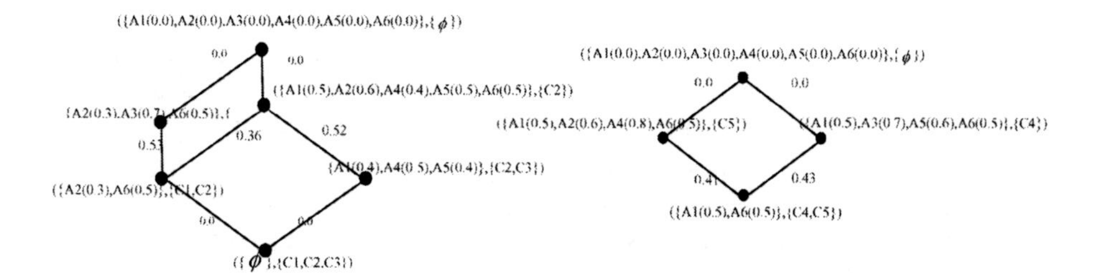


**Fig. 2.** The fuzzy concept lattices of the context in the Table 1

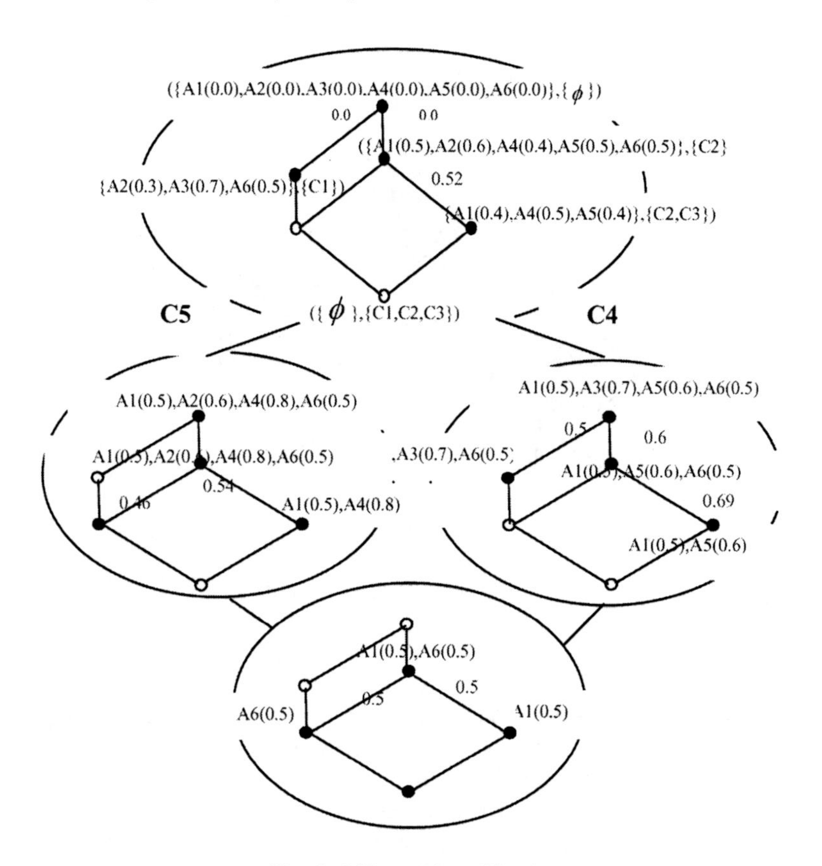


**Fig. 3.** A Fuzzy Nested Lattice

This very simple sorting procedure gives us for each many-valued attribute the distribution of the objects in the line diagram of the chosen fuzzy scale. Usually, we are interested in the interaction between two or more fuzzy many-valued attributes. This interaction can be visualized using the so-called fuzzy nested line diagrams. It is used for visualizing larger fuzzy concept lattices, and combining fuzzy conceptual scales on-line. Fig. 3 shows the fuzzy nested lattice constructed from Fig.2.

In this fuzzy nested line diagram, we are interested to see for each concepts of diagram represented in Fig.2 how its students are distributed in the fuzzy scale surface. We blow up each circle of fuzzy line diagram of Fig. 2 and insert the fuzzy line diagram of the surface fuzzy scale. Hence, Fig. 3 represents all pairs $(c,d)$ of concepts $c$ from the first and concepts $d$ from the second fuzzy lattice. This structure is called the direct product of the two given fuzzy lattices.

From the fuzzy nested lattice, we can draw a nice usual fuzzy lattice of the same fuzzy context. This illustrated in Fig. 4.

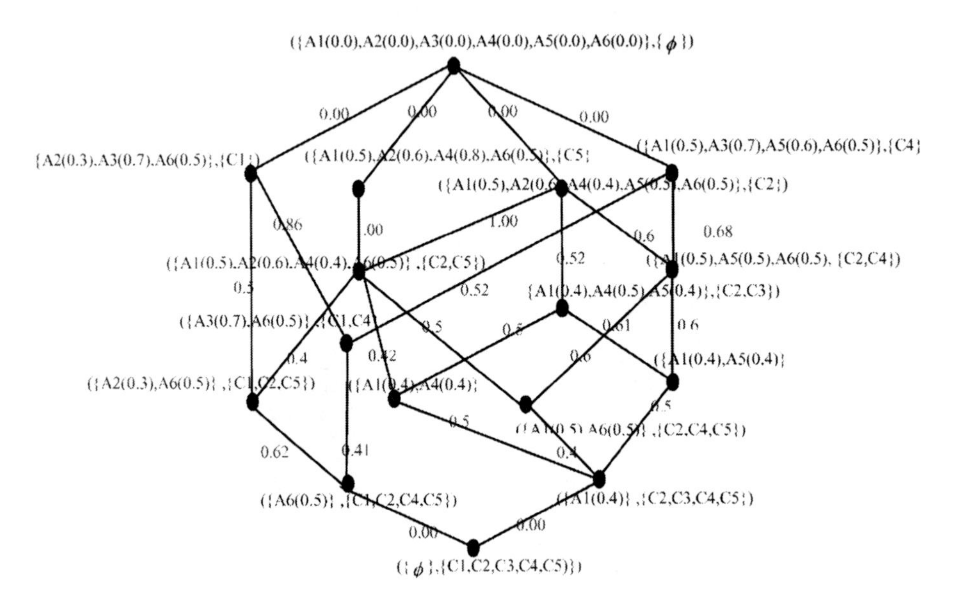


Fig. 4. A Fuzzy Clusters Lattice: FCL

### 3.3 Quality Evaluation

In general, the evaluation is based on a clustering validity index. The general principle of these indices consists on minimizing the compactness within a cluster and maximizing the separation between clusters.

Because they disregard lack of considering the theoretical characterization of the overlapping phenomenon, they often yield questionable results for cases involving overlapping clusters.

To cure this problem, we propose a new process of quality clustering evaluation. We give firstly an interpretation of the generated clusters and then study the quality. It consists of selection of characteristics in a given data set.

From fig. 4, we can deduce the possible overlapping between the various clusters. Let $V = \{v_j : j = 1,\ldots,C\} \subset R^M$ a set of $C$ clusters generated from the dataset $X = \{x_i : i = 1,\ldots,N\} \subset R^M$

We define a distance function $D$ as follows:

$$D : V \rightarrow |R^{+}$$
$$(v_j, v_k) \rightarrow d$$

$d$ is the weight of the arc connecting $v_j$ with $v_k$ in FCL. We note $v_j \Re v_k$ if $\exists$ $d / D(v_j, v_k) = d$

The following properties are required:

-If $v_j \Re v_k$ and $v_k \Re v_i$ then $v_j \Re v_i$.

-If $v_j \Re v_k$ then $v_j$ and $v_k$ overlapped.

We study the overlapping phenomenon in the case of deducing the overlapping rate.

These properties enabled us to deduce the overlapping between different clusters. They will be used in the quality evaluation process. So, we define the separation degree and the overlapping rate. These measures form the quality index which will judge if two clusters must be merged or not.

Let $V = \{v_j : j = 1,..., C\} \subset R^M$ a set of $C$ clusters generated from the dataset $X = \{x_i : i = 1,..., N\} \subset R^M$, $v_j \Re v_k$ and $v_k \Re v_i$ having respectively $D(v_j, v_k) = d_{jk}$ and $D(v_k, v_i) = d_{ki}$ as similarity between concepts $S(C_j, C_k)$ and $S(C_k, C_j)$. We can deduce that $D(v_j, v_i) = d_{ji} = d_{jk} + d_{ki} = S(C_k, C_i)$.

The separation $Sep$ is given by equation 1:

$$Sep = \sum_{j=1, k \neq j}^{C} S(C_j, C_k) \quad (2)$$

In general, when $Sep$ is large, the $j^{th}$ and $k^{th}$ clusters are well separated.

For example, $D(v_1, \{v_1, v_4\}) = 0.86$ and $D(\{v_1, v_4\}, v_4) = 0.52$ imply $D(v_1, v_4) = 0.86 + 0.52 = 1.38$.

We can calculate the overlapping rate, noted $Overl$, as the ratio between the number of extensions of sub-concepts, noted $Extension(Sub-Concepts)$, and the number of extensions of super-concepts, noted $Extension(Super-Concepts)$. This rate is given by equation 2.

$$Overl = \frac{\sum Extension(Sub-Concepts)}{\sum Extension(Super-Concepts)} \quad (3)$$

In general, a definition for the overlap rate implements the following principle: 1) the overlap tends to decrease $(\rightarrow 0)$ as the two components become more separated, 2) the overlap rate increases $(\rightarrow 1)$ as the two components become more strongly overlapped.

Once these requirements are met, we can evaluate the quality of clusters while basing on separation degree and the overlapping rate. We noted $Ind\_Quality$ as the quality index for a given clustering.

$$Overl = \frac{Overl}{Sep} \quad (4)$$

The clusters to be merged into one cluster are those which must maximizing the overlapping rate and minimizing the separation degree. So, a large value of $Ind\_Quality$ imply that the clusters must be merged into one.

## 4 Conclusion

Validity indices measure the goodness of the clustering result. A clustering is considered good if it optimizes two conflicting criteria. One of these is related to within-class scattering (the compactness), which needs to be minimized; the other to between-class scattering (the separation), which needs to be maximized. Most existing indices give good results for data sets with well separated clusters, but usually fail for complex data sets, for example, data sets with overlapping clusters.

This motivated our search for a new quality evaluation process based on Fuzzy FCA (FFCA). It consists of three steps. The first step consists of analyse the clustering results. To do this, we have proposed the fuzzy conceptual scaling notion. The second step consists of visualization. The FFCA has been proposed. It bases itself on a Fuzzy Clusters Lattice (FCL) which includes the similarity distances between different concepts in the FCL. The third step consists of evaluation the quality of generated clusters. We have defined a formal separation degree and an overlapping rate. We have defined a quality index while basing on the separation degree and the overlapping rate. The large value if this index means that the clusters must be merged into one cluster.

Future work will focus on the applicability of our quality evaluation process formal to test clustering algorithms in a more controlled way.

## References

1. Menard, M., Eboueya, M.: Extreme physical information and objective function in fuzzy clustering. Fuzzy Sets and Systems 128, 285–303 (2002)
2. Bezdek, J.C.: Pattern Recognition in Handbook of Fuzzy Computation, ch. F6. IOP Publishing Ltd, Bristol (1998)
3. Sun, H.: A theory on distinguishing overlapping components in mixture models, Research Report, DMI, University of Sherbrooke, No 345 (Novenber 2003)
4. Sassi, M., Grissa Touzi, A., Ounelli, H.: Two Levels of Extensions of Validity Function Based Fuzzy Clustering. In: The 4th International Multiconference on Computer Science & Information Technology (CSIT 2006), Amman-Jordan (April 5-7, 2006)
5. Priss, U.: Formal Concept Analysis in Information Science, Annual Review of Information Science and Technology (ARIST), Preview, vol. 40 (2006)

6. Ganter, B., Wille, R.: Formal Concept Analysis: Mathematical Foundations. Springer, Heidelberg (1999)
7. Valtxhev, P., Missaoui, P., Godin, R.: Formal Concept analysis for Knowledge Discovery and Data Mining: The New Challenges. In: Eklund, P.W. (ed.) ICFCA 2004. LNCS (LNAI), vol. 2961, Springer, Heidelberg (2004)
8. Ganter, B., Wille, R.: Formal Concept Analysis: mathematical foundations. Springer, Heidelberg (1999)
9. Vogt, F., Wille, R.: TOSCANA - a graphical tool for analyzing and exploring data. In: Tamassia, R., Tollis, I.G. (eds.) Graph Drawing, pp. 193–205. Springer, Heidelberg (1994)
10. Zadeh, L.A.: Fuzzy sets. Information and Control 8, 338–353 (1965)

# Comparing Clustering Algorithms and Their Influence on the Evolution of Labeled Clusters

Rene Schult

Otto-von-Guericke-University Magdeburg
Institute of Technical and Business Information Systems

**Abstract.** We study the influence of different clustering algorithms on cluster evolution monitoring in data streams. The capturing and interpretation of cluster change delivers indicators on the evolution of the underlying population. For text stream monitoring, the clusters can be summarized into *topics*, so that cluster monitoring provides insights on the data and decline of thematic subjects over time. However, such insights should always be taken with a grain of salt: The quality of the clusters has a decisive impact on the observed changes. In the simplest case, cluster change across the stream may be due to the low quality of the original cluster than to a drift in the population belonging to this cluster. We show our framework *ThemeFinder* for topic evolution monitoring in streams and compare the influence to the quality of two very different cluster algorithms. After an evaluation of different cluster algorithms with external and internal quality measures, we use the center based bisecting k-means algorithm and the density-based DBScan algorithm. Our results show that the influence is relatively high and show that different clustering algorithms results allow to draw conclusion to the evaluation of the other cluster algorithm. Our experiments were done on a subarchive of the ACM library.

## 1 Introduction

Stream clustering considers two different perspectives: On the one hand, one is focused on grouping multiple independent streams of signals (e.g. signals from multiple sensors) into clusters and adapting or monitoring these clusters while the senders continue delivering signals. On the other hand, data of complex structures or texts can arrive as a single stream, the contents of which are grouped into clusters according to similarity; cluster monitoring is then devoted to identify changes in these clusters, as the data flow continues. In this paper, we take the second approach on stream clustering and cluster monitoring into account and study the impact of different clustering algorithms on cluster evolution. For monitoring clusters evolution we focus on monitor labels of clusters.

As a case study, we consider the ACM archive[1], which uses the ACM taxonomy for keyword assignment, categorization and browsing as well. A stream of documents is added to the archive and is ranged to the existing taxonomy. This taxonomy could be considered as the labels/topic of the clusters of the documents under this taxonomy

[1] http://portal.acm.org/ccs.cfm

R. Wagner, N. Revell, and G. Pernul (Eds.): DEXA 2007, LNCS 4653, pp. 650–659, 2007.

point. The ACM taxonomy has been expanded with subjects like "data mining" and "image databases" under the existing subject "database applications" to assign the document to the appropriate taxonomy. All existing documents in "database applications" are not automaticially assigned to the new created subcategories like "data mining" or "image databases". If a knowledge seeker is interested on early advances on data mining, he or she has to go through the whole subarchive on database applications. Although keyword-based search is available, the appropriate keywords for search on data mining in the early nineties are likely to be different from those of today. A retrospective *re-categorization* of the documents or at least the discovery of the keywords characterizing them is needed.

We propose the monitoring of cluster evolution by focusing on monitor cluster labels. We are aiming to find persistent labels of clusters or at least some changes at the labels in order to find the real cluster evolution. We have set up our experiments under the assumption that the terminology of the document archive changes over long time. Thus, we should be able to detect cluster changes over time. In our case study of the ACM archive, we have looked for evidence on the emergence of themes like "data mining" and for words that correspond to this theme like subtopics of this from the time it emerged till today. To detect temporal trends, we have also used a non accumulated archive in [11]. Here our goal is to find out the influence of the used cluster algorithm to the quality of detected long term trends. We have discovered evolution of the detecting trends, but we have tried to minimize the changes of the used feature space.

We have applied text clustering in an evolving feature space. It must be stressed that a classification is not appropriate for this problem: Classification (even adaptive classification) requires a labeled dataset. Here, the challenge lays more in identifying documents that adhere to a yet unknown subject, with these documents having been assigned to some more generic class label. Moreover, themes consists of words in a feature space that must be adapted as the language of the documents' authors evolves. In [11] and in [10] we present an algorithm, the *ThemeFinder* , to detect such "themes" at cummulated and non-cummulated datasets and show it with experiments on clustering the data with a bisecting k-means algorithm. Now we will show the influence of the used clustering algorithm to the functionality and the quality of the *ThemeFinder* and to the quality of the detected cluster evolutions.

In the next section, we have discussed relevant research. In section 3, we first have short present the *ThemeFinder* and the used clustering algorithms. Section 4 we have shown our new experiments with the ACM archive. The last section concludes our work.

## 2 Related Work

The subjects of Topic Detection and Topic Tracking are defined in [1], where the five tasks of TDT are enlisted. As stated in that book, TDT concentrates on the detection and tracking of *stories* (a "topic" is a story) and encompasses the tasks of (1) story segmentation, (2) first story detection, (3) cluster detection, (4) tracking and (5) story link detection. However in this paper we address the influence of different clustering algorithms on the quality of label monitoring that is not the focus of TDT works.

Moringa et al. [8] and Wang and McCallum [12] present different methods for detecting topics over time, but in [8] only persistent topics over the whole time are detected and in [12] they have the assumption that topics never change over time, which differ from our point of view.

In [5] the authors compare hierarchical algorithms with K-means and bisecting K-means. First the authors compare different hierarchical algorithms to select the algorithm which produces the best quality. After that, they compare the selected hierarchical (UPGMA) with both K-means algorithms. The result of this comparison is the following: the bisecting K-means shows the same or better quality as other algorithms. One reason for this result could be the nature of text documents. The authors used the entropy and the "Overall Similarity" to measure the quality of the clustering algorithm. The comparison was made with a different number of clusters. We use these results as input for our evaluation of different cluster algorithms. We exclude the incremental versions of the k-means algorithm, developed by Lin et al in [6] or from Pham et al in [9], and the online spherical k-means algorithm presented by Zhong in [13] from the evaluations point because we will not mix incremental and not incremental algorithms.

We decided also to use the DBScan algorithm from Ester [7] because it is a fundamental different cluster algorithm to k-means. So we have two very different algorithms for evaluating the impact of clusterer on the monitoring process.

## 3 *ThemeFinder* on an Accummulated Document Collection

In this section we shortly summarise our model of the *ThemeFinder* in order to provide a better understanding of our algorithm and for understanding the experiments and the results. For a more detailed presentation we refer to [10].

The *ThemeFinder* takes a text collection as input, which consist of documents over several time periods $t_i$. A document is described as a vector of words derived from a feature space. We do not observe documents in their entirety, but concentrate on title, keywords and a limited number of sentences (e.g. from the document's abstract), assuming that these fragments are particularly designed to disseminate the content to the reader in a compact way[2].

In each period $t_i$, the document set $D_i$ contains the documents of the period $t_{i-1}$ and the documents which have been inserted in the archive during the period $t_i$. The feature space is the set of the $n$ "dominant" words, which we define as the words with the highest TF$\times$IDF values. For each time period $t_i$ we define the "*period-specific feature space*" as the set of $n$ dominant words of the document set of the period. For some documents a small value of $n$ could produce null-vectors.

The notions of label and thematic cluster reflect the insights of [4] on concept indexing and of [2] on latent semantic indexing: Both studies agree, that the importance of a component can be derived from the weight it receives in the analysis. We define the *label for the clusters* consisting of all words of the actual feature space. In it the fraction of documents in the cluster containing the word, divided by the number of documents in the cluster have to exceed a threshold. It means that the dominant words in the cluster

[2] PageRank of Google also checks only a small part of the document, including header, preample etc.

become the label, in case the support of the word passes over a threshold. All clusters, whose label is not empty, are called *thematic clusters*.

In order to deal with clustering quality assessment, *ThemeFinder* will define a clustering as being *good*, if the number of thematic clusters in it is no less than a threshold. Hence we call it a good clustering. The feature space used for the clustering is good enough and doesn't have be updated if it results in a good clustering. It is implicitly assumed that the period-specific feature space always delivers a good clustering of the document set, since it reflects exactly the dominant features of the documents. Only in case the feature space for the actual period is not sufficient it will be changed to the periodic specific feature space.

Based on these clarified fundamentals we are now briefly explaining the procedure of *ThemeFinder*. Starting with the establishment of a clustering at period $t_1$ for the document set $D_1$ using the period-specific feature space, the first set of labels is derived over $\zeta(D_1, FS_1)$. For each subsequent period $t_i, i \geq 2$, the algorithm first builds the clustering $\zeta_i$ over $D_i$ using the feature space of the previous period(s) (originally $FS_1$). *ThemeFinder* initially verifies, if the used feature space is a good feature space for this period. If not, the feature space is replaced by the current period-specific feature space and the actual period is re-clustered. In case a good clustering is achieved, *ThemeFinder* compares this to the clustering of the previous period and identifies pairs of similar clusters. The function $best\_match(\cdot)$ returns for each cluster in period $t_{i-1}$ the cluster of period $i$ with the most similar label, or the empty set if no such cluster exists.

## 4 Experimenting with the ACM Archive

**The Dataset.** As a case study we used the ACM archive and particularly the subsections of H2.8, named "database applications", which we downloaded from the ACM library. We considered the documents accumulated until December 2004. We will use the term "ACM subarchive"for this collection hereafter.

The ACM taxonomy below H2.8 has been *gradually* expanded with 5 specific topics. Once made available to the authors, they have been used as keywords. The topic "image databases" appears already in the first period of observations ($\leq$ 1994), the topic "data mining" first appears in 1995, "spatial databases and GIS" in 1996, while "scientific databases" and "statistical databases" are used since 1997. The parent topic "database applications" is further not used, giving a total of 5 "ACM topics". So we decide to use only documents from 1996 till 2004. The distribution of documents is shown in Table 1.

**Table 1.** Number of documents in the ACM subarchive "database applications"

| Period | 1996 | 1997 | 1998 | 1999 | 2000 | 2001 | 2002 | 2003 | 2004 |
|---|---|---|---|---|---|---|---|---|---|
| numbers | 89 | 150 | 369 | 675 | 1155 | 1634 | 2338 | 3371 | 4434 |

For our case study, we considered the title and keywords of each document, removing the ACM topic to which the document was assigned. We did not consider abstracts, because many documents did not have an abstract and those having one could otherwise bias the feature space contents. The document excerpts were vectorised as described before. For the first partition, its period-specific feature space was used. For subsequent partitions, the selection of the feature space was governed by the heuristics of *ThemeFinder*, as we will see in the follow up. As preprocessing tasks we used basic NLP preprocessing and stopword removal and vectorisation with TFxIDF weighting.

The goal of our case study is to show the influence of the clustering algorithm for the functionality of *ThemeFinder*. First we start with the bisecting k-means algorithm, as a result from [5] and try to find out the best cluster number. We produced clusterings with $k = \{2...10\}$ over the whole dataset and calculate different quality indexes. We used the Normalized Mutual Information (NMI), the purity index and the rand index. The NMI measure has the best value at $k = 5$, the purity and the Rand index have at different $k$ the same, best value, but both include $k = 5$, see table 2. So we decide to use $k = 5$ for the experiments.

**Table 2.** External quality measures at different k

| | 2 | 3 | 4 | 5 | 6 | 7 | 8 | 9 | 10 |
|---|---|---|---|---|---|---|---|---|---|
| NMI measure | 0,09 | 0,17 | 0,16 | 0,19 | 0,15 | 0,17 | 0,16 | 0,18 | 0,17 |
| purity index | 0,58 | 0,62 | 0,63 | 0,66 | 0,64 | 0,65 | 0,64 | 0,66 | 0,66 |
| rand index | 0,46 | 0,56 | 0,61 | 0,62 | 0,61 | 0,62 | 0,62 | 0,62 | 0,62 |

Secondly we want to check if a change of the clustering algorithm leads to an increasing of the quality of the clustering. For good measure the quality of the clustering we used the same indexes as before. We clustered the dataset with the normal k-means, a spherical k-means and a fuzzy c-means algorithm as partitional algorithms. We used the Non-negative Matrix Factorization (NMF) technique that at first minimises the Kullback-Leibler divergence and second minimises squared Euclidean distance. We also tried a kernel k-means clustering algorithm. As result of all the measures, it can be said that the best clustering is produced by the bisecting k-means algorithm, see table 3.

**Table 3.** External quality measures at different algorithms (k=5)

| | bisecting KM | KM | spherical KM | fuzzy CM | NMF(K-L) | NMF(en) | kernel KM |
|---|---|---|---|---|---|---|---|
| NMI measure | 0,19 | 0,14 | 0,15 | 0,05 | 0,04 | 0,11 | 0,15 |
| purity index | 0,66 | 0,62 | 0,64 | 0,58 | 0,58 | 0,61 | 0,63 |
| rand index | 0,62 | 0,59 | 0,59 | 0,41 | 0,57 | 0,57 | 0,59 |

In contrast to the nature of the bisecting k-means algorithm, we decide to used the density-based DBScan algorithm from [3].

**Number of themes.** The number of themes that our algorithm can find, depends on the number of thematic clusters it finds in each period of observation. Hence, we have varied the value of $k$ for bisecting k-means. Since bisecting k-means generates one bucket-cluster, in which all otherwise dissimilar vectors are put together, a clustering can contain at most $k-1$ thematic clusters. Based on the relatively high value of the threshold for labels $(0,60)$, we have set the threshold for cluster matching from different periods so that a clustering over a given feature space is sufficient if three clusters are thematic clusters. As explained before, in case a clustering contains less thematic clusters, the feature space is replaced by the period-specific feature space.

### 4.1 Using *ThemeFinder* at the Clustering Results

The new ACM topics in the subarchive indicate that the ACM taxonomy designers have responded to emerging research threads. These threads are associated with a drift in the frequent terms in the documents: new research areas use new terms. A simple way of detecting such a drift is by clustering the documents and check whether the thematic clusters degenerate. So we first have checked whether the anticipated themes could be found without using *ThemeFinder*.

A high number of feature space changes is not desirable, because it is apt to features of short-term popularity and prohibits a long-term observation of the clusters.

**Bisecting K-Means Clustering Results.** For the bisecting k-means clustering we used the time periods from "1996" to "2004". So the first time period was "1996" and the feature space of this period was used for the first clustering. In case three clusters should be minimum thematic clusters, a change of the feature space is only needed for 2 periods. The same holds true for the threshold of only 2 thematic clusters, which is less restrictive. Although the value of threshold for thematic clusters is too small (4 thematic clusters) for generalisation, this experiment indicates that the selected value of 2 thematic clusters for matching is appropriate for those experiments. We found that no label persists across all periods. However, there are several quite interesting themes: the label {datum,mine} qualifies as theme because it is present at 4 periods, while the label {retrieval,image,base} persists in 4 non-consecutive periods. If we allow that a label may change by at most one word, then {retrieval,image} with the additional word "base" becomes a very stable theme, appearing for the last 5 time periods. This theme obviously refers to "image retrieval", a subcategory of image databases that emerges in 1997, disappears for a short time and then becomes stable from 2000 on.

The emergence and evolution of labels associated to data mining is also very interesting. The first cluster of period 1996 contains the words "discovery", "knowledge" and "datum" (data) in all documents, the word "pattern" is also very frequent. With the period-specific feature space of 1997, the cluster on data mining becomes separated under the label {datum,discovery}. The words "knowledge" and "discovery" persist in the next three periods. The label {datum,discovery,knowledge} is present at 3 periods, the "knowledge discovery [from] data". Starting from 1998, the label {datum,mining} becomes present; the two sibling labels {datum,mine} and {datum,mining} finally absorb the older label {datum,discovery,knowledge} and the new theme for "data mining" becomes a very stable label.

An explaination of the sibling labels {datum,mine} and {datum,mining} is necessary here. They are an artefact of the linguistic preprocessor, which (correctly) distinguishes between "mining" and "mine". Since the documents of the ACM subarchive are quite unlikely to refer to explosives, though, we can assume that all appearances of "mine" refer to data mining. We intend to remove the artefact in future implementations. For the time being, however, the artefact causes either distinct clusters (as in 2001) or cannibalisation – none of the two words is adequately frequent to appear in a label. We suspect that this is the cause of the uninformative label "datum" that appears in the last three periods. This is further indicated by the juxtaposition of the cluster labeled "datum" to the ACM categories: 64% of its members refer to data mining.

**DBScan Clustering Results.** First we describe our experiments to find out good values for the input parameters for the DBScan, the *eps* and the $minPts$ value.

*Experiments with different eps and* $minPts$ *values:* For clustering the data with the DBScan algorithm we made experiments with the dataset from period "1996" and tried to change the values for *eps* and $minPts$, which gave the size of the eps-region and the minimum number of other points in this region, in order to build a new cluster. We tried short values for both and short values for one of them and for the other a bigger value. On this experiments every time we get only one small cluster and a big number of noise documents. One reason could be the relative small number of documents at period "1996". Thats why we made similar experiments with the period "2000" to find out good values at this period and if the problem with the good clustering is caused by the number of documents. In figure 1(a) we show the results of the cluster numbers at different *eps* and different $minPts$ where we used the feature space of period 1996 to make a better comparision with the bisecting k-means clustering. The clustering results looks very similar every time. We get a relative big number of noise point, which are not in any cluster and get different numbers of very small clusters.

*Experiments from period "2000":* For the experiments with DBScan we decided to start at period "2000". But if we set the period "2000" as start period we had to use the feature space of the period "2000" too as start feature space and so we make short experiments with this feature space and dataset to check the decision and find out a good start value for *eps* and $minPts$. The results are shown in figure 1(b). From this diagram it shows clearly that we used $eps = 0.5$ and $minPts = 10$ for the following clustering experiments, so the smallest clusters have minimum 11 members.

We limited the feature space size to the most 100 used words for this experiments. At the cluster results we see that we have many "noise" data points, which are not a member of a cluster. But with increased size of the data set the relative size of the "noise" data points decrease. We concentrate only on the clusters founded by the algorithm, and on the labels of the clusters. We recognised that we could not produce a label for only one founded cluster. After a detailed view to this cluster we see the reason. This cluster members are all documents which are represented by a null-vector by the used feature space. So it was acceptable that this cluster had no label.

The following table 4 shows the number of clusters of each period and the number of refound clusters of the previous period via *ThemeFinder*.

It can clearly be seen that we have no matched clusters at period "2000" with a previous period because "2000" is the start period. We found all clusters of the previous period again in the next period with the exception at period "2004" where we did not found again only one of the 27 clusters of period "2003". Another effect of found again nearly all clusters at the next period is that the labeled clusters at one period are persistent over future periods, for example the 3 labels of period 2000 are also present on

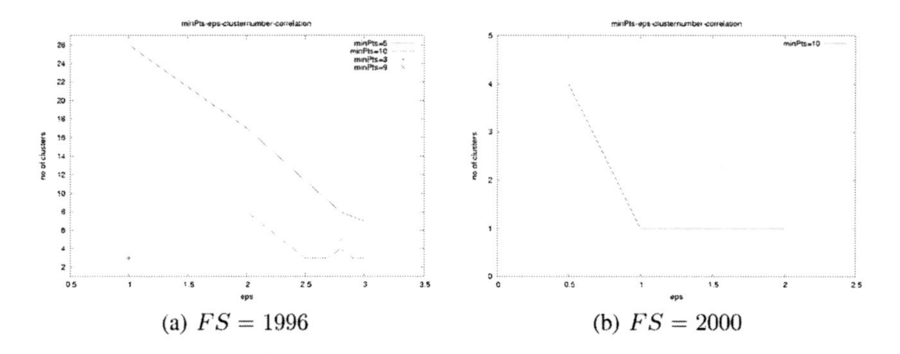


(a) $FS = 1996$ (b) $FS = 2000$

**Fig. 1.** Cluster numbers at different eps and minPts

**Table 4.** Results of matched clusters with the DBScan

| period | 2000 | 2001 | 2002 | 2003 | 2004 |
|---|---|---|---|---|---|
| no.of clusters | 3 | 6 | 12 | 27 | 41 |
| matched clusters | | 3 | 6 | 12 | 26 |

period 2004. It could be mentioned that every new period new clusters showed up because at earlier periods there were not enough documents about the topic of the new clusters. So the relative size of the noise decrease.

**Comparison of the *ThemeFinder* for different clustering algorithms.** Both algorithm found labels with our method for label creation and label monitoring with the *ThemeFinder*. A comparision of the labels is not easy because the number of labels at the bisecting k-means clustering results is static and with the DBScan variable. We have 4 labels at a maximum at the bisecting k-means results and found 2 of the labels at the periods "2000" and "2001" again at the labels at the clustering results from the DBScan algorithm. For the periods from "2002" and later we found 3 labels actually at both label sets.

As second both cluster algorithm produce the dominant clusters from the dominant group. The number of documents from the "data mining" subgroup becomes more and more dominant with time. If we evaluate the cluster labels to the subgroups by calculate the percentage of documents in the cluster from which subgroup they originate, we found 2 labeled clusters from the "data mining" subgroup with the bisecting k-means. With the DBScan algorithm all clusters at period "2000" and "2001" are from the subgroup "data mining" and in the following periods the percentage of labeled clusters from this group is very high.

In figure 2 we show the number of clusters where the most documents ($> 51\%$) come from one subgroup. The shortcuts at the lines stands for the names of the subgroups, data mining (dm), spatial databases (sp), image databases (im), statistical databases (st) and scientific databases (sc). The difference between the number of clusters at each period from this figure and table 4 arises because we have some clusters which have a

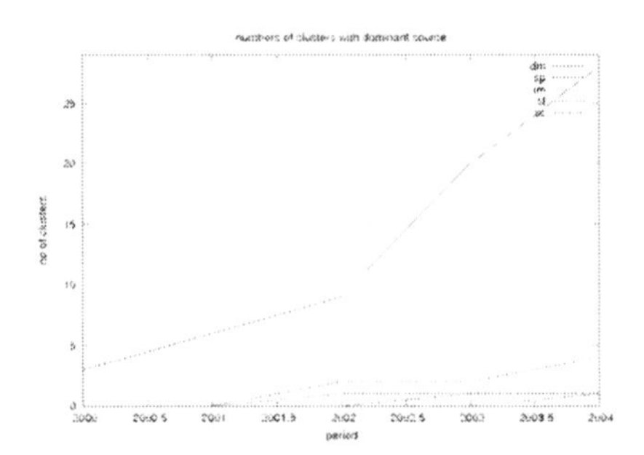


**Fig. 2.** Clusters with dominant source subgroup

mixed composition from the several subgroups and here we list only the clusters where the majority come from one group.

We also see too that changes are be found at the labels by monitoring the labels with the *ThemeFinder*. But the difference between both clustering algorithms is the kind of changes which we found. At the clustering with the bisecting k-means the label of a cluster changes during the time. In contrast to this, with the DBScan, the labels of a cluster are very stable but the number of clusters increase and so from one to an other period new labels will be created and show new topics.

We present ahead the results of the *ThemeFinder* ones with the bisecting k-means and ones with the DBScan clustering algorithm. As a main result, we see that *ThemeFinder* is useful at both algorithms to detect and monitor the topics of the clustering. The big size of noise data as one result of the DBScan shows us, that the data used is very noisy and so it could be one reason for rather negative clustering results with the bisecting k-means. This is already known as one disadvantage of this algorithm.

Another result is, that we can see the change of topics in the data, when monitoring the labels at both algorithm results. The difference is, at the bisecting k-means we see it on the change at the labels itself and at the DBScan we see it on the labels of the new created clusters from one to next period.

We have also seen, that the number of adjusting the feature space is very small. At bisecting k-means it was for 9 periods only two times and for the 5 periods with the DBScan zero times.

## 5 Conclusions and Outlook

We have seen that our *ThemeFinder* is able to monitoring label/topic evolutions at both cluster result sets, one from a bisecting k-means algorithm and one from the DBScan algorithm and so can monitor cluster evolutions. This underlines our design point of the *ThemeFinder* to be cluster algorithm independent and work only on the results of the clustering and has an influence only to the used feature space. We also see the influence of the clustering algorithms to the results of label monitoring and to monitor this labels with *ThemeFinder*. As a next step we will try experiments with different label mechanism, to see the influence of the label mechanisms to the monitoring results with *ThemeFinder*.

## References


1. Allan, J.: Introduction to Topic Detection and Tracking. Kluwer Academic Publishers, Dordrecht (2002)
2. Deerwester, S., Dumais, S.T., Furnas, G.W., Landauer, T.K., Harshman, R.: Indexing by Latent Semantic Analysis. Journal of the American Society of Information Science 44(6), 391–407 (1990)
3. Ester, M., Sander, J.: Knowledge Discovery in Databases. Techniken und Anwendungen. Springer, Heidelberg (2000)
4. Karypis, G., Han, E.-H.(S): Fast Supervised Dimensionality Reduction Algorithm with Apllications to Document Categorization & Retrieval. In: Proceedings of CIKM-00, pp. 12–19. ACM Press, New York (2000)
5. Karypis, G., Steinbach, M., Kumar, V.: A comparison of document clustering techniques. In: TextMining Workshop at KDD2000 (May 2000)
6. Lin, J., Vlachos, M., Keogh, E., Gunopulos, D.: Iterative Incremental Clustering of Time Series. In: Bertino, E., Christodoulakis, S., Plexousakis, D., Christophides, V., Koubarakis, M., Böhm, K., Ferrari, E. (eds.) EDBT 2004. LNCS, vol. 2992, Springer, Heidelberg (2004)
7. Ester, M., Kriegel, H.-P., Sander, J., Xu, X.: A density-based algorithm for discovering clusters in large spatial databases with noise. In: Proc. 2nd int. Conf. on Knowledge Discovery and Data Mining (KDD 96), Portland, Oregon, AAAI Press, Stanford (1996)
8. Moringa, S., Yamanishi, K.: Tracking Dynamics of Topic Trends Using a Finite Mixture Model. In: Kohavi, R., Gehrke, J., DuMouchel, W., Ghosh, J. (eds.) Proceedings of the 2004 ACM SIGKDD international conference on Knowledge discovery and data mining, pp. 811–816. ACM Press, New York (2004)
9. Pham, D.T., Dimov, S.S., Nguyen, C.D.: An Incremental K-means algorithm. In: Proceedings of the Institution of Mechanical Engineers, Part C: Journal of Mechanical Engineering Science, vol. 218, pp. 783–795 (2004)
10. Schult, R., Spiliopoulou, M.: Discovering emerging topics in unlabelled text collections. In: Manolopoulos, Y., Pokorný, J., Sellis, T. (eds.) ADBIS 2006. LNCS, vol. 4152, pp. 353–366. Springer, Heidelberg (2006)
11. Schult, R., Spiliopoulou, M.: Expanding the Taxonomies of Bibliographic Archives with Persistent Long-Term Themes. In: Procedings of the 21th Annual ACM Symposium on Applied Computing (SAC'06), ACM Press, New York (2006)
12. Wang, X., McCallum, A.: Topics over time: a non-markov continuous-time model of topical trends. In: Procedings of KDD06, Philadelphia, Pennsylvania, ACM Press, New York (2006)
13. Zhong, S.: Efficient streaming text clustering. Neural Networks 18(5-6), 790–798 (2005)

# Journey to the Centre of the Star: Various Ways of Finding Star Centers in Star Clustering

Derry Tanti Wijaya and Stéphane Bressan*

School of Computing, National University of Singapore, 3 Science Drive 2, Singapore
derrytan@comp.nus.edu.sg, steph@nus.edu.sg

**Abstract.** The Star algorithm is an effective and efficient algorithm for graph clustering. We propose a series of novel, yet simple, metrics for the selection of Star centers in the Star algorithm and its variants. We empirically study the performance of off-line, standard and extended, and on-line versions of the Star algorithm adapted to the various metrics and show that one of the proposed metrics outperforms all others in both effectiveness and efficiency of clustering. We empirically study the sensitivity of the metrics to the threshold value of the algorithm and show improvement with respect to this aspect too.

## 1 Introduction

Clustering is the task of grouping similar objects. In a graph, when vertices represent objects and (possibly weighted) edges represent similarity among objects, clustering is the separation of dense from sparse regions where clusters are the dense regions. A vector space clustering naturally translates into a graph clustering problem for a dense graph in which vertices correspond to vectors and pairs of vertices are connected with an edge whose weight is the cosine of the corresponding vectors. The graph is a clique [1]. In 1998, Aslam et al. [2] proposed the Star algorithm for graph clustering and gave a complete presentation in 2004 in [3], in which they proposed both an off-line and an on-line version and studied analytically and empirically the properties and performance of their algorithms. Star algorithm replaces the computation of vertex covering of a graph by cliques by a very simple computation of dense sub-graphs: lower weight edges of the graph are first pruned; then vertices with higher degree are chosen in turn as Star centers while vertices connected to a center become satellites. The algorithm terminates when every vertex is either a center or a satellite; each center and its satellites forming one cluster. Two critical elements in this algorithm are the threshold value for the pruning of edges and the metrics for selecting Star centers. In their paper [2], Aslam et al. derived the expected similarity between two satellite vertices in a cluster as a function of similarities between satellites and their star center and the threshold value. They show empirically that this theoretical lower bound on the expected similarity is a good estimate of the actual similarity. Yet the metrics they

* This work was funded by the National University of Singapore ARG project R-252-000-285-112.

R. Wagner, N. Revell, and G. Pernul (Eds.): DEXA 2007, LNCS 4653, pp. 660–670, 2007.

propose for selecting star centers: the degree of a vertex; does not leverage this finding. Our work is motivated by the suspicion that degree is not the best possible metrics for selecting Star centers. A metric that maximizes intra cluster similarity, i.e. average similarity among all pairs of vertices in the cluster, should improve performance. We propose a series of novel, yet simple, metrics for selecting Star centers that take into account the similarity between vertices (i.e. weight of edges). In particular we propose average metrics which, as we argue analytically, maximizes intra-cluster similarity. We empirically study the performance of off-line and on-line versions of Star algorithm adapted to the various proposed metrics. We show that average metrics outperforms all others in both effectiveness and efficiency. Our contribution is the presentation of metrics and their comprehensive comparative performance analysis with real world and standard corpora for the task of clustering documents.

## 2 Background and Related Works

### 2.1 Vector and Graph Clustering

Algorithms for vector and graph clustering can be grouped into: partitioning, hierarchical, and graph algorithms. Partitioning algorithms, such as K-means [4] divide vertices into K clusters by defining K centroids and associate each vertex to nearest centroid. The algorithm iteratively recalculates the position of K centroids until the centroids no longer move. Hierarchical algorithms [5] can be categorized into agglomerative and divisive. The agglomerative method treats each vertex as a separate cluster, iteratively merges clusters that have smallest distance from one another until all clusters are grouped into one, yielding a hierarchical tree of clusters. Divisive method is the reverse of agglomerative that starts with all objects in one cluster and divides them into smaller clusters. Markov Clustering (MCL) is a graph clustering algorithm based on simulation of (stochastic) flow/random walks in graphs [6]. The Star clustering algorithm replaces the NP-complete computation of a vertex-cover by cliques by the greedy, simple and inexpensive computation of star shaped dense sub graphs. Star clustering does not require the indication of an a priori number of clusters. It also allows the clusters to overlap. Star clustering analytically guarantees a lower bound on the similarity between objects in each cluster and computes more accurate clusters than either the single [7] or average [8] link hierarchical clustering.

### 2.2 Star Clustering and Extended Star Clustering

To produce reliable document clusters of similarity $\sigma$ (i.e. clusters where documents have pair-wise similarities of at least $\sigma$, where $\sigma$ is a user-defined threshold), Star algorithm starts by representing the document collection by its $\sigma$-similarity graph where vertices correspond to documents and there is an undirected edge from vertex $v_i$ to vertex $v_j$ if their cosine similarity in a vector space [1, 9] is greater than or equal to $\sigma$. Star clustering formalizes clustering by performing a minimum clique vertex cover with maximal cliques on this $\sigma$-similarity graph [10]. Since covering by cliques is an NP-complete problem [11, 12], Star clustering approximates a clique cover greedily by dense sub-graphs that are star shaped, consisting of a single Star center and its satellite vertices, where there exist edges between the Star center and each satellite vertex. Star clustering guarantees pair-wise similarity of at least $\sigma$ between the Star and each of its satellites. However, it does not guarantee such similarity between satellite vertices. By investigating the geometry of the vector space model, Aslam et al. derive a lower bound on the similarity between satellite vertices and predict that the pair-wise similarity between satellite vertices in a Star-shaped sub-graph is high. Together with empirical evidence [2], Aslam et al. conclude that covering $\sigma$-similarity graph with Star-shaped sub-graphs is an accurate method for clustering a set of documents.

Each vertex v in Star clustering has a data structure containing v.degree: the degree of v, v.adj: the list of v's adjacent vertices, v.marked: a bit denoting if v is already in a cluster, and v.center: a bit denoting if v is a Star center. The off-line Star algorithm (for static data) sorts vertices by degree. Then, it scans the sorted vertices from highest to lowest degree as a greedy search for Star centers. Only vertices that are not yet in a cluster can be Star centers. Once a new Star center v is selected, v.center and v.marked bits are set, and for all vertices w adjacent to v (i.e. w $\in$ v.adj), w.marked is set. Only one scan of the sorted vertices is needed to determine all Star centers. Upon termination when all vertices have their marked bits set, these conditions must be met: (1) the set of Star centers are the Star cover of the graph, (2) a Star center is not adjacent to any other Star center, and (3) every satellite vertex is adjacent to at least one center vertex of equal or higher degree. The algorithm has a run time of $\Theta$ (V + E$\sigma$) where V is the set of vertices and E$\sigma$ edges in the $\sigma$-similarity graph G$\sigma$ [2].

Star algorithm has some drawbacks that the Extended Star algorithm [13] by Gil et al. proposes to solve. The first drawback [13] is the Star covers (hence the clusters produced) are not unique. When there are several vertices of the same highest degree, the algorithm arbitrarily chooses one as Star. The second drawback is because no two Star centers can be adjacent to one another; the algorithm can produce illogical clusters [13]. The extended Star algorithm addresses these issues by choosing Star centers independently of document order. It uses complement degree of a vertex v, CD(v), which is the degree of v only taking into account its adjacent vertices not yet included in any cluster: CD (v) = | v.adj \ Clu |; where Clu is the set of vertices already clustered. Extended Star algorithm also considers a new notion of Star center. A vertex v is considered a Star if it has at least an adjacent vertex x with less or equal degree than v, and x satisfies these conditions: (1) x has no adjacent Star, or (2) the highest degree of Stars adjacent to x is not greater than v's degree. Extended Star algorithm has two versions: unrestricted and restricted. In the restricted version, another condition is imposed: only unmarked vertices: vertices not yet included in any cluster can be Star centers. Extended Star processes vertices independently of document order. If more than one vertex with the same highest degree exists, the extended Star selects all of them as Star centers. Extended Star algorithm allows two Stars to be adjacent to one another as long as they satisfy the required conditions to be Star centers.

On-line Star algorithm [2] supports insertion and deletion of vertices from the graph. When a new vertex is inserted into (or deleted from) the graph, new Stars may need to be created and existing Stars may need to be destroyed. On-line Star algorithm maintains a queue containing all satellite vertices that have the possibility of being 'promoted' into Star centers. As long as these vertices are indeed satellites, the existing Star cover is correct. The vertices in the queue are processed in order of their degree (from highest to lowest). When an enqueued satellite is promoted to Star center, one or more existing Stars may be destroyed; creating new satellites that have the possibility of being promoted. These satellites are put into the queue and the process repeats. Our discussion in the next section is, to some extent, orthogonal to the improvement of the extended Star algorithm or to the extensions of the on-line version. Our algorithms retain the logic of the off-line and on-line extended and original Star algorithms but aim within reasonable complexity at improving the performance of off-line, on-line, extended and original Star algorithms by maximizing the 'goodness' of the greedy vertex cover through novel yet simple metrics for selecting Star centers.

# 3 Finding Star Centers

Our work is motivated by the suspicion that degree is not the best metrics for selecting Star centers. We believe Star centers should be selected by metrics that take into consideration the weight of the edges in order to maximize intra-cluster similarity.

## 3.1 Markov Stationary Distributions

Star centers are 'important' or 'representative' vertices. Intuitively, it seems that maximum flow vertices as used in Markov clustering should be good candidate Star centers. This idea is similar to the one used in Google's page Rank algorithm [14] for quantifying the importance of a Web page. The first metrics that we propose is therefore the probability of reaching a vertex after a random walk, i.e. in the stationary distribution of the Markov matrix. A Markov matrix $A=[a_{ij}]$ is a square matrix such that: $\forall i, \forall j\ 0 \leq a_{ij} \leq 1$ , and $\forall i\ \Sigma_{\forall j}\ a_{ij} \leq 1$ (or $\forall j\ \Sigma_{\forall i}\ a_{ij} \leq 1$). The adjacency matrix of a weighted graph is or can be normalized into a symmetric Markov matrix. We represent the similarity graph (without threshold) G = (V, E) as an adjacency matrix A where each entry aij is the normalized similarity between vertex i and j in G: aij = sim (i, j) / $\Sigma_{\forall x \in V}$ sim (i, x). A is therefore a sub-stochastic matrix. Each entry $a_{ij}$ represents the probability of passing from i to j traversing one edge during a random walk in G. The matrix A×A or $A^2$, represents probabilities of two edge transversals. At infinity, the Markov chain $A^k$ converges. We call A* the fix point of the product of A with itself. The sum of the stationary values is the probability to end up on a given vertex after a random walk. For a given vertex represented by row and column i in the adjacency matrix A of the graph, it is the value $\Sigma_{\forall j}\ b_{ij}$ in the matrix $A^*= [b_{ij}]$. Because A is a Markov matrix A* can be computed directly as follows: $A^* = (I-A)^{-1}$.

Incorporating this metrics in Star algorithm, we sort vertices by the sum of their stationary values and pick unmarked vertex with highest value to be the new Star center. In the remainder we refer to Star algorithm in which Star centers are determined using sum of stationary values as Star-markov: the off-line star algorithm with Markov stationary distribution metrics. The computation of A* is relatively expensive (it is accounted for in experimental results presented in section 4). For this reason, we do not consider this metrics for extended and on-line versions of the algorithm; neither do we investigate in this paper technique for on-line approximation of the value.

### 3.2 Lower Bound, Average and Sum

In their derivation of expected similarity between satellite vertices in a Star cluster [2], Aslam et al. show that the similarity $\cos(\gamma_{i,j})$ between two satellite vertices $v_i$ and $v_j$ in a Star cluster is such that:

$$\cos(\gamma_{i,j}) \geq \cos(\alpha_i)\cos(\alpha_j) + (\sigma / \sigma + 1)\sin(\alpha_i)\sin(\alpha_j)\,. \quad (1)$$

Where $\cos(\alpha_i)$ is the similarity between the Star center v and satellite $v_i$ and $\cos(\alpha_j)$ is the similarity between the Star center v and satellite $v_j$. They show that the right hand side of inequality (1) above is a good estimate of its left hand side. Hence it can be used to estimate the average intra-cluster similarity. For a cluster of n vertices and center v, the average intra-cluster similarity is therefore: $(\Sigma_{(vi,\, vj) \in v.adj \times v.adj} (\cos(\gamma_{i,j}))) / n^2 \geq ((\Sigma_{vi \in v.adj} \cos(\alpha_i))^2 + (\sigma / \sigma + 1)(\Sigma_{vi \in v.adj} \sin(\alpha_i))^2)/n^2$; where $\gamma_{i,j}$ is the angle between vertices $v_i$ and $v_j$ and $\alpha_i$ is the angle between v and vertex $v_i$ and where v.adj is the set of vertices adjacent to v in $G\sigma$ (i.e. vertices in the cluster). This is computed on the pruned graph (i.e. the $\sigma$-similarity graph). Based on this average intra-cluster similarity, we derive our first metrics: for each vertex v in $G\sigma$, we let: $lb(v) = ((\Sigma_{vi \in v.adj} \cos(\alpha_i))^2 + (\sigma / \sigma + 1)(\Sigma_{vi \in v.adj} \sin(\alpha_i))^2)/n^2$. We call this metrics **lb(v)**, the lower bound. This is the theoretical lower bound on the actual average intra-cluster similarity when v is Star center and v.adj are its satellites. The lower bound metrics is slightly expensive to compute because it uses both cosine and sine. We consider two additional simpler metrics and empirically evaluate whether they constitute good approximations of lb(v). The metrics are computed on the pruned graph. Namely, for each vertex v in $G\sigma$, we let: $ave(v) = \Sigma_{\forall vi \in v.adj} \cos(\alpha_i) / degree(v)$ and $sum(v) = \Sigma_{\forall vi \in v.adj} \cos(\alpha_i)$; where $\alpha_i$ is the angle between v and vertex $v_i$. We call the metrics **ave(v)** and **sum(v)** the average and sum metrics, respectively. Notice that the ave(v) metrics is the square root of the first term of the lb(v) metrics. lb(v) grows together with ave(v). Therefore ave(v) should be a criteria equivalent to lb(v) for the selection of Star centers. Incorporating these metrics in off-line and on-line Star algorithms, we sort vertices by sum and average and pick unmarked vertex with highest sum and average to be the new Star center, respectively. In the remainder we refer to Star algorithm in which Star centers are determined using the sum and average as the Star-sum and Star-ave, respectively. We incorporate the lower bound, average and sum metrics in the original and extended Star algorithms by using lb(v), sum(v) and ave(v) in the place of degree(v) and by defining and using complement lower bound **CL(v)**, complement sum **CS(v)**, and complement average **CA(v)** in the place of complement degree, CD(v), respectively. We define $CL(v) = ((\Sigma_{vi \in v.adj \setminus Clu} \cos(\alpha_i))^2 + (\sigma / \sigma + 1)(\Sigma_{vi \in v.adj \setminus Clu} \sin(\alpha_i))^2)/n^2$, $CS(v) = \Sigma_{\forall vi \in v.adj \setminus Clu} \cos(\alpha_i)$, $CA(v) = \Sigma_{\forall vi \in v.adj \setminus Clu} \cos(\alpha_i) / CD(v)$, where Clu is the set of vertices already clustered.

We integrate the above metrics in the Star algorithm and its variants to produce the following extensions: (1) Star-lb: the off-line star algorithm with lb(v) metrics; (2) Star-sum: the off-line star algorithm with sum(v) metrics; (3) Star-ave: the off-line star algorithm with ave(v) metrics; (4) Star-markov: the off-line star algorithm with Markov stationary distribution metrics; (5) Star-extended-sum-(r): the off-line restricted extended star algorithm with sum(v) metrics; (6) Star-extended-ave-(r): the off-line restricted extended star algorithm with ave(v) metrics; (7) Star-extended-sum-(u): the off-line unrestricted extended star algorithm with sum(v) metrics; (8) Star-extended-ave-(u): the off-line unrestricted extended star algorithm with ave(v) metrics; (9) Star-online-sum: the on-line star algorithm with sum(v) metrics; (10) Star-online-ave: the on-line star algorithm with ave(v) metrics. For the sake of simplicity and concision, we only show results for lower bound in the original Star algorithm in the performance analysis section (we do not evaluate Star-extended-lb-(u) and Star-extended-lb-(r), the off-line unrestricted and restricted extended star with lower bound metrics, nor Star-online-lb: the on-line star algorithm with lower bound metrics).

# 4 Experiments

In order to evaluate the proposed metrics, we compare the performance of our extensions with the original off-line and on-line Star clustering algorithms and with the restricted and unrestricted extended Star clustering algorithms. We use data from Reuters-21578 [15], TIPSTER–AP [16] and our original collection: Google. The Reuters-21578 collection contains 21,578 documents that appeared in Reuter's newswire in 1987. The TIPSTER–AP collection contains AP newswire from the TIPSTER collection. Our original collection: Google contains news documents obtained from Google News website [17] in December 2006. Each collection is divided into several sub-collections. By default and unless otherwise specified, we set the value of threshold $\sigma$ to be the average similarity of documents in the given sub-collection. We measure effectiveness (recall, r, precision, p, and F1 measure, $F1 = (2 * p * r) / (p + r)$), efficiency (running time) and sensitivity to $\sigma$. In each experiment, for each topic, we return the cluster which "best" approximates the topic, i.e. cluster that produces maximum F1with respect to the topic: $topic(i) = max_j \{F1(i, j)\}$; where $F1(i, j)$ is the F1 measure of the cluster number j with respect to topic i. The weighted average of F1 measure for a sub-collection is calculated as: $F1 = \Sigma (n_i/S) * F1(i, topic (i))$; for $0 \leq i \leq N$; where N is the number of topics in the sub-collection, $n_i$ is the number of documents belonging to topic i in the sub-collection, and $S = \Sigma n_i$; for $0 \leq i \leq N$. For each sub-collection, we calculate the weighted-average of precision, recall and F1-measure produced by each algorithm. We then present the average results over each collection.

## 4.1 Performance of Off-Line Algorithms

We empirically evaluate the effectiveness (precision, recall, F1 measure) and efficiency (time) of our proposed off-line algorithms: Star-markov, Star-lb, Star-sum, Star-ave and compare their performance with the original Star algorithm: Star and its variant: Star-random that picks star centers randomly. The results reported for Star-random are average results for various seeds that gives us a base line for comparison.

In Fig. 1 we see that Star-lb and Star-ave achieve the best F1 values on all collections. This is not surprising because as we argued, lb(v) metrics maximizes intra-cluster similarity. The fact that our Star-ave achieves comparable F1 to Star-lb is evidence that average is a sufficient metric for selecting star centers. In Fig. 1, based on

F1: when compared to original Star, our proposed algorithms: Star-ave and Star-markov by far outperform the original Star on all collections. An interesting note is that Star-random performs comparably to original Star when threshold σ is the average similarity of documents in the collection. This further proves our suspicion that degree may not be the best metric. In Fig. 2 we see that our proposed algorithms with the exception of Star-markov perform as efficiently as original Star. Star-markov takes longer time as it uses matrix calculation to compute random walk. As expected, Star-lb takes the longest time due to its expensive computation.

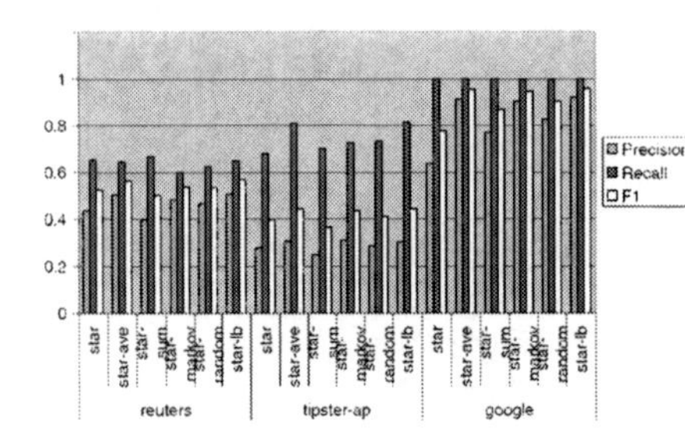


Fig. 1. Effectiveness of off-line Star

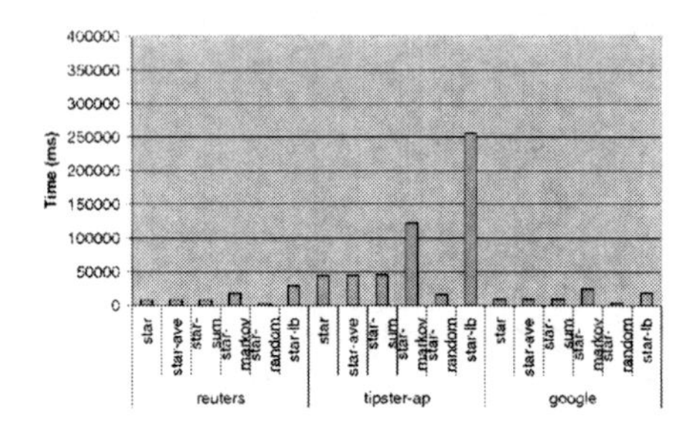


Fig. 2. Efficiency of off-line Star

From Fig. 1 and Fig. 2 we can conclude that (1) since Star-random is more efficient and achieves comparable F1 to original Star, using degree to pick stars may not be the best metric, (2) since Star-ave is more efficient and achieves comparable F1 to Star-lb; Star-ave can be used as a good approximation to Star-lb to maximize the resulting intra-cluster similarity.

## 4.2 Order of Stars

We empirically demonstrate that Star-ave indeed approximates Star-lb better than other algorithms by a similar choice of star centers.

In Fig. 3 and 4 we present the order in which the algorithms choose their star centers on TIPSTER-AP and Reuters collections (similar trend is observed on Google

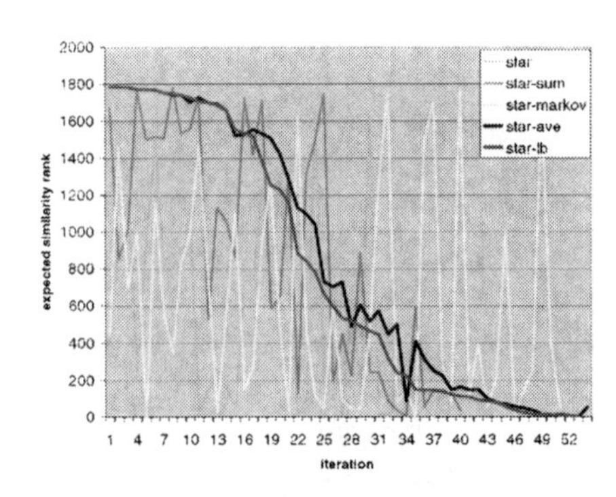


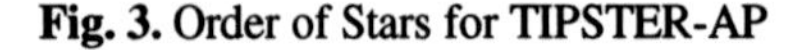

Fig. 3. Order of Stars for TIPSTER-AP

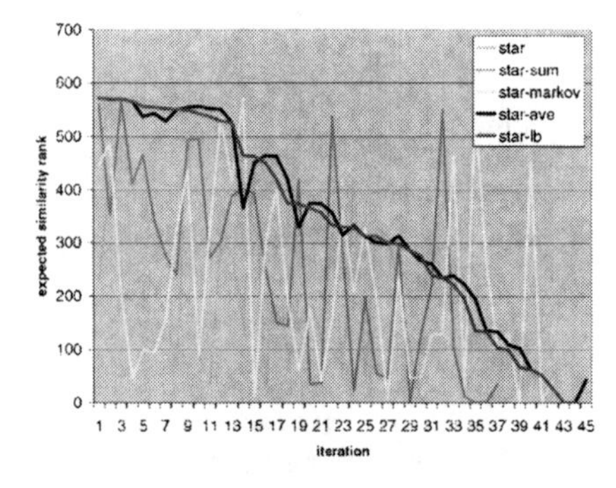


Fig. 4. Order of Stars for Reuters

collection): from the first star center to the $n^{th}$ star center and where star centers are ranked by their expected intra-cluster similarity (computed from equation 1) from highest to lowest. Star-lb chooses star centers in the order of their expected similarity rank, from highest to lowest. In Fig. 3 and 4, we see that Star-ave chooses star centers in an order similar to Star-lb. This is not the case of the other algorithms. Therefore picking star centers in descending order of expected intra-cluster similarity can be approximated simply by picking star center in descending order of average similarity with its adjacent vertices.

### 4.3 Performance of Off-Line Algorithms at Different Threshold (σ)

Star clustering has one parameter σ. With a good choice of σ, just enough edges are removed from the graph to disconnect sparsely connected dense sub-graphs. Removing too few edges will group these sparsely connected sub-graphs together; producing high recall but low precision clusters. Removing too many edges will break these dense sub-graphs into smaller, perhaps not-so-meaningful components; producing low recall but high precision clusters. Fig. 5 illustrates the empirical performance evaluation of our proposed off-line algorithms at different threshold values σ on Reuters collection (similar observation is found on TIPSTER-AP and Google collections). For a similarity graph G (V, E), we represent σ as a fraction (s) of the average edge weight ($E_{mean}$) in G, i.e. $\sigma = E_{min} + (s * E_{mean})$; where $E_{min}$ is the minimum edge weight in G.

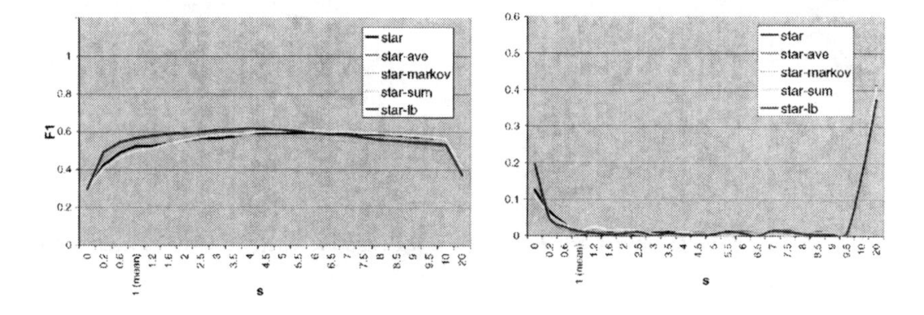


**Fig. 5.** Performance of off-line algorithms with varying σ on Reuters data

In Fig. 5, we see that Star-ave and Star-markov converge to a maximum F1 at a lower threshold than the original Star. Star-ave and Star-markov are able to 'spot' sparsely connected dense sub-graphs and produce reliable vertex cover even when there are few edges removed from the graph. On the contrary, the original Star algorithm needs to remove more edges from the graph before it is able to produce reliable clusters. The maximum F1 value of Star-ave is also higher than the maximum of the original Star. In Fig. 5 we see that the F1 values of Star-ave coincide closely with the F1 values of Star-lb at all thresholds. This is further evidence that Star-ave can be used to approximate Star-lb at varying thresholds. In Fig. 5, we see that the F1 gradient (the absolute change in F1 value with each change in threshold) is smaller (some approaching zero) for Star-ave and Star-markov as compared to the original Star. This small gradient means the value of F1 does not change/fluctuate much with each change in threshold. The smaller F1 gradient means that Star-ave and Star-markov are less sensitive to the change in threshold as compared to the original Star.

From Fig. 5 we can conclude that: (1) Based on maximum F1 value: our proposed algorithm Star-ave outperforms the original Star, (2) our proposed algorithms: Star-ave and Star-markov are able to produce reliable clusters even at a lower threshold where there are only fewer edges removed from the graph; (3) F1 values of Star-ave coincide closely with F1 values of Star-lb at all thresholds; this is further evidence that Star-ave can approximate Star-lb at any given threshold value.

### 4.4 Performance of Off-Line Extended Star Algorithms

We present the results of the experiment that incorporates our idea into the extended Star. In Fig. 6 and 7, we present effectiveness and efficiency comparison between: our algorithm: Star-ave that has performed the best so far, the original restricted extended star: Star-extended-(r), our algorithms: Star-extended-ave-(r), and Star-extended-sum-(r). Due to space constraints we do not present the results of comparison with the unrestricted version of extended star in which we observe similar findings.

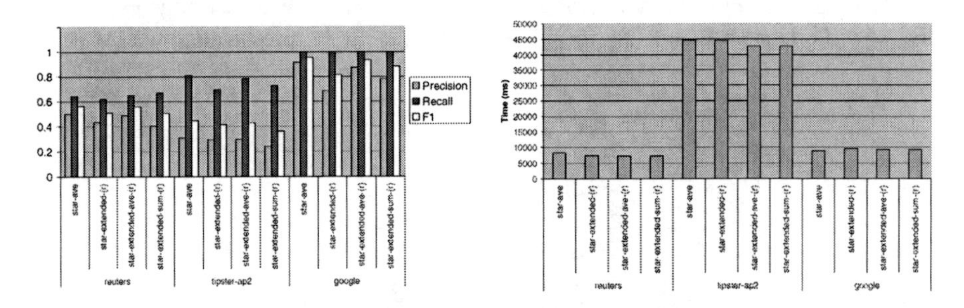


Fig. 6. Effectiveness of restricted extended Star

Fig. 7. Efficiency of restricted extended Star

In Fig. 6, based on F1 values: we see that our proposed algorithm: Star-ave; outperforms Star-extended-(r) on all collections. This is despite the fact that Star-ave uses only very simple idea to incorporate into the original Star. Our proposed algorithm: Star-extended-ave-(r) that incorporates the idea of complement-ave metric to the extended Star algorithms improves the performance of Star-extended-(r). In Fig. 7, in terms of efficiency: our proposed algorithms perform comparably or faster than the extended Star; with the exception on TIPSTER-AP collection (in which Star-ave takes longer time). We believe this difference in time could be because Star-ave picks different Star centers from the extended Star. We also see that incorporating the idea of complement-ave and complement-sum to extended Star does not reduce its original efficiency on all collections. From Fig. 6 and 7, we can conclude that: (1) our proposed algorithm: Star-ave obtain higher F1 values than the extended Star on all collections, (2) incorporating the idea of complement-ave metric to the extended Star improves its F1 without affecting its efficiency on all collections.

### 4.5 Performance of On-Line Algorithms

Fig. 8 and 9 illustrate the effectiveness and efficiency of the on-line algorithms: the original on-line Star: Star-online, and our on-line algorithms, Star-online-ave and Star-online-sum. We compare the performance of these algorithms with a randomized version of the algorithm: Star-online-random that picks star centers randomly.

Star-online-ave outperforms Star-online on all collections in terms of F1 (cf. Fig. 8). Both Star-online-ave and Star-online-sum perform better on all collections than Star-Random. They are more efficient (cf. Fig. 9) than Star-online on TIPSTER-AP data and comparable on other collections. From Fig. 8 and 9, we can conclude that Star-online-ave achieves higher F1 than Star-online. However, due to the fact that Star-online-ave may pick different star centers than Star-online; its efficiency may be affected.

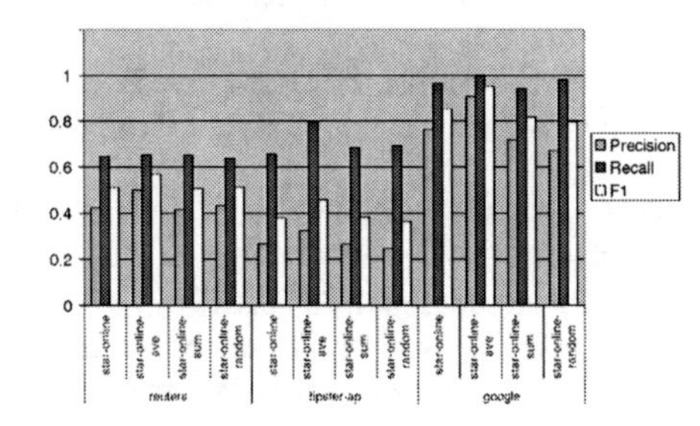


Fig. 8. Effectiveness of on-line algorithms

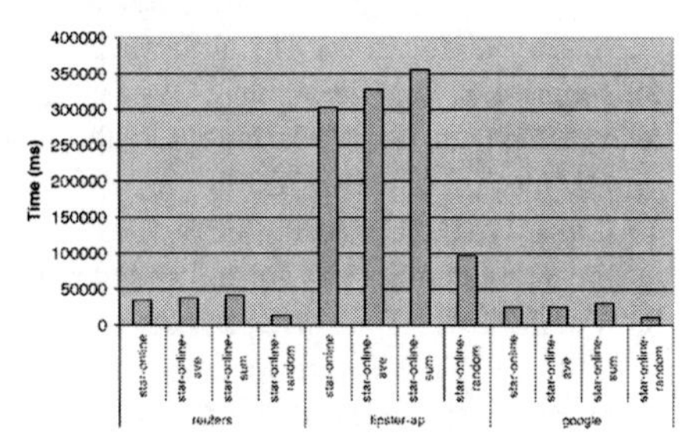


Fig. 9. Efficiency of on-line algorithms

## 5 Conclusion

We suspected that the metrics used for selecting star centers in Star clustering is not optimal. The theoretical argument was presented in the original papers of Star clustering but was not exploited. We therefore proposed various new metrics for selecting star centers: Markov metrics that find vertices of maximum flow; and metrics estimating and commensurating to the maximum intra-cluster similarity (lower bound, average and sum). We empirically studied the performance of off-line star, extended restricted and unrestricted Star and on-line Star with these different metrics. Our results confirm our conjecture: selecting star centers based on degree (as proposed by the original algorithm inventors) performs almost as poorly as a random selection. One needs to use a metrics that maximizes intra-cluster similarity such as the lower bound metrics. While it indeed yields the best results, it is expensive to compute. The average metrics is a fast and good approximation of the lower bound metrics in all variants of Star algorithm: (1) Star-ave yields up to 4.53% improvement of F1 with a 19.1% improvement on precision and 1.77% on recall; (2) Star-extended-ave-(r) yields up to 14.65% improvement of F1 with a 27.2% improvement on precision and 0.25% on recall; (3) Star-extended-ave-(u) yields up to 138% improvement of F1 with a 102% improvement on precision and 3.4% on recall; (4) Star-online-ave yields up to an outstanding 20.81% improvement of F1 with a 20.87% improvement on precision and 20.67% on recall. We notice that, since intra-cluster similarity is maximized, it is precision that is mostly improved. We can therefore propose Star-online-ave as a very efficient and very effective graph clustering algorithm.

## References


1. Salton, G.: Automatic Text Processing: the transformation, analysis, and retrieval of information by computer. Addison-Wesley (1989)
2. Aslam, J., Pelekhov, K., Rus, D.: Static and Dynamic Information Organization with Star Clusters. In Proceedings of the 1998 Conference on Information Knowledge Management, Baltimore, MD (1998)
3. 3. Aslam, J., Pelekhov, K., Rus, D.: The Star Clustering Algorithm. In Journal of Graph Algorithms and Applications, 8(1) 95–129 (2004)
4. MacQueen, J. B.: Some Methods for classification and Analysis of Multivariate Observations. Proceedings of 5th Berkeley Symposium on Mathematical Statistics and Probability. Berkeley, University of California Press, 1:281-297 (1967)
5. Johnson, S. C.: Hierarchical Clustering Schemes. Psychometrika, 2:241-254 (1967)
6. van Dongen, Stijn Marinus: Graph clustering by flow simulation - Tekst. - Proefschrift Universiteit Utrecht (2000)
7. Croft, W. B.: Clustering large files of documents using the single-link method. Journal of the American Society for Information Science, 189-195 (November 1977)
8. Voorhees, E.: The cluster hypothesis revisited. In Proceedings of the 8th SIGIR, 95-104
9. Salton, G.: The Smart document retrieval project. In Proceedings of the Fourteenth Annual International ACM/SIGIR Conference on Research and Development in Information Retrieval, 356-358
10. Karp, R.: Reducibility among combinatorial problems. Computer Computations, 85–104, Plenum Press, NY (1972)
11. Lund, C., Yannakakis, M.: On the hardness of approximating minimization problems. Journal of the ACM 41, 1960-981 (1994)
12. Press W., Flannery B., Teukolsky S., Vetterling W.: Numerical Recipes in C: The Art of Scientific Computing, Cambridge University Press (1988)
13. García, R.J. Gil, Contelles, J.M. Badía, Porrata, A. Pons: Extended Star Clustering Algorithm. Proc. Of CIARP'03, LNCS 2905, 480-487 (2003)
14. Brin Sergey, Page Lawrence: The anatomy of a large-scale hypertextual Web search engine. Proceedings of the seventh international conference on World Wide Web 7, 107-117 (1998)
15. http://www.daviddlewis.com/resources/testcollections/reuters21578/ (visited on December 2006)
16. http://trec.nist.gov/data.html (visited on December 2006)
17. Google News (http://news.google.com.sg)

# Improving Semantic Query Answering

Norbert Kottmann and Thomas Studer

Institut für Informatik und angewandte Mathematik,
Universität Bern, Neubrückstrasse 10, CH-3012 Bern, Switzerland
{kottmann,tstuder}@iam.unibe.ch

**Abstract.** The retrieval problem is one of the main reasoning tasks for knowledge base systems. Given a knowledge base $\mathsf{K}$ and a concept $C$, the retrieval problem consists of finding all individuals $a$ for which $\mathsf{K}$ logically entails $C(a)$. We present an approach to answer retrieval queries over (a restriction of) OWL ontologies. Our solution is based on reducing the retrieval problem to a problem of evaluating an SQL query over a database constructed from the original knowledge base. We provide complete answers to retrieval problems. Still, our system performs very well as is shown by a standard benchmark.

## 1 Introduction

Over the last decade, ontologies left the realm of academia and became an important technology in many domains. However, in order to be of practical use for full-fledged applications, tools and techniques that can deal with huge amounts of (ontological) information are needed.

Relational databases are one of the well-established cornerstones for systems managing large data loads. In this paper, we present a method to solve the ontological retrieval problem based on a relational database system. Our implementation shows that this provides an efficient and scalable solution for the retrieval problem.

An ontology defines the terms used to describe and represent an area of knowledge [7]. It consists of the definitions for the basic concepts of a domain and their relationships. These definitions and relations are formulated in a so-called ontology language which should be not only understandable for humans but also machine readable, hence supporting automatic knowledge processing. The W3C defined the ontology language OWL for applications in the semantic web. However, OWL also became the language of choice for many other applications in the area of knowledge representation and reasoning.

One of the main reasoning tasks for knowledge base systems is the so-called *retrieval problem* [2]. Let us illustrate this problem by an example.

Assume a zoological ontology defines the following:

(1) A carnivore is an animal that eats only animals.
(2) A lion is a carnivore.
(3) A lion eats gnus.

R. Wagner, N. Revell, and G. Pernul (Eds.): DEXA 2007, LNCS 4653, pp. 671–679, 2007.

Further, assume that this ontology has been loaded into a knowledge base system. This system will answer the query *show me all animals* as follows:

1. A lion is an animal since a lion is a carnivore (2) and every carnivore is an animal (1).
2. A gnu is an animal since a lion is a carnivore (2), everything that is eaten by a carnivore is an animal (1), and gnus are eaten by lions (3).

The abstract definition of the retrieval problem reads as follows: given a knowledge base $\mathsf{K}$ and a concept description $C$, find all individuals $a$ such that $\mathsf{K}$ logically entails $C(a)$. That is given $\mathsf{K}$ and $C$, look for all individuals $a$ such $C(a)$ is a logical consequence of $\mathsf{K}$. There is a trivial algorithm for this problem, namely to test for each individual occurring in $\mathsf{K}$ whether it satisfies the concept $C$ or not. This approach has the advantage that it provides (almost) complete reasoning for quite expressive knowledge representation languages. However, if large data sets have to be treated, then for efficiency reasons, one may need to turn to another approach.

It is possible to extend relational database systems to support storing and querying OWL [4] data as follows: when data is loaded into the database, the system precomputes the subsumption hierarchy and stores also the statements inferred from this hierarchy. Prominent projects following this approach are DLDB [10] and its successor HAWK[1]. Queries to the knowledge base can then be translated to standard SQL queries that are evaluated over the relational representation of the knowledge base. This has the advantage that all the query optimization techniques provided by relational database systems can be used and it becomes possible to work with huge datasets.

However, DLDB and HAWK often do not give complete answers to queries. We overcome this problem by identifying the description logic pos-$\mathcal{ALE}$ which is the positive fragment of $\mathcal{ALE}$ with transitive and inverse roles. Based on pos-$\mathcal{ALE}$, we present an extension of relational database systems to support OWL retrieval queries which is sound and complete with respect to pos-$\mathcal{ALE}$.

The language of pos-$\mathcal{ALE}$ provides enough expressive power for many applications. It also suffices for the LUBM benchmark for OWL knowledge base systems [5]. We evaluate our system with this standard benchmark and compare our results with HAWK. The main observation is that our systems performs very well although we provide complete pos-$\mathcal{ALE}$ reasoning. For many queries we are even faster that HAWK (which is often not complete).

Another approach for querying ontologies with the use of an SQL engine has been presented in [1,3]. There, the DL-Lite family of description logics is introduced. These languages are also well suited for the translation of description logic queries into SQL queries. However, DL-Lite languages are quite different from pos-$\mathcal{ALE}$. On one hand, pos-$\mathcal{ALE}$ features value restrictions and transitive roles which are both not included in DL-Lite. DL-Lite, on the other hand, supports functional restrictions and conjunctive queries which cannot be treated in

[1] `http://swat.cse.lehigh.edu/downloads/`

pos-$\mathcal{ALE}$. Because of all these differences, we could not compare our approach with a DL-Lite system.

## 2 DL to DB Mapping

The concepts of pos-$\mathcal{ALE}$ are given as follows, where $A$ is used for an *atomic concept*, $S$ is an *atomic role*, $R, T$ denote *roles*, and $C, D$ stand for *concept descriptions*:

$$
\begin{array}{rll}
C, D \rightarrow & A \mid & \text{(atomic concept)} \\
& \top \mid & \text{(top)} \\
& C \sqcap D \mid & \text{(conjunction)} \\
& \forall R.C \mid & \text{(value restriction)} \\
& \exists R.C & \text{(full existential quantification)} \\
R \rightarrow & S \mid & \text{(atomic role)} \\
& S^{-} & \text{(inverse role).}
\end{array}
$$

Additionally, we consider a set $\mathsf{R}^{+}$ of transitive roles. A TBox $\mathsf{T}$ contains concept inclusions $C \sqsubseteq D$ as well as role inclusions $R \sqsubseteq T$. An ABox $\mathsf{A}$ contains concept assertions $C(a)$ and role assertions $R(a, b)$. A knowledge base $\mathsf{K}$ is the union of a TBox and an ABox.

We make use of the standard semantics for description logics [2]. Accordingly, we write $\mathsf{K} \models C(a)$ if every model of $\mathsf{K}$ is a model of $C(a)$.

Our aim is to build a *completion* $\mathsf{A}^{*}$ of the ABox $\mathsf{A}$ such that is possible to answer arbitrary pos-$\mathcal{ALE}$ retrieval queries by only querying *atomic* concept and roles in $\mathsf{A}^{*}$. Assume $\mathsf{DB}_{\mathsf{K}}$ is the subset of all atomic concept and role assertion of such a completed ABox (stemming from an initial knowledge base $\mathsf{K}$). Then we write $\mathsf{DB}_{\mathsf{K}} \models_{\mathsf{DB}} C(a)$ if $a$ is in the answer to the retrieval query $C$ when it is evaluated over $\mathsf{DB}_{\mathsf{K}}$. This evaluation is inductively defined as follows.

1. $\mathsf{DB}_{\mathsf{K}} \models_{\mathsf{DB}} A(a)$ if $A(a) \in \mathsf{DB}_{\mathsf{K}}$
2. $\mathsf{DB}_{\mathsf{K}} \models_{\mathsf{DB}} R(a, b)$ if $R(a, b) \in \mathsf{DB}_{\mathsf{K}}$
3. $\mathsf{DB}_{\mathsf{K}} \models_{\mathsf{DB}} C \sqcap D(a)$ if $\mathsf{DB}_{\mathsf{K}} \models_{\mathsf{DB}} C(a)$ and $\mathsf{DB}_{\mathsf{K}} \models_{\mathsf{DB}} D(a)$
4. $\mathsf{DB}_{\mathsf{K}} \models_{\mathsf{DB}} \forall R.C(a)$ if $\mathsf{DB}_{\mathsf{K}} \models_{\mathsf{DB}} C(\forall_{R,a})$
5. $\mathsf{DB}_{\mathsf{K}} \models_{\mathsf{DB}} \exists R.C(a)$ if there exists a constant $b$ with $\mathsf{DB}_{\mathsf{K}} \models_{\mathsf{DB}} C(b)$ and $\mathsf{DB}_{\mathsf{K}} \models_{\mathsf{DB}} R(a, b)$

The constants $\forall_{R,a}$ are special individual terms introduced in the completion process in order to correctly answer queries which involve value restrictions.

*Remark 1.* The above definition makes it possible to formulate retrieval queries over $\mathsf{DB}_{\mathsf{K}}$ using standard SQL.

Before we can perform the completion algorithm which computes the relational representation of a knowledge base $\mathsf{K}$, we have to normalize $\mathsf{K}$, that is replace every occurrence of $\forall R.(C \sqcap D)$ with $\forall R.C \sqcap \forall R.D$.

Then, the *precompletion* $\mathsf{A}'$ of $\mathsf{A}$ is built by applying the following rules to an ABox $\mathsf{A}$ until a fixed point is reached. Of course, these rule are reminiscent of the tableau construction for description logics with transitive and inverse roles, see for instance [8].

1. $\top(a) \in \mathsf{X}$ if $a$ occurs in $\mathsf{X}$
2. $C(a) \in \mathsf{X}$ and $D(a) \in \mathsf{X}$ if $C \sqcap D(a) \in \mathsf{X}$
3. $C(x) \in \mathsf{X}$ and $R(a,x) \in \mathsf{X}$ for a new $x$ if $\exists R.C(a) \in \mathsf{X}$ and no such $x$ exists yet
4. $C(a) \in \mathsf{X}$ if $\forall R.C(b) \in \mathsf{X}$ and $R(b,a) \in \mathsf{X}$ for some $b$
5. $C(\forall_{R,a}) \in \mathsf{X}$ if $\forall R.C(a) \in \mathsf{X}$
6. $C(\forall_{R,a}) \in \mathsf{X}$ if $\forall R.C(b) \in \mathsf{X}$, $R(b,a) \in \mathsf{X}$, and $R \in \mathsf{R}^+$
7. $R(a,b) \in \mathsf{X}$ if $T(a,b) \in \mathsf{X}$ and $T \sqsubseteq R \in \mathsf{T}$
8. $R^-(a,b) \in \mathsf{X}$ if $R(b,a) \in \mathsf{X}$ where we set $(R^-)^- := R$
9. $R(a,c) \in \mathsf{X}$ if $R \in \mathsf{R}^+$, $R(a,b) \in \mathsf{X}$, and $R(b,c) \in \mathsf{X}$ for some $b$

The only part of $\mathsf{K}$ that is not taken into account in the build up of the precompletion are the concept inclusions present in the TBox. In order to treat them properly, we have to apply the following algorithm.

**Algorithm 1.** Procedure for building the completion of an ABox

**Input:** ABox $\mathsf{X}$ and TBox $\mathsf{T}$
**Output:** Completion of the ABox

$\mathsf{Y} \longleftarrow \emptyset$
**repeat**
  $\mathsf{X} \longleftarrow \mathsf{X} \cup \mathsf{Y}$
  $\mathsf{X} \longleftarrow$ precompletion of $\mathsf{X}$
  $\mathsf{Y} \longleftarrow \{C(a)$ : there exists a concept $D$ such that $\mathsf{DB}_\mathsf{X} \models D(a)$ and $D \sqsubseteq C \in \mathsf{T}\}$
**until** $\mathsf{Y} = \emptyset$
**return** $\mathsf{X}$

That is starting from an initial ABox $\mathsf{A}_1$ we build the precompletion $\mathsf{A'}_1$. Then we add all assertions implied by inclusion axioms yielding an ABox $\mathsf{A}_2$. We have to precomplete this ABox resulting in $\mathsf{A'}_2$. Again, the inclusion axioms may imply new assertions which gives us an ABox $\mathsf{A}_3$. This process eventually stops which provides the completion $\mathsf{A}^*$ of $\mathsf{A}_1$

*Example 1.* Consider the zoological ontology given in the introduction. We have

$$\mathsf{T} := \{\mathsf{Carnivore} \sqsubseteq \mathsf{Animal} \sqcap \forall \mathsf{Eats.Animal}\}$$

and

$$\mathsf{A} := \{\mathsf{Carnivore(lion)}, \mathsf{Eats(lion, gnu)}\}.$$

Let $\mathsf{A}_1 := \mathsf{A}$. Building the precompletion of $\mathsf{A}_1$ does not give any new assertions. The second step deals with the concept hierarchy. That yields

$$\mathsf{A}_2 = \mathsf{A}_1 \cup \{\mathsf{Animal} \sqcap \forall \mathsf{Eats.Animal(lion)}\}.$$

Building the precompletion of $\mathsf{A}_2$ gives us

$$\mathsf{Animal(lion)}, \forall \mathsf{Eats.Animal(lion)}, \mathsf{Animal(gnu)} \in \mathsf{A'}_2.$$

Since no new individuals have been added to Carnivore, the second step dealing with the concept hierarchy does not result in additional assertions. Thus, we reached a fixed point and we have $\mathsf{A'}_2 = \mathsf{A}^*$.

If we start from a finite knowledge base $\mathsf{K}$, then by a cardinality argument we easily can see that a fixed point is reached after finitely many steps.

**Theorem 1.** *If $\mathsf{K}$ is a finite knowledge base, then also its completion $\mathsf{K}^*$ is finite.*

*Proof.* First observe, that only finitely many new individual constants are needed in the course of the inductive built up of $\mathsf{K}^*$. Consider the case for value restrictions. There, it may be that a constant of the form $\forall_{R,\forall_{\dots,\forall_{T,a}}}$ is introduced. However the role depth of such a new constant can at most be the role depth of the original knowledge base $\mathsf{K}$. The same goes for the case of existential quantification. Therefore, only finitely many new constants have to be introduced. Moreover, individual constants are only added to subconcepts of concepts occurring in $\mathsf{K}$ and there are only finitely many such subconcepts. Hence, the fixed point will be reached after finitely many stages. □

The database instance $\mathsf{DB}_\mathsf{K}$ of a knowledge base $\mathsf{K}$ consists of all atomic concept assertions $A(a)$ and role assertions $R(a, b)$ of the completion $\mathsf{K}^*$.

Due to the interplay of $\forall$ and $\exists$, the size of $\mathsf{DB}_\mathsf{K}$ can be exponential in the size of $\mathsf{K}$ [2]. However, in many practical applications this blow-up does not happen. For instance, our evaluation shows that in the LUBM benchmark the database size grows only linearly in the size of the original knowledge base.

In the sequel we show that query evaluation over $\mathsf{DB}_\mathsf{K}$ is sound and complete.

**Theorem 2 (Completeness).** *Let $\mathsf{K}$ be a knowledge base, $C$ be a concept description, and $a$ be an individual constant. We have that*

$$\mathsf{K} \models C(a) \Longrightarrow \mathsf{DB}_\mathsf{K} \models_{\mathsf{DB}} C(a).$$

*Proof.* Assume $\mathsf{DB}_\mathsf{K} \not\models_{\mathsf{DB}} C(a)$. Completeness is easily established by constructing a canonical counter model $\mathcal{M}$ with $\mathcal{M} \models \mathsf{K}$ and $\mathcal{M} \not\models C(a)$. The only point that needs a bit of attention is the case when $C$ is a value restriction $\forall R.D$. In this case we have to observe that it is always possible in pos-$\mathcal{ALE}$ to extend a given interpretation of $R$ to falsify $\forall R.D(a)$. □

In order to prove soundness we need some auxiliary definitions.

**Definition 1.** *The $*$ unfolding of a concept assertion is given by:*

1. $C(a)^* := C(a)$ *if $a$ is not of the form $\forall_{R,b}$,*
2. $C(a)^* := (\forall R.C(b))^*$ *if $a = \forall_{R,b}$.*

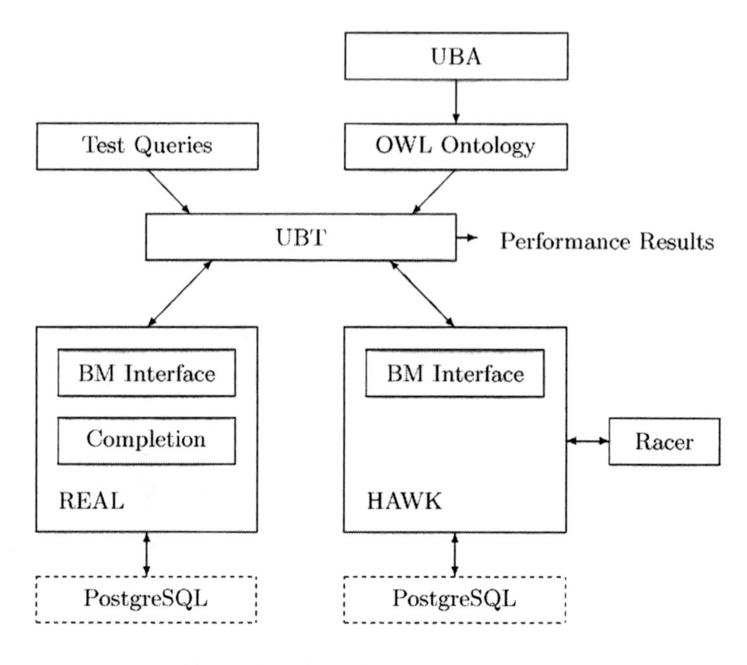


**Fig. 1.** Benchmark setup

**Definition 2.** *The basis of an individual constant is given by:*

1. $\mathsf{basis}(a) := \mathsf{basis}(b)$ *if* $a = \forall_{R,b}$,
2. $\mathsf{basis}(a) := a$ *otherwise.*

**Theorem 3.** *Let* $\mathsf{K}$ *be a knowledge base,* $C$ *be a concept description, and* $a$ *be an individual constant such that* $\mathsf{basis}(a)$ *occurs in* $\mathsf{K}$*. We have that*

$$\mathsf{DB}_\mathsf{K} \models_\mathsf{DB} C(a) \Longrightarrow \mathsf{K} \models C(a)^*.$$

*Proof.* By induction on the structure of $C$. We show only the case when $C$ is $\forall R.D$. Then $\mathsf{DB}_\mathsf{K} \models_\mathsf{DB} D(\forall_{R,a})$. By the induction hypothesis we get $\mathsf{K} \models D(\forall_{R,a})^*$. That is $\mathsf{K} \models \forall R.D(a)^*$ by the previous definition. □

As a corollary we obtain:

**Corollary 1 (Soundness).** *Let* $\mathsf{K}$ *be a knowledge base,* $C$ *be a concept description, and* $a$ *be an individual constant. We have that*

$$\mathsf{DB}_\mathsf{K} \models_\mathsf{DB} C(a) \Longrightarrow \mathsf{K} \models C(a).$$

## 3 Evaluation

To show the applicability of our approach, we developed a system called REAL based on pos-$\mathcal{ALE}$ and the completion procedure presented above, see [9]. We

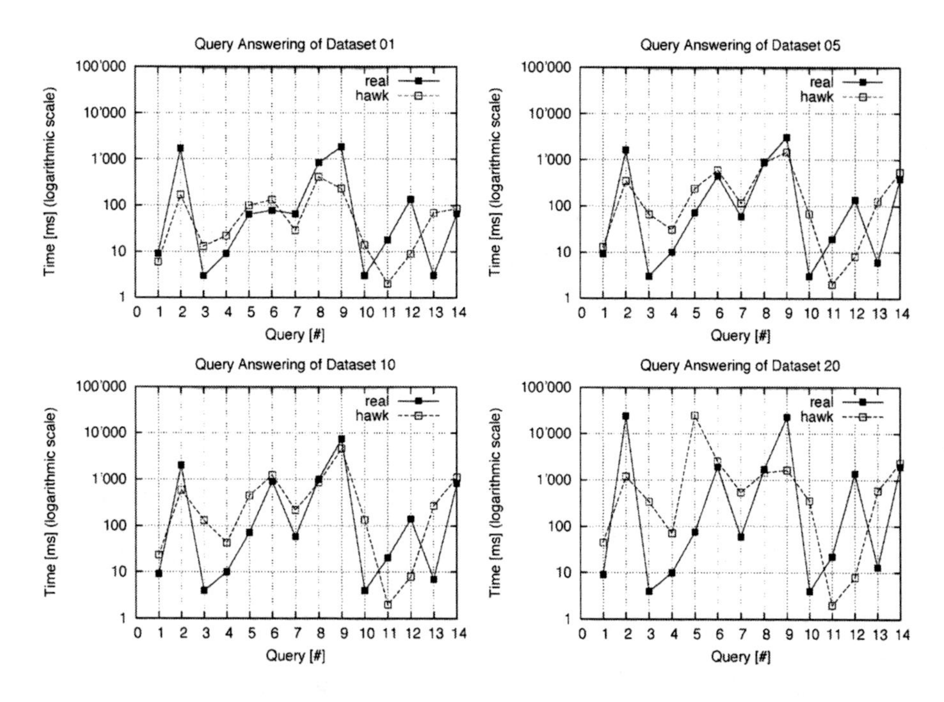


**Fig. 2.** All datasets with the response time of each query

evaluated our implementation with the Lehigh University Benchmark [5]. This is a standard benchmark for expressive semantic web knowledge base systems. We compared the performance of our system with that of HAWK which also follows a completion approach. We used a 3 GHz Pentium 4, 2 GB RAM, and a PostgreSQL DB to run the tests.

The overall setup of the benchmark is shown in Figure 2.The benchmark system contains

1. an OWL ontology modeling a university domain,
2. a data generator (UBA) creating datasets (ABoxes) of different size,
3. a performance test application (UBT) which runs the benchmark, and
4. a set of test queries.

The system under consideration have to provide a benchmark interface (BM) which is called by the test application. Note that HAWK uses an external reasoner (RacerPro, see [6]) for some initial computations in the loading process of an ontology. The detailed settings and all results of the evaluation can be found in [9].

We tested four different datasets called 01, 05, 10, and 20. The smallest dataset (01) contains about 100'000 triples which equals a file size of 8 MB whereas the largest dataset (20) counts more than 2'700'000 triples with a total size of 234 MB.

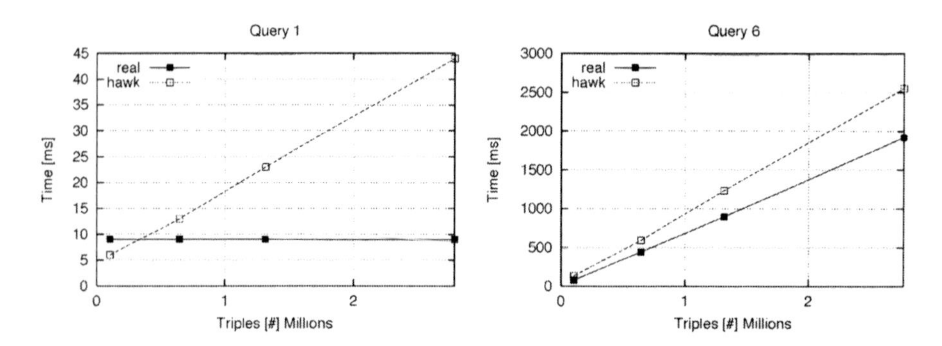


**Fig. 3.** Response time of queries 1 and 6 in relation to the number of triples in the database

The benchmark consists of 14 queries which we issued against the four datasets in both systems. First note that our system provides complete answers to all queries whereas HAWK only provides complete answers to the queries 1,2,3,4, and 14. For all datasets, the answer times to the queries are shown in Figure 2. We find that although we provide complete answers to all queries, our system often perform even better than HAWK.

Our system also scales up very well. For many queries we have linear (sometimes almost constant) behavior of the answering time with respect to the number of triples. See Figure 3 for two typical examples.

For the queries 2, 9, and 12 we need much more time than HAWK and our system does not show a good scaling behavior. These queries need a lot of joins and we have not yet found an optimal database configuration to better support such queries in our approach. However note that HAWK does not provide complete answers to 9 and 12. Still query 2 shows that there is room for improvement.

## 4 Concluding Remarks

We identified pos-$\mathcal{ALE}$, a description logic which can easily be represented in a relational setting. This leads to an extension of relational databases which supports semantic web queries. So far, such extensions often did not provide complete answers to retrieval problems. Our relational representation provides sound and complete query answering with respect to pos-$\mathcal{ALE}$. The evaluation of our implementation with the LUBM benchmark showed that our approach is suitable for practical applications. In particular, it exhibits good scaling properties. However, the tests also showed that we still need better support for queries that involve many join operations.

Scalability is not the only feature which makes our approach valuable for practical applications. The use of a classical database system has the additional advantage that all the features provided by the database may be employed. For example, access control for the ontological system can be implemented based on the privileges and rights system the database provides.

## Acknowledgments

We would like to thank Yuanbo Guo for his support on LUBM.

## References

1. Acciarri, A., Calvanese, D., De Giacomo, G., Lembo, D., Lenzerini, M., Palmieri, M., Rosati, R.: Quonto: Querying ontologies. In: Proc. of the 20th Nat. Conf. on Artificial Intelligence (AAAI 2005), pp. 1670–1671 (2005)
2. Baader, F., Calvanese, D., McGuinness, D., Nardi, D., Patel-Schneider, P.: The Description Logic Handbook. Cambridge (2003)
3. Calvanese, D., De Giacomo, G., Lembo, D., Lenzerini, M., Rosati, R.: DL-Lite: Tractable description logics for ontologies. In: Proc. of the 20th Nat. Conf. on Artificial Intelligence (AAAI 2005), pp. 602–607 (2005)
4. Dean, M., Schreiber, G.: OWL web ontology language reference (2004)
5. Guo, Y., Pan, Z., Heflin, J.: LUBM: A benchmark for OWL knowledge base systems. Journal of Web Semantics 3, 158–182 (2005)
6. Haarslev, V., Möller, R.: Racer system description. In: Goré, R.P., Leitsch, A., Nipkow, T. (eds.) IJCAR 2001. LNCS (LNAI), vol. 2083, pp. 701–705. Springer, Heidelberg (2001)
7. Heflin, J.: OWL web ontology language use cases and requirements (2004), Available at `http://www.w3c.org/TR/webont-req/`
8. Horrocks, I., Sattler, U.: A description logic with transitive and inverse roles and role hierarchies. Journal of Logic and Computation 9(3), 385–410 (1999)
9. Kottmann, N.: Description logic query answering with relational databases. Master's thesis, University of Bern (2006)
10. Pan, Z., Heflin, J.: DLDB: Extending relational databases to support semantic web queries. In: Fensel, D., Sycara, K.P., Mylopoulos, J. (eds.) ISWC 2003. LNCS, vol. 2870, pp. 109–113. Springer, Heidelberg (2003)

# A Method for Determining Ontology-Based Semantic Relevance

Tuukka Ruotsalo and Eero Hyvönen

Semantic Computing Research Group (SeCo)
Helsinki University of Technology (TKK), Laboratory of Media Technology
University of Helsinki, Department of Computer Science
firstname.lastname@tkk.fi
http://www.seco.tkk.fi/

**Abstract.** The semantic web is based on ontologies and metadata that indexes resources using ontologies. This indexing is called annotation. Ontology based information retrieval is an operation that matches the relevance of an annotation or a user generated query against an ontology-based knowledge-base. Typically systems utilising ontology-based knowledge-bases are semantic portals that provide search facilities over the annotations. Handling large answer sets require effective methods to rank the search results based on relevance to the query or annotation. A method for determining such relevance is a pre-requisite for effective ontology-based information retrieval. This paper presents a method for determining relevance between two annotations. The method considers essential features of domain ontologies and RDF(S) languages to support determining this relevance. As a novel use case, the method was used to implement a knowledge-based recommendation system. A user study showing promising results was conducted.

## 1 Introduction

The semantic web [4] promotes the use of explicit background knowledge (metadata) to manage diverse resources. Metadata has a defined meaning in terms of a domain ontology that provides a shared conceptualisation of the domain of discourse [9]. Resources are indexed using metadata schemas and values from domain ontologies. Resources indexed with ontological values are called annotations. While the research of the logical structure of the ontologies and metadata schemas has gained much popularity in the past years, the methods for information retrieval have mainly concentrated on strict boolean querying of a knowledge-base rather than assessing relevance for the annotations in the knowledge-base. Good examples can be found in a field of semantic portals [14], that provide search facilities to access the data. Many of the portals so far have utilised search facilities that are based on Boolean queries or facet-based search [2,16]. To enable effective information retrieval, methods for ranking and clustering the search results are a necessity.

The relevance determination problem has an important background in text retrieval where document-term matrices are used to calculate similarity of the documents [5]. Good results have been achieved with *tf-idf* weighting of the feature vectors [19,3].

R. Wagner, N. Revell, and G. Pernul (Eds.): DEXA 2007, LNCS 4653, pp. 680–688, 2007.

The majority of current research in ontology-based information retrieval has focused on crisp logic with intelligent user interfaces to formulate the query [16]. These have been further developed to support fuzzy logic [10]. Determining structural similarity has been investigated in SimRank [13], SimFusion [20], AKTiveRank [1] and Swoogle [8]. SimRank measures similarity of structural contexts, but concentrates only on graph theoretical model instead of feature vectors. SimFusion considers object features, but does not bind the features to ontologies. The Swoogle search engine uses *term rank* and *onto rank* algorithms to provide the relevance to predict correct ontology and instances for terms and concepts. However, Swoogle concentrates on matching classes and terms to ontologies, but does not consider the mutual relevance of annotations. AKTiveRank uses semantic similarity and the alignment of the terms in separate ontologies as a ranking principle.

In this paper we present a method that calculates the mutual relevance of annotations. Unlike SimRank, SimFusion, Swoogle and AKTiveRank we concentrate on the relevance of the annotations based on the underlying domain ontology. We extend the *tf-idf* [19,3] method by considering essential features of the domain ontologies and RDF(S) languages. The method can be used for numerous applications such as knowledge-based recommendation [15,7], information retrieval and clustering [3]. As a novel use case we present a recommendation system implemented with a real-life dataset. Finally, we show initial empirical results from a user test that support the method.

## 2 Knowledge Representation on the Semantic Web

### 2.1 Representation of Annotations

The Semantic web contains metadata about resources. This metadata is called annotations. Annotations are formulated with a RDF(S) [6] language, where each statement about the resource is given as an *annotation triple*. A set of annotation triples describe a resource $x$.

An annotation is a set of triples $E = \{< subject, predicate, object >\}$, where at least one $subject \equiv x$ (i.e. it is connected to the resource that is being annotated). Triples that have a resource as the object value are called *ontological elements*; triples with literal values are called *literal elements*. In this paper, we are concerned with ontological relevance and therefore only ontological elements are considered.

In addition to the triples, RDF Schema language (RDFS) defines a schema for RDF. RDFS separates classes and instances. For example a resource *GeorgeWBush* could be defined to be an instance of the classes *Person* and *President*. In RDFS classes can be defined as subsumption hierarchies. For example class *President* could be defined to be a *subClassOf* a class *PoliticalRole*.

We next present requirements for a method by which the ontology-based semantic relevance between resources can be determined.

### 2.2 Requirements for Annotation Relevance Calculation

To fully support the data model behind RDF(S), the following criteria must be taken into account by the method determining the relevance:

1. **Classes and instances.** A typical approach in knowledge representation is to separate classes and instances. For example, when annotating a web page with the resource *'GeorgeWBush'*, the particular instance of a class, say *'Politician'*, is referred to. Any other annotation stating something about the same resource *'GeorgeWBush'* would also refer to this particular instance. The instance sharing approach leads to undisputed benefits because the resources have a unique identifier. However, if the instance is commonly referred in the knowledge-base it may over-dominate traditional retrieval methods.
2. **Subsumption.** Concepts in the domain ontologies are typically ordered in subsumption hierarchies. For example, if *'GeorgeWBush'* is an instance of the class *'Politician'* this could be subsumed by the class *'Person'*. It is clear that the fact that it is implicitly known that *'GeorgeWBush'* is also related to class *'Person'* has to be taken into a consideration by information retrieval methods, but intuitively with less relevance than the class *'Politician'*, since persons may also be non-political.
3. **Part-of relations.** In addition to subsumption, many domain ontologies introduce relations to support the theory of parts and wholes (part-of). For example, if a resource is annotated with the instance *'New York City'*, it may be relevant in the scope of *'New York State'* due to the part-of-relation between the resources, although the notion of the state does not subsume the notion of the city. Part-of relations are useful in information retrieval, but require separate handling from subsumption relations [17].

Next we present the method for determining ontology-based annotation relevance where these requirements are taken into consideration.

## 3 A Method for Determining Semantic Relevance of Annotations

Consider resources $x$ and $y$. The ontological relevance $r$ of a resource $y$, when a resource $x$ is given, is defined by the quadripartite relation $S$

$$S \subset \{< x, y, r, e > | x \in C, y \in C, r = ar(ann(x), ann(y)) \in [0, 1], e \text{ is a literal}\},$$

where $C$ is the set of resources, $ar$ is a real valued function *annotation relevance* expressing how relevant $y$ is given $x$, $ann$ is a function returning annotation triples for a resource, and $e$ is a literal explanation of why $y$ is relevant given $x$. A tuple $< x, y, r, e > \in S$ intuitively means that "$x$ is related to item $y$ by relevance $r$ because of $e$". For example:

*<GeorgeWBush, WhiteHouse, 0.8, "George W. Bush workingIn Whitehouse">*

The relevance relation can be used in semantic recommending: it provides the set of explained recommendations for each content item $x$. In addition, the relevance relation could be used for clustering or as a search base of its own if the end-user is interested in finding relations between resources instead of resources themselves.

Below, we first present a method for computing the annotation relevance $r = ar(ann(x), ann(y))$ for resources $x$ and $y$, and then discuss how to provide the explanation $e$.

A widely used method for determining the relevance of a document with respect to a keyword $k$ is *tf-idf* [3]. Here the relevance $r_k$ of a document $d$ with respect to $k$ is the product of term frequency (*tf*) and inverse document frequency (*idf*) $r_k = tf \times idf$. The term frequency $tf = n_k/n_d$ is the number of occurrences of $k$ in $d$ divided by the number $n_d$ of terms in $d$. The inverse document frequency is $idf(d) = log\frac{N}{N_k}$ where $N$ is the number of documents and $N_k$ is the number of documents in which $k$ appears. Intuitively, *tf-idf* determines relevance based on two components: *tf* indicates how relevant $k$ is w.r.t. $d$ and *idf* lessens the relevance for terms that are commonly used in the whole document set.

Our case is different from the classical text document retrieval in the following ways. First, the document set is a set of ontological annotations. The *tf* component cannot be based on term frequency as in *tf-idf*. Second, we will not search for relevant documents with respect to a key word, but try to find semantically related ontological annotations. To account for these differences, we devised *idf*-like measures *inverse class factor, inverse instance factor*, and *inverse triple factor* that account for the global usage of classes, individuals and triples in the annotations.

**Definition 1 (inverse class factor).** *The inverse class factor $icf(c)$ for a class $c$ is $icf(c) = log\frac{N}{N_c}$, where N is the total number of instances of all classes used in the annotations, and $N_c$ is the number of instances of the class c.*

Intuitively, *icf(c)* is higher for annotation instances whose class are rarely used in annotation.

**Definition 2 (inverse instance factor).** *The inverse instance factor $iif(i)$ for an instance $i$ is $iif(i) = log\frac{I}{n}$, where I is the total number of instances shared by the annotations, and $n$ is the number of usage of the instance $i$.*

This measure takes into account the fact that instances can be shared by the annotations. The idea of using *iif(i)* will be to lessen relevance of content items that share same instances, when such instances are commonly used.

In order to define the inverse triple factor we first define the predicate $cmatch(x, y)$ for matching two instances and $pmactch(p, q)$ for matching two properties (rdf predicates). Let $cl(x)$ denote the class of instance $x$, $sp(c)$ denote the set of superclasses of class $c$, and $pr(p)$ denote the set of super properties of property $p$. Then:

$$cmatch(x, y) = \text{true}, \text{ if } x = y \text{ or } cl(x) \in sp(cl(y))$$
$$pmatch(p, q) = \text{true}, \text{ if } p = q \text{ or } p \in pr(q).$$

**Definition 3 (inverse triple factor).** *The inverse triple factor $itf(t)$ for a triple $t =< s, p, o >$ is: $itf(t) = log\frac{T}{N}$, where T is the total number of annotation triples $< s', p', o' >$, such that $cmatch(s', s)$ and $pmatch(p', p)$ and $cmatch(o', o)$ hold, and N is the total number of annotation triples.*

In addition, a measure is needed to determine the relevance between two instances based on the class membership, part-of, and an instance equivalence relations in the domain ontology:

**Definition 4 (ontological instance relevance).**
*The ontological instance relevance of instance $y$ given instance $x$ is*

$$oir(x,y) = \begin{cases} iif(y) \times icf(cl(y)) \text{ if } cmatch(x,y) \\ 0.70 \times iif(y) \times icf(cl(y)) \text{ if } partOf(y,x) \text{ or } partOf(cl(y), cl(x)), \\ 0 \text{ otherwise} \end{cases}$$

*where $partOf(x,y)$ is true if $x$ is a part of $y$. If two first cases match at the same time, the maximum value is selected.*

The ontological instance relevance is given by the product of the inverse instance factor and inverse class factor. The similarity can be calculated if the instance between annotation objects is shared or the class membership of the target instance is in the transitive closure of the class membership of the source instance. In addition the target instance or target class membership can be connected to the source instance or to the source class membership with a part-of relation. We have used 0.70 as the part-of multiplier based on extensive empirical tests by Rodriquez and Egenhofer [17]. Intuitively, $oir(x,y)$ is high, i.e. $y$ is relevant for $x$, when $y$ and $sp(cl(y))$ are rarely used in annotations, and $x$ and $y$ are related by hyponymy, meronymy, or equivalence.

The ontological triple relevance can now be defined.

**Definition 5 (ontological triple relevance).** *The ontological relevance of a triple $y =< s,p,o >$, given a triple $x =< s',p',o' >$ and assuming that $pmatch(p,p')$ holds, is:*

$$otr(x,y) = oir(s,s') + oir(o,o') + itf(y).$$

Intuitively, $otr(x,y)$ is high, i.e. $y$ is relevant for $x$, when the subject and object of $y$ are relevant given the subject and object of $x$, respectively, and $y$ is rarely used.

Finally the annotation relevance is the sum of the ontological triple relevances for the annotations.

**Definition 6 (annotation relevance).** *The annotation relevance $ar(A,B)$ of an annotation A, given an annotation B, is*

$$ar(A,B) = \frac{\sum_{a \in A, b \in B} otr(a,b)}{n_t},$$

*where $n_t$ is a number of triples in a target annotation used as a normalisation factor.*

When determining the values $ar(x,y)$, the explanation literal $e$ can be formulated based on the labels of the matching triples.

## 4 Implementation and Evaluation

The method presented above has been implemented in the recommendation system of the CULTURESAMPO prototype portal [11]. A user study was conducted to evaluate how well the method predicted the ranking of the resources compared to opinions of the users. In information retrieval systems, the users usually want to see just ten to twenty documents, and if these do not correspond to the information need of the users the search is re-adjusted [3]. This is why in practical applications, such as knowledge-based recommendation, the ranking of the documents is a crucial task.

A user study was conducted to evaluate the method. The hypothesis tested was: does the ranking performed by the annotation relevance method correspond with the end-user's opinion of the ranking. In practice this means testing whether the users liked more the recommended resources ranked higher (target documents) based on a source resource (source document) they were looking at.

The most obvious way to measure this is to calculate the correlation between the ordering of the items made by the method and by the user. Based on a preliminary user test, it turned out that the ordering of the documents was difficult for the users. However, the users indicated that it was rather easy to classify the documents into two groups: the highly relevant and less relevant. Therefore, this simple ranking dichotomy was used in the test.

### 4.1 Dataset

The dataset used contained annotations of three different resources: images of museum items, images of photographs and images of paintings. These were annotated by domain experts in Finnish museums. The General Finnish Ontology (YSO) [12] was used as a domain ontology. This domain-ontology consists of more than 23.000 classes organised in subsumption and part-of hierarchies. Seven documents were randomly selected as source documents representing the source resources: two images of museum items, three images of photographs and two images of paintings. Ten target recommendations were then calculated for each source document representing the target resources, which resulted to a set of 70 target documents. The calculation was performed against a knowledge-base that contained annotations of nearly 10.000 resources of before-mentioned types.

The five top-ranked recommendation documents given by our method were considered the *higher relevance group*. The other five, the *lower relevance group*, were a sample of the lower half of the ranking based on the median relevance. To exclude highly non-relevant recommendations its was required that the source document and its target recommendation should share at least two triples.

### 4.2 Test Setting

Figure 1 illustrates the user interface showing a page about a photograph in a dataset. The pages were printed without the recommendations that can be seen on the right side of the figure. A card sorting experiment was conducted based on the printed pages [18]. Seven test subjects were first asked to classify the recommendations according to the seven source documents, based on the metadata and the image. After this the subjects were asked to formulate the higher relevance group and the lower relevance group for each source document. In both tests the test subjects were able to leave a target recommendation document out, if the they felt that it was not relevant given any source document.

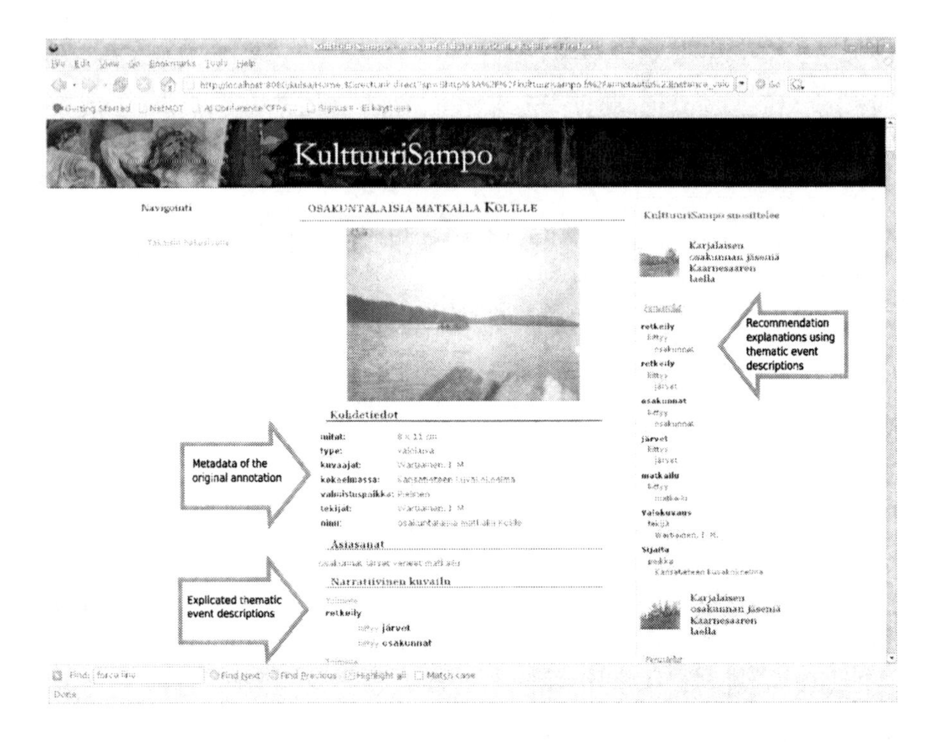


**Fig. 1.** CULTURES AMPO user interface showing a photograph, its metadata, and semantic recommendation links

### 4.3 Results

The right source document for a recommendation was found in 71 per cent (%) of the cases. Test subjects then classified the items in the higher relevance group correctly in 82% of the cases. From the documents that were classified under the wrong group, 23% were in the higher relevance group and in 77% in the lower relevance group. The share of the target recommendation documents that the test subjects were unable to classify into either the high or low relevance group, 5% were in higher relevance group, and 95% in the lower relevance group.

### 4.4 Conclusions of Empirical Evaluation

These results show that the method predicted the relevance very well for a first attempt: only 5% of the recommendations were left out as non-relevant. All of the recommendations left out were from the lower relevance group. Only 18% of the recommendations were classified wrongly in the higher relevance group (including the items that still belonged to the lower relevance group). Only 3% of the recommendations were classified under the wrong source item in the higher relevance group.

## 5 Conclusions

Previous work in knowledge-based recommendation and object relevance has focused on measuring the similarity of feature vectors [7,15], where similarity measures are used to calculate nearest neighbours from a vector space. This approach works well in a single domain, where features can be predefined and weights assessed for features. The *tf-idf* methods have shown promising results in measuring the relevance in information retrieval from natural language documents [3].

Our work extends such measures by adopting the ontology and the annotation triples as a source for feature matching. In terms of *tf-idf*, we have extended the *idf* component to consider three essential features of ontology-based systems, namely separation of classes and instances, support for subsumption and support for part-of relations. We have implemented the method in the semantic portal CULTURESAMPO. In addition, we have conducted a user study that gives preliminary empirical evidence of the value of the approach.

## Acknowledgements

This research is part of the National Finnish Ontology Project (FinnONTO) 2003-2007[1], funded mainly by The National Technology Agency (Tekes) and a consortium of 37 companies and public organisations.

## References

1. Alani, H., Brewster, C.: Ontologies and knowledge bases: Ontology ranking based on the analysis of concept structures. In: Proceedings of the 3rd international conference on Knowledge capture K-CAP 2005 (2005)
2. Athanasis, N., Christophides, V., Kotzinos, D.: Generating on the?y queries for the semantic web: The ics-forth graphical rql interface. In: Proceedings of the Third International Semantic Web Conference, pp. 486–501 (2004)
3. Baeza-Yates, R., Ribeiro-Neto, B.: Modern Information Retrieval. Addison-Wesley, ACM Press, New York (1999)
4. Berners-Lee, T., Hendler, J., Lassila, O.: The semantic web. Scientific American 284(5), 34–43 (2001)
5. Berry, M.: Survey of Text Mining Clustering, Classification, and Retrieval. Springer, Heidelberg (2004)
6. Brickley, D., Guha, R.V.: RDF Vocabulary Description Language 1.0: RDF Schema W3C Recommendation 10 February 2004. Recommendation, World Wide Web Consortium (February 2004)
7. Burke, R.: Knowledge-based Recommender Systems. In: A. Kent (ed.) Encyclopedia of Library and Information Systems. vol. 69, Supplement 32. Marcel Dekker (2000)
8. Ding, L., Finin, T., Joshi, A., Pan, R., Cost, R.S., Peng, Y., Reddivari, P., Doshi, V., Sachs, J.: Swoogle: a search and metadata engine for the semantic web. In: Proceedings of the thirteenth ACM international conference on Information and knowledge management, pp. 652–659 (2004)

[1] http://www.seco.tkk.fi/projects/finnonto/

9. Klein, H.K., Hirschheim, R., Lyytinen, K.: Information Systems Development and Data Modeling: Conceptual and Philosophical Foundations. Cambridge University Press, Cambridge (1995)
10. Holi, M., Hyvönen, E.: Fuzzy view-based semantic search. In: Proceedings of the 1st Asian Semantic Web Conference (ASWC2006), Beijing, Springer, Heidelberg (2006)
11. Hyvönen, E., Ruotsalo, T., Häggström, T., Salminen, M., Junnila, M., Virkkilä, M., Haaramo, M., Mäkelä, E., Kauppinen, T., Viljanen, K.: Culturesampo–finnish culture on the semantic web: The vision and first results. In: Developments in Artificial Intelligence and the Semantic Web - Proceedings of the 12th Finnish AI Conference STeP 2006, October 26-27 (2006)
12. Hyvönen, E., Valo, A., Komulainen, V., Seppälä, K., Kauppinen, T., Ruotsalo, T., Salminen, M., Ylisalmi, A.: Finnish national ontologies for the Semantic Web - towards a content and service infrastructure. In: Proceedings of International Conference on Dublin Core an Meltadata Applications (DC 2005) (November 2005)
13. Jeh, G., Widom, J.: Simrank: A measure of structural-context similarity. In: Proceedings of the Eighth ACM SIGKDD International Conference on Knowledge Discovery and Data Mining, pp. 538–543 (2002)
14. Maedche, A., Staab, S., Stojanovic, N., Struder, R., Sure, Y.: Semantic portal — the SEAL approach. MIT Press, Cambridge (2003)
15. McSherry, D.: A generalized approach to similarity-based retrieval in recommender systems. Artificial Intelligence Review 18, 309–341 (2002)
16. Mäkelä, E., Hyvönen, E., Saarela, S.: Ontogator — a semantic view-based search engine service for web applications. In: Cruz, I., Decker, S., Allemang, D., Preist, C., Schwabe, D., Mika, P., Uschold, M., Aroyo, L. (eds.) ISWC 2006. LNCS, vol. 4273, Springer, Heidelberg (2006)
17. Rodriquez, A., Egenhofer, M.: An asymmetric and context-dependent similarity measure. International Journal of Geographical Information Science 18(3), 229–256 (2004)
18. Rugg, G., McGeorge, P.: The sorting techniques: a tutorial paper on card sorts, picture sorts and item sorts. Expert Systems 14(2), 80–93 (1997)
19. Salton, G., Buckley, C.: Term weighting approaches in automatic text retrieval. Technical report tr87-881, Cornell University Ithaca, NY (1987)
20. Xi, W., Fox, E.A., Fan, W., Zhang, B., Chen, Z., Yan, J., Zhuang, D.: Simfusion: measuring similarity using unified relationship matrix. In: Proceedings of the 28th annual international ACM SIGIR conference on Research and development in information retrieval, pp. 130–137 (2005)

# Semantic Grouping of Social Networks in P2P Database Settings*

Verena Kantere, Dimitrios Tsoumakos, and Timos Sellis

School of Electr. and Comp. Engineering, National Technical University of Athens
{vkante, dtsouma, timos}@dbnet.ece.ntua.gr

**Abstract.** Social network structures map network links to semantic relations between participants in order to assist in efficient resource discovery and information exchange. In this work, we propose a scheme that automates the process of creating schema synopses from semantic clusters of peers which own autonomous relational databases. The resulting mediated schemas can be used as global interfaces for relevant queries. As our experimental evaluations show, this method increases both the quality and the quantity of the retrieved answers and allows for faster discovery of semantic groups by joining peers.

## 1 Introduction

In the variety of P2P applications that have been proposed, Peer Data Management Systems (PDMSs) (e.g., [6, 19]) hold a leading role in sharing semantically rich information. In a PDMS, each peer is an autonomous source that has a local schema. Sources store and manage their data locally, revealing part of their schemas to the rest of the peers. Due to the lack of global schema, they express and answer queries based on their local schema. Peers also perform local coordination with their *acquaintees*, i.e., their one-hop neighbors in the overlay. During the acquaintance procedure, the two peers exchange information about their local schemas and create mediating mappings semi-automatically [9]. The establishment of an acquaintance implies an agreement for the performance of data coordination between the acquaintees based on the respective schema mapping. However, peers do not have to conform to any kind of data or schema transformation to establish acquaintances with other peers and participate in the system. The common procedure for query processing in such a system is the propagation of the query on paths of bounded depth in the overlay. At each routing step, the query is rewritten to the schema of its new host based on the respective acquaintance mappings. A query may have to be rewritten several times from peer to peer till it reaches nodes that are able to answer it sufficiently in terms of quality but also quantity.

In such systems, in order to enable efficient data sharing between heterogeneous sources, the properties of *social networks* [21] are usually applied: Just as humans direct their queries either to personal acquaintances or other knowledgeable individuals,

* This work has been funded by the project PENED 2003. The project is cofinanced 75% of public expenditure through EC - European Social Fund, 25% of public expenditure through Ministry of Development - General Secretariat of Research and Technology and through private sector, under measure 8.3 of OPERATIONAL PROGRAMME "COMPETITIVENESS" in the 3rd Community Support Programme.

R. Wagner, N. Revell, and G. Pernul (Eds.): DEXA 2007, LNCS 4653, pp. 689–699, 2007.

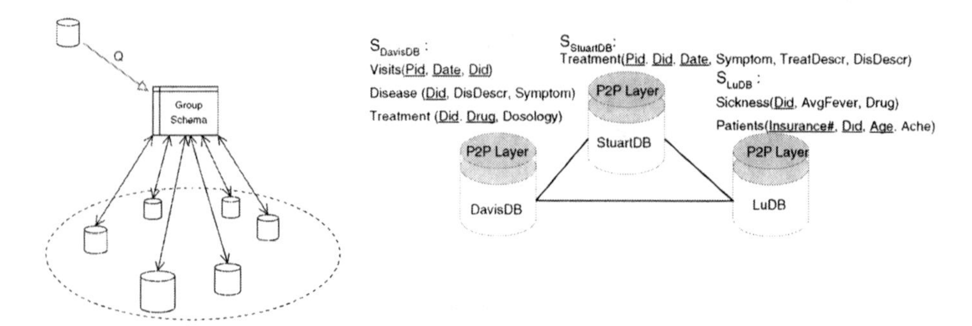


**Fig. 1.** Query directed towards a group schema

**Fig. 2.** Part of a P2P system with peer-databases from the health environment

peers try to identify other participants in the overlay with interests that match theirs. Similar to human social networks, social networking services such as MySpace [15], Orkut [17], etc, form virtual communities, with each participant setting her own characteristics and interests. Their goal is to allow members to form relationships through communication with other members and sharing of common interests. Extending this paradigm, computing social networks consist of a mesh of interconnected nodes (peers). Initially, each of the nodes is connected to a random subset of peers. Gradually, nodes get acquainted with each other, with the new connections indicating semantic proximity. Although not explicitly stated, there has been considerable work to apply the principle of semantic grouping and routing in order to improve performance in distributed systems (e.g., [2, 16, 10, 18, 1], etc).

Assuming a social network organization in a PDMS, an interesting question is how to automatically create a synopsis of the common interests of a group of semantically related nodes. This will be a mediating schema representative of the group along with its mappings with the local databases. Queries can then be expressed on this mediated schema (see Figure 1). This functionality is desirable for multiple reasons: First, it allows queries to be directed to a single, authoritative schema. Second, it actively expedites the acquaintance between semantically related peers. Finally, it minimizes human involvement in the process of creating/updating the group schema. Until now, nodes have been organized by means of a human-guided process (usually by one or more administrators and application experts) into groups of peers that store semantically related data. The administrator, using schema matching tools as well as domain knowledge, initiates and maintains these synopses. This approach requires manual work, extensive peer coordination and repetition of this process each time the group changes.

As a motivating example, envision a P2P system where the participating peers are databases of private doctors of various specialties, diagnostic laboratories and databases of hospitals. Figure 2 depicts a small part of this system, where the peer databases (or else, pDBMSs) are: DavisDB - the database of the private doctor Dr. Davis, LuDB - the database of pediatrist Dr Lu and StuartDB - the database of the pharmacist, Mr Stuart. A P2P layer, responsible for all data exchange of a peer with its acquaintees, sits on top of each database. Among others, the P2P layer is responsible for the creation and maintenance of mappings of local schemas during the establishment of acquaintances towards the line of [9]. Moreover, each peer owns a query rewriting and a query-schema matching mechanism. The schemas of the databases are shown in Figure 2.

We would like to automatically produce a merged schema for all three peers of our example, semantically relevant to their local schemas. Such a merged schema could be the following:

Disease/Sickness(Did, DisDescr, Symptom, Drug)
Visits/Patients(Pid/Insurance#, Did, Date, Age, Ache)
Treatment(Did,Drug, Dosology)

Obviously, in the merged schema we would like alternative names for relations or attributes (separated by '/' above). We would also like the merged schema to contain relations or attributes according to their frequency in the set of local schemas. For example, the attribute *Patients.AvgFever* is not present, possibly because the respective concept is not considered to be frequent in the set of local schemas.

In this paper, we describe a mechanism that operates on a semantically clustered PDMS and automatically creates relational schemas that are representative of the existing clusters. Given the semantic neighborhoods, our system can initiate the creation of a mediating schema $\mathcal{S}_{\mathcal{G}}$ that summarizes the semantics of the participating database schemas. It is created by the gradual merging of peer schemas along the path followed by the process. We call *interest* or *semantic groups* the semantic clusters that exist in social networks operating on PDMSs; moreover, we call *group schema* the inferred schema of the group $\mathcal{S}_{\mathcal{G}}$. $\mathcal{S}_{\mathcal{G}}$ holds mappings with each of the peers involved in its creation and functions as a point of contact for all incoming queries, whether from inside or outside the semantic neighborhood. Thus, requesters of information need only maintain mappings and evaluate queries against one schema, instead of multiple ones. Our experimental evaluation shows that our group creation process increases both the accuracy and the number of answers compared to individually propagating and answering queries in an unstructured PDMS.

## 2 Interest Group Creation

Our goal in creating a group schema is to represent the semantic clusters in a social network using a distributed process that iteratively merges local schemas into the final group schema that preserves their most frequent semantics.

In the following, we assume a PDMS with a social-network organization of peers, i.e., semantically relevant pDBMSs are acquainted or close in the overlay. This can be achieved either manually or using one of the proposed schemes (e.g., [12, 16]). Finally, we also assume that peer mappings between acquaintees are of the widely-known GAV/LAV/GLAV form [6, 13] and peer schemas are relational, (i.e., the only internal mappings are foreign key constraints). Moreover, peers do not carry semantic information about their schemas and mappings.

### 2.1 Group Inference

In this section, we describe the process through which a group schema emerges from a set of clustered nodes in our system. The group-creation procedure (or group *inference*) comprises the following steps:

- Initialization: Who and when initiates the group schema inference
- Propagation: How does the process advance among peers of the same group
- Termination and Refinement: When is the process over/reiterated

**Initialization:** The nature of our application requires that the group inference is performed in a distributed manner, without global coordination. Peers should be able to start the process that creates the respective schema with minimum message exchange. In our system, each member of the social group is eligible to initiate the inference process. Nevertheless, such groups may consist of numerous participants resulting in very frequent collisions among competing initiators. Hence, we only allow *active* members to become the initiators of the process. This is enforced by a system-wide parameter that defines the minimum number of queries posed in the most recent time frame. Intuitively, more active peers have a better knowledge of the social network and the schemas of the other participants through the answers they receive.

The initiator's local schema becomes a point-of-reference regarding the inferred one. Thus, the peer schemas considered for the formation of the group schema should not differ semantically a lot from the schema of the initiator. Specifically, we require that the participating local schemas should be at least $t$-similar to the initiator's schema: $t$ is a parameter that mainly determines how specialized (only peers very similar to the initiator considered) or general (a broad collection of peers participate in the process) the inferred schema will be. The initiator peer is called the *originator* of the group, its schema is the *origin* of the group schema and the maximum similarity distance between the origin and the peer schemas that participate in the group schema inference is the *semantic radius* of the group. The following function calculates the *directed* semantic similarity, *SS*, of two relational schemas:

$$SS(S,T) = \frac{\sum_i \sum_j w_{ij} Mapped_T(SR_{ij})}{\sum_i \sum_j w_{ij} SR_{ij}}$$

In the above function, $S$ is the source schema and $T$ is the target schema. *SS* calculates the portion of $S$'s attributes (*SR*) that are mapped on $T$, with the indices $i, j$ referring to the $j^{th}$ attribute of the $i^{th}$ relation. Also, $w_{ij} > 1$ for attributes that belong to relation keys and $w_{ij} = 1$ otherwise. Obviously, $SS(S,T) \neq SS(T,S)$ in general. *SS* achieves to measure semantic similarity because it takes into consideration the mapping of *concepts* beyond their structural interpretations on the schema level. In our setting we define a distinct concept of a schema $S$ to be each element $R.A$, where $A$ is an attribute of relation $R$ of schema $S$[1]. Moreover, since *SS* ignores the schema structure, it is very easily calculated.

**Propagation:** The initiator $I$ (with schema $S_I$) of the inference process initializes the group schema to its own and creates a stack $ST(I)$ with its acquaintees that are part of

[1] For more details on concepts, see [12].

the cluster. Specifically, $ST(I) = \{A_1, A_2, ..., A_m\}$ is an ordered set of elements $A_j = \{P_j, SS(S_I, S_{P_j})\}$, where $P_j$ is a peer with schema $S_{P_j}$. Elements $A_j$ refer to the $I$'s most similar acquaintees: $SS(S_I, S_{P_j}) \geq t$, $j = 1, .., m$ and $SS(S_I, S_{P_j}) \geq SS(S_I, S_{P_{j+1}})$, $j = 1, .., m-1$. The initiator propagates the inference procedure to the first peer on the stack. The latter is supposed to merge its own schema with the group schema it receives according to the merging procedure described in the section 2.2. Every peer $P$ on the network path of the inference process determines its acquaintees $P_j$ for which $SS(S_I, S_{P_j}) \geq t$, adds the respective pair $P_j$, $SS(S_I, S_{P_j})$, to $ST(I)$ and orders it. Any peer $P$ on the inference process path calculates $SS(S_I, S_{P_j})$ indirectly, as the product: $SS(S_I, S_P) \cdot SS(S'_P, S_{P_j})$, where $S'_P$ is the part of $S_P$ mapped on $S_I$. Essentially, $SS(S_I, S_P)$ aims to measure how much of the semantics of $S_I$ can be found on schema $S_P$, independently of other semantics that the latter captures. The only way to measure this (without automatic matching) is through the chain of mappings of $S_I$ all the way to $S_P$. Thus, the value of $SS(S_I, S_P)$ depends on the path followed by the inference process and fails to consider concepts that exist both in $S_I$ and $S_P$ but not in the schemas of intermediate nodes.

In order for this formula to produce a satisfactory result, the existing clustering in the social network should assure that the similarity between local schemas decreases with the hop-distance of the respective peers in the overlay. Therefore, schemas that are considered later in the process will have lower similarity than previously considered ones. Moreover, if a peer $P$ already in $ST(I)$ is considered for addition, the entry with the highest $SS(S_I, S_P)$ value is kept.

Even though the participation or not of peers in the inference process is judged by a part of their schemas, their whole schema contributes to the inferred group schema (see subsection 2.2). Intuitively, the goal of the inference process is to produce a schema that represents semantics encapsulated in the cluster. In order to determine the cluster's semantic borders we use the semantics of the initiator as reference. This way, the process is safe from producing a schema too broad or distorted from the interests of the initiator.

**Termination:** As aforementioned, the group inference procedure ends when the stack of participating peers becomes empty. However, if too many peers own schemas very similar to the originator's schema or the similarity threshold $t$ is too small (i.e., the semantic radius of the inferred group is big), then it may be the case that the stack is provided at each step with a lot of new entries. Thus, the inference procedure is prolonged taking into account a big number of peers. After a certain number of iterations, there is usually no point of considering more peer schemas in the inference procedure, because they do not alter the schema significantly. In order to expedite the inference process and reduce the exchanged messages, we add a limit to the maximum number of encountered peer schemas, *MaxP*, as a termination condition. *MaxP* is not a TTL condition, since successive hops are not always on the same path; *MaxP* refers to the total number of participating peers and not just the peers on one path.

## 2.2 Schema Merging Algorithm

The goal of the merging procedure is to produce a schema that represents the semantics of the majority of the peers that belong to the respective cluster. This is achieved gradually by merging the schemas of peers on consecutive steps of the path that the merging procedure follows. We need a merging procedure that preserves the most popular concepts of the respective peer schemas and produces a schema representative of almost all the source schemas. Thus, we require a merging procedure that performs high compression before throwing away schema elements (i.e., relations or attributes). Finally, we require that merging is only based on available information on the peers, i.e., it solely exploits the peer schemas and the peer mappings. Each mapping is considered to be a set of 1-1 correspondences between attributes that hold with an optional set of value and join constraints on some attributes (see [12] for details).

The schema merging procedure is designed with respect to the following dictations:

- D1 Fewer relations with more attributes are preferred to more relations with fewer attributes
- D2 The semantic relevance of two relations is proportional to the number of correspondences between their sets of attributes
- D3 If the keys of two relations are mapped thoroughly, both relations are considered to be projections of the same relation with the same key
- D4 The key of a merged relation consists of the keys of both relations that are merged
- D5 If two attributes are merged and at least one of them is a key, then the merged attribute is part of the key of the merged relation
- D6 Correspondences that involve the same attribute imply that all involved attributes are semantically equivalent
- D7 Correspondences that are based on any value constraints are considered valid only under certain conditions and never produce merged attributes.
- D8 There are two pre-specified constants that represent the maximum number of relations that the schema of the interest group is allowed to have and the maximum number of attributes per relation

Briefly, the schema merging procedure produces the interest group schema but also a set of internal mappings and a dictionary. The internal mappings are the peer mappings that were not consumed in the successive schema merges. These hold additional syntactic and implicit semantic information for the group schema elements; thus, they can be very helpful to peers that would like to join the group and create mappings to their local schema. Moreover, this set of mappings includes all mappings with value constraints met during the merging procedure. Such mappings cannot be consumed: the involved relations/attributes cannot be merged, since they are mapped under certain conditions (the value constraints). Furthermore, the merged schema has alternative keywords for the same element that result from the merged mapping correspondences. These alternatives are entered in the dictionary that accompanies the group schema. The dictionary is then used to identify semantic similarity between a group and a new node and also assist in the creation of mappings if so desired.

The algorithm first merges relations that share the same key and then those that do not. In the latter case, priority is given to relations that share most of their attributes. Additional criteria in order to break ties can be based on whether the corresponding attributes are parts of the relation keys, or whether unmapped attributes are parts of the relation keys. Nevertheless, refining the algorithm based on additional criteria is future work. At the end of the schema merging procedure, i.e., when all relevant peer schemas

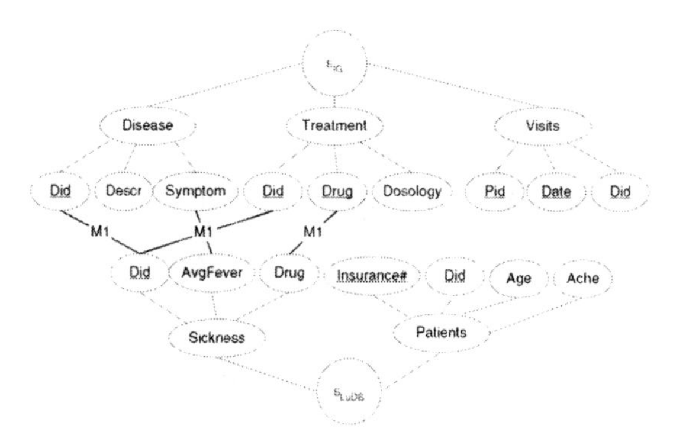


**Fig. 3.** $S_{IG}$ is initialized to $S_{DavisDB}$ and there is mapping $M1$ between $S_{IG}$ and $S_{LuDB}$

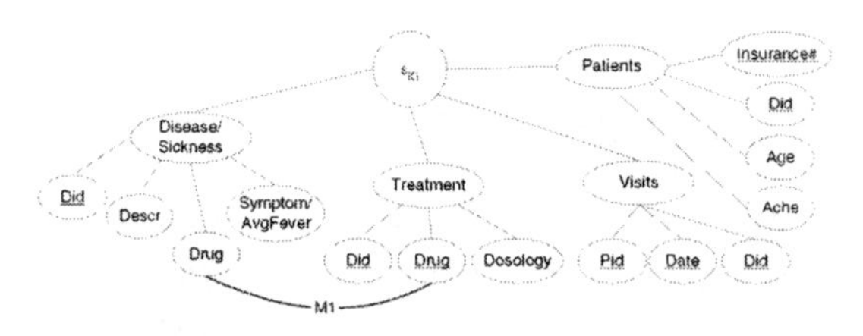


**Fig. 4.** Relations Disease and Sickness of Figure 3 are merged

have been merged, relations and relation attributes that have been rarely met during the procedure can be dropped. In order to do this, we need to keep a counter for each of them during the merging. For a thorough analysis and presentation of the merging algorithm the reader is referred to [11]. We present a simple merging example.

**Example:** Assume the pDBMSs of the motivating example in Section 1. The schemas of DavisDB and LuDB are presented in Figure 2; assume that the databases have the following mapping:

$M1_{LuDB\_DavisDB}$:

```
Disease (Did, _, Symptom), Treatment (Did, Drug, _):-
Sickness(Did, AvgFever, Drug),
```

where the correspondences Symptom = AvgFever and Disease = Sickness are implied and '_' is introduced for attributes that are not needed.

As shown in Figure 3, there are three correspondences that are encapsulated in mapping $M1$. We assume that the peer of Dr Davis initializes the schema merge. Thus, the group schema $S_{IG}$ is initialized to $S_{DavisDB}$. First, all relations of $S_{LuDB}$ are added to $S_{IG}$. Relations *Disease* and *Sickness* are merged in one (Figure 4), since they share the same key; thus, attributes *Symptom* and *AvgFever* are merged. The correspondence $Disease/Sickness.Drug = Treatment.Drug$ is kept as an internal one. Also, the dictionary $D$ is enriched with correspondences $Disease = Sickness$ and $Symptom = AvgFever$; actually the schema keeps one name for each relation or attribute from the alternative ones. At the end of the schema merging procedure we propose that the schema

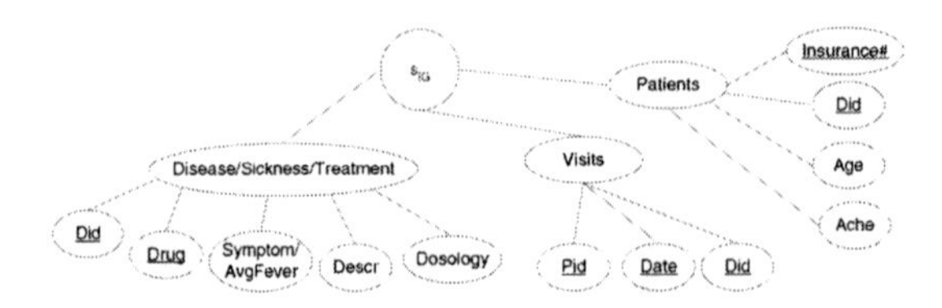


**Fig. 5.** Relations Disease/Sickness and Treatment of Figure 4 are merged

keeps for relation and attribute names the most common ones encountered during the procedure. Relations *Disease/Sickness* and *Treatment* are merged (Figure 5), since they are the only ones related with a mapping. Now there is one attribute named 'Drug' and it is part of the relation key, even though just one of the attributes that where merged was a key. Additional iterations can merge relations based on foreign key constraints, since no other internal mappings exist.

## 3 Performance Evaluation

To evaluate the performance of the proposed group inference procedure, we use a message-level simulator that implements it over an unstructured overlay of semantically clustered nodes. The clustering is performed using the *GrouPeer* system [12]. In GrouPeer, peers decide to add new (and abolish old) one-hop neighbors in the overlay (acquaintees) according to the accuracy of the answers they receive from remote peers. This is measured using a function that tries to capture the semantic similarity between rewritten versions of a query. Specifically, requesters (i.e., peers that pose queries) accumulate correct and erroneous mappings with remote peers through a learning procedure. Based on these mappings, they decide to become acquainted with peers that store information similar to their interests. The result is an effective semantic clustering of the overlay, where the accuracy of query rewritings and answers is a lot higher compared to the unclustered overlay (for details see [12]).

We compare the query evaluation performed by GrouPeer with the evaluation that utilizes the inferred groups on the overlay. When the first group is created, we direct relevant queries to the inferred schema. The basic performance metrics are the average *accuracy* of answers to the original queries (i.e., the similarity of the rewritten query that is answered over the original one), as well as the number of nodes that provide an answer. Similarity is calculated by a formula presented in [12] that identifies erroneous or not-preserved correspondences in mappings, which degrade the complete and perfect rewriting. To identify the gains of our grouping approach, we present the percentile increase/decrease in accuracy and number of answers compared to GrouPeer's clustering as these are measured on the *first* created group. Participants of the group hold mappings with the group schema; thus, when the query is rewritten to the group schema, the successive rewritings through the chain of mappings are avoided. Non-members create mappings with relevant group schemas.

We present results for 1,000-node random graphs (an adequate number of participants regarding our motivating application) with average node degrees around 4, created by the *BRITE* [14] topology generator. Results are averaged over 20 graphs of the

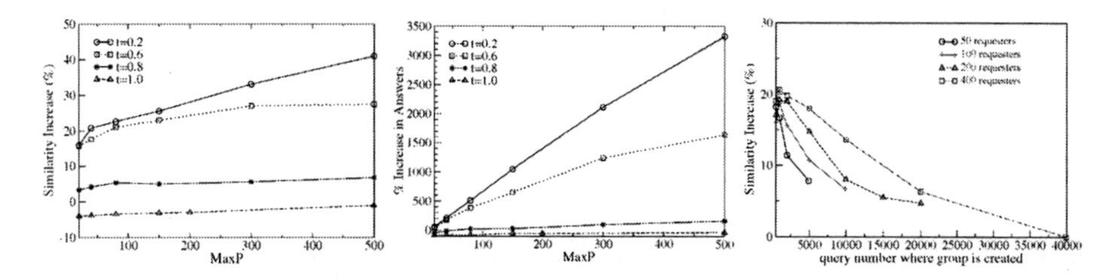


**Fig. 6.** % Increase in answer similarity over variable MaxP and t

**Fig. 7.** % Increase in number of answers over variable MaxP and t

**Fig. 8.** % Increase in answer similarity over variable group creation time

same type and size, with multiple runs in each. Results using power-law topologies constructed by Inet-3.0 [8] with the same number of peers are qualitatively similar.

For the schemas stored at each node, we use an initial schema whose relations and attributes are uniformly distributed at the nodes. The initial schema comprises of 5 relations and 33 attributes. Seven attributes are keys with a total of 11 correspondences between them. Each peer stores 10 table columns (attributes) on average. Queries are formed on a single or multiple tables if applicable (join queries). The maximum size of the inferred schema is always in the order of the size of the initial schema used to produce the local ones during start-up.

First, we vary the maximum group size limit, $MaxP$, as well as the minimum similarity of participating peers to the initiator node, $t$. Figures 6 and 7 show the obtained results for 100 requesters and maximum 100 queries each. As $t$ increases, the group becomes more specialized and less general. In contrast, small similarity values produce groups too general that incorporate many concepts foreign to the initiator. This results in *specialized* groups (i.e., high value of $t$) that receive fewer queries, while more "general" ones receive more but cannot answer them all satisfactorily. As the graphs show, there exists a point where grouping ceases to increase its relative gains to clustering.

Both metrics increase as $MaxP$ increases. This is reasonable since more nodes can participate and produce results. Very specialized grouping causes significantly less populated groups, which in turn affects the number of returned answers. As groups get more general (around $t = 0.6$), an improvement of 13-23% in accuracy is achieved, while the gains in replies are 40-900%. As $t$ decreases, the gains in accuracy decrease but more results are generated. These curves show that a $t$ value of around 0.65 with the group initiator and $MaxP = 80$ achieve good results without too much generalization. These will be our default values for the rest of this discussion.

Next, we try to determine the quality of the created group based on the quality of the semantic clustering. Figure 8 show the percentile improvement in the similarity of answers when the first group is created at various points (i.e., number of queries) in the clustering process. The results show a decrease in the relative gains in accuracy which is due to the improvement of clustering with time. What is important is that groups that are allowed to be created as soon as possible (which would be the common case) show about 20% more accurate answers and return about three times more results compared to clustering, even though the inference procedure is performed on a less optimally clustered overlay. The clustering process is expedited with more active requesters, which suits the purposes of grouping. An extended experimental study is presented in [11].

## 4 Related Work

There exist several interesting research efforts that have discussed about semantics and semantic clustering of peers. The work in [3] is one of the first to consider semantics in P2P systems and suggest the construction of semantic overlay networks, i.e., SONs. Various other researchers have attempted to go beyond the a priori static formulation of SONs: the work in [18] suggests the dynamic construction of the interest-based shortcuts in order for peers to route queries to nodes that are more likely to answer them. Inspired by this work, the authors of [20] and [7] exploit implicit approaches for discovering semantic proximity based on the history of query answering and the least recently used nodes. In the same spirit, the work in [4] presents preliminary results about the clustering of the workload on the popular e-Donkey and Kazaa systems.

Finally, Bibster [5] is a project that exploits ontologies in order to enable P2P sharing of bibliographic data. Ontologies are used for importing data, formulating and routing queries and processing answers. Peers advertise their expertise and learn through ontologies about peers with similar data and interests.

## 5 Summary

In this paper we have described a method to automatically create schemas in order to characterize semantic clusters in PDMSs. Our scheme operates on clustered unstructured P2P overlays. By iteratively merging relevant peer schemas and maintaining only the most frequent common characteristics, we provide a schema representative of the cluster. Group schemas can be used in order to increase both query performance and the volume of returned data. Our experimental evaluations confirm these observations in a detailed comparison with the GrouPeer system.

## References

[1] Aberer, K., Cudre-Mauroux, P., Hauswirth, M., Van Pelt, T.: Gridvine:Building internet-scale semantic overlay networks. In: International Semantic Web Conference (2004)

[2] Cohen, E., Fiat, A., Kaplan, H.: Associative search in peer to peer networks: Harnessing latent semantics. In: INFOCOM (2003)

[3] Crespo, A., Garcia-Molina, H.: Semantic Overlay Networks for P2P Systems. In: Technical Report (2003)

[4] Le Fessant, F., Handurukande, S., Kermarrec, A.-M., Massoulie, L.: Clustering in Peer-to-Peer File Sharing WorkLoads. In: IPTPS (2004)

[5] Haase, P., Schnizler, B., Broekstra, J., Ehrig, M., van Harmelen, F., Menken, M., Mika, P., Plechawski, M., Pyszlak, P., Siebes, R., Staab, S., Tempich, C.: Bibster - a semantics-based bibliographic peer-to-peer system. In: Journal of Web Semantics (2005)

[6] Halevy, A., Ives, Z., Suciu, D., Tatarinov, I.: Schema Mediation in Peer Data Management Systems. In: ICDE (2003)

[7] Handurukande, S., Kermarrec, A.-M., Le Fessant, F., Massoulie, L.: Exploiting Semantic Clustering in the eDonkey P2P Network. In: ACM SIGOPS, ACM Press, New York (2004)
[8] Jin, C., Chen, Q., Jamin, S.: Inet: Internet Topology Generator. Technical Report CSE-TR443-00, Department of EECS, University of Michigan (2000)
[9] Kantere, V., Kiringa, I., Mylopoulos, J., Kementsientidis, A., Arenas, M.: Coordinating P2P Databases Using ECA Rules. In: DBISP2P (2003)
[10] Kantere, V., Tsoumakos, D., Roussopoulos, N.: Querying Structured Data in an Unstructured P2P System. In: WIDM (2004)
[11] Kantere, V., Tsoumakos, D., Sellis, T.: Semantic Grouping of Social Networks in P2P Database Settings. Technical Report TR-2007-2, National Technical Un. of Athens (2007), http://www.dbnet.ece.ntua.gr/pubs/uploads/TR-2007-2
[12] Kantere, V., Tsoumakos, D., Sellis, T., Roussopoulos, N.: GrouPeer: Dynamic Clustering of P2P Databases. Technical Report TR-2006-4, National Technical Un. of Athens (2006), http://www.dbnet.ece.ntua.gr/pubs/uploads/TR-2006-4
[13] Levy, A.Y.: Answering Queries Using Views: A Survey. VLDB Journal (2001)
[14] Medina, A., Lakhina, A., Matta, I., Byers, J.: BRITE: An Approach to Universal Topology Generation. In: MASCOTS (2001)
[15] MySpace website: http://www.myspace.com
[16] Ooi, B., Shu, Y., Tan, K.L., Zhou, A.Y.: PeerDB: A P2P-based System for Distributed Data Sharing. In: ICDE (2003)
[17] Orkut website: http://www.orkut.com
[18] Sripanidkulchai, K., Maggs, B., Zhang, H.: Efficient Content Location Using Interest-Based Locality in Peer-to-Peer Systems. In: INFOCOM (2003)
[19] Tatarinov, I., Halevy, A.: Efficient Query Reformulation in Peer-Data Management Systems. In: SIGMOD (2004)
[20] Voulgaris, S., Kermarrec, A.-M., Massoulie, L., van Steen, M.: Exploiting Semantic Proximity in Peer-to-Peer Content Searching. In: FTDCS (2004)
[21] Wang, F., Moreno, Y., Sun, Y.: Structure of peer-to-peer social networks. Physical Review E, 73 (2006)

# Benchmarking RDF Production Tools

Martin Svihla and Ivan Jelinek

Czech Technical University in Prague,
Karlovo namesti 13, Praha 2, Czech republic
{svihlm1, jelinek}@fel.cvut.cz
http://webing.felk.cvut.cz

**Abstract.** Since a big part of web content is stored in relational databases (RDB) there are several approaches for generating of semantic web metadata from RDB. In our previous work we designed a novel approach for RDB to RDF data transformation. This paper describes experimental comparison of our system with several means of RDF production. We benchmarked both systems for the RDB to RDF transformation and native RDF repositories. The test results show a good performance of our system but also bring a new look at the effectiveness of the RDF production.

## 1 Introduction

It is widely accepted fact that the growth of the semantic web is dependent on the mass creation of metadata that will cover current web resources. Since the most of web content is backed by relational databases our previous work is focused on the transformation of relational data into RDF metadata, which is based on mapping between a relational database schema and existing RDF-S ontology. We have recently proposed *METAmorphoses* [1] – a new data transformation model based on two layers. This model is designed with regard to its performance, robustness and usability. Our work stands on the theoretical foundations of the semantic web technologies, but we also took into the account practical issues while developing the formal model. In this paper we briefly introduce our approach and compare our system with several other approaches for the RDB to RDF transformation as well as with some native RDF repositories built over relational databases. Our approach appears to be the fastest solution for RDF production between selected systems but results shows much more general conclusions – it appears that a well designed RDF production directly from RDB can be faster than querying the native RDF repositories.

## 2 METAmorphoses

The *METAmorphoses* processor[1] is a data transformation tool developed in our previous work [1]. It transforms relational data to RDF according to mapping from a relational schema into an existing ontology. The transformation process

[1] Available at http://metamorphoses.sourceforge.net/

R. Wagner, N. Revell, and G. Pernul (Eds.): DEXA 2007, LNCS 4653, pp. 700–709, 2007.

has two steps: (i) mapping from a particular RDB schema into an existing RDF-S ontology and (ii) creating RDF documents from relational data based on the mapping from the first step. Thus the model has two layers (figure 1). The mapping between a database schema and ontology consists of mapping elements and is processed in the mapping layer. A mapping element addresses relational concepts by SQL queries. The mapping is used in the template layer, which processes templates – XML-based documents for querying RDB. Since templates have a form of RDF and uses SQL from mapping elements to fetch data from RDB, they can be considered as a RDF view to RDB. METAmorphoses supports all RDF and RDF-S features and is relational complete. We designed the system so that it uses no RDF API, which is not neccessary for a data transformation and can slow a performance of the system. This fact together with the template user interface instead of RDF query language is a novel contribution to the RDB to RDF data transformation. The system is implemented in Java language.

The scope of this paper does not allow us to describe the system in detail. The complete description is available in [1].

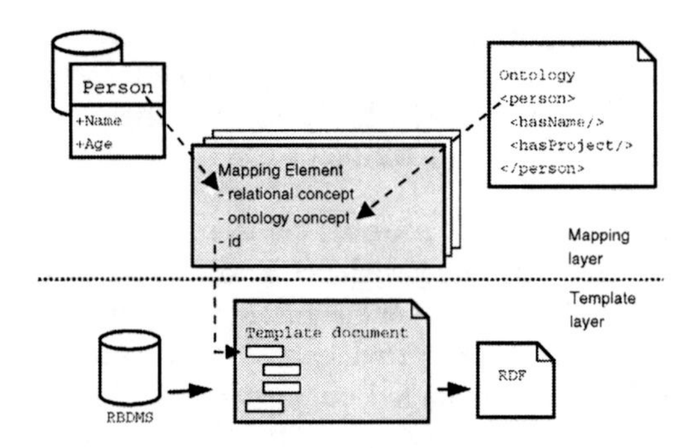


**Fig. 1.** Two-layer data transformation in METAmorphoses

## 3 Experiment Overview

### 3.1 Compared Systems

We compared 5 different systems in our experiments – 3 tools for the RDB to RDF transformation (*METAmorphoses*, *D2RQ* and *SquirrelRDF*) and 2 native RDF repositories with RDB back-end (*Jena* and *Sesame1*). Moreover, we performed 2 different tasks with D2RQ and Jena in the most of tests – we queried dataset with both SPARQL and graph API.

**METAmorphoses v0.2.5** is briefly described above. To perform tests we created a schema mapping between the relational schema of the experimental dataset and RDF-S ontology and then queried the dataset using our template documents as queries.

**D2RQ v0.5** [2] is a plug-in for the Jena [4], which uses mappings to rewrite Jena API calls to SQL queries and passes query results as RDF triples up to the higher layers of the Jena framework. Using D2RQ mapping it is possible to access relational database as a virtual RDF graph via classical Jena API. This way the relational database can be queried by SPARQL [8] or `find(s p o)` functions and a result is an RDF. When testing D2RQ, we performed two separate experiments: one with `find(s p o)` functions and another with SPARQL. We run D2RQ in *Jena v2.5.1* in these tests.

**SquirrelRDF** [3] is a tool which allows relational databases to be queried using SPARQL. It provides a tool that creates just a rough mapping for a database schema (its *the naïve RDB to RDF mapping*, described in [9], which does not consider an ontology) and a set of different SPARQL interfaces. Result of the SPARQL query over RDB is RDF. SquirrelRDF requires *Jena v2.4* and we used its API to perform SPARQL queries in our experiments.

**Jena v2.5.1 (persistent DB model)** [4] is a Java framework for building semantic web applications. It provides a programmatic environment for RDF, RDF-S, SPARQL and includes a rule-based inference engine. Jena can also store RDF data persistently in relational databases. We stored testing dataset in the persistent storage (backed by MySQL RDBMS) and then performed exactly the same experiments as in the case of D2RQ.

**Sesame v1.2.6 (persistent DB model)** [6] is an open source Java framework for storing, querying and reasoning with RDF and RDF-S. It can be used as a database for RDF and RDF-S, or as a Java library for applications that need to work with RDF internally. Sesame provides also relational storage for RDF data (so called *RDBMS-Sail*). We uploaded our testing RDF dataset to the Sesame persistent datastore (backed by MySQL RDBMS) and queried it with SeRQL (the internal query language of Sesame) in our experiments.

### 3.2 Experiment Methodology

To compare the tools listed above we used micro-benchmarks. The measured aspect was a time of an RDF production on a given query. Each test task consisted of *(i) preliminary phase*, where the source data, query engine and query were prepared and *(ii) measured phase*, where the query was executed and resulting RDF was written to standard output in the RDF/XML syntax. We decided to measure also RDF output because the tested aspect is the RDF production. According to granularity of the benchmarking tool (10ms) and the very short time of a query execution, we executed each query 100 times in a measured phase of each task. A warm-up was executed before each measured task in order to avoid JVM performance unbalance. A test consisted of the 5 same tasks executed in a row and its result was an arithmetic mean computed of the 5 task times.

### 3.3 Testing Dataset

The dataset used for the benchmarks is an XML dump of the DBLP computer science bibliography [7]. XML was converted into an SQL database dump and into an RDF representation[2]. The relational version of the dataset consists of 6 tables (`InProceeding`, `Person`, `Proceeding`, `Publisher`, `RelationPersonIn Proceeding`, `Series`) without indexes and contains 881,876 records. These relational data were stored in a MySQL database and used while testing *META-morphoses*, *D2RQ* and *SquirrelRDF*. The RDF representation of DBLP contains 1,608,344 statements and it was loaded into relational backends of *Jena2* and *Sesame1* to test them. To obtain more granular data, we created the tables `Proceeding500` and `Proceeding1500`, which contains the first 500 and 1500 records respectively from the table `Proceeding`. These data were also added to the RDF version of the dataset.

### 3.4 Testing Environment

The tests ran on Intel Pentium M processor 1400MHz with 1536 MB of RAM. The operating system was Linux (i386) with kernel version 2.6.12. Java Virtual Machine was the one implementated by Sun Microsystems Inc., version 1.5.0_01-b08. The RDBMS for storing data was MySQL server 5.0.30-Debian_1. All tests were performed within a simple Java benchmarking framework called JBench[3], which provides an easy way to compare Java algorithms for speed. The CPU timer based on the native JVM profiling API was used to obtain more accurate times. This timer reports the actual CPU time spent executing code in the test case thread rather than the *wall-clock* time that is affected by CPU load. The granularity of the timer was 10ms.

## 4 Experiments and Results

Testing queries are described in SPARQL formal terminology even their form vary between systems. In case of SPARQL and SeRQL queries we used *CONSTRUCT* form so that the result was a graph - as well as in the case of META-morphoses templates or Jena Graph API. To compare various aspects of the RDF production, we divided our tests into three groups. In these groups we tested RDF production according to a *(i)* result size, *(ii)* query graph pattern complexity and *(iii)* query condition complexity.

### 4.1 Experiments with the Result Size

In this test set we performed very simple query, based on the following general graph pattern:

$$(?s \ < rdf:type > < particular\_RDF-S\_class\_URI >)$$

[2] Thanks to Richard Cyganiak (FU Berlin) for providing us the converted datasets.
[3] Available at http://www.yoda.arachsys.com/java/jbench/

We applied this query on RDF-S classes with different amount of RDF individuals and we compared the behaviour of the tools according to the size of resulting RDF graph. We performed 5 tests in this group - we stepwise selected all resources with type `Series`, `Publisher`, `Proceeding500`, `Proceeding1500` and `Proceeding`. The specific SPARQL queries for these tests with number of triples in the resulting graphs are in the table 1, results are listed in the table 2.

**Table 1.** Tests with the result size: queries

| **Test no.** | **Query** | **Result triples** |
|---|---|---|
| 1.1 | CONSTRUCT * WHERE {?r rdf:type d:Series.} | 24 |
| 1.2 | CONSTRUCT * WHERE {?r rdf:type d:Publisher.} | 64 |
| 1.3 | CONSTRUCT * WHERE {?r rdf:type d:Proceeding500.} | 500 |
| 1.4 | CONSTRUCT * WHERE {?r rdf:type d:Proceeding1500.} | 1500 |
| 1.5 | CONSTRUCT * WHERE {?r rdf:type d:Proceeding.} | 3007 |

**Table 2.** Tests with the result size: results (times in ms)

| | **Test no.** (number of result triples) | | | | |
|---|---|---|---|---|---|
| **System** | **1.1** (20) | **1.2** (64) | **1.3** (500) | **1.4** (1500) | **1.5** (3007) |
| METAmorphoses | *84* | *230* | *1840* | *5692* | 13124 |
| SquirrelRDF | 830 | 1314 | 5180 | 16228 | 42504 |
| D2RQ SPARQL | 522 | 1332 | 7730 | 25530 | 45704 |
| Jena SPARQL | 482 | 1444 | 8296 | 27478 | 49968 |
| D2RQL Graph API | 366 | 1134 | 6434 | 20922 | 38253 |
| Jena Graph API | 368 | 1028 | 6902 | 22874 | 42588 |
| Sesame1 SeRQL | 194 | 423 | 2144 | 6624 | *12826* |

## 4.2 Experiments with the Graph Pattern Complexity

Test queries from this group consist of one graph pattern matching condition, which identifies exactly one RDF resource:

$$(?s\ < my\_ontology : hasTitle >\ "TITLE"\hat{}\hat{}xsd : string)$$

These queries differ in amount of resources and literals linked by graph patterns in the query. The size of resulting RDF graph does not differ too much in these queries so that we can compare tools according to complexity of the query graph pattern. The graph patterns are depicted on the figure 4.3, test results are listed in the table 3.

## 4.3 Experiments with the Query Condition Complexity

The tests in the third test set have a very simple graph pattern and they reffer to individuals from only one RDF-S class. These tests differ in number and

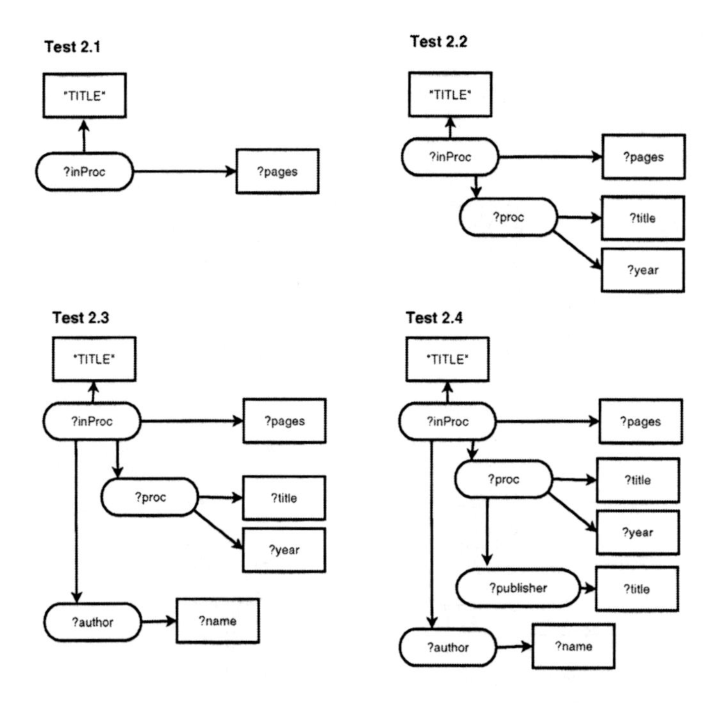


**Fig. 2.** Tests with the graph pattern complexity: graph patterns

**Table 3.** Tests with the graph pattern complexity: results (times in ms)

| System | **Test no.** (number of result triples) | | | |
|---|---|---|---|---|
| | **2.1** (2) | **2.2** (4) | **2.3** (6) | **2.4** (8) |
| METAmorphoses | *28* | *76* | *110* | *124* |
| SquirrelRDF | 640 | 678 | 768 | 808 |
| D2RQ SPARQL | 252 | 426 | 674 | 850 |
| Jena SPARQL | 212 | 360 | 456 | 552 |
| D2RQL Graph API | 106 | 224 | 434 | 506 |
| Jena Graph API | 94 | 150 | 204 | 262 |
| Sesame1 SeRQL | 100 | 198 | 272 | 324 |

type of conditions in the query. We performed these tests only with 5 systems – we ommited Jena and D2RQ graph API because this API does not allow straightforward queries with more conditions.

In SPARQL, there are two ways how to restrict possible solutions of a query: graph pattern matching and constraining values. This test set contains 4 tests that combines these conditions in various ways. The size of a resulting RDF is very small so that tests are focused on the query algorithm performance.

The test queries are depicted on the figure 3. The first query (test 3.1) contains a single graph pattern matching condition (similar to queries from the second test set) and the resulting graph contains 8 triples. The second query (test 3.2) adds one graph pattern matching condition to the first query and fetches 2 triples from the dataset. The query in the test 3.3 uses a condition with constraining value to obtain the same result as the test 3.2. The last query (test 3.4) combines conditions from test 3.1 and 3.3. The test times are in the table 4.

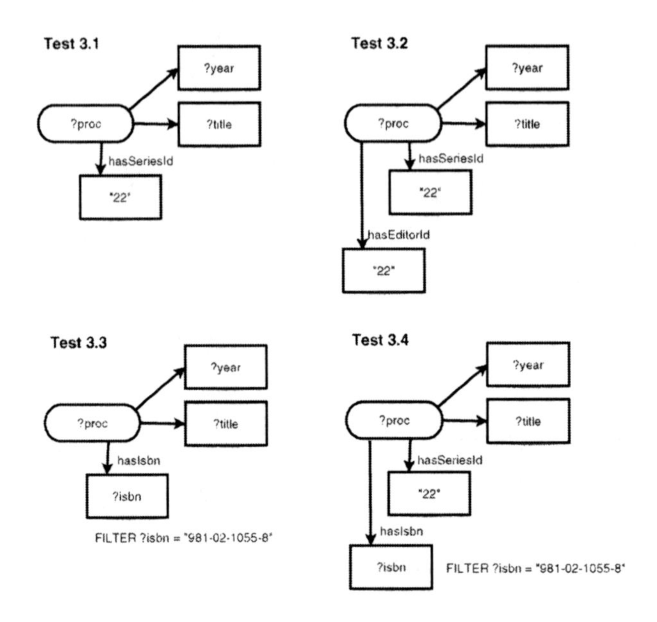


**Fig. 3.** Tests with the query condition complexity: graph patterns

**Table 4.** Tests with the query condition complexity: results (times in ms)

| System | **Test no.** (number of result triples) | | | |
|---|---|---|---|---|
| | **3.1** (8) | **3.2** (2) | **3.3** (2) | **3.4** (2) |
| METAmorphoses | *78* | *36* | *30* | *32* |
| SquirrelRDF | 636 | 582 | 9670 | 598 |
| D2RQ SPARQL | 618 | 336 | 17480 | 360 |
| Jena SPARQL | 544 | 240 | 30794 | 232 |
| Sesame1 SeRQL | 238 | 124 | 110 | 126 |

## 5 Discussion on Results

The test results show that our approach (METAmorphoses) was the fastest one almost in all tests (except test 1.5, where Sesame1 had slightly better performace).

In the first test set all systems have approximatelly linear computation performace. This relation between the result size and performance is illustrated in the table 5, which contains average times for producing 1 triple (100 times). It is obvious that these times are almost the same for each system, but vary between systems. Interesting is also the test 1.1, where the *time-for-one-triple* index is slightly higher for the most systems. We reason that this is caused by a *starting phase* of the query execution, which does not depend on the result size and is obvious at the query with a small resulting RDF (24 triples). The relatively shortest starting phase appears with METAmorphoses, the longest is with SquirrelRDF (more than 2 times higher than in other tests). Considering this starting phase there is no point to compute the time-for-one-triple index in the second and third test set – a resulting RDF is very small in these tests (2-8 triples).

The METAmorphoses was the fastest system also in the second test set. It keeps its high performance and small growth of the test time with the growing graph pattern complexity. The METAmorphoses has the best performance also in all tests in the third test set 4. Sesame1 is a little bit slower but similarity of its results with those of METAmorphoses is interesting. On the other hand, other three systems are much more slower. It is obvious in the test 3.3, where result times are very high. This is probably caused with non-optimized constraint value handling in the Jena SPARQL query engine, which is used by all these systems.

**Table 5.** Time (in ms) for producing one triple (100 times) - based on the first test set

| | **Test no.** (number of result triples) | | | | |
|---|---|---|---|---|---|
| **System** | **1.1** (20) | **1.2** (64) | **1.3** (500) | **1.4** (1500) | **1.5** (3007) |
| METAmorphoses | 3,5 | 2,67 | 3,68 | 3,79 | 4,36 |
| SquirrelRDF | 34,58 | 15,28 | 10,36 | 10,82 | 14,14 |
| D2RQ SPARQL | 21,75 | 15,49 | 15,46 | 17,02 | 15,2 |
| Jena SPARQL | 20,08 | 16,79 | 16,59 | 18,32 | 16,62 |
| D2RQL Graph API | 15,25 | 13,19 | 12,87 | 13,95 | 12,72 |
| Jena Graph API | 15,33 | 11,95 | 13,8 | 15,25 | 14,16 |
| Sesame1 SeRQL | 8,08 | 4,92 | 4,29 | 4,42 | 4,27 |

An interesting observation is that all systems based on Jena (SquirrelRDF, D2RQ and Jena itself) have very similar results, especially in the first and third test set. We explain this by the same algorithms for a resulting RDF graph composition (in the first test set) and SPARQL query execution (in the third test set). This means all solutions build above Jena shares all its advantages and disadvantages and are limited by its performance. Sesame1 and METAmorphoses had considerably different (and usually much better) test performance. Sesame1 is obviously optimized for querying big amounts of data and METAmorphoses was designed to be a *high performance* data transformation tool. According to the test results, we can say that our concepts implemented in METAmorphoses shows higher performance compared to other tested data transformation systems (D2RQ and SquirrelRDF). Our assumption that using RDF API is performance limitation was correct – our system is faster than those with RDF API.

METAmorphoses is also faster than tested native RDF persistent storages (persistent DB model in Jena and Sesame1). This is very interesting point. We proved that if one needs just to publish relational data in RDF, there is no need to migrate RDB to RDF repository and query this repository. On-the-fly data transformation (using METAmorphoses) can be done faster than queries over native RDF repository.

We did not measured RAM footprint of tested systems. However, METAmorphoses does not build RDF graph in a memory (it is a stream data transformation processor) thus its memory consumption does not depend on a size of a resulting RDF. All other tested systems first create resulting graph in memory and then serialize it, which means that their RAM footprint depends on a size of resulting RDF graph.

## 6 Related Work

There are not many similar comparison experiments for RDF production tools because the lack of a common query language and access method make benchmarking RDF stores a time consuming task (as mentioned in [5]). However, several attemps are described in [5], [11] or [10]. Due to different metodologies and tested systems it is very difficult to compare results, but our performance comparison can be considered as one of the most complex due to the number of tested systems and performed tests, too.

## 7 Conclusion

In this work we performed three test sets focused on computational performance to compare our ideas implemented in METAmorphoses with other RDB to RDF transformation tools (D2RQ and SquirrelRDF) and native RDF stores with RDB back-end (Jena and Sesame1). METAmorphoses had the best performance in the most tests (12 out of 13) and also other performance aspects discused in the section 5 were better. We proved that our system of data transformation has higher performance than other similar tools as well as native RDF repositories.

The main contribution of this paper is that we showed the on-the fly data transformation can be faster than queries over native RDF repository – thus it is not neccessary to migrate relational data to RDF repositories in order to publish them as RDF.

## Acknowledgements

This research has been supported by MSMT under research program no. 6840770014 and by the grant of the Czech Grant Agency no. 201/06/0648.

## References


1. Svihla, M., Jelinek, I.: The Database to RDF Mapping Model for an Easy Semantic Extending of Dynamic Web Sites. In: Proceedings of IADIS International Conference WWW/Internet, Lisbon, Portugal (2005)
2. Seaborne, A., Bizer, C.: D2RQ – Treating Non-RDF Databases as Virtual RDF Graphs. In: McIlraith, S.A., Plexousakis, D., van Harmelen, F. (eds.) ISWC 2004. LNCS, vol. 3298, Springer, Heidelberg (2004)
3. Steer, D.: SquirrelRDF, `http://jena.sourceforge.net/SquirrelRDF/`
4. Jeremy, J., et al.: Jena: implementing the semantic web recommendations. In: Proceedings of the 13th international World Wide Web conference on Alternate track papers & posters, New York (2004)
5. Harth, A., Decker, S.: Optimized index structures for querying RDF from the Web. In: Proceedings of LA-WEB (2005)
6. Broekstra, J., Kampman, A., van Harmelen, F.: Sesame: A Generic Architecture for Storing and Querying RDF and RDF Schema. In: Proceedings of the First International Semantic Web Conference, Sardinia, Italy (2002)
7. Ley, M.: DBLP Bibliography, `http://www.informatik.uni-trier.de/~ley/db/`
8. Prud'hommeaux, E., Seaborne, A.: SPARQL Query Language for RDF. W3C Recommendation (February 2005)
9. Beckett, D., Grant, J.: Semantic Web Scalability and Storage: Mapping Semantic Web Data with RDBMSes. SWAD-Europe deliverable (February 2003)
10. Streatfield, M., Glaser, H.: Report on Summer Internship Work For the AKT Project: Benchmarking RDF Triplestores. Technical Report. Electronics and Computer Science, University of Southampton (November 2005)
11. Cyganiak, R.: Benchmarking D2RQ v0.2. Technical Report. Freie Universität Berlin, Germany (June 2004)

# Creating Learning Objects and Learning Sequence on the Basis of Semantic Networks

Przemysław Korytkowski[1] and Katarzyna Sikora[2]

Szczecin University of Technology, Faculty of Computer Science and Information Technology, ul. Zołnierska 49, 71-210 Szczecin, Poland
pkorytkowski@wi.ps.pl, ksikora@wi.ps.pl

**Abstract.** With the growing popularity of distance courses the demand for good didactic materials also increases. However, preparing such material is a complicated task, often done in an inproper way, by simply transforming traditional teaching/learning material into a digital form. In such case, one of the most important aspects of learning – the presence of a teacher, is not considered. The following paper presents a method that allows creating didactic materials for distance learning that could compensate for the lack of direct contact with a teacher in the learning process.

## 1 Introduction

The main goal of education, that every organisation should aim at, is not only spreading information or raw knowledge, but creating competency. According to [3], competency is a set of certain characteristics owned and used by a person to achieve desired results. Knowledge by itself, without the ability to use it, does not lead to achieving any level of competency in a given domain. Therefore, nowadays teaching should not only mean sharing information, but rather presenting a way of thinking, a way of manipulating knowledge.

Although the popularity of distance courses increases, it still happens very often that materials meant for distance learning are simply a digitalised version of didactic materials meant for a traditional teaching/learning process. This means that material created for being presented by a teacher is being presented without his/her participation, causing a significant decrease in the quality of learning. It is connected to the fact that the way of reasoning in a given domain is usually presented by the teacher during class, didactic materials include only raw information about the subject.

In this paper, authors present a method of preparing didactic materials for distance learning in such a way, that allows including in them the teacher's way of thinking.

The presented method was succesfully used as a basis for creating learning material and sequence for teaching the course of „Modelling and simulation of production processes". In this paper a part of the discrete-event simulation domain serves as an example for presenting the procedure.

R. Wagner, N. Revell, and G. Pernul (Eds.): DEXA 2007, LNCS 4653, pp. 710–719, 2007.

## 2 Developing Semantic Network of the Chosen Scope of Material

The first step in creating a distance learning course is defining the scope of knowledge that has to be mastered during the course. This scope should be presented in the form of a knowledge representation model.

In order to ensure reflecting the teacher's way of thinking, semantic network was chosen as the best mean of representing domain knoweldge. It consists of two basic types of elements [7]:

- concepts – allow unambigous recognition of objects and classes of objects belonging to a given domain;
- relations – bind together different concepts, reflecting relationships and connections that occur between them and associations that link them.

Semantic networks of the same part of domain knowledge may differ when created by different experts, as for each them direct and indirect connections may occur between different concepts. Hence, a semantic network represents the way of thinking and reasoning of a given expert in a certain domain. Therefore, using it is the first step to ensuring such a level of education that will lead to creating in the student's mind his/her own thinking schemes, what is necessary to achieve competency in a given area of knowledge.

The type of relations used is very important for the level of unambiguousness of the semantic network. Many approaches, for example the depth matrix presented in [11], assume that for a proper representation of a concept out of the known relations [9] two: IS A and PART OF are enough. However, in case of a semantic network this might cause an ambiguous interpretation. Therefore, the method presented in this article assumes using relations that represent three basic types of connecting information, according to [4]: IS A – generalisation, PART OF – aggregation, MEMBER OF – association, and relation HAS – describing a characteristic of an object or, eventually, a possesion relationship.

A semantic network represents a certain fragment of a domain. However, very often just one domain, or it's part, is not enough to represent the whole subject. In such cases several semantic networks are created and then joined together. The joining requires a concepts mapping process.

Mapping can be defined as the process of identifying concepts and relations being approxiamtely equal [2]. This process can be described as follows [10]: with two given sets of concepts, $S_a$ and $S_b$, mapping one set of concepts onto another means that for each element of set $S_a$ a matching element (or elements) with the same or similar semantic in set $S_b$ should be found, and vice-versa.

We can distinguish two layers of mapping. The first one is connected to the way that a concept is understood in a given context. In this case, according to [5], the following types of mapping can be defined:

- concept-to-concept mapping – two concepts include the same type of information,

- attribute-to-attribute mapping – two concepts posses attributes that can be mapped,
- attribute-to-concept mapping – this way can be defined as searching for an instance of a certain object.

The second layer refers to the amount of information included in both concepts. The following types of equivalence can be defined [2]:

- partial equivalence – one concept becomes superior to the other,
- exact equivalence – concepts are identical,
- inexact equivalence – the definitions of concepts overlap, there is a certain common area,
- single-to-multiple equivalence – any of the above occuring for more than two concepts, e.g. `A is partially equivalent to B AND C`; logical operators `AND`, `OR` and negation are used.

The process of mapping is a complicated one, especially since finding equivalent concepts can be truely difficult due to [2]:

- different use of concepts in individual domains,
- different meaning scope of the concept: in one domain it can be considered generally, while in the other in a detailed way,
- different semantic, usually caused by different classification, e.g. „museum" can mean an organisation in one domain and a building in another.

## 3 Creating a Distance Learning Course

The scope of material, prepared in the form of a semantic network, is the basis for developing a distance learning course. However, first of all the structure of such a course should be specified.

According to current distance learning standards, the most popular of which is SCORM [1], each course can be divided into knowledge modules, called Learning Objects (LO), arranged in a proper order to create a learning sequence. Therefore, designing such modules and creating a proper sequence makes up the basis for developing didactic materials for distance learning.

The module structure assumes dividing knowledge embedded in the course into pieces small enough to be mastered in one learning session, and at the same time big enough for their content to be meaningful by itself. It is a very important aspect, because unful mastering of knowledge can cause a need to go back to lessons already completed, what can significantly discourage from further learning.

We can distinguish three basic aspects that have to be considered when deciding about the size (capacity) of LO:

- reuse possibility – small elements are easier to use again;
- time required to master the content – the smaller the size, the faster the material can be mastered;

– the amount of information that has to be presented in one session to enable understanding all the content included in the LO;

The educational process assumes spreading knowledge in such a way, that it will be permanently accumulated in the memory of the student. It can be achieved only if all the information and knowledge is repeated enough times [6]. The neccessity to refresh information is an important aspect, as it means that neither the course, nor a single knowledge module can consist of only new knowledge. Information should be used repeatedly and connected to the knowledge already obtained, only then it will be remembered permanently.

For that reason, during one learning session too many new concepts should not be introduced. According to [8] the optimal number of concepts in one LO should be between five and seven. For the purpose of the method being presented, number five was chosen as the basic capacity of LO. Not always an even division of the domain knowledge is possible, therefore the possibility of a deviation of one is assumed.

### 3.1 Marking LO Out of the Semantic Network

In order to mark out of the semantic network concepts that should be included in individual knowledge modules, several assumptions should be made:

– concepts not connected by a direct or indirect relation should not be placed in one LO;
– choosing concepts for individual LOs should begin at the lowest level of hierarchy in the network;
– very often one concept posseses many outgoing relations, in case when those relations are of different types and including all of them in one LO is not possible, it is necessary to decide which relation is more important for represanting the concept.

Considering the essenciality of considering the type of relation for a proper choice of concepts that should be placed in individual learning units, creating a hierarchy of importance is necessary. The hierarchy is as follows:

1. is a (generalisation),
2. part of (aggregation),
3. has (characteristic, possesion),
4. member of (association).

The best representation of knowledge is achieved when in one LO as many concepts connected with the same type of relation as possible are placed. That allows understanding a given concept and its related concepts in a comprehensive way.

Before marking the LOs out of the semantic network it is necessary to make several transformations and formalisations.

First of all, the semantic network will be presented as a list of concepts, where concept is understood as:

$$P = \{Def, Lev, R_{in}, R_{out}, LO\},$$

where:

$Def$ is the definition of the concept,
$Lev$ the level in hierarchy, where the concept is placed,
$R_{in}$ the set of relations entering the concept,
$R_{out}$ the set of relations leaving the concept, and
$LO$ is the list of knowledge modules that the concept belongs to.

All sets of relations are ordered according to the presented hierarchy of importance, then according to their length, beginning with the shortest one. The length is understood as the distance between the level in hierarchy where the concept the relation leaves from is placed and the level where the concept the relation enters can be found.

Relation is defined in the following way:

$$R = \{Lev_k, P_k, P_p, L, LO\},$$

where:

$Lev_k$ is the level in hierarchy, where the end of relation is found, in other words, the level where the concept the relation enters is placed,
$P_k$ the ending concept of the relation, similarly
$P_p$ the starting concept,
$L$ the length relation, and
$LO$ the number of the learning module the relation was assigned to.

Before beginning to create the set of learning units it is necessary to make a so called list of free relations ($LFR$). At first, it will include all relations in the semantic network. Relations in the list will be ordered according to the following three criterions:

1. location in the semantic network hierarchy (the lowest level has the highest priority),
2. type of relation (according to the presented hierarchy),
3. length of relation.

Regarding to LO as a set of concepts we can consider its capacity as the cardinal number of this set and mark it as: $O = cardLO = 5$. Number five was chosen for the purpose of presenting the method, however, it can be adjusted by the expert controlling the process of marking out knowledge modules, therefore it is possible for a bit smaller or bigger LOs to exist.

For transforming the semantic network into a set of Learning Object the following algorithm is proposed:

1. $Rel$ := first relation from $LFR$;
2. (a) IF ($P_k(Rel) \in LO_i$ AND $P_p(Rel) \in LO_i$) THEN
   $LO_i := LO_i \cup Rel$;
   $LFR := LFR - Rel$;
   GO TO 1;

(b) ELSE
$NLO := new\ LO$;
$NLO := NLO \cup P_k$;
$P1 := P_k(Rel)$;
3. $Rel :=$ first free relation $\in R_{in}(P1)$;
4. $P2 := P_p(Rel)$;
(a) IF ($card\ R_{out}(P2) > 1$)
IF ($O - card\ NLO < card$ {free relations $\in R_{out}(P2)$})
GO TO 3;
ELSE
$NLO := NLO \cup$ each free relation $\in R_{out}(P2)$ and its $P_k$;
(b) $NLO := NLO \cup P2 \cup Rel$;
(c) $LFR := LFR - Rel$;
(d) IF ($card\ NLO = O$) THEN ask for expert's opinion before GO TO 1;
(e) ELSE
IF ($card$ {free relations $\in R_{in}(P1)$} $= \emptyset$)
$Rel :=$ first relation from $R_{in}(P1) \in NLO$;
$P1 := P_p(Rel)$;
GO TO 3;

### 3.2 Learning Sequence Creation

There are several possible approaches to organising LOs in a learning sequence. The first one assumes presenting modules according to their position in hierarchy, introducing the ones placed at the highest level first and then moving to lower levels. However, this approach does not consider relationships between individual LOs. A sequence based on connections linking the learning units is much more effective in teaching, especially if we consider the fact that new information is remembered through associations with old knowledge.

Requirements and abilities of a student should be considered when creating the learning sequence. Through evaluating the student's knowledge of a given domain and domains that should also be known, it is possible to decide which parts of the course have to be emphasised, which ones should be introduced at the beginning and which ones can be introduced later. Evaluating the level of student's knowledge can be done with the use of different kinds of control tests. The knowledge mastered to the lowest degree should be introduced first, so that it could be as fast as possible, permanently and properly remembered.

However, the process of establishing the sequence is not entirely flexible, a part of the sequence – connected to the relationships between LOs, has to be maintained for proper understanding of information. It can be changed only if tests show that the student has mastered all the knowledge included in a given LO. In such case this LO should not be included in the sequence at all.

The entire procedure of creating learning objects and learning sequence on the basis of semantic networks was presented in Fig.1.

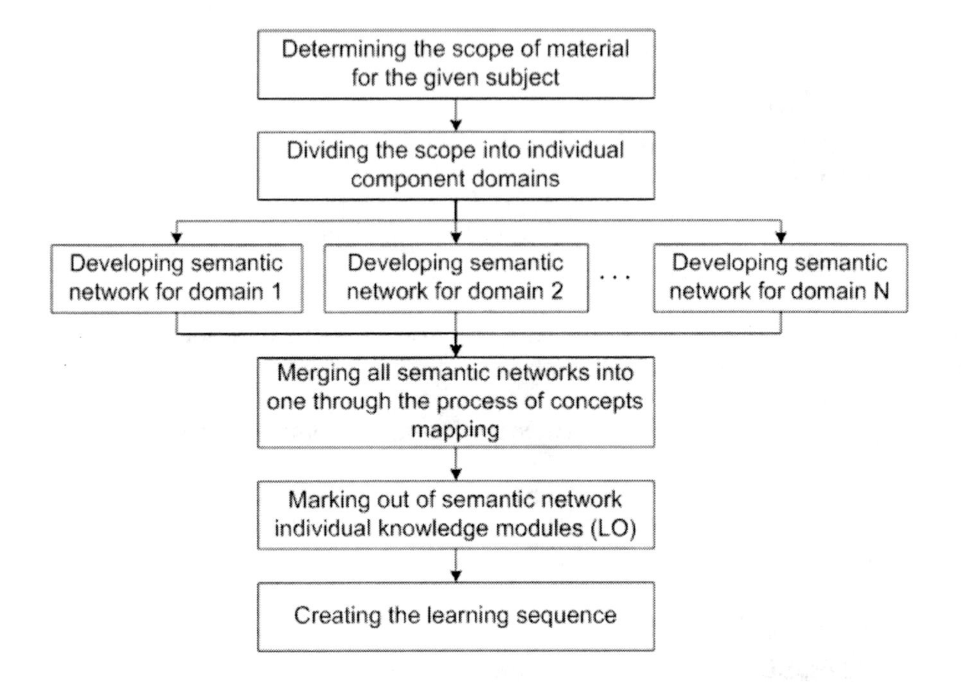


**Fig. 1.** Procedure of creating a distance learning course on the basis of a semantic network

## 4 Example of Using the Method

Let's illustrate the procedure described above on the example of the course of „Modelling and simulation of production processes" which is given for Production Engineering and Management students. It is a one-semester course, the aim of which is to create competency regarding understanding the functioning of production systems and their optimalization with the use of discrete-event simulation.

As the scope of the entire course is quite wide and includes knowledge from more than one domain, in order to facilitate understanding of the proposed method it was restricted to just a part of one knowledge domain – discrete-event simulation.

The entire procedure begins with analysing the semantic network (Fig.2), created in cooperation with the domain expert.

According to the presented procedure, first of all the list of free relations is created. Relation IS A with the end at the lowest level is going to be considered first. The concept at its end („state of system") is added to the newly created LO1, like the relation itself and the concept it leaves from („variable"). The relation is removed from the list of free relations. Because the concidered concept has no more entering relations we are moving to a higher level in hierarchy.

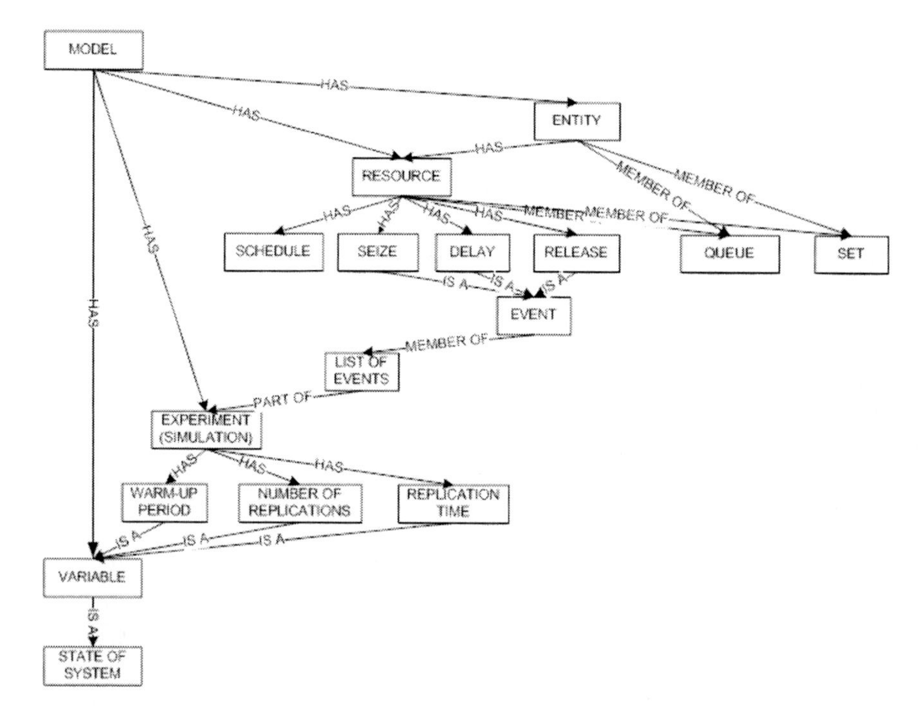


**Fig. 2.** Semantic network of a part of the modeling and simulation domain

At the higher level we can choose between four relations, three of them are of IS A kind and one is a HAS relation. Because the choice is made according to the hierarchy of importance of the relation type, three IS A relations and the concepts at their beginnings will be added to LO1. This way LO1 is filled.

Relation HAS joining the concepts of „variable" and „model" is next on the list. „Model" has an un-empty set of outgoing relations, fortunately, all of them can be included in one LO, thus creating LO2.

The next LO is created beginning with relation HAS between „warm-up period" and „experiment". Further proceeding according to the algorithm leads to creating the following six LOs:

- *LO1* = state of system, variable, warm-up period, number of replications, replication time,
- *LO2* = variable, model, entity, resource, experiment,
- *LO3* = warm-up period, experiment, number of replications, replication time, list of events,
- *LO4* = list of events, event, seize, delay, release,
- *LO5* = schedule, resource, seize, delay, release,
- *LO6* = queue, resource, entity, set.

The modules themselves are not enough to create a distance learning course. It is also necessary to create a sequence in which these modules will be presented

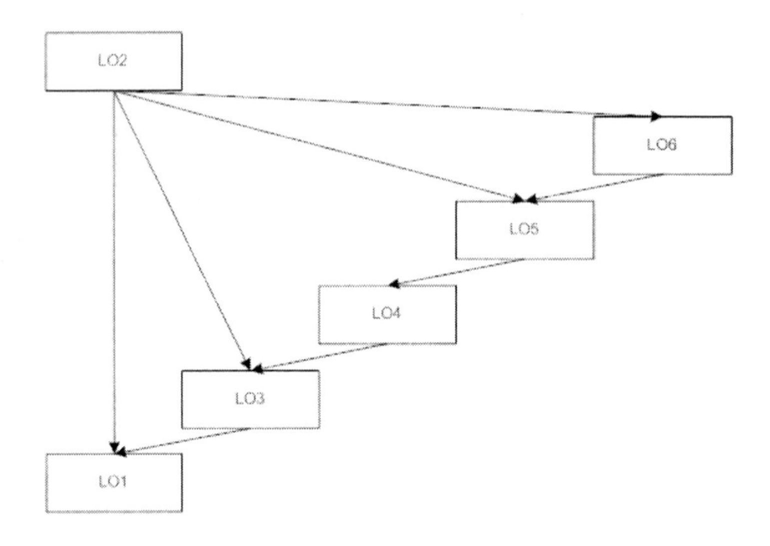


**Fig. 3.** Knowledge modules hierarchy

to the user. In order to do so, first of all, the knowledge modules marked out of the semantic network have to be ordered. Connections between the created LOs can be seen in Fig.3.

Because there are connections between LOs from all the levels, they have to be introduced according to hierarchy. Therefore, the sequence for this course will be as follows: $LO2 \rightarrow LO6 \rightarrow LO5 \rightarrow LO4 \rightarrow LO3 \rightarrow LO1$.

Criterions of creating the learning sequence can be set by the creator of the course, but they can be just as well left to the user of the course.

## 5 Conclusion

Developing didactic materials for distance learning is not a trivial matter. That is caused mainly by the fact that those materials have to be able to succesfully replace the teacher. To achieve such an effect it is best to use semantic network in developing the course, as it not only presents the facts themselves but also the rules of connecting them. The relations are what best reflects the teacher's way of thinking, what enables recognising, developing and creating one's own schemes of reasoning in a given domain.

The algorithm of transforming a semantic network into a set of Learning Object presented in this paper assumes creating the course beginning with the concept that requires knowing all other concepts in order to be understood. Thanks to this, identifying concepts that have to be introduced first, before moving to the ones at lower levels of the semantic network hierarchy, is much easier. Moreover, all the time the semantic of relations is respected, it also significantly influences the process of creating knowledge modules. The paper also proposes a hierarchy of importance of the types of relations.

The modules created according to the presented method can be used for creating a learning sequence adapted to the needs of a certain student. Maintaining the semantic of relations represented in the network allows giving priorities to individual LOs in the sequence not only on the basis of their placement in hierarchy. Considering the type of relations connecting concepts shared with modules appearing at higher levels in the hierarchy gives the possibility to better adjust the learning sequence within individual levels.

## References


1. Advanced Distributed Learning: Sharable Content Object Reference Model (SCORM), 2nd edn. Overview. ADL (2004)
2. Doerr, M.: Semantic Problems of Thesaurus Mapping. Journal of Digital Information, Article No. 52 (2001)
3. Dubois, D.D., Rothwell, W.J., et al.: Competency-Based Human Resource Management. Davies-Black Publishing (2004)
4. Goldstein, R.C., Storey, V.C.: Data abstractions: Why and how. Data & Knowledge Engineering 29, 293–311 (1999)
5. Maier, A., et al.: Integration with Ontologies. In: Conference Paper WM2003 (2003)
6. Maruszewski, T.: Cognitive psychology (in Polish). Biblioteka Myśli Semiotycznej. Znak - J ezyk - Rzeczywistość. Warszawa (1996)
7. Mulawka, J.J.: Expert Systems (in Polish). WNT. Warszawa (1996)
8. Różewski, P.: Method of developing an information system of knowledge representation and sjaring for distance learning (in Polish). PhD thesis. Politechnika Szczecińska. Szczecin (2004)
9. Storey, V.C.: Understanding Semantic Relationships. VLDB Journal 2, 455–488 (1993)
10. Xiaomeng, S., Gulla, J.A.: Semantic Enrichment for Ontology Mapping. NLDB, 217–228 (2004)
11. Zaikine, O., Kushtina, E., Róewski, P.: Model and algorithm of the conceptual scheme formation for knowledge domain in distance learning. European Journal of Operational Research 175(3), 1379–1399 (2006)

# SQORE-Based Ontology Retrieval System

Rachanee Ungrangsi[1], Chutiporn Anutariya[1], and Vilas Wuwongse[2]

[1] School of Technology, Shinawatra University
99 Moo 10 Bangtoey, Samkok, Pathum Thani, 12160 Thailand
{rachanee, chutiporn}@shinawatra.ac.th
[2] School of Engineering and Technology, Asian Institute of Technology
P.O. Box 4, Klong Luang, Pathum Thani, 12120 Thailand
vw@cs.ait.ac.th

**Abstract.** Most existing ontology search engines primarily base their search mechanisms solely on keyword matching. Users, therefore, are not well equipped with expressive means to structurally and semantically describe their ontology needs. We propose an ontology retrieval system based on *SQORE*, a novel framework that enables precise formulation of a *semantic query* in order to best capture a user's ontology requirements and rank the resulting ontologies based on their conceptual closeness to the given query. This paper develops a prototype system, investigates the effectiveness of SQORE framework, and compares its experimental results with the ones obtained from well-known ontology search engines. This investigative analysis indicates that SQORE system yields better results and better rankings.

## 1 Introduction

Ontology is a key technology that supports the increasing need in knowledge sharing and *Semantic Web* [5] application integration. In general, creating an ontology from scratch is often a difficult and time-consuming task, whereas reusing and modifying an existing one is preferable. However, as the number of publicly available ontologies grows, it becomes more difficult to find a suitable ontology that meets a user's needs. Therefore, an ontology retrieval system is demanded to enable ontology developers and users to find and compare existing ontologies, reuse complete or partial ones.

Existing ontology search engines, such as *Swoogle* [9], *OntoSearch* [18], *OntoSearch2* [13] and *OntoKhoj* [14], mainly base their approaches solely on search terms which cannot sufficiently capture the structural and semantic information about the desired domain concepts and relations. Semantics of a query and ontologies are thus ignored and not taken into account when executing a query, hence resulting in a demand for a semantic-based retrieval system, which can significantly improve the retrieval performance.

To tackle these problems, this paper presents and develops an effective ontology retrieval system, namely *SQORE* (*Semantic Query Ontology Retrieval Framework*) [2], which enables users to precisely and structurally formulate their ontology requirements in terms of a *semantic query*, containing not only the desired classes and properties, but also their relations and restrictions. Each query is evaluated by

R. Wagner, N. Revell, and G. Pernul (Eds.): DEXA 2007, LNCS 4653, pp. 720–729, 2007.

considering the semantic closeness between the query itself and an ontology in the database by means of SQORE's *semantic similarity measure*.

In order to evaluate and demonstrate SQORE's effectiveness, a prototype system available at http://ict.shinawatra.ac.th:8080/sqore is developed. Comprehensive user-based experiments are performed to analyze its precision and recall on a set of real-world ontologies. Furthermore, the comparison of the search results returned by SQORE and other existing systems is conducted, which shows that SQORE can yield better search results and better rankings.

The paper is organized as follows. Sect. 2 reviews related works. Sect. 3 informally introduces SQORE, and Sect. 4 illustrates the approach via an example. Sect. 5 explains the prototype system. Sect. 6 discusses the conducted experiments and the obtained results, and followed by conclusions and future work in Sect. 7.

## 2 Related Work

This section reviews recently-emerging ontology retrieval approaches, which can be classified into the following three groups:

- *WordNet-based approaches:* Hwang et al. [10] has proposed an approach to domain ontology retrieval based on concepts of search terms. It adopts *WordNet* [11] to determine semantic relations of terms and the concepts of searched ontology and then uses *Jaccard similarity* [16] for measuring similarity between query concepts and ontology concepts. However, this approach relies on only keywords, while lacking a capability to capture the structural and semantic information about the desirable concepts and relations.
- *PageRank-based approaches:* By employing the PageRank algorithm of Google [6], the most popular Web search engine, the two ontology search engines, namely, Swoogle [9], and OntoKhoj [14], have implemented their PageRank-like search algorithms based on ontology referral network. OntoKhoj focuses only on ontologies only whereas Swoogle supports searching and querying of both ontologies and Semantic Web databases. This approach is currently inefficient due to the lack of links among ontologies on the Web. Furthermore, it ignores semantic relations between query terms and concept/property terms in ontologies. For instance, it cannot determine that Academician and Faculty_Member are synonym; hence an ontology containing the concept Academician will not be the answer of the query searching for Faculty_Member.
- *Structure-based approaches:* Another well-known ontology search engine is OntoSearch [18] which allows users to search, evaluate and browse the ontologies based on several criteria. It has been enhanced by employing *AKTiveRank* [1] as metrics for ontology ranking based on the taxonomic structure information such as class names, shortest paths, linking density and positions of focused classes in the ontology. Recently, OntoSearch2 [13], the successor of OntoSearch, has been developed to search and query both ontologies and their associated data-sets using *SPARQL* [15] query language. Each query is evaluated by a *DL-Lite* [7] inference engine. Although this approach considers the semantics of a query and an ontology, when evaluating a query, it still lacks a capability to determine semantic relations (e.g., synonym, hyponym) between

two given terms. Users may have to learn specific query syntax to be able to compose queries. Moreover, since DL-Lite is a sub-language of OWL DL, it could not support OWL ontologies and ontological constructs that exceed its expressiveness such as owl:TransitiveProperty, owl:SymmetricProperty and owl:InverseProperty.

## 3 SQORE: Semantic Query Based Ontology Retrieval

*SQORE* [2] employs *XML Declarative Description* (*XDD*) *theory* [3, 17] as its theoretical foundation for modeling ontology databases and evaluating semantic queries, which does not only facilitate ontology matching and retrieval, but also support reasoning capability to enhance the matching results. Moreover, it also enables the use of a semantic lexical database, such as *WordNet* [11], for determining semantic relation between two given terms. Thus, the retrieval performance (precision and recall) can be significantly improved when compared to a conventional keyword search. To rank the relevant ontologies, SQORE employs *similarity score* by focusing on their conceptual closeness to the formulated semantic query.

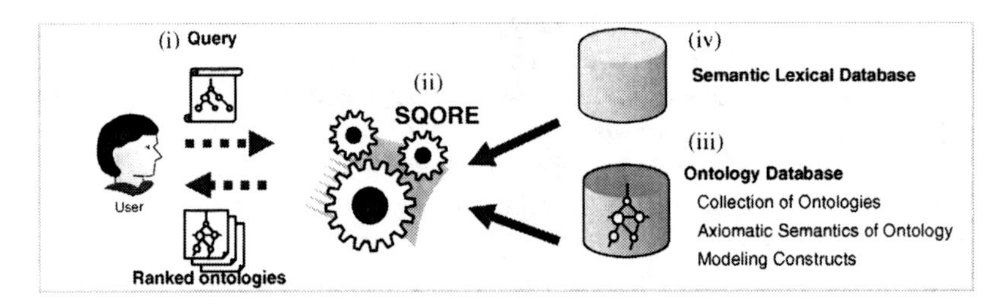


**Fig. 1.** SQORE System Architecture

Fig. 1 illustrates the system architecture of SQORE which comprises four major components: i) a *semantic query*, ii) a *retrieval engine*, iii) an *ontology database*, and iv) a *semantic lexical database*. In essence, the system works as follows: First, a user formulates and submits a semantic query which precisely captures his/her ontology requirements. The system then executes such query by semantically evaluating it against the ontology database, which comprises a collection of ontologies and a set of rules defining ontology axiomatic semantics. By incorporating these rules, implicit information about classes/properties in a query and an ontology can be derived, and hence enabling semantic query evaluation. Furthermore, when class/property names defined in a query and an ontology do not *exactly match* (=), four possibilities occur: i) *equivalence* ($\equiv$): the two terms are synonym, ii) *more general* ($\supseteq$): the query term is broader, iii) *less general* ($\subseteq$): the ontology term is broader, and iv) *unknown* ($\neq$): the relation is unknown. To support this, a referenced lexical database, such as *WordNet* [11], is employed in order to determine their appropriate semantic relation. Finally, the system computes the *semantic similarity score* between a given query and an ontology in the collection, which ranges from 0 (strong dissimilarity) to 1 (strong similarity), and returns as the answer the list of ranked ontologies.

In SQORE, there are four measures used for calculating similarity scores, as follows:

- **Element Similarity Score ($SS_E$):** The similarity score of any two given elements $x$ and $y$, denoted by $SS_E(x, y)$, depends on their semantic relation determined by the referenced lexical database as explained earlier. For any two given restrictions $r(a_1,b_1)$ and $r(a_2,b_2)$, their similarity is equal to the product of $a_1$-$a_2$ similarity score and $b_1$-$b_2$ similarity score i.e., $SS_E(a_1, a_2)*SS_E(b_1, b_2)$. When $x$ and $y$ do not belong to the same type, for instance $x$ is a class name and $y$ a property name, their similarity score is undefined.
- **Best Similarity Score ($SS_B$):** Based on the element similarity score $SS_E$, $SS_B(x,O)$ represents the similarity between a given element $x$ of a query and an ontology $O$ by finding the highest similarity score between $x$ and each element $y$ that is semantically defined by $O$. In other words, the element $y$ in $O$ that is most similar to $x$, will be used for measuring the closeness between $x$ and $O$.
- **Satisfaction Score of Mandatory conditions ($SS_M$) and Optional conditions ($SS_O$):** In SQORE, a semantic query comprises mandatory conditions and optional conditions. If an ontology semantically satisfies all mandatory conditions of a given query, then that ontology will be included in the answer. Optional conditions, on the other hand, are useful for expressing additional means for measuring the extent of closeness between the ontology and the query.
- **Query-Ontology Similarity Score (SS):** This similarity score represents the semantic closeness between a query and an ontology, which is measured by the satisfaction degree of the ontology with respect to the mandatory and optional conditions of the query.

In addition, SQORE defines several weight factors, as follow:

- **Semantic relation weight factors** ($w_=$, $w_\equiv$, $w_\supseteq$, $w_\subseteq$, $w_\neq$) quantify the similarity between two elements based on the discovered semantic relations.
- **Mandatory-condition weight factor** ($w_M$) indicates how important the mandatory conditions are, and hence $1-w_M$ becomes the weight for the optional conditions.

In practice, these weights are configured as default settings of the system which can be redefined by a user. For more details of the measures, the readers are referred to Reference [2].

## 4 An Example: Step by Step

This section illustrates how SQORE works by means of an example. Assume that an application developer wants to search for an ontology that meets the following conditions:

- It ***must*** contain concepts about Student and Professor.
- It ***may*** have telephone property, the domain of which is the concepts Student and Professor.

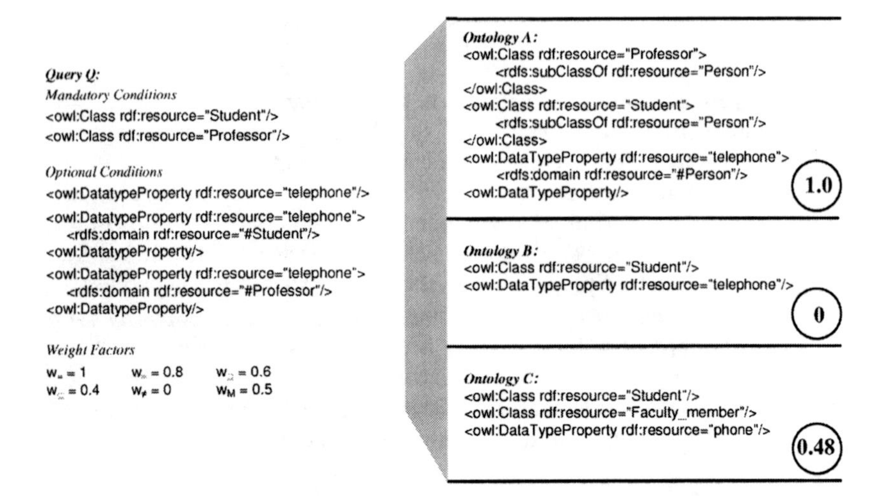


**Fig. 2.** An example of SQORE evaluation

Fig. 2 demonstrates how to formulate such a semantic query and shows how SQORE evaluates the query against three different ontologies and compute their similarity scores with the weight factors $w_{=}$=1, $w_{\equiv}$=0.8, $w_{\sqsupseteq}$=0.6, $w_{\sqsubseteq}$=0.4, $w_{\neq}$=0 and $w_M$ =0.5.

In this example, Query $Q$ searches for ontologies that must contain both Student and Professor classes. Ontology $A$ is obviously one of the answer due to the exact matches. Ontology $C$ is also an answer because Faculty_member is a hypernym of Professor. The similarity score of Ontology $B$ is equal to 0 since it fails to meet one of the mandatory requirements, i.e., it does not have the concept Professor defined. Therefore, Ontology $B$ will not be included in the answer list.

In addition, Query $Q$ has optional conditions which specify that telephone is a DatatypeProperty, the domain of which is Student and/or Professor. By incorporating inference mechanisms, one can see that Ontology $A$ satisfies this requirement because it has telephone DatatypeProperty with Person as the domain, and since both Student and Professor are defined as subClassOf Person, one can derive that Student and Professor are also the domain of telephone. On the other hand, Ontology $C$ contains a DatatypeProperty named phone, which is a synonym of telephone; thus $C$ partially satisfies the optional conditions.

As a result, SQORE returns two Ontologies $A$ and $C$ as the query's answer. The similarity score of Ontology $A$ is 1.0 since it perfectly satisfies all requirements whereas that of Ontology $C$ is 0.48.

## 5 SQORE Prototype System

In the implementation of SQORE, the main component is a Java Servlet that receives a semantic query from a user. The query consists of both mandatory conditions and optional conditions to search for concepts, properties and relations. When a query is submitted, SQORE starts evaluating the query against ontologies in the database and returns the results including the ontology URIs, their similarity scores and their ranks. WordNet [11] and XET inference engine [4] are employed during query evaluation to semantically match the query with ontologies and compute their similarity scores. The user can formulate a query using OWL syntax and can also specify his/her preferable weight factors, such as $w_{=}$, $w_{\equiv}$, $w_{\supseteq}$, $w_{\subseteq}$, $w_{\neq}$ and $w_M$.

SQORE system development is in progress with continuing enhancement. Currently, it allows user to query by using owl:Class, owl:ObjectProperty, owl:DatatypeProperty, rdfs:subClassOf, rdfs:domain, rdfs:range, owl:onProperty, owl:someValuesFrom and owl:allValuesFrom. Its ontology database comprises a large number of OWL ontologies related to the University domain.

A complete support of all OWL/RDFS modeling constructs as well as an insertion of more ontologies in various domains are part of SQORE's development plan. In addition, important NLP techniques that are often used in information retrieval systems, such as word stemming, tokenization, stop word elimination, will be applied in order to further improve the precision and recall of the system. Fig. 3 shows the current SQORE interface and sample search results.

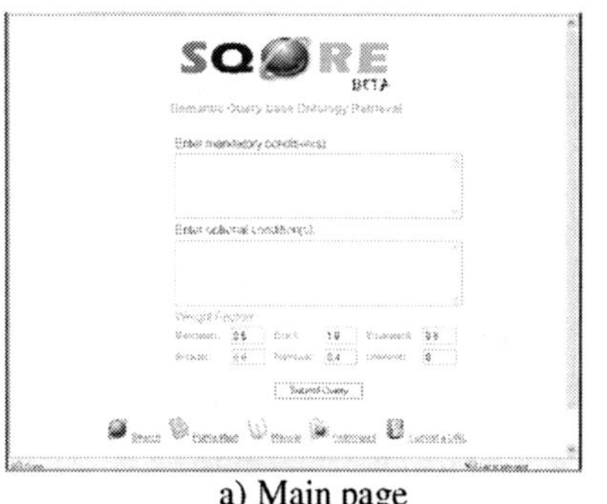

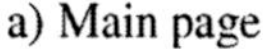
a) Main page

b) Sample search results

**Fig. 3.** SQORE Interface

# 6 Experiment

This section describes the design and methodology of the conducted experiment, and then reports the results of running SQORE over a set of existing ontologies with a set of test queries for measuring its precision and recall. Finally, the ranking result obtained from SQORE is compared with those recommended by users and by existing ontology search systems. However, OntoKhoj [14] and OntoSearch [18] seem to be currently inactive and are no longer accessible from the Web, while OntoSearch2 [13] is not specifically developed as an ontology retrieval system, but focuses mainly on datasets query. Therefore, Swoogle [9] is the only system used in the experiment.

## 6.1 Experimental Design

A user-based experiment was conducted to evaluate SQORE system. There were 10 users in the experiment and all of them are familiar with ontology technologies and

*Protègè* [12] ontology editor. The users were presented a general scenario, a set of questions and a set of existing ontologies. In the experiment, they need to find an OWL ontology that provides schema about people and their relationships in the university domain such as contact information, position, research interest, publication, etc.

The experiment employed top ten OWL ontologies about University domain, retrieved from Swoogle as listed in Table 1. Users were asked to generate a small set of keywords/queries, evaluate a set of queries against the given set of ontologies and also rank them based on the application requirements. The weight factors for computing similarity score in SQORE for this experiment were as follows: $w_{=}=1.0$, $w_{\equiv}=0.8$, $w_{\sqsupseteq}=0.6$, $w_{\sqsubseteq}=0.4$, $w_{\neq}=0$ and $w_M=0.5$.

**Table 1.** Top ten OWL ontologies about University domain retrieved from Swoogle and used in the Experiment

| Ranking | Ontology | URL |
|---|---|---|
| 1 | A | http://swrc.ontoware.org/ontology |
| 2 | B | http://www.aktors.org/ontology/portal |
| 3 | C | http://annotation.semanticweb.org/iswc/iswc.owl |
| 4 | D | http://www.csd.abdn.ac.uk/~cmckenzi/playpen/rdf/akt_ontology_LITE.owl |
| 5 | E | http://protege.stanford.edu/plugins/owl/owl-library/ka.owl |
| 6 | F | http://www.architexturez.in/+/--c--/caad.3.0.rdf.owl |
| 7 | G | http://www.mindswap.org/2004/SSSW04/aktive-portal-ontology-latest.owl |
| 8 | H | http://www.lehigh.edu/~zhp2/2004/0401/univ-bench.owl |
| 9 | I | http://ontologies.isx.com/onts/saturn/2004/10/core.owl |
| 10 | J | http://ontologies.isx.com/onts/2005/02/isxbusinessmgmtont.owl |

### 6.2 Precision vs. Recall

To evaluate SQORE's retrieval performance, a set of test queries and the ten given ontologies are presented to five expert users for evaluating the ontology-query relevance. Then, for each query, the set of these manually determined relevant ontologies are compared with the results suggested by SQORE and Swoogle. The precision values of each query at each recall level are computed and averaged. Finally, the precision values of all test queries are calculated and represented as a precision-recall graph as shown in Fig. 4.

As depicted by Fig. 4, SQORE has higher precision values over Swoogle at all recall levels, which implies that it can search and retrieve practically well and has better relevance ranking metrics. However, at very high recall levels (> 0.8), the precision dramatically decreases because the system fails to retrieve ontologies that have required classes or properties, but appear in different word-forms. For example, it yet fails to match the properties Supervises, Supervised, Supervising, isSupervisedBy and Supervise. Therefore, its performance can be significantly improved by applying additional NLP techniques used in information retrieval systems, such as eliminating the stop word list, word stemming, tokenization, etc.

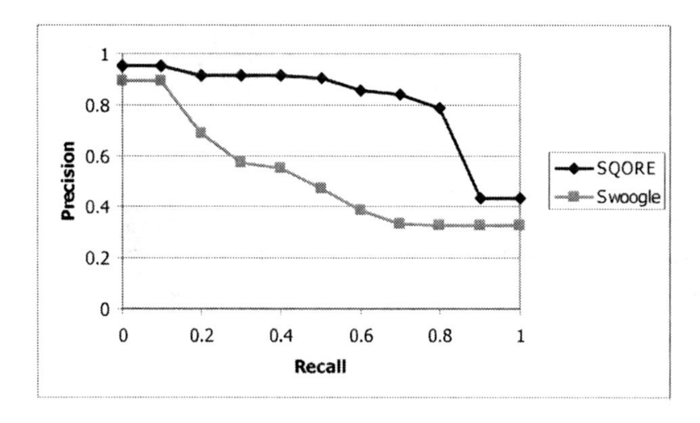


**Fig. 4.** Average Recall-Precision Curves

## 6.3 Rank Evaluations

In the experiment, the users were asked to give 16 keywords that represent the application requirements such as main concepts and relationships for searching an existing ontology. We selected only keywords that were proposed by more than half of the users in order to formulate a test query as presented in Table 2.

**Table 2.** Top Keywords by users

| Top Keywords | Ratio |
|---|---|
| Student. Course | 7/10 |
| Professor, University, Department, Faculty | 6/10 |

In order to compare the user ranking with that of Swoogle and SQORE, the users were requested to rank the ten ontologies based on how much they fit the application requirements. Then, all the top keywords, given by Table 2 and related by the OR condition, were submitted to Swoogle and SQORE. Table 3a shows the average user ranking compared to those obtained from Swoogle and SQORE.

Ontology *A* is ranked the first in both of the user and Swoogle rankings. On the other hand, SQORE ranks Ontology *H* the first, and Ontology *A* the third. When manually evaluating the submitted query with both Ontologies *A* and *H*, we observe that Ontology *H* satisfies all requirements in the query, while Ontology *A* fails to satisfy some keyword requirements; for instance, it does not have the keywords Professor and Faculty. We suspect that since those ten ontologies used in the experiments are rather huge in size, it is possible that the user rankings might contain some mistakes, and the ranking computed by SQORE is more accurate and can eliminate human errors.

*Pearson Correlation Coefficient (PCC)* [8] is then employed to measure the similarity between the user ranking and the system ranking. If the calculated PCC value is closer to 1, it indicates a stronger linear relationship between the two rankings. Table 3b shows that the PCC value of SQORE-user ranking is 0.765, while that of Swoogle-user ranking is 0.474. Therefore, one can draw that the ranking returned by SQORE is much closer to the user ranking than Swoogle.

**Table 3.** Results from User Evaluations

a. Ranks given by users, Swoogle and SQORE

| Ontology | User Ranking | Swoogle | SQORE |
|---|---|---|---|
| A | 1 | 1 | 3 |
| B | 6 | 2 | 5 |
| C | 6 | 3 | 2 |
| D | 8 | 4 | 7 |
| E | 2 | 5 | 3 |
| F | 3 | 6 | 5 |
| G | 5 | 7 | 8 |
| H | 3 | 8 | 1 |
| I | 9 | 9 | 9 |
| J | 10 | 10 | 10 |

b. Pearson Correlation Coefficient for SQORE and Swoogle wrt. user ranking

| System | Pearson Correlation Coefficient |
|---|---|
| SQORE | 0.765 |
| Swoogle | 0.474 |

## 7 Conclusions and Future Work

This paper has presented an ontology retrieval system based on SQORE framework. It enables a user to precisely and structurally formulate their ontology requirements, which include not only the desired class and property names, but also their relations and restrictions. The system not only facilitates ontology matching and retrieval, but also supports reasoning capability and incorporates semantics of keyword terms to enhance the matching results. In other words, during the query evaluation, a query's semantics together with an ontology's semantics are also taken into account in order to correctly and semantically match them. Comprehensive user-based experiments have been conducted and the results have shown that SQORE system can offer better quality of results and rankings than Swoogle, a widely-used existing ontology retrieval system.

SQORE system development is in progress with continuing enhancement. More features will be added to the system in order to improve the quality of search results and facilitate user search activities. In addition, since it is possible that none of the existing ontologies in the database satisfies all user requirements, the next goal of SQORE is to find an efficient algorithm that recommends a set of relevant ontologies such that their integration will lead to the solution for users with minimum ontology modification efforts.

## References

1. Alani, H., Brewster, C.: Metrics for Ranking Ontologies. In: Proc. of 4th Int'l EON Workshop, 15th Int'l WWW Conf., Edinburgh (2006)
2. Anutariya, C., Ungrangsi, R., Wuwongse, V.: SQORE: a Framework for Semantic Query based Ontology Retrieval. In: DASFAA07. LNCS, vol. 4443, pp. 924–929. Springer, Heidelberg (2007)
3. Anutariya, C., Wuwongse, V., Akama, K.: XML Declarative Description with First-Order Logical Constraints. Computational Intelligence 21(2), 130–156 (2005)
4. Anutariya, C., Wuwongse, V., Wattanapailin, V.: An Equivalent-Transformation-Based XML Rule Language. In: Proc. Int'l Workshop on Rule Markup Languages for Business Rules in the Semantic Web (CEUR Workshop Proc.), Sardinia, Italy, vol. 60 (2002)
5. Berners-Lee, T., Handler, J., Lassila, O.: The Semantic Web, Scientific American (May 2001)
6. Brin, S., Page, L.: The anatomy of a large-scale hyper-textual web search engine. In: Proc. of 7th Int'l WWW Conf., Brisbane, Australia (1998)
7. Clavanese, D., et al.: DL-Lite: Tractable Description Logics for Ontologies. In: Proc. of the 20th National Conference on Artificial Intelligence, pp. 602–607. Pittsburgh (2005)
8. Conover, W.J.: Practical Non-Parametric Statistics, 2nd edn. John Wiley and Sons, Chichester (1980)
9. Ding, L., Finin, T., Joshi, A., Pan, R., Scott Cost, R., Peng, Y., Reddivari, P., Doshi, V., Sachs, J.: Swoogle: a search and metadata engine for the semantic web. In: Proc. 13th ACM Int'l Conf. Information and Knowledge Management, DC, pp. 8–13. ACM Press, New York (2004)
10. Hwang, M., Kong, H., Kim, P.: The Design of the Ontology Retrieval System on the Web. In: Proc. 8th Int'l Conf. Advanced Communication Technology (ICACT2006) (2006)
11. Miller, A.: WordNet: A lexical database for English. Communications of the ACM 38(11) (1995)
12. Noy, N.F., Sintek, M., Decker, S., Crubezy, M., Fergerson, R.W., Musen, M.A.: Creating semantic web contents with protege-2000. IEEE Intelligent Systems, pp. 60–71 (March/April 2001)
13. Pan, J.Z., Thomas, E., Sleeman, D.: ONTOSEARCH2: Searching and Querying Web Ontologies. In: Proc. of the IADIS International Conference WWW/Internet (2006)
14. Patel, C., Supekar, K., Lee, Y., Park, E.K.: OntoKhoj: a semantic web portal for ontology searching, ranking and classification. In: Proc. 5th ACM Int'l Workshop on Web Information and Data Management, Louisiana, (November 07-08, 2003)
15. Prud'hommeaux, E., Seaborne, A.: SPARQL Query Language for RDF. W3C, Working Draft (2004), http://www.w3.org/TR/rdf-sparql-query/
16. Sokal, R.R., Sneath, P.H.A.: Principles of numerical taxonomy, San Francisco (1963)
17. Wuwongse, V., Anutariya, C., Akama, K., Nantajeewarawat, E.: XML Declarative Description (XDD): A Language for the Semantic Web. IEEE Intelligent Systems 16(3), 54–65 (2001)
18. Zhang, Y., Vasconcelos, W., Sleeman, D.: Ontosearch: An ontology search engine. In: Proc. 24th SGAI In'l. Conf. on Innovative Techniques and Applications of Artificial Intelligence, Cambridge (2004)

# Crawling the Web with OntoDir

Antonio Picariello and Antonio M. Rinaldi*

Universitá di Napoli Federico II - Dipartimento di Informatica e Sistemistica
80125 Via Claudio, 21 - Napoli, Italy
{picus,amrinald}@unina.it

**Abstract.** Managing large amount of information on the internet needs more efficient and effective methods and techniques for mining and representing information. The use of ontologies for knowledge representation has had a fast increase in the last years: in fact the use of a common and formal representation of knowledge allows a more accurate analysis of a number of documents content, in several contexts. One of these challenging applications is the Web: the World Wide Web, in fact, has nowadays those kinds of requirements which are hard to satisfy, especially when one considers a complex scenario as the Semantic Web. In this paper we present a methodology for automatic topic annotation of Web pages. We describe an algorithm for words disambiguation using an apposite metric for measuring the semantic relatedness and we show a technique which allows to detect the topic of the analyzed document by means of ontologies extracted from a knowledge base. The strategy is implemented in a system where these information are taken into account to build a topic hierarchy automatically created and not a priori defined. Experimental results are presented and discussed in order to measure the effectiveness of our approach.

## 1 Introduction

The extremely rapid growth of information on the internet requires of novel approaches to help users during their information searches. There are several ways to aid users in this task and in the last years new techniques for mining the Web have been proposed. One of these approaches is based on the creation of catalogues in which the Web pages are arranged into categories using methods and techniques from the field of information filtering and retrieval [14]. The navigation across categories has a great impact on the perception of the user information needs satisfaction [4]. The large amount of information on the Web makes impossible to manually classify data contained in tons of Web pages [9]. Thus, several methods for automatic Web pages classification have been presented in the last years. The aim of this paper is to define a framework for automatic Web page classification using hierarchical categories dynamically built. The approach presented in this paper is based on ontologies. This choice depends on some considerations derived from the definition of ontology and the specific application field. A formal definition of ontology is proposed in [6] *"An ontology can be defined as a formal, explicit specification of a shared conceptualization"*; in it we can find some useful terms

* Authors wish to thank the student Antonluca Paruolo for the precious contribution to the development of the system.

R. Wagner, N. Revell, and G. Pernul (Eds.): DEXA 2007, LNCS 4653, pp. 730–739, 2007.

to explain our approach: *conceptualization* is referred to an abstract model of specified reality in which the component concepts are identified; *explicit* means that the type of concepts used and the constraints on them are well defined; *formal* is referred to the ontology propriety of being "machine-readable"; *shared* is about the propriety that an ontology captures the consensual knowledge, accepted to a group of person, not only to a single one. On the other hand there are new views for mining the Web such as the one in the Semantic Web vision [2]. Therefore the ontological aspects of information can be used to really get in cooperation computers and people. In our approach we use a general knowledge base for extracting domain ontology to define the Web pages topic. In the proposed framework ontologies are not used only for indexing the terms in the documents but also for computing the topic detection step. The proposed techniques are implemented in a system for automatic Web page classification using hierarchical categories dynamically created by means of a knowledge base. For this purpose we implemented a Semantic Web Crawler called OntoDir to perform the necessary tasks. The Crawler is not limited to explore the Web but it has a specific module to annotate a Web page using a Word Sense Disambiguation and a Topic Detection task based on innovative metric and algorithms.

The paper is organized as follows: in section 2 some related works are presented; the OntoDir system architecture is drawn in section 3; the proposed method and the Word Sense Disambiguation and the Topic Detection algorithms and metrics are discussed in section 4; experimental results and conclusions are presented in section 5 and 6 respectively.

## 2 Related Works

Web page organization and classification is a hot topic in the field of Information Retrieval and Representation (IRR) and a number of works about it have been presented using various approaches. These methods includes: Web summarization-based classification, fuzzy similarity, natural language parsing for Web page classification and clustering, text classification approach using supervised neural networks, machine learning methods, *k*NN model-based and fuzzy classifiers. In [15] is presented an approach for automatic classification based on ontologies used to express the meaning of relationships contained in Web documents. The classification method is based on the similarities of documents already categorized by ontologies using information extracted from the documents. The proposed document classification technique does not involve any learning processes. A probabilistic approach for Web page classification is described in [16] where the authors propose a dynamic and hierarchical classification system that is capable of adding new categories as required, organizing the Web pages into a tree structure, and classifying Web pages by searching through only one path of the tree structure. In [17] is proposed a fuzzy classification method for Web pages. It is used for reducing the complexity related to the large amount of number of keywords of Web pages. The method is based on fuzzy learning and parallel feature selection where the fuzzy learning is adopted to increase the accuracy, while parallel feature selection based on weighted similarity is used to decrease the dimension of the features and inhibit the learning parameters. A genetic K-means approach is presented in [13]. The authors adopt this technique to evolve the weights associated with the keywords which characterize each class instead of evolving the centroid population. The approach is used to implement a hierarchical automated Web page classifier. LiveClassifier [7] is a system that automatically train classifiers through Web corpora based on user-defined topic hierarchies. The system is based on the assumption that the Web offers an inexhaustible data source for almost all subjects. Therefore LiveClassifier uses Web search result pages as the corpus source than it exploits the structural information inherent in the topic hierarchy to train the classifier and create key terms to amend the insufficiency of the topic hierarchy. In [8] a subject-oriented Web information classification system is presented. The Web pages are collected and classified into several subjects using text preprocessing, index, inverted files and vector space distance algorithm. The subjects are defined using a classification prototype built in according to the users requirements. Calado et al. [3] use a hybrid approach considering link-based and content-based methods for Web documents classification. The authors evaluate four different measures of subject similarity, derived from the Web link structure and by means of a bayesian network model, they combine these measures with the results obtained by traditional content-based classifiers to improve the results. A hierarchical structure to classify Web content is presented in [5]. The proposed method uses support vector machine (SVM) classifier and the authors use the hierarchical structure to train a different second-level classifier and to combine scores from the top and second-level models using different combination rules.

## 3 The System Architecture

So far we have described both the complexity of a Web classification tasks and its actual emergent utility for a variety of applications in the Web domain. For this reason, we provide a novel and effective general IIR system that can be suitably and simple used in a number of application domains. Let us quickly describe our system architecture at a glance. In Figure 1, we can consider the several modules we have proposed to solve the basic Web page classification steps. In other words, we imagine that our system is an intelligent *Crawler* exploring the Web. This crawler is not built for a single and specific problem, but it has been thought as a general engine: so we first provide a user with a *Crawler Interface* for setting the crawler's search parameters, such as a Web page address to analyze, a single IP or an IP range, for scanning a number of Web sites, the number of exploration levels in a Web site. Once retrieved a single page, we need to download its information content for our subsequent analysis: this is the *Fetcher* job, which downloads the information contents from the several links and it explores the Web considering the external Web sites linked to the actual Web site. All the extracted information is thus stored in the Web Page Data Base. Note that this data base is designed and implemented considering the Web page structure, HTML tags and metatags. The Web Page Data Base is used by the *Web Page Pre-Processor*. In fact, in order to analyze the real information content of a page, we have to "clear" the page, collecting only useful information from a semantic point of view. For this reason, we have to eliminate HTML tags and to use appropriate morphological text functions to get into basic form the terms extracted from the analyzed text. At the end of this module, pre-processed Web pages are stored in the *Web Text Repository*. The core of our system is the *Annotator* module. In fact, its purpose is that of furnishing a meaning to a certain text, so effectively annotating the given text according to a number of topics. To best understand the task of the Annotator component, we just think of a module which try to associate to a list of basic topic the considered Web page. We note that it is not simple to do that: in fact, natural language is imprecise and full of ambiguity also at the word level: a single word may have, in fact, several meanings (*polysemy*) that need to be associated to a context. In order to do that, the module has two main components: the *Word Sense Disambiguator* (WSD) and the *Topic Detector* (TD). The first module tries to set the right sense to the analyzed word due to the polysemy property; the second one, after the effective WSD task, performs a topic detection over the considered page, identifying the page argument. The algorithms at the base of those tasks will be described in the next section and make use of a general knowledge base, i.e. WordNet [12]. Even if WordNet has several lacks in some conceptual domains, it is one of the most used linguistic resources in the research community. The detected topic is arranged in a directory schema built by means of the hierarchical WordNet structure.

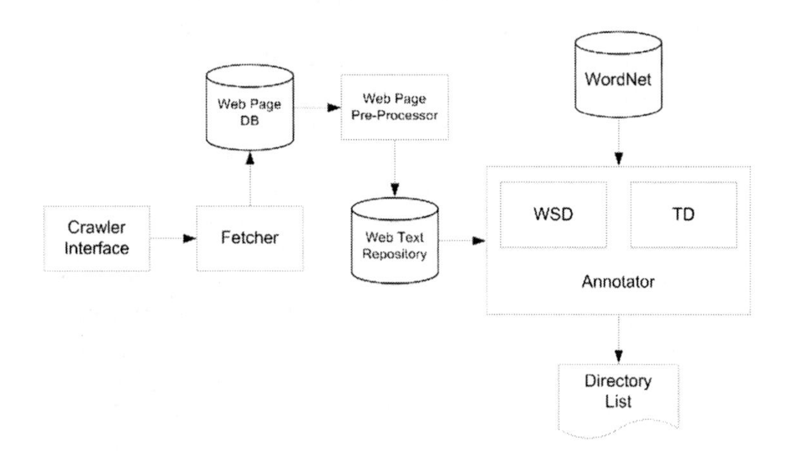


**Fig. 1.** System Architecture

## 4 The Proposed Method

In the previous section we have had a look over the whole proposed system. It is not difficult to recognize that several steps in this challenging process are nowadays well established and some research and also commercial products are available. The core of our system is the Annotator module, particularly what we have called the Word Sense Disambiguation and Topic Detection tasks. Because these steps are accomplished using WordNet as previously described, before reporting and detailing each phase, it's useful to introduce some consideration about the WordNet structure, so we can better understand our novel algorithm. All information in WordNet is arranged using linguistic properties. The basic unit is the synset a logic set of words related through the synonymy property. Each synset is a concept in WordNet. All the synsets are related to the others by pointers that represent linguistic properties. Two kinds of relations are represented: *lexical* and *semantic*; lexical relations hold between word forms while semantic relations hold between word meanings.

### 4.1 Word Sense Disambiguation Step

In the Word Sense Disambiguation step, the system assigns the right sense to the terms in the analyzed page due to the polysemy property. To perform this task we must analyze the context where the considered term is located; each sense term is compared with all the senses of the others terms. The similarity between terms is computed using an ad hoc metric that we will discuss in the next, in order to measure their semantic relatedness. The sense with the best score is chosen as representative of the considered term. We need a similarity function, that may be implemented using a suitable metric in order to measure a semantic relatedness between words. At the end of this process, we can easily identify what is the sense which has the best score, so that it can be chosen as the representative sense of the considered term. In our approach we propose a metric to measure the semantic relatedness among terms in a document introducing a novel technique to calculate the paths between terms and we also consider a component to take into account the weight of a single term in the document itself. The paths between terms are computed by means of a semantic network dynamically built from the first common subsumer (i.e. the first common ancestor) of the considered term senses. We can have more concepts related to a term due the polysemy property; therefore we consider the combination of all possible senses of the analyzed terms. In a previous work [1], the authors have just proposed an innovative algorithm to build dynamically a semantic network (DSN: Dynamic Semantic Network) starting from WordNet. Briefly the DSN is built starting from the synset that represents the concept $S_i$. We then consider all the component synsets and construct a hierarchy, only based on the hyponymy property; the last level of our hierarchy corresponds to the last level of WordNet one. After this step we enrich our hierarchy considering all the other kinds of relationships in WordNet. Based on these relations we can add other terms in the hierarchy obtaining an highly connected semantic network. An example of DSN is shown in Figure 2 where we can see its complexity 2(a) and structure 2(b); they are music and car DSNs respectively. The labeling of the graph is obtained using different colors.

We are now in a position to introduce the similarity metric. Our assumption is that we are considering the intersection among DSN and the retrieved documents, leaving out stop words. First of all we assign to the properties, represented by arcs between the nodes of the DSN, a weight $\sigma_i$, in order to express the strength of the relation. The weights are real numbers in the [0,1] interval and their values are defined by experiments. To calculate the relevance of a term we assign a weight to each one in the DSN considering the polysemy property, that can be considered as a measure of the ambiguity in the use of a word, if it can assume several senses.

Thus we define as *centrality* of the term $i$ as:

$$\varpi(i) = \frac{1}{poly(i)} \tag{1}$$

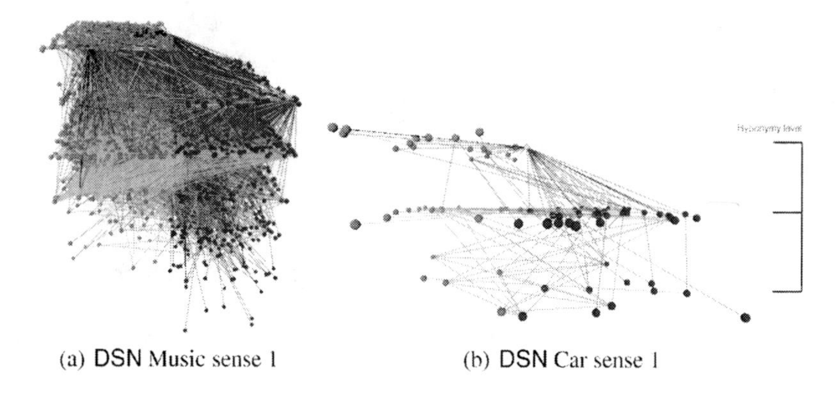

(a) DSN Music sense 1 (b) DSN Car sense 1

**Fig. 2.** A DSN example

*poly(i)* being the polysemy (number of senses) of $i$. As an example, the word *music* has five senses in WordNet so the probability that it is used to express a specific meaning is equal to $1/5$. We argue that those words have only one meaning strongly characterize the expressed concept. The metric has two types of information; one in order to take into account syntactic information based on the concepts of document word centrality and another one in order to consider the semantic component calculated on relevant couple of words in the document.

About the first contribute, the *syntactic-semantic grade* (SSG), we can define the relevance of a document in a domain represented by a DSN as the sum of its terms centrality:

$$\mathsf{SSG}(\nu) = \sum_{i=1}^{n} \varpi(i) \tag{2}$$

$n$ being the number of terms in the document $\nu$.

The other contribution is based on a combination of the path length ($l$) between pairs of terms and the depth ($d$) of their subsumer, expressed as the number of hops. The correlation between the terms is the semantic relatedness and it is computed through a nonlinear function. The choice of a nonlinear function to express the semantic relatedness between terms derives from several considerations. The values of path length and depth, based on their definition, may range from 0 to infinity, while relatedness between two terms should be expressed as a number in the $[0, 1]$ interval. In particular, when the path length decreases toward 0, the relatedness should monotonically increase toward 1, while it should monotonically decrease toward 0 when path length goes to infinity. We need a scaling effect w.r.t. the depth, because words in the upper levels of a semantic hierarchy express more general concepts than the words in a lower level. We use a non linear function for scaling down the contribution of subsumers in a upper level and scaling up those in a lower one.

Given two words $w_1$ and $w_2$, the length $l$ of the path between $w_1$ and $w_2$ is computed using the DSN and it is defined as:

$$l(w_1, w_2) = \min_j \sum_{i=1}^{h_j(w_1,w_2)} \frac{1}{\sigma_i} \quad (3)$$

$j$ spanning over all the paths between $w_1$ and $w_2$, $h_j(w_1, w_2)$ being the number of hops in the $j$-th path and $\sigma_i$ being the weight assigned to the i-th hop in the $j$-th path in respect to the hop linguistic property; the weights are experimentally set. Using this formula we find the *best path* between two words because we consider not only a geometric distance (number of hops) but also a *logic proximity*, i.e. the kind of properties between words. The depth $d$ of the subsumer of $w_1$ and $w_2$ is also computed using WordNet. To this aim only the hyponymy and hyperonymy relations (i.e. the IS-A hierarchy) are considered. $d(w_1, w_2)$ is computed as the number of hops from the subsumer of $w_1$ and $w_2$ to the root of the hierarchy. Given the above considerations, we selected an exponential function, that satisfies the previously discussed constraints.

We can now introduce the definition of *Semantic Grade* (SeG), that extends a metric proposed in [11]:

$$\mathsf{SeG}(\nu) = \sum_{(w_i,w_j)} e^{-\alpha \cdot l(w_i,w_j)} \frac{e^{\beta \cdot d(w_i,w_j)} - e^{-\beta \cdot d(w_i,w_j)}}{e^{\beta \cdot d(w_i,w_j)} + e^{-\beta \cdot d(w_i,w_j)}} \quad (4)$$

$(w_i, w_j)$ being a pairs of words in the document $\nu$, $\alpha \geq 0$ and $\beta > 0$ being two scaling parameters whose values have been defined by experiments.

The final grade is the sum of the Syntactic-Semantic Grade and the Semantic Grade.

### 4.2 Topic Detection Step

In the Topic Detection step the system gets a synset which represents the set of synsets detected in the word sense disambiguation step. This synset called `Topic Synset`, is representative of the page topic. We consider every concepts previously found in the Word Sense Disambiguation step and the system builds a DSN starting from the related synset. Therefore we calculate the intersection among the DSN from $S_i$ with each DSNs from all single concepts. We argue that the number of common concepts between the $\mathsf{DSN}_{S_i}$ and the other DSNs is the representation grade of the considered concept with respect to the whole Web page: this measure is called `Sense Coverage`. We take into account the concepts specialization also using a scaling factor get from the depth of the synset from the related WordNet root. After those considerations, the `Topic Synset` will be the synset having the best trade-off between the Sense Coverage and the Sense Depth form the correspondent WordNet root:

$$\mathsf{TopicSynset} = \max(\mathsf{depth}(\mathsf{S_i}) * \mathsf{Cover}(\mathsf{DSN}_{\mathsf{S_i}})) \quad (5)$$

If the intersection set is empty, the system does not return any classification at all.

The directory structure, in which the analyzed page is arranged, is obtained from the path started from the `Topic Synset` to the correspondent WordNet root.

This path is calculated using the hyperonymy property. An example of this path using XML is in the next figure where there is a comparison with the Yahoo! directory.

```
<web-page>
  <Url>
    http://dialspace.dial.pipex.com/agarman/jaguar.htm
  </Url>
  <Yahoo_Category>
    Science/Biology/Zoology/Animals_Insects_and_Pets/
      Mammals/Cats/Wild_Cats/Jaguars
  </Yahoo_Category>
  <Category>
    entity/object/living_thing/organism/animal/chordate
        /vertebrate/mammal/placental/carnivore
        /feline/big_cat/jaguar/
  </Category>
  <Topic_Description>
    a large spotted feline of tropical America similar
        to the leopard
  </Topic_Description>
  <Keywords>
    jaguar,panther,Panthera_onca,Felis_onca
  </Keywords>
</web-page>
```

This XML file is used to annotate the considered Web page; in addition to the directory paths we have information about the Url, a description of page topic using the WordNet gloss of the Topic Synset and some keywords derived from the synonyms of the Topic Synset.

## 5 Experimental Results

In order to have a reliable evaluation of our system and methods, we used a standard test set collection of Web documents. Note that to the best of our knowledge, no general Web page test set collections are available for our propose, therefore we choose to use the 20 Newsgroups test collection [10]. In particular we analyze all the categories because they refer to very different subjects and we use this heterogeneity to obtain a general and formal evaluation of our system, algorithms and our knowledge base. In fact WordNet is a general ontology with a different specialization on conceptual domains. We think that this test set is a good example of a real scenario and the task performed by our system could be a step before an opinion analysis or other emergent application on the Web. As reported in the 20 Newsgroups documentation, some of the newsgroups are very closely related to each other (e.g. *comp.sys.ibm.pc.hardware / comp.sys.mac.hardware*), while others are highly unrelated (e.g m*isc.forsale / soc.religion.christian*). We calculate what we call "local precision" simply considering precision of each category. Structural relation among categories in the 20 Newsgroups defined by the talk topics have been found during the experimental test. We defined different classes of annotated documents with respect to the system analysis output. The evaluation of the experimental results has been performed by human experts following the strategy afterward described. The *Right Classification* class is referred to the annotations fitting with the relevance assessment given to the 20 Newsgroups categories and subjects; in the *Wrong Classification* class are all the documents with an erroneous analysis; with the label *General Classification* we suggest the categories too much general to satisfy the user needs but with a right beginning root path; *None Classification* is the tag for those documents with none annotation. In Figure 3 are shown the experimental results where the number of documents for each classes are on the y-axis and the categories are on the x-axis. From the documents in the None Classification class we found very few or none words and the DSNs built from them have no concepts in common. On the other hand annotations in the General Classification class are obtained from documents about general arguments within no specific concepts. In a general point of view we notice that for all the categories the lack of specific terms in WordNet lead the system to an incorrect classification giving a misrepresent analysis. We also investigated about the document topics. In fact we have a good detection for some topics; for some other ones, we have a low accuracy. We argue that this is due to the considered general dictionary WordNet and to the fact that such kind of dictionary has small sized ontologies that can be extracted for specific conceptual domain. This idea is also supported by the analysis of log system, where we noticed many DSNs and several matching terms in the document with good precision.

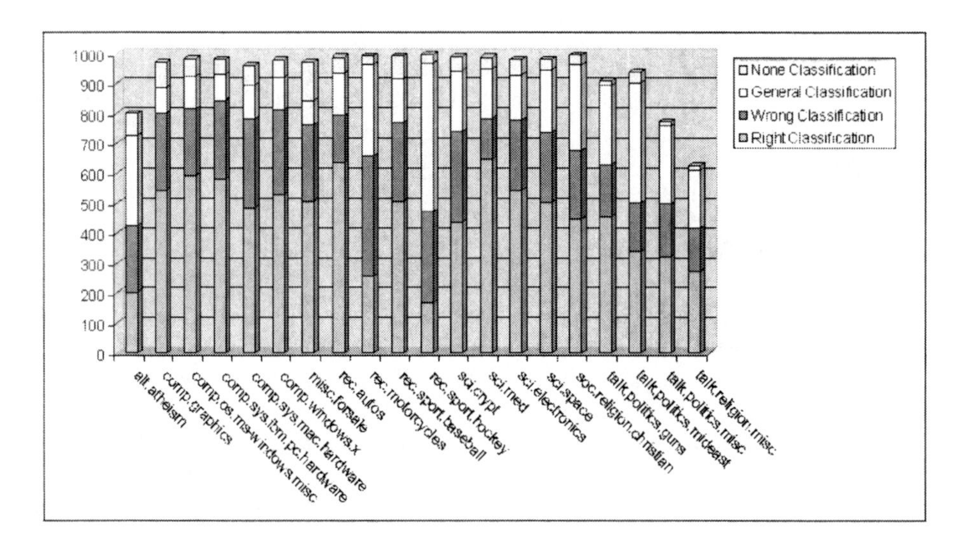


**Fig. 3.** Topic Detection Accuracy

# 6 Conclusions

In this paper we used ontologies in the Web page classification process. We use a novel approach to classify Web documents in order to create a directory structure not a priori defined but dynamically built using a general knowledge base. Our method involves two tasks to perform a word sense disambiguation and a topic detection. Novel algorithms and metrics are proposed and the preliminary results are very promising. Currently we are investigating about the use of other types of data, such as images, in order to obtain a more accurate classification; we are studying ontology merging techniques to increase the ontological definition of a conceptual domain. Furthermore, we are planning to successively compare our method with the ones proposed in the literature.

## References

1. Albanese, M., Picariello, A., Rinaldi, A.M.: A semantic search engine for web information retrieval: an approach based on dynamic semantic networks. In: ACM SIGIR Semantic Web and Information Retrieval Workshop (SWIR 2004), pp. 25–29. ACM Press, New York (2004)
2. Berners-Lee, T., Hendler, J., Lassila, O.: The semantic web: A new form of web content that is meaningful to computers will unleash a revolution of new possibilities. Scientific American 284(5), 28–37 (2001)
3. Calado, P., Cristo, M., Moura, E., Ziviani, N., Ribeiro-Neto, B., Gonçalves, M.A.: Combining link-based and content-based methods for web document classification. In: CIKM '03: Proceedings of the twelfth international conference on Information and knowledge management, pp. 394–401. ACM Press, New York (2003)
4. Chen, H., Dumais, S.: Bringing order to the web: automatically categorizing search results. In: CHI '00: Proceedings of the SIGCHI conference on Human factors in computing systems, pp. 145–152. ACM Press, New York (2000)
5. Dumais, S., Chen, H.: Hierarchical classification of web content. In: SIGIR '00: Proceedings of the 23rd annual international ACM SIGIR conference on Research and development in information retrieval, pp. 256–263. ACM Press, New York (2000)
6. Gruber, T.R.: A translation approach to portable ontology specifications. Knowl. Acquis. 5(2), 199–220 (1993)
7. Huang, C.-C., Chuang, S.-L., Chien, L.-F.: Liveclassifier: creating hierarchical text classifiers through web corpora. In: WWW '04: Proceedings of the 13th international conference on World Wide Web, pp. 184–192. ACM Press, New York (2004)
8. Huang, Y., Wang, Q., Yang, J., Ding, Q.: The design and implementation of a subject-oriented web information classification system. In: Proceedings of the 9th International Conference on Computer Supported Cooperative Work in Design, vol. 2, pp. 836–840 (2005)
9. Jackson, M., Burden, P.: WWLib-TNG - new directions in search engine technology. IEE Informatics Colloquium Lost in the Web - navigation on the Internet, pp. 10/1–10/8 (1999)
10. Lang, K.: Newsweeder: Learning to filter netnews. In: Proceedings of the Twelfth International Conference on Machine Learning, pp. 331–339 (1995)
11. Li, Y., Bandar, Z., McLean, D.: An approach for measuring semantic similarity between words using multiple information sources. IEEE Trans. Knowl. Data Eng. 15(4), 871–882 (2003)
12. Miller, G.A.: Wordnet: a lexical database for english. Commun. ACM 38(11), 39–41 (1995)
13. Qi, D., Sun, B.: A genetic k-means approaches for automated web page classification. In: IRI, pp. 241–246 (2004)
14. Salton, G.: Automatic Text Processing: The Transformation, Analysis, and Retrieval of Information by Computer. Addison-Wesley, London (1989)
15. Song, M.-H., Lim, S.-Y., Kang, D.-J., Lee, S.-J.: Automatic classification of web pages based on the concept of domain ontology. In: Proceeding of the 12th Asia-Pacific Software Engineering Conference (APSEC '05), Taipei, Taiwan, pages CD–ROM (2005)
16. Xiaogang, P., Choi, B.: Automatic web page classification in a dynamic and hierarchical way. In: Proceeding of the IEEE International Conference on Data Mining (ICDM '02), Maebashi City, Japan, pp. 386–393 (2002)
17. Zhang, M.-Y., Lu, Z.-D.: A fuzzy classification based on feature selection for web pages. In: WI '04: Proceedings of the IEEE/WIC/ACM International Conference on Web Intelligence (WI'04), pp. 469–472. IEEE Computer Society Press, Washington (2004)

# Extracting Sequential Nuggets of Knowledge

Christine Froidevaux[1,*], Frédérique Lisacek[2], and Bastien Rance[1]

[1] LRI; Univ. Paris-Sud, CNRS UMR 8623; F-91405 Orsay, France
chris@lri.fr, bastien.rance@lri.fr
[2] Proteome Informatics Group, Swiss Institute of Bioinformatics, Geneva, Switzerland
frederique.lisacek@isb-sib.ch

**Abstract.** We present the notion of sequential association rule and introduce Sequential Nuggets of Knowledge as sequential association rules with possible low support and good quality, which may be highly relevant to scientific knowledge discovery. Then we propose the algorithm SNK that mines some interesting subset of sequential nuggets of knowledge and apply it to an example of molecular biology. Unexpected nuggets that are produced may help scientists refine a rough preliminary classification. A first implementation in Java is freely available on the web[1].

## 1 Introduction

Mining the collection of records in a large database to find out association rules is a classical problem introduced by [1] that has received a great deal of attention. Association rules are expressions of the form $A \rightarrow B$, where $A$ and $B$ are disjoint itemsets. Frequent sequential patterns mining was introduced in [2] in the case where the data stored in the database are relative to behavioural facts that occur over time as a refinement of frequent pattern mining that accommodates ordered items. It is an active research field in data mining that is applied in various domains including, among others, analysis of customer shopping sequences, web usage mining, medical processes, DNA sequences.

In this paper, we introduce the notion of sequential association rule which is based on the notion of interestingness measure. Unlike common approaches, we are only interested in producing rules whose consequent belongs to some predefined set of items (target items), disjoint from the set of the items present in the antecedent. We want to detect tight associations between antecedents of rules and their consequent rather than rules with high support. Thus as in [14], we also search for significant rare data that co-occur in relatively high association with the specific data. Namely discovering close dependencies between facts that almost always co-occur is informative, even if these facts are not frequent in the database. In contrast, associations with large support cannot be surprising since they are relative to a large part of the objects ([3], [8]). Unexpected associations are interesting because they may reveal an aspect of the data that needs further study [7].

* To whom correspondence should be addressed.
[1] http://www.lri.fr/~rance/SNK/

R. Wagner, N. Revell, and G. Pernul (Eds.): DEXA 2007, LNCS 4653, pp. 740–750, 2007.

We determine the relevance of a rule merely by its value for some *interestingness measure*. We will consider several interestingness measures because not all measures are equally good at capturing the dependencies between the facts and no measure is better than others in all cases [12]. Then we introduce *Sequential Nuggets of Knowledge* as sequential association rules that may have a low support in the database but are highly relevant for some interestingness measure. Finally, not all Sequential Nuggets of Knowledge, but only the maximal ones are searched for. The rational is to reduce the number and the length of rules, assuming that such rules correspond in some way to a typical signature of the objects, that is, represent concise characteristics of the studied objects. Moreover they are easier to analyse for human experts.

Maximal Sequential Nuggets of Knowledge could be used for example to improve the organisation of a web site. Given the log (list of tuples <IP address, date, visited web page>) of visitors to our university web site, IP addresses could be used to identify different profiles of users: e.g. students of our university, researchers from other universities, visitors from the remainder of the world. If we could discover typical signatures for each profile, we would improve our web site organisation by adding hyperlinks between different pages and would simplify the navigation for the users.

In this paper, we present the algorithm SNK which calculates the most general Sequential Nuggets of Knowledge and illustrate its use in the domain of molecular biology, more specifically, in the perspective of protein functional classification. Sequential Nuggets of Knowledge express context-sensitive sequential constraints that are mostly verified in a sub-class of objects as opposed to another sub-class. This approach is particularly interesting in biology.

The remainder of the paper is organised as follows. In section 2 we introduce the fundamental concepts underlying the notion of Sequential Nuggets of Knowledge. We present and study the algorithm SNK (section 3) that computes these nuggets. We show in (section 4) how this algorithm is useful in an example of SNK application in the domain of molecular biology. We report related work and conclude by discussing our results and giving some perspectives (section 5).

## 2 Basic Concepts

### 2.1 Definitions

We aim at discovering dependencies between the descriptions of objects in terms of sequences of items in relation with some specific target item. We denote by $IDT$ the set of identifiers of the objects and by $T$ the set of the target items. Let $I$ be the set of all *items* (boolean attributes). The sets $I$ and $T$ are supposed to be disjoint. An *itemset* is any subset of $I$.

The following notion of sequence is borrowed from [2]. A *sequence* $s$ on $I$ is an ordered list of itemsets, denoted by $\langle E_1, E_2, ..., E_l \rangle$, where $E_i \subseteq I, 1 \leq i \leq l$. Note that an itemset can have multiple occurrences in a sequence.

The *size* of a sequence $s$ is the number of itemsets in $s$ and is written $|s|$. A sequence $s = \langle E_1, E_2, ..., E_n \rangle$ is called a *subsequence* of another sequence

$s' = \langle F_1, F_2, ..., F_m \rangle$, denoted $s \sqsubseteq s'$, if and only if there exist integers $j_1, ..., j_n$, such that $1 \leq j_1 < j_2 < ... < j_n \leq m$ and $E_1 \subseteq F_{j_1}$, $E_2 \subseteq F_{j_2}$, ... , $E_n \subseteq F_{j_n}$, where $\subseteq$ denotes the classical inclusion between sets. We will say that $s'$ *contains* $s$. If $s$ and $s'$ are distinct sequences such that $s \sqsubseteq s'$, we will write $s \sqsubset s'$.

Let $s = \langle E_1, E_2, ..., E_n \rangle$ and $s' = \langle F_1, F_2, ..., F_m \rangle$ be two sequences on $I$. We will denote by $s \cdot s'$ the sequence resulting from the concatenation of the two sequences: $s \cdot s' = \langle E_1, E_2, ..., E_n, F_1, F_2, ..., F_m \rangle$.

We define a *categorised sequence database* as a set $CSD$ of tuples $\langle sid, s, tg \rangle$, $sid \in IDT$, $tg \in T$, where $sid$ is the object identifier, $s$ the sequence of itemsets from $I$ describing it and $tg$ the target item associated to it. A tuple $\langle sid, s, tg \rangle$ is said to *contain* a sequence $s'$ if and only if $s'$ is a subsequence of $s$.

**Running example:**

$CSD =$

| id | seq | target |
|---|---|---|
| $\{\alpha_1 = \langle id_1,$ | $\langle a, b, f, c, e, f, g \rangle$ | $, tg_1 \rangle,$ |
| $\alpha_2 = \langle id_2,$ | $\langle a, e, b, h, c, f, g \rangle$ | $, tg_1 \rangle,$ |
| $\alpha_3 = \langle id_3,$ | $\langle c, e, a, b, e, g, f \rangle$ | $, tg_2 \rangle,$ |
| $\alpha_4 = \langle id_4,$ | $\langle c, e, a, b, e, g, f, a, e, b, f, d \rangle$ | $, tg_2 \rangle\}$ |

In $CSD$ the sequence $\langle b, e, f \rangle$ is a subsequence of $\langle a, b, f, c, e, f, g \rangle$ and $\alpha_1$ contains the sequence $\langle b, e, f \rangle$. In this example all the itemsets are singletons denoted by their unique element, which is not required in the general definition.

We introduce the notion of sequential association rule as a combination of classical association rules and sequential patterns. Formally, a *sequential association rule* $r$ on $CSD$ is an implication of the form $ANT \rightarrow CONS$, where $ANT$ is a sequence of itemsets from $I$ and $CONS$ an element of $T$. We call $ANT$ (resp. $CONS$) the *antecedent* (resp. *consequent*) of $r$ and write $ant(r)$ (resp. $cons(r)$).

The *support* of a sequential association rule $r$ in a database $CSD$ is defined as the number of tuples of $CSD$ that contain both its antecedent and its consequent. Formally we have: $support_{CSD}(ANT \rightarrow CONS) =$
$|\{\langle sid, s, tg \rangle \in CSD \text{ s.t. } (ANT \sqsubseteq s) \wedge (CONS = tg)\}|$.
Note that the items in $ANT$ need not be consecutive in $s$, in order to be supported by the tuple.

**Example:** $support_{CSD}(\langle a, b, f \rangle \rightarrow tg_1) = 2$
The *confidence* of a sequential association rule $r$ in the database $CSD$ indicates amongst all the tuples of $CSD$ containing its antecedent the fraction in which its consequent appears. $conf_{CSD}(ANT \rightarrow CONS) =$

$$\frac{|\{\langle sid, s, tg \rangle \in CSD \text{ s.t. } (ANT \sqsubseteq s) \wedge (CONS = tg)\}|}{|\{\langle sid, s, tg \rangle \in CSD \text{ s.t. } ANT \sqsubseteq s\}|}$$

**Example:** $conf_{CSD}(\langle a, b, f \rangle \rightarrow tg_1) = 0.5$; $conf_{CSD}(\langle a, b, f, g \rangle \rightarrow tg_1) = 1$.
A sequential association rule $r_1$ is said to *contain* another rule $r_2$, written ($r_2 \preceq r_1$), if and only if $cons(r_1) = cons(r_2)$ and $ant(r_2) \sqsubseteq ant(r_1)$. We also say that $r_2$ is *more general* than $r_1$. If $r_1 \neq r_2$ and $r_2 \preceq r_1$ we will write $r_2 \prec r_1$.

We now focus on the main notion of this paper, namely *Sequential Nuggets of Knowledge.* We introduce them as sequential association rules with possible low support but with high quality. Minimal support is required in order not to discover strong associations that involve only a few objects, which may come from noise.

A *sequential nugget of knowledge* is defined as a sequential association rule $r$ in $CSD$ such that its support is no less than some threshold and its interestingness measure value (cf. section 2.2) is no less than to some other threshold.

In the applications we have foreseen, objects are merely described by sequences of items, so that sequences of itemsets are unnecessarily complicated. Therefore, in the remainder of the paper, we will consider only sequences where itemsets have a single item. The definition of subsequence can be rewritten in a simpler form where inclusion is replaced by equality.

## 2.2 Interestingness Measures

Identifying sequences of variables that are strongly correlated and building relevant rules with those variables is a challenging task. Interestingness measures help to estimate the importance of a rule: they can be used for pruning low utility rules, or ranking and selecting interesting rules. Selecting a good measure allows to reduce time and space costs during the mining process ([12], [7]). As pointed earlier, all the interestingness measures do not capture the same kind of association. For example, using a support-confidence approach, a rule $ANT \rightarrow CONS$ may be considered as important, even if $CONS$ is often found without $ANT$. In our work we mainly studied, besides confidence, another measure which is well adapted to our data, Zhang's measure as it takes into consideration the counter-examples [16].

[8] and [7] suggest a number of key properties to be examined for selecting the right measure that best suits the data. Note that while support satisfies anti-monotonicity (if $r \preceq r'$ then $support_{CSD}(r') \leq support_{CSD}(r)$), not all interestingness measures satisfy monotonicity (if a rule is considered to be relevant any of its specialisations is relevant too).

## 2.3 Postfix-Projection

The method proposed for mining sequential nuggets of knowledge follows the approach of [11] for sequential patterns. We recursively project the initial categorised sequential database into a set of smaller categorised sequential databases, thus generating projected databases by growing prefixes.

Let $CSD$ be a categorised sequential database, $\alpha = \langle sid_1, \langle e_1...e_n \rangle, c_1 \rangle$ a tuple of $CSD$ and $s' = \langle e'_1...e'_m \rangle$ a sequence with $m \leq n$. $s'$ is called a *prefix* of $\alpha$ if and only if $\forall i, 1 \leq i \leq m$ , $e'_i = e_i$.

**Example (continued):** The sequence $\langle a, b, f \rangle$ is a prefix of $\alpha_1$.
Let $\alpha = \langle sid, s, tg \rangle$ be a tuple of $CSD$. We denote *id*, *seq* and *target* the methods which return respectively the identifier, the sequence and the target of $\alpha$: $id(\alpha) = sid$, $seq(\alpha) = s$ and $target(\alpha) = tg$.

The notion of $s'$-projection corresponds to the longest subsequence having $s'$ as a prefix. Let $\alpha$ be a tuple and $s'$ be a sequence such that $s' \sqsubseteq seq(\alpha)$. A tuple $\alpha' = \langle id(\alpha'), seq(\alpha'), target(\alpha') \rangle$ is the *$s'$-projection* of $\alpha$ if and only if (1) $id(\alpha') = id(\alpha)$, (2) $seq(\alpha') \sqsubseteq seq(\alpha)$, (3) $target(\alpha') = target(\alpha)$, (4) $s'$ is *a prefix* of $\alpha'$ and (5) $\nexists \alpha''$ a tuple s.t. $seq(\alpha') \sqsubset seq(\alpha'')$ and $seq(\alpha'') \sqsubseteq seq(\alpha)$ and $s'$ is a prefix of $\alpha''$.

Note that with such a definition only the subsequence of $seq(\alpha)$ prefixed with the first occurrence of $s'$ should be considered for $\alpha'$.

**Example (continued):**
$\langle id_1, \langle a, b, f, c, e, f, g \rangle, tg_1 \rangle$ is an abf-projection of $\alpha_1$, while $\langle id_1, \langle a, b, f, g \rangle, tg_1 \rangle$ is not because (5) is not satisfied. Similarly, $\langle id_4, \langle a, b, f, a, e, b, f, d \rangle, tg_2 \rangle$ is an abf-projection of $\alpha_4$, while $\langle id_4, \langle a, b, f, d \rangle, tg_2 \rangle$ is not because of (5).

The $s'$-projection of $\alpha$, if it exists (i.e. if $s'$ can be a prefix of a tuple whose sequence is contained in $\alpha$) is unique. It is *the* $s'$-projection of $\alpha$.

Let $\alpha$ be a tuple of $CSD$ and let $s = \langle e_1, ..., e_n \rangle$ be a sequence on I. Let $\alpha' = \langle id_1, \langle e_1, ..., e_n, e_{n+1}, ..., e_{n+p} \rangle, tg_1 \rangle$ be the $s$-projection of $\alpha$, where $s$ is a prefix of $\alpha'$. Then $\gamma = \langle id_1, \langle e_{n+1}, ..., e_{n+p} \rangle, tg_1 \rangle$ is the *$s$-postfix* of $\alpha'$. If $p > 0$, then the $s$-postfix has a sequence of size $> 0$: it is said to be not empty and is denoted by $\alpha/s$. Note that $\gamma$ satisfies: $seq(\alpha') = s \cdot seq(\gamma)$.

The *$s$-projected database*, denoted by $s - postfix(CSD)$, is defined as follows: $s$−postfix$(CSD) = \{(\alpha/s),\ \alpha \in CSD\}$

**Running example :**

$abf$−postfix$(CSD)$ =

| id | seq | target |
|---|---|---|
| $\{\langle id_1,$ | $\langle c, e, f, g \rangle$ | $, tg_1 \rangle,$ |
| $\langle id_2,$ | $\langle g \rangle$ | $, tg_1 \rangle,$ |
| $\langle id_4,$ | $\langle a, e, b, f, d \rangle$ | $, tg_2 \rangle\}$ |

The recursive principle of our algorithm is based on the following property:

**Property 1:**
Let $CSD$ be a categorised database. Let $s_1$ and $s_2$ be any sequences on $I$, and let $r$ be any sequential association rule. Then:

(i) $s_2$−postfix($s_1$−postfix$(CSD)$) = $s_1 \cdot s_2$−postfix$(CSD)$
(ii) $\text{support}_{s_1.s_2-\text{postfix}(CSD)}(r) = \text{support}_{CSD}((s_1 \cdot s_2 \cdot ant(r)) \rightarrow cons(r))$
(iii) $\text{support}_{CSD}(r) \geq \text{support}_{s_1-\text{postfix}(CSD)}(r)$.

# 3 SNK Algorithm

## 3.1 Specification and Pseudo-code

Now we present SNK, an algorithm which mines the most general sequential nuggets of knowledge from a categorised sequential database, given some thresholds specified by the user.

**SNK method**
**Parameters:**
In: $CSD$ a categorised sequential database; $min\_supp$ a support threshold; $IM$ an interestingness measure; $min\_meas$ an IM value threshold;
Out: $RESULTS$ the set of the most general Sequential Nuggets of Knowledge;
**Method used**: SNKrec;
**Begin**
$RESULTS = \emptyset$; $ST$ = the set of all target items of $T$ present in $CSD$;
**Foreach** $y$ in $ST$ **do**
 //sequential nuggets of knowledge targeted on $y$ are searched for
 $S_y$ = the set of all tuples of $CSD$ having $y$ as a target;
 SNKrec($S_y$,$y$,$min\_supp$,$IM$,$min\_meas$,$\langle\rangle$,$RESULTS$) **endfor end_SNK;**

**SNKrec method**
// generates rules $r$ of the form $(p \cdot x) \rightarrow y$, where $x$ is any item occurring in $S$ and $p$ the prefix used; updates $RESULTS$ with $r$ in order to get only the most general sequential nuggets of knowledge; calls recursively itself on the $x$-projected database of $S$ if $r$ has good support but bad interestingness measure value
**Parameters:**
In: $S$ a set of tuples having $y$ as a target; $min\_supp$, $IM$, $min\_meas$;
$p$ the sequence used as a prefix;
In/Out: $RESULTS$ a set of Sequential Nuggets of Knowledge s.t. $\not\exists r_1, r_2 \in RESULTS$ with $r_1 \prec r_2$;
**Methods used:**
add_rule; //add_rule($r$,$RES$) adds rule $r$ to $RES$ unless if $r$ is less general than or equal to some rule in $RES$ and removes from $RES$ any rule that is less general than $r$.
measure; // measure$_{IM,CSD}$(r) evaluates the value of $r$ for $IM$ in $CSD$
support: // support$_S$(r) evaluates the support of $r$ in $S$
**Begin** $SI$ = the set of all items of $I$ occurring in elements of $S$;
**Foreach** $x$ in $SI$ **do**
 **if** support$_S(x \rightarrow y) \geq min_supp$ **then**
  **if** measure$_{IM,CSD}((p \cdot x) \rightarrow y) \geq min_meas$ **then**
   $RESULTS$ = add_rule($(p \cdot x) \rightarrow y$,$RESULTS$)
  **else if** $x$-postfix(S) $\neq \emptyset$ **then**
   SNKrec($x$-postfix(S),$y$,$min\_supp$,$IM$,$min\_meas$,$p \cdot x$,$RESULTS$)
**endifendifendifendfor end_SNKrec;**

**Running example:**
Let $min\_supp = 2$, $IM$ = confidence, $min\_meas = 1$. SNK yields the set of all the maximal sequential nuggets of knowledge:
$RESULTS = \{\langle e, e\rangle \rightarrow tg_2, \langle e, a\rangle \rightarrow tg_2, \langle c, b\rangle \rightarrow tg_2, \langle c, a\rangle \rightarrow tg_2, \langle g, f\rangle \rightarrow tg_2, \langle b, c\rangle \rightarrow tg_1, \langle f, g\rangle \rightarrow tg_1, \langle a, c\rangle \rightarrow tg_1\}$.

### 3.2 Properties of SNK

First the algorithm is sound and complete w.r.t its specification [6]. Formally:

**Theorem 2.** Let $CSD$ be a categorised sequential database, $IM$ an interestingness measure, $min\_supp$ a support threshold and $min\_meas$ an interestingness measure threshold for $IM$. Then:
SNK returns exactly all the most general sequential association rules $r$ on $CSD$ that satisfy $supp_{CSD}(r) \geq min\_supp$ and $meas_{IM,CSD}(r) \geq min\_meas$.

The time complexity of SNK is related to the number of target items, and for each target item, to the number of recursive calls of SNKrec. The worst case for SNKrec occurs when all the rules generated have good support but bad measure, leading to a maximal number of recursive calls. Each call requires a calculation of support and of $IM$ measure, and involves either the cost of a postfix-projection or that of the add_rule method. With our depth-first search approach all the projected databases need not be stored in memory and they can be built independently. The analysis shows (see [6] for details) that the theoretical time complexity is high in the worst case. However, in practice, for the applications foreseen, the SNK algorithm remains efficient because the size of the projected databases decreases very quickly.

SNK allows to discover rules describing regularities in a sequential data set. Moreover, SNK provides the user with a parameterisation process for adapting the tool to specific needs. The user can select among a dozen measures the measure that best fits his application field (by default confidence is selected) [7]. A bootstrap mode is also available, where SNK is run on a categorised sequential database resampled from the original database as an input in order to check the consistency of the generated rules. In the data mining mode, SNK runs in about 3 seconds for mining sequential nuggets of knowledge for 760 tuples (described by sequences of size less than 17 where the set of items has about 35 distinct elements), 6 seconds for 1200 tuples. SNK is fully implemented in Java and the web Applet is freely available on SNK website (http://www.lri.fr/~rance/SNK/).

## 4 Example

We show how SNK can be useful through the study of a family of bacterial proteins. Each protein is described by its sequence of motifs (we call "motif" a functional or well conserved part of the amino acid sequence). We consider the Phospholipase D (PLD) family of proteins which are present in all species from virus to eukaryote, and involved in many cell processes. These proteins are grouped together simply because they carry the PLDc motif repeated once. They also contain a wide range of other motifs. In [10], a surprising regularity concerning the C-terminal part of proteins was reported. More precisely, the distance between the end of the second PLDc motif and the C-terminal end of the protein (rightmost) was shown to correlate with the known functions of the proteins. Consequently, proteins could be grouped into classes using this distance as a classification criterion. In the remainder of this section we will refer to the length of this region as the *C-terminal length* (this length is either: 40, 60, 72, 82, 100). Each class is then functionally consistent. Using SNK we have investigated a possible relationship between module architecture, C-terminal length and function. We have considered all bacterial proteins of the UniProtKB database [4] which contain two PLDc motifs. The corresponding set of proteins showed a variety of motif combinations involving other protein family signatures as well as so-called "low complexity regions" (poorly informative sequences [13]). The total number of proteins is 676. We first considered the possible existence of a link between low-complexity regions and C-terminal length. In this first test, proteins were described as successions of PLDc motifs and low complexity regions. We studied a set of proteins containing all the PLD proteins with C-terminal length from classes "72" and "82" using Zhang's interestingness measure. SNK was performed with a very low support threshold (*min_supp*=15) and with a good measure threshold (*min_meas*=0.8). Among the 20 most general sequential nuggets of knowledge obtained, 3 rules were especially interesting. In the rules presented below, *lc* denotes low-complexity region and the values between brackets are respectively support and Zhang's measure values.

`(1a) lc,PLDc,PLDc -> 82, (273,0.80),`
`(1b) lc,lc,PLDc -> 82, (174,0.68),`
`(1c) PLDc,PLDc,lc -> 72, (136,0.90)`

The sequential association rules returned by SNK are high quality rules. Rule (1a) and (1c) highlight the importance of the order between modules in the assignment to a class. The location of low complexity regions is closely linked to the C-terminal length. Depending on whether *lc* is in front of or behind the double PLDc motif, the conclusion of the rule is one or the other class. Simple association rules could not have expressed such a clear distinction.

In a second test information about protein family signatures as compiled in both Pfam-A [5] and Pfam-B databases was added. Amongst the rules generated with *min_supp*=7 and *min_meas*=0.9,

`(2a) PLDc,Pfam-B_115,Pfam-B_2786 -> 40, (7,0.90)` and
`(2b) PLDc,Pfam-B_115,Pfam-B_6054 -> 40, (7,0.93)`

are the only rules where PLDc precedes Pfam-B_115 and therefore appear to characterise class 40. In all other rules where the two entities occur Pfam-B_115 precedes PLDc. A complementary test was performed with the same initial data set but taking the protein function as a target for SNK (either diacyltransferase, cardiosyntase, transphosphatidylase or unspecialised phospholipase D). Amongst the 9 rules (*min_supp*=7, *min_meas*=0.9), one strongly corresponds to the cardiosyntase function:

`(3a) Pfam-B_1038,lc,Pfam-B_115,Pfam-B_2786 -> cardio, (7,1.00)`

This rule appears quite similar to one of the rules generated (same thresholds) for the length criterion

`(2c) lc,Pfam-B_115,lc,Pfam-B_2786 -> 60, (15,0.94)`

Likewise, `(3b) lc,Pfam-B_5151 -> diacyltransferase, (7,0.91)` strongly corresponds to the diacyltransferase function while `(3c) Pfam-B_5151 -> 72, (53,1.00)` was previously generated for the length criterion.

This generalises the correlation suggested in [10] between length 60 and 72 respectively and the cardiosyntase and diacyltransferase functions. Other rules generated with the protein function as a target are potentially misleading due to inconsistencies of the automated assignment of function in these proteins. We are currently testing the possibility of correcting mistakes using rules generated with the length criterion.

## 5 Related Work and Discussion

In this paper, we have proposed a definition of sequential association rules and introduced sequential nuggets of knowledge. Those definitions are based on the works presented in [11], but unlike classical sequential pattern mining, our approach focuses on rules with predefined targets as consequents. We have designed SNK, an algorithm based on a pattern-growth strategy (as PrefixSpan [11]) to generate the most general sequential nuggets of knowledge using an interestingness measure that evaluates the pertinence of a rule. Other efficient works have been proposed for sequential pattern mining. SPADE [15] is as fast as PrefixSpan but uses a bitmap structure which is better adapted to the study of very long sequences but less suitable for short sequences. [9] had proposed a method to generate sequential association rules, but is based on an *a priori*-like strategy with two steps, a candidate test step and a candidate generation step. This approach generates many unnecessary candidates that our pattern-growth approach avoids.

Sequential nuggets of knowledge are defined by a good interestingness measure value. SNK offers the choice between a dozen of interestingness measures. The choice of a suitable measure for a given application domain can be guided by the examination of criteria described in [7] and in [12]. On the other hand, [8] proposes a statistical bootstrap-based method to assess the significance of a measure (thus avoiding false discoveries) that could be used with SNK. A first implementation of SNK is freely available on the web with some other functionalities.

Finally we have presented an example in biology involving the PLDc family of proteins. The link between C-terminal length of a PLDc protein and its function was investigated. Let us recall that a protein function usually corresponds to a specific sequence of structural units. Most studies take into account the combinatorial aspect of the structural composition of proteins. We showed that the identification of sequential constraints could lead to a refinement of the functional classification of proteins. As a result, a large class grouped upon one rough criterion can be subdivided into sub-classes upon explicit and informative distinctive traits. We are currently testing the possibility of using the rules discovered as a way of automatically correcting mistakes.

We also envisage to use our algorithm in other applications, e.g. on web logs, and to extend it by adding non-sequential items in the antecedent of a rule. In that way, it could take into account more expressive descriptions of objects. Since the projected databases can be considered independently, we also plan to develop a distributed version for a cluster of PC thereby drastically speeding up SNK.

## Acknowledgement

Authors are very grateful to Céline Arnaud for her great help for the implementation of SNK applet. This work was supported in part by the French ACI IMPBio grant RAFALE.

## References

[1] Agrawal, R., Imielinski, T., Swami, A.N.: Mining Association Rules between Sets of Items in Large Databases. In: Proc. of the 1993 ACM SIGMOD International Conference on Management of Data, pp. 207–216 (1993)

[2] Agrawal, R., Srikant, R.: Mining sequential patterns. In: Proc. Eleventh International Conference on Data Engineering, pp. 3–14 (1995)

[3] Azé, J., Kodratoff, Y.: A study of the Effect of Noisy Data in Rule Extraction Systems. In: Proc. of the Sixteenth European Meeting on Cybernetics and Systems Research (EMCSR'02) (2), pp. 781–786 (2002)

[4] Bairoch, A., Apweiler, R., Wu, C.H., Barker, W.C., Boeckmann, B., Ferro, S., Gasteiger, E., Huang, H., Lopez, R., Magrane, M., Martin, M.J., Natale, D.A., O'Donovan, C., Redaschi, N., Yeh, L.S.: The Universal Protein Resource (UniProt), Nucleic Acids Res. 33, D154–159 (2005)

[5] Finn, R.D., Mistry, J., Schuster-Backler, B., Griffiths-Jones, S., Hollich, V., Lassmann, T., Moxon, S., Marshall, M., Khanna, A., Durbin, R., Eddy, S.R., Sonnhammer, E.L.L., Bateman, A.: Pfam: clans, web tools and services. Nucleic Acids Research, Database Issue 34, D247–D251 (2006)

[6] Froidevaux, C., Lisacek, F., Rance, B.: Mining sequential nuggets of knowledge UPS-LRI, Technical report (to appear)

[7] Geng, L., Hamilton, H.J.: Interestingness Measures for Data Mining: A Survey. ACM Computing surveys 38(3), Article 9 (2006)

[8] Lallich, S., Teytaud, O., Prudhomme, E.: Association rule interestingness: measure and statistical validation. In: Guillet, F., Hamilton, H.J. (eds.) Quality Measures in data Mining, Springer, Heidelberg (to appear, 2006)

[9] Masseglia, F., Tanasa, D., Trousse, B.: Web Usage Mining: Sequential Pattern Extraction with a Very Low Support. In: Yu, J.X., Lin, X., Lu, H., Zhang, Y. (eds.) APWeb 2004. LNCS, vol. 3007, pp. 513–522. Springer, Heidelberg (2004)

[10] Nikitin, F., Rance, B., Itoh, M., Kanehisa, M., Lisacek, F.: Using Protein Motif Combinations to Update KEGG Pathway Maps and Orthologue Tables. Genome Informatics 2, 266–275 (2004)

[11] Pei, J., Han, J., Mortazavi-Asl, B., Wang, J., Pinto, H., Chen, Q., Dayal, U., Hsu, M.-C.: Mining Sequential Patterns by Pattern-Growth: The PrefixSpan Approach. IEEE Transactions on Knowledge and Data Engineering 16, 1424–1440 (2004)

[12] Tan, P.N., Kumar, V., Srivastava, J.: Selecting the Right Interestingness Measure for Association Patterns. In: SIGKDD'02 (2002)

[13] Wootton, J.C., Federhen, S.: Statistics of local complexity in amino acid sequences and sequence databases. Comput. Chem. 17, 149–163 (1993)

[14] Yun, H., Ha, D., Hwang, B., Ryu, K.H.: Mining association rules on significant rare data using relative support. The Journal of Systems and Software 67, 181–191 (2003)
[15] Zaki, M.J.: SPADE: An Efficient Algorithm for Mining Frequent Sequences. In: Fisher, D. (ed.) Machine Learning Journal, special issue on Unsupervised Learning, vol. 42, pp. 31–60 (2001)
[16] Zhang, T.: Association Rules. In: Terano, T., Chen, A.L.P. (eds.) PAKDD 2000. LNCS, vol. 1805, pp. 245–256. Springer, Heidelberg (2000)

# Identifying Rare Classes with Sparse Training Data

Mingwu Zhang, Wei Jiang, Chris Clifton, and Sunil Prabhakar

Department of Computer Science, Purdue University
West Lafayette, IN 47907-2107, USA
{mzhang2, wjiang, clifton, sunil}@cs.purdue.edu

**Abstract.** Building models and learning patterns from a collection of data are essential tasks for decision making and dissemination of knowledge. One of the common tools to extract knowledge is to build a classifier. However, when the training dataset is sparse, it is difficult to build an accurate classifier. This is especially true in biological science, as biological data are hard to produce and error-prone. Through empirical results, this paper shows challenges in building an accurate classifier with a sparse biological training dataset. Our findings indicate the inadequacies in well known classification techniques. Although certain clustering techniques, such as seeded k-Means, show some promise, there are still spaces for further improvement. In addition, we propose a novel idea that could be used to produce more balanced classifier when training data samples are very limited.

## 1 Introduction

With the explosion of data, data mining techniques gain much attention for their promise in building models and learning patterns from a collection of data. These tasks are essential for decision making and dissemination of knowledge in many areas. Well-known learning techniques such as association rule mining, classification and clustering have been successfully applied in many applications.

Recently, biological science has emerged as a challenging area to apply data mining techniques. One common problem in this field is that given a dataset of which a small fraction has class labels, we need to identify class labels for the other data items [1]. To solve this problem, we can either use supervised (e.g., classification) or unsupervised (e.g., clustering) learning methods. To apply classification techniques, the data with class labels can be treated as a training dataset, and a classifier can be constructed from it. Then, the classifier is used to predict class labels for the rest of unlabeled data items. On the other hand, clustering techniques can also be adopted to achieve this task. For example, assume the total number of class labels in the dataset is known. The dataset can be clustered first, then each unlabeled data is assigned to the majority class label in its cluster. (Thereafter, we term the set of data with class labels as a training dataset and the rest of data as unlabeled data or dataset.)

R. Wagner, N. Revell, and G. Pernul (Eds.): DEXA 2007, LNCS 4653, pp. 751–760, 2007.

Although these techniques can be applied directly, different techniques produce various results. Therefore, how to choose the best method or design a suitable method for a specific domain is challenging. Before making any decision, we first need to understand the characteristics of biological data. Generally speaking, collecting biological data requires major efforts and years of research, and biological data are noisy and error-prone. Thus, it is very likely that the training dataset are *sparse*: either the size of the training dataset is small or the training dataset contains incomplete class information.

For example, in cell wall genomics research, the mutants of cell wall synthesis are extremely valuable to study the genes responsible for biosynthesis of the cell wall and the genes that regulate the cell wall biosynthesis pathways. Traditional experimental methods to find the mutants are time consuming and labor intensive. Although techniques such as Fourier Transform InfraRed microspectroscopy (FTIR) followed by Principle Component Analysis (PCA) and Linear Discriminant Analysis (LDA) has been successfully applied to rapidly identify mutants [1], one common challenge biologists faced is the fact that certain mutants might not have visually abnormal or known phenotypes. In other words, there may not exist any training data for these mutants even though they are detectable. These mutants that do not have known phenotypes are very valuable to biologists because their mutations may be in the regulatory component of the cell wall biosynthetic pathways. In addition, these unidentified mutants could be much less common than the known mutants. The problem appears when the training dataset has many data samples for common mutants but very few or none for rare mutants that are potentially important. Consequently, classifiers built on this kind of sparse training dataset are biased toward the common mutants and could be useless in identifying rare mutants. It would be a great loss for biological science if these potentially valuable mutants cannot be discovered.

Another issue that has not been addressed in data mining community is that the training data may not be reliable and contain errors. Some biological experiments (e.g., Yeast 2-hybrid Assay, Mass Spectrometry) are known to produce a large number of false positives. If the results of these experiments are used as the training data for supervised learning, the classifier could be defective because it is built on unreliable training data. Another source of uncertainty comes from the computational methods extensively used in bioinformatics. With the development of high throughput experiment techniques, biologists more and more rely on data mining and machine learning methods to rapidly and automatically process the data. For example, Swiss-Prot is a curated protein function database [2]. To alleviate the intense labor of manually curating protein function annotations, scientists explore using decision tree to predict the functions of the protein sequences [3]. The function annotations in the Swiss-Prot database are used as training data. As Peter Karp points out in [4], some function annotations are computed using computational methods such as BLAST [5] and may not be reliable. Because of the complexity and inherent uncertainty of the biological data, collected biological data samples are very unlikely to be complete and accurate. Therefore, when training data samples are sparse, developing learning techniques that can discover rare important classes and tolerate the noise in the training data has great value.

These examples highlight the nature of biological data in that the training dataset portion is sparse and some rare data may have great value in biological research. Under our problem domain, classification techniques generally perform better than clustering techniques if sufficient and unbiased training data are available. When training data are sparse, the computed classifier is likely biased. We expect that such a classifier is likely to ignore rare class labels. This could lead to potential loss in research. Therefore, with sparse training data, clustering techniques could be the better option in assigning more correct class labels without ignoring rare class items.

Through empirical results, this paper shows challenges faced by biologists in building an accurate classifier with a sparse training dataset. Our findings indicate that when the training dataset is sparse, well known classification techniques are inadequate in producing accurate classifiers. Using them to discover rare classes is almost impossible. Semi-supervised clustering techniques, such as seeded k-means, show some promise in identifying rare classes, but there are still spaces for further improvement. Based on these observations, we also propose a novel idea that could be used to identify rare classes when training data samples are very limited. The paper is organized as follows: Section 2 presents a brief overview of related works. Section 3 provides empirical results showing inadequacies of common classification techniques to detect rare classes when the training dataset is sparse. Section 4 proposes a novel idea in hopes that better learning techniques can be designed to produce unbiased classifiers. Section 5 concludes the paper with lessons learned and future research directions.

## 2 Related Work

Machine learning and data mining methods can be classified into supervised and unsupervised learnings. Supervised learning requires a training dataset while unsupervised learning does not. Lately, semi-supervised learning has gained increasing attention [6,7] because semi-supervised learning promises the advantage of both supervised and unsupervised methods. In particular, semi-supervised clustering tries to use a small number of labeled data to guide the clustering process. By incorporating the domain knowledge in the clustering process, one hopes that the result of semi-supervised clustering will be better than totally ignoring this information. In [6], unlike the traditional k-means algorithm, instead of using random seed, the initial seeds are the mean of each class of the labeled data.

However, this approach cannot be applied directly to the problem presented in this paper because they assume that every class labels are included in the training dataset. In section 3, we leverage this work and show how to choose the seeds when training data contain incomplete information. A related problem often emerging in biological application is Single Class Classification (SCC). In [8], SCC is defined as distinguishing one class of the data from the universal set of multiple classes. In our problem domain, because we would like to identify rare classes from multiple classes, without training data for the rare classes, single-class approaches cannot be applied.

## 3 Seeded k-Means vs. Classification Techniques

Here, we show that when training data is sparse, well-known classification techniques rarely produce accurate and unbiased classifiers. We also point out that with careful choice of seeds, seeded k-means (SkM) performs better in classifying unknown data instances and identifying rare class instances with a sparse training dataset. By sparse training data, we mean that either the size of the training dataset is small or the training dataset contains incomplete class informations due to errors occurred during data collection process. For the rest of this section, we distinguish these two cases and present our findings independently.

The experiments are done using two datasets, Ecoli and Yeast datasets from UC Irvine Machine Learning Repository [9]. Ecoli dataset contains 336 instances, 7 numeric attributes and 8 classes: cp (143), im (77), pp (52), imU (35), om (20), omL (5), imL (2) and imS (2) (the number in parentheses is the number of instances belonging to that class). The Yeast dataset contains 1462 instances, 8 attributes and 10 classes: CYT (463), NUC (429), MIT (244), ME3 (163), ME2 (51), ME1 (44), EXC (37), VAC (30), POX (20) and ERL (5).

As stated in Section 1, different techniques produce various results. Our experiments focus on three commonly used methods: decision tree (C4.5) [10], k-nearest neighbor (kNN) [11,12], and seeded k-means (SkM) [6]. Based on our own observations, the generic k-means did not produce better results than SkM. Hence, we only show SkM's results. In addition, when there are missing class labels in the training data, SkM cannot be used directly because the number of seeds it picks is equal to the number of distinct class labels in the training dataset. To get around this issue, we propose two variations of SkM: $SkM^r$ and $SkM^d$. When there are missing class labels, $SkM^r$ chooses the rest of cluster centers randomly (as with the basic k-means) and $SkM^d$ chooses the rest of cluster centers by picking the seed with largest Euclidean distance to the chosen cluster centers, randomly choosing the seed if there are multiple candidates.

Both C4.5 and kNN were used in [13] to classify Ecoli and Yeast datasets, where it was reported that the two classification techniques are most effective for these datasets. We choose the same k values (for kNN) as those used in [13]. First, the training data and test data are generated using Weka [14] to create the stratified n-fold cross-validation. Since we are interested in the situation when little training data is available, we use one fold of data as the training data to build the classifier and n-1 folds of data to test the classifier. In order to fairly compare the clustering techniques with the classifiers, only the test data are used to estimate the accuracy (or precision) and confusion matrices.

Fig. 1 shows the results for Ecoli dataset. Fig. 1 (a) is related to the original Ecoli dataset, and it presents the accuracy changes with the number of data samples varying from 16 to 168 (or 5% to 50%). Each error bar indicates maximum, minimum and average values. Note that the label J48 in the fig-

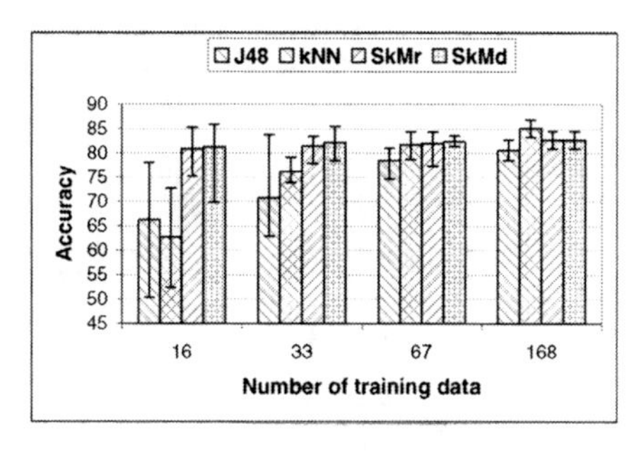


(a) Ecoli

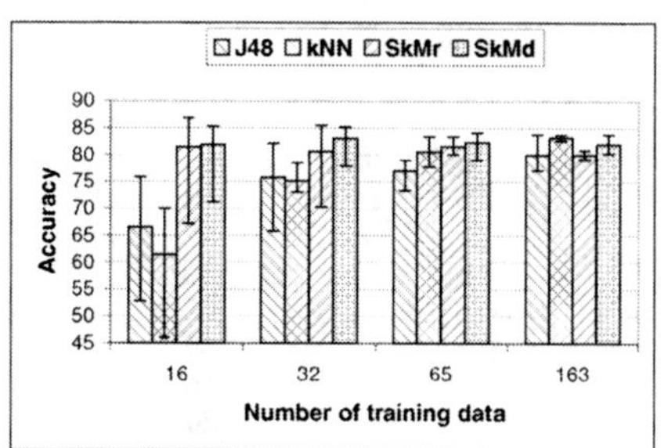


(b) Ecoli with missing classes

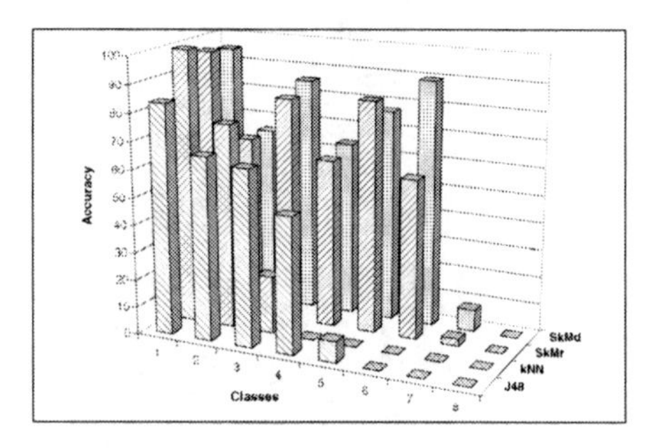


(c) Ecoli 5% training data

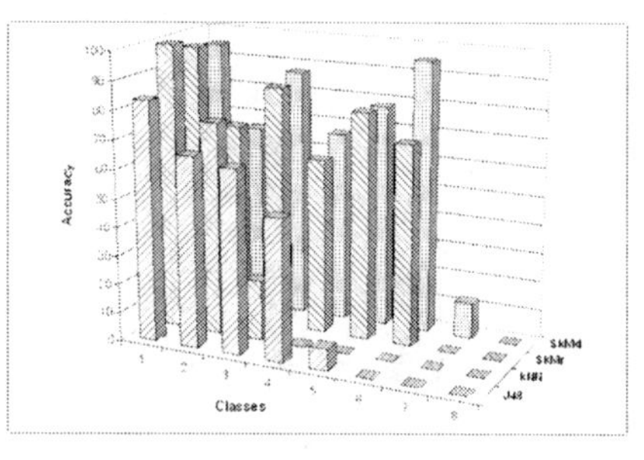

(d) Ecoli 5% training data with missing classes

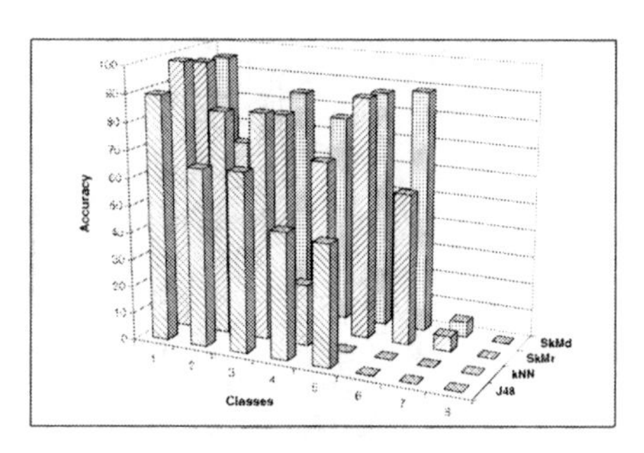


(e) Ecoli 10% training data

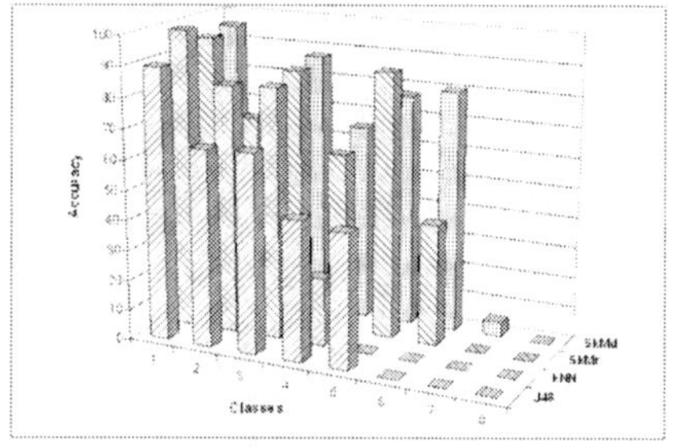

(f) Ecoli 10% training data with missing classes

**Fig. 1.** Ecoli dataset: Accuracy is computed based on test data only. Subfigures (a) and (b) indicate overall accuracy of each algorithm, and the error bars indicate maximum, minimum and average values across n-1 folds of test data. The rest of subfigures indicate overall accuracy regarding each individual class.

ure indicates a Java implementation of C4.5. The figure shows when the number of training data is small, C4.5 and kNN perform worse than $SkM^d$ and $SkM^r$. As

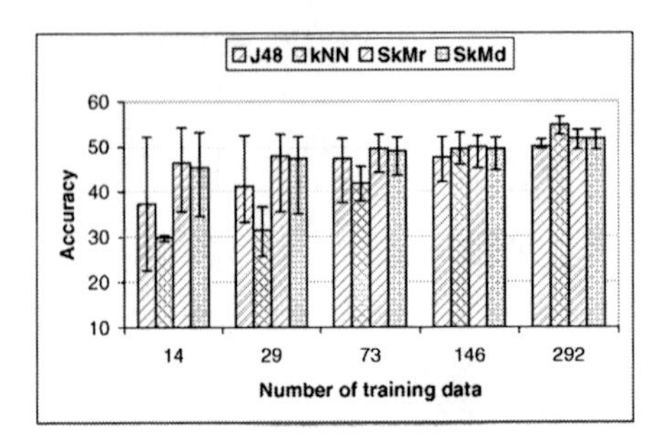


(a) Yeast

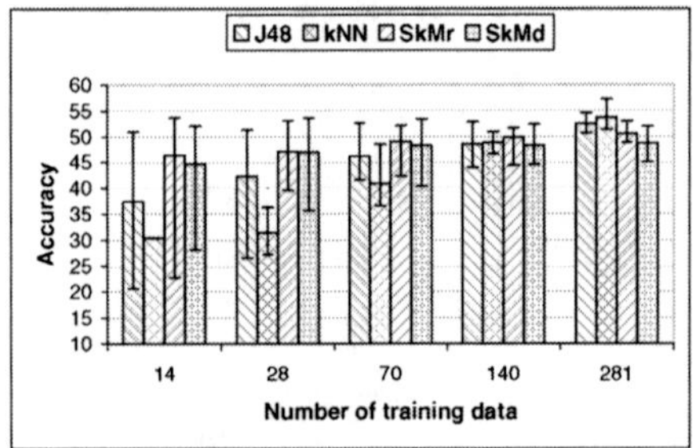


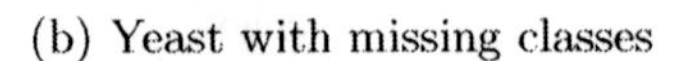

(b) Yeast with missing classes

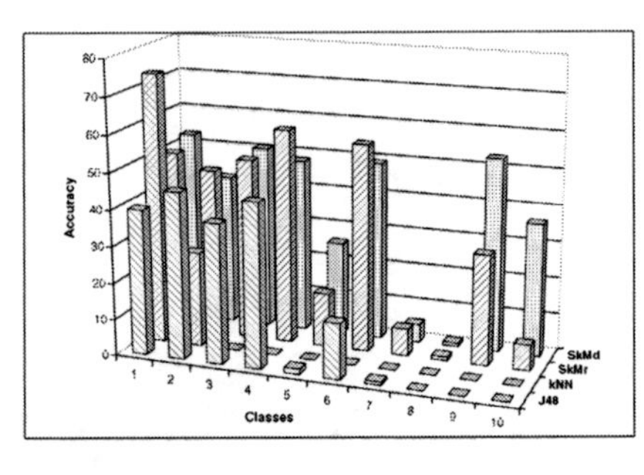


(c) Yeast 1% training data

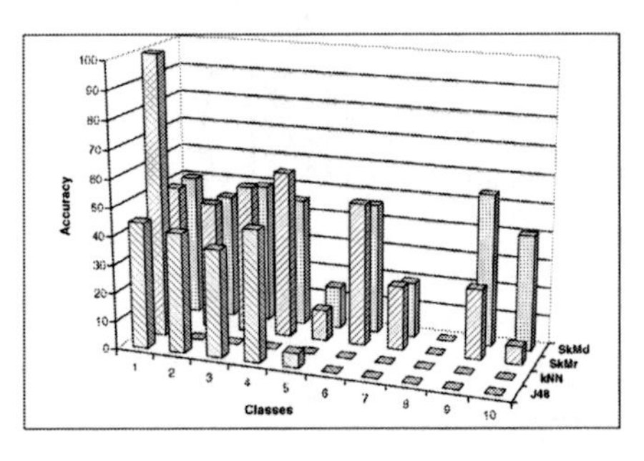


(d) Yeast 1% training data with missing classes

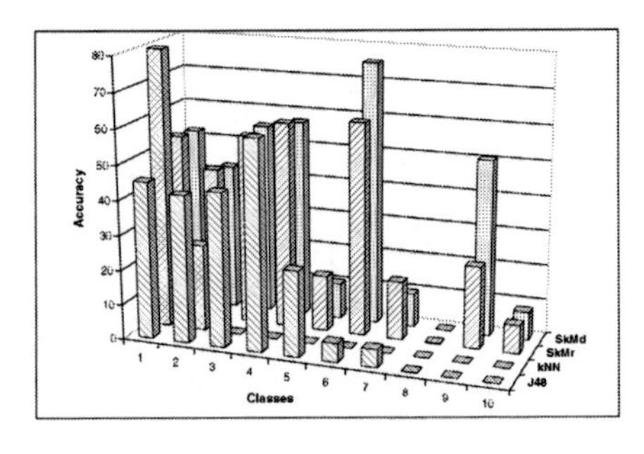


(e) Yeast 2% training data

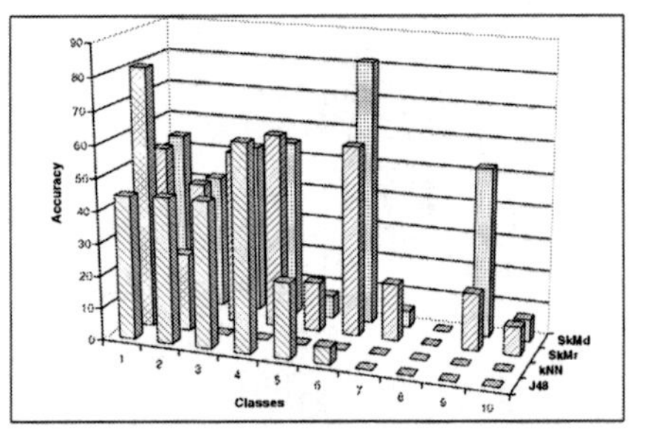


(f) Yeast 2% training data with missing classes

**Fig. 2.** Yeast dataset: Accuracy is computed based on test data only. Subfigures (a) and (b) indicate overall accuracy of each algorithm, and the error bars indicate maximum, minimum and average values across n-1 folds of test data. The rest of subfigures indicate overall accuracy regarding each individual class. In addtion, subfigures (d) and (f) are related to the case where class 8 (VAC), class 9 (POX) and class 10 (ERL) have been removed from the training data.

the number of the training data increases, the classifiers outperform both $SkM^d$ and $SkM^r$. This confirms that when training data are adequate, classification techniques are well suited for the tasks. Since $SkM^d$ performs better than $SkM^r$, we can be certain that the choice of seeds does make a difference in the outcome.

Fig. 1 (c) and (e) present the accuracy in each class with 5% and 10% of the training data respectively. The figures reveal that the classifiers fail to discover the rare classes, such as class 6 (omL), class 7 (imL), and class 8 (imS), while $SkM^d$ and $SkM^r$ successfully identify class 6 and class 7. For class 5 (om), $SkM^d$ and $SkM^r$ substantially perform better than the two classifiers. In particularly, the $SkM^d$ outperforms $SkM^r$ in rare classes. Since $SkM^d$ chooses the seeds that are furthest away from the known seeds, it has a higher chance of picking a seed that is close to the true center of the rare class. $SkM^r$ chooses a random seed for the missing class, this random seed could be actually in a known cluster and is a bad seed for the missing class. The reason that the classifiers does not perform well is that when the training data is sparse, the training dataset contains none or few data items belonging to rare classes.

In order to test the performance when the training dataset does not contain some rare classes, we remove the most scarce classes from the dataset. In the case of Ecoli dataset, the data of three classes 6, 7 and 8 are removed. The training and test data are generated as described before. Then the rare classes are added back to the test data. Fig. 1 (b)[1] shows that the classifiers perform better as the number of training data increases. Fig. 1 (d) and (f) show that the classifiers fail to discover any rare classes as expected even when the size of the training dataset increases. We conclude that $SkM^d$ and $SkM^r$ outperform classifiers for the rare classes and that $SkM^d$ outperforms $SkM^r$ in rare classes. Fig. 2 shows the results for Yeast dataset indicating similar trends as Ecoli dataset. Particularly, Fig. 2 (d) and (f) show that $SkM^d$ outperforms $SkM^r$ in class 9 (POX) and class 10 (ERL) but fails to identify class 8 (VAL). In [13], even when the whole dataset was used to construct kNN classifier, VAL could not be identified. Thus, we suspect that the training data related to class 8 are too similar to other classes.

## 4 Entropy-Based Semi-supervised Learning

As discussed in Section 3, when the training dataset is sparse (less than 70 data samples), seeded k-means can classify data more accurate than C4.5 and kNN algorithms. In addition, it can also identify more instances that belong to rare classes. Seeded k-means only utilizes the labeled instances at the initial stage of the algorithm, so here we propose a novel semi-supervised approach that uses labeled instances during each execution round to make a more reasonable and logical choice when assigning a data instance to clusters. We term this new approach as entropy k-means (EkM).

[1] Note that the numbers indicating the sizes of training data are slightly different between the top two sub-figures. This is because the underlying datasets are modified slightly to fit our experiments.

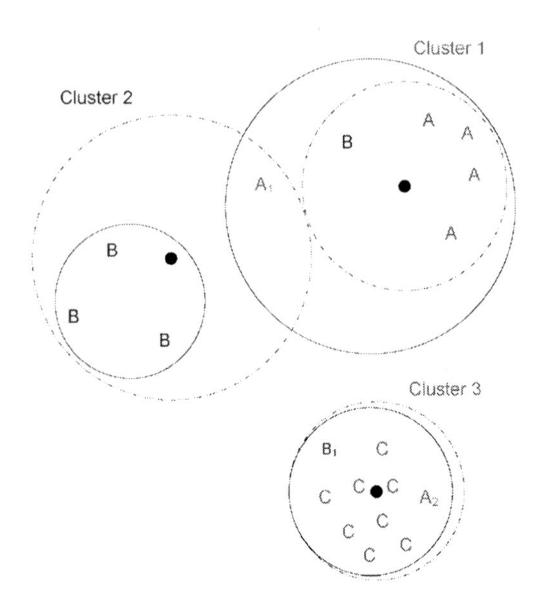


**Fig. 3.**

The intuition behind the EkM is shown in Figure 3. In this example, the broken line circle represents the clusters before one iteration and solid line circle represents the clusters after the iteration. The distance from the $A_1$ [2] to the center of cluster 1 is more than that of cluster 2. Under the k-means algorithm, $A_1$ will be assigned to cluster 2. However, since most data in cluster 2 have class label B and most data in cluster 1 have class label A, it is apparently more reasonable to assign $A_1$ to cluster 1. Because class labels of some data are known, we want the data having the same class labels stay in the same cluster as much as possible. Having this goal in mind, we incorporate entropy into our decision making process. EkM works as follows: given a small number of data items having class labels, EkM decides a cluster for a data item based on a *score* metric that combines both the distance-based similarity metric and the entropy of the cluster to which the item is added. The metric is defined as:

$$s_i^j = p \cdot D_i^j + q \cdot E_i^j \tag{1}$$

$$E_i^j = \sum_{m=1}^{n} -P_{mi}^j \lg P_{mi}^j \tag{2}$$

In Equation 1, $s_i^j$ represents the *score* of data item $t_i$ related to the center of cluster $C_j$, $D_i^j$ is the Euclidean distance between $C_j$'s center and $t_i$; $E_i^j$ is the entropy of $C_j$ when $t_i$ is added. If $t_i$ is labeled, $E_i^j$ is calculated, otherwise the entropy does not change. In Equation 2, $P_{mi}^j$ is the probability a given class

[2] $A_1$ has class label $A$. The subscription is used to identify data element.

has been assigned to cluster $C_j$. $n$ is the number of distinct classes in $C_j$. The lowest $s_i^j$ value indicates $t_i$ is assigned to $C_j$ during current iteration, and $p, q$ are coefficients to adjust the weight between distance and entropy. This *score* function determines if the distances to the centers of different clusters are similar, the cluster having lower entropy will win. Nevertheless, if the distance to a cluster is very small and adding the data item to this cluster will increase its entropy, this could indicate that the label actually is not correct. Consequently, EkM will allow this item to go into a different cluster with the correct class label. We expect that EkM will perform better in classifying and identifying rare classes.

The distance and entropy can offset each other. Fig. 3 shows a situation where distance could dominate entropy. $A_2$ and $B_1$ are assigned to cluster 3. Although we would like the data items having the same labels stay together, the distance from $A_2$ to cluster 1 is much larger than the distance to cluster 3. The same is true for $B_1$. In this case, the distance is too large to overcome and entropy has little impact on the score metric; as a result, $B_1$ and $A_2$ should be assigned to cluster 3 if we believe Euclidean distance is a good measure of similarity between objects. On the other hand, if the distance from $A_1$ to cluster 1 is close enough to that from $A_1$ to cluster 2, the entropy will guide $A_1$ to cluster 1.

Our work is still in preliminary stage, but it did show some promise on certain datasets. Several issues need to be solved before giving a full evaluation of EkM: using the *score* metric, convergence is not guaranteed because EkM is no longer an EM based algorithm. Also, how to decide the values for $p$ and $q$ (in equation 1) is another challenge. Our experiments conducted on Ecoli and Yeast datasets show the magnitudes of distance and entropy are very similar, so we set $p = q = 0.5$. In other words, distance and entropy have the same weight in calculation of the *score* metric. In general, we think $p$ and $q$ are dataset dependent. Furthermore, overfitting could cause potential problems in identifying rare classes when using classification techniques. Since our proposed approach is a combination of classification and clustering, our approach might be less likely to cause overfitting. We will investigate this issue extensively in the future.

Another issue is that using this as an enhancement on k-means only affects the center of the cluster. In reality, clusters may have different sizes or shapes; using the (limited) class data to adjust size/shape of clusters as well as the center would have even greater promise. We have started with a k-means basis due to the success of k-means clustering in our problem domain, but K-means algorithm assumes that K is known in advance, which may not be true for biological applications. (e.g., the number of types of mutants are not known). Density-based and hierarchical clustering algorithms are more suitable. We believe the entropy-based idea can be used to guide density-based or hierarchical clustering as well. The difficulty is to avoid over-reliance on the known data (leading to the same problem of not recognizing rare classes that standard classifiers face), while still getting full benefit. The simplicity of k-means makes this less of a problem; further research is needed to see how this can affect other techniques.

## 5 Conclusion / Future Work

We have shown that when rare classes have few instances or are completely missing in the training data, classification techniques using this training data perform poorly to identify rare classes. We also showed seeded k-means can be adopted in our problem domain, but the choice of seeds makes a difference. In the future, we will systematically and theoretically investigate the best ways to choose these seeds. Under the semi-supervised learning framework, we proposed a novel idea that incorporates entropy into the *score* metric to guide the clustering process. The preliminary results show some promise in identifying rare classes, and we will thoroughly investigate this idea and apply it to a real application in cell wall genomics. Since many clusters in biological data do not have a spherical shape, we will extend this idea into density-based clustering techniques.

## References

1. Chen, L., Carpita, N., Reiter, W., Wilson, R., Jeffries, C., McCann, M.: A rapid method to screen for cell-wall mutants using discriminant analysis of fourier transform infrared spectra. The plant Journal 16(3), 385–392 (1998)
2. Apweiler, R., Bairoch, A., Wu, C.H., Barker, W.C., Boeckmann, B., Ferro, S., Gasteiger, E., Huang, H., Lopez, R., Magrane, M., Martin, M.J., Natale, D.A., O'Donovan, C., Redaschi, N., Yeh, L.L.: Uniprot: the universal protein knowledgebase. Nucleic Acids Research 32, D115–D119 (2004)
3. Kretschmann, E., Fleischmann, W., Apweiler, R.: Automatic rule generation for protein annotation with the c4.5 data mining algorithm applied on swiss-prot. Bioinformatics 17(10), 920–926 (2001)
4. Karp, P.D.: What we do not know about sequence analysis and sequence databases. BioInformatics 14(9), 753–754 (1998)
5. Altschul, S.F., Madden, T.L., Schaffer, A.A., Zhang, J., Zhang, Z., Miller, W., Lipman, D.: Gapped blast and psi-blast: a new generation of protein database search programs. Nucleic Acids Res. 25(17), 3389–3402 (1997)
6. Basu, S., Banerjee, A., Mooney, R.J.: Semi-supervised clustering by seeding. In: ICML, pp. 27–34 (2002)
7. Bilenko, M., Basu, S., Mooney, R.J.: Integrating constraints and metric learning in semi-supervised clustering. In: ICML (2004)
8. Yu, H.: Svmc: Single-class classification with support vector machines. In: IJCAI, pp. 567–574 (2003)
9. Blake, C., Merz, C.: UCI repository of machine learning databases (1998)
10. Quinlan, J.R.: C4.5: Programs for Machine Learning. Morgan Kaufmann Publishers, San Francisco (1993)
11. Duda, R., Hart, P.E.: Pattern Classification and Scene Analysis. John Wiley & Sons, Chichester (1973)
12. Fukunaga, K.: Introduction to Statistical Pattern Recognition. Academic Press, San Diego (1990)
13. Horton, P., Nakai, K.: Better prediction of protein cellular localization sites with the *k* nearest neighbors classifier. In: Proc Int Conf Intell Syst Mol Biol., pp. 147–152 (1997)
14. Witten, I.H., Frank, E.: Data Mining: Practical Machine Learning Tools and Techniques with Java Implementations. Morgan Kaufmann, San Fransisco (1999)

# Clustering-Based K-Anonymisation Algorithms

Grigorios Loukides and Jianhua Shao

School of Computer Science
Cardiff University
Cardiff CF24 3AA, UK
{G.Loukides, J.Shao}@cs.cf.ac.uk

**Abstract.** K-anonymisation is an approach to protecting private information contained within a dataset. Many k-anonymisation methods have been proposed recently and one class of such methods are clustering-based. These methods are able to achieve high quality anonymisations and thus have a great application potential. However, existing clustering-based techniques use different quality measures and employ different data grouping strategies, and their comparative quality and performance are unclear. In this paper, we present and experimentally evaluate a family of clustering-based k-anonymisation algorithms in terms of data utility, privacy protection and processing efficiency.

## 1 Introduction

Advances in the Internet and data storage technologies have resulted in an increasing amount of data being produced and stored. Often, the collected data is used in studies such as healthcare research, business analysis and lifestyle surveys. However, if sensitive information contained within the data is not suitably protected, individuals' privacy, such as medical history and consumer preferences, may be revealed. Unfortunately, simply removing unique identifiers (e.g. names or credit card numbers) from data is not enough, as individuals can still be identified through a combination of non-unique attributes (often called quasi-identifiers or QIDs) such as age and postcode [14].

In response, k-anonymisation, a conceptually simple but powerful technique, has been proposed to address this issue. It works by organising data into groups of at least $k$ tuples and modifying the tuples in each group so that they share the same value in the set of QIDs. This makes individual identification through QIDs difficult, hence protects individuals' privacy. Many methods for k-anonymising data have been proposed [2,9,8,15] and one class of such methods are clustering-based [13,3,11,16,4]. These methods work by first grouping data into clusters using a quality measure and then generalise the data in each group separately to achieve k-anonymity. It has been shown that clustering-based methods are able to produce high quality anonymisations. However, existing clustering-based techniques use different quality measures and employ different data grouping strategies, and their comparative quality and performance are unclear.

Motivated by this, we study clustering algorithms in this paper by comparing them in terms of three criteria that we believe are essential to gauging the quality

R. Wagner, N. Revell, and G. Pernul (Eds.): DEXA 2007, LNCS 4653, pp. 761–771, 2007.

of a k-anonymisation and its derivation. These are *data utility* (the extent of information loss as a result of anonymisation), *data protection* (the extent to which individual identification is prevented), and *processing efficiency* (how fast k-anonymisations may be derived). Several representative clustering strategies are analysed and compared in terms of these criteria, which provide insight into the strength and weakness of the existing clustering-based algorithms.

The paper is organised as follows. Section 2 discusses optimality criteria for k-anonymisation. Section 3 presents several clustering-based k-anonymisation strategies, which are experimentally evaluated in Section 4. Finally, we conclude in Section 5.

## 2 Preliminaries

In this section, we first formally define k-anonymisation and then discuss optimality criteria for k-anonymising data.

**Definition 1 (K-anonymisation).** *K-anonymisation is the process in which a table* $T(A_1, \ldots, A_d)$*, where* $A_j, j = 1, \ldots, q$ *are quasi-identifiers (QIDs) and* $A_m, m = (q+1), \ldots, d$ *are sensitive attributes (SAs), is partitioned into groups* $\{g_1, \ldots, g_h\}$ *s.t.* $|g_i| \geq k, 1 \leq i \leq h$ *and tuples in* $g_i$ *have the same values in every* $A_j$*.*

Consider Table 2, for example. The values contained in *Age* and *Postcode* (QIDs) are anonymised in such a way that each tuple is made identical to at least other two tuples. Thus, Table 2 is 3-anonymised w.r.t. $\{Age, Postcode\}$. It is typical that values are anonymised by replacing specific QID values with more general ones, such as ranges (for interval attributes) or sets of values (for discrete attributes). For instance, the original values in *Age* in Table 1 have been replaced by [20-45] and [25-50], and the values in *Postcode* by {NW,SO} in Table 2. Obviously, this process incurs some information loss, but makes identification of private information difficult: an individual can not be associated with a specific value in *Disease* (SA) with a probability greater than $1/k$.

**Table 1.** Original data

| Age | Postcode | Disease |
|---|---|---|
| 20 | NW | HIV |
| 45 | SO | Cancer |
| 25 | NW | HIV |
| 21 | NW | HIV |
| 47 | SO | Cancer |
| 50 | SO | Cancer |

**Table 2.** An anonymisation of Table 1

| Age | Postcode | Disease |
|---|---|---|
| [20-45] | {NW,SO} | HIV |
| [20-45] | {NW,SO} | HIV |
| [20-45] | {NW,SO} | Cancer |
| [25-50] | {NW,SO} | HIV |
| [25-50] | {NW,SO} | Cancer |
| [25-50] | {NW,SO} | Cancer |

**Table 3.** Another anonymisation of Table 1

| Age | Postcode | Disease |
|---|---|---|
| [20-25] | NW | HIV |
| [20-25] | NW | HIV |
| [20-25] | NW | HIV |
| [45-50] | SO | Cancer |
| [45-50] | SO | Cancer |
| [45-50] | SO | Cancer |

Many different anonymisations are possible for a given table [4]. Thus, optimality criteria should be considered. First, anonymised data should remain useful for data analysis. Measuring information loss incurred by k-anonymisation is one way to capture data utility. For example, *Discernability Metric* (DM) [2] measures information loss as the sum of squared size of each anonymised group. Intuitively, a larger anonymised group may indicate more information loss, as more tuples are made indistinguishable. Alternatively, the size of the range or set used in recoding original values in a QID (i.e. the distance between its two furthest values) can be used to quantify information loss. For instance, Loss Metric [8], Ambiguity Metric [13], Information Loss Metric [15] and Usefulness Metric (US) [11] quantify data utility using the distances between the two furthest values in all QIDs of a group (i.e. group extent). Intuitively, groups that contain close QID values are preferable, as they tend to incur less information loss in anonymising data. Finally, how well anonymised data supports an intended task can also be used to indicate data utility [8,10,5]. A well-known measure specifically constructed for evaluating anonymised data intended for classification is the *Classification Metric* (CM) proposed in [8]. It measures the accuracy of a classifier built on anonymised data by counting the number of tuples whose class label is different from that of the majority of tuples in their group, as the labels of these tuples are not retained in the group after anonymisation.

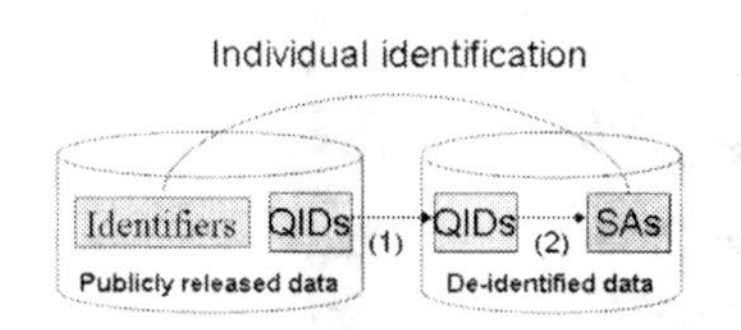


**Fig. 1.** Individual identification through linkage of data

Second, anonymised data should prevent individual identification. As discussed before, individuals can be identified when de-identified data is linked to publicly available data using QIDs (illustrated by link (1) in Figure 1). K-anonymisation attempts to weaken link (1) by grouping tuples together. However, the possibility of identification through linking data is not completely eliminated. For instance, when there is a strong correlation between anonymised QIDs and SAs (link (2) in Figure 1), sensitive information can be inferred in presence of some background knowledge [12]. In order to see this, observe that knowing that somebody aged 20 is included in Table 3, one can infer that this person suffers from HIV. In contrast, such an inference cannot be easily made in Table 2, as each group has 2 "well-represented" values in *Disease*. Machanavajjhala et al. [12] proposed *l-diversity* to measure protection based on how "diverse" SA values are, using the frequency of unordered distinct SA values in each group. Recently, protection measures that can be applied on attributes with ordered domains have been proposed as well [15,10,11]. In [11], for example, a measure called Protection Metric (PR) was proposed, which is expressed as the average distance of SA values in anonymised groups.

Third, anonymisations should be derived efficiently. Many formulations of the k-anonymisation problem have been shown to be NP-hard [9,3,16] and heuristic methods are typically employed. Many methods often differ substantially in terms of processing efficiency, as shown in our experiments.

## 3 Clustering-Based k-Anonymisation Methods

Clustering is a well-established technique for data analysis that attempts to divide data into groups of similar objects. However, the goal of clustering in the case of k-anonymisation is not just to find groups of similar data as pursued by traditional methods, but also requires that at least $k$ tuples are contained in each group and data remains "useful" after being anonymised. Many well-known methods for clustering are not applicable in this context and alternative approaches have been proposed [3,13,11,5,16,4]. They typically work as follows. A seed tuple is chosen and a cluster is built around the seed by adding tuples into the cluster until a stopping criterion is met. Deciding which tuples should be inserted into a cluster is determined by some optimality measures (e.g. one of those discussed in Section 2). This process is iteratively repeated until every tuple of the dataset is clustered. Then, tuples in each group are anonymised separately. Thus, a clustering method for k-anonymisation can be seen as built around three main components: seed selection, similarity measurement and stopping criterion.

**Seed selection.** Unlike many traditional clustering methods which select all seeds prior to forming clusters, clustering-based k-anonymisation methods tend to select one seed [13,1,3,16,11] or a pair of seeds [4] at a time. Seeds can either be chosen randomly [13,1,11] or using a furthest-first selection strategy [4,16,3]. That is, the most dissimilar tuple from the last selected seed [4,16] or from the last tuple added into a cluster [3] is selected as seed.

**Similarity measurement.** A key component in clustering-based algorithms for k-anonymisation is the measurement of similarity between the tuples in a cluster and candidate tuples that are being examined for insertion into the clusters. There are two different ways to define similarity. One is to compute similarity between the whole cluster and each candidate tuple (full linkage) [3,16]. Alternatively, a single tuple of a cluster (cluster representative) can be used instead of the whole cluster in similarity measurement. This tuple can be either a random tuple (single linkage) [11,1] or the cluster centroid (centroid linkage). Centroids can be constructed by averaging interval values or by using the median for discrete ones [4].

**Stopping criterion.** Many clustering-based methods employ a size-based stopping criterion that restricts the maximum cluster size, so that clusters of nearly $k$ tuples [3,16] are created, based on the intuition that large clusters do not help data utility. Alternatively, a quality-based stopping criterion is suggested in [11] that allows a cluster to be extended only when its quality does not exceed a user-specified threshold.

**Discussion.** We now comment on the clustering strategies in terms of runtime performance and quality.

Selecting cluster seeds randomly is clearly more efficient than using a furthest-first seed selection strategy, which requires quadratic time to the cardinality of the dataset. For example, a heuristic furthest-first seed selection strategy [4] finds a pair of seeds $t, t'$ in two steps. First, it finds the most dissimilar tuple $t$ from the most central tuple of the dataset (i.e. the centroid tuple of the whole dataset) and then the most dissimilar tuple $t'$ from $t$. This can find a pair of seeds can be found in linear time to the cardinality of the dataset. However, the number of seeds can be comparable to the cardinality of the dataset when $k$ is small, making furthest-first seed selection strategy expensive.

All linkage strategies have a quadratic time complexity to the cardinality of the dataset, since all candidate tuples are checked for insertion into a cluster every time a cluster is extended. However, sorting pairwise distances between the cluster representative and candidate tuples can speed up single linkage strategy. This is because a cluster can be formed by retrieving the $|c| - 1$ most similar tuples to a cluster representative (for single and centroid linkage strategies) in $O(|c| - 1)$ time ($|c|$ is the cluster size), and sorting requires log-linear time to $n$, the cardinality of the dataset. Thus, the complexity of single linkage clustering becomes $O(\frac{n^2}{|c|-1} log(n))$ , which is smaller than $O(n^2)$, as typically $|c| > log(n)$.

Furthermore, the quality-based stopping criterion has a higher computational cost compared to that of the size-based stopping criterion. In fact, that computational cost may not be negligible in practice. For example, assuming that evaluating quality of a cluster needs accessing all tuples currently in this cluster after each tuple insertion and that $\frac{n}{k}$ clusters comprised of exactly $k$ tuples have been formed, the cost of the quality-based criterion is $O(\frac{n}{k} \times (1 + \ldots + k)) \approx O(n \times k)$. Thus, the quality-based stopping criterion can be expensive when $k$ is large.

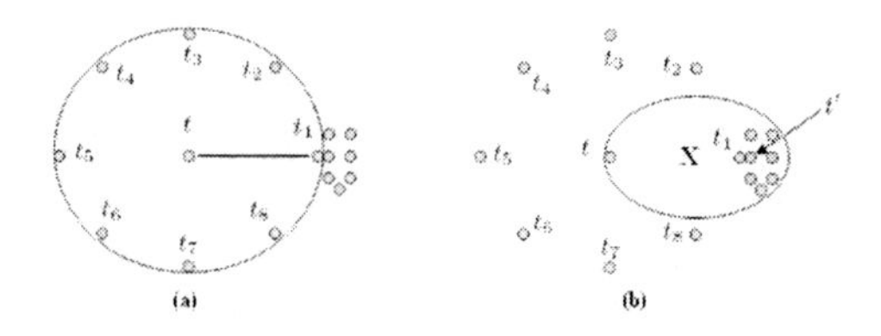


**Fig. 2.** The effect of linkage strategy on clustering quality

Linkage strategies also affect the quality of clustering. Representing a cluster by using only one tuple for example, may degrade the quality of clustering. Consider Figure 2(a) where single linkage is used. The representative tuple is denoted with $t$ and a size-based stopping criterion that creates clusters of 9 tuples is used. In this case, tuples $t_1$ to $t_8$ are added into the cluster (one at a time), as they are the closest to $t$. On the contrary, when full or centroid linkage is used, tuple $t'$ is added to the cluster after choosing $t_1$. This is because $t'$ becomes closer to the cluster (for full linkage) or to the centroid of the cluster (for centroid linkage). The cluster centroid is depicted with X in Figure 2(b). Subsequent iterations of the full or centroid linkage strategies result in creating the cluster depicted in Figure 2(b). This cluster is preferred, as it has a smaller extent than that of the cluster shown in Figure 2(a). However, in the case of large or arbitrary shaped clusters, it is difficult to represent a cluster using only one tuple as a centroid [6].

In addition to linkage strategy, seed selection does play an important role in generating a good clustering for k-anonymisation as well. Randomly selected seeds for example, may end up being close together, resulting in clusters with large extents. In order to see this, observe Figure 3(a), where close seeds (depicted as encircled points) resulted in the two clusters shown with different symbols, when single linkage is used. Unlike single linkage, full and centroid linkage strategies are less susceptible to bad seeds, as closeness to a candidate tuple is determined by the shape of the cluster, which changes when tuples are added into the cluster. Figure 3(b) depicts the clusters generated by centroid linkage. Observe that the position of the centroid (denoted with X) moved away from the centroid in the first two iterations. Thus, the created clusters are more separated than those of Figure 3(a). On the other hand, a furthest-first seed selection strategy achieves two compact and well separated clusters as shown in Figure 3(c), where the previously described heuristic was used for seed selection (seeds are depicted as encircled points).

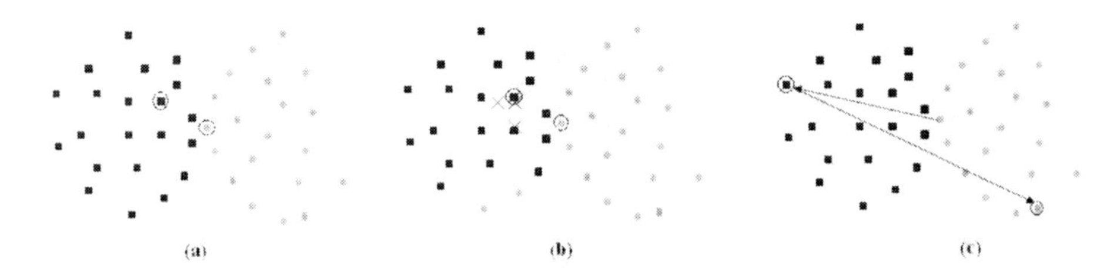


**Fig. 3.** The effect of seed selection strategy on clustering quality

Finally, the stopping criterion can also affect the quality of anonymisations. Intuitively, creating compact, equal-sized clusters using a size-based stopping criterion can help data utility. However, this depends on how anonymised data is used. For example, equal-sized clusters may degrade classification accuracy [5]. The quality-based stopping criterion is more flexible, as it is expressed in terms of optimality measures. For instance, special requirements implied by a security policy or application, which can be expressed as the maximum distance between values in a group, can be modelled.

## 4 Experimental Evaluation of Clustering Strategies

In this section we experimentally evaluate clustering-based k-anonymisation methods. For our experiments, we have implemented two well-known clustering-based algorithms called Greedy Clustering [11] and K-Members [3], which combine some of the strategies described in Section 3. For convenience, we refer to the first algorithm as TSR (Threshold-based stopping, Single linkage and Random seed selection). This algorithm is expected to achieve anonymisations of low quality (due to random seed selection and single-linkage) but will be very efficient (due to single linkage). We also refer to K-Members as SFF (Size-based stopping, Full linkage and Furthest-first seed selection). SFF allows us to examine the effect that furthest-first seed selection and full linkage strategies have on the quality and efficiency. Due to these two strategies, SFF is expected to achieve anonymisations of high quality but will be less efficient than TSR. Finally, we developed SCF (Size-based stopping Centroid linkage Furthest-first seed selection). SCF differs from SFF only in the employed linkage strategy, thus it allows us to compare the centroid and full linkage strategies. As discussed in Section 3, centroid linkage can be worse than full linkage in terms of quality but it is more efficient. Thus, we expect SCF to lie in between TSR and SFF in terms of data quality and processing efficiency. We have used a heuristic [11] that captures information loss using the size of ranges or sets used in recoding QID values for all three algorithms. Different heuristics [3,16] can also be used but for sake of space we do not report these experiments in this paper. Furthermore, we compare these clustering-based algorithms to Mondrian [9], a very effective non-clustering based algorithm, which works in a way reminiscent to kd-tree construction. It uses a search strategy which recursively splits a group of tuples at the median value of the QID that has the largest range of values, until the resultant partitions contain at least $k$ but no more than $2k-1$ tuples.

We have used two datasets in our experiments. The first dataset is a projection of the Adults dataset [7] on 6 attributes: Age, Race, Marital Status, Salary, Education and Occupation. This dataset has 30718 tuples after removing tuples with missing values. We treat Education and Occupation as SAs. We also used synthetic data, generated from a standard normal distribution. This dataset has 8000 tuples and 6 attributes, two of which are discrete. We treat one interval and one discrete attribute as SAs. All the algorithms were implemented in Java and ran on a Pentium-D 3GHz machine with 1 GB of RAM under Windows XP.

**Quality evaluation.** We evaluated the quality of k-anonymisations in terms of three utility measures applying the algorithms on the Adults dataset. First, we used the DM measure. As illustrated in Figure 4, both SFF and SCF achieve very good results with respect to DM compared to that of TSR. This is because the size-based stopping criterion used by SFF and SCF creates small anonymised groups of no more than $2k-1$ tuples. Mondrian also achieves a good result for that reason. In contrast, the quality-based stopping criterion (expressed as a maximum allowed US value) used in TSR does not pose any restriction in terms of cluster size and thus creates larger clusters. We then used the US measure to evaluate data utility. The result is shown in Figure 5. As expected, SFF achieved the best result due to the full linkage strategy. In order to see this, observe that the US values achieved by SFF are much better than those achieved by SCF (SFF and SCF differ only in the linkage strategy). The US values achieved by SCF, TSR and Mondrian were comparable for small $k$'s (up to $k=15$), while for larger $k$'s TSR outperformed both SCF and Mondrian. This is because clusters cannot

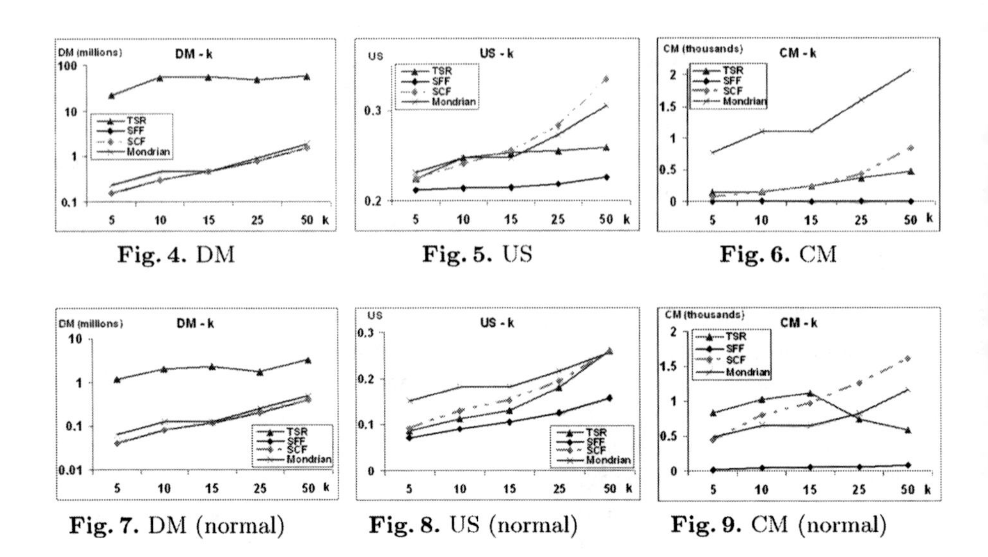


**Fig. 4.** DM **Fig. 5.** US **Fig. 6.** CM

**Fig. 7.** DM (normal) **Fig. 8.** US (normal) **Fig. 9.** CM (normal)

be accurately represented by centroids for large $k$'s and thus SCF performed as poorly as Mondrian did. Finally, we used the CM measure, retaining *Salary* as the target attribute for classification. As illustrated in Figure 6, SFF continues to perform better (lower CM values are preferred) than both SCF and TSR due to the full linkage strategy. However, both SCF and TSR outperform Mondrian. This is because Mondrian hardly splits data using *Salary* (0.4%-13% of the total cuts), and as a result groups tend to have a lot of different values in this attribute.

We repeated the same experiments with the synthetic dataset. The result for DM illustrated in Figure 7 is similar to that of Figure 4, as TSR created large groups due to the quality-based stopping criterion. Evaluating data utility using US, we observed that again SFF outperforms both TSR and SCF, similarly to when real data was used. However, all clustering-based algorithms perform significantly better than the Mondrian. The reason is that when data is skewed, the median-splitting criterion employed by Mondrian creates elongated clusters that have very high (bad) US values. Indeed, when data skewness is reduced (i.e. by increasing the standard deviation of the normal distribution) the gap between clustering-based algorithms and Mondrian becomes smaller. Due to space limitation, we do not report the full result of this experiment. The result for CM is illustrated in Figure 9. Again, SFF outperforms both TSR and SCF, due to the full linkage strategy. Observe that Mondrian also performs reasonably well. This is because, most splits were done on the attribute used for evaluating CM. Thus, the resultant groups were similar with respect to this attribute.

We then studied the protection using the entropy l-diversity [12] and PR [11] measures. For the Adults dataset, we used *Occupation* for computing l-diversity. Figure 10 depicts the result. It is easy to see that TSR achiever better protection for all values of $k$. This is because the quality-based threshold employed by TSR created large groups with many diverse values in the sensitive attribute in this experiment. Note however, that large groups cannot always guarantee protec-

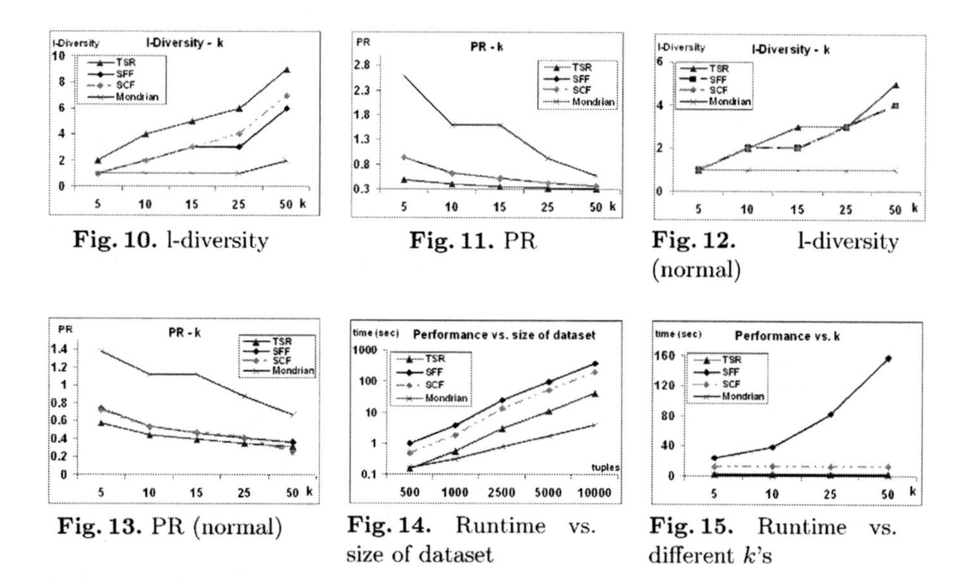


**Fig. 10.** l-diversity

**Fig. 11.** PR

**Fig. 12.** l-diversity (normal)

**Fig. 13.** PR (normal)

**Fig. 14.** Runtime vs. size of dataset

**Fig. 15.** Runtime vs. different $k$'s

tion and special attention in handling sensitive attributes is required [12,15]. In contrast, Mondrian, SFF and SCF achieve a very low level of l-diversity particularly for small $k$'s. For instance, Mondrian creates groups which do not offer any protection (i.e. they contain the same value in *Occupation*) for all $k$ values up to 25. As for PR, the algorithms achieved a similar result illustrated in Figure 11 (low values in PR are preferred). Our results indicate that protection can in practice be unacceptably low, especially when optimising towards data utility and QIDs are strongly correlated to the SAs as in this dataset. Finally, we evaluated data protection using synthetic data. As can be seen in Figures 12 and 13, all algorithms achieve a low level of protection as measured by l-diversity and PR respectively, although there is no correlation between QIDs and SAs in this dataset. This is because many tuples have similar SA values due to large data skewness and thus groups that offer low level of protection are formed.

**Performance evaluation.** We also evaluated the runtime performance of the algorithms. First, we used random samples of the Adults dataset with a cardinality $n$ ranging from 500 to 10000 fixing k to 5. As illustrated in Figure 14, Mondrian is faster by at least one order of magnitude than all clustering-based algorithms. This is because, in contrast to clustering-based algorithms which have a quadratic time complexity, the complexity of Mondrian is log-linear with respect to the cardinality of the dataset. TSR is the most efficient, SFF the least and SCF is lying in between of the two. These results confirm our theoretical analysis in Section 3. Second, we tested clustering-based algorithms fixing $n$ to 2500 and varying $k$. As can be seen in Figure 15, SFF is sensitive to $k$ being more expensive than TSR and SCF when a large $k$ is used. This again confirms the computational overhead that the full linkage brings. Furthermore, SCF is slower than TSR due to the additional overhead imposed by the centroid computation, while TSR and SCF are not significantly affected by $k$. This is because, we sorted pairwise distances between cluster representatives and candidate tuples prior to forming clusters, as discussed in Section 3, and centroid computation was not significantly affected by $k$.

## 5 Conclusions

Recent research has recognised the power of clustering-based algorithms for creating high quality anonymisations when compared to alternative methods. However, the superiority of certain clustering strategies and their performance under different quality measures had not been clearly shown. In this paper, we performed a comparative study of different strategies employed by well-known clustering techniques and demonstrated their effectiveness and efficiency using both real world and synthetic data. We showed that each of the three key components of a clustering-based algorithm (seed selection, similarity measurement and stopping criterion) can have significant impact on both quality and performance. In particular, furthest-first seed selection and full linkage strategy tend to be good for data utility, whereas random seed selection and single linkage enhance processing efficiency. However, combining furthest-first seed selection with centroid linkage can offer a good balance between data utility and performance as long as $k$ is small. Finally, we observed that the quality-based stopping criterion is flexible but can be expensive to compute, and many clustering-based algorithms [16,13] with the exception of [11,3] optimise data utility but may not sufficiently protect from individual identification.

## References


1. Aggarwal, C.C., Yu, P.S.: A condensation approach to privacy preserving data mining. In: Bertino, E., Christodoulakis, S., Plexousakis, D., Christophides, V., Koubarakis, M., Böhm, K., Ferrari, E. (eds.) EDBT 2004. LNCS, vol. 2992, pp. 183–199. Springer, Heidelberg (2004)
2. Bayardo, R.J., Agrawal, R.: Data privacy through optimal k-anonymization. In: ICDE '05, pp. 217–228 (2005)
3. Byun, J., Kamra, A., Bertino, E., Li, N.: Efficient k-anonymization using clustering techniques. In: DASFAA '07, pp. 188–200 (2007)
4. Domingo-Ferrer, J., Torra, V.: Ordinal, continuous and heterogeneous k-anonymity through microaggregation. Data Mining and Knowledge Discovery 11(2), 195–212 (2005)
5. Fung, B.C.M., Wang, K., Yu, P.S.: Top-down specialization for information and privacy preservation. In: ICDE, pp. 205–216 (2005)
6. Guha, S., Rastogi, R., Shim, K.: ROCK: A robust clustering algorithm for categorical attributes. Information Systems 25(5), 345–366 (2000)
7. Hettich, S., Merz, C.J.: Uci repository of machine learning databases (1998)
8. Iyengar, V.S.: Transforming data to satisfy privacy constraints. In: KDD '02, pp. 279–288 (2002)
9. LeFevre, K., DeWitt, D.J., Ramakrishnan, R.: Mondrian multidimensional k-anonymity. In: ICDE '06, p. 25 (2006)

10. LeFevre, K., DeWitt, D.J., Ramakrishnan, R.: Workload-aware anonymization. In: KDD '06, pp. 277–286 (2006)
11. Loukides, G., Shao, J.: Capturing data usefulness and privacy protection in k-anonymisation. In: SAC '07, pp. 370–374 (2007)
12. Machanavajjhala, A., Gehrke, J., Kifer, D., Venkitasubramaniam, M.: l-diversity: Privacy beyond k-anonymity. In: ICDE '06, p. 24 (2006)
13. Nergiz, M.E., Clifton, C.: Thoughts on k-anonymization. In: 22nd International Conference on Data Engineering Workshops (ICDEW'06), p. 96 (2006)
14. Sweeney, L.: k-anonymity: a model for protecting privacy. International Journal on Uncertainty, Fuzziness and Knowledge-based Systems 10, 557–570 (2002)
15. Xiao, X., Tao, Y.: Personalized privacy preservation. In: SIGMOD '06, pp. 229–240 (2006)
16. Xu, J., Wang, W., Pei, J., Wang, X., Shi, B., Fu, A.W-C.: Utility-based anonymization using local recoding. In: KDD '06, pp. 785–790 (2006)

# Investigation of Semantic Similarity as a Tool for Comparative Genomics

Danielle Welter[1], W. Alexander Gray[1], and Peter Kille[2]

[1] School of Computer Science, Cardiff University, Cardiff, UK
[2] School of Biosciences, Cardiff University, Cardiff, UK
D.N.Welter@cs.cardiff.ac.uk, W.A.Gray@cs.cf.ac.uk, kille@cf.ac.uk

**Abstract.** The project sets out to investigate the concept of semantic similarity between individual and collections of gene products based on functional descriptors such as "Gene Ontology" (GO) annotations. Different existing concepts for quantifying semantic similarity are implemented into a basic "Semantic Similarity Calculator" and the resulting tool applied to assess its utility in different biological contexts. It is discussed what kinds of problems were encountered during the implementation of the prototype, and how these problems were addressed, or are planned to be addressed in the future. An overview over future work is given.

## 1 Introduction

### 1.1 The Gene Ontology

The last decade has seen an exponential increase in the availability of biological data such as DNA sequences, protein sequences and gene expression data. In addition to the raw experimental data, there are annotations that put the data into context, i.e. by describing the function of proteins, mutations associated with certain genes, conditions associated with mutations and other factors. In the light of this wealth of information, the need has arisen to process information computationally. However, most of the annotation data is available in formats that are readily accessible to humans but do not readily lend themselves to computer processing.[1, 2, 3, 4]

The understanding gained from this new data has revealed that the core mechanisms of life are common to all eukaryotic organisms, from simple yeast cells to complex mammals.[2] However, each research field or species group has its own conventions for annotating and storing data due to historical research origins. This has led to the problem that, although biological data is readily available in the public domain, its distribution through a wide variety of heterogeneous data sources makes its computational analysis very difficult.

Changing and homogenizing all the existing knowledge bases would however be a Herculean task that would most likely cause more confusion than it would contribute to solving this problem. The development of a common controlled vocabulary that can be used to annotate the different data sources with the same terms represents a much better solution. To this end, in 1998, the model organism databases Flybase, Mouse

R. Wagner, N. Revell, and G. Pernul (Eds.): DEXA 2007, LNCS 4653, pp. 772–779, 2007.

Genome Informatics and the *Saccharomyces* Genome Database began a collaborative project, the Gene Ontology (GO) Consortium.[2] The Consortium's mission statement is *"to produce a structured, precisely defined, common, controlled vocabulary for describing the roles of genes and gene products in any organism."*[2, 5]

Since the early days of the GO, the number of member databases has more than quadrupled. A complete list of members can be found under [6].

The GO actually consists of three orthogonal ontologies, rather than a single hierarchy, as one might expect at first. The three ontologies or "sub-ontologies" are:

- "molecular function", which refers to the biochemical activity of a gene product;
- "biological process", which is the larger context or pathway that a gene product contributes to, and
- "cellular component", which refers to the location within the cell at which the gene product is active.[7]

Each ontology is structure as a directed acyclic graph (DAG), i.e. each parent can have more than one child, and each child can have more than one parent [7], but circular relationships between terms are not allowed [8]. Relationships between terms belong to one of the followings two types:

- "is-a" links between parent and child, which make up the majority of links;
- "part-of" links between part and whole.[7]

The Gene Ontology is updated daily and currently comprises over 22500 terms. About 96% currently have definitions.[5]

### 1.2 Semantic Similarity

A number of similarity measures have been exploited in molecular biology to compare gene products, such as DNA sequence similarity, amino acid sequence similarity, and protein family similarity (based on the presence of certain 3D structures in proteins). During the last few years, another form of similarity has become of interest: comparing individual or groups of gene products based on their "meaning", i.e. their functional annotation ("*what they do, where they do it and how they do it*" [9]). This concept is termed "semantic similarity", and has been employed for decades in domains such as artificial intelligence, psychology and natural language processing.[10].

In 2002, P.W. Lord et al.[3] were the first to publish a study on the use of semantic similarity measures in the context of the Gene Ontology. They used the method described by Resnik (1995)[10] to quantify the semantic relationship between different proteins based on the information content of a common ancestor of two proteins.

The information content of a concept c is defined as

"*negative the log likelihood*, **-ln p(c)**"[10],

where p(c) is the probability of a concept c occurring in a taxonomy. This probability is determined as:

$$p(c) = \frac{freq(c)}{N} \tag{1}$$

where

- $freq(c) = \sum_{n \in concepts(c)} total(n)$;
- concepts(c) is the set of concepts that are descendants of c;
- total(n) is the number of occurrences of term n in the corpus;
- N is the total number of terms in the corpus.

As two terms may have more than one common ancestor, the most meaningful of those ancestors should be considered. This is generally the first common ancestor, i.e. the ancestor with the smallest p(c). Lord et al. called this the

*"probability of the minimum subsumer"* [3]

$$p_{ms} = \min\{p(c)\} \quad (2)$$

Therefore, similarity between concepts $c_1$ and $c_2$ according to Resnik [10] is given by

$$similarity(c_1, c_2) = -\ln p_{ms}(c) \quad (3)$$

In 2003, Lord's group published another paper [4] in which they also investigated the information content approach of Lin [11], which differs from the Resnik method insofar that it also considers the information content of the query terms themselves, as well as a distance-based approach proposed by Jiang and Conrath [12].

Similarity according to Lin is given, in equation (4), as

$$similarity(c_1, c_2) = \frac{2 * \ln p_{ms}(c)}{\ln p(c_1) + \ln p(c_2)} \quad (4)$$

while Jiang's approach in equation (5) represents the semantic distance between two terms, which is the inverse of the semantic similarity.

$$distance(c_1, c_2) = -2\ln p_{ms}(c) - (\ln p(c_1) + \ln p(c_2)) \quad (5)$$

The group's findings [3, 4] revealed a significant correlation between semantic similarity and sequence similarity, i.e. two gene products that were strongly semantically similar were generally found to have a high degree of sequence similarity. Out of the three approaches to quantifying semantic similarity, none was found to significantly outperform the other two, although each approach presented some advantage that the other two lacked. The Resnik approach was found to give identical similarity scores when comparing a term $\beta$'s parent $\alpha$ with some other term $\gamma$, as when comparing $\beta$ and $\gamma$ directly because it does not take into account the query terms, but only their common ancestor. This problem did not occur with the Lin approach, which takes into account the information content of both query terms and their common ancestor. On the other hand, Lin is bounded between 0 and 1, and only really performs well if a large corpus is underlying the calculations. As GO is constantly growing, this is not a problem in this context. The Jiang approach scored the lowest correlation for the "molecular function" aspect, but like Lin's approach, it does have the benefit of considering the information content of query terms and shared parents. Considering these findings, it was decided to use all three measures in the present project. However, we propose that it might be of greater interest for researchers to find semantically similar gene products from different species than only consider gene products within the human species.

## 2 The Semantic Similarity Calculator

### 2.1 Design and Implementation

An application was implemented to compare a single given gene product from one species with all annotated products from a different species to determine whether this approach might allow a better analysis and interpretation of biological data than the approach that only compares pairs of given gene products.[9]

In a first phase, the nematode worm *C.elegans* [13] and the mouse *M. musculus* [14] were chosen as ideal model organism. The program prototype was designed in a way to ensure that addition of further species could be done easily and without the need to rewrite much code. In the near future, the fruit fly *D.melanogaster* [15], as well as the human will be added to list of species available for semantic comparison. It is also possible to use this software to do comparisons within a species.

Lord et al. [4] stated that the correlation between the three different ontologies of GO was of little significance, i.e. the orthogonal ontologies are largely independent of each other. Although we considered that there might be some cases in which researchers would want to look for high semantic similarity scores across all three ontologies, especially when comparing gene products from two or more different species, it was decided to follow Lord's conclusion at this stage of our investigation and only use one the sub-ontologies per query.

The implementation of the semantic similarity calculator was done in the Java programming language, and the GO data and species-specific gene product annotations were downloaded and stored in a MySQL database. Any kind of database could have been used for the species-specific annotations as these can be downloaded from the GO download website [5, 16] as flat files, but the full GO database is only available in two formats, namely GO RDF-XML and MySQL, and only the latter contains the full database, including electronically inferred annotations (IEA evidence code). Connection between the program and the database occurred through the use of JDBC.

After entering a query term, the GO terms associated with that query term for the ontology selected are returned to the user. The user has to decide which term the query is run on, as it was decided that this was the best way to ensure the query is run in the desired context of a researcher. The selection of a single GO term for comparison is based on the assumption that a researcher wants to find semantically similar terms based on a specific characteristic of the query term. Another, equally valid assumption states that a protein has all its functions at the same time, which would be expressed by computing the average similarity of all the GO terms annotated to the query protein. This approach will be investigated in the future.

Once the query GO term has been determined, the program retrieves all the gene products of the comparison species annotated with one or more terms from the right ontology. It then goes through the list term by term. For each term, the common ancestor between that term and the query term is determined using a complex retrieval statement that was directly taken from the GO help website and only slightly modified to fit its intended purpose. The common ancestor term is compared to a list of GO terms that are judged to be too general, i.e. not informative enough to produce any significant results. If the common ancestor term corresponds to any of the terms in that list, the term is disregarded, and the program passes to the next comparison term.

If the common ancestor is appropriate, the different probabilities (depending on the calculation approach selected by the user) are calculated. This involves several MySQL queries, the results of which are returned to the program for processing. After the semantic similarity coefficient has been calculated, the program moves on to the next term.

Overall, there are quite a large number of database accesses in the program. At the time of design, this was not considered to be a problem as the aim was to investigate whether the approach was promising.

## 2.2 Testing

Initial tests of the semantic similarity calculator revealed runtimes in the order of 24 to 36 hours, due to the very large amount of comparisons that need to be done and to the limitation of the technology being used. Adding indexes to the species tables, which did not have indexes before this, speeded the program up 100fold, giving run times of between 15 minutes and 3 to 4 hours. The slower queries were however still deemed to be a problem as end users would not want to wait for such lengths of time for their results. The reason for the excessive runtimes was found to be in the access to the database server, where a bottleneck situation occurred. This problem was provisionally remedied by moving the database to a local machine instead of accessing it over a network, which reduced the run time by almost half.

This can however not be a permanent solution. The semantic similarity calculator is expected to eventually be available to all members of the scientific community, and should ideally even work through a web interface. In this situation, downloading a multi-megabyte database to a local machine is highly impractical.

Another option, which was being worked on at the time of submission of this paper, was to limit the number of times the database server needs to be accessed by either retrieving large amounts of data at a time, or performing some of the data processing within the database, rather than returning the data to the program for handling.

To ensure that the different methods had been implemented correctly and that the prototype generated biologically meaningful results, a three-step testing strategy was designed:

- Stage 1 involves testing gene products from a well-documented pathway, such as the TCA cycle [1], known to be conserved in mouse and nematode worm
- Stage 2 involves testing gene products from a well-documented pathway that is known not to be fully conserved in the two species, such as fatty acid metabolism [1]
- Stage 3 involves testing the gene products of a pathway that is known to be completely unconserved between mouse and worm.

As this project is a work-in-progress, only stage 1 of this testing strategy has been undertaken at the time of writing to determine whether the approach is promising. Evaluation of the results of this phase showed that the expected gene products scored the highest similarity coefficients for the Resnik and Lin approaches, and in each of the three ontologies. A sample result set is shown in table 2, with the mouse terms used to run the query, and their chosen GO terms in table 1. Only results obtained through the Lin approach are shown as these are the easiest to evaluate because they are bounded between 0 and 1.

As the Jiang approach is distance-based, i.e. the closer two terms are in the hierarchy, the more semantically similar they are, it would be expected that the terms with the highest similarity would obtain the lowest score values. Initial test with the semantic similarity calculator revealed results which followed the opposite trend, i.e. the most similar terms had the highest scores. It was determined that this was due to an error in the implementation for the Jiang approach, although the exact nature of this error has not yet been identified at this time. Until the problem with the Jiang approach has been resolved, the measure will be excluded from further tests.

From the data shown in tables 1 and 2, a particularly noteworthy observation can be made in the annotations of the mouse gene products for the "Cellular Component" ontology: one term is annotated with "cytoplasm" (GO:0005737), while the other is annotated with "mitochondrion" (GO:0005739), even though the two gene products are actually involved in the same biological pathway, the TCA cycle. In fact,

**Table 1.** Query mouse gene products, with the selected GO term for each ontology

| *Mouse Gene Product* | *GO Ontology* | *GO Term* | *Definition* |
|---|---|---|---|
| MGI:87879 | MF | GO:0003994 | Aconitate hydratase activity |
| | CC | GO:0005737 | cytoplasm |
| | BP | GO:0006099 | Tricarboxylic acid cycle |
| MGI:88529 | MF | GO:0004108 | Citrate (Si)-synthase activity |
| | CC | GO:0005739 | mitochondrion |
| | BP | GO:0006099 | Tricarboxylic acid cycle |

"mitochondrion" is a child of the term "cytoplasm", i.e. it is more detailed than its parent. The same situation exists for mouse gene product "MGI:87879" and its worm equivalents: the worm gene products are annotated with the more detailed "mitochondrion", leading to a seemingly lower similarity score of 0.811 rather than 1 because of the lack of detail in the mouse term annotation.

These examples illustrate how the level of detail with which gene products are annotated varies greatly. This is an issue in the whole area of bioinformatics: any biological data is annotated at a level of detail known to the annotator. Annotations are not set in stone, and will become more detailed as new knowledge is gained. In many cases, evaluation by a human is required to determine whether or not a result reflects a less than ideal annotation or is caused by a wrong assignment of an annotation.

**Table 2.** Worm gene products expected to correspond to each query mouse gene product. Note that some worm gene products are annotated with more than one GO term from a given ontology. This can be because a protein may have the different functions, or be involved in more than one biological process, or because new knowledge about a protein was gained and a more detailed annotation was made to reflect this.

| *Mouse Gene Product* | *Expected Worm Gene Products* | *MF* | | *CC* | | *BP* | |
|---|---|---|---|---|---|---|---|
| | | *GO Term* | *Score* | *GO Term* | *Score* | *GO Term* | *Score* |
| MGI:87879 | WP:CE25005 | GO:0003994 | 1 | GO:0005739 | 0.811 | GO:0006099 | 1 |
| | | GO:0016836 | 0.853 | | | | |
| | WP:CE30144 | GO:0003994 | 1 | GO:0005739 | 0.811 | GO:0006099 | 1 |
| | | GO:0016836 | 0.853 | | | | |
| | WP:CE32436 | GO:0003994 | 1 | GO:0005739 | 0.811 | GO:0006099 | 1 |
| | | GO:0016836 | 0.853 | | | | |
| | WP:CE03812 | GO:0003994 | 1 | n/a | n/a | n/a | n/a |
| | | GO:0016836 | 0.853 | | | | |
| MGI:88529 | WP:CE000513 | GO:0004108 | 1 | GO:0005737 | 0.812 | GO:0006099 | 1 |
| | | GO:0046912 | 0.916 | | | GO:0006092 | 0.878 |

# 3 Conclusions and Future Work

The results obtained from the semantic similarity calculator prototype have demonstrated that the semantic similarity concept originally investigated by Lord et al. [3, 4] shows great potential when applied to compare similarity across species. Even though the implemented prototype is still far from being utilizable as an end user tool, proof of concept has been delivered by the results that were as expected. Ongoing testing will show whether this promising start is also going to be successful for more complex, and less well-studied, and therefore less predictable, gene products. A number of modifications of the initial prototype are expected to be necessary before it can function as an end user tool. In particular, the advantages and disadvantages of evaluating single-term similarity versus average similarity require further evaluation.

In addition to the approach used so far, i.e. the comparison of a single gene product on a list of gene products, it is planned to allow comparison of a whole list of gene products, for example obtained from gene expression experiments, on both inter- and intra-species level. It is also planned to combine the semantic similarity measure with other existing measures such as sequence similarity, family similarity and over- and under-representation of terms in gene datasets. This is expected to allow more efficient and diversified analysis and interpretation of biological data than existing approaches that only look at one of these aspects.

On a computational level, the challenge of this project is to study the interoperation of data in different databases that do not necessarily follow identical standards, in order to integrate these different sources into one integrated search and calculation process. This ties in with current research at Cardiff University, which has been developing a system that uses as "Soft Link Method" (SLM) [17] to enable different approaches to statistical linkage of bioinformatics data sources to be run through a single system. One option to resolve the run time problems encountered during the early phases of this project is to investigate the use of GRID enabled resources, in particular the Cortex 2 Grid-enabled Pattern Matching Engine at the Welsh e-Science

Centre. This is expected to enable hardware matching of the annotations which will speed up this process.

It is expected that this project will find new and improved ways for comparative genomic analysis to allow novel insights into a number of aspects that are currently at the forefront of advanced biological research. Computational prediction of results will be made more efficient, allowing scientists to avoid the waste of precious time and resources on "wet" experiments by accurately evaluating which experiments are worth undertaking.

## References


1. Berg, J.M., Tymoczko, J.L., Stryer, L.: Biochemistry. W. H. Freeman and Co., New York (2002)
2. The Gene Ontology Consortium: Gene Ontology: Tool for the Unification of Biology. Nature Genet. 25, 25–29 (2000)
3. Lord, P.W., Stevens, R.D., Brass, A., Goble, C.A.: Investigating Semantic Similarity Measures across the Gene Ontology: The Relationship between Sequence and Annotation. Bioinformatics 19, 1275–1283 (2003)
4. Lord, P.W., Stevens, R.D., Brass, A., Goble, C.A.: Semantic Similarity Measures as Tools for Exploring the Gene Ontology. In: Proc. Pacific Symp. Biocomputing, vol. 8, pp. 601–612 (2003)
5. Gene Ontology general documentation, http://www.geneontology.org/doc/GO.doc.html
6. The GO Consortium, http://www.geneontology.org/GO.consortiumlist.shtml
7. The Gene Ontology Consortium. Creating the Gene Ontology resource: Design and implementation. Genome Res. 11, 1425–1433 (2001)
8. Sevilla, J.L., Segura, V., Podhorski, A., Guruceaga, E., Mato, J.M., Martinez-Cruz, L.A., Corrales, F.J., Rubio, A.: Correlation between Gene Expression and GO Semantic Similarity. IEEE TCBB 2(4), 330–338 (2005)
9. Welter, D.: Development of a System for Comparison of Gene Products through Semantic Similarity based on their Gene Ontology Annotation. Thesis (MSc), Cardiff University (2006)
10. Resnik, P.: Using Information Content to Evaluate Semantic Similarity in a Taxonomy. In: Proc. 14th Int'l Joint Conf. Artificial Intelligence, pp. 448–453 (1995)
11. Lin, D.: An Information-Theoretic Definition of Similarity. In: Proc.15th Int'l Conf. Machine Learning, pp. 296–304 (1998)
12. Jiang, J.J., Conrath, D.W.: Semantic Similarity Based on Corpus Statistics and Lexical Taxonomy. In: Proc. Int'l Conf. Research in Computational Linguistics, ROCLING X (1997)
13. Wormbase website release WS170, date 12/02/2007, http://www.wormbase.org
14. Eppig, J.T., Bult, C.J., Kadin, J.A., Richardson, J.E., Blake, J.A., the members of the Mouse Genome Database Group: The Mouse Genome Database (MGD): from genes to mice—a community resource for mouse biology. Nucleic Acids Res. 2005 33, D471–D475 (2005)
15. Grumbling, G., Strelets, V., The FlyBase Consortium,: FlyBase: anatomical data, images and queries. Nucleic Acids Research 34, D484–D488 (2006)
16. The Gene Ontology Consortium. The Gene Ontology (GO) Project in 2006. Nucleic Acids Res. 34, D322–D326 (2006)
17. Al-Daihani, B., Gray, A., et al.: Bioinformatics data source integration based on Semantic Relationships across species. In: Dalkilic, M.M., Kim, S., Yang, J. (eds.) VDMB 2006. LNCS (LNBI), vol. 4316, Springer, Heidelberg (2006)

# On Estimating the Scale of National Deep Web*

Denis Shestakov and Tapio Salakoski

Turku Centre for Computer Science,
University of Turku, Turku, Finland-20520
denis.shestakov@utu.fi

**Abstract.** With the advances in web technologies, more and more information on the Web is contained in dynamically-generated web pages. Among several types of web "dynamism" the most important one is the case when web pages are generated as results of queries submitted via search web forms to databases available online. These pages constitute the portion of the Web known as deep Web. The existing estimates of the deep Web are predominantly based on study of English deep web sites. The key parameters of other-than-English segments of the deep Web were not investigated so far. Thus, currently known characteristics of the deep Web may be biased, especially owing to a steady increase in non-English web content. In this paper, we survey the part of the deep Web consisting of dynamic pages in one particular national domain. The estimation of the national deep Web is performed using the proposed sampling techniques. We report our observations and findings based on the experiments conducted in summer 2005.

## 1 Introduction

With the advances in web technologies, more and more information on the Web is contained in dynamically-generated web pages. The "dynamism" apparently improves the interactivity of web pages but, at the same time, leads to ignoring a huge number of dynamic pages by the current-day web crawlers (like, for instance, `google.com`) due to crawlers' limited abilities in retrieving and indexing dynamic web data. Among several types of web dynamism the most important one is the case when web pages are generated as results of queries submitted via search web forms to databases available online. These pages constitute the portion of the Web known as *deep Web* [2] and often also referred to as *hidden* or *invisible Web*. The data in the deep Web is hidden behind search forms which are the only access points to myriads of databases on the Web.

Recent study [3] has estimated the total number of online databases for April 2004 as around 450,000. Since current web search engines cannot effectively query networked databases, the information contained in these hundred thousands of repositories is mostly invisible to web crawlers and hence hidden from users.

This paper surveys databases on one specific national segment of the Web. The survey is based on our experiments for the scale of the national deep Web

* This work was partially supported by Yandex LLC (grant number 102104).

R. Wagner, N. Revell, and G. Pernul (Eds.): DEXA 2007, LNCS 4653, pp. 780–789, 2007.

conducted in summer 2005. This work extends our study [10], which has not been reported in English, by including additional experiments and providing slightly updated results.

To our knowledge, this survey is the first attempt to consider the specific national segment of the deep Web. The known deep Web characterization efforts (see Section 2) have predominantly concentrated on study of English deep web sites and, therefore, the estimates of the deep Web obtained in these works may be biased, especially owing to a steady increase in non-English web content. The national deep Web was studied on the example of the Russian segment of the Web (called *Runet* hereafter in this paper). There were several reasons to choose exactly the Russian part of the deep Web. Firstly, since Russian is written using the Cyrillic alphabet, which is non-Latin, one can expect that Runet is considerably more separated from the entire Web than, say, the German segment (where Latin alphabet is used). Secondly, we had an access to the data set provided by Yandex, a Russian web search engine (see Section 3.2 and Appendix C). And last but not least, having Russian as one of the author's mother tongue language was essential due to many web sites in Runet need to be manually inspected.

The rest of the paper is organized as follows. We start in Section 2 by reviewing the related work. Section 3 reports our experimental results. Section 4 discusses our findings and, finally, Section 5 concludes the paper.

## 2 Related Work

Several studies on the characterization of the indexable Web space of various national domains have been published. The review work [1] surveys several reports on national Web domains, discusses a methodology to present these kinds of reports, and presents a side-by-side comparison of their results. At the same time, national domains of the deep Web have not been studied so far and, hence, their characteristics can be only hypothesized.

There are two key works devoted to the characterization of the entire deep Web. The first one is a highly cited study [2] where *overlap analysis* [5] approach was used. Several well-known estimates of the deep Web for March 2000 have been reported in this paper. Particularly, it has been estimated that there were 43,000-96,000 deep web sites in the deep Web at that time (ultimate but disputable estimate in this study is even 200,000 deep sites for the year 2001).

The second survey [3] is based on the experiments performed in April 2004. In this work, the scale of the deep Web has been measured using the *random sampling of IP addresses* method, which was originally applied to the characterization of the entire Web [7]. To reduce uncertainty when identifying deep web resources Chang et al. distinguished three related notions[1] for accessing the deep Web - a *deep web site*, a *web database*, and a *web interface*. Among the findings obtained are the total number of deep web sites, web databases, and web interfaces, which were estimated as 307,000, 450,000 and 1,258,000 correspondingly.

[1] See [3] for more details.

There are solid grounds for supposing that the estimates obtained in the aforementioned surveys [2,3] are actually lower-bound because web databases in, at least, several national segments of the Web were for the most part ignored. For instance, the semi-automatic process of identifying a web interface to a web database in [3] consists of automatic extraction of web forms from all considered web pages, automatic removal of forms, which are definitely non-searchable, and finally manual inspection of the rest, potentially searchable forms, by human experts. Due to the limited number of experts (presumably just the authors of [3] were those experts) and, hence, the limited number of languages they were able to work with one can expect that a certain number of web interfaces in unknown (to experts) languages has not been taken into account. Notwithstanding that the approach used in [2] did not require multilingual skills from people involved in the study, we still argue that some number of non-English deep web sites has not been counted in this report as well. Indeed, the results produced by the overlap analysis technique depend significantly on "quality" of sources used in pairwise comparisons. The deep-Web directories considered by Bergman are mainly for English-speaking web users and, thus, omitted a number of national deep web sites. This makes the overlap analysis imperfect under the circumstances and suggests that the Bergman's estimate for the total number of deep web sites is a lower bound. In this way, we consider our survey as an attempt to supplement and refine the results presented in [2,3] by studying online databases in one particular national segment of the Web.

In our work, we adopted the random sampling of **IP** addresses (*rsIP*) method to our needs. Unlike [3] we noted an essential drawback of the rsIP method leading to underestimating of parameters of interest and suggested a way to correct the estimates produced by the rsIP. Additionally, we proposed a new technique for the deep Web characterization, the ***stratified random sampling of hosts*** (called *srsh* further on) method.

## 3 Estimation of Russian Deep Web Scale

In June 2005 and in August 2005 we performed a series of experiments to estimate the number of deep web sites in Runet. We used two techniques: the rsIP method and the srsh method. The experiments themselves and the results for each method are described in the following subsections.

### 3.1 Random Sampling of IP Addresses

We extracted all ranges of IP addresses used by Russian networks from the IP-Country database [6]. There were totally around 10.5 millions of IPs at the time of June 2005, $N = 10.5 \times 10^6$. Then, $n = 10.5 \times 10^4$ unique IP addresses (1% of the total number) were randomly selected and scanned for active web servers (tools we used for that are mentioned in Appendix A). We detected

1,379 machines with web servers running on port 80. For each of these machines the corresponding hostnames were resolved[2] based on a machine's IP address. Next step was crawling each host to depth three[3]. While crawling we checked if links point to pages located on hosts on the same IP. To not violate the sampling procedure (i.e., study only those IP addresses which are in the sample) we ignored any page returned by a server on another IP. The automatic analysis of retrieved pages performed by our script in Perl was started after that. All pages which do not contain web forms and pages which do contain forms, but those forms that are not interfaces to databases (i.e., forms for site search, navigation, login, registration, subscription, polling, posting, etc.) were excluded. In order to consider just unique search forms pages with duplicated forms were removed as well. Finally, we manually inspected the rest of pages and identified totally $x = 33$ deep web sites. It should be noted that unlike [3] we counted only the number of deep web sites. The number of web databases accessible via found deep web sites as well as the number of interfaces to each particular database were not counted since we did not have a consistent and reliable procedure to detect how many web databases are accessible via particular site. The typical case here (not faced in this sample though) is to define how many databases are accessed via a site with two searchable forms - one form for searching new cars while another for searching used ones. Both variants, namely two databases for used and new cars exist in this case or it is just one combined database, are admissible. Nevertheless, according to our non-formal database detection, 5 of 33 deep web sites found had interfaces to two databases, which gives us 38 web databases in the sample in total.

The estimate for the total number of deep web sites is $\widehat{D_{rsIP}} = \frac{33\times10.5\times10^6}{10.5\times10^4} = 3300$. An approximate 95% confidence interval[4] for $\widehat{D_{rsIP}}$ is given by the following formula: $\widehat{D_{rsIP}} \pm 1.96\sqrt{\frac{N(N-n)(1-p)p}{n-1}}$, where $p = \frac{x}{n}$ (see Chapter 5 in [11]). Thus, the total number of deep web sites estimated by the rsIP method is 3300±1120.

To our knowledge, there are four factors which were not taken into account in the rsIP experiment and, thus, we can expect that the obtained estimate $\widehat{D_{rsIP}}$ is biased from the true value. Among four sources of bias the most significant one is the virtual hosting. A recent analysis [12] of all second-level domains in the .RU zone conducted in March 2006 has shown that there are, in average, 7.5 web sites on one IP address. Unfortunately, even with the help of advanced tools for reverse IP lookup (see Appendix B) there is not much guarantee that all hostnames related to a particular IP address would be resolved correctly. This means that during the experiment we certainly overlooked a number of sites, some of which are apparently deep web sites.

[2] The first hostname is always the IP address itself in a string format. Others hostnames are non-empty values returned by *gethostbyaddr* function.

[3] See [3] for discussion on crawling depth value.

[4] This interval contains the true value of the estimated parameter with 95 percent certainty.

Next essential factor is the assignment of multiple IP addresses to only one web site. A web site mapped to several IPs is more likely to be selected for the sample than a site with one IP address. Therefore, our rsIP estimate should be expected to be greater than the true value. Similar to the virtual hosting factor, there is no guarantee in detecting all web sites with multiple IPs. We checked all 33 identified deep web sites for multiple IPs by resolving their IP addresses to corresponding hostnames and then resolving those hostnames back to their corresponding IPs (the same technique as in [7]). Though no multiple IP addresses for any of these sites were detected by this procedure, we are are not confident whether every deep web site in the sample is accessible only via one IP address. In any case, sites on multiple IPs are less common than sites sharing the same IP address and, hence, we believe that the virtual hosting's impact on the estimate should exceed one-site-on-multiple-IPs factor influence.

Third unconsidered factor is the exclusion of web servers running on ports other than 80 (default port for web servers). In our experiment we did not detect web servers that are not on port 80 and, obviously, missed a number of servers that may host deep web sites. However, the number of deep resources on non-default port numbers seems to be negligible since using non-default port numbers for web servers is not a widespread practice.

While previous three factors are about how well we are able to detect deep web sites in the sample the IP geodistribution factor concerns how well the whole IP pool covers the object of study, Runet in our case. Recall that our pool of IP addresses is all IPs assigned to the Russian Federation. Since web hosting is not restricted to geographical borders one can expect that a number of Runet web sites are hosted outside Russia. Analysis of all second-level domains in the .RU zone [12] revealed that, indeed, this is the case and approximately 10.5% of all studied domains are hosted on IPs outside the Russian Federation. Although only second-level RU-domains were investigated in [12] we suppose that the found distribution (89.5% of web servers are in Russia and the rest is outside) is applicable to all domains related to Runet[5]. This allows us to make a correction to our rsIP estimate. Under the fact that our sample was selected from the population which covers just around 90% of Runet, we updated the estimate and, finally, got that there are approximately **3650±1250** (rounded to the nearest 50) **deep web sites in Runet**.

## 3.2 Stratified Random Sampling of Hosts

In this experiment we used the data set "Hostgraph" (its description is given in Appendix C) provided by Yandex, a Russian search engine. All hosts indexed by Yandex were extracted from the Hostgraph. Besides, "host citation index", i.e. the number of incoming links for each host, was calculated. To improve method's accuracy, the shortening procedure was applied to the list of extracted hosts:

[5] The geodistribution of third and higher-level domains in the .RU zone should be almost the same. The distribution of second and higher-level domains in other than .RU zone (i.e., those ending with .com, .net, and so on) may differ but their fraction in all Runet domains is not so significant, around 25% according to Appendix C.

- Web sites which are certainly not deep web sites were removed from the list. In particular, we removed all sites on free web hostings known to us.
- We grouped all host by their second and third-level domain names (for example, all hosts of the form **.something.ru* are in the same group). The largest groups of hosts were checked and groups with sites leading to the same web databases were removed. As an example, we eliminated all hosts of the form **.mp3gate.ru* (except the host *www.mp3gate.ru*) since the same web database is available via any of these hosts.

We stopped the procedure at the total list of $N = 299,241$ hosts. At the beginning, we decided to check our assumption that the proportion of deep web sites among highly cited sites is higher than among less cited sites. If so, applying stratified random sampling technique to our data would be more preferable than using simple random sampling.

**Table 1.** Results of the preliminary stratified random sampling

| Stratum(k) | $N_k$ | $n_k^*$ | $d_k^*$ |
|---|---|---|---|
| 1 | 49,900 | 50 | 7 |
| 2 | 52,100 | 50 | 2 |
| 3 | 197,240 | 100 | 1 |

To examine the assumption we divided the list of hosts into three strata according to the number of incoming links for each host. The first stratum contained the most cited $N_1 = 49,900$ hosts, less cited $N_2 = 52,100$ hosts were in the second stratum, and the rest, $N_3 = 197,240$ hosts, was assigned to the third stratum (the strata sizes are rounded to the nearest ten). Then, $n_1^* = 50$, $n_2^* = 50$, and $n_3^* = 100$ unique hosts were randomly selected from each stratum correspondingly. Similar to the rsIP method, each of the selected hosts were crawled to a depth three. While crawling we checked if links point to pages located on the same host or on other hosts not mentioned in the total list of hosts. To meet the conditions of the sampling procedure all pages on hosts, which are in the list, were ignored. After that, the procedure becomes identical to the rsIP one (see Section 3.1). The results, number of deep web sites $d_k^*$ detected in each stratum, are shown in Table 1.

It is easy to see from Table 1 that our assumption is correct and, indeed, the probability of having a web interface to a web database is higher for highly cited hosts than for less cited ones. Thus, for reliable estimation of the total number of deep web sites in Runet we decided to use the stratified random sampling approach with the same division into strata.

We selected $n_1 = 294$ (including $n_1^*$ hosts already studied in the preliminary sampling), $n_2 = 174$ (including $n_2^*$ hosts), and $n_3 = 400$ (including $n_3^*$ hosts) unique hosts for each of the three strata correspondingly. The process of analyzing totally 668[6] sampled hosts and identifying deep web sites was the

[6] 200 of 868 hosts were already studied.

**Table 2.** Results of the srsh experiment

| Stratum(k) | $N_k$ | $n_k$ | $d_k$ | $\widehat{D_k}$ | $D_{k,cor}$ | Duplication |
|---|---|---|---|---|---|---|
| 1 | 49,900 | 294 | 35 | 5940 | 5600 | 10 of 35 resources have one or more duplicates: 4 duplicates in $stratum_1$, 1 in $stratum_2$, and 7 in $stratum_3$ |
| 2 | 52,100 | 174 | 8 | 2395 | 2090 | 2 of 8 resources have one duplicate: 1 duplicate in $stratum_2$ and 1 in $stratum_3$ |
| 3 | 197,240 | 400 | 3 | 1480 | 100 | 0 of 3 resources have duplicates |
| **Total** | 299,240 | 868 | 46 | 9815 | 7790 | 12 of 46 resources have duplicates |

same as in the preliminary sampling of hosts. Our findings are summarized in Table 2, where $d_k$ and $\widehat{D_k} = N_k \frac{d_k}{n_k}$ are the number of deep web sites in the sample from stratum $k$ and the estimated total number of deep web sites in stratum $k$ correspondingly.

The estimate for the total number of deep web sites is $\widehat{D_{srsh}} = \sum_{k=1}^{3} \widehat{D_k} = 9815$. An approximate 95% confidence interval for $\widehat{D_{srsh}}$ is given by the following formula: $\widehat{D_{srsh}} \pm 1.96\sqrt{\sum_{k=1}^{3} \frac{N_k(N_k-n_k)(1-p_k)p_k}{n_k-1}}$, where $p_k = \frac{d_k}{n_k}$ (see Chapter 11 in [11]). In this way, the total number of deep web sites estimated by the srsh method is 9815±2970. This estimate is not a final one, however. There are two factors which have to be considered in order to correct $\widehat{D_{srsh}}$.

The most significant source of bias in the srsh experiment is the host duplication problem. It is rather common that a web site can be accessed using more than one hostname. Naturally, several hostnames per one deep web site is typical as well. By using tools described in Appendix B (resolving a hostname of interest to an IP address and then reverse resolving IP to a set of hostnames) and inspecting our list of hosts manually we were able to identify that 12 of 46 deep web sites are accessible via more than one hostname. The distribution of duplicates among different strata is given in Table 2 in the "*Duplication*" column. The correction to be done is pretty straightforward: the existence of one duplicate for a particular host means that this host is twice more likely to be in the sample than a host without any duplicate. Thus, the corrected estimate for the first stratum is $D_{1,cor} = \frac{N_1}{n_1} \times (31 + \frac{4}{2}) = 5600$. $D_{1,cor} \times \frac{1}{35}$ and $D_{1,cor} \times \frac{7}{35}$ deep resources should be excluded from the estimates for second and third stratum correspondingly since they were already counted in the estimate for the first stratum. Similarly, we obtained the corrected estimates for the second and third stratum: $D_{2,cor} = \frac{N_2}{n_2} \times (7 + \frac{1}{2}) - D_{1,cor} \times \frac{1}{35} = 2090$ and $D_{3,cor} = \widehat{D_3} - D_{1,cor} \times \frac{7}{35} - D_{2,cor} \times \frac{1}{8} = 100$ deep web sites respectively.

The second factor, similar to the geodistribution factor in the rsrIP experiment, is how well the list of hosts we studied covers Runet. The list we worked with contained all hosts indexed by Yandex, a Runet search engine, at the time of February 2005. Recent study [9] has shown that Yandex has one of the best coverage of Runet among the largest web crawlers indexing Russian part of the Web. In this way, one can expect that our list represented Runet with sufficient accuracy. More importantly, we believe that the only way for a web database content to be available for web users is having at least one web interface located on a page indexed by a search engine. Otherwise, not only data in this database is hidden but also its whole existence is completely unknown to anyone. Therefore, according to our point of view, the population of hosts used in the srsh experiment is essentially complete for purposes of detecting deep web sites. It should be noted however that since the "Hostgraph" data was created in February 2005 and our experiments were performed in June and August 2005 those deep web sites which appeared mainly in spring 2005 were not counted.

To sum up, due to the fact that at least one of four deep web sites is accessible via more than one hostname we corrected $\widehat{D_{srsh}}$ and obtained that the **total number of deep web sites in Runet** is around **7800±2350** (rounded to the nearest 50).

### 3.3 Subject Distribution of Web Databases

We manually categorized 79 deep web sites sampled in the rsIP and srsh experiments into ten subject categories: Libraries (*lib*), Online Stores (*shop*), Auto (*auto*), Business (*biz*), Address Search (*addr*), Law&Goverment (*law*), People Search (*pe*), Travel (*tra*), Health (*he*), and Science (*sci*). Note that more than one category may be assigned to a deep web site - for instance, nearly half of deep web sites assigned to the "Auto" category were also placed into "Online Stores" since these sites were auto parts&accessories online stores. Figure 1 shows the distribution of deep web databases over the categories. The particular observation we made is that almost 90% of deep resources in the category "Online Stores" (13 of 15 sites) have a navigational access to their data, i.e. these sites have not only one or several web search forms but also have a browse interface which allows a user to reach the necessary data from a web database via a series of links. In this way, such resources cannot be considered as entirely "hidden" from search engines since their content may be accessed by following hyperlinks only.

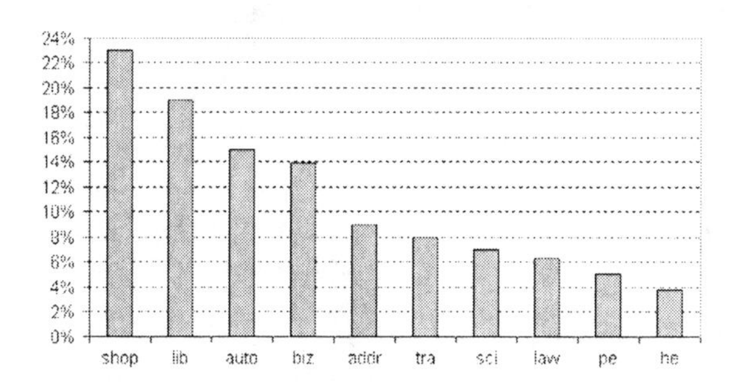


**Fig. 1.** Distribution of deep web sites over subject category

## 4 Discussion

The experiments provided us with two estimates for the total number of deep web sites in Runet: 3650±1250 as estimated by the rsIP method, and 7800±2350 as estimated by the srsh method.

In fact, there is no contradiction between the estimates since the rsIP method should give us a lower-bound estimate due to the virtual hosting factor (see Section 3.1) while the srsh method should result in an upper-bound estimate because of the host duplication problem (see Section 3.2). In any case, it is unquestionable that a 95% confidence interval for the total number of deep web sites in Runet is (2400,10150), that is, the **scale of the Russian deep Web is on the order of $10^3$ resources**. We also believe that the estimate obtained by the srsh method is closer to the true number than the estimate by the rsIP method because the host duplication factor was at least partially addressed in the srsh experiment (duplicates were identified and then the corresponding correction was done) while the influence of virtual hosting on the rsIP estimate was just mentioned as very important but not measured quantitatively.

The quick and indirect attempt to correct the rsIP estimate is to reconstruct the process of detecting deep web sites for specifically designed list of IPs. In order to build such a list, we took 46 deep web sites detected by the srsh method and resolved their hostnames to the list of IP addresses. "Ideal" rsIP method should detect 46 deep web resources in this list, non-ideal rsIP method detects less due to shortcomings of IP-to-host resolving procedure. Our rsIP method was able to detect 25 (54%) deep resources, and the rest, 21 (46%) of deep web sites, were not detected[7]. Thus, we can expect that around 46% deep resources were missed in our rsIP experiment and, hence, the corrected rsIP estimate is 6800±2300. The intervals for the rsIP and srsh estimates, (4500,9100) and (5450,10150) correspondingly, are well overlapping, and their intersection, namely (5450,9100) or approximately (rounded to the nearest 100) **7300±1800**, is our final estimate for the **total number of deep web sites in Runet**.

It is interesting to compare the number of deep resources in Runet and in the entire Web. The number of deep web sites resources in the entire Web estimated by Chang et al. [3] is 307,000[8] for April 2004 while our estimate for Runet obtained by the same method as in [3] is 3650±1250 for summer 2005. The comparison suggests that in terms of the number of deep web sites the Russian deep Web is approximately the hundredth part of the entire deep Web. This roughly coincides with the portion of Russian web sites in the Web - survey [8] indicated that one percent of public sites in the Web in 1999 as well as in 2002 were in Russian.

[7] In more than half cases no hostnames were resolved from an IP address and then the only URL to use `http://IP_address` returned just an error page or a web server default page.

[8] No confidence intervals were mentioned in [3] but it is easy to calculate them from their data: particularly, a 95% confidence interval for the number of deep web sites is 307,000±54,000. A 99% confidence intervals have also been specified in more recent work of the same authors [4].

## 5 Conclusion

This paper presented our survey of web databases on one specific national segment of the Web. The national deep Web was studied on the example of the Russian segment of the Web. Based on the proposed sampling techniques we estimated the total number of deep web sites in Runet as 7300±1800. The comparison with results of [3] showed that in terms of the number of deep web resources the Russian deep Web is approximately the hundredth part of the entire deep Web. Additionally, we demonstrated that the proportion of deep web sites among highly cited web sites is higher than among less cited sites.

*Notes and Comments.* We would like to thank Yandex LLC for providing us with the data set "Hostgraph". Appendices can be downloaded at:

`http://denshe.googlepages.com/shestakov_dexa07_appendices.pdf`.

## References

1. Baeza-Yates, R., Castillo, C., Efthimiadis, E.: Characterization of national Web domains. TOIT 7(2) (2007)
2. Bergman, M.: The deep Web: surfacing hidden value. Journal of Electronic Publishing 7(1) (2001)
3. Chang, K., He, B., Li, C., Patel, M., Zhang, Z.: Structured databases on the web: observations and implications. SIGMOD Rec. 33(3), 61–70 (2004)
4. He, B., Patel, M., Zhang, Z., Chang, K.: Accessing the deep web. CACM 50(5), 94–101 (2007)
5. Lawrence, S., Giles, C.: Searching the World Wide Web. Science 280(5360), 98–1000 (1998)
6. MaxMind GeoIP Country® Database. URL: `http://www.maxmind.com/app/country`
7. O'Neill, E., McClain, P., Lavoie, B.: A methodology for sampling the World Wide Web. Annual Review of OCLC Research 1997 (1997)
8. O'Neill, P., Lavoie, B., Bennett, R.: Trends in the evolution of the public Web. D-Lib Magazine 9(4) (2003)
9. Segalovich, I., Zelenkov, Y., Nagornov, D.: Methods for comparative analysis of modern search systems and Runet size determination. In: Proc. of RCDL'06, [in Russian] (2006)
10. Shestakov, D., Vorontsova, N.: Characterization of Russian deep Web. In: Proc. of Yandex Research Contest 2005, pp. 320–341 [In Russian] (2005)
11. Thompson, S.: Sampling. John Wiley&Sons, New York (1992)
12. Tutubalin, A., Gagin, A., Lipka, V.: Black quadrate: Runet in March (2006) [In Russian] (2006), `http://www.rukv.ru/analytics-200603.html`

# Mining the Web for Appearance Description

Shun Hattori, Taro Tezuka, and Katsumi Tanaka

Department of Social Informatics, Graduate School of Informatics, Kyoto University
Yoshida-Honmachi, Sakyo, Kyoto 606-8501, Japan
{hattori, tezuka, tanaka}@dl.kuis.kyoto-u.ac.jp

**Abstract.** This paper presents a method to extract appearance descriptions for a given set of objects. Conversion between an object name and its appearance descriptions is useful for various applications, such as searching for an unknown object, memory recall support, and car/walk navigation. The method is based on text mining applied to Web search results. Using a manually constructed dictionary of visual modifiers, our system obtains a set of pairs of a visual modifier and a component/class for a given name of object, which best describe its appearance. The experimental results have demonstrated the effectiveness of our method in discovering appearance descriptions of various types of objects.

## 1 Introduction

In recent years, there has been an exponentially growing amount of information available on the Internet, especially the World Wide Web. With the improvement and maintenance of mobile computing environments, we have been able to access information anywhere at any time in our daily lives. Information retrieval systems such as Web search engines, which we must often pass through to access necessary information, have increased their significances or more.

In utilizing such a large information source as the Web for our daily activities, converting a name of an object to its appearance description and vice versa sometimes becomes necessary for us. For example, let us suppose that the user sees an object which she does not know the name at all or exactly and wishes to retrieve web pages about the object. Using conventional search engines such as Google, the user will have difficulty in finding relevant web pages by submitting only the appearance descriptions of the object as a query. Although conventional search engines are efficient in finding relevant web pages when the name of the object is given, they are far from satisfactory in retrieving information when the name of the object is not given.

On the other hand, there are situations where the user wants to get the appearance description by using the name of the target object as a clue. For example, in an auditory car/walk navigation system, the user wants to know the appearance description of the landmark, rather than its name, especially when she is not well acquainted with the area.

These conversions between the name of an object and its appearance description have many applications, as described below in more detail.

R. Wagner, N. Revell, and G. Pernul (Eds.): DEXA 2007, LNCS 4653, pp. 790–800, 2007.

**Object Name Search by Appearance Description**

Web search engines are powerful tools for finding information when the name of the object is given. On the other hand, if the user does not know the name of the object, she has to put in much effort to identify it.

There are cases where the user encounters an unknown species of bird, insect, or plant when she is out in a field, and wants to know its name and obtain more information about it. Such situations have become very common with the advent of mobile devices that enable access to the Web at any moment.

The user also forgets the name of an object, such as a do-it-yourself tool that she rarely uses. Unless the user can remember the name of the tool, she cannot obtain further information about it by doing a Web search.

Search engines for children also need to incorporate this type of function, because children have smaller vocabularies than adults and can only describe many objects by their appearance descriptions [1].

**Appearance Description Search by Object Name**

There are cases where the user wants to know the appearance of a certain object. For example, in a car/walk navigation system, presenting the appearance descriptions of a landmark is more helpful to the driver/walker, especially when the navigation is performed by using auditory media.

By just submitting the name of the object as a query to an image search engine, the user can obtain relevant images. Converting the object name to an appearance description by surveying them, however, is not a difficulty task for humans, but the task is hard for computational systems.

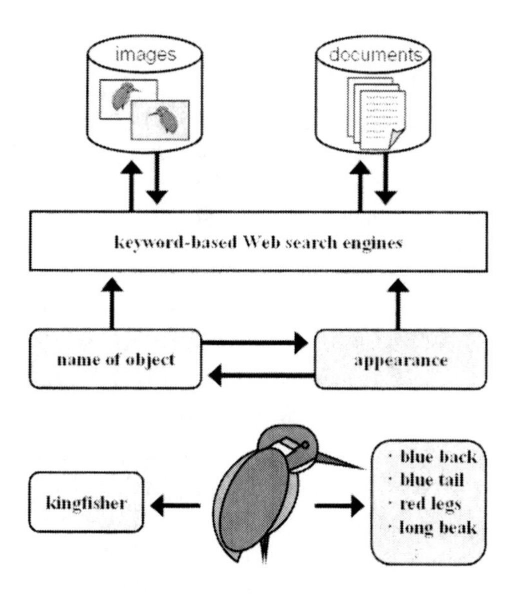


**Fig. 1.** Conversion between object names and appearance descriptions

One way to cope with these problems is to manually construct a dictionary that stores relationships between names of objects and their appearance descriptions. There are already many good databases for certain domains. There is, however, still a vast amount of information yet to be utilized on the Web. By mining the Web, we can construct much larger databases for various domains.

The aim of this paper is to present a method that automatically constructs a database of correspondences between names of objects and their appearance descriptions. We accomplish this task by finding typical appearance descriptors and components of the target object by Web mining. Our proposed method involves immensely time consuming calculations, so the method should be implemented as a process of constructing a database before any search is performed.

The remainder of the paper is organized as follows. Section 2 presents some work related to this research. Section 3 presents our method to discover appearance descriptions for names of objects and its implementation. Section 4 describes the experimental results. Finally, we conclude the paper in Section 5.

## 2 Related Work

### 2.1 Visual Description Mining

Tezuka et al. has developed a method for extracting object names and their "visual descriptions" from large document collections, such as the Web and encyclopedias, by applying knowledge mining and dependency analysis [2]. Their method was mainly targeted on extracting descriptions of geographic objects, such as buildings and landmarks, and developing search and navigation support applications in geographic space.

### 2.2 Visualness of Adjectives

Yanai *et al.* has proposed the concept of "visualness" of word concepts, especially adjectives [3]. By applying image processing to text-annotated images by Google Image Search, they obtained values that indicate how appropriate each adjective is for describing an image's visual characteristics. Their definition of visualness is based on the calculation of decision trees. Adjectives that could be used to divide images into characteristic groups have higher visualness. Their method was not applied to nouns and other parts of speech, which could have visualness as well. The vocabulary of visual modifiers could be automatically built by visualness.

### 2.3 Object Identification by Photographs

There are some online services where the user takes a picture of an unknown object, submits it to the system through a network, and the system identifies the name of the object based on image analysis [4,5]. Such systems often identify the name of an unknown object in a picture by measuring the similarity with the corpus of pictures labeled by names of known objects. However, the precisions of systems for object identification, especially general object identification, are not sufficiently high at this moment [6,7,8,9].

# 3 Extracting Appearance Descriptions for General Objects from Very Large Corpus of Documents

In this section, we present our method to discover appearance descriptions for each name of object by mining a very large corpus of documents such as the Web. First, we give an overview of our method.

## 3.1 Overview

We have formalized the most simplified model of an *appearance* description for a general object, as a triplet of the name of *object* itself, a name of its *component* or *class*, and a *visual modifier* describing its visual characteristic:

$$appearance = (object,\ \ visual\text{-}modifier,\ \ component/class)$$

For example, "A kingfisher has a short blue tail and a long bill." could be simplified to three triplets of an object, a visual modifier, and its modifying component of the object, (kingfisher, short, tail), (kingfisher, blue, tail) and (kingfisher, long, bill). Meanwhile, the sentence "A kingfisher is a small bird." could be simplified to one triplet of an object, a visual modifier, and its modifying class of the object, (kingfisher, small, bird). This model is based on the observation that an object is usually perceived as an aggregation of components with specific visual characteristics. If the visual modifier describes the whole of the object itself, such as in the case of "A kingfisher is colorful.", then the class would be the name of the object itself, that is, the simplified triplet would be (kingfisher, colorful, kingfisher).

The goal of our method presented in this paper is to collect a set of pairs (*visual*, *component/class*) that correctly describes the appearance of the given object $o$. We call these pairs, *V-C pairs*,

$$o \Longrightarrow \{(v_1, c_1), (v_2, c_2), ..., (v_n, c_n)\}.$$

Moreover, we also aim to rank these *V-C pairs* in the order of weight $w_i$, which indicates the suitability of each *V-C pair* as an appearance description of the target object $o$,

$$o \Longrightarrow \{(v_1, c_1, w_1), (v_2, c_2, w_2), ..., (v_n, c_n, w_n)\}.$$

When a name of object $o \in O$ is inputted, our method processes the following four steps and then outputs its appearance as several *V-C pairs* ordered by their weights. In the remainder of this section, we describe these steps in detail.

**Step 0: Constructing General Dictionary of Visual Modifiers**

**Step 1: Collecting Components/Classes of Object for V-Modifiers**

**Step 2: Ranking of V-C Pairs Using Sampled Documents**

**Step 3: Filtering of V-C Pairs Using Whole Documents**

### 3.2 Dictionary of Visual Modifiers

We have manually collected a set of visual modifiers as the basic data set for our method. It consists of 617 words that describe color, shape, size, and surface material of objects. The composition of the set is shown in Table 1.

**Table 1.** Set of Visual Modifiers

| Type | Number | Examples |
|---|---|---|
| Color | 192 | aeruginous, amber, amethyst, antique, apricot, ..., wheat, white, wine, wisteria, yellow |
| Shape | 143 | antisymmetric, aquiline, arc, asymmetric, ..., vertical, wavy, wedge-shaped, winding |
| Texture | 119 | abrasive, allover, argyle, banded, belted, ..., veined, velvety, watermarked, wet, zebra-stripe |
| Size | 53 | abundant, average, big, bold, brief, broad, ..., thick, thin, tiny tremendous, trivial, vast, wide |
| Surface material | 110 | acrylic, adobe, alloy, aluminum, asphalt, bamboo, ..., vinyl, waxy, wire, wood, wooden, woolly |
| **Total** | 617 | |

The dictionary contains many words (maybe too many) referring to size, such as short/long, small/big, high/low, many, much, few, and little. Although they are important for describing the appearance in some occasions, they have also caused some noise in the results.

### 3.3 Collection of Components/Classes

To collect the names of components/classes for a target object, first, our system crawls web pages described about only the object, by submitting the name of the object as a query to Google Web Search [10], which is a common Web search engine. To increase the accuracy of our method, the system only retrieves web pages that contain the name of the object $o$ in the title, by submitting not `["o"]` but `[intitle:"o"]` as a query to Google. Henceforth, we use $D(o)$ as the set of crawled documents for each name of object $o$ obtained from $D$, the set of all documents of a corpus such as the Web.

Next, the parser scans through the collected web pages and finds phrases that contain a visual modifier in the dictionary. Words that immediately follow a visual modifier in the crawled web documents are considered as candidates for component/class name of the target object. There are, however, many irrelevant words on the candidate list. We apply the following ranking technique to refine the results.

### 3.4 Ranking of V-C Pairs Using Sampled Documents

After obtaining the *V-C pairs*, our system evaluates their weights to offer the users them ranked according to their significance for the target object. We present three methods to weight the *V-C pairs*.

**Method 1**

This method is a very simple approach that evaluates each weight of a *V-C pair*, $(v_i, c_i)$, for the target object, $o$, by the number of web documents in $D(o)$ that contains the phrase "$v_i$ $c_i$", that is, $v_i$ immediately followed by $c_i$:

$$\text{weight}_o^1(v, c) := \text{df}_o(\text{“}v\ \ c\text{”}),$$

where $\text{df}_o(\text{“}p\text{”})$ stands for the number of web documents within $D(o)$ that contain the phrase $p$. Because this method considers a word that frequently appears after a visual modifier to be a component/class automatically, it is vulnerable to a compound word that starts with a visual modifier but is not an appearance description, such as "high school" and "yellow pages." Therefore, the below Method 3 is proposed to cope with this problem.

**Method 2**

This method is a more refined approach that evaluates the significance of *V-C pair*, $(v_i, c_i)$, for the target object, $o$, in the following manner:

$$\text{weight}_o^2(v, c) := \text{weight}_o^1(v, c) \cdot \text{weight}_o(c),$$

$$\text{weight}_o(c) := \sum_{v_i \in V} f_o(v_i, c),$$

$$f_o(v_i, c) := \begin{cases} 1 & \texttt{if } \text{df}_o(\text{“}v_i\ \ c\text{”}) > t_2, \\ 0 & \texttt{otherwise.} \end{cases}$$

In the formula, $t_2$ is a threshold value. In this paper, our system sets $t_2 = 1$. $f_o(v_i, c)$ is a boolean function that indicates whether or not there is a meaningful co-occurrence as the phrase "$v_i$ $c$" in the crawled web documents $D(o)$ for the target object $o$, and $\text{weight}_o(c)$ means that the number of variations of visual modifiers that have a meaningful co-occurrence with a candidate component/class $c$ of the object in $D(o)$.

**Method 3:**

This method filters problematical *V-C pairs*, as described in Method 1, almost completely. However, the method also filters too many acceptable *V-C pairs* for appearance descriptions of the target object, because the number of the crawled web documents $D(o)$ is at most 1000 and they do not often include the phrase "$c$ is/are $v$" even when $(v, c)$ is acceptable. Therefore, our system does not adopt this method.

$$\text{weight}_o^3(v, c) := \begin{cases} \text{weight}_o^2(v, c) & \texttt{if } \dfrac{\text{df}_o(\text{“}c \text{ is/are } v\text{”})}{\text{df}_o(\text{“}v\ \ c\text{”})} > 0, \\ 0 & \texttt{otherwise.} \end{cases}$$

### 3.5 Filtering of V-C Pairs Using the Whole Documents

After weighting the *V-C pairs* by using the sampled documents $D(o)$ for the target object $o$, our system filters them by using all the documents of a corpus such as the Web to offer the better results to the users.

**Method 4:**

This method is similar to Method 3 with regard to the below-mentioned fundamental, but it does not filter too many acceptable *V-C pairs* for the appearance descriptions of the target object unlike Method 3 because the number of all the documents of a corpus is much greater than the number of the crawled web documents $D(o)$. Of course, problematical *V-C pairs*, as described in Method 1, are filtered out almost completely:

$$\mathrm{weight}_o^4(v,c) := \begin{cases} \mathrm{weight}_o^2(v,c) & \texttt{if } \dfrac{\mathrm{df}(\text{“}c \text{ is/are } v\text{”})}{\mathrm{df}(\text{“}v\ c\text{”})} > t_4, \\ 0 & \texttt{otherwise.} \end{cases}$$

where $t_4$ is a threshold value set to $10^{-4}$ in our system. df(“$p$”) stands for the number of documents that contains the phrase $p$ within $D$, the set of all the documents of a corpus, not just the sampled documents $D(o)$.
The formula is based on an observation for a set phrase “$v$ $c$” that df(“$v$ $c$”) is too great but df(“$c$ is/are $v$”) is too small or nearly equal to 0. For example, in the case of (kingfisher, red, legs), both “legs are red” and “red legs” appear with a high frequency in all the documents and maybe also in $D$(“kingfisher”). Therefore, this *V-C pair* (red, legs) is considered as an appearance description in general. On the other hand, although the phrase “high school” appears at a certain high rate in $D$(“kingfisher”), this *V-C pair* (high, school) is not considered as an appearance description in general because of low frequency of the phrase “school is high”, and thus should not always be considered as an appearance description of the object “kingfisher”.

**Method 5:**

This method is an approach like conventional tf·idf methods, that weights a *V-C pair* $(v, c)$ based on the proportion of the local connectedness to the global connectedness in the form of phrase “$v$ $c$”:

$$\mathrm{weight}_o^5(v,c) := \begin{cases} \mathrm{weight}_o^2(v,c) & \texttt{if } \dfrac{\mathrm{connect}_o(v,c)}{\mathrm{connect}(v,c)} > t_5, \\ 0 & \texttt{otherwise.} \end{cases}$$

$$\mathrm{connect}_o(v,c) = \frac{\mathrm{df}_o(\text{“}v\ c\text{”})^2}{\mathrm{df}_o(\text{“}v\text{”}) \cdot \mathrm{df}_o(\text{“}c\text{”})},$$

$$\mathrm{connect}(v,c) = \frac{\mathrm{df}(\text{“}v\ c\text{”})^2}{\mathrm{df}(\text{“}v\text{”}) \cdot \mathrm{df}(\text{“}c\text{”})}.$$

However, our system in this paper does not adopt this method because its evaluation task requires submitting numerous queries to a web search engine such as Google to evaluate each local connectedness more accurately.

## 4 Experiment

In this section, we justify our method by using the above-defined weight to extract appearance descriptions for a name of a target object from the Web.

We performed experiments on a set of object names, consisting of four typical categories that the users may encounter in their daily lives. Each set has five objects, as indicated in the following list.

**Landmarks:** Big Ben, Leaning Tower of Pisa, Statue of Liberty, Taj Mahal, Tokyo Tower
**Birds:** Jungle Myna, Kingfisher, Shoebill, Snowy Owl, Sun Conure
**Flowers:** Edelweiss, Japanese Cherry, Lavender, Lily of the Valley, Sunflowers
**Products:** InterCityExpress, PS3, TGV, ThinkPad, Wii

Table 2 illustrates the results of an experiment performed on the object, "edelweiss" (a name of flower). We have applied the methods described in the previous section, to obtain appearance descriptions appropriate for "edelweiss". *V-C pairs* in bold font in Table 2 indicate correct answers for the target object, where authors have checked manually whether or not each *V-C pair* is acceptable for appearance descriptions for it. The results suggest that Method 2 is substantially better than Method 1. Filtering based on Method 4 also improves the result, by giving higher ranks to the correct answers. On the other hand, the correct answers that were ranked too low by Method 2 did not become sufficiently high or were filtered out as a result of using Method 4.

Fig. 2 to 5 compares our defined weights on the top $k$ average precision for each category. Fig. 6 compares our defined weights on the top $k$ average precision

**Table 2.** For an object $o=$ "edelweiss" (as a name of flower)

| Method 1 | | | | Method 2 | | | | Method 4 | | | |
|---|---|---|---|---|---|---|---|---|---|---|---|
| | v | c | $w_o^1$ | | v | c | $w_o^2$ | | v | c | $w_o^4$ |
| 1 | few | steps | 18 | 1 | beautiful | mountain | 65 | 1 | beautiful | mountain | 65 |
| 2 | snow | report | 17 | 2 | rock | climbing | 48 | 2 | **small** | **flower** | 45 |
| 3 | dark | matter | 15 | 3 | **small** | **flower** | 45 | 3 | **yellow** | **flower** | 30 |
| 4 | high | quality | 14 | 4 | arc | climbing | 44 | 3 | big | mountain | 30 |
| 4 | yellow | pages | 14 | 5 | **yellow** | **flower** | 30 | 5 | dry | rope | 27 |
| 6 | snow | may | 13 | 5 | big | mountain | 30 | 6 | **white** | **flower** | 25 |
| 6 | beautiful | mountain | 13 | 7 | yellow | pages | 28 | 7 | wide | selection | 20 |
| 8 | old | world | 12 | 8 | dry | rope | 27 | 8 | few | steps | 18 |
| 8 | great | deals | 12 | 9 | **white** | **flower** | 25 | 9 | snow | mountain | 15 |
| 8 | rock | climbing | 12 | 9 | red | mountain | 25 | 9 | dark | matter | 15 |
| ⋮ | ⋮ | ⋮ | ⋮ | ⋮ | ⋮ | ⋮ | ⋮ | ⋮ | ⋮ | ⋮ | ⋮ |
| 17 | **small** | **flower** | 9 | 72 | **silver** | **star** | 6 | 67 | **small** | **heads** | 5 |
| 29 | **yellow** | **flower** | 6 | 72 | **cream** | **edelweiss** | 6 | 67 | **beautiful** | **flower** | 5 |
| 42 | **white** | **flower** | 5 | 98 | **small** | **heads** | 5 | - | **silver** | **star** | 0 |
| 42 | **small** | **heads** | 5 | 98 | **beautiful** | **flower** | 5 | - | **cream** | **edelweiss** | 0 |

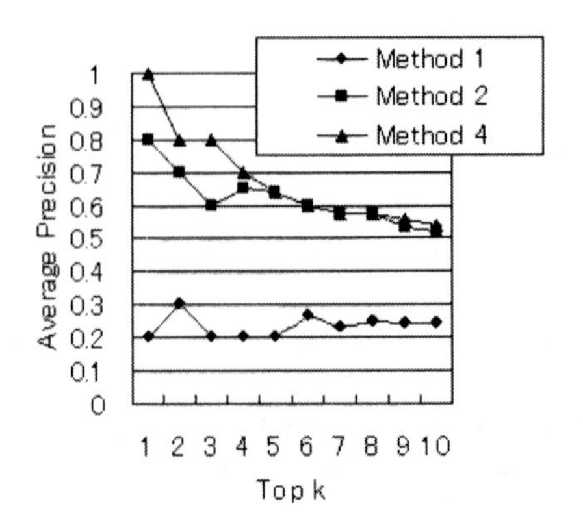


**Fig. 2.** For five landmarks

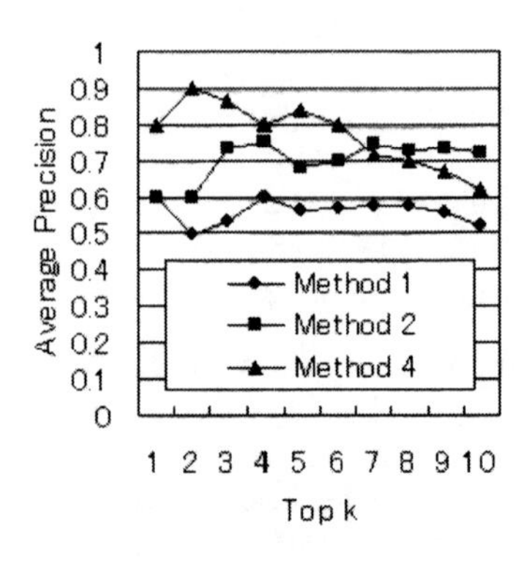


**Fig. 3.** For five birds

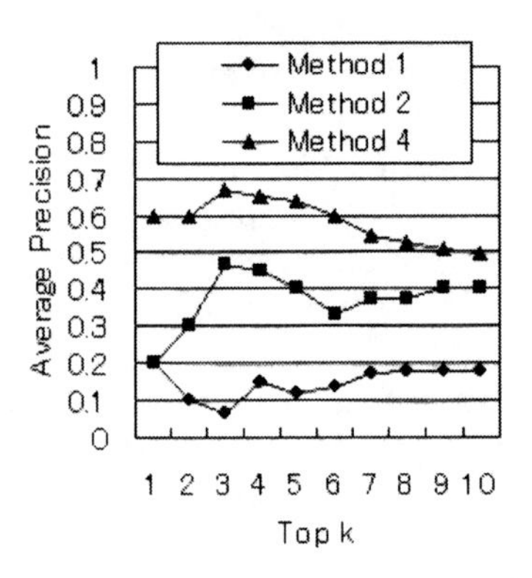


**Fig. 4.** For five flowers

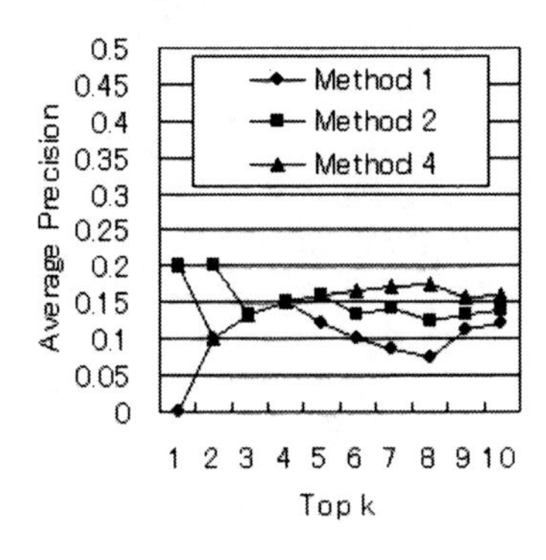


**Fig. 5.** For five products

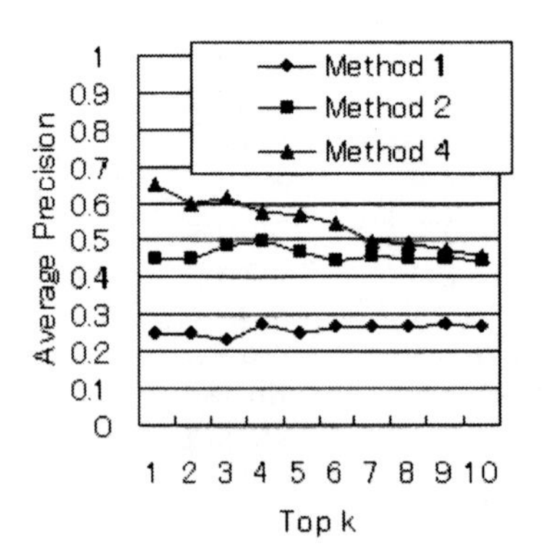


**Fig. 6.** For twenty general objects

in total for all categories. The graphs illustrate that the results obtained after applying Method 4 are substantially better than the previous results. This is especially significant for highly ranked results (with low $k$).

## 5 Conclusion and Future Work

We proposed a method to extract appearance descriptions for a given set of objects to build a database that stores relationships between the name of an object and its appearance descriptions and allows users to convert between them bi-directionally. Our system mines a very large corpus of documents, such as the Web, for a set of *V-C pairs* of a visual modifier and a component/class that best describe the appearance of the given object by using a manually constructed dictionary of visual modifiers. The experimental results have showed the effectiveness of our method in discovering appearance descriptions for various types of objects. We plan to improve the method by using the lexical relations of hyponymy and meronymy in a thesaurus for component and class names of a target object, and also challenge to automatically build a dictionary of visual modifiers.

**Acknowledgments.** This work was supported by a MEXT Grant for "Development of Fundamental Software Technologies for Digital Archives", Software Technologies for Search and Integration across Heterogeneous-Media Archives (Project Leader: Katsumi Tanaka), and a MEXT Grant-in-Aid for Scientific Research on Priority Areas "Cyber Infrastructure for the Information-explosion Era", Planning Research "Contents Fusion and Seamless Search for Information Explosion" (Project Leader: Katsumi Tanaka, A01-00-02, Grant#: 18049041).

## References

1. Nakaoka, M., Shirota, Y., Tanaka, K.: Web information retrieval using ontology for children based on their lifestyle. In: Proceedings of the First International Special Workshop on Databases for Next Generation Researchers (SWOD'05) in conjunction with ICDE'05 (2005)
2. Tezuka, T., Tanaka, K.: Visual description conversion for enhancing search engines and navigational systems. In: Zhou, X., Li, J., Shen, H.T., Kitsuregawa, M., Zhang, Y. (eds.) APWeb 2006. LNCS, vol. 3841, pp. 955–960. Springer, Heidelberg (2006)
3. Yanai, K., Barnard, K.: Image region entropy: A measure of "visualness" of web images associated with one concept. In: Proceedings of ACM International Conference on Multimedia 2005 (MM'05), pp. 420–423 (2005)
4. NTT-IT Corporation: Magicfinder/o (2007), `http://www.ntt-it.co.jp/goods/bcj/MagicFinder/o/pro_index_j.html`
5. TechIndex Corporation: PicLin (2007), `http://www.piclin.jp/`
6. Rue, H., Hurn, M.A.: Bayesian object identification. Biometrika 86(3), 649–660 (1999)
7. Duygulu, P., Barnard, K., de Freitas, N., Forsyth, D.: Object recognition as machine translation: Learning a lexicons for a fixed image vocabulary. In: Heyden, A., Sparr, G., Nielsen, M., Johansen, P. (eds.) ECCV 2002. LNCS, vol. 2353, pp. 97–112. Springer, Heidelberg (2002)

8. Barnard, K., Duygulu, P., Forsyth, D.A.: Recognition as translating images into text. In: Proceedings of Internet Imaging IV, SPIE, vol. 5018, pp. 168–178 (2003)
9. Barnard, K., Duygulu, P., Forsyth, D.A., de Freitas, N., Blei, D.M., Jordan, M.I.: Matching words and pictures. Journal of Machine Learning Research 3, 1107–1135 (2003)
10. Google Corporation: Google Web Search (2007), `http://www.google.com/`

# Rerank-by-Example: Efficient Browsing of Web Search Results

Takehiro Yamamoto, Satoshi Nakamura, and Katsumi Tanaka

Department of Social Informatics, Graduate School of Informatics,
Kyoto University
Yoshida-Honmachi, Sakyo, Kyoto 606-8501 Japan
{tyamamot,nakamura,tanaka}@dl.kuis.kyoto-u.ac.jp

**Abstract.** The conventional Web search has two problems. The first is that users' search intentions are diverse. The second is that search engines return a huge number of search results which are not ordered correctly. These problems decrease the accuracy of Web searches. To solve these problems, in our past work, we proposed a reranking system based on the user's search intentions whereby the user edits a part of the search results and the editing operations are propagated to all of the results to rerank them. In this paper, we propose methods of reranking Web search results that depend on the user's delete and emphasis operations. Then, we describe their evaluation. In addition, we propose a method to support deletion and emphasis by using Tag-Clouds.

**Keywords:** edit-and-propagate, tag-cloud, reranking, user-interface.

## 1 Introduction

A variety of ranking algorithms have been proposed and implemented. They are used in many Web search engines such as Yahoo! and Google. However, they sometimes do not return search results that adequately meet the user's needs as they are affected by two big problems in Web searches.

The first problem is that users' search intentions are diverse. For example, if a product named "*A*" is used as a query, one user might be seeking reviews of "*A*", whereas another user might be seeking information on how to buy "*A*". Moreover, the search intentions might be diverse even if the same user uses the same query. If a Web search engine returns the search results that do not satisfy the user's intention, the precision of the search results is considered low, because the user has to check more search results or modify the query. Instead, if the user could inform the user's search intention to the system while browsing the search results, the system might be able to return good search results for the user.

The second problem is that Web search engines return a huge number of Web search results, but the user usually checks merely the top 5 or 10 results. Therefore, almost all Web search results are not checked by the users. Moreover, there has been widespread usage of search engine optimization (SEO) techniques recently. Many companies and site owners spend much money on SEO in order to raise the search

R. Wagner, N. Revell, and G. Pernul (Eds.): DEXA 2007, LNCS 4653, pp. 801–810, 2007.

rank order of their Web sites. As a result, currently, high ranked search results are often the result of applying SEO techniques and the accuracy and reliability of Web search results have decreased. This problem stems from the fact that it is burdensome for the user to check low ranked search results. Therefore, if we introduce reranking functions that the user can use easily, the user will be able to check a lot of good Web search results efficiently and the impact of SEO techniques will become weaker.

To solve these problems, in our past work, we proposed and implemented a reranking system based on edit-and-propagate operations [1]. In this work the system enables the user to edit any portion of a browsed page of Web search results at any time while the user is searching. Our system detects the user's search intention from the editing operations. For example, if the user deletes a part of the search results, our system guesses that "*this user does not want this kind of the result*". If the user emphasizes a part of the search results, our system guesses that "*this user wants more of this kind of result*". Our system propagates the users' search intention based on their editing operations to all search results in order to rerank them. In this way, the user can easily obtain optimized search results.

In this paper, we propose the three reranking methods of Web search results when the user performs the delete/emphasis operation to rerank them. Then, we provide the experimental results of these reranking methods. In addition, in order to perform delete/emphasis operations more effectively for the user, we propose and implement *MultiTagCloud*, which are generated by using Web search results.

## 2 Edit-and-Propagate Operations for Reranking

Our system reranks search results according to the type of operations users want to edit and the target that the user edits. The flow is as follows:

1. The user inputs a query to our system.
2. The system sends to the query to Web search engine.
3. The system receives the search results and shows them to the user.
4. The user checks the search results ranked by the search engine.
5. The user edits a portion of the search results by using editing operations.
6. The system detects the user's editing operations and guesses the user's search intention. Then the system reranks the whole search results according to this guess and shows the reranked search results to the user.
7. If the user is not satisfied with the reranked search results, go to 4.

The editing operations are as follows:

- **Deletion:** Deletion is an operation that indicates what types of search results the user does not want to obtain from the system.
- **Emphasis:** The role of an emphasis operation is opposite to the delete operation. Emphasis operation indicates what types of search results the user wants to obtain.

Search engines usually return title, snippet and URL as one of the search results to the user. If a user deletes/emphasizes a keyword or sentence from a title or snippet of a search result, the system guesses that the user wants to degrade/upgrade some results which include the deleted/emphasized text. The system also guesses that the user wants to degrade/upgrade some results which include the topic about the deleted/emphasized text. For example, while browsing the Web search results of "*Katsumi Tanaka*", if a user deleted "*pianist*" from a search result, the system guesses that the user has the intention to degrade the search results which are related to the topic about Katsumi Tanaka who is a pianist.

If a user deletes a URL or a part of a URL, the system guesses that the user wants to degrade the results which include deleted strings. If the user does not need the results whose domain is JP, he/she might delete ".jp" from a URL of the search results. If the user deletes a search result item, the system guesses that he/she wants to degrade the search results which are similar to the deleted item.

Figure 1 shows the implementation of our system when the user submits "*kyoto*" as the query. For this query, some of the high ranked search results are related to the Kyoto Protocol. If the user does not need information about the Kyoto Protocol, the user deletes "*protocol*" from the search result. After deleting a keyword, the system guesses the user's intention and degrades the search results which include "*protocol*" in their titles or snippets. Then, the user receives reranked search results whose high ranked ones are related to Kyoto City or Kyoto University.

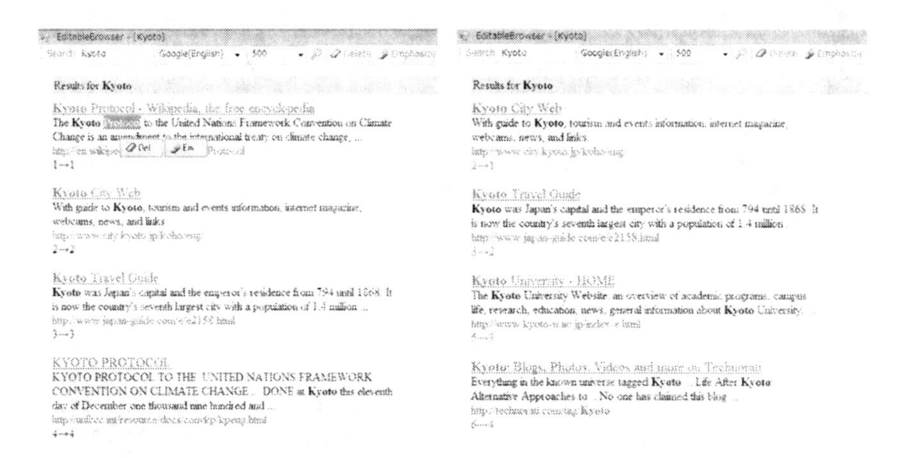


**Fig. 1.** Examples of the results produced by our system: The search results are reranked after deleting a keyword

## 3 Reranking Methods

Our goal is to estimate the user's intention when he/she uses editing operations and to reflect this intention in the search results page. For this purpose, we propose methods of reranking Web search results by deleting or emphasizing a keyword.

When the user inputs a query to our system, the system obtains the top $N$ search results $r_1$, $r_2$, …, $r_N$ ($r_i$ is the $i_{th}$ search result item). Then, the system initializes the score of each search result item as:

$$Score(r_i) = N - i \quad (1)$$

(This formula allocates scores to search results depending on their search ranks.)

If the user deletes/emphasizes a keyword, the system re-calculates the scores of search result items and shows the search results in order of their scores to the user.

### 3.1 Selection Keyword Method

When the user deletes or emphasizes the term *t*, this method degrades or upgrades the search results including *t*. Our system calculates the score of $r_i$ when it includes *t* in its title or snippet according to Equation 2.

$$Score(r_i)_{new} = Score(r_i)_{last} + type \times N \quad (2)$$

$$type = \begin{cases} 1 & \text{if edit operation is emphasis} \\ -1 & \text{if edit operation is deletion} \end{cases}$$

### 3.2 Extended Keyword Method

When the user inputs query like "*Katsumi Tanaka*" in order to seek information about Katsumi Tanaka who is a professor, he/she receives the search results which include not only a professor, but also a pianist, a poet, and so on. If a user deletes the term "*pianist*" in the search result, the system guesses his/her search intention as "the user wants to degrade the results about '*Katsumi Tanaka who is a pianist*' ". So when the user deletes "*pianist*", it is better for the user to degrade the search results that not only include "*pianist*" but also include terms like "*piano*" or "*concert*" and so on. To accomplish this, we propose two methods called "*inner extended keyword*" and "*outer extended keyword*".

If the user deleted/emphasized term $t$, both methods obtain terms $w_1, w_2, \ldots, w_k$ as extended keywords. Then, for each term $t, w_1, w_2, \ldots, w_k$, the system calculates the score of the search result items which include these terms in their titles or snippets according to Equation 2. *R(q)* denotes the set of search result items of *q*. *W(q)* denotes the set of terms which appear in *R(q)*. *TF(w, R(q))* denotes the term frequency of $w \in W(q)$ in *R(q)*. In this work, by using Japanese morphological analyzer ChaSen [3], we extract proper nouns and general nouns as terms appear in the search results.

- **Inner Extended Keyword Method:** To obtain extended keywords, this method uses the term frequency of search results that include the term (we call this set of search result items as *R'(q)*). For each term $w \in W(q)$, the system calculates *TF(w, R'(q))* and extracts the top *k* terms in order of this value as extended keywords.
- **Outer Extended Keyword Method:** This method uses the difference of term frequency between *R(q AND t)* and *R(q)*. For each term $w \in W(q)$, the system calculates *TF(w,R(q AND t)) – TF(w,R(q))*. Then the system extracts the top *k* terms in order of this value as extended keywords.

For example, when the user deletes "*pianist*" in the search results of "*Katsumi Tanaka*", the inner extended keyword method degrades the search results which include *"homepage"*, *"information"*, *"profile"*, or *"concert"*. On the other hand, the outer extended keyword method degrades the search results which include *"profile"*, *"concert"*, *"lesson"*, or *"piano"* (if the system uses Google, $N=500$ and $k=4$).

## 4 Experiment

To evaluate our reranking methods, we investigated how the search results are reranked by deleting or emphasizing keywords using the reranking methods in Section 3. We prepared 15 search tasks and also prepared the correct answer sets for each search task.

In the keyword deletion task, the user inputs the prepared search query and submits the query to the search engine using our system. Then, the user checks the search results sequentially after receiving the search results. If the user reaches an unmatched search result, the user selects the keyword and deletes it. If the system detects the user's deletion, the system reranks the search results and shows the reranked search results to the user. Then the user also rechecks the search results from top to bottom. The user does this three times.

After each deletion, we calculated the precision of the ranking and the number of correct search results which the system incorrectly degraded. We compared the *selection keyword method* (*SKM*), *inner extended keyword method* (*IEKM*), and *outer extended keyword method* (*OEKM*) in this task.

In the keyword emphasis task, the user inputs the prepared search query and submits the query to the search engine using our system. Then, he/she checks the search results sequentially after receiving the search results. If the user finds a keyword which the user thinks is relevant, the user selects the keyword and emphasizes it. If the system detects the user's emphasis operation, the system reranks the search results and shows the reranked search results to the user. Then the user rechecks the search results from top to bottom. This action is performed once, because the keyword emphasis task is so effective that after second emphasis almost all the top 20 reranked search results become correct.

After each emphasis, we calculated the precision of the ranking. We compared the *SKM*, the *IEKM*, and the *OEKM* in this task.

If the reranked all top 20 search results are correct, the user stops the operation.

Figure 2 shows the average precisions of the top $K$ ($K$=1, 2, ..., 20) in the search results of each deletion. Figure 3 shows average precisions of each emphasis. Figure 4 shows average initial order of lowest original matched search results in top $K$ by using *SKM*. This figure shows how low rank of the matched search results are upgraded as the user performs the emphasis operation. Table 1 shows the numbers of matched search results which are removed from the top 20 by the deletion. Number of total deletions means the total number of search results which are removed from the top 20 by three delete operations. Number of error deletions means the number of the matched search results which are removed from the top 20 by three deletions. Error ratio means the ratio of the total deletion number and the error number.

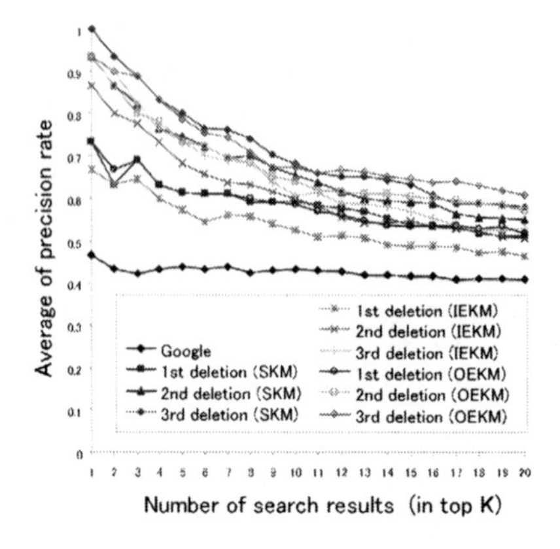


**Fig. 2.** Average precision obtained after each deletion

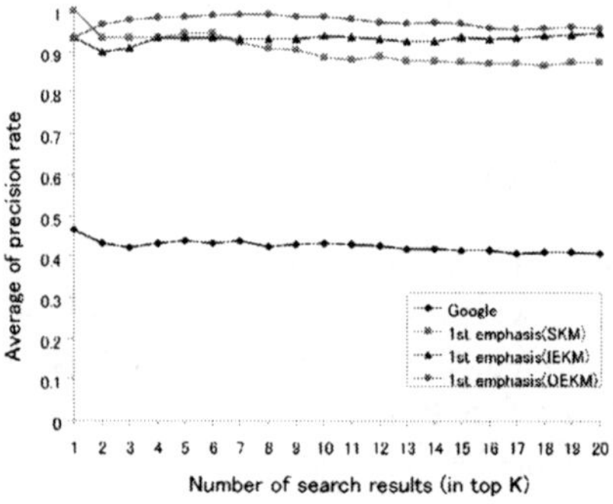

**Fig. 3.** Average precision obtained after each emphasis

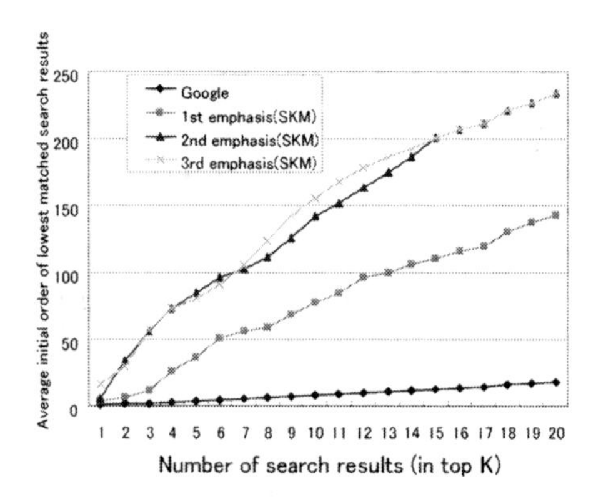


**Fig. 4.** Average initial order of lowest matched search results in top *K*

**Table 1.** Ratio of matched search results removed from the top 20

| | # of total deletions | # of error deletions | Error ratio |
|---|---|---|---|
| *SKM* | 127 | 2 | 1.57% |
| *IEKM* | 191 | 20 | 10.47% |
| *OEKM* | 208 | 35 | 16.83% |

### 4.1 Discussion

Figure 2 shows that the number of relevant search results in the top 20 increases after each delete operation in all reranking methods. We note that the effect of the first deletion is biggest and the effect of the deletion becomes smaller as the user repeats deletions.

The effect of the *IEKM* is smallest and the difference between the effectiveness of the *OEKM* and that of the *SKM* was relatively small. In particular, if we assume that users usually check only the top 10 search results, both reranking methods have the same effectiveness. As shown in Table 1, the main reason for this is that although the two extended keyword methods degrade more search results than the *SKM* does, they also degrade more relevant search results. As shown in Section 3.2, when the user deletes "*pianist*" from the search results of "*Katsumi Tanaka*", the *OEKM* degrades the search results which contain terms related to "*pianist*". In contrast, the *IEKM* degrades the search results which contain "*information*". The search results which are related to the Katsumi Tanaka who is a professor have frequent co-occurrences with "*information*". Therefore, when the system degrades the search results which contain "*information*", many relevant search results are degraded. The *OEKM* also obtains popular teams or terms which appear in multiple topics although the *OEKM* obtains fewer such terms than the *IEKM*. We need to solve this problem to increase the effect of the two extended keyword methods.

The effect of emphasis operations is much bigger than that of delete operations. We note that the first emphasis operation is effective, by which all top 20 search results are relevant in more than two-thirds of the queries.

In the emphasis operation task, the *OEKM* outperformed the *SKM* and the *IEKM*. In the delete operation task, the fact that the extended keyword methods often obtain popular terms like "*information*", which are related to multiple topics, lowers their accuracy. In the emphasis task, however, if the term which the user emphasized directly is related to relevant search results, this problem does not occur. We think the extended keyword methods outperformed the *SKM* because of this reason.

As shown in Figure 4, our system can pick up very low ranked relevant search results as the user performs the emphasis.

# 5 Supporting Deletion and Emphasis by Using MultiTagCloud

A tag-cloud is used as a visual depiction of tags. In a tag-cloud, more frequently used tags or more popular tags are shown bigger or more emphasized. The widely known use of tag-clouds is made by flickr [4]. Flickr displays many kinds of tags that users tag to photos by using a tag-cloud. One of the biggest merits of tag-clouds is that users can instantly check the popular tags and easily look over many tags.

We therefore applied this interface model to the Web search results page in order to support the use of the delete and emphasis operations for users.

## 5.1 Method of Generating MultiTagCloud

The general tag-cloud is a list of tags that users tagged. Our tag-cloud is a list of keywords that appear in Web search results. Moreover, in order to visualize how the contents of the Web search results are distributed on the page, we split search results into several groups, and for each of the groups we generate a tag-cloud. The system shows these tag-clouds simultaneously as a *MultiTagCloud*. The process of creating a *MultiTagCloud* is as follows:

1. The system splits a set of the search result items $R=\{r_1, r_2, \dots, r_N\}$ in page $p$, ordered by their scores, into $n$ groups. (We call this split group $R'=\{R_1, R_2, \dots, R_n\}$)
2. For each $R_i \in R'$, we generate a tag-cloud according to the following steps.
   (a) By doing a morphological analysis on the titles and the snippets of the search result items in $R_i$, the system obtains the set of keywords $T_i$, and calculates the term frequency of keyword $T \in T_i$

(b) The system finds the top $k$ frequent keywords, and then adds them to a tag-cloud in appearance order in $R_i$.
(c) When the term frequency of a keyword in a tag-cloud exceeds the threshold $\theta_i$ ($i$=1, 2, 3, … ), the system modifies the font size of the keyword in it.

3. The system simultaneously shows all the tag-clouds as a *MultiTagCloud.*

## 5.2 Example of System Output

Figure 5 shows an example of the output produced by our system. The normal search results are displayed on the left side of the system. The *MultiTagCloud* is displayed on the right side of the system. In Figure 5, the system displays the top 500 search results, with *MultiTagCloud* consisting of tag-clouds of the 1st to 100th, 101st to 200th, …, 401st to 500th search results. For each tag-cloud, the system displays keywords whose term frequency is within the top 30. The users can use editing operations on any of the keywords in the tag-clouds as well as those in search results. Therefore, the user can rerank search results by deleting or emphasizing any keywords in the *MultiTagCloud.* According to the user's operations, the system reranks the search results using the methods in Section 3, and then the system shows the reranked search results and the re-generated *MultiTagCloud* (right of Figure 5).

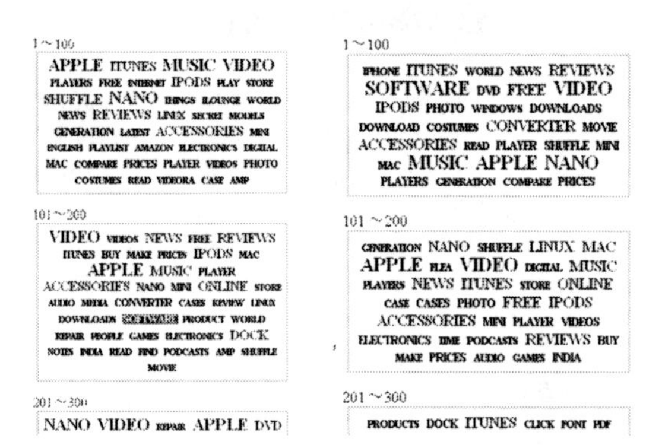


**Fig. 5.** Example of *MultiTagCloud.* The *MultiTagCloud* is re-generated depending on the reranked search results (Right).

## 5.3 Effectiveness of MultiTagCloud

By showing *MultiTagCloud* with the search results, the user can easily and quickly understand what keywords are influential.

We can visually understand that if we delete a keyword displayed in a large font, many search results will be reranked to the bottom and that if we emphasize a keyword displayed in a small font, a few sparse search results will be gathered and reranked to the top. For these reasons, showing tag-clouds enables users to know how to modify the search results when they delete or emphasize a keyword. Users can more easily find keywords they need or do not need in a tag-cloud than in the conventional search results. Moreover, by using a *MultiTagCloud*, the user can also see the propagation of keywords. In Figure 5, when the user submits "*iPod*" and then emphasizes "*software*" in a *MultiTagCloud*, keywords related to the "*software*" such as "*iPhone*", "*iTunes*", "*Free*", and "*Download*" come to appear at the top of the *Multi-Tag-Cloud*.

# 6 Related Work

Broder [2] and Rose and Levinson [8] have classified the goals of a user searching the Web into three categories: *navigational*, *informational*, and *transactional/resource*. There are also many studies on automatic classification of user goals by using queries [5, 6]. These works focused on the recognition of the users' search intentions by analyzing their queries. However, it is difficult to recognize such intentions just by analyzing the inputted queries because even if the same user uses the same query in a Web search, the user's search intention might depend on the time and situation. Our goal is to estimate the user's search intention through interactions between the user and the system, and to reflect this intention in the Web search results page.

Yahoo! Mindset [11] allows the user to specify his/her search intentions to the system. It weights each Web page as 'more research' or 'more shopping'. The user can specify that his/her search intention is research or shopping by using a slide bar; then the system re-ranks the search results according to the weight. However, in this system, the user cannot re-rank the search intentions without prepared factors.

There are many reranking methods. Relevance feedback [9] is the most popular retrieval system to retrieve documents using the user's feedback. Relevance feedback is based on a vector space model [10]. Non-relevance feedback [7] is a retrieval method for documents. It uses only non-relevant documents to find the target documents from a large data set of documents. We think that we can apply both of the relevance feedback and non-relevance feedback to optimizing searched results.

# 7 Conclusion and Future Work

We proposed reranking methods for users to delete or emphasize keywords and did detailed evaluation tests. We showed the effectiveness of edit-and-propagate operations for reranking Web search results. In the future, we plan to compare the effect of our reranking methods with that of a query expansion which uses Boolean operators such as AND and NOT. We also plan to devise more advanced reranking methods because reranking methods proposed in this paper are simple algorithms.

Moreover, we proposed the use of the *MultiTagCloud* in order to support the delete and emphasis operations.

In this work, we only implemented reranking by keyword deletion and emphasis. In the future, however, we plan to implement reranking by deletion and emphasis of search results item or sentence in a snippet. In addition, we would like to introduce other operations such as drag-and-drop, and replace.

In this paper, we only focused on reranking. However, we think that the user may have many other intentions in browsing search results pages. For example, the user may want to modify a snippet which is generated by the search engine in order to check another part of the original content which does not appear in the snippet. Users might also want to compare multiple targets using Web searches. We will attempt to support such a comparison by introducing editing operations.

Moreover, we can apply edit-and-propagate operations to browse other types of Web pages. For example, while browsing a bulletin board, users might want to delete improper messages or spam in order to increase readability, or emphasize a term in order to check topics quickly. We will develop such an editable browser for many kinds of Web pages.

## Acknowledgements

This work was supported in part by MEXT Grant-in-Aid for Scientific Research on Priority Areas: "Cyber Infrastructure for the Information-explosion Era", "Contents Fusion and Seamless Search for Information Explosion" (Project Leader: Katsumi Tanaka, A01-00-02, Grant#: 18049041), and by "Design and Development of Advanced IT Research Platform for Information" (Project Leader: Jun Adachi, Y00-01, Grant#: 18049073), and by Grant-in-Aid for Young Scientists (B) "Content Manipulation and Browsing by Reversible Display" (Leader: Satoshi Nakamura, Grant#: 18700129).

## References

1. Yamamoto, T., Nakamura, S., Tanaka, K.: An Editable Browser for Reranking Web Search Results. In: Proceedings of the Third International Special Workshop on Databases for Next-Generation Researchers (2007)
2. Broder, A.: A taxonomy of web search. ACM SIGIR Forum 36(2), 3–10 (2002)
3. Morphological analyzer: ChaSen. http://chasen.naist.jp/hiki/ChaSen/
4. flickr. http://www.flickr.com/
5. Kang, I.H., Kim, G.C.: Query type classification for web document retrieval. In: Proceedings of SIGIR2006, pp. 64–71 (2006)
6. Lee, U., Liu, Z., Cho, J.: Automatic identification of user goals in Web search. In: Proceedings of WWW2005, pp. 391–400 (2005)
7. Onoda, T., Murata, H., Yamada, S.: Non-Relevance Feedback Document Retrieval Based on One Class SVM and SVDD. In: 2006 IEEE World Congress on Computational Intelligence, pp. 2191–2198 (2006)
8. Rose, D.E., Levinson, D.: Understanding user goals in web search. In: Proceedings of the WWW2004, pp. 13–19 (2004)
9. Salton, G.: The SMART Retrieval System Experiments in Automatic Document Processing. pp. 312–323 (1971)
10. Salton, G., McGill, M.J.: Introduction to Modern Information Retrieval. McGraw-Hill, New York (1986)
11. Yahoo! Mindset. http://mindset.research.yahoo.com/

# Computing Geographical Serving Area Based on Search Logs and Website Categorization*

Qi Zhang[2], Xing Xie[1], Lee Wang[1], Lihua Yue[2], and Wei-Ying Ma[1]

[1] Microsoft Research Asia, 5F, Sigma Building, No.49, Zhichun Road. Beijing, 100080, P.R. China
{xingx, wyma}@microsoft.com

[2] Department of CS, University of Science and Technology of China, Hefei, 230027, P.R. China
wizard@mail.ustc.edu.cn, llyue@ustc.edu.cn

**Abstract.** Knowing the geographical serving area of web resources is very important for many web applications. Here serving area stands for the geographical distribution of online users who are interested in a given web site. In this paper, we proposed a set of novel methods to detect the serving area of web resources by analyzing search engine logs. We use the search logs to detect serving area in two ways. First, we extracted the user IP locations to generate the geographical distribution of users who had the same interests in a web site. Second, query terms input by users were considered as the user knowledge about a web site. To increase the confidence and to cover new sites for use in real-time applications, we also proposed a categorization system for local web sites. A novel method for detecting the serving area was proposed based on categorizing the web content. For each category, a radius was assigned according to previous logs. In our experiments, we evaluated all these three algorithms. From the results, we found that the approach based on query terms was superior to that based on IP locations, since search queries for local sites tended to include location words while the IP locations were sometimes erroneous. The approach based on categorization was efficient for sites of known categories and were useful for small sites without sufficient number of query logs.

**Keywords:** Location-based web application, serving area; serving radius, web classification.

## 1 Introduction

More and more web applications such as local search and local advertisement take geographical dimension of web resources into consideration. In this paper, we are interested in a special characteristic, "serving area", which can be regarded as the expected geographical distribution of online users who are interested in a given web site. For instance, in a search engine, giving a query of "*pizza seattle*", the search

* This work was done when the first author was visiting Microsoft Research Asia.

R. Wagner, N. Revell, and G. Pernul (Eds.): DEXA 2007, LNCS 4653, pp. 811–822, 2007.

engine should better return pizza related web sites whose serving area is *Seattle*. The serving area of a web resource can be different from the street address of the entity who owns that web resource.

Many research works have been carried out to detect the "geographical scope" of web resources. They are mainly based on analyzing web content and hyperlink structures. Geographical names, postal codes, telephone numbers and a number of other features are extracted from the web content to help get the geographical scope of a web page or a web site [1][2][3][4][5][6][16][17][18]. The underlying assumption is that if a web resource does have a non-global (thus local) geographical scope, it will be more likely to contain the location names or other named entities covered by the geographical scope. The geographical scope of a web resource can be used as an approximation to its serving area. However, geographical scope is different from serving area in that it describes content, not user. For example, www.newzealand.com has a clear geographical scope of New Zealand, but it will interest global users. Its serving area, therefore, should be global.

Link analysis has been widely used to measure the relevance among web documents. Two documents are considered relevant when one links to another. In location based web applications, hyperlinks can be used to measure the geographical relevance among web resources [1]. Hyperlinks are indirect hints for the serving area of a web resource, since they indicate the interests of web authors of those sites link to that web resource. However, information brought by links is usually incomplete and inaccurate. For new or less popular web sites, there are few linked resources that can be used to help the serving area detection.

In this paper, we propose and study a novel method for detecting serving areas by analyzing web search logs, which are direct hints of end user interests. A web content classification based serving area detection algorithm is also proposed for web sites without sufficient logs. Experiments on large samples of real world data are carried out to evaluate the performance of our algorithms.

The rest of the paper is organized as follows: Section 2 surveys the related work. In section 3, we describe the concept of serving area. Section 4 provides the details of our proposed algorithms. We give the experimental results in section 5 and in section 6 we conclude our work.

## 2 Related Work

Much work has been done to improve the accuracy of extracting web locations. The progress in Named Entity Recognition (NER) has helped a lot to the location detection problem [5]. By using various geographical cues (such as geographical names, IPs and hyperlinks, etc.) and developing effective computation approaches, the precision of extracted geographical scope has been greatly enhanced.

Ding [1] proposed an algorithm to detect the geographical scope of a web resource based on analyzing resource links and resource content. They first analyzed the geographical locations of the hosts linking to the site in question. Then they exploited the web content by taking an NER step. In the algorithm, they defined two measures: *power* for measuring interest and *spread* for measuring uniformity. Locations with significant interest (large *power*) and smooth distribution (large *spread*) were considered as the geographical scope.

Amitay [5] presented a system named web-a-where which employed a gazetteer based approach. A geographical focus was assigned to each web page after identifying and processing all geographical occurrences in these pages.

In [6], the authors classified locations of web resources into three types: provider location, content location, and serving location. Provider location is defined as the physical location of the entity that owns the web resource. Content location is the geographical location that the content of the web resource talks about. Serving location is the geographical scope that a web resource reaches. In our paper, the concept of serving area is somewhat similar to the serving location proposed in [6], but our definition is more user-oriented. In [6], page links were used to calculate the serving location, while search logs are studied here.

There exist quite a few commercial local search engines such as Google local search [19] and MSN local search [20]. Most of them get geographical information from Yellow Pages or manual classification of web resources. The geographical information got from Yellow Pages is usually a geo-point, for example, when we search for "*pizza redmond*" on a local search engine, one of the results might be a pizza restaurant "*Third Place Pizza and Sandwich*" whose street address is "*509 Jackson St Seattle, WA 98104.*" However, we still don't know its serving area. Will people who live in Redmond (about 30 miles from Seattle) be interested in this restaurant? Does the serving radius of this pizza restaurant cover Redmond? Current approaches can hardly give us an accurate answer. In the following sections, for a local web site, we propose a novel approach to analyze search logs to understand more about its serving area. Before going to the algorithm details, we will first describe the concept of serving area.

## 3 Definitions

In this paper, we use the following definitions:

**Serving area:** The geographical distribution of users who are interested in a certain web site. Serving area of a web resource can be also seen as the geographical area that this web resource intends to reach.

**Serving radius:** It represents the size of the geographical serving area of a web resource. Serving radius can be continuous, like miles, or discrete levels, like city level or state level.

**Provider location:** We take the same definition as in [6]. Provider location in this paper can be seen as the street address where the owner of the web resource locates.

## 4 Serving Area Detection

### 4.1 Computing Serving Area by Analyzing User IP Locations

In search logs, the relationship between user locations and clicked URLs can be estimated by analyzing the collection of user IP locations. In our algorithm, we use two measures: *weight* and *spread*. *Weight* is used to measure the percentage of users in a certain location who are interested in a web site. *Spread* of a certain location is used to measure the uniformity of *weight* in its child locations on an administrative hierarchy. The user's interest here is regarded as the number of clicks on a web site URL in search logs. The more clicks on the URL, the higher interests the users put on.

*Weight* is defined as follows:

$$Weight(w,l) = \frac{Click(w,l) / Population(l)}{Click(w,Parent(l)) / Population(Parent(l))} \tag{1}$$

Where *Click(w, l)* is the number of clicks on web resource *w* by people in location *l*. *Population (l)* is the population of location *l*. *Parent(l)* is the parent location of *l* on an administrative hierarchy.

*Spread* is defined as same as that in Ding [1] and the entropy definition is chosen for the best performance based on their results:

$$Spread(w,l) = \frac{-\sum_{i=1}^{n} \frac{Weight(w,l_i)}{\sum_{j=1}^{n} Weight(w,l_j)} \times \log\left(\frac{Weight(w,l_i)}{\sum_{j=1}^{n} Weight(w,l_j)}\right)}{\log n} \tag{2}$$

Where $n$ is the number of children of location $l$. $l_i$ or $l_j$ is a direct children of $l$.
Once *weight* and *spread* are computed, user logs can be used to detect the serving area of a web site:

1 Map all user IPs to locations.
2 Map all the locations got from step 1 onto a geographical hierarchy, where location nodes distribute on different geographical levels such as country, state or city.
3 Travel the geographical hierarchy down from the root. For each node, *weight* and *spread* values will be calculated and the node will be pruned if its *spread* or *weight* values do not exceed given thresholds. Otherwise we continue the traveling to its offspring nodes if there are any. When the algorithm stops, the nodes where we stop at constitute the serving area.

## 4.2 Computing Serving Area by Analyzing Query Terms

When a user wants to find a local web resource, he or she is very likely to input a location term in the query. For instance, a user will input a query "*pizza seattle*" or "*seattle pizza*" if he or she wants to find some pizza related sites in *Seattle*.

From search logs, we can build up a relationship between query terms and user clicked URLs. If we get all the query terms which lead to clicks on URLs in a specific domain, we can then detect the geographical distribution by analyzing the location information in these query terms.

For example, we have extracted a sample list of query terms in whose search results users clicked one of the URLs in web site *Lombardi's* (www.lombardispizza.com). We list all these queries below (the number after each query term is the number of its occurrences in the log)

| | | | | | |
|---|---|---|---|---|---|
| *Lombardi pizza* | *3* | *lombardi spring "street nyc"* | *1* | *lombardi's pizza new york* | *2* |
| *big sauage pizza.com* | *1* | *lombardi's* | *1* | *lombardi's new york* | *1* |
| *lombardis* | *1* | *lombardi's pizza nyc* | *1* | *Lombardi pizza in new york city* | *6* |

As one can see that the percentage of queries which contain *New York* or *NYC* is 11/17 = 0.647. By only looking at the query terms, we conclude that *Lombardi's* has a strong relationship with *New York City*. Actually, *Lombardi's* is a very famous pizzeria in *New York City*.

The query terms are often short, so it is more difficult to analyze query terms than to analyze web pages. There are several difficulties that we need to address:

1. Geo/geo and geo/non-geo disambiguation.
   Due to the shortness of query terms, there are usually only one or two candidate geo-names in a query. Disambiguation approaches used in web content analysis can be hardly applied here. In the above example, we found that many query terms contain "new york" or "ny" only. It is therefore difficult to know whether it stands for *New York State* or *New York City*.
2. Recognizing location abbreviations.
   From the above example, we find that users like to input abbreviations instead of full names when they are using search engines. In the example, "ny" and "nyc" are used to represent *New York City*. It is difficult to recognize all these abbreviations.
3. Lack of query terms.
   For new or less popular web sites, there will be fewer query terms associated with them. The lack of query terms will affect the confidence of location information extraction.

In our algorithm, we solve geo/geo and geo/non-geo ambiguities by looking at query context. The query context here is the query terms input by other users. Using the example of *Lombardi's*, if we see query terms like "*lombardi's new york*", we don't know whether "*new york*" here stands for *New York City* or *New York State*, we will go forward to look at other query terms. From other query terms, we found that users have explicitly stated *New York City* instead of *New York State*, like "*lombardi's pizza nyc*" and "*Lombardi pizza in new york city*". This information is what we call query context. Finally, we know that "*new york*" here is more likely to represent *New York City* than *New York State*. For the other two difficulties, we leave them for future work.

The serving area detecting includes three steps:

1. Given a web site, extract all the related query terms and build a query document.
2. Run a content location detection algorithm on the query document. Any content location detecting algorithms can be applied here, such as the algorithm proposed in [6]. The query context will be considered in the algorithm.

3. The content location computed from step 2 is regarded as the serving area of the web site. The result can be seen as the user knowledge about the location information of the web site.

The disadvantage of this approach, as mentioned above, is that the number of query terms might be insufficient for obtaining reasonable results.

### 4.3 Computing Serving Area by Web Content Classification

In this section, we propose a web content classification based algorithm which extracts serving radius and provider location from the web resource, then combines them to obtain the serving area.

As defined previously, serving radius can be numeric or levels. In real applications, it is hard to calculate or manually assign a numeric radius to each web resource. Therefore, here we use coarse levels to represent the radius for practicability. Serving radius of a local web site is defined as three levels:

**City Level:** The serving area of the web resource is within the city where it locates.
**State Level:** The serving area of the web resource is beyond its city but within its state.
**Country Level:** The serving area of the web resource covers the entire country.

First we group web sites into categories. Then we either calculate or assign (from common knowledge or user study) a radius for each of them. The radius can be calculated by averaging the radius of sites with sufficient search logs. In this approach, all the web resources which fall in a certain category will have the same radius. For example, businesses such as *banking service* and *restaurant* are local services whose users are usually within a city, while businesses such as *airport* are state-level service, since not all cities have local *airport*. There are also country level services such as *software development*.

Note that not all categories can be assigned with a fixed level, for example, the category of *government* can be in all levels. This is because the serving radius of web sites in this category can be either a city, a state or a country. To solve this problem, we add a new level: *hybrid level*. If a web site is in a hybrid level category, we regard its content location as its serving area directly, because no unified radius can be assigned.

Web content classification is an extension to text classification, which involves a training phase and a testing phase. During the training phase, a set of web pages or sites with known category labels is used to train a classifier. During the testing phase, the trained classifier is used to classify new web content. For classifying web sites, there are already many research works [7][8][9][10][14] in the literature. We choose *Supported Vector Machine (SVM)* [12][13][15] since it has been proved to be an efficient algorithm in many applications.

Provider location can be acquired from existing business databases such as Yellow Pages or other commercial contact information databases. In many cases, provider locations can also be found in the home pages or contact pages.

Now we conclude our algorithm as three steps:

1. Classify web sites into categories according to their business types.
2. Assign each category a serving radius according to search logs or common knowledge. Radius here can be either numeric, like miles, or levels.
3. Compute provider locations of the web resources. We refer to the algorithm proposed in [6]. If the provider location is not available, content location will be regarded as an approximation.

The main difficulties for this approach are the assignment of radius for different categories and the definition of categories. It is difficult and time-consuming to propose a complete categorization scheme for every local site.

# 5 Experiments

## 5.1 Gazetteers and Search Logs

The gazetteers we used are extracted from various sources on the Web [11][21][22][23]. There are six attributes for each location: name, population, longitude, latitude, postal code and area code. Our experiments focus on the geographical scope of United States. Therefore, the gazetteers just cover all the cities and states in US.

The log used in our experiments is a 30-day collection of search logs from a famous search engine. We define web resources in the unit of web sites. Each URL will be converted to the corresponding domain. Besides, a commercial IP to location database is used in our experiments, which contains the mapping relationship between IP addresses and corresponding geographical locations.

## 5.2 Results for Using User IP Locations

We first evaluated the performance of detecting serving area of web resources from user IP locations. We chose 535 USA governmental sites whose top domains are *.gov* as the test set. The sites were selected so that they have more than 200 clicks in the search log. The governmental sites can be easily labeled. For example, the serving area of *Wisconsin Government Home Page* (www.wisconsin.gov) is *Wisconsin state*. We manually labeled all the test sites with correct serving areas in advance.

As described in the previous section, thresholds for spread and weight were used to prune the traveling on the geographical tree. Fig. 1 shows the impact of Ts (threshold for spread) on the final performance. As we can see, F–measure reaches the highest point 0.87 when Ts = 0.5. Recall and F-measure will drop dramatically when Ts is larger than 0.8. In the following experiments, we will fix *Ts* = 0.5.

The performance of the algorithms is also greatly affected by the number of clicks available in the log. If there are more clicks for a web site, more precise results are expected to be obtained. From Fig. 2, we can clearly see that with the increasing of click count, F-measure increases quickly. In addition, the click count has a much bigger impact on recall than precision.

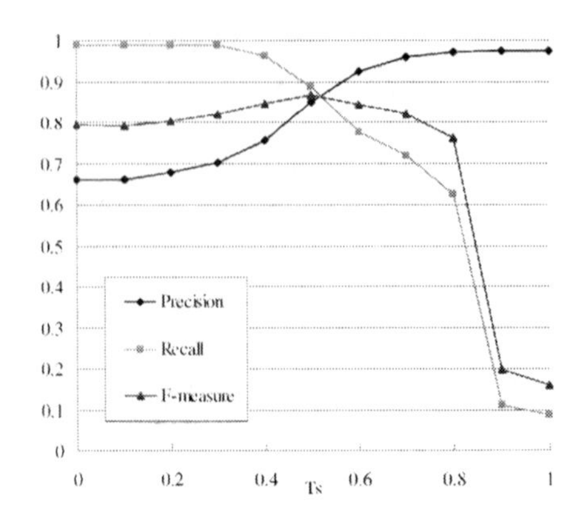


Fig. 1. Impact of Ts on computing serving area from IP locations

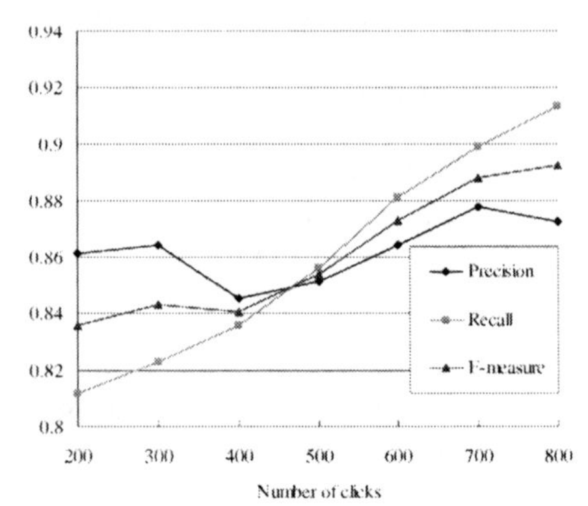


Fig. 2. Impact of click count on computing serving area from IP locations

### 5.3 Results for Using Query Terms

As shown in the example of section 4.2, users tend to input location names in queries when searching local information. In this section, we will test the performance of using query terms for serving area detection.

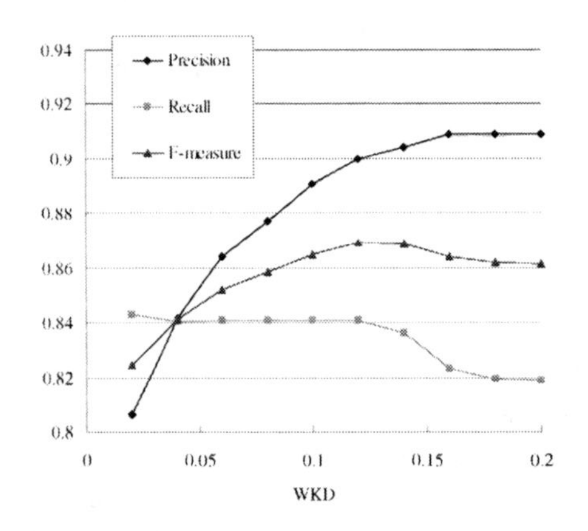


**Fig. 3.** Impact of *Well-Know Degree* (*WKD*) on computing serving area from query terms

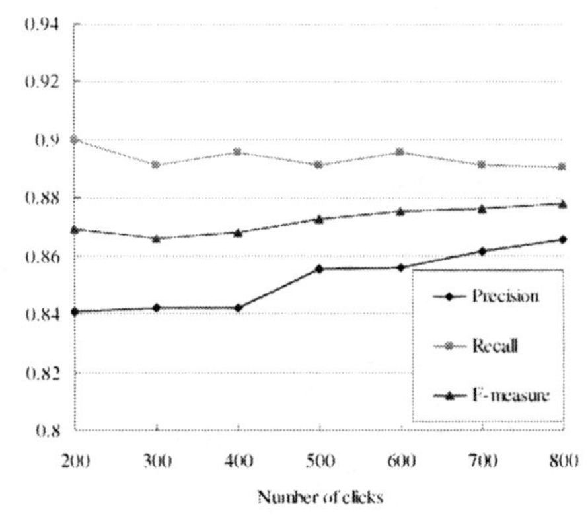


**Fig. 4.** Impact of click count on computing serving area from query terms

We use the same test set of 535 USA government sites as in the previous sub-section. According to our log, every log item will have a user IP and a query. For each web site, we do not group identical queries together. Therefore, the number of queries equals to the number of user IPs here.

When analyzing the collection of query items, we try to study how the number of query terms that contain location *l* affects the performance. In other words, how many people who think the web site relates to *l* will make us believe it is truly serving *l*?

We define a *Well-Known Degree* (*WKD*) in the algorithm. When *WKD* is 0.05, it means a web site is serving location *l* only if more than 5% of query terms contain *l*. Figure 3 shows the impact of *WKD* on the performance of our algorithm. From the figure, we find that when *WKD* is 0.12, F-measure reaches the best value of 0.87. In the following experiments, we will fix *WKD=0.12*.

Fig. 4 shows the performance of computing serving area from query terms. Comparing Fig. 4 and Fig. 2, we found that the performance of using query terms is more stable than that of using IP locations. The F-measure of query term based approach is better than that of IP location based approach when the click count is less than 600. The main reason here is that IP locations are usually not very precise. Therefore, in most cases, query term information will be considered superior to user IPs.

### 5.4 Results for Web Content Classification

In this experiment, we randomly chose 3162 web sites from ten categories in the *Open Directory Project (ODP)* [24] which is the largest human-edited directory on the Web. The names of these ten categories are shown in the first column of Table 1. Each web site has already been annotated with location information by *ODP* editors. This location information is regarded as the actual serving area of these web sites.

In the experiments, support vector machine (SVM) with a linear kernel is used as the classifier. Our data set is divided into two parts: training set (70%) and test set (30%). The performance of classification on our data is listed in Table 1. In the table, *P*, *R* and *F* stand for precision, recall and F-measure respectively. *Train* and *Test* columns represent the number of sites included in training set and test set.

**Table 1.** Performance of web site classification

| Category Name | Radius Level | Train | Test | P | R | F |
|---|---|---|---|---|---|---|
| Limousines and shuttles | City level | 93 | 39 | 0.97 | 0.85 | 0.90 |
| Airport | State level | 102 | 43 | 0.95 | 0.86 | 0.90 |
| Government | Hybrid level | 180 | 76 | 0.96 | 0.87 | 0.91 |
| Apartment rentals | City level | 131 | 55 | 0.80 | 0.67 | 0.73 |
| Hardware retailers | Country level | 340 | 144 | 0.83 | 0.85 | 0.84 |
| Real estate residential | City level | 350 | 149 | 0.95 | 0.95 | 0.95 |
| Software development | Country level | 341 | 146 | 0.87 | 0.86 | 0.87 |
| Software retailers | Country level | 99 | 42 | 0.73 | 0.45 | 0.56 |
| Zoos and aquariums | State level | 93 | 39 | 0.90 | 0.90 | 0.90 |
| Restaurants and bars | City level | 150 | 64 | 0.98 | 0.89 | 0.93 |
| Average | | - | - | 0.90 | 0.82 | 0.86 |

Table 1 shows that the SVM classifier is efficient for most of the categories (0.86 for average F-measure). There are two categories which have pretty low F-measure, *software retailer* and *apartment rentals*. The serving radius levels for each category are also shown in Table 1 (In the second column).

Due to the lack of address strings in many web sites and the difficulties in recognizing whether an address string is the provider location, we use content location to approximate the provider location in our experiments. Serving area of web resources can be computed by combining serving radius and provider location. The final performance of our algorithm is shown in Table 2.

In Table 2, we compare our algorithm with a straightforward algorithm that directly uses content location detection [6] results as the serving area. As we can see, the results of categorization based algorithm (average F-measure is 0.86) are much better than that of the content only algorithm (average F-measure is 0.61), especially when the serving radius is a state or a country. The reason here can be illustrated by an example. Suppose a web site is in the category of *software development*. The company who owns this site locates in *California*. It is very likely to have *California* as its content location by content analysis. In our algorithm, we will recognize this site as covering the whole USA, because it's serving radius is defined to be country level. On the contrary, if we use content location as the serving area directly, we will mistake *California* as its serving area, which is incorrect for this software company.

**Table 2.** Performance comparison of our algorithm with the content location only algorithm

| Category Name | F-measure(our algorithm) | F-measure(content only) |
|---|---|---|
| Limousines and shuttles | 0.70 | 0.70 |
| Airport | 0.94 | 0.60 |
| Government | 0.98 | 0.98 |
| Apartment rentals | 0.71 | 0.71 |
| Hardware retailers | 1 | 0.37 |
| Real estate residential | 0.72 | 0.72 |
| Software development | 0.59 | 0.59 |
| Software retailers | 1 | 0.44 |
| Zoos and aquariums | 1 | 0.41 |
| Restaurants and bars | 0.93 | 0.59 |
| Average | 0.86 | 0.61 |

## 6 Conclusion

In this paper, we studied the serving area of web resources, which stands for the geographical distribution of their potential users. Knowing the serving area is important to improve the performance of certain web applications such as local search and local advertisement.

Experimental results showed that all the algorithms we proposed worked well while the query term based algorithm was more effective than the IP location based approach.

For web resource without sufficient logs, the performance of the classification based algorithm was much better than that of a content location only algorithm.

## References

1. Ding, J., Gravano, L., Shivakumar, N.: Computing geographical scopes of web resource. In: 26th International Conference on Very Large Data Bases (VLDB'00), Cairo, Egypt (September 2000)
2. Buyukkokten, O., Cho, J., Garcia-Molina, H., Gravano, L., Shivakumar, N.: Exploiting geographical location information of web pages. In: ACM SIGMOD Workshop on the Web and Databases 1999 (WebDB'99), Philadelphia (June 1999)
3. Yokoji, S., Takahashi, K., Miura, N.: Kokono search: a location based search engine. In: 10th International World Wide Web Conference (WWW01), Hong Kong (May 2001)
4. Kosala, R., Blocakeel, H.: Web mining research: a survey. In: 6th ACM SIGKDD International Conference on Knowledge Discovery and Data Mining (KDD'00), Boston (August 2000)
5. Amitay, E., Har'El, N., Sivan, R., Soffer, A.: Web-a-where: geotagging web content. In: Proceedings of the 27th SIGIR, pp. 273–280 (2004)
6. Wang, C., Xie, X., Wang, L., Lu, Y., Ma, W.-Y.: Detecting Geographic Locations from Web Resources. In: The 2nd Internatinal Workshop on Geographic Information Retrieval (GIR 2005), ACM Fourteenth Conference on Information and Knowledge Management (CIKM 2005), Bremen (October 2005)
7. Dumais, S., Chen, H.: Hierarchical classification of web content. In: Proceeding of SIGIR-00, 23rd ACM International Conference on Research and Development in Information Retrieval, Athens, Greece, pp. 256–263. ACM Press, New York (2000)
8. Glover, E.J., Tsioutsiouliklis, K., Lawrence, S., Pennock, D.M., Flake, G.W.: Using web structure for classifying and describing web pages. In: Proceedings of the Eleventh International Conference on World Wide Web, pp. 562–569. ACM Press, New York (2002)
9. Yang, Y., Slattery, S., Ghani, R.: A study of approaches to hypertext categorization. Journal of Intelligent Information Systems
10. Gravano, L., Hatzivassiloglou, V., Lichtenstein, R.: Categorizing web queries according to geographical locality. In: 12th ACM Conference on Information and Knowledge Management (CIKM'03), New Orleans (November 2003)
11. CITY-DATA.COM. http://www.city-data.com
12. Burges, C.J.C.: A tutorial on support vector machines for pattern recognition. Data Mining and Knowledge Discovery 2(2), 121–167 (1998)
13. Platt, J.: Fast training of support vector machines using sequential minimal optimization. In: Schölkopf, B., Burges, C.J.C., Smola, A.J. (eds.) Advances in Kernel Methods—Support Vector Learning, pp. 185–208. MIT Press, Cambridge (1999)
14. Chakrabarti, S., Dom, B., Indyk, P.: Enhanced hypertext categorization using hyperlinks. In: Proceedings of the 1998 ACM SIGMOD international conference on Management of data (1998)
15. Hearst, M.A.: Trends and controversies: support vector machines. IEEE Intelligent Systems 13(4), 18–28 (1998)
16. Hill, L.L., Frew, J., Zheng, Q.: Geographic names: the implementation of a gazetteer in a georeferenced digital library. Digital Library, 5(1) (January 1999)

17. Iko, P., Takahiko, S., Katsumi, T., Masaru, K.: User behavior analysis of location aware search engine. In: 3rd International Conference on Mobile Data Management (MDM'02), Singapore (January 2002)
18. McCurley, K.S.: Geographical mapping and navigation of the web. In: 10th International World Wide Web Conference (WWW01), Hong Kong (May 2001)
19. Google Local Search. http://www.google.com/local
20. MSN Local Search. http://search.msn.com/local
21. Geographic Names Information System (GNIS). http://geonames.usgs.gov/
22. North American Numbering Plan. http://sd.wareonearth.com/ phil/npanxx
23. USPS – The United States Postal Services. http://www.usps.com
24. Open Directory Project. http://dmoz.org/

# A General Framework to Implement Topological Relations on Composite Regions

Magali Duboisset[1], François Pinet[1], Myoung-Ah Kang[2], and Michel Schneider[2]

[1] Cemagref, Clermont Ferrand, France
{magali.duboisset,francois.pinet}@cemagref.fr
[2] Laboratory of Computer Science, Modeling and System Optimisation (LIMOS)
Blaise Pascal University, Clermont Ferrand, France
{kang,schneider}@isima.fr

**Abstract.** Many GIS (Geographic Information Systems) handle composite geometries, i.e. geometries made of the union of simple shapes. Recent works show that relations between composite regions can be modelled with the well-known 9-Intersection Method (9IM). In this case, each relation is represented by a matrix. The proposed paper presents a general method to deduce the topological relations between the "parts" of regions from the matrix representation. Thus relations between composite regions could be easily implemented.

## 1 Introduction

The specification of topological relations on regions composed of several parts remains difficult. Existing methods very often propose a particular semantics for topological predicates applied on this type of regions [2][3][4][7][8].

**Using 9-Intersection Method to model relations between composite regions.** As presented in [1] and [8], a relation between two composite regions can be represented by a matrix in using the well-known 9-Intersection Method [5][6]. This model constitutes an excellent mean to distinguish an interesting number of possible topological relations; in using 9IM, the authors of [1] and [8] presents 33 possible topological relations between two composite regions with or without holes.

To illustrate the model let us consider the following 9IM matrix $R$ which represents a topological relation between two composite regions $A$ and $B$.

$$R = \begin{array}{c} \\ A^\circ \\ \partial A \\ A^- \end{array} \begin{array}{c} \begin{array}{ccc} B^\circ & \partial B & B^- \end{array} \\ \begin{pmatrix} 1 & 0 & 1 \\ 1 & 0 & 1 \\ 1 & 1 & 1 \end{pmatrix} \end{array}$$

The 9IM matrix represents the intersections between the interiors, the boundaries and the exteriors of 2 spatial objects. The result of these intersections might be empty (0) or not (1). The interior, the boundary and the exterior of a spatial object $O$ are

R. Wagner, N. Revell, and G. Pernul (Eds.): DEXA 2007, LNCS 4653, pp. 823–833, 2007.

respectively denoted by $O°$, $\partial O$ and $O^{-}$. The interior, the boundary and the exterior of the composite region *A* are respectively the union of the interiors, the union of the boundaries and the intersection of the exteriors of all parts of *A*.

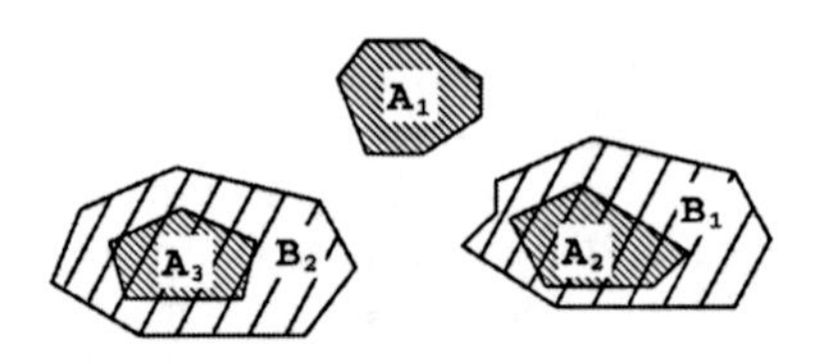


**Fig. 1.** Drawing of a spatial configuration corresponding to matrix *R*. There are two composite regions *A* and *B* made of several simple parts.

Figure 1 illustrates a possible spatial scene corresponding to the matrix *R*. In this example, the composite region *A* is composed of 3 parts (i.e. 3 simple regions, $A_1$, $A_2$ and $A_3$) and the composite region *B* is composed of 2 parts (i.e. 2 simple regions, $B_1$ and $B_2$). According to the matrix *R*, the interiors of *A* and *B* must intersect each other (cf. the value of the cell in first line first column); this is the case in figure 1 because the interior of at least one part of *A* intersects the interior of at least one part of *B*. In the matrix *R*, the boundary of *A* must intersect the interior of *B*; this is the case in figure 1 because the boundary of at least one part of *A* intersects the interior of at least one part of *B*, etc.

**Deducing the topological relations between the "parts".** The relations between each part of two composite regions do not appear explicitly in the definition of the 9IM matrix. In the previous example, the relations "disjoint" and "inside/contains" between the parts are not "explicit" in the matrix (see the drawing in figure 1 – for instance "$A_3$ is inside $B_2$"). This is due to the fact that a 9IM matrix represents the intersections between the interior, the boundary and the exterior of the whole of each spatial object involved in the relation.

Thus, our objective is to propose a transformation method to deduce the basic topological relations between parts, from each 9IM matrix representing a relation between 2 composite regions. More precisely, we suggest a logical expression semantically equivalent to a 9IM matrix; each element in this expression involves only topological relations between parts of composite regions. In this paper, we will consider only composite regions without holes.

For example, our method produces from the matrix *R* (previously presented) the following output. This is a logical expression presenting the possible relations between the parts of *A* and *B*:

$$(\exists i \in 1..n,\ \exists j \in 1..m \mid \langle A_i,\ \text{inside},\ B_j \rangle) \wedge$$
$$(\exists k \in 1..n,\ \forall l \in 1..m \mid \langle A_k,\ \text{disjoint},\ B_l \rangle) \wedge$$
$$(\forall r \in 1..n,\ \forall s \in 1..m,\ \langle A_r,\ \text{inside},\ B_s \rangle \vee \langle A_r,\ \text{disjoint},\ B_s \rangle)$$

In this expression, $A_i$ and $B_j$ are simple regions i.e. parts of the composite regions *A* and *B* involved in the relation. **This logical expression and the matrix R have the**

**same semantics; they represent the same topological relation.** The drawing of figure 1 verifies this logical expression.

**Implementing relations between the composite regions.** The main interest of this logical expression is to allow a systematic coding of procedures which check if a topological relation between two composite regions is true or false. We show that these procedures can be coded in any language supporting an implementation of the 8 Egenhofer's basic relations (disjoint, contains, inside, equal, meet, covers, coveredBy, overlap) which can exist between parts.

For example, at present, no DBMS' SQL implements all the topological relations proposed in [1][8] but the Oracle Spatial SQL supports the 8 Egenhofer's basic relations cited before. Thus, as illustrated in this paper, it becomes possible to check relations between composite regions (without holes) in Oracle Spatial SQL. In this case, we can benefit of the optimization already developed in the implementation of these 8 basic relations as well as the associated existing functionalities (e.g. support of different spatial coordinate systems). The programmers do not need to develop new geographic functions to check relations on composite regions.

The remainder of the paper is organized as follows: section 2 introduces the 9-intersection method (9IM). Section 3 presents our transformation method, which translates 9IM matrixes into logic-based expressions. Section 4 is dedicated to the mapping to Oracle Spatial SQL queries. Lastly, section 5 draws some conclusions and discusses future works.

## 2 Overview of 9IM

A well-known model for specifying topological relations is the 9-intersection model (9IM) [5][6]. It provides a methodology to characterize topological relations between two spatial objects.

In 9IM, each topological relation is represented by a matrix. This matrix presents the intersections of boundary, interior and exterior of two spatial objects. The result of these 9 intersections might be empty (0) or not (1). The interior, the boundary and the exterior of a spatial object $O$ are respectively denoted by $O°$, $\partial O$ and $O^{-}$. Thus, each topological relation between two spatial objects $A$ and $B$ is represented by a 3x3 matrix whose coefficients correspond to the results of the intersections between $A°$, $\partial A$, $A^{-}$ and $B°$, $\partial B$, $B^{-}$ as shown in figure 2.

$$M = \begin{pmatrix} A° \cap B° \neq \varnothing & A° \cap \partial B \neq \varnothing & A° \cap B^{-} \neq \varnothing \\ \partial A \cap B° \neq \varnothing & \partial A \cap \partial B \neq \varnothing & \partial A \cap B^{-} \neq \varnothing \\ A^{-} \cap B° \neq \varnothing & A^{-} \cap \partial B \neq \varnothing & A^{-} \cap B^{-} \neq \varnothing \end{pmatrix}$$

**Fig. 2.** Matrix M characterizing a topological relation between two spatial objects A and B [6]

In theory, there are $2^9 = 512$ matrixes. However, some of them are inconsistent; they cannot be drawn in a 2 dimensional space.

For two simple regions, 8 meaningful configurations have been identified which lead to the 8 predicates illustrated in figure 3.

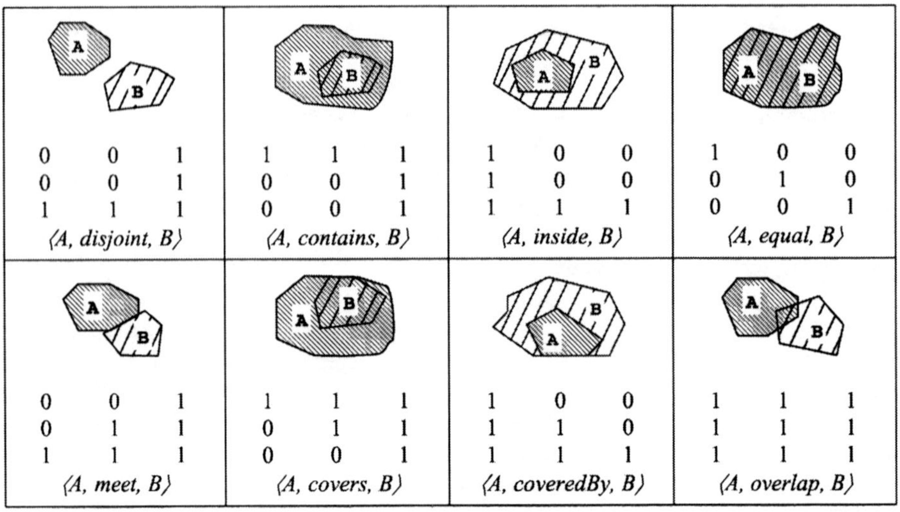

| | | | |
|---|---|---|---|
| 0 0 1<br>0 0 1<br>1 1 1 | 1 1 1<br>0 0 1<br>0 0 1 | 1 0 0<br>1 0 0<br>1 1 1 | 1 0 0<br>0 1 0<br>0 0 1 |
| *⟨A, disjoint, B⟩* | *⟨A, contains, B⟩* | *⟨A, inside, B⟩* | *⟨A, equal, B⟩* |
| 0 0 1<br>0 1 1<br>1 1 1 | 1 1 1<br>0 1 1<br>0 0 1 | 1 0 0<br>1 1 0<br>1 1 1 | 1 1 1<br>1 1 1<br>1 1 1 |
| *⟨A, meet, B⟩* | *⟨A, covers, B⟩* | *⟨A, coveredBy, B⟩* | *⟨A, overlap, B⟩* |

**Fig. 3.** 8 Egenhofer's basic topological relations between two simple regions [6]

In [1] and [8], the authors extend this model to complex geographic objects. This paper only focuses on composite regions without holes defined as follows.

**Definition 1.** Region abstract model.
A simple region is a closed connected point set without holes in a 2-dimensional space $R^2$. A composite region is a set $CR = \{R_1, ..., R_i, ..., R_n\}$ where $R_i$ is a simple region also called "part" of $CR$.

We define that: $\forall i \neq j$, $R_i^\circ \cap R_j^\circ = \varnothing$ and $\partial R_i \cap \partial R_j = \varnothing$ to avoid the cases where two or more parts form something similar to a hole. □

The interior, the boundary and the exterior of a composite region are respectively the union of the interiors, the union of the boundaries and the intersection of the exteriors of all its parts. The 9IM matrixes applied on composite regions have the same interpretation as the one presented above; the matrix coefficients are the results of the intersections of boundaries, interiors and exteriors of two composite regions without taking into account their number of parts. Therefore, the relations between each part of two composite regions do not appear explicitly in the definition of the matrix.

In [1], the authors enumerated all the possible matrixes for composite regions; they identified 33 topological relations between composite regions (with and without holes).

## 3 Mapping Theorem

In this paper, we only consider the topological relations between composite regions (CR) without holes. Our proposal of mapping theorem (theorem 1) returns the "logical expression" equivalent to a 9IM matrix; they both represent the same relation. In the proposed logical expressions, quantifiers and topological operators on simple geometries are combined (see the example in section 1). This logical form highlights the set of topological relations between the parts of the two considered CR.

In this section, we will show that all the topological relations between CR without holes defined in 9IM could be rewritten in a logical form.

We first identify the set of matrixes that characterize relations between CR with holes among the 33 matrixes of [1]. These matrixes represent relations that imply at least one region with one or several holes.

**Definition 2.** 9IM matrix characterizing topological relations between composite regions with holes.

Let *A* and *B* be two non-empty composite regions. Two conditions imply that at least *A* or *B* contains holes. They can be written as follows:

i) $(A^{\circ} \cap B^{-} = \neg\varnothing) \wedge (\delta A \cap B^{-} = \varnothing)$, and,
ii) $(A^{-} \cap B^{\circ} = \neg\varnothing) \wedge (A^{-} \cap \delta B = \varnothing)$. □

The matrix patterns of these 2 cases are given in table 1. From this characterization it results that among the 33 matrixes of [1], 17 involve a composite region with holes.

**Table 1.** Matrix patterns of topological relations between composite regions with holes. $x \in \{0;1\}$.

| Case (i) | | | Case (ii) | | |
|---|---|---|---|---|---|
| x | x | 1 | x | x | x |
| x | x | 0 | x | x | x |
| x | x | x | 1 | 0 | x |

**Corollary 1.** 9IM matrix characterizing topological relations between composite regions without holes.

All relations between 2 non-empty composite regions without holes can be represented by the 16 remaining matrixes. If the 2 regions *A* and *B* are without holes, the interior of *A* cannot be outside *B* while the boundary of *A* is not outside *B*, and reciprocally. For the following, we note *M* the set of these 16 matrixes. □

We will show that the 16 matrixes of this set *M* can be rewritten into a logical form. First, it appears that each of these 16 matrixes can be defined as a logical combination of topological relations between the parts of the composite regions involved in the relation. In these combinations each topological relation between 2 parts corresponds to a "factorization matrix".

We define the concept of factorization matrix.

**Definition 3.** Factorization matrixes.
The factorization matrixes are 9IM matrixes that represent topological relations between parts of composite regions.

The set of factorization matrixes is $F = \{fContains, fInside, fEqual, fMeet, fCovers, fCoveredBy, fOverlap, fDisjoint1st, fDisjoint2nd\}$. □

The 9IM matrixes that correspond to the factorization matrixes are given in table 2.

**Table 2.** Factorization matrixes

| | | |
|---|---|---|
| 1 1 1<br>0 0 1<br>0 0 1<br>⟨A, fContains, B⟩ | 1 0 0<br>1 0 0<br>1 1 1<br>⟨A, fInside, B⟩ | 1 0 0<br>0 1 0<br>0 0 1<br>⟨A, fEqual, B⟩ |
| 0 0 1<br>0 1 1<br>1 1 1<br>⟨A, fMeet, B⟩ | 1 1 1<br>0 1 1<br>0 0 1<br>⟨A, fCovers, B⟩ | 1 0 0<br>1 1 0<br>1 1 1<br>⟨A, fCoveredBy, B⟩ |
| 1 1 1<br>1 1 1<br>1 1 1<br>⟨A, fOverlap, B⟩ | 0 0 1<br>0 0 1<br>0 0 1<br>⟨A, fDisjoint1st, B⟩ | 0 0 0<br>0 0 0<br>1 1 1<br>⟨A, fDisjoint2nd, B⟩ |

There are 2 cases for the disjoint relation: *fDisjoint1st* and *fDisjoint2nd*. Indeed, the disjoint relation applied to composite regions' parts is not always commutative. One part of *A* can be disjoint of each part of *B*, while no parts of *B* are disjoint from *A*. ⟨*A, fDisjoint1st, B*⟩ means that at least one part of *A* is disjoint from every part of *B*. Reciprocally, ⟨*A, fDisjoint2nd, B*⟩ means that at least one part of *B* is disjoint from every part of *A*.

We now introduce theorem 1 which details how a topological relation between two composite regions can be expressed into a logical combination of factorization matrixes.

**Theorem 1**
Let *R* be a matrix of the set *M* defined in corollary 1. Let $c_i$ be a boolean and $b_i$ a factorization matrix of the set *F*:

$$R = \sum_{i \in \{1..9\}} c_i b_i$$

Notice that $c_8$ cannot be the only non-null coefficient, as well as $c_9$ as they cannot exist between simple geometries. □

**Sketch.** The interior, the boundary and the exterior of a composite region are respectively the union of the interiors, the union of the boundaries and the intersection of the exteriors of all its parts. Thus, a matrix specifying a relation between 2 composite regions is equal to the logical union of factorization matrixes, i.e. the Egenhofer matrixes that represent the topological relations between composite regions' parts. □

**Example 1**

We recall the matrix $R$ presented in section 1.

$$R = \begin{bmatrix} 1 & 0 & 1 \\ 1 & 0 & 1 \\ 1 & 1 & 1 \end{bmatrix}$$

We have to determine all the booleans from $c_1$ to $c_9$ such that:

$R = c_1.fContains + c_2.fInside + c_3.fEqual + c_4.fMeet + c_5.fCovers + c_6.fCoveredBy + c_7.fOverlap + c_8.fDisjoint1st + c_9.fDisjoint2nd$

More explicitly we have :

$$\begin{bmatrix} 1 & 0 & 1 \\ 1 & 0 & 1 \\ 1 & 1 & 1 \end{bmatrix} = c_1. \begin{bmatrix} 1 & 1 & 1 \\ 0 & 0 & 1 \\ 0 & 0 & 1 \end{bmatrix} + c_2. \begin{bmatrix} 1 & 0 & 0 \\ 1 & 0 & 0 \\ 1 & 1 & 1 \end{bmatrix} + c_3. \begin{bmatrix} 1 & 0 & 0 \\ 0 & 1 & 0 \\ 0 & 0 & 1 \end{bmatrix} + c_4. \begin{bmatrix} 0 & 0 & 1 \\ 0 & 1 & 1 \\ 1 & 1 & 1 \end{bmatrix} + c_5. \begin{bmatrix} 1 & 1 & 1 \\ 0 & 1 & 1 \\ 0 & 0 & 1 \end{bmatrix}$$

$$+ c_6. \begin{bmatrix} 1 & 0 & 0 \\ 1 & 1 & 0 \\ 1 & 1 & 1 \end{bmatrix} + c_7. \begin{bmatrix} 1 & 1 & 1 \\ 1 & 1 & 1 \\ 1 & 1 & 1 \end{bmatrix} + c_8. \begin{bmatrix} 0 & 0 & 1 \\ 0 & 0 & 1 \\ 0 & 0 & 1 \end{bmatrix} + c_9. \begin{bmatrix} 0 & 0 & 0 \\ 0 & 0 & 0 \\ 1 & 1 & 1 \end{bmatrix}$$

It leads to the following system of boolean sums:

($e_{11}$) $c_1 + c_2 + c_3 + c_5 + c_6 + c_7 = 1$
($e_{12}$) $c_1 + c_5 + c_7 = 0$
($e_{13}$) $c_1 + c_4 + c_5 + c_7 + c_8 = 1$
($e_{21}$) $c_2 + c_6 + c_7 = 1$
($e_{22}$) $c_3 + c_4 + c_5 + c_6 + c_7 = 0$
($e_{23}$) $c_1 + c_4 + c_5 + c_7 + c_8 = 1$
($e_{31}$) $c_2 + c_4 + c_6 + c_7 + c_9 = 1$
($e_{32}$) $c_2 + c_4 + c_6 + c_7 + c_9 = 1$
($e_{33}$) $c_1 + c_2 + c_3 + c_4 + c_5 + c_6 + c_7 + c_8 + c_9 = 1$

The equations ($e_{13}$) and ($e_{23}$) are equivalent so as ($e_{31}$) and ($e_{32}$).
The equation ($e_{12}$) implies $c_1 = c_5 = c_7 = 0$.
The equation ($e_{22}$) implies $c_3 = c_4 = c_5 = c_6 = c_7 = 0$.

It remains:

($e_{11}$) $c_2 = 1$
($e_{13}$) $c_8 = 1$
($e_{21}$) $c_2 = 1$
($e_{31}$) $c_2 + c_9 = 1$
($e_{33}$) $c_2 + c_8 + c_9 = 1$
($e_4$) $c_1 = c_3 = c_4 = c_5 = c_6 = c_7 = 0$

($e_{11}$) and ($e_{13}$) imply that $c_2 = 1$ and $c_8 = 1$.

The final solution of this system is:

$$(c_2=1 \text{ and } c_8=1) \text{ and } (c_1=c_3=c_4=c_5=c_6=c_7=0).$$

The value of $c_9$ is not significant as in ($e_{31}$) and ($e_{33}$), $c_2$ is never null. □

The next definition provides a logical expression for each of the factorization matrices. Since a factorization matrix can be expressed into a logical form, the overall combination can be also expressed into a logical form.

**Definition 4**
A factorization matrix that corresponds to one of the first seven Egenhofer relations (i.e. *fContains*, *fInside*, *fEqual*, *fMeet*, *fCovers*, *fCoveredBy*, *fOverlap*), can be expressed into the following logical expression:

$\langle A, fTopologicalRel, B\rangle \rightarrow \exists r\in 1..n, \exists s\in 1..m \mid \langle A_r, \texttt{topologicalRel}, B_s\rangle$

where $A_r$ and $B_s$ are parts of the composite regions *A* and *B*, and *topologicalRel* is an Egenhofer relation of figure 3.

The matrixes *fDisjoint1st* and *fDisjoint2nd* can be expressed as follows:

$\langle A, fDisjoint1st, B\rangle \rightarrow \exists r\in 1..n, \forall s\in 1..m \mid \langle A_r, \texttt{disjoint}, B_s\rangle$
$\langle A, fDisjoint2nd, B\rangle \rightarrow \forall r\in 1..n, \exists s\in 1..m \mid \langle A_r, \texttt{disjoint}, B_s\rangle$ □

In order to derive the logical expression of the total combination we must distinguish 2 parts in this combination: the first part with non-null coefficients and the second part with null coefficients. The mapping of these 2 parts into a logical form is given by theorem 2.

**Theorem 2**
*Logical expression of the first part (with non-null coefficients)*

Let $N_1$ be the set of *i* such as the coefficient $c_i$ is not null in the result of the application of theorem 1. The first part of the final logical form is the logical intersection of the factorization matrixes expressed in a logical form (cf. definition 4).

*Logical expression of the second part (with null coefficients)*

Let $N_2$ be the set of *i* such as the coefficient $c_i$ is null in the result of the application of theorem 1. The second part of the final logical form is then:

$$\forall r\in 1..n, \exists s\in 1..m \mid \bigvee_{i\in\{1..9\}\setminus N2} \langle A_r, \texttt{topologicalRel}_i, B_s\rangle$$

where $A_r$ and $B_s$ are parts of the composite regions *A* and *B*, and $topologicalRel_i$ is a relation of figure 3. *fDisjoint1st* and *fDisjoint2nd* are associated to *disjoint.* □

Thus, the topological relations that can exist between parts of the composite regions are, on the one hand, those corresponding to the non-null coefficients and, on the other hand, those that do not appear among the null-coefficients. The logical form could be seen as the logical intersection of 2 parts: mapping of non-null coefficients ∧ mapping of null coefficients.

The second part of the logical form is the union of each factorization matrixes that do not appear among the null coefficients, applied on all parts of the 2 composite regions. It corresponds to the union of each factorization matrixes that could be applied between parts of the 2 composite regions involved in the relation.

**Example 2.** Let us continue with example 1. The first part of the result, defined from the non-null coefficients, is ($c_2$=1 and $c_8$=1) leads to:

$$(\exists i \in 1..n,\ \exists j \in 1..m \mid \langle A_i,\ \text{inside},\ B_j \rangle) \wedge (\exists k \in 1..n,\ \forall l \in 1..m \mid \langle A_k,\ \text{disjoint},\ B_l \rangle)$$

The second part of the result, defined from the null coefficients,

($c_1 = c_3 = c_4 = c_5 = c_6 = c_7 = 0$) is rewritten as follows:

$$(\forall r \in 1..n,\ \forall s \in 1..m \mid \langle A_r,\ \text{inside},\ B_s \rangle \vee \langle A_r,\ \text{disjoint},\ B_s \rangle)$$

Thus, the final logical expression is the intersection of these 2 expression parts:

$$(\exists i \in 1..n,\ \exists j \in 1..m \mid \langle A_i,\ \text{inside},\ B_j \rangle) \wedge$$
$$(\exists k \in 1..n,\ \forall l \in 1..m \mid \langle A_k,\ \text{disjoint},\ B_l \rangle) \wedge$$
$$(\forall r \in 1..n,\ \forall s \in 1..m \mid \langle A_r,\ \text{inside},\ B_s \rangle \vee \langle A_r,\ \text{disjoint},\ B_s \rangle)$$

where $A_i$, $B_j$, $A_k$, $B_l$, $A_r$ and $B_s$ are parts of the composite regions $A$ and $B$ involved in the relation specified by $R$. One possible spatial configuration of 2 composite regions checking the topological relation $R$ is exemplified in figure 1. □

# 4 Example of Implementation

This section illustrates how the generated logical expressions can be used to obtain an implementation of procedures for checking topological relations on composite regions. Indeed, the produced logical expressions provide all the conditions that must be checked between the parts of the composite regions. These conditions can be checked in different programming or query languages.

Thus, a major interest of our work is to propose a generic mechanism to obtain a direct implementation of procedures checking topological relations between composite regions in languages and systems supporting the 8 Egenhofer's basic relations. Thus, we can benefit of the optimizations already developed in the implementation of these 8 relations.

To illustrate this mechanism we consider the matrix $R$ of example 1. Our goal is to elaborate the procedure which checks if the relation $R$ between two CR is true or false. Parts of the 2 composite regions $A$ and $B$ are stored in a relational table (named TCR). The SQL query which checks the relation can be easily derived from the logical expression obtained in example 2. The query is composed of 3 subqueries connected with an intersect operator. It returns a row containing a 'TRUE' record if the relation is true, or no row if it is false.

```
-- (∃i∈1..n, ∃j∈1..m | ⟨Ai, inside, Bj⟩):
(select distinct 'TRUE' from TCR TA, TCR TB where
   TA.CR_ID='A' and TB.CR_ID='B' and
   MDSYS.SDO_RELATE(TA.PART_GEO, TB.PART_GEO,
                         'mask=INSIDE querytype=WINDOW')='TRUE')
intersect
--(∃k∈1..n, ∀l∈1..m | ⟨Ak, disjoint, Bl⟩):
(select distinct 'TRUE' from TCR TA, TCR TB where
   TA.CR_ID='A' and TB.CR_ID='B' and
   MDSYS.SDO_RELATE(TA.PART_GEO,TB.PART_GEO,
                         'mask=DISJOINT querytype=WINDOW')='TRUE'
   GROUP BY TA.part_id
   having count(*) = (select count (*) from TCR where CR_ID= 'B'))
intersect
-- (∀r∈1..n, ∀s∈1..m, ⟨Ar, inside, Bs⟩ ∨ ⟨Ar, disjoint, Bs⟩):
(select distinct 'TRUE' from tcr where
   (select count(*) from TCR TA, TCR TB
    where TA.CR_ID='A' and TB.CR_ID='B')
   = (select count(*) from TCR TA, TCR TB where
      TA.CR_ID='A' and TB.CR_ID='B' and
      MDSYS.SDO_RELATE(TA.PART_GEO,TB.PART_GEO,
      'mask=DISJOINT+INSIDE querytype=WINDOW')='TRUE'))
```

## 5 Conclusion and Perspectives

A well-known method for specifying topological relations is the 9-intersection model (9IM). In the context of composite regions, each topological relation can be modelled by a 9IM matrix. However, the topological relations between each part do not appear explicitly in the definition of the matrix. The proposed paper offers the possibility to deduce the relations between parts from a 9IM matrix representing a relation between 2 composite regions without holes.

As presented in section 4, the proposed approach can facilitate the implementation of operations that check the relations on composite regions in different languages supporting the 8 Egenhofer relations. In this case, we can benefit, in a transparent manner, of the optimization already developed in the implementation of these 8 relations as well as the associated existing functionalities (e.g. support of different spatial coordinate systems). In using these already developed spatial operations, the programmers or the database users do not need to develop other geographic functions to check relations on composite regions. Thus our approach could be very useful to help the implementation of the integrity subsystem of a GIS.

This paper focuses on topological relations between composite regions without holes. In the future, we will try to generalize the proposed approach to consider topological relations between different kinds of composite geometries (e.g. points, lines, regions with holes) and between heterogeneous set of geometries. The optimization of the implementation is another important issue.

## References

[1] Behr, T., Schneider, M.: Topological Relationships of Complex Points and Complex Regions. In: Kunii, H.S., Jajodia, S., Sølvberg, A. (eds.) ER 2001. LNCS, vol. 2224, pp. 56–69. Springer, Heidelberg (2001)

[2] Claramunt, C.: Extending Ladkin's Algebra on Non-convex Intervals towards an Algebra on Union-of Regions. In: GeoInformatica Proc. of the Int. ACM Symposium on Advances in Geographic Information Systems, USA, pp. 9–14 (2000)

[3] Clementini, E., Di Felice, P., Califano, G.: Composite Regions in Topological Queries. Information Systems 20(7), 579–594 (1995)
[4] Clementini, E., Di Felice, P., Oosterom, P.: A Small Set of Formal Topological Relationships For End-User Interaction. In: Abel, D.J., Ooi, B.-C. (eds.) SSD 1993. LNCS, vol. 692, pp. 277–295. Springer, Heidelberg (1993)
[5] Egenhofer, M., Franzosa, R.: Point-Set Topological Spatial Relations. Int. Journal of Geographical Information Systems 5(2), 161–174 (1991)
[6] Egenhofer, M., Herring, Categorizing, J.: Binary topological relationships between regions, lines, and points in geographic databases. Technical Report, 1992. Department of Surveying Engineering, University of Maine, Orono, ME. http://www.cs.umn.edu/Research/shashi-group/CS8715/MSD11_egenhofer_herring.pdf
[7] Schneider, M.: Implementing Topological Predicates for Complex Regions. In: Symposium on Geospatial Theory, Processing and Applications, Ottawa (2002)
[8] Schneider, M., Behr, T.: Topological Relationships between Complex Spatial Objects. ACM Transactions on Database Systems (TODS) 31, 39–81 (2006)

# Active Adjustment: An Approach for Improving the Performance of the TPR*-Tree

Sang-Wook Kim[1], Min-Hee Jang[1], and Sungchae Lim[2]

[1] Department of Information and Communications, Hanyang University, Korea
wook@hanyang.ac.kr, zzmini@ihanyang.ac.kr
[2] Department of Computer Science, Dongduk Women's University, Korea
sclim@dongduk.ac.kr

**Abstract.** The TPR*-tree is most popularly accepted as an index structure for processing *future-time queries* in moving object databases. In the TPR*-tree, the future locations of moving objects are predicted based on the CBR(Conservative Bounding Rectangle). Since the areas predicted from CBRs tend to grow rapidly over time, CBRs thus enlarged lead to serious performance degradation in query processing. Against the problem, we propose a novel method to adjust CBRs to be tight, thereby improving the performance of query processing. Our method examines whether the adjustment of a CBR is necessary when accessing a leaf node for processing a user query. Thus, it does not incur extra disk I/Os in this examination. Also, in order to make a correct decision, we devise a cost model that considers the I/O overhead for the CBR adjustment and the performance gain in the future-time owing to the CBR adjustment. With the cost model, we can prevent unusual expansions of BRs even when updates on nodes are infrequent and also avoid unnecessary execution of the CBR adjustment. For performance evaluation, we conducted a variety of experiments. The results show that our method improves the performance of the original TPR*-tree significantly.

## 1 Introduction

The recent advances of technologies in mobile communications and global positioning systems have increased people's attentions to an effective use of location information on the objects that move in 2-dimensional space. *Moving objects* usually send their current positions to a central server in a periodic fashion. A database that stores the time-varying information of numerous objects' positions is called a *moving object database* [10].

The *future-time query* in a moving ojbect database is to predict moving objects' movements in the future-time [7]. The system answering the future-time queries can be used for the applications such as location-based services, traffic information services, and air traffic controls [2]. A typical example of future-time queries seems to be "Retrieve all the vehicles that will pass over the Golden Gate Bridge at 1 pm".

Many studies have been done to develop efficient index structures such as the VCR-tree [5], the TPR-tree [6], and the TPR*-tree [8] suitable for the process-

R. Wagner, N. Revell, and G. Pernul (Eds.): DEXA 2007, LNCS 4653, pp. 834–843, 2007.

ing of future-time queries [4]. Among them, the TPR*-tree has been reported to provide the best query performance by solving some shortcomings of the previous TPR-tree. The TPR*-tree basically adopts the data structure of the R*-tree [1] and uses the notion of a conservative bounding rectangle (CBR) to predict the future-time positions of the moving objects indexed in a node. The CBR is represented with a minimum bounding rectangle (MBR) [1] enclosing all the moving objects within it and a velocity vector expressing their maximum and minimum speeds along X and Y axes. To predict the future-time position of a moving object, we apply the velocity vector to its MBR and thus get a rectangular region wherein the object may move. Since the predicted region obtained from a CBR expands in a rapid and continuous rate, the TPR*-tree usually contains large dead space [1] in its index space, which causes serious performance degradation in query processing.

In this paper, we address a novel CBR adjustment method that can significantly reduce the size of dead space in CBRs. In our method, we make the CBR adjustment performed in query processing times. Our CBR adjustment comes from the idea that the I/O cost can be reduced if we use the CBR data read for query processing. If we cache the nodes accessed during query processing on buffer memory, the cached nodes can be used without extra I/O overhead thereafter. From that, we can avoid the I/O cost paid for the CBR adjustment. In our research, we call our CBR adjustment by the query processor the *active CBR adjustment* (ACA). As a result, the proposed method is able to continuously detect the unusual growing of the BRs, thereby efficiently reducing dead space. Also, in order to make a correct decision, we devise a cost model that considers the I/O overhead for the CBR adjustment and the performance gain in the future-time owing to the CBR adjustment. With the cost model, we can prevent unusual expansions of BRs even when updates on nodes are infrequent and also avoid unnecessary execution of the CBR adjustment. To show the superiority of our method, we conduct extensive experiments. The results show that our method provides significant performance improvement in query processing compared to the original TPR*-tree.

## 2 Related Work

### 2.1 TPR-Tree

The TPR-tree [6], which has been devised based on the R*-tree [1], predicts the future-time positions of moving objects by storing the current position and velocity of each object at a specific time point. To express the time-varying positions of moving objects, the TPR-tree uses the notion of the *conservative bounding rectangle(CBR)*.

A CBR is composed of two sorts of data: the *minimum bounding rectangle(MBR)* and a velocity vector. The MBR is a rectangle encompassing a group of moving objects indexed in a node, and the velocity vector represents the maximum and minimum speeds of the moving objects within the MBR along X and Y axes. If it is necessary to predict the future time position of a moving object, then the velocity vector is applied to the corresponding MBR in order to compute a rectangle covering the predictable positions of the queried object. Such a predicted rectangle is called the *bounding rectangle(BR)*. The CBR of a leaf node is to express the BR for the moving objects in that node, and the CBR of a non-leaf node is used to represent the BR covering the BRs of its child nodes. To process a user query $Q$ with a future prediction time $t$, the TPR-tree computes the BRs for time $t$ by expanding the CBRs saved in the root node and recursively searches down the sub-trees whose BRs overlap with the target query region of $Q$ [1, 6].

### 2.2 TPR*-Tree

The data structure and the query processing algorithm of the TPR*-tree [8] are very similar to those of the TPR-tree. A difference is found in their insertion algorithm. The TPR*-tree employs an insertion algorithm that considers how much the BR will sweep the index space from the insertion time to a specific future-time. For instance, consider two different time points of $t_1$ and $t_2$ ($t_1 < t_2$). The *sweeping region* of a BR from $t_1$ to $t_2$ is defined to be an index space area that is swept by the BR expanding during the time interval ($t_2$-$t_1$).

The insertion algorithm searches down the TPR*-tree for a leaf node, by recursively choosing the child pointers to the sub-trees where its insertion will occur. During the downward search, it chooses its insertion paths so that the overall sweeping region remains smallest. Consequently, the TPR*-tree can provide better performance in processing future-time queries because its compactness of CBRs.

## 3 The Proposed Method

### 3.1 Motivation

Because the CBR of the TPR*-tree stores only the maximum and minimum speeds of moving objects in the MBR, the BR predicted from the CBR enlarges in a rapid and continuous rate. Such rapid growth of the BR leads to huge dead space and thus causes large overlaps among nodes' BRs as time goes on.

In Figure 1(a), a CBR is created to contain the objects of $O_1$ and $O_2$ at the initial time 0 and it is denoted by rectangle $N$. Note that, at the initial time, the BR is identical to the MBR of the CBR. If those objects move at their velocities then their real positions at time 1 are enclosed by a minimum rectangle of $N''$ as in Figure 1(b). On the other hand, the BR for objects $O_1$ and $O_2$ at time 1 becomes larger than the minimum rectangle $N''$. Since the BR is expanded according to CBR's maximum and minimum speeds, it usually gets larger than is necessary. The BR is denoted by $N'$ in Figure 1(b) and the dead space in this case is the difference between $N'$ and $N''$

To prevent continuous BR's growth, the TPR-tree updates the CBR data in such a way that it covers the positions of objects more tightly, whenever any object pertaining to the CBR changes its velocity or location information. By

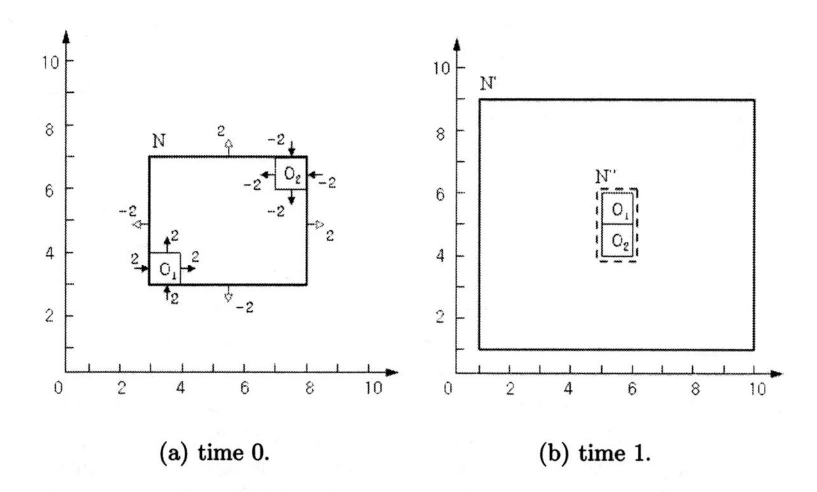


(a) time 0. (b) time 1.

**Fig. 1.** An example: CBR adjustment is executed at time 1

updating the CBR data, the TPR-tree can reduce the areas of the dead space. We call such a CBR update the *CBR adjustment*.

As known from Figure 1(b), the BR $N'$ contains large dead space inside. If any of objects $O_1$ and $O_2$ changes its location information and thus a node update occurs, then the TPR*-tree will perform the CBR adjustment along the tree path from that leaf node to the root in a bottom-up fashion. In the case of Figure 1(b), a CBR adjustment can reduce the BR into $N''$ if an update on the node arises at time 1. The reduction of the BR improves the query processing time by decreasing dead space and overlaps among BRs.

The CBR adjustment is allowable only at the update time in the original TPR*-tree. Therefore, the CBR adjustment does not happen for a long period if moving objects are infrequently updated. Such a problem mainly comes from the *passive* characteristic of the CBR adjustment in the original TPR*-tree.

### 3.2 Basic Strategy

For exposition, let us consider a situation where a user query reaches a leaf node $N$ of a TPR*-tree for its query processing. Let $T_{u_j}$ and $T_{u_{j+1}}$ be the time points when the $j$-th and the $(j+1)$-th updates arise on $N$, respectively. If the user query is the $i$-th one accessing $N$ within the interval of $[T_{u_j}, T_{u_{j+1}}]$, we denote the query by $Q_{j,i}$. That is, $Q_{j,i}$ is the $i$-th user query accessing $N$ after the $j$-th update has occurred on $N$. Also, we let $T_{q_{j,i}}$ be the access time of $Q_{j,i}$ to $N$. If $k$ user queries has reached $N$ in the time of $[T_{u_j}, T_{u_{j+1}}]$, then we have a timing sequence as below.

$$T_{u_j} < T_{q_{j,1}} < T_{q_{j,2}} < T_{q_{j,3}} < \cdots < T_{q_{j,k}} < T_{u_{j+1}}$$

Because the original TPR*-tree can do its CBR adjustment only at the update times, user queries at $T_{q_{j,i}}$ $(1 \leq i \leq k)$ tend to view the constantly growing BR of $N$ since there is no CBR adjustment on $N$ by time $T_{u_{j+1}}$. If the time interval to $T_{u_{j+1}}$ is a large one, the dead space in $N$ also can be large. Meanwhile, if we can do a CBR adjustment on $N$ within $[T_{u_j}, T_{u_{j+1}}]$, then we can avoid the performance degradation by reducing the number of user queries unnecessarily reaching $N$.

We allow the query processor, which is reaching $N$ at $T_{q_{j,i}}$, to perform a CBR adjustment, if needed. Here, the necessity depends on the result of benefit prediction regarding the CBR adjustment on $N$. Since the benefit prediction requires the involved CBR data, disk I/Os are needed to read nodes in general. Fortunately, the processor can access those CBR data without such disk I/Os since the processor has cached those nodes on its buffer memory during its downward search. Using the caching mechanism, the query arriving at $N$ can do the benefit prediction without additional node reads. If the result of the benefit prediction indicates that the CBR adjustment will favorably affect the future query performance, then the query processor actively initiates a CBR adjustment. In this paper, we call this an *active CBR adjustment* (ACA). We note that the ACA is performed in query processing.

Although our benefit prediction does not incur any I/O overhead, the execution of the ACA requires disk writes for updating the changed CBR on the search path. Therefore, while making the benefit prediction, we should take into account the I/O cost for an ACA execution as well as the expected performance gains of future-time's query processing. For this, we develop a cost model that evaluates the performance trade-off based on some probabilistic assumptions.

Our cost model used for benefit prediction is mainly based on the notion of the sweeping region. Consider a situation where we are processing a future-time query $Q$. In processing $Q$, the query processor reads every node having any overlap between the BR of that node and the target query region of $Q$. Therefore, the probability of the access to node $N$ by a specific time becomes proportional to the sweep region size of $N$ by that time if query regions are randomly selected in the index space [8]. Since the reduced sweeping region decreases the probability of unnecessary node reads in query processing, we can improve the query performance by properly executing the ACAs.

For example, consider the time point $T_{q_{j,i}}$ when the $i$-th user query reads the leaf node $N$ after the update time of $T_{u_j}$. Let $SR$ be the size of the sweeping region of $N$ from $T_{q_{j,i}}$ to $T_{u_{j+1}}$. Also, let $SR'$ be a new size of the reduced sweeping region of $N$ in case the CBR of $N$ is adjusted at time $T_{q_{j,i}}$. In this case, the difference $(SR - SR')$ determines the performance enhancement for other user queries that will arrive at $N$ in $[T_{q_{j,i}}, T_{u_{j+1}}]$. That is, a greater difference entails more enhancement in the future-time's query processing. From this observation, we use the reduced size of sweeping regions to estimate the amount of performance enhancement. We refer to the profit from the ACA as the *CAB*(*CBR Adjustment Benefit*). If the amount of the CAB is greater than I/O overheads for executing the ACA, we make the ACA initiated at time $T_{q_{j,i}}$.

The query frequency is a major factor in the computation of the CAB. Since the CAB is the sum of benefits of the user queries issued in $[T_{q_{j,i}}, T_{u_{j+1}}]$, we have to take into account the future-time's queries that will reach $N$ at time $t$ ($T_{q_{j,i}} < t < T_{u_{j+1}}$). The number of such future-time's queries can be obtained by multiplying the query frequency with the length of interval $[T_{q_{j,i}}, T_{u_{j+1}}]$.

Since we cannot exactly forecast the update time of $T_{u_{j+1}}$ in advance, a probabilistic approach is taken instead. That is, we compute the average update period of a leaf node, which can be obtained from dividing the average update period of a moving object by the average number of objects in a leaf node. If we let $P_u$ be the average update period of a leaf node, we can predict the time of $T_{u_{j+1}}$ as the time point of $T_{u_j} + P_u$. In our research, we use a time stamp (TS) field for saving the expected update time of $T_{u_j} + P_u$. Each leaf node has a TS field and the field is looked up in the time of CAB computation. From the notations above, we can have a formula for the CAB computation as follows.

$$CAB(T_{q_{j,i}}, T_{u_j} + P_u) = \left(\frac{SR(T_{q_{j,i}}, T_{u_j} + P_u) - SR'(T_{q_{j,i}}, T_{u_j} + P_u)}{2}\right) \times Q_{freq} \times (T_{q_{j,i}}, T_{u_j} + P_u)$$

In the formula, $CAB(T_{q_{j,i}}, T_{u_j} + P_u)$ denotes the CAB that seems obtainable by the next update time if the ACA is executed at time $T_{q_{j,i}}$. $SR(T_{q_{j,i}}, T_{u_j} + P_u)$ denotes the size of the sweeping regions that the non-adjusted CBR will sweep by the next update time, and $SR'(T_{q_{j,i}}, T_{u_j} + P_u)$ does the size of the reduced sweeping regions in case an ACA is executed at $T_{q_{j,i}}$. $Q_{freq}$ denotes the average query frequency of our TPR*-tree, and its multiplication with $(T_{q_{j,i}}, T_{u_j} + P_u)$ yields the average number of queries in the future time. By multiplying $(SR - SR')$ with the number of queries, we can estimate the average number of queries that will not access node $N$ owing to the reduced sweeping region. Note that the estimated number implies the number of reduced disk accesses to $N$ caused by the ACA execution. As it is not possible to predict the exact size of sweeping regions viewed by future-time's queries, we divide $SR - SR'$ by 2 to get the average size of sweeping regions for the CAB computation.

Although the CAB can be computed without extra I/Os using cached CBR data, the execution of the ACA incurs additional disk writes. To avoid the unprofitable ACA executions, we use the benefit prediction as below.

$$CAB(T_{q_{j,i}}, T_{u_j} + P_u) > H - 1 \qquad (Condition\ i)$$

The ACA is profitable if (Condition i) holds. In other words, if (Condition i) is satisfied, the ACA is allowable. Here, $H$ is the height of the TPR*-tree and $H - 1$ acts as the maximum number of node writes for an ACA execution. Note that, since the root node resides on the memory buffer in most cases, we do not count its node write. Therefore, $H - 1$ is the number of disk writes that are needed in the presence of worst-case upward propagation of CBR updates.

Although there is no I/O overhead for the CAB computation in (Condition i), there exists a CPU cost. Because the query processor may perform such computation every time, the overall CPU cost could be a performance bottleneck with a large volume of user queries. Against this, we check (Condition ii), which is much cheaper than (Condition i) in computation, to avoid unnecessary computations of CAB. That is, we compute (Condition i) only when (Condition ii) below holds.

$$T_{q_{j,i}} + epsilon < T_{u_j} + P_u \qquad (Condition\ ii)$$

This is because the ACA is not that advantageous if the remaining time to $T_{u_j} + P_u$ is too short. The *epslilon* is a parameter variable. As its value, we can use the time taken to read $H$ nodes. Since the remaining time should be enough to read $H$ nodes to meet (Condition i), that time can be used as a minimum value of *epsilon*.

# 4 Performance Evaluation

## 4.1 Experimental Environment

We generated the datasets for our experiments using the GSTD [9], a data generator widely used in many previous researches for performance evaluation [3]. With the GSTD, we generated 100,000 moving objects in such a way that each object has a random speed in the range of [0, 70]. Those objects are made to move around within the normalized 2-dimensional space whose size is 10,000 by 10,000. In the space, an object is represented as a point, and its initial position follows one of uniform, skewed, and Gaussian distribution.

Parameters can be flexibly altered in order to reflect the various characteristics of moving objects and user queries. Table 1 gives such parameters. While any parameter in Table 1 is being varied for different experimental environments, other parameters are fixed as the boldface pivot values.

**Table 1.** Simulation parameters and their values

| parameters | parameter values |
|---|---|
| update frequency (per moving object) | 20, **50**, 100, 150 |
| query frequency (per unit time) | 20, **40**, 60, 80, 100 |
| size of target query regions | 0.01%, **0.16%**, 0.64%, 2.56% |
| average speed of moving objects | 30, **50**, 70, 90 |
| future prediction time point | 20, 40, **60**, 80, 100 |

We used two performance measures, i.e., the average number of node accesses and the average response time for a hundred of future-time queries. The experiments was performed on a Windows 2000 server equipped with a Pentium 4 processor of 4.23 GHz and 512 MB of main memory.

## 4.2 Results and Analyses

In this section, we present performance comparisons between our method and the original TPR*-tree. Due to space limitations, we omit the experiment results from Gaussian distribution because the results are very similar to those of uniform and skewed distribution. Figure 2 shows how the query frequency affects the query performance. The results show that our method provides a better performance, and the performance gains get greater as the query frequency increases. The average number of node accesses is reduced in our method by up to 34% in uniform distribution and by up to 37% in skewed distribution. The average response time is also improved in our method by up to 26% and 28% in uniform and skewed distribution, respectively.

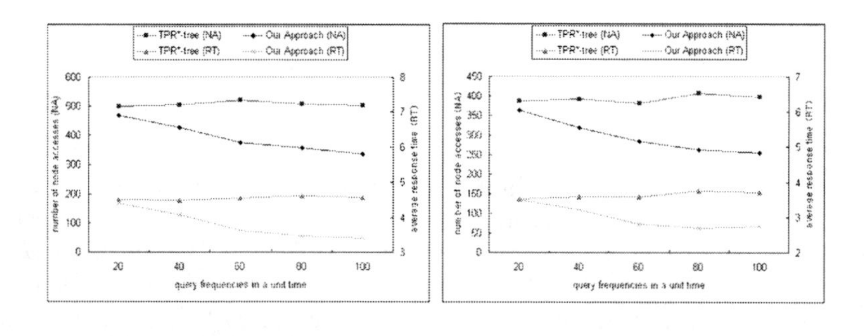


(a) Uniform distribution.

(b) Skewed distribution.

**Fig. 2.** Performance comparisons with respect to the query frequency

In Figure 2, the performance gain increases until the query frequency does not exceed 80 per unit time. After that point, the performance gain comes to be nearly constant. That means that the occurrence frequency of the ACA becomes constant above a high point of the query frequency. This is because the ACA is not executed when it seems unprofitable.

Also, we examined performance gains with respect to the update frequency of the moving objects. The results are given in Figure 3, where our method consistently outperforms the original TRR*-tree. Our performance gain increases as the update period gets longer, i.e., the update frequency gets lower. In our method, the average number of node accesses is reduced by up to 39% in uniform distribution and by up to 43% in skewed distribution. The response time is also improved by up to 34% and 37% in uniform and skewed distribution, respectively. Since the original TPR*-tree can execute the CBR adjustment only when the update operation occurs on a node, the prolonged update period entails infrequent CBR adjustments. On the other hand, the proposed method can initiate the ACA when user queries are being processed. From this, our method provides a better performance even while the update period gets longer. On the contrary as in Figure 3, our method does not provide a performance gain in the presence of a quite short update period of less than 10 unit time. However, since such a short update period is not common in reality, we can say that the proposed method have a better performance than the original TPR*-tree in almost cases.

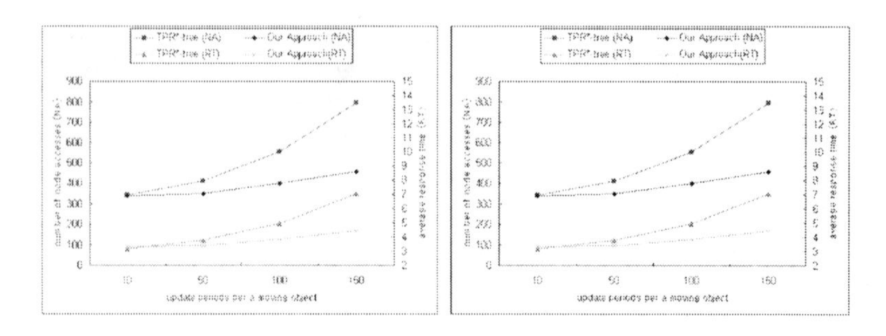


(a) Uniform distribution. (b) Skewed distribution.

**Fig. 3.** Performance comparisons with respect to different update periods

Lastly, in Figure 4, we examined how the performance varies as the future prediction time point goes farther. From the figure, we can see that our method reduces the number of node accesses by up to 15% and 16% in uniform and skewed distribution, respectively. As to the response time, the performance improves by up to 18% in uniform distribution and by up to 16% in skewed distribution. As to a target query region of Figure 4, the future prediction time is also a less dominant factor. However, since user queries with farther prediction times require greater numbers of node accesses, the proposed method can improve the query performance greatly.

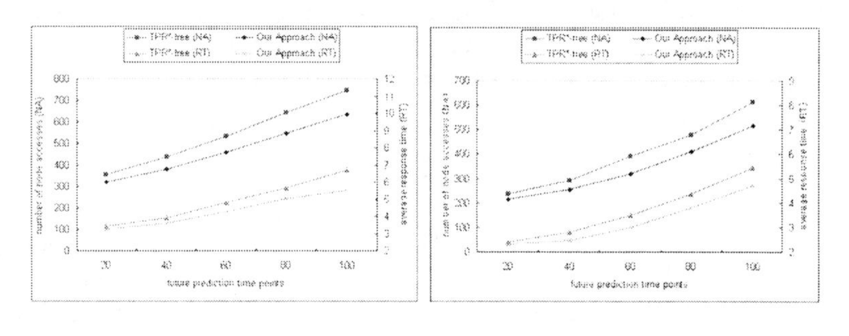


(a) Uniform distribution. (b) Skewed distribution.

**Fig. 4.** Performance comparisons with respect to varying future prediction time points

## 5 Conclusions

The TPR*-tree for processing future-time queries is apt to suffer from performance degradation because of quickly growing dead space and overlapping areas among BRs over time. Against the problem, we have proposed a new method capable of executing the CBR adjustment during query processing. In our method, the query processor can do the CBR adjustment in an active manner. Since the query processor can cache the CBR data of its search path on the buffer memory, we allow the query processor to perform the active CBR adjustment (ACA) in a quite efficient way. For the ACA mechanism to be useful, we have also proposed a cost model that assesses both of the performance benefit to be obtained during the future-time's query processing and the I/O cost for an ACA execution. To show the superiority of the proposed method, we have conducted various performance experiments. The results reveal that our method outperforms the original TPR*-tree significantly in most cases.

## References


[1] Beckmann, N., et al.: The R*-tree: An Efficient and Robust Access Method for Points and Rectangles. In: Proc. ACM Int'l. Conf. on Management of Data(ACM SIGMOD), pp. 322–331. ACM Press, New York (1990)

[2] Lee, D.L., Xu, J., Zheng, B., Lee, W.C.: Data Management in Location-Dependent Information Services. IEEE Pervasive Computing 1(3), 65–72 (2002)

[3] Lin, B., Su, J.: On Bulk Loading TPR-Tree. In: Proc. IEEE Int'l. Conf. on Mobile Data Management, pp. 395–406 (2004)

[4] Mokbel, M.F., Ghanem, T.M., Aref, W.G.: Spatio-Temporal Access Methods. Bulletin of the IEEE Computer Society Technical Committee on Data Engineering 26(2), 40–49 (2003)

[5] Prabhakar, S., et al.: Query Indexing and Velocity Constrained Indexing: Scalable Techniques for Continuous Queries on Moving Objects. IEEE Trans. on Computers 51(10), 1124–1140 (2002)

[6] Saltenis, S., et al.: Indexing the Positions of Continuously Moving Objects. In: Proc. ACM Int'l. Conf. on Management of Data(ACM SIGMOD), pp. 331–342. ACM Press, New York (2000)

[7] Sistla, A.P., et al.: Modeling and Querying Moving Objects. In: Proc. IEEE Int'l. Conf. on Data Engineering (ICDE), pp. 422–432 (1997)

[8] Tao, Y., Papadias, D., Sun, J.: The TPR*-Tree: An Optimized Spatio-Temporal Access Method for Predictive Queries. In: Proc. Int'l. Conf. on Very Large Data Bases (VLDB), pp. 790–801 (2003)

[9] Theodoridis, Y., Silva, R., Nascimento, M.: On the Generation of Spatiotemporal Datasets. In: Proc. Int'l. Symp. on Spatial Databases, pp. 147–164 (1999)

[10] Wolfson, O., Xu, B., Chamberlain, S., Jiang, L.: Moving Objects Databases: Issues and Solutions. In: Proc. Int'l. Conf. on Scientific and Statistical Database Management (SSDBM), pp. 111–122 (1998)

# Performance Oriented Schema Matching

Khalid Saleem[1], Zohra Bellahsene[1], and Ela Hunt[2]

[1] LIRMM - UMR 5506 CNRS University Montpellier 2,
161 Rue Ada, F-34392 Montpellier
{saleem, bella}@lirmm.fr
[2] Department of Computer Science, ETH Zurich, CH-8092
hunt@inf.ethz.ch

**Abstract.** Semantic matching of schemas in heterogeneous data sharing systems is time consuming and error prone. Existing mapping tools employ semi-automatic techniques for mapping two schemas at a time. In a large-scale scenario, where data sharing involves a large number of data sources, such techniques are not suitable. We present a new robust mapping method which creates a mediated schema tree from a large set of input XML schema trees and defines mappings from the contributing schema to the mediated schema. The result is an almost automatic technique giving good performance with approximate semantic match quality. Our method uses node ranks calculated by pre-order traversal. It combines tree mining with semantic label clustering which minimizes the target search space and improves performance, thus making the algorithm suitable for large scale data sharing. We report on experiments with up to 80 schemas containing 83,770 nodes, with our prototype implementation taking 587 seconds to match and merge them to create a mediated schema and to return mappings from input schemas to the mediated schema.

## 1 Introduction

Schema matching relies on discovering correspondences between similar elements of two schemas. Several different types of schema matching tools [8,9] have been studied, demonstrating their benefit in different scenarios. In data integration schema matching is of central importance [1]. The need for information integration arises in data warehousing, OLAP, data mashups [6], and work flows. Omnipresence of XML as a data exchange format on the web and the presence of metadata available in that format force us to focus on schema matching, and on matching for XML schemas in particular.

We consider schemas to be rooted, labeled trees. This supports the computation of contextual semantics in the tree hierarchy. The contextual aspect is exploited by tree-mining, making it feasible to use almost automated approximate schema matching [4] and integration in a large-scale scenario. The individual semantics of node labels have their own importance. We utilize linguistic matchers, based on tokenisation, and synonym and abbreviation tables, to extract the

R. Wagner, N. Revell, and G. Pernul (Eds.): DEXA 2007, LNCS 4653, pp. 844–853, 2007.

concepts hidden within them. The use of synonym and abbreviation tables is considered to be a form of user intervention.

Tree mining techniques extract similar sub tree patterns from a large set of trees and predict possible extensions of these patterns. A pattern starts with one node and is incrementally augmented. There are different techniques [11] which mine rooted, labeled, embedded or induced, ordered or unordered sub-trees. The function of tree mining is to find sub-tree patterns that are frequent in the given set of trees, which is similar to schema matching that tries to find similar concepts among a set of schema trees.

**Our Contributions**
a) Matching, merging and the creation of a mediated schema with semantically approximate mappings, in one algorithm which has good performance.
b) Use of tokenisation, abbreviation and synonym matching of label tokens, intuitively supporting the clustering of similar labels to minimize the search space.
c) Extension of a tree mining data structure [11] to schema matching, using ancestor/ descendant properties for quality contextual semantic matching.
d) Ability to produce element level [8] simple 1:1 mappings, and complex mappings, including 1:n(leaf mapped to non-leaf) and n:1(non-leaf mapped to leaf).
e) Experiments with real XML schema instances (OAGIS, XCBL[1]) showing performance appropriate for a large scale scenario.
f) Quality evaluation based on *precision*, showing that our method is reliable.

The remainder of the paper is organized as follows. Section 2 presents the related work in large-scale schema integration. Section 3 defines the concepts. In Section 4 we describe our approach, Performance Oriented Schema Matching, along with a running example. Section 5 presents the experimental evaluation. Section 6 gives a discussion, outlines future work and concludes.

## 2 Related Work

Nearly all schema-matching systems compare two schemas at a time and aim for quality matching but require significant human intervention. Several surveys [3,8,9] shed light on this aspect and show that extending the matching to data integration is time consuming and limited in scope. Large scale schema matching has been investigated in the web interface schema integration [5,10] using data mining. Matching of two large bio-medical taxonomies [3,7] was demonstrated using COMA++ [3] and Protoplasm [2].

There are numerous issues in the semantic integration of a large number of schemas. For example, Semantic Web, by definition, offers a large-scale dynamic environment where individual service providers are independent. In such a situation the mappings can never be exact, rather they are approximate [4,5].

Performance is an open issue in schema matching [3,8,9]. The complexity of the schema matching task is proportional to the size of participating schemas

[1] OAGIS : http://www.openapplications.org/, XCBL : http://www.xcbl.org/

and the number of match algorithms employed, i.e. $O(nma)$, where $n$ and $m$ are node counts in the source and target schema and $a$ is the number of algorithms applied. The quality of mappings depends on the type and number of matching algorithms and their combination strategy [2,3].

One of the most recent matching and merging tool is Coma++ [3] which produces quality matches. Coma is a composite matcher which can reuse previous mappings. It uses user defined synonym and abbreviation tables, along with other matchers. Coma can map large schemas with the help of user input. The user can identify fragments of the schema to be mapped. This option is intended to manage the namespace/ include characteristics of XML schemas. However, human intervention in the schema mapping and merging process is needed. Systems like Coma, which produce mappings and no integrated schema, do not support automated data integration suitable for application environments with hundreds of schemas.

Semantically, a match between two nodes can be either an equivalence or a partial equivalence. In a partial match, the similarity is partial, e.g., an element Name = 'John M. Brown' in source schema is partially matched to LastName='Brown' and FirstName ='John' in the target, because Name also contains the MiddleInitial='M'. If there are several possible matchings of the source element to the mediated schema, the best/most correct match can be selected (manual in [2,3]). The choice can depend upon match quality confidence computed at run time [2,3,8,9]. We use a hybrid approach which automatically selects the best match and performs the binary integration of schemas using the ladder technique [1]. Our method caters both for the quality as well as performance in large scale scenarios, using domain specific linguistic matching (synonym and abbreviation oracles), clustering, and tree mining.

## 3 Preliminaries

Semantic matching requires the comparison of concepts which are structured as schema elements. Node labels of schema elements are considered to be concepts and each element's contextual placement in the schema enhances the semantics of the concept. For example, in Fig. 1 Sb, the elements writer/**name** and publisher/**name** have similar labels but their contexts are different, which makes them conceptually disjoint. In an XML tree, the combination of the element label and the structural placement of the element produces the concept.

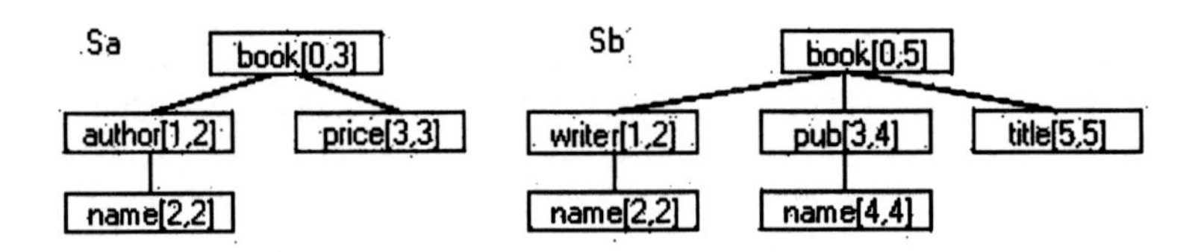


**Fig. 1.** Input Schema Trees $S_a$ and $S_b$

**Def. 1 (Schema Tree).** A *schema tree* is a rooted, labeled tree [11]. A schema tree, S=(V,E), consists of V={0,1,...,n}, a set of nodes, and E={(x,y) | x,y ∈ V}, a set of edges. One distinguished node r ∈ V is designated the root, and for all x ∈ V, there is a unique path from r to x. Further, *lab*:V → L is a labeling function mapping nodes to labels in L=$\{l_1, l_2, \ldots\}$, and *vt*:L → $\mathcal{P}(V)$ is a function which returns for each label $l_i$∈ L a set of nodes $V_i$ ⊆V with labels similar to $l_i$.

**Def. 2 (Node Scope).** Nodes x ∈ V are numbered according to their position in the pre-order traversal of the tree S (where the root is numbered 0, and x is numbered X). Let T(x) denote the sub-tree rooted at x, and let y be the rightmost leaf (or highest numbered descendant) under x, numbered Y. Then the scope of x is defined as scope(x)=[X,Y]. Intuitively, scope(x) is the range of nodes under x, and includes x itself (Fig. 1). The count of nodes in T(x) is Y-X+1.

**Def. 3 (Label Semantics).** Label semantics corresponds to the conceptual meaning of the label (irrespective of context). It is a composition of concepts attached to the tokens making up the label. $C_l : l \rightarrow (C(t_1), \ldots, C(t_m))$ where $m$ is the number of tokens making up the label.

**Def. 4 (Node Semantics).** Node semantics for x ∈ V, $C_x$, combines the semantics of the node label $C_{l_x}$ with its contextual placement in the tree, TreeContext($x$), as follows [11]: $C_x : x \rightarrow (C_{l_x}, TreeContext(x))$. TreeContext of a node x is its scope (Def. 2).

### 3.1 Scope Properties

Scope properties describe the contextual placement of a node [11]. Property testing involves simple integer comparisons.

**Unary Properties**, given a node x with scope [X,Y]
**Pr. 1:** Leaf Node(x): X=Y, **Pr. 2**: Non-Leaf Node(x): X < Y.
**Binary Properties**, given x [X,Y], xd[Xd,Yd], xa[Xa,Ya], and xr[Xr,Yr]
**Pr. 3:** Descendant (x,xd), xd is a descendant of x: Xd>X ∧ Yd≤Y
**Pr. 4:** DescendantLeaf (x,xd) combines Pr. 1 and 3:
Xd>X ∧ Yd≤Y ∧ Xd=Yd
**Pr. 5:** Ancestor (x,xa), complements Pr. 3, xa is ancestor of x: Xa<X ∧ Ya≥Y
**Pr. 6:** RightHandSideNode (x,xr) with non-overlapping scope, xr is RHSNode of x: Xr>Y.

**Example 3.1:** In Fig. 1 Sa, Pr. 1 for node **price**[3,3] defines it as a leaf. Pr. 2 for **author**[1,2] states that it is a non-leaf (an inner node).

**Example 3.2:** The task is to find a mapping for Sa tree node author/**name** in Sb. In Fig. 1 Sb there are two nodes called **name**: [2,2] and [4,4]. Given synonymy between words **author** and **writer**, and top down traversal, Sa **author**[1,2] is already mapped to Sb **writer** [1,2], we perform the descendant node check on nodes [2,2] and [4,4] with respect to **writer**[1,2]. Node [2,2] is a descendant of [1,2], using Pr. 3, and node [4,4] is not a descendant of [1,2], thus author/**name** is mapped to writer/**name**.

# 4 Our Approach

We assume that only schema trees are available as input. Our method accepts a set of schema trees and outputs the mediated schema tree and the corresponding mappings.

**Def. 5 (Semantic Mediation)**
INPUT: A set of schema trees $SSet$=$\{S_1$,$S_2 \ldots S_u\}$.
OUTPUTS:

a) A mediated schema tree $S_m$, which is a composition of all distinct concepts in $SSet$. $S_m = P^u_{i=1}(S_i)$, $P(S_i) = \{C_1 \ \rho \ C_2 \ \rho \ldots C_n\}$ includes all distinct concepts in $S_i$ (Def. 4). $P$ is the composition function and $\rho$ denotes the composition operator.
b) A set of mappings $M = \{M_1, M_2, \ldots M_w\}$ from the concepts of input schema trees to the concepts in the mediated schema.

The mediated schema tree $S_m$ is a composition of all nodes representing distinct concepts in the set of schemas. During the integration process if a node is not present in $S_m$, a new edge $e'$ is created in $S_m$ and a node is added to it.

## 4.1 Assumptions

We make the following assumptions, valid in domain specific schema integration (extended from [10]).
a) Schemas in the same domain contain the same domain concepts, but differ in structure and concept naming.
b) We select the input schema with the highest number of nodes as the initial mediated schema. Since each node represents a concept, this covers the maximum number of concepts. This choice minimizes the addition of new concepts (nodes not present in the mediated schema) to the mediated schema and improves performance.
c) Only one type of matching between two labels is possible. For example, author is a synonym of writer.
d) In one schema, different labels for the same concept are rarely present.
e) A node from the input schema is only matched to the set (cluster) of similarly labeled nodes present in the mediated schema.

## 4.2 Example of Schema Integration

We developed an algorithm which works in three steps. First, we perform *pre-mapping*. Schema trees are input to the system as a stream of XML and the node number and parent for each node, **node scope**, schema size, and schema depth are calculated. A listing of nodes and of distinct labels for each tree is constructed.

Next, a *linguistic matcher* identifies semantically similar node labels. The user can set the level of similarity of labels as A) Label String Equivalence, B) Label Token Set Equivalence (using abbreviation table), or C) Label Synonym Token Set Equivalence (synonym table). The matcher derives the meaning for each individual token and combines these meanings to form a label concept. Similar labels are clustered. Since each input node corresponds to its label object, this intuitively forms **clusters of similarly labeled nodes** within the group of schemas to be merged.

**Table 1.** Before Node Mapping

a. List of labels, ordered alphabetically

| 0 | 1 | 2 | 3 | 4 | 5 | 6 | 7 | 8 |
|---|---|---|---|---|---|---|---|---|
| author | book | name | name | price | pub | title | writer | ROOT |

b. Input Schema Nodes' Matrix : Row 1 is Sa and Row 2 is Sb

| | | | | | | | | |
|---|---|---|---|---|---|---|---|---|
| 1,2,0 | 0,3,-1 | 2,2,1 | | 3,3,0 | | | | |
| | 0,5,-1 | 2,2,1 | 4,4,3 | | 3,4,0 | 5,5,0 | 1,2,0 | |

c. Initial Mediated Schema, Sm, renumbered after adding ROOT to Sb

| | | | | | | | | |
|---|---|---|---|---|---|---|---|---|
| | 1,6,0 | 3,3,2 | 5,5,4 | | 4,5,1 | 6,6,1 | 2,3,1 | 0,6,-1 |

*Column entries show node scope and parent

**Example 4.1:** Consider labels "POShipDate" and "PurchaseOrderDeliverDate". In the abbreviation table PO stands for *purchase order* and in the synonym table 'deliver'='ship'. This implies that the two labels are similar.

In the *integration and mapping part*, we first select the input schema tree with the highest number of nodes and designate it as the initial mediated schema (Section 4.1). Next, we take each schema in turn and merge it with the mediated schema, following the **binary ladder technique** highlighted in [1]. This requires matching, mapping and merging. Concepts from input schemas are mapped to the mediated schema.

The algorithm traverses the input schema depth-first, mapping parents before siblings. If a new concept is found, with no match in the mediated schema, a new concept node is created and added to the mediated schema as the right most child of the node in the mediated schema, to which the parent of current node is mapped. This new node is used as the target node in the mapping (Def. 5). The algorithm combines node label similarity and contextual positioning, calculated using properties defined in Section 3.1. Our example uses the two schemas shown in Fig. 1 where $S_a$ and $S_b$ are shown with information calculated during *pre-mapping.* A list of labels created in this traversal is shown in Table 1a. Nodes 2

**Table 2.** After Node Mapping

a. Labels List

| 0 | 1 | 2 | 3 | 4 | 5 | 6 | 7 | 8 |
|---|---|---|---|---|---|---|---|---|
| author | book | name | name | price | pub | title | writer | ROOT |

b. Mapping Matrix : Row 1 is Sa and Row 2 is Sb

| | | | | | | | | |
|---|---|---|---|---|---|---|---|---|
| 1,2,0,**7** | 0,3,-1,**1** | 2,2,1,**2** | | 3,3,0,**4** | | | | |
| | 0,5,-1,**1** | 2,2,1,**2** | 4,4,3,**3** | | 3,4,0,**5** | 5,5,0,**6** | 1,2,0,**7** | |

c. Final Mediated Schema

| | | | | | | | | |
|---|---|---|---|---|---|---|---|---|
| | 1,7,0, **1.0,2.0** | 3,3,2, **1.2,2.2** | 5,5,4, **2.4** | 7,7,1, **1.3** | 4,5,1, **2.3** | 6,6,1, **2.5** | 2,3,1, **1.1,2.1** | 0,7,-1 |

*Column entries show node scope, parent and **mapping**

and 3, with the same label 'name' but different parents (author and publisher) are shown to be disjoint. The last label is a new label, ROOT, created by our algorithm.

Table 1b shows a matrix of size $um$, where $u$ is the number of schemas and $m$ the total number of distinct labels in all schemas (the length of the label list). A matrix row represents an input schema tree. Each non-null entry contains the node scope and parent node number. Each node is placed in the column which holds its label.

The larger schema tree $S_b$, see Fig. 1, is selected as the initial mediated schema $S_m$. ROOT is added to $S_m$, and the nodes are renumbered to reflect this. A list of size m (Table 1c) holds $S_m$, assuming the same column order as in Table 1b.

The node mapping algorithm takes the data structures in Table 1 as input, and produces mappings shown in Table 2b, and the integrated schema in Table 2c. In the process, the input schema $S_a$ is mapped to mediated schema $S_m$. The mapping is read as the column number (Table 2b **mapping**) of node in the mediated schema. Saving mappings as column number gives us the flexibility to add new nodes to $S_m$, without disturbing the previous mappings. Scope values of some existing nodes are affected, cf. Table 1c and Table 2c, because of addition of new nodes (identified by Pr. 5 or 6; scope values adjusted accordingly), but column numbers of all previous nodes remain the same. Thus, intuitively, none of the existing mappings are affected.

Node mapping for input schema tree $S_a$ (Table 1b row 1) starts from the label 'book' with scope [0,3]. As it is a root node, with only one similar node 'book' [1,6] in $S_m$, mapping **1** is added in column 1 in $S_a$ row, shown in bold in Table 2b, for node $S_a[0,3]$. This is now recorded in the mediated schema, see Table 2c, as **1.0** i.e., node 0 in Schema 1 i.e., $S_a$ mapped to node 0 in $S_m$. Next node to map is $S_a.author[1,2]$, similar to $S_m.writer[2,3]$. Both nodes are internal and the ancestor check returns true since parent nodes of both are already mapped. The resulting mapping for label 0 is **7**. For node 2 with label 'name', there are two possibilities, nodes attached to label 2 (col. 2) and label 3 (col. 3). Descendant(name,writer) is true for node in column 2 and false for 3 by Pr. 3 (Example 3.2). Hence **2** is the correct map. The last node in $S_a$ is price[3,3]. There is no match in $S_m$, so a new node is added to $S_m$, as an entry in the column with label 'price' in the mediated schema list (Table 2c). This node is created as the right most sibling of node in the mediated tree to which the parent node of current input node is mapped, i.e. 'book'. The scope and parent node link are adjusted for the new node and its ancestors, and a mapping is created from the input node to this newly created node.

**Algorithm complexity:** Given a set of input schemas S= $\{S_1, S_2, \ldots S_u\}$, we select as the mediated schema the schema tree with the highest number of nodes, max(N($S_i$)) where N($S_i$)returns the number of nodes in a schema. We match each node of each input schema to the mediated schema. The number of input schema nodes $N_t$ is given by $N_t = \sum_{i=1}^{u} N(S_i)$. Therefore the complexity of Node Mapper algorithm is O($N_t N(S_m)$). This is quadratic in the size of schema set that is to be integrated. Our experiments confirm this complexity.

## 5 Experimental Evaluation

The experiments were performed on a PC, Pentium 4-M, 1.80 GHz, 768 MB RAM, running Windows XP, and Java 1.5. Three sets of schemas from different domains were used, with *ISN*, Integrated Schema Nodes, giving the schema size of the largest integrated schema.

1. Books: 176 synthetic schemas; Avg./Max/Min nodes 8/14/5, ISN 23
2. OAGIS: 80 real schemas; Avg./Max/Min nodes 1047/3519/26, ISN 70191
3. XCBL: 44 real schemas; Avg./Max/Min nodes 1678/4578/4, ISN 4803

**Performance** was evaluated in three label similarity scenarios: A) Label String Equivalence, B) Label Token Set Equivalence, and C) Label Synonym Token Set Equivalence. Figures 2 and 3 demonstrate the performance of our

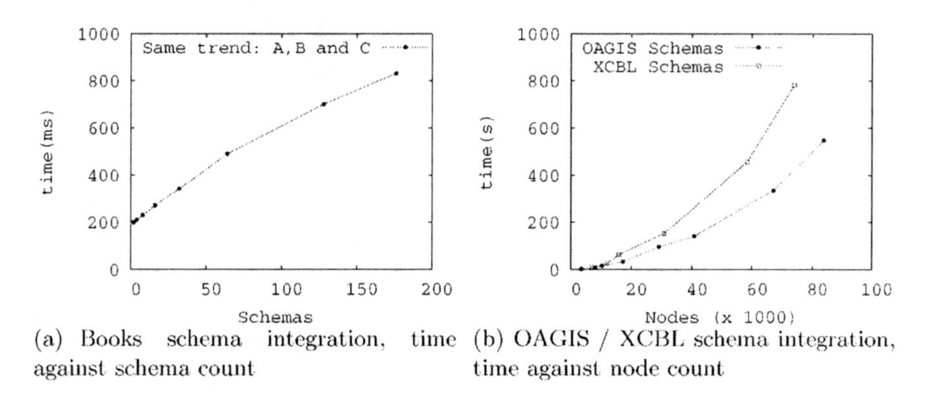


(a) Books schema integration, time against schema count

(b) OAGIS / XCBL schema integration, time against node count

**Fig. 2.** Small synthetic schemas are matched faster than complex real life schemas

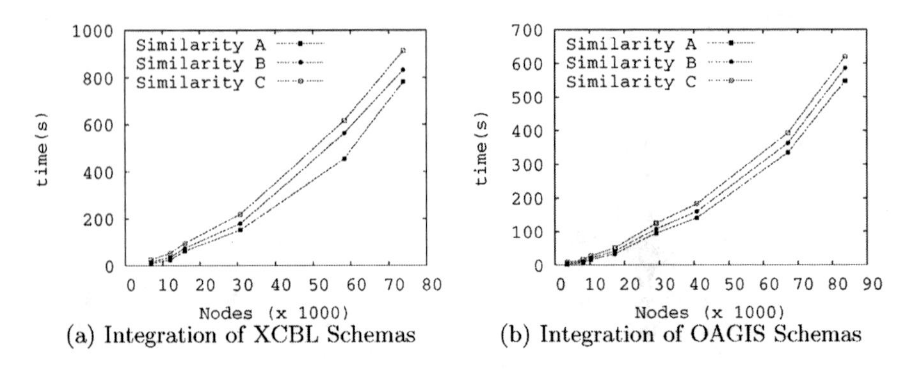

(a) Integration of XCBL Schemas (b) Integration of OAGIS Schemas

**Fig. 3.** Influence of matchers A, B, C on schema integration time

method. Our experiments show that the execution time reflects the number of schemas to be integrated, and appears to be at worst quadratic in the number of nodes compared.

Fig. 2a shows a comparison of three kinds of matching: A, B, and C for sets of 2, 4, 8, 16, 32, 64, 128, 176 Books schemas. There is no difference in the performance of various matchers, which is possibly due to the fact that synthetic schemas vary little in their labels. Fig. 2b shows time in seconds for Domains 2 and 3. Fig. 3a shows the time (s) against the number of nodes processed, for the three similarity methods for XCBL. XCBL schemas are slower to match than OAGIS schemas, see Fig. 3b. This is due to the higher average number of nodes in XCLB schemas. It takes less than 600 seconds to match 80 OAGIS schemas, while 44 XCLB schemas require more than 800 seconds.

**Table 3.** Quality Evaluation. A, B, C are the label similarity levels, and schemas are at www.lirmm.fr/~saleem/matching/schemas.

| **Domain** | PurchaseOrd | | Books | | OAGIS | | XCBL | |
|---|---|---|---|---|---|---|---|---|
| **Schema** | S1 | Sm | S1 | Sm | S1 | Sm | S1 | Sm |
| **Size** | 14 | 18 | 8 | 15 | 26 | 34 | 647 | 743 |
| **Precision A/B/C** | 0.29/0.36/1 | | 0.5/0.5/1 | | 0.77 | | 0.96 | |

Since there is no established schema integration benchmark for a large scale scenario, including both schemas and mappings, it is impossible to carry out a full **quality comparison**. We evaluate the quality of our solution by looking at a random schema pair in the set and counting the number of correctly placed nodes in the integrated schema, *correctlyPlacedNodes/ allPlacedNodes*, as our algorithm will always add a node if it cannot find a match. We establish the precision measure by manual inspection of schemas, see Table 3 for results. Since our method takes the larger schema as the initial mediated schema, the smaller schema is integrated into the larger. Real domains schemas follow the same namespaces, with no abbreviations and synonym applicability, as established by manual inspection. Thus showing no variance in quality for the three label similarity cases. Synthetic domain schemas show fluctuation because of the abbreviated and synonym labels incorporated manually to study the algorithm.

## 6 Conclusions

We have introduced a novel technique based on tree mining, for schema matching, integrating and mapping of a large set of schemas. We have investigated its scalability with respect to time performance, in the context of approximate mapping. The experimental results demonstrate that our approach scales to hundreds of schemas and thousands of nodes. The linguistic matching of node labels uses tokenisation, abbreviations and synonyms. The matching strategy is hybrid, and optimized for schemas in tree format. Our algorithm provides an almost automated solution to the large scale mediation problem.

Our results point to significant future work. We are planning to investigate the application of our approach in P2P architectures, and enhancements to linguistic matching. Another issue for the future is a benchmark for schema mapping evaluation in a large scale schema integration scenario. To further benefit from tree mining, we are going to use it to identify co-relationships between sub-trees within a forest of schema trees, which will help in identifying subsumptions and overlaps, for the discovery of n:m complex mappings.

## References

1. Batini, C., Lenzerini, M., Navathe, S.B.: A comparitive analysis of methodologies for database schema integration. ACM Computing Surveys 18(4), 323–364 (1986)
2. Bernstein, P.A., Melnik, S., Petropoulos, M., Quix, C.: Industrial-strength schema matching. SIGMOD Record 33(4), 38–43 (2004)
3. Do, H.-H., Rahm, E.: Matching large schemas: Approaches and evaluation. Information Systems 32(6), 857–885 (2007)
4. Doan, A., Madhavan, J., Dhamankar, R., Domingos, P., Halevy, A.Y.: Learning to match ontologies on the semantic web. VLDB J. 12(4), 303–319 (2003)
5. He, B., Chang, K.C.-C., Han, J.: Discovering complex matchings across web query interfaces: a correlation mining approach. In: KDD, pp. 148–157 (2004)
6. Jhingran, A.: Enterprise information mashups: Integrating information, simply - keynote address. In: VLDB (2006)
7. Mork, P., Bernstein, P.A.: Adapting a generic match algorithm to align ontologies of human anatomy. In: ICDE (2004)
8. Rahm, E., Bernstein, P.A.: A survey of approaches to automatic schema matching. VLDB J. 10(4), 334–350 (2001)
9. Shvaiko, P., Euzenat, J.: A survey of schema-based matching approaches. J. Data Semantics IV, 146–171 (2005)
10. Su, W., Wang, J., Lochovsky, F.: Holistic query interface matching using parallel schema matching. In: ICDE (2006)
11. Zaki, M.J.: Efficiently mining frequent embedded unordered trees. Fundamenta Informaticae 65, 1–20 (2005)

# Preference-Based Integration of Relational Databases into a Description Logic

Olivier Curé[1] and Florent Jochaud[2]

[1] S3IS, Université Paris Est, Marne-la-Vallée, France
ocure@univ-mlv.fr
[2] University of Bayreuth, Bayreuth, Germany
florent.jochaud@uni-bayreuth.de

**Abstract.** This paper aims to bridge the gap between legacy databases and knowledge bases in the context of the Semantic Web. Such approaches may facilitate the design of ontologies and thus accelerate the adoption of the next generation Web. We have implemented a mapping-based system where sources are relational databases and the target is a Description Logics based knowledge base. The cornerstone of this approach is the consideration that the Description Logics ABox is a view of the relational database. In this solution, the target is materialized and a global-as-view approach is adopted. In order to deal with possible cases of inconsistencies, we support the setting of preferences over the views of the mapping.

## 1 Introduction

The future of the Semantic Web partly depends on the availability of efficient ontological engineering tools. In this paper, we concentrate on the following set of ontological engineering's tasks: creation of the ontology schema from schemas of relational databases and automatic instantiation of the knowledge base from tuples of relational databases.

In the context of practical and expressive ontologies, these tasks are considered complex, time-consuming and financially expensive because they require a collaboration between knowledge engineers and domain experts. Thus approaches aiming to facilitate the design of ontologies may be valuable for organizations willing to implement Semantic Web applications.

The DataBase Ontology Mapping (DBOM) system we present processes a declarative mapping to create and instantiate a knowledge base from multiple data sources. This choice is motivated by our need to semantically enrich the data contained in relational databases and fulfils a need to implement practical and efficient inference enabled services which are based on application-dependent ontologies. A materialized approach is adopted, meaning that our system computes the extensions of the concepts and relationships in the target schema by replicating the data at the sources. This materialization is motivated by the fact that some sources may not be accessible on-demand at application runtime. This computation is performed using a global-as-view (GAV) approach which

R. Wagner, N. Revell, and G. Pernul (Eds.): DEXA 2007, LNCS 4653, pp. 854–863, 2007.

requires that the target schema is expressed in terms of the data sources [13]. In this work, we consider that data stored at the sources are always locally consistent, meaning that they respect their set of integrity constraints.

Anyhow, the integration of these autonomous and consistent sources may result in an inconsistent knowledge base. This is due to violations of integrity constraints specified on the target schema. In the related domain of data integration, two approaches are proposed to deal with inconsistent data: (i) a procedural approach which is based on domain-specific transformation and cleaning, (ii) a declarative approach. In this last approach, information integration semantics given in terms of *repairs* is usually proposed to solve inconsistency. In the context of sound interpretation of the mapping, repairs are provided by means of insertions and deletions [2] of tuples over the inconsistent target. In [16], the author also allows tuple updates as a repair primitive.

Our method to deal with inconsistent data adopts a declarative approach which exploits information on view preferences. In this solution, we consider that all sources do not have the same level of reliability and we enable the mapping designer to set confidence values over the views of the mapping. Using these information at mapping processing time, there is a motivation to prefer values coming from a view with respect to data retrieved from another view.

The system we propose tackles very expressive ontologies, those enabling the expression of general logical constraints [9]. The implementation support the serialization of the knowledge base resulting from the integration in the OWL ontology language [6].

This paper is organized as follows. In Section 2, we introduce the DBOM system with a comparison with data exchange and integration then we present the syntax and semantics of the framework. This section also introduces a solution to the impedance mismatch problem encountered in such integration environments. In Section 3, we outline our view preference approach. Section 4 presents an implementation of the DBOM system and evaluates its usefulness in the context of a concrete medical informatics application. Finally, section 5 concludes and presents ongoing work on the DBOM framework.

## 2 Framework

In this section, we define a general framework for the DBOM system and contrast this approach with the comparison of data exchange and integration provided in [7]:

- as in both data exchange and integration, the source schema is given and the mapping is a set of formulas constructed by a human expert.
- as in data integration [13], the target schema, i.e. the intensional knowledge of a Description Logic (DL), is constructed from the processing of the source schema given a mapping.
- as in data exchange [11], the target instances are materialized, while they are usually virtualized in the case of data integration.

The main contribution of this system is to enable the migration of data stored in multiple database sources to a DL knowledge base [1]. This approach tackles the fundamental aspect of impedance mismatch, i.e. database store values while DL based ontologies represent objects. In this section, we present the syntax and the semantics of the underlying framework, and a mapping-based solution that allows specifying how to transform data into objects.

### 2.1 Syntax

We can define our system as follows: $\mathcal{D} = (\mathcal{S}, \mathcal{K}, \mathcal{M})$, where:

- $\mathcal{S}$ is a set of relational database source schemas, $\{S_1, .., S_n\}$. A signature for $S_i$ is $(R_i, IC_i)$ with $R_i$ a set of relations and $IC_i$ a set of integrity constraints that we assume are locally satisfied by $S_i$.
- $\mathcal{K}$ is the (target) ontology schema formalized in an DL knowledge base. Such a knowledge base generally comprises two components $\langle \mathcal{T}, \mathcal{A} \rangle$ where (i) $\mathcal{T}$ is called a TBox which contains intensional knowledge in the form of a terminology, and (ii) $\mathcal{A}$ is called an ABox which contains extensional knowledge that is specific to the individuals of the domain of discourse [1].
- $\mathcal{M}$ is the mapping between $\mathcal{S}$ and $\mathcal{K}$. This mapping is represented as a set of GAV assertions in which views, i.e. queries, expressed over elements of $\mathcal{S}$ are put in correspondence to elements of the ontology $\mathcal{K}$. Queries in the mapping are conjunctive queries, i.e. conjunctive queries of the form

$$\{x_1, .., x_n | \exists y_1, .., y_m conj(x_1, .., x_n, y_1, .., y_m)\}$$

  where *conj* is a conjunction of atoms, whose predicate symbols are relation names from $\mathcal{S}$, $x_1, .., x_n$ are free variables of the query, and n is the arity of the query. We omit $\exists y_1, .., y_m$ when clear from the context.

In order to understand the transformations involved in the resolution of the impedance mismatch problem, it is important to introduce some notions on the target ontology.

In a DL we are defining concepts and properties which can respectively be considered as unary and binary predicates. Thus the representation of n-ary relations is not directly possible and requires the reification of relationships [1]. This aspect is supported in our system but compel from the mapping designer some extra modeling efforts.

We refer to 'members' of the mapping as the set of concepts and properties of the TBox. We distinguish between object properties which relate individuals to individuals and datatype properties which relate individuals to typed values.

In our framework, we make the distinction between concrete and abstract members. The meaning associated to these terms refers to the ability to instantiate a member via the processing of a mapping file. The comprehension of concrete and abstract members is relatively straightforward as it is equivalent to the assumption made in object-oriented programming. Thus instances (individuals) can be created for a concrete concept and a concrete object property can relate two existing individuals. These instantiations are undergone using the answers retrieved from the processing of the conjunctive queries associated with each concrete member. Thus the head of mapping assertions are formed of atomic concrete members. Abstract members cannot be explicitly instantiated from processing the mapping since they are not allowed in the head of mapping assertions. The objective of abstract members is to support member hierarchies which are henerally represented as trees. It's obvious from this presentation that leaves in such trees must be defined as concrete memebers.

This framework's mapping language allows specifying associations between variables involved in the head of mapping assertions and user defined datatype properties.

In the case of concrete concepts, each variable involved is mapped to a datatype property where the domain of this property must correspond to the given concrete concept or to one if its super concepts. The value retrieved from the execution of the conjunctive query is then associated to the range of this datatype property.

Finally, in the case of concrete object properties, the domain and the range must correspond to individuals already existing in the ABox. This aspect requires an ordering in the processing of the mapping: datatype properties are first created, then concepts are designed and instantiated, finally object properties can relate existing individuals. It is important to stress that this task ordering modeling approach is generally adopted for ontology editors, e.g. Protg, and is thus not considered as a constraint for end-users.

This aspect implies that ABox individuals can be accessed from values identifying source tuples, i.e. primary keys. Thus our system must support a form of relation between the object identifiers of the ABox individuals and the tuples from source relations. This aspect is dealt with a special property *dbom:id* which identifies which are the datatype properties involved in a concept specification that are related by the mapping $\mathcal{M}$ to primary keys in relations of $\mathcal{S}$.

*Example 1.* Let $\mathcal{D}^1 = (\mathcal{S}^1, \mathcal{K}^1, \mathcal{M}^1)$ be a DBOM integration system where $\mathcal{S}^1$ consists of a single source with three relations. Relation *drug* is of arity 3 and contains information about drugs with their codes, names and prices. Relation *ephMRA*, arity 2, contains codes and names of the Anatomical Classification, i.e. a standard that represents a subjective method of grouping certain pharmaceutical products, proposed by the European Pharmaceutical Market Research Association© . Finally, relation ephDrug, of arity 2, proposes relationships between drug codes and codes of the Anatomical Classification.

The ontology schema is made of the following members. The concrete concepts *Drug* and *EphMRA*. The datatype properties *hasDrugCode*, *hasDrugName*, *hasDrugPrice* whose domains must be instances of the *Drug* concept and ranges are respectively a drug code, a drug name and a drug price. The *hasEphCode* and *hasEphName* are datatype properties with instances of the *EphMRA* concept as domain and respectively EphMRA code and name as range. Finally, we need a concrete object property, namely *hasEphMRA*, to relate instances of the *Drug* concept to instances of the *EphMRA* concept.

The mapping $\mathcal{M}^1$ is defined by:

$$\begin{array}{c}(1)\ Drug(x,y,z) \leftarrow drug(x,y,z)\\ (2)\ EphMRA(x,y) \leftarrow ephMRA(x,y)\\ (3)\ hasEphMRA(x,y) \leftarrow ephDrug(x,y)\end{array}$$

This mapping is completed by the following:

- in assertion (1) of $\mathcal{M}^1$, the x,y and z of $Drug(x, y, z)$ are related respectively to *hasDrugCode*, *hasDrugName* and *hasDrugPrice*. The *hasDrugCode* property is related to *dbom:id*.
- in assertion (2) of $\mathcal{M}^1$, the x and y of $EphMRA(x, y)$ are related respectively to *hasEphCode* and *hasEphName*. The *hasEphCode* property is related to *dbom:id*.
- finally, in assertion (3) of $\mathcal{M}^1$, the x and y of $hasEphMRA(x, y)$ correspond respectively to a *Drug* individual identified by x and a *EphMRA* individual identified by y.

### 2.2 Semantics

DL TBoxes and relational schemas are interpreted according to standard first-order semantics. They distinguish the legal structures, i.e. structures that satisfy all axioms, from the illegal structures, i.e. the structures that violate some of them. Thus we can use a first-order semantics with the domain of interpretation being a fixed denumerable set of elements $\Delta$, and every such element is denoted uniquely by a constant symbol in $\Gamma$. In this setting, constants in $\Gamma$ act as *standard names* [14].

In order to specify the semantics of $\mathcal{D}$, we first have to consider a set of data at the sources, and we need to specify which are the data that satisfy the target schema with respect to such data at the sources. We call $\mathcal{C}$ a *source model* for $\mathcal{D}$.

Starting from this specification of a *source model*, we can define the information content of the target $\mathcal{K}$. From now on, any interpretation over $\Delta$ of the symbols used in $\mathcal{K}$ is called a *target interpretation* for $\mathcal{D}$.

**Definition 1.** *Let $\mathcal{D} = \mathcal{S}, \mathcal{K}, \mathcal{M}$ be a DBOM system, $\mathcal{C}$ be a source model for $\mathcal{D}$, a target interpretation $\mathcal{B}$ for $\mathcal{D}$ is a model for $\mathcal{D}$ with respect to $\mathcal{C}$ if the following holds:*

- *$\mathcal{B}$ is a model of $\mathcal{K}$, i.e. $\mathcal{B} \models \mathcal{K}$;*
- *$\mathcal{B}$ satifies the mapping $\mathcal{M}$ with respect to $\mathcal{C}$.*

The notion of $\mathcal{B} \models \mathcal{K}$ requires some attention since schema statements in DL are interpreted differently to similar statements in a relational database setting. These differences can be explained studying the restriction and constraint notions. Basically, a restriction restrains the number of models of a logical theory and allow inference of additional information. This is the approach generally adopted to reason with DLs and in particular in OWL. However, a constraint specifies conditions which may not be violated by an interpretation of the logical theory. So constraints do not allow inference of additional information but can be used to check the information with respect to certain conditions. This is the approach usually adopted by databases.

Ideally, we would like to exploit both restrictions and constraints in order to integrate consistently the sources in the knowledge base. So, in addition to the DL checking whether $\mathcal{B}$ satisfies $\mathcal{K}$, we need to adopt a database-like constraint approach. In order to fulfill such an approach, we introduce the Unique Name Assumption (UNA), i.e. requiring each object instance to be interpreted as a different individual. After this constraint checking step, UNA is relaxed in order to free DL knowledge bases from this restrictive aspect.

The introduction of UNA is suported by the built-in *dbom:id* property. This property is considered as an object property and needs to satisfy the following axioms:

$$(4)\ \forall x, y_1, y_2(\neg dbom : id(x, y_1) \lor \neg dbom : id(x, y_2) \lor y_1 = y_2)$$
$$(5)\ \forall x, y_1, y_2(\neg dbom : id(y_1, x) \lor \neg dbom : id(y_2, x) \lor y_1 = y_2)$$

These constraints state that the domain of an object property is identified by a single range (4) and that a range identifies a single domain (5). The implementation of the DBOM system, the *dbom:id* property is defined in terms of cardinality constraints of OWL properties: owl:functionalProperty, i.e. equivalent to (4) and owl:inverseFunctionalProperty, i.e. equivalent to (5). Using this approach, we are still able to design ontologies in the decidable fragment of OWL, namely OWL DL.

The notion of $\mathcal{B}$ satisfying the mapping $\mathcal{M}$ with respect to $\mathcal{C}$ depends on the interpretation of the mapping assertions. In the context of our solution, we consider the GAV mappings to be sound, i.e. data that it is possible to retrieve from the sources through the mapping are assumed to be a subset of the intended data of the corresponding target schema [13]. In this case, there may be more than one legal knowledge base that satisfies the mapping $\mathcal{M}$ with respect to $\mathcal{C}$.

It is generally considered that DL knowledge bases can be understood as incomplete databases [15]. And there is a general agreement that in the context of incomplete databases, the 'correct' answers are the *certain answers*, that is, answers that occur in the intersection of all 'possible' databases. We adopt this notion for query answering in the DBOM system and use the results that the evaluation of conjunctive queries on an arbitrarily chosen universal solution gives precisely the set of certain answers and that universal solution are the only solutions that have this property [7].

*Example 2.* We consider the $\mathcal{D}^1$ system from Example 1 and let $\mathcal{C}^1$ be a source model for $\mathcal{D}^1$ such that the set of facts holding in $\mathcal{C}^1$ is as follows:

{drug(33316809,Nodex,1.69), ephMRA(R5D1,'Plain antitussives'),
ephDrug(3316809,'R5D1')}

The representation of the ABox of $\mathcal{K}^1$ is now proposed as a graph:

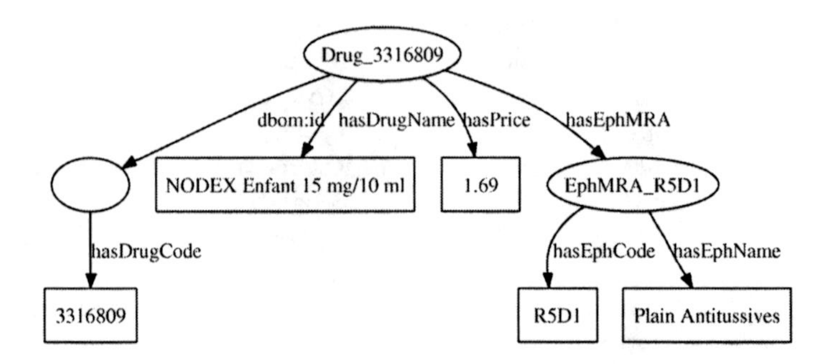


**Fig. 1.** Graph representation of $\mathcal{K}^1$'s ABox

# 3 Enriching the Framework with Preferences over the Views

According to the semantics proposed in Section 2, it may be the case that data retrieved from the sources cannot satisfy both the target ontology schema and the mapping. Example 3 underlines such a violation where data coming from different sources are mutually inconsistent.

*Example 3.* Let $\mathcal{D}^2 = (\mathcal{S}^2, \mathcal{K}^2, \mathcal{M}^2)$ be a DBOM system where $\mathcal{S}^2$ enriches $\mathcal{S}^1$ with a new relation *drug'* of arity 4 and which contains drug codes, names, prices and ephMRA codes. Let $\mathcal{C}^2$ be a source model for $\mathcal{D}^2$ such that the set of facts holding in $\mathcal{C}^2$ is as follows:

{drug(3316809,Nodex,1.69), drug'(3316809,Nodex,2.19,R5D2), ephMRA(R5D1,'Plain antitussives'), ephDrug(3316809,'R5D1')}

The ontology schema $\mathcal{K}^2$ is identical to $\mathcal{K}^1$ and the mapping $\mathcal{M}^2$ is defined as:

$$(6)\ Drug(x,y,z) \leftarrow drug(x,y,z)$$
$$(7)\ Drug(x,y,z) \leftarrow drug'(x,y,z,w)$$
$$(8)\ EphMRA(x,y) \leftarrow ephMRA(x,y)$$
$$(9)\ hasEphMRA(x,y) \leftarrow ephDrug(x,y)$$
$$(10)\ hasEphMRA(x,y) \leftarrow drug'(x,v,w,y)$$

The processing of $\mathcal{M}^2$ violates the axioms associated with the *dbom:id* property as a Drug individual (the Nodex drug) may be created from assertions (6) and (7). In such a situation, we propose to correct inconsistencies using preferences over views of the mapping $\mathcal{M}$.

Preferences have some of their origins in decision theory where they support complex, multifactorial decision processes [8]. Preferences have also motivated researches in the field of databases starting with [12]. In [10,3], the authors distinguish between quantitative and qualitative approaches to preferences. In the quantitative approach, a preference is associated with an atomic scoring function. This approach usually restricts the approach to total orderings of result tuples. The qualitative approach is more general than the quantitative as it proposes partial ordering of results.

**Definition 2.** *In the $\mathcal{D}$ system, a preference function is a function that maps views of concrete ontology members to a score between 0 and 1. A score value of 0 is set by default and corresponds to indifference from the user.*

We now present the reasons for a quantitative approach in the $\mathcal{D}$ system:

- it responds effectively to our need to correct inconsistencies when a given individual, identified by a given key, may be generated from several mapping assertions. This is the case in $\mathcal{D}^2$ where the *Nodex* drug may be either produced from assertion (6) or (7). The resulting individuals would be characterized by a relation with EphMRA code R5D1 (respectively R5D2) and with a price of 1.69 (respectively 2.19) euros.
- the task of setting preferences is only required for ontology members defined by different source views. Thus the total order aspect of this quantitative approach does not make the preference setting task more restrictive.
- in practice, the task of setting preferences to views of a given concrete members may not be complex for a mapping designer. User responsible for the design of mappings generally know the sources well and are able to tell which source is more reliable to others.

*Example 4.* The mapping of Example 3, $\mathcal{M}^2$ can now be enriched with preferences:

$$(11)\ Drug(x,y,z) \leftarrow drug(x,y,z),\ \text{pref=1}$$
$$(12)\ Drug(x,y,z) \leftarrow drug'(x,y,z,w)$$
$$(13)\ EphMRA(x,y) \leftarrow ephMRA(x,y)$$
$$(14)\ hasEphMRA(x,y) \leftarrow ephDrug(x,y),\ \text{pref=1}$$
$$(15)\ hasEphMRA(x,y) \leftarrow drug'(x,v,w,y)$$

Given this mapping and its preferences, Drug individuals coming from the execution of assertion (11) are preferred to those of assertion (12). The same mechanism is adopted for assertion (14) over assertion (15) for the *hasEphMRA* object property. The preferences on assertions (12) and (15) are assumed to be 0 according to the indifference default assumption. Finally, assertion (13) does not need a preference value as it is the only assertion in $\mathcal{M}^2$ for the EphMRA concept. Thus the processing of mapping $\mathcal{D}^2$ with source model $\mathcal{C}^2$ results in the same graph as Example 1, namely Figure 1.

## 4 Implementation

In order to propose a user-friendly and efficient environment for the design of schema mappings, we have implemented a Protégé plug-in [4] which enables via interactions with relational database schemas and SQL queries (i) the creation from scratch of an ontology and (ii) the enrichment of a given ontology. In this

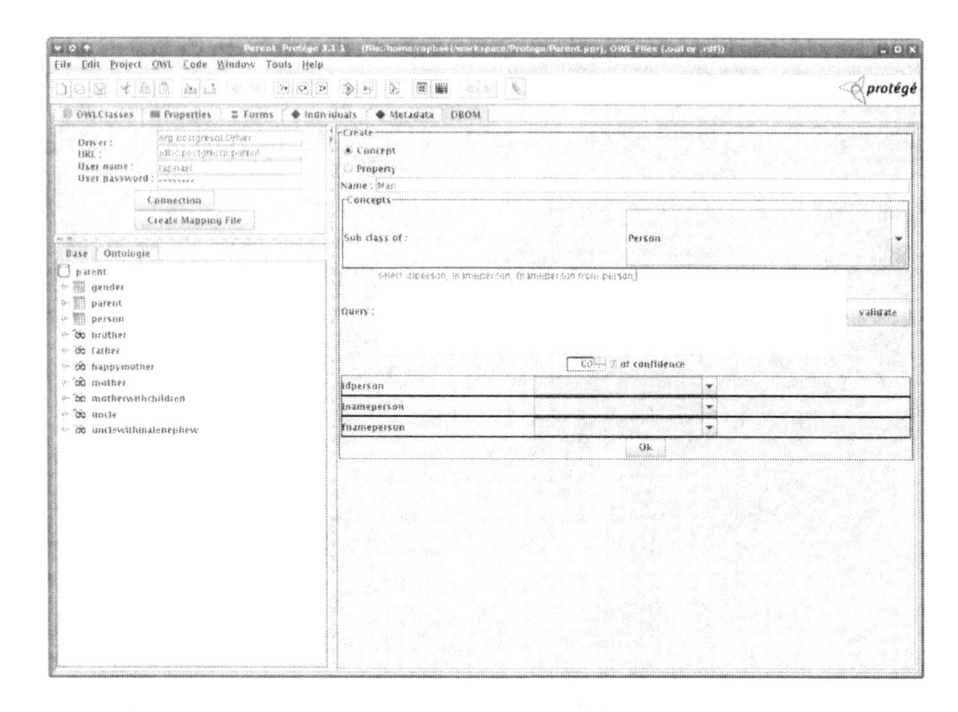


**Fig. 2.** Protégé and the DBOM plug-in

context, the definition of a mapping corresponds to associate SQL queries to DL concepts and roles (Figure 2) and to enrich these elements with assertions using the standards OWL Protégé tabs, i.e. OWLClasses and Properties.

In this plug-in, the validation of a mapping schema implies the recording of (i) the mapping file which is serialized in an XML file, (ii) the DL TBox which is recorded using the Protégé API, and (iii) the DL ABox which is the result of processing the mapping in a DBOM system. This plug-in is implemented in Java and uses Hewlett-Packard's Jena API.

We evaluated the DBOM system and its Protégé plug-in on a drug related medical informatics application. Our approach enabled design and instantiation of an OWL DL knowledge base of the drug domain which now supports inferences on a self-medication application [5]. The readiness of the plug-in was appreciated by all project members, i.e. computer scientists and health care professionals, and enabled to integrate drug related data from more than ten sources (AFSAPPS, EphMRA, Self, etc...).

## 5 Conclusion

In this paper, we have introduced and formalized a solution to enable the migration of data stored in multiple data sources to a DL knowledge base. Because data from the sources may not be retrieved and shared on-demand at query time, we have opted for a materialization of the migrated data.

In order to deal with possible cases of inconsistencies, we support the setting of preferences over the views of the mapping. This approach is particularly useful when several sources are required to cover a domain, e.g. the french drug domain.

We believe that the approach presented in this work can be extended in several ways. First, we can refine the notion of preference at the view level to preference at the attribute level over the views. Using this approach of preferences, it will be easier to design fine-grained ABoxes.

## References


1. Baader, F., Calvanese, D., McGuinness, D., Nardi, D., Patel-Schneider, P.: The Description Logic Handbook: Theory, Implementation and Applications. Cambridge University Press, Cambridge (2003)
2. Bertossi, L., Chomicki, J.: Query Answering in Inconsistent Databases. In: Chomicki, J., Saake, G., van der Meyden, R. (eds.) Logics for emerging applications of databases(Chapter in book), Springer, Heidelberg (2003)
3. Chomicki, J.: Preference Formulas in Relational Queries. ACM Trans. Database System 28, 427–466 (2003)
4. Curé, O., Squelbut, R.: Integrating data into an OWL Knowledge Base via the DBOM Protégé plug-in. In: Proc. of the 9th International Protégé conference (2006)
5. Curé, O.: XIMSA: eXtended Interactive Multimedia System for Auto-medication IEEE Computer-Based Medical System Symposium (CBMS), pp. 570–575 (2004)
6. Dean, M., Schreiber, G.: OWL Web Ontology Language Reference. 2004. W3C Recommendation (February 10, 2004)
7. Fagin, R., Kolaitis, P., Miller, R., Popa, L.: Data exchange: semantics and query answering. In: Calvanese, D., Lenzerini, M., Motwani, R. (eds.) ICDT 2003. LNCS, vol. 2572, pp. 207–224. Springer, Heidelberg (2002)
8. Fishburn, P.C.: Utility Theory for Decision Making. Wiley, Chichester (1970)
9. Gómez-Pérez, A., Fernández-López, M., Corcho, O.: Ontological Engineering. Springer, Heidelberg (2003)
10. Kieβling, W.: Foundations of Preferences in Database Systems. In: Proc. of the 28th International Conference on Very Large Data Bases (VLDB). pp. 311–322 (2002)
11. Kolaitis, P.: Schema mappings, data exchange, and metadata management. In: Proc. ACM Symposium on Principles of Database Systems (PODS), pp. 61–75. ACM Press, New York (2005)
12. Lacroix, M., Lavency, P.: Preferences: Putting More Knowledge into Queries. In: Proc. 13th International Conference on Very Large Data Bases (VLDB). pp. 217–225 (1987)
13. Lenzerini, M.: Data integration: a theoretical perspective. In: Proc. ACM Symposium on Principles of Database Systems (PODS 02), pp. 233–246 (2002)
14. Levesque, H., Lakemeyer, G.: The Logic of Knowledge Bases. MIT Press, Cambridge (2001)
15. Levy, A.Y.: Obtaining Complete Answers from Incomplete Databases. In: Proc. VLDB'96. pp. 402–412 (1996)
16. Wijsen, j.: Condensed representation of database repairs for consistent query answering. In: Calvanese, D., Lenzerini, M., Motwani, R. (eds.) ICDT 2003. LNCS, vol. 2572, pp. 378–393. Springer, Heidelberg (2002)

# A Context-Based Approach for the Discovery of Complex Matches Between Database Sources

Youssef Bououlid Idrissi and Julie Vachon

Department of Computer Science and Operational Research
University of Montreal, CP 6128, succ. Centre-Ville
Montreal, Quebec, H3C 3J7, Canada
{bououlii,vachon}@iro.umontreal.ca
http://www.iro.umontreal.ca

**Abstract.** The elaboration of semantic matching between hetero geneous data sources is a fundamental step in the design of data sharing applications. This task is tedious and often error prone if handled manually. Therefore, many systems have been developed for its automation. But, the majority of them focus on the problem of finding simple (*one-to-one*) matching. This is likely due to the fact that complex (*many-to-many*) matching raises a far more difficult problem since the search space of concept combinations can be tremendously large. This article presents INDIGO, a system which can compute complex matching by taking into account data sources' context. First, it enriches data sources with complex concepts extracted from their respective development artifacts. It then computes a mapping between the two data sources thus enhanced.

**Keywords:** complex semantic matching, context analysis, multi-strategy.

## 1 Introduction

Semantic matching[1] consists in finding semantically meaningful relationships across data stored in heterogeneous sources. When done manually, this task can prove to be very tedious and error prone [4]. To date, many systems [1,5] have addressed the automation of this stage. However, most solutions confine themselves to simple matching (*one-to-one*) (e.g. *postal_code* → *zip_code*) although complex matching (*many-to-many*) (e.g. *concat(street,city)* → *address*) is frequently required in practice. The little work addressing complex matching can be explained by the greater complexity of finding complex matches than of discovering simple ones. In fact, while the search space of matching candidates is finite in the case of simple matching (n.b. it is indeed limited by the product of the number of concepts in data sources), it can be extremely large in the case of complex matching. Indeed, one can imagine an important number of operators (*concat*, +, *, ...) to combine concepts of a data source, therefore giving rise to many possible combinations. To cope with this challenging problem, this article

[1] Also called *semantic alignment*, mapping or simply *matching*.

R. Wagner, N. Revell, and G. Pernul (Eds.): DEXA 2007, LNCS 4653, pp. 864–873, 2007.

introduces INDIGO[2], a system that avoids searching such large spaces of possible concept combinations. Rather, it implements an innovative solution, based on the exploration of the data sources' *informational context*, to identify semantically pertinent combinations worth to be matched. The informational context of a data source is composed of all the available textual and formal artifacts documenting, specifying, implementing this data source. It therefore conceals precious supplementary information which can provide useful insights about the semantics of data sources' concepts.

### 1.1 Informational Context

INDIGO distinguishes two main sets of documents in the informational context (cf. [2] for details). The first set, called the descriptive context, gathers all the available data source specification and documentation files produced during the different development stages (e.g. requirement description). These are usually written in some informal or semi-formal style. The second set is called the operational context. It is composed of formal artifacts such as programs, forms or XML files. In formal settings, significant combinations of concepts are more easily located (e.g. they can be found in formulas, function declarations, etc.). This is one of the reason why INDIGO favors the exploration of the operational context for the extraction of complex concepts.

### 1.2 Complex Concept Mining

Each complex concept extracted from the operational context is to be represented by a triplet <*name*, *dataType*, *conceptComb*> whose components respectively denote the name of the complex concept, its data type and a combination of concepts to which it is associated (e.g. <*totalprice*, *float*, *price*(1+taxes)*>). To search for complex concepts, INDIGO relies on a set of context analyzers each specialized to work over a particular category of artifacts (e.g. programs, forms, etc.). Each analyzer applies a set of heuristic extraction rules to mine complex concepts. Once extracted, complex concepts are added to data sources as new candidates for the matching phase. In INDIGO, the enhancement of a data source with information gathered from its context is called *data source enrichment*. INDIGO was used for the semantic matching of four database schemas taken from four open-source e-commerce applications: *Java Pet Store* [9], *eStore* [12], *PetMarket* [10] and *PetShopDNG* [11]. In all cases, complete source code files were freely available and could thus be used for our experiments.

The rest of the paper is organized as follows. Section 2 surveys recent work on complex semantic matching. Section 3 describes the current implementation of INDIGO's *Context Analyzer* and *Mapper* modules. Explanations focus on specific modules required for complex concept mining and matching. Experimental results showing INDIGO's performance in computing complex matches are presented and commented in Section 4. Concluding remarks and comments on future work are given in Section 5.

---

[2] INteroperabilty and Data InteGratiOn.

## 2 Related Work

To this day, most tools and approaches dealing with semantic matching have solely addressed the "one-to-one" mapping aspect of the problem. To the best of our knowledge, the only work directly tackling complex matching is described in [6], [7] and [8]. An overview of these respective approaches is given below.

In [6], authors advocate an approach which maps data sources to some intermediary handcrafted domain ontology in order to facilitate the discovery of complex matches. This domain ontology defines generic complex concepts along with their relationships (e.g. aggregation, generalization or specialization) to other concepts. This approach globally performs well but requires the manual development of a domain ontology for each specific application.

In [7], a framework called **DCM** is described which exploits co-occurrence patterns across database query interfaces over the Internet to discover complex semantic relationships. Even if it performs with good accuracy, **DCM** does not provide still an efficient solution to the complex matching problem. In fact, many complex combinations (e.g. $total = (1 + taxeRate) * unitPrice$) have few chance from being identified through co-occurrences in query interfaces.

Finally, authors in [8] presents the **iMAP** system which discovers complex concepts by searching the space of all the possible concept combinations. To make the search effective the system relies on several search modules, each one exploring a subspace composed of a single type of combinations (e.g. textual or arithmetic combinations subspaces). Besides, **iMAP** uses an iterative technique called Beam Search in order to control the search for matching combinations. Although supporting an interesting approach, **iMAP** still presents some limitations. For instance, **iMAP**'s iterative Beam Search algorithm may prevents relevant combinations from being found since it limits the maximal number of computation iterations.

Many other works [1,5] have addressed the general semantic matching problem. INDIGO belongs to the category of systems relying on multiple strategic matchers [8]. It distinguishes itself by taking into account the informational context of data sources in its alignment process and by proposing a particularly flexible hierarchical matching architecture. Moreover, it can generate both simple and complex matchings quite efficiently, with no need for specific domain considerations. From this point of view, INDIGO surpasses many limitations of existing matching systems. Nevertheless, the fact of totally relying on data sources' context to find combinations gives INDIGO some limitations. In fact, this makes INDIGO dependent of the data sources' contexts availability in practice. In the other hand, even when available, not always that contexts contain all relevant combinations. What prevents INDIGO from identifying all complex matches.

## 3 INDIGO's Architecture

To handle both context analysis and semantic matching, INDIGO has an architecture composed of two main modules: a *Context Analyzer* and a *Mapper* module.

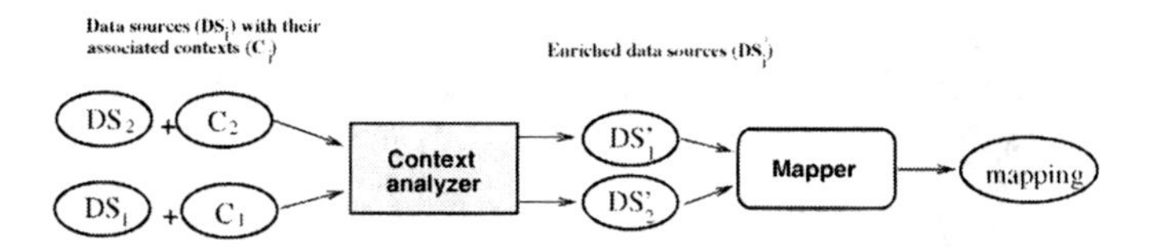


**Fig. 1.** INDIGO's matching process

Figure 1 shows the intervention of these two modules within INDIGO's general matching process. The *Context Analyzer* module takes the data sources to be matched along with their related contexts and proceeds to their enrichment before delivering them to the *Mapper* module for their effective matching.

### 3.1 *Context Analyzer*

The *Context Analyzer* comprises two main modules, each being specialized in a specific type of concept extraction and enrichment. 1) The *Concept name collector* [3] explores the descriptive context of a data source to find (simple) concept names which can be related, by vicinity in a phrase for example, to the ones found in the data source's schema. 2) The Complex concept extractor analyzes the operational context to extract complex concepts. In both cases, the new concepts found are integrated to respective data sources to enrich the semantics of schemas' concepts. These two analyzer modules are supervised by a head meta-analyzer which coordinates their respective tasks and is in charge of data source enrichment. Figure 2a) shows the current architecture of our *Context Analyzer*. Basic analyzers composing the Complex concept extractor are depicted by white boxes. INDIGO's basic analyzers currently deal with forms, programs and SQL requests.

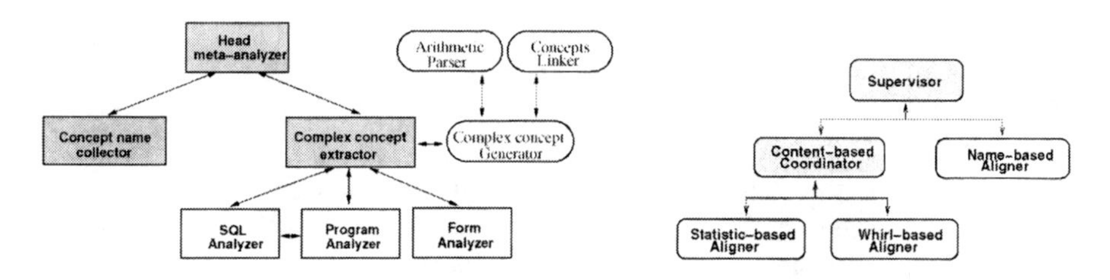


**Fig. 2.** a) A *Context Analyzer* and b) a *Mapper* module

Each analyzing module implements some set of heuristic rules. These rules can readily be modified or extended to answer the specific needs of the documents being analyzed. All modules within the *Context Analyzer* are therefore extensible and can be adapted to deal with different kinds of data sources' contexts. The sequel discusses the use of heuristic extraction rules as well as the implementation of *basic* and *meta* analyzers involved in complex matching.

**Basic Analyzers.** Basic analyzers are responsible for the effective mining of the operational context. They rely on heuristic rules for the extraction of complex concepts. An extraction rule has the following shape:

$$ruleName :: SP_1 || SP_2 ... || SP_n \rightarrow extractionAction$$

The left part is a disjunction of syntactic patterns (noted SP) that basic analyzers try to match[3] when parsing a document. A SP is a regular expression that can contain pattern variables *name*, *type*, $exp_1$, $exp_2$, ... $exp_n$. When a basic analyzer recognizes one of the SPs appearing on the left-hand side of a rule, pattern variables are assigned values (by pattern matching) and the corresponding right-hand side action of the rule is executed. This action builds a complex concept $< name, type, concept\_combinaison >$ using the pattern variables' values. As an example, let's consider the heuristics according to which an accessor method within a class is likely to contain a relevant concept combination in its *return* statement. This heuristics is actually implemented by our *Program analyzer* using an extraction rule

- whose left-hand side contains a set of $SP_i$ $(1 \leq i \leq n)$ recognizing accessor methods (e.g. in Java: $SP_1$ = {`public` *type* `get`*name* * `return` $exp_1$}, or in C#: $SP_2$ = {`public` *type name* { `get` * `return` $exp_1$}).
- whose right-hand side is a function returning a complex concept <*name*, *type*, $exp_1$> for the matched $SP_i$.

```
public string fullName {
        get {
                if (firstName.Length>0)
                        return firstName + ' ' + lastName;
                else
                        return lastName;
        }
}

public double getTotalCost() {
        return quantity * unitCost;
}
```

**Fig. 3.** Part of a C# program of the *eStore* context and part of java program of the *PetStore* context

Each basic analyzer applies its own set of heuristic extraction rules over each of the artifacts it is assigned. Figure 3 shows two program extracts, one in Java and the other in C#. When applied to these extracts, the heuristic rule presented above produces these two complex concepts: <*fullname*, *string*, *firstname*+' '+*lastname*> and <*TotalCost*, *double*, *quantity***unitCost*>. The following paragraphs give an overview of the heuristic rules implemented by each of our current basic analyzers.

*Program analyzer.* As just said, a program analyzer is responsible for parsing files containing source code. In addition to the rule about accessors and update methods, this analyzer implements another heuristics concerning class

[3] N.b. This kind of text matching is called "pattern matching."

constructors. We agree that a constructor's role, in a class $C$, generally consists in building a composite object of type $C$ (e.g. *creditCard*) using the components given in parameter (e.g. *cardNumber, expiryDate, cardType*). The aggregation of these parameters is therefore likely to form a complex concept for the concept represented by $C$.

*Form analyzer.* A *form analyzer* is a module specialized in the discovery of complex concepts emerging from forms. Its strategy consists in exploiting the intuitive semantics associated to tables found in forms. As an example, let's consider part of a form used in the *eStore* application. It presents a tabular structure composed of two columns. The first column contains the word *Street* while the second column is composed of two input fields, respectively labeled *Address1Label* and *Address2Label*, aligned one above the other. The form analyzer recognizes this typical tabular pattern. The concept appearing in the first column is interpreted as a composite concept defined by the concatenation of the two components appearing in the next column.

*SQL analyzer.* A SQL analyzer is specialized in the parsing of SQL requests. Its extraction rules search SELECT statement queries for clauses containing expressions which are likely to be concept combinations, like in this case: `SELECT street, concat(address_line_1, address_line_2), FROM Person.` If no combination is found in a SELECT query, a new complex concept is nevertheless generated by simply concatenating all the concepts found in the clauses composing this query.

**Meta-Analyzers.** Each meta-analyzer is in charge of a set of artifacts composing the informational context. Its role essentially consists in classifying these artifacts and assigning each of them to a relevant child. To do so, it applies heuristics like checking file name extensions or parsing file internal structures. Here follows a short description of meta-analyzers implemented in INDIGO. Given the scope of this paper, the *Concept name collector* is not be presented here since it is not directly involved in complex matching.

*Head meta-analyzer.* The meta-analyzer module at the head of the *Context Analyzer* is in charge of the *Concept name collector* and the *Complex concept extractor* coordination. It enhances data sources with the simple and complex concepts respectively delivered by these two modules. For complex concepts, the enrichment step not only requires the name of the enriching concepts but also the values[4] associated to them. These values are assessed by querying the database using SQL SELECT statements.

*Complex concept extractor.* The Complex concept extractor coordinates actions of the three modules respectively specialized in the extraction of complex concepts from programs, forms and SQL requests. In addition, it relies on an internal module, called *complex concept generator* (shown on Fig.2), to perform the two following tasks ensuring the consistency of data source enrichment:

[4] Concepts are more likely to be similar when they share similar values.

- *Validation and harmonization:* When a complex concept has a numeric type, an arithmetic parser is used to analyze its concept combination expression. If any problem is encountered during the parsing, the complex concept is rejected. As for string complex concepts, their corresponding concept combination is usually given as a list of concepts separated by commas or '+' operators. Each concept is checked to be a simple literal word or a program constant. A concatenation expression, suiting the format of the data source to be enriched, is then created from these concepts.
- *Concept linkage:* Concept names found in complex concepts provided by basic analyzers must be replaced by corresponding concepts of the data source. This conformity relation is checked using the *JaroWinkler* similarity metric. A similarity threshold is arbitrarily fixed to a high value (e.g. fixed to 0.85) to solely retain combinations of concepts which can each be matched to some concept of the data source. As soon as a complex concept comprises a concept name which can not be matched, it is rejected.

### 3.2 Mapper Module

The current architecture of our *Mapper* implementation is shown on Figure 2b). In addition to the supervisor, it comprises three aligners and one coordinator hierarchically organized. An overview of each module is given below.

*Name-based aligner.* A name-based aligner proposes matches between concepts having similar names. Name similarity is measured using the *JaroWinkler* lexical similarity metric [5].

*Whirl-based aligner.* Given a source concept $s$, the whirl-based aligner tries to identify the target concept $t$ whose instance values best matches the ones of $s$. It uses an adapted version of a so-called *WHIRL* technique which is a sort of nearest neighbor classification algorithm developed by Cohen and Hirsh [13].

*Statistic-based aligner.* The statistic-based aligner also compares concepts' contents. The content of a concept is represented by a normalized vector describing seven characteristics: three data type status bits (*string*, *numeric* and *date*) and four statistical information (*minimum*, *maximum*, *average* and *variance values*). The similarity between two concepts is given by the euclidean distance separating their respective characteristic vectors.

*Content-based coordinator.* The content-based coordinator combines matches proposed by its subordinate *whirl-based* and *statistic-based* aligners.

## 4 Experimental Results

Experiments were conducted over four open-source systems related to the domain of *pet sale*. Our main objective was to get a first evaluation of INDIGO's performance with regards to complex concept mining and complex matching.

### 4.1 Context Mining Performance

Two measures, respectively called *cohesion* and *relevance*, were defined to evaluate INDIGO's *Context Analyzer* performance.

- *Cohesion.* A complex concept is said to be cohesive if it presents a semantically sound combination of concepts. For instance, a complex concept which is associated to concept combination *concat(first_name, unit_price)* can't be considered cohesive. Indeed, the concatenation of a person's first name with the unit price of a product does not convey any commonly accepted meaning. The cohesion measure indicates the percentage of extracted complex concepts which are cohesive.
- *Relevance.* A complex concept is relevant if it appears in the reference mapping. Given two data sources to be matched, the relevance measure thus computes the proportion of complex concepts in the reference mapping that were effectively mined by the *Context Analyzer.*

**Table 1.** Performance results of the *Context Analyzer* mining applications' context for complex concepts

| Appli-cation | tables | concepts | complex concepts | extracted complex concepts total | cohesive | relevant | Performance cohesion | relevance |
|---|---|---|---|---|---|---|---|---|
| PetStore | 13 | 86 | 7 | 40 | 32 | 5 | 80% | 71% |
| eStore | 15 | 73 | 3 | 23 | 18 | 3 | 78% | 100% |
| PetShop | 5 | 41 | 2 | 8 | 7 | 2 | 87% | 100% |
| PetMarket | 11 | 71 | 3 | 11 | 8 | 2 | 73% | 67% |
| total | 44 | 271 | 15 | 82 | 65 | 12 | 79% | **80%** |

To improve performances and eliminate noisy side effects, the *Context Analyzer* resorts to two filters. The first one (introduced in Section 3.1) consists in keeping only the combinations for which all concepts have been successfully linked to a data source with a similarity exceeding a predefined threshold. As for the second filter, it simply rejects those excessively large combinations whose number of concepts exceeds a certain threshold (n.b. lengthy combinations are likely to lack cohesion). Nevertheless, it is worth noting that in the specific case of *concat* expressions, large combinations can be separated into shorter ones. This is done by grouping adjacent concept names that share a same prefix or suffix. For instance, the following expression *concat(street1, street2, zip, city, state, country)* can be divided into the two following ones *concat(street1, street2)* and *concat(zip, city, state, country)*. The reason for this separation is to retrieve a larger number of cohesive complex concepts that would otherwise be eliminated by the second filter. Table 1 shows the results of our experiments with complex concept mining. The two filters' thresholds were respectively fixed to 0.8 (similarity) and 5 (nb of concepts). Before the application of filters, we observed that 277 complex concepts were initially extracted from the *PetStore*'s context which recorded 1123 documents. In contrast, 48 complex concepts were discovered in the *eStore* application whose context contained only 72 documents. This constitutes an interesting result in itself since it suggests that the efficiency of complex concepts discovery may essentially depend, not on the quantity, but on the relevance of the data source's context artifacts. As indicated in Table 1, when filters were applied, the *Context Analyzer* discovered a total of 82 complex concepts among which 79% were cohesive. As for relevant complex concepts, 12 were discovered on a maximal total of 15, for a relevance evaluation of 80%. It is important to underline the fact that our *Context Analyzer* was not designed for the specific analysis of the given applications. Genericity and extensibility are fundamental qualities of the *Context Analyzer*'s architecture. Indeed analyzer modules are designed to deal with a general "category" of documents rather than with a specific file format. For instance, the implementation proposes a single *Program analyzer* supporting all C#, Java and coldfusion programs. With only five heuristic extraction rules being applied, the *Context Analyzer* already achieves good mining performances as shown by results in Table 1.

**Table 2.** Performance results for the mapping

| | global matching | | | | | | complex matching | | |
|---|---|---|---|---|---|---|---|---|---|
| | $prec_1$ | $rec_1$ | $f\text{-}m_1$ | $prec_2$ | $rec_2$ | $f\text{-}m_2$ | prec | rec | f-m |
| PetStore/eStore | 70% | 56% | 62% | 87% | 49% | 63% | 100% | 100% | 100% |
| PetStore/PetShop | 67% | 62% | 64% | 68% | 66% | 67% | 100% | 40% | 57% |
| PetStore/PetMarket | 54% | 84% | 66% | 55% | 86% | 67% | 100% | 50% | 67% |
| eStore/PetShop | 83% | 64% | 73% | 87% | 68% | 77% | 100% | 100% | 100% |
| eStore/PetMarket | 67% | 69% | 68% | 68% | 70% | 69% | 100% | 100% | 100% |
| PetShop/PetMarket | 72% | 79% | 75% | 76% | 79% | 78% | 100% | 100% | 100% |
| Average | 69% | 69% | 69% | 73% | 69% | 70% | 100% | 82% | **87%** |

N.b. **prec**: precision; **rec**: recall; **f-m**: f-measure; $_1$: without c.c. enrichment i.e. simple matching only; $_2$: with c.c. enrichment i.e. simple and complex matching;

## 4.2 Complex Matching Evaluation

For global matching runs (i.e. computing both simple and complex matchings), experiments were executed with and without taking into account the enrichment of data sources with complex concepts. For complex matching runs, experiments necessarily required data sources to be previously enriched with complex concepts. Among the complex concepts that were mined, only cohesive ones were used. The performance of each computed alignment was evaluated using the *f-measure*. Table 2 shows that the *Mapper* module performed well with an average *f-measure* of 87% for the discovery of complex matches. We can see that enrichment with complex concepts does not work against single concept matching ($prec_1 \leq prec_2 \rightarrow$ no loss due to undesirable noise effects). Our expectations are that concept enrichment should contribute positively to "global" semantic matching. We therefore plan to extend enrichment to both single and complex concepts and study its effect on single and complex matching.

## 5 Conclusion

We proposed INDIGO, a innovative solution for the discovery of complex matches between database sources. Avoiding to search the unbounded space of possible concept combinations, INDIGO discovers complex concepts by searching through data sources' artifacts. Newly discovered complex concepts are added to data sources as new matching candidates for complex matching. The use of INDIGO for the matching of four open-source e-commerce applications emphasizes the pertinence of this approach. Experiments showed that INDIGO could perform very well on complex concept extraction by discovering 80% of relevant complex concepts. Moreover, its efficiency for complex matching was stressed by an *f-measure* of 87%.

## References

1. Rahm, E., Bernstein, P.A.: A Survey of Approaches to Automatic Schema Matching. VLDB Journal 10(4), 334–350 (2001)
2. Bououlid, I.Y., Vachon, J.: Context Analysis for Semantic Mapping of Data Sources Using a Multi-Strategy Machine Learning Approach. In: Proc. of the International Conf. on Enterprise Information Systems (ICEIS05), Miami, pp. 445–448 (2005)
3. Bououlid, I.Y., Vachon, J.: A Context-Based Approach for Linguistic Matching. In: Proc. of the International Conf. on Software and Data Technologies (ICSOFT07), Barcelona, Spain (2007)
4. Li, W.S., Clifton, C.: Semantic Integration in Heterogeneous Databases Using Neural Networks. In: Proc. of the 20th Conf. on Very Large Databases (VLDB), pp. 1–12 (1994)
5. Euzenat, J., et al.: State of the Art on Ontology Alignment. Part of a research project funded by the IST Program of the Commission of the European Communities, project number IST-2004-507482. Knowledge Web Consortium (2004)
6. Xu, L., Embley, D.: Using domain ontologies to discover direct and indirect matches for schema elements. In: Proc. of the Semantic Integration Workshop (2003)
7. He, B., Chang, K.C.-C., Han, J.: Discovering complex matchings across web query interfaces: A correlation mining approach. In: Proc. of the SIGKDD conf. (2004)
8. Dhamankar, R., Lee, Y., Doan, A., Halevy, A., Domingos, P.: iMAP: Discovering Complex Semantic Matches between Database Schemas. In: Proc. of the ACM SIGMOD Conference on Management of Data, pp. 383–394. ACM Press, New York (2004)
9. Sun Microsystems (2005), `http://java.sun.com/developer/releases/petstore/`
10. Adobe (2007), `http://www.adobe.com/devnet/blueprint/`
11. DotNetGuru.org (2003), `http://www.dotnetguru.org/modules.php`
12. McUmber, R.: Developing pet store using rup and xde. Web Site (2003)
13. Cohen, W., Hirsh, H.: Joins that Generalize: Text Classification using Whirl. In: Proc. of the Fourth Int. Conf. on Knowledge Discovery and Data Mining (1998)

# Ontology Modularization for Knowledge Selection: Experiments and Evaluations*

Mathieu d'Aquin[1], Anne Schlicht[2], Heiner Stuckenschmidt[2], and Marta Sabou[1]

[1] Knowledge Media Institute (KMi), The Open University, Milton Keynes, UK
{m.daquin, r.m.sabou}@open.ac.uk
[2] University of Mannheim, Germany
{anne, heiner}@informatik.uni-mannheim.de

**Abstract.** Problems with large monolithical ontologies in terms of reusability, scalability and maintenance have led to an increasing interest in modularization techniques for ontologies. Currently, existing work suffers from the fact that the notion of modularization is not as well understood in the context of ontologies as it is in software engineering. In this paper, we experiment on applying state-of-the-art tools for ontology modularization in the context of a concrete application: the automatic selection of knowledge components to be used for Web page annotation and semantic browsing. We conclude that, in a broader context, an evaluation framework is required to guide the choice of a modularization tool, in accordance with the requirements of the considered application.

**Keywords:** Ontology modularization, partitioning, module extraction.

## 1 Introduction

Modularization is a crucial task to allow ontology reuse and exploitation on the Semantic Web. The notion of modularization comes from Software Engineering where it refers to a way of designing software in a clear, well structured way that supports maintenance and reusability. From an ontology engineering perspective, modularization should be considered as a way to structure ontologies, meaning that the construction of a large ontology should be based on the combination of self-contained, independent and reusable knowledge components. In reality, even if they implicitly relate several sub-domains, most of the ontologies are not structured in a modular way. Therefore, in order to facilitate the management and the exploitation of such ontologies, *ontology modularization techniques* are required to identify and extract significant modules in existing ontologies.

While there is a clear need for modularization, there are no well-defined and broadly accepted definitions of modularity for ontologies. Several approaches have been recently proposed to extract modules from ontologies, each of them implementing its own intuition about what a module should contain and what

* This work is partially funded by the Open Knowledge (IST-FF6-027253) and NeOn projects (IST-FF6-027595), and partially supported by the German Science Foundation under contract STU 266/1 as part of the Emmy-Noether Program.

R. Wagner, N. Revell, and G. Pernul (Eds.): DEXA 2007, LNCS 4653, pp. 874–883, 2007.

should be its qualities, generally without making this intuition explicit. This lack of consensus and of clarity hinders the application of these techniques in concrete scenarios, leading to difficulties in choosing the appropriate one. Moreover, to our knowledge, no other study has focused on the evaluation and comparison of ontology modularization techniques.

Our hypothesis is that there is no universal way to modularize an ontology and that the choice of a particular technique should be guided by the requirements of the considered application. We believe that modularization criteria should be defined in terms of the applications for which the modules are catered. For this reason, we detail in this paper some experiments conducted with several ontology modularization tools on a particular application: the selection of relevant knowledge components from online available ontologies. The goal is to characterize the requirements of this particular application using criteria from the literature on ontology modularization, and thus, to analyze the results of existing ontology modularization techniques regarding these requirements. In this way, we aim at better understanding the fundamental assumptions underlying the current modularization techniques. This work can be seen as a first step towards a broader framework, guiding application developers in choosing the appropriate technique and the designers of techniques in further developments.

The paper is structured as follows. Section 2 briefly describes the concrete scenario in which we apply modularization techniques. Section 3 and Section 4 respectively overview ontology modularization techniques and evaluation criteria that have been proposed in the literature. In Section 5 we evaluate, using the considered criteria, the results of the application of modularization techniques on our case-study. We conclude in Section 6 on the need for a comprehensive evaluation framework for ontology modularizations.

## 2 A Case-Study for Modularization: The Knowledge Selection Scenario

Knowledge selection has been described in [1] as the process of selecting the relevant knowledge components from online available ontologies and has been in particular applied to the Magpie application. Magpie [2] is a Semantic Web browser, available as a browser plugin, in which instances of ontology classes are identified in the current Web page and highlighted with the color associated to each class. In our current work we are extending Magpie towards open semantic browsing in which the employed ontologies are automatically selected and combined from online ontologies. As such, the user is relieved from manually choosing a suitable ontology every time he wishes to browse new content. Such an extension relies on mechanisms that not only dynamically select appropriate ontologies from the Web, but also extract from these ontologies the relevant and useful parts to describe classes in the current Web page.

Our previous work and experiences in ontology selection [3] made it clear that modularization may play a crucial role in complementing the current selection techniques. Indeed, selection often returns large ontologies that are virtually

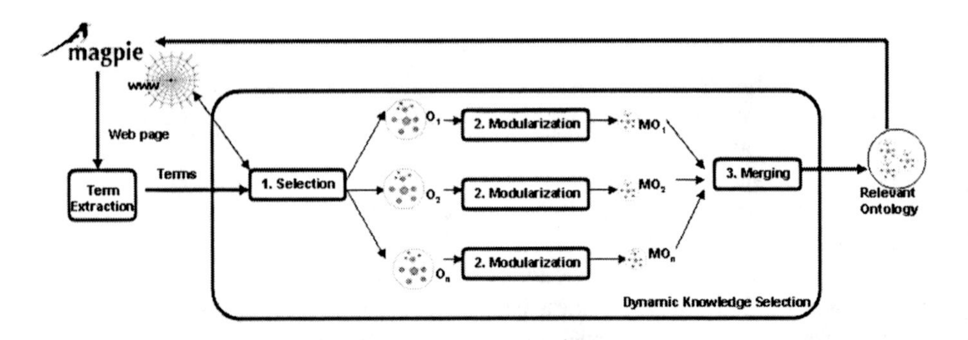


**Fig. 1.** The knowledge selection process and its use for semantic browsing with Magpie

useless for a tool such as Magpie which only visualises a relatively small number of classes at a time. What is needed instead is that the selection process returns a part (module) of the ontology that defines the relevant set of terms. These considerations justify the need to extend selection techniques with modularization capabilities. In Figure 1 we depict the three major generic steps of the *knowledge selection* process that integrates ontology selection, modularization and merging. We focus in this paper on applying existing techniques for the second step of this process: ontology modularization.

## 3 Modularization Techniques

We consider an ontology $O$ as a set of axioms (subclass, equivalence, instantiation, etc.) and the signature $Sig(O)$ of an ontology $O$ as the set of entity names occurring in the axioms of $O$, i.e. its vocabulary.

In the following, we deal with several approaches for ontology modularization, having different assumptions about the definition of an ontology module. Therefore, we define an ontology module in a very general way as a part of an ontology: a module $M_i(O)$ of an ontology $O$ is a set of axioms, such that $Sig(M_i(O)) \subseteq Sig(O)$.

Two different approaches have been considered for the modularization of existing ontologies: ontology partitioning, which divides an ontology into a set of modules, and module extraction, which reduces an ontology to a module focusing on a given set of elements.

### 3.1 Ontology Partitioning Approaches

The task of partitioning an ontology is the process of splitting up the set of axioms into a set of modules $\{M_1, \cdots, M_k\}$ such that each $M_i$ is an ontology and the union of all modules is semantically equivalent to the original ontology $O$. There are several approaches for ontology partitioning that have been developed for different purposes. We have chosen to consider only available techniques that are sufficiently stable:

**PATO** refers to a standalone application described in [4]. The goal of this approach is to support maintenance and use of very large ontologies by providing the possibility to individually inspect smaller parts of the ontology. The algorithm operates with a number of parameters that can be used to tune the result to the requirements of a given application.

**SWOOP** refers to the partitioning functionality included in the SWOOP ontology editor and described in [5]. This tool partitions an ontology into a set of modules connected by $\varepsilon$-connections. It aims at preserving the completeness of local reasoning within all created modules. This requirement is supposed to make the approach suitable for supporting selective use and reuse since every module can be exploited independently of the others.

### 3.2 Module Extraction Approaches

The task of module extraction consists in reducing an ontology to the sub-part, the module, that covers a particular sub-vocabulary. This task has been called segmentation in [6] and traversal view extraction in [7]. More precisely, given an ontology $O$ and a set $SV \subseteq Sig(O)$ of terms from the ontology, a module extraction mechanism returns a module $M_{SV}$, supposed to be the relevant part of $O$ that covers the sub-vocabulary $SV$ ($Sig(M_{SV}) \supseteq SV$). Techniques for module extraction often rely on the so-called *traversal approach*: starting from the elements of the input sub-vocabulary, relations in the ontology are recursively "traversed" to gather related elements to be included in the module.

Two module extraction tools are considered here:

**KMi** refers to a standalone application developed at the Knowledge Media Institute (KMi) of the Open University, for the purpose of the knowledge selection scenario, as described in [1]. The input sub-vocabulary can contain either classes, properties, or individuals. The mechanism is fully automatized, is designed to work with different kinds of ontologies (from simple taxonomies to rich and complex OWL ontologies) and relies on inferences during the modularization process.

**Prompt** refers to the module extraction feature of the Prompt toolkit, integrated as a plugin of the Protégé ontology editor, as described in [7]. This approach recursively follows the properties around a selected class of the ontology, until a given distance is reached. The user can exclude certain properties in order to adapt the result to the needs of the application.

It is worth mentioning that the technique described in [6] is also freely available, but can only be used on the Galen ontology in its current state.

## 4 Evaluation Criteria for Modularization

In the previous section, we have briefly presented a number of different approaches for ontology partitioning and module extraction. In this section, we take a closer look at different criteria for evaluating either the modules resulting from the application of a modularization technique, or the system implementing this technique.

### 4.1 Evaluating the Result of Modularization

In [8], the authors describe a set of criteria based on the structure of the modularized ontology and that have been designed to trade-off maintainability as well as efficiency of reasoning in a distributed system, using distributed modules.

*Size.* Despite its evident simplicity, the relative size of a module (number of classes and properties) is among the most important indicators of the efficiency of a modularization technique. Indeed, the size of a module has a strong influence on its maintainability and on the robustness of the applications relying on it.

*Redundancy.* Allowing the modules of a partition to overlap is a common way of improving efficiency and robustness. On the other hand, having to deal with redundant information increases the maintenance effort.

*Connectedness.* The independence of a set of modules resulting from a partitioning technique can be estimated by looking at the degree of interconnectedness of the generated modules. A modularized ontology can be depicted as a graph, where the axioms are nodes and edges connect every two axioms that share a symbol. The connectedness of a module is then evaluated on the basis of the number of edges it shares with other modules.

*Distance.* It is worth to measure how the terms described in a module *move closer to each other* compared to the original ontology, as an indication of the simplification of the structure of the module. This *intra-module distance* is computed by counting the number of relations in the shortest path from one entity to the other. An *inter-module distance*, counting the number of modules that have to be considered to relate two entities, can also be envisaged, as a way to to characterize the communication effort caused by the partition of an ontology.

Several authors also defined criteria for evaluating ontology modules, in general focusing on the logical and formal aspects of modularizations (see e.g., [9]). Logical criteria are of particular importance when the modules resulting of the modularization techniques are intended to be used in reasoning mechanisms, but should not be emphasized in our case-study, which focuses on a human interpretation of the module.

### 4.2 Evaluating the Modularization Tool

In [1] the authors focus on the use of modularization for a particular application. This leads to the definition of several criteria, most of them characterizing the adequacy of the design of a modularization tool with respect to constraints introduced by the application.

*Assumptions on the ontology.* Most of the existing approaches rely on some assumptions. For example, those described in [5] and [6] are explicitly made to work on OWL ontologies, whereas [4] can be used either on RDF or OWL but only exploits RDF features.

*Level of user interaction.* In many systems the required user entries are limited to the inputs of the algorithm. In certain cases, some numerical parameters can be required [4] or some additional procedures can be manually (de)activated [6]. The technique in [7] has been integrated in the Protégé ontology editor to support knowledge reuse during the building of a new ontology. In this case, modularization is an interactive process where the user has the possibility to extend the current module by choosing a new starting point for the traversal algorithm among the *boundary classes* of the module.

*Performance.* Most of the papers concerning modularization techniques do not give any indication about the performance of the employed method (with the noticeable exception of [6]). Performance is a particularly important element to be considered when using a modularization technique for the purpose of an application. Different applications may have different requirements, depending on whether the modularization is intended to be used dynamically, at run-time, or as a "batch" process.

## 5 Experiments

In the scenario described in Section 2, modularization is integrated in a fully automatic process, manipulating automatically selected online ontologies for the purpose of annotation in Magpie. In this section, we simulate the process of knowledge selection on two examples, using four different techniques, in order to evaluate and compare their results[1]. The purpose is to characterize the requirements of this particular scenario using the criteria defined in Section 4, and to show how modularization techniques respond to the selected experiments regarding these requirements.

As already described in [1], it is quite obvious that module extraction techniques fit better in the considered scenario than partitioning tools. Indeed, we want to obtain *one* module covering the set of keywords used for the selection of the ontology and constituting a *sub-vocabulary* of this ontology. However, the result of partitioning techniques can also be used by selecting the set of generated modules that cover the considered terms. The criteria are then evaluated on this set of modules as grouped together by union. Furthermore, we primarily focus on the criteria that appear to be relevant in our scenario: application related criteria (Section 4.2), the size, and the intra-module distance (Section 4.1).

### 5.1 Considered Ontologies

We consider two examples, originally described in the context of ontology selection in [3], where the goal is to obtain an ontology module for the annotation of news stories. We simulate the scenario described in Section 2 by manually

[1] Actual results are available at `http://webrum.uni-mannheim.de/math/lski/Modularization`

extracting relevant keywords in these stories, using ontology selection tools[2] to retrieve ontologies covering these terms, and then applying modularization techniques on these ontologies (steps 1 and 2 in figure 1).

In the first example, we consider the case where we want to annotate the news stories available on the KMi website[3]. We used the keywords *Student*, *Researcher*, and *University* to select ontologies to be modularized, and obtain three ontologies covering these terms:

**ISWC:** `http://annotation.semanticweb.org/iswc/iswc.owl`
**KA:** `http://protege.stanford.edu/plugins/owl/owl-library/ka.owl`
**Portal:** `http://www.aktors.org/ontology/portal`

It is worth mentioning that this example is designed to be simple: we have chosen a well covered domain and obtained three well defined OWL ontologies of small sizes (33 to 169 classes).

The second example was used in [3] to illustrate the difficulties encountered by ontology selection algorithms. Consequently, it also introduces more difficulties for the modularization techniques, in particular because of the variety of the retrieved ontologies in terms of size and quality. It is based on the following news snippet:

*"The Queen will be 80 on 21 April and she is celebrating her birthday with a family dinner hosted by Prince Charles at Windsor Castle"*[4]

Using the keywords *Queen*, *Birthday* and *Dinner*, we obtained the following ontologies, covering (sometimes only partially) this set of terms:

**OntoSem:** `http://morpheus.cs.umbc.edu/aks1/ontosem.owl`
**TAP:** `http://athena.ics.forth.gr:9090/RDF/VRP/Examples/tap.rdf`
**Mid-Level:** `http://reliant.teknowledge.com/DAML/Mid-level-ontology.owl`, covering only the terms *Queen* and *Birthday*

Compared to Example 1, the ontologies used in Example 2 are bigger (from 1835 classes in **Mid-Level** to 7596 in **OntoSem**). Moreover, they contain different levels of descriptions. For example, **OntoSem** is a big, complex OWL ontology containing a lot of properties (about 600), whereas TAP is simple RDFS taxonomy without any properties. In that sense, we use Example 1 to assess basic characteristics of the modularization techniques and then, rely on Example 2 to show how these characteristics are influenced by the properties of the ontologies.

### 5.2 Results for Example 1

Running the four modularization techniques on the three ontologies of the first example allowed us to test how they behave on simple, but yet practical real word examples.

[2] in particular Watson (`http://watson.kmi.open.ac.uk`).
[3] `http://news.kmi.open.ac.uk/`
[4] `http://news.billinge.com/1/hi/entertainment/4820796.stm`

Concerning the *level of user interaction*, **SWOOP** is fully automatic and does not need any parameters besides the input ontology. As a module extraction tool, **KMi** requires, in addition to the source ontology, a set of terms from the signature of the ontology, defining the sub-vocabulary to be covered by the module. This sub-vocabulary corresponds to the initial terms used for selecting the ontology: *Researcher*, *Student* and *University*. **Pato** has to be fine tuned with several parameters, depending on the ontology and on the requirements of the application. Here, it has been configured in such a way that modularizations in which the considered terms are in the same module are preferred. **Prompt** is an interactive mechanism, in which the user is involved in each step of the process. In particular, the class to be covered and the property to traverse have to be manually selected, requiring that the user has a good insight of the content of the ontology, can easily navigate in it, and that he understands the modularization mechanism. When using **Prompt**, we manually included the input terms and tried to obtain an (intuitively) good module, without going too deep in the configuration. Note that, since the system crashed at the early stage of the process, we did not manage to obtain results for the **KA** ontology with **Prompt**.

Concerning *performance*, apart from **Prompt** for which this criteria is irrelevant, each tool has only taken a few seconds or less on these *small ontologies*. Experiences in Example 2 should give us a better insight on this criteria and on the way techniques behave on different and larger ontologies.

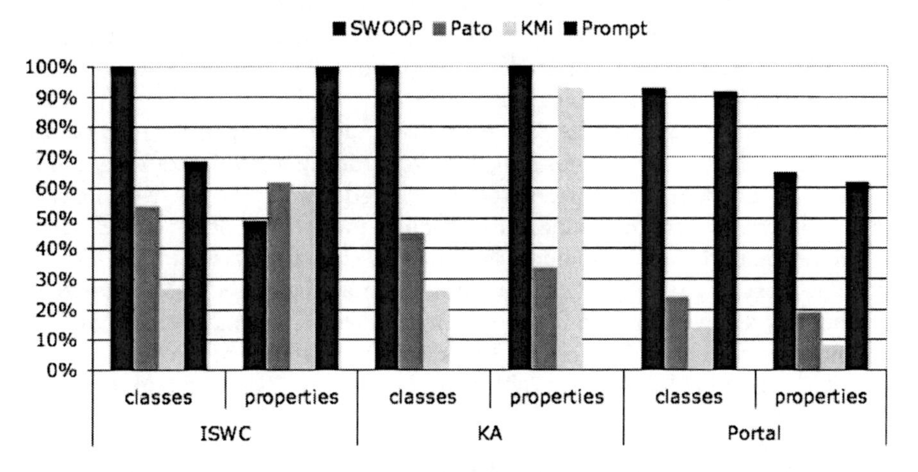


**Fig. 2.** Relative size of the resulting modules for the first example

Figure 2 shows the *size* of the resulting modules for each system in terms of number of classes and properties. It can be easily remarked that **SWOOP** generally generates very large modules, containing 100% of the classes for two of the three ontologies, and an important proportion of the properties: in most of the cases, **SWOOP** generates one module with almost the same content as the original ontology. The tool developed in **KMi** is focused on generating modules with a small number of classes (the smallest), so that the ontology hierarchy would be easy to visualize. It nevertheless includes a large proportion of the properties, in order to keep the definition of the included classes intact. **Pato** is optimized to give an appropriate size. It generally operates an important reduction of the size of the ontology.

The **KMi** tool relies on mechanisms that "take shortcuts" in the class hierarchy[5] for reducing the size of the module. It is thus the only one that potentially reduces the *intra-module distance* between the considered terms. For example, in the **Portal** ontology, by eliminating an intermediary class between *Researcher* and *Person*, **KMi** has reduced the distance between *Researcher* and *Student*, while keeping a well formed structure for the module.

### 5.3 Results for Example 2

The second example concerns larger ontologies, with more heterogeneous levels of description. For example, **TAP** contains around 5500 classes, but no property or individual, whereas **Mid-Level** relies on almost 200 properties and is populated with more than 650 individuals for less than 2000 classes. These elements obviously have an important impact on the *performance* of the modularization techniques: in the worst cases (**Pato** and **KMi** on **TAP**), it takes several minutes to get a modularization and none of the tested techniques can be used at run-time for such ontologies.

Moreover, some of the techniques are not designed to take into account such big and heterogeneous ontologies. It is particularly hard for the user to handle the process of module extraction in **Prompt** when having to deal with several thousands of classes and hundreds of properties. We also did not manage to partition the **OntoSem** ontology using **Pato**.

Finally, concerning the *size* of the resulting modules, the difference between **SWOOP** and other techniques is even more significant in this example. Indeed, because of the poor structure of the considered ontologies (restricted uses of OWL constructs, few or insufficiently defined properties), **KMi** and **Pato** result in particularly small modules (less than 10 classes), whereas **SWOOP** still includes most of the content of the ontology in a single module. Therefore, regarding the requirement about the *assumption on the ontology*, this shows that techniques are highly influenced by the inherent properties of the ontology to be modularized and that, in general, they *assume* a high level of description.

## 6 Conclusion and Discussion: Towards a Benchmark for Modularization Techniques

There is currently an important growth in interest concerning modularization techniques for ontologies, as more ontology designers and users become aware of the difficulty of reusing, exploiting and maintaining big, monolithic ontologies. The considered notion of modularity comes from software engineering, but,

[5] Instead of including all the super-classes of the included classes, it only considers classes that relate these entities: their common super-classes.

unfortunately, it is not yet as well understood and used in the context of ontology design as it is for software development. Different techniques implicitly rely on different assumptions about modularity in ontologies and these different *intuitions* require to be made explicit.

This paper reports on preliminary steps towards the characterization of ontology modularization techniques. We reviewed existing modularization tools as well as criteria for evaluating different aspects of a modularization, and used them on a particular scenario: the automatic selection of knowledge components for the annotation of Web pages. The main conclusion of these experiments is that the evaluation of a modularization (technique) is a difficult and subjective task that requires a formal, well described framework – a *benchmark* – taking into account the requirements of applications. Such a framework would be useful in two ways: first for application developers, it would provide a guide for choosing the appropriate modularization technique, and second, for the developers of modularization techniques, it would give directions in which techniques can be improved with respect to particular scenarios. The definition of this evaluation framework requires to build an adequate, well understood dataset for benchmarking and to improve the definition of the criteria for evaluation, in particular to allow the expression of requirements concerning subjective notions like the *quality of the module.*

## References


1. d'Aquin, M., Sabou, M., Motta, E.: Modularization: a Key for the Dynamic Selection of Relevant Knowledge Components. In: Proc. of the ISWC 2006 Workshop on Modular Ontologies (2006)
2. Dzbor, M., Domingue, J., Motta, E.: Magpie - towards a semantic web browser. In: Fensel, D., Sycara, K.P., Mylopoulos, J. (eds.) ISWC 2003. LNCS, vol. 2870, Springer, Heidelberg (2003)
3. Sabou, M., Lopez, V., Motta, E.: Ontology Selection on the Real Semantic Web: How to Cover the Queens Birthday Dinner? In: Proc. of the European Knowledge Acquisition Workshop (EKAW), Podebrady, Czech Republic (2006)
4. Stuckenschmidt, J., Klein, M.: Structure-Based Partitioning of Large Concept Hierarchies. In: McIlraith, S.A., Plexousakis, D., van Harmelen, F. (eds.) ISWC 2004. LNCS, vol. 3298, Springer, Heidelberg (2004)
5. Cuenca Grau, B., Parsia, B., Sirin, E., Kalyanpur, A.: Automatic Partitioning of OWL Ontologies Using E-Connections. In: Proc. of Description Logic Workshop (DL) (2005)
6. Seidenberg, J., Rector, A.: Web Ontology Segmentation: Analysis, Classification and Use. In: Proc. of the World Wide Web Conference (WWW) (2006)
7. Noy, N., Musen, M.: Specifying Ontology Views by Traversal. In: McIlraith, S.A., Plexousakis, D., van Harmelen, F. (eds.) ISWC 2004. LNCS, vol. 3298, Springer, Heidelberg (2004)
8. Schlicht, A., Stuckenschmidt, H.: Towards Structural Criteria for Ontology Modularization. In: Proc. of the ISWC 2006 Workshop on Modular Ontologies (2006)
9. Cuenca Grau, B., Horrocks, I., Kazakov, Y., Sattler, U.: A Logical Framework for Modularity of Ontologies. In: Proc. of the International Joint Conference on Artificial Intelligence, IJCAI (2007)

# The Role of Knowledge in Design Problems

Zdenek Zdrahal

Knowledge Media Institute, The Open University, UK

**Abstract.** The paper presents design as an example of ill-structured problems. Properties of ill-structured problems are discussed. French's model of design processes is analysed. The role of domain knowledge as a means for structuring the problem space is explained. Design process can be viewed as a sequence of problem re-representations gradually reducing problem indeterminacy. The results are demonstrated on pilot applications developed as a part of the Clockwork project.

## 1 Introduction

In this paper we discuss two issues related to knowledge sharing and reuse in engineering design. First, we show why domain knowledge is an essential component of problem solving in design. We will argue that knowledge provides the necessary structure for problem spaces. Then we will interpret the process of designing tasks as a sequence of problem re-representation where each subsequent problem specification is better structured. The described topics motivated the CEC funded Clockwork project whose objectives included the support for reuse of design knowledge.

## 2 Design as an Ill-Structured Problem

In mid sixties, when studying planning tasks, Rittel and Webber [18] noticed that most problems do not follow the linear "waterfall" model consisting of data analysis, problem specification and problem solving. In particular, problem specification and problem solving are mutually intertwined - problem cannot be completely specified without committing solution to the concrete problem solving method and the problem solving method cannot be selected without a complete problem specification. Similar characteristics were also observed in other domains such as architecture, urban planning or design. The problematic domains very often included social context. Solving this class of problems is difficult because at least a conceptual or tentative solution must be assumed as a part of problem specification. Rittel and Webber call these problems "wicked" to distinguish from standard problems which they call "tame". According to Rittel and Webber [18] p.161, the class of wicked problems is characterised by the following ten properties:

1. There is no definitive formulation of a wicked problem. The information needed for understanding the problem and formulating its specification depends on the idea of how to solve it.

R. Wagner, N. Revell, and G. Pernul (Eds.): DEXA 2007, LNCS 4653, pp. 884–894, 2007.

2. Wicked problems have no stopping rules. There is no criterion or test proving that the solution has been found.
3. Solutions to wicked problems are not true-or-false, but good-or-bad. When solving mathematical equations, the result is either correct or incorrect i.e. true or false. Such a dichotomy does not apply to wicked problems. Solutions could be good, bad, better, worse or perhaps only good enough.
4. There is no immediate and no ultimate test of a solution to a wicked problem. Solutions may produce unexpected consequences that are not obvious when the solution attempt is made. These even might not be immediate but they would emerge sometime in the future.
5. Every solution to a wicked problem is a "one-shot operation". There is no opportunity to undo already implemented solutions.
6. Wicked problems do not have an exhaustive set of potential solutions nor is there a well-described set of permissible operations that may be included into the problem solving plan.
7. Every wicked problem unique.
8. Every wicked problem can be considered as a symptom of another problem. Wicked problems may create a causal chain.
9. The existence of a discrepancy representing a wicked problem can be explained in numerous ways. The explanation depends on the "world view" of the problem solver.
10. The wicked problem solver has no right to be wrong. This point is especially important in social policy planning where the ultimate objective is not to find the truth but to improve the current state of public affairs.

These characteristics of wicked problems were formulated for social policy planning, but similar properties were also identified in other areas including design [2], [3]. For example, Schön [20], p.79 describes how the "designer shapes the situation and the situation 'talks back'", or that problem "naming, framing and, if needed, re-framing" is a way of dealing with the lack of definitive problem formulation.

For engineering design not all characteristics described above are equally important. The concept of "wicked problem" emerged from studying problems with social connotation where social criteria play important part of the problem domain. Planning city infrastructure, proposing solutions to the problems of poverty, education, crime or health care are typical examples that manifest wickedness. Conklin even explains that social complexity of design problems relates to the "measure" of their wickedness [3]. However, we claim that social context does not have the same impact in every domain. While it is certainly the key issue in policy planning, there are areas, where it is less important. This becomes obvious when we try to apply the above mentioned characteristics in the context of engineering design. When a planner tries to resolve a societal problem it is usually a one-shot operation. The solution cannot be withdrawn because it creates permanent or long lasting changes. Also, the solution must not be completely wrong because there is too much at stake. However, in engineering design unsuccessful trials that fail to solve the problem frequently exist. Thomas Alva

Edison was a pioneer of designing by trial and error and yet he patented over 1000 original designs addressing problems that certainly satisfied multiple properties of wickedness. Trials and errors are acceptable if the situation is reversible and return is not too expensive. The cost depends on the feedback loop through which the solution is evaluated. It is probably impossible to completely rebuild already constructed road system, because it would be too costly. It is certainly impossible to correct a medical therapy if the patient has died. But it is easy to redesign breaks for the bicycle, especially if the bicycle has not yet been put on the market. Conklin [3] demonstrates the properties of wicked problems on designing a new car with the aims of improving side-impact safety. In his example, the success of a car is measured by the response of the market. When a new model is launched, the customers will either accept or reject the product, but the innovative project cannot be turned back and the preparation of manufacturing technology cannot be undone. In this sense car design is a one-shot operation. However, there is a number of counterexamples when manufacturers had to recall hundreds of thousands of cars to correct design errors. The "second shot" was certainly very expensive, but it has been done. If the evaluation loop is short and cheap, the problem solving allows proposing, evaluating and rejecting alternatives. Returning to the car example, there are various legislations, national and international regulations that constrain possible designs before they reach the market, even before a prototype hits the road. Within these regulations, design by educated trial and error is a common practice. Yet these design problems retain the majority of characteristics of wickedness. Though design problems are presented as unique, similarities with other problems play an important role. How would otherwise designers develop their expertise if problems were always the "universe of one" [19]? It is well known that designers often acquire their inspiration from already existing cases. Schön [20] argues that in professions, unique problems are common, but they are often approached by adapting and combining known solutions.

Simon analyses similar class of problems but without emphasising the social context [21], [22]. He introduces the concepts of ill-structured and well-structured problems with a meaning similar to wicked and tame problems. In analogy to the way used by Alan Turing who formalised the intuitive concept of algorithm in terms of Turing machine, Simon associated well-structured problems with the General Problem Solver (GPS) [15]. This formalisation makes it possible to introduce the concepts like state space, states, state transitions and structures. Well-structured problems are defined by the following conditions (adapted from [21]):

1. There is a definite criterion for testing proposed solutions
2. There is a problem space in which it is possible to represent the initial problem state, the goal state and all other reachable states.
3. Attainable state transitions can be represented in a problem space.
4. Any knowledge that the problem solver can acquire can be represented in a problem space

5. If the problem depends on input from the external world, then the state transitions reflect accurately the laws of nature that govern the external world.
6. All these conditions are practically computable.

The problems that do not satisfy one or more of these conditions are called ill-structured. The correspondence with many of characteristics of wicked problems is straightforward. It makes sense to consider the measure to which the real problem satisfies these conditions. We can then position the problem on a scale with ill-structured problems at one end and well-structured on the other one.

When solving a problem each method has associated preconditions of its applicability. Since ill-structured problems are not fully specified these preconditions are impossible to evaluate. The preconditions specify what kind of information about the problem space is needed and, consequently what is the quality of the solution delivered by the method. For example, solutions can be optimal globally, locally or only "good enough", methods can guarantee that if a solution exists it will be found, they may search a large state space or progress directly to the solution, they may convergence slowly or quickly. There are many other measures that assess the quality of solution. In general, the more a priori information about the problem space is available, the better results are delivered by the method [16]. If the problem is well structured, i.e. the problem space is well described and the quality criterion is known, it is possible to apply "strong" problem solving methods, such as optimisation algorithms. However, since design is an ill-structured problem, generally applicable strong problem solving methods do not exist. But why is design an ill-structured problem? Buchanan [2], p.15 argues that the reason is that "design has no specific subject matter of its own apart from what the designer conceives it to be". It means that designer has to bring his/her domain knowledge to structure design tasks. This observation has important consequences for knowledge management in design processes.

Imagine an engineering problem of designing a bridge over a river. The problem can be certainly regarded as ill-structured. Possible solutions include a beam bridge, an arch bridge, a suspension bridge or perhaps even a pontoon bridge. There are many other bridge types, each of them having its pros and cons. By making the decision and selecting one type, the designer brings into the design process the corresponding part of physics and provides structure for the problem space. If the designer opts for a suspension bridge, he/she will have to express the problem in terms of concepts and laws of theory of elasticity. If he/she decides to construct a pontoon bridge, he/she will use theory of hydrostatics and hydrodynamics. The progress in problem solving makes it possible to improve the problem specification (e.g. in the case of the suspension bridge the task of calculating the strength of cables is a part of the additional problem specification). Problem solving is intertwined with problem specification as the problem acquires structure and moves from ill-structured towards well-structured end of the scale.

For solving wicked problems Kunz and Rittel [13] proposed the method known as Issue Based Information Systems (IBIS), that allows the designers to uncover conflicting criteria in the reasoning process, capture the argumentation, and arrive at an acceptable solution. A number of variations of IBIS-based methods focused on design problem solving are discussed in [14]. Though these methods successfully capture design rationale, they provide only limited support to converting the ill-structured design problem into a well-structured one because they not offer any means for integrating new domain knowledge with the existing problem specification. Mismatch of expectations and real benefits of IBIS-like methods are analysed in [12].

## 3 Models of Design Processes

In engineering, various models of design processes have been proposed. Typically, design starts with a problem analysis and ends with a detailed specification for the artefact. At the beginning, the problem is indefinite, its specification is incomplete and therefore, the problem space is ill-structured. At the end the problem space is already well-structured and the artefact is fully specified at the required level of detail. In order to structure the problem space, new knowledge need to be brought into the design process. The model of design process elaborated by French [9] is frequently used in engineering applications. This model is shown in figure 1 (a). French's model divides the design process into a number of distinct stages. They are: problem analysis, conceptual design, embodiment of schemes and detailed design. In the first stage, the designer analyses and sets the problem. The remaining three stages synthesize the solution. French admits that this model is one of many possible, however since it captures well our intuitive understanding of the design process, it has been adopted by many other authors, e.g. [5], [11], [23], [8]. We argue that French's model also well describes how the designer gradually commits the solution to the selected engineering domain. This can be manifested by the increasing use of domain knowledge and the decreasing indefiniteness of problem specification as the design process progresses through consecutive stages. This is highlighted in figure 1(b). Model in figure 1(a) describes the design process as a sequence of steps with possible feedback (progress in vertical direction). However, each step is also a design process on its own. These "horizontal" processes are shown in figure 1(b). For example, conceptual design is a process which produces a conceptual solution to the problem in terms of schemes or other conceptual objects. Similarly, embodiment of schemes and detailing are design processes that provide the conceptual solution with more detailed structures and elaborate final details.

Each stage produces a complete solution to the design problem, but the solutions differ by the coarseness of detail due to different indeterminacy of problem specification. This does not imply that the solution at the higher design stage must be finalised before the lower stage starts. The most important decisions are made in early stages. In final stages, the problem is already well-structured and strong problem solving methods, such as optimisation procedures are likely to be found. Conceptual design can be very creative but provides only limited guidance for problem solving (Newell's weak methods) because the problem is

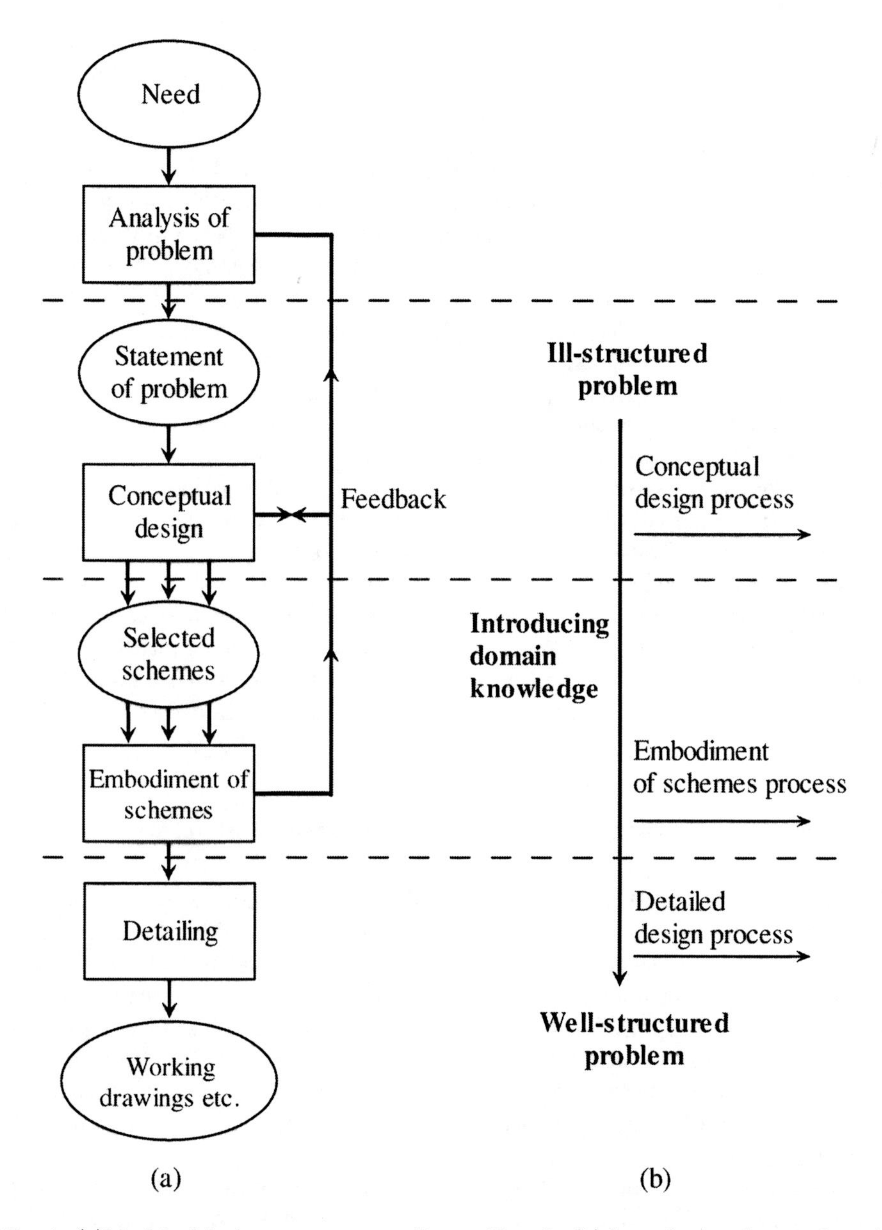


**Fig. 1.** (a)Model of design process according to French, (b) Introducing domain knowledge into process

ill-structure and there is no commitment to a specific domain. Goel [11], p. 131 characterises conceptual design as follows: " Generation and exploration of alternatives is facilitated by the abstract nature of information being considered, a low degree of commitment to generated ideas, the coarseness of detail and a large number of lateral transformation. A lateral transformation is one where movement is from one idea to a slightly different idea, rather than a more detailed version of the same idea. Lateral transformations are necessary for widening the problem space and the exploration and development of kernel ideas". In general, design decisions made during early design stages direct the solution towards a specific domain and make the corresponding domain knowledge available for the later design stages. This principle applies throughout the design process. Domain knowledge available at each stage affects the repertoire of applicable problem solving methods. Problem solving methods typically used in early design stages differ from those applicable later. In engineering practice, creativity is often restricted to selecting and exploring a few most promising alternatives using case-based or IBIS-like problem solving methods. This was reported by Ball, Ormerod and Morley [1] who observed that designers solve new problems by looking for analogies with old ones. Moreover, they showed how designers' experience changes their reasoning processes: Experts develop and use more abstract knowledge schemas while novices tend to draw the analogies from specific cases. Cross [7] arrives at similar conclusions claiming that the experienced designers tend to keep design ill-structured in order to have more opportunities for creative thinking while novices quickly moves towards the first possible solution. Goel [11] demonstrates that different parts of brain are specialised in solving different design stages. Right hemisphere is predominantly used for ill-structured problems i.e. for conceptual design, while left hemisphere supports mental activities needed for solving well-structured problems, i.e. methods used in detailed design.

## 4 Solving Design by Problem Re-representation

Design process based on French's model can be viewed as problem re-representation. First, a tentative conceptual model is built from the initial problem specification . The choice of conceptual objects and relations is based on designer's assumptions about the expected solution. Conceptual solution is further elaborated by specifying the structure and components to be used for implementing. A new set of assumptions is needed, now at a different level of abstraction. Finally, the parameters of selected components are specified. Their calculation may require a few additional assumptions. In general, design decisions have the form of assumptions.

Problem re-representation is also used in a compilation and linking of computer programs. The algorithm represented by a source code, say in the C language, is re-represented step by step until the absolute machine code is achieved. However, there is a difference: when compiling a C program, all necessary information is already contained in the source code. The problem is well-structured from the very beginning and there is no need for supplying additional information into the compilation/linking process.

In design, on the other hand, the problem is initially ill-structured. Each re-representation step converts the results of the previous stage but require also additional assumptions about the solution in next step. These assumptions add new information into the design process. For example, when constructing a model of car suspension, the designer may assume that the elasticity of the body can be omitted. At the current stage of problem solving, this cannot be proved or disproved but the design must make tentative decision, otherwise the process cannot continue. Design decisions are based on formal and experiential knowledge of the designer. Capturing designer's most important and innovative decision and making them available in the future was one of the major objectives of the Clockwork project described in the following section.

## 5 The Clockwork Project

Structuring design as problem re-representation was the basis of the CEC funded project "Creating Learning Organisations with Contextualised Knowledge-Rich Work Artifacts" (Clockwork). The project objectives included the development of methodology and support tools for sharing and reuse of design knowledge. The support tools were integrated in a configurable web-based toolkit, called the Clockwork Knowledge Manager (CKM). The main focus of the project was on modelling and simulation models of dynamic systems and therefore, the first pilot application was from this area. Since the development of simulation models has all features of design, the Clockwork approach can be applied to other areas of engineering design. This was tested by the second pilot application in the area of designing industrial thermal technologies. Both pilot applications were developed as instantiations of CKM. In Clockwork, design stages are called *worlds*. Each world has associated world objects and relations specified in terms of domain ontologies.

Clockwork methodology provides a number of extensions on top of French's model. For example, the world representation can be also used for stages and environments that are not part of the French's model. Since it might be convenient to collect post-design data about the designed artefact, the CKM application can be extended by Production, Installation or Maintenance worlds. The collected data may serve to inform future designs. Similarly, in modelling and simulation applications, the task can be to simulate a real object. In CKM, all objects to be simulated could be described in a Real world. Example of four worlds used in the modelling and simulation application is shown in figure 2. In this appli-

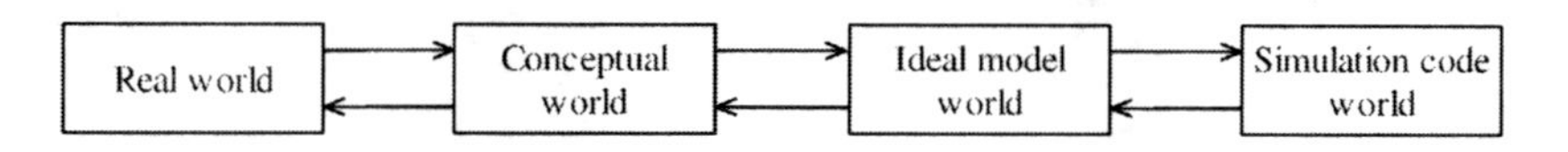


**Fig. 2.** Four world model of modelling and simulation

cation the problem to be investigated by computer simulation is introduced in Real world. For example, imagine that the task of designing a car suspension that guarantees passenger comfort within certain speed range on standard types of road. Though concepts like passenger comfort and road type are likely to be defined by international, national or manufacturer's norms, the problem is clearly ill-defined. There is a number of possible conceptualisations of the problem. For example, the suspension system can be modelled using the concept of a "quartercar" (model of a single wheel), quartercar models can be connected by a set of rigid or elastic bodies, the whole suspension can be modelled using finite elements etc. Each conceptual solution can be further elaborated, for example models can be built using ideal physical models, bond graphs, etc. These will be eventually converted into simulation code. Most important decisions for the final outcome are the assumptions justifying conceptualisation and modelling approaches. Re-representation from model world to simulation world usually does not require any additional design-related assumptions because the problem is already well structured. For this reason, simulation packages often allow the user to build the model in a graphical environment and the simulation code is uploaded automatically. CKM toolkit not only supports the integration of domain ontologies with the design process, but also capturing design decisions, their assumptions and consequences at more detailed level. In order to minimise additional tedious work of designer, the Clockwork methodology assumes that only important or non-trivial knowledge will be recorded. The Clockwork therefore does not build a complete mapping between adjacent worlds. An example of four world environment for modelling and simulation applications is shown in figure 3. CKM provides two mechanisms for representing case specific knowledge. Formal knowledge is represented by instances of ontology classes. In Clockwork, these instances are called *semantic indexes.* The relations between semantic indexes of adjacent worlds are parts of problem re-representation. Informal knowledge has the form of textual annotations and can be associated with transitions between whole models or with individual semantic indexes. CKM in-

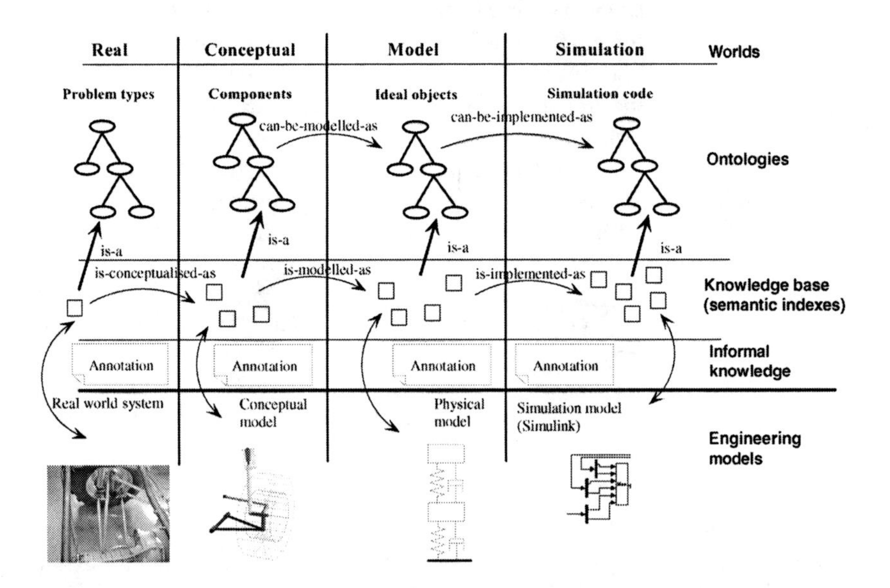


**Fig. 3.** Modelling and Simulation application

tegrates formal and informal knowledge with standard work representation such as technical drawings, engineering diagrams, conceptual graphs or graphical environments. The application to thermal machinery design and production was also developed by instantiating CKM. The configuration includes worlds shown in figure 4. The design-related worlds follow from the French's model. The post

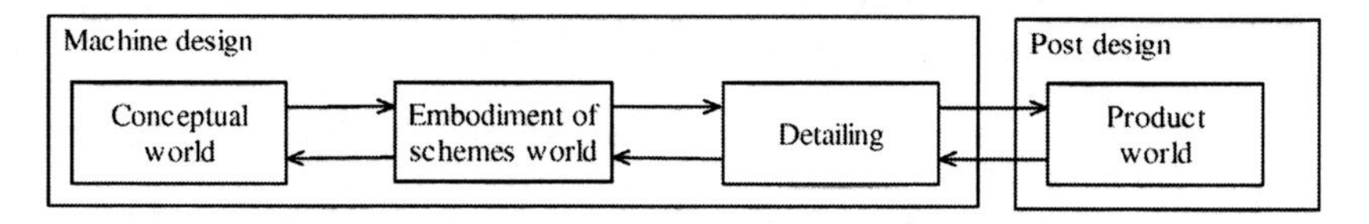


**Fig. 4.** Product design and post design worlds

design worlds collect data related to the machine installation and maintenance. The company design team uses these data for informing future design of machines with a similar specification.

## 6 Conclusions

Design is a human activity which aims at converting the current state of the world into a new, preferred one. Since we usually do not specify in advance how to reached the desirable outcome, the problem is ill-structured. Designers need their expertise to overcome the problem of indeterminacy and to interpret the problem. We have presented a methodology and a tool that supports knowledge reuse and sharing between designers. This overview could not present the problem in a full complexity. For example, we did not discuss the role of tacit knowledge in design, see [4], [17] or sharing design knowledge in settings where participants must collaborate but also compete on the same market (cautious knowledge sharing). These issues were also addressed by the Clockwork project The following persons significantly contributed to the Clockwork project: Paul Mulholland of The Open University, Michael Valasek of the Czech Technical University and Ansgar Bernardi of DFKI.

## References

1. Ball, L.J., Ormerod, T.C., Morley, N.J.: Spontaneous analogising in engineering design: a comparative analysis of experts and novices. Design Studies 25, 495–508 (2004)
2. Buchanan, R.: Wicked Problems in Design Thinking. Design Issues 8(2), 5–12 (1992)
3. Conklin, J.: Wicked Problems and Social Complexity. In: Book Conklin J. Dialogue Mapping: Building Shared Understanding of Wicked Problems, John Wiley and Sons Ltd, Chichester (2005)

4. Cook, S.D.N., Brown, J.S.: Bridging Epistemology: The Generative Dance Between Organizational Knowledge and Organizational Knowing. Organization Science 10(4), 381–400 (1999)
5. Cross, N.: Engineering Design Methods. In: Strategies for Product Design, 2nd edn., John Wiley & Sons, Chichester (1994)
6. Cross, N.: Descriptive model of creative design: application to an example. Design Studies 18, 427–440 (1997)
7. Cross, N.: Expertise in design: an overview. Design Studies 25, 427–441 (2004)
8. Dym, C.L.: Engineering Design, A Synthesis of Views. Cambridge University Press, Cambridge (1994)
9. French, M.J.: Engineering Design: The Conceptual Stage. Heinemann Educational Books, London (1971)
10. Goel, V., Pirolli, P.: The Structure of Design Spaces. Cognitive Science 16, 395–429 (1992)
11. Goel, V.: Cognitive Role of Ill-Structured Representations in Preliminary Design. In: Gero, J.S., Tversky, B. (eds.) Visual and Spatial Reasoning in Design, Key Centre of Design Computing and Cognition, pp. 131–145. University of Sydney, Sydney (1999)
12. Isenmann, S., Reuter, W.D.: IBIS - a convincing concept ... but a lousy instrument? In: Proceedings of the Conference on Designing interactive Systems: Processes, Practices, Methods, and Techniques, pp. 163–172. ACM Press, New York (1997)
13. Kunz, W., Rittel, H.W.J.: Issues as Elements of Information Systems. Working Paper WP131. July 1970, reprinted May 1979. Inst. Urban and Regional Development., Univ. Calif., Berkeley (1970)
14. Louritas, P., Loucopoulos, P.: A Generic Model for Reflective Design. ACM Transactions on Software Engineering and Methodology 9(2), 199–237 (2000)
15. Newell, A., Simon, H.A.: GPS, a program that simulates human thought. In: Computers and Thought, pp. 279–293. MIT Press, Cambridge (1995)
16. Newell, A.: Artificial Intelligence and the Concept of Mind. In: Schank, R.C., Colby, K.M. (eds.) Computer Models of Thought and Language, pp. 1–60. W.H. reeman and Comp., San Francisco (1973)
17. Nonaka, I.: The Knowledge-Creating Company. In: Harvard Business Review on Knowledge Management, pp. 21–45. Harvard Business School Press, Boston (1998)
18. Rittel, H.W.J., Webber, M.M.: Dilemmas in a General Theory of Planning. Policy Sciences 4, 155–169 (1973)
19. Schön, D.A.: Designing: Rules, types and worlds. Design Studies. 9(3), 181–190 (1988)
20. Schön, D.A.: The Reflective Practitioner. Ashgate Publishing Ltd., Aldershot, England (1991)
21. Simon, H.A.: The Structure of Ill Structured Problems. Artificial Intelligence 4, 181–201 (1973)
22. Simon, H.A.: The Science of the Artificial, 3rd edn. MIT Press, Cambridge (1996)
23. Tansley, D.S.W., Hayball, C.C.: Knowledge-Based System Analysis and Design. Prentice-Hall, Englewood Cliffs (1993)

# e-Infrastructures

Wolfgang Gentzsch

D-Grid, RENCI, and Duke University

**Abstract.** In the last decades, the Internet and the World Wide Web have evolved into a new infrastructure for science, business, and the public. Driven by the need to better cope with recent trends and developments caused by globalization, complexity, and the grand challenges, we are refining and enhancing this infrastructure with powerful new tools for communication, collaboration, computation and the huge amounts of resulting data and knowledge. Researchers and business people alike are more and more able to easily access the tools, the data and the IT resources they need to solve their applications and to increase knowledge, via Grid and Service Oriented Architectures, and the technologies and tools to build them. This presentation will highlight Grids, SOA, and Web 2.0 and how they relate and complement, and the benefits they bring to scientists, businesses, and our whole society, supported by a number of use cases from research, industry, and public community.

## 1 Introduction

In the last decades, the Internet and the World Wide Web have evolved into a new infrastructure for science, business, and the public. Driven by the need to better cope with recent trends and developments caused by globalization, complexity, and the grand challenges, recently, we are refining and enhancing this infrastructure with powerful new tools for communication, collaboration, computation and the huge amounts of resulting data and knowledge. Researchers and business people alike are more and more able to remotely and easily access the tools, the data and the IT resources they need to solve their applications and to increase knowledge, via Grid and Service Oriented Architectures, and the technologies and tools to build them. While the Web offers easy access to mostly static information, the Grid adds another fundamental layer to the Internet, by enabling direct access to and use of underlying resources, such as computers, storage, scientific instruments and experiments, sensors, applications, data, and middleware services. Based on widely accepted grid and web services standards, resources communicate with each other and deliver results as services to the user. These resources are part of a service-oriented architecture, called OGSA, the Open Grid Services Architecture, [2]. For the past several years, early adopters in research and industry have been building and operating prototypes of grids for global communities, virtual organizations, and within enterprises.

R. Wagner, N. Revell, and G. Pernul (Eds.): DEXA 2007, LNCS 4653, pp. 895–904, 2007.

## 2 Benefits of Grid Computing for Research and Industry

Grid infrastructures provide a wide spectrum of benefits, [3], [4]: transparent access to and better utilization of resources; almost infinite compute and storage capacity; flexibility, adaptability and automation through dynamic and concerted interoperation of networked resources [5]; cost reduction through utility model; higher quality of products designed and developed via grid tools; shorter time-to-market; and more. This grid revolution is already well underway in scientific and engineering organizations with high demand of computing and data processing, mostly as prototype grids, and a few already in full production, see e.g. [6] - [14]. But also (and especially ) for those research departments and businesses which cannot afford powerful supercomputing resources grid computing is of great benefit.

An outstanding example for grid computing benefits in research is the data collection and processing infrastructure for the high-energy physics experiment at the European research center CERN in Geneva, the so-called Large Hadron Collider, LHC, [15], which will be operational from early 2007. Over 5000 physicists world-wide will analyze the collision events of the largest particle accelerator in the world, resulting in petabytes of data per year, which will be filtered, sorted and stored into digital repositories and accessed by scientists. This LHC Grid is built in four tiers, with tier zero being the compute servers at CERN, tier one the national research centers, tier two the servers in the local research centers and universities, and finally the desktops of the researchers.

Besides the obvious benefits for the researchers, grid technology has great benefits also for the industry, [16]. In an era of increasing dynamics, shrinking distances, and global competition, those organizations are in an advantageous position which have access either to natural or to highly specialized resources, on demand, in an efficient and effective way. Countries like Germany for example don't have enough natural resources; thus, competition has to be strengthened via specialization, e.g. an excellent education for everybody, use of highly modern tools and machines, optimized development and production processes, and highly efficient communication and sales processes. Here, grid technology can provide great benefit. It enables engineers to access any IT resource (computer, software, applications, data, etc) in an easy and efficient way, to simulate any process and any product (and even the whole product life cycle) in virtual reality before it is build, resulting in higher quality, increased functionality, and cost and risk reduction. Grid technology helps to adjust an enterprise's IT structures to real business requirements (and not vice versa). For example, global companies will be able to decompose their highly complex processes into modular components of a workflow which can be distributed around the globe such that on-demand access to suitable workforce and resources is assured, productivity increased, and cost reduced. Application of grid technology in these processes, guarantees seamless integration of and communication among all distributed components and provides transparent and secure access to sensitive company information and other proprietary assets, world-wide.

## 3 Grid Business and Services

From a bird's eye view, the business model for grid services will be similar to those for electrical power, water, or telephony: our payments will be based on widely agreed billing units which include cost for computers, storage, software, applications, work, electrical power, and square footage for the equipment.

In the future, we will see many different grid-based services and an army of new service providers. For example, providers offering compute cycles or storage, such as Amazon's Elastic Computing Cloud (EC2, [17]) or its Simple Storage Service (S3, [18]). But also 'Application Service Providers' offering a specific application service to engineering firms, accessible via browser. At the end of each month, we will receive an invoice from our favorite service provider, broken down into the services which we received from the different providers, very similar to our Telekom invoice today.

The Web will be the platform for many of our future grid-based businesses. We will surf to the service provider's Web portal, login and set up a personal account. Service providers will offer any services, securely, on demand, with highest quality, at reasonable cost, according to the Service Level Agreement (SLA) negotiated with the customer.

Grid technology will also revolutionize society. Let's look at education, for example. On one hand, our knowledge is increasing exponentially - e.g. in bioinformatics it's doubling every 12 months -, on the other hand, schools can't keep pace with this exponential development, especially in the natural sciences, [19]. Grid technologies will become the fundament for novel teaching and learning tools such as virtual laboratories which enable children (and teachers) to interactively experience the secrets of nature, engineering and society. This will dramatically increase their motivation, creativity and knowledge.

All this will take a few more years to happen. Firstly, we have to 'grid-enable' our data, our applications, our knowledge repositories. We need security technologies which guarantee that one's identity can't be stolen and that confidential data can be stored in highly secure containers if needed. This requires close collaboration among computer scientists, researchers, engineers, and businesses. Now that we share specific resources because it's more efficient and fosters communication and collaboration, we have to make sure that we only pay for what WE use and that our computational results remain confidential. Good news is that thousands of experts in research and industry are working very hard on solving these problems, in hundreds of grid projects such as the ones presented herein, and on Grid standards in the Open Grid Forum, OGF [20].

## 4 Case Study: The German D-Grid Initiative

In 2003, German scientists and scientific organizations started the D-Grid initiative [21], jointly publishing a strategic paper in July 2003. This paper examined the status and consequences of grid technology on scientific research in Germany and recommended a long-term strategic grid research and development initiative. This resulted in the German e-Science Initiative founded by the German

Ministry for Research and Education (BMBF) in March 2004, together with a call for proposals in the areas of Grid Computing, e-Learning, and Knowledge Management. In November 2004, the BMBF presented the vision of a new quality of digital scientific infrastructure which will enables our globally connected scientists to collaborate on an international basis; exchange information, documents and publications about their research work in real time; and guarantee efficiency and stability even with huge amounts of data from measurements, laboratories and computational results.

The e-Science Initiative and the first phase of D-Grid started on September 1, 2005. BMBF is funding over 100 German research organizations with 100 Million Euro over the next 5 years. For the first 3-year phase of D-Grid, financial support is approximately 25 Million Euro. The goal is to design, build and operate a network of distributed, integrated and virtualized high-performance resources and related services to enable the processing of large amounts of scientific data and information. The Ministry for Research and Education is funding the assembling, set-up and operation in several overlapping stages:

1. D-Grid 1, 2005-2008: IT services for scientists, designed and developed by the 'early adopters' of the computer science community. This global services infrastructure will be tested and used by so-called Community Grids in the areas of high-energy physics, astrophysics, medicine and life sciences, earth sciences (e.g. climate), engineering sciences, and scientific libraries.
2. D-Grid 2, 2007-2009: IT services for scientists, industry, and business, including new applications in chemistry, biology, drug design, economy, visualization of data, and so on. Grid services providers will offer basic services to these users.

D-Grid 3 (around 2008- 2010) will extend the grid infrastructure with a business and a knowledge management layer, and adding several virtual competence centers, encourage global service-oriented architectures in the industry, and use this grid infrastructure for the benefit of our whole society, as discussed in chapter 5.

D-Grid consists of the DGI Infrastructure project, [22], and (currently) the following seven Community Grid projects:

- AstroGrid-D (Astronomy)
- C3-Grid (Earth Sciences)
- HEP Grid (High-Energy Physics)
- InGrid (Engineering)
- MediGrid (Medical Research)
- TextGrid (Scientific Libraries, Humanities)
- WISENT (Knowledge Network Energy Meteorology)

Short-term goal of D-Grid is to build a core grid infrastructure for the German scientific community, until the end of 2006. Then, first test and benchmark computations will be performed by the Community Grids, to provide technology feedback to DGI. Then, climate researchers of the C3-Grid, for example, will be able to predict climate changes faster and more accurately than before, to inform governments about potential environmental measures. Similarly, astrophysicists will be able to access and use radio-telescopes and supercomputers remotely via the grid, which they wouldn't be able to access otherwise, resulting in novel quality of research and the resulting data.

## 4.1 The D-Grid Infrastructure Project

Scientists in the D-Grid Infrastructure project DGI are developing and implementing a set of basic grid middleware services which will be offered to the other Community Grids. For example, services include access to large amounts of data distributed in the grid, the management of virtual organizations, monitoring and accounting. So far, a core-grid infrastructure has been built for the community grids for testing, experimentation, and production. High-level services will be developed which guarantee security, reliable data access and transfer, and fair-use policies for computing resources. This core-grid infrastructure will then be further developed into a reliable, generic, long-term production platform which can be enhanced in a scalable and seamless way, such as the addition of new resources and services, distributed applications and data, and automated "on demand" provisioning of a support infrastructure.

An important aspect in every grid is security, especially with the industry expected to join soon, such as automotive and aerospace. Therefore, an important DGI work package is "Authentication and Authorization" [23]. It's obviously important to know that a user is really the one she pretends to be, and that she is authorized to access and use the requested resources and information. While enterprise grids are mostly operating behind firewalls, global community grids use security technology like VOMS, Virtual Organization Membership Service, [24]. However, building and managing so-called Certificate Authorities is still a very cumbersome activity to date.

The following D-Grid DGI infrastructure services are available for the current and D-Grid 2 community projects, at the end of 2006:

- The core D-Grid infrastructure offers central grid services. New resources can be easily integrated in the help-desk and monitoring system, allowing central control of resources to guarantee sustainable grid operation.
- DGI offers several grid middleware packages (gLite, Globus und Unicore) and data management systems (SRB, dCache und OGSA-DAI). A support infrastructure helps new communities and "Virtual Organizations" (VOs) with the installation and integration of new grid resources via a central Information Portal ("Point of Information"). In addition, software tools for managing VOs are offered, based on the VOMS and Shibboleth [25] systems.
- Monitoring und Accounting prototypes for distributed grid resources exist, as well as an early concept for billing in D-Grid.
- DGI offers consulting for new Grid Communities in all technical aspects of network and security, e.g. firewalls in grid environments, alternative network protocols, and CERT (Computer Emergency Response Team).

- DGI partners operate "Registration Authorities" to support simple application of internationally accepted Grid Certificates from DFN (German Research Network organization) and GridKA (Grid Project Karlsruhe). DGI partners support new members to build their own "Registration Authorities".
- Core D-Grid is offering resources for testing, via middleware systems (gLite, Globus and UNICORE). The Portal Framework Gridsphere serves as the user interface. Within the D-Grid environment the dCache system takes care of the administration of large amount of scientific data.

### 4.2 Community Grid Projects

- The High-Energy Physics (HEP, [25]) Grid community is developing applications and components for evaluating terabytes of data from large high-energy physics experiments, including the Large Hadron Collider at CERN.
- ASTRO Grid [26] combines research institutions in astronomy and astrophysics into a single, nationwide virtual organization for distributed collaboration and integration of distributed astronomical data archives, instruments and experiments.
- MEDI Grid [27] represents the medical and bio-informatics community in Germany. It focuses on application scenarios in medical image processing, bioinformatics, and clinical research, and their interaction.
- C3-Grid [28] for the Collaborative Climate Community has the goal to develop a highly proficient grid-based research platform for the German earth-system research community to efficiently access and analyze distributed, high-volume scientific data from earth-system modeling and observation.
- For InGrid [29], the Industry applications grid project, a grid framework will be developed to enable modeling, optimization, and simulation of engineering applications from areas such as foundry technology, metal forming, groundwater flows, turbine simulation, and fluid-structure interaction.
- The TextGrid project [30] is developing tools and standard interfaces for publication software, modules for scientific text processing and editing, and administration and access to distributed data and tools on the grid.
- WISENT [31] is developing tools and methods in the area of energy meteorology to accurately forecast energy usage to be matched with energy provisioning on demand.

In the past, organizations strived to successfully collaborate within their own communities. But in D-Grid, for the first time ever, all these different communities are working together on a single, inter-community grid middleware platform to share computing resources, middleware tools, applications, and expertise. This will result in an IT infrastructure which is interoperable with other international grids, scalable and extensible for more community grids in the future, and available for all of our scientists for national and international collaboration.

## 5 The Future: What Comes After the Grid Projects?

How will the Internet evolve under the influence of these new grid technologies? It will certainly take another few years until we see the next-generation Internet which allows access to compute resources and services as easily as the access to billions of Web sites today. For this to happen we have to continue to improve the new e-infrastructure in projects such as the ones mentioned herein, to fully benefit from the availability of vast amount of resources and services in a transparent way. In my interviews, I have collected a few thoughts on a potential roadmap for research, industry and society to achieve this goal:

Research:

- Development of user-friendly and automated grid infrastructure building blocks with standard interfaces to easily build local and special grids (e.g. campus grids in universities) and global grids for international research projects, to collaboratively use computers, storage, applications, and data resources distributed in the Internet.
- Adaptation of application software for grid infrastructure and services, in areas like physics, chemistry, biology, weather, climate, environment, bioinformatics, medicine, aero- and fluid mechanics, oil and gas, economy, finance, and so on.
- Participation and contribution to standardization organizations, e.g. OGF [20], OASIS [32], W3C [33], and to European organizations such as ESFRI [34] and e-IRG [35].
- Development of training material and organization of training courses to learn how to build, operate and use grid infrastructures.
- Encourage independent service providers of grid resource and applications, develop operational and accounting models, utility computing, and service level agreements.
- Integration of local, national, community grids into international grid infrastructures.
- Overcome mental, legal and regulatory barriers, via case studies, demonstrators, and pilot projects.

Industry and Business:

- Development of new enterprise IT infrastructures based on OGSA (Open Grid Services Architecture) and SOA (Service Oriented Architecture), with SLOs (Service Level Objectives) and SLAs (Service Level Agreements) to mapping business processes to resource and application usage in an enterprise.
- Global enterprise grids to network all resources of globally distributed subsidiaries and branches, and for seamless integration of companies after merger or acquisition.
- Close collaboration with research to efficiently transfer reliable global grid technology to the industry.

- Partner grids for close collaboration with business partners and suppliers, to optimize distributed product development, complex workflows for multi-disciplinary processes and applications, productivity and quality improvement through global "Six Sigma" processes.
- Sensor Grids und Wireless Grids, to enable communication and interaction of electronic devices e.g. for safety reasons in airplanes, cars, bridges, skyscrapers, etc.
- Development of local and global training grids to support active and interactive, flexible and dynamic education of enterprise personnel.

Society:

- Development of grids for the masses, in areas such as healthcare (illness, fitness, sensor-based monitoring of bodily functions), leisure (multi-player games, digital entertainment, sports), education (life-long learning, school grids, digital interactive laboratories), and work (Internet-based courses, online training, global teamwork, collaboratories).
- Starting with pilot projects in these areas, partnering with end-users (consumers), application service providers, and resource providers.
- Grid resources and services for education in schools, universities, and in enterprises. Integration of grid resources and application simulations into existing curricula to dramatically improve motivation and creativity of the learners (and the teachers).
- Development of personal digital assistants including technology and service infrastructure for the mass market.
- Integration of these new applications for the masses into user-friendly web portals.

In the near future, on an enhanced Internet, all kinds of service providers will offer their services for computing, data, applications, and many more. On an enhanced World Wide Web, via secure Web Portals, we will access grid components like Lego building blocks, which enable us to dynamically build grids 'on the fly', according to our specific needs. We will rent or lease the resources required and pay for what we use or on a subscription basis. We still might have our own resources, to fulfill a certain basic need, or for highly proprietary applications and data, which can be extended in a seamless way with resources from service providers, available on the grid. But, as already said, this will still take a few years.

As with any new infrastructure, development and deployment of the next Internet generation will require vision and endurance. We have to work continuously on strategic, long-term projects on a national or international scale, which demand collaboration of research and industry on complex inter-disciplinary projects, and which will enable and improve the tools of our scientists, business people, and educators and strengthen our position in the international competition.

## References


1. O'Reilly, T.: What Is Web 2.0, Design Patterns and Business Models for the Next Generation of Software, http://www.oreillynet.com/pub/a/oreilly/tim/news/2005/09/30/what-is-web-20.html
2. OGSA Open Grid Services Architecture, http://www.globus.org/ogsa/
3. Foster, I., Kesselman, C.: The GRID: Blueprint for a new Computing Infrastructure, 1st edn. Morgan Kauffman Publishers, San Francisco (1999) 2nd (edn.) 2003
4. WS-RF Web Services Resource Framework, http://www.globus.org/wsrf/
5. Dini, P., Gentzsch, W., Potts, M., Clemm, A., Yousif, M., Polze, A.: Internet, Grid, Self-Adaptability and Beyond: Are We Ready?. In: Proc. 2nd Intl. Workshop on Self-Adaptive & Autonomic Computing Systems, Zaragoza, Spain, August 30- September 03 (2004), www.dcl.hpi.uni-potsdam.de/papers/papers/134_SAACS_Panel_II_v3.0.pdf
6. TeraGrid, www.teragrid.org
7. NAREGI Japanese national grid project, www.naregi.org/index_e.html
8. APAC Australian Partnership for Advanced Computing, www.apac.edu.au
9. Website of CEC funded European grid projects, www.cordis.lu/ist/results
10. EGEE, Enabling Grids for e-Science, http://eu-egee.org/
11. DEISA, Distributed European Infrastructure for Supercomputing Applications, www.deisa.org/index.php
12. Large Hadron Collider Computing Grid Project, LCG, http://lcg.web.cern.ch/LCG/
13. UK e-Science Programme, www.rcuk.ac.uk/escience/
14. Gentzsch, W.: Grid Computing in Research and Business. In: International Supercomputing Conference, Heidelberg (2005), www.isc2005.org/download/cp/gentzsch.pdf
15. LHC Large Hadron Collider, http://lhc-new-homepage.web.cern.ch
16. Gentzsch, W.: Enterprise Resource Management: Applications in Research and Industry. In: Foster, I., Kesselman, C. (eds.) Grid II: Blueprint for a new computing infrastructure, Morgan Kaufmann Publisher, San Francisco (2003)
17. Amazon Elastic Computing Cloud, www.amazon.com/gp/browse.html?node=201590011
18. Amazon Simple Storage Service, www.amazon.com/gp/browse.html?node=16427261
19. IAETE School grids panel, www.iaete.org/soapbox/summary.cfm
20. Open Grid Forum standardization organization, www.ogf.org
21. German D-Grid Initiative, https://www.d-grid.de/index.php?id=1\&L=1
22. D-Grid DGI Project, https://www.d-grid.de/index.php?id=61\&L=1
23. Authentifizierung im D-Grid (in German), www.d-grid.de/fileadmin/dgrid_document/Dokumente/vorschlagspapier-authz_v2.pdf
24. Thesenpapier zum VO Management in D-Grid (in German), www.d-grid.de/fileadmin/dgrid_document/Dokumente/VOMS-Thesenpapier.pdf
25. HEP-Grid, www.d-grid.de/index.php?id=44
26. Astro-Grid, www.d-grid.de/index.php?id=45, http://www.gac-grid.org/
27. Medi-Grid, www.d-grid.de/index.php?id=42, http://www.medigrid.de/
28. C3-Grid, www.d-grid.de/index.php?id=46, http://www.c3-grid.de/
29. InGrid, www.d-grid.de/index.php?id=43, http://www.ingrid-info.de/index.php

30. Text-Grid, www.d-grid.de/index.php?id=167, http://www.textgrid.de/
31. WISENT Energy Meteorology, http://www.offis.de/projekte/projekt_e.php?id=181&bereich=bi
32. OASIS: Organization for the Advancement of Structured Information Standards, http://www.oasis-open.org/
33. W3C: The World Wide Web Consortium, http://www.w3.org/
34. ESFRI: European Strategy Forum on Research Infrastructures, http://cordis.europa.eu/esfri/
35. e-IRG: e-Infrastructure Reflection Group, http://www.e-irg.org/

# Author Index

Adaikkalavan, Raman 369
Alonso, Miguel A. 529
Aly, Robin 98
Amagasa, Toshiyuki 298, 414
Anutariya, Chutiporn 720
Apostolou, Dimitris 213
Atay, Mustafa 603

Bača, Radim 1
Bao, Zhifeng 130
Bastien, Rance 740
Bayer, Rudolf 277
Belkhatir, Mohammed 392
Bellahsene, Zohra 844
Bellatreche, Ladjel 479
Berrabah, D. 593
Bertino, Elisa 434
Bhatnagar, Vasudha 629
Bhowmick, Sourav S. 617
Binder, Walter 172
Bonifati, Angela 539
Böttcher, Stefan 424
Boufarès, F. 593
Boukhalfa, Kamel 479
Bououlid Idrissi, Youssef 864
Brando, Carmen 254
Bressan, Stéphane 233, 660

Cabanac, Guillaume 202
Carneiro Filho, Heraldo J.A. 141
Casali, Alain 572
Chakravarthy, Sharma 369
Chang, Shi-Kuo 509
Charhad, Mbarek 392
Che, Dunren 87
Chebotko, Artem 603
Chen, Yangjun 243
Chevalier, Max 202
Choi, Byung-Uk 404
Chrisment, Claude 202
Christine, Froidevaux 740
Cicchetti, Rosine 572
Clifton, Chris 751
Constantinescu, Ion 172

Curé, Olivier 854
Cuzzocrea, Alfredo 539

d'Aquin, Mathieu 874
d'Orazio, Laurent 162
Denneulin, Yves 162
Deufemia, Vincenzo 509
Deveaux, Jean-Paul 109
Do, Tai T. 264, 445
dos Santos Mello, Ronaldo 13, 65
Draheim, Dirk 519
Duboisset, Magali 823

El-Mahgary, Sami 489
El Sayed, Ahmad 54
Elmongui, Hicham G. 434

Faltings, Boi 172
Farfán, Fernando 75
Fazzinga, Bettina 287
Fegaras, Leonidas 551
Flesca, Sergio 287
Fotouhi, Farshad 603
Fousteris, N. 23
Frédérique, Lisacek 740

Gao, Jun 562
Gentzsch, Wolfgang 895
Gergatsoulis, M. 23
Gómez-Rodríguez, Carlos 529
Goncalves, Marlene 254, 469
Gonçalves, Rodrigo 13
González, Vanessa 254
Gray, W. Alexander 772
Gu, Jie 339
Güneş, Salih 45
Guo, Hang 223

Hacid, Hakim 54
Hattori, Shun 790
He, Weimin 551
Helmer, Sven 98
Himsl, Melanie 519
Hose, Katja 308

Hou, Wen-Chi 87
Hristidis, Vagelis 75
Hua, Kien A. 264, 445
Hunt, Ela 844
Hyvönen, Eero 680

Jabornig, Daniel 519
Jang, Min-Hee 834
Jelinek, Ivan 700
Jeong, Seungdo 404
Jiang, Wei 751
Jiang, Zhewei 87
Jin, Xiaoming 339
Jochaud, Florent 854
Jose, Joemon M. 380
Jouanot, Fabrice 162
Julien, Christine 202

Kambur, Dalen 182
Kang, Myoung-Ah 823
Kantere, Verena 689
Kaur, Sharanjit 629
Kille, Peter 772
Kim, Sang-Wook 404, 834
Kirchberg, Markus 319
Kitagawa, Hiroyuki 298, 414
Klampanos, Iraklis A. 380
Korytkowski, Przemysław 710
Kottmann, Norbert 671
Krátký, Michal 1
Küng, Josef 519

Labbé, Cyril 162
Lakhal, Lotfi 572
Lee, Mong-Li 233
Leithner, Werner 519
Leonardi, Erwin 617
Levine, David 551
Li, Zhanhuai 151
Lim, Sungchae 834
Lin, Dan 434
Ling, Tok Wang 130
Liu, Fuyu 264, 445
Liu, Jun 499
Loukides, Grigorios 761
Lu, Shiyong 603
Lu, Yan Sheng 499
Luo, Cheng 87

Ma, Wei-Ying 811
Machado, Javam C. 141

Meersman, Robert 34
Mentzas, Gregoris 213
Moerkotte, Guido 98
Mohania, Mukesh 479
Murphy, John 182

Nakamura, Satoshi 801
Nedjar, Sébastien 572
Neumann, Thomas 98, 329

Ooi, Beng Chin 434
Ounelli, Habib 639

Papailiou, Niki 213
Pernici, Barbara 64
Philippi, Hans 459
Picariello, Antonio 730
Pinet, François 823
Polat, Kemal 45
Polese, Giuseppe 509
Prabhakar, Sunil 751
Pradhan, Sujeet 192
Prinzie, Anita 349
Pugliese, Andrea 287

Rangaswami, Raju 75
Rau-Chaplin, Andrew 109
Regner, Peter 519
Rezende Rodrigues, Khaue 65
Rinaldi, Antonio M. 730
Roantree, Mark 182
Roncancio, Claudia 162
Rugui, Yao 151
Ruotsalo, Tuukka 680

Sabou, Marta 874
Salakoski, Tapio 780
Saleem, Khalid 844
Sassi, Minyar 639
Sattler, Kai-Uwe 308
Schlicht, Anne 874
Schneider, Michel 823
Schult, Rene 650
Şekerci, Ramazan 45
Sellis, Timos 689
Shao, Jianhua 761
Shestakov, Denis 780
Sikora, Katarzyna 710
Snášel, Václav 1
Soisalon-Soininen, Eljas 489

Sousa, Flávio R.C. 141
Speer, Jayson 319
Spycher, Samuel 172
Stavrakas, Y. 23
Steinmetz, Rita 424
Stuckenschmidt, Heiner 874
Studer, Thomas 671
Sun, Chong 499
Svihla, Martin 700

Tanaka, Katsumi 790, 801
Tang, Yan 34
Tezuka, Taro 790
Tineo, Leonid 469
Tok, Wee Hyong 233
Tosun, Ali Şaman 120
Touzi, Amel Grissa 639
Tsoumakos, Dimitrios 689
Tzitzikas, Yannis 582

Ungrangsi, Rachanee 720

Vacca, Mario 509
Vachon, Julie 864
Valentin, Olivier 162
Van den Poel, Dirk 349
Vilares, Jesús 529

Wang, Lee 811
Wang, Tengjiao 562
Watanabe, Yousuke 414
Welter, Danielle 772
Wen, Lianzi 298
Widjanarko, Klarinda G. 617
Wiesinger, Thomas 519
Wijaya, Derry Tanti 660
Wuwongse, Vilas 720

Xie, Xing 811
Xu, Juan 151
Xu, Liang 130

Yamada, Shinichi 414
Yamamoto, Takehiro 801
Yang, Dongqing 562
Yanlong, Wang 151
Yasukawa, Michiko 359
Yokoo, Hidetoshi 359
Yue, Lihua 811

Zdrahal, Zdenek 884
Zeh, Norbert 109
Zhang, Jun 223
Zhang, Mingwu 751
Zhang, Qi 811
Zhou, Chong 499
Zhou, Lizhu 223
Zhu, Qiang 87
Zighed, Djamel 54

# Lecture Notes in Computer Science

For information about Vols. 1–4578

please contact your bookseller or Springer

Vol. 4720: B. Konev, F. Wolter (Eds.), Frontiers of Combining Systems. X, 2283 pages. 2007. (Sublibrary LNAI).

Vol. 4708: L. Kučera, A. Kučera (Eds.), Mathematical Foundations of Computer Science 2007. XVIII, 764 pages. 2007.

Vol. 4707: O. Gervasi, M.L. Gavrilova (Eds.), Computational Science and Its Applications – ICCSA 2007, Part III. XXIV, 1205 pages. 2007.

Vol. 4706: O. Gervasi, M.L. Gavrilova (Eds.), Computational Science and Its Applications – ICCSA 2007, Part II. XXIII, 1129 pages. 2007.

Vol. 4705: O. Gervasi, M.L. Gavrilova (Eds.), Computational Science and Its Applications – ICCSA 2007, Part I. XLIV, 1169 pages. 2007.

Vol. 4703: L. Caires, V.T. Vasconcelos (Eds.), CONCUR 2007 – Concurrency Theory. XIII, 507 pages. 2007.

Vol. 4697: L. Choi, Y. Paek, S. Cho (Eds.), Advances in Computer Systems Architecture. XIII, 400 pages. 2007.

Vol. 4685: D.J. Veit, J. Altmann (Eds.), Grid Economics and Business Models. XII, 201 pages. 2007.

Vol. 4683: L. Kang, Y. Liu, S. Zeng (Eds.), Intelligence Computation and Applications. XVII, 663 pages. 2007.

Vol. 4682: D.-S. Huang, L. Heutte, M. Loog (Eds.), Advanced Intelligent Computing Theories and Applications. XXVII, 1373 pages. 2007. (Sublibrary LNAI).

Vol. 4681: D.-S. Huang, L. Heutte, M. Loog (Eds.), Advanced Intelligent Computing Theories and Applications. XXVI, 1379 pages. 2007.

Vol. 4679: A.L. Yuille, S.-C. Zhu, D. Cremers, Y. Wang (Eds.), Energy Minimization Methods in Computer Vision and Pattern Recognition. XII, 494 pages. 2007.

Vol. 4678: J. Blanc-Talon, W. Philips, D. Popescu, P. Scheunders (Eds.), Advanced Concepts for Intelligent Vision Systems. XXIII, 1100 pages. 2007.

Vol. 4673: W.G. Kropatsch, M. Kampel, A. Hanbury (Eds.), Computer Analysis of Images and Patterns. XX, 1006 pages. 2007.

Vol. 4671: V. Malyshkin (Ed.), Parallel Computing Technologies. XIV, 635 pages. 2007.

Vol. 4660: S. Džeroski, J. Todorovski (Eds.), Computational Discovery of Scientific Knowledge. X, 327 pages. 2007. (Sublibrary LNAI).

Vol. 4659: V. Mařík, V. Vyatkin, A.W. Colombo (Eds.), Holonic and Multi-Agent Systems for Manufacturing. VIII, 456 pages. 2007. (Sublibrary LNAI).

Vol. 4658: T. Enokido, L. Barolli, M. Takizawa (Eds.), Network-Based Information Systems. XIII, 544 pages. 2007.

Vol. 4657: C. Lambrinoudakis, G. Pernul, A M. Tjoa (Eds.), Trust and Privacy in Digital Business. XIII, 291 pages. 2007.

Vol. 4656: M.A. Wimmer, J. Scholl, Å. Grönlund (Eds.), Electronic Government. XIV, 450 pages. 2007.

Vol. 4655: G. Psaila, R. Wagner (Eds.), E-Commerce and Web Technologies. VII, 229 pages. 2007.

Vol. 4654: I.Y. Song, J. Eder, T.M. Nguyen (Eds.), Data Warehousing and Knowledge Discovery. XVI, 482 pages. 2007.

Vol. 4653: R. Wagner, N. Revell, G. Pernul (Eds.), Database and Expert Systems Applications. XXII, 907 pages. 2007.

Vol. 4651: F. Azevedo, P. Barahona, F. Fages, F. Rossi (Eds.), Recent Advances in Constraints. VIII, 185 pages. 2007. (Sublibrary LNAI).

Vol. 4649: V. Diekert, M.V. Volkov, A. Voronkov (Eds.), Computer Science – Theory and Applications. XIII, 420 pages. 2007.

Vol. 4647: R. Martin, M. Sabin, J. Winkler (Eds.), Mathematics of Surfaces XII. IX, 509 pages. 2007.

Vol. 4645: R. Giancarlo, S. Hannenhalli (Eds.), Algorithms in Bioinformatics. XIII, 432 pages. 2007. (Sublibrary LNBI).

Vol. 4644: N. Azemard, L. Svensson (Eds.), Integrated Circuit and System Design. XIV, 583 pages. 2007.

Vol. 4643: M.-F. Sagot, M.E.M.T. Walter (Eds.), Advances in Bioinformatics and Computational Biology. XII, 177 pages. 2007. (Sublibrary LNBI).

Vol. 4642: S.-W. Lee, S.Z. Li (Eds.), Advances in Biometrics. XX, 1216 pages. 2007.

Vol. 4641: A.-M. Kermarrec, L. Bougé, T. Priol (Eds.), Euro-Par 2007 Parallel Processing. XXVII, 974 pages. 2007.

Vol. 4639: E. Csuhaj-Varjú, Z. Ésik (Eds.), Fundamentals of Computation Theory. XIV, 508 pages. 2007.

Vol. 4638: T. Stützle, M. Birattari, H.H. Hoos (Eds.), Engineering Stochastic Local Search Algorithms. X, 223 pages. 2007.

Vol. 4637: C. Kruegel, R. Lippmann, A. Clark (Eds.), Recent Advances in Intrusion Detection. XII, 337 pages. 2007.

Vol. 4635: B. Kokinov, D.C. Richardson, T.R. Roth-Berghofer, L. Vieu (Eds.), Modeling and Using Context. XIV, 574 pages. 2007. (Sublibrary LNAI).

Vol. 4634: H.R. Nielson, G. Filé (Eds.), Static Analysis. XI, 469 pages. 2007.

Vol. 4633: M. Kamel, A. Campilho (Eds.), Image Analysis and Recognition. XII, 1312 pages. 2007.

Vol. 4632: R. Alhajj, H. Gao, X. Li, J. Li, O.R. Zaïane (Eds.), Advanced Data Mining and Applications. XV, 634 pages. 2007. (Sublibrary LNAI).

Vol. 4628: L.N. de Castro, F.J. Von Zuben, H. Knidel (Eds.), Artificial Immune Systems. XII, 438 pages. 2007.

Vol. 4627: M. Charikar, K. Jansen, O. Reingold, J.D.P. Rolim (Eds.), Approximation, Randomization, and Combinatorial Optimization. XII, 626 pages. 2007.

Vol. 4626: R.O. Weber, M.M. Richter (Eds.), Case-Based Reasoning Research and Development. XIII, 534 pages. 2007. (Sublibrary LNAI).

Vol. 4624: T. Mossakowski, U. Montanari, M. Haveraaen (Eds.), Algebra and Coalgebra in Computer Science. XI, 463 pages. 2007.

Vol. 4622: A. Menezes (Ed.), Advances in Cryptology - CRYPTO 2007. XIV, 631 pages. 2007.

Vol. 4619: F. Dehne, J.-R. Sack, N. Zeh (Eds.), Algorithms and Data Structures. XVI, 662 pages. 2007.

Vol. 4618: S.G. Akl, C.S. Calude, M.J. Dinneen, G. Rozenberg, H.T. Wareham (Eds.), Unconventional Computation. X, 243 pages. 2007.

Vol. 4617: V. Torra, Y. Narukawa, Y. Yoshida (Eds.), Modeling Decisions for Artificial Intelligence. XII, 502 pages. 2007. (Sublibrary LNAI).

Vol. 4616: A. Dress, Y. Xu, B. Zhu (Eds.), Combinatorial Optimization and Applications. XI, 390 pages. 2007.

Vol. 4615: R. de Lemos, C. Gacek, A. Romanovsky (Eds.), Architecting Dependable Systems IV. XIV, 435 pages. 2007.

Vol. 4613: F.P. Preparata, Q. Fang (Eds.), Frontiers in Algorithmics. XI, 348 pages. 2007.

Vol. 4612: I. Miguel, W. Ruml (Eds.), Abstraction, Reformulation, and Approximation. XI, 418 pages. 2007. (Sublibrary LNAI).

Vol. 4611: J. Indulska, J. Ma, L.T. Yang, T. Ungerer, J. Cao (Eds.), Ubiquitous Intelligence and Computing. XXIII, 1257 pages. 2007.

Vol. 4610: B. Xiao, L.T. Yang, J. Ma, C. Muller-Schloer, Y. Hua (Eds.), Autonomic and Trusted Computing. XVIII, 571 pages. 2007.

Vol. 4609: E. Ernst (Ed.), ECOOP 2007 – Object-Oriented Programming. XIII, 625 pages. 2007.

Vol. 4608: H.W. Schmidt, I. Crnkovic, G.T. Heineman, J.A. Stafford (Eds.), Component-Based Software Engineering. XII, 283 pages. 2007.

Vol. 4607: L. Baresi, P. Fraternali, G.-J. Houben (Eds.), Web Engineering. XVI, 576 pages. 2007.

Vol. 4606: A. Pras, M. van Sinderen (Eds.), Dependable and Adaptable Networks and Services. XIV, 149 pages. 2007.

Vol. 4605: D. Papadias, D. Zhang, G. Kollios (Eds.), Advances in Spatial and Temporal Databases. X, 479 pages. 2007.

Vol. 4604: U. Priss, S. Polovina, R. Hill (Eds.), Conceptual Structures: Knowledge Architectures for Smart Applications. XII, 514 pages. 2007. (Sublibrary LNAI).

Vol. 4603: F. Pfenning (Ed.), Automated Deduction – CADE-21. XII, 522 pages. 2007. (Sublibrary LNAI).

Vol. 4602: S. Barker, G.-J. Ahn (Eds.), Data and Applications Security XXI. X, 291 pages. 2007.

Vol. 4600: H. Comon-Lundh, C. Kirchner, H. Kirchner (Eds.), Rewriting, Computation and Proof. XVI, 273 pages. 2007.

Vol. 4599: S. Vassiliadis, M. Berekovic, T.D. Hämäläinen (Eds.), Embedded Computer Systems: Architectures, Modeling, and Simulation. XVIII, 466 pages. 2007.

Vol. 4598: G. Lin (Ed.), Computing and Combinatorics. XII, 570 pages. 2007.

Vol. 4597: P. Perner (Ed.), Advances in Data Mining. XI, 353 pages. 2007. (Sublibrary LNAI).

Vol. 4596: L. Arge, C. Cachin, T. Jurdziński, A. Tarlecki (Eds.), Automata, Languages and Programming. XVII, 953 pages. 2007.

Vol. 4595: D. Bošnački, S. Edelkamp (Eds.), Model Checking Software. X, 285 pages. 2007.

Vol. 4594: R. Bellazzi, A. Abu-Hanna, J. Hunter (Eds.), Artificial Intelligence in Medicine. XVI, 509 pages. 2007. (Sublibrary LNAI).

Vol. 4593: A. Biryukov (Ed.), Fast Software Encryption. XI, 467 pages. 2007.

Vol. 4592: Z. Kedad, N. Lammari, E. Métais, F. Meziane, Y. Rezgui (Eds.), Natural Language Processing and Information Systems. XIV, 442 pages. 2007.

Vol. 4591: J. Davies, J. Gibbons (Eds.), Integrated Formal Methods. IX, 660 pages. 2007.

Vol. 4590: W. Damm, H. Hermanns (Eds.), Computer Aided Verification. XV, 562 pages. 2007.

Vol. 4589: J. Münch, P. Abrahamsson (Eds.), Product-Focused Software Process Improvement. XII, 414 pages. 2007.

Vol. 4588: T. Harju, J. Karhumäki, A. Lepistö (Eds.), Developments in Language Theory. XI, 423 pages. 2007.

Vol. 4587: R. Cooper, J. Kennedy (Eds.), Data Management. XIII, 259 pages. 2007.

Vol. 4586: J. Pieprzyk, H. Ghodosi, E. Dawson (Eds.), Information Security and Privacy. XIV, 476 pages. 2007.

Vol. 4585: M. Kryszkiewicz, J.F. Peters, H. Rybinski, A. Skowron (Eds.), Rough Sets and Intelligent Systems Paradigms. XIX, 836 pages. 2007. (Sublibrary LNAI).

Vol. 4584: N. Karssemeijer, B. Lelieveldt (Eds.), Information Processing in Medical Imaging. XX, 777 pages. 2007.

Vol. 4583: S.R. Della Rocca (Ed.), Typed Lambda Calculi and Applications. X, 397 pages. 2007.

Vol. 4582: J. Lopez, P. Samarati, J.L. Ferrer (Eds.), Public Key Infrastructure. XI, 375 pages. 2007.

Vol. 4581: A. Petrenko, M. Veanes, J. Tretmans, W. Grieskamp (Eds.), Testing of Software and Communicating Systems. XII, 379 pages. 2007.

Vol. 4580: B. Ma, K. Zhang (Eds.), Combinatorial Pattern Matching. XII, 366 pages. 2007.

Vol. 4579: B. M. Hämmerli, R. Sommer (Eds.), Detection of Intrusions and Malware, and Vulnerability Assessment. X, 251 pages. 2007.

CPSIA information can be obtained at www.ICGtesting.com
Printed in the USA
LVOW09s0746120616

492225LV00002B/10/P